JS Miller
Goshen College
Biol. Dept.

Developmental
Biology

FIFTH EDITION

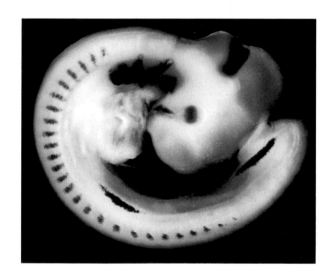

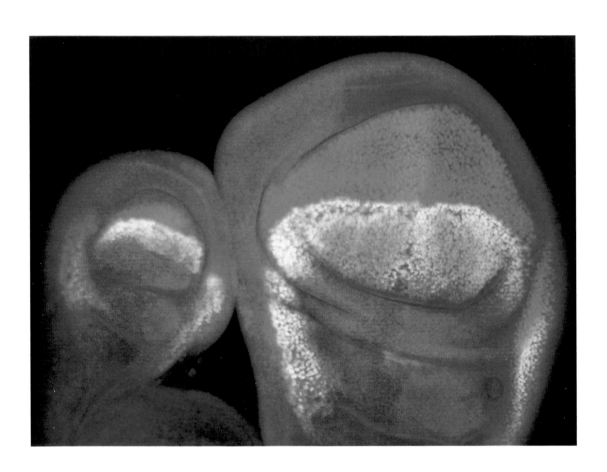

Developmental Biology

FIFTH EDITION

Scott F. Gilbert

Swarthmore College

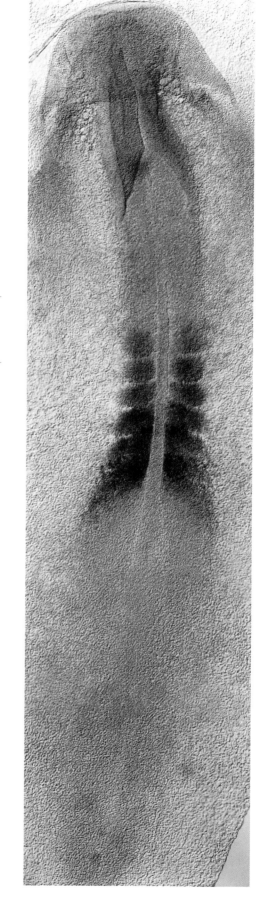

SINAUER ASSOCIATES, INC.
PUBLISHERS
Sunderland, Massachusetts

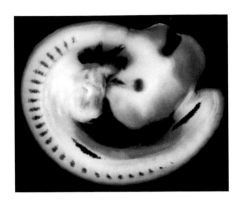

The cover

COVER PHOTOGRAPH: The mRNA for Fibroblast Growth Factor-8 can be detected by wholemount in situ hybridization using chemically labeled RNA that is complementary to this message. In this 3-day chick embryo, the *Fgf8* message is found in the most distal ectoderm of the limb buds, in the boundary between the midbrain and hindbrain, in the somites, in the branchial arches of the neck, and in the developing tail. FGF8 is important for several developmental processes, and it plays critical roles in the outgrowth of the limbs and the patterning of the developing brain. Chapters 3, 7 and 18. (Photograph courtesy of E. Laufer, C.-Y. Yeo and C. Tabin.)

BACK COVER PHOTOGRAPH: Photograph of a Day 20–21 chicken embryo at the pipping and prehatching stage. Note the prominent peridermal covering at the tip of the beak (egg-tooth), used by the chick to make holes in the eggshell, which has become thinner and more brittle as a consequence of mineral utilization by the embryo for its growing skeleton. This developmental stage marks the transition of the embryo into an air-breathing chick. Chapters 1 and 5. (Photograph from the *International Poultry Journal,* courtesy of R. Tuan.)

The title pages

LEFT PAGE: Gene expression generates boundaries in *Drosophila* imaginal discs. The large and small discs within the fly larva form the adult wing and haltere, respectively. At this stage, Apterous protein (red) is expressed only in the dorsal compartments; the Cubitus interruptus protein (blue) marks the anterior (but not the posterior) compartments (a line forming this boundary can be seen). The green staining (from the Vestigial protein) in the interior demarcates the boundary between the free limb and the hinge linking it to the thoracic wall. Chapter 19. (Photograph courtesy of J. Williams, S. Paddock and S. Carroll.)

RIGHT PAGE: Expression of the *paraxis* gene in the 6-somite chick embryo. Wholemount in situ hybridization using a digoxygenin-labeled RNA complementary to a portion of the chick *paraxis* message shows the expression of this gene during somite formation. The Paraxis protein is important in establishing the structure of these mesodermal clusters. Chapters 2 and 9. (Photographic montage courtesy of R. Tuan.)

Developmental Biology, FIFTH EDITION

Copyright © 1997 by Sinauer Associates, Inc.
All rights reserved.
This book may not be reproduced in whole or in part without permission from the publisher. For information or to order address:
Sinauer Associates, Inc., P. O. Box 407,
Sunderland, Massachusetts 01375-0407
U.S.A.
Fax: 413-549-1118
E-mail: publish@sinauer.com

**Library of Congress
Cataloging-in-Publication Data**

Gilbert, Scott F., 1949–
　Developmental biology/by Scott F. Gilbert.
—5th ed.
　　　p.　　cm.
　Includes bibliographical references and index.
　ISBN 0-87893-244-5 (hc)
　1. Embryology.　2. Developmental biology.
I. Title.
QL955.G48　1997　　　　97-6182
571.8—dc21　　　　　　　CIP

Printed in U.S.A.
6 5 4 3 2 1

To Daniel, Sarah, and David

Table of Contents

PART III Mechanisms of Cellular Differentiation

PART IV Specification of Cell Fate and the Embryonic Axes

Autonomous cell specification by cytoplasmic determinants 505 — 13

The genetics of axis specification in Drosophila 543 — 14

PART V Cellular Interactions During Organ Formation

Preface

The last years of the twentieth century find developmental biology returning to the position it held a century earlier: the discipline that unifies the studies of heredity, evolution, and physiology. In 1896, E. B. Wilson's first edition of *The Cell in Development and Inheritance* announced "the wonderful truth that a single cell may contain within its microscopic compass the sum-total of the heritage of the species." Today, developmental biology is in the forefront of this study of our natural heritage. In its *molecular* aspects, it touches physical chemistry in its probing of the biochemical mechanisms by which different proteins are made in different cells from the same genome. It is also at the lead of *evolutionary* studies that seek to understand how macroevolutionary changes have occurred. It has recently opened a new area of *ecological* developmental biology wherein environmental changes are seen to create alterations in organismal development. During the past three years, developmental biology has also expanded into *medicine*, merging with clinical genetics to create a revitalized science of human embryology, a science that has already become important in explaining congenital malformations.

The fifth edition of *Developmental Biology* has been revised and rewritten to reflect these ongoing revolutions. There have been four major changes in the structure of the book since the last edition. First, it has become impossible to discuss the fundamental principles of embryology without knowledge of *gene activity* or *signal transduction* pathways. Therefore, this information has been brought into the introductory section of the book so that cellular interactions such as fertilization and induction can be appreciated on the molecular as well as on the morphological levels.

Second, new interest in the effects of the environment on normal and abnormal development has led to a new chapter. Chapter 21, "Environmental Regulation of Animal Development," concerns the ways by which the environment effects the phenotype of the organism. Interest in environmental protection and in controversies surrounding possibly teratogenic pollutants has forced a new awareness of the influences that the environment plays in normal and abnormal development. Indeed, developmental biologists may soon find themselves at the forefront of ecological conservation movements. The first four editions of this book sought to integrate the molecular, cellular, and organismal approaches to developmental biology; this edition adds the ecological dimension.

Third, this edition places new emphasis on the roles of paracrine factors in development. Not only are the signal transduction studies placed in the introductory section of the book, but Part V of the Fifth Edition begins with an overview of the fibroblast growth factor, TGF-β, Wnt, and Hedgehog families of growth and differentiation factors.

Fourth, this book is connected to a website wherein students and faculty can find more material on many selected topics. Such material includes (1) details of experiments that are too specialized to put into the textbook, (2) historical information about particular areas of developmental biology and the personalities involved, (3) medical implications of particular developmental phenomena, (4) debates or commentaries on issues relevant to the field, and (5) updates of the text material in this increasingly rapidly growing area of biology. Movies and taped interviews are included and this feature will be expanded as the technology makes them easier to use. This website is also connected to other websites and can be used to enrich one's perspective about what is happening in animal development. The presence of a website allows me to keep directing this book to the people for whom it was originally intended: upper-level undergraduates and introductory graduate students. It also has helped me keep the book from becoming a substitute for standing weights.

It was Roux's vision that developmental biology would "sometime constitute the common basis of all other biological disciplines and, in continued symbiosis with these disciplines, play a prominent part in the solutions of the problems of life." These were bold, even arrogant words one hundred years ago; today, they express a widely held assumption. Development integrates all areas of biology and plays the crucial role of relating genotype to phenotype. Development can be studied using any organism and at any level of organization, from molecules to phyla.

As the field continues to expand and deepen, a word of warning is called for: developmental biology cannot be taught or learned in a single semester. This text is an attempt to provide each person with sufficient material for their course, but an instructor need not feel guilty for not assigning every chapter, and students need not feel deprived if they have not read every chapter. This is the beginning of the path, not its conclusion.

How to use the website

One can enter the website through its homepage [http://zygote.swarthmore.edu/index.html] or through its table of chapter files located at [http://zygote.swarthmore.edu/info.html]. Alternatively, we have placed specific access addresses throughout the book wherever a relevant entry exists at the time of publication. These addresses all begin with [http://zygote.swarthmore.edu/] and are followed by the code given in the textbook. Thus, the location specified on page 20 of the textbook is:

http://zygote.swarthmore.edu/intro2.html

More locations are being added to the website, and these can be accessed by entering the chapter files. In addition, by clicking on the "Other Files" button on the bottom of each chapter file, connections to other websites will be facilitated. Have fun.

Acknowledgments

This edition, like its earlier incarnations, owes a great deal to the suggestions and criticisms of the students in my developmental biology and developmental genetics classes. The extremely supportive staff and faculty of Swarthmore College have also played major roles in producing this book, and science librarians E. Horikawa and M. Spencer are due special thanks for keeping recent volumes from being sent to the bindery while I was writing the book. The scientists who reviewed these chapters provided enormous help in both the technical accuracy of the chapters and in suggestions for further work. These investigators include S. Carroll, J. Cebra-Thomas, E. M. De Robertis, S. DiNardo, E. Eicher, C. Emerson, D. J. Grunwald, G. Grunwald, M. Hollyday, L. A. Jaffe, W. Katz, R. Keller, K. Kemphues, D. Kirk, G. Martin, H. F. Nijhout, D. Page, R. Schultz, C. Stern, S. Tilghman, R. Tuan, and M. Wickens. I also want to thank several scientists who went out of their way to help make this edition better by reading specific portions of the chapters. They include M. Bronner-Fraser, J. Fallon, N. M. Le Douarin, E. McCloud, J. Opitz, K. Sainio, H. Sariola, I. Thesleff, and T. Valente. If I left anyone out, please forgive me. Needless to say, the final editorial judgments were my responsibility. My special thanks to Judy Cebra-Thomas who not only advised me on certain chapters but whose excellent help during my sabbatical leave allowed me to finish this book. Thanks also to the scientists and philosophers, especially C. van der Weele, R. Amundson, L. Nyhart, R. Burian, H. F. Nijhout, A. F. Sterling, K. Smith, and A. I. Tauber, who participated in the developmental biology workshops of the International Society for the History, Philosophy, and Social Studies of Biology. Some of the best constructive critiques of this textbook have come from these people.

Andy Sinauer has yet again managed to gather the same remarkable people around this project, and it has been a privilege to work with them. My thanks to him and to editors Nan Sinauer and Carol Wigg, production coordinator Chris Small, artists John Woolsey and Gary Welch, designer Susan Schmidler, copy editor Janet Greenblatt, and layout artist Janice Holabird. The editorial skills of Tinsley Davis are greatly appreciated. Because publishing deadlines must be met and other work gets put aside, I have to thank my family for once again allowing me to get away with this. In particular, this book could never have been completed were it not for the encouragement of my wife, Anne Raunio, who, as an obstetrician, enjoys the more practical side of developmental biology. My thanks to you all.

SCOTT F. GILBERT
MARCH 1, 1997

An Introduction to Developmental Biology

I

An introduction to animal development

1

Nature is always the same, and yet its appearance is always changing. It is our business as artists to convey the thrill of nature's permanence along with the elements and the appearances of all its changes.
PAUL CEZANNE (ca. 1900)

Happy is the person who is able to discern the causes of things.
VIRGIL (37 B.C.E.)

THE CONCEPT OF AN EMBRYO is a staggering one, and forming an embryo is the hardest thing you will ever have to do. To become an embryo, you had to build yourself from a single cell. You had to respire before you had lungs, digest before you had a gut, build bones when you were pulpy, and form orderly arrays of neurons before you knew how to think. One of the critical differences between you and a machine is that the machine is never called on to function until after it is built. Every animal has to function as it builds itself.

The scope of developmental biology

For animals and plants, the sole way of getting from egg to adult is by developing an embryo. The embryo mediates between genotype and phenotype, between the inherited genes and the adult organism. Whereas most of biology studies *adult* structure and function, developmental biology finds the *transient* stages more interesting. Developmental biology is a science of becoming, a science of process. To say that a mayfly lives but one day is meaningless to a developmental biologist. It may only be an adult for a day, but it spends the other 364 days as an embryo and larva.

The questions asked by developmental biologists are often questions about becoming rather than about being. To say that XX mammals are usually females and XY mammals are usually males does not explain sex determination to a developmental biologist. The developmental biologist wants to know *how* the XX genotype produces a female and *how* the XY genotype produces a male. Similarly, a geneticist may ask how globin genes are transmitted from one generation to the other, and a physiologist might ask about the function of globin in the body. But the developmental biologist asks how it is that the globin genes become expressed only in red blood cells and how they become active only at certain times in development. (We do not know the answers yet.)

Developmental biology is a great science for people who want to integrate different levels of biology. We can take a problem and study it on the molecular and chemical levels (e.g., How are globin genes transcribed, and how do the factors activating their transcription interact with one another on the DNA?), on the cellular and tissue levels (e.g., Which cells are able to make globin, and how does globin mRNA leave the nucleus?), on the organ and organ system levels (e.g., How do the capillaries form in each tissue, and how are they instructed to branch and connect?), and even at the ecological and evolutionary levels (e.g., How do differences in globin gene activation enable oxygen to flow from mother to fetus, and how do environmental factors trigger the differentiation of more red blood cells?). Developmental biologists can study any organism and any cell type.

Developmental biology is one of the fastest growing and most exciting fields in biology. Part of the excitement comes from its subject matter, for we are just beginning to understand the molecular mechanisms of animal development. Another part of the excitement comes from the unifying role that developmental biology is assuming in the biological sciences. Developmental biology is creating a framework that integrates molecular biology, physiology, cell biology, anatomy, cancer research, neurobiology, immunology, ecology, and evolutionary biology. The study of development has become essential for understanding any other area of biology.

The problems of developmental biology

Development accomplishes two major functions: it generates cellular diversity and order within each generation, and it ensures the continuity of life from one generation to the next. Thus, there are two fundamental questions in developmental biology: How does the fertilized egg give rise to the adult body, and how does that adult body produce yet another body? Each species has its own answers, but some generalizations can be made. Traditionally, these questions have been subdivided into four general problems of developmental biology:

- *The problem of differentiation.* A single cell, the fertilized egg, gives rise to hundreds of different cell types—muscle cells, epidermal cells, neurons, lymphocytes, blood cells, fat cells, and so on. This generation of cellular diversity is called **differentiation.** Since each cell of the body contains the same set of genes, we need to understand how this same set of genetic instructions can produce different types of cells.
- *The problem of morphogenesis.* Our differentiated cells are not randomly distributed. Rather, they are organized into intricate tissues and organs. These organs are arranged in a given way: the fingers at the tip of our hands, not in the middle; the eyes in our heads, not in our toes or gut. This creation of ordered form is called **morphogenesis.** How do cells organize themselves and form the correct arrangements?
- *The problem of growth.* We are bigger than the egg, but how did the cells know when to stop dividing? If each cell in our face were to undergo just one more cell division, we would be considered horribly malformed. If each cell of our arms underwent just one more round of cell division, we could tie our shoelaces without bending over.
- *The problem of reproduction.* The sperm and egg are very specialized cells. Only they can transmit the instructions to make an organism from one generation to the next. How are these cells set apart to form the next generation, and what are the instructions in the nucleus and cytoplasm that allow them to function this way?

Recently, a fifth problem has been reemphasized:

- *The problem of evolution.* Evolution involves inherited changes in development. When we say that today's one-toed horse had a five-toed ancestor, we are saying that changes in the development of cartilage and muscles occurred over many generations in the embryos of the horse's ancestors. How do changes in development create new body forms? Which heritable changes are possible, given the constraints imposed by the necessity of the organism to survive as it develops?

The stages of animal development

According to Aristotle, the first major embryologist known to history, science begins with wonder: "It is owing to wonder that people began to philosophize, and wonder remains the beginning of knowledge." The development of an animal from an egg has been a source of wonder throughout human history. The simple procedure of cracking open a chick egg on each successive day of its three-week incubation provides a remarkable experience as a thin band of cells is seen to give rise to an entire bird. Aristotle performed this procedure and noted the formation of the major organs. Anyone can wonder at this remarkable—yet commonplace—phenomenon, but it is the scientist who seeks to discover how development actually occurs. And rather than dissipating wonder, new understanding increases it.

Multicellular organisms do not spring forth fully formed. Rather, they arise by a relatively slow process of progressive change that we call **development.** In nearly all cases, the development of a multicellular organism begins with a single cell—the fertilized egg, or **zygote,** which divides mitotically to produce all the cells of the body. The study of animal development has traditionally been called **embryology,** referring to the fact that between fertilization and birth the developing organism is known as an **embryo.** But development does not stop at birth, or even at adulthood. Most organisms never stop developing. Each day we replace over a gram of skin cells (the older cells being sloughed off as we move), and our bone marrow sustains the development of millions of new erythrocytes every minute of our lives. Therefore, in recent years it has become customary to speak of **developmental biology** as the discipline that studies embryonic and other developmental processes.

The major features of animal development are illustrated in Figure 1.1. The life of a new individual is initiated by the fusion of genetic material from the two **gametes**—the sperm and the egg. This fusion, called **fertilization,** stimulates the egg to begin development. The subsequent stages of development are collectively called **embryogenesis.** Throughout the animal kingdom an incredible variety of embryonic types exist, but most patterns of embryogenesis comprise variations on four themes:

1. Immediately following fertilization, cleavage occurs. **Cleavage** is a series of extremely rapid mitotic divisions wherein the enormous volume of zygote cytoplasm is divided into numerous smaller cells. These cells are called **blastomeres,** and by the end of cleavage, they generally form a sphere known as a **blastula.**
2. After the rate of mitotic division has slowed down, the blastomeres undergo dramatic movements wherein they change their positions relative to one another. This series of extensive cell rearrangements is called **gastrulation.** As a result of gastrulation, the typical embryo contains three cell regions, called **germ layers.*** The outer layer, the **ectoderm**, produces the cells of the epidermis and the nervous sys-

---

*From the Latin *germen,* meaning "bud" or "sprout" (the same root as in the word *germination*). The names of the three germ layers are from the Greek: ectoderm from *ektos* ("outside") plus *derma* ("skin"); mesoderm from *mesos* ("middle"); and endoderm from *endon* ("within").

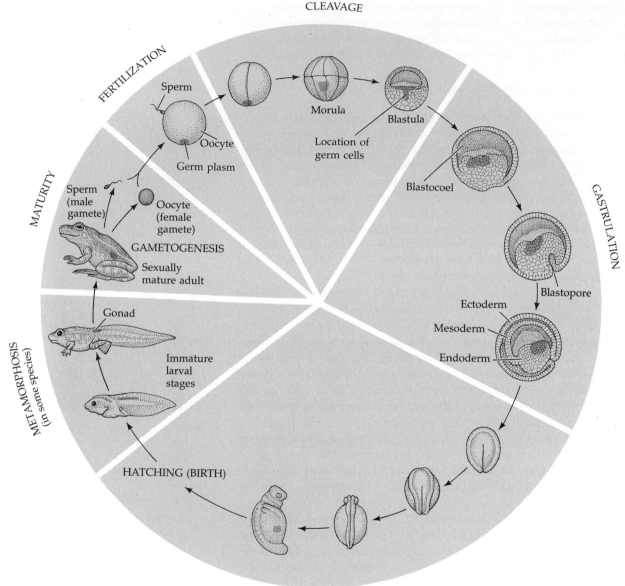

CLEAVAGE

FERTILIZATION

Sperm

Morula

Oocyte

Germ plasm

Blastula

Location of
germ cells

MATURITY

Sperm
(male
gamete)

Oocyte
(female
gamete)

GAMETOGENESIS

Sexually
mature adult

Blastocoel

GASTRULATION

Blastopore

Ectoderm

Mesoderm

Endoderm

Gonad

Immature
larval
stages

METAMORPHOSIS
(in some species)

HATCHING (BIRTH)

ORGANOGENESIS

Figure 1.1
Developmental history of a representative animal, a frog. The stages from fertilization through hatching (birth) are
known collectively as embryogenesis.
The region set aside for producing germ
cells is shown in color. Gametogenesis,
which is complete in the sexually
mature adult, begins at different times
during development, depending on the
species.

tem; the inner layer, the **endoderm,** produces the lining of the digestive tube and its associated organs (pancreas, liver, lungs, etc.); and
the middle layer, the **mesoderm,** gives rise to several organs (heart,
kidney, gonads), connective tissues (bone, muscles, tendons, blood
vessels), and the blood cells.

3. Once the three germ layers are established, the cells interact with one
   another and rearrange themselves to produce tissues and organs.
   This process is called **organogenesis.** (In vertebrates, organogenesis
   is initiated when a series of cellular interactions causes the mid-dor-
   sal ectodermal cells to form the neural tube. This tube will become
   the brain and spinal cord.) Many organs contain cells from more than
   one germ layer, and it is not unusual for the outside of an organ to be
   derived from one layer and the inside from another. Also during

organogenesis, certain cells undergo long migrations from their place of origin to their final location. These migrating cells include the precursors of blood cells, lymph cells, pigment cells, and gametes. Most of the bones of our face are derived from cells that have migrated ventrally from the dorsal region of the head.

4. As seen in Figure 1.1, in many species, a specialized portion of egg cytoplasm gives rise to cells that are the precursors of the gametes. These cells are called **germ cells,** and they are set aside for their reproductive function. All the other cells of the body are called **somatic cells.** This separation of somatic cells (which give rise to the individual body) and germ cells (which contribute to the formation of a new generation) is often one of the first differentiations to occur during animal development. The germ cells eventually migrate to the gonads, where they differentiate into gametes. The development of gametes, called **gametogenesis,** is usually not completed until the organism has become physically mature. At maturity, the gametes may be released and participate in fertilization to begin a new embryo. The adult organism eventually undergoes senescence and dies.

Our eukaryotic heritage

Organisms are divided into two major groups, depending on whether their cells possess a nuclear envelope. The **prokaryotes** (from the Greek *karyon,* meaning "nucleus"), which include the archaebacteria and the eubacteria, lack a true nucleus. The **eukaryotes,** which include protists, animals, plants, and fungi, have a well-formed nuclear envelope surrounding their chromosomes. This fundamental difference between eukaryotes and prokaryotes influences how these two groups arrange and utilize their genetic material. In both groups, the inherited information needed for development and metabolism is encoded in the DNA sequences of the chromosomes. The prokaryotic chromosome is generally a small, circular double helix of DNA consisting of approximately 1 million base pairs. The eukaryotic cell usually has several chromosomes, and the simplest eukaryotic protists have over 10 times the amount of DNA found in the most complex prokaryotes. Moreover, the structure of a eukaryotic gene is more complex than that of a prokaryotic gene. The amino acid sequence of a prokaryotic protein is a direct reflection of the DNA sequence in the chromosome. The protein-coding DNA of a eukaryotic gene, however, is usually divided up such that the complete amino acid sequence of a protein is derived from discontinuous segments of DNA (Figure 1.2). The intervening DNA often contains sequences that are involved with regulating the time and place that the gene is activated.

Eukaryotic chromosomes also are very different from prokaryotic chromosomes. Eukaryotic DNA is wrapped around specific protein complexes called **nucleosomes** that are composed of **histone proteins.** These nucleosomes organize the DNA into compact structures and are important in regulating which genes become expressed in which cells. In bacteria, there are no histones. Moreover, eukaryotic cells undergo mitosis, wherein the nuclear envelope breaks down and the replicated chromosomes are equally divided between the daughter cells (Figure 1.3). In prokaryotes, cell division is not mitotic; no mitotic spindle develops, and there is no nuclear envelope to break down. Rather, the daughter chromosomes remain attached to adjacent points on the cell membrane. These attachment points are separated by the growth of the cell membrane between them, eventually placing the chromosomes into different daughter cells.

Figure 1.2
Summary of steps by which proteins are synthesized from DNA. (A) Prokaryotic (bacterial) gene expression. The coding regions of DNA are co-linear with the protein product. (B) Eukaryotic gene expression The genes are discontinuous, and a nuclear envelope separates the DNA from the cytoplasm.

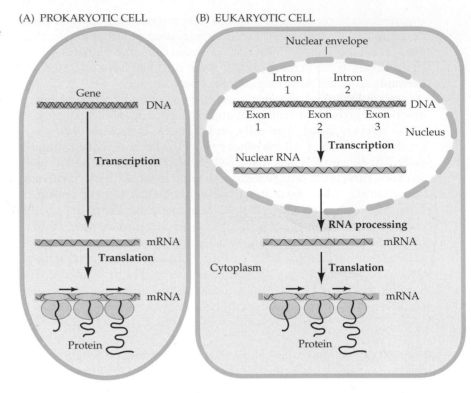

Prokaryotes and eukaryotes have different mechanisms of gene regulation. In both prokaryotes and eukaryotes, DNA is transcribed by enzymes called RNA polymerases to make RNA. When messenger RNA (mRNA) is produced in prokaryotes, it is immediately translated into a protein while the other end of it is still being transcribed from the DNA (Figure 1.4). Thus, in prokaryotes, transcription and translation are simultaneous and coordinated events. But the existence of the nuclear envelope in eukaryotes provides the opportunity for an entirely new type of cell regulation. The ribosomes, which are responsible for translation, are on one side of the nuclear envelope, and the DNA and the RNA polymerases needed for transcription are on the other side. In between transcription and translation, the transcribed RNA must be processed so that it can pass through the nuclear envelope. By regulating which mRNAs can pass into the cytoplasm, the cell is able to select which of the newly synthesized messages will be translated. Thus, a new level of complexity has been added, one that is extremely important for the developing organism.

Development among the unicellular eukaryotes

All multicellular eukaryotic organisms have evolved from unicellular protists. It is in these protists that the basic features of development first appeared. Simple eukaryotes give us our first examples of the nucleus directing morphogenesis, the use of the cell surface to mediate cooperation between individual cells, and the first occurrences of sexual reproduction.

Control of Developmental Morphogenesis in Acetabularia

A century ago, it had not yet been proved that the nucleus contained hereditary or developmental information. Some of the best evidence for this theory came from studies in which unicellular organisms were fragmented into nu-

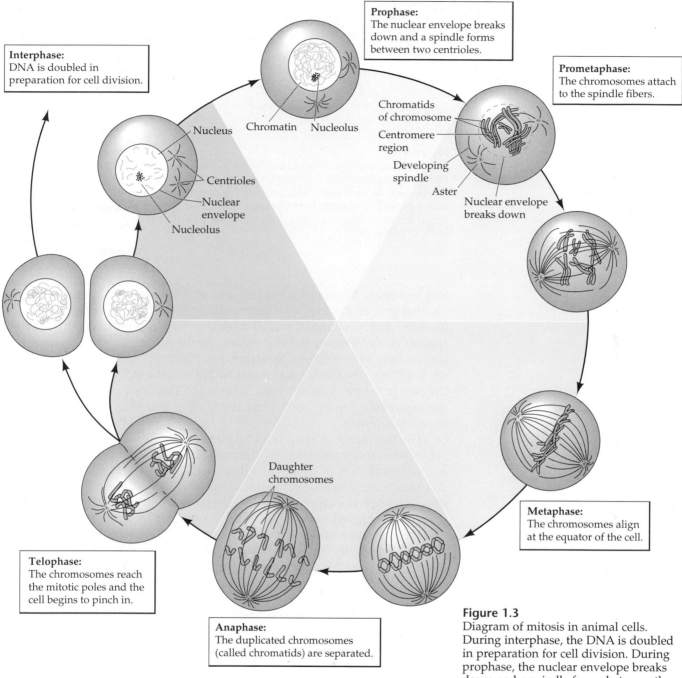

Interphase:
DNA is doubled in preparation for cell division.

Prophase:
The nuclear envelope breaks down and a spindle forms between two centrioles.

Prometaphase:
The chromosomes attach to the spindle fibers.

Nucleus

Chromatin   Nucleolus

Chromatids of chromosome

Centromere region

Developing spindle

Aster

Nuclear envelope breaks down

Centrioles

Nuclear envelope

Nucleolus

Metaphase:
The chromosomes align at the equator of the cell.

Daughter chromosomes

Telophase:
The chromosomes reach the mitotic poles and the cell begins to pinch in.

Anaphase:
The duplicated chromosomes (called chromatids) are separated.

Figure 1.3
Diagram of mitosis in animal cells. During interphase, the DNA is doubled in preparation for cell division. During prophase, the nuclear envelope breaks down and a spindle forms between the two centrioles. At metaphase, the chromosomes align at the equator of the cell, and as anaphase begins, the duplicated chromosomes (each duplicate is called a chromatid) are separated. At telophase, the chromosomes reach the mitotic poles and the cell begins to pinch in. At each pole are the same number and type of chromosomes as were present in the cell before it divided.

cleate and anucleate pieces (reviewed in Wilson, 1896). When various protists were cut into fragments, nearly all the pieces died. However, the fragments containing nuclei were able to live and to regenerate entire complex cellular structures (Figure 1.5).

Nuclear control of cell morphogenesis and the interaction of nucleus and cytoplasm are beautifully demonstrated in studies of *Acetabularia*. This enormous single cell (2–4 cm long) consists of three parts: a cap, a stalk, and a rhizoid (Figure 1.6A). The rhizoid is located at the base of the cell and holds it to the substrate. The single nucleus of the cell resides within the rhizoid. The size of *Acetabularia* and the location of its nucleus allow investiga-

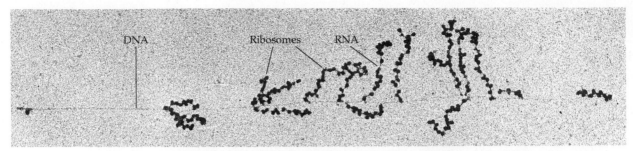

Figure 1.4
Concurrent transcription and translation in prokaryotes. A portion of *Escherichia coli* DNA runs horizontally across this electron micrograph. Messenger RNA transcripts can be seen on both sides. Ribosomes have attached to the mRNA and are synthesizing proteins (which cannot be seen). The size of the mRNA can be seen increasing in length from left to right, indicating the direction of transcription. (Courtesy of O. L. Miller, Jr.)

tors to remove the nucleus from one cell and replace it with a nucleus from another cell. In the 1930s, J. Hämmerling took advantage of these unique features and exchanged nuclei between two morphologically distinct species, *A. mediterranea* and *A. crenulata*. As the photographs show, these two species have very different cap structures. Hämmerling found that when the nucleus from one species was transplanted into the stalk of another species, the newly formed cap eventually assumed the form associated with the donor nucleus (Figure 1.6B). Thus, the nucleus was seen to control *Acetabularia* development.

The formation of a cap is a complex morphogenic event involving the synthesis of numerous proteins, the products of which must be accumulated in a certain portion of the cell and then assembled into complex, species-specific structures. The transplanted nucleus does indeed direct the synthesis of its species-specific cap, but it takes several weeks to do so. Moreover, if the nucleus is removed from an *Acetabularia* cell early in development, before it first forms a cap, a normal cap is formed weeks later, even though the organism will eventually die. These studies suggest that (1) the nucleus contains information specifying the type of cap produced (i.e., it contains the genetic information that specifies the proteins required for the production of a certain type of cap), and (2) material containing this information enters the cytoplasm long before cap production occurs. This information in the cytoplasm is not used for several weeks.

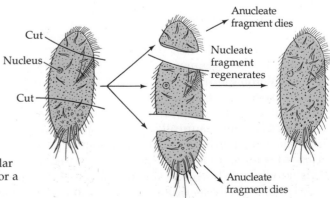

Figure 1.5
Regeneration of the nucleate fragment of the unicellular protist *Stylonychia*. The anucleate fragments survive for a time but finally die.

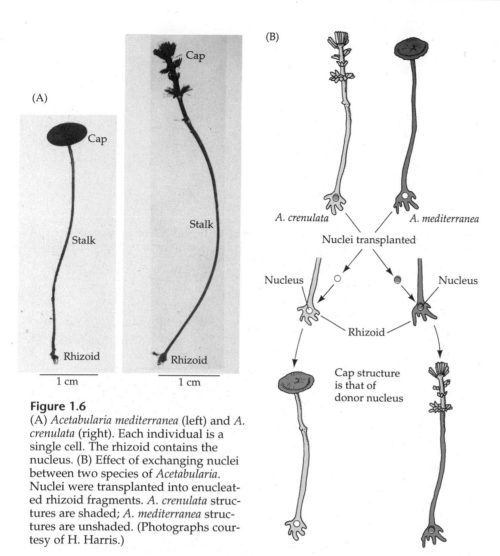

Figure 1.6
(A) *Acetabularia mediterranea* (left) and *A. crenulata* (right). Each individual is a single cell. The rhizoid contains the nucleus. (B) Effect of exchanging nuclei between two species of *Acetabularia*. Nuclei were transplanted into enucleated rhizoid fragments. *A. crenulata* structures are shaded; *A. mediterranea* structures are unshaded. (Photographs courtesy of H. Harris.)

One current hypothesis proposed to explain these observations is that the nucleus synthesizes a stable mRNA that lies dormant in the cytoplasm until the time of cap formation. This hypothesis is supported by an observation that Hämmerling published in 1934. Hämmerling fractionated young *Acetabularia* into several parts (Figure 1.7). The portion with the nucleus eventually formed a new cap as expected; so did the apical tip of the stalk. However, the intermediate portion of the stalk did not form a cap. Thus, Hämmerling postulated (nearly 30 years before the existence of mRNA was known) that the instructions for cap formation originated in the nucleus and were somehow stored in a dormant form near the tip of the stalk. Many years later, Kloppstech and Schweiger (1975) established that nucleus-derived mRNA does accumulate in this region. Ribonuclease, an enzyme that cleaves RNA, completely inhibits cap formation when added to the seawater in which *Acetabularia* is growing. In enucleated cells, this effect is permanent; once the RNA is destroyed, no cap formation can occur. In nucleated cells, however, a new cap can form after the ribonuclease is washed away, presumably because new mRNA is then made by the nucleus. Garcia and Dazy (1986) have also shown that protein synthesis is especially active in the apex of *Acetabularia*.

It is clear from the preceding discussion that nuclear transcription plays an important role in the formation of the *Acetabularia* cap. But note that the

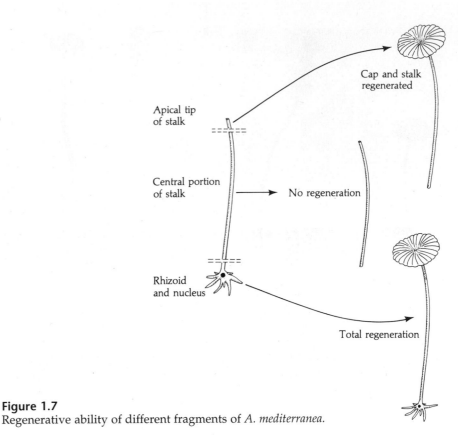

Figure 1.7
Regenerative ability of different fragments of *A. mediterranea.*

cytoplasm also plays an essential role in cap formation. The mRNAs are not translated for weeks, even though they are in the cytoplasm. Something in the cytoplasm controls when the message is utilized. Hence, the expression of the cap is controlled not only by nuclear transcription but also by the **translational control** of the cytoplasmic RNA. In this unicellular organism, "development" is controlled at both the transcriptional and translational levels.

Differentiation in the Amoeboflagellate Naegleria

One of the most remarkable cases of protist "differentiation" is that of *Naegleria gruberi.* This organism occupies a special place in protist taxonomy because it can change its form from that of an amoeba to that of a flagellate (Figure 1.8). During most of its life cycle, *N. gruberi* is a typical amoeba, feeding on soil bacteria and dividing by fission. However, when the bacteria are diluted (either by rainwater or by water added in an experiment), each *N. gruberi* rapidly develops a streamlined body shape and two long anterior flagella, which it uses to find regions of more abundant bacteria. Thus, instead of having several differentiated cell types in one organism, this one cell has different cell structures and biochemistry at different times of its life.

Differentiation into the flagellate form occurs in about an hour (Figure 1.9). During this time, the amoeba has to create centrioles to serve as the basal bodies (microtubule-organizing centers) of the flagella, as well as to create the flagella themselves. Basal bodies and flagella are made from many proteins, the most abundant of which is **tubulin.** The tubulin molecules are organized into microtubules, and the microtubules are further organized into an arrangement that permits flagellar movement. Fulton and Walsh (1980) showed that the tubulin for the *Naegleria* flagella does not exist in the amoeba stage. It is made de novo ("from scratch"), starting with new tran-

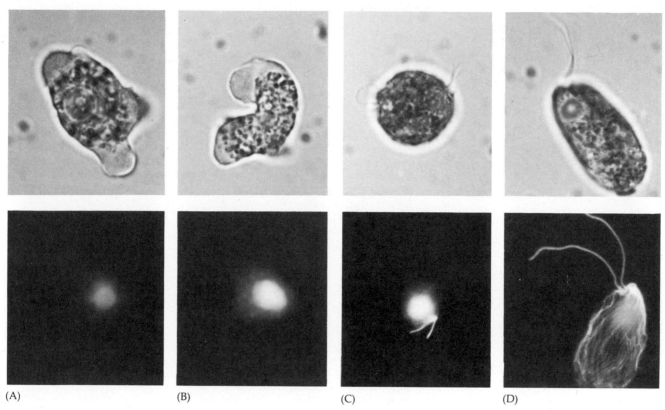

(A)   (B)   (C)   (D)

Figure 1.8
Transformation from amoeboid to flagellated state in *Naegleria gruberi*. Top row stained with Lugol's iodine; bottom row stained with fluorescent antibody to the tubulin protein of the microtubules. Transformation is initiated by rinsing the food (bacteria) from the *Naegleria* colony. (A) 0 minutes. (B) 25 minutes, showing new tubulin synthesis. (C) 70 minutes, showing emergence of visible flagella. (D) 120 minutes, showing mature flagella and streamlined body shape. (From Walsh, 1984, courtesy of C. Walsh.)

scription from the nucleus. To show this, the investigators manipulated transcription at various stages with actinomycin D, an antibiotic drug that selectively inhibits RNA synthesis. When added before the dilution of the food supply, this antibiotic prevents tubulin synthesis. However, if the actinomycin D is added 20 minutes after dilution, tubulin is still made at the normal time (about 30 minutes later). Therefore, it appears that the mRNA for tubulin is made during the first 20 minutes after dilution and is used shortly thereafter. This interpretation was confirmed when it was shown that mRNA extracted from amoebae does not contain any detectable messages for flagellar tubulin, whereas mRNA extracted from differentiating cells contains a great many such messages (Walsh, 1984).

Here, then, is an excellent example of the **transcriptional control** of a development process: the *Naegleria* nucleus responds to environmental changes by synthesizing the mRNA for flagellar tubulin. We also see another process that remains extremely important in the development of all other animals and plants, namely, the assembly of tubulin molecules to produce a flagellum. This arrangement, whereby tubulin is polymerized into microtubules and the microtubules assembled into an ordered array, is seen throughout nature. In mammals, it is evident in the sperm flagellum and in the cilia of the spinal cord and respiratory tract. Moreover, tubulin alone does not make a flagellum. There are around 300 other proteins in each flagellum, and flagellar movement depends on the proper orientation of these proteins with respect to each other. So even cellular processes have their own "morphogenesis" based on molecular interactions between the parts of the proteins. Such **posttranslational control,** whereby a protein is not functional until it is linked with other molecules, will be discussed more fully later. We see, then, that development in unicellular eukaryotes can be controlled at the transcriptional, translational, and posttranslational levels.

Figure 1.9
Differentiation of the flagellate phenotype in *Naegleria*. Amoebae that had been growing in the bacteria-enriched medium are washed free of bacteria at time 0. By 80 minutes, nearly the entire population has developed flagella. (After Fulton, 1977.)

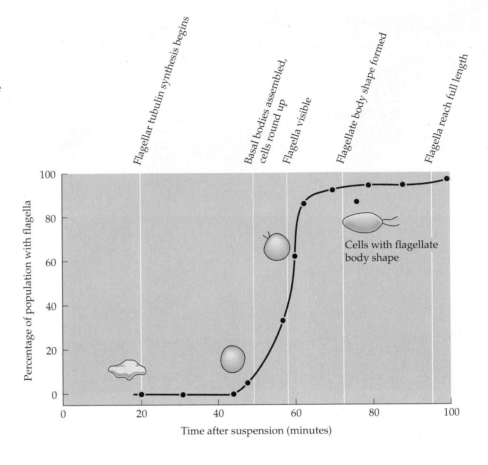

Figure 1.10
Sex in bacteria. Some bacterial cells are covered with numerous appendages (pili) and are capable of transmitting genes to a recipient cell (lacking pili) through a sex pilus. In this figure, the sex pilus is highlighted by viral particles that bind specifically to that structure. (Courtesy of C. C. Brinton, Jr. and J. Carnahan.)

The Origins of Sexual Reproduction

Sexual reproduction is another invention of the protists that has had a profound effect on more complex organisms. It should be noted that sex and reproduction are two distinct and separable processes. *Reproduction* involves the creation of new individuals. *Sex* involves the combining of genes from two different individuals into new arrangements. Reproduction in the absence of sex is characteristic of organisms that reproduce by fission; there is no sorting of genes when an amoeba divides or when a hydra buds off cells to form a new colony. Sex without reproduction is also common among unicellular organisms. Bacteria are able to transmit genes from one individual to another by means of sex pili (Figure 1.10). This transmission is separate from reproduction. Protists are also able to reassort genes without reproduction. Paramecia, for instance, reproduce by fission, but sex is accomplished by **conjugation.** When two paramecia join together, they link their oral apparatuses and form a cytoplasmic connection through which they can exchange genetic material (Figure 1.11). Each macronucleus (which controls the metabolism of the organism) degenerates while each micronucleus undergoes meiosis to produce eight haploid micronuclei, of which all but one degenerate. The remaining micronucleus divides once more to form a stationary micronucleus and a migratory micronucleus. Each migratory micronucleus crosses the cytoplasmic bridge and fuses with ("fertilizes") the stationary micronucleus, thereby creating a new diploid nucleus in each cell. This diploid nucleus then divides mitotically to give rise to a new micronucleus and a new macronucleus as the two partners disengage. Therefore, no reproduction has occurred, only sex.

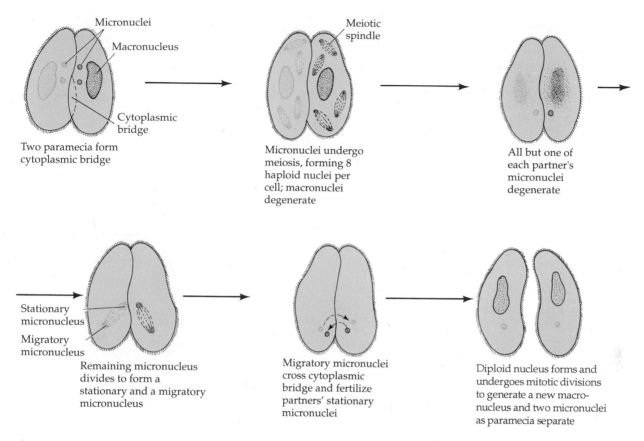

Micronuclei
Macronucleus
Cytoplasmic bridge

Two paramecia form cytoplasmic bridge

Meiotic spindle

Micronuclei undergo meiosis, forming 8 haploid nuclei per cell; macronuclei degenerate

All but one of each partner's micronuclei degenerate

Stationary micronucleus
Migratory micronucleus

Remaining micronucleus divides to form a stationary and a migratory micronucleus

Migratory micronuclei cross cytoplasmic bridge and fertilize partners' stationary micronuclei

Diploid nucleus forms and undergoes mitotic divisions to generate a new macronucleus and two micronuclei as paramecia separate

Figure 1.11
Conjugation across a cytoplasmic bridge in paramecia. Two paramecia can exchange genetic material, leaving each with genes that differ from those with which they started. (After Strickberger, 1985.)

The union of these two distinct processes, sex and reproduction, into **sexual reproduction** is seen in unicellular eukaryotes. Figure 1.12 shows the life cycle of *Chlamydomonas*. This organism is usually haploid, having just one copy of each chromosome (like a mammalian gamete). The individuals of each species, however, are divided into two mating groups: *plus* and *minus*. When these meet, they join their cytoplasms, and their nuclei fuse to form a diploid zygote. This zygote is the only diploid cell in the life cycle, and it eventually undergoes meiosis to form four new *Chlamydomonas* cells. Here is sexual reproduction, for chromosomes are reassorted during the meiotic divisions and more individuals are formed. Note that in this protist type of sexual reproduction, the gametes are morphologically identical; the distinction between sperm and egg has not yet been made.

In evolving sexual reproduction, two important advances had to be achieved. The first is the mechanism of **meiosis** (Figure 1.13), whereby the diploid complement of chromosomes is reduced to the haploid state (discussed in detail in Chapter 22.) The other advance is the mechanism whereby the two different mating types recognize each other. In *Chlamydomonas*, recognition occurs first on the flagellar membranes (Figure 1.14; Bergman et al., 1975; Goodenough and Weiss, 1975). The agglutination of flagella enables specific regions of the cell membranes to come together. These specialized sectors contain mating-type-specific components that enable the cytoplasms to fuse. Following agglutination, the *plus* individuals initiate the

Figure 1.12
Sexual reproduction in *Chlamydomonas*. Two strains, both haploid, can reproduce asexually when separate. Under certain conditions, the two strains can unite to produce a diploid cell that can undergo meiosis to form four new haploid organisms. (After Strickberger, 1985.)

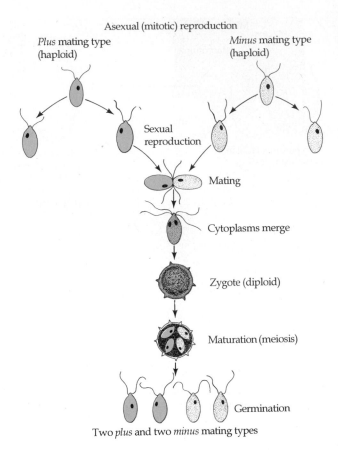

Figure 1.13
Summary of meiosis. The DNA and associated proteins replicate during interphase. During prophase, the nuclear envelope breaks down and homologous chromosomes (each chromosome being double, with the chromatids joined at the centromere) align in pairs. Chromosomal rearrangements can occur between the four homologous chromatids at this time. After the first metaphase, the two original homologous chromosomes are segregated into different cells. During the second division the centromere splits, thereby leaving each new cell with one copy of each chromosome.

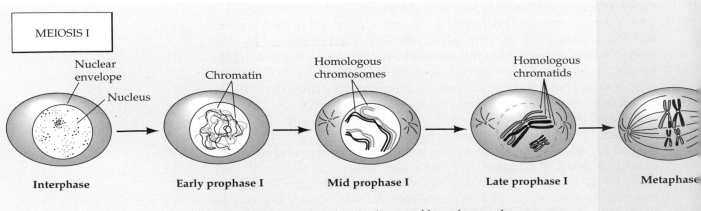

(A)

(B)

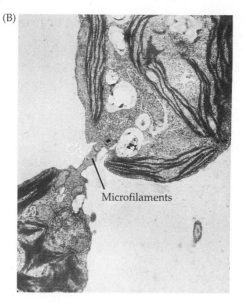

Microfilaments

Figure 1.14
Two-step recognition in mating *Chlamydomonas*. (A) Scanning electron micrograph (7000×) of mating pair. The interacting flagella twist about each other, adhering at the tips (arrows). (B) Transmission electron micrograph (20,000×) of a cytoplasmic bridge connecting the two organisms. The microfilaments extend from the donor (lower) cell to the recipient (upper) cell. (From Goodenough and Weiss, 1975, and Bergman et al., 1975; by permission of U. Goodenough.)

fusion by extending a **fertilization tube.** This tube contacts and fuses with a specific site on the *minus* individual. Interestingly, the mechanism used to extend this tube—the polymerization of the protein **actin**—is also used to extend processes of sea urchin eggs and sperm. In Chapter 4, we will see that the recognition and fusion of sperm and egg occur in a manner amazingly similar to that of these protists.

Unicellular eukaryotes appear to have the basic elements of the developmental processes that characterize the more complex organisms: cellular synthesis is controlled by transcriptional, translational, and posttranslational regulation; there is a mechanism for processing RNA through the nuclear membrane; the structures of individual genes and chromosomes are as they will be throughout eukaryotic evolution; mitosis and meiosis are perfected; and sexual reproduction exists, involving cooperation between individual cells. Such intercellular cooperation becomes even more important with the evolution of multicellular organisms.

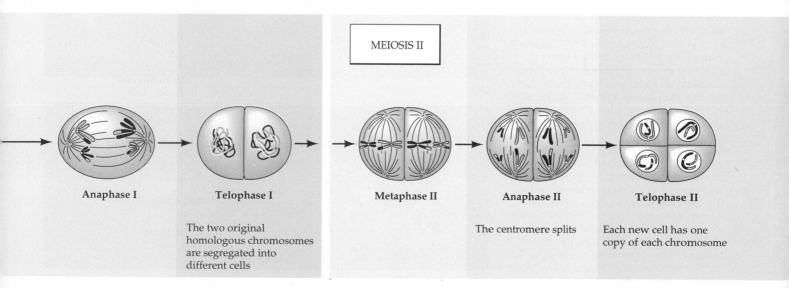

MEIOSIS II

Anaphase I

Telophase I

The two original homologous chromosomes are segregated into different cells

Metaphase II

Anaphase II

The centromere splits

Telophase II

Each new cell has one copy of each chromosome

Colonial eukaryotes: The evolution of differentiation

One of evolution's most important experiments was the creation of multicellular organisms. There appear to be several paths by which single cells evolved multicellular arrangements; we will discuss only two of them here (see Chapter 23 for a fuller discussion). The first path involves the orderly division of the reproductive cell and the subsequent differentiation of its progeny into different cell types. This path to multicellularity can be seen in a remarkable series of multicellular organisms collectively referred to as the family Volvocaceae, or the volvocaceans.

The Volvocaceans

The simpler organisms among the volvocaceans are ordered assemblies of numerous cells, each resembling the unicellular protist *Chlamydomonas*. A single organism of the volvocacean genus *Gonium* (Figure 1.15) for example, consists of a flat plate of 4 to 16 cells, each with its own flagellum. In a related genus, *Pandorina*, the 16 cells form a sphere; and in *Eudorina*, the sphere contains 32 or 64 cells arranged in a regular pattern. In these organisms, then, a very important developmental principle has been worked out: *the ordered division of one cell to generate a number of cells that are organized in a predictable fashion.* As occurs in most animal embryos, the cell divisions by which a single volvocacean cell produces an organism of 4 to 64 cells occur in very rapid sequence and in the absence of cell growth.

The next two genera of the volvocacean series exhibit another important principle of development: *the differentiation of cell types within an individual organism.* The reproductive cells become differentiated from the somatic cells. In all the genera mentioned earler, every cell can, and normally does, produce a complete new organism by mitosis (Figure 1.16A,B). In the genera *Pleodorina* and *Volvox*, however, relatively few cells can reproduce. In *Pleodorina californica*, the cells in the anterior region are restricted to a somatic func-

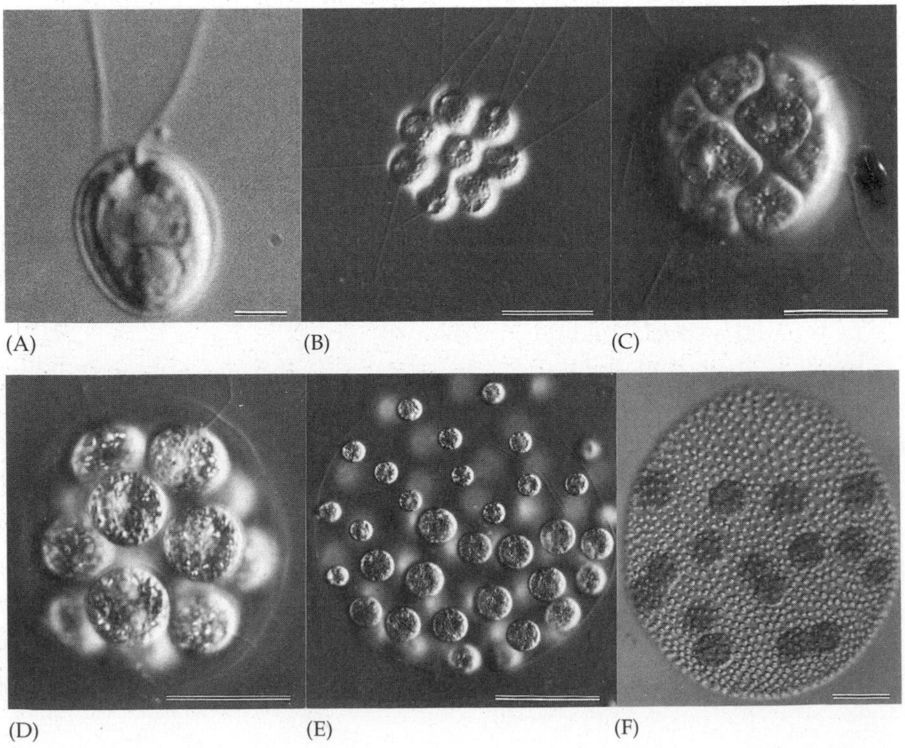

Figure 1.15
Representatives of the order Volvocales. (A) The unicellular protist *Chlamydomonas reinhardtii*. (B) *Gonium pectorale*, with eight *Chlamydomonas*-like cells in a convex disc. (C) *Pandorina morum*. (D) *Eudorina elegans*. (E) *Pleodorina californica*. Here, all 64 cells are originally similar, but the posterior ones dedifferentiate and redifferentiate as asexual, reproductive cells called gonidia, while the anterior cells remain small and biflagellate, like *Chlamydomonas*. (F) *Volvox carteri*. Here, cells destined to become gonidia are set aside early in development and never have somatic characteristics. The smaller, somatic, cells resemble *Chlamydomonas*. All but *Chlamydomonas* are members of the family Volvocaceae. Complexity increases from the single-celled *Chlamydomonas* to the multicellular *Volvox*. Bar in A is 5 μm; B–D, 25 μm; E, F, 50 μm. (Courtesy of D. Kirk.)

(A)   (B)   (C)

(D)   (E)   (F)

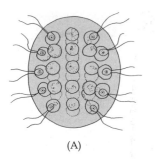
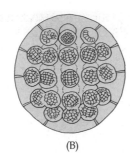
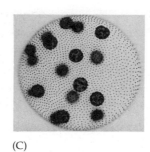

(A)   (B)   (C)

Figure 1.16
Asexual reproduction in volvocaceans.
(A) Mature colony of *Eudorina elegans*.
(B) Each of the *E. elegans* cells divides
and produces a new colony. (C) Mature
Volvox carteri. Most of the cells are incapable of reproduction. Germ cells (gonidia) have begun dividing into new
organisms. (A and B after Hartmann,
1921; C from Kirk et al., 1982, courtesy
of D. Kirk.)

tion; only those cells on the posterior side can reproduce. In *P. californica*, a
colony usually has 128 or 64 cells, and the ratio of the number of somatic
cells to the number of reproductive cells is usually 3:5. Thus, a 128-cell
colony typically has 48 somatic cells, and a 64-cell colony has 24.

In *Volvox*, almost all the cells are somatic, and very few of the cells are
able to produce new individuals. In some species of *Volvox*, reproductive
cells, as in *Pleodorina*, are derived from cells that originally look and function
like somatic cells before they enlarge and divide to form new progeny. However, in other members of the genus, such as *V. carteri*, there is a complete division of labor: the reproductive cells that will create the next generation are
set aside during the division of the reproductive cells that are forming a new
individual. The reproductive cells never develop functional flagella and
never contribute to motility or other somatic functions of the individual; they
are entirely specialized for reproduction. Thus, although the simpler volvocaceans may be thought of as colonial organisms (because each cell is capable
of independent existence and of perpetuating the species), in *V. carteri* we
have a truly multicellular organism with two distinct and interdependent cell
types (somatic and reproductive), both of which are required for perpetua-
tion of the species (Figure 1.16C). Although not all animals set aside the reproductive cells from the somatic cells (and plants hardly ever do), this separation of germ cells from somatic cells early in development is characteristic
of many animal phyla and will be discussed in more detail in Chapter 13.

Although all the volvocaceans, including their unicellular relative
Chlamydomonas, reproduce predominantly by asexual means, they are also
capable of sexual reproduction. This involves the production and fusion of
haploid gametes. In many species of *Chlamydomonas*, including the one illustrated in Figure 1.12, sexual reproduction is **isogamous,** since the haploid
gametes that meet are similar in size, structure, and motility. However, in
other species of *Chlamydomonas*—as well as many species of colonial volvocaceans—swimming gametes of very different sizes are produced by the different mating types. This is called **heterogamy.** But the larger volvocaceans
have evolved a specialized form of heterogamy, called **oogamy,** which involves the production of large, relatively immotile eggs by one mating type
and small, motile sperm by the other (see Sidelights & Speculations). Here
we see one gamete specialized for the retention of nutritional and developmental resources and the other gamete specialized for the transport of nuclei. Thus, the volvocaceans include the simplest organisms that have distinguishable male and female members of the species and that have distinct
developmental pathways for the production of eggs or sperm. In all the
volvocaceans, the fertilization reaction resembles that of *Chlamydomonas* in
that it results in the production of a dormant diploid zygote that is capable
of surviving harsh environmental conditions. When conditions allow the zygotes to germinate, they first undergo meiosis to produce haploid offspring
of the two different mating types in equal numbers. [other.html#intro1]

Sex and Individuality in Volvox

SIMPLE AS IT IS, *Volvox* shares many features that characterize the life cycles and developmental histories of much more complex organisms, including ourselves. As already mentioned, *Volvox* is among the simplest organisms to exhibit a division of labor between two completely different cell types. As a consequence of this, it is among the simplest organisms to include death as a regular, genetically programmed part of its life history.

Death and Differentiation

Unicellular organisms that reproduce by simple cell division, such as amoebae, are potentially immortal. The amoeba you see today under the microscope has no dead ancestors! When an amoeba divides, neither of the two resulting cells can be considered either ancestor or off-spring; they are siblings. Death comes to an amoeba only if it is eaten or meets with a fatal accident; and when it does, the dead cell leaves no offspring.

Death becomes an essential part of life, however, for any multicellular organism that establishes a division of labor between somatic (body) cells and germ (reproductive) cells. Consider the life history of *Volvox carteri* when it is reproducing asexually (Figure 1.17). Each asexual adult is a spheroid containing some 2000 small, biflagellated somatic cells along its periphery and about 16 large, asexual reproductive cells, called **gonidia,** toward one end of the interior. When mature, each gonidium divides rapidly 11 or 12 times. Certain of these divisions are asymmetrical and produce the 16 large cells that will become a new set of gonidia. At the end of cleavage, all the cells that will be present in an adult have been produced from each gonidium. But the embryo is "inside out": its gonidia are on the outside and the flagella of its somatic cells are pointing toward the interior of the hollow sphere of cells. This predicament is corrected by a process called inversion, in which the embryo turns itself right side out by a set of cell movements that resemble the gastrulation movements of animal embryos (Figure 1.18). Clusters of bottle-shaped cells open a hole at one end of the embryo by producing tension on the interconnected cell sheet (Figure 1.19).

Figure 1.17   Asexual reproduction in *V. carteri*. When reproductive cells (gonidia) are mature, they enter a cleavage-like stage of embryonic development to produce juveniles within the adult. Through a series of cell movements resembling gastrulation, the embryonic volvox inverts and is eventually released from the parent. The somatic cells of the parent, lacking the gonidia, undergo senescence and die, while the juvenile colonies mature. The entire sexual cycle takes two days. (After Kirk, 1988.)

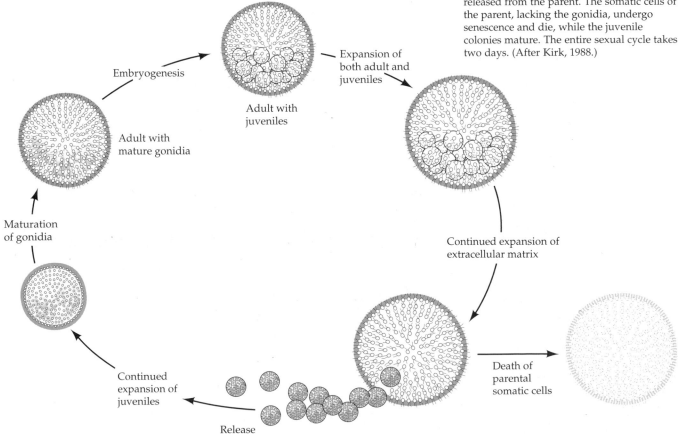

Embryogenesis

Expansion of both adult and juveniles

Adult with juveniles

Adult with mature gonidia

Maturation of gonidia

Continued expansion of extracellular matrix

Continued expansion of juveniles

Release of juveniles

Death of parental somatic cells

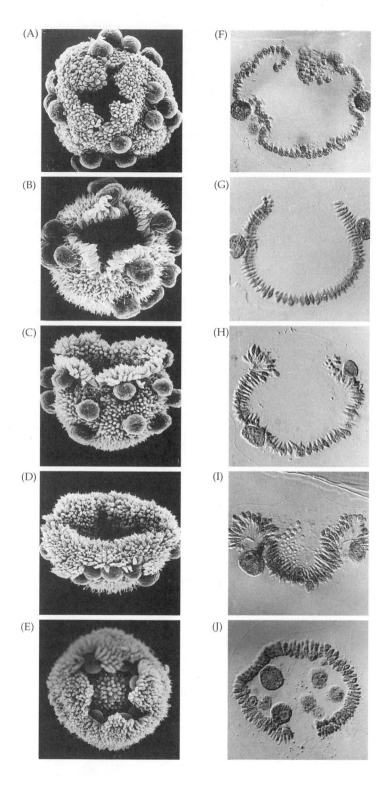

(A)

(B)

(C)

(D)

(E)

(F)

(G)

(H)

(I)

(J)

Figure 1.18   Inversion of asexually produced embryos of *V. carteri.* A–E are scanning electron micrographs of whole embryos. F–J are sagittal sections through the center of the embryo, visualized by differential interference microscopy. Before inversion, the embryo is a hollow sphere of connected cells. When cells change their shape, a hole (the phialopore) opens at the apex of the embryo (A, B, F, G). Cells then curl around and rejoin at the bottom (C–E, H–J). (From Kirk et al., 1982, courtesy of D. Kirk.)

their own offspring, all the nutrients that they had stored during life. "Thus emerges," notes David Kirk, "one of the great themes of life on planet Earth: 'Some die that others may live.'"

In *V. carteri,* a specific gene* that plays a central role in regulating cell death has been identified (Kirk, 1988). In laboratory strains possessing mutations of this gene, somatic cells abandon their suicidal ways, gain the ability to reproduce asexually, and become potentially immortal (Figure 1.20). The fact that such mutants have never been found in

---

*This gene (*regA*) has now been cloned and found to encode a protein that acts to repress (directly or indirectly) all the genes whose products are required for cells to develop as gonidia. Loss-of-function mutations would prevent the protein from acting, and the cells would be able to become gonidia (D. Kirk, personal communication).

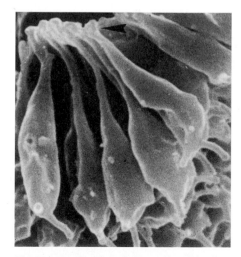

Figure 1.19   "Bottle cells" near the opening of the phialopore. These cells remain tightly interconnected through cytoplasmic bridges near their elongated apices, thereby creating the tension that causes the curvature of the interconnected cell sheet. (From Kirk et al., 1982, courtesy of D. Kirk.)

The embryo everts through this hole and then closes it up. About a day after this is done, the juvenile colonies are enzymatically released from the parent and swim away.

What happens to the somatic cells of the "parent" *Volvox* now that its young have "left home"? Having produced offspring and being incapable of further reproduction, these somatic cells die. Actually, they commit suicide, synthesizing a set of proteins that cause the death and dissolution of the cells that make these proteins (Pommerville and Kochert, 1982). Moreover, in this death, the cells release for the use of others, including

(A)

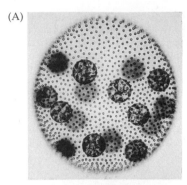

(B)

Figure 1.20 Mutation of a single gene (called *somatic regenerator A*) abolishes programmed cell death in *V. carteri*. The newly hatched volvox carrying this mutation (A) is indistinguishable from the wild-type spheroid. However, shortly before the time when the somatic cells of the wild-type spheroids begin to die, the somatic cells of this mutant redifferentiate as gonidia (B). Eventually, every cell of the mutant will divide to form (regenerate) a new spheroid that will repeat this potentially immortal developmental cycle.

nature indicates that cell death most likely plays an important role in the survival of *V. carteri* under natural conditions. [intro2.html]

Enter Sex

Although *V. carteri* reproduces asexually much of the time, in nature it reproduces sexually once each year. When it does, one generation of individuals passes away, and a new and genetically different generation is produced. The naturalist Joseph Wood Krutch (1956) put it more poetically:

> *The amoeba and the paramecium are potentially immortal…But for* Volvox, *death seems to be as inevitable as it is in a mouse or in a man.* Volvox *must die as Leeuwenhoek saw it die because it had children and is no longer needed. When its time comes it drops quietly to the bottom and joins its ancestors. As Hegner, the Johns Hopkins zoologist, once wrote, "This is the first advent of inevitable natural death in the animal kingdom and all for the sake of sex." And he asked: "Is it worth it?"*

For *Volvox carteri*, it most assuredly is worth it. *V. carteri* lives in shallow temporary ponds that fill with spring rains but dry out in the heat of late summer. During most of that time, *V. carteri* swims about, reproducing asexually. These asexual volvoxes would die in minutes once the pond dried out, but *V. carteri* is able to survive by turning sexual shortly before the pond dries up, producing dormant zygotes that survive the heat and drought of late summer and the cold of winter. When rain fills the ponds in spring, the zygotes break their dormancy and hatch out a new generation of individuals to reproduce asexually until the pond is about to dry up once more. How do these simple organisms predict the coming of adverse conditions with sufficient accuracy to produce a sexual generation just in time, year after year?

The stimulus for switching from the asexual to the sexual mode of reproduction in *V. carteri* is known to be a 30-kDa sexual inducer protein. This protein is so powerful that concentrations as low as 6×10^{-17} cause gonidia to undergo a modified pattern of embryonic development that results in the production of eggs or sperm, depending on the genetic sex of the individual (Sumper et al., 1993). The sperm are released and swim to a female,

Figure 1.21 Sexual reproduction in *V. carteri*. Males and females are indistinguishable in their asexual phase. When the sexual inducer protein is present, the gonidia of both mating types undergo a modified embryogenesis that leads to the formation of eggs in the females and sperm in the males. When the gametes are mature, sperm packets (containing 64 or 128 sperm each) are released and swim to the females. Upon reaching the female, the sperm packet breaks up into individual sperm, which can fertilize the eggs. The resulting zygote has tough cell walls that can resist drying, heat, and cold. When spring rains cause the zygote to germinate, it undergoes meiosis to produce haploid males and females that reproduce asexually until heat induces the sexual cycle again.

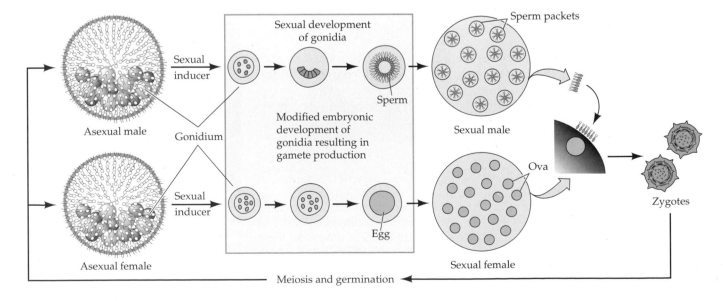

where they fertilize eggs to produce the dormant zygotes (Figure 1.21).

What is the source of this sexual inducer protein? Kirk and Kirk (1986) discovered that the sexual cycle could be initiated by heating dishes of *V. carteri* to temperatures that might be expected in a shallow pond in late summer. When this was done, the somatic cells of the asexual volvoxes produced the sexual inducer protein. Since the amount of sexual inducer protein secreted by one individual is sufficient to initiate sexual development in over 500 million asexual volvoxes, a single inducing volvox can convert the entire pond to sexuality. This discovery explained an observation made nearly 90 years ago that "in the full blaze of Nebraska sunlight, *Volvox* is able to appear, multiply, and riot in sexual reproduction in pools of rainwater of scarcely a fortnight's duration" (Powers, 1908). Thus, in temporary ponds formed by spring rains and dried up by summer's heat, *Volvox* has found a means of survival: it uses the heat to induce the formation of sexual individuals whose mating produces zygotes capable of surviving conditions that kill the adult organism. We see, too, that development is critically linked to the ecosystem in which the organism has adapted to survive. ■

Differentiation and Morphogenesis in Dictyostelium

THE LIFE CYCLE OF *DICTYOSTELIUM*. Another type of multicellular organization derived from unicellular organisms is found in *Dictyostelium discoideum*.* The life cycle of this fascinating organism is illustrated in Figure 1.22. In its vegetative cycle, solitary haploid amoebae (called **myxamoebae** or "social amoebae" to distinguish them from amoeba species that always remain solitary) live on decaying logs, eating bacteria and reproducing by binary fission. When they have exhausted their food supply, tens of thousands of these amoebae join together to form moving streams of cells that converge at a central point. Here they pile atop one another to produce a conical mound called the tight aggregate. Subsequently, a tip arises at the top of this mound, and the mound bends over to produce the migrating slug (with the tip at the front). The slug (often given the more dignified title of pseudoplasmodium or **grex**) is usually 2–4 mm long and is encased in a slimy sheath. The grex begins to migrate (if the environment is dark and moist) with its anterior tip slightly raised. When the grex reaches an illuminated area, migration ceases, and the grex differentiates into a **fruiting body** composed of spore cells and a stalk. The *anterior* cells, representing 15–20 percent of the entire cellular population, form the tubed **stalk.** The stalk begins as some of the central anterior cells, the **prestalk cells,** begin secreting an extracellular coat and extending a tube through the grex. As the prestalk cells differentiate, they form vacuoles and enlarge, lifting up the mass of **prespore cells** that had been in the posterior four-fifths of the grex (Jermyn and Williams, 1991). The stalk cells die, but the posterior cells, elevated above the stalk, become **spore cells.** These spore cells disperse, each one becoming a new myxamoeba.

In addition to this asexual cycle, there is a possibility for sex in *Dictyostelium.* Two amoebae can fuse to create a giant cell, which digests all the other cells of the aggregate. When it has eaten all its neighbors, it encysts itself in a thick wall and undergoes meiotic and mitotic divisions; eventually, new myxamoebae are liberated.

Dictyostelium has been a wonderful experimental organism for developmental biologists, because initially identical cells are differentiated into one of two alternative cell types, spore and stalk. It is also an organism wherein individual cells come together to form a cohesive structure composed of differentiated cell types, akin to tissue formation in more complex organisms. The aggregation of thousands of amoebae into a single organism is an in-

*Though colloquially called a "cellular slime mold," *Dictyostelium* is not a mold (such as *Neurospora*), nor is it consistently slimy. It is perhaps best to think of it as a social amoeba.

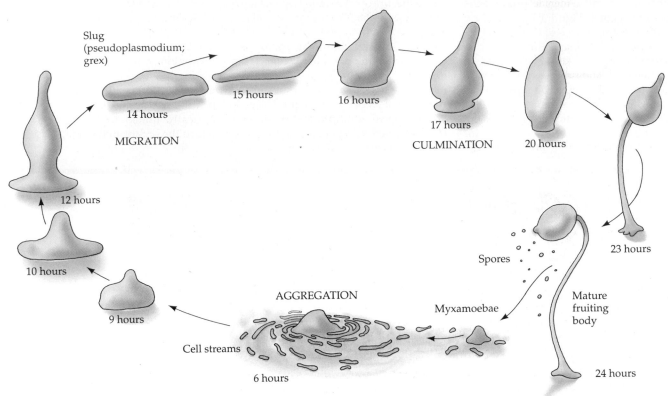

Slug (pseudoplasmodium; grex)

14 hours

15 hours

16 hours

17 hours

20 hours

MIGRATION

CULMINATION

12 hours

10 hours

9 hours

23 hours

AGGREGATION

Spores

Myxamoebae

Mature fruiting body

Cell streams

6 hours

24 hours

Figure 1.22
Life history of *Dictyostelium discoideum*. Haploid spores give rise to myxamoebae, which can reproduce asexually to form more haploid myxamoebae. As the food supply diminishes, aggregation occurs at central points, and a migrating pseudoplasmodium is formed. Eventually, it stops moving and forms a fruiting body that releases more spores. The numbers refer to hours since nutrient dilution began the developmental sequence.

credible feat of organization and invites experimentation to answer questions about the mechanisms involved.

AGGREGATION OF *DICTYOSTELIUM* CELLS. The first question is, What causes the amoebae to aggregate? Time-lapse microcinematography has shown that no directed movement occurs during the first 4–5 hours following nutrient starvation. During the next 5 hours, however, the cells are seen moving at about 20 μm/min for 100 seconds. This movement ceases for about 4 minutes, then resumes. Although the movement is directed toward a central point, it is not a simple radial movement. Rather, cells join with each other to form streams; the streams converge into larger streams, and eventually all streams merge at the center. Bonner (1947) and Shaffer (1953) showed that this movement is due to **chemotaxis:** the cells are guided to aggregation centers by a soluble substance. This substance was later identified as **cyclic adenosine 3′,5′-monophosphate (cAMP)** (Konijn et al., 1967; Bonner et al., 1969), the chemical structure of which is shown in Figure 1.23A.

Aggregation is initiated as each of the cells begins to synthesize cAMP. There are no "dominant" cells that begin the secretion or control the others. Rather, the sites of aggregation are determined by the distribution of amoebae (Keller and Segal, 1970; Tyson and Murray, 1989). Neighboring cells respond to cAMP in two ways: they initiate a movement toward the cAMP pulse, and they release cAMP of their own (Robertson et al., 1972; Shaffer,

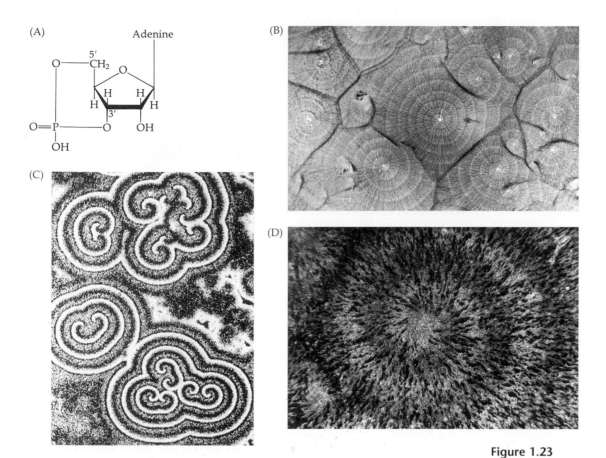

(A) Adenine

(C)

(B)

(D)

Figure 1.23
Chemotaxis of *Dictyostelium* amoebae due to spiral waves of cAMP. (A) Chemical structure of cAMP. (B) Visualization of several cAMP "waves" in the medium. Central cells secrete cAMP at regular intervals, and each secretion diffuses outward as a concentric wave. Waves are charted by saturating filter paper with radioactive cAMP and placing it on an aggregating colony. The cAMP from the secreting cells dilutes the radioactive cAMP. When the radioactivity on the paper is recorded (by placing it over X-ray film), the regions of high cAMP concentration in the culture appear lighter than those of low cAMP concentration. (C, D) Spiral waves of amoebae moving toward the initial source of cAMP. (C) This digitally processed dark-field photomicrograph shows about 10^7 cells. Because moving and nonmoving cells scatter light differently, the photograph reflects cell movement. The bright bands are composed of elongated migrating cells; the dark bands are cells that have stopped moving and have rounded up. (D) As cells form streams, the spiral of movement can still be seen moving toward the center. (B from Tomchick and Devreotes, 1981, courtesy of P. Devreotes; C and D from Siegert and Weijer, 1989, courtesy of F. Siegert.)

1975). After this, the cell is unresponsive to further cAMP pulses for several minutes. The result is a rotating spiral wave of cAMP that is propagated throughout the population of cells (Figure 1.23B–D). As each wave arrives, the cells take another step toward the center.*

The differentiation of individual amoebae into either stalk (somatic) or spore (reproductive) cells is a complex matter. Raper (1940) and Bonner (1957) have demonstrated that the anterior cells normally become stalk, while the remaining, posterior, cells are usually destined to form spores. However, surgically removing the anterior part of the slug does not abolish the ability of the grex to form a stalk. Rather, the cells that now find themselves at the anterior end following the surgery (and which originally had been destined to produce spores) now form the stalk (Raper, 1940). Somehow a decision is made so that whichever cells are anterior become stalk cells and whichever are posterior become spores. This ability of cells to change their developmental fates according to their location within the whole organism and thereby

---

*The biochemistry of this reaction involves a receptor that binds cAMP. When this binding occurs, specific gene transcription takes place, motility toward the source of the cAMP is initiated, and adenyl cyclase enzymes (which synthesize cAMP from ATP) are activated. The newly formed cAMP activates its own receptors as well as those of its neighbors. The cells in the area remain insensitive to new waves of cAMP until the bound cAMP is removed from the receptors by another cell-surface enzyme, phosphodiesterase (Johnson et al., 1989). The mathematics of such oscillation reactions predict that the diffusion of cAMP would initially be circular. However, as cAMP interacts with the cells that receive and propagate the signal, the cells that receive the front part of the wave begin to migrate at a different rate than the cells behind them. The result is a rotating spiral of cAMP and migration as seen in Figure 1.23. Interestingly, the same mathematical formulas predict the behavior of certain chemical reactions and the formation of new stars in rotating spiral galaxies (Tyson and Murray, 1989).

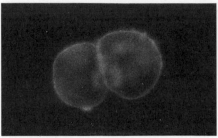

Figure 1.24
Dictyostelium cells synthesize an adhesive, 24-kDa glycoprotein shortly after nutrient starvation. *Dictyostelium* cells were stained with a fluorescent antibody that binds to the 24-kDa glycoprotein and were then observed under ultraviolet light. This protein is not seen on amoebae that have just stopped dividing. However, as shown here—10 hours after cell division has ceased—individual amoebae are seen to have this protein in their cell membranes and are capable of adhering together. (Courtesy of W. Loomis.)

compensate for missing parts is called **regulation.** We will see this phenomenon in many embryos, including those of mammals.

CELL ADHESION MOLECULES IN *DICTYOSTELIUM.*   How do these individual cells stick together to form a cohesive organism? This is the same problem that embryonic cells face, and the solution that evolved in the protists is the same one used by embryos: developmentally regulated **cell adhesion molecules.**

While growing mitotically on bacteria, *Dictyostelium* cells do not adhere to one another. However, once cell division stops, the cells become increasingly adhesive, reaching a plateau of maximum cohesiveness around 8 hours after starvation. The initial cell-cell adhesion is mediated by a 24,000-Da (24-kDa) glycoprotein that is absent in growing cells but is seen shortly thereafter (Figure 1.24; Knecht et al, 1987; Loomis, 1988). This protein is synthesized from newly transcribed mRNA and becomes localized in the cell membranes of the myxamoebae. If these cells are treated with antibodies that bind to and mask this protein, the cells will not stick to each other and all subsequent development ceases.

Once this initial aggregation has occurred, it is stabilized by a second cell adhesion molecule. This 80-kDa glycoprotein is also synthesized during the aggregation phase. If it is defective or absent in the cells, small slugs will form, and their fruiting bodies will be only about one-third the normal size. Thus, the second cell adhesion system seems to be needed for retaining a large enough number of cells to form large fruiting bodies (Müller and Gerisch, 1978; Loomis, 1988). In addition, a third cell adhesion system is activated late in development, while the slug is migrating. The protein or group of proteins that mediates the third system may exist only on prespore cells and may be responsible for separating the prespore cells from the prestalk cells (Loomis, personal communication). Thus, *Dictyostelium* has evolved three developmentally regulated systems of cell-cell adhesion that are necessary for the morphogenesis of individual cells into a coherent organism. As we will see in subsequent chapters, metazoan cells also use cell adhesion molecules to form the tissues and organs of the embryo.

Dictyostelium is a "part-time multicellular organism" that does not form many cell types (Kay et al., 1989), and the more complex multicellular organisms do not form by the aggregation of formerly independent cells. Nevertheless, many of the principles of development demonstrated by this "simple" organism also appear in embryos of more complex phyla. The ability of individual cells to sense a chemical gradient (as in the amoeba's response to cAMP) is very important for cell migration and morphogenesis during animal development. Moreover, the role of cell-surface proteins for cell cohesiveness is seen throughout the animal kingdom, and differentiation-inducing molecules are just beginning to be isolated in metazoan organisms.

Sidelights & Speculations

Evidence and Antibodies

BIOLOGY, like any other science, does not deal with Facts; rather, it deals with evidence. Several types of evidence will be presented in this book, and they are not equivalent in strength. As an example, we will use the analysis of cell adhesion in *Dictyostelium*. The first, and weakest, type of evidence is *correlative evidence*. Here, correlations are made between two or more events, and there is an inference that one event causes the other. As we have seen, fluorescently labeled antibodies to a certain 24-kDa glycoprotein do not label dividing vegetative cells, but they *do* find this protein in myxamoeba cell membranes soon after the cells stop dividing and become competent to aggregate (see Figure 1.24). Thus, there is a correlation between the presence of this cell membrane glycoprotein and the ability to aggregate.

Correlative evidence gives a starting point to investigations, but one cannot say with certainty that one event causes the other based solely on correlations. Although one might infer that the synthesis of this protein caused the adhesion of the cells, it is also possible that cell adhesion caused the cells to synthesize this new glycoprotein or that cell adhesion and the synthesis of the 24-kDa glycoprotein are separate events initiated by the same underlying cause. The simultaneous occurrence of the two events could even be coincidental, the events having no relationship to each other.*

How, then, does one get beyond mere correlation? In the study of cell adhesion in *Dictyostelium*, the next step was to use those same antibodies to block the adhesion of myxamoebae. Using a technique pioneered by Gerisch's laboratory (Beug et al., 1970), Knecht and co-workers (1987) took the antibodies that bound this 24-kDa glycoprotein and isolated their antigen-binding sites (the portions of the antibody molecule that actually recognize the antigen). This was necessary because the whole antibody molecule contains two antigen-binding sites and would therefore artificially crosslink and agglutinate the myxamoebae. When these antigen-binding fragments (called Fab Fragments) were added to the aggregation-competent cells, the cells could not aggregate. The antibody fragments inhibited the cells' adhering together, presumably by binding to the 24-kDa glycoprotein and blocking its function. This type of evidence is called *loss-of-function evidence*. While stronger than correlative evidence, it still does not make other inferences impossible. For instance, perhaps the antibodies killed the cell (as might have been the case if the 24-kDa glycoprotein were a critical transport channel). This would also stop the cells from adhering. Or perhaps the 24-kDa glycoprotein has nothing to do with adhesion itself but is necessary for the *real* adhesive molecule to function (such as by stabilizing membrane proteins in

general). In this case, blocking the glycoprotein would similarly cause the inhibition of cell aggregation. Thus, loss-of-function evidence must be bolstered by many controls demonstrating that the agents causing the loss of function specifically knock out the particular function and nothing else.

The strongest type of evidence is *gain-of-function evidence*. Here, the initiation of the first event causes the second event to happen even in instances where neither event usually occurs. Recently, da Silva and Klein (1990) and Faix and co-workers (1990) have obtained such evidence to show that the 80-kDa glycoprotein is an adhesive molecule. They isolated the gene for the 80-kDa protein and modified the gene in a way that would cause it to be expressed all the time. They then placed it back into well-fed, vegetatively growing myxamoebae that do not usually express this protein and that are not usually able to adhere to each other. The presence of this protein on the cell membrane of these dividing cells was confirmed by antibody labeling. Moreover, such cells now adhered to one another even in the vegetative stages, when they normally do not. Thus, they had gained an adhesive function solely upon expressing this particular glycoprotein on their cell surfaces. This gain-of-function evidence is more convincing than other types of analysis. Similar experiments have recently been performed on mammalian cells (see Chapter 3) to demonstrate the presence of particular cell adhesion molecules in the developing embryo. ∎

---

*In a tongue-in-cheek letter spoofing such correlative inferences, Sies (1988) demonstrated a remarkably good correlation between the number of storks seen in West Germany from 1965 to 1980 and the number of babies born during those same years.

DIFFERENTIATION IN *DICTYOSTELIUM*. Differentiation into stalk cell or spore cell reflects one of the major phenomena of embryogenesis: the cell's selection of a developmental pathway. Cells often select a particular developmental fate when alternatives are available. A particular cell in a vertebrate embryo, for instance, can become either an epidermal skin cell or a neuron. In *Dictyostelium*, we see a simple dichotomous decision, because only two cell types are possible. How is it that a given cell becomes a stalk cell or a spore cell? Although the details are not fully known, a cell's fate appears to be regulated by certain diffusible molecules. The two major candidates are **differentiation-inducing factor (DIF)** and cAMP. DIF appears to be necessary for stalk cell differentiation. This factor, like the sex-inducing factor of *Volvox*, is effective at very low concentrations ($10^{-10}M$); and, like the *Volvox* protein, it appears to induce the differentiation of a particular type of cell. When added to isolated amoebae or even to prespore (posterior) cells, it causes them to form stalk cells. The synthesis of this low-molecular-weight lipid is genetically regulated, for there are mutant strains of *Dictyostelium* that form only spore precursors and no stalk cells. When DIF is added to these mutant cultures, stalk cells are able to differentiate (Kay and Jermyn, 1983; Morris et al., 1987), and new prestalk-specific mRNAs are seen in the cell cytoplasm (Williams et al., 1987). While the mechanisms by which DIF induces 20 percent of the grex cells to become stalk tissue are still controversial (see Early et al., 1995), DIF may act by releasing calcium ions from intracellular compartments within the cell (Shaulsky and Loomis, 1995). [other.html#intro3]

Although DIF stimulates amoebae to become prestalk cells, the differentiation of *prespore* cells is most likely controlled by the continuing pulses of cAMP. High concentrations of cAMP initiate the expression of prespore-specific mRNAs in aggregated amoebae. Moreover, when slugs are placed in a medium containing an enzyme that destroys extracellular cAMP, the prespore cells lose their differentiated characteristics (Figure 1.25; Schaap and van Driel, 1985; Wang et al., 1988a,b).

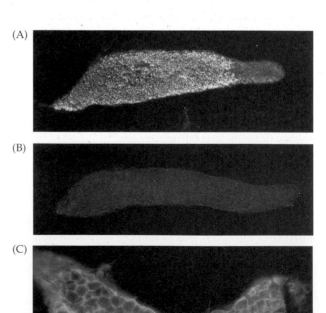

(A)

(B)

(C)

Figure 1.25
Chemicals controlling differentiation in *Dictyostelium*. (A) and (B) show the effects of placing *Dictyostelium* slugs into a medium containing enzymes that destroy extracellular cAMP. (A) Control grex stained for the presence of a prespore-specific protein (white regions). (B) Similar grex stained after the treatment with cAMP-degrading enzymes. No prespore-specific product is seen. (C) Higher magnification of a slug treated with DIF (in the absence of ammonia). The stain used here binds to the cellulose wall of the stalk cells. All cells of the grex have become stalk cells. (A and B from Wang et al., 1988a; C from Wang and Schaap, 1989; courtesy of the authors.)

How the Grex Knows Which End Is Up

IF ALL THE AMOEBAE of the grex start out equal, how can cells in the posterior four-fifths of the slug differentiate into spore cells while equivalent cells in the anterior fifth become stalk cells? The answer may lie in the observation that all the original cells are *not* equal. Amoebae that become starved during the early portion of their cell cycle tend to move to the anterior portion of the slug, while amoebae starved toward the end of their cell cycle tend to remain in the posterior (McDonald and Durston, 1984; Weijer et al., 1984). This work has been confirmed and extended by Ohmori and Maeda (1987), who showed that cells starved in the latter part of the cell cycle respond differently to cAMP and show much higher adhesivity than cells starved just after mitosis. Williams and co-workers (1989) have found that prespore and prestalk cells can be distinguished in early aggregates and that they are randomly distributed throughout these hemispherical mounds. Thus, the *tendencies* toward certain fates have been established even before the grex starts migrating. Within each aggregate, most of the prestalk cells actively migrate to the anterior, while prespore cells remain in what becomes the posterior region of the grex. This migration is probably due to repeated pulses of cAMP that are still emanating from the apical tip of the aggregate. These pulses are chemotactic for prestalk cells but not for prespore cells, so they draw the prestalk cells to the tip of the aggregate (Matsukuma and Durs-

ton, 1979; Mee et al., 1986; Siegert and Weijer, 1991; Takeuchi, 1991). Cyclic AMP, then, is seen to play several different roles in *Dictyostelium* development. It aggregates the cells together, it induces prespore cell differentiation, and it directs the migration of prestalk cells to the anterior of the aggregate.

Once the aggregate is complete, it topples over onto its side and forms the migrating grex. Most of the prestalk cells are in the anterior 20 percent of the grex, but there are also some scattered prestalk cells throughout the posterior. Prestalk cells can be distinguished by their secretion of extracellular matrix protein A into the spaces between their cells. Within the center of the anterior portion of the grex, another group of prestalk cells begin secreting a second new protein (extracellular matrix protein B) into their extracellular matrix. These central cells are called prestalk B (pstB) cells, while the majority of the prestalk cells are denoted as prestalk A (pstA) cells (Figure 1.26). Another group of prestalk cells, the pstO cells, are scattered sparsely throughout the prespore cells, and they migrate more slowly toward the anterior. When the grex finds itself in sunlight, it ceases migrating and undergoes the final differentiation into spores and a stalk. During this process (called **culmination**), the grex sits on one end so that the rearguard cells become its base. Some pstA cells migrate onto the central tube of pstB cells, and as they contact the central tube, they differentiate into pstB cells by synthesiz-

ing new extracellular matrix components. The new cells are added to the anterior region of the tube and force the tube farther down into the culminating structure. This tube differentiates to become the stalk. At the same time, the pstA cells that had been in the posterior region of the grex migrate to the boundaries of the prespore region and differentiate into the spore case and the basal disc (Williams and Jermyn, 1991; Harwood et al., 1992). Eventually, the spores are lifted 2 mm off the ground, from which point they can be dispersed by the wind or by a passing animal.

The trigger for culmination appears to be sunlight or low humidity. Recent experiments suggest that these two factors cause the diffusion of ammonia from the slug. Ammonia is copiously produced by

Figure 1.26 Regulation of stalk cell differentiation during the culmination phase of *Dictyostelium* growth. Schematic representation of cell migration shows that prespore and prestalk cells are usually mixed in the early aggregate stage, but sort out so that most of the prestalk cells are at the anterior of the grex. The prestalk A cells constitute most of the anterior of the grex, with some similar cells in the posterior. Prestalk B cells are seen in the center of the anterior portion of the grex. In the early culmination stages, the prestalk cells in the posterior migrate to form the basal disc and cups of the spore sac; and the anterior prestalk A cells migrate toward the center and become prestalk B cells. This extends the stalk until it lifts the spore case off the surface. (After Harwood et al., 1992.)

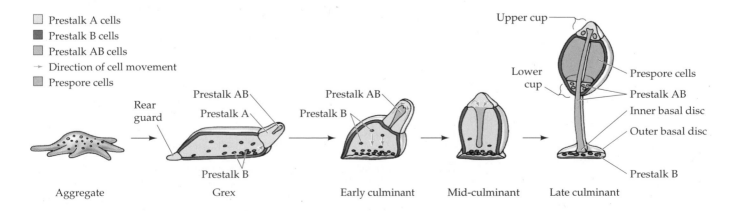

- ☐ Prestalk A cells
- ■ Prestalk B cells
- ▨ Prestalk AB cells
- → Direction of cell movement
- ▨ Prespore cells

Rear guard · Prestalk AB · Prestalk A · Prestalk B · Prestalk AB · Prestalk B · Upper cup · Lower cup · Prespore cells · Prestalk AB · Inner basal disc · Outer basal disc · Prestalk B

Aggregate · Grex · Early culminant · Mid-culminant · Late culminant

migrating slugs and represses culmination. Whenever ammonia is depleted (either naturally or experimentally), culmination begins (Schindler and Sussman, 1977; Newell and Ross, 1982; Bonner et al., 1985). Ammonia inhibits the conversion of the pstA cells into pstB cells and prohibits further stalk formation (Gross et al., 1983; Wang et al., 1990). Bonner and co-workers (1985) have suggested that because light causes more rapid diffusion of ammonia, it removes the inhibitor and thereby allows culmination to proceed.

Ammonia appears to inhibit stalk production in at least two ways. First, it inhibits the action of DIF (Wang and Schaap, 1989). Second, it inhibits the production of cAMP in the prestalk cells (Schindler and Sussman, 1977; Harwood et al., 1992). This cAMP is needed to activate the enzyme cAMP-dependent protein kinase (PKA). Prestalk cells carrying nonfunctional PKA cannot phosphorylate certain proteins. These cells do not migrate into the central anterior region, nor do they differentiate into stalk cells (Firtel and Chapman, 1990; Harwood et al., 1992). The data suggest that when PKA is activated, it phosphorylates a repressor that had been inhibiting the stalk differentiation genes from being expressed. In its phosphorylated state, the repressor is inactive. Therefore, once cAMP levels are elevated (by the removal of ammonia), PKA can inactivate the inhibitor of the stalk-forming genes (Figure 1.27). [intro.4html] ∎

Figure 1.27 Hypothesis for the coordinated initiation of culmination and the differentiation of stalk cells in *Dictyostelium*. Sunlight dissipates the ammonia in the anterior of the grex, enabling more cAMP to be produced inside the prestalk cells. The higher concentrations of cAMP activate PKA, which phosphorylates an inhibitor of stalk gene expression. The phosphorylated inhibitor no longer can inhibit the stalk-specific genes. The sequence by which spore formation is inhibited is not as clear. (Based on models of Bonner et al., 1985, and Harwood et al., 1992.)

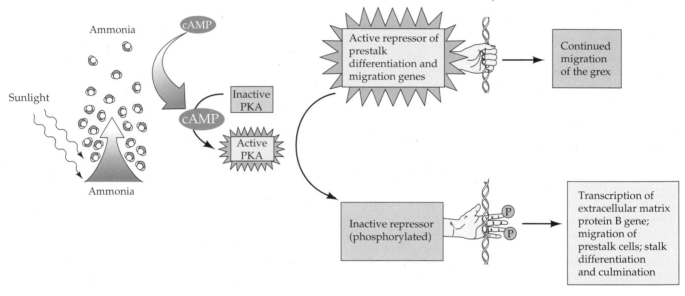

Developmental patterns among the metazoans

Since the remainder of this book concerns the development of **metazoans**—multicellular animals that pass through embryonic stages of development—we will present an overview of their developmental patterns.* The final chapter of the text will discuss these patterns in more detail. Figure 1.28 illustrates the major evolutionary trends of metazoan development. The most striking observation is that life has not evolved in a straight line; rather, there are several branching evolutionary paths. We can see that most of the species of metazoans belong to one of two major branches of animals: protostomes and deuterostomes.

---

*Plants undergo equally complex and fascinating patterns of embryonic and post-embryonic development. However, plant development differs significantly from that of animals, and to have included a comprehensive treatment of plant development would have doubled the length of this text. Therefore, the decision was made to focus this text on the development of animals. For an overview, see Singer, 1997.

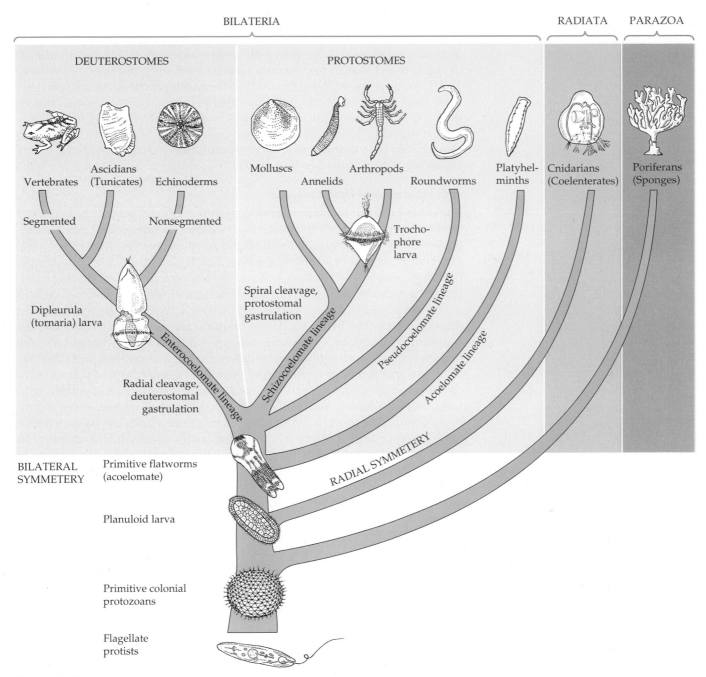

Figure 1.28
Major evolutionary divergences in extant animals. (Other models are possible, but the general schemes are all similar to the one shown here.)

The Porifera

The colonial protists are thought to have given rise to at least two groups of metazoans, both of which pass through embryonic stages of development. One of these groups is the Porifera (sponges). These animals develop in a manner so different from that of any other animal group that some taxonomists do not consider them metazoans at all (and call them "parazoans"). A sponge has three major types of somatic cells, but one of these, the **archeocyte,** can differentiate into all the other cell types in the body. The cells of a sponge passed through a sieve can regenerate new sponges from individual cells. Moreover, in some instances, such reaggregation is species-specific: if individual sponge cells from two different species are mixed together, each of the sponges that re-forms contains cells from only one species (Wilson, 1907). In these cases, it is thought that the motile archeocytes collect cells from their own species and not from others (Turner, 1978). Sponges contain no mesoderm, so there are no true organ systems in the Porifera; nor do they have a digestive tube or circulatory system, nerves, or muscles. Thus, even though they pass through an embryonic and a larval stage, sponges are very unlike most metazoans (see Fell, 1997).

Protostomes and Deuterostomes

The other group of metazoans arising from the colonial protists is characterized by the presence of three germ layers during development. Some members of this group constitute the **Radiata,** so called because they have radial symmetry, like that of a tube or a wheel. The Radiata include the cnidarians (jellyfish, corals, and hydras) and ctenophores (comb jellies). In these animals, the mesoderm is rudimentary, consisting of sparsely scattered cells in a gelatinous matrix. Most metazoans, however, have bilateral symmetry and thus constitute the **Bilateria.** These bilateral phyla are classified as either flatworms, protostomes, or deuterostomes. All Bilateria are thought to have descended from a primitive type of flatworm. These flatworms were the first to have a true mesoderm (although it was not hollowed out to form a body cavity), and they are thought to have resembled the larvae of certain contemporary coelenterates. While the flatworms are acoelomate (having no body cavity), the roundworms (and rotifers) have a body cavity distinctive from all other animals since it is not lined with mesoderm. The majority of phyla are coelomate, that is, they possess a mesoderm-lined body cavity.

The differences in the two coelomate divisions of the Bilateria are illustrated in Figure 1.29. **Protostomes** (from the Greek, meaning "mouth first"), which include the mollusc, arthropod, and worm phyla, are so called because the mouth is formed first, at or near the opening to the gut, which is produced during gastrulation. The anus forms later at another location. The body cavity of these animals forms from the hollowing out of a previously solid cord of mesodermal cells. The other great division of the Bilateria is the **deuterostome** lineage. Phyla in this division include chordates and echinoderms. Although it may seem strange to classify humans and horses in the same group as starfish and sea urchins, certain embryological features stress this kinship. First, in deuterostomes (from the Greek, meaning "mouth second"), the mouth opening is formed after the anal opening. Also, whereas protostomes generally form their body cavities by hollowing out a solid mesodermal block (schizocoelous formation of the body cavity), most deuterostomes form their body cavities from mesodermal pouches extending from the gut (enterocoelous formation of the body cavity). It should be mentioned that there are many exceptions to these generalizations.

Protostomes and deuterostomes differ in the way they undergo cleavage. In most deuterostomes, the blastomeres are perpendicular or parallel to each other. This is called **radial cleavage.** Protostomes, on the other hand, have a wide variety of cleavage types. Many species form blastulae composed of cells that are at acute angles to the polar axis of the embryo. Thus, they are said to undergo **spiral cleavage.** Furthermore, the cleavage-stage blastomeres of most deuterostomes have a greater ability to regulate development than do those of protostomes. If a single blastomere is removed from a four-cell sea urchin or mouse embryo, that blastomere will develop

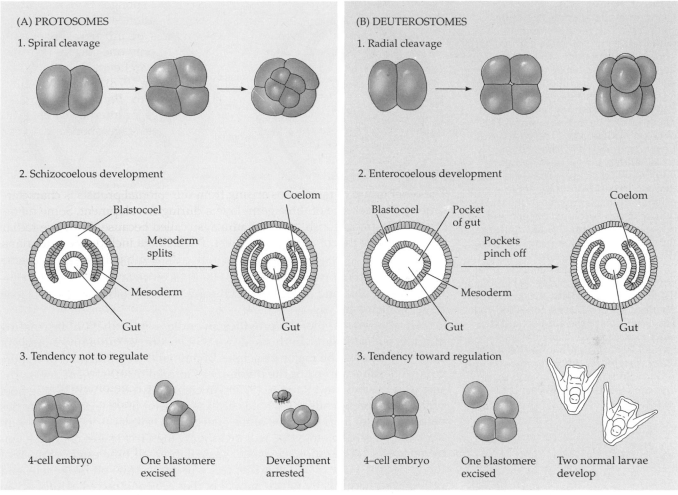

(A) PROTOSOMES

1. Spiral cleavage

2. Schizocoelous development

Blastocoel

Mesoderm
splits

Mesoderm

Gut

Coelom

Gut

3. Tendency not to regulate

4-cell embryo

One blastomere
excised

Development
arrested

(B) DEUTEROSTOMES

1. Radial cleavage

2. Enterocoelous development

Blastocoel

Pocket
of gut

Pockets
pinch off

Mesoderm

Gut

Coelom

Gut

3. Tendency toward regulation

4–cell embryo

One blastomere
excised

Two normal larvae
develop

Figure 1.29
Principal tendencies of the protostomes and deuterostomes. Exceptions to all these general tendencies have evolved secondarily in certain members of each group. (Most vertebrates, for instance, do not have a strictly enterocoelous formation of the body cavity; and the embryos of certain deuterostomes such as tunicates do not undergo regulation when blastomeres are taken from them.)

into an entire organism, and the remaining three-quarters of the embryo will also develop normally. However, if the same operation were performed on a snail or worm embryo, both the single blastomere and the remaining ones develop into partial embryos—each lacking what was formed from the other.

The evolution of organisms depends on inherited changes in their development. One of the greatest evolutionary advances—the **amniote egg**—occurred among the deuterostomes. This type of egg, exemplified by that of a chicken (Figure 1.30), is thought to have originated in the amphibian ancestors of reptiles about 255 million hears ago. The amniote egg allowed vertebrates to roam on land, far from existing ponds. Whereas most amphibians must return to water to breed and enable their eggs to develop, the amniote egg carries its own water and food supplies. The egg is fertilized internally and contains **yolk** to nourish the developing embryo. Moreover, it contains four sacs: the **yolk sac,** which stores the nutritive proteins, the **amnion,** which contains the fluid bathing the embryo; the **allantois,** in which waste materials from embryonic metabolism collect; and the **chorion,** which interacts with the outside environment, selectively allowing materials to reach the embryo. The entire structure is encased in a shell that allows the diffusion of oxygen but is hard enough to protect the embryo from environmen-

Figure 1.30
Diagram of the amniote egg of the chick, showing the development of membranes enfolding the embryo. (A) Three-day incubation. The extraembryonic mesoderm extends from the embryo to provide blood vessels to and from the various regions outside the embryo. (B) Seven-day incubation. The origin of the membranes will be detailed in Chapter 9. The yolk is eventually surrounded by the yolk sac, which allows the entry of nutrients into the blood vessels. The chorion is derived in part from the ectoderm and extends from the embryo to the shell (where it will exchange oxygen and carbon dioxide and obtain calcium from the shell). The amnion provides the fluid medium in which the embryo grows, and the allantois collects nitrogenous wastes that would be dangerous to the embryo. Eventually, the endoderm becomes the gut and encircles the yolk. The evolution of the amnion and the other extraembryonic membranes formed a great dividing line between those vertebrates whose reproduction is tied to water (anamniotes) and those that can reproduce on dry land (amniotes).

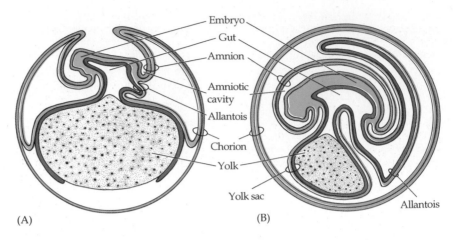

(A)   (B)

tal assaults. A similar development of egg casings enabled arthropods to be the first terrestrial invertebrates. Thus, the final crossing of the boundary between water and land occurred with the modification of the earliest stage in development—the egg.

Developmental biology provides an endless assortment of fascinating animals and problems. In this text, we will encounter but a tiny sample of them to illustrate the major principles of animal development. (For a more comprehensive survey of the diversity of animal development across the phyla, see Gilbert and Raunio, 1997.) We are merely observing the tide pool within our reach while the whole ocean of developmental phenomena lies before us. After a brief outline of the genetic and cellular principles relevant to developmental biology, we will investigate the early stages of animal embryogenesis: fertilization, cleavage, gastrulation, and the establishment of the vertebrate body plan. Later chapters will concentrate on the genetic and cellular mechanisms by which animal bodies are constructed. Although an attempt has been made to survey the important variations throughout the animal kingdom, a certain deuterostome chauvinism may be apparent.

LITERATURE CITED

Bergman, K., Goodenough, U. W., Goodenough, D. A., Jawitz, J. and Martin, H. 1975. Gametic differentiation in *Chlamydomonas reinhardtii*. II. Flagellar membranes and the agglutination reaction. *J. Cell Biol.* 67: 606–622.

Beug, H., Gerisch, G., Kempff, S., Riedel, V. and Cremer, G. 1970. Specific inhibition of cell contact formation in *Dictyostelium* by univalent antibodies. *Exp. Cell Res.* 63: 147–158.

Bonner, J. T. 1947. Evidence for the formation of cell aggregates by chemotaxis in the development of the slime mold *Dictyostelium discoideum*. *J. Exp. Zool.* 106: 1–26.

Bonner, J. T. 1957. A theory of the control of differentiation in the cellular slime molds. *Q. Rev. Biol.* 32: 232–246.

Bonner, J. T., Berkley, D. S., Hall, E. M., Konijn, T. M., Mason, J. W., O'Keefe, G. and

Wolfe, P. B. 1969. Acrasin, acrasinase and the sensitivity to acrasin in *Dictyostelium discoideum*. *Dev. Biol.* 20: 72–87.

Bonner, J., Hay, A. and John, D. 1985. pH affects fruiting and slug orientation in *Dictyostelium discoideum*. *J. Embryol. Exp. Morphol.* 87: 207–213.

da Silva, A. M. and Klein, C. 1990. Cell adhesion transformed *D. discoideum* cells: Expression of gp80 and its biochemical characterization. *Dev. Biol.* 140: 139–148.

Early, A., Abe, T. and Williams, J. 1995. Evidence for positional differentiation of prestalk cells and for a morphogenetic gradient in *Dictyostelium*. *Cell* 83: 91–99.

Faix, J., Gerisch, G. and Noegel, A. A. 1990. Constitutive overexpression of the contact A glycoprotein enables growth-phase cells of *Dictyostelium discoideum* to aggregate. *EMBO J.* 9: 2709–2716.

Fell, P. 1997. Porifera, the sponges. *In* S. F. Gilbert and A. M. Raunio (eds.), *Embryology: Constructing the Organism*. Sinauer Associates, Sunderland, MA.

Firtel, R. A. and Chapman, A. L. 1990. A role for cAMP-dependent protein kinase A in early *Dictyostelium* development. *Genes Dev.* 4: 18–28.

Fulton, C. 1977. Cell differentiation in *Naegleria gruberi*. *Annu. Rev. Microbiol.* 31: 597–629.

Fulton, C. and Walsh, C. 1980. Cell differentiation and flagellar elongation in *Naegleria gruberi*: Dependence on transcription and translation. *J. Cell Biol.* 85: 346–360.

Garcia, E. and Dazy, A.-C. 1986. Spatial distribution of poly(A)+ RNA and protein synthesis in *Acetabularia mediterranea*. *Biol. Cell* 58: 23–29.

Gilbert, S. F. and Raunio, A. M. 1997. *Embryology: Constructing the Organism*. Sinauer Associates, Sunderland, MA.

Goodenough, U. W. and Weiss, R. L. 1975. Gametic differentiation in *Chlamydomonas reinhardtii*. III. Cell wall lysis and microfilament associated mating structure activation in wild-type and mutant strains. *J. Cell Biol.* 67: 623–637.

Gross, J., Bradbury, J., Kay, R. and Peacey, M. 1983. Intracellular pH and the control of cell differentiation in *Dictyostelium discoideum*. *Nature* 303: 244–245.

Hämmerling, J. 1934. Über formbildended Substanzen bei *Acetabularia mediterranea*, ihre räumliche und zeitliche Verteilung und ihre Herkunft. *Wilhelm Roux Arch. Entwicklungsmech. Org.* 131: 1–81.

Hartmann, M. 1921. Die dauernd agame Zucht von *Eudorina elegans*, experimentelle Beiträge zum Befruchtungs-und Todproblem. *Arch. Protistk.* 43: 223–286.

Harwood, A. J., Hopper, N. A., Simon, M.-N., Driscoll, D. M., Veron, M. and Williams, J. G. 1992. Culmination in *Dictyostelium* is regulated by the cAMP-dependent protein kinase. *Cell* 69: 615–624.

Jermyn, K. A. and Williams, J. 1991. An analysis of culmination in *Dictyostelium* using prestalk and stalk-specific cell autonomous markers. *Development* 111: 779–787.

Johnson, R. L., Gundersen, R., Lilly, P., Pitt, G. S., Pupillo, M., Sun, T. J., Vaughan, R. A. and Devreotes, P. N. 1989. G-protein-linked signal transduction systems control development in *Dictyostelium*. *Development* [Suppl.]: 75–80.

Kay, R. R. and Jermyn, K. A. 1983. A possible morphogen controlling differentiation in *Dictyostelium*. *Nature* 303: 242–244.

Kay, R. R., Berks, M. and Traynor, D. 1989. Morphogen hunting in *Dictyostelium*. *Development* [Suppl.]: 81–90.

Keller, E. F. and Segal, L. A. 1970. Initiation of slime mold aggregation viewed as an instability. *J. Theoret. Biol.* 26: 399–415.

Kirk, D. L. 1988. The ontogeny and phylogeny of cellular differentiation in *Volvox*. *Trends Genet.* 4: 32–36.

Kirk, D. L. and Kirk, M. M. 1986. Heat shock elicits production of sexual inducer in *Volvox*. *Science* 231: 51–54.

Kirk, D. L., Viamontes, G. I., Green, K. J. and Bryant, J. L. Jr. 1982. Integrated morphogenetic behavior of cell sheets: *Volvox* as a model. *In* S. Subtelny and P. B. Green (eds.), *Developmental Order: Its Origin and Regulation*. Alan R. Liss, New York, pp. 247–274.

Kloppstech, K. and Schweiger, H. G. 1975. Polyadenylated RNA from *Acetabularia*. *Differentiation* 4: 115–123.

Knecht, D. A., Fuller, D. and Loomis, W. F. 1987. Surface glycoprotein gp24 involved in early adhesion of *Dictyostelium discoideum*. *Dev. Biol.* 121: 277–283.

Konijn, T. M., van der Meene, J. G. C., Bonner, J. T. and Barkley, D. S. 1967. The acrasin activity of adenosine -3′,5′-cyclic phosphate. *Proc. Natl. Acad. Sci. USA* 58: 1152–1154.

Krutch, J. W. 1956. *The Great Chain of Life*. Houghton Mifflin, Boston. [pp. 28–29]

Loomis, W. F. 1988. Cell-cell adhesion in *Dictyostelium discoideum*. *Dev. Genet.* 9: 549–559.

Matsukuma, S. and Durston, A. J. 1979. Chemotactic cell sorting in *Dictyostelium* discoideum. *J. Embryol. Exp. Morphol.* 50: 243–251.

McDonald, S. A. and Durston, A. J. 1984. The cell cycle and sorting out behaviour in *Dictyostelium discoideum*. *J. Cell Sci.* 66: 196–204.

Mee, J. D., Tortolo, D. M. and Coukell, M. B. 1986. Chemotaxis-associated properties of separated prestalk and prespore cells of *Dictyostelium discoideum*. *Biochem. Cell Biol.* 64: 722–732.

Morris, H. R., Taylor, G. W., Masento, M. S., Jermyn, K. A. and Kay, R. R. 1987. Chemical structure of the morphogen differentiation-inducing factor from *Dictyostelium discoideum*. *Nature* 328: 811–814.

Müller, K. and Gerisch, G. 1978. A specific glycoprotein as the target of adhesion blocking Fab in aggregating *Dictyostelium* cells. *Nature* 274: 445–447.

Newell, P. and Ross, F. 1982. Genetic analysis of the slug stage of *Dictyostelium* discoideum. *J. Genet. Microbiol.* 128: 1639–1652.

Ohmori, T. and Maeda, Y. 1987. The developmental fate of *Dictyostelium discoideum* cells depends greatly on the cell-cycle position at the onset of starvation. *Cell Differ.* 22: 11–18.

Pommerville, J. and Kochert, G. 1982. Effects of senescence on somatic cell physiology in the green alga *Volvox carteri*. *Exp. Cell Res.* 14: 39–45.

Powers, J. H. 1908. Further studies on *Volvox*, with description of three new species. *Trans. Am. Micros. Soc.* 28: 141–175.

Raper, K. B. 1940. Pseudoplasmodium formation and organization in *Dictyostelium discoideum*. *J. Elisha Mitchell Sci. Soc.* 56: 241–282.

Robertson, A., Drage, D. J. and Cohen, M. H. 1972. Control of aggregation in *Dictyostelium discoideum* by an external periodic pulse of cyclic adenosine monophosphate. *Science* 175: 333–335.

Schaap, P. and van Driel, R. 1985. The induction of post-aggregative differentiation in *Dictyostelium discoideum* by cAMP. Evidence for involvement of the cell surface cAMP receptor. *Exp. Cell Res.* 159: 388–398.

Schindler, J. and Sussman, M. 1977. Ammonia determines the choice of morphogenetic pathways in *Dictyostelium discoideum*. *J. Mol. Biol.* 116: 161–169.

Shaffer, B. M. 1953. Aggregation in cellular slime molds: In vitro isolation of acrasin. *Nature* 171: 975.

Shaffer, B. M. 1975. Secretion of cyclic AMP induced by cyclic AMP in the cellular slime mold *Dictyostelium discoideum*. *Nature* 255: 549–552.

Shaulsky, G. and Loomis, W. F. 1995. Mitochondrial DNA replication but no nuclear DNA replication during development. *Proc. Natl. Acad. Sci. USA* 92: 5660–5663.

Siegert, F. and Weijer, C. J. 1989. Digital image processing of optical density wave propagation in *Dictyostelium discoideum* and analysis of the effects of caffeine and ammonia. *J. Cell Sci.* 93: 325–335.

Siegert, F. and Weijer, C. J. 1991. Analysis of optical density wave propagation and cell movement in the cellular slime mould *Disctyostelium discoideum*. *Physica D* 49: 224–232.

Sies, H. 1988. A new parameter for sex education. *Nature* 332: 495.

Singer, S. 1997. Plant development. *In* S. F. Gilbert and A. M. Raunio (eds.), *Embryology: Constructing the Organism*. Sinauer Associates, Sunderland, MA.

Strickberger, M. W. 1985. *Genetics*, 3rd Ed. Macmillan, New York.

Sumper, M., Berg, E., Wenzl, S. and Godl, K. 1993. How a sex pheromone might act at a concentration below 10^{-16} M. *EMBO J.* 12: 831–836.

Takeuchi, I. 1991. Cell sorting and pattern formation in *Dictyostelium discoideum*. *In* J. Gerhart (ed.), *Cell-Cell Interactions in Early Development*. Wiley-Liss, New York, pp. 249–259.

Tomchick, K. J. and Devreotes, P. N. 1981. Adenosine 3′,5′ monophosphate waves in *Dictyostelium discoideum*: A demonstration by isotope dilution-fluorography. *Science* 212: 443–446.

Turner, R. S. Jr. 1978. Sponge cell adhesions. *In* D. R. Garrod (ed.), *Specificity of Embryological Interactions*. Chapman and Hall, London, pp. 199–232.

Tyson, J. J. and Murray, J. D. 1989. Cyclic AMP waves during aggregation of *Dictyostelium* amoebae. *Development* 106: 421–426.

Walsh, C. 1984. Synthesis and assembly of the cytoskeleton of *Naegleria gruberi* flagellates. *J. Cell Biol.* 98: 449–456.

Wang, M. and Schaap, P. 1989. Ammonia depletion and DIF trigger stalk cell differentiation in intact *Dictyostelium discoideum* slugs. *Development* 105: 569–574.

Wang, M., van Driel, R. and Schaap, P. 1988a. Cyclic AMP-phosphodiesterase induces dedifferentiation of prespore cells in *Dictyostelium discoideum* slugs: Evidence that cyclic AMP is the morphogenetic signal for prespore differentiation. *Development* 103: 611–618.

Wang, M., Aerts, R. J., Spek, W. and Schaap, P. 1988b. Cell cycle phase in *Dictyostelium discoideum* is correlated with the expression of cyclic AMP production, detection and degradation. Involvement of cAMP signalling in cell sorting. *Dev. Biol.* 125: 410–416.

Wang, M., Roelfsema, J. H., Williams, J. G. and Schaap, P. 1990. Cytoplasmic acidification facilitates but does not mediate DIF-induced prestalk gene expression in *Dictyostelium discoideum*. *Dev. Biol.* 140: 182–188.

Weijer, C. J., Duschl, G. and David, C. N. 1984. Dependence of cell-type proportioning and sorting on cell cycle phase in *Dictyostelium discoideum*. *Exp. Cell Res.* 70: 133–145.

Williams, J. G. and Jermyn, K. A. 1991. Cell sorting and positional differentiation during *Dictyostelium* morphogenesis. *In* J. Gerhart (ed.), *Cell-Cell Interactions in Early Development*. Wiley-Liss, New York, pp. 261–272.

Williams, J. G., Ceccarelli, A., McRobbie, S., Mahbubani, H., Kay, R. R., Early, A., Berks, M. and Jermyn, K. A. 1987. Direct induction of *Dictyostelium* pre-stalk gene expression of DIF provides evidence that DIF is a morphogen. *Cell* 49: 185–192.

Williams, J. G., Duffy, K. T., Lane, D. P., McRobbie, S. J., Harwood, A. J., Traynor, D., Kay, R. R. and Jermyn, K. A. 1989. Origins of the prestalk-prespore pattern in *Dictyostelium* development. *Cell* 59: 1157–1163.

Wilson, E. B. 1896. *The Cell in Development and Inheritance*. Macmillan, New York.

Wilson, H. V. 1907. On some phenomena of coalescence and regeneration in sponges. *J. Exp. Zool.* 5: 245–258.

Genes and development: Introduction and techniques

2

What we would like to know is whether structure is directly determined by DNA-encoded information laid down in the egg...the extent to which structure can be reduced to information.
JONATHAN BARD (1990)

The secrets that engage me—that sweep me away—are generally secrets of inheritance: how the pear seed becomes a pear tree, for instance, rather than a polar bear.
CYNTHIA OZICK (1989)

"BETWEEN THE CHARACTERS that furnish the data for the theory, and the postulated genes, to which the characters are referred, lies the whole field of embryonic development." Here Thomas Hunt Morgan (1926) was noting that the only way to get from genotype to phenotype is through developmental processes. In the early twentieth century, embryology and genetics were not considered separate sciences. They diverged in the 1920s, when Morgan redefined genetics as the science studying the *transmission* of traits, as opposed to embryology, the science studying the *expression* of these traits. Within the past decade, however, the techniques of molecular biology have effected a rapprochement of embryology and genetics. In fact, the two fields have again become linked to a degree that necessitates an early discussion of molecular genetics in this text. Problems in animal development that could not be addressed a decade ago are now being solved by a set of techniques involving nucleic acid synthesis and hybridization. This chapter seeks to place these new techniques within the context of the ongoing dialogues between genetics and embryology.

The embryological origins of the gene theory

Nucleus or Cytoplasm: Which Controls Heredity?

Mendel called them *Formbildungelementen,* form-building elements; we call them genes. It is in Mendel's term, however, that we see how closely intertwined were the concepts of inheritance and development in the nineteenth century. Mendel's observations, however, did not indicate where these hereditary elements existed in the cell or how they came to be expressed. The gene theory that was to become the cornerstone of modern genetics originated from a controversy within the field of embryology. In the late 1800s, a group of scientists began to study, for its own intrinsic value, how fertilized eggs give rise to adult organisms. Two young American embryologists, Edmund Beecher Wilson and Thomas Hunt Morgan (Figure 2.1), became part of this group of "physiological embryologists," and each became a partisan in the controversy over which of the two compartments of the fertilized egg—the nucleus or the cytoplasm—controls inheritance.

(A)

(B)

Figure 2.1
(A) E. B. Wilson (1856–1939; shown here around 1899), an embryologist whose work on early embryology and sex determination greatly advanced the chromosomal hypotheses of development. (Wilson was also acknowledged to be among the best amateur cellists in the country.) (B) Thomas Hunt Morgan (1866–1945), who brought the gene theory out of embryology. This photo—taken in 1915, as the basic elements of the gene theory were coming together—shows Morgan using a hand lens to sort flies. (A courtesy of W. N. Timmins; B courtesy of G. Allen.)

When Morgan and Wilson entered into this debate, the dispute was already a lively one. One school, associated with Oskar Hertwig, Wilhelm Roux, and Theodor Boveri, proposed that the chromosomes of the *nucleus* contained the form-building elements. This group was challenged by Eduard Pflüger, T. L. W. Bischoff, Wilhelm His, and their colleagues, who believed that no preformed structures could cause such enormous changes during development; rather, that the inherited patterns of development were caused by the creation of new molecules from the interacting gamete *cytoplasms.* Morgan allied himself with this latter group and obtained data that he interpreted as being consistent with the cytoplasmic model of inheritance. In his most crucial experiment, he removed cytoplasm from the newly fertilized ctenophore (comb jelly) egg. Morgan (1897) reported:

> *Here, although the entire segmentation nucleus is present, yet by loss of cytoplasm, defects are produced in the embryos…There seems to be no escape from the conclusion that in the cytoplasm and not in the nucleus lies the differentiating power of the early stages of development.*

Wilson, meanwhile, became a major proponent of the view that the form-building elements were on the nuclear chromosomes. He stated this view forcefully in his book *The Cell in Development and Inheritance* (1896), pointing to the necessary presence of the nucleus for protozoan regeneration (see Chapter 1):

> *This fact establishes the presumption that the nucleus is, if not the seat of the formative energy, at least the controlling factor in that energy and hence the controlling factor in inheritance. This presumption becomes a certainty when we turn to the facts of maturation [meiosis], fertilization, and cell-division. All those converge to the conclusion that the chromatin is the most essential element in development.*

Wilson (1895) did not shrink from the consequences of this conclusion:*

> *Now, chromatin is known to be closely similar, if not identical with, a substance known as nuclein . . . which analysis shows to be a tolerably definite chemical composed of a nucleic acid (a complex organic acid rich in phosphorus) and albumin. And thus we reach the remarkable conclusion that inheritance may, perhaps, be effected by the physical transmission of a particular chemical compound from parent to offspring.*

Wilson thought that the organ-forming material Morgan had removed from the cytoplasm of ctenophore eggs had already been secreted there by the nuclear chromosomes (Wilson, 1894, 1904). To Wilson (1905), "The protoplasmic stuffs appear to be only the immediate means or the efficient cause of differentiation, and we still seek its primary determination in the causes that lie more deeply."

Some of the major support for the chromosomal hypothesis of inheritance was coming from the embryological studies of Theodor Boveri (Figure 2.2A), a researcher at the Naples Zoological Station. Boveri fertilized sea urchin eggs with large concentrations of their sperm and obtained eggs that had been fertilized by two sperm. At first cleavage, these eggs formed four mitotic poles and divided the egg into four cells instead of two (see Chapter 4). Boveri then separated the blastomeres and demonstrated that each cell developed abnormally and in different ways as a result of each of the cells having different *types* of chromosomes. Thus, Boveri claimed that each chromosome had an individual nature and controlled different vital processes.

The X Chromosome as Bridge Between Genes and Development

Adding to Boveri's evidence, E. B. Wilson (1905) and Nettie Stevens (1905a,b) demonstrated a critical correlation between nuclear chromosomes and organismal development. Stevens (Figure 2.2B), a former student of Morgan's, showed that in 92 species of insects (and one primitive chordate), females had two sex-specific chromosomes in each nucleus (XX), while males had only one X chromosome (XY or XO). It appeared that a nuclear structure, the X chromosome, was controlling sexual development.[†] Morgan disagreed

---

*Notice that Wilson is writing about form-building units in chromatin in *1896*—before the rediscovery of Mendel's paper or the founding of the gene theory. For further analysis of the interactions between Morgan and Wilson that led to the gene theory, see Gilbert (1978, 1987) and Allen (1986).

[†]Wilson was one of Morgan's closest friends, and Morgan considered Stevens his best graduate student at that time. Both were against Morgan on this issue. Even though they disagreed, Morgan wholeheartedly supported Stevens's request for research funds, saying that her qualifications were the best possible. Wilson wrote an equally laudatory letter of support, even though she would be a rival researcher (see Brush, 1978).

(A)

(B)

Figure 2.2
Chromosomal uniqueness was shown by Boveri and Stevens. (A) Theodor Boveri (1862–1915) whose work, Wilson (1918) said, "accomplished the actual amalgamation between cytology, embryology, and genetics—a biological achievement which…is not second to any of our time." This photograph was taken in 1908, when Boveri's chromosomal and embryological studies were at their zenith. (B) Nettie M. Stevens (1861–1912), who trained with both Boveri and Morgan, seen here in 1904 when she was a postdoctoral student pursuing the research that correlated the number of X chromosomes with sexual development. (A from Baltzer, 1967; B courtesy of the Carnegie Institute of Washington.)

with their interpretation that the chromosomes actually determined sex. Rather, he viewed the assortment of chromosomes as a secondary sexual characteristic that was controlled by some cytoplasmic sex-determining substance. [gene1.html]

Morgan's "conversion" to the chromosomal hypothesis came after he obtained data contrary to his theories (see Allen, 1978; Gilbert, 1978; Lederman, 1989). While breeding *Drosophila* for a series of experiments on evolution, Morgan began to obtain several mutations that correlated with sex. (As Morgan would soon show, X-linked mutations appear sooner than mutations on other chromosomes, because defects on the X chromosome are not masked by the homologous chromosome in the male.) In 1910, Morgan showed that the traits for both sex and white eye color are correlated in some way to the presence of a particular X chromosome; but he avoided calling them physically linked. By 1911, however, Morgan had shown that factors regulating eye color, body color, wing shape, and sex all segregate together with the X chromosome, and he began to view the genes as being physically linked to one another on the chromosomes. The embryologist Morgan had shown that nuclear chromosomes are responsible for the development of inherited characters.

The split between embryology and genetics

Morgan's evidence provided a material basis for the concept of the gene. Genetics had mostly been an empirical science of animal and plant breeding; Morgan gave it a scientific foundation. Driven by the urge for progress in animal and plant (and human) breeding and by the ability of geneticists to quickly obtain mathematically verifiable and concrete results, genetics soon became the predominant biological science in the United States (see Allen, 1986; Sapp, 1987; Paul and Kimmelman, 1988). In the 1930s, genetics became its own discipline, developing its own vocabulary, journals, societies, favored organisms, professorships, and rules of evidence. Hostility between embryology and genetics also emerged. Geneticists believed that the embryologists were old-fashioned and that development would be completely explained as the result of gene expression. As Richard Goldschmidt (1938) proclaimed, "Development is, of course, the orderly production of pattern, and therefore after all, genes control pattern." If embryologists were not going to look at embryogenesis in terms of gene activity, the geneticists would.

Conversely, the embryologists regarded the geneticists as irrelevant and uninformed. Embryologists such as Frank Lillie (1927), Ross Granville Harrison (1937), Hans Spemann (1938), and Ernest E. Just (1939) (Figure 2.3) claimed that there could be no genetic theory of development until at least three major challenges had been met by the geneticists:

1. Geneticists had to explain how chromosomes—which were thought to be *identical* in every cell of the organism—direct *different* and *changing* types of cell cytoplasms.
2. Almost all the genes known at the time affected the final modeling steps (eye color, bristle shape, wing veination). Geneticists had to provide evidence that genes control the early stages of embryogenesis. As Just said (quoted in Harrison, 1937), embryologists were interested in how a fly forms its back, not in the number of bristles on its back.
3. Geneticists had to explain phenomena such as sex determination in certain invertebrates (and vertebrates such as reptiles), where the environment determines sexual phenotype.

(A)

(B)

Figure 2.3

Embryologists attempted to keep genetics from "taking over" their field in the 1930s. (A) Frank Lillie headed the Marine Biology Laboratory at Woods Hole and was a leader in fertilization research and reproductive endocrinology. (B) Hans Spemann (left) and Ross Harrison (right) perfected transplantation operations to discover when the body and limb axes are determined. They argued that geneticists had no mechanism for explaining how the same nuclear genes could create different cell types during development. (C) Ernest E. Just made critical discoveries on fertilization. He spurned genetics and emphasized the role of the cell membrane in determining the fates of cells. (A courtesy of V. Hamburger; B courtesy of T. Horder; C courtesy of the Marine Biology Laboratory, Woods Hole.)

(C)

The debate became quite vehement. In rhetoric reflecting the political anxieties of the late 1930s, Harrison (1937) warned:

> *Now that the necessity of relating the data of genetics to embryology is generally recognized and the* Wanderlust *of geneticists is beginning to urge them in our direction, it may not be inappropriate to point out a danger of this threatened invasion. The prestige of success enjoyed by the gene theory might easily become a hindrance to the understanding of development by directing our attention solely to the genom, whereas cell movements, differentiation, and in fact all of developmental processes are actually effected by cytoplasm. Already we have theories that refer the processes of development to gene action and regard the whole performance as no more than the realization of the potencies of genes. Such theories are altogether too one-sided.*

Until geneticists could demonstrate the existence of inherited variants during early development and until geneticists had a well-documented theory for how the same chromosomes could produce different cell types, embryologists generally felt no need to ground their embryology in gene action. [gene2.html]

Early attempts at developmental genetics

Some scientists, however, felt that neither embryology nor genetics was complete without the other. There were several attempts to synthesize the two disciplines, but the first successful reintegration of genetics and embryology came in the late 1930s from two embryologists, Salome Gluecksohn-

Schoenheimer (now S. Gluecksohn Waelsch) and Conrad Hal Waddington. Both were trained in European embryology and had learned genetics in the United States from Morgan's students. Gluecksohn-Schoenheimer and Waddington attempted to find mutations that affected early development and to find the processes these genes affected. Gluecksohn-Schoenheimer (1938, 1940) showed that mutations in the *Brachyury* genes of the mouse caused the aberrant development of the posterior portion of the embryo, and she traced the effects of these mutant genes to defects in the axial mesoderm that would normally have helped induce the dorsal axis.* Moreover, Gluecksohn-Schoenheimer (1938) reflected that in working with mice, it was not possible to do what experimental embryologists should be doing— changing a structure during its development and seeing what the consequences of that operation were. Rather, a new type of scientist was called for: the **developmental geneticist:**

> *While the experimental embryologist carries out a certain experiment and then studies its results, the developmental geneticist first has to study the course of development (that is, the results of the developmental disturbance) and then sometimes draw conclusions on the nature of the "experiment" carried out by the gene.*

At the same time, Waddington (1939) isolated several genes that caused wing malformations in the fruit fly, *Drosophila*. He, too, analyzed them in terms of how these genes might affect the developmental primordia that give rise to these structures. The *Drosophila* wing, he correctly claimed, "appears favorable for investigations on the developmental action of genes." Thus, one of the main objections of embryologists to the genetic model of development—that genes appear to be working only on the final modeling of the embryo and not on its major outlines—was countered. [gene3.html]

Evidence for genomic equivalence

There still remained the other big objection to a genetically based embryology: How could nuclear genes direct development when the genes were the same in every cell type? This genomic equivalence was not so much proved as assumed (because each cell is the mitotic descendant of the fertilized egg), and one of the first problems of developmental genetics was to determine whether each cell of an organism had the same genome as every other cell.

Metaplasia

The first evidence for genomic equivalence came after World War II, from embryologists studying the regeneration of excised tissues. The study of salamander eye regeneration has demonstrated that even adult *differentiated* cells can retain their potential to produce other cell types. Therefore, the genes for these other cell types' products must still be present, though not normally expressed. In the salamander, removal of the neural retina promotes its regeneration from the pigmented retina, and a new lens can be

---

*Gluecksohn-Schoenheimer's observations took 60 years to be confirmed by DNA hybridization. However, when the *T*-locus gene was cloned and its expression detected by the in situ hybridization technique (discussed later in this chapter), Wilkinson and co-workers (1990) found that "the expression of the *T* gene has a direct role in the early events of mesoderm formation and in the morphogenesis of the notochord." While a comprehensive history of early developmental genetics needs to be written, more information on its turbulent origins can be found in Oppenheimer, 1981; Sander, 1986; Gilbert, 1988, 1991, 1996; Burian et al., 1991; Harwood, 1993; Keller, 1995; and Morange, 1996).

formed from the cells of the dorsal iris. The regeneration of lens tissue from the iris (called Wolffian regeneration after the person who first observed it, in 1894) has been intensively studied. Yamada and his colleagues (Yamada, 1966; Dumont and Yamada, 1972) have found that after the removal of a lens, a series of events leads to the production of a new lens from the iris (Figure 2.4). The nuclei of the dorsal side of the iris begin to synthesize enormous amounts of ribosomes, their DNA replicates, and a series of mitotic divisions ensues. The pigmented iris cells then begin to dedifferentiate by throwing out their melanosomes (the pigmented granules that give the eye its color; these melanosomes get ingested by macrophages that enter the wound site). The dorsal iris cells continue to divide, forming a globe of dedifferentiated tissue in the region of the removed lens. These cells then start synthesizing the differentiated products of *lens* cells, the crystallin proteins. These proteins are made in the same order as in normal lens development. Once a new lens has formed, the cells on the dorsal side of the iris cease mitosis.

These events are not the normal route by which the vertebrate lens is formed. As we will detail later, the lens normally develops from a layer of head epithelial cells induced by the underlying retinal precursor cells. The formation of the lens by the differentiated cells of the iris represents **metaplasia** (or **transdifferentiation**), the transformation of one differentiated cell type into another (Okada, 1991). The salamander iris, then, has not lost any of the genes that are used to differentiate the cells of the lens.

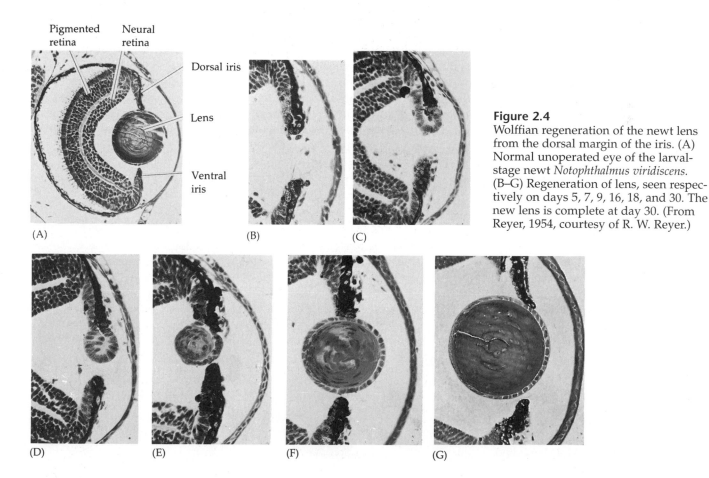

Figure 2.4
Wolffian regeneration of the newt lens from the dorsal margin of the iris. (A) Normal unoperated eye of the larval-stage newt *Notophthalmus viridiscens.* (B–G) Regeneration of lens, seen respectively on days 5, 7, 9, 16, 18, and 30. The new lens is complete at day 30. (From Reyer, 1954, courtesy of R. W. Reyer.)

Amphibian Cloning: The Restriction of Nuclear Potency

The ultimate test of whether or not the nucleus from a differentiated cell has undergone any irreversible functional restriction would be to have that nucleus generate every other type of differentiated cell in the body. If each nucleus were identical to the zygote nucleus, each cell's nucleus should be capable of directing the entire development of the organism when transplanted into an activated enucleated egg. Before such an experiment could be done, however, three techniques had to be perfected: (1) a method for enucleating host eggs without destroying them; (2) a method for isolating intact donor nuclei; and (3) a method for transferring such nuclei into the egg without damaging either the nucleus or the oocyte.

These techniques were developed in the 1950s by Robert Briggs and Thomas King. First, they combined the enucleation of the egg with its activation. When an oocyte from the leopard frog (*Rana pipiens*) is pricked with a clean glass needle, the egg undergoes all the cytological and biochemical changes associated with fertilization. The internal cytoplasmic rearrangements of fertilization occur, and the completion of meiosis takes place near the animal pole of the cell. This meiotic spindle can easily be located as it pushes away the pigment granules at the animal pole, and puncturing the oocyte at this site causes the spindle and its chromosomes to flow outside the egg (Figure 2.5). The host egg is now considered both activated (fertilization reactions necessary to initiate development have been completed) and enucleated. The transfer of a nucleus into the egg is accomplished by disrupting a donor cell and transferring the released nucleus into the oocyte through a micropipette. Some cytoplasm accompanies the nucleus to its new home, but the ratio of donor to recipient cytoplasm is only $1:10^5$, and the donor cytoplasm does not seem to affect the outcome of the experiments. In 1952, Briggs and King demonstrated that blastula cell nuclei could direct the development of complete tadpoles when transferred into the oocyte cytoplasm.

What happens when nuclei from more advanced stages are transferred into activated enucleated oocytes? The results of King and Briggs (1956) are outlined in Figure 2.6. Whereas most blastula nuclei could produce entire tadpoles, there was a dramatic decrease in the ability of nuclei from later

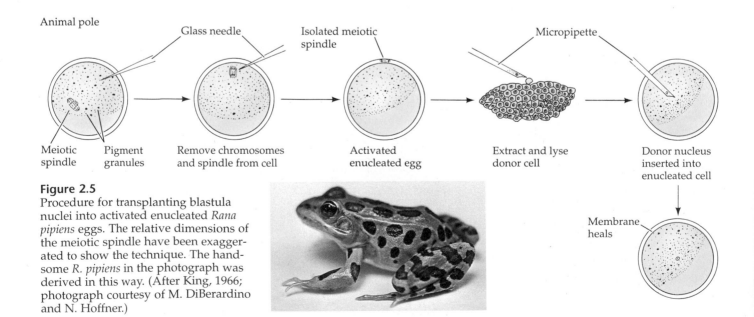

Figure 2.5
Procedure for transplanting blastula nuclei into activated enucleated *Rana pipiens* eggs. The relative dimensions of the meiotic spindle have been exaggerated to show the technique. The handsome *R. pipiens* in the photograph was derived in this way. (After King, 1966; photograph courtesy of M. DiBerardino and N. Hoffner.)

Animal pole

Glass needle

Isolated meiotic spindle

Micropipette

Meiotic spindle   Pigment granules

Remove chromosomes and spindle from cell

Activated enucleated egg

Extract and lyse donor cell

Donor nucleus inserted into enucleated cell

Membrane heals

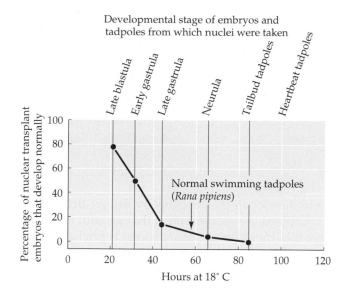

Developmental stage of embryos and
tadpoles from which nuclei were taken

Figure 2.6
Graph of successful nuclear transplants as a function of the developmental age of the nucleus. The abscissa represents the stage at which the donor nucleus (from *R. pipiens*) was isolated and inserted into the activated enucleated oocyte. The ordinate shows the percentage of those transplants capable of producing blastulae that could then direct development to the swimming tadpole stage. (After McKinnell, 1978.)

stages to direct development to the tadpole stage. When nuclei from the *somatic cells* of tailbud-stage tadpoles were used as donors, normal development did not occur. However, *germ cell* nuclei from tailbud-stage tadpoles (which eventually will give rise to a complete organism after fertilization) were capable of directing normal development in 40 percent of the blastulae that developed (Smith, 1956). Thus, somatic cells appear to lose their ability to direct complete development as they become determined and differentiated, and the progressive restriction of nuclear potency during development appears to be the general rule. It is possible that some differentiated cell nuclei differ from others.

Amphibian Cloning: The Pluripotency of Somatic Cells

John Gurdon and his colleagues, using slightly different methods of nuclear transplantation on the frog *Xenopus*, have obtained results suggesting that the nuclei of some differentiated cells may remain totipotent. Gurdon, too, found a progressive loss of potency with increased development, although *Xenopus* cells retained their potencies for a longer period of development (Plate 1). The exceptions to this rule, however, proved very interesting. Gurdon had transferred nuclei from the intestinal endoderm of feeding *Xenopus* tadpoles into activated enucleated eggs. These donor nuclei contained a genetic marker (one nucleolus per cell instead of the usual two) that distinguished them from host nuclei. Out of 726 nuclei transferred, only 10 (1.4 percent) promoted development to the feeding tadpole stage. Serial transplantation (which involved placing an intestinal nucleus into an egg and, when the egg had become a blastula, transferring the nuclei of the blastula cells into several more eggs) increased the yield to 7 percent (Gurdon, 1962). In some instances nuclei from tadpole intestinal cells were capable of generating all the cell lineages—neurons, blood cells, nerves, and so forth—of a living tadpole. Moreover, seven of these tadpoles (from two original nuclei) metamorphosed into fertile adult frogs (Gurdon and Uehlinger, 1966); these nuclei were totipotent (Figure 2.7).

King and his colleagues criticized these experiments, pointing out that (1) not enough precaution was taken to make certain that primordial germ cells which can migrate through to the gut were not used as sources of nuclei, and (2) the intestinal epithelial cell of such a young tadpole may not

Figure 2.7
Procedure used to obtain mature frogs from the intestinal nuclei of *Xenopus* tadpoles. The wild-type egg (2 nucleoli per nucleus; 2-nu) is irradiated to destroy the maternal chromosomes, and an intestinal nucleus from a marked (1-nu) tadpole is inserted. In some cases, there is no division; in some cases, the embryo is arrested in development; but in other cases, an entire, new frog is formed and has a 1-nu genotype. (After Gurdon, 1968, 1977.)

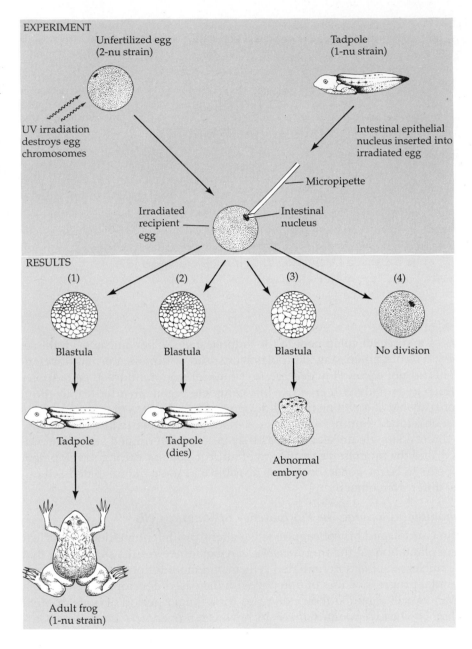

qualify as a truly differentiated cell type because such cells of feeding tadpoles still contain yolk platelets (DiBerardino and King, 1967; McKinnell, 1978; Briggs, 1979). To answer these criticisms, Gurdon and his colleagues cultured epithelial cells from adult frog foot webbing. These cells were shown to be differentiated; each of them contained keratin, the characteristic protein of adult skin cells. When nuclei from these cells were transferred into activated, enucleated *Xenopus* oocytes, none of the first-generation transfers progressed further than the formation of the neural tube shortly after gastrulation. By serial transplantation, however, numerous tadpoles were generated (Gurdon et al., 1975). Although these tadpoles all died prior to feeding, a single differentiated cell nucleus still retained incredible potencies. A single nucleus derived from an adult frog red blood cell (which neither replicates nor synthesizes mRNA) can undergo over 100 divisions after being transplanted into an activated oocyte and still retain the ability to generate swimming tadpoles (Orr et al., 1986; DiBerardino, 1989). Although DiBerardino (1987) has observed that "to date, no nucleus of a documented specialized cell nor of an adult cell has yet been shown to be totipotent," such a nucleus can still instruct the formation of all the organs of a swimming tadpole.

Some of the differences between the results of Briggs's laboratory and Gurdon's may involve differences in developmental physiology between *Rana* and *Xenopus* frogs. When transferring a nucleus of a differentiated cell into oocyte cytoplasm, one is asking the nucleus to revert back to physiological conditions it is not used to. The cleavage nuclei of frogs divide at a rapid rate, whereas some differentiated cell nuclei divide rarely, if at all. Failure to replicate DNA rapidly can lead to chromosomal breakage, and such chromosomal abnormalities have been seen in many cells of the cloned tadpoles. Sally Hennen (1970) has shown that the developmental success of donor nuclei can be increased by treating such nuclei with spermine and by cooling the egg to give the nucleus time to adapt to the egg cytoplasm. Spermine is thought to remove histones from chromatin and may "reset" the activity of the nuclei. When endoderm nuclei from *Rana pipiens* tailbud-stage tadpoles were treated in this fashion, 62 percent of those nuclei that initiated normal development went on to generate normal tadpoles. In control animals, none of the nuclei succeeded in generating such tadpoles. Thus, the genes for the development of the entire tadpole do not appear to be lost in the tadpole endoderm cells.

We can look at these amphibian cloning experiments in two ways. First, we can recognize a general restriction of potency concomitant with development. Second, we can readily see that the differentiated cell genome is remarkably potent in its ability to produce all the cell types of the amphibian tadpole. In other words, even if there is debate over the *totipotency* of such nuclei, there is little doubt that they are extremely *pluripotent*. Certainly, many unused genes in skin or blood cells can be reactivated to produce the nerves, stomach, or heart of a swimming tadpole. Thus, each nucleus in the body contains most (if not all) of the same genes.

Sidelights & Speculations

Cloning Mammals for Fun and Profit

CLONING HUMAN BEINGS from previously differentiated cells seems to be the goal of newspaper editors and novelists. It should be obvious from the preceding discussion, however, that cloning a fully developed individual from differentiated cells is a formidable undertaking. Even in amphibians, the nuclei of differentiated adult cells have not been able to generate adult animals when they are placed into activated enucleated eggs.

Moreover, even if adult frogs could be generated from differentiated cell nuclei, this ability could not be extrapolated to human cells. Besides the ethical and technical difficulties working with the human organism, human oocyte cytoplasm may not be responsive to signals from the nucleus of an advanced-stage cell. Still, nuclear transplantation has been accomplished in mice by removing the sperm and egg pronuclei (haploid nuclei) from one zygote and replacing them with pronuclei from other zygotes (Figure 2.8; McGrath and Solter, 1983).

These reconstructed zygotes begin dividing and are then implanted into the uterus. The resulting mice display the phenotype of the donor nucleus.

Whereas over 90 percent of enucleated mouse zygotes receiving pronuclei from other zygotes develop successfully to the blastocyst (blastula) stage, not a single embryo (out of 81) developed even this far when nuclei from 4-cell embryos were transferred into the enucleated zygotes (McGrath and Solter, 1984). Similarly, nuclei from 8-cell

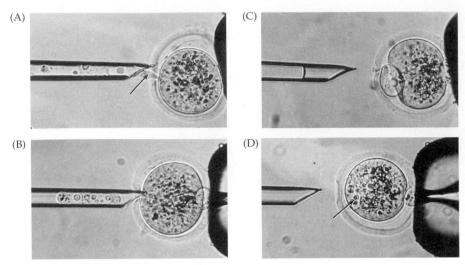

Figure 2.8 Procedure for transferring nuclei into an activated enucleated mammalian egg. A single-cell embryo, incubated in colcemid and cytochalasin to relax the cytoskeleton, is held on a suction pipette. The sperm-derived and egg-derived haploid nuclei have not yet come together. The enucleation pipette pierces the zona pellucida (the protein coat around the egg) and sucks up adjacent cell membrane and the area of the cell containing the pronuclei. (A) The enucleation pipette is then withdrawn, and the pronuclei-containing cytoplasm is removed from the egg. The cell membrane is not broken; the continuity of the membrane-bounded cytoplasm is indicated by the arrow. (B) The cell membrane forms a vesicle around the pronuclei within the enucleation pipette. (C) This vesicle is mixed with Sendai virus (which induces the fusion of cell membranes) and is inserted into the space between the zona pellucida and another enucleated egg. (D) The Sendai virus mediates the fusion of the enucleated egg and the membrane-bounded pronuclei, allowing the pronuclei (arrow) to enter the cell. (From McGrath and Solter, 1983, courtesy of the authors.)

the same, however, and mammalian species differ greatly in the time of gene activation and implantation into the uterus. Using modifications of the preceding technique, Willadsen (1986) has produced full-term lambs from the transplanted nuclei of 8-cell-stage blastomeres, and nuclei from the preimplantation embryos of cattle, pigs, and rabbits have been able to direct full development when transplanted into enucleated and activated oocytes (Prather et al., 1987; Stice and Robl, 1988; Prather et al., 1989; Willadsen, 1989). However, the nuclei in all these cases of mammalian cloning came from preimplantation embryos. A recent report (Wilmut et al., 1997) shows that it is possible to clone a sheep from an adult mammary gland cell nucleus. In general, no genes appear to be lost or destroyed in going from egg to adult. This ability to clone may have important agricultural and legal consequences (Prather, 1991). [gene4.html]

Plant Cloning

Only in plants are the nuclei of differentiated cells of adult organisms readily seen to be capable of directing the development of another adult organism. This ability has been demonstrated dramatically in cells from carrots and tobacco. In 1958, F. C. Steward and his colleagues established a procedure by which the differentiated tissue of carrot roots could give rise to an entire new plant (Figure 2.9). Small pieces of the phloem tissue are isolated from the carrot and rotated in large flasks containing coconut milk. This fluid (which is actually the liquid

embryos and the inner cell mass (the blastomeres that form the embryo rather than the placenta*) would not support development. Unlike nuclei from sea urchins and amphibians, the nuclei of

early mouse blastomeres (whose cells are known to be totipotent) do not support full development. These experiments probably fail because blastomere nuclei cannot function properly in zygote cytoplasm. Thus, the cloning of Elvis Presley from fully differentiated cells is not something we should count on.

Not all mammalian blastomeres are

*Each blastomere of the inner cell mass is totipotent in that it retains the ability to form any cell type in the body. This ability allows the possibility of twins.

Figure 2.9 Steward's experiment demonstrating the totipotency of carrot phloem cells.

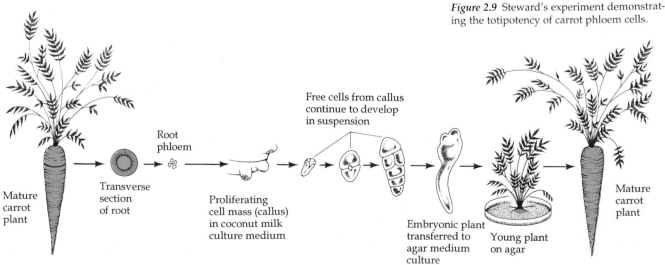

Mature carrot plant → Transverse section of root → Root phloem → Proliferating cell mass (callus) in coconut milk culture medium → Free cells from callus continue to develop in suspension → Embryonic plant transferred to agar medium culture → Young plant on agar → Mature carrot plant

endosperm of the coconut seed) contains the factors and nutrients necessary for plant growth and the hormones required for plant differentiation. Under these conditions, the tissues proliferate and form a disorganized mass of tissue called a **callus.** Continued rotation causes the shearing away of individual cells from the callus and into suspension. These individual cells give rise to rootlike nodules of cells that continue to grow as long as they remain in suspension. When these nodules are placed into a medium solidified with agar, the rest of the plant is able to develop, ultimately forming a complete, fertile carrot plant (Steward et al., 1964; Steward, 1970).

But plants and animals develop differently; the vegetative propagation of plants by cutting (i.e., portions of the plant that, when nourished, regenerate the missing parts) is a common agricultural practice. Moreover, in contrast to amphibians and mammals (in which germ cells are set off as a distinctive cell lineage early in development), plants normally derive their gametes from somatic cells. It is not overly surprising, then, that a single plant cell can differentiate into other cell types and form a genetically identical clone (from the Greek *klon*, meaning "twig"). ■

Of E. coli *and elephants: The operon model*

In most cases studied, the genome is the same from cell to cell within the organism. The genes for globin proteins can be found in skin cells, and the genes for skin keratins can be found in brain neurons. But that still leaves unanswered the other big question posed by the embryologists: If the nucleus of each cell in an organism has the same genes, how can these genes cause the cells to become different?* Shortly after World War II, many biologists agreed that

> the widest gap, still to be filled, between two fields of research in biology, is probably the one between genetics and embryology. It is the repeatedly stated and thus far unsolved problem of understanding how cells with identical genomes may become differentiated, of acquiring the property of manufacturing molecules with new or, at least, different specific patterns or configurations.

Interestingly, this quotation comes from Jacques Monod (1947), a microbial geneticist who had been working on adaptive enzyme synthesis. Adaptive enzymes are those proteins that, although not usually synthesized by a strain of bacteria or yeast, become synthesized when the microorganisms encounter a novel substrate. One example is the ability of the bacterium *Escherichia coli* to synthesize β-galactosidase and other lactose-digesting enzymes when it encounters lactose. If lactose is absent in the cytoplasm, these enzymes are not synthesized. Upon the introduction of lactose into the cytoplasm, this set of new enzymes appears. In microbes, at least, the same genome can produce two functionally different cytoplasmic states, depending on the presence of absence of a particular compound (in this case, lactose). Monod hypothesized that the phenomenon of enzymatic adaptation could be a solution to the problem of how identical genomes can synthesize different "specific" molecules.

Monod was not the only scientist who felt that unicellular microbes might explain multicellular differentiation. Microbiologist Sol Spiegelman (1947) claimed that embryology was being impeded by its own terminology. First, the problem of differentiation had to change from being seen as a structural property of tissues to being seen as a biochemical property of in-

---

*The big exception to this rule of gene constancy—the immunoglobulin genes—is discussed in Chapter 10. Each cell has all the immunoglobulin gene subunits, but in lymphocytes, some of these subunits get rearranged or even deleted from the genome. The third challenge—the explanation of how the environment can direct development—followed readily once the general explanation for differential gene expression was established. As we will see, the operon model demonstrated how an environmental substance could effect differential gene expression.

dividual cells. Differentiation was to be seen not in terms of tissue anatomy but "as the controlled production of unique enzyme patterns." This redefinition would focus attention on "the relationship between the genes in the nucleus and the properties of the cytoplasm." Second, the synthesis of adaptive enzymes in the presence of their substrate was to be discussed as an "induction." This is the technical term in embryology for the ability of one cell to produce a substance that influences the differentiation of another cell. Similarly, the molecular agent responsible for this induction was to be called "the inducer." Thus, Spiegelman saw an essential similarity between the induction of new cell types in the embryo and the induction of new enzymes in the microorganism. [gene5.html]

By the late 1950s, a number of researchers believed that microbes were an excellent (and easily studied) model for embryonic differentiation. In fact, many microbial geneticists explicitly linked inducible enzymes to embryological concepts. The geneticists felt the extrapolation valid, and they appealed to the unity of nature and the ultimately simple rules that they expected to find. As Monod claimed (see Judson, 1979), if you understand the bacterium, you understand the elephant. Many embryologists, however, remained skeptical of the extrapolation from bacteria to embryos, and they emphasized the complexity of development and the diversity of embryological performances.

By 1961, Jacob and Monod had synthesized the data on β-galactosidase induction into the operon model. This model postulates that the small inducer molecule causes the transcription of different genes in *E. coli* (Figure 2.10). In inductive systems, a gene-encoded repressor protein binds at the operator site adjacent to the structural genes. This prevents RNA polymerase from binding to the promoter site and initiating transcription. If the inducer is present, it binds to and alters the conformation of the repressor protein so that it cannot bind to the operator. Thus, the gene becomes able to transcribe mRNA, which can then be translated into proteins. In this way, the same genome can synthesize different enzymes, depending on whether or not the inducer is present. In a major 1961 article, Jacob and Monod emphasized that operon-like control mechanisms may be a universal part of gene regulation. They linked their results to "the fundamental problem of chemical embryology [which] is to understand why tissue cells do not express, all the time, all the potencies inherent in their genome."

The operon model of development was brought into embryology texts immediately by researchers who had been looking for a synthesis of genetics and embryology. Waddington's 1962 book, *New Patterns in Genetics and Development,* begins with a chapter relating the Jacob and Monod operon model to the control of gene expression in amphibian development. Waddington especially liked this model because it meant that genes are not only active, but reactive. They respond to changes in the cytoplasm. Waddington viewed the genes and the cytoplasm as mutually interacting with each other. This view was also celebrated in *Heredity and Development* (1963), John Moore's synthesis of embryology and genetics, which concludes:

> *A generation ago, few embryologists or geneticists would have predicted that a synthesis of their fields would be made possible by studies on the bacterium* Escherichia coli. *But this microscopic creature, with no embryology of its own, has shown a way. A decade from now it may be difficult to distinguish between a geneticist and an embryologist, as they advance their science beyond what each might independently achieve.*

(A) The *lac* operon

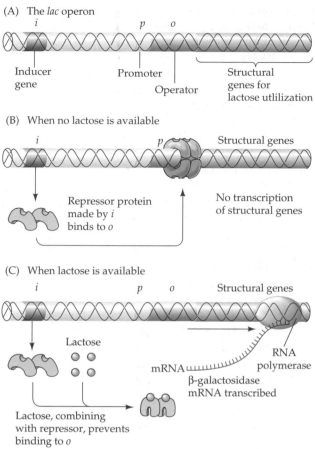

Inducer gene

Promoter

Operator

Structural genes for lactose utilization

(B) When no lactose is available

Structural genes

Repressor protein made by *i* binds to *o*

No transcription of structural genes

(C) When lactose is available

Structural genes

Lactose

mRNA

RNA polymerase

β-galactosidase mRNA transcribed

Lactose, combining with repressor, prevents binding to *o*

(D) The lactose repressor is soluble

Structural genes

Wild-type *i* gene can make repressor for both chromosomes which binds to *o* in the absence of lactose

Structural genes

Figure 2.10
Differential gene regulation in *E. coli.* (A–C) In the wild-type, inducible state, no β-galactosidase RNA is transcribed unless lactose is present. (B) When no lactose is available, a repressor protein, made by gene *i,* binds to the operator site (*o*), inhibiting transcription by RNA polymerase from the promoter (*p*). (C) When the inducer, lactose, is present, it combines with the repressor protein, changing the repressor's shape. The altered repressor protein can no longer bind to the operator DNA, and transcription ensues. (D) The soluble nature of this repressor protein is shown in studies on mutant *E. coli.* When haploid bacterial cells with a nonfunctional inducer gene (i^-) are made partially diploid with the wild-type *i* gene (i^+), wild-type repressor is manufactured and is able to make the original β-galactosidase gene inducible.

Differential RNA synthesis

The desired unification did not come as quickly as Moore had hoped. However, based on the embryological evidence for genomic equivalence and the operon model of *E. coli,* a consensus emerged in the 1960s that cells regulate development through differential gene expression. Because bacteria were the models for such activity, expression usually meant the transcription of mRNA. The three postulates of differential gene expression were as follows:

1. Every cell nucleus contains the complete genome established in the fertilized egg. In molecular terms, the DNAs of all differentiated cells are identical.
2. The unused genes in differentiated cells are not destroyed or mutated, and they retain the potential for being expressed.
3. Only a small percentage of the genome is being expressed in each cell, and a portion of the RNA synthesized is specific for that cell type.

The first two postulates have already been discussed. The third postulate—that only a small portion of the genome is active in making tissue-specific products—was first tested in insect larvae. Upon hatching, an insect larva has two distinct cell populations. About 10,000 cells form the larval tissue. Most of these larval cells have **polytene chromosomes.** Such chromosomes undergo DNA replication in the absence of mitosis and therefore contain 512 (2^9), 1024 (2^{10}), or even more parallel DNA double helices instead of just one (Figure 2.11; Plate 31). These cells do not undergo mitosis, and they grow by expanding to about 150 times their original volume. During metamorphosis, these cells die and are replaced by the nonpolytene diploid cells clustered in certain regions of the larva (see Chapter 19). Beermann (1952) showed that the banding patterns of polytene chromosomes were identical throughout the larva and that no loss or addition of any chromosomal region was seen when different cell types were compared (Figure 2.12). However, Beermann, studying the larval gall midge *Chironomus,* and Becker (1959), studying

Figure 2.11
Polytene chromosomes. (A) Polytene chromosomes from the salivary gland cells of *Drosophila melanogaster.* The four chromosomes are connected at their centromere regions, forming a dense chromocenter. The structural genes for alcohol dehydrogenase (ADH), aldehyde oxidase (Aldox), and octanol dehydrogenase (ODH) have been mapped to the assigned positions on these chromosomes. (B) Electron micrograph of a small region of a *Drosophila* polytene chromosome. The bands (dark) are highly condensed compared with the interband (lighter) regions. (A from Ursprung et al., 1968, courtesy of H. Ursprung; B from Burkholder, 1976, courtesy of G. D. Burkholder.)

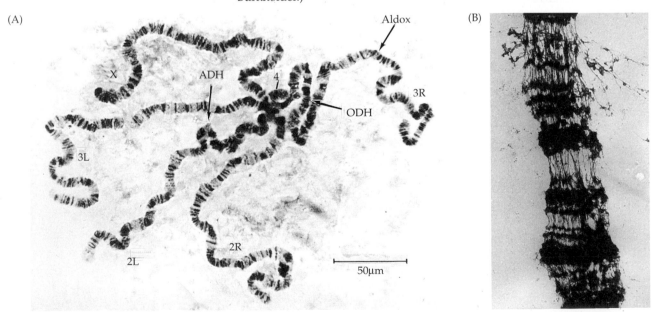

Drosophila, found that there were regions of the chromosomes that were "puffed out." These puffs appeared in different places on the chromosomes in different tissues, and their appearances changed with the development of these cells (Figure 2.13). Furthermore, certain puffs could be induced or inhibited by certain physiological changes caused by heat or hormones (Clever, 1966; Ashburner, 1972; Ashburner and Berondes, 1978).

Beermann (1961) provided evidence that these puffs represent a local loosening of the polytene chromosomes (Figure 2.14) and that they are sites of active RNA synthesis. He found two interbreeding species of *Chironomus* that differed: one produced a major salivary protein, whereas the other did not (Figure 2.15). The producers had a large puff (Balbiani ring) at a certain band in the polytene chromosomes in cells of the larval salivary gland; that puff was absent in the nonproducers. When producer was mated with nonproducer, the hybrid larva produced an intermediate amount of salivary protein. When two hybrid flies were mated, the ability to produce salivary protein segregated in proper Mendelian fashion (1 high producer:2 intermediate producers:1 nonproducer). Moreover, whereas high producers were found to have two puffs (one on each homologous chromosome), the intermediate producers had only one puff. The nonproducers lacked any puffs at

(A)

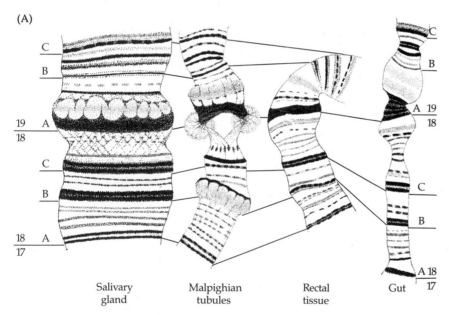

Salivary gland   Malpighian tubules   Rectal tissue   Gut

(B)

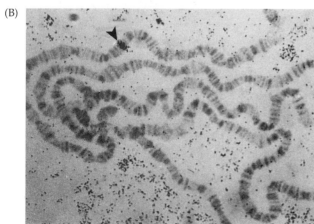

Figure 2.12
Genomic identity in polytene chromosomes. (A) A region of the polytene chromosome set of the midge *Chironomus tentans.* Note the constancy of band number in different tissues. (B) Hybridization of a yolk protein mRNA to the polytene chromosome of a larval *Drosophila* salivary gland. The dark grains (arrow) show where the radioactive yolk protein message bound to the chromosomes. Note that the gene for the yolk protein is present in the salivary gland chromosomes, even though yolk protein is not synthesized there. (A After Beermann, 1952; B from Barnett et al., 1980; photograph courtesy of P. C. Wensink.)

Figure 2.13
Puffing sequence of a portion of chromosome 3 in the larval *Drosophila melanogaster* salivary gland. (A,B) 110-hour larva. (C) 115-hour larva. (D,E) Prepupal stage (4 hours apart). Note the puffing and regression of bands 74EF and 75B. Other bands (71DE, 78D) puff later, but most do not puff at all during this time. (Courtesy of M. Ashburner.)

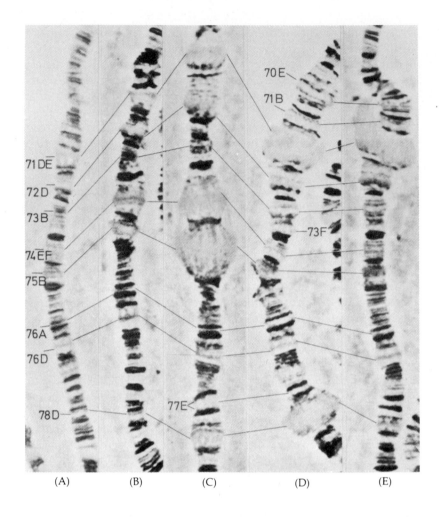

(A)   (B)   (C)   (D)   (E)

these sites. Beermann concluded that the genetic information needed for the synthesis of this salivary protein is present in this distal chromosome band and that the synthesis of that product depends on that band's becoming a puff region.

Further proof that chromosomal puffs make mRNA comes from studies of Balbiani ring 2 (BR2) puffs in *Chironomus tentans.* Because of its exceptionally large size, BR2 can be isolated by microdissection, and its products can

Figure 2.14
Proximal end of chromosome 4 from the salivary gland of *Chironomus pallidivitatus,* showing the enormous puff, BR2. (A) Phase-contrast photomicrograph of stained preparation showing extended puff on the polytene chromosome. (B) Diagram of the BR2 region undergoing puffing. (A from Grossbach, 1973, courtesy of U. Grossbach; B after Beermann, 1963.)

(A)

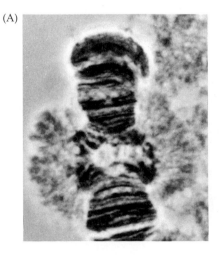

(B)

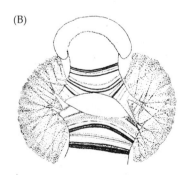

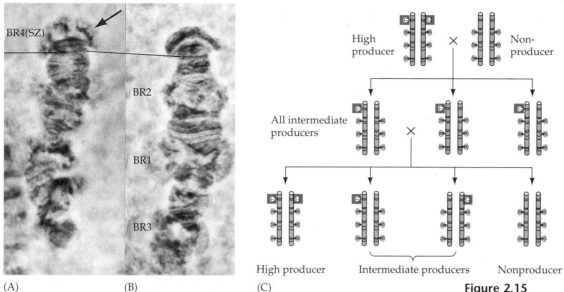

(A) (B) (C)

Figure 2.15
Correlation of puffing patterns with specialized functions in salivary gland cells of *Chironomus pallidivitatus*. (A) Chromosome 4 from a salivary gland cell producing a granular secretion and showing an additional Balbiani ring [4(SZ)]. (B) Chromosome 4 from a salivary cell, showing only Balbiani rings 1, 2, and 3 (BR1, BR2, BR3). (C) Genetic evidence that the synthesis of a major salivary protein is dependent on the formation of BR4(SZ) puffs. Larvae with high amounts of granular secretions have salivary gland cells with BR4(SZ) puffs on both chromosomes 4 (color), whereas larvae without these secretions have no such puffs. Intermediate producers have only one chromosome 4 with a puffed BR4(SZ) region in each salivary cell making the secretion. (A and B from Beermann, 1961, courtesy of W. Beermann.)

be analyzed by autoradiography (Lambert and Daneholt, 1975). Figure 2.16A,B shows the isolation of BR2 from chromosome 4 of *C. tentans*. BR2 transcription was demonstrated by incubating isolated salivary glands with radioactive RNA precursors. Radioactive RNA could then be extracted from the BR2 portion of the dissected chromosome (Lambert, 1972). This RNA was found to be exceptionally large—about 50,000 bases. This large radioactive RNA segment hybridized specifically to the BR2 region of the chromosome, showing that the puffed DNA—and no other locus—had been actively transcribing it (Figure 2.16C). This same RNA could also be isolated from protein-synthesizing polysomes, indicating that it is active in protein synthesis (Wieslander and Daneholt, 1977). Thus, an RNA transcribed from a specific band of DNA that puffs in the larval salivary gland is later seen to

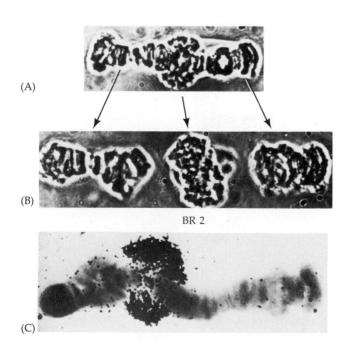

(A)

(B)

BR 2

(C)

Figure 2.16
(A,B) Isolation of BR2 region of *Chironomus tentans* by micromanipulation. The intact chromosome 4 can be divided into three regions, one containing BR2. (C) Transcription from the BR2 region of chromosome 4 of *C. tentans* salivary gland cells, as shown by an in situ autoradiograph after BR2 RNA was hybridized to the chromosome preparation. (A and B from Lambert and Daneholt, 1975; C from Lambert, 1972; photographs courtesy of B. Lambert.)

be making proteins on cytoplasmic ribosomes. Therefore, the puffs on salivary chromosomes are actively making mRNA. In cells that synthesize this protein, the gene is activated; in cells that do not use this protein, the gene remains repressed.

Nucleic acid hybridization

Few genes could be analyzed like those on *Chironomus* polytene chromosome puffs. And while these puff genes were active in cells (such as those of the salivary gland) that had already differentiated, they were not the genes that actually caused the cells to differentiate. To find and analyze the genes that *are* responsible for embryonic development, new techniques had to be perfected.

Most techniques of eukaryotic gene analysis are based on **nucleic acid hybridization.** This technique involves annealing single-stranded pieces of RNA and DNA to allow complementary strands to form double-stranded hybrids. For example, if DNA is cut into small pieces and each piece dissociated into two single strands—**denatured**—each strand in the solution should find and reunite with its complementary partner, given sufficient time. The conditions of renaturation must be such that specific binding between complementary strands is maintained while nonspecific matchings are dissociated. This is usually achieved by varying the temperature or the ionic conditions in the solution in which renaturation is taking place (Wetmur and Davidson, 1968). Similarly, *RNA* synthesized from a particular region of DNA would be expected to bind to the strand from which it was transcribed (Figure 2.17). Thus, RNA is expected to hybridize specifically with a gene that encodes it. To measure this hybridization, one of the nucleic acid strands (the *probe*) is usually labeled by the incorporation of radioactive nucleotides. One technical problem that originally plagued nucleic acid hybridization studies was the difficulty of getting enough radioactivity into the RNA molecule. This problem is circumvented by isolating the RNA and making a **complementary DNA (cDNA)** copy in the presence of radioactive precursors. This can be done in a test tube containing the RNA, a short stretch of DNA (called a **primer**), radioactive DNA precursors, and the viral

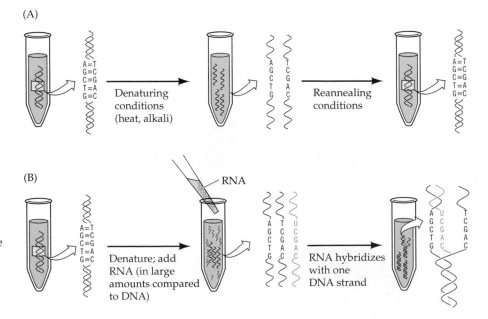

(A)

Denaturing conditions (heat, alkali)

Reannealing conditions

(B)

RNA

Denature; add RNA (in large amounts compared to DNA)

RNA hybridizes with one DNA strand

Figure 2.17
Nucleic acid hybridization. (A) If the DNA helix is separated into two strands, the strands should reanneal, given the appropriate ionic conditions and time. (B) Similarly, if DNA is separated into its two strands, RNA should be able to bind to the genes that encode it. If present in sufficiently large amounts compared with the DNA, the RNA will replace one of the DNA strands in this region.

enzyme **reverse transcriptase.** This enzyme is capable of making DNA from an RNA template (Figure 2.18). Because the DNA is synthesized in vitro, one need not worry about the dilution of the radioactive precursors. Furthermore, the cDNA can hybridize with both the gene that produced the RNA (albeit the other strand) and the RNA itself, making it extremely useful in detecting small amounts of specific RNAs. [other.html#gene6]

Cloning from genomic DNA

As early as 1904, Theodor Boveri despaired that the techniques of his time might never allow him to study how genes create embryos. A particular type of gene amplification technique was needed:

> *For it is not cell nuclei, not even individual chromosomes, but certain parts of certain chromosomes from certain cells that must be isolated and collected in enormous quantities for analysis; that would be the precondition for placing the chemist in such a position as would allow him to analyse [the hereditary material] more minutely than the morphologists.*

However, since the 1970s, nucleic acid hybridization has enabled developmental biologists to do just what Boveri wanted: to isolate and amplify specific regions of the chromosome. The main technique for isolating and amplifying individual genes is called **gene cloning.** The first step in this process involves cutting nuclear DNA into discrete pieces by incubating the DNA with a **restriction endonuclease** (more commonly called a **restriction enzyme**). These endonucleases are usually bacterial enzymes that recognize specific sequences of DNA and cleave the DNA at these sites (Table 2.1; Nathans and Smith, 1975). For example, when human DNA is incubated with the enzyme *Bam*HI (from *Bacillus amyloliquifaciens* strain H), the DNA is cleaved at every site where the sequence GGATCC occurs. The products are variously sized pieces of DNA, all ending with G on one end and GATCC on the other (Figure 2.19). These pieces are often called **restriction fragments.**

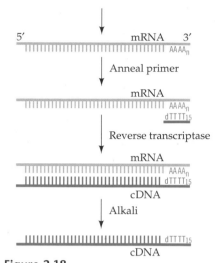

Figure 2.18
Method for preparing complementary DNA (cDNA). Most mRNA contains a long stretch of adenosine residues (AAA$_n$) at the 3′ end of the message (to be discussed in Chapter 12); therefore, investigators anneal a primer consisting of 15 deoxythymidine residues (dT$_{15}$) to the 3′ end of the message. Reverse transcriptase then transcribes a complementary DNA strand, starting at the dT$_{15}$ primer. The cDNA can be isolated by raising the pH of the solution, thereby denaturing the double-stranded hybrid and cleaving the RNA.

Table 2.1 Commonly used restriction enzymes

Enzyme site[a]	Derivation	Recognition and cleavage
*Eco*RI	*Escherichia coli*	G▾A A T T C C T T A A▴G
*Bam*HI	*Bacillus amyloliquifaciens*	G▾G A T C▴C C C T A G G
*Hind*III	*Haemophilus influenzae*	A▾A G C T T T T C G A▴A
*Sal*I	*Streptomyces albus*	G▾T C G A C C A G C T▴G
*Sma*I	*Serratia marcescens*	C C C▾G G G G G G▴C C C
*Hha*I	*Haemophilus haemolyticus*	G C G▾C C▴G C G
*Hae*III	*Haemophilus aegyptius*	G G▾C C C C▴G G
*Alu*I	*Arthrobacter luteus*	A G▾C T T C▴G A

[a]All restriction enzyme recognition sites have a center of symmetry. The double-stranded sequence read in one direction is identical to the sequence read backward in the other direction.

The next step in gene cloning is to incorporate these restriction fragments into **cloning vectors.** These vectors are usually circular DNA molecules that replicate in bacterial cells independently of the bacterial chromosome. Either drug-resistant plasmids or specially modified viruses (which are especially useful for cloning large DNA fragments) are used. For instance, a vector can be constructed to have only one *Bam*HI-sensitive site. This vector can be opened by incubating it with that restriction enzyme. After being opened, it can be mixed with the *Bam*HI-fragmented human DNA. In numerous cases, the cut DNA pieces will become incorporated into these vectors (because their ends are complementary to the vector's open ends), and the pieces can be joined covalently by placing them in a solution containing the enzyme **DNA ligase.** The whole process yields bacterial plasmids that each contain a single piece of human DNA. These are called recombinant plasmids or, usually, **recombinant DNA** (Cohen et al., 1973; Blattner et al., 1978).

The plasmid illustrated in Figure 2.19 is pUC18, a cloning vector often used by molecular biologists (Vierra and Messing, 1982). It contains (1) a drug-resistance gene, Ap^R, which makes the bacterium immune to ampicillin and allows researchers to select for those bacteria that have incorporated a plasmid; (2) an origin of DNA replication that enables the plasmid to replicate hundreds of times in each bacterium; and (3) a polylinker, a short, artificial stretch of DNA that contains the restriction enzyme sites for several of these endonucleases. The polylinker resides within a *lacZ* gene that encodes *E. coli* β-galactosidase. The polylinker is short enough (and has the correct number of base pairs) so that it does not interfere with the enzymatic activity of the β-galactosidase. The cloning procedure begins when the restriction fragments of the nuclear DNA are mixed with the opened pUC18 plasmids and are then ligated shut. The putative recombinant plasmids made in this manner are then incubated with ampicillin-sensitive *E. coli* cells that lack a β-galactosidase gene. Even though the bacteria and the plasmids are mixed together under conditions that encourage the bacteria to take in plasmids, not every bacterium incorporates a plasmid. To screen for those bacteria that have incorporated plasmids, the treated *E. coli* cells are grown on agar containing ampicillin. Only those bacteria that have incorporated a plasmid (with its dominant ampicillin-resistance gene) survive.

But not every plasmid has incorporated a foreign gene, because it is possible for the "sticky ends" of the restriction enzyme site to renature with themselves. To distinguish bacterial colonies that have incorporated foreign DNA from those that have not, the agar also contains a dye called X-gal. This compound is colorless, but when acted upon by β-galactosidase, it forms a blue precipitate.* Thus, if a plasmid has not incorporated a restriction fragment into its restriction enzyme site in the polylinker, the β-galactosidase (*lacZ*) gene is functional, and the resulting β-galactosidase turns the dye blue. The result is the appearance of "blue colonies." However, if the plasmid has taken up a DNA fragment, the β-galactosidase gene is destroyed by the insertion. These bacteria will not turn the dye blue; they produce colorless colonies on the agar.

Colorless colonies are then screened for the presence of the particular gene. Cells from each of these colonies are placed on a paper-thin nitrocellulose or nylon filter. When these cells are lysed, their DNA gets stuck on the filter. Next, the DNA strands are separated by heating, and the filter is incu-

---

*The dye is 5-bromo-4-chloroindole, and it is blue unless complexed with a molecule such as galactose. The β-galactosidase encoded by the plasmid gene cleaves the galactose off the dye, allowing the dye to achieve its blue conformation.

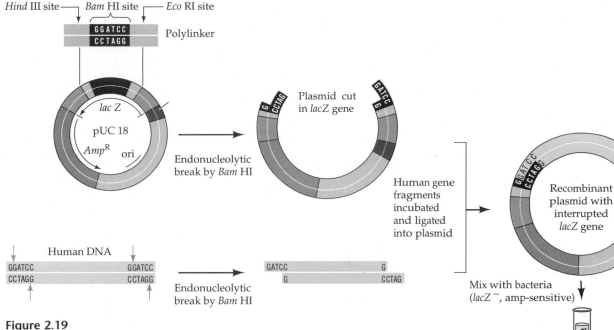

Hind III site — Bam HI site — Eco RI site

GGATCC
CCTAGG

Polylinker

lac Z

pUC 18

Amp^R ori

Endonucleolytic
break by Bam HI

Plasmid cut
in lacZ gene

G CCTAG
GGATCC

Human gene
fragments
incubated
and ligated
into plasmid

Recombinant
plasmid with
interrupted
lacZ gene

Human DNA

GGATCC   GGATCC
CCTAGG   CCTAGG

Endonucleolytic
break by Bam HI

GATCC   G
G   CCTAG

Figure 2.19
The general protocol for cloning DNA, using as an example the insertion of a
human DNA sequence into a plasmid with one *Bam*HI-sensitive site.

bated in a solution containing the radioactive RNA (or its cDNA copy) of the
gene one wishes to clone. (In some cases, the sequence of the mRNA or gene
is not yet known, and one has to guess the sequence from the amino acid sequence of the protein.) If a plasmid contains that gene, its DNA should be on
the filter, and only that DNA should be able to bind the radioactive RNA or
cDNA probe. Therefore, only those areas will be radioactive. The radioactivity of these regions is detected by **autoradiography.** Sensitive X-ray film is
placed over the treated paper. The high-energy electrons emitted by the radioactive RNA sensitize the silver grains in the film, causing them to turn
dark when the film is developed. Eventually, a black spot is produced over
each colony containing the recombinant plasmid carrying that particular
gene (see Figure 2.19). This colony is then isolated and grown, producing
billions of bacteria, each containing hundreds of identical recombinant plasmids.

The recombinant plasmids can be separated from the *E. coli* chromosome by centrifugation, and incubating the plasmid DNA with *Bam*HI releases the foreign DNA fragment that contains the gene. This fragment can
then be separated from the plasmid DNA, so the investigator has micro-
grams of purified DNA sequences containing a specific gene. Although this
procedure sounds very logical and easy, the number of colonies that must be
screened is often astronomical. The number of random fragments that must
be cloned to obtain the gene we want gets larger with the increasing complexity of the organism's genome.* To detect a particular gene from a mammalian genome, millions of individual clones must be screened.

*Complexity refers to the number of different types of genes within a nucleus. Although mil-
lions of clones must be screened, about 100,000 colonies can now be screened on a single plate.
Another common way of screening the clones is to use a plasmid that has its restriction enzyme
site near a strong bacterial promoter (such as the one for β-galactosidase). The bacteria will transcribe the cDNA and translate it into protein. After the bacterial colonies are lysed on filter
paper, the proteins stick to the paper and can be found by antibodies directed against that protein. This is called expression cloning, and the plasmids are referred to as expression vectors.

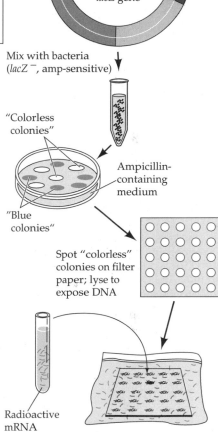

Mix with bacteria
(lacZ^−, amp-sensitive)

"Colorless
colonies"

"Blue
colonies"

Ampicillin-
containing
medium

Spot "colorless"
colonies on filter
paper; lyse to
expose DNA

Radioactive
mRNA

Filter paper incubated with radioactive
mRNA from gene to be cloned

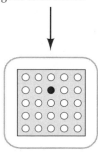

Prepare autoradiograph to show those
bacterial clones with a DNA fragment that
hybridized to the radioactive DNA

DNA hybridization: Within and across species

Clones can be screened by any radioactive stretch of nucleotides. Therefore, the genes cloned from one organism can be probed with radioactive cDNAs derived from the mRNAs of another species. One of the most exciting findings of modern developmental biology has been that genes used for specific developmental processes in one organism may be used for similar processes in other organisms. *Drosophila* has been critical in the discovery of these genes. Starting with Morgan, these genes have been mapped, and in the 1960s, E. B. Lewis confirmed that some of these genes are responsible for the formation of basic body parts (see Chapter 14). One of these, *Antennapedia*, is a gene whose protein product is essential for inhibiting head structures from forming in the thorax. If the gene is missing, antennae grow where the legs should be. If the gene is expressed in the head (as it is in a particular mutant), the fly develops an extra set of legs coming out of its eye sockets (see Figure 14.28). Could such a gene exist in vertebrates?

Evidence for such genes in vertebrates came first from **DNA blots,** sometimes called **Southern blots** after their inventor, E. M. Southern (1975). DNA from numerous vertebrate and invertebrate organisms was treated with a restriction enzyme, and the resulting DNA fragments were separated on an electrophoresis gel. The mixtures of fragments were placed into slots on one side of a gel, and an electric current was passed through the gel. The negatively charged DNA fragments migrated toward the positive pole, the smaller fragments moving faster than the larger ones.* However, hybridization cannot be done inside a gel; the DNA must be transferred to a flat surface, and this is done by blotting. After denaturing the DNA strands in alkali, investigators returned the gel to a neutral pH, then placed it on wet filter paper atop a plastic support (Figure 2.20; McGinnis et al., 1984; Holland and Hogan, 1986). Nitrocellulose paper (capable of binding single-stranded DNA) was placed directly over the gel and covered with multiple layers of dry paper towels. The filter paper beneath the gel extended into a trough of high-ionic-strength buffer. The buffer traveled through the gel up through the nitrocellulose filter and into the towels. The DNA was brought up through the gel by this flow of buffer, but it was stopped by the nitrocel-

Figure 2.20
Southern blotting. DNA is treated with restriction enzymes, and the resulting restriction fragments of DNA are placed in a gel. After the fragments of DNA are separated on the gel by electrophoresis, the DNA is denatured into single strands. The gel is then placed on a support on top of a filter paper saturated with high-ionic-strength buffer. Nitrocellulose paper or a nylon filter is placed atop the filter, and towels are placed atop the filter. The transfer buffer makes its way through the gel, nitrocellulose paper, and towels by capillary action, taking the DNA with it. The single-stranded DNA is stopped by the nitrocellulose paper. The positions of the DNA in the paper directly reflect the positions of the DNA fragments in the gel.

*Given the same charge-to-mass ratio, smaller fragments obtain a faster velocity than larger ones when propelled by the same energy. This is a function of the kinetic energy equation, $E = \frac{1}{2}mv^2$. Solving for velocity, we find that it is inversely proportional to the square root of the mass.

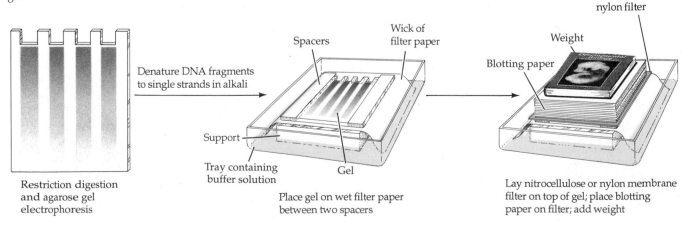

Restriction digestion and agarose gel electrophoresis

Denature DNA fragments to single strands in alkali

Spacers

Wick of filter paper

Support

Tray containing buffer solution

Gel

Place gel on wet filter paper between two spacers

Nitrocellulose or nylon filter

Weight

Blotting paper

Lay nitrocellulose or nylon membrane filter on top of gel; place blotting paper on filter; add weight

Figure 2.21
Southern blots of various organisms' DNA using a radioactive probe from the *Antennapedia* gene of *Drosophila melanogaster*. Because we do not expect the sequences between such diverged species to be perfectly identical, the stringency of the hybridization is lowered by changing the salt conditions. (Such low-stringency blots across phyla are colloquially refered to as "zoo blots," for obvious reasons.) Autoradiography shows that *Drosophila* genes contain several portions that are like *Antennapedia* genes in structure and that many organisms contain several genes that will hybridize this radioactive gene fragment, suggesting that *Antennapedia*-like genes exist in these organisms. The numbers beside the blots indicate size of bands, in kilobases. (From McGinnis et al., 1984, courtesy of W. McGinnis.)

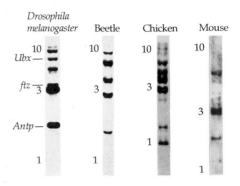

lulose filter; thus, the DNA was transferred from the gel to the nitrocellulose paper. After baking the DNA fragments onto the nitrocellulose paper (otherwise they would have come off), the DNA fragments were incubated with radioactive cDNA from a portion of the *Drosophila Antennapedia* gene. An autoradiogram of the nitrocellulose paper showed where the radioactive DNA had found its match. The results from these experiments (Figure 2.21) showed that even vertebrates (mice, humans, and chicks) have genes that hybridize to these sequences. This radioactive section of the *Antennapedia* gene was then used to screen a genomic library of DNA clones derived from the genome of these different species. As we will see in Chapter 16, investigators found clones containing genes that resemble *Antennapedia,* and these genes were revealed to be extremely important in the formation of the vertebrate body axis.

DNA sequencing

Sequence data can tell us the structure of the encoded protein and can identify regulatory DNA sequences that certain genes have in common. The simplicity of the Sanger "dideoxy" sequencing technique (Sanger et al., 1977) has made it a standard procedure in many molecular biology laboratories. One starts with the vector carrying the cloned gene and isolates a single strand of the circular DNA (Figure 2.22). One then anneals a radioactive primer of DNA (about 20 base pairs) complementary to the vector DNA immediately 3′ to the cloned gene. (Because these vector sequences are known, oligonucleotide primers can be readily synthesized or purchased commercially.) The primer has a free 3′ end to which more nucleotides can be added. One places the primed DNA and all four deoxyribonucleoside triphosphates into four test tubes. Each of the test tubes contains the polymerizing subunit of DNA polymerase and a different *di*deoxynucleoside triphosphate: one tube contains dideoxy-G, one tube contains dideoxy-A, and so forth. The structures of the deoxynucleotides and the dideoxynucleotides are shown in Figure 2.23. Whereas a deoxyribonucleotide has no hydroxyl (OH) group on the 2′ carbon of its sugar, a dideoxyribonucleotide lacks hydroxyl groups on both the 2′ and 3′ carbons. So even though a dideoxyribonucleotide can be bound to a growing chain of DNA by DNA polymerase, it stops the chain's growth because, lacking a 3′ hydroxyl group, no new nucleotide can bind to it. Thus, when the DNA polymerase is synthesizing DNA from the primer, the new DNA will be complementary to the cloned gene. In the tube with dideoxy-A, however, every time the polymerase puts an A into the growing chain, there is a chance that the dideoxy-A will be placed there instead of the deoxy-A. If this happens, the chain stops. Similarly, in the tube with dideoxy-G, the chain has the potential to stop every time a G is inserted. (The process has been likened to a Greek folk dance in which some small

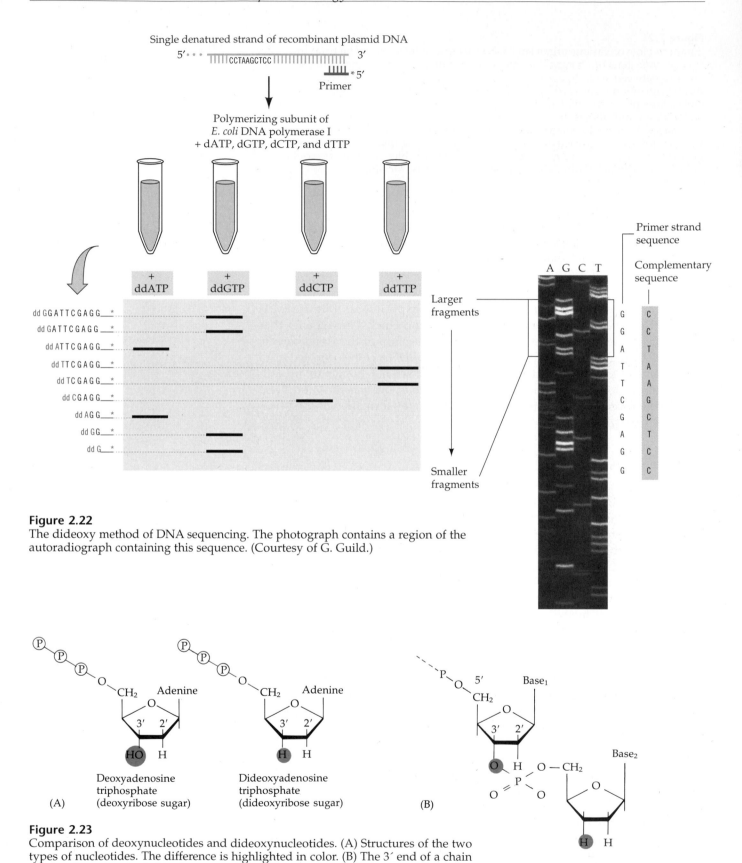

Figure 2.22
The dideoxy method of DNA sequencing. The photograph contains a region of the autoradiograph containing this sequence. (Courtesy of G. Guild.)

Figure 2.23
Comparison of deoxynucleotides and dideoxynucleotides. (A) Structures of the two types of nucleotides. The difference is highlighted in color. (B) The 3′ end of a chain that has been terminated by incorporation of dideoxynucleotide because it has no free 3′ hydroxyl group for further DNA polymerization.

percentage of the potential dancers have one arm in a sling.) Because there are millions of chains being made in each tube, each tube will contain a population of chains, some stopped at the first possible site, some at the last, and some at sites in between. The tube with dideoxy-A, for instance, will contain chains of different discrete lengths, each ending at an A residue. The resulting radioactive DNA fragments are separated by electrophoresis. The result is a "ladder" of fragments wherein each "rung" is a nucleotide sequence of a different length. By reading up the ladders, one obtains the DNA sequence complementary to that of the cloned gene.

Analyzing mRNA through cDNA libraries

Now we can return to the specificity of mRNA transcription: Can we isolate populations of mRNA that characterize certain cell types and are absent in all others? To find these RNAs, we can "clone" the mRNA from different types of cells and compare them. As shown in Figure 2.24A, this is done by taking the messenger RNAs from a cell or tissue and converting them into complementary DNA strands. By taking the procedure a step further (with the aid of DNA polymerase and S1 nuclease), we can change this population of single-stranded cDNA into a population of double-stranded cDNA pieces. These strands of DNA can be inserted into plasmids by adding the appropriate "ends" onto them with DNA ligase. Appending a GATCC/G fragment onto the blunt ends of such a DNA piece creates an artificial *Bam*HI restriction cut and enables the piece to be inserted into a virus or plasmid cut with that enzyme (Figure 2.24B).

Such collections of clones derived from mRNAs are often called **libraries.** Thus, we can have a 16-day embryonic mouse liver library, representing all the genes active in making embryonic liver proteins. We can also have a *Xenopus* vegetal oocyte library, representing messages present only in a particular part of that cell. Genes cloned in this manner are very important because they lack introns. When added to bacterial cells, these genes can be transcribed and then translated into the proteins they encode.

Libraries have been extremely useful in studying development, as seen in the efforts of Wessel and co-workers (1989) to detect differences in the RNAs in different parts of the gastrulating sea urchin embryo. To find endoderm-specific mRNAs in sea urchins, Wessel and co-workers prepared a cDNA library from gastrulating embryos. The mRNA of these samples (most of the RNA of eukaryotic cells is ribosomal) was isolated by running the samples through oligo-dT beads that capture the poly(A) tails of the messages (see legend to Figure 2.19). Then the mRNA population was converted into a cDNA population by using reverse transcriptase (see Figure 2.24A). By using *E. coli* polymerase I, the single-stranded cDNA was then made double-stranded. Next, commercially available *Eco*RI "ends" were ligated onto the double-stranded cDNAs. This made them clonable into vectors that were cut with *Eco*RI restriction enzyme. The DNA was then mixed with the arms of a genetically modified λ phage (see Figure 2.24B). This phage is so constructed that when grown in a petri dish, the phages that have incorporated the DNA (and thus destroyed the β-galactosidase gene) produce colorless plaques (Figure 2.24C). In this way, approximately 4 million recombinant phages were generated, each containing a cDNA representing an mRNA molecule.

The next steps involved screening the recombinant phages. Which ones might represent mRNAs found in endoderm and not in the other cell layers?

(A) Preparation of clonable cDNA

(B) Insertion of double-stranded cDNA into viral vector (bacteriophage λ)

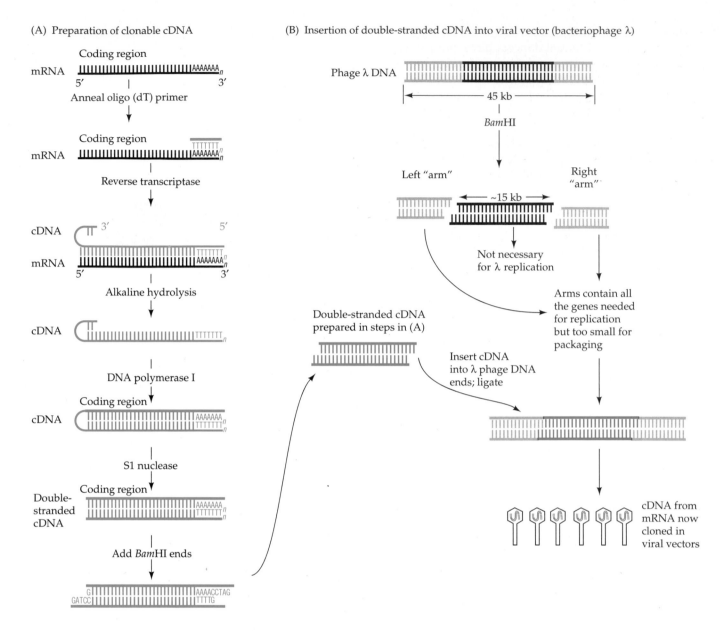

Wessel and his colleagues isolated the mRNA populations from mesoderm, ectoderm, and endoderm. They then made labeled cDNAs from each of the mRNA populations using radioactive precursors. They now had three collections of radioactive cDNA molecules, each representing the mRNA population from one of the three germ layers.

The recombinant phages representing the mRNAs of the gastrulating sea urchin embryo were grown, and samples of numerous colonies—each containing thousands of phages—were placed on two nitrocellulose filters (Figure 2.24D). These samples were then placed in alkaline solutions to lyse the phages and make the DNA single-stranded. One of these filter papers was incubated with radioactive cDNA made from the total mRNA of the *endoderm*; the other paper was incubated with radioactive probes to both *mesoderm and ectoderm*. The filters were then washed to remove any unhybridized radioactive cDNA, dried, and exposed on X-ray film. If an mRNA were present in the endoderm but not in either the ectoderm or mesoderm, the recombinant DNA made from that message should bind radioactive cDNA from the endoderm but should not find an mRNA anywhere else. As a re-

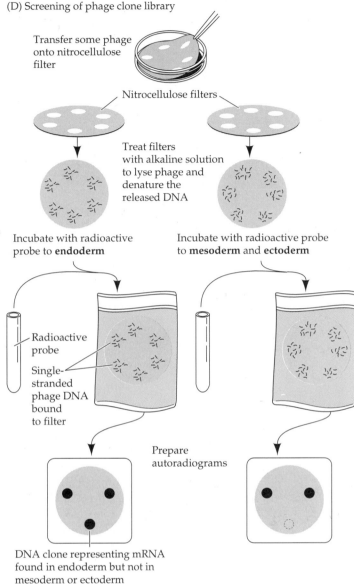

(C) Preparing library of phage clones

Hybrid phage

Add to lawn of *E. coli* cells

Phage infection of *E. coli*

Lysis

Plaque

Lawn of *E. coli* bacteria

Zone of lysis indicating phage clone

(D) Screening of phage clone library

Transfer some phage onto nitrocellulose filter

Nitrocellulose filters

Treat filters with alkaline solution to lyse phage and denature the released DNA

Incubate with radioactive probe to **endoderm**

Incubate with radioactive probe to **mesoderm** and **ectoderm**

Radioactive probe

Single-stranded phage DNA bound to filter

Prepare autoradiograms

DNA clone representing mRNA found in endoderm but not in mesoderm or ectoderm

Figure 2.24
Protocol used to make cDNA libraries. (A) Messenger RNA is isolated and made into cDNA. This cDNA is made double-stranded, and restriction fragment ends are added. (B) The cDNA "genes" can then be inserted into specially modified vectors, in this case bacteriophages. (C) Phages containing the recombinant DNA will lyse *E. coli,* forming plaques. Biochemical techniques can distinguish plaques of recombinant phages from those that lack the inserted gene. (D) The plaques are transferred to nitrocellulose paper and treated with alkali to lyse the phages and denature the DNA in place. These filters are then incubated in radioactive probes (usually cDNA) from a tissue. For the differential cDNA library screening discussed in the text, the same phage library was screened with radioactive probes from two different tissues, allowing the researchers to look for an mRNA that would be found in one type of tissue but not in the other.

sult, that spot of recombinant DNA from the endoderm should be radioactive (since it bound radioactive cDNA from the endoderm), but the same clone should not be radioactive when exposed to ectodermal or mesodermal mRNA. This was found to be the case. One recombinant phage in particular only bound radioactive cDNA made from endodermal mRNA; hence, it represented an mRNA found in the endoderm and not in the mesoderm or ectoderm. The phage containing this gene can now be grown in large quantities and characterized.

RNA localization techniques

In Situ Hybridization

The process of **in situ hybridization,** developed by Mary Lou Pardue and Joseph Gall (1970), allows researchers to visualize the positions of specific nucleic acids within cells and tissues. After a particular clone is identified as being interesting (e.g., the endoderm-specific clone just mentioned), it is grown in large quantities, and the cloned gene is isolated by treating the recombinant vector with restriction enzymes. This gene is then made single-stranded and radioactive. When the radioactive cDNA is added to appropriately fixed cells on a microscope slide, the radioactive cDNA binds only where the complementary mRNA is present. After the unbound cDNA is washed off, the slide is covered with a transparent photographic emulsion for autoradiography. The resulting spots, which are directly above the places where the radioactive cDNA had bound, appear black when viewed directly or white when viewed under dark-field illumination. Thus, we can visualize those cells (or even regions within cells) that have accumulated a specific type of mRNA. Figure 2.25A,B shows in situ hybridization using the cDNA found to be specific for endodermal cells. The cDNA finds mRNAs only in the endoderm of the early sea urchin gastrula. As gastrulation continues, the cDNA (and thus the mRNA) becomes localized even more precisely—to the hindgut-to-midgut region of the endodermal tube.

Working with radioactive probes and emulsions necessitates the use of finely sliced microscopic sections. A more recent technique for in situ hybridization utilizes probes that bind colored reagents. Thus, scientists can look at entire organs (and organisms) without sectioning them, thereby observing large regions of gene expression. Figure 2.25C shows an in situ hybridization performed on a wholemount of a 10.5-day mouse embryo. The probe recognizes mRNA encoded by the *Brachyury* gene (discussed on page 40) and which synthesizes a protein necessary for the production of mesodermal cells in the posterior of the mouse embryo.

Northern Blots

We can also determine the temporal and spatial expression of RNAs by running an **RNA blot** (often called a **Northern blot**). Whereas Southern blots transfer *DNA* fragments from gel to paper, Northern blots (not named after their inventor) transfer *RNA* from gel to paper in the same way. An investigator can extract messenger RNA from embryos at various stages of their development and electrophorese these RNAs side by side on a gel. After transferring the separated RNA (by blotting) to nitrocellulose paper or nylon membrane, the researcher incubates the RNA-containing paper in a solution containing a radioactive single-stranded DNA fragment from a particular gene. This DNA sticks only to regions where the complementary RNA is located. Thus, if the mRNA for that gene is present at a certain embryonic stage, the radioactive DNA binds to it and can be detected by autoradiography. Autoradiographs of this type, in which several stages are compared simultaneously, are called **developmental Northern blots.** Figure 2.26A shows a developmental Northern blot for the expression of the endoderm-specific gene during sea urchin development. We can see that the mRNA for this endodermal protein is first synthesized during the mesenchyme blastula stage and is continually made during the rest of development. The Northern blot in Figure 2.26B shows that the accumulation of this mRNA in the prism stage is restricted to the endoderm (Wessel et al., 1989). In situ hybridization and Northern blots provide the best evidence for differential RNA transcription in both space and time. Transcription of certain genes can be tissue-specific and time-specific.

The timing of transcription for several genes can be visualized by **dot blots.** For instance, Sargent and Dawid (1983) have isolated from *Xenopus* gastrulae the mRNA that was not present in the egg. To do this, they ex-

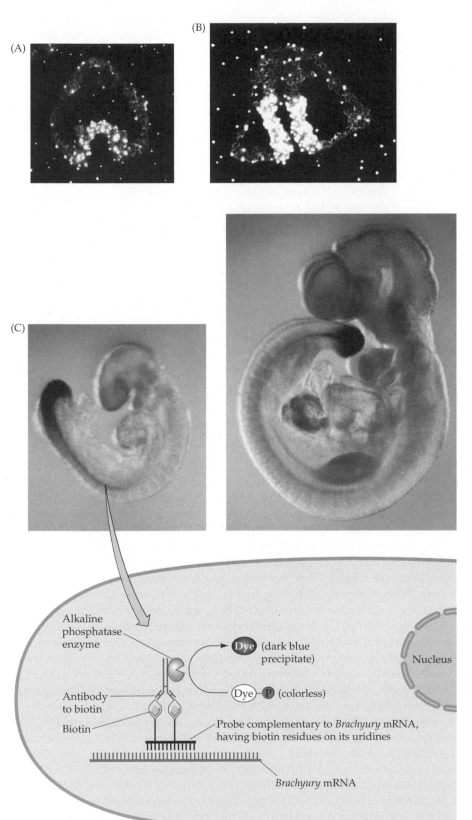

Figure 2.25
In situ hybridization. (A,B) Dark-field photomicrographs of in situ hybridization, showing the localization of an endoderm-specific mRNA in sea urchin embryos. The radioactive cDNA used as a probe was prepared from the cloned gene made from endoderm-specific mRNA (see Figure 2.24). This radioactive cDNA binds to mRNA in the endoderm of early-gastrula sea urchins (A) and in the mid- and hindgut endoderm of late-gastrula sea urchins (B). (C) Wholemount in situ hybridization of a 9.5-day and 10.5-day mouse embryo stained for *Brachyury* mRNA. This message is transcribed in those cells forming new mesoderm, and in this stage, it is found in the posterior portion of the embryo. Fixed embryos were incubated in a probe to *Brachyury* mRNA (the antisense strand, complementary to the mRNA), which had been synthesized using biotinylated uridine. After washing away all the probe that had not bound to *Brachyury* mRNA (and inactivating any endogenous alkaline phosphatase activity in the embryo), the embryo was treated with antibodies to biotin. These antibodies had been joined to alkaline phosphatase enzymes. Staining for the presence of alkaline phosphatase thereby enables one to determine the location of a specific mRNA. For color pictures of wholemount in situ hybridization, see Plates 22, 23 and 25. (A and B from Wessel et al., 1989, courtesy of G. Wessel; C from the author's laboratory.)

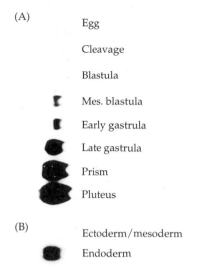

(A)

Egg

Cleavage

Blastula

Mes. blastula

Early gastrula

Late gastrula

Prism

Pluteus

(B)

Ectoderm/mesoderm

Endoderm

Figure 2.26
Northern blots for an endoderm-specific gene of the sea urchin *Lytechinus variegatus*. (A) Developmental Northern blot showing stage-specific accumulation of the mRNA from this gene. Total mRNA (10 μg per stage) was electrophoresed into an agarose gel. The gel was blotted onto treated paper, and the mRNAs were bound to the paper. The paper was then incubated with radioactive cDNA from the endoderm-specific clone. This mRNA was found to be synthesized during the mesenchyme blastula stage and increased throughout development. (B) Northern blot at the prism stage, showing that the mRNA is present in the endoderm (with some mesoderm attached) but not in the ectoderm. Total RNA from the endoderm was electrophoresed (lane 2) next to total mRNA from the rest of the sea urchin (lane 1). Binding of the radioactive cDNA detected mRNA only in the endoderm. (From Wessel et al., 1989, courtesy of G. Wessel.)

tracted the gastrula mRNA and made cDNA copies of these messages. They mixed the gastrula cDNAs with large quantities of oocyte mRNA. If the oocyte mRNA hybridized with gastrula cDNA, then the cDNA was derived from an mRNA present in both oocyte and gastrula stages. These double-stranded hybrid molecules were removed by filtration, thereby leaving a population of gastrula-specific cDNAs. These cDNAs were then made double-stranded (by DNA polymerase) and inserted into cloning vehicles. This technique is called **subtraction cloning.** Like the dual screening of cDNA libraries, subtraction cloning generates a stage-specific set of clones whose mRNA is found in some stages but not others or in some tissues but not others (Figure 2.27).

Sargent and Dawid then took embryos from the zygote through the tailbud tadpole stage and separately isolated their RNAs. The RNAs were spotted onto nitrocellulose filters directly (with no prior gel electrophoresis) so that each filter had RNAs from all stages. After the RNAs were baked onto the filter, single-stranded DNA derived from a particular "gastrula" clone was made radioactive and incubated with the filters. If a gene was being transcribed at a given stage, the radioactive cDNA from that gene would find its complement in the mRNAs from that stage on the filter. The nonbound cDNA was then washed off and the binding of radioactive cDNA observed by autoradiography. The dot blots in Figure 2.28 show the time courses of expression for 17 genes that are active at various stages of gastrulation. None of them is expressed earlier than the midblastula transition at 7 hours. Some genes (*DG64, DG39*) are expressed immediately afterward, whereas other genes (*DG72, DG81*) begin to be transcribed at midgastrula, about 7 hours later. Some genes (*DG76, DG81*) appear to "stay on" once activated, while the activity of other genes (*DG56, DG21*) is far more transient.

Finding rare messages by the polymerase chain reaction

The polymerase chain reaction (PCR) is a method of in vitro cloning that can generate enormous quantities of a specific DNA fragment from a small amount of starting material (Saiki et al., 1985). It can be used for cloning a specific gene or for determining if a specific gene is actively transcribing RNA in a particular organ or cell type. The standard methods of cloning use living microorganisms to amplify recombinant DNA. PCR, however, can amplify a single DNA molecule several millionfold in a few hours, and it does it in a test tube. This technique has been extremely useful in cases where there is very little nucleic acid to study. Preimplantation mouse embryos, for instance, have very little mRNA, and we cannot obtain millions of such embryos to study. If we wanted to know whether the preimplantation mouse embryo contained the mRNA for a particular protein, it would be very difficult to find out using standard cloning methods. We would have to lyse thousands of mouse embryos to get enough mRNA. However, the PCR technique allows us to find this message in a few embryos by specifically amplifying only that message millions of times (Rappolee et al., 1988).

The use of PCR for finding rare mRNAs is illustrated in Figure 2.29. The mRNAs from a group of cells are purified and converted into cDNA by reverse transcriptase. By using DNA polymerase and S1 nuclease, the population of single-stranded DNAs is then made into double-stranded DNAs. Next we target a specific DNA for amplification. To do this, we separate the DNA double helices and add to them two small oligonucleotide primers that are complementary to a portion of the message being looked for. If the oligonucleotides recognize sequences in the DNA, then the mRNA was originally present. The oligonucleotides have been made so that they hybridize

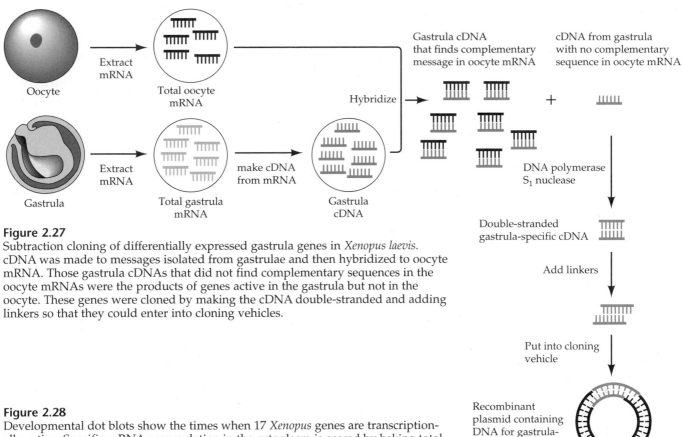

Figure 2.27
Subtraction cloning of differentially expressed gastrula genes in *Xenopus laevis*. cDNA was made to messages isolated from gastrulae and then hybridized to oocyte mRNA. Those gastrula cDNAs that did not find complementary sequences in the oocyte mRNAs were the products of genes active in the gastrula but not in the oocyte. These genes were cloned by making the cDNA double-stranded and adding linkers so that they could enter into cloning vehicles.

Figure 2.28
Developmental dot blots show the times when 17 *Xenopus* genes are transcriptionally active. Specific mRNA accumulation in the cytoplasm is scored by baking total mRNA from gene embryonic stages onto nitrocellulose paper and incubating that strip of paper with radiolabeled DNA derived from a gastrula-specific cDNA clone. By placing these strips together, we get a time course for the activity of the specific genes. The r5 line represents a ribosomal RNA control that should always be present. (From Jamrich et al., 1985, courtesy of I. Dawid and M. Sargent.)

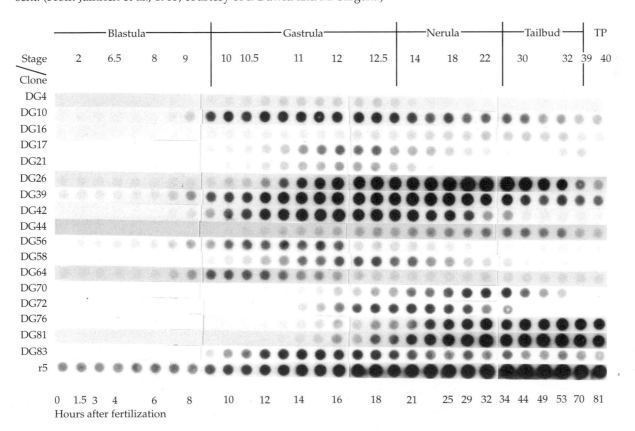

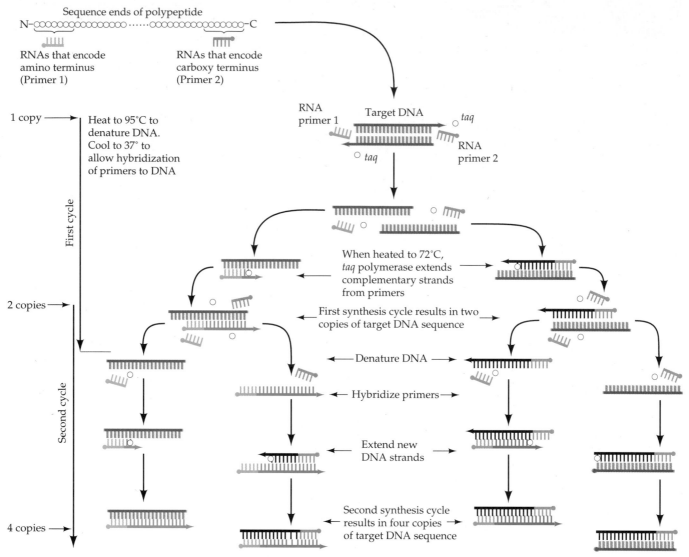

Figure 2.29

Protocol for the polymerase chain reaction (PCR). For determining whether a particular type of mRNA is present, all the mRNA is converted to double-stranded DNA by reverse transcriptase and DNA polymerase. This DNA is denatured, and two sets of primers are added. If the specific sequence is present, the primers will hybridize to its opposite ends. (Specific primers are made based on the sequence we are looking for. If we know only the sequence of the protein encoded by the message, a set of different primers is made, each possibly being complementary to the DNA.) Using thermostable DNA polymerase from *T. aquaticus*, each DNA strand synthesizes its complement. These strands are then denatured and the primers are hybridized to them, starting the cycle again. In this way, the number of new strands having the sequence between the two primers increases exponentially.

to opposite strands and opposite sides of the targeted sequence. (If we are trying to isolate the gene or mRNA for a specific protein of known sequence, these flanking regions can be prepared by synthesizing oligonucleotides that would encode the amino end of the protein and oligonucleotides complementary to those that would encode the carboxyl end of the protein.) The 3′ ends of these primers face each other, so that replication is through the target DNA. Once the first primer has hybridized, the DNA polymerase can synthesize a new strand.

This is not normal *E. coli* DNA polymerase, however; it is DNA polymerase from bacteria such as *Thermus aquaticus* or *Thermococcus litoralis*. These bacteria live in hot springs (such as those in Yellowstone National Park) or in submarine thermal vents, where the temperature reaches nearly 90°C. These DNA polymerases can withstand temperatures near boiling; PCR takes advantage of this evolutionary adaptation. Once the second strand is made, it is heat-denatured from its complement. The second primer is added, and now both strands can synthesize new DNA. Repeated cycles of denaturation and synthesis will amplify this region of DNA in geometric fashion. After 20 such rounds, that specific region has been amplified 2^{20} (a little more than a million) times. When run on electrophoresis, the presence of such an amplified fragment is easily detected. This shows that the original mRNA having this sequence was present in the sample. (This could be confirmed by Southern blots, as in Figure 2.30.) Moreover, we can take these amplified copies and place them into cloning vectors and clone them.

Determining the function of a gene: Transgenic cells and organisms

Techniques of Inserting New DNA into a Cell

While it is important to know the sequence of a gene and its temporospatial pattern of expression, what's really crucial is the function of that gene during development. Recent techniques have enabled us to study gene function by taking certain genes into and out of embryonic cells. Cloned pieces of DNA can be isolated, modified (if so desired), and placed into cells by several means. One very direct technique is *microinjection,* in which a solution containing the cloned gene is injected very carefully into the nucleus of a cell (Capecchi, 1980). This is an especially useful technique for injecting genes into newly fertilized eggs, since the haploid nuclei of the sperm and egg are relatively large (Figure 2.31). In *transfection,* DNA is incorporated directly into the cell by incubating it in a particular solution that makes the cells drink it in. The chances of such a DNA fragment being incorporated into the chromosomes are relatively small, so the DNA is usually mixed with another gene that enables those rare cells that incorporate them to survive under culture conditions that will kill all the other cells (Perucho et al., 1980; Robins et al., 1981).

Another technique is *electroporation,* in which a high-voltage pulse "pushes" the DNA into the cells. A more "natural" technique to get genes into cells is to put the cloned gene into a *transposable element* or retroviral vector. These are naturally occurring mobile regions of DNA that can integrate into the genome. Retroviruses are RNA-containing viruses. Within a host cell, they make a DNA copy of themselves (using their own reverse transcriptase); the copy becomes double-stranded and integrates into a host chromosome. The integration is accomplished as a result of two identical sequences (long terminal repeats) at the ends of the retroviral DNA. Retroviral vectors are made by removing the viral packaging genes (needed for the exit of viruses from the cell) from the center of a mouse retrovirus. This extraction creates a vacant site where other genes can be placed. By using the appropriate restriction enzymes, researchers can remove genes from a cloned phage or plasmid and reinsert the gene into the retroviral vectors. These retroviral vectors infect mouse cells with an efficiency approaching 100 percent. In *Drosophila*, new genes can be carried into a fly via P elements. These DNA sequences are naturally occurring transposable elements that can integrate like viruses into any region of the *Drosophila* genome. Moreover, they

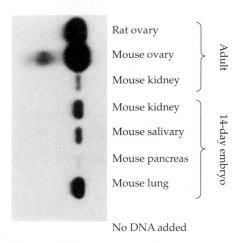

Figure 2.30
Evidence from PCR for the synthesis of a growth factor, activin, from embryonic mouse organs. The mRNA from these organs was converted into DNA and amplified through 20 replication cycles. The DNA was electrophoresed and Southern blotted using a radioactive probe to a portion of the activin gene. Activin mRNA was found in the adult mouse ovary (as expected) and also in several embryonic organs. The possible role of activin in these organs will be discussed in Chapter 17. (Courtesy of O. Ritvos.)

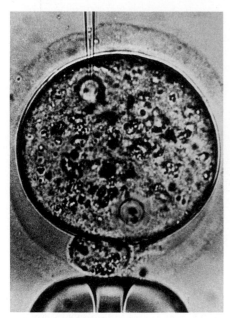

Figure 2.31
Injection of DNA (from cloned genes) into a nucleus (in this case, a pronucleus of a mouse egg). (From Wagner et al., 1981, courtesy of T. E. Wagner.)

can be isolated, and cloned genes can be inserted into the center of the P element. When the recombined P element is injected into a *Drosophila* oocyte, it can integrate into the DNA and provide the embryo with the new gene (Spradling and Rubin, 1982).

Chimeric Mice

The techniques described have recently been used to transfer genes into every cell of the mouse embryo (Figure 2.32). During mouse development, there is a stage when only two cell types are present: the outer cells, which will form the fetal portion of the placenta, and the inner cells, which will give rise to the embryo itself. These inner cells are called **embryonic stem cells** because each of them can, if isolated, generate all the cells of the embryo (Gardner, 1968; Moustafa and Brinster, 1972). These cells can be isolated from a mouse embryo and grown in culture. Once in culture, they can be treated as described so that they will incorporate new DNA. The new embryonic stem cell (the entire cell, not just the DNA) can then be injected into another early-stage mouse embryo, and the treated stem cell will integrate into the host embryo. The result is a chimeric mouse.* Some of its cells are derived from the host embryonic stem cells, but some portion of its cells are also derived from the treated stem cell. If the treated cells have become part of the germ line of the mouse, some of its gametes will be derived from the donor cell. When mated with a wild-type mouse, some of the progeny will therefore carry one copy of the inserted gene. When these heterozygote progeny are mated, about 25 percent of the resulting embryos will carry two copies of the inserted gene in every cell of its body (Gossler et al., 1986). Thus, in three generations—the chimeric mouse, the heterozygous mouse, and the homozygous mouse—a gene that had been cloned from some other individual is now present in both copies of the chromosomes within the mouse genome. Mice with stable genes derived from other individuals are called **transgenic mice.** These strains have been particularly useful in determining the functions of regulatory regions that flank the genes.

Gene-targeting ("Knockout") Experiments

The analysis of early mammalian embryos has long been hindered by our inability to breed and select mutations that affect early embryonic development. This block has been circumvented by the techniques of gene targeting (or, as it is sometimes called, gene "knockout"). These techniques are similar to those that generate transgenic mice, but instead of adding genes, gene targeting replaces wild-type alleles with mutant ones. Chisaka and Capecchi (1991) used gene targeting to find the function of the *Hoxa-3* gene in the development of the mouse. *Hoxa-3* is similar to several genes of *Drosophila* that have been found to control segment-specific gene expression in the early embryo, and the protein encoded by the *Hoxa-3* gene binds to DNA, just like its *Drosophila* counterparts. Could *Hoxa-3* similarly regulate spatially specific gene expression in mammals? Chisaka and Capecchi isolated the *Hoxa-3* gene, cut it with a restriction enzyme, and inserted a gene for neomycin resistance into that site (Figure 2.33). In other words, they mutated the *Hoxa-3* gene by inserting into it a large piece of DNA that contained a neomycin-resistance gene, destroying the ability of the Hoxa-3 protein to bind to DNA.

---

*It is critical to note the difference between a chimera and a hybrid. A hybrid results from the union of two different *genomes* within the same cell; the offspring of an *AA* genotype parent and an *aa* genotype parent is an *Aa* hybrid. A chimera results when *cells* of different genetic constitution appear in the same organism. The term is apt: it refers to a mythical beast with lion's head, goat's body, and serpent's tail.

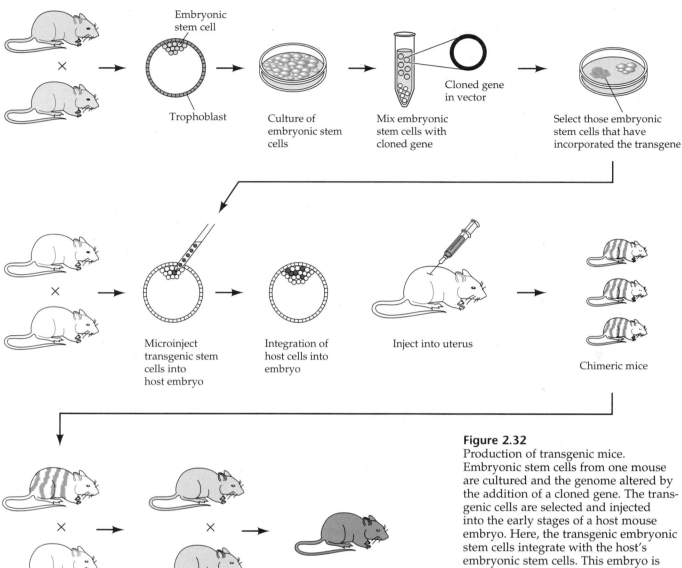

Embryonic
stem cell

Trophoblast

Culture of
embryonic stem
cells

Mix embryonic
stem cells with
cloned gene

Cloned gene
in vector

Select those embryonic
stem cells that have
incorporated the transgene

Microinject
transgenic stem
cells into
host embryo

Integration of
host cells into
embryo

Inject into uterus

Chimeric mice

Heterozygous
transgenic mice

Homozygous
transgenic mouse

Figure 2.32
Production of transgenic mice.
Embryonic stem cells from one mouse
are cultured and the genome altered by
the addition of a cloned gene. The transgenic cells are selected and injected
into the early stages of a host mouse
embryo. Here, the transgenic embryonic
stem cells integrate with the host's
embryonic stem cells. This embryo is
placed into the uterus of a pregnant
mouse and develops into a chimeric
mouse. If donor stem cells contributed
to the germ line, and the chimeric
mouse is crossed with a wild-type
mouse, some of the progeny will be heterozygous for the added allele. By mating heterozygotes, a strain of mice can
be generated that is homozygous for the
added allele. This would be a transgenic
strain. The added gene (the *transgene*)
can be from any eukaryotic source.

These mutant *Hoxa-3* genes were electroporated into embryonic stem cells
that were sensitive to neomycin. Once inside the nucleus of these cells, the
mutated *Hoxa-3* gene replaced a normal allele of *Hoxa-3* by a process called
homologous recombination. Here the enzymes involved in DNA repair and
replication incorporate the mutant gene in the place of the normal copy. It's
a rare event, but such cells can be selected by growing the stem cells in
neomycin. Most of the cells die in the drug, but the ones that acquired resistance from the incorporated gene survive. The resulting cells have one normal *Hoxa-3* gene and one mutated *Hoxa-3* gene. The heterozygous stem cells
are then microinjected into a mouse blastocyst and integrate into the cells of
the embryo. The resulting mouse is a chimera composed of wild-type cells
from the host embryo and heterozygous *Hoxa-3* cells from the donor stem
cells. The chimeras are mated to wild-type mice, and if some of the donor

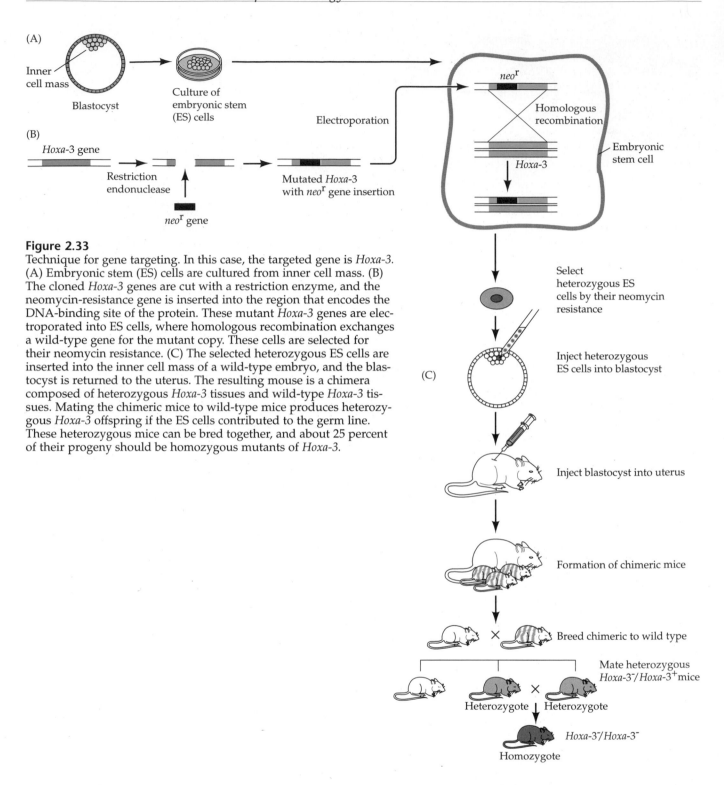

Figure 2.33
Technique for gene targeting. In this case, the targeted gene is *Hoxa-3*.
(A) Embryonic stem (ES) cells are cultured from inner cell mass. (B)
The cloned *Hoxa-3* genes are cut with a restriction enzyme, and the
neomycin-resistance gene is inserted into the region that encodes the
DNA-binding site of the protein. These mutant *Hoxa-3* genes are electroporated into ES cells, where homologous recombination exchanges
a wild-type gene for the mutant copy. These cells are selected for
their neomycin resistance. (C) The selected heterozygous ES cells are
inserted into the inner cell mass of a wild-type embryo, and the blastocyst is returned to the uterus. The resulting mouse is a chimera
composed of heterozygous *Hoxa-3* tissues and wild-type *Hoxa-3* tissues. Mating the chimeric mice to wild-type mice produces heterozygous *Hoxa-3* offspring if the ES cells contributed to the germ line.
These heterozygous mice can be bred together, and about 25 percent
of their progeny should be homozygous mutants of *Hoxa-3*.

cells became integrated into the germ cell lineage, some of the progeny will
be heterozygous for the *Hoxa-3* gene. These heterozygous mice can be bred
with each other, and about 25 percent of their progeny are expected to carry
two copies of the mutated *Hoxa-3* gene. These homozygous mutant mice
lack thyroid, parathyroid, and thymus glands! In this way, gene targeting
can be used to analyze the roles of particular genes during mammalian development. [gene7.html]

Determining the function of a message: Antisense RNA

Another method for determining the function of a gene is to make "antisense" copies of its message. Antisense messages can be generated by taking cloned cDNA and recloning it *backward* next to a strong bacterial promoter in another vector. The bacterial promoter will initiate transcription of the message "in the wrong direction" when it is incubated with RNA polymerase and nucleotide triphosphates. In so doing, it synthesizes a transcript that is complementary to the natural one (Figure 2.34A). This complementary transcript is called antisense RNA because it is the reverse of the original message. When large amounts of antisense RNA are injected or transfected into cells containing the normal mRNA from this gene, the antisense RNA binds to the normal message; the resulting double-stranded polypeptide is degraded. (Cells have enzymes to digest double-stranded nucleic acids in the cytoplasm.) This causes a functional depletion of the message, just as if there were a deletion mutation for that gene.

These results were seen when antisense RNA was made to the *Krüppel* gene of *Drosophila. Krüppel* is critical for forming the thorax and abdomen of the fly. If this gene is absent, fly larvae die because they lack thoracic and anterior abdominal segments (Figure 2.34B); a similar situation is created when large amounts of antisense RNA against the *Krüppel* message are injected into early fly embryos (Rosenberg et al., 1985). Antisense RNA allows developmental biologists to determine the function of genes during development and analyze the action of genes in animals that would otherwise be inaccessible for genetic analysis.

Reinvestigation of old problems with new methods

The union of embryology and molecular biology is giving developmental biologists a new appreciation of how genes work to construct the organism. We are in the middle of a revolution in our understanding of development, and one of the first major successes of all this cloning and sequencing has been a new "anatomy" of the eukaryotic gene. Although we will be describing gene structure in more detail in Chapter 10, it is important to realize that eukaryotic genes that encode proteins have several regulatory sites (Figure 2.35). One site, the **promoter,** is located directly upstream (before the start) of the gene and is the site where RNA polymerase binds. Located somewhere

Figure 2.34
Production of antisense RNA to examine the roles of genes in development. (A) Production of antisense message (in this case, to the *Krüppel* gene of *Drosophila*) by placing the cloned cDNA fragment for the *Krüppel* message between two strong promoters. The promoters are in opposite orientation with respect to the *Krüppel* cDNA. In this case, the T3 promoter is in normal orientation and the T7 promoter is reversed. These promoters recognize different RNA polymerases (from the T3 and T7 bacteriophages, respectively). T3 polymerase enables the transcription of "sense" mRNA, whereas T7 polymerase transcribes antisense transcripts. (B) Result of injecting *Krüppel* antisense message into the early embryo (syncytial blastoderm stage) of *Drosophila* before normal *Krüppel* message is produced. The central figure is the wild-type embryo just prior to hatching. Above it is the mutant caused by the lack of *Krüppel* genes. Below it is a wild-type embryo that had been injected with *Krüppel* antisense message in the early-embryo stage. Both the mutation and the antisense-treated embryos lack their thoracic and anterior abdominal segments. (B after Rosenberg et al., 1985.)

(A)

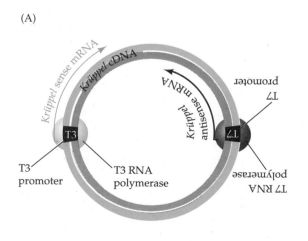

(B) *Krüppel* mutant embryo

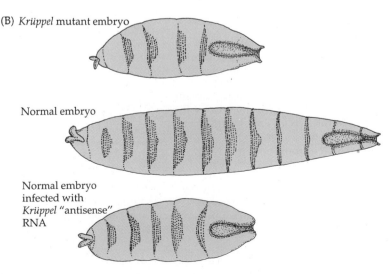

Normal embryo

Normal embryo infected with *Krüppel* "antisense" RNA

near the gene (either downstream, upstream, or in an intron within the gene) is a second region called the **enhancer.** Protein factors that bind to the enhancer enable it to interact with the promoter to enable the RNA polymerase to transcribe the gene. Whereas some promoters (such as those used by products involved in the cell's general metabolism) do not need to be activated by enhancers, most genes involved in development are activated at specific times and in specific cells. These genes need to be activated by factors that bind to the enhancer and promoter. As we will see in Chapter 10, the binding of different transcription factors to the promoters and enhancers of specific genes is one of the mechanisms that control the production of different proteins from identical genomes. One example is the activation of the gene for ZP3.

As we will detail in Chapter 4, ZP3 is the principal sperm-binding protein of the mouse egg surface. It is a glycoprotein synthesized by the oocyte as it matures in the ovum (Roller et al., 1989). A Northern blot shows that the mRNA for this glycoprotein is made only in growing oocytes and cannot be detected in any other cell type (Figure 2.36). What enables this gene to become active only in the oocytes? Lira and co-workers (1990) have isolated the gene for ZP3, determined its sequence, and found a promoter site 28 base pairs upstream from the site where transcription of the gene is initiated. They hypothesized that the sequences conferring oocyte-specific activation might exist even farther upstream from the gene. So they used restriction enzymes to isolate the DNA from the 5′ upstream region (150 base pairs long) and fused it to the gene for firefly luciferinase. (Needless to say, this light-producing enzyme is not found in mice. It is being used here as a "reporter gene" to monitor where the upstream DNA may cause its expression.) The newly constructed gene containing the upstream region of the *ZP3* gene attached to the *luciferinase* structural gene was injected into mouse zygotes to create transgenic mice that carried the *ZP3* regulatory region-*luciferinase* gene in each nucleus. In female transgenic mice, in situ hybridization located *luciferinase* mRNA in only one cell type: the oocyte (Figure 2.37). Thus, the 150-base-pair sequence of DNA was necessary and sufficient for activating the gene (any gene!) in the oocyte. Within this 150-base-pair region (at 99 to 86 base pairs upstream from the *ZP3* structural gene) there is a sequence 5′-GATAA-3′, which was found to bind a protein called OSP-1. OSP-1 is found solely in maturing oocytes, and it activates the *ZP3* gene by binding to this DNA sequence in the promoter. It appears, then, that ZP3 is made in oocytes because oocytes have OSP-1 protein, which binds to certain sequences of DNA that are part of its promoter (Schickler et al., 1992). Research is now under way to see how the OSP-1-encoding gene is regulated.

Figure 2.35
Basic structure of a developmentally regulated gene. The promoter of most genes encoding proteins is found at the 5′ (upstream) end of the gene. The enhancer is often even farther upstream, but it can also be within an intron of the gene or at the 3′ end. Proteins that bind to the promoter and enhancers interact to regulate the transcription of the gene. (In the ZP3 example, the OSP-1 site, GATAA, is located in the promoter, about 95 base pairs upstream from the transcription initiation site. An estrogen-sensitive enhancer site is found in the first intron of the *ZP3* gene.)

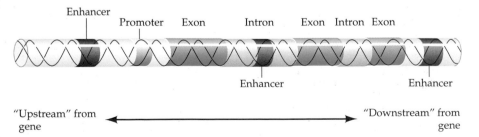

"Upstream" from gene
"Downstream" from gene

Figure 2.36
Northern blot of *ZP3* mRNA accumulation in the mouse. RNA from various tissues (10 μg per lane) and oocytes (125 ng) was electrophoresed and blotted onto nitrocellulose paper. A radioactively labeled fragment of the *ZP3* gene was used to probe the mRNA. The ZP3 message was found only in the ovary, especially within the oocytes. (From Roller et al., 1989, courtesy of P. Wassarman.)

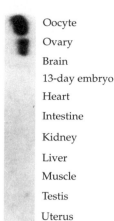

Oocyte

Ovary

Brain

13-day embryo

Heart

Intestine

Kidney

Liver

Muscle

Testis

Uterus

A conclusion and a caveat

After nearly a century, we are beginning to understand how cells regulate the differential expression of their genes so that different genes can become active in different cells. This knowledge is helping to explain how inherited information is utilized to construct the basic body plans and specific cell types of developing organisms.

A word of warning, however. Lest the celebratory tone of this chapter give you the impression that development is solely a function of gene activity, you might recall from Chapter 1 that the distinction between stalk and spore (*Dictyostelium*), amoeboid and flagellate state (*Naegleria*), and sexual and asexual gonidia (*Volvox*) is determined by *environment*. In later chapters (especially Chapter 21), we will see other examples of environmental control of development: temperature-dependent sex determination in reptiles, diet-dependent development in insects, and the experience-dependent differentiation of neurons and lymphocytes in mammals. In these cases, the organism inherits the *ability* to respond to environmental cues, but there is no predicting the phenotype from the genotype.

(A)   (B)

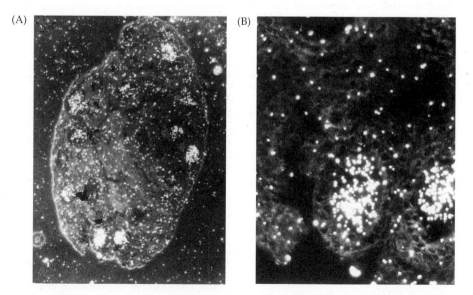

Figure 2.37
In situ hybridization of reporter gene *luciferinase* expression when *luciferinase* was attached to the promoter of the *ZP3* gene. Radioactive probe was against the *luciferinase* message, which appeared where it was expressed under the direction of the ZP3 promoter. (A) View of entire ovary (60×). (B) Higher magification (160×) of two of the ovarian follicles containing maturing oocytes. (From Lira et al., 1990, courtesy of P. Wassarman.)

LITERATURE CITED

Allen, G. E. 1978. *Thomas Hunt Morgan: The Man and His Science.* Princeton University Press, Princeton, NJ.

Allen, G. E. 1986. T. H. Morgan and the split between embryology and genetics, 1910–1935. *In* T. J. Horder, J. A. Witkowski and C. C. Wylie (eds.), *A History of Embryology.* Cambridge University Press, New York, pp. 113–146.

Ashburner, M. 1972. Patterns of puffing activity in the salivary glands of *Drosophila.* VI. Induction by ecdysone in salivary glands of *D. melanogaster* cultured in vitro. *Chromosoma* 38: 255–281.

Ashburner, M. and Berondes, H. D. 1978. Puffing of polytene chromosomes. In *The Genetics and Biology of Drosophila,* Vol. 2B. Academic Press, New York, pp. 316–395.

Baltzer, F. 1967. *Theodor Boveri: Life and Work of a Great Biologist.* (Trans. D. Rudnick.) University of California Press, Berkeley.

Barnett, T., Pachl, C., Gergen, J. P. and Wensink, P. C. 1980. The isolation and characterization of *Drosophila* yolk protein genes. *Cell* 21: 729–738.

Becker, H. J. 1959. Die Puffs der Speicheldrüsenchromosomen von *Drosophila melanogaster.* I. Beobachtungen zum Verhalten des Puffmusters im Normalstamm und bei zwei Mutanten, giant- und lethal-giant Larvae. *Chromosoma* 10: 654–678.

Beermann, W. 1952. Chromomerenkonstanz und spezifische Modifikationen der Chromosomenstruktur in der Entwicklung und Organdifferenzierung von *Chironomus tentans. Chromosoma* 5: 139–198.

Beermann, W. 1961. Ein Balbiani-ring als Locus einer Speicheldrüsen-Mutation. *Chromosoma* 12: 1–25.

Beermann, W. 1963. Cytological aspects of information transfer in cellular differentiation. *Am. Zool.* 3: 23–28.

Blattner, F. R. and eight others. 1978. Cloning human fetal g-globin and mouse a-type globin DNA: Preparation and screening of shotgun collections. *Science* 202: 1279–1283.

Boveri, T. 1904. *Ergebnisse über die Konstitution der chromatischen Substanz des Zelkerns.* Gustav Fisher, Jena. [p. 123]

Briggs, R. 1979. Genetics of cell type determination. *Int. Rev. Cytol.* [Suppl.] 9: 107–127.

Briggs, R. and King, T. J. 1952. Transplantation of living nuclei from blastula cells into enucleated frogs' eggs. *Proc. Natl. Acad. Sci. USA* 38: 455–463.

Brush, S. 1978. Nettie Stevens and the discovery of sex determination. *Isis* 69: 132–172.

Burian, R., Gayon, J. and Zallen, D. T. 1991. Boris Ephrussi and the synthesis of genetics and embryology. *In* S. Gilbert (ed.), *A Conceptual History of Modern Embryology.* Plenum, New York, pp. 207–227.

Burkholder, G. D. 1976. Whole mount electron microscopy of polytene chromosome from *Drosophila melanogaster. Can. J. Genet. Cytol.* 18: 67–77.

Capecchi, M. R. 1980. High efficiency transformation by direct microinjection of DNA into cultured mammalian cells. *Cell* 22: 479–488.

Chisaka, O. and Capecchi, M. R. 1991. Regionally restricted developmental defects resulting from targeted disruption of the homeobox gene *hox 1.5. Nature* 350: 473–479.

Clever, U. 1966. Induction and repression of a puff in *Chironomus tentans. Dev. Biol.* 14: 421–438.

Cohen, S. N., Chang, A. C. Y., Boyer, H. W. and Helling, R. B. 1973. Construction of biologically functional bacterial plasmids in vitro. *Proc. Natl. Acad. Sci. USA* 70: 3240–3244.

DiBerardino, M. A. 1987. Genomic potential of differentiated cells analyzed by nuclear transplantation. *Am. Zool.* 27: 623–644.

DiBerardino, M. A. 1989. Genomic activation in differentiated somatic cells. *In* M. A. DiBerardino and L. D. Etkin (eds.), *Developmental Biology: A Comprehensive Synthesis.* Plenum, New York, pp. 175–198.

DiBerardino, M. A. and King, T. J. 1967. Development and cellular differentiation of neural nuclear transplants of known karyotypes. *Dev. Biol.* 15: 102–128.

Dumont, J. N. and Yamada, T. 1972. Dedifferentiation of iris epithelial cells. *Dev. Biol.* 29: 385–401.

Gardner, R. L. 1968. Mouse chimeras obtained by the injection of cells into the blastocyst. *Nature* 220: 596–597.

Gilbert, S. F. 1978. The embryological origins of the gene theory. *J. Hist. Biol.* 11: 307–351.

Gilbert, S. F. 1987. In friendly disagreement: Wilson, Morgan, and the embryological origins of the gene theory. *Am. Zool.* 27: 797–806.

Gilbert, S. F. 1988. Cellular politics: Ernest Everett Just, Richard B. Goldschmidt, and the attempts to reconcile embryology and genetics. *In* R. Rainger, K. R. Benson and J. Maienschein (eds.), *The American Development of Biology.* University of Pennsylvania Press, Philadelphia, pp. 311–346.

Gilbert, S. F. 1991. Induction and the origins of developmental genetics. *In* S. Gilbert (ed.), *A Conceptual History of Modern Embryology.* Plenum, New York, pp. 181–206.

Gilbert, S.F. 1996. Enzyme adaptation and the entrance of molecular biology into embryology. *In* S. Sarkar (ed.), *The Molecular Philosophy and History of Molecular Biology: New Perspectives.* Kluwer Academic Publishers, Dordrecht, pp. 101–123.

Glueksohn-Schoenheimer, S. 1938. The development of two tailless mutants in the house mouse. *Genetics* 23: 573–584.

Glueksohn-Schoenheimer, S. 1940. The effect of an early lethal (t°) in the house mouse. *Genetics* 25: 391–400.

Goldschmidt, R. B. 1938. *Physiological Genetics.* McGraw-Hill, New York. [p. 1]

Gossler, A., Doetschman, T., Korn, R., Serfling, E. and Kemler, R. 1986. Transgenesis by means of blastocyst-derived stem cell lines. *Proc. Natl. Acad. Sci. USA* 83: 9065–9069.

Grossbach, U. 1973. Chromosome puffs and gene expressions in polytene cells. *Cold Spring Harbor Symp. Quant. Biol.* 38: 619–627.

Gurdon, J. B. 1962. The developmental capacity of nuclei taken from intestinal epithelial cells of feeding tadpoles. *J. Embryol. Exp. Morphol.* 10: 622–640.

Gurdon, J. B. 1968. Transplanted nuclei and cell differentiation. *Sci. Am.* 219(6): 24–35.

Gurdon, J. B. 1977. Egg cytoplasm and gene control in development. *Proc. R. Soc. Lond.* [B] 198: 211–247.

Gurdon, J. B. and Uehlinger, V. 1966. "Fertile" intestinal nuclei. *Nature* 210: 1240–1241.

Gurdon, J. B., Laskey, R. A. and Reeves, O. R. 1975. The developmental capacity of nuclei transplanted from keratinized cells of adult frogs. *J. Embryol. Exp. Morphol.* 34: 93–112.

Harrison, R. G. 1937. Embryology and its relations. *Science* 85: 369–374.

Harwood, J. 1993. *Styles of Scientific Thought: The German Genetics Community 1900–1933.* The University of Chicago Press, Chicago.

Hennen, S. 1970. Influence of spermine and reduced temperature on the ability of transplanted nuclei to promote normal development in eggs of *Rana pipiens. Proc. Natl. Acad. Sci. USA* 66: 630–637.

Holland, P. W. H. and Hogan, B. L. M. 1986. Phylogenetic distribution of *Antennapedia*-like homeoboxes. *Nature* 321: 251–253.

Jacob, F. and Monod, J. 1961. Genetic regulatory mechanisms in the synthesis of proteins. *J. Mol. Biol.* 3: 318–356.

Jamrich, J., Sargent, T. D. and Dawid, I. 1985. Altered morphogenesis and its effects on gene activity in *Xenopus laevis* embryos. *Cold Spring Harbor Symp. Quant. Biol.* 50: 31–35.

Judson, H. F. 1979. *The Eighth Day of Creation.* Simon & Schuster, New York.

Just, E. E. 1939. *The Biology of the Cell Surface.* Blakiston, Philadelphia.

Keller, E. F. 1995. *Refiguring Life: Metaphors of Twentieth-Century Biology.* Colorado University Press.

King, T. J. 1966. Nuclear transplantation in amphibia. *Methods Cell Physiol.* 2: 1–36.

King, T. J. and Briggs, R. 1956. Serial transplantation of embryonic nuclei. *Cold Spring Harbor Symp. Quant. Biol.* 21: 271–289.

Lambert, B. 1972. Repeated DNA sequences in a Balbiani ring. *J. Mol. Biol.* 72: 65–75.

Lambert, B. and Daneholt, B. 1975. Microanalysis of RNA from defined cellular components. *Methods Cell Biol.* 10: 17–47.

Lederman, M. 1989. Research note: Genes on chromosomes: The conversion of Thomas Hunt Morgan. *J. Hist. Biol.* 22: 163–176.

Lillie, F. R. 1927. The gene and the ontogenetic process. *Science* 64: 361–368.

Lira, A. A., Kinloch, R. A., Mortillo, S. and Wassarman, P. A. 1990. An upstream region of the mouse *ZP3* gene directs expression of firefly luciferinase specifically to growing oocytes in transgenic mice. *Proc. Natl. Acad. Sci. USA* 87: 7215–7219.

McGinnis, W., Garber, R. L., Wirz, J. Kurioiwa, A. and Gehring, W. J. 1984. A homologous protein-coding sequence in *Drosophila* homeotic genes and its conservation in other metazoans. *Cell* 37: 403–408.

McGrath, J. and Solter, D. 1983. Nuclear transplantation in the mouse embryo by microsurgery and cell fusion. *Science* 220: 1300–1302.

McGrath, J. and Solter, D. 1984. Inability of mouse blastomere nuclei transferred to enucleated zygotes to support development in vitro. *Science* 226: 1317–1319.

McKinnell, R. G. 1978. *Cloning: Nuclear Transplantation in Amphibia.* University of Minnesota Press, Minneapolis.

Monod, J. 1947. The phenomenon of enzymatic adaptation and its bearing on problems of genetics and cellular differentiation. *Growth Symp.* 11: 223–289.

Moore, J. A. 1963. *Heredity and Development.* Oxford University Press, Oxford. [p. 236]

Morange, M. 1996. Construction of the developmental gene concept. The crucial years: 1960-1980. *Biol. Zent. bl.* 115: 132–138.

Morgan, T. H. 1897. *The Frog's Egg.* Macmillan, New York. [p. 135]

Morgan, T. H. 1926. *The Theory of the Gene.* Yale University Press, New Haven.

Moustafa, L. A. and Brinster, R. L. 1972. Induced chimaerism by transplanting embryonic cells into mouse blastocysts. *J. Exp. Zool.* 181: 193–202.

Nathans, D. and Smith, H. O. 1975. Restriction endonucleases in the analysis and restructuring of DNA molecules. *Annu. Rev. Biochem.* 44: 273–293.

Okada, T. S. 1991. *Transdifferentiation.* Oxford University Press, New York.

Oppenheimer, J. M. 1981. Walter Landauer and developmental genetics. *In* S. Subtelny and U. K. Abbott (eds.), *Levels of Genetic Control in Development.* Alan R. Liss, New York, pp. 1–13.

Orr, N. H., DiBerardino, M. A. and McKinnell, R. G. 1986. The genome of frog erythrocytes displays centuplicate replications. *Proc. Natl. Acad. Sci. USA* 83: 1369–1373.

Pardue, M. L. and Gall, J. G. 1970. Chromosomal localization of mouse satellite DNA. *Science* 168: 1356–1358.

Paul, D. B. and Kimmelman, B. A. 1988. Mendel in America: Theory and practice, 1900–1919. *In* R. Rainger, K. R. Benson and J. Maienschein (eds.), *The American Development of Biology.* University of Pennsylvania Press, Philadelphia, pp. 281–310.

Perucho, M., Hanahan, D. and Wigler, M. 1980. Genetic and physical linkage of exogenous sequences in transformed cells. *Cell* 22: 309–317.

Prather, R. S. 1991. Nuclear transplantation and embryo cloning in mammals. *Int. Lab. Animal Res. News* 33: 62–68.

Prather, R. S., Barnes, F. L., Sims, M. M., Robl, J. M., Eyestone, W. H. and First, N. L. 1987. Nuclear transplantation in the bovine embryo: Assessment of donor nuclei and recipient oocyte. *Biol. Reprod.* 37: 859–866.

Prather, R. S., Sims, M. M. and First, N. L. 1989. Nuclear transplantation in early porcine embryos. *Biol. Reprod.* 41: 414–418.

Rappolee, D. A., Brenner, C. A., Schultz, R., Mark, D. and Werb, Z. 1988. Developmental expression of *PDGF, TGF-α* and *TGF-β* genes in preimplantation mouse embryos. *Science* 241: 1823–1825.

Reyer, R. W. 1954. Regeneration in the lens in the amphibian eye. *Q. Rev. Biol.* 29: 1–46.

Robins, D. M., Ripley, S., Henderson, A. S. and Axel, R. 1981. Transforming DNA integrates into the host chromosome. *Cell* 23: 29–39.

Roller, R. J., Kinloch, R. A., Hiraoka, B. Y., Li, S. S.-L. and Wassarman, P. M. 1989. Gene expression during mammalian oogenesis and early embryogenesis: Quantification of three messenger RNAs abundant in fully grown mouse oocytes. *Development* 106: 251–261.

Rosenberg, U. B., Preiss, A., Seifert, E., Jäckle, H. and Knipple, D. C. 1985. Production of phenocopies by *Krüppel* antisense RNA injection into *Drosophila* embryos. *Nature* 313: 703–706.

Saiki, R. K., Scharf, S., Faloona, F., Mullis, K. B., Horn, G. T., Erlich, H. A. and Arnheim, N. 1985. Enzymatic amplification of β-globin genomic sequences and restriction site analysis for diagnosis of sickle cell anemia. *Science* 230: 1350–1354.

Sander, K. 1986. The role of genes in ontogenesisevolving concepts from 1883 to 1983 as perceived by an insect embryologist. *In* T. J. Horder, J. A. Witkowski and C. C. Wylie (eds.), *A History of Embryology.* Cambridge University Press, New York, pp. 363–395.

Sanger, F., Nicklen, S. and Coulson, A. R. 1977. DNA sequencing with chain-terminating inhibitors. *Proc. Natl. Acad. Sci. USA* 74: 5463–5467.

Sapp, J. 1987. *Beyond the Gene: Cytoplasm Inheritance and the Struggle for Authority in Genetics.* Oxford Universtiy Press.

Sargent, T. D. and Dawid, I. 1983. Differential gene expression in the gastrula of *Xenopus laevis. Science* 222: 135–139.

Schickler, M., Lira, S., Kinloch, R. A. and Wassarman, P. A. 1992. A mouse oocyte-specific protein that binds to a region of mZP3 promoter responsible for oocyte-specific mZP3 gene expression. *Mol. Cell Biol.* 122: 120–127.

Smith, L. D. 1956. Transplantation of the nuclei of primordial germ cells into enucleated eggs of *Rana pipiens. Proc. Natl. Acad. Sci. USA* 54: 101–107.

Southern, E. M. 1975. Detection of specific sequences among DNA fragments separated by gel electrophoresis. *J. Mol. Biol.* 98: 503–517.

Spemann, H. 1938. *Embryonic Development and Induction.* Yale University Press, New Haven.

Spiegelman, S. 1947. Differentiation as the controlled production of unique enzymatic patterns. *In* J. F. Danielli and R. Brown (eds.), *Growth in Relation to Differentiation and Morphogenesis.* Cambridge University Press, Cambridge, p. 287.

Spradling, A. C. and Rubin, G. M. 1982. Transposition of cloned P elements into *Drosophila* germ line chromosomes. *Science* 218: 341–347.

Stevens, N. M. 1905a. A study of the germ cells of *Aphis rosae* and *Aphis oenotherae. J. Exp. Zool.* 2: 371–405; 507–545.

Stevens, N. M. 1905b. *Studies in Spermatogenesis with Especial Reference to the "Accessory Chromosome."* Carnegie Institute of Washington, Washington, D.C.

Steward, F. C. 1970. From cultured cells to whole plants: The induction and control of their growth and morphogenesis. *Proc. R. Soc. Lond.* [B] 175: 1–30.

Steward, F. C., Mapes, M. O. and Smith, J. 1958. Growth and organized development of cultured cells. I. Growth and division of freely suspended cells. *Am. J. Bot.* 45: 693–703.

Steward, F. C., Mapes, M. O., Kent, A. E. and Holsten, R. D. 1964. Growth and development of cultured plant cells. *Science* 143: 20–27.

Stice, S. J. and Robl, J. M. 1988. Nuclear reprogramming in nuclear transplant rabbit embryos. *Biol. Reprod.* 39: 657–664.

Ursprung, H., Smith, K. D., Sofer, W. H. and Sullivan, D. T. 1968. Assay systems for the study of gene function. *Science* 160: 1075–1081.

Vierra, J. and Messing, J. 1982. The pUC plasmids, an M13mp7-derived system for insertion mutagenesis and sequencing with synthetic universal primers. *Gene* 19: 259–268.

Waddington, C. H. 1939. Preliminary notes on the development of wings in normal and mutant strains of *Drosophila*. *Proc. Natl. Acad. Sci. USA* 25: 299–307.

Waddington, C. H. 1962. *New Patterns in Genetics and Development*. Columbia University Press, New York, pp. 14–36.

Wagner, T. E., Hoppe, P., Jollick, J. D., Scholl, D. R., Hodinka, R. L. and Gault, J. B. 1981. Microinjection of rabbit β-globin gene into zygotes and its subsequent expression in adult mice and offspring. *Proc. Natl. Acad. Sci. USA* 78: 6376–6380.

Wieslander, L. and Daneholt, B. 1977. Demonstration of Balbiani ring RNA sequence in polysomes. *J. Cell Biol.* 73: 260–264.

Wessel, G. M., Goldberg, L., Lennarz, W. J. and Klein, W. H. 1989. Gastrulation in the sea urchin is accompanied by the accumulation of an endoderm-specific mRNA. *Dev. Biol.* 136: 526–538.

Wetmur, J. G. and Davidson, N. 1968. Kinetics of renaturation of DNA. *J. Mol. Biol.* 31: 349–370.

Wilkinson, D. G., Bhatt, S. and Herrmann, B. G. 1990. Expression pattern of the mouse *T* gene and its role in mesoderm formation. *Nature* 343: 657–659.

Willadsen, S. M. 1986. Nuclear transplantation in sheep embryos. *Nature* 320: 63–65.

Willadsen, S. M. 1989. Cloning of sheep and cow embryos. *Genome* 31: 956–962.

Wilmut, I., Schnieke, A. E., McWhir, J., Kind, A. J. and Campbell, K. H. S. 1997. Viable offspring from fetal and adult mammalian cells. *Nature* 385: 810–813.

Wilson, E. B. 1894. The mosaic theory of development. *Biol. Lect. Marine Biol. Lab. Woods Hole* 2: 1–14.

Wilson, E. B. 1895. *An Atlas of the Fertilization and Karyogenesis of the Ovum*. Macmillan, New York. [p. 4]

Wilson, E. B. 1896. *The Cell in Development and Inheritance*. Macmillan, New York. [p. 262]

Wilson, E. B. 1904. Experimental studies on germinal localization. I. The germ regions in the egg of *Dentalium*. *J. Exp. Zool.* 1: 1–72.

Wilson, E. B. 1905. The chromosomes in relation to the determination of sex in insects. *Science* 22: 500–502.

Yamada, T. 1966. Control of tissue specificity: The pattern of cellular synthetic activities in tissue transformation. *Am. Zool.* 6: 21–31.

The cellular basis of morphogenesis: Differential cell affinity

3

But nature is not atomized. Its patterning is inherent and primary, and the order underlying beauty is demonstrably there; what is more, the human mind can perceive it only because it is itself part and parcel of that order.
PAUL WEISS (1960)

I am fearfully and wonderfully made.
PSALM 139 (ca. 500 B.C.E.)

A BODY IS NOT merely a collection of randomly distributed cell types. Development involves not only the *differentiation* of cells, but also their *morphogenesis* into multicellular arrangements such as tissues and organs. When we observe the detailed anatomy of a tissue such as the neural retina, we see an intricate and precise arrangement of many different types of cells. In this chapter, we will introduce the ways by which the cells of the developing embryo change to create the functional organs of the body. There are four major questions that form the framework for discussions of morphogenesis:

- *How are tissues formed from cells?* How do neural retina cells stick to other neural retina cells and not become integrated into the pigmented retina or iris cells next to them? How are the various cell types within the neural retina (the three distinct layers of photoreceptors, bipolar neurons, and ganglion cells) arranged so that the retina is functional?

- *How are organs constructed from tissues?* The retinal cells of the eye form behind the cornea and lens at a precise distance. The retina would be useless if it formed behind a bone or somewhere else where the lens could not focus light on to it. Moreover, the neurons from the retina must enter the brain to innervate the regions of the brain cortex that analyze the visual information. All these connections must be precisely ordered.

- *How do migrating cells reach their destinations, and how do organs form in particular locations?* Eyes develop in the head and nowhere else. What stops an eye from forming in other parts of the body if all the cells have the same genetic potential? In some cases, such as the precursors of our pigment cells, germ cells, and adrenal gland, the cells have to migrate long distances to reach their final destinations. How are cells instructed to travel along certain routes and to stop when they reach a particular region of the body?

- *How do organs and their cells grow, and how is their growth coordinated throughout development?* The cells of the eye must grow together, and the cells of the retina are hardly ever seen to divide after birth. Our intestine, however, is constantly shedding cells and regenerating new cells, and yet its mitotic rate is carefully regulated. If it were to generate more cells than it sloughed off, it could produce cancerous outgrowths. If it produced fewer cells than it sloughed off, the intestine could not digest food. What controls these differences in growth rate?

All these questions concern aspects of cell behavior. There are two main groups of cells in the embryo: **epithelial cells,** which are tightly connected to one another in sheets or tubes, and **mesenchymal cells,** which are unconnected to each other and which operate as individual units. Morphogenesis is brought about through a limited repertoire of cellular processes in these two classes of cells: (1) the direction and number of cell divisions; (2) cell shape changes; (3) cell movement; (4) cell growth; (5) cell death; and (6) changes in the composition of the cell membrane and extracellular matrix. How these processes are accomplished can differ between mesenchymal and epithelial cells (Figure 3.1).

There also appear to be two major ways by which cells communicate with one another to effect morphogenesis. The first is through **diffusible substances** that are made by one type of cell and that change the behavior of other types of cells. These substances include hormones, growth factors, and morphogens; each will be detailed in subsequent chapters. The second method involves **contact** between the surfaces of adjacent cells. Cells can selectively recognize other cells, adhering to some and migrating over others. The molecular events mediating the selective recognition of cells and their formation into tissues and organs occur at the cell surface. Whereas the dominant paradigm of developmental genetics is differential gene expression, the dominant paradigm of morphogenesis involves **differential cell affinities.** These affinities can be for the surfaces of other cells or for the molecules of the extracellular matrix secreted by the cells. In this chapter, we will see how adjacent cell surfaces interact during development to localize cells in their proper sites within tissues and organs.

Differential cell affinity

Just as there was a struggle to show the importance of genes in development, there was also a debate over whether the cell surface plays a role in forming the embryo. The cell surface looks pretty much the same in all cell types, and many early investigators thought that the cell surface was not even a living part of the cell. Observations of fertilization and early embryonic development made by E. E. Just (1939) suggested that the cell surface differed between cell types, but the modern analysis of morphogenesis begins with the experiments of Townes and Holtfreter in 1955. Taking advantage of the discovery that amphibian tissues become dissociated into single cells when placed in alkaline solutions, they prepared single-cell suspensions from each of the three amphibian germ layers soon after the neural tube had formed. Two or more of these single-cell suspensions could be combined in various ways, and when the pH was normalized, the cells adhered to each other, forming aggregates on agar-coated petri dishes. By using embryos from species having cells of different sizes and colors, Townes and Holtfreter were able to follow the behavior of the recombined cells (Figure 3.2).

Figure 3.1   ▶
Summary of major morphogenetic processes in mesenchymal cells and epithelial cells.

PROCESS	ACTION	MORPHOLOGY	EXAMPLE

MESENCHYMAL CELLS

PROCESS	ACTION	MORPHOLOGY	EXAMPLE
Condensation	Mesenchyme becomes epithelium	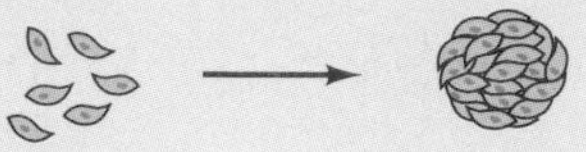	Cartilage mesenchyme
Cell division	Mitosis to produce more cells (hyperplasia)	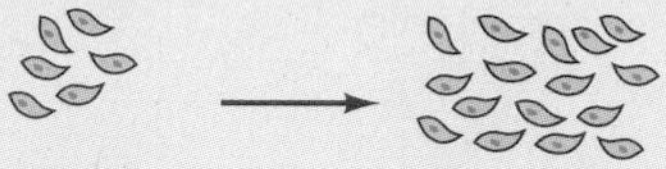	Limb mesenchyme
Cell death	Cells die	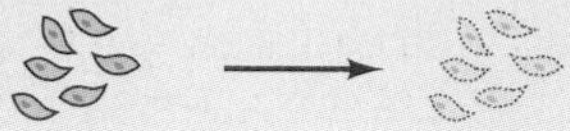	Interdigital mesenchyme
Migration	Cells move at particular times and places	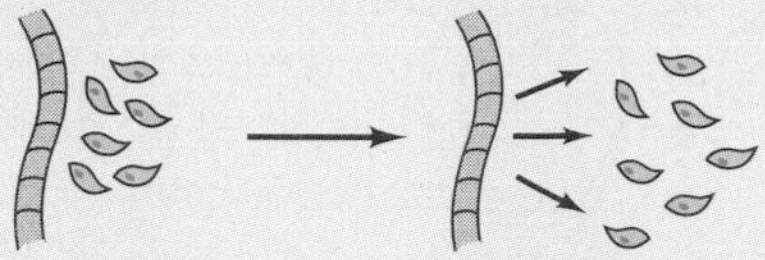	Heart mesenchyme
Matrix secretion and degradation	Synthesis or removal of extracellular layer	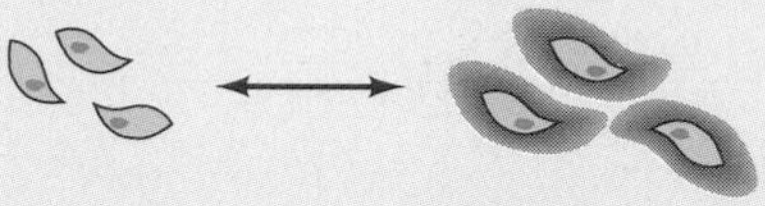	Cartilage mesenchyme
Growth	Cells get larger (hypertrophy)		Fat cells

EPITHELIAL CELLS

PROCESS	ACTION	MORPHOLOGY	EXAMPLE
Dispersal	Epithelium ⟶ Mesenchyme (entire structure)	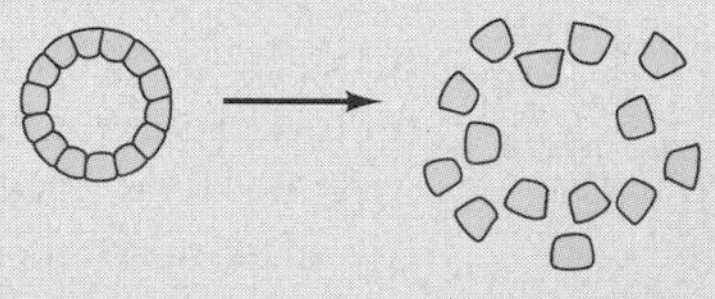	Müllerian duct degeneration
Delamination	Epithelium ⟶ Mesenchyme (part of structure)	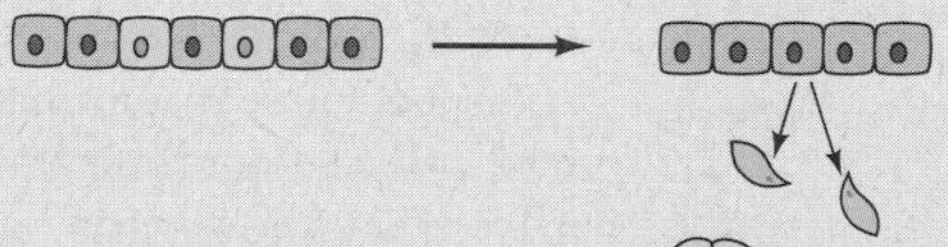	Chick hypoblast
Shape change or growth	Cells remain attached as morphology is altered	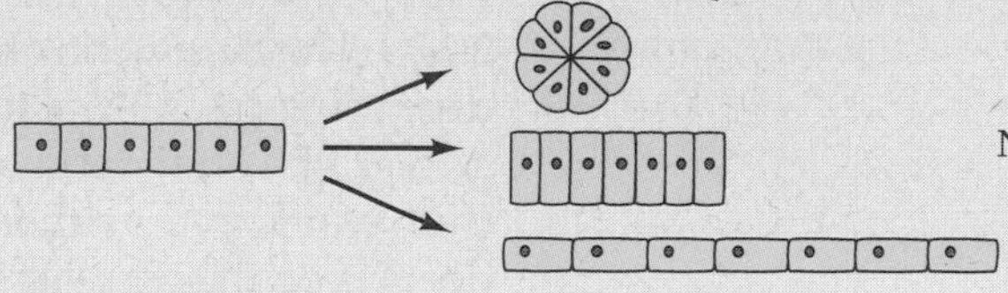	Neurulation
Cell migration (intercalation)	Rows of epithelia merge to form fewer rows	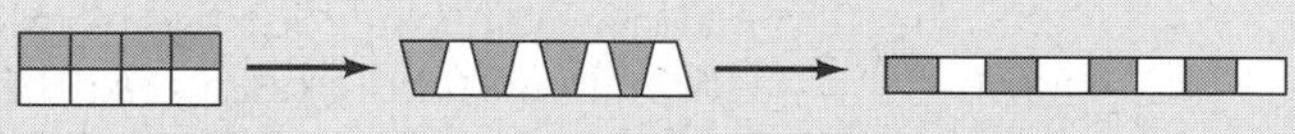	Vertebrate gastrulation
Cell division	Mitosis within row or other direction		Vertebrate gastrulation
Matrix secretion and degradation	Synthesis or removal of extracellular layer	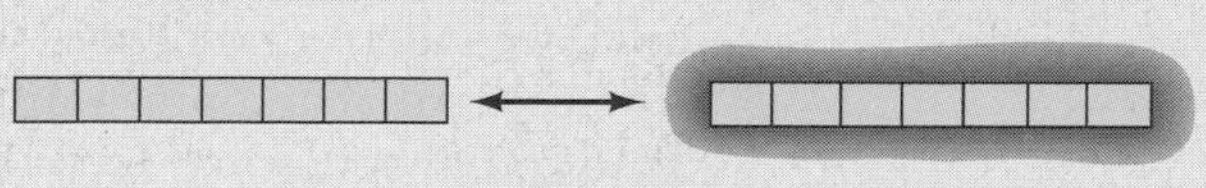	Vertebrate organ formation
Migration	Formation of free edges	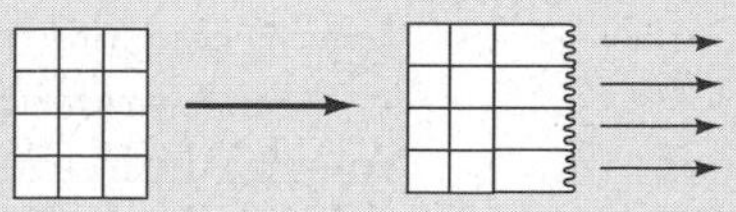	Chick ectoderm

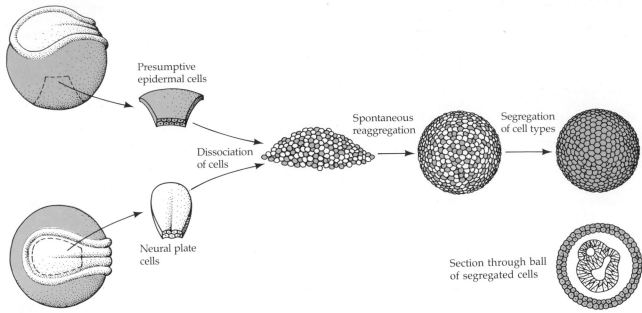

Figure 3.2
Reaggregation of cells from amphibian neurulae. Presumptive epidermal cells from pigmented embryos and neural plate cells from unpigmented embryos are dissociated and mixed together. The cells reaggregate such that one type (here, the presumptive epidermis) covers the other. (Modified from Townes and Holtfreter, 1955.)

The results of their experiments were striking. First, they found that reaggregated cells become spatially segregated. That is, instead of remaining mixed, each cell type sorts out into its own region. Thus, when epidermal (ectodermal) and mesodermal cells are brought together to form a mixed aggregate, the epidermal cells are found at the periphery of the aggregate and the mesodermal cells are found inside. In no case do the recombined cells remain randomly mixed, and in most cases, one tissue type completely envelops the other.

Second, the researchers found that the final positions of the reaggregated cells reflect their embryonic positions. The mesoderm migrates centrally to the epidermis, adhering to the inner epidermal surface (Figure 3.3A). The mesoderm also migrates centrally with respect to the gut or endoderm (Figure 3.3B). However, when the three germ layers are mixed together, the endoderm separates from the ectoderm and mesoderm and is then enveloped by them (Figure 3.3C). In its final configuration, the ectoderm is on the periphery, the endoderm is internal, and the mesoderm lies in the region between them. Holtfreter interpreted this in terms of selective affinity. The inner surface of the ectoderm has a positive affinity for mesodermal cells and a negative affinity for the endoderm, while the mesoderm has positive affinities for both ectodermal and endodermal cells. The mimicry of normal embryonic structure by cell aggregates is also seen in the recombination of epidermis and neural plate cells (Figure 3.3D). The presumptive epidermal cells migrate to the periphery as before; the neural plate cells migrate inward, forming a structure reminiscent of the neural tube. When axial mesoderm (notochord) cells are added to the suspension of presumptive epidermal and presumptive neural cells, the cell segregation results in an external epidermal layer, a centrally located neural tissue, and a layer of mesodermal tissue between them (Figure 3.3E). Somehow, the cells are able to sort out into their proper embryological positions.

Such preferential affinities have also been noted by Boucaut (1974), who injected individual cells from specific germ layers back into the cavity of amphibian gastrulae. He found that these cells migrate to their appropriate germ layer. Endodermal cells find positions in the host endoderm, whereas ecto-

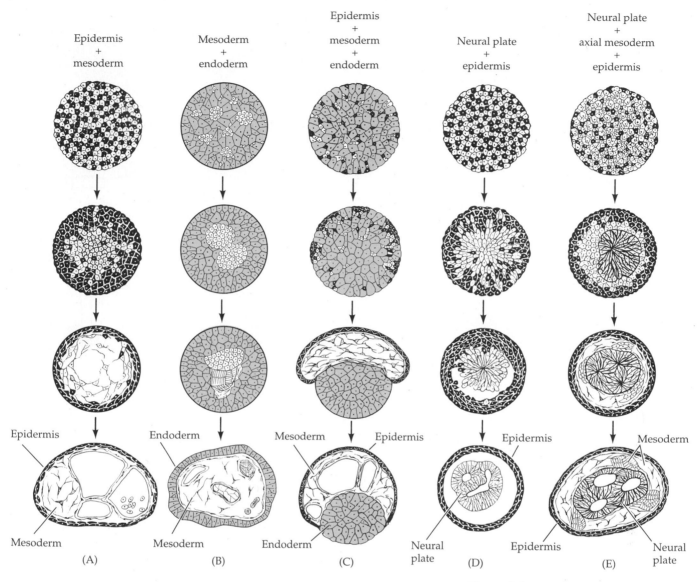

Epidermis
+
mesoderm

Mesoderm
+
endoderm

Epidermis
+
mesoderm
+
endoderm

Neural plate
+
epidermis

Neural plate
+
axial mesoderm
+
epidermis

Epidermis

Mesoderm

(A)

Endoderm

Mesoderm

(B)

Mesoderm — Epidermis

Endoderm

(C)

Epidermis

Neural
plate

(D)

Mesoderm

Epidermis

Neural
plate

(E)

Figure 3.3
Sorting out and reorganization of embryonic spatial relationships in aggregates of embryonic amphibian cells. (Modified from Townes and Holtfreter, 1955.)

dermal cells are found only in host ectoderm. Thus, selective affinity appears to be important for imparting positional information to embryonic cells.

The third conclusion of Holtfreter and his colleagues was that selective affinities change during development. This should be expected, because embryonic cells do not retain a single stable relationship with other cells. For development to occur, cells must interact differently with other cell populations at specific times. Such changes in cellular affinity were dramatically confirmed by Trinkaus (1963), who showed a clear correlation between adhesive changes in vitro and changes in embryonic cell behavior. More recently, the experiments of Fink and McClay (1985) demonstrated this behavior during sea urchin development. In the blastula, all cells seem to have the same affinity for one another. Each cell also has high affinity for the extracellular matrix (hyaline layer) covering the embryo and a low affinity for the proteins inside the embryonic cavity (blastocoel). However, at the onset of gastrulation, a specific group of cells at the vegetal end of the blastula lose their affinity for neighboring cells and for the external extracellular matrix while simultaneously acquiring affinity for the protein fibrils lining the blastocoel (Figure

Figure 3.4
Summary of the changes in cell adhesion of skeletal precursor cells (boxed). (A) In the sea urchin blastula, each cell has high affinity for its neighbors and for its substrate, the hyaline layer. (B) As development progresses, changes in the cell surface of these cells cause them to weaken their affinities for surrounding cells and the hyaline layer and to increase their affinities for the proteins inside the blastocoel cavity. The result is that these cells migrate into the blastocoel (arrows) and will form the skeleton.

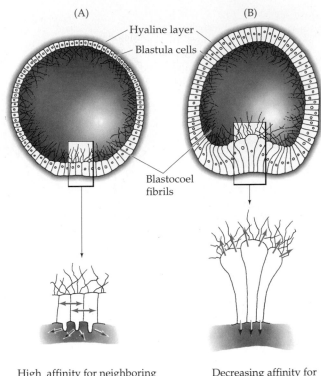

High affinity for neighboring cells and hyaline layer

Decreasing affinity for neighboring cells and hyaline layer. Increased affinity for blastocoel fibrils

3.4). These changes in affinities cause these cells to release their contacts with their neighboring cells and to migrate inside the blastocoel, where they will form the larval skeleton. When they begin to form this skeleton, their adhesion properties will have to change again. These cells, which had been "antisocial" toward each other since their ingression into the blastocoel, must now adhere together to form the rudiments of the skeletal ring. These changes in adhesion are temporally specific and are also specific to the skeletal precursor cells (McClay and Ettensohn, 1987). Such changes in cell affinity are extremely important in the processes of morphogenesis.

The reconstruction of aggregates from later embryos of birds and mammals was accomplished by the use of the protease trypsin to dissociate the cells from one another (Moscona, 1952). When the resulting single cells were mixed together in a flask and swirled so that the shear force would break any nonspecific adhesions, the cells sorted out according to their cell type. In so doing, they reconstructed the organization of the original tissue (Moscona, 1961; Giudice, 1962). Figure 3.5 shows the "reconstruction" of skin tissue from a 15-day embryonic mouse. The skin cells are separated by proteolytic enzymes and then aggregated in a rotary culture. The epidermal cells migrate to the periphery, and the dermal cells migrate toward the center. By 72 hours, the epidermis has been reconstituted, a keratin layer has formed, and hair follicles are seen in the dermal region. Such reconstruction of complex tissues from single cells is called **histotypic aggregation.**

The Thermodynamic Model of Cell Interactions

Cells, then, do not sort randomly but can actively move to create tissue organization. What forces direct cell movement during morphogenesis? In 1964,

Figure 3.5
Reconstruction of skin from a suspension of skin cells from a 15-day embryonic mouse. (A) Section through the embryonic skin, showing epidermis, dermis, and primary hair follicle. (B) Suspension of single skin cells from both the dermis and the epidermis. (C) Aggregates after 24 hours. (D) Section through an aggregate, showing migration of epidermal cells to the periphery. (E) Further differentiation of aggregates (72 hours), showing reconstituted epidermis and dermis, complete with hair follicles and keratinized layer. (From Monroy and Moscona, 1979, courtesy of A. Moscona.)

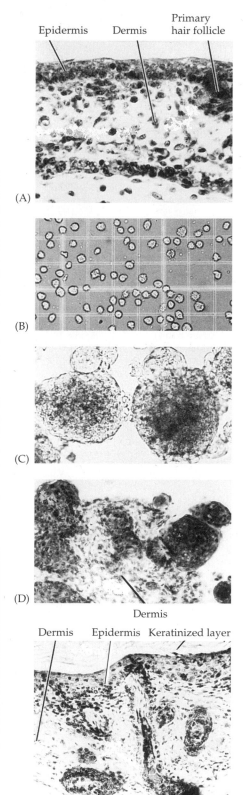

Malcolm Steinberg proposed a model that explained the directions of cell sorting on thermodynamic principles. Using cells derived from trypsinized embryonic tissues, Steinberg showed that certain cell types always migrate centrally when combined with some cell types, but migrate peripherally when combined with others. Figure 3.6 illustrates the interactions between cultures of pigmented retina cells and neural retina cells. When single-cell suspensions of these two cell types are mixed together, they form aggregates of randomly arranged cells. However, after several hours, the pigmented retina cells are no longer seen on the periphery of the aggregates, and by two days, two distinct layers are seen, with the pigmented retina lying internal to the neural retina cells. The same type of interactions can also be seen when spherical aggregates of tissues are placed in contact with each other. One of the tissues eventually envelops the other, and the final topography is independent of the starting positions (Figure 3.7).

Moreover, such interactions form a hierarchy (Steinberg, 1970). If the final position of one cell type, A, is internal to a second cell type, B, and the final position of B is internal to a third cell type, C, then the final position of A will always be internal to C. For example, pigmented retina cells migrate internally to neural retina cells, and heart cells migrate centrally to pigmented retina. Therefore, heart cells migrate internally to neural retina cells. This observation led Steinberg to propose that the mixed cells interact to

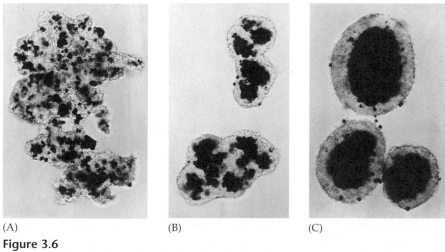

(A)   (B)   (C)

Figure 3.6
Aggregates formed by mixing 7-day-old chick embryo neural retina (unpigmented) cells with pigmented retina (dark) cells. (A) 5 hours after the single-cell suspensions are mixed, aggregates of randomly distributed cells are seen. (B) At 19 hours, the pigmented retina cells are no longer seen on the periphery. (C) By two days, a great majority of the pigmented retina cells are located in a central internal mass, surrounded by the neural retina cells. (The scattered pigmented cells are probably dead cells.) (From Armstrong, 1989, courtesy of P. B. Armstrong.)

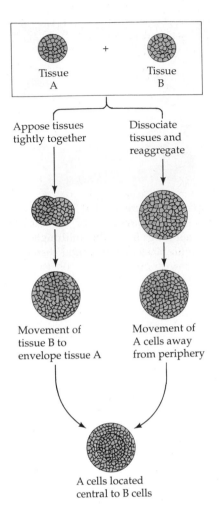

Tissue A + Tissue B

Appose tissues tightly together

Dissociate tissues and reaggregate

Movement of tissue B to envelope tissue A

Movement of A cells away from periphery

A cells located central to B cells

Figure 3.7
Spreading of one cell type over another. The final position of aggregates composed of two tissue types is independent of their initial position. An identical final condition is formed whether the tissues are made into single-cell suspensions and then reaggregated or the two tissues are kept intact and brought into contact. (After Armstrong, 1989.)

form an aggregate with the smallest interfacial free energy (Figure 3.8). In other words, the cells rearrange themselves into the most thermodynamically stable pattern. If cell types A and B have different strengths of adhesion, and if the strength of A-A connections is greater than the strength of A-B or B-B connections, sorting will occur, with the A cells becoming central. If the strength of A-A connections is less than or equal to the strength of A-B connections, then the aggregate will remain as a random mix of cells. Finally, if the strength of A-A connections is far greater than the strength of A-B connections—in other words, A and B cells show essentially no adhesivity toward one another—then A cells and B cells will form separate aggregates.

All that is needed for sorting out to occur is that cells differ in the strengths of their adhesion. In the simplest form of this model, all cells could have the same type of "glue" distributed on the cell surface. The amount of this cell-surface product or the cellular architecture that allows the substance to be differentially concentrated could cause a difference in the number of stable contacts made between cell types. Alternatively, the thermodynamic differences could be caused by several types of adhesion molecules. This thermodynamic model is called the **differential adhesion hypothesis.** In this hypothesis, the early embryo can be viewed as existing in an equilibrium state until some change in gene activity changes the cell-surface molecules. The movements that occur seek to restore the cells to a new equilibrium configuration.

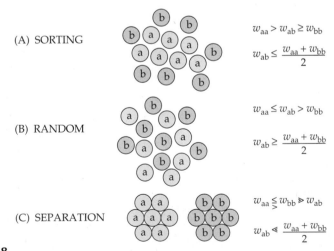

(A) SORTING

$$w_{aa} > w_{ab} \geq w_{bb}$$

$$w_{ab} \leq \frac{w_{aa} + w_{bb}}{2}$$

(B) RANDOM

$$w_{aa} \leq w_{ab} > w_{bb}$$

$$w_{ab} \geq \frac{w_{aa} + w_{bb}}{2}$$

(C) SEPARATION

$$w_{aa} \lesseqgtr w_{bb} \gg w_{ab}$$

$$w_{ab} \lll \frac{w_{aa} + w_{bb}}{2}$$

Figure 3.8
Sorting out as a process tending toward the maximum thermodynamic stability. (A) Sorting out occurs when the average strength of adhesions between different types of cells (w_{ab}) is less than the average of homotypic (A-A or B-B) adhesive strengths (w_{aa}, w_{bb}). The more adhesive cells becomes centrally located. (B) If the strength of the A-B adhesions is greater than or equal to the average of the homotypic adhesions, no sorting will occur, because the system has already achieved a thermodynamic equilibrium, and the mixture of cell types will be random. (C) If the A-B bonds are much weaker than the average of the homotypic adhesions, complete separation will ensue, as is characteristic of oil and water.

Evidence for the Thermodynamic Model

EVIDENCE for the differential adhesion hypothesis has recently emerged from research aimed at answering two questions: (1) Can the sorting-out phenomenon be explained by cell-adhesion-generated surface tensions, and (2) does such sorting out actually occur during development?

Foty and colleagues in Steinberg's laboratory (1994) have analyzed the interfacial surface tensions of various embryonic tissues. They compressed tissue samples between glass plates of a tensiometer and measured the tissues' surface tensions as their ability to restore themselves to their original spheroidal shapes. In this way, the surface tension of each tissue could be calculated in dynes per centimeter. Foty and his co-workers (1996) found a complete correlation between the surface tension of the tissue and whether it sorted out to the center or the periphery of a mixed aggregate. Tissues with the higher surface tension always became internal when mixed with tissues having a lower surface tension. It appears that sorting out can be explained solely by the interfacial surface tensions of the apposing cells. [cell1.html]

Until recently, it was extremely difficult to design an experiment to test this model of cell sorting in vivo; but evidence for this hypothesis is emerging from studies of limb regeneration in salamanders. Here, the more proximal (close to the body) tissue will envelop the more distal tissues (Nardi and Stocum, 1983).

Salamander limbs have certain remarkable attributes. When a forelimb is amputated at the *upper arm,* the remaining stump forms a dedifferentiated mass of cells at its tip (**the regeneration blastema**), which divides and differentiates to form a new limb. The new limb tissue starts at the amputation site, in this case forming the remainder of the limb from the upper arm downward. When the forelimb is amputated at the *wrist,* a similar regeneration blastema forms. However, it does not start making upper arm tissue, elbow, and ulna. Rather, it "knows" its location and regenerates only the wrist and digits.

How is this "positional memory" stored? Nardi and Stocum (1983) demonstrated that when two salamander limb blastemas from the same level of origin are placed together, they fuse, but neither tissue surrounds the other (Figure 3.9). However, when the blastemas are from different levels, the more proximal (closer to the body) blastema surrounds the more distal one. It appears, then, that the adhesive properties of these cells form a gradient along the proximodistal axis; these properties are greatest at the wrist and lowest at the upper arm.

Crawford and Stocum (1988) were able to relate this in vitro sorting of cells to processes in the living, regenerating limb. Blastemas from the wrist, elbow, or upper arm were grafted into the blastema-stump junction of a hindlimb regenerating from the midthigh. The forelimb blastemas migrated distally to the corresponding level of the host hindlimb and then regenerated a new structure (Figure 3.10). The *upper arm* blastema immediately regenerated a complete arm from the midthigh level; the *elbow* blastema moved to the level of the knee and then formed the remainder of the arm from this point onward; and the *wrist* blastema was displaced to the end of the regenerating hindlimb, where it formed a wrist adjacent to the tarsus of the foot. These data suggest that the hierarchies of cellular sorting out seen in vitro reflect differences that are used by the body in constructing new organs in vivo. ∎

Figure 3.9 Sorting when blastemas from the same or different forelimb levels are brought together in culture. (One member of each pair was marked with tritium to distinguish it from the other.) After three days in culture, the aggregates were fixed and sectioned. Blastemas from the same level fused in a straight line. When blastemas were from different levels, the proximal blastema could be seen attempting to surround the more distal cells. (From Nardi and Stocum, 1983, courtesy of D. Stocum.)

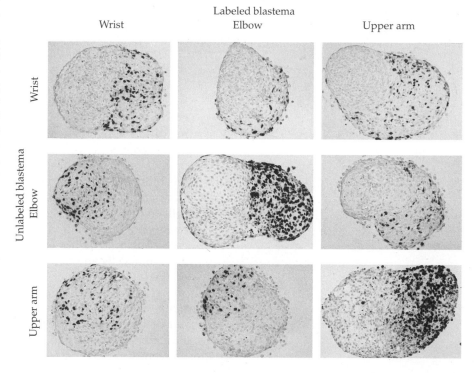

Labeled blastema

Wrist · Elbow · Upper arm

Unlabeled blastema: Wrist · Elbow · Upper arm

Figure 3.10 Sorting out in vivo, whereby regenerating forelimb blastemas (color) grafted into the midthigh blastemas (gray) are displaced to the corresponding region of the regenerating hindlimb (wrist to tarsus; elbow to knee; upper arm to midthigh) and initiate forelimb formation from that point distally. (After Crawford and Stocum, 1988.)

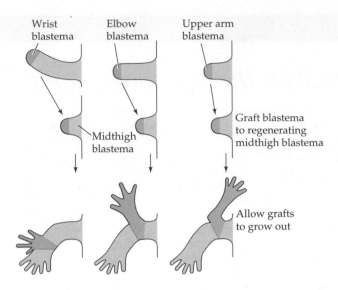

The molecular basis of cell-cell adhesions

The Classes of Cell Adhesion Molecules

The formation of tissues and organs is mediated by events occurring at the cell surfaces of adjacent cells. The cell surface includes the plasma membrane, the molecules directly beneath the membrane and associated with it, and the molecules found in the extracellular spaces. Eukaryotic cells are surrounded by a complex molecular border called the **plasma** (or cell) **membrane.** The plasma membrane is a fluid lipid bilayer that contains proteins capable of interacting with the outside environment. Certain proteins have their active sites pointing outward, toward other cells, and there are three classes of cell membrane molecules (mostly proteins) that are particularly involved in controlling specific interactions with other cells (Edelman and Thiery, 1985):

- *Cell adhesion molecules.* These proteins are involved in cell-cell adhesion. They can unite cells into epithelial sheets and condense mesenchymal cells into cohesive aggregates. They are critical in separating different tissues from one another.
- *Cell junction molecules.* These molecules provide communication pathways between the cytoplasm of adjacent cells and provide permeability barriers and mechanical strength to epithelial sheets.
- *Substrate adhesion molecules.* These molecules are involved in binding cells to their extracellular matrices. They include components of the extracellular matrix and the cell-surface receptors for these molecules. Substrate adhesion molecules permit the movement of mesenchymal cells and neurons and allow the spatial separation of epithelial sheets.

The local patterns of expression of these cell-surface molecules are thought to provide a major link between the one-dimensional genetic code and the three-dimensional organism. By modulating the appearance of these molecules, the genetic potential can become manifest in the mechanical processes of morphogenesis.

Monoclonal Antibodies and Reverse Genetics

Monitoring Cell Membrane Changes Through Monoclonal Antibodies

The expression of membrane components changes in time and space. Different cell types display different cell-surface components, and these components change as the cell develops. Such tissue-specific membrane components are often recognized by antisera and are therefore called **differentiation antigens** (Boyse and Old, 1969). Specific differentiation antigens can now be identified by **monoclonal antibodies** (Figure 3.11). These antibodies are usually made by injecting foreign cells into mice (or mouse cells of one strain into mice of another strain). The mouse B lymphocytes will begin producing antibodies against each foreign component on these cells, with each B lymphocyte producing a single type of antibody. These lymphocytes are then "immortalized" by fusing them with cultured B lymphocyte tumor cells (myelomas) that have been mutated so that (1) they can no longer synthesize their own antibodies and (2) they lack the purine salvage enzyme hypoxanthine phosphoribosyltransferase (HPRT). This latter alteration means that the myeloma cells can only make purine nucleotides de novo and cannot utilize purines from the culture medium. After fusion, the cells are grown in a medium containing aminopterin, a drug that inhibits the de novo purine synthetic pathway. Thus, unfused myeloma cells die of purine starvation. They cannot make purine nucleotides using the HPRT-mediated salvage pathway, and the aminopterin blocks the de novo pathway as well. Normal B lymphocytes do not divide in culture, so they, too, die. The fused product of the B lymphocyte and the myeloma cell—a **hybridoma**—proliferates, having the purine salvage enzyme from the B lymphocyte and the growth properties from the tumor. Moreover, each of these hybridomas secretes the specific antibody of the B lymphocyte. The medium in which the hybridomas are growing is then tested

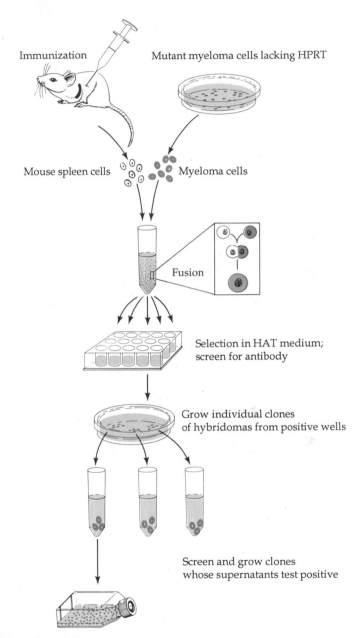

Immunization

Mutant myeloma cells lacking HPRT

Mouse spleen cells

Myeloma cells

Fusion

Selection in HAT medium; screen for antibody

Grow individual clones of hybridomas from positive wells

Screen and grow clones whose supernatants test positive

Figure 3.11 Protocol for making monoclonal antibodies. Spleen cells from an immunized mouse are fused with mutated myeloma cells lacking the enzyme HPRT. Cells are grown in a medium containing hypoxanthine, aminopterin, and thymidine (HAT). Unfused myeloma cells cannot grow in this medium because aminopterin blocks the only way they have of making purine nucleotides. B cells die in this medium even though they contain an enzyme (HPRT) that would allow them to utilize the hypoxanthine placed in the medium. The fused cells (hybridomas) grow and divide. The wells in which the hybridomas grow are screened for the presence of the effective antibody, and the cells from positive wells are plated at densities low enough to allow individual cells to give rise to discrete clones. These clones are isolated and screened for the effective antibody. Such an antibody is monoclonal. The hybridomas producing this antibody can be grown and frozen. (After Yelton and Scharff, 1980.)

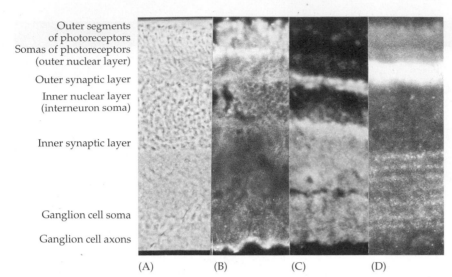

Outer segments
of photoreceptors
Somas of photoreceptors
(outer nuclear layer)

Outer synaptic layer

Inner nuclear layer
(interneuron soma)

Inner synaptic layer

Ganglion cell soma

Ganglion cell axons

(A)   (B)   (C)   (D)

Figure 3.12 Cell-surface specificity in the chick neural retina. (A) Phase-contrast photograph of a section through newly hatched chick neural retina. (B) Retinal section stained with a fluorescent monoclonal antibody that recognizes retinal (but not other neuronal) cells. (C) Retinal section stained with fluorescent monoclonal antibody that recognizes neuronal processes but not cell bodies in the retina. (D) Retinal section stained with fluorescent monoclonal antibody that recognizes antigens on a subset of nerve cell processes in the outer and inner synaptic layers. (Courtesy of G. Grunwald.)

for antibodies that binds to the original population of foreign cells. Such antibodies, having a single B lymphocyte as its original source, is called a **monoclonal antibody.** Monoclonal antibodies can be produced in enormous amounts and can recognize antigens (proteins, lipids, or carbohydrates) that are only weakly expressed (Köhler and Milstein, 1975).

Monoclonal antibodies directed against specific types of cells have uncovered numerous differentiation antigens appearing at different times and places during development. Figure 3.12 shows different cell-surface molecules at different *spatial* layers of the newly hatched chick neural retina. Each of the monoclonal antibodies recognizes a different molecule in the cell membrane. As is evident from this composite photograph, the membranes of all the cells of the neural retina are not the same. In fact, regions of the same cell membrane can differ; for example, the membranes of the axons and the membranes of the nerve soma contain some different molecules. Figure 3.13 shows *temporal* changes in the cell membrane of a single *Drosophila* epithelial cell as it develops into a retinal photoreceptor. Monoclonal antibodies were obtained after mice had been injected with homogenates of *Drosophila* head tissue, and a panel of antibodies was tested on

the cells of the larval eye imaginal disc that were differentiating into eye structures. As soon as the undifferentiated epithelial cells of the disc show neuronal properties, they express the 22C10 antigen. This antigen is also found in other neuronal cell types. Shortly thereafter, though, the cell begins to express another cell membrane molecule, the 24B10 antigen. This molecule is seen only in those neurons destined to become photoreceptors. At subseqent stages (about 80 hours later), the 21A6 antigen

becomes expressed on certain regions of the maturing photoreceptors, and another antigen, 28H9, is characteristic of the terminally differentiated retinal photoreceptor (Zipursky et al., 1984). Thus, cell membranes of different cell types contain different molecules, and these molecules can change during the maturation of the particular cell.

From Protein to Gene

Since differentiation antigens are often proteins whose expression is regulated in time and space, and since these changes often correlate with specific morphological transitions (as shown in Figure 3.13), it would be interesting to know how their genes are regulated. For example, an understanding of how the 24B10 protein becomes expressed might provide clues to the genetic mechanisms of neuronal diversity. How do we accomplish this "reverse genetics," going from protein to gene?

First, we bind the monoclonal antibody to resinous beads and pass retinal homogenates through a column of such material (Figure 3.14). (This is called an immunoaffinity column.) The antibody binds only to the antigen it originally recognized, and the bead-bound protein is eluted from the beads (by ionic solutions) and run on an electrophoresis gel to separate it from any possible contaminants. The region of the gel containing this protein is cut from the gel, and the protein is eluted from the gel matrix and partially sequenced. Radioactive oligonucleotides that would bind to a DNA

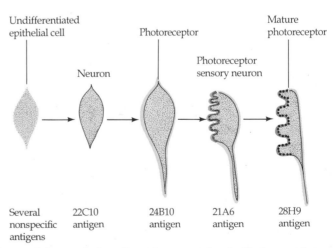

Undifferentiated epithelial cell	Neuron	Photoreceptor	Photoreceptor sensory neuron	Mature photoreceptor
Several nonspecific antigens	22C10 antigen	24B10 antigen	21A6 antigen	28H9 antigen

Figure 3.13 Temporal changes in the cell membrane correlated with the morphogenesis of a *Drosophila* retinal photoreceptor cell. As differentiation proceeds, different antigens become expressed on the cell membrane. (After Venkatesh et al., 1985.)

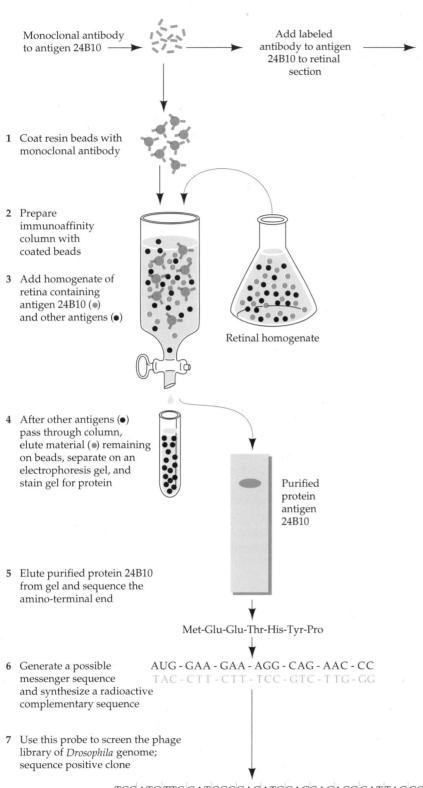

Monoclonal antibody to antigen 24B10 → → Add labeled antibody to antigen 24B10 to retinal section →

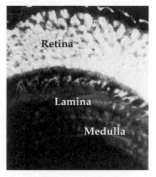

Localization of 24B10 by fluorescein-labeled monoclonal antibody

1 Coat resin beads with monoclonal antibody

2 Prepare immunoaffinity column with coated beads

3 Add homogenate of retina containing antigen 24B10 (●) and other antigens (●)

Retinal homogenate

4 After other antigens (●) pass through column, elute material (●) remaining on beads, separate on an electrophoresis gel, and stain gel for protein

Purified protein antigen 24B10

5 Elute purified protein 24B10 from gel and sequence the amino-terminal end

Met-Glu-Glu-Thr-His-Tyr-Pro

6 Generate a possible messenger sequence and synthesize a radioactive complementary sequence

AUG - GAA - GAA - AGG - CAG - AAC - CC
TAC - CTT - CTT - TCC - GTC - TTG - GG

7 Use this probe to screen the phage library of *Drosophila* genome; sequence positive clone

TCC	ATG	TTC	GAT	CGC	GAG	ATG	GAG	GAG	ACG	CAT	TAC	CCG	CCC	TGC	ACC	TAC	AAC	GTG	ATG	TGC
Ser	Met	Phe	Asp	Arg	Glu	Met	Glu	Glu	Thr	His	Tyr	Pro	Pro	Cys	Thr	Tyr	Asn	Val	Met	Cys

Expected sequence

8 Isolate and characterize gene

Figure 3.14 Protocol for finding the gene that encodes a protein identified by a monoclonal antibody. The oligonucleotide decoded from the protein structure need not be a perfect match with the actual sequence. (After Venkatesh et al., 1985; photograph courtesy of S. Benzer.)

sequence capable of encoding such a protein are then synthesized. In the case of 24B10, these radioactive probes were used to screen a library of recombinant DNA clones containing regions of the *Drosophila* genome. The *Drosophila* DNA of each positive clone was sequenced to see if it matched the sequence of the original protein isolated by the monoclonal antibody. In this way, we can go from a rare protein identified by a monoclonal antibody to a specific piece of genomic DNA (Zipursky et al., 1984; Venkatesh et al., 1985). ■

Cell adhesion molecules

Identifying Cell Adhesion Molecules and Their Role in Development

The sorting-out studies of Holtfreter and Steinberg did not identify the molecules involved in the differential cell adhesions. Roth (1968; Roth et al., 1971) demonstrated that different cell types display selective cell adhesion independent of cell sorting. He modified the rotary aggregation assay by incubating ^{3}H-labeled cartilage cells and ^{14}C-labeled hepatocytes in a rotating solution containing small aggregates of unlabeled cartilage cells. By measuring the ^{14}C- and ^{3}H-labeled cells in these aggregates, he demonstrated that the cartilage aggregates specifically picked up cartilage cells. Similar experiments extended this finding to liver and muscle cells (Figure 3.15). These studies demonstrated that different cell types could use different adhesion molecules.

The next task is to identify the molecules mediating cell adhesion and to discover how they accomplish this feat. Several **cell adhesion molecules** (**CAMs**) have been identified and are grouped into two general categories: the **cadherins,** whose cell-adhesive properties depend on calcium ions, and the **immunoglobulin superfamily CAMs,** whose cell-binding domains resemble those of antibody molecules. Table 3.1 lists some recently discovered CAMs.

Cadherins

Calcium ions are often necessary for cell adhesion. The ions stabilize the adhesive conformations of certain cell-surface proteins called cadherins. Cadherins are critical for establishing and maintaining intercellular connections, and they appear to be crucial to the spatial segregation of cells and to the organization of animal form (Takeichi, 1987). Cadherins interact with other cadherins on adjacent cells, and they are anchored into the cell by a complex of proteins called **catenins** (Figure 3.16). The cadherin-catenin complex forms the classical adherans junctions that connect epithelial cells together. Moreover, since the catenins bind to the actin cytoskeleton, they integrate the epithelial cells together into a mechanical unit. In vertebrate embryos, four major cadherin classes have been identified:

Figure 3.15
Specificity of cell-cell attachment. Collecting aggregates, each consisting of one type of cell, are placed in a rotating culture containing single cells of both the same (isotypic) and different (heterotypic) types. The single isotypic and heterotypic cells were previously labeled with different radioactive isotopes. After 6 hours, the aggregates are collected, washed, and counted for both isotypic and heterotypic cells that adhered to the aggregate as shown in the table. (Data from Roth, 1968.)

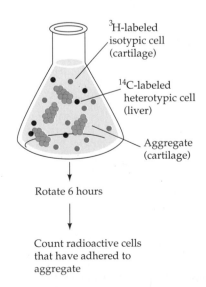

^{3}H-labeled isotypic cell (cartilage)

^{14}C-labeled heterotypic cell (liver)

Aggregate (cartilage)

Rotate 6 hours

Count radioactive cells that have adhered to aggregate

Count of radioactive cells that have adhered to aggregate

Aggregate type	Labeled single cells in suspension*		
	Cartilage	Liver	Pectoral muscle
Cartilage	100	6	48
Liver	10	100	0
Pectoral muscle	38	49	100

*Percentage of mean number of cells collected by isotopic aggregates.

Table 3.1 General classification of major cell adhesion molecules (CAMs)

Class	CAM	Cell types
Cadherins (calcium-dependent)	N-cadherin (a.k.a. A-CAM) P-cadherin E-cadherin (a.k.a. L-CAM, uvomorulin)	Nerve, kidney, lens, heart Placenta, epithelia Epithelia, mouse blastula
Immunoglobulin superfamily CAMs (calcium-independent)	N-CAM Ng-CAM (a.k.a. L1, NILE) Neurofascin Cell-CAM LFA-1 CD4 glycoprotein (HIV receptor)	Muscle, nerve, kidney Glia, neurons *Drosophila* neurons Hepatocytes Lymphocytes T-inducer cells

- **E-cadherin** (epithelial cadherin, also called uvomorulin and L-CAM) is expressed on all early mammalian embryonic cells, even at the one-cell stage. Later, this molecule is restricted to epithelial tissues of embryos and adults.
- **P-cadherin** (placental cadherin) appears to be expressed primarily on those placental cells of the mammalian embryo that contact the uterine wall (the trophoblast cells) and the uterine wall epithelium itself (Nose and Takeichi, 1986). It is possible that P-cadherin facilitates the connection of the trophoblast with the uterus, since P-cadherin on the uterine cells is seen to contact P-cadherin on the trophoblastic cells of mouse embryos (Kadokawa et al., 1989).

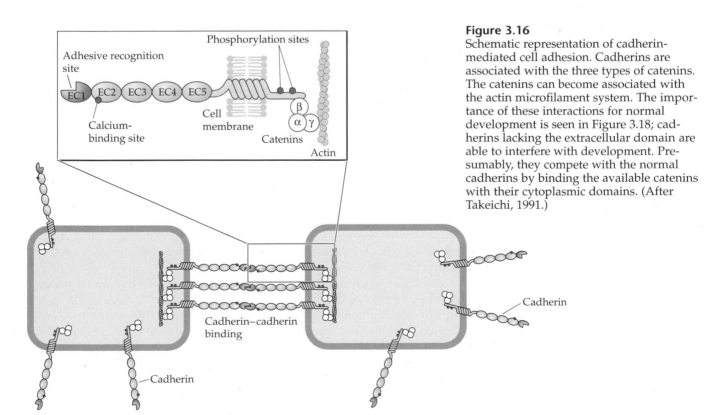

Figure 3.16
Schematic representation of cadherin-mediated cell adhesion. Cadherins are associated with the three types of catenins. The catenins can become associated with the actin microfilament system. The importance of these interactions for normal development is seen in Figure 3.18; cadherins lacking the extracellular domain are able to interfere with development. Presumably, they compete with the normal cadherins by binding the available catenins with their cytoplasmic domains. (After Takeichi, 1991.)

(A)

(B)

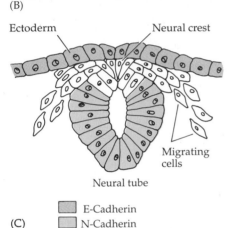

(C)

| | E-Cadherin |
| | N-Cadherin |

Figure 3.17
Localization of two different cadherins during the formation of the mouse neural tube. Double immunofluorescent staining was used to localize E-cadherin (A) and N-cadherin (B) in the same transverse section of an 8.5-day embryonic mouse hindbrain. Antibodies to E-cadherin were labeled with one type of fluorescent dye (which fluoresces under one set of wavelengths), while antibodies to N-cadherin were marked with a second type of fluorescent dye (which emits its color at other wavelengths). Photographs taken at the different wavelengths reveal that the outer ectoderm expresses predominantly E-cadherin, while the invaginating neural plate ceases E-cadherin expression and instead expresses N-cadherin. (C) When the neural tube has formed, it expresses N-cadherin, the epidermis expresses E-cadherin, and the neural crest cells between them express neither.
(Photographs by K. Shimamura and H. Matsunami, courtesy of M. Takeichi; C after Rutishauser, 1988.)

- **N-cadherin** (neural cadherin) is first seen on the mesodermal cells in the gastrulating embryo as they lose their E-cadherin expression. It is highly expressed on the cells of the developing central nervous system (Figure 3.17; Hatta and Takeichi, 1986).
- **EP-cadherin** (C-cadherin) is critical for maintaining cell adhesion between the blastomeres of the *Xenopus* blastula and is required for the normal movements of gastrulation (Figure 3.18; Heasman et al., 1994; Lee and Gumbiner, 1995).

Cadherins adhere cells together by binding to the same type of cadherin on another cell. Thus, cells with E-cadherin stick to other cells having E-cadherin, and they will sort out from cells containing N-cadherin in their membranes. This is called **homophilic binding.** Cells expressing N-cadherin readily sort out from N-cadherin-negative cells in vitro, and univalent antibodies against cadherins will convert a three-dimensional histotypic aggregate of cells into a monolayer (Takeichi et al., 1979). Moreover, when activated E-cadherin genes are transfected into and expressed in cultured mouse fibroblasts (which usually do not express this protein), E-cadherin is seen on their cell surfaces, and the treated fibroblasts become tightly connected to each other (Nagafuchi et al., 1987). In fact, these cells begin acting like epithelial cells.

Cadherin expression is often correlated with aggregation and dispersion. Neural crest cells (which are at the dorsalmost portion of the neural tube) initially express N-cadherin. Then, as they begin migrating away from the neural tube as individual cells (to form pigment cells, sensory neurons, and other cell types), they lose their N-cadherin expression (see Figure 3.17; see also Chapter 7). However, when the migrating cells reach their destination and begin to aggregate together to form nerve ganglia, they re-express N-cadherin (Hatta et al., 1987).

Differential cadherin expression can also explain the homotypic sorting-out data presented earlier. As we discussed, Roth and co-workers showed that liver cells tend to collect liver cells and that retinal cells collect other retinal cells. Takeichi (1987) demonstrated that the retinal cells express N-cadherin and the liver cells express E-cadherin and that the sorting out would be expected because of this difference in cadherin expression. He also suggested that the observations of Townes and Holtfreter may likewise be explained by differential cadherin expression. Support for this idea came from studies in which different cadherin genes were transfected into mouse fibroblasts that do not usually express either cadherin type. Fibroblasts expressing E-cadherin adhered to other E-cadherin-bearing fibroblasts, while P-cadherin fibroblasts stuck to other fibroblasts expressing P-cadherin. Moreover, when embryonic lung tissue was dissociated and allowed to recombine in the presence of untreated or E-cadherin-bearing fibroblasts, the E-cadherin-expressing fibroblasts became integrated into the epithelial lung tubules (which express E-cadherin), while the untreated fibroblasts became associated with the mesenchymal cells (which do not express any cadherin) (Nose et al., 1988).

All these experiments have been performed on cells in culture. Recently, however, in vivo studies have shown that cadherins may be critical in sorting-out phenomena occurring within the embryo. When the mRNA for chicken N-cadherin is injected into one of the two first-cleavage blastomeres of the *Xenopus* frog embryo, N-cadherin is often expressed on cells that would normally lack it. The embryos that express extra N-cadherin are often characterized by clumps of cells and thickened tissue layers. Normally, the neural tube (which expresses N-cadherin) separates from the cells that would become the epidermis (which express E-cadherin). However, in em-

Ectoderm
Neural crest
Migrating cells
Neural tube

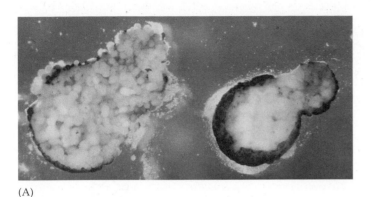

(A)

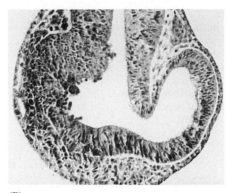

(B)

Figure 3.18
Importance of cadherins for maintaining cohesion between developing cells. (A) When oocytes are injected with an antisense oligonucleotide against a maternally inherited cadherin message, the inner cells disperse when the animal cap is removed. In control embryos (right), the inner cells remain together. (B) At the four-cell stage, the blastomeres that form the left side of the frog are injected with an mRNA for N-cadherin that lacks the extracellular region of the cadherin. During neurulation the cells with the mutant protein do not form a coherent layer. (A after Heasman et al., 1994; B from Kintner et al., 1992; photographs courtesy of J. Heasman and C. Kintner.)

bryos in which both the epidermis and the neural tube express this extra N-cadherin, the neural tube does not separate from the epidermis (Detrick et al., 1990; Fujimori et al., 1990). Thus, the cadherins are probably playing a major role in the assortment of cells into tissues. [cell2.html]

Immunoglobulin Superfamily CAMs

As we discussed in Chapter 1, antibodies were first used to identify cell adhesion molecules in *Dictyostelium*. Gerisch and colleagues (Beug et al., 1970) prepared antibodies against *Dictyostelium* and then chemically split the antibodies so that only their monovalent antigen-binding regions—the Fab fragments—remained. (The divalent antibodies had to be split, for if they remained divalent, they might artificially join the cells together, and the effect could not be measured.) This led to the discovery of the 80-kDa glycoprotein that mediates cell-cell adhesion during slime mold aggregation. The same approach was used to study embryonic cell adhesion by Edelman and his colleagues (Brackenbury et al., 1977) and led to the isolation of **neural cell adhesion molecule (N-CAM)**. [cell3.html]

N-CAM is a member of a class of CAMs that do not need calcium ions and that share a similar structure (Figure 3.19). This structure, with its extracellular globular domains held in place by disulfide bonds, resembles that of the immunoglobulin molecule, and it is likely that the immunoglobulins are derived from this group of CAMs (Williams and Barclay, 1988; Lander, 1989). Thus, these glycoproteins are called the immunoglobulin superfamily CAMs.*

The immunoglobulin superfamily CAMs may play an important role in the development of the nervous system. N-CAM is needed for the proper attachment of axons to target muscle cells (Covault and Sanes, 1986; Tosney et al., 1986). Moreover, N-CAM appears to be critical in bundling ("fasciculating") axons together so that they travel as a unit. Antibodies to N-CAM can

---

*The designation *superfamily* is often given because the different classes of immunoglobulin molecules themselves constitute a "family." These other members of the superfamily have structures resembling immunoglobulins, but are not exactly "close" family.

Figure 3.19
Immunoglobulin superfamily adhesion molecules. (A) Three members of the immunoglobulin superfamily. The membrane-bound form of IgM molecule has two heavy chains, each with five domains, and two light chains, each with two domains. N-CAM is a single polypeptide chain with five domains. Its anchor into the membrane can be either a transmembrane amino acid sequence in the protein or a lipid. L1 is a transmembrane protein with six globular domains. The insect cell adhesion molecules fasciclin II and neuroglian resemble N-CAM and L1, respectively. (B) Model for adhesion of Ig superfamily CAMs.

(A)

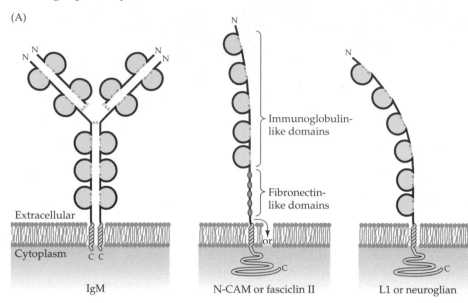

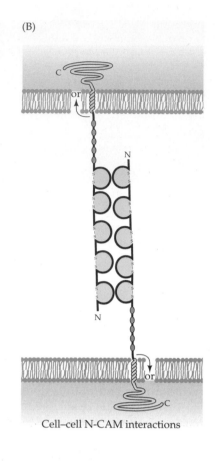

(B)

| IgM | N-CAM or fasciclin II | L1 or neuroglian | Cell–cell N-CAM interactions |

break these attachments, allowing axons to wander (Fraser et al., 1988; Landmesser et al., 1988). A similar situation appears to occur in insects, where immunoglobulin superfamily CAMs called **fasciclins** (Figure 3.20) aid the migration of the axons (Harrelson and Goodman, 1988). L1 is needed for the outgrowth of certain axons (Lemmon et al., 1989), and mutations of L1 in humans cause a spectrum of anomalies characterized by hydrocephalus, mental retardation, and the inability to control limb movements (Vits et al., 1994).

Differential CAM expression is critical at boundaries between two groups of cells. At such places, the body segregates different cells into differ-

Figure 3.20
Fasciclin expression in the developing grasshopper nervous system. (A) Scaffold of fasciculated axons in a grasshopper embryo as seen by Nomarski microscopy. A com and P com are the anterior and posterior commissures whose axons traverse the segment; ISN is the intersegmental neuron, and con is a connective neuron. (B,C) Embryonic nervous system as in (A), but stained with monoclonal antibodies made to the cell-surface fasciclin glycoproteins. The antibody on (B) recognizes a subset of axons in the anterior and posterior commissures, while the antibody in (C) binds to a membrane glycoprotein of most longitudinal axon fascicles. The arrows show the same locations on (B) and (C). Note that the antibody stains only a portion of each axon. (From Bastiani et al., 1987, courtesy of C. Goodman.)

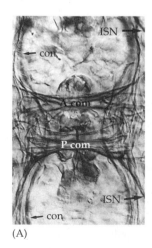

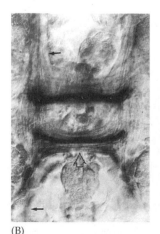

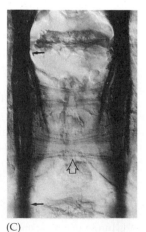

(A)

(B)

(C)

Figure 3.21
Distribution of different CAMs at tissue boundaries. As the mesodermal cells collect to induce the feather bud in the ectoderm, the newly aggregated mesenchyme cells express N-CAM (A), while the ectodermal cells express E-cadherin (B) in their respective cell membranes. (From Chuong and Edelman, 1985a, courtesy of G. Edelman.)

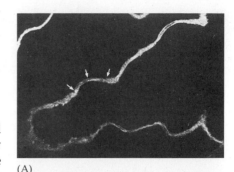
(A)

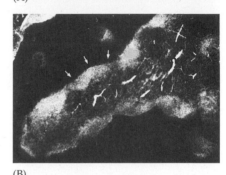
(B)

ent regions. Notochord cells do not enter into the neural tube, nor do dermal cells trespass into the epidermis. Such segregation may be accomplished by the adjacent populations' having different CAMs. For instance, feathers are induced when mesodermally derived mesenchymal cells condense together to form a ball of cells directly beneath the epidermis of the chicken skin. The ectodermal cells are linked together by E-cadherin, while the previously CAM-negative mesenchymal cells begin to express N-CAM and collect to form an aggregate (Figure 3.21). Throughout the developing feather, different groups of cells become separated from one another as a result of their ability to express N-CAM, E-cadherin, or both proteins (Chuong and Edelman, 1985a,b).

Cell junctional molecules: Gap junction proteins

Gap junctions are specialized intercellular regions where adjacent cells are 15–40 nm apart. Fine connections serve as communication channels between the adjacent cells (Figure 3.22A,B). Cells so linked are said to be "coupled,"

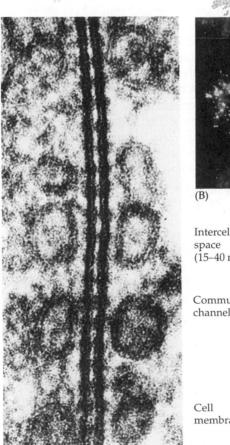
(A)

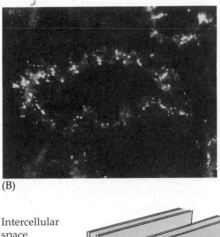
(B)

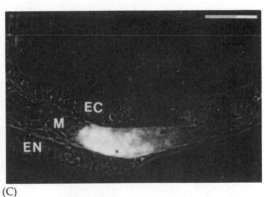
(C)

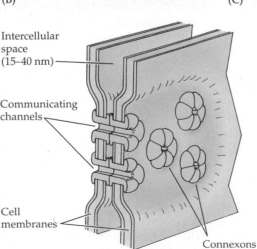
Intercellular
space
(15–40 nm)

Communicating
channels

Cell
membranes

Connexons

(D)

Figure 3.22
Gap junction proteins. (A) Electron micrograph of a row of gap junctions connecting two apposed cells. (B) Fluorescence micrograph of gap junctions in a 17-day embryonic mouse kidney tubule. (C) Compartments formed by gap junction proteins between cells that communicate with one another. This compartment in the mouse gastrula can be seen by injecting Lucifer Yellow dye into one cell and seeing it transferred to a small group of cells. (D) Subunit structure of the gap junction. (A from Peracchia and Dulhunty, 1976, courtesy of C. Peracchia; B from Sainio et al., 1992, courtesy of K. Sainio; C from Kalimi and Lo, 1988, courtesy of C. Lo; D after Darnell et al., 1986.)

and small molecules (MW < 1500) and ions can freely pass from one cell to another. In most embryos, at least some of the early blastomeres are connected by gap junctions, thereby enabling ions and small soluble molecules to pass readily between them. The ability of cells to form gap junctions with some cells and not with others creates physiological "compartments" within the developing embryo (Figure 3.22C).

The importance of gap junctions in development has been demonstrated in amphibian and mammalian embryos (Warner et al., 1984). When antibodies to gap junction proteins were microinjected into one specific cell of an eight-cell *Xenopus* blastula, the progeny of that cell, which are usually coupled through gap junctions, could no longer pass ions or small molecules from cell to cell. Moreover, the tadpoles that resulted from such treated blastulae showed defects specifically relating to the developmental fate of the injected cell (Figure 3.23). The progeny of such a cell did not die, but they were unable to undergo their normal development (Warner et al., 1984). In the mouse embryo, the first eight blastomeres are connected to one another by gap junctions. Although loosely associated with one another, these eight cells move together to form a compacted embryo. If compaction is inhibited by antibodies against gap junction proteins, further development ceases. The treated blastomeres continue to divide, but compaction fails to occur (Lo and Gilula, 1979; Lee et al., 1987). If antisense RNA to gap junction messages is injected into *one* of the blastomeres of a normal mouse embryo, that cell will not form gap junctions and will not be included in the embryo (Bevilacqua et al., 1989).

The gap junction channels are made of **connexin** proteins. In each cell, six identical connexins in the membrane group together to form a transmembrane channel containing a central pore. The gap junction complex of one cell connects to the gap junction complex of another cell, enabling the cytoplasms of both cells to be joined (Figure 3.22D). There are about a dozen types of connexins, and some may be regulated by cadherins. Jongen and co-workers (1991) found that when cells are coupled by E-cadherin, gap junction-mediated communication between the cells depends on cadherin function. Evidence suggests that cadherin works both by keeping the cells in contact and by modifying the connexin protein. The different types of connexin proteins have separate, but overlapping, roles in normal development. For example, the connexin-43 gap junction protein is found throughout the developing mouse embryo, in nearly every tissue. However, if the connexin-43 genes are knocked out by gene targeting, the mouse embryo will still develop. It appears that the functions of the connexin-43 protein can be taken over by other connexin proteins. But shortly after birth, these mice take gasping breaths, turn bluish, and die. Autopsies of these mice show that the right ventricle—the chamber that pumps blood through the pulmonary artery to the lungs—is filled with tissue that occludes the chamber and obstructs blood flow (Reaume et al., 1995) Although loss of the connexin-43 protein can be compensated for in many tissues, it appears to be critical for normal heart development. [cell4.html]

The cell membrane, then, has several mechanisms by which it can form attachments with other cell membranes. It can use immunoglobulin super-

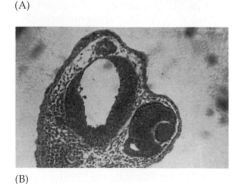

(A)

(B)

Figure 3.23
Developmental effects of gap junctions. Section through *Xenopus* tadpole in which one of the blastomeres at the eight-cell stage had been injected with (A) a control antibody or (B) antibody against gap junction protein. The side formed by the injected blastomere lacks its eye and has abnormal brain morphology. (From Warner et al., 1984, courtesy of A. E. Warner.)

family CAMs, calcium-dependent CAMs, and junctional proteins. This does not exhaust its repertoire. As we already mentioned, the cell can also bind to specific components of the extracellular matrix. It is to these components that we turn our attention.

The molecular basis of cell-substrate affinity

Differential Substrate Affinity

The migration of cells, like the migration of birds and monarch butterflies, depends on sensing when to start migration, when to cease migration, and what route to follow. There are numerous cues that the environment can give cells, but the major cues appear to involve substances on the extracellular matrix. The **differential substrate affinity hypothesis** postulates that different cells recognize different molecules in various extracellular matrices. Each migrating cell type prefers certain combinations of matrix molecules to others, and these matrix molecules tell the cells when to migrate and where. Weiss (1945) and Tyler (1946) suggested that cells can sometimes interact with their substrata through lock-and-key interactions between the cell membrane and the extracellular matrix. The relationship between the cell membrane protein and the matrix molecule would resemble that between enzyme and substrate or antibody and antigen. During the past decade, this type of interaction has been shown to be extremely important for cell migration. [cell5.html]

The Extracellular Matrix

The **extracellular matrix** consists of macromolecules secreted by the cells into their immediate environment. These molecules interact to form insoluble networks that can play several roles during development. In some instances, they separate two adjacent groups of cells and prevent any interactions. In other cases, the extracellular matrix may serve as the substrate on which cells migrate, or it may even induce differentiation in certain cell types. One type of matrix is shown in Figure 3.24. Here, a sheet of epithelial cells is adjacent to a layer of loose mesenchymal tissue. The epithelial cells have formed a tight extracellular layer called the **basal lamina;** the mesenchymal cells secrete a loose **reticular lamina.** Together, these layers constitute the **basement membrane** of the epithelial cell sheet. There are three major components of most extracellular matrices: collagen, proteoglycans, and the large glycoproteins that are called substrate adhesion molecules (Table 3.2).

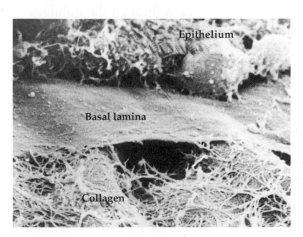

Figure 3.24
Location and formation of extracellular matrix in the chick embryo. The scanning electron micrograph shows extracellular matrix at the junction of the epithelial cells (above) and mesenchymal cells (below). The epithelial cells synthesize a tight glycoprotein-based lamina, while the mesenchymal cells secrete reticular lamina made primarily of collagen. (Courtesy of R. L. Trelsted.)

Table 3.2 Main constituents of the extracellular matrix

Mesenchymal extracellular matrix	Basal laminae of epithelial cells

COLLAGENS
 	
Long, thin molecules (most common is type I; types II, III, and V-XIII are also found) that assemble to form fibrils, usually 60–70 nm in diameter. Collagens provide strength and stability to tissues.
 	
MATRIX PROTEOGLYCANS
 	
Made up of proteins and repeating disaccharides (glycosaminoglycans). Glycosaminoglycans include hyaluronic acid, a very large (10^8 Da) molecule that binds large amounts of water. Sulfated proteoglycans each comprise a linear core protein to which are attached chains of one or more of the sulfated glycosaminoglycans (chondroitin, heperan, keratan, and dermatan sulfate).
 	
   Proteoglycans stimulate and modulate cell movement; their range suggests they may have other properties that are not yet known.
 	
SUBSTRATE ADHESION MOLECULES
 	
Molecules to which cells make the adhesions that allow them to move. They include fibronectin, chondronectin, and tenascin.
	COLLAGEN IV
	
	The major structural component of the basal lamina. Unlike the other collagens, its fibrils are like fine "chicken wire" and assemble into a feltlike substrate.
	
	MATRIX PROTEOGLYCANS
	
	Hyaluronic acid and sulfated proteoglycans, often present in basal laminae. Their presence may facilitate the passage of secretory products through the lamina.
	
	SUBSTRATE ADHESION MOLECULES
	
	Laminin, the major functional component of the basal lamina. A glycoprotein trimer with adhesion sites for the cell membrane, collagen IV, and glycosaminoglycans.
	
	   Basal laminae may contain fibronectin, tenascin, nidogen, and other adhesive glycoproteins.

Source: Adapted from Bard, 1990.

COLLAGEN.   **Collagen** is a family of glycoproteins containing a large percentage of glycine and proline residues. Because collagen is the major structural support of almost every animal organ, it constitutes nearly half the total body protein. There are numerous types of collagen serving special functions. Type I collagen, found in the extracellular matrices of skin, tendons, and bones, makes up 90 percent of the collagen in the body. Type II collagen is most evident as the secretion of cartilage cells, but it is also found in the notochord and in the vitreous body of the eye. Type III collagen is most evident in blood vessels, and type IV is found in basal laminae produced by epithelial cells (Vuorio, 1986). Other types of collagen are found throughout the body, especially in cartilage. Collagen is important for the formation of the basal lamina, and it is also implicated in the branching of epithelial tubules in the salivary glands, lungs, and other organs. [cell6.html]

PROTEOGLYCANS.   These are specific types of glycoproteins in which (1) the weight of the carbohydrate residues far exceeds that of the protein and (2) the carbohydrates are linear chains composed of repeating disaccharides. Usually, one of the sugars of the disaccharide has an amino group, so the repeating unit is called a **glycosaminoglycan** (**GAG**). Table 3.3 lists the common glycosaminoglycans; the basic proteoglycan structure is illustrated in Figure 3.25. The interconnection of protein and carbohydrate forms a weblike matrix, and in many motile cell types, the proteoglycan surrounds the cells and is thought to mediate against their coming together (Figure 3.26). The consistency of the extracellular matrix depends on the ratio of collagen to proteoglycan. Cartilage, which has a high percentage of proteoglycans, is soft, whereas tendons, which are predominantly collagen fibers, are tough. In basal laminae, proteoglycans predominate, forming a molecular sieve in addition to providing structural support.

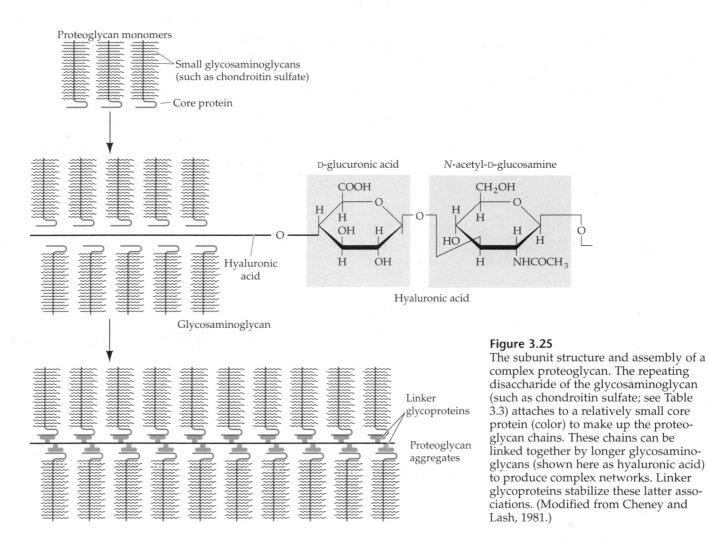

Figure 3.25
The subunit structure and assembly of a complex proteoglycan. The repeating disaccharide of the glycosaminoglycan (such as chondroitin sulfate; see Table 3.3) attaches to a relatively small core protein (color) to make up the proteoglycan chains. These chains can be linked together by longer glycosaminoglycans (shown here as hyaluronic acid) to produce complex networks. Linker glycoproteins stabilize these latter associations. (Modified from Cheney and Lash, 1981.)

Table 3.3 Repeating disaccharide units of the most common glycosaminoglycans of matrix proteoglycans

Glycosaminoglycan	Repeating disaccharide unit[a]	Distribution
Hyaluronic acid	Glucuronic acid-N-acetylglucosamine	Connective tissues, bone, vitreous body
Chondroitin sulfate	Glucuronic acid-N-acetylgalactosamine sulfate	Cartilage, cornea, arteries
Dermatan sulfate	[Glucuronic or iduronic acid]-N-acetylgalactosamine sulfate	Skin, heart, blood vessels
Keratan sulfate	Galactose-N-acetylglucosamine sulfate	Cartilage, cornea
Heparan sulfate	[Glucuronic or iduronic acid]-N-acetylglucosamine sulfate	Lung, arteries, cell surfaces

[a]These are the typical repeating units of these glycosaminoglycans. However, regions of each GAG can have slightly modified saccharides.

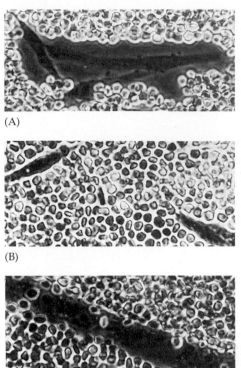

(A)

(B)

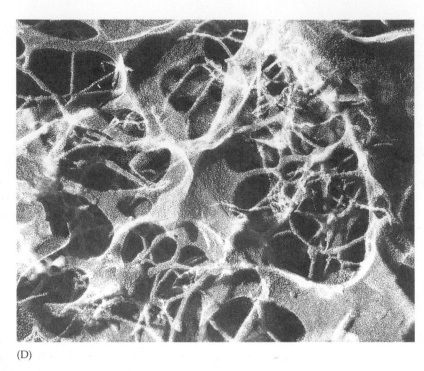

(D)

(C)

Figure 3.26
The proteoglycan coat surrounding mobile cells. (A) Hyaluronidate coat surrounds chick myoblasts. Myoblasts in culture exclude small particles (in this case, fixed red blood cells) for a significant distance from the cell border. (B) When the myoblasts are treated with hyaluronidase (which dissolves hyaluronic acid), this extracellular coat vanishes. (C) The coat also vanishes as the myoblasts cease dividing and join together as they differentiate. (D) Electron micrograph of hyaluronidate in aqueous solution shows a branching fibrillar network. (A–C from Orkin et al., 1985, courtesy of B. Toole; D from Hadler et al., 1982, courtesy of N. M. Hadler.)

Proteoglycans are also important in mediating the connections between adjacent tissues of an organ. Here they draw loose cells together to form an epithelial sheet* (San Antonio et al., 1987; Thesleff et al., 1989; Vainio et al., 1989; Bernfield and Sanderson, 1990). In some cases, proteoglycans secreted by one cell type are essential for the growth of neighboring cells. Axons of the dorsal root ganglia have heparan sulfate proteoglycan on some of their cell-surface proteins, and the removal of this proteoglycan prevents the associated Schwann cells from proliferating around them (Ratner et al., 1985). One of the ways that the glycosaminoglycan chains of the proteoglycans can function is to retain and present **growth factors** to cellular receptors. Growth factors are hormone-like proteins that regulate mitosis or differentiation when they bind to particular cells. However, the cell receptor for the growth factor often does not bind the factor with great affinity. Rather, the growth factor is first bound by the proteoglycan's carbohydrates, and this concentrates the growth factor so that it can bind to its receptor (Massagué, 1991; Yayon et al., 1991).

EXTRACELLULAR GLYCOPROTEINS. Extracellular matrices contain a variety of other specialized molecules, such as fibronectin, laminin, and tenascin. These large glycoproteins probably are responsible for organizing the collagen, proteoglycan, and cells into an ordered structure. **Fibronectin** is a very large (460-kDa) glycoprotein dimer synthesized by fibroblasts, chondrocytes, endothelial cells, macrophages, and certain epithelial cells (such as hepatocytes and amniocytes). One function of fibronectin is to serve as a general adhesive molecule, linking cells to substrates such as collagen and proteo-

*Heparan sulfate proteoglycans are thought to bring together chondrocytes, the cartilage-producing cells. Excessive levels of glucose, however, inhibit the synthesis of the core protein of the proteoglycan, thus inhibiting cartilage formation. Leonard and co-workers (1989) propose this as a possible mechanism for the skeletal problems frequently seen in children born to severely diabetic mothers.

glycans. Fibronectin also organizes the extracellular matrix by having several distinct binding sites, and interaction with the appropriate molecules results in the proper alignment of cells with their extracellular matrix (Figure 3.27).

As we will see in later chapters, fibronectin also has an important role in cell migration. The "roads" over which certain migrating cells travel are paved with this protein. Mesodermal cell migration during gastrulation is seen on the fibronectin surfaces of many species, and the movement of these cells ceases when fibronectin is locally removed. In chick embryos, the precursors of the heart, the precardiac cells, migrate on fibronectin to travel from the lateral sides of the embryo to the midline. If chick embryos are injected with antibodies to fibronectin, the precardiac cells fail to migrate to the midline and two separate hearts develop. Fluorescent antibodies to fibronectin have demonstrated a gradient of fibronectin in the path of migration between the endoderm and the mesoderm. If this region is cut and rotated, the heart cells follow the gradient to new positions away from the midline (Linask and Lash, 1988a,b). Thus, fibronectin appears to play a major role in the migration of the precardiac cells to the embryonic midline. Other cell types, such as the germ cell precursors of frog embryos, also travel over cells that secrete fibronectin onto their surfaces (Heasman et al., 1981).

Laminin is a major component of basal laminae. It is made of three peptide chains, and, like fibronectin, it can bind to collagen, glycosaminoglycans, and cells. The collagen bound by laminin is type IV (specific for basal lamina), and the cell-binding region of laminin recognizes chiefly epithelial cells and neurons. The adhesion of epithelial cells to laminin (which they sit upon and utilize) is much greater than the affinity of mesenchymal cells for fibronectin (which they have to bind to and release if they are to migrate). Like fibronectin, laminin plays a role in assembling the extracellular matrix, promoting cell adhesion and growth, changing cell shape, and permitting cell migration (Hakamori et al., 1984).

Not all the large extracellular glycoproteins promote cell adhesion. **Tenascin** (also called **cytotactin**) resembles fibronectin for about half the

Figure 3.27
Fibronectin in the developing chick embryo. (A) Fluorescent antibodies to fibronectin show that the fibronectin deposition in the 24-hour chick embryo lies along the basal lamina of many organs. (B) Structure and binding domains of fibronectin. The rectangles represent protease-resistant domains. The fibroblast-cell-binding domain consists of two units, the RGD site and the high-affinity site, both of which are essential for cell binding. Avian neural crest cells have another site that is necessary for their motility on a fibronectin substrate. Other regions of fibronectin enable it to bind to collagen, heparin,* and other molecules of the extracellular matrix. (A courtesy of J. Lash; B after Dufour et al., 1988.)

*Heparin is a portion of the heparin proteoglycan secreted by mast cells and basophils. *Heparan* and *heparan sulfate* are names given to similar glycosaminoglycans found in the extracellular matrix or on the cell surface. It is presumed that the binding sites for heparin are also those of heparan sulfate (Bernfield and Sanderson, 1990).

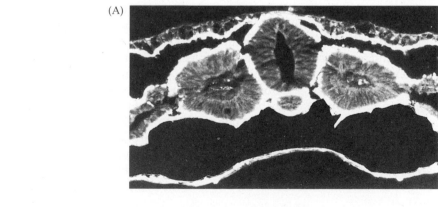

(A)

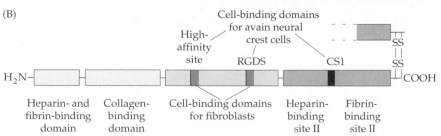

(B)

H₂N— Heparin- and fibrin-binding domain — Collagen-binding domain — High-affinity site — Cell-binding domains for fibroblasts — RGDS — Cell-binding domains for avain neural crest cells — Heparin-binding site II — CS1 — Fibrin-binding site II — SS SS —COOH

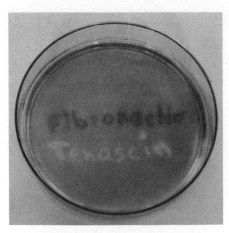

Figure 3.28
Inhibition of cell adhesion by tenascin. Fibronectin and tenascin were both plated onto a tissue culture dish in the form of letters. Fibroblasts were added to the dish and allowed to adhere and migrate. The result shows that fibronectin was a much preferred substrate over tissue culture plastic, whereas cells did not adhere to or migrate well upon tenascin. (Courtesy of M. Chiquet.)

length of the molecule, and it is found transiently in several extracellular matrices during embryonic development. However, different cells react in different ways to tenascin. Some cells stick to it, while other cells round up and detach from tenascin (Figure 3.28; Spring et al., 1989). Different relative amounts of fibronectin and tenascin may generate substrates of various degrees of adhesiveness. In addition, tenascin appears to increase the synthesis and secretion of proteases from the cells that are on it (Werb et al., 1990). Both these characteristics may be important in generating paths for cell migration and in remodeling the extracellular matrix during development (Tan et al., 1987; Bronner-Fraser, 1988; Wehrle and Chiquet, 1990).

Cell Receptors for the Extracellular Matrix Molecules

INTEGRINS.   The ability of a cell to bind these adhesive glycoproteins depends on its expressing a cell membrane receptor for the cell-binding site of these large molecules. The main fibronectin receptors have been purified by using monoclonal antibodies that block the attachment of cells to fibronectin (Chen et al., 1985; Knudsen et al., 1985). The fibronectin receptor complex was found not only to bind fibronectin on the *outside* of the cell, but also to bind cytoskeletal proteins on the *inside* of the cell. Thus, the fibronectin receptor complex appears to span the cell membrane and unite two types of matrices. On the inside of the cell, it serves as an anchorage site for the actin microfilaments that move the cell; on the outside of the cell, it binds to the fibronectin of the extracellular matrix (Figure 3.29). Horwitz and co-workers (1986; Tamkun et al., 1986) have called this family of receptor proteins **integrins** because they integrate the extracellular and intracellular scaffolds, allowing them to work together. Integrin proteins have been found to span the cell membrane of numerous cell types. On the extracellular side, integrin binds to the sequence arginine-glycine-aspartate (RGD) of several adhesive proteins in extracellular matrices, including vitronectin (found in the basal lamina of the eye), fibronectin, and laminin (Ruoslahti and Pierschbacher, 1987). On the cytoplasmic side, integrin binds to **talin** and **α-actinin,** two proteins that connect to actin microfilaments. This dual binding enables the cell to move by contracting the actin microfilaments against the fixed extracellular matrix (see Wang et al., 1993). Different cell types can have different types of integrin molecules with different affinities for extracellular matrix molecules (Hemler et al., 1987; Hemler, 1990). Each integrin protein has two distinct subunits, α and β, and different binary combinations of α and β subunits enable the integrin to bind to particular extracellular molecules. For instance, α2β1 binds to collagen and laminin, while α4β1 binds solely to fibronectin.

While both α and β subunits of integrin are needed for binding to fibronectin or laminin, only the β subunit connects with the internal cytoskeleton. During migration, the bonds linking the β subunit of integrin to the cytoskeleton can be continually made and broken by a protease that cleaves talin and is specifically localized to the cell membrane sites where integrin binds to the substratum. It is possible that this protease cleaves the bridge between the fibronectin receptor and the cytoskeleton (Beckerle et al., 1987).

The importance of integrins for morphogenesis is dramatically illustrated during *Drosophila* embryogenesis. Like vertebrate integrins, *Drosophila* integrins are composed of α and β subunits that span the cell membrane. In the two known *Drosophila* integrins, the β subunits are identical, but the α subunits differ. These two integrins often work together to effect tissue and cell adhesion during development. In the development of the *Drosophila* wing, two epithelial sheets are brought together. The PS1 integrin is found on the basal surface of the presumptive *dorsal* wing epithelium, while the

(A)

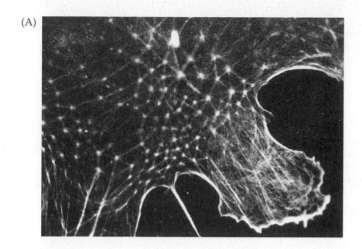

(B)

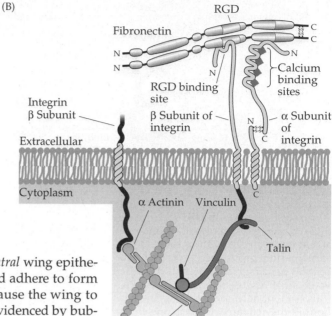

Figure 3.29
Dual function of integrin in binding to extracellular matrices and internal cytoskeleton. (A) Indirect immunofluorescence staining the actin microfilaments of a cell extending a lamellipodium. The actin fibers radiate from the ordered cytoskeletal lattice into the lamellipodium. (B) Speculative diagram relating the binding of cytoskeleton to the extracellular matrix through the integrin molecule. (A from Lazarides, 1976, courtesy of E. Lazarides; B after Luna and Hitt, 1992.)

PS2 integrin is on the upper surface of the presumptive *ventral* wing epithelium. During metamorphosis, these two epithelia meet and adhere to form the two-layered wing blade. Mutations in the integrins cause the wing to have regions where the two wing epithelia come apart, as evidenced by bubbles between the two blades (Brower and Jaffe, 1989; Wilcox et al., 1989). Some of the mutants of *Drosophila* integrins are lethal, because integrin is needed to attach muscles to the epidermis and the gut wall. In the lethal mutation *lethal (1) myospheroid*, there is a deficiency in the genes encoding the β subunit of the *Drosophila* integrins. In the absence of this subunit, neither integrin is formed, and the somatic muscles are contracted into spheres that lack attachments to the body wall or gut (Leptin et al., 1989).

Integrins are not the only molecules capable of binding to laminin and fibronectin. While the integrin receptor binds to an RGD sequence in the A chain of laminin, another laminin receptor protein in the cell membrane binds to a different sequence (YIGSR) in the B1 chain (Graf et al., 1987; Yow et al., 1988). The receptors have different affinities for laminin, and these might be important in their function (Horwitz et al., 1985). The a3β1 integrin of fibroblasts, for example, has a relatively low affinity for laminin ($K^d = 10^{-6}$ M), whereas the affinity for laminin of a laminin receptor from epithelial cells is much higher ($K^d = 2 \times 10^{-9}$ M). The receptor that is used may be important in allowing the cells to use laminin either as a basement membrane (in which case the affinity of the receptor would be high) or as a substrate for migration (in which lower-affinity receptors would be used).

GLYCOSYLTRANSFERASES. Another set of proteins that can adhere cells to extracellular matrix proteins are the cell-surface **glycosyltransferases.** These membrane-bound enzymes are routinely found in the endoplasmic reticulum and Golgi vesicles, where they are responsible for adding sugar residues onto peptides to make glycoproteins. There are numerous glycosyltransferases, each one specific for a given sugar and some showing substrate specificity as well. Thus, a *galactosyl*transferase is an enzyme capable of transferring *galactose* from an activated donor molecule (UDP-galactose) to an acceptor. There may be several galactosyltransferases with affinities for different acceptor molecules.

Galactosyltransferases are functional cell membrane enzymes, and their adhesion to the extracellular matrix represents a "frustrated" catalysis (Figure 3.30). The enzyme needs two substrates for catalysis to occur, the acceptor carbohydrate and the activated sugar. The membrane glycosyltransferases recognize the carbohydrate acceptors on the extracellular matrix

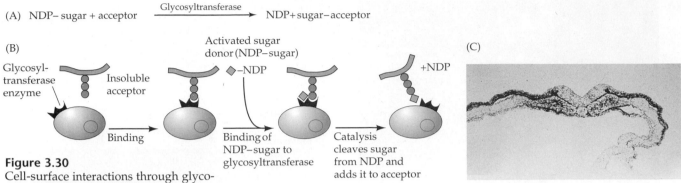

Figure 3.30
Cell-surface interactions through glycosyltransferases. (A) The standard glycosyltransferase reaction, in which a sugar is transferred from a nucleoside diphosphate carrier to an acceptor. (B) Interaction between glycosyltransferases and carbohydrate group (acceptor) on extracellular matrix glycoprotein. If the activated sugar is absent, adhesion results (this is thought to occur during fertilization). If the activated sugar is present in low amounts, migration is permitted. (C) Staining for cell-surface galactosyltransferase by incubating microscopic sections of a 10-somite chick embryo with UDP-[^{3}H]galactose. Insoluble radioactivity seen by autoradiography shows that this radioactive sugar was transferred to the cell surfaces, especially of the migrating mesoderm cells. (A and B modified from Pierce et al., 1980; C from Shur, 1977a, courtesy of B. Shur.)

proteins such as laminin. This causes adhesion to occur. When the second substrate appears, these adhesions can be broken by the catalysis. In some instances (such as fertilization in mice, where galactosyltransferase on the sperm cell membrane interacts with carbohydrate components of the extracellular matrix secreted by the egg), adhesion is critical and no catalysis occurs. In migrating cells, both adhesion and catalysis are seen (Toole, 1976; Shur, 1977a,b; Turley and Roth, 1979; Eckstein and Shur, 1989).

Differential Adhesion Resulting from Multiple Adhesion Systems

Although we have been discussing adhesion systems as separate units, the morphogenetic processes of cell-cell interaction are probably brought about by combinations of cell adhesion molecules. For example, the initial attachment of the mouse embryo to the uterine wall appears to be mediated through several adhesion systems. First, the outer cells of the embryo (the trophoblastic cells) have receptors for the collagen and heparan sulfate proteoglycans of the uterine endometrium, and interference with this binding can impede implantation (Farach et al., 1987; Carson et al., 1988, 1993). Second, Dutt and co-workers (1987) have shown that the trophoblast cells can also adhere to uterine cells through cell-surface glycosyltransferases. Third, Kadokawa and co-workers (1989) have shown P- and E-cadherin on both the trophoblast and the uterine tissue at the site of implantation. Thus, cells may have many adhesive systems that enable them to bind and/or migrate over specific substrates.

Multiple systems also are used by cells to remodel tissues by digestion. For example, when mammalian embryos embed in the uterus, they digest their way through the uterine epithelium and through its basement membrane of laminin, fibronectin, and type IV collagen (Behrendtsen et al., 1992). Bone growth, tadpole tail regression, and branched organ formation (such as in the salivary gland, kidney, and lung) also require the breakdown of basement membranes. This controlled degradation of extracellular matrix molecules is accomplished by a set of enzymes collectively called **matrix-degrading metalloproteinases** (Matrisian, 1992; Sato et al., 1994). Some of these enzymes are bound to the cell membrane, while others are secreted directly from the cells onto the matrix to be dissolved. These metalloproteinases include (1) collagenases that digest types I, II, and III collagens; (2) gelatinases that digest elastin and collagens IV and V; and (3) stromelysins that digest proteoglycans, fibronectin, and laminin. The activation of the metalloproteinase genes is accomplished coordinately, and several metalloproteinases interact to amplify the intensity of the digestive enzymes (Figure 3.31). Soon after the metalloproteinases are activated, the cells activate the genes for in-

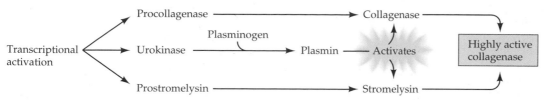

Figure 3.31
Cascade of membrane metalloproteinase activation. Urokinase is a plasminogen activator, cleaving plasminogen into plasmin. Plasmin activates the precursor forms of the stromelysins and collagenases to provide a highly active mix of enzymes capable of digesting extracellular matrices. (After Matrisian, 1992.)

hibitors of these proteins. The controlled production and degradation of the extracellular matrix is an essential part of normal development.

Receptor molecules and signal transduction pathways

The fates of cells are often determined by interactions at their cell surfaces, where a receptor molecule meets its complementary ligand. But how does some interaction on the cell surface cause specific genes to be transcribed inside the nucleus? The pathways between the cell membrane and the genome are called **signal transduction pathways.** Several types of pathways have been discovered, and we will mention some of the major ones here. As you will see, these appear to be variations on a common theme. The theme is quite elegant: Each receptor spans the membrane and has an extracellular region, a transmembrane region, and a cytoplasmic region. When the receptor binds its ligand on its extracellular region, its shape changes, and the cytoplasmic portion of the molecule now has enzymatic activity. This activity is usually a kinase activity that can use ATP to phosphorylate proteins, including itself. The active receptor can now catalyze reactions that phosphorylate other proteins, and eventually, the phosphorylation activates a transcription factor that had formerly been dormant. This transcription factor can now activate (or repress) a new set of genes. The ligand initiating the reaction can be bound on a cell or extracellular matrix or it can be a diffusible molecule. When the diffusible molecule is coming from the blood, it is considered an **endocrine** signal. If the signal comes from nearby cells diffusing from one cell to another, it is called a **paracrine** signal.

The JAK-STAT Pathway

In Chapter 2, we discussed a set of transcription factors that were inactive until a signal from another cell phosphorylated them. These transcription factors are the **STAT** (signal transducers and activators of transcription) proteins (Ihle, 1995, 1996). STATs are phosphorylated by the active form of the **JAK** family of kinases. The JAK-STAT pathway is extremely important in the differentiation of blood cells and in the activation of the casein gene during milk production (Briscoe et al., 1994; Groner and Gouilleux, 1995). In these instances, a certain differentiation factor binds to its membrane-spanning receptors, causing them to dimerize (Figure 3.32). A JAK protein is bound to each of the receptors (in their respective cytoplasmic regions), and these JAK proteins are now brought together, where they can phosphorylate the receptors at several sites. The receptors are now activated and have their own protein kinase activity. They can phosphorylate particular inactive STATs and cause them to dimerize. These dimers are the active form of the STAT, and

Figure 3.32
The JAK-STAT pathway—in this case, the casein gene activation pathway activated by prolactin. The casein gene is activated during the last (lactogenic) phase of mammary gland development, and its signal is the secretion of prolactin, a 9-amino-acid peptide from the anterior pituitary gland. Prolactin causes the dimerization of the prolactin receptors in the mammary duct epithelial cells. A particular JAK protein (Jak2) is "hitched" onto these receptors. When the receptors are dimerized, the JAK proteins phosphorylate each other and the neighboring receptors, activating the dormant kinase activity of these receptors. The activated receptors add a phosphate group onto a tyrosine residue (Y) of a particular STAT protein (in this case, Stat5). This allows it to dimerize, be translocated into the nucleus, and bind to particular regions of DNA. In combinations with other transcription factors (which presumably have been waiting for its arrival), the STAT protein activates transcription of the casein gene. GR is the glucocorticoid receptor, OCT1 is a general transcription factor, and TBP is the set of proteins responsible for binding RNA polymerase. (For details, see Groner and Gouilleux, 1995.)

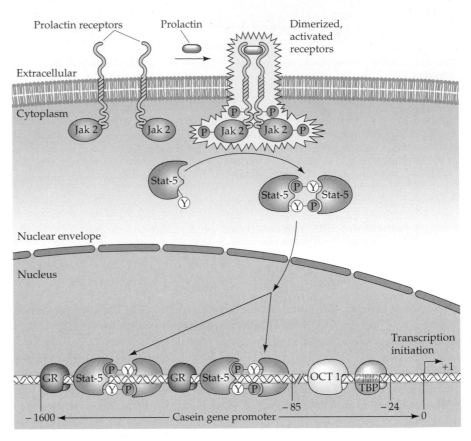

they are translocated into the nucleus, where they bind to specific regions of DNA.

The RTK-Ras Pathway

The RTK-Ras signal transduction pathway was one of the first pathways to unite various areas of developmental biology. Researchers studying *Drosophila* eyes, nematode vulvae, and human cancers all found that they were studying the same genes. The RTK-Ras pathway begins at the cell surface, where a **receptor tyrosine kinase** (**RTK**) binds its specific ligand. Ligands that bind to RTKs include the fibroblast growth factors, epidermal growth factors, and platelet-derived growth factors. The receptor tyrosine kinase spans the membrane, and when it binds its ligand, it undergoes a conformational change that enables it to dimerize. These dimers have a latent kinase activity that becomes activated by the conformational change, and the receptors phosphorylate each other on particular tyrosine residues. Thus, the binding of the ligand to the receptor causes the **autophosphorylation** of the cytoplasmic domain of the receptor.

The phosphorylated tyrosine on the receptor is then recognized by an adaptor protein (Figure 3.33)—specifically, the phosphorylated tyrosines are recognized by a portion of the adaptor protein called the SH2 domain. The adaptor proteins serve as a bridge that links the phosphorylated receptor kinase to a powerful intracellular signaling system. While binding to the phosphorylated receptor kinase through its SH2 domain, the adaptor protein uses its SH3 domain to regulate the activator of a **Ras G protein.** Normally, the wild-type Ras protein is in an inactive, GDP-binding state. When activated by the ligand-bound receptor, it exchanges a phosphate from another GTP to transform the bound GDP into GTP. This catalysis is aided by the **guanine**

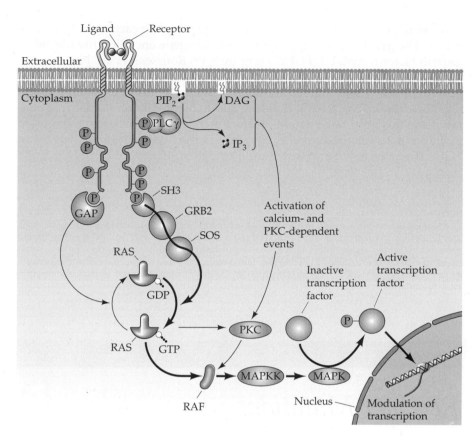

Figure 3.33
The widely used RTK-Ras pathway. The receptor tyrosine kinase is dimerized by the ligand. This causes the autophosphorylation of the receptor. The SH3 protein recognizes the phosphotyrosines and activates the intermediate proteins (GRB2 and SOS), which activate the Ras G protein by allowing the phosphorylation of the GDP moiety of Ras. At the same time, the GAP protein stimulates the hydrolysis of this phosphate bond. The active Ras activates protein kinase C (PKC), which in turn phosphorylates a series of kinases. Eventually, the activated MAP kinase alters gene expression by phosphorylating certain transcription factors (which can then enter the nucleus to change the types of genes transcribed) and certain translation factors (which alter the level of protein synthesis). In many cases, this pathway is reinforced by the release of calcium ions.

nucleotide exchange factor. The GTP-bound Ras is the active form of the protein that transmits the signal. After delivering the signal, the GTP on the Ras protein is hydrolyzed into GDP. This catalysis is greatly stimulated by the normal complexing of the Ras protein to the **GTPase-activating protein** (**GAP**). This 120-kDa protein increases its GTP-hydrolyzing activity over 100-fold (Trahey and McCormick, 1987; Gibbs et al., 1988), which returns the Ras to its inactive state. Indeed, mutations in the *RAS* gene account for a large proportion of human tumors (Shih and Weinberg, 1982), and the mutations of the *RAS* gene that make it oncogenic all inhibit the binding of the GAP protein. Without GAP protein, Ras protein cannot catalyze GTP well and so remains in its active configuration (Cales et al., 1988; McCormick, 1989).

The active Ras protein associates with a kinase called Raf. The Ras protein places the inactive Raf protein onto the cell membrane, where it becomes active (Leevers et al., 1994; Stokoe et al., 1994). The Raf protein is called a MAP-kinase-kinase-kinase (MAPKKK). (MAP stands for mitosis-associated protein, but is now thought to be a larger set of transcription factors.) The MAPKKK phosphorylates the MAPKK, which can then phosphorylate the MAP kinase. This last kinase can phosphorylate the transcription factors that specify cell fate or proliferation. In *Drosophila* eyes, for instance, the cascade is thought to activate the Sina (Sevenless-in-Absentia) transcription factor, whose presence is necessary for the differentiation of photoreceptor 7 (Carthew and Rubin, 1990; Dickson et al., 1992).

As we will see later in this book, this pathway is critical in numerous developmental processes. In humans, mutations in this pathway generate the most common forms of dwarfism, including **achondroplasia,** occurring in about 1 out of 50,000 births. Here, the torso and head undergo normal growth, but the arms and legs are proximally shortened. The deficiency resides in the minimal proliferation of the growth plate cartilage of the long bones. The genetic lesion appears to be in the gene encoding the fibroblast growth factor receptor 3 (FGFR3) (Figure 9.19; Rousseau et al., 1994; Shiang et al., 1994). This gene is expressed in the developing cartilage cells in the growth plates of the long bones. When activated by an FGF, the FGFR3 signals the chondrocyte to stop dividing and to begin differentiating. The mutations in this gene cause a gain-of-function phenotype, wherein the mutant FGFR3 is active constitutively (i.e. without the need to be activated by an FGF)* (Deng et al., 1996; Webster and Donoghue, 1996). [cell7.html]

*Names can be dangerously misleading. Many compounds do more than one thing in a cell, and what they do depends on the context of the cell. Certain "growth factors" can also inhibit growth, and some "transcription factors" can be utilized to inhibit transcription. Indeed, some transcription factors can also be used to regulate translation. Here we see that cell adhesion molecules may also be used for signal transduction. Cellular proteins do not respect our disciplinary boundaries.

Sidelights & Speculations

Dominant Negative Receptor Mutations

THE FUNCTIONAL SIGNIFICANCE of a ligand molecule can be assessed by deleting its receptor. One way of doing that is to create **dominant negative receptor mutations.** This type of experiment will work when dimerization is critical for the function of the receptor. The active FGF receptors, for instance, are dimers of two identical molecules embedded in the cell membrane. The dominant negative mutant will not form an active dimer, even with a wild-type partner. Therefore, when present in high enough concentrations, the mutant receptor can compete with normal FGF receptors and prevent normal FGF receptor proteins from being activated. This can occur by natural mutation or it can be induced. Amaya and co-workers (1991) injected the mRNA for a mutant form of an FGF receptor into two-cell *Xenopus* embryos. Such blastulae are unable to respond to FGF (Figure 3.34). In this experiment, the embryos that lacked functional FGF receptors had dramatically reduced amounts of posterior and lateral mesoderm (Plate 3). ∎

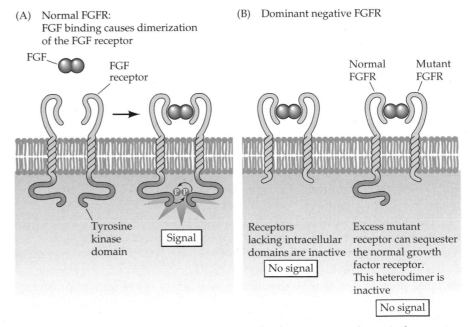

(A)   Normal FGFR:
FGF binding causes dimerization of the FGF receptor

(B)   Dominant negative FGFR

Figure 3.34   The dominant negative receptor assay for the importance of a particular receptor. The FGF receptor (FGFR) is a transmembrane RTK. (A) When FGF dimers bind to the extracellular portion of these receptors, it causes them to dimerize, and the two protein kinase domains phosphorylate each other. When phosphorylated, they initiate a signal through the cytoplasm. (B) The dominant negative receptor lacks the protein kinase domain. When it binds FGF, it creates an inactive dimer, even if the other partner is wild-type. Thus, the effect of FGF is not transmitted into the cell.

The Inositol Phosphate Pathway

Sometimes, the transduction of a cell-surface signal causes a great many changes, and alterations in gene expression consitute only a small subset of what the signal does. The activation of the **inositol phosphate pathway** drastically changes the physiology of the cell by releasing calcium ions from the endoplasmic reticulum. This pathway is extremely important in both sperm and egg activation, both of which require an increase in the concentration of intracellular calcium ions.

The pathway can have two initiation points (Figure 3.35; Berridge, 1993; Shilling et al., 1994). One initiation point is the receptor tyrosine kinase,

Figure 3.35
The inositol phosphate pathway. (A) The phospholipase C reaction, splitting PIP_2 into DAG and IP_3. (B) This reaction can be initiated at two major points on the cell membrane. First, the pathway is initiated when a G-protein-linked transmembrane receptor is activated by binding ligand. This activation causes the binding of GTP to the heteromeric G protein and its dissociation into active subunits. These subunits activate phospholipase C enzymes (PLC), which can catalyze the formation of DAG and IP_3. Second, this pathway can be activated by the RTK pathway. IP_3 can bind to a receptor to release calcium ions from the endoplasmic reticulum. Meanwhile, DAG (in the presence of the released calcium ions) activates protein kinase C. This protein kinase stimulates the sodium/hydrogen transporter to exchange cellular hydrogen ions for extracellular sodium ions, thereby leading to the increase in pH.

(A)

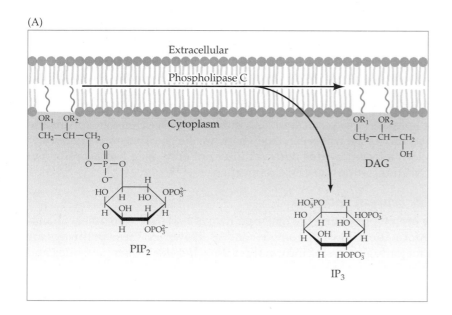

(B)

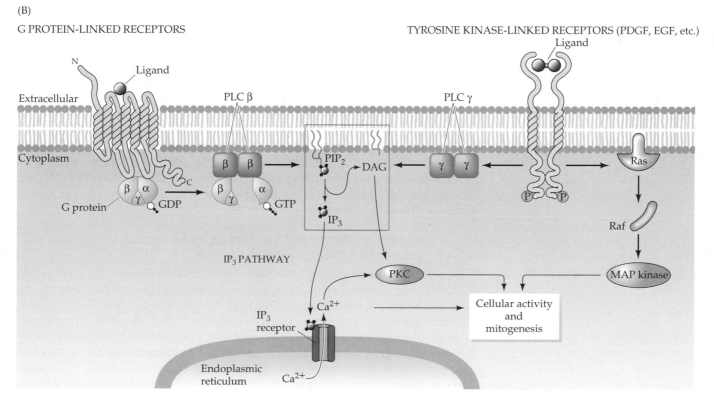

mentioned earlier. In addition to activating the Ras G protein, the activated tyrosine kinases can interact with a type of **phospholipase C** enzyme (PLC-γ1, which also has an SH2 domain that recognizes the autophosphorylated tyrosines). Phospholipase C can catalyze the hydrolysis of phosphatidylinositol 4,5-bisphosphate (**PIP₂**) into two **second messengers:** inositol 1,4,5-trisphosphate (**IP₃**) and diacylglycerol (**DAG**). IP₃ is able to open calcium ion channels of the endoplasmic reticulum, releasing a large store of calcium ions into the cytoplasm. DAG activates **protein kinase C**, which in turn activates the protein pump that exchanges sodium ions for hydrogen ions (Swann and Whitaker, 1986; Nishizuka, 1986). The result is the elevation of intracellular calcium ions and the increase of intracellular pH.

A second initiation point is another class of receptors, sometimes called the serpentine receptors, because they have seven transmembrane domains and "snake" through the cell membrane. These receptors are connected to another type of G protein, the **heteromeric G protein**. When the ligand binds to its receptor, the receptor activates the G protein. This activation dissociates the G protein into its subunits, which are then able to activate another set of phospholipase C, namely, PLC-β1 and PLC-β2. These two types of phospholipase C can also split PIP₂ into inositol 1,4,5-trisphosphate and diacylglycerol. As we will see in later chapters, the changes in hydrogen and calcium ions effected by this pathway alter not only gene transcription but also mRNA translation and DNA replication.

Cross-Talk between Pathways

We have been representing the major pathways as if they were linear chains, where information flows in a single conduit. However, these pathways are just the major highways of information flow. Between them, avenues and streets connect one pathway with another. (This may be why there are so many steps between the cell surface and the nucleus. Each step is a potential regulatory point and a potential cross-talk intersection.) This cross-talking can be seen in Figure 3.35, where two pathways reinforce each other. We must also remember that a cell has numerous receptors and is constantly receiving many signals simultaneously. In some cases, gene transcription requires two signals. This is seen during lymphocyte development, where two signals are needed, each one producing one of the two peptides of a transcription factor needed for the production of interleukin 2 (IL-2, also known as T cell growth factor). One factor, c-Fos, is produced by the binding of the T cell receptor to the antigen (Figure 3.36). This activates the Ras cascade, creating a transcription factor, Elk-1, that activates the c-*fos* gene to synthesize c-Fos. The second signal comes from the B7 glycoprotein on the surface of the cell presenting the antigen. This signal activates a second cascade of kinases, eventually producing c-Jun. The two peptides, c-Fos and c-Jun, can make the AP-1 protein, a transcription factor that binds to the IL-2 enhancer and activates its expression (Li et al., 1996).

The Extracellular Matrix and Cell Surface as Sources of Developmentally Critical Signals

Bissell and colleagues (1982; Martins-Green and Bissell, 1995) have proposed that the extracellular matrix is capable of inducing specific gene expression in developing tissues, especially those of the liver and mammary gland, where the induction of specific transcription factors depends on cell-substrate binding (Liu et al., 1991; Streuli et al., 1991; Notenboom et al., 1996). Often, the presence of bound integrin prevents the activation of genes that specify cell death (Brooks et al., 1994; Montgomery et al., 1994). Therefore, the extracellular matrix is an important source of signals that can be trans-

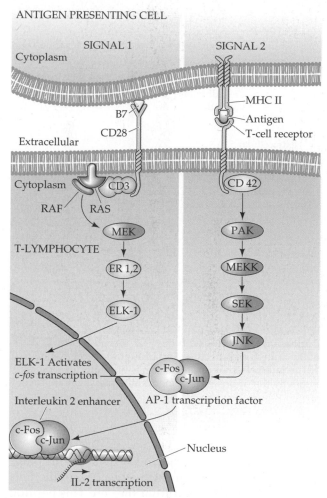

Figure 3.36

Two signals are needed to effect the differentiation of the T lymphocyte. The first signal comes from the receptors that bind the antigen that is presented on the cell surface of the B cell or macrophage. The second signal comes from binding the CD28 protein to the B7 protein that is on the surface of the antigen-presenting cell. The first signal directs the synthesis of one subunit of the AP-1 transcription factor. The second signal directs the synthesis of the other subunit. The two subunits, c-fos and c-jun, form the AP-1 transcription factor that can activate T-cell-specific enhancers such as that regulating interleukin 2 production.

duced into the nucleus to give specific gene expression. Recent studies have shown that the binding of integrins to the extracellular matrix can stimulate the RTK-Ras pathway, and so can the interaction of the cell with a neighboring cell's L1, N-CAM, and cadherins (Bixby et al., 1994; Williams et al., 1994a; Clark and Brugge, 1995). Cadherins (even soluble ones) appear to be able to dimerize FGF receptors just like the normal FGF ligands, causing the calcium ion release, transcriptional activation, and developmental phenomena characteristic of the cell's FGF responses (Figure 3.37; Williams et al., 1994b; Doherty et al., 1995). Cross-talk is almost certain to happen when the cell adhesion molecules are also signal transducers.

Reciprocal Interactions at the Cell Surface

When two cells interact during development, both cells are usually changed. This reciprocal induction is mediated by interactions at the cell membrane.

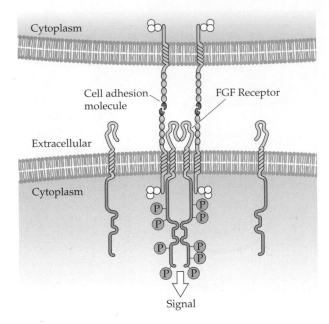

Figure 3.37
Possible interactions of cell adhesion molecules with the FGF receptors. The FGF receptors may be "hijacked" by adhesion molecules and brought together. They may be brought together by the interaction of opposite adhesion molecules, or the "crosslinking" of FGF receptors by the opposing cell membrane may activate their kinase domains.

One widely used pathway is the Wingless-Hedgehog system. In this pathway, two cells (or groups of cells) are adjacent to each other. One of the cells produces the Hedgehog protein and secretes it from the cell. This peptide acts on the neighboring cell, causing this neighbor to produce the Wingless (Wnt) protein. The Wingless protein is itself secreted and binds to the neighboring cell, stimulating it to continue the synthesis of Hedgehog. The result is the stabilization of a border wherein the tissue on one side secretes Hedgehog protein, while the tissue on the other side produces Wingless. This border is critical in the production of segments and appendages in *Drosophila* as well as brain subdivision and limbs in mammals (Figure 3.38; Ingham, 1994; Niswander et al., 1994; Wilder and Perrimon, 1995). [cell8.html]

The cell surface is an extremely important place for developmental interactions. These include the differential adhesion of one cell to others, the differential adhesion of a cell type to an extracellular matrix, and the communication of signals for cell differentiation and division. In 1782, the French essayist Denis Diderot posed the question of morphogenesis in the fevered dream of a noted physicist. This character could imagine that the body was formed from myriads of "tiny sensitive bodies" that collected together to form an aggregate, but he could not envision how this aggregate could become an animal. Recent studies have shown that this ordering is due to the molecules on the surfaces of these cells. In subsequent chapters, we will look more closely at some of these morphogenetic interactions. We are now at the stage where we can begin our study of early embryogenesis and see the integration of the organismal, genetic, and cellular processes of animal development.

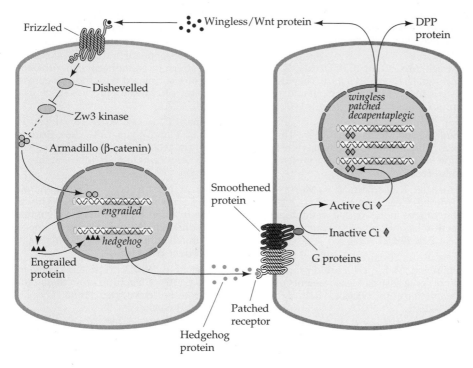

Figure 3.38
Reciprocal interactions between cells in the wingless-hedgehog pathway in *Drosophila*. Wingless protein is secreted from one cell and diffuses only a small distance. Its neighboring cell binds the Wingless protein, causing it to activate the Disheveled protein, which blocks the inhibiting action of Zeste-white-3 upon the Armadillo protein (a catenin). The active Armadillo protein enables the cell to transcribe the *hedgehog* (*hh*) gene. This protein is secreted and bound by the neighboring cell. Binding the Hedgehog protein causes this cell to transcribe the *wingless* gene and secrete that protein.

LITERATURE CITED

Ayama, E., Musci. T. J. and Kirschner, M. W. 1991. Expression of a dominant negative mutant of the FGF receptor disrupts mesoderm formation in *Xenopus* embryos. *Cell* 66: 257–270.

Armstrong, P. B. 1989. Cell sorting out: The self-assembly of tissues in vitro. *CRC Crit. Rev. Biochem. Mol. Biol.* 24: 119–149.

Bastiani, M. J., Harrelson, A. L., Snow, P. M. and Goodman, C. S. 1987. Expression of fasciclin I and II glycoproteins on subsets of axon pathways during neuronal development in the grasshopper. *Cell* 48: 745–755.

Beckerle, M. C., Burridge, K., DeMartino, G. N. and Croall, D. E. 1987. Colocalization of calcium-dependent protease II and one of its substrates and sites of adhesion. *Cell* 51: 569–577.

Behrendsten, O., Alexander, C. M. and Werb, Z. 1992. Metalloproteinases mediate extracellular matrix degradation by cells from mouse blastocyst outgrowths. *Development* 114: 447–456.

Bernfield, M. and Sanderson, D. 1990. Syndecan, a morphogenetically regulated cell surface proteoglycan that binds extracellular matrix and growth factors. *Philos. Trans. R. Soc. Lond.* [A] 327: 171–186.

Berridge, M. J. 1993. Inositol triphosphate and calcium signalling. *Nature* 361: 315–325.

Beug, H., Gerisch, G., Kempff, S., Riedel, V. and Cremer, G. 1970. Specific inhibition of cell contact formation in *Dictyostelium* by univalent antibodies. *Exp. Cell Res.* 63: 147–158.

Bevilacqua, A., Loch-Caruso, R. and Erickson, R. P. 1989. Abnormal development and dye coupling produced by antisense RNA to gap junction protein in mouse preimplantation embryos. *Proc. Natl. Acad. Sci. USA* 86: 5444–5448.

Bissell, M. J., Hall, H. G. and Parry, G. 1982. How does the extracellular matrix direct gene expression? *J. Theoret. Biol.* 99: 31–68.

Bixby, J. L., Grunwald, G. B. and Bookman, R. J. 1994. Ca⁺⁺ influx and neurite growth in response to purified N-cadherin and laminin. *J. Cell Biol.* 127: 1461–1475.

Boucaut, J. C. 1974. Étude autoradiographique de la distribution de cellules embryonnaires isolées, transplantées dans le blastocèle chez *Pleurodeles waltii* Michah (Amphibien, Urodele). *Ann. Embryol. Morphol.* 7: 7–50.

Boyse, E. A. and Old, L. J. 1969. Some aspects of normal and abnormal cell surface genetics. *Annu. Rev. Genet.* 3: 269–289.

Brackenbury, R., Thiery, J.-P., Rutishauser, U. and Edelman, G. M. 1977. Adhesion among neural cells of the chick embryo. I. Immunological assay for molecules involved in cell-cell binding. *J. Biol. Chem.* 252: 6835–6840.

Briscoe, J., Guschin, D., and Müller, M. 1994. Just another signalling pathway. *Curr. Biol.* 4: 1033–1035.

Bronner-Fraser, M. 1988. Distribution of tenascin during cranial neural crest development in the chick. *J. Neurosci. Res.* 21: 135–147.

Brooks, P. C. Montgomery, A. M. P., Rosenfeld, M., Reisfeld, R. A., Hu, T., Klier, G. and Cheresh, D. A. 1994. Integrin α v β 3 antagonists promote tumor regression by inducing apoptosis of angiogenic blood vessels. *Cell* 79: 1157–1164.

Brower, D. L. and Jaffe, S. M. 1989. Requirement for integrins during *Drosophila* wing development. *Nature* 342: 285–287.

Cales, C., Hancock, J. F., Marshall, C. J. and Hall, A. 1988. The cytoplasmic protein GAP is implicated as a target for regulation by the *ras* gene product. *Nature* 332: 548–551.

Carson, D. D., Tang, J.-P. and Gay, S. 1988. Collagens support embryo attachment and outgrowth in vitro: Effects of the Arg-Gly-Asp sequence. *Dev. Biol.* 127: 368–375.

Carson, D. D., Tang, J.-P. and Julian, J. 1993. Heparan sulfate proteoglycan (perlecan) expression by mouse embryos during acquisition of attachment competence. *Dev. Biol.* 155: 97–106.

Carthew, R.W. and Rubin, G.M. 1990. *Seven-in-absentia*, a gene required for the specification of R7 cell fate in the *Drosophila* eye. *Cell* 63: 561–577

Chen, W. T., Hasegawa, E., Hasegawa, T., Weinstock, C. and Yamada, K. M. 1985. Development of cell-surface linkage complexes in cultured fibroblasts. *J. Cell Biol.* 100: 1103–1114.

Cheney, C. M. and Lash, J. W. 1981. Diversification within embryonic chick somites: Differential response to notochord. *Dev. Biol.* 81: 288–298.

Chuong, C.-M. and Edelman, G. M. 1985a. Expression of cell adhesion molecules in embryonic induction. I. Morphogenesis of nestling feathers. *J. Cell Biol.* 101: 1009–1026.

Chuong, C.-M. and Edelman, G. M. 1985b. Expression of cell adhesion molecules in embryonic induction. II. Morphogenesis of adult feathers. *J. Cell Biol.* 101: 1027–1043.

Clark, E. A. and Brugge, J. S. 1995. Integrin and signal transduction pathways: The road taken. *Science* 268: 233–239.

Covault, J. and Sanes, J. R. 1986. Distribution of N-CAM in synaptic and extrasynaptic portions of developing and adult skeletal muscle. *J. Cell Biol.* 102: 716–730.

Crawford, K. and Stocum, D. L. 1988. Retinoic acid coordinately proximalizes regenerate pattern and blastema differential affinity in axolotl limbs. *Development* 102: 687–698.

Darnell, J., Lodish, H. and Baltimore, D. 1986. *Molecular Cell Biology.* Scientific American Books, New York.

Deng, C., Wynshaw-Boris, A., Zhou, F., Kuo, A. and Leder, P. 1996. Fibroblast growth factor receptor-3 is a negative regulator of bone growth. *Cell* 84: 911–921.

Detrick, R. J., Dickey, D. and Kintner, C. R. 1990. The effects of N-cadherin misexpression on morphogenesis in *Xenopus* embryos. *Neuron* 4: 493–506.

Dickson, B., Sprenger, F., Morrison, D. and Hafen, E. 1992. Raf functions downstream of Ras1 in the Sevenless signal transduction pathway. *Nature* 360: 600–603.

Diderot, D. 1782. *D'Alembert's Dream.* Reprinted in J. Barzun and R. H. Bowen (eds.), *Rameau's Nephew and Other Works* (1956). Doubleday, Garden City, NY. [p. 114]

Doherty, P., Williams, E. and Walsh, F. S. 1995. A soluble chimeric form of the L1 glycoprotein stimulates neurite outgrowth. *Neuron* 14: 57–66.

Dufour, S., Duband, J.-L., Humphries, M. J., Obara, M., Yamada, K. M. and Thiery, J. P. 1988. Attachment, spreading and locomotion of avian neural crest cells are mediated by multiple adhesion sites on fibronectin molecules. *EMBO J.* 7: 2661–2671.

Dutt, A., Tang, T.-P. and Carson, D. D. 1987. Lactosaminoglycans are involved in uterine epithelial cell adhesion in vitro. *Dev. Biol.* 119: 27–37.

Eckstein, D. J. and Shur, B. D. 1989. Laminin induces the stable expression of surface glycosyltransferases on lamellipodia of migrating cells. *J. Cell Biol.* 108: 2507–2517.

Edelman, G. M. and Thiery, J.-P. 1985. *The Cell in Contact: Adhesions and Junctions as Morphogenic Determinants.* Wiley, New York.

Farach, M. C., Tang, J. P., Decker, G. L. and Carson, D. D. 1987. Heparin/heparan sulfate is involved in attachment and spreading of mouse embryos in vitro. *Dev. Biol.* 123: 401–410.

Fink, R. and McClay, D. R. 1985. Three cell recognition changes accompany the ingression of sea urchin primary mesenchyme cells. *Dev. Biol.* 107: 66–74.

Fraser, S., E., Carhart, M. S., Murray, B. A., Chuong, C.-M. and Edelman, G. E. 1988. Alterations in the *Xenopus* retinotectal projection by antibodies to *Xenopus* N-CAM. *Dev. Biol.* 129: 217–230.

Foty, R. A., Forgacs, G., Pfleger, C. M. and Steinberg, M. S. 1994. Liquid properties of embryonic tissues: Measurements of interfacial tensions. *Physic. Rev. Lett.* 72: 2298–2301.

Foty, R. A., Pfleger, C. M., Forgacs, G. and Steinberg, M. S. 1996. Surface tensions of embryonic tissues predict their mutual envelopment behavior. *Development* 122: 1611–1620.

Fujimori, T., Miyatani, S. and Takeichi, M. 1990. Ectopic expression of N-cadherin perturbs histogenesis in *Xenopus* embryos. *Development* 110: 97–104.

Gibbs, J. B., Scaber, M. D., Allard, W. J., Sigal, I. S. and Scolnick, E. M. 1988. Purification of *ras* GTPase activating protein from bovine brain. *Proc. Natl. Acad. Sci. USA* 85: 5026–5030.

Giudice, G. 1962. Restitution of whole larvae from disaggregated cells of sea urchin embryos. *Dev. Biol.* 5: 402–411.

Graf, J., Ogle, R. C., Robey, F. A., Sasaki, M., Martin, G. R., Yamada, Y. and Kleinman, H. K. 1987. A pentapeptide from the laminin B1 chain mediates cell adhesion and binds to the 67000 laminin receptor. *Biochemistry* 26: 6896–6900.

Groner, B. and Gouilleux, F. 1995. Prolactin-mediated gene activation in mammary epithelial cells. *Curr. Opin. Genet. Dev.* 5: 587–594.

Hadler, N. M., Dourmash, R. R., Nermut, M. V. and Williams, L. D. 1982. Ultrastructure of a hyaluronic acid matrix. *Proc. Natl. Acad. Sci. USA* 79: 307–309.

Hakamori, S., Fukuda, M., Sekiguchi, K. and Carter, W. G. 1984. Fibronectin, laminin, and other extracellular glycoproteins. *In* K. A. Picz and A. H. Reddi (eds.), *Extracellular Matrix Biochemistry.* Elsevier, New York, pp. 229–275.

Harrelson, A. L. and Goodman, C. S. 1988. Growth cone guidance in insects: Fasciclin-II is a member of the immunoglobulin superfamily. *Science* 242: 700–708.

Hatta, K. and Takeichi, M. 1986. Expression of N-cadherin adhesion molecules associated with early morphogenetic events in chick development. *Nature* 320: 447–449.

Hatta, K. Takagi, S., Fujisawa, H. and Takeichi, M. 1987. Spatial and temporal expression pattern of N-cadherin cell adhesion molecules correlated with morphogenetic processes of chicken embryos. *Dev. Biol.* 120: 215–227.

Heasman, J., Hines, R. D., Swan, A. P., Thomas, V. and Wylie, C. C. 1981. Primordial germ cells of *Xenopus* embryos: The role of fibronectin in their adhesion during migration. *Cell* 27: 437–447.

Heasman, J., Ginsberg, D., Goldstone, K., Pratt, T., Yoshidanaro, C., and Wylie, C. 1994. A functional test for maternally inherited cadherin in Xenopus shows its importance in cell adhesion at the blastula stage. *Development* 120: 49–57.

Hemler, M. E. 1990. VLA proteins in the integrin family: Structures, functions, and their role on leukocytes. *Annu. Rev. Immunol.* 8: 365–400.

Hemler, M. E., Huang, C. and Schwartz, L. 1987. The VLA protein family. Characterization of five distinct cell surface heterodimers each with a common 130,000 molecular weight β subunit. *J. Biol. Chem.* 262: 3300–3309.

Horwitz, A., Duggan, K., Greggs, R., Decker, C. and Buck, C. 1985. The cell substrate attachment (CSAT) antigen has properties of a receptor for laminin and fibronectin. *J. Cell Biol.* 101: 2134–2144.

Horwitz, A., Duggan, K., Buck, C., Beckerle, M. C. and Burridge, K. 1986. Interaction of plasma membrane fibronectin receptor with talin-a actinin transmembrane linkage. *Nature* 320: 531–533.

Ingham, P. W. 1994. Hedgehog points the way. *Curr. Biol.* 4: 1–4.

Jongen, W, M. F. and seven others. 1991. Regulation of connexin 43-mediated gap junction intercellular communication by Ca^{2+} in mouse epidermal cells is controlled by E-cadherin. *J. Cell Biol.* 114: 545–555.

Ihle, J. N. 1995. Cytokine receptor signalling. *Nature* 377: 591–594.

Ihle, J. N. 1996. STATs: Signal transducers and activators of transcription. *Cell* 84: 331–334.

Just, E. E. 1939. *The Biology of the Cell Surface.* Blackiston, Philadelphia.

Kadokawa, Y., Fuketa, I., Nose, A., Takeichi, M. and Nakatsuji, N. 1989. Expression of E- and P-cadherin in mouse embryos and uteri during the periimplantation period. *Dev. Growth Diff.* 31: 23–30.

Kalimi, G. H. and Lo, C. 1988. Communication compartments in the gastrulating mouse embryo. *J. Cell Biol.* 107: 241–255.

Kintner, C. 1992. Regulation of embryonic cell adhesion by the cadherin cytoplasm domain. *Cell* 69: 225–236.

Knudsen, K., Horwitz, A. F. and Buck, C. 1985. A monoclonal antibody identifies a glycoprotein complex involved in cell-substratum adhesion. *Exp. Cell Res.* 157: 218–226.

Köhler, G. and Milstein, C. 1975. Continuous cultures of fused cells secreting antibody of predefined specificity. *Nature* 256: 495–497.

Lander, A. D. 1989. Understanding the molecules of neural cell contacts: Emerging patterns of structure and function. *Trends Neurosci.* 12: 189–195.

Landmesser, L., Dahm, L., Schultz, K. and Rutishauser, U. 1988. Distinct roles for adhesion molecules during innervation of embryonic chick muscle. *Dev. Biol.* 130: 645–670.

Lazarides, E. 1976. Actin, a actinin, and tropomyosin interaction in the structural organization of actin filaments in nonmuscle cells. *J. Cell Biol.* 68: 202–219.

Lee, C.-H. and Gumbiner, B. M. 1995. Disruption of gastrulation movements in *Xenopus* by a dominant-negative mutant for C-cadherin. *Dev. Biol.* 171: 363–373.

Lee, S., Gilula, N. B. and Warner, A. E. 1987. Gap junctional communication and compaction during preimplantation stages of mouse development. *Cell* 51: 851–860.

Leevers, S. J., Paterson, H. F., and Marshall, C. J. 1994. Requirement for Ras in Raf activation is overcome by targeting Raf to the plasma membrane. *Nature* 369: 411–414.

Lemmon, V., Farr, K. L., and Lagenauer, C. 1989). L1-mediated axon growth occurs via homophilic binding mechanism. *Neuron* 2: 1597–1603.

Leonard, C. M., Bergman, M., Frenz, D. A., Macreery, L. A. and Newman, S. A. 1989. Abnormal ambient glucose levels inhibit proteoglycan core protein gene expression and reduce proteoglycan accumulation during chondrogenesis: Possible mechanism for teratogenic effects of maternal diabetes. *Proc. Natl. Acad. Sci. USA* 86: 10113–10117.

Leptin, M., Bogaert, T., Lehmann, R. and Wilcox, M. 1989. The function of PS integrins during *Drosophila* embryogenesis. *Cell* 56: 401–408.

Li, W., Whaley, C. D., Mondino, A. and Mueller, D. L. 1996. Blocked signal transduction to the ERK and JNK protein kinases in anergic CD4[+] T cells. *Science* 271: 1272–1274.

Linask, K. L. and Lash, J. W. 1988a. A role for fibronectin in the migration of avian precardiac cells. I. Dose-dependent effects of fibronectin antibody. *Dev. Biol.* 129: 315–323.

Linask, K. L. and Lash, J. W. 1988b. A role for fibronectin in the migration of avian precardiac cells. II. Rotation of the heart-forming region during different stages and its effects. *Dev. Biol.* 129: 324–329.

Liu, J-K., Di Persio, M. C., and Zaret, K. S. 1991. Extracellular signals that regulate liver transcription factors during hepatic differentiation *in vitro. Mol. Cell Biol.* 11: 773–784.

Lo, C. and Gilula, N. B. 1979. Gap junctional communication in the preimplantation mouse embryo. *Cell* 18: 399–409.

Luna, E. J. and Hitt, A. L. 1992. Cytoskeleton-plasma membrane interactions. *Science* 258: 955–964.

Martins-Green, M. and Bissell, M. J. 1995. Cell-ECM interactions in development. *Semin. Dev. Biol.* 6: 149–159.

Massagué, J. 1991. A helping hand from proteoglycans. *Curr. Biol.* 1: 117–119.

Matrisian, L. M. 1992. The matrix-degrading metalloproteinases. *BioEssays* 14: 455–463.

McClay, D. R. and Ettensohn, C. A. 1987. Cell recognition during sea urchin gastrulation. *In* W. F. Loomis (ed.), *Genetic Regulation of Development.* Alan R. Liss, New York, pp. 111–128.

McCormick, F. 1989. *ras* GTPase activating protein: Signal transmitter and signal terminator. *Cell* 56: 5–8.

Monroy, A. and Moscona, A. A. 1979. *Introductory Concepts in Developmental Biology.* University of Chicago Press, Chicago.

Montgomery, A. M. P., Reisfeld, R. A., and Cheresh, D. A. 1994. Integrin α v β 3 rescues melanoma cells from apoptosis in a three-dimensional dermal collagen. *Proc. Natl. Acad. Sci. USA* 91: 8856–8860.

Moscona, A. A. 1952. Cell suspension from organ rudiments of chick embryos. *Exp. Cell Res.* 3: 535–539.

Moscona, A. A. 1961. Rotation-mediated histogenetic aggregation of dissociated cells: A quantifiable approach to cell interaction in vitro. *Exp. Cell Res.* 22: 455–475.

Nagafuchi, A., Shirayoshi, Y., Okazaki, K., Yasuda, K. and Takeichi, M. 1987. Transformation of cell adhesion properties of exogenously introduced E-cadherin cDNA. *Nature* 329: 341–343.

Nardi, J. B. and Stocum, D. L. 1983. Surface properties of regenerating limb cells: Evidence for gradation along the proximodistal axis. *Differentiation* 25: 27–31.

Nishizuka, Y. 1986. Studies and perspectives of protein kinase C. *Science* 233: 305–312.

Niswander, L., Jeffrey, S., Martin, G. R. and Tickle, C. 1994. A positive feedback loop coordinates growth and patterning in the vertebrate limb. *Nature* 371, 609–612.

Nose, A. and Takeichi, M. 1986. A novel cadherin adhesion molecule: Its expression patterns associated with implantation and organogenesis of mouse embryos. *J. Cell Biol.* 103: 2649–2658.

Nose, A., Nagafuchi, A. and Takeichi, M. 1988. Expressed recombinant cadherins mediate cell sorting in model systems. *Cell* 54: 993–1001.

Notenboom, R. G. E., de Poer, P. A. J., Moorman, A. F. M. and Lamers, W. H. 1996. The establishment of the hepatic architecture is a prerequisite for the development of a lobular pattern of gene expression. *Development* 122: 321–332.

Orkin, R. W., Knudson, W. and Toole, P. T. 1985. Loss of hyaluronidate-dependent coat during myoblast fusion. *Dev. Biol.* 107: 527–530.

Peracchia, C. and Dulhunty, A. F. 1976. Low resistance junctions in crayfish: Structural changes with functional uncoupling. *J. Cell Biol.* 70: 419–439.

Pierce, M., Turley, E. A. and Roth, S. 1980. Cell surface glycosyltransferase activities. *Int. Rev. Cytol.* 65: 1–47.

Ratner, N., Bunge, R. P. and Glaser, L. 1985. A neuronal cell surface heparan sulfate proteoglycan is required for dorsal root ganglion neuron stimulation of Schwann cell proliferation. *J. Cell Biol.* 101: 744–754.

Reaume, A. G. and eight others. 1995. Cardiac malformation in neonatal mice lacking connexin43. *Science* 267: 1831–1834.

Roth, S. 1968. Studies on intracellular adhesive selectivity. *Dev. Biol.* 18: 602–631.

Roth, S., McGuire, E. J. and Roseman, S. 1971. An assay for intercellular adhesive specificity. *J. Cell Biol.* 51: 525–535.

Rousseau, F. and seven others. 1994. Mutations in the gene encoding fibroblast growth factor receptor-3 in achondroplasia. *Nature* 371: 252–254.

Ruoslahti, E. and Pierschbacher, M. D. 1987. New perspectives in cell adhesion: RGD and integrins. *Science* 238: 491–497.

Rutishauser, U., Acheson, A., Hall, A., Mann, D. M. and Sunshine, J. 1988. The neural cell adhesion molecule (N-CAM) as a regulator of cell-cell interactions. *Science* 240: 53–57.

Sainio, K., Gilbert, S. F., Lehtonen, E., Nishi, M., Kumar, N. M., Gilula, N. B. and Saxén, L. 1992. Differential expression of gap junction mRNAs and proteins in the developing murine kidney and in experimentally induced nephric mesenchymes. *Development* 115: 827–837.

San Antonio, J. D., Winston, B. M. and Tuan, R. S. 1987. Regulation of chondrogenesis by heparan sulfate and structurally related glycosaminoglycans. *Dev. Biol.* 123: 17–24.

Sato H., Takino, T., Okada, Y., Cao, J., Shinagawa, A., Yamamoto, E. and Seiki, M. 1994. A matric metalloproteinase expressed on the surface of invasive tuomour cells. *Nature* 370: 61–65.

Shiang, R. and seven others. 1994. Mutations in the transmembrane domain of FGFR3 cause the most common genetic form of dwarfism, achondroplasia. *Cell* 78: 335–342.

Shih, C. and Weinberg, R. A. 1982. Isolation of a transforming sequence from a human bladder carcinoma cell line. *Cell* 29: 161–169.

Shilling, F. M., Carroll, D. J., Muslin, A. J., Escobodo, J. A., Williams, L. T. and Jaffe, L. A. 1994. Evidence for both tyrosine kinase and G-protein-coupled pathways leading to starfish egg activation. *Dev. Biol.* 162: 590–599.

Shur, B. D. 1977a. Cell surface glycosyltransferases in gastrulating chick embryos. I. Temporally and spatially specific patterns of four endogenous glycosyltransferase activities. *Dev. Biol.* 58: 23–29.

Shur, B. D. 1977b. Cell surface glycosyltransferases in gastrulating chick embryos. II. Biochemical evidence for a surface localization of endogenous glycosyltransferase activities. *Dev. Biol.* 58: 40–55.

Spring, J., Beck, K. and Chiquet-Ehrismann, R. 1989. Two contrary functions of tenascin: Dissection of the active sites by recombinant tenascin fragments. *Cell* 59: 325–334.

Steinberg, M. S. 1964. The problem of adhesive selectivity in cellular interactions. *In* M. Locke (ed.), *Cellular Membranes in Development*. Academic Press, New York, pp. 321–434.

Steinberg, M. S. 1970. Does differential adhesion govern self-assembly processes in histogenesis? Equilibrium configurations and the emergence of a hierarchy among populations of embryonic cells. *J. Exp. Zool.* 173: 395–434.

Stokoe, D., Macdonald, S. G., Cadwallader, K., Symons, M. and Hancock, J. F. 1994. Activation of raf as well as recruitment to the plasma membrane. *Science* 264: 1463–1467.

Streuli, C. H., Bailey, N. and Bissell, M. J. 1991. Control of mammary epithelial differentiation: Basement membrane induces tissue specific gene expression in the absence of cell-cell interactions and morphological polarity. *J. Cell Biol.* 115: 1383–1395.

Swann, K. and Whitaker, M. 1986. The part played by inositol trisphosphate and calcium in the propagation of the fertilization wave in sea urchin eggs. *J. Cell Biol.* 103: 2333–2342.

Takeichi, M. 1987. Cadherins: A molecular family essential for selective cell-cell adhesion and animal morphogenesis. *Trends Genet.* 3: 213–217.

Takeichi, M. 1991. Cadherin cell adhesion receptors as a morphogenetic regulator. *Science* 251: 1451–1455.

Takeichi, M., Ozaki, H. S., Tokunaga, K. and Okada, T. S. 1979. Experimental manipulation of cell surface to affect cellular recognition mechanisms. *Dev. Biol.* 70: 195–205.

Tamkun, J. W., DeSimone, D. W., Fonda, D., Patel, R. S., Buck, C, Horwitz, A. F. and Hynes, R. O. 1986. Structure of integrin, a glycoprotein involved in transmembrane linkage between fibronectin and actin. *Cell* 46: 271–282.

Tan, S.-S., Crossin, K. L., Hoffman, S. and Edelman, G. M. 1987. Asymmetric expression of somites of cytotactin and its proteoglycan ligand is correlated with neural crest cell migration. *Proc. Natl. Acad. Sci. USA* 84: 7977–7981.

Thesleff, I., Vainio, S. and Jalkanen, M. 1989. Cell-matrix interactions in tooth development. *Int. J. Dev. Biol.* 33: 91–95.

Toole, B. P. 1976. Morphogenetic role of glycosaminoglycans (acid mucopolysaccharides) in brain and other tissues. *In* S. H. Barondes (ed.), *Neuronal Recognition*. Plenum, New York, pp. 276–329.

Tosney, K. W., Watanabe, M., Landmesser, L. and Rutishauser, U. 1986. The distribution of N-CAM in the chick hindlimb during axon outgrowth and synaptogenesis. *Dev. Biol.* 114: 468–481.

Townes, P. L. and Holtfreter, J. 1955. Directed movements and selective adhesion of embryonic amphibian cells. *J. Exp. Zool.* 128: 53–120.

Trahey, M. and McCormick, F. 1987. A cytoplasmic protein stimulates normal N-*ras* p21 GTPase, but does not affect oncogenic mutants. *Science* 238: 542–544.

Trinkaus, J. P. 1963. The cellular basis of *Fundulus* epiboly. Adhesivity of blastula and gastrula cells in culture. *Dev. Biol.* 7: 513–532.

Turley, E. A. and Roth, S. 1979. Spontaneous glycosylation of glycosaminoglycan substrates by adherent fibroblasts. *Cell* 17: 109–115.

Tyler, A. 1946. An auto-antibody concept of cell structure, growth, and differentiation. *Growth* 10 (Symposium 6): 7–19.

Vainio, S., Jalkanen, M., Lehtonen, E. and Bernfield, M. 1989. Epithelial-mesenchymal interactions regulate stage-specific expression of a cell surface proteoglycan, syndecan, in the development kidney. *Dev. Biol.* 134: 382–391.

Venkatesh, T. R., Zipursky, S. L. and Benzer, S. 1985. Molecular analysis of the development of the compound eye in *Drosophila*. *Trends Neurosci.* 8: 251–257.

Vits, L., Van Camp, G., Couke, P., Wilson, G., Schrander-Stumpel, C., Schwarz, C. and Willems, P. J. 1994. MASA syndrome is due to mutations in the L1CAM gene. *Nature Genet.* 7: 408–413.

Vuorio, E. 1986. Connective tissue diseases: Mutations of collagen genes. *Ann. Clin. Res.* 18: 234–241.

Wang, N., Butler, J. P. and Ingber, D. E. 1993. Mechanotransduction across the cell surface and through the cytoskeleton. *Science* 260: 1124–1127.

Warner, A. E., Guthrie, S. C. and Gilula, N. B. 1984. Antibodies to gap junctional protein selectively disrupt junctional communication in the early amphibian embryo. *Nature* 311: 127–131.

Webster, M. K. and Donoghue, D. J. 1996. Constitutive activation of fibroblast growth factor receptor 3 by the transmembrane domain point mutation found in achondroplasia. *EMBO J.* 15: 520–527.

Wehrle, B. and Chiquet, M. 1990. Tenascin is accumulated along developing peripheral nerves and allows neurite outgrowth *in vitro*. *Development* 110: 401–415.

Weiss, P. 1945. Experiments on cell and axon orientation in vitro: The role of colloidal exudates in tissue organization. *J. Exp. Zool.* 100: 353–386.

Werb, Z., Tremble, P. and Damsky, C. H. 1990. Regulation of extracellular matrix degradation by cell-extracellular matrix interaction. *Cell Differ. Dev.* 32: 299–306.

Wilcox, M., DiAntonio, A. and Leptin, M. 1989. The functions of the PS integrins in *Drosophila* wing morphogenesis. *Development* 107: 891–897.

Wilder, E. L. and Perrimon, N. 1995. Dual functions of *wingless* in the *Drosophila* leg imaginal disc. *Development* 121: 477–488.

Williams, A. F. and Barclay, A. N. 1988. The immunoglobulin superfamily: Domains for cell surface recognition. *Annu. Rev. Immunol.* 6: 381–405.

Williams, E. J., Walsch, F. S. and Doherty, P. 1994a. Tyrosine kinase inhibitors can differentially inhibit integrin-dependent and CAM-dependent neurite outgrowth. *J. Cell Biol.* 124: 1029–1037.

Williams, E. J., Furness, J., Walsh, F. S. and Doherty, P. 1994b. Activation of the FGF receptor underlies neuriote outgrowth stimulated by L1, N-CAM, and N-cadherin. *Neuron* 13: 583–594.

Yayon, A., Klagsbrun, M., Esko, J. D., Leder, P. and Ornitz, D. M. 1991. Cell surface heparin-like molecules are required for binding of basic fibroblast growth factor to its high affinity receptor. *Cell* 64: 841–849.

Yelton, D. E. and Scharff, M. D. 1980. Monoclonal antibodies. *Am. Sci.* 68: 510–516.

Yow, H., Wong, J. M., Chen, H. S., Lee, C., Steel, G. D. Jr. and Chen, L. B. 1988. Increased mRNA expression of a laminin-binding protein in human colon carcinoma: Complete sequence of a full-length cDNA encoding the protein. *Proc. Natl. Acad. Sci. USA* 85: 6394–6398.

Zipursky, S. L., Venkatesh, T. R., Teplow, D. B. and Benzer, S. 1984. Neuronal development in the *Drosophila* retina: Monoclonal antibodies as molecular probes. *Cell* 36: 15–26.

Patterns of Development

II

Fertilization: Beginning a new organism

4

Urge and urge and urge,
Always the procreant urge of the world.
Out of the dimness opposite equals advance,
Always substance and increase, always sex,
Always a knit of identity, always distinction,
Always a breed of life.
WALT WHITMAN (1855)

The final aim of all love intrigues, be they comic or tragic, is really of more importance than all other ends in human life. What it turns upon is nothing less than the composition of the next generation.
A. SCHOPENHAUER
(QUOTED BY C. DARWIN, 1871)

FERTILIZATION is the process whereby two sex cells (gametes) fuse together to create a new individual with genetic potentials derived from both parents. Fertilization, then, accomplishes two separate activities: sex (the combining of genes derived from the two parents) and reproduction (the creation of new organisms). Thus, the first function of fertilization is to transmit genes from parent to offspring, and the second function is to initiate in the egg cytoplasm those reactions that permit development to proceed.

Although the details of fertilization vary from species to species, the events of conception generally consist of four major activities:

- *Contact and recognition between sperm and egg.* In most cases, this ensures that the sperm and egg are of the same species.
- *Regulation of sperm entry into the egg.* Only one sperm can ultimately fertilize the egg. This is usually accomplished by allowing only one sperm to enter the egg and inhibiting any others from entering.
- *Fusion of the genetic material of sperm and egg.*
- *Activation of egg metabolism to start development.*

Structure of the gametes

A complex dialogue exists between egg and sperm. The egg activates the sperm metabolism that is essential for fertilization, and the sperm reciprocates by activating the egg metabolism needed for the onset of development. But before we investigate these aspects of fertilization, we need to consider the structures of the sperm and egg—the two cell types specialized for fertilization.

Sperm

It is only within the past century that the sperm's role in fertilization has been known. Anton van Leeuwenhoek, the Dutch microscopist who co-discovered sperm in 1678, first believed them to be parasitic animals living within the semen (hence the term *spermatozoa*, meaning "sperm animals"). He originally assumed that they had nothing at all to do with reproducing the organism in which they were found, but he later came to believe that each sperm contained a preformed embryo. Leeuwenhoek (1685) wrote that

121

Figure 4.1
The human infant preformed in the sperm, as depicted by Nicolas Hartsoeker (1694).

sperm were seeds (both *sperma* and *semen* mean "seed") and that the female merely provided the nutrient soil into which the seeds were planted. In this, he was returning to a notion of procreation promulgated by Aristotle 2000 years earlier. Try as he could, Leeuwenhoek was continually disappointed in his attempts to find the preformed embryo within the spermatozoa. Nicolas Hartsoeker, the other co-discoverer of sperm, drew a picture of what he hoped to find: a preformed human ("homunculus") within the human sperm (Figure 4.1). This belief that the sperm contained the entire embryonic organism never gained much acceptance, as it implied an enormous waste of potential life. Most investigators regarded the sperm as unimportant. (See Pinto-Correia, 1997, for details of this remarkable story.) [fert1.html]

The first evidence suggesting the importance of sperm in reproduction came from a series of experiments performed by Lazzaro Spallanzani in the late 1700s. Spallanzani demonstrated that filtered toad semen devoid of sperm would not fertilize eggs. He concluded, however, that the viscous fluid retained by the filter paper, and not the sperm, was the agent of fertilization. He, too, felt that the spermatic "animals" were parasitic.

The combination of better microscopic lenses and the cell theory led to a new appreciation of spermatic function. In 1824, J. L. Prevost and J. B. Dumas claimed that sperm were not parasites but rather the active agents of fertilization. They noted the universal existence of sperm in sexually mature males and their absence in immature and aged individuals. These observations, coupled with the known absence of spermatozoa in the sterile mule, convinced them that "there exists an intimate relation between their presence in the organs and the fecundating capacity of the animal." They proposed that the sperm entered the egg and contributed materially to the next generation.

These claims were largely disregarded until the 1840s, when A. von Kolliker described the formation of sperm from cells within the adult testes. He ridiculed the idea that the semen could be normal and yet support an enormous number of parasites. Even so, von Kolliker denied that there was any physical contact between sperm and egg. He believed that the sperm excited the egg to develop, much as a magnet communicates its presence to iron. It was only in 1876 that Oscar Hertwig and Herman Fol independently demonstrated sperm entry into the egg and the union of those cells' nuclei. Hertwig had sought an organism suitable for detailed microscopic observations, and he found that the Mediterranean sea urchin, *Toxopneustes lividus,* was perfect. Not only was it common throughout the region and sexually mature throughout most of the year, but its eggs were available in large numbers and were transparent even at high magnifications. After mixing sperm and egg suspensions together, Hertwig repeatedly observed a sperm entering an egg and saw the two nuclei unite. He also noted that only one sperm was seen to enter each egg and that all the nuclei of the embryo were derived from the fused nucleus created at fertilization. Fol made similar observations and detailed the mechanism of sperm entry. Fertilization was at last recognized as the union of sperm and egg, and the union of sea urchin gametes remains one of the best-studied examples of fertilization. [fert2.html]

Each sperm consists of a haploid nucleus, a propulsion system to move the nucleus, and a sac of enzymes that enable the nucleus to enter the egg. Most of the sperm's cytoplasm is eliminated during maturation, leaving only certain organelles that are modified for spermatic function (Figure 4.2). During the course of sperm maturation, the haploid nucleus becomes very streamlined, and its DNA becomes tightly compressed. In front of this compressed haploid nucleus lies the **acrosomal vesicle,** which is derived from the Golgi apparatus and contains enzymes that digest proteins and complex sugars; thus, it can be considered a modified secretory vesicle. These stored enzymes are used to lyse the outer coverings of the egg. In many species, such as sea urchins, a region of globular actin molecules lies between the nucleus and the acrosomal vesicle. These proteins are used to extend a finger-like process during the early stages of fertilization. In sea urchins and several other species, recognition between sperm and egg involves molecules on this **acrosomal process.** Together, the acrosome and nucleus constitute the head of the sperm.

The means by which sperm are propelled varies according to how the species has adapted to environmental conditions. In some species (such as the parasitic roundworm *Ascaris*), the sperm travel by the amoeboid motion of lamellipodial extensions of the cell membrane. In most species, however, each sperm is able to travel long distances by whipping its **flagellum.** Flagella are complex structures. The major motor portion of the flagellum is called the **axoneme.** An axoneme is formed by the microtubules emanating from the centriole at the base of the sperm nucleus (Figures 4.2 and 4.3). The core of the axoneme consists of two central microtubules surrounded by a row of nine doublet microtubules. Actually, only one microtubule of each doublet is complete, having 13 protofilaments; the other is C-shaped and has only 11 protofilaments (Figure 4.3B). A three-dimensional model of a complete microtubule is shown in Figure 4.3C. Here we can see the 13 interconnected protofilaments, which are made exclusively of the dimeric protein tubulin.

Although tubulin is the basis for the structure of the flagellum, other proteins are also critical for flagellar function. The force for sperm propulsion is provided by **dynein,** a protein that is attached to the microtubules (Figure 4.3B). Dynein hydrolyzes molecules of ATP and can convert the re-

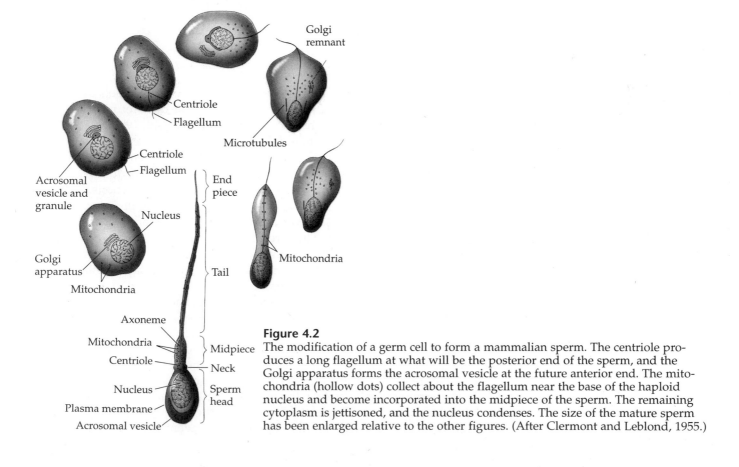

Figure 4.2
The modification of a germ cell to form a mammalian sperm. The centriole produces a long flagellum at what will be the posterior end of the sperm, and the Golgi apparatus forms the acrosomal vesicle at the future anterior end. The mitochondria (hollow dots) collect about the flagellum near the base of the haploid nucleus and become incorporated into the midpiece of the sperm. The remaining cytoplasm is jettisoned, and the nucleus condenses. The size of the mature sperm has been enlarged relative to the other figures. (After Clermont and Leblond, 1955.)

Figure 4.3
The motile apparatus of the sperm. (A) Cross section of the flagellum of a mammalian spermatozoon, showing the central axoneme and the external fibers. (B) Interpretive diagram of the axoneme, showing the "9 + 2" arrangement of the microtubules and other flagellar components. The schematic diagram shows the association of tubulin protofilaments into a microtubule doublet. The first ("A") portion of the doublet is a normal microtubule comprising 13 protofilaments. The second ("B") portion of the doublet contains only 11 (occasionally 10) protofilaments. (C) A three-dimensional model of the "A" microtubule. The α- and β-tubulin subunits are similar but not identical, and the microtubule can change size by polymerizing or depolymerizing tubulin subunits at either end. (A courtesy of D. M. Phillips; B after De Robertis et al., 1975, and Tilney et al., 1973; C from Amos and Klug, 1974, courtesy of the authors.)

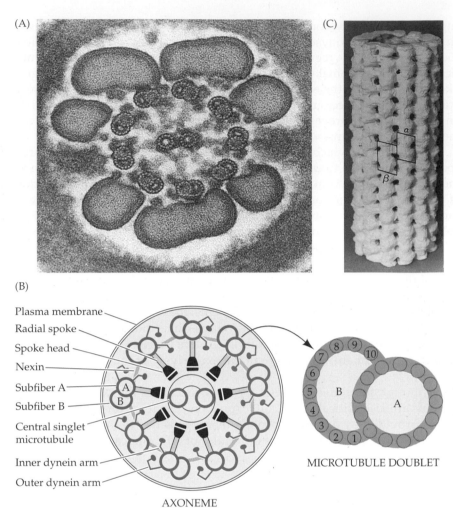

(A)

(C)

(B)

Plasma membrane
Radial spoke
Spoke head
Nexin
Subfiber A
Subfiber B
Central singlet microtubule
Inner dynein arm
Outer dynein arm

A
B

AXONEME

8 9 10 7 6 5 4 3 2 1
B
A

MICROTUBULE DOUBLET

leased chemical energy into the mechanical energy that propels the sperm. This energy allows the active sliding of the outer doublet microtubules, causing the flagellum to bend (Ogawa et al., 1977; Asai, 1996). The importance of dynein can be seen in individuals with the genetic syndrome called Kartagener triad. These individuals lack dynein on all their ciliated and flagellated cells, rendering these structures immotile. Males with this disease are sterile (immotile sperm), are susceptible to bronchial infections (immotile respiratory cilia), and have a 50 percent chance of having their heart on the right side of their body (Afzelius, 1976). Another important flagellar protein appears to be histone H1. This protein is usually seen inside the nucleus, where it folds the chromatin into tight clusters. However, Multigner and colleagues (1992) found that this same protein stabilizes the flagellar microtubules so that they do not disassemble.

The "9 + 2" microtubule arrangement with the dynein arms has been conserved in axonemes throughout the eukaryotic kingdoms, suggesting that this arrangement is extremely well suited for transmitting energy for movement. The energy to whip the flagellum and thereby propel the sperm comes from rings of mitochondria located in the neck region of the sperm (see Figure 4.2). In many species (notably mammals), a layer of dense fibers has interposed itself between the mitochondrial sheath and axoneme. This fiber layer stiffens the sperm tail. Because the thickness of this layer decreases toward the tip, the fibers probably prevent the sperm head from being whipped around too suddenly. Thus, the sperm has undergone extensive modification for the transmission of its nucleus to the egg.

The differentiation of the sperm is not completed in the testes, however. After being expelled into the lumen of the seminiferous tubules, the sperm are stored in the epididymis, where they acquire the ability to move. This motility is achieved through changes in the ATP-generating system (possibly through modification of dynein) as well as changes in the plasma membrane that enable the membrane to become more fluid (Yanagimachi, 1994). The sperm released during ejaculation are able to move, yet they do not yet have the capacity to bind to and fertilize the egg. These final stages of sperm maturation (called *capacitation*) don't occur until the sperm has been inside the female reproductive tract for a certain period of time.

The Egg

All the material necessary for the beginning of growth and development must be stored in the mature egg (the **ovum**). Whereas the sperm has eliminated most of its cytoplasm, the developing egg (called the **oocyte** before it is haploid) not only conserves its material but is actively involved in accumulating more. It either synthesizes or absorbs proteins, such as yolk, that act as food reservoirs for the developing embryo. Thus, birds' eggs are enormous single cells that have become swollen with their accumulated yolk. Even eggs with relatively sparse yolk are comparatively large. The volume of a sea urchin egg is about $2 \times 10^{-4} \ \mu m^3$, more than 10,000 times the volume of the sperm. The depiction of the sea urchin egg and sperm in Figure 4.4 shows their relative sizes, as well as the various components of the mature egg. So, while sperm and egg have equal haploid nuclear components, the egg also has a remarkable cytoplasmic storehouse that it has accumulated during its maturation. This cytoplasmic trove includes proteins, RNAs, protective chemicals, and morphogenetic factors:*

- *Proteins.* It will be a long while before the embryo is able to feed itself or obtain food from its mother. The early embryonic cells need some storable supply of energy and amino acids. In many species, this is accomplished by accumulating yolk proteins in the egg. Many of the yolk proteins are made in other organs (liver, fat body) and travel through the maternal blood to the egg.
- *Ribosomes and tRNA.* The early embryo needs to make many of its own proteins, and in some species, there is a burst of protein synthesis soon after fertilization. Protein synthesis is accomplished by ribosomes and tRNA, which preexist in the egg. The developing egg has special mechanisms to synthesize ribosomes, and certain amphibian oocytes produce as many as 10^{12} ribosomes during their meiotic prophase.
- *Messenger RNA.* In most organisms, the messages for proteins made during early development are already packaged in the oocyte. It is estimated that the eggs of sea urchins contain 25,000 to 50,000 different types of mRNA. This mRNA, however, remains dormant until after fertilization (see Chapter 12).
- *Morphogenetic factors.* These molecules direct the differentiation of cells into certain cell types. They appear to be localized in different regions of the egg and become segregated into different cells during cleavage (see Chapter 13).

*The contents of the egg vary greatly from species to species. The synthesis and placement of these materials will be addressed in Chapter 22, when we discuss the differentiation of germ cells.

Figure 4.4
Structure of the sea urchin egg during fertilization. (After Epel, 1977.)

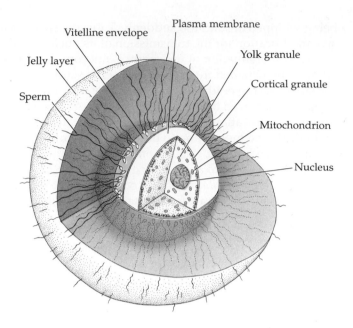

- *Protective chemicals.* The embryo cannot run away from predators or move to a safer environment, so it must come equipped to deal with these factors. Many eggs contain ultraviolet filters and DNA repair enzymes that protect them from sunlight; some eggs contain molecules that potential predators find distasteful; and the yolk of bird eggs even contain antibodies. [fert3.html]

Within this enormous volume of cytoplasm resides a large nucleus. In some species (e.g., sea urchins), the nucleus is already haploid at the time of fertilization. In other species (including many worms and most mammals), the egg nucleus is still diploid, and the sperm enters before the meiotic divisions are completed. The stage of the egg nucleus at the time of sperm entry is illustrated in Figure 4.5.

Enclosing the cytoplasm is the egg **plasma membrane.** This membrane must regulate the flow of certain ions during fertilization and must be capable of fusing with the sperm plasma membrane. Above the plasma membrane is the **vitelline envelope** (Figure 4.6). The major components of this envelope form a fibrous mat above the egg. This mat is supplemented by the extensions of membrane glycoproteins from the plasma membrane and by proteinaceous vitelline posts that adhere the mat to the membrane (Mozingo and Chandler, 1991). The vitelline envelope is essential for the species-specific binding of sperm. In mammals, the vitelline envelope is a separate and thick extracellular matrix called the **zona pellucida.** The mammalian egg is also surrounded by a layer of cells, the **cumulus cells** (Figure 4.7). The cumulus layer represents ovarian follicular cells that were nurturing the egg at the time of its release from the ovary. Mammalian sperm have to get past these cells to fertilize the egg.*

Lying immediately beneath the plasma membrane of the egg is a thin shell (about 5 μm) of gel-like cytoplasm called the **cortex.** The cytoplasm in this region is stiffer than the internal cytoplasm and contains high concentrations of globular actin molecules. During fertilization, these actin molecules

*In mammalian species, the extracellular coverings of the egg are divided into two regions: the zona pellucida and the cumulus. The term *corona radiata* refers to those follicle cells immediately adjacent to the zona pellucida, and they are the innermost cells of the cumulus.

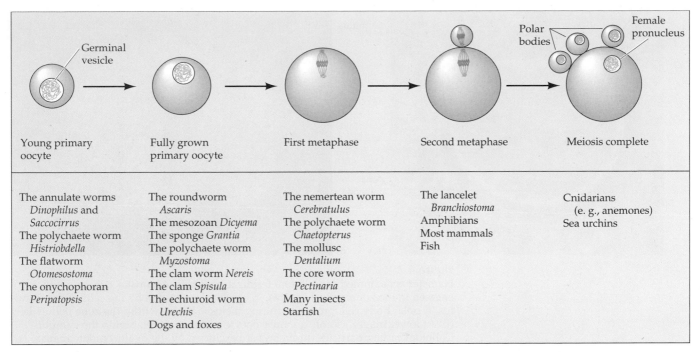

Young primary oocyte	Fully grown primary oocyte	First metaphase	Second metaphase	Meiosis complete
The annulate worms *Dinophilus* and *Saccocirrus* The polychaete worm *Histriobdella* The flatworm *Otomesostoma* The onychophoran *Peripatopsis*	The roundworm *Ascaris* The mesozoan *Dicyema* The sponge *Grantia* The polychaete worm *Myzostoma* The clam worm *Nereis* The clam *Spisula* The echiuroid worm *Urechis* Dogs and foxes	The nemertean worm *Cerebratulus* The polychaete worm *Chaetopterus* The mollusc *Dentalium* The core worm *Pectinaria* Many insects Starfish	The lancelet *Branchiostoma* Amphibians Most mammals Fish	Cnidarians (e. g., anemones) Sea urchins

Figure 4.5
Stages of egg maturation at the time of sperm entry in different animals. (After Austin, 1965.)

polymerize to form long cables of actin known as **microfilaments.** Microfilaments are necessary for cell division, and they also are used to extend the egg surface into the microvilli, which aid sperm entry into the cell (see Figure 4.6; also see Figure 4.19). Also within this cortex are the **cortical granules** (see Figures 4.4 and 4.6). These membrane-bound structures are homologous

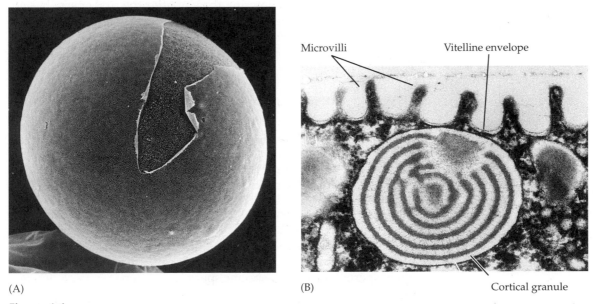

Figure 4.6
The sea urchin egg cell surface. (A) Scanning electron micrograph of an egg before fertilization. The plasma membrane is exposed where the vitelline envelope has been torn. (B) Transmission electron micrograph of an unfertilized egg, showing microvilli and plasma membrane, which are closely covered by the vitelline envelope. A cortical granule lies directly beneath the plasma membrane of the egg. (From Schroeder, 1979, courtesy of T. E. Schroeder.)

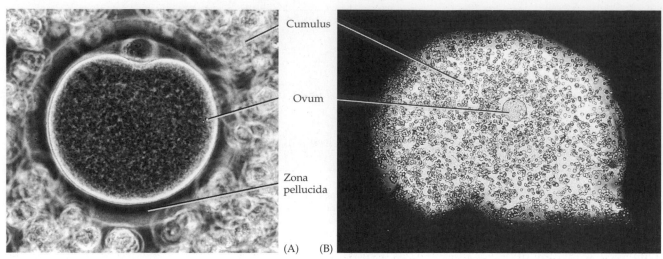

Figure 4.7
Hamster eggs immediately before fertilization. (A) The hamster egg, or ovum, is encased in the zona pellucida. This, in turn, is surrounded by the cells of the cumulus. A polar body cell, produced during meiosis, is also within the zona pellucida. (B) At lower magnification, a mouse oocyte is shown in relation to the cumulus. Colloidal carbon particles (India ink) are excluded by the hyaluronidate matrix. (Courtesy of R. Yanagimachi.)

to the acrosomal vesicle of the sperm, being Golgi-derived organelles containing proteolytic enzymes. However, whereas each sperm contains one acrosomal vesicle, each sea urchin egg contains approximately 15,000 cortical granules. Moreover, in addition to containing the digestive enzymes, the cortical granules also contain mucopolysaccharides, adhesive glycoproteins, and **hyalin protein.** The enzymes and mucopolysaccharides are active in preventing other sperm from entering the egg after the first sperm has entered, and the hyalin and adhesive glycoproteins surround the early embryo and provide support for the cleavage-stage blastomeres.

Many types of eggs also have an **egg jelly** outside their vitelline envelope (see Figure 4.4). This glycoprotein meshwork can have numerous functions, but mostly it is used to either attract or activate sperm. The egg, then, is a cell specialized for receiving sperm and initiating development.

Recognition of egg and sperm: Action at a distance

Many marine organisms release their gametes into the environment. This environment may be as small as a tide pool or as large as an ocean. Moreover, this environment is shared with other species that may shed their sex cells at the same time. These organisms are faced with two problems: (1) How can sperm and eggs meet in such a dilute concentration, and (2) what mechanism prevents starfish sperm from trying to fertilize sea urchin eggs? Two major mechanisms have evolved to solve these difficulties: species-specific attraction of sperm and species-specific sperm activation.

Sperm Attraction

Species-specific sperm attraction (a type of chemotaxis) has been documented in numerous species, including cnidarians, molluscs, echinoderms, and urochordates (Miller, 1985; Yoshida et al., 1993). In 1978, Miller demonstrated that the eggs of the cnidarian *Orthopyxis caliculata* not only secrete a chemotactic factor but also regulate the timing of its release. Developing

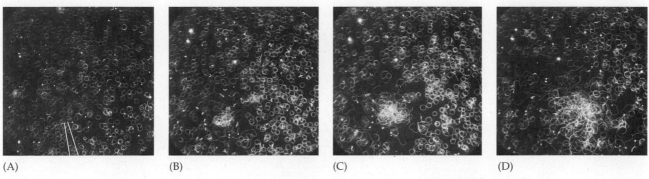

(A)   (B)   (C)   (D)

Figure 4.8
Sperm chemotaxis in *Arbacia*. One nano-liter of a 10-n*M* solution of resact is injected into a 20-µl drop of sperm suspension. The position of the micropipette is indicated in (A). (A) A 1-second photographic exposure showing sperm swimming in tight circles before the addition of resact. (B–D) Similar 1-second exposures showing migration of sperm to the center of the resact gradient 20, 40, and 90 seconds after injection. (From Ward et al., 1985, courtesy of V. D. Vacquier.)

oocytes at various stages in their maturation were fixed onto microscope slides, and sperm were released at a certain distance from the eggs. Miller found that when sperm were added to oocytes that had not yet completed their second meiotic division, there was no attraction of sperm to eggs. However, after the second meiotic division was finished and the eggs were ready to be fertilized, the sperm migrated toward them. Thus, these oocytes control not only the type of sperm they attract but also the time at which they attract them.

The mechanisms for chemotaxis are different in other species (see Metz, 1978; Ward and Kopf, 1993). One such chemotactic molecule, a 14-amino-acid peptide called **resact** has been isolated from the egg jelly of the sea urchin *Arbacia punctulata* (Ward et al., 1985). Resact diffuses readily in seawater and has a profound effect at very low concentrations when added to a suspension of *Arbacia* sperm (Figure 4.8). When a drop of seawater containing *Arbacia* sperm is placed on a microscope slide, the sperm generally swim in circles about 50 µm in diameter. Within seconds after a minute amount of resact is introduced into the drop, sperm migrate into the region of the injection and congregate there. As resact continues to diffuse from the area of injection, more sperm are recruited into the growing cluster. Resact is specific for *A. punctulata* and does not attract sperm of other species. *A. punctulata* sperm bind resact to receptors in their cell membranes (Ramarao and Garbers, 1985; Bentley et al., 1986) and can swim up a concentration gradient of this compound until they reach the egg.

Resact also acts as a **sperm-activating peptide.** These peptides (over 70 of which have been isolated from different species of sea urchins) cause dramatic and immediate increases in sperm motility and oxygen consumption (Hardy et al., 1994). The receptor for resact is a transmembrane protein. When it binds resact on the extracellular side, the resact causes a conformational change that activates the receptor's guanylate cyclase activity on the cytoplasmic side. This increases the egg's concentration of cyclic GMP (Shimomura et al., 1986), which appears to activate the dynein ATPase that stimulates tail beating in the sperm (Cook and Babcock, 1993).

Sperm Activation: The Acrosome Reaction in Sea Urchins

A second interaction between sperm and egg involves the activation of sperm by egg jelly. In most marine invertebrates, this **acrosome reaction** has two components: the fusion of the acrosomal vesicle with the sperm plasma membrane (an exocytosis that results in the release of the contents of the acrosomal vesicle) and the extension of the acrosomal process (Figure 4.9; Colwin and Colwin, 1963). The acrosome reaction can be initiated by soluble egg jelly, by the egg jelly surrounding the egg, or even by contact with the egg itself in certain species. It can also be activated artificially by increasing the calcium concentration of seawater.

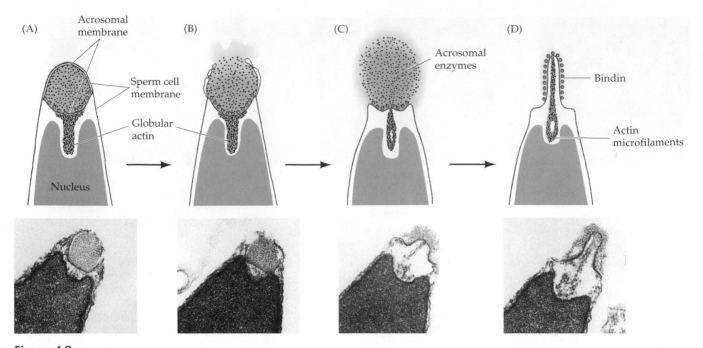

Figure 4.9
Acrosome reaction in echinoderm sperm. (A–C) The portion of the acrosomal membrane lying directly beneath the sperm cell membrane fuses with the cell membrane to release the contents of the acrosomal vesicle. (D) As the actin molecules assemble to produce microfilaments, the acrosomal process is extended outward. Actual photographs of the acrosome reaction in sea urchin sperm are shown below. (After Summers and Hylander, 1974; photographs courtesy of G. L. Decker and W. J. Lennarz.)

In sea urchins, contact with egg jelly causes the exocytosis of the acrosomal vesicle and release of protein-digesting enzymes that can digest a path through the jelly coat to the egg surface (Dan, 1967; Franklin, 1970; Levine et al., 1978). The sequence of these events is outlined in Figure 4.9. The acrosome reaction is thought to be initiated by a protein-bound oligosaccharide in the egg jelly that allows calcium to enter into the sperm head (SeGall and Lennarz, 1979; Schackmann and Shapiro, 1981; Keller and Vacquier, 1994a,b). The exocytosis of the acrosomal vesicle is caused by the calcium-mediated fusion of the acrosomal membrane with the adjacent sperm plasma membrane (Figures 4.9 and 4.10). This exocytosis enables the acrosomal vesicle to release its contents at the head of the sperm.*

The second part of the acrosome reaction involves the extension of the acrosomal process (see Figure 4.9). This protrusion arises from the polymerization of globular actin molecules into actin filaments (Tilney et al., 1978). Exposure of sea urchin sperm to egg jelly also causes a rapid utilization of ATP and a 50 percent increase in mitochondrial respiration. The energy generated is used primarily for flagellar motility (Tombes and Shapiro, 1985).

The egg jelly factors that initiate the acrosome reaction in sea urchins are often highly specific. The sperm of sea urchins *Arbacia punctulata* and *Strongylocentrotus drobachiensis* will react only with jelly of their own eggs. However, *S. purpuratus* sperm can also be activated by *Lytechinus variegatus* (but not *A. punctulata*) egg jelly (Summers and Hylander, 1975). Therefore, egg jelly may provide species-specific recognition in some species but not in others.

*Such exocytotic reactions are seen in the release of insulin from pancreatic cells and in the release of neurotransmitters from synaptic terminals. In all cases, there is a calcium-mediated fusion between the secretory vesicle and the cell membrane. Indeed, the similarity of acrosomal vesicle exocytosis and synaptic vesicle exocytosis may actually be quite deep. Recent studies of acrosome reactions in sea urchins and mammals (Florman et al., 1992; González-Martínez et al., 1992) suggest that when the receptors for the sperm-activating ligands bind these molecules, they cause a depolarization of the membrane that would open voltage-dependent calcium ion channels in a manner reminiscent of synaptic transmission. The proteins that dock the cortical granules to the cell membrane also appear to be homologous to those used in the axon tip (Bi et al., 1995).

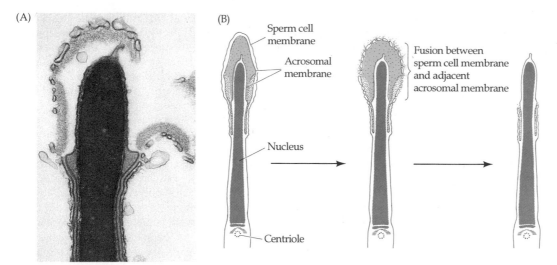

Figure 4.10
Acrosome reaction in hamster sperm. (A) Transmission electron micrograph of hamster sperm undergoing the acrosome reaction. The acrosomal membrane can be seen to form vesicles. (B) Interpretive diagram of electron micrographs showing the fusion of the acrosomal and cell membranes in the sperm head. (A from Meizel, 1984, courtesy of S. Meizel; B after Yanagimachi and Noda, 1970.)

Sidelights & Speculations

Action at a Distance: Mammalian Gametes

IT IS VERY DIFFICULT to study the interactions that might be occurring between mammalian gametes prior to sperm-egg contact. One obvious reason for this is that mammalian fertilization occurs inside the oviducts of the female. While it is relatively easy to mimic the conditions surrounding sea urchin fertilization (using either natural or artificial seawater), we do not yet know the components of the various natural environments that the mammalian sperm encounter as they travel to the egg. A second reason for this difficulty is that the sperm population ejaculated into the female is probably very heterogeneous, containing spermatozoa at different stages of maturation. Of the 280×10^6 human sperm normally ejaculated into the vagina, only about 200 reach the ampullary region of the oviduct, where fertilization takes place (Ralt et al., 1991). Since fewer than 1 in 10,000 sperm get close to the egg, it is difficult to assay those molecules that might enable the sperm to swim toward the egg and become activated. There is a great deal of controversy concerning the translocation of mammalian sperm to the oviduct, the capacitation and hyperactivation reactions that appear necessary in some species' sperm to bind with the egg, and the possibility that the egg may be attracting the sperm through chemotaxis.

Translocation and Capacitation
The reproductive tract of female mammals plays a very active role in the mammalian fertilization process. While sperm motility is needed for mouse sperm to encounter the egg once it is in the oviduct, sperm motility is probably a minor factor in getting the sperm into the oviduct in the first place. Sperm is found in the oviduct of mice, hamsters, guinea pigs, cows, and humans within 30 minutes of sperm deposition, a time "too short to have been attained by even the most Olympian sperm relying on their own flagellar power" (Storey, 1995).

Rather, the sperm appear to be transported to the oviduct by the muscular activity of the uterus.

Newly ejaculated mammalian sperm are unable to undergo the acrosome reaction without residing for some amount of time in the female reproductive tract (Chang, 1951; Austin, 1952). This requirement for **capacitation** varies from species to species (Gwatkin, 1976) and can be mimicked in vitro by incubating sperm in tissue culture media (containing calcium ions, bicarbonate, and serum albumin) or in fluid from the oviducts. Sperm that are not capacitated are "held up" in the cumulus matrix and so do not reach the egg (Austin, 1960; Corselli and Talbot, 1987).

The molecular changes that account for capacitation are still unknown (see Yanigamachi, 1994), but there are four sets of molecular changes that may be important. First, the sperm cell membrane may be altered by changing its lipid composition. The concentration of cholesterol in sperm is lowered during sperm capacitation in several species (Davis, 1981), and two proteins found both in serum and in the female reproductive tract (albumin and lipid transfer protein 1) have been found to remove cholesterol from human sperm (Langlais et al., 1988; Ravnik et al., 1992). Second, particular proteins or carbohydrates on the sperm surface are lost during capacitation (Poirier and Jackson, 1981; Lopez et al., 1985; Wilson and Oliphant, 1987). It is possible that the moieties lost during capacitation had been blocking the recognition sites for the proteins that bind to the zona pellucida. Third, certain proteins are phosphorylated in a cyclic-AMP-dependent pathway. Cyclic AMP can artificially induce competence through the cAMP-dependent protein kinase (PKA), and PKA is needed both for the acquisition of competence and the phosphorylation of tyrosine kinases. It is possible that the female reproductive tract stimulates sperm adenylyl cyclase to make more cAMP and that the cAMP activates the protein kinase, which starts a phosphorylation cascade ending in the phosphorylation and activation of the proteins involved in binding the sperm to the zona pellucida and mediating the exocytosis of the acrosomal vesicle (Leyton and Saling, 1989a; Visconti et al., 1995a,b). Fourth, the membrane potential of the sperm is dramatically lowered (from about -30 mV to about -50 mV (Zeng et al., 1995). However, it is still uncertain if these events are independent of one another and to what extent each of them causes sperm capacitation.

Hyperactivation and Chemotaxis

The different regions of the female reproductive tract may secrete different, regionally specific molecules. These factors may influence sperm motility as well as capacitation. For instance, when sperm of certain mammals (especially hamsters, guinea pigs, and some strains of mice) pass from the uterus into the oviducts, they become "hyperactivated," swimming at higher velocities and generating greater force than before. Suarez and coworkers (1991) have shown that while this behavior is not conducive for traveling in low-viscosity fluids, it appears to be extremely well suited for linear sperm movement in the viscous fluid that sperm might encounter in the oviduct.

In addition to increasing the activity of sperm, soluble factors in the oviduct may also provide the directional component of sperm movement. There has been speculation that the ovum (or, more likely, the ovarian follicle in which the egg had developed) may be secreting chemotactic substances that might attract the sperm toward the egg during the last stages of its migration (see Hunter, 1989). Ralt and colleagues (1991) tested this hypothesis using follicular fluid from human follicles whose eggs were being used for in vitro fertilization. Performing an experiment similar to the one described earlier with sea urchins, they microinjected a drop of follicular fluid into a larger drop of sperm suspension. When they did this, some of the sperm changed their direction of movement to migrate toward the source of follicular fluid. Microinjection of other solutions did not have this effect. These

studies did not rule out the possibility that the effect was due to a general stimulation of sperm movement or metabolism. However, these investigations uncovered a fascinating correlation: the fluid from only about half the follicles tested showed a chemotactic effect, and in nearly every case, the egg was fertilizable if, and only if, the fluid showed chemotactic ability (P < .0001). It is possible, then, that like certain invertebrate eggs, the human egg secretes a chemotactic factor only when it is capable of fertilization.

It should be noted that "the race is not always to the swiftest." Although some sperm can reach the ampullary region of the oviduct (where fertilization occurs) within a half hour after intercourse, those sperm may have little chance of fertilizing the egg. Wilcox and colleagues (1995) found that nearly all human pregnancies result from sexual intercourse during a six-day period *ending* on the day of ovulation. This means that the fertilizing sperm could take as long as six days to make the journey. Eisenbach (1995) has proposed a hypothesis wherein capacitation is a transient event, and sperm are given a relatively brief window of competence in which they can successfully fertilize the egg. As the sperm reach the ampulla, they acquire competence, but if they stay around too long, they lose it. Sperm may also have different survival rates, depending on their location within the reproductive tract, and this may allow some sperm to arrive late but with better chance of success than those that have arrived days earlier. ■

Recognition of egg and sperm: Contact of gametes

Species-specific Recognition in Sea Urchins

Once the sea urchin sperm has penetrated the egg jelly, the acrosomal process of the sperm contacts the vitelline envelope of the egg (Figure 4.11). A major species-specific recognition step occurs at this point. The acrosomal protein mediating this recognition is called **bindin.** In 1977, Vacquier and coworkers isolated this nonsoluble 30,500-Da protein from the acrosome of *Strongylocentrotus purpuratus.* This protein is capable of binding to dejellied eggs from *S. purpuratus* (Figure 4.12; Vacquier and Moy, 1977). Further, its interaction with eggs is relatively species-specific (Glabe and Vacquier, 1977;

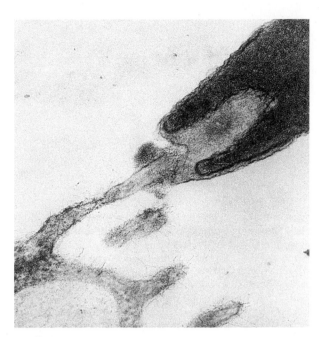

Figure 4.11
Contact of a sea urchin sperm acrosomal process with an egg microvillus. (From Epel, 1977, courtesy of F. D. Collins and D. Epel.)

Figure 4.12
Species-specific agglutination of dejellied eggs by bindin. (A) Agglutination was promoted by adding 212 µg of bindin to a plastic well containing 0.25 ml of a 2 percent (volume to volume) suspension of eggs. After 2–5 minutes of gentle shaking, the wells were photographed. Each bindin only bound to its own eggs. (B) Fluorescence photomicrograph of *S. purpuratus* eggs bound together by fluorescein-labeled *S. purpuratus* bindin particles. Bindin particles were invariably present at the places where two eggs came together. (A based on photographs of Glabe and Vacquier, 1977; B from Glabe and Lennarz, 1979, courtesy of the authors.)

Glabe and Lennarz, 1979); bindin isolated from the acrosomes of *S. purpuratus* agglutinates its own dejellied eggs, but not those of *Arbacia punctulata*. Using immunological techniques, Moy and Vacquier (1979) demonstrated that bindin is located specifically on the acrosomal process, exactly where it should be for sperm-egg recognition (Figure 4.13).

Biochemical studies have shown that the bindins of closely related sea urchin species are indeed different. This finding implies the existence of species-specific bindin receptors on the vitelline envelope. Such receptors

(A)

SPERM BINDIN

(B)

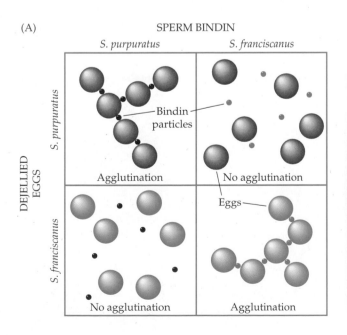

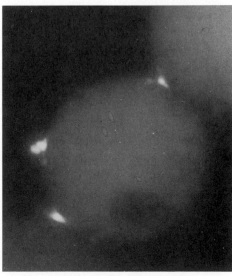

(A)

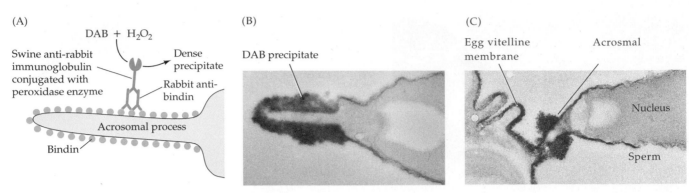

(B)

DAB precipitate

(C)

Egg vitelline membrane   Acrosmal

Nucleus

Sperm

Figure 4.13
Localization of bindin on the acrosomal process. (A) Immunochemical localization technique places a rabbit antibody wherever bindin is exposed. Rabbit antibodies were made to the bindin protein, and these antibodies were incubated with sperm that had undergone the acrosome reaction. If bindin were present, the rabbit antibodies would remain bound to the sperm. After any unbound antibody was washed off, the sperm were treated with *swine* antibodies that could bind to *rabbit* antibodies. These swine antibodies had been covalently linked to peroxidase enzymes. In such a fashion, peroxidase molecules were placed wherever bindin was present. Peroxidase catalyzes the formation of a dark precipitate from diaminobenzidine (DAB) and hydrogen peroxide. Thus, this precipitate will form only where bindin is present. (B) Localization of bindin to the acrosomal process after the acrosome reaction (33,200×). (C) Localization of bindin to the acrosomal process at the junction of the sperm and the egg. (B and C from Moy and Vacquier, 1979, courtesy of V. D. Vacquier.)

were also suggested by the experiments of Vacquier and Payne (1973), who saturated sea urchin eggs with sperm. As seen in Figure 4.14A, sperm binding does not occur over the entire egg surface. Even at saturating numbers of sperm (approximately 1500), there appears to be room on the ovum for more

(A)

Figure 4.14
Bindin receptors on the egg. (A) Scanning electron micrograph of sea urchin sperm bound to the vitelline envelope of an egg. (B) Binding of *S. purpuratus* sperm to polystyrene beads that have been coated with purified bindin receptor protein. (A courtesy of C. Glabe, L. Perez, and W. J. Lennarz; B from Foltz et al., 1993.)

(B)

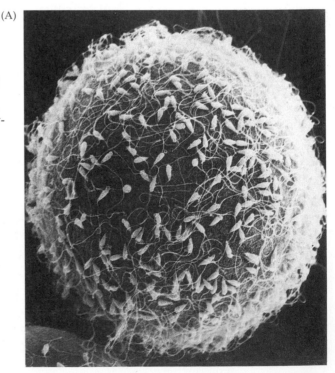

sperm heads, implying a limiting number of sperm-binding sites. A large glycoprotein complex from the vitelline envelopes of sea urchin eggs has been isolated and shown to bind radioactively labeled bindin in a species-specific manner (Glabe and Vacquier, 1978; Rossignol et al., 1984). This glycoprotein is also able to compete with eggs for the sperm of the same species. That is, if *S. purpuratus* sperm are mixed with the bindin receptor from *S. purpuratus* vitelline envelopes, the sperm bind to it and will not fertilize the eggs. The isolated bindin receptor from *S. purpuratus,* however, does not interfere with the fertilization of other related sea urchins. This bindin receptor is a transmembrane glycoprotein of nearly 1300 amino acids (Foltz et al., 1993). The bindin-binding region of this protein extends into the extracellular space and probably becomes a component of the vitelline envelope. These bindin receptors are aggregated into complexes, and hundreds of these complexes are probably needed to tether the sperm to the egg (Figure 4.14B). Thus, species-specific recognition of sea urchin gametes occurs at the levels of sperm attraction, sperm activation, and sperm adhesion to the egg surface. [fert4.html]

Gamete Binding and Recognition in Mammals

ZP3: THE SPERM-BINDING PROTEIN OF THE MOUSE ZONA PELLUCIDA. The zona pellucida in mammals plays a role analogous to that of the vitelline envelope in invertebrates. This glycoprotein matrix is synthesized and secreted by the growing oocyte, and it plays two major roles during fertilization: it binds the sperm, and it initiates the acrosome reaction *after* the sperm is bound (Saling et al., 1979; Florman and Storey, 1982; Cherr et al., 1986). The binding of sperm to the zona is relatively, but not absolutely, species-specific (species specificity should not be a major problem when fertilization occurs internally), and the binding of mouse sperm to the mouse zona can be inhibited by first incubating the sperm with zona glycoproteins. Bleil and Wassarman (1980, 1986, 1988) have isolated from the mouse zona pellucida an 83-kDa glycoprotein, **ZP3,** that is the active competitor in this inhibition assay. The other two zona glycoproteins, ZP1 and ZP2, failed to compete for sperm binding (Figure 4.15). Moreover, radiolabeled ZP3 bound to the heads of mouse sperm with intact acrosomes. Thus, ZP3 is the specific glycoprotein in the zona pellucida to which the mouse sperm bind. ZP3 also initiates the acrosome reaction after sperm have bound to it. The mouse sperm can

Figure 4.15
Binding of sperm to the zona pellucida. (A) Inhibition assay showing the specific decrease of mouse sperm binding to zonae pellucidae when sperm and zonae are incubated with increasingly large amounts of glycoprotein ZP3. The importance of the carbohydrate portion of ZP3 is also indicated by this figure. (B) Binding of radioactively labeled ZP3 to capacitated mouse sperm. (A after Bleil and Wassarman, 1980, and Florman and Wassarman, 1985; B from Bleil and Wassarman, 1986, courtesy of the authors.)

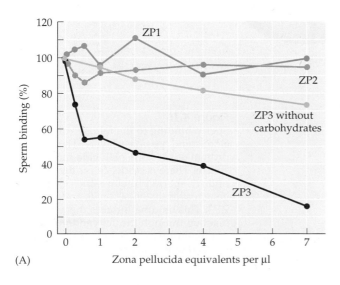

(A)

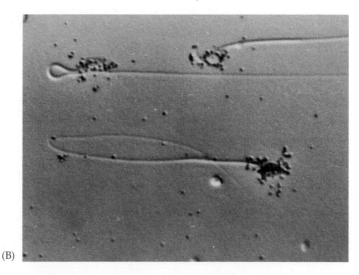

(B)

thereby concentrate its proteolytic enzymes directly at the point of attachment at the zona pellucida.

The molecular mechanism by which the zona pellucida and the mammalian sperm recognize each other is presently being studied. The current hypothesis of mammalian gamete binding postulates a set of proteins on the sperm capable of recognizing specific carbohydrate regions on the egg zona ZP3 (Florman et al., 1984; Florman and Wassarman, 1985; Wassarman, 1987; Saling, 1989). Removal of these threonine- or serine-linked carbohydrate groups abolishes the ability of ZP3 to bind sperm.

SPERM-ZONA ADHESION PROTEINS. Mouse sperm do not "bore into" the zona. Rather, the sperm approach parallel to the plane of the zona surface and are actively tethered there (Baltz et al., 1988). How is the zona able to bind and retain these wriggling sperm? It appears that ZP3 can bind to at least three adhesive proteins on the sperm cell membrane, and thousands of these sites might be needed to keep these two cells from coming apart. There is significant controversy concerning whether all three proteins on the sperm are needed to bind to the zona and what their respective functions might be (see Figure 4.16; Snell and White, 1996). It appears that each of these proteins may play specific, but somewhat overlapping, roles in sperm-egg adhesion and the acrosome reaction. These three proteins are the galactose-binding protein, the galactosyltransferase, and the zona receptor kinase.

THE 56-KDA GALACTOSE-BINDING PROTEIN (SP56). One critical zona-binding protein of the sperm appears to be a protein that specifically binds to the galactose residues of ZP3. Bleil and Wassarman (1980) have shown that one of the critical carbohydrates of the ZP3 glycoprotein is a terminal galactose group. If this terminal galactose is removed or chemically modified, sperm-binding activity is lost. These researchers later isolated this protein by binding ZP3

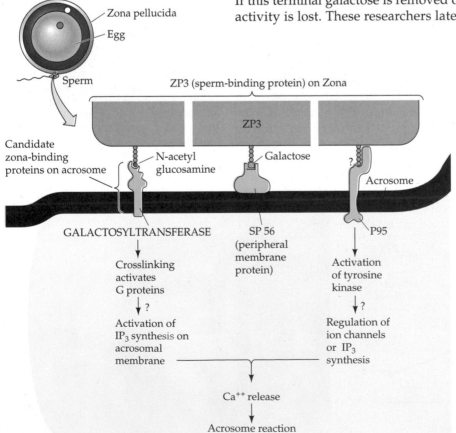

Figure 4.16
Sperm binding to the mouse zona pellucida: some possible players. The ZP3 protein of the zona pellucida binds sperm. There is evidence for the binding of three sperm proteins—the cell surface galactosyltransferase, sp56, and P95—to ZP3. This binding induces the acrosome reaction through the activation of calcium ion influx. The details have yet to be worked out. (After Snell and White, 1996.)

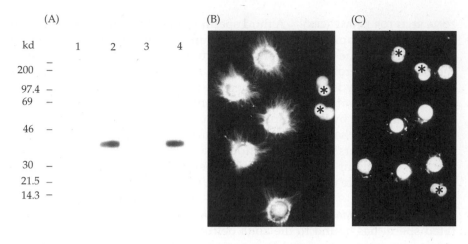

Figure 4.17
Purified sp56 binds to the zona pelluci-da and inhibits sperm binding to mouse eggs. (A) Binding of sp56 to zona pellu-cida of unfertilized eggs. Lane 1 is the result of lysing unfertilized eggs, run-ning the extracted proteins on a gel, blotting the gel, and probing for the presence of sp56 with labeled antibody. No sp56 is seen. Lane 2 shows the posi-tive result when the unfertilized egg is preincubated with sp56, indicating that sp56 binds to the eggs. Lane 3 shows the negative results when sp56 was added to two-cell embryos. Lane 4 shows the control when purified sp56 is run on the gel. (The antibody recognizes the nonreduced form of sp56, which runs at 40 kDa.) (B) Sperm binding nor-mally to unfertilized mouse eggs (about 76 sperm per egg). The two-cell embryos (labeled here by asterisks) are internal controls showing no binding. (C) In the presence of sp56, sperm were inhibited from binding to the zona. (From Bookbinder et al., 1995, courtesy of J. D. Bleil.)

covalently to beads and passing the proteins isolated from mouse sperm membranes over them in a column (Bleil and Wassarman, 1990). Most of the proteins passed through the column; but one, a 56-kDa peptide, bound to the ZP3-coated beads. It did not bind to ZP2-coated beads in a similar exper-iment. This protein was found to be exposed in the sperm membrane, and it bound to galactose residues, strongly suggesting that it is a sperm receptor binding to terminal galactose moiety on the ZP3 glycoprotein. Purified sp56 binds to the zona pellucida of unfertilized (but not fertilized) eggs and blocks sperm-egg binding (Figure 4.17; Bookbinder et al., 1995).

GALACTOSYLTRANSFERASE. The second sperm protein that appears to be important for sperm-zona binding is a sperm cell membrane glycosyltrans-ferase enzyme. Shur's laboratory has shown that this receptor for the zona is an enzyme that recognizes the sugar *N*-acetylglucosamine on ZP3 (Shur and Hall, 1982a,b; Lopez et al., 1985; Miller et al., 1992). This enzyme, *N*-acetyl-glucosamine:galactosyltransferase, is embedded in the sperm plasma mem-brane, directly above the acrosome, with its active site pointed outward. The enzymatic function of this 60-kDa enzyme would be to catalyze the addition of a galactose sugar (from UDP-galactose) onto a carbohydrate chain termi-nating with an *N*-acetylglucosamine sugar (see Chapter 3). However, there are no UDP-galactose residues in the female reproductive tract. Although the enzyme can bind to the *N*-acetylglucosamine residues of the zona pro-teins exactly as any enzyme would bind to a substrate, it cannot catalyze the reaction because the second reactant is missing. Therefore, the enzymes (on the sperm) remain bound to their substrate (on the zona).

If this hypothesis were correct, we would expect that sperm-egg binding could be inhibited either by inhibiting the enzyme or by adding the second reactant, UDP-galactose. This is exactly what Shur and co-workers found to be the case. Sperm-zona binding was blocked by (1) addition of UDP-galac-tose, (2) removal of the *N*-acetylglucosamine residues from ZP3, (3) addition of antibodies that blocked the activity of the galactosyltransferase, and (4) placement of excess galactosyltransferases in the medium (the excess en-zymes would bind to the zona and inhibit the sperm from binding) (Lopez, et al., 1985; Shur and Neely, 1988). Moreover, mouse sperm cell membranes will transfer a sugar from UDP-galactose specifically to ZP3 (Miller et al., 1992). Thus, the sperm surface galactosyltransferase appears to recognize a carbohydrate group on the ZP3 protein of the mouse zona pellucida. The ag-gregation of these galactosyltransferases causes the activation of a G protein that may be important in initiating the acrosome reaction (Gong et al., 1995).

ZONA RECEPTOR KINASE (ZRK).   A third sperm protein that binds to the mouse zona pellucida appears to be a 95-kDa transmembrane protein with two functional sites. The extracellular site specifically binds ZP3, while the intracellular site has tyrosine kinase enzymatic activity (Leyton et al., 1992). This enzymatic activity is stimulated when the protein binds ZP3. This implies that the 95-kDa protein is a receptor tyrosine kinase, and it may initiate the acrosome reaction by the phosphorylation of its target proteins (see Chapter 3). Human sperm have a similar protein, and human ZP3 stimulates kinase activity. Moreover, synthetic peptides that mimic the extracellular (ZP3-binding) domain of this protein inhibit sperm binding to the human zona, suggesting possible uses as a contraceptive (Burks et al., 1995).

INDUCTION OF THE MAMMALIAN ACROSOME REACTION BY ZP3.   Once the capacitated sperm has bound to the zona pellucida, how does the mammalian acrosome reaction take place? The reaction is induced by the protein portion of ZP3 (Endo et al., 1987; Leyton and Saling, 1989a), and ZP3 appears to work by crosslinking the receptors for it on the sperm membrane. This crosslinking opens the calcium ion channels to increase the concentration of calcium in the sperm (Leyton and Saling, 1992b). The mechanism by which ZP3 induces the opening of the calcium channels and the subsequent exocytosis of the acrosome remains controversial, but it may involve the IP_3 pathway (Florman, 1994; Suarez and Dai, 1995). [fert5.html]

SECONDARY BINDING OF SPERM TO ZONA PELLUCIDA.   During the acrosome reaction, the anterior portion of the sperm plasma membrane is shed from the sperm (see Figure 4.10). This is where the ZP3-binding proteins are located, and yet the sperm must still remain bound to the zona in order to lyse a path through it. In mice, it appears that the *secondary* binding to the zona is accomplished by proteins in the inner acrosomal membrane that bind specifically to ZP2 (Bleil et al., 1988). Whereas acrosome-intact sperm will not bind to the ZP2 glycoprotein, acrosome-reacted sperm will. Moreover, antibodies against the ZP2 protein will not prevent the binding of acrosome-intact sperm to the zona, but will inhibit the attachment of acrosome-reacted sperm. The structure of the zona consists of repeating units of ZP3 and ZP2, occasionally crosslinked by ZP1 (Figure 4.18). It appears that the acrosome-reacted sperm transfer their binding from ZP3 to the adjacent ZP2 molecules. After a mouse sperm has entered the egg, the egg cortical granules release their contents. One of the proteins released by these granules is a protease that specifically alters ZP2 (Moller and Wassarman, 1989). This inhibits other acrosome-reacted sperm from moving further toward the egg.

It is not known which of the mouse sperm proteins bind to ZP2. In porcine sperm, secondary zona binding appears to be mediated by proacrosin. Proacrosin becomes the protease acrosin that has long been known to be involved in digesting the zona pellucida. However, proacrosin is also a fucose-binding protein that maintains the connection between acrosome-reacted sperm and the zona pellucida (Jones et al., 1988). It is possible that proacrosin binds to the zona and is then converted into the active enzyme that locally digests the zona pellucida.

In guinea pigs, secondary binding to the zona is thought to be mediated by the protein PH-20. Moreover, when this inner acrosomal membrane protein was injected into male or female guinea pigs, 100 percent of them became sterile for several months (Primakoff et al., 1988). The blood sera of these sterile guinea pigs had extremely high concentrations of antibodies to PH-20. The antiserum from guinea pigs sterilized by injections of PH-20 not only bound specifically to this protein, but also blocked sperm-zona adhe-

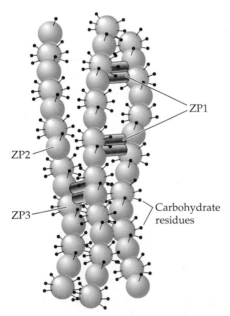

Figure 4.18
Diagram of the fibrillar structure of the mouse zona pellucida. The major strands of the zona are composed of repeating dimers of proteins ZP2 and ZP3. These strands are occasionally crosslinked together by ZP1, forming a meshlike network. (After Wassarman, 1989.)

sion in vitro. The contraceptive effect lasted several months, after which fertility was restored. These animals had been temporarily sterilized by these antibodies. The human analogue of the PH-20 protein is not yet known, but certain sperm antigens display a similar pattern of localization in the sperm. Similarly, the human zona proteins and their functions are not as clearly established as those in the mouse. Nevertheless, these experiments show that the principle of immunological contraception is well founded.

Gamete fusion and the prevention of polyspermy

Fusion Between Egg and Sperm Cell Membranes

Recognition of sperm by the vitelline envelope or zona is followed by the lysis of that portion of the envelope or zona in the region of the sperm head (Colwin and Colwin, 1960; Epel, 1980). This lysis is followed by the fusion of the sperm cell membrane with the cell membrane of the egg.

The entry of a sperm into a sea urchin egg is illustrated in Figure 4.19. The egg surface is covered with small extensions called microvilli; sperm-egg fusion appears to cause the polymerization of actin and the extension of several microvilli to form the **fertilization cone** (Summers et al., 1975; Schatten and Schatten, 1980, 1983). Homology between the egg and the sperm is

(A)

(B)

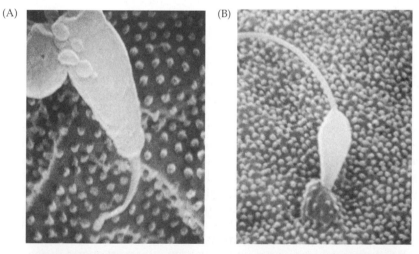

Figure 4.19
Scanning electron micrographs of the entry of sperm into sea urchin eggs. (A) Contact of sperm head with egg microvillus through the acrosomal process. (B) Formation of fertilization cone. (C) Internalization of sperm within the egg. (D) Transmission electron micrograph of sperm internalization through the fertilization cone. (A–C from Schatten and Mazia, 1976, courtesy of G. Schatten; D courtesy of F. J. Longo.)

(C)

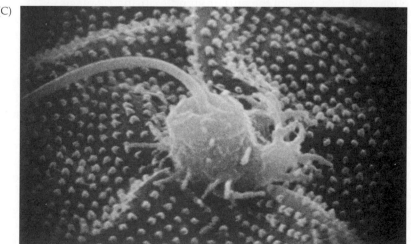

(D)

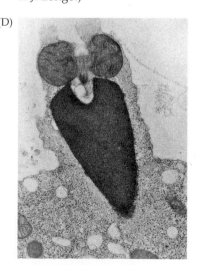

again demonstrated, because the transitory fertilization cone, like the acrosomal process, appears to be extended by the polymerization of actin. The sperm and egg membranes join together, and material from the sperm cell membrane can later be found on the egg membrane (Gundersen et al., 1986). The sperm nucleus and tail pass through the cytoplasmic bridge, which is widened by the actin polymerization. Yanagimachi and Noda (1970) have shown a similar process to occur during the fusion of mammalian gametes (Figure 4.20).

In the sea urchin, all regions of the egg are capable of fusing with sperm; in several other species, there are specialized regions of the membrane for sperm recognition and fusion (Vacquier, 1979). Fusion is an active process, often mediated by specific "fusogenic" proteins. Proteins such as the HA protein of influenza virus and the F protein of Sendai virus are known to promote cell fusion, and it is possible that bindin is also such a protein. Glabe (1985) has shown that sea urchin bindin will cause phospholipid vesicles to fuse together and that, like the viral fusogenic proteins, bindin contains a long stretch of hydrophobic amino acids near its amino terminus. In abalones, the lysin that dissolves the vitelline envelope has also been found to have fusogenic activity (Hong and Vacquier, 1986).

The **fertilin proteins** in the plasma membrane of mammalian sperm are essential for sperm-egg membrane fusion (Primakoff et al., 1987; Blobel et al., 1992; Myles et al., 1994). Mouse fertilin has a hydrophobic region similar to that of viral fusogenic proteins, and it has a sequence that suggests that it binds an integrin in the egg cell membrane. Current evidence suggests that mouse fertilin binds to the $\alpha6\beta1$ integrin protein on the egg plasma membrane, and it is hypothesized that the hydrophobic region of the fertilin might then mediate the union of the two membranes (Almeida et al., 1995). When the membranes are fused, the sperm nucleus, mitochondria, centriole, and flagellum can enter the egg.

Prevention of Polyspermy

As soon as one sperm has entered the egg, the fusibility of the egg membrane, which was so necessary to get the sperm inside the egg, becomes a dangerous liability. In sea urchins, as in most animals studied, any sperm that enters the egg can provide a haploid nucleus and a centriole to the egg. In normal **monospermy,** where only one sperm enters the egg, a haploid sperm nucleus and a haploid egg nucleus combine to form the diploid nucleus of the fertilized egg (zygote), thus restoring the chromosome number appropriate for the species. The centriole, coming from the sperm, will divide to form the two poles of the mitotic spindle during cleavage.

The entrance of multiple sperm—**polyspermy**—leads to disastrous consequences in most organisms. In the sea urchin, fertilization by two sperm results in a triploid nucleus, where each chromosome is represented three times rather than twice. Worse, since the sperm centriole divides to form the two poles of a mitotic apparatus, instead of a bipolar mitotic spindle separating the chromosomes into two cells, the triploid chromosomes would be divided into as many as four cells. Because there is no mechanism to ensure that each of the four cells receives the proper number and type of chromosomes, the chromosomes would be apportioned unequally. Some cells would receive extra copies of certain chromosomes and other cells would lack them. Theodor Boveri demonstrated in 1902 that such cells either die or develop abnormally (Figure 4.21). [fert6.html]

Species have evolved ways to prevent the union of more than two haploid nuclei. The most common way is to prevent the entry of more than one sperm into the egg. The sea urchin egg has two mechanisms to avoid

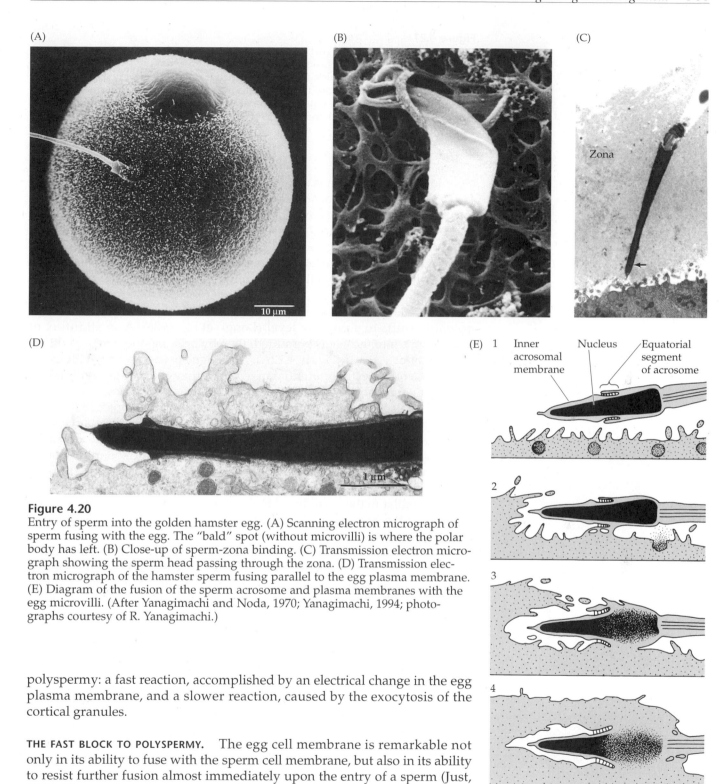

Figure 4.20
Entry of sperm into the golden hamster egg. (A) Scanning electron micrograph of sperm fusing with the egg. The "bald" spot (without microvilli) is where the polar body has left. (B) Close-up of sperm-zona binding. (C) Transmission electron micrograph showing the sperm head passing through the zona. (D) Transmission electron micrograph of the hamster sperm fusing parallel to the egg plasma membrane. (E) Diagram of the fusion of the sperm acrosome and plasma membranes with the egg microvilli. (After Yanagimachi and Noda, 1970; Yanagimachi, 1994; photographs courtesy of R. Yanagimachi.)

polyspermy: a fast reaction, accomplished by an electrical change in the egg plasma membrane, and a slower reaction, caused by the exocytosis of the cortical granules.

THE FAST BLOCK TO POLYSPERMY.   The egg cell membrane is remarkable not only in its ability to fuse with the sperm cell membrane, but also in its ability to resist further fusion almost immediately upon the entry of a sperm (Just, 1919). Thus, only one sperm enters the egg cytoplasm.

The **fast block to polyspermy** is achieved by changing the electrical potential of the egg membrane. The cell membrane provides a selective barrier between the egg cytoplasm and the outside environment, and the ionic concentration of the egg differs greatly from that of its surroundings. This concentration difference is especially true for sodium and potassium ions. Seawater has a particularly high sodium ion concentration, whereas the egg cytoplasm has relatively little sodium. The reverse is the case with potas-

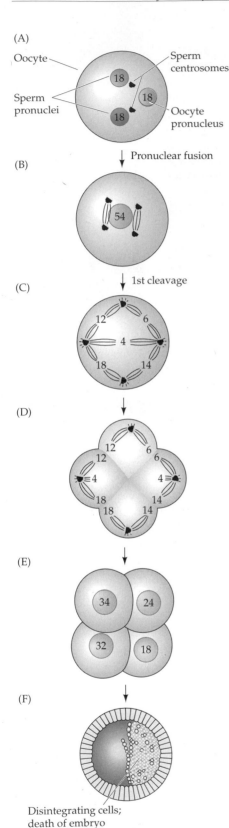

(A)

Oocyte

Sperm
centrosomes

Sperm
pronuclei

Oocyte
pronucleus

↓ Pronuclear fusion

(B)

(C) ↓ 1st cleavage

(D)

(E)

(F)

Disintegrating cells;
death of embryo

Figure 4.21
Aberrant development in a dispermic sea urchin egg. (A) Fusion of three haploid nuclei, each containing 18 chromosomes, and the division of the two sperm centrioles to form four mitotic poles. (B,C) The 54 chromosomes randomly assort on the four spindles. (D) At anaphase of the first division, the duplicated chromosomes are pulled to the four poles. (E) Four cells containing different numbers and types of chromosomes are formed, thereby causing (F) the early death of the embryo. (After Boveri, 1907.)

sium ions. This condition is maintained by the cell membrane, which steadfastly inhibits the entry of sodium into the oocyte and prevents potassium ions from leaking into the environment. If we insert an electrode into the egg and place a second electrode outside the oocyte, we can measure the constant potential difference across the egg plasma membrane. This **resting membrane potential** is generally about 70 mV, usually expressed as –70 mV because the inside of the cell is negatively charged with respect to the exterior. [fert7.html]

Within 1–3 seconds after the binding of the first sperm, the membrane potential shifts to a positive level (Longo et al., 1986). A small influx of sodium ions into the egg is permitted, thereby bringing the potential difference to about +20 mV (Figure 4.22A). Although sperm can fuse with membranes having a resting potential of –70 mV, they cannot fuse with membranes having a positive resting potential. It is not known how the binding or entry of a sperm signals the opening of the sodium channels, but Gould and Stephano (1987, 1991) have provided what may be an important clue to understanding this process. They isolated from *Urechis* (a marine echiuroid worm) sperm an acrosomal protein that is able to open the sodium channels of unfertilized *Urechis* eggs. Moreover, when these eggs are exposed to this protein, the rate of sodium influx and the resulting membrane potential shift are very similar to those produced by live sperm. The opening of the sodium channels in the egg appears to be caused by the binding of the sperm to the egg.

Jaffe and her co-workers have found that polyspermy can be induced if eggs are artificially supplied with an electric current that keeps their membrane potential negative. Conversely, fertilization can be prevented entirely by artificially keeping the membrane potential of eggs positive (Jaffe, 1976). The fast block to polyspermy can also be circumvented by lowering the concentration of sodium ions in the water (Figure 4.22B–D). If the sodium ions are not sufficient to cause the positive shift in membrane potential, polyspermy occurs (Gould-Somero et al., 1979; Jaffe, 1980). It is not known how the differences in membrane potential act on the sperm to block secondary fertilization. Most likely, the sperm carry a voltage-sensitive component (possibly a positively charged fusogenic protein), and the insertion of this component into the egg cell membrane is probably regulated by the electrical charge across the membrane (Iwao and Jaffe, 1989). An electrical block to polyspermy also occurs in frogs (Cross and Elinson, 1980), but probably not in most mammals (Jaffe and Cross, 1983).

THE SLOW BLOCK TO POLYSPERMY. Sea urchin eggs (and many others) have a second mechanism to ensure that multiple sperm do not enter the egg cytoplasm (Just, 1919). The fast block to polyspermy is transient, and the membrane potential of the sea urchin egg remains positive for only about a minute. This brief potential shift is not sufficient to permanently prevent polyspermy, for Carroll and Epel (1975) have demonstrated that polyspermy

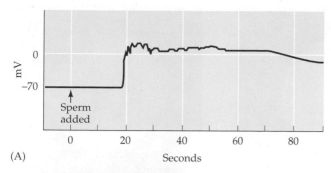

(A)

Figure 4.22
Membrane potential of sea urchin eggs before and after fertilization. (A) Before the addition of sperm, the potential difference across the egg cell membrane is about –70 mV. Within 1–3 seconds after the fertilizing sperm contacts the egg, the potential shifts in a positive direction. (B) Control eggs developing in 490 mM Na$^+$. (C) Polyspermy in eggs fertilized in 120 mM Na$^+$ (choline was substituted for sodium). The *Lytechinus* eggs were photographed during the first cleavage. (D) Table showing the rise of polyspermy with decreasing sodium ion concentration. (From Jaffe, 1980, photographs courtesy of L. A. Jaffe.)

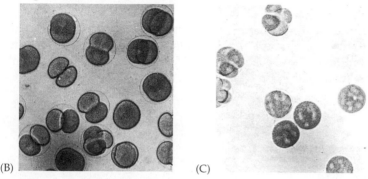

(B)   (C)

[Na$^+$](mM)	Percentage of polyspermic eggs
490	22
360	26
120	97
50	100

(D)

can still occur if the sperm bound to the vitelline envelope are not somehow removed. This removal is accomplished by the **cortical granule reaction,** a slower, mechanical block to polyspermy that becomes active about 1 minute after the first successful sperm-egg attachment.

Directly beneath the sea urchin egg membrane are 15,000 cortical granules, each about 1 μm in diameter (see Figure 4.6B). Upon sperm entry, these cortical granules fuse with the egg plasma membrane and release their contents into the space between the cell membrane and the fibrous mat of vitelline envelope proteins. There are several proteins associated with this cortical granule exocytosis. The first are the proteases. These enzymes dissolve the vitelline posts that connect the vitelline envelope proteins to the cell membrane, and they clip off the bindin receptor and any sperm attached to it (Vacquier et al., 1973; Glabe and Vacquier, 1978). Other proteins, mucopolysaccharides released by the cortical granules, produce an osmotic gradient that causes water to rush into the space between the cell membrane and the envelope, causing the vitelline envelope to expand and become the **fertilization envelope** (Figures 4.23 and 4.24). A third protein product of the cortical granules, a peroxidase enzyme, hardens the fertilization envelope by crosslinking tyrosine residues on adjacent proteins (Foerder and Shapiro, 1977; Mozingo and Chandler, 1991). As shown in Figure 4.23, the fertilization envelope starts to form at the site of sperm entry and continues its expansion around the egg. As this envelope forms, sperm are released. This process starts about 20 seconds after sperm attachment and is complete by the end of the first minute of fertilization. Finally, a fourth cortical granule protein, hyalin, forms a coating around the egg (Hylander and Summers, 1982). The cell extends elongated microvilli whose tips attach to this **hyaline layer.** This hyaline layer provides support for the blastomeres during cleavage.

In mammals, the cortical granule reaction does not create a fertilization envelope, but the effect is the same. Released enzymes modify the zona pellucida sperm receptors such that they can no longer bind sperm (Bleil and

Figure 4.23
Formation of the fertilization envelope and removal of excess sperm. Sperm were added to sea urchin eggs, and the suspension was fixed in formaldehyde to prevent further reactions. (A) At 10 seconds after sperm addition, sperm are seen surrounding the egg. (B,C) At 25 and 35 seconds after insemination, a fertilization envelope forms around the egg, starting at the point of sperm entry. (D) The fertilization envelope is complete, and excess sperm are removed. (From Vacquier and Payne, 1973, courtesy of V. D. Vacquier.)

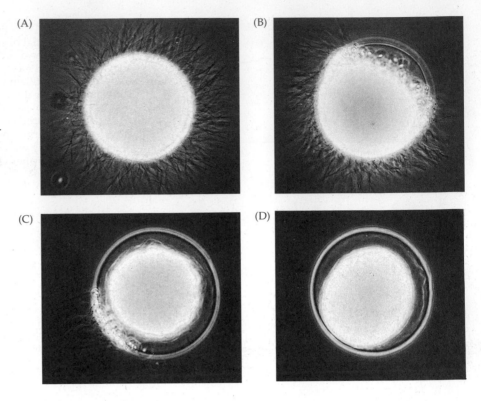

Wassarman, 1980). This modification process is called the **zona reaction.** During the zona reaction, both ZP3 and ZP2 are modified. Florman and Wassarman (1985) have proposed that the cortical granules of mouse eggs contain an enzyme that clips off the terminal sugar residues of ZP3, thereby releasing bound sperm from the zona and preventing the attachment of other sperm. Cortical granules of mouse eggs contain *N*-acetylglucosaminidase enzymes capable of cleaving *N*-acetylglucosamine from ZP3 carbohydrate chains. Miller and co-workers (1992, 1993) have demonstrated that after fertilization, the *N*-acetylglucosamine residue is removed, and ZP3 will not serve as a substrate for the binding of sperm galactosyltransferase. ZP2 is clipped by the cortical granule proteases and loses its ability to bind sperm as well (Moller and Wassarman, 1989). Thus, sperm can no longer initiate or maintain their binding to the zona pellucida and are rapidly shed.

CALCIUM AS THE INITIATOR OF THE CORTICAL GRANULE REACTION. The mechanism for the cortical granule reaction is similar to that of the acrosome reaction. Upon fertilization, the intracellular calcium ion concentration of the egg increases greatly. In the elevated concentration of intracellular free calcium ions, the cortical granule membranes fuse with the egg plasma membrane, thereby causing the exocytosis of their contents (see Figure 4.24). Following the fusion of the cortical granules about the point of sperm entry, a wave of cortical granule exocytosis propagates around the cortex to the opposite side of the egg.

The release of calcium from intracellular storage can be monitored visually by calcium-activated luminescent dyes such as aequorin (isolated from luminescent jellyfish) or fluorescent dyes such as fura-2. These dyes emit light when they bind free calcium ions. Eggs are injected with the dye and then fertilized. Plate 12 shows the striking wave of calcium release that propagates across the sea urchin egg. Starting at the point of sperm entry, a

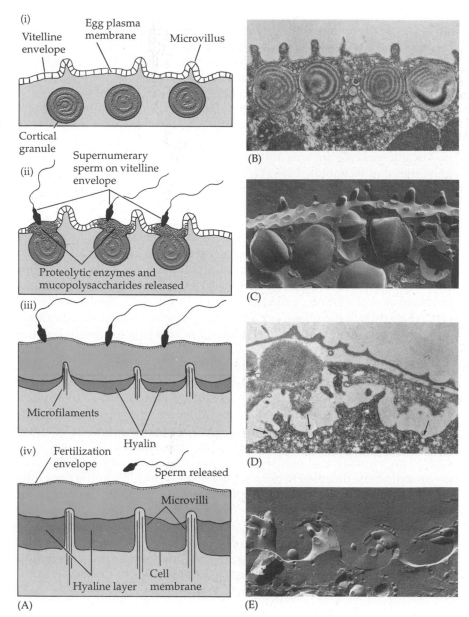

(i)

Vitelline envelope · Egg plasma membrane · Microvillus

Cortical granule

(ii) Supernumerary sperm on vitelline envelope

Proteolytic enzymes and mucopolysaccharides released

(iii)

Microfilaments

Hyalin

(iv) Fertilization envelope

Sperm released

Microvilli

Hyaline layer · Cell membrane

(A)

(B)

(C)

(D)

(E)

Figure 4.24
Cortical granule exocytosis. (A) Schematic diagram showing the events leading to the formation of the fertilization envelope and the hyaline layer. As cortical granules undergo exocytosis, they release proteases that cleave the proteins linking the vitelline envelope to the cell membrane. Mucopolysaccharides released by the cortical granules form an osmotic gradient, thereby causing water to enter and swell the space between the vitelline envelope and the cell membrane. Other enzymes released from the cortical granules harden the vitelline envelope (now the fertilization envelope) and release sperm bound to it. (B,C) Transmission and scanning electron micrographs of the cortex of an unfertilized sea urchin egg. (D,E) Transmission and scanning electron micrographs of the same region of a recently fertilized egg, showing the raising of the fertilization envelope and the points at which the cortical granules fused with the plasma membrane of the egg (arrows in D). (A after Austin, 1965; B–E from Chandler and Heuser, 1979, courtesy of D. E. Chandler.)

band of light traverses the cell (Steinhardt et al., 1977; Gilkey et al., 1978; Hafner et al., 1988). As these photographs document, the calcium ions do not merely diffuse across the egg from the point of sperm entry. Rather, the release of calcium ions starts at one end of the cell and proceeds to the other end. The mechanism for this wave of calcium ion release is discussed shortly (see Sidelights & Speculations, page 147). The entire release of calcium ions is complete in roughly 30 seconds in sea urchin eggs, and the free calcium ions are resequestered shortly after they are released. If two sperm enter the egg cytoplasm, calcium ion release can be seen starting at the two separate points on the cell surface (Hafner et al., 1988).

Several experiments have demonstrated that calcium ions are directly responsible for propagating the cortical reaction and that the calcium ions are stored within the egg itself. The drug A23187 is a calcium ionophore that transports calcium ions across membranes, allowing these cations to traverse otherwise impermeable barriers. Placing unfertilized sea urchin eggs into seawater containing A23187 causes the cortical granule reaction and the elevation of the fertilization envelope. Moreover, this reaction occurs in the absence of any calcium ions in the seawater. Therefore, A23187 causes the release of calcium ions already sequestered in organelles within the egg (Chambers et al., 1974; Steinhardt and Epel, 1974). Further studies (Hollinger and Schuetz, 1976; Fulton and Whittingham, 1978; Hamaguchi and Hiramoto, 1981; Kline, 1988) have shown that calcium ions will initiate cortical granule reactions when injected into sea urchin, mouse, and frog eggs.

The internal calcium ions are stored in the endoplasmic reticulum of the egg (Eisen and Reynolds, 1985; Terasaki and Sardet, 1991). In sea urchins and frogs, whose eggs undergo a cortical granule reaction, this reticulum is pronounced in the cortex and surrounds the cortical granules (Figure 4.25; Gardiner and Grey, 1983; Luttmer and Longo, 1985). In the frog *Xenopus*, the cortical endoplasmic reticulum becomes 10-fold more abundant during the maturation of the egg and disappears locally within a minute after the wave of exocytosis occurs in any particular region of the cortex. Jaffe (1983) likens this calcium-sequestering cortical endoplasmic reticulum to the sarcoplasmic reticulum of skeletal and cardiac muscle. Once initiated, the release of calcium is self-propagating. Free calcium is able to release sequestered calcium from its storage sites, thus causing a wave of calcium ion release and cortical granule exocytosis.

Variations in polyspermy-preventing strategies exist throughout nature. In mammals, polyspermy is minimized by the small number of sperm that reach the site of fertilization (Braden and Austin, 1954). The block to polyspermy in hamsters appears to be controlled solely by the release of sperm-binding sites on the zona pellucida (Miyazaki and Igusa, 1981; Jaffe and Gould, 1985). Rabbits, however, rely completely on a membrane-level block to polyspermy, and nobody will argue with their success. Lastly, certain animals have defenses to polyspermy about which we know very little. In the yolky eggs of certain birds, reptiles, and salamanders, several sperm actually do enter the egg cytoplasm. In some unknown way, all but one of these sperm are induced to disintegrate in the cytoplasm after the fusion of the egg pronucleus with one of the sperm pronuclei (Ginzburg, 1985; Elinson, 1986). Whatever the mechanism, only one haploid sperm nucleus is allowed to fuse with the haploid nucleus of the egg.

Figure 4.25
Endoplasmic reticulum surrounding cortical granule in sea urchin eggs. (A) Endoplasmic reticulum has been stained with osmium-zinc iodide to allow visualization by transmission electron microscopy. The cortical granule is seen to be surrounded by the endoplasmic reticulum.(B) Picture of an entire egg stained with fluorescent antibodies to the calcium-dependent calcium release channels. The antibodies show these channels in the cortical endoplasmic reticulum. (A From Luttmer and Longo, 1985, courtesy of S. Luttmer; B from McPherson et al., 1992, courtesy of F. J. Longo.)

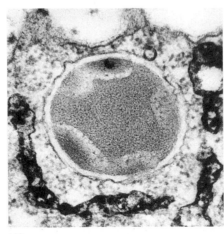

(A)

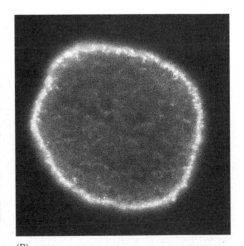

(B)

Sidelights & Speculations

The Activation of Gamete Metabolism

IF CALCIUM ION RELEASE is necessary for the activation of the oocyte, how does the sperm cause it to occur? We do not really know. As one investigator (Berridge, 1993) stated, "Just how the sperm triggers the explosive release of calcium in the egg is still something of a mystery." Recent data suggest that the production of **inositol 1,4,5-trisphosphate (IP₃)** is the primary mechanism for releasing calcium ions from their intracellular storage.

Injected IP₃ can release sequestered calcium ions in eggs and in many other cell types (Swann and Whitaker, 1986; Berridge 1993), and increased concentrations of intracellular IP₃ are seen within 10 seconds after sea urchin eggs are fertilized (Ciapa and Whitaker, 1986). The release of calcium ions and the cortical granule reaction quickly follow the formation or injection of IP₃ (Whitaker and Irvine, 1984; Busa et al., 1985). Moreover, these IP₃-mediated effects can be thwarted by preinjecting the egg with calcium-chelating agents (Turner et al., 1986), thereby confirming that IP₃ stimulates the release of stored calcium in the egg.

IP₃-responsive calcium channels have been found in the egg endoplasmic reticulum. The IP₃ formed at the site of sperm entry is thought to bind to these IP₃ receptors in the endoplasmic reticulum, effecting a local release of calcium (Ferris et al., 1989; Furuichi et al., 1989; Terasaki and Sardet, 1991). Once released, the calcium ions can diffuse directly or they can facilitate the release of more calcium ions from calcium-sensitive receptors located in the cortical endoplasmic reticulum (McPherson et al., 1992). The binding of calcium ions to these receptors releases more calcium, and this released calcium can continue the wave by binding to more receptors, and so on. Mohri and colleagues (1995) have shown that the IP₃-released calcium is both necessary and sufficient for the calcium release. This wave of calcium ions is propagated throughout the cell, starting at the point of sperm entry; and the cortical granules,

which will fuse with the cell membrane in the presence of high calcium concentrations, respond in a wave of exocytosis that follows the calcium ions.

IP₃ is similarly found to release calcium ions in vertebrate eggs, and blocking the IP₃ receptor in hamster eggs prevents the release of calcium at fertilization. As with sea urchins, waves of IP₃ are thought to mediate calcium release from sites within the endoplasmic reticulum (Lechleiter and Clapham, 1992; Miyazaki et al., 1992; Ayabe et al., 1995). Xu and colleagues (1994) have found that blocking the IP₃-mediated calcium release blocks every aspect of sperm-induced egg activation, including cortical granule exocytosis, mRNA recruitment, and cell cycle resumption.

The question then becomes, What initiates the production of IP₃? There are two pathways that appear to stimulate the release of calcium once sperm has bound to a receptor in the egg plasma membrane. The first pathway, that of the **G-protein-linked receptor**, is widely known to release calcium ions in muscle contraction, cell growth, hormonal secretion, sensory perception, and neurotransmitter release (Berridge, 1993). The second pathway is the **receptor tyrosine kinase** cascade that also is used to signal cell proliferation and differentiation. As presented in Chapter 3, the G-protein-linked receptor pathway is initiated by the binding of an extracellular ligand (such as acetylcholine or serotonin) to a transmembrane receptor protein. On the inside of the plasma membrane, this receptor is connected to a **trimeric G protein.** When the ligand binds to its receptor, the receptor activates the G protein (see Figures 3.33 and 3.35), causing the dissociation of the G protein into its subunits, which are then able to activate a set of enzymes called **phospholipase C.** Phospholipase C catalyzes the hydrolysis of phosphatidylinositol 4,5-bisphosphate (**PIP₂**) into two **second messengers:** inositol 1,4,5-trisphosphate (**IP₃**) and diacylglycerol (**DAG**). IP₃ is able to

open calcium ion channels, and DAG is able to stimulate those proton exchange proteins that enable the efflux of hydrogen ions from the cells. The result is the elevation of intracellular calcium ions and the increase of intracellular pH.

The "second" second messenger, DAG, is thought to activate the membrane enzyme **protein kinase C (PKC)**. PKC is translocated from the cytosol to the egg plasma membrane shortly after fertilization, and it is thought to be responsible for activating the protein that exchanges sodium ions for hydrogen ions (Swann and Whitaker, 1986; Nishizuka, 1986; Shen and Burgart, 1986; Olds et al., 1995). Blocking the activity of protein kinase C in sea urchin eggs inhibits the alkalinization of the cytoplasm seen during normal fertilization (Shen and Buck, 1990). The Na+/H+ exchange protein also needs calcium ions for activity. Thus, both DAG and IP₃ are involved in the developmental activities of the egg. The key regulatory step is the activation of phospholipase C, the enzyme that produces these two compounds. Jaffe and her co-workers have found G proteins in sea urchin and frog eggs, and when they injected G protein activators into eggs, the cortical granules underwent exocytosis in the absence of sperm (Turner et al., 1986, 1987; Kline et al., 1991). This activation was inhibited when calcium chelators such as EGTA were added. [fert8.html]

It appears, then, that a G protein may be involved in regulating the release of sequestered calcium ions and the exocytosis of the cortical granules. There are several ways that this might be done. First, the binding of sperm to a receptor in the egg plasma membrane might change the conformation of the receptor such that it activates the G protein and initiates the cascade (Figure 4.26A). This is a model seen in the study by Kline and co-workers (1988, 1991), who hypothesized that if a G protein mediated the fertilization events by being activated by a sperm-binding receptor, then that same G protein might be activated by a *neurotransmitter* if the egg contained a neurotransmitter receptor that could activate the G protein. They injected mRNA for the serotonin receptor or the acetylcholine receptor into frog eggs. These cell-surface receptors were synthesized by the egg and were seen on the

egg cell membrane. In these instances, the eggs could be "fertilized" by the neurotransmitters serotonin and acetylcholine, and the cortical reaction was observed. Similar experiments have shown that when neurotransmitters activate the G protein–IP$_3$ pathway in mouse oocytes, fertilization events are induced (Williams et al., 1992; Moore et al., 1993).

However, the G-protein-linked receptor cascade is not the only pathway capable of generating IP$_3$ (see Chapter 3). Recent evidence (Moore et al., 1994;

Shilling et al., 1994; Yim et al., 1994) demonstrates that activating receptor tyrosine kinases also produce IP$_3$ and activate the calcium wave and the cortical granule reaction (Figure 4.26B). When the mRNA for a receptor tyrosine kinase, the receptor for platelet-derived growth factor (PDGF), was injected into starfish oocytes, the PDGF receptor was synthesized and incorporated into their cell membranes. When the oocytes matured and PDGF was added to the seawater in which the eggs were incubating, the eggs underwent a rise in

Figure 4.26 Possible mechanisms of egg activation. (A) G-protein-mediated phosphatidylinositol pathway. (B) Receptor tyrosine kinase (RTK) pathway. (C) Cytoplasmic tyrosine kinase pathway. (D) Pathway wherein activated G protein or tyrosine kinase on sperm membrane activates egg pathways. (E) Soluble activator pathways.

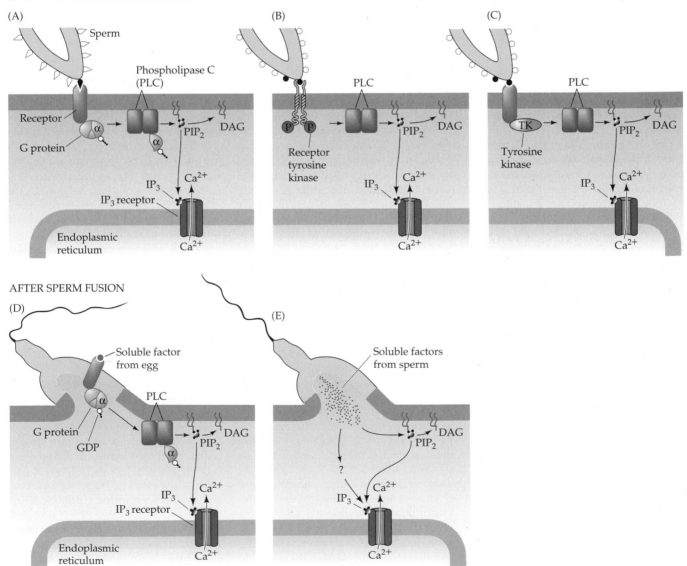

intracellular free calcium ions, cortical granule exocytosis, and DNA synthesis. Some of them developed into larvae. If the mRNA contained a point mutation such that the receptor could not interact with phospholipase C, then no such fertilization reactions occurred when PDGF was added (Shilling et al., 1994). Thus, the G-protein-linked receptor pathway and the receptor tyrosine kinase pathway both appear able to activate phospholipase C, create IP_3, and induce the calcium flux in the egg. The bindin receptor gives no real clue as to how activation takes place, since it resembles no other transmembrane protein. However, within 5 seconds of binding bindin, it becomes phosphorylated on one of its cytoplasmic tyrosine residues (Abassi and Foltz, 1994). This suggests that the bound bindin receptor might interact with a cytoplasmic tyrosine kinase such as those that mediate calcium release during the activation of T cells (Figure 4.26C; Hall et al., 1993).

Another possibility is that the activation of the IP_3 pathway is not caused by the *binding* of the sperm and egg but by the *fusion* of the sperm and egg membranes. McCulloh and Chambers (1992) have electrophysiological evidence that sea urchin egg activation does not occur until *after* sperm and egg cytoplasms are joined. They suggest that the egg-activating components are located on the sperm cell membrane or in the cytoplasm. It is even possible that when the fusion of gamete membranes occurs, the sperm membrane G proteins or tyrosine kinases (activated by the egg jelly to initiate the acrosome reaction) might activate the polyphosphoinositide cascade for calci-um release in the egg. (In this scenario, shown in Figure 4.26D, bindin merely binds the egg or perhaps causes the phosphorylation of proteins needed later in development).

Still another possibility is that the agent active in releasing the bound calcium comes from the sperm cytosol. Parrington and colleagues (1996) have isolated a 33-kDa protein called oscillin that is localized in the sparse cytoplasm of the mammalian sperm head (Figure 4.26E). Microinjection of this protein into mouse eggs can initiate calcium release, but other parameters of egg activation (cortical granule exocytosis, mRNA recruitment, and cell cycle resumption) are not seen. It is not known what role this protein may have in the physiology of egg activation. ■

Activation of egg metabolism

Although fertilization is often depicted as merely the means to merge two haploid nuclei, it has an equally important role in initiating the processes that begin development. These events happen in the cytoplasm and occur without the involvement of the nuclei.*

The mature sea urchin egg is a metabolically sluggish cell that is reactivated by the sperm. This activation is merely a stimulus, however; it sets into action a preprogrammed set of metabolic events. The responses of the egg to the sperm can be divided into "early" responses that occur within seconds of the cortical reaction and "late" responses that take place several minutes after fertilization begins (Table 4.1).

Early Responses

Contact between sea urchin sperm and egg activates the two major blocks to polyspermy: the fast block, initiated by sodium influx into the cell, and the slow block, initiated by the intracellular release of calcium ions. The activation of all eggs appears to depend on an increase in the concentration of free calcium ions within the egg. In protostomes such as snails and worms, at least part of the calcium usually enters the egg from outside. In deuterostomes such as fish, frogs, sea urchins, and mammals, the activation is accomplished by the release of calcium ions from the endoplasmic reticulum, resulting in a wave of calcium ions sweeping across the egg (Jaffe, 1983; Terasaki and Sardet, 1991).

---

*In certain salamanders, this developmental function of fertilization has been totally divorced from the genetic function. The silver salamander (*Ambystoma platineum*) is a hybrid subspecies consisting solely of females. Each female produces an egg with an unreduced chromosome number. This egg, however, cannot develop on its own, so the silver salamander mates with the male Jefferson salamander (*A. jeffersonianum*). The sperm from the male Jefferson salamander only stimulates the egg's development; it does not contribute genetic material (Uzzell, 1964). For details of this complex mechanism of procreation, see Bogart et al., 1989.

Table 4.1 Events of sea urchin fertilization

Event	Approximate time postinsemination[a]
Sperm-egg binding	0 seconds
Fertilization potential rise (fast block to polyspermy)	within 1 sec
Sperm-egg membrane fusion	within 6 sec
Calcium increase first detected	6 sec
Cortical vesicle exocytosis (slow block to polyspermy)	15–60 sec
Activation of NAD kinase	starts at 1 min
Increase in NADH and NADPH	starts at 1 min
Increase in O_2 consumption	starts at 1 min
Sperm entry	1–2 min
Acid efflux	1–5 min
Increase in pH (remains high)	1–5 min
Sperm chromatin decondensation	2–12 min
Sperm nucleus migration to egg center	2–12 min
Egg nucleus migration to sperm nucleus	5–10 min
Activation of protein synthesis	starts at 5–10 min
Activation of amino acid transport	starts at 5–10 min
Initiation of DNA synthesis	20–40 min
Mitosis	60–80 min
First cleavage	85–95 min

Main Sources: Whitaker and Steinhardt, 1985; Mohri et al., 1995.
[a]Approximate times based on data from *S. purpuratus* (15–17°C), *L. pictus* (16–18°C), *A. punctulata* (18–20°C), and *L. variegatus* (22–24°C). The timing of events within the first minute is best known for *Lytechinus variegatus*, so times are listed for that species.

This release of calcium ions is essential for activating the development of the embryo. If the calcium-chelating chemical EGTA is injected into the sea urchin egg, there is no exocytosis of cortical granules, no fertilization potential changes, no sperm decondensation, and no reinitiation of cell division (Kline, 1988). Conversely, eggs can be activated artificially in the absence of sperm by procedures that release free calcium into the oocyte. Steinhardt and Epel (1974) found that micromolar amounts of calcium ionophore A23187 elicit most of the egg responses characteristic of a normally fertilized egg. The elevation of the fertilization envelope, the rise of intracellular pH, the burst of oxygen utilization, and the increases in protein and DNA synthesis are all generated in their proper order. Moreover, this activation takes place in the total absence of calcium ions in the seawater. In most of these cases, development ceases before the first mitosis because the eggs are still haploid and lack the sperm centriole needed for division.

This calcium release activates a series of metabolic reactions (Figure 4.27). One of these is the activation of the enzyme NAD^+ kinase, which converts NAD^+ to $NADP^+$ (Epel et al., 1981). This change may have important consequences for lipid metabolism, since $NADP^+$ (but not NAD^+) can be used as a coenzyme for lipid biosynthesis. Thus, the switch from NAD^+ to $NADP^+$ may be important in the construction of the many new cell membranes required during cleavage. Another effect of this change involves oxygen consumption. A burst of oxygen reduction is seen during fertilization, and much of this "respiratory burst" is used to crosslink the fertilization membrane. The enzyme responsible for this reduction of oxygen (to hydrogen peroxide) is also NADPH-dependent (Heinecke and Shapiro, 1989).

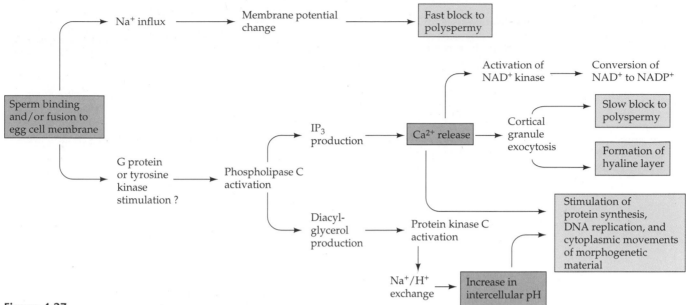

Figure 4.27
Model of possible mechanism of egg activation in the sea urchin. (After Epel, 1980, and L. A. Jaffe, personal communication.)

Lastly, NADPH helps regenerate glutathione and ovothiols that may be crucial for scavenging free radicals that could otherwise damage the DNA of the egg and early embryo (Mead and Epel, 1995).

Late Responses

Shortly after the calcium ion levels rise, the intracellular pH also increases. It is thought that these two ionic conditions (higher $[Ca^{2+}]$, lower $[H^+]$) act in concert to give the full spectrum of fertilization events, including protein synthesis and DNA synthesis (Winkler et al., 1980; Whitaker and Steinhardt, 1982). The rise in intracellular pH begins with a second influx of sodium ions, causing a 1:1 exchange between sodium ions from the seawater and hydrogen ions from the egg.* This loss of hydrogen ions causes the pH to rise from 6.8 to 7.2 and brings about enormous changes in egg physiology (Shen and Steinhardt, 1978).

The late responses of fertilization brought about by these ionic changes include the activation of DNA synthesis and protein synthesis. The burst of protein synthesis usually occurs within several minutes after sperm entry and does not depend on new messenger RNA synthesis (Figure 4.28). Rather, new protein synthesis utilizes mRNAs already present in the oocyte cytoplasm. (Much more will be said about this in Chapter 12.) These include mRNAs that encode proteins such as histones, tubulins, actins, and morphogenetic factors that are utilized during early development. Such a burst of protein synthesis can be induced by artificially raising the pH of the cytoplasm by ammonium ions (Winkler et al., 1980). Conversely, agents that block the rise in pH inhibit late fertilization events such as DNA and protein synthesis. When newly fertilized eggs are placed into solutions containing low concentrations of sodium ions and amiloride (a drug that inhibits

---

*Again, species-to-species variation is rampant. In the much smaller egg of the mouse, there is no elevation of pH after fertilization. Similarly in the mouse, there is no dramatic increase in protein synthesis immediately following fertilization.

Figure 4.28
Burst of protein synthesis at fertilization uses mRNAs stored in the oocyte cytoplasm. (A) Protein synthesis in embryos of the sea urchin *Arbacia punctulata* fertilized in the presence or absence of actinomycin D, an inhibitor of transcription. For the first few hours, protein synthesis occurs without any new transcription from the zygote or embryo nuclei. A second burst of protein synthesis occurs during midblastula stages, and this represents translation of newly transcribed messages (and therefore is not seen in the embryos growing in actinomycin). (B) Increase in the percentage of ribosomes recruited into polysomes during the first hours of sea urchin development, especially during the first cell cycle. (A after Gross et al., 1964; B after Humphreys, 1971.)

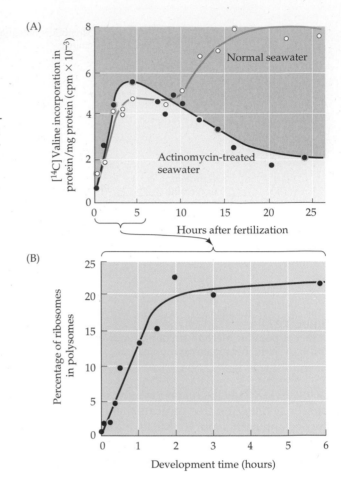

sodium-hydrogen exchange), protein synthesis fails to occur, the movements of the egg and sperm pronuclei are prevented, and cell division fails to take place (Dube et al., 1985).

Fusion of the genetic material

In sea urchins, the sperm nucleus enters the egg perpendicular to the egg surface. After fusion of the sperm and egg membranes, the sperm nucleus and its centriole separate from the mitochondria and the flagellum. The mitochondria and the flagellum disintegrate inside the egg, so very few, if any, sperm-derived mitochondria are found in developing or adult organisms (Dawid and Blackler, 1972; Giles et al., 1980). In mice, it is estimated that only 1 out of every 10,000 mitochondria are sperm-derived (Gyllensten et al., 1991). Thus, although each gamete contributes a haploid genome to the zygote, the mitochondrial genome is transmitted primarily from the maternal parent. Conversely, in almost all animals studied (the mouse being the major exception), the centrosome needed to produce the mitotic spindle of the subsequent divisions is derived from the sperm centriole (Sluder et al., 1989, 1993).

The egg nucleus, once it is haploid, is called the **female pronucleus.** Within the egg cytoplasm, the sperm nucleus decondenses to form the **male pronucleus.** Once inside the egg, the male pronucleus undergoes a dramatic transformation. The pronuclear envelope vesiculates into small packets, thereby exposing the compact sperm chromatin to the egg cytoplasm (Longo

(A)

(B)

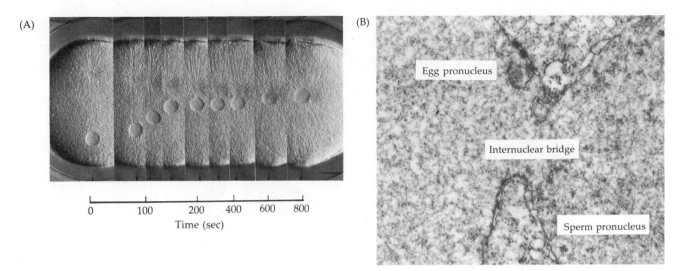

Time (sec)

Egg pronucleus

Internuclear bridge

Sperm pronucleus

Figure 4.29
Nuclear events in the fertilization of the sea urchin. (A) Migration of the egg pronucleus and the sperm pronucleus in an egg of *Clypeaster japonicus.* The sperm pronucleus is surrounded by its aster of microtubules. (B) Fusion of pronuclei in the sea urchin egg. (A from Hamaguchi and Hiramoto, 1980, courtesy of the authors; B courtesy of F. J. Longo.)

and Kunkle, 1978). The proteins holding the sperm chromatin in its condensed, inactive state are exchanged for proteins derived from the egg cytoplasm. This exchange permits the **decondensation** of the sperm chromatin. In sea urchins, decondensation appears to be initiated by the phosphorylation of two sperm-specific histones that bind tightly to the DNA. This process begins when the sperm comes into contact with a glycoprotein in the egg jelly that elevates the level of cAMP-dependent protein kinase activity. (Such cAMP-dependent protein kinases were mentioned in Chapter 1). These protein kinases phosphorylate several of the basic residues of the sperm-specific histones and thereby interfere with their binding to DNA (Garbers et al., 1980, 1983; Porter and Vacquier, 1986). This loosening is thought to facilitate the replacement of the sperm-specific histones by other histones that have been stored in the oocyte cytoplasm (Poccia et al., 1981; Green and Poccia, 1985). Once decondensed, the DNA can begin transcription and replication. [fert9.html]

After the sea urchin sperm enters the egg cytoplasm, the male pronucleus rotates 180° so that the sperm centriole is between the sperm pronucleus and the egg pronucleus. The sperm centriole then acts as a microtubule organizing center, extending its own microtubules and integrating them with egg microtubules to form an aster.* These microtubules extend throughout the egg, contact the female pronucleus, and bring the two pronuclei toward each other (Hamaguchi and Hiramoto, 1980; Bestor and Schatten, 1981). Their fusion forms the diploid **zygote nucleus** (Figure 4.29). The initiation of DNA synthesis can occur either in the pronuclear stage (during migration) or after the formation of the zygote nucleus.

In mammals, the process of pronuclear migration takes about 12 hours, compared with less than 1 hour in the sea urchin. The mammalian sperm enters almost tangentially to the surface of the egg rather than approaching it perpendicularly, and it fuses with numerous microvilli (see Figure 4.20). The mammalian sperm nucleus also breaks down as its chromatin decon-

*When Oscar Hertwig observed this radial array of sperm asters forming in his newly fertilized sea urchin eggs, he called it "the sun in the egg," and he thought it the happy indication of a successful fertilization (Hertwig, 1877). More recently, Simerly and co-workers (1994) have found that certain types of human male infertility are due to defects in the centrosome's ability to form these microtubular asters. This deficiency causes the failure of pronuclear migration and the cessation of further development.

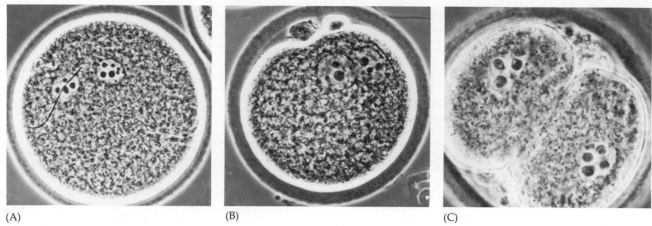

(A) (B) (C)

Figure 4.30
Pronuclear movement in the hamster. (A) Sperm entry into the cell and swelling of the sperm pronucleus. The sperm tail can be seen within the egg. (B) Apposition of sperm and egg pronuclei. (C) Two-cell stage, showing two equally sized cells with well-defined nuclei. Debris in the perivitelline space is the degenerating polar bodies. (From Bavister, 1980, courtesy of B. D. Bavister.)

denses and is then reconstructed by coalescing vesicles. The DNA of the sperm nucleus is bound by basic proteins called protamines, and these nuclear proteins are tightly compacted through disulfide bonds. Once in the egg, glutathione reduces these disulfide bonds and allows the uncoiling of the sperm chromatin (Calvin and Bedford, 1971; Kvist et al., 1980; Perreault et al., 1988). The mammalian male pronucleus enlarges while the oocyte nucleus completes its second meiotic division (Figure 4.30A). The centrosome accompanying the male pronucleus produces its asters (largely from proteins stored in the oocyte) and contacts the female pronucleus. Then each pronucleus migrates toward the other, replicating its DNA as it travels. Upon meeting, the two nuclear envelopes break down (Figure 4.30B). However, instead of producing a common zygote nucleus (as happens in sea urchin fertilization), the chromatin condenses into chromosomes that orient themselves on a common mitotic spindle. Thus, a true diploid nucleus in mammals is first seen not in the zygote, but at the two-cell stage (Figure 4.30C). [fert10.html]

Sidelights & Speculations

The Nonequivalence of Mammalian Pronuclei

I T IS GENERALLY ASSUMED that males and females carry equivalent haploid genomes. Indeed, one of the fundamental tenets of Mendelian genetics is that genes derived from the sperm are functionally equivalent to those derived from the egg. However, recent studies show that in mammals, the sperm-derived genome and the egg-derived genome may be functionally different and play complementary roles during certain stages of development. The first evidence for this nonequivalence came from studies of a human tumor called a

hydatidiform mole. These tumors resemble placental tissue. A majority of such moles have been shown to arise from a haploid sperm fertilizing an egg in which the female pronucleus is absent. After entering the egg, the sperm chromosomes duplicate themselves, thereby restoring the diploid chromosome number. Thus, the entire genome is derived from the sperm (Jacobs et al., 1980; Ohama et al., 1981). Here we see a situation in which the cells survive, divide, and have a normal chromosome number, but development is abnormal.

Instead of forming an embryo, the egg becomes a mass of placenta-like cells. Normal development does not occur when the entire genome comes from the male parent.

Evidence for the nonequivalence of mammalian pronuclei also comes from attempts to get ova to develop in the absence of sperm. The ability to develop an embryo without spermatic contribution is called **parthenogenesis** (from the Greek, meaning "virgin birth"). The eggs of many invertebrates and some vertebrates are capable of developing normally in the absence of sperm if the egg is artificially activated. In these situations, sperm's contribution to development seems dispensable. Mammals, however, do not exhibit parthenogenesis. Placing mouse oocytes into a culture medium that artificially activates the ooyte while supressing the formation of the second polar body produces diploid mouse eggs whose inheritance is derived from the egg alone (Kaufman et al., 1977). These cells divide to form embryos with spinal cords, muscles, skeletons, and organs, including beating hearts. However, development does not continue, and by day 10 or 11 (halfway through the mouse's gestation), profound differences are observed between normal and parthenogenetic embryos, with the parthenogenetic embryos deteriorating and becoming grossly disorganized (Figure 4.31). Neither human nor mouse development can be completed solely with egg-derived chromosomes.

The hypothesis that male and female pronuclei are different also gains support from pronuclear transplantation experiments (Surani and Barton, 1983; Surani et al., 1986; McGrath and Solter, 1984). Either male or female pronuclei of recently fertilized mouse eggs can be removed and added to other recently fertilized eggs. (The two pronuclei can be distinguished at this stage, because the female pronucleus is the one beneath the

Table 4.2 Pronuclear transplantation experiments

Class of reconstructed zygotes	Operation	Number of successful transplants	Number of progeny surviving
Bimaternal		339	0
Bipaternal		328	0
Control		348	18

Source: McGrath and Solter, 1984.

polar bodies.) Thus, zygotes with two male or two female pronuclei can be constructed. Although embryonic cleavage occurs, neither of these types of eggs develops to birth, whereas some control eggs (containing one male pronucleus and one female pronucleus from different zygotes) undergoing such transplantation develop normally (Table 4.2). Moreover, the bimaternal or bipaternal embryos cease development at the same time as the parthenogenetic mice. Thus, although the two pronuclei are equivalent in many animals, in mammals there are important functional differences between them.

The reason for these embryonic deaths is that in some cells, only the maternally derived allele of certain

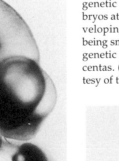

Figure 4.31 (A) Control and (B) parthenogenetic (two female pronuclei) mouse embryos at 11 days' gestation. The mice were developing in the same female. In addition to being smaller and deteriorating, the parthenogenetic embryos also had much smaller placentas. (From Surani and Barton, 1983, courtesy of the authors.)

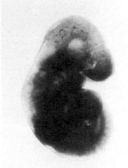

genes is active, while in other cells, only the paternally derived allele of those genes is functional. (In most genes, of course, the male-derived and female-derived alleles are equivalent and are activated to the same degree in every cell. We are dealing here with exceptions to that Mendelian rule.) For instance, insulin-like growth factor II (IGF-II) promotes the growth of embryonic and fetal organs. In embryonic mice, the *paternally* derived allele of IGF-II is active throughout the embryo, whereas the *maternally* derived allele is usually inactive (except in a few neural cells). Thus, if a mouse inherits a mutant IGF-II allele from its mother, the mouse will develop to a normal size (since the maternally derived allele is not expressed); but if the same mutant allele is inherited from its father, the mouse will have stunted growth (DeChiara et al., 1991). The opposite pattern of allele expression is found for one of the *receptors* of IGF-II. Here, the *paternal* gene for the receptor is poorly transcribed, while the *maternal* allele is active (Barlow et al., 1991). The differences between the active and inactive alleles are thought to be caused by modifications of DNA that occur differently in the egg and sperm nuclei, and these will be discussed further in Chapter 11. Because certain developmentally important genes are active only if they come from the sperm and other such genes are active only if they come from the egg, maternal and paternal pronuclei are both necessary for the completion of mammalian development. ■

Rearrangement of egg cytoplasm

Fertilization can initiate radical displacements of the egg's cytoplasmic materials. These rearrangements of oocyte cytoplasm are often crucial for cell differentiation later in development. As we will see in Chapters 13 and 14, the cytoplasm of the egg frequently contains **morphogenetic determinants** that become segregated into specific cells during cleavage. These determinants ultimately lead to the activation or repression of specific genes and thereby confer certain properties to the cells that incorporate them. The correct spatial arrangement of these determinants is crucial for proper development.

In some species, the rearrangement of these determinants into their required orientation can be visualized because cytoplasmic pigment granules are present. One such example is the egg of the tunicate *Styela partita* (Conklin, 1905). The unfertilized egg of this animal is depicted in Figure 4.32A. A central gray cytoplasm is enveloped by a cortical layer containing yellow lipid inclusions. During meiosis, the breakdown of the nucleus releases a clear substance that accumulates in the animal (upper) hemisphere of the egg. Within 5 minutes of sperm entry, the inner clear and cortical yellow cytoplasms migrate into the vegetal (lower) hemisphere of the egg. As the male pronucleus migrates from the vegetal pole to the equator of the cell along the future posterior side of the embryo, the lipid inclusions migrate with it. This migration forms a **yellow crescent,** extending from the vegetal pole to the equator (Figure 4.32B–D), and brings the yellow cytoplasm into the area where muscle cells will later form in the tunicate larva. The movement of these cytoplasmic regions depends on microtubules that are generated by the sperm centriole and by the wave of calcium ions that contract the animal pole cytoplasm (Sawada and Schatten, 1989; Speksnijder et al., 1990; Roegiers et al., 1995).

Cytoplasmic movement is also seen in amphibian eggs. In frogs, a single sperm can enter anywhere on the animal hemisphere; when it does, it changes the cytoplasmic pattern of the egg. Originally, the egg is radially symmetrical about the animal-vegetal axis. After sperm entry, however, the cortical (outer) cytoplasm shifts about 30° toward the point of sperm entry, relative to the inner cytoplasm (Manes and Elinson, 1980; Vincent et al., 1986). In some frogs (such as *Rana*), one region of the egg that had formerly been covered by the dark cortical cytoplasm of the animal hemisphere is now exposed (Figure 4.33). This underlying cytoplasm, located near the equator on the side opposite the point of sperm entry, contains diffuse pigment granules and therefore appears gray. Thus, this region has been referred to as the **gray crescent** (Roux, 1887; Ancel and Vintenberger, 1948). As

Figure 4.32
Cytoplasmic rearrangement in the egg of the tunicate *Styela partita.* (A) Before fertilization, yellow cortical cytoplasm surrounds gray yolky cytoplasm. (B) After sperm entry, the yellow cortical cytoplasm and the clear cytoplasm derived from the breakdown of the oocyte nucleus stream vegetally toward the sperm. (C) As the sperm pronucleus migrates animally toward the egg pronucleus, the yellow and clear cytoplasms move with it. (D) The final positions of the clear and yellow cytoplasms mark the locations where the cells give rise to the mesenchyme and muscles, respectively. (After Conklin, 1905.)

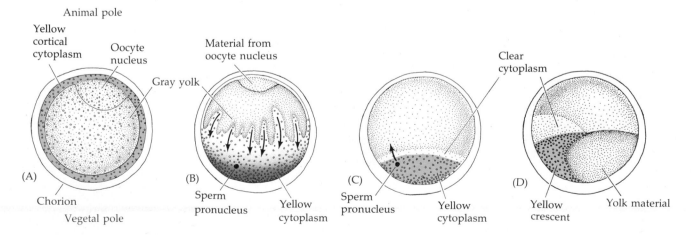

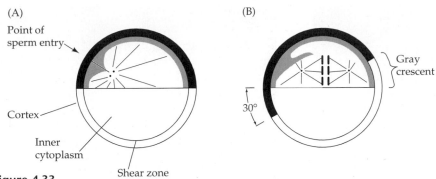

(A)

Point of
sperm entry

Cortex

Inner
cytoplasm

Shear zone

(B)

Gray
crescent

30°

Figure 4.33
Reorganization of cytoplasm in the newly fertilized frog egg. (A) Schematic cross section of an egg midway through the first cleavage cycle. The egg has radial symmetry about its animal-vegetal axis. The sperm has entered at one side, and the sperm nucleus is migrating inward. The cortex is represented like that of *Rana,* with a heavily pigmented animal hemisphere and a transparent vegetal hemisphere. (B) About 80 percent of the way into first cleavage, the cortical cytoplasm rotates 30° relative to the internal cytoplasm. This rotation is important in that gastrulation will begin in that region opposite the point of sperm entry where the greatest displacement of cytoplasm occurs. (After Gerhart et al., 1989.)

we will see in subsequent chapters, the gray crescent marks the region where gastrulation is initiated in amphibian embryos.

In frogs such as *Xenopus,* where no gray crescent is seen, we can still observe the rotation of the cortical cytoplasm relative to the internal, subcortical cytoplasm. This movement was demonstrated by Vincent and his colleagues (1986). These investigators imprinted a hexagonal grid of dye (Nile blue) onto the cytoplasm beneath the cortex while applying another type of dye (a fluorescein-bound lectin) to the egg surface. When the egg was held in one position by embedding it in gelatin, the Nile blue dots were seen to rotate 30° relative to the fluorescent lectin spots (Figure 4.34). In normal, unembedded eggs, the egg surface is thought to rotate while the subcortical cytoplasm, heavy with yolk platelets, remains stabilized by gravity.

The motor for these cytoplasmic movements in amphibian eggs appears to be an array of parallel microtubules that lie between the cortical and inner cytoplasm of the vegetal hemisphere parallel to the direction of cytoplasmic rotation. These microtubular tracks are first seen immediately before the rotation commences, and they disappear when rotation ceases (Figure 4.35; Elinson and Rowning, 1988). Treating the egg with colchicine or ultraviolet radiation at the beginning of rotation stops the formation of these microtubules, thereby inhibiting the cytoplasmic rotations. Using antibodies that bind to the microtubules, Houliston and Elinson (1991a) found that these tracks are formed from sperm- and egg-derived microtubules and that the sperm centriole directs the polymerization of these microtubules so that they grow into the vegetal region of the egg. Upon reaching the vegetal cortex, these microtubules angle away from the point of sperm entry, toward the vegetal pole. The off-center position of the sperm centriole as it initiates microtubule polymerization provides the directionality to the rotation. The motive force for the rotation is possibly provided by the ATPase **kinesin.** Like dynein and myosin, kinesin is able to attach to fibers and produce energy through ATP hydrolysis. This ATPase is located on the vegetal microtubules and the membranes of the cortical endoplasmic reticulum (Houliston and Elinson, 1991b).

The movement of the cortical cytoplasm with respect to the inner cytoplasm causes profound movement within the inner cytoplasm. Danilchik

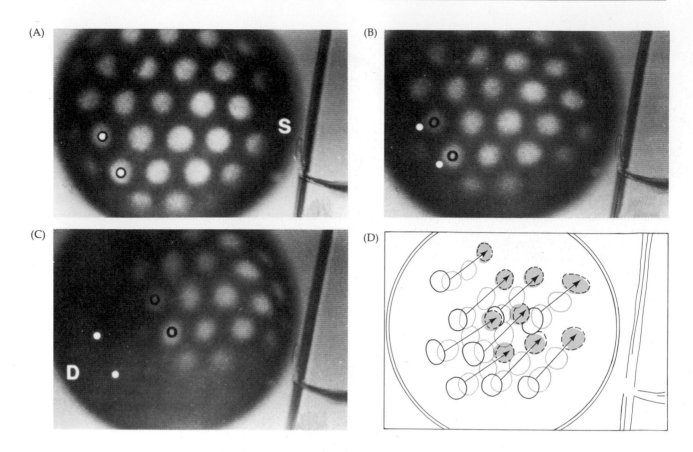

Figure 4.34
The rotation of subcortical cytoplasm relative to the cell surface cytoplasm. (A) A recently fertilized egg was imprinted with a hexagonal grid of Nile Blue dye (which stains the lipids in the yolk platelets). The egg was embedded in gelatin, and the original positions of some of the dots were marked on the cell surface with fluorescein (circles in A). The point of sperm entry is marked with an S. (B,C) As the first cell cycle progressed, the dots of the subcortical cytoplasm shifted roughly 30° with respect to the outer immobilized surface of the egg. The site on the egg marking the future dorsal surface of the embryo is marked with a D. (D) Summary of these movements in the vegetal (lower) region of the egg. (From Vincent et al., 1986, photographs courtesy of J. C. Gerhart.)

and Denegre (1991) have labeled yolk platelets with Nile blue and watched their movement by fluorescent microscopy (the bound dye fluoresces red). During the middle part of the first cell cycle, a mass of central egg cytoplasm flows from the presumptive ventral (belly) to the future dorsal (back) side of the embryo (Plate 7). By the end of first division, the cytoplasm of the prospective dorsal side of the embryo is distinctly different from that of the prospective vegetal side. What had been a radially symmetrical embryo is now a bilaterally symmetrical embryo.

As we will see in Chapters 6 and 15, these cytoplasmic movements initiate a cascade of events that determine the dorsal-ventral axis of the frog. Indeed, the parallel microtubules that allow these rearrangements appear to stretch along the future dorsal-ventral axis (Klag and Ubbels, 1975; Gerhart et al., 1983).

Preparation for Cleavage

The increase in levels of intracellular free calcium ions also sets in motion the apparatus for cell division. The mechanisms by which cleavage is initi-

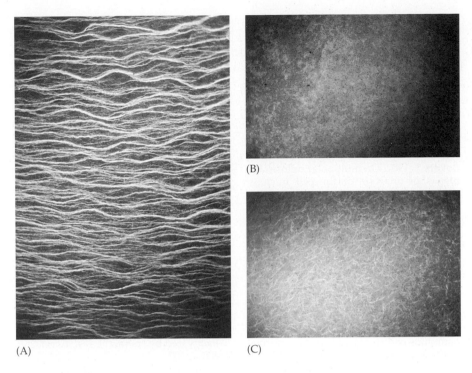

Figure 4.35
Parallel arrays of microtubules extend along the vegetal hemisphere along the future dorsal-ventral axis. (A) Parallel arrays of microtubules seen in the second half of the first cell cycle by fluorescent antibodies to tubulin. (B) Prior to cytoplasmic rotation (about midway through the first cell cycle), no array of microtubules can be seen. (C) At the end of cytoplasmic rotation, the microtubules depolymerize. (From Elinson and Rowning, 1988, courtesy of R. Elinson.)

ated probably differ between species, depending on the stage of meiosis at which fertilization occurs. However, in all species studied, the rhythm of cell divisions is regulated by the synthesis and degradation of **cyclin.** Cyclin keeps the cells in metaphase, and the breakdown of cyclin enables the cells to return to interphase. In addition to their other activities, calcium ions also appear to initiate the degradation of cyclin (Watanabe et al., 1991). Once the cyclin is degraded, the cycles of cell division can begin anew.

Cleavage has a special relationship to these cytoplasmic regions. In tunicate embryos, the first cleavage bisects the egg into mirror-image duplicates. From that stage on, every division on one side of the cleavage furrow has a mirror-image division on the opposite side. Similarly, the gray crescent is bisected by the first cleavage furrow in amphibian eggs. Thus, the position of the first cleavage is not random, but tends to be specified by the point of sperm entry and the subsequent rotation of egg cytoplasm. The coordination of cleavage plane and cytoplasmic rearrangements is probably mediated through the microtubules of the sperm aster (Manes et al., 1978; Gerhart et al., 1981; Elinson, 1985).

Toward the end of the first cell cycle, then, the cytoplasm is rearranged, the pronuclei have met, DNA is replicating, and new proteins are being translated. The stage is set for the development of a multicellular organism. [fert11.html], [other.html#fert13]

LITERATURE CITED

Abassi, Y. A. and Foltz, K. R. 1994. Tyrosine phosphorylation of the sperm receptor at fertilization. *Dev. Biol.* 164: 430–443.

Afzelius, B. A. 1976. A human syndrome caused by immotile cilia. *Science* 193: 317–319.

Almeida, E. A. C. and ten others. 1995. Mouse egg integrin α6β1 functions as a sperm receptor. *Cell* 81: 1095–1104.

Amos, L. A. and Klug, A. 1974. Arrangement of subunits in flagellar microtubules. *J. Cell Sci.* 14: 523–549.

Ancel, P. and Vintenberger, P. 1948. Recherches sur le determinisme de la symmetrie bilatérale dans l'oeuf des amphibiens. *Bull. Biol. Fr. Belg.* [Suppl.] 31: 1–182.

Asai, D. J. 1996. Functional and molecular diversity of dynein heavy chains. *Semin. Cell Dev. Biol.* 7: 311–320.

Austin, C. R. 1952. The "capacitation" of mammalian sperm. *Nature* 170: 326.

Austin, C. R. 1960. Capacitation and the release of hyaluronidase. *J. Reprod. Fert.* 1: 310–311.

Austin, C. R. 1965. *Fertilization.* Prentice-Hall, Englewood Cliffs, New Jersey.

Ayabe, T., Kopf, G. S. and Schultz, R. M. 1995. Regulation of mouse egg activation: presence of ryanodine receptors and effects of microinjected ryanodine and cyclic ADP ribose on uninseminated and inseminated eggs. *Development* 121: 2233–2244.

Baltz, J. M., Katz, D. F. and Cone, R. A. 1988. The mechanics of sperm-egg interaction at the zona pellucida. *Biophys. J.* 54: 643–654.

Barlow, D. P., Stöger, R., Herrmann, B. G., Saito, K. and Schweifer, N. 1991. The mouse insulin-like growth factor type-2 receptor is imprinted and closely linked to the *Tme* locus. *Nature* 349: 84–87.

Bavister, B. D. 1980. Recent progress in the study of early events in mammalian fertilization. *Dev. Growth Differ.* 22: 385–402.

Bentley, J. K., Shimomura, H. and Garbers, D. L. 1986. Retention of a functional resact receptor in isolated sperm plasma membranes. *Cell* 45: 281–288.

Berridge, M. J. 1993. Inositol triphosphate and calcium signalling. *Nature* 361: 315–325.

Bestor, T. M. and Schatten, G. 1981. Anti-tubulin immunofluorescence microscopy of microtubules present during the pronuclear movements of sea urchin fertilization. *Dev. Biol.* 88: 80–91.

Bi, G.-Q., Alderton, J.M. and Steinhardt, R.A. 1995. Calcium-mediated exocytosis is required for cell membrane resealing. *J. Cell Biol.* 131: 1747–1758.

Bleil, J. D. and Wassarman, P. M. 1980. Mammalian sperm and egg interaction: Identification of a glycoprotein in mouse-egg zonae pellucidae possessing receptor activity for sperm. *Cell* 20: 873–882.

Bleil, J. D. and Wassarman, P. M. 1986. Autoradiographic visualization of the mouse egg's sperm receptor bound to sperm. *J. Cell Biol.* 102: 1363–1371.

Bleil, J. D. and Wassarman, P. M. 1988. Galactose at the nonreducing terminus of O-linked oligosaccharides of mouse egg zona pellucida glycoprotein ZP3 is essential for the glycoprotein's sperm receptor activity. *Proc. Natl. Acad. Sci. USA* 85: 6778–6782.

Bleil, J. D. and Wassarman, P. M. 1990. Identification of a ZP3-binding protein on acrosome-intact mouse sperm by photoaffinity crosslinking. *Proc. Natl. Acad. Sci. USA* 87: 5563–5567.

Bleil, J. D., Greve, J. M. and Wassarman, P. M. 1988. Identification of a secondary sperm receptor in the mouse egg zona pellucida: Role in maintenance of binding of acrosome-reacted sperm to eggs. *Dev. Biol.* 28: 376–385.

Blobel, C. P, Wolfsberg, T. G., Turck, C. W., Myles, D. G., Primakoff, P. and White, J. M.

1992. A potential fusion peptide and an integrin domain in a protein active in sperm-egg fusion. *Nature* 356: 248–251.

Bogart, J. P., Elinson, R. P. and Licht, L. E. 1989. Temperature and sperm incorporation in polyploid salamanders. *Science* 246: 1032–1034.

Bookbinder, L. H., Cheng, A., and Bleil, J. D. 1995. Tissue- and species-specific expression of sp56, a mouse sperm fertilization protein. *Science* 269: 86–87.

Boveri, T. 1902. On multipolar mitosis as a means of analysis of the cell nucleus. (Translated by S. Gluecksohn-Waelsch.) *In* B. H. Willier and J. M. Oppenheimer (eds.), *Foundations of Experimental Embryology.* Hafner, New York, 1974.

Boveri, T. 1907. Zellenstudien VI. Die Entwicklung dispermer Seeigeleier. Ein Beiträge zur Befruchtungslehre und zur Theorie des Kernes. *Jena Z. Naturwiss.* 43: 1–292.

Braden, A. W. H. and Austin, C. R. 1954. The number of sperm about the eggs in mammals and its significance for normal fertilization. *Aust. J. Biol. Sci.* 7: 543–551.

Burks, D. J., Carballada, R., Moore, H. D. M. and Saling, P. M. 1995. Interaction of a tyrosine kinase from human sperm with the zona pellucida at fertilization. *Science* 269: 83–86.

Busa, W. B., Ferguson, J. E., Joseph, S. K., Williamson, J. R. and Nuccitelli, R. 1985. Activation of frog (*Xenopus laevis*) eggs by inositol triphosphate. I. Characterization of Ca^{2+} release from intracellular stores. *J. Cell Biol.* 100: 677–682.

Calvin, H. I. and Bedford, J. M. 1971. Formation of disulfide bonds in the nucleus and accessory structures of mammalian spermatozoa during maturation in the epididymis. *J. Reprod. Fertil.* [Suppl.] 13: 65–75.

Carroll, E. J. and Epel, D. 1975. Isolation and biological activity of the proteases released by sea urchin eggs following fertilization. *Dev. Biol.* 44: 22–32.

Chambers, E. L., Pressman, B. C. and Rose, B. 1974. The activity of sea urchin eggs by the divalent ionophores A23187 and X-537A. *Biochem. Biophys. Res. Commun.* 60: 126–132.

Chandler, D. E. and Heuser, J. 1979. Membrane fusion during secretion: Cortical granule exocytosis in sea urchin eggs as studied by quick-freezing and freeze fracture. *J. Cell Biol.* 83: 91–108.

Chang, M. C. 1951. Fertilizing capacity of spermatozoa deposited into the fallopian tubes. *Nature* 168: 697–698.

Cherr, G. N., Lambert, H., Meizel, S. and Katz, D. F. 1986. In vitro studies of the golden hamster sperm acrosomal reaction: Completion on zona pellucida and induction by homologous zonae pellucidae. *Dev. Biol.* 114: 119–131.

Ciapa, B. and Whitaker, M. 1986. Two phases of inositol polyphosphate and diacylglycerol production at fertilization. *FEBS Lett.* 195: 347–351.

Clermont, Y. and Leblond, C. P. 1955. Spermiogenesis of man, monkey, and other animals as shown by the "periodic acid-Schiff" technique. *Am. J. Anat.* 96: 229–253.

Colwin, A. L. and Colwin, L. H. 1963. Role of the gamete membranes in fertilization in *Saccoglossus kowalevskii* (Enteropneustra). I. The acrosome reaction and its changes in early stages of fertilization. *J. Cell Biol.* 19: 477–500.

Colwin, L. H. and Colwin, A. L. 1960. Formation of sperm entry holes in the vitelline membrane of *Hydroides hexagonis* (Annelida) and evidence of their lytic origin. *J. Biophys. Biochem. Cytol.* 7: 315–320.

Conklin, E. G. 1905. The orientation and cell-lineage of the ascidian egg. *J. Acad. Nat. Sci. Phila.* 13: 5–119.

Cook, S. P. and Babcock, D. F. 1993. Selective modulation by cGMP of the K$^+$ channel activated by speract. *J. Biol. Chem.* 268: 22402–22407.

Corselli, J. and Talbot, P. 1987. In vivo penetration of hamster oocyte-cumulus complexes using physiological numbers of sperm. *Dev. Biol.* 122: 227–242.

Cross, N. L. and Elinson, R. P. C. 1980. A fast block to polyspermy in frogs mediated by changes in the membrane potential. *Dev. Biol.* 75: 187–198.

Dan, J. C. 1967. Acrosome reaction and lysins. *In* C. B. Metz and A. Monroy (eds.), *Fertilization*, Vol. 1. Academic Press, New York, pp. 237–367.

Danilchik, M. V. and Denegre, J. M. 1991. Deep cytoplasmic rearrangements during early development in *Xenopus laevis. Development* 111: 845–856.

Davis, B. K. 1981. Timing of fertilization in mammals: Sperm cholesterol/phospholipid ratio as determinant of capacitation interval. *Proc. Natl. Acad. Sci. USA* 78: 7560–7564.

Dawid, I. B. and Blackler, A. W. 1972. Maternal and cytoplasmic inheritance of mitochondria in *Xenopus. Dev. Biol.* 29: 152–161.

DeChiara, T. M., Robertson, E. J. and Efstradiatis, A. 1991. Parental imprinting of the mouse insulin-like growth factor II gene. *Cell* 64: 849–859.

De Robertis, E. D. P., Saez, F. A. and De Robertis, E. M. F. 1975. *Cell Biology*, 6th Ed., Saunders, Philadelphia.

Dube, F., Schmidt, T., Johnson, C. H. and Epel, D. 1985. The hierarchy of requirements for an elevated pH during early development of sea urchin embryos. *Cell* 40: 657–666.

Eisen, A. and Reynolds, G. T. 1985. Sources and sinks for the calcium release during fertilization of single sea urchin eggs. *J. Cell Biol.* 100: 1522–1527.

Eisenbach, M. 1995. Sperm changes enabling fertilization in mammals. *Curr. Opin. Endocrinol. Diabetes* 2: 468–475.

Elinson, R. P. 1985. Changes in levels of polymeric tubulin associated with activation and dorsoventral polarization of the frog egg. *Dev. Biol.* 109: 224–233.

Elinson, R. P. 1986. Fertilization in amphibians: The ancestry of the block to polyspermy. *Int. Rev. Cytol.* 101: 59–100.

Elinson, R. P. and Rowning, B. 1988. A transient array of parallel microtubules in frog eggs: Potential tracks for a cytoplasmic rotation that specifies the dorso-ventral axis. *Dev. Biol.* 128: 185–197.

Endo, Y. G., Kopf, G. S. and Schultz, R. M. 1987. Effects of phorbol ester on mouse eggs: Dissociation of sperm receptor activity from acrosome reaction-inducing activity of the mouse zona pellucida protein, ZP3. *Dev. Biol.* 123: 574–577.

Epel, D. 1977. The program of fertilization. *Sci. Am.* 237(5): 128–138.

Epel, D. 1980. Fertilization. *Endeavour N.S.* 4: 26–31.

Epel, D., Patton, C., Wallace, R. W. and Cheung, W. Y. 1981. Calmodulin activates NAD kinase of sea urchin eggs: An early response. *Cell* 23: 543–549.

Ferris, C. D., Huganir, R. L., Supattapone, S. and Snyder, S. H. 1989. Purified inositol 1,4,5-trisphosphate receptor mediates calcium flux in reconstituted lipid vesicles. *Nature* 342: 87–89.

Florman, H. M. 1995. Sequential focal and global elevations of sperm intracellular Ca^{2+} are initiated by the zona pellucida during acrosomal exocytosis. *Dev. Biol.* 165: 152–164.

Florman, H. M. and Storey, B. T. 1982. Mouse gamete interactions: The zona pellucida is the site of the acrosome reaction leading to fertilization in vitro. *Dev. Biol.* 91: 121–130.

Florman, H. M. and Wassarman, P. M. 1985. O-linked oligosaccharides of mouse egg ZP3 account for its sperm receptor activity. *Cell* 41: 313–324.

Florman, H. M., Bechtol, K. B. and Wassarman, P. M. 1984. Enzymatic dissection of the functions of the mouse egg's receptor for sperm. *Dev. Biol.* 106: 243–255.

Florman, H, M., Corron, M. E., Kim, T. D.-H. and Babcock, D. F. 1992. Activation of voltage-dependent calcium channels of mammalian sperm is required for zona pellucida-induced acrosomal exocytosis. *Dev. Biol.* 152: 304–314.

Foerder, C. A. and Shapiro, B. M. 1977. Release of ovoperoxidase from sea urchin eggs hardens fertilization membrane with tyrosine crosslinks. *Proc. Natl. Acad. Sci. USA* 74: 4214–4218.

Fol, H. 1877. Sur le commencement de l'hémogénie chez divers animaux. *Arch. Zool. Exp. Gén.* 6: 145–169.

Foltz, K. R., Partin, J. S. and Lennarz, W. J. 1993. Sea urchin egg receptor for sperm: Sequence similarity of binding domain to hsp70. *Science* 259: 1421–1425.

Franklin, L. E. 1970. Fertilization and the role of the acrosomal reaction in non-mammals. *Biol. Reprod.* [Suppl.] 2: 159–176.

Fulton, B. P. and Whittingham, D. G. 1978. Activation of mammalian oocytes by intracellular injection of calcium. *Nature* 273: 149–151.

Furuichi, T., Yoshikawa, S., Miyawaki, A., Wada, K., Maeda, N. and Mikoshiba, K. 1989. Primary structure and functional expression of the inositol 1,4,5-trisphosphate-binding protein P400. *Nature* 342: 32–38.

Garbers, D. L., Tubb, D. J. and Kopf, G. S. 1980. Regulation of sea urchin sperm cAMP-dependent protein kinases by an egg associated factor. *Biol. Reprod.* 22: 526–532.

Garbers, D. L., Kopf, G. S., Tubb, D. J. and Olson, G. 1983. Elevation of sperm adenosine 3':5'-monophosphate concentrations by a fucose sulfate-rich complex associated with eggs. I. Structural characterization. *Biol. Reprod.* 29: 1211–1220.

Gardiner, D. M. and Grey, R. D. 1983. Membrane junctions in *Xenopus* eggs: Their distribution suggests a role in calcium regulation. *J. Cell Biol.* 96: 1159–1163.

Gerhart, J., Ubbels, G., Black, S., Hara, K. and Kirschner, M. 1981. A reinvestigation of the role of the grey crescent in axis formation in *Xenopus laevis*. *Nature* 292: 511–516.

Gerhart, J., Black, S., Gimlich, R. and Scharf, S. 1983. Control of polarity in the amphibian egg. *In* W. R. Jeffery and R. A. Raff (eds.), *Time, Space, and Pattern in Embryonic Development*. Alan R. Liss, New York, pp. 261–286.

Gerhart, J., Danilchik, M., Doniach, T., Roberts, S., Rowning, B. and Stewart, R. 1989. Cortical rotation of the *Xenopus* egg: Consequences for the anterioposterior pattern of embryonic dorsal development. *Development 1989* [Suppl.]: 37-51.

Giles, R. E., Blanc, H., Cann, H. M. and Wallace, D. C. 1980. Maternal inheritance of human mitochondrial DNA. *Proc. Natl. Acad. Sci. USA* 77: 6715–6719.

Gilkey, J. C., Jaffe, L. F., Ridgway, E. G. and Reynolds, G. T. 1978. A free calcium wave traverses the activating egg of the medaka, *Oryzias latipes. J. Cell Biol.* 76: 448–466.

Ginzburg, A. S. 1985. Phylogenetic changes in the type of fertilization. *In* J. Mlíkovsky and V. J. A. Novák (eds.), *Evolution and Morphogenesis*. Academia, Prague, pp. 459–466.

Glabe, C. G. 1985. Interaction of the sperm adhesive protein, bindin, with phospholipid vesicles. II. Bindin induces the fusion of mixed-phase vesicles that contain phosphatidylcholine and phosphatidylserine in vitro. *J. Cell Biol.* 100: 800–806.

Glabe, C. G. and Lennarz, W. J. 1979. Species-specific sperm adhesion in sea urchins: A quantitative investigation of bindin-mediated egg agglutination. *J. Cell Biol.* 83: 595–604.

Glabe, C. G. and Vacquier, V. D. 1977. Species-specific agglutination of eggs by bindin isolated from sea urchin sperm. *Nature* 267: 836–838.

Glabe, C. G. and Vacquier, V. D. 1978. Egg surface glycoprotein receptor for sea urchin sperm bindin. *Proc. Natl. Acad. Sci. USA* 75: 881–885.

Gong, X., Dubois, D.H., Miller, D. J., and Shur, B. D. 1995. Activation of a G protein complex by aggregation of β-1,4-galactosyltransferase on the surface of sperm. *Science* 269: 1718–1721.

González-Martínez, M. T., Guerrero, A., Morales, E., de la Torre, L. and Darszon, A. 1992. A depolarization can trigger Ca^{2+} uptake and the acrosome reaction when preceded by a hyperpolarization in *L. pictus* sea urchin sperm. *Dev. Biol.* 150: 193–202.

Gould, M. and Stephano, J. L. 1987. Electrical response of eggs to acrosomal protein similar to those induced by sperm. *Science* 235:1654–1656.

Gould, M. and Stephano, J. L. 1991. Peptides from sperm acrosomal protein that activate development. *Dev. Biol.* 146: 509–518.

Gould-Somero, M., Jaffe, L. A. and Holland, L. Z. 1979. Electrically mediated fast polyspermy block in eggs of the marine worm, *Urechis caupo. J. Cell Biol.* 82: 426–440.

Green, G. R. and Poccia, E. L. 1985. Phosphorylation of sea urchin sperm H1 and H2B histones precedes chromatin decondensation and H1 exchange during pronuclear formation. *Dev. Biol.* 108: 235–245.

Gross, P. R., Malkin, L. I., and Moyer, W. 1964. Templates for the first proteins of embryonic development. *Proc. Natl. Acad. Sci. USA* 51: 407–414.

Gundersen, G. G., Medill, L. and Shapiro, B. M. 1986. Sperm surface proteins are incorporated into the egg membrane and cytoplasm after fertilization. *Dev. Biol.* 113: 207–217.

Gwatkin, R. B. L. 1976. Fertilization. *In* G. Poste and G. L. Nicolson (eds.), *The Cell Surface in Animal Embryogenesis and Development*. Elsevier North-Holland, New York, pp. 1–53.

Gyllensten, U., Wharton, D., Josefson, A. and Wilson, A. 1991. Paternal inheritance of mitochondrial DNA in mice. *Nature* 352: 255–258.

Hafner, M., Petzelt, C., Nobiling, R., Pawley, J. B., Kramp, D. and Schatten, G. 1988. Wave of free calcium at fertilization in the sea urchin egg visualized with Fura-2. *Cell Motil. Cytoskel.* 9: 271–277.

Hall, C. G., Sancho, J., and Terhorst, C. 1993. Reconstitution of T cell receptor ζ-mediated calcium mobilization in nonlymphoid cells. *Science* 261: 915–918.

Hamaguchi, M. S. and Hiramoto, Y. 1980. Fertilization process in the heart-urchin, *Clypaester japonicus*, observed with a differential interference microscope. *Dev. Growth Differ.* 22: 517–530.

Hamaguchi, M. S. and Hiramoto, Y. 1981. Activation of sea urchin eggs by microinjection of calcium buffers. *Exp. Cell Res.* 134: 171–179.

Hardy, D. M., Harumi, T. and Garbers, D. L. 1994. Sea urchin sperm receptors for egg peptides. *Sem. Dev. Biol.* 5: 217–224.

Hartsoeker, N. 1694. *Essai de dioptrique.* Paris.

Heinecke, J. W. and Shapiro, B. M. 1989. Respiratory oxygen burst of fertilization. *Proc. Natl. Acad. Sci. USA* 86: 1259–1263.

Hertwig, O. 1877. Beiträge zur Kenntniss der Bildung, Befruchtung, und Theilung des theirischen Eies. *Morphol. Jahr.* 1: 347–452.

Hollinger, T. G. and Schuetz, A. W. 1976. "Cleavage" and cortical granule breakdown in *Rana pipiens* oocytes induced by direct microinjection of calcium. *J. Cell Biol.* 71: 395–401.

Hong, K. and Vacquier, V. D. 1986. Fusion of liposomes induced by a cationic protein from the acrosomal granule of abalone spermatozoa. *Biochemistry* 25: 543–549.

Houliston, E. and Elinson, R. P. 1991a. Patterns of microtubule polymerization relating to cortical rotation in *Xenopus laevis* eggs. *Development* 112: 107–117.

Houliston, E. and Elinson, R. P. 1991b. Evidence for the involvement of microtubules, endoplasmic reticulum, and kinesin in cortical rotation of fertilized frog eggs. *J. Cell Biol.* 114: 1017–1028.

Humphreys, T. 1971. Measurements of messenger RNA entering polysomes upon fertilization in sea urchins. *Dev. Biol.* 26: 201–208.

Hunter, R. H. F. 1989. Ovarian programming of gamete progression and maturation in the female genital tract. *Zool. J. Linn. Soc.* 95: 117–124.

Hylander, B. L. and Summers, R. G. 1982. An ultrastructural and immunocytochemical localization of hyaline in the sea urchin egg. *Dev. Biol.* 93: 368–380.

Iwao, Y. and Jaffe, L. A. 1989. Evidence that the voltage-dependent component in the fertilization porcess is contributed by the sperm. *Dev. Biol.* 134: 446–451.

Jacobs, P. A., Wilson, C. M., Sprenkle, J. A., Rosenshein, N. B. and Migeon, B. R. 1980. Mechanism of origin of complete hydatidiform moles. *Nature* 286: 714–716.

Jaffe, L. A. 1976. Fast block to polyspermy in sea urchins is electrically mediated. *Nature* 261: 68–71.

Jaffe, L. A. 1980. Electrical polyspermy block in sea urchins: Nicotine and low sodium experiments. *Dev. Growth Differ.* 22: 503–507.

Jaffe, L. A. 1996. Egg membranes during fertilization. *In* S. G. Schultz et al. (eds.), *Molecular Biology of Membrane Transport Disorders.* Plenum, NY, pp. 367–378.

Jaffe, L. A. and Cross, N. L. 1983. Electrical properties of vertebrate oocyte membranes. *Biol. Reprod.* 30: 50–54.

Jaffe, L. A. and Gould, M. 1985. Polyspermy-preventing mechanisms. *Biol. Fert.* 3: 223–250.

Jaffe, L. F. 1983. Sources of calcium in egg activation: A review and hypothesis. *Dev. Biol.* 99: 265–276.

Jones, R., Brown, C. R. and Lancaster, R. T. 1988. Carbohydrate-binding properties of boar sperm proacrosin and assessment of its role in sperm-egg recognition and adhesion during fertilization. *Development* 102: 781–792.

Just, E. E. 1919. The fertilization reaction in *Echinarachnius parma*. *Biol. Bull.* 36: 1–10.

Kaufman, M. H., Barton, S.C. and Surani, M. A. H. 1977. Normal postimplantation development of mouse parthenogenetic embryos to the forelimb bud stage. *Nature* 265: 53–55.

Keller, S. H. and Vacquier, V. D. 1994a. N-linked oligosaccharides of sea urchin egg jelly induces the sperm acrosome reaction. *Dev. Growth Differ.* 36: 551–556.

Keller, S. H. and Vacquier, V. D. 1994b. The isolation of acrosome-reaction-inducing glycoproteins from sea urchin egg jelly. *Dev. Biol.* 162: 304–312.

Klag, J. J. and Ubbels, G. A. 1975. Regional morphological and cytochemical differentiation of the fertilized egg of *Discoglossus pictus* (Anura). *Differentiation* 3: 15-20.

Kline, D. 1988. Calcium-dependent events at fertilization of the frog egg: Injection of a calcium buffer blocks ion channel opening, exocytosis, and formation of pronuclei. *Dev. Biol.* 126: 346–361.

Kline, D., Simoncini, L., Mandel, G., Maue, R., Kado, R. T. and Jaffe. L. A. 1988. Fertilization events induced by neurotransmitters after injection of mRNA into *Xenopus* eggs. *Science* 241: 464–467.

Kline, D, Kopf, G., Muncy, L. F. and Jaffe, L. A. 1991. Evidence for the involvement of a pertussis toxin-insensitive G-protein in egg activation of the frog *Xenopus laevis*. *Dev. Biol.* 143: 218–229.

Kvist, U., Afzelius, B. A. and Nilsson, L. 1980. The intrinsic mechanism of chromatin decondensation and its activation in human spermatozoa. *Dev. Growth Differ.* 22: 543–554.

Langlais, J., Kan, F. W. K., Granger, L., Raymond, L., Bleau, G. and Roberts, K. D. 1988. Identification of sterol acceptors that stimulate cholesterol efflux from human spermatozoa during in vitro capacitation. *Gamete Res.* 20: 185–201.

Lechleiter, J. D. and Clapham, D. E. 1992. Molecular mechanisms of intracellular calcium excitability in *X. laevis* oocytes. *Cell* 69: 283–294.

Leeuwenhoek, A. van. 1685. Letter to the Royal Society of London. Quoted in E. G. Ruestow, 1983, Images and ideas: Leeuwenhoek's perception of the spermatozoa. *J. Hist. Biol.* 16: 185–224.

Levine, A. E., Walsh, K. A. and Fodor, E. J. B. 1978. Evidence of an acrosin-like enzyme in sea urchin sperm. *Dev. Biol.* 63: 299–306.

Leyton, L. and Saling, P. 1989a. 95 kd sperm proteins bind ZP3 and serve as tyrosine kinase substrates in response to zona binding. *Cell* 57: 1123–1130.

Leyton, L. and Saling, P. 1989b. Evidence that aggregation of mouse sperm receptors by ZP3 triggers the acrosome reaction. *J. Cell Biol.* 108: 2163–2168.

Leyton, L., Leguen, P., Bunch, D. and Saling, P. M. 1992. Regulation of mouse gametic interaction by a sperm tyrosine kinase. *Proc. Natl. Acad. Sci. USA* 93: 1164–1169.

Longo, F. J. 1986. Surface changes at fertilization: Integration of sea urchin (*Arbacia punctulata*) sperm and oocyte plasma membranes. *Dev. Biol.* 116: 143–159.

Longo, F. J. and Kunkle, M. 1978. Transformation of sperm nuclei upon insemination. *In* A. A. Moscona and A. Monroy (eds.), *Current Topics in Developmental Biology*, Vol. 12. Academic Press, New York, pp. 149–184.

Longo, F. J., Lynn, J. W., McCulloh, D. H. and Chambers, E. L. 1986. Correlative ultrastructural and electrophysiological studies of sperm-egg interactions of the sea urchin *Lytechinus variegatus*. *Dev. Biol.* 118: 155–166.

Lopez, L. C., Bayna, E. M., Litoff, D., Shaper, N. L., Shaper, J. H. and Shur, B. D. 1985. Receptor function of mouse sperm surface galactosyltransferase during fertilization. *J. Cell Biol.* 101: 1501–1510.

Luttmer, S. and Longo, F. J. 1985. Ultrastructural and morphometric observations of cortical endoplasmic reticulum in *Arbacia, Spisula*, and mouse eggs. *Dev. Growth Differ.* 27: 349–359.

Manes, M.E. and Elinson, R.P. 1980. Ultraviolet light inhibits gray crescent formation in the frog egg. *Wilhelm Roux Arch. Dev. Biol.* 189: 73–76.

Manes, M. E., Elinson, R. P. and Barbieri, F. D. 1978. Formation of the amphibian gray crescent: Effects of colchicine and cytochalasin-B. *Wilhelm Roux Arch. Dev. Biol.* 185: 99–104.

McCulloh, D. H. and Chambers, E. L. 1992. Fusion of membranes during fertilization. *J. Gen. Physiol.* 99: 137–175.

McGrath, J. and Solter, D. 1984. Completion of mouse embryogenesis requires both the maternal and paternal genome. *Cell* 37: 179–183.

McPherson, S. M., McPherson, P. S., Mathews, L., Campbell, K. P. and Longo, F. J. 1992. Cortical localization of a calcium release channel in sea urchin eggs. *J. Cell Biol.* 116: 111–1121.

Mead, K. S. and Epel, D. 1995. Beakers and breakers: how fertisation in the laboratory differs from fertisation in nature. *Zygote* 3: 95–99.

Meizel, S. 1984. The importance of hydrolytic enzymes to an exocytotic event, the mammalian sperm acrosome reaction. *Biol. Rev.* 59: 125–157.

Metz, C. B. 1978. Sperm and egg receptors involved in fertilization. *Curr. Top. Dev. Biol.* 12: 107–148.

Miller, D. J., Macek, M. B. and Shur, B. D. 1992. Complementarity between sperm surface β-1,4-galactosyltransferase and egg-coat ZP3 mediates sperm-egg binding. *Nature* 357: 589–593.

Miller, D.L., Gong, X., Decker, G. and Shur, B. D. 1993. Egg cortical granule N-acetylglucosaminidase is required for the mouse zona block to polyspermy. *J. Cell Biol.* 123: 1431–1440.

Miller, R. L. 1978. Site-specific agglutination and the timed release of a sperm chemoattractant by the egg of the leptomedusan, *Orthopyxis caliculata. J. Exp. Zool.* 205: 385–392.

Miller, R. L. 1985. Sperm chemo-orientation in the metazoa. *In* C. B. Metz, Jr. and A. Monroy (eds.), *Biology of Fertilization*, Vol. 2. Academic Press, New York, pp. 275–337.

Miyazaki, S. and Igusa, Y. 1981. Fertilization potential in golden hamster eggs consists of recurring hyperpolarizations. *Nature* 290: 702–704.

Miyazaki, S.-I., Yuzaki, M., Nakada, K., Shirakawa, H., Nakanishi, S., Nakade, S. and Mikoshiba, K. 1992. Block of Ca^{2+} wave and Ca^{2+} oscillation by antibody to the inositol 1,4,5-trisphosphate receptor in fertilized hamster eggs. *Science* 257: 251–255.

Mohri, T., Ivonnet, P. I. and Chambers, E. L. 1995. Effect of sperm-induced activation current and increase of cytosolic Ca^{2+} by agents that modify the mobilization of $[Ca^{2+}]_i$ I. Heparin and pentosan polysulfate. *Dev. Biol.* 172: 139–157.

Moller, C. C. and Wassarman, P. M. 1989. Characterization of a proteinase that cleaves zona pellucida glycoprotein ZP2 following activation of mouse eggs. *Dev. Biol.* 132: 103–112.

Moore, G. D., Kopf, G. S. and Schultz, R. M. 1993. Complete mouse egg activation in the absence of sperm by stimulation of an exogenous G protein-coupled receptor. *Dev. Biol.* 159: 669–678.

Moore, G. D., Ayabe, T., Visconti, P. E., Schultz, R. M. and Kopf, G. 1994. Roles of heteromeric and monomeric G proteins in sperm-induced activation of mouse eggs. *Development* 120: 3313–3323.

Moy, G. W. and Vacquier, V. D. 1979. Immunoperoxidase localization of bindin during the adhesion of sperm to sea urchin eggs. *Curr. Top. Dev. Biol.* 13: 31–44.

Mozingo, N. M. and Chandler, D. E. 1991. Evidence for the existence of two assembly domains within the sea urchin fertilization envelope. *Dev. Biol.* 146: 148–157.

Multigner, L., Gagnon, J., Dorsselaer, A. van and Job, D. 1992. Stabilization of sea urchin flagellar microtubules by histone H1. *Nature* 360: 33–39.

Myles, D. G., Kimmel, L. H., Blobel, C. P., White, J. M. and Primakoff, P. 1994. Identification of a binding site in the disintegrin domain of fertilin required for sperm-egg fusion. *Proc. Natl. Acad. Sci. USA* 91: 4195–4198.

Nishizuka, Y. 1986. Studies and perspectives of protein kinase C. *Science* 233: 305–312.

Ogawa, K., Mohri, T. and Mohri, H. 1977. Identification of dynein as the outer arms of sea urchin sperm axonomes. *Proc. Natl. Acad. Sci. USA* 74: 5006–5010.

Ohama, K., Kajii, T., Okamoto, E., Fukada, Y., Imaizumi, K., Tsukahara, M., Kobayashi, K. and Hagiwara, K. 1981. Dispermic origin of XY hydatidiform moles. *Nature* 292: 551–552.

Olds, J. L. and thirteen others. 1995. Imaging protein kinase C activation in living sea urchin eggs after fertilization. *Dev. Biol.* 172: 675–682.

Parrington, J., Swann, K., Shevchenko, V. I., Sesay, A. K. and Lai, F. A. 1996. Calcium oscillations in mammalian eggs triggered by a soluble sperm protein. *Nature* 379: 364–368.

Perreault, S. D., Barbee, R. R., and Slott, V. L. 1988. Importance of glutathione in the acquisition and maintainance of sperm nuclear decondensing activity in maturing hamster oocytes. *Dev. Biol.* 125: 181–186.

Pinto-Correia, C. 1997. *The Ovary of Eve: Eggs and Sperm and Preformation.* University of Chicago Press, Chicago.

Poccia, D., Salik, J. and Krystal, G. 1981. Transitions in histone variants of the male pronucleus following fertilization and evidence for a maternal store of cleavage-stage histones in the sea urchin egg. *Dev. Biol.* 82: 287–296.

Poirier, G. R. and Jackson, J. 1981. Isolation and characterization of two proteinase inhibitors from the male reproductive tract of mice. *Gamete Res.* 4: 555–569.

Porter, D. C. and Vacquier, V. D. 1986. Phosphorylation of sperm histone H1 is induced by the egg jelly layer in the sea urchin *Strongylocentrotus purpuratus. Dev. Biol.* 116: 203–212.

Prevost, J. L. and Dumas, J. B. 1824. Deuxieme mémoire sur la génération. *Ann. Sci. Nat.* 2: 129–149.

Primakoff, P., Hyatt, H. and Tredick-Kline, J. 1987. Identification and purification of a sperm cell surface protein with a potential role in sperm-egg membrane fusion. *J. Cell Biol.* 104: 141–149.

Primakoff, P., Lathrop, W., Woolman, L., Cowan, A. and Myles, D. 1988. Fully effective contraception in male and female guinea pigs immunized with the sperm protein PH-20. *Nature* 335: 543–546.

Ralt, D. and eight others. 1991. Sperm attraction to a follicular factor(s) correlates with human egg fertilizability. *Proc. Natl. Acad. Sci. USA* 88: 2840–2844.

Ramarao, C. S. and Garbers, D. L. 1985. Receptor-mediated regulation of guanylate cyclase activity in spermatozoa. *J. Biol. Chem.* 260: 8390–8396.

Ravnik, S. E., Zarutskie, P. W. and Muller, C. H. 1992. Purification and characterization of a human follicular fluid lipid transfer protein that stimulates human sperm capacitation. *Biol. Reprod.* 47: 1126–1133.

Roegiers, F., McDougall, A. and Sardet, C. 1995. The sperm entry point defines the orientation of the calcium-induced contraction wave that directs the first phase of cytoplasmic reorganization in the ascidian egg. *Development* 121: 3457–3466.

Rossignol, D. P., Earles, B. J., Decker, G. L. and Lennarz, W. J. 1984. Characterization of the sperm receptor on the surface of eggs of *Strongylocentrotus purpuratus. Dev. Biol.* 104: 308–321.

Roux, W. 1887. Beiträge zur Entwicklungsmechanik des Embryo. *Arch. Mikrosk. Anat.* 29: 157–212.

Saling, P. M. 1989. Mammalian sperm interaction with extracellular matrices of the egg. *Oxford Rev. Reprod. Biol.* 11: 339–388.

Saling, P. M., Sowinski, J. and Storey, B. T. 1979. An ultrastructural study of epididymal mouse spermatozoa binding to zonae pellucida in vitro: Sequential relationship to acrosome reaction. *J. Exp. Zool.* 209: 229–238.

Sawada, T. and Schatten, G. 1989. Effects of cytoskeletal inhibitors on ooplasmic segregation and microtubule organization during fertilization and early development in the ascidian *Molgula occidentalis. Dev. Biol.* 132: 331–342.

Schackmann, R. W. and Shapiro, B. M. 1981. A partial sequence of ionic changes associated with the acrosome reaction of *Strongylocentrotus purpuratus. Dev. Biol.* 81: 145–154.

Schatten, G. and Mazia, D. 1976. The penetration of the spermatozoan through the sea urchin egg surface at fertilization: Observations from the outside on whole eggs and from the inside on isolated surfaces. *Exp. Cell Res.* 98: 325–337.

Schatten, G. and Schatten, H. 1983. The energetic egg. *The Sciences* 23(5): 28–35.

Schatten, H. and Schatten, G. 1980. Surface activity at the plasma membrane during sperm incorporation and its cytochalasin-B sensitivity: Scanning electron micrography and time-lapse video microscopy during fertilization of the sea urchin *Lytechinus variegatus. Dev. Biol.* 78: 435–449.

Schroeder, T. E. 1979. Surface area change at fertilization: Resorption of the mosaic membrane. *Dev. Biol.* 70: 306–326.

SeGall, G. K. and Lennarz, W. J. 1979. Chemical characterization of the component of the jelly coat from sea urchin eggs responsible for induction of the acrosome reaction. *Dev. Biol.* 71: 33–48.

Shen, S. S. and Buck, W. R. 1990. A synthestic peptide of the pseudosubstrate domain of protein kinase C blocks cytoplasmic alakalinization during activation of the sea urchin egg. *Dev. Biol.* 140: 272–280.

Shen, S. S. and Burgart, L. J. 1986. 1,2-Diacylglycerols mimic phorbol 12-myristate 13-acetate activation of the sea urchin egg. *J. Cell. Physiol.* 127: 330–340.

Shen, S. S. and Steinhardt, R. A. 1978. Direct measurement of intracellular pH during metabolic depression of the sea urchin egg. *Nature* 272: 253–254.

Shilling, F. M., Carroll, D. J., Muslin, A. J., Escobodo, J. A., Williams, L. T. and Jaffe, L. A. 1994. Evidence for both tyrosine kinase and G-protein-coupled pathways leading to starfish egg activation. *Dev. Biol.* 162: 590–599.

Shimomura, H., Dangott, L. J. and Garbers, D. L. 1986. Covalent coupling of a resact analogue to guanylate cyclase. *J. Biol. Chem.* 261: 15778–15782.

Shur, B. D. and Hall, N. G. 1982a. Sperm surface galactosyltransferase activities during in vitro capacitation. *J. Cell Biol.* 95: 567–573.

Shur, B. D. and Hall, N. G. 1982b. A role for mouse sperm surface galactosyltransferase in sperm binding for the egg zona pellucida. *J. Cell Biol.* 95: 574–579.

Shur, B. D. and Neely, C. A. 1988. Plasma membrane association, purification, and partial characterization of mouse sperm β 1,4-galactosyltransferase. *J. Biol. Chem.* 263: 17706–17714.

Simerly, C. and ten others. 1994. The paternal inheritance of the centrosome, the cell's microtubule-organizing center, in humans, and the implications for infertility. *Nat. Med.* 1: 1–10.

Sluder, G., Miller, F. J., Lewis, K., K., Davison, E. D. and Reider, C. L. 1989. Centrosome inheritance in starfish zygotes: Selective loss of the maternal centrosome after fertilization. *Dev. Biol.* 131: 567–579.

Sluder, G., Miller, F. J. and Lewis, K. 1993. Centrosome inheritance in starfish zygotes II. Selective suppression of the maternal centrosome during meiosis. *Dev. Biol.* 155: 58–67.

Snell, W. J. and White, J. M. 1996. The molecules of mammalian fertilization. *Cell* 85: 629–637.

Speksnijder, J. E., Sardet, C. and Jaffe, L. F. 1990. The activation wave of calcium in the ascidian egg and its role in ooplasmic segregation. *J. Cell Biol.* 110: 1589–1598.

Steinhardt, R. A. and Epel, D. 1974. Activation of sea urchin eggs by a calcium ionophore. *Proc. Natl. Acad. Sci. USA* 71: 1915–1919.

Steinhardt, R., Zucker, R. and Schatten, G. 1977. Intracellular calcium release at fertilization in the sea urchin egg. *Dev. Biol.* 58: 185–196.

Storey, B. T. 1995. Interactions between gametes leading to fertilization: The sperm's eye view. *Reprod. Fert. Dev.* 7: 927–942.

Suarez, S. S. and Dai, X. 1995. Intracellular calcium reaches different levels of elevation in hyperactivated and acrosome-reacted hamster sperm. *Mol. Reprod. Dev.* 42: 325–333.

Suarez, S. S., Katz, D. F., Owen, D. H., Andrew, J. B. and Powell, R. L. 1991. Evidence for the function of hyperactivated motility in sperm. *Biol. Reprod.* 44: 375–381.

Summers, R. G. and Hylander, B. L. 1974. An ultrastructural analysis of early fertilization in the sand dollar, *Echinarachnius parma. Cell Tiss. Res.* 150: 343–368.

Summers, R. G. and Hylander, B. L. 1975. Species-specificity of acrosome reaction and primary gamete binding in echinoids. *Exp. Cell Res.* 96: 63–68.

Summers, R. G., Hylander, B. L., Colwin, L. H. and Colwin, A. L. 1975. The functional anatomy of the echinoderm spermatozoon and its interaction with the egg at fertilization. *Am. Zool.* 15: 523–551.

Surani, M. A. H. and Barton, S. C. 1983. Development of gynogenetic eggs in the mouse: Implications for parthenogenetic embryos. *Science* 222: 1034–1036.

Surani, M. A. H., Barton, S. C. and Norris, M. L. 1986. Nuclear transplantation in the mouse: Hereditable differences between parental genomes after activation of the embryonic genome. *Cell* 45: 127–136.

Swann, K. and Whitaker, M. 1986. The part played by inositol trisphosphate and calcium in the propagation of the fertilization wave in sea urchin eggs. *J. Cell Biol.* 103: 2333–2342.

Terasaki, M. and Sardet, C. 1991. Demonstration of calcium uptake and release by sea urchin egg cortical endoplasmic reticulum. *J. Cell Biol.* 115: 1031–1037.

Tilney, L. G., Bryan, J., Bush, D. J., Fujiwara, K., Mooseker, M. S., Murphy, D. B. and Snyder, D. H. 1973. Microtubules: Evidence for 13 protofilaments. *J. Cell Biol.* 59: 267–275.

Tilney, L. G., Kiehart, D. P., Sardet, C. and Tilney, M. 1978. Polymerization of actin. IV. Role of Ca^{2+} and H^+ in the assembly of actin and in membrane fusion in the acrosomal reaction of echinoderm sperm. *J. Cell Biol.* 77: 536–560.

Tombes, R. M. and Shapiro, B. M. 1985. Metabolite channeling: A phosphocreatine shuttle to mediate high-energy phosphate transport between sperm mitochondrion and tail. *Cell* 41: 325–334.

Turner, P. R., Jaffe, L. A. and Fein, A. 1986. Regulation of cortical vesicle exocytosis in sea urchin eggs by inositol 1,4,5-trisphosphate and GTP-binding protein. *J. Cell Biol.* 102: 70–76.

Turner, P. R., Jaffe, L. A. and Primakoff, P. 1987. A cholera toxin-sensitive G-protein stimulates exocytosis in sea urchin eggs. *Dev. Biol.* 120: 577–583.

Uzzell, T. M. 1964. Relations of the diploid and triploid species of the *Ambystoma jeffersonianum* complex. *Copeia* 1964: 257–300.

Vacquier, V. D. 1979. The interaction of sea urchin gametes during fertilization. *Am. Zool.* 19: 839–849.

Vacquier, V. D. and Moy, G. W. 1977. Isolation of bindin: The protein responsible for adhesion of sperm to sea urchin eggs. *Proc. Natl. Acad. Sci. USA* 74: 2456–2460.

Vacquier, V. D. and Payne, J. E. 1973. Methods for quantitating sea urchin sperm in egg binding. *Exp. Cell Res.* 82: 227–235.

Vacquier, V. D., Tegner, M. J. and Epel, D. 1973. Protease release from sea urchin eggs at fertilization alters the vitelline layer and aids in preventing polyspermy. *Exp. Cell Res.* 80: 111–119.

Vincent, J. P., Oster, G. F. and Gerhart, J. C. 1986. Kinematics of gray crescent formation in *Xenopus* eggs. The displacement of subcortical cytoplasm relative to the egg surface. *Dev. Biol.* 113: 484–500.

Visconti, P. E., Bailey, J. L., Moore, G. D., Pan, D., Olds-Clarke, P. and Kopf, G. S. 1995a. Capacitation of mouse spermatozoa I. Correlation between capacitation state and protein tyrosine phosphorylation. *Development* 121: 1129–1137.

Visconti, P. E. and seven others. 1995b. Capacitation of mouse spermatozoa II. Protein tyrosine phosphorylation and capacitation are regulated by a cAMP-dependent pathway. *Development* 121: 1139–1150.

von Kolliker, A. 1841. *Beiträge zur Kenntnis der Geschlectverhältnisse und der Samenflüssigkeit wirbelloser Thiere, nebst einem Versuch Über Wesen und die Bedeutung der sogenannten Samenthiere.* Berlin.

Ward, C. R. and Kopf, G. S. 1993. Molecular events mediating sperm activation. *Dev. Biol.* 158: 9–34.

Ward, G. E., Brokaw, C. J., Garbers, D. L. and Vacquier, V. D. 1985. Chemotaxis of *Arbacia punctulata* spermatozoa to resact, a peptide from the egg jelly layer. *J. Cell Biol.* 101: 2324–2329.

Wassarman, P. 1987. The biology and chemistry of fertilization. *Science* 235: 553–560.

Wassarman, P. M. 1989. Fertilization in mammals. *Sci. Am.* 256(6): 78–84.

Watanabe, N., Hunt, T., Ikawa, Y. and Sagata, N. 1991. Independent inactivation of MPF and cytostatic factor (Mos) upon fertilization of *Xenopus* eggs. *Nature* 352: 247–249.

Whitaker, M. and Irvine, R. F. 1984. Inositol 1,4,5-triphosphate microinjection activates sea urchin eggs. *Nature* 312: 636–639.

Whitaker, M. and Steinhardt, R. 1982. Ionic regulation of egg activation. *Q. Rev. Biophys.* 15: 593–666.

Wilcox, A. J., Weinberg, C. R. and Baird, D. D. 1995. Timing of sexual intercourse in relation to ovulation: Effects on the probability of conception, survival of pregnancy, and the sex of the baby. *N. Engl. J. Med.* 333: 1517–121.

Williams, C. J., Schultz, R. M. and Kopf, G. S. 1992. Role of G proteins in mouse egg activation: Stimulatory effects of acetylcholine on the ZP2 to ZP2f conversion and pronuclear formation in eggs expressing a functional m1 muscarinic receptor. *Dev. Biol.* 151: 288–296.

Wilson, W. L. and Oliphant, G. 1987. Isolation and biochemical characterization of the subunits of the rabbit sperm acrosome stabilizing factor. *Biol. Reprod.* 37: 159–169.

Winkler, M. M., Steinhardt, R. A., Grainger, J. L. and Minning, L. 1980. Dual ionic controls for the activation of protein synthesis at fertilization. *Nature* 287: 558–560.

Xu, Z., Kopf, G. S. and Schultz, R. M. 1994. Involvement of inositol-1,4,5-trisphosphate-mediated Ca^{2+} release in early and late events of mouse egg activation. *Development* 120: 1851–1859.

Yanagimachi, R. 1994. Mammalian fertilization. *In* E. Knobil and J. D. Neill (eds.), *The Physiology of Reproduction*, 2nd Ed. Raven Press, NY.

Yanagimachi, R. and Noda, Y. D. 1970. Electron microscope studies of sperm incorporation into the golden hamster egg. *Am. J. Anat.* 128: 429–462.

Yim, D. L., Opresko, L. K., Wiley, H. S. and Nuccitelli, R. 1994. Highly polarized EGF receptor tyrosine kinase activity initiates egg activation in *Xenopus*. *Dev. Biol.* 162: 41–55.

Yoshida, M., Inabar, K. and Morisawa, M. 1993. Sperm chemotaxis during the process of fertilization in the ascidians *Ciona savignyi* and *Ciona intestinalis*. *Dev. Biol.* 157: 497–506.

Zeng, Y., Clark, E. N. and Florman, H. M. 1995. Sperm membrane potential: Hyperpolarization during capacitation regulates zona pellucida-dependent acrosomal secretion. *Dev. Biol.* 171: 554–563.

Cleavage: Creating multicellularity

5

To our limited intelligence, it would seem a simple task to divide a nucleus into equal parts. The cell, manifestly, entertains a very different opinion.
E. B. WILSON (1923)

One must show the greatest respect towards any thing that increases exponentially, no matter how small.
GARRETT HARDIN (1968)

REMARKABLE AS IT IS, fertilization is but the initiating step in development. The zygote, with its new genetic potential and its new arrangement of cytoplasm, now begins the production of a multicellular organism. In all animal species known, this begins by a process called **cleavage,** a series of mitotic divisions whereby the enormous volume of egg cytoplasm is divided into numerous smaller, nucleated cells. These cleavage-stage cells are called **blastomeres.**

In most species (mammals being the chief exception), the rate of cell division and the placement of the blastomeres with respect to one another is completely under the control of the proteins and mRNAs stored in the oocyte by the mother. The zygotic genome, transmitted by mitosis to all the other cells, does not function in early-cleavage embryos. Few, if any, mRNAs are made until relatively late in cleavage, and the embryo can divide properly even when chemicals are used to inhibit transcription. Also, in most species, there is no net increase in embryonic volume during cleavage. This differs from most cases of cell proliferation, in which there is a period of cell growth between mitoses: a cell expands to nearly twice its volume, then divides. This growth produces a net increase in the total volume of cells while maintaining a relatively constant ratio of nuclear to cytoplasmic volume. During embryonic cleavage, however, cytoplasmic volume does not increase. Rather, the enormous volume of zygote cytoplasm is divided into increasingly smaller cells. First the egg is divided in half, then quarters, then eighths, and so forth. This division of egg cytoplasm without increasing its volume is accomplished by abolishing the growth period between divisions, while the cleavage of nuclei occurs at a rapid rate never seen again (not even in tumor cells). A frog egg, for example, can divide into 37,000 cells in just 43 hours. Mitosis in cleavage-stage *Drosophila* occurs every 10 minutes for over 2 hours and in just 12 hours forms some 50,000 cells. This increase in cell number can be appreciated by comparing cleavage with other states of development. Figure 5.1 shows the logarithm of cell number in a frog embryo plotted against the time of development (Sze, 1953). It illustrates a sharp discontinuity between cleavage and gastrulation.

One consequence of this rapid division is that the ratio of cytoplasmic to nuclear volume gets increasingly smaller as cleavage progresses. In many types of embryos, the decrease in the cytoplasmic-to-nuclear volume ratio is

Figure 5.1
Formation of new cells during the early development of the frog *Rana pipiens*. (After Sze, 1953.)

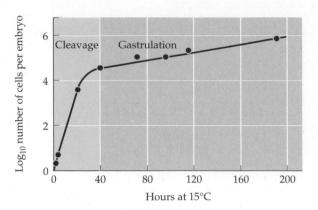

crucial in timing the activation of certain genes. For example, in the frog *Xenopus laevis*, transcription of new messages is not activated until after 12 divisions. At that time, the rate of cleavage decreases, the blastomeres become motile, and nuclear genes begin to be transcribed. It is thought that some factor in the egg is being titrated by the newly made chromatin, because the time of this transition can be changed by experimentally altering the ratio of chromatin to cytoplasm in the cell (Newport and Kirschner, 1982a,b). Thus, cleavage begins soon after fertilization and ends shortly after the stage when the embryo achieves a new balance between nucleus and cytoplasm.

■ PATTERNS OF EMBRYONIC CLEAVAGE

Cleavage is a very well coordinated process and is under genetic regulation. The pattern of embryonic cleavage particular to a species is determined by two major parameters: (1) the amount and distribution of yolk protein within the cytoplasm and (2) those factors in the egg cytoplasm influencing the angle of the mitotic spindle and the timing of its formation.

The amount and distribution of yolk determines where cleavage can occur and the relative size of the blastomeres. When one pole of the egg is relatively yolk-free, the cellular divisions occur there at a faster rate than at the opposite pole. The yolk-rich pole is referred to as the **vegetal pole;** the yolk concentration in the **animal pole** is relatively low. The zygote nucleus is frequently displaced toward the animal pole. In general, yolk inhibits cleavage. Table 5.1 provides a classification of cleavage types and shows the influ-

Table 5.1 Classification of cleavage types

Cleavage pattern	Position of yolk	Cleavage symmetry	Representative animals
Holoblastic (complete cleavage)	Isolecithal (oligolecithal) (sparse, evenly distributed yolk)	Radial	Echinoderms, *Amphioxus*
		Spiral	Most molluscs, annelids, flatworms, and roundworms
		Bilateral	Ascidians
		Rotational	Mammals
	Mesolecithal (moderately telolecithal)	Radial	Amphibians
Meroblastic (incomplete cleavage)	Telolecithal (dense yolk concentrated at one end of egg)	Bilateral	Cephalopod molluscs
		Discoidal	Reptiles, fishes, birds
	Centrolecithal (yolk concentrated in center of egg)	Superficial	Most arthropods

ence of yolk on the cleavage symmetry and pattern. In zygotes with relatively little yolk (*isolecithal* and *mesolecithal* eggs), cleavage is **holoblastic,** meaning that the cleavage furrow extends through the entire egg. Zygotes containing large accumulations of yolk protein undergo **meroblastic** cleavage, wherein only a portion of the cytoplasm is cleaved. The cleavage furrow does not penetrate into the yolky portion of the cytoplasm. Meroblastic cleavage can be **discoidal,** as in birds' eggs, or **superficial,** as in insect zygotes, depending on whether the yolk deposit is located to one side (*telolecithal*) or in the center of the cytoplasm (*centrolecithal*), respectively.

Yolk is an evolutionary adaptation that enables an embryo to develop in the absence of an external food source. Animals developing without large yolk concentrations, such as sea urchins, usually form a larval stage fairly rapidly. This larval stage can feed itself, and development then continues in this free-swimming larva. Mammalian embryos, which also lack large quantities of yolk, adopt another strategy: the placenta. As we will see, the first differentiation of the mammalian embryo sets aside the cells that will form the placenta. This organ supplies food and oxygen for the embryo during its long gestation.

At the other extreme are the eggs of insects, fishes, reptiles, and birds. Most of their cell volumes are yolk. The yolk must be sufficient to nourish these animals, since they develop without a larval stage or placental attachment. The correlation between heavy yolk concentration and lack of larval forms is seen in certain species of frogs. Certain tropical frogs, such as *Eleutherodactylus* and *Arthroleptella*, lack a tadpole stage. Rather, they provision their eggs with an enormous yolk concentration (Lutz, 1947). The eggs do not need to be laid in water because the tadpole stage has been eliminated. (This will be discussed further in Chapter 19.)

However, the yolk is just one factor influencing a species' pattern of cleavage. There are also inherited patterns of cell division that are superimposed upon the restraints of the yolk. This can readily be seen in isolecithal eggs, in which very little yolk is present. In the absence of a large concentration of yolk, four major cleavage types can be observed: radial holoblastic, spiral holoblastic, bilateral holoblastic, and rotational holoblastic cleavage.

Radial holoblastic cleavage

Radial holoblastic cleavage is the simplest form of cleavage to understand. The furrows in this type of cleavage are oriented parallel to and perpendicular to the animal-vegetal axis of the egg. This type of cleavage is characteristic of echinoderms and the protochordate *Amphioxus,* as well as frogs and salamanders.

The Sea Cucumber, Synapta

The cleavage pattern of the sea cucumber, *Synapta digita,* is illustrated in Figure 5.2. After the union of the pronuclei, the axis of the first mitotic spindle is formed perpendicular to the animal-vegetal axis of the egg. Therefore, the first cleavage furrow passes directly through the animal and vegetal poles, creating two equal-sized daughter cells. This cleavage is said to be **meridional** because it passes through the two poles like a meridian on a globe. The cleavage furrows of the second cleavage are at right angles to the furrow of the first cleavage but are still perpendicular to the animal-vegetal axis of the egg. The two cleavage furrows appear simultaneously in both blastomeres and also pass through the two poles. Thus, the first two divisions are both meridional and perpendicular to each other. The third division is **equatorial:** the mitotic spindles of each blastomere are now positioned parallel to the

Figure 5.2
Holoblastic cleavage in the echinoderm *Synapta digita,* leading to the formation of a hollow blastula, as shown in the cutaway view in the last panel. (After Saunders, 1982.)

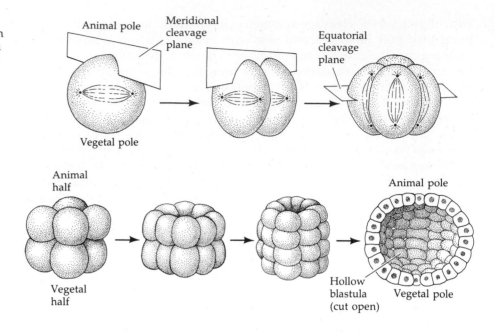

animal-vegetal axis, and the resulting cleavage furrow separates the two poles from each other, dividing the embryo into eight equal blastomeres. Each blastomere in the animal half of the embryo is now directly above a blastomere of the vegetal half.

The fourth division is again meridional, producing two tiers of 8 cells each, while the fifth division is equatorial, producing four tiers of 8 cells each. Successive divisions produce 64-, 128-, and 256-cell embryos, with meridional divisions alternating with equatorial divisions. The resulting embryo consists of blastomeres arranged in horizontal rows along a central cavity. At both poles of the embryo, blastomeres move toward each other to create a hollow sphere composed of a single cell layer. This hollow sphere is called the **blastula,** and the central cavity is referred to as the **blastocoel.** At any time during the cleavage of *Synapta,* an embryo bisected through any meridian produces two mirror-image halves. This type of symmetry is characteristic of a sphere or cylinder and is called radial symmetry. Thus, *Synapta* is said to have radial holoblastic cleavage.

Sea Urchins

Sea urchins also exhibit radial holoblastic cleavage, but with some important modifications. The first and second cleavages are similar to those of *Synapta;* both are meridional and are perpendicular to each other. Similarly, the third cleavage is equatorial, separating the two poles from one another (Figure 5.3). In the fourth cleavage, however, events are very different. The four cells of the animal tier divide meridionally into eight blastomeres, each with the same volume. These cells are called **mesomeres.** The vegetal tier, however, undergoes an *unequal* equatorial cleavage to produce four large cells, the **macromeres,** and four smaller **micromeres** at the vegetal pole (Figure 5.4; Summers et al., 1993). As the 16-cell embryo cleaves, the eight mesomeres divide to produce two "animal" tiers, an_1 and an_2, one staggered above the other. The macromeres divide meridionally, forming a tier of eight cells below an_2. The micromeres also divide, producing a small cluster beneath the larger tier. All the cleavage furrows of the sixth division are equatorial; and the seventh cleavage is meridional, producing a 128-cell blastula.

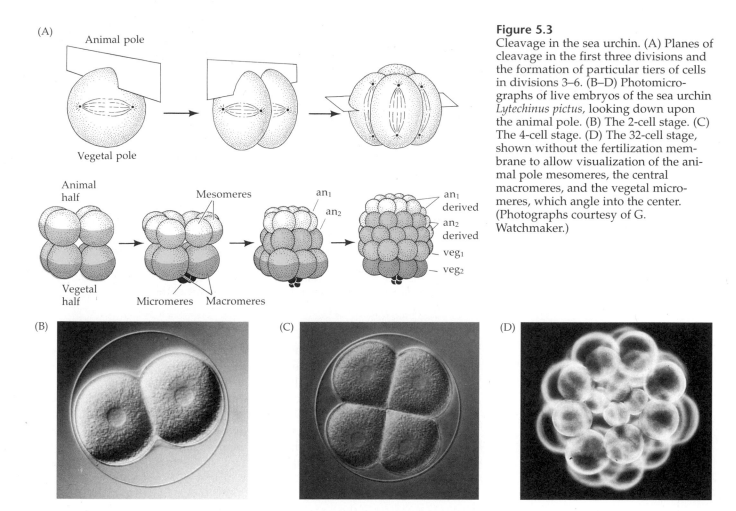

(A)

Animal pole

Vegetal pole

Animal
half

Mesomeres

Vegetal
half

Micromeres   Macromeres

an₁
an₂

an₁
derived

an₂
derived

veg₁

veg₂

(B)

(C)

(D)

Figure 5.3
Cleavage in the sea urchin. (A) Planes of cleavage in the first three divisions and the formation of particular tiers of cells in divisions 3–6. (B–D) Photomicrographs of live embryos of the sea urchin *Lytechinus pictus*, looking down upon the animal pole. (B) The 2-cell stage. (C) The 4-cell stage. (D) The 32-cell stage, shown without the fertilization membrane to allow visualization of the animal pole mesomeres, the central macromeres, and the vegetal micromeres, which angle into the center. (Photographs courtesy of G. Watchmaker.)

In 1939, Sven Hörstadius performed a simple experiment demonstrating that the timing and placement of each sea urchin cleavage is independent of preexisting cleavages. He showed that if he inhibited the first one, two, or three cleavages by shaking the eggs or placing them into hypotonic seawater, the unequal (fourth) cleavage division that forms the micromeres would still occur at the appropriate time. Thus, Hörstadius concluded that there are three factors that determine cleavage in the 8-cell embryo: (1) there are "pro-

(A)   (B)

Figure 5.4
Formation of the micromeres during the fourth division of sea urchin embryos. The vegetal poles of the embryos are viewed from below. (A) The location and orientation of the mitotic spindle at the bottom portion of the vegetal cells is shown by viewing the living embryo with polarized light. (B) The cleavage through these asymmetrically placed spindles has produced micromeres and macromeres. (From Inoué, 1982, courtesy of S. Inoué.)

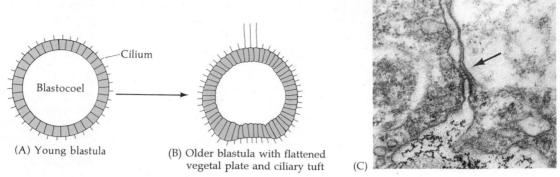

Figure 5.5
Sea urchin blastulae. (A) Sketch of a control cross-section through an early sea urchin blastula, showing a single layer of rounded cells surrounding a large blastocoel. (B) As division continues, the cells of the late blastula show differences in shape as the vegetal plate cells elongate. (C) Tight junctions (arrow) forming between cells of a 1024-cell echinoderm blastula. (A and B after Giudice, 1973; C from Dan-Sohkawa and Fujisawa, 1980, courtesy of the authors.)

gressive changes in the cytoplasm which cause spindles formed after a certain time after fertilization to lie in a certain direction"; (2) there must be micromere-forming material in the vegetal cytoplasm; and (3) there must be some mechanism by which the micromere-forming material is activated at the correct time (Hörstadius, 1973).

The blastula stage of sea urchin development begins at the 128-cell stage. Here the cells form a hollow sphere surrounding a central blastocoel (Figure 5.5A,B). By this time, all the cells are the same size, the micromeres having slowed down their cell division and cleaving less frequently. Every cell is in contact with the proteinaceous fluid of the blastocoel and with the hyaline layer within the fertilization envelope. During this time, contacts between the cells are tightened. Dan-Sohkawa and Fujisawa (1980) analyzed this process in starfish embryos and showed that the closure of the hollow sphere is contemporaneous with the formation of tight junctions between the blastomeres. These junctions unite the loosely connected cells into an epithelial tissue and seal off the blastocoel from the outside environment (Figure 5.5C). As the cells continue to divide, the cell layer expands and thins out. During this period, the blastula remains one cell layer thick.

Two theories have been offered to explain the concomitant cell proliferation and blastocoel formation. Dan (1960) hypothesized that the motive force of this expansion is the influx of water into the blastocoel cavity. As the blastomeres secrete proteins into the blastocoel, the blastocoel fluid becomes syrupy. This blastocoel sap absorbs large quantities of water by osmosis, thereby swelling and putting pressure on the blastomeres to expand outward. This pressure would also align the long axis of each cell so that division would never be inward toward the blastocoel. This would create further expansion by having the population oriented in one plane only. Wolpert and Gustafson (1961) and Wolpert and Mercer (1963) proposed that pressure from the blastocoel is not needed to get this effect. They emphasized the role of differential adhesiveness of the cells to one another and to the hyaline layer. They found that as long as the cells remain strongly attached to the hyaline layer, the cells have no alternative but to expand. This expansion creates the blastula rather than the other way around. Certainly, the hyaline layer is critical for blastocoel expansion, and if the adhesion of cells to the hyaline layer is inhibited by antibodies to hyalin, then blastocoel expansion ceases (Adelson and Humphreys, 1988). A recent review (Ettensohn and Ingersoll, 1992) concludes that it is likely that both these mechanisms expand the blastocoel. During early cleavage, the adhesion to the hyaline layer appears to be the most important factor, while at later stages, the osmotic pressure also seems to play a role.

The blastula cells develop cilia on their outer surfaces (Figure 5.6), thereby causing the blastula to rotate within the fertilization envelope. Soon after, the cells of the animal half of the embryo synthesize and secrete a **hatching enzyme** that enables them to digest the fertilization membrane (Lepage et al., 1992). The embryo is now a free-swimming **hatched blastula.**

Amphibians

Cleavage in most frog and salamander embryos is radially symmetrical and holoblastic, just like echinoderm cleavage. The amphibian egg, however, contains much more yolk. This yolk, which is concentrated in the vegetal hemisphere, is an impediment to cleavage. Thus, the first division begins at the animal pole and slowly extends down into the vegetal region (Figure 5.7). In the axolotl salamander, the cleavage furrow extends through the animal hemisphere at a rate close to 1 mm/min. The cleavage furrow bisects the gray crescent and then slows down to a mere 0.02–0.03 mm/min as it approaches the vegetal pole (Hara, 1977).

Figure 5.8A is a scanning electron micrograph showing the first cleavage in a frog egg. One can see the folds in the cleavage furrow and the difference between the furrows in the animal and vegetal hemispheres. Figure 5.8B shows that while the first cleavage furrow is still trying to cleave the yolky cytoplasm of the vegetal hemisphere, the second cleavage has already started near the animal pole. This cleavage is at right angles to the first one and is also meridional. The third cleavage, as expected, is equatorial. However, because of the vegetally placed yolk, this cleavage furrow in amphibian eggs is much closer to the animal pole. It divides the frog embryo into four small animal blastomeres (micromeres) and four large blastomeres (macromeres) in the vegetal region. This unequal holoblastic cleavage establishes two major embryonic regions: a rapidly dividing region of micromeres near the animal pole and a more slowly dividing macromere area (Figure 5.8C). As cleavage progresses, the animal region becomes packed with numerous small cells, while the vegetal region contains only a relatively small number of large, yolk-laden macromeres (see Figure 5.7).

Amphibian embryos containing 16 to 64 cells are commonly called **morulae** (from the Latin, meaning "mulberry," whose shape they vaguely resemble). At the 128-cell stage, the blastocoel becomes apparent, and the

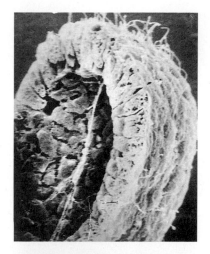

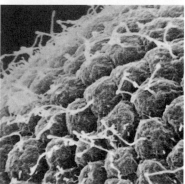

Figure 5.6
Ciliated blastula cells. Each cell develops a single cilium. (Courtesy of W. J. Humphreys.)

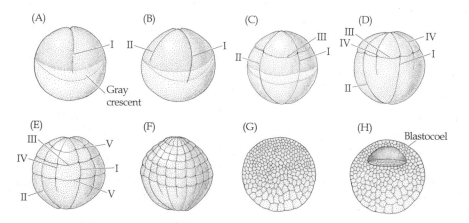

Figure 5.7
Cleavage of a frog egg. Cleavage furrows, designated by Roman numerals, are numbered in order of appearance. (A,B) The vegetal yolk impedes the cleavage such that the second division begins in the animal region of the egg before the first division has divided the vegetal cytoplasm. (C) The third division is displaced toward the animal pole. (D–H) The vegetal hemisphere ultimately contains longer and fewer blastomeres than the animal half. (After Carlson, 1981.)

(A)

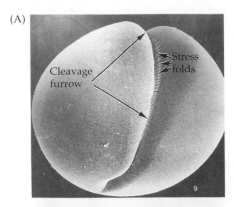

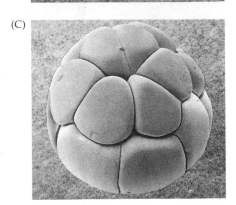

(B)

(C)

Figure 5.8
Scanning electron micrographs of the cleavage of a frog egg. (A) First cleavage. (B) Second cleavage (4 cells). (C) Fourth cleavage (16 cells), showing the size discrepancy between the animal and vegetal cells after arising from the third division. (A from Beams and Kessel, 1976, courtesy of the authors; B and C courtesy of L. Biedler.)

embryo is considered a blastula. Actually, the formation of the blastocoel has been traced back to the very first cleavage furrow. Kalt (1971) demonstrated that in the frog *Xenopus laevis,* the first cleavage furrow widens in the animal hemisphere to create a small intercellular cavity that is sealed off from the outside by tight intercellular junctions (Figure 5.9). This cavity expands during subsequent cleavages to become the blastocoel.

The blastocoel probably serves two major functions in frog embryos: (1) it is a cavity that permits cell migration during gastrulation, and (2) it prevents the cells beneath it from interacting prematurely with the cells above it. When Nieuwkoop (1973) took embryonic newt cells from the roof of the blastocoel and placed them next to the yolky vegetal cells from the base of the blastocoel, these animal cells became mesoderm tissue instead of ectoderm. Because mesodermal tissue is normally formed from those animal cells that are adjacent to the endoderm precursors, it seems plausible that the vegetal cells influence adjacent cells to differentiate into mesodermal tissues. Thus, the blastocoel appears to prevent the contact of endoderm with those cells fated to give rise to the skin and nerves.

While these cells are dividing, numerous cell adhesion molecules keep the blastomeres together. One of the most important of these molecules is **EP-cadherin**. The mRNA for this protein is supplied in the oocyte cytoplasm, and if this message is destroyed (by injecting into the oocyte antisense oligonucleotides complementary to this mRNA), the EP-cadherin is not made, and the adhesion between the blastomeres is dramatically reduced (Heasman et al., 1994). This results in the obliteration of the blastocoel (Figure 5.10).

Figure 5.9
Formation of the blastocoel in a frog egg. (A) First cleavage plane, showing a small cleft, which later develops into the blastocoel. (B) 8-cell embryo showing a small blastocoel (arrow) at the junction of the three cleavage planes. (From Kalt, 1971, courtesy of M. R. Kalt.)

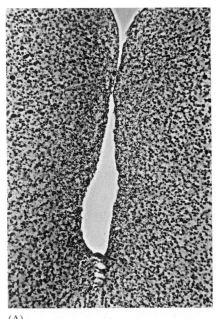

(A)

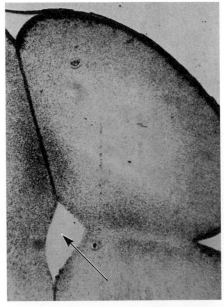

(B)

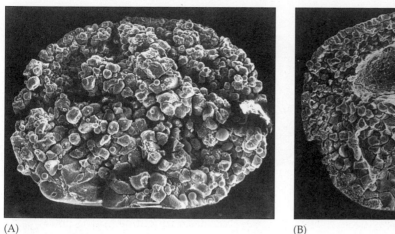

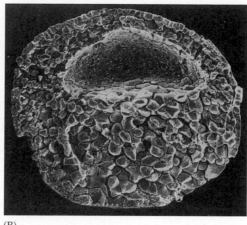

(A)   (B)

Figure 5.10
Depletion of EP-cadherin mRNA in the *Xenopus* oocyte, resulting in the loss of adhesion between blastomeres and the obliteration of the blastocoel. Antisense oligonucleotides complementary to the EP-cadherin message were injected into the one-cell embryo, preventing the expression of EP-cadherin. (A) The blastocoel is obliterated in EP-cadherin depleted embryos, but (B) not in controls. (From Heasman et al., 1994; photographs courtesy of J. Heasman.)

Spiral holoblastic cleavage

Spiral cleavage is characteristic of annelid worms, turbellarian flatworms, nemertean worms, and all molluscs except cephalopods. It differs from radial cleavage in numerous ways. First, the eggs do not divide in parallel or perpendicular orientations to the animal-vegetal axis of the egg; rather, cleavage is at oblique angles, forming the "spiral" arrangement of daughter blastomeres. Second, the cells touch each other at more places than do those of radially cleaving embryos. In fact, they take the most thermodynamically stable packing orientation, much like that of adjacent soap bubbles (Figure 5.11). Third, spirally cleaving embryos usually undergo fewer divisions before they begin gastrulation, making it possible to know the fate of each cell of the blastula. When the fates of the individual cells from annelid, flatworm, and mollusc embryos were compared, the same cells were seen in the same places, and their general fates were identical (Wilson, 1898). The blastulae so produced have no blastocoel and are called **stereoblastulae.**

Figures 5.12 and 5.13 depict the cleavage of mollusc embryos. The first two cleavages are nearly meridional, producing four large macromeres (labeled A, B, C, and D). In many species, the blastomeres are different sizes (D being the largest), a characteristic that allows them to be individually identified. In each successive cleavage, each macromere buds off a small micromere at its animal pole. Each successive quartet of micromeres is displaced to the right or to the left of its sister macromere, creating the spiral relationship characteristic of the cleavage. Looking down on the embryo from the animal pole, the upper ends of the mitotic spindle appear to alternate clockwise and counterclockwise. This causes alternate micromeres to form obliquely to the left and to the right of their macromere. At the third cleavage, the A macromere gives rise to two daughter cells, macromere 1A and micromere 1a. The B, C, and D cells behave similarly, producing the first quartet of micromeres. In most species, the micromeres are to the right of their macromeres (looking down on the animal pole), an arrangement indicating a dextral (as opposed to sinistral) spiral. At the fourth cleavage, the

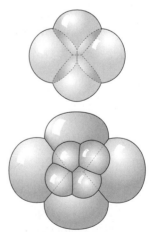

Figure 5.11
Diagram showing the arrangement of four and eight soap bubbles in a slightly concave dish. The thermodynamic arrangement maximizes contact and is very reminiscent of spirally cleaving embryos. (After Morgan, 1927.)

(A) View from animal pole

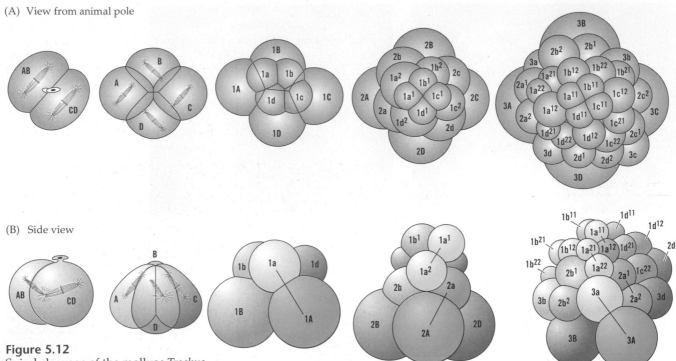

(B) Side view

Figure 5.12
Spiral cleavage of the mollusc *Trochus* viewed (A) from the animal pole and (B) from one side. In (B), the cells derived from the A blastomere are in color. The mitotic spindles, sketched in the early stages, divide the cells unequally and at an angle to the vertical and horizontal axes.

1A macromere divides to form macromere 2A and micromere 2a; and micromere 1a divides to form two more micromeres, $1a^1$ and $1a^2$. Further cleavage yields blastomeres 3A and 3a from the 2A macromere; and micromeres such as $1a^2$ divide to produce cells such as $1a^{21}$ and $1a^{22}$.

The orientation of the cleavage plane to the left or to the right is controlled by cytoplasmic factors within the oocyte. This was discovered by analyzing mutations of snail coiling. Some snails have their coils opening to the right of their shells, whereas other snails have their coils opening to the left. Usually, the rotation of coiling is the same for all members of a given species. Occasionally, though, mutants are found. For instance, in the species in which the coils open on the right, some individuals will be found with coils that open on the left. Crampton (1894) analyzed the embryos of such aberrant snails and found that their early cleavage differed from the norm.

Figure 5.13
Spiral cleavage of the snail *Ilyanassa.* The D blastomere is larger than the others, allowing the identification of each cell. Cleavage is dextral. (A) 8-cell stage. PB is polar body. (B) Mid-fourth cleavage; the macromeres have already divided into large and small spirally oriented cells. (From Craig and Morrill, 1986, courtesy of the authors.)

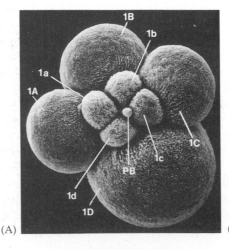

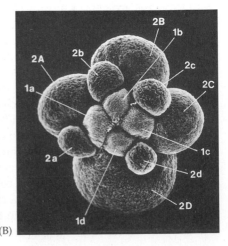

(A) Left-handed coiling

(B) Right-handed coiling

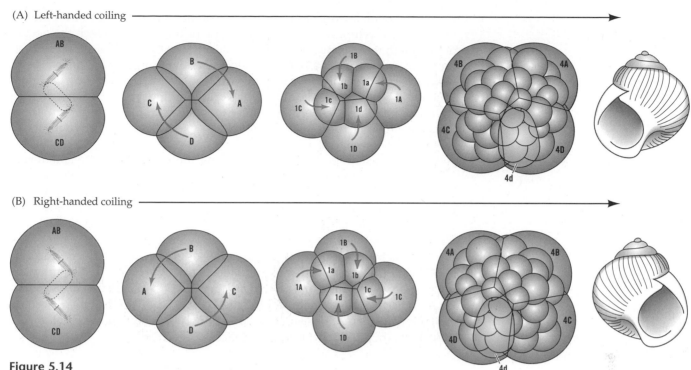

Figure 5.14
Looking down on the animal pole of right-handed and left-handed snails. The origin of right-handed and left-handed snail coiling can be traced to the orientation of the mitotic spindle at second cleavage. The left-handed (A) and right-handed (B) snails develop as mirror images of each other. (After Morgan, 1927.)

The orientation of the cells after second cleavage was different (Figure 5.14) owing to a different orientation of the mitotic apparatus in the sinistrally coiling snails. All subsequent divisions in left-coiling embryos are mirror images of those of dextrally coiling embryos. In Figure 5.14, one can see that the position of the 4d blastomere (which is extremely important, as its progeny will form the mesodermal organs) is different in the two types of spiraling embryos. Eventually, the two snails are formed with their bodies on different sides of the coil opening.

The direction of snail shell coiling is controlled by a single pair of genes (Sturtevant, 1923; Boycott et al., 1930). In the snail *Limnaea peregra*, most individuals are dextrally coiled. Rare mutants exhibiting left-handed coiling were found and mated with wild-type snails. These matings showed that there is a "right-handed" allele *D*, which is dominant to the "left-handed" allele *d*. However, the direction of cleavage is determined not by the genotype of the developing snail but by the genotype of the snail's mother. A *dd* female snail can produce only sinistrally coiling offspring, even when the offspring's genotype is *Dd*. A *Dd* individual will coil either left or right, depending on the genome of its mother. The matings produce a chart like this:

		Genotype	Phenotype
DD♀ × dd♂	→	*Dd*	All right-coiling
DD♂ × dd♀	→	*Dd*	All left-coiling
Dd × Dd	→	1*DD*:2*Dd*:1*dd*	All right-coiling

The genetic factors involved in snail coiling are brought to the embryo in the oocyte cytoplasm. It is the genotype of the *ovary* in which the oocyte develops that determines which orientation the cleavage will take. When Freeman and Lundelius (1982) injected a small amount of cytoplasm from dextrally coiling snails into the eggs of *dd* mothers, the resulting embryos coiled to the right. Cytoplasm from sinistrally coiling snails did not affect the right-coiling embryos. This confirmed the view that the wild-type mothers were placing a factor into their eggs that was absent or defective in the *dd* mothers. [cleave1.html]

Another exciting discovery concerning molluscan cleavage is that certain blastomeres communicate with each other. In molluscs with equal-sized blastomeres at the 4-cell stage,* the determination of which cell will give rise to the mesodermal precursor cell is accomplished between the fifth and sixth cleavages. At this time, the 3D macromere extends inward and contacts the micromeres of the animal pole. Without this contact, the 4d cell given off by the 3D macromere does not produce mesoderm (van den Biggelaar and Guerrier, 1979). By injecting low-molecular weight-dyes, de Laat and co-workers (1980) demonstrated that at the time of contact (and not before), small molecules are capable of diffusing between the 3D macromere and the central micromeres. Transmission electron microscopy shows that at this time, gap junctions appear on the surfaces of these cells.

*Don't worry about mollusc embryos with unequal-sized blastomeres at the 4-cell stage. We will deal with them in Chapter 16.

Sidelights & Speculations

Adaptation by Modifying Embryonic Cleavage

EVOLUTION is caused by the hereditary alteration of embryonic development. Sometimes we are able to identify a modification of embryogenesis that has enabled the organism to survive in an otherwise inhospitable environment. One such modification, discovered by Frank Lillie in 1898, is brought about by altering the typical pattern of spiral cleavage in the unionid family of clams.

Unlike most clams, *Unio* and its relatives live in swift-flowing streams. Streams create a problem for the dispersal of larvae: because the adults are sedentary, the free-swimming larvae would always be carried downstream by the current. These clams, however, have solved this problem by effecting two changes in their development. The first alters embryonic cleavage. In the typical cleavage of molluscs, either all the macromeres are equal in size or the 2D blastomere is the largest cell at that embryonic stage. However, the division of this *Unio* is such that the 2d blastomere gets the largest amount of cytoplasm (Figure 5.15). This cell divides to produce most of the larval structures, including a gland capable of producing a massive shell. These larvae (called *glochidia*) resemble tiny bear traps; they have sensi-

Figure 5.15 Formation of glochidium larvae by the modification of spiral cleavage. After the 8-cell embryo is formed (A), placement of the mitotic spindle causes most of the D cytoplasm to enter the 2d blastomere (B). This large 2d blastomere divides (C) to eventually give rise to the large "bear-trap" shell of the larva (D). (After Raff and Kaufman, 1983.)

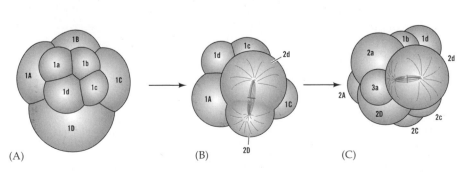

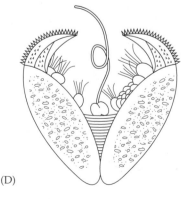

(A)　　　(B)　　　(C)　　　(D)

tive hairs that cause the valves of the shell to snap shut when they are touched by the gills or fins of a wandering fish. They "hitchhike" with the fish until they are ready to drop off and metamorphose into adult clams. In this manner, they can spread upstream.

In some species, glochidia are released from the female's brood pouch and merely wait for a fish to come wandering by. Some other species, such as *Lampsilis ventricosa,* have increased the chances of its larvae finding a fish by yet another modification of their development (Welsh, 1969). Many clams develop a thin mantle that flaps around the shell and surrounds the brood pouch. In some unionids, the shape of the brood pouch (marsupium) and the undulations of the

Figure 5.16   Phony fish atop the unionid clam *Lampsilis ventricosa.* The "fish" is actually the brood pouch and mantle of the clam. (Photograph courtesy of J. H. Welsh.)

mantle mimic the shape and swimming behavior of a minnow. To make the deception all the better, they develop a black "eyespot" on one end and a flaring "tail" on the other. The "fish" seen in Figure 5.16 is not a fish at all, but the brood pouch and mantle of the clam beneath it. When a predatory fish is lured within range, the clam discharges the glochidia from the brood pouch. Thus, the modification of existing developmental patterns has permitted unionid clams to survive in difficult environments. ■

Bilateral Holoblastic Cleavage

Bilateral holoblastic cleavage is found primarily in ascidians (tunicates). Figure 5.17 shows the cleavage pattern of a tunicate, *Styela partita.* The most striking phenomenon in this type of cleavage is that the first cleavage plane establishes the only plane of symmetry in the embryo, separating the embryo into its future right and left sides. Each successive division orients itself to this plane of symmetry, and the half-embryo formed on one side of the first cleavage is the mirror image of the half-embryo on the other side. The second cleavage is meridional, like the first division; but unlike the first division, it does not pass through the center of the egg. Rather, it creates two large anterior cells (the A and D blastomeres) and two smaller posterior cells (blastomeres B and C). Each side now has a large and a small blastomere. During the next three divisions, differences in cell size and shape highlight the bilateral symmetry of these embryos. At the 32-cell stage, a small blastocoel is formed and gastrulation begins.

As mentioned in Chapter 4, certain tunicates (including *S. partita*) contain colored cytoplasmic regions. During cleavage, these plasms become partitioned into different cells. Moreover, the type of cytoplasm the cells receive determines their eventual fates. Cells receiving clear cytoplasm become ectoderm; those containing yellow cytoplasm give rise to mesodermal cells; the cells that incorporate the slate-gray inclusions become endoderm;

Figure 5.17
Bilateral symmetry in a tunicate egg. (A) Uncleaved egg, showing the fates of various cytoplasmic regions. (B) 8-cell embryo, showing the blastomeres and the fates of various cells. It can be viewed as two 4-cell halves; from here on, each division on the right side of the embryo has a mirror-image division on the left. (C,D) Views of later embryos from the vegetal pole. (The regions of cytoplasm destined to form particular organs are labeled in A and are coded by color throughout the diagram.) (After Balinsky, 1981.)

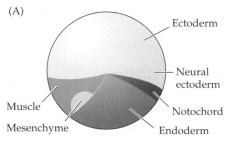

(A)

Ectoderm

Neural ectoderm

Muscle

Mesenchyme

Notochord

Endoderm

Vegetal pole

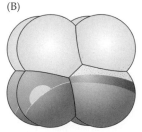

(B)

Vegetal pole

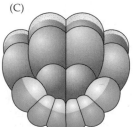

(C)

View from vegetal pole

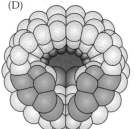

(D)

and the light gray cells become the neural tube and notochord. These colored plasms are localized bilaterally around the plane of symmetry, so they are bisected by the first cleavage furrow into the right and left halves of the embryo. The second cleavage causes the prospective mesoderm to lie in the two posterior cells, while the prospective neural and chordamesoderm are to be formed from the two anterior cells. The third division further partitions these cytoplasmic regions such that the mesoderm-forming cells are confined to the two vegetal posterior blastomeres, and the chordamesoderm cells are likewise restricted to the two vegetal anterior cells. The fate of each cell of the early *Styela* embryo has been followed and will be discussed in detail in Chapter 13.

Rotational holoblastic cleavage

It is not surprising that mammalian cleavage has been the most difficult to study. Mammalian eggs are among the smallest in the animal kingdom, making them hard to manipulate experimentally. The human zygote, for instance, is only 100 μm in diameter—barely visible to the eye and less than one-thousandth the volume of a *Xenopus* egg. Also, mammalian zygotes are not produced in numbers comparable to sea urchin or frog embryos. Usually, fewer than 10 eggs are ovulated by a female at a given time, so it is difficult to obtain enough material for biochemical studies. As a final hurdle, the development of mammalian embryos is accomplished within another organism rather than in the external environment. Only recently has it been possible to duplicate some of these internal conditions and observe development in vitro.

With all these difficulties, knowledge of mammalian cleavage was worth waiting for, as mammalian cleavage turned out to be strikingly different from most other patterns of embryonic cell division. The mammalian oocyte is released from the ovary and swept by the fimbriae into the oviduct (Figure 5.18). Fertilization occurs in the ampulla of the oviduct, a region close to the ovary. Meiosis is completed at this time, and first cleavage begins about a day later. Cleavages in mammalian eggs are among the slowest in the animal kingdom—about 12–24 hours apart. Meanwhile, the cilia in the oviduct push the embryo toward the uterus; the first cleavages occur along this journey.

There are several features of mammalian cleavage that distinguish it from other cleavage types. The first feature is the relative slowness of the divisions. The second fundamental difference is the unique orientation of

Figure 5.18
Development of a human embryo from fertilization to implantation. Compaction in human embryos takes place at day 4, when it is at the 10-cell stage. The egg "hatches" from the zona upon reaching the uterus, and the zona probably prevents the cleaving cells from sticking to the oviduct rather than traveling to the uterus. (After Tuchmann-Duplessis et al., 1972.)

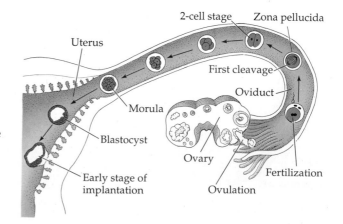

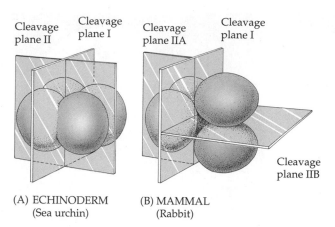

Cleavage plane II Cleavage plane I Cleavage plane IIA Cleavage plane I

Cleavage plane IIB

(A) ECHINODERM
(Sea urchin)

(B) MAMMAL
(Rabbit)

Figure 5.19
Comparison of early cleavage in (A) echinoderms (radial cleavage) and (B) mammals (rotational cleavage). Nematodes also have a rotational form of cleavage, but they do not form the blastocyst structure characteristic of mammals. Details of nematode cleavage will be given in Chapter 13. (After Gulyas, 1975.)

mammalian blastomeres with relation to one another. The first cleavage is a normal meridional division; however, in the second cleavage one of the two blastomeres divides meridionally and the other divides equatorially (Figure 5.19). This type of cleavage is called **rotational cleavage** (Gulyas, 1975).

The third major difference between mammalian cleavage and that of most other embryos is the marked asynchrony of early division. Mammalian blastomeres do not all divide at the same time. Thus, mammalian embryos do not increase evenly from 2- to 4- to 8-cell stages, but frequently contain odd numbers of cells. Also, unlike almost all other animal genomes, the mammalian genome is activated during early cleavage, and the genome produces the proteins necessary for cleavage to occur. In the mouse and goat, the switch from maternal to zygotic control occurs at the 2-cell stage (Piko and Clegg, 1982; Prather, 1989). [cleave2.html]

Compaction

Perhaps the most crucial difference between mammalian cleavage and all other types involves the phenomenon of **compaction.** As seen in Figure 5.20, mammalian blastomeres through the 8-cell stage form a loose arrangement with plenty of space between them. Following the third cleavage, however,

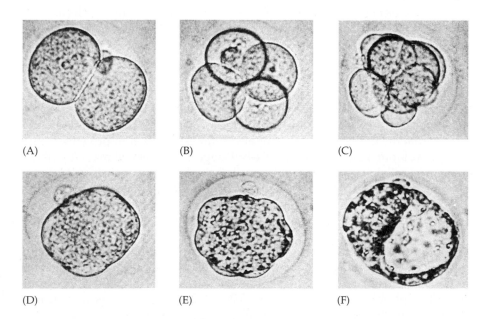

(A) (B) (C)

(D) (E) (F)

Figure 5.20
The cleavage of a single mouse embryo in vitro. (A) 2-cell stage. (B) 4-cell stage. (C) Early 8-cell stage. (D) Compacted 8-cell stage. (E) Morula. (F) Blastocyst. (From Mulnard, 1967, courtesy of J. G. Mulnard.)

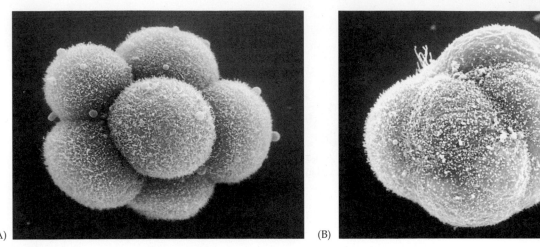

(A)   (B)

Figure 5.21
Scanning electron micrograph of (A) uncompacted and (B) compacted 8-cell mouse embryos. (Courtesy of C. Ziomek.)

the blastomeres undergo a spectacular change in their behavior. They suddenly huddle together, maximizing their contact with the other blastomeres and forming a compact ball of cells (Figures 5.20C,D and 5.21). This tightly packed arrangement is stabilized by tight junctions that form between the outside cells of the ball, sealing off the inside of the sphere (Figure 5.22). The cells within the sphere form gap junctions, thereby enabling small molecules and ions to pass between the cells.

The cells of the compacted embryo divide to produce a 16-cell morula. This morula consists of a small group of internal cells surrounded by a larger group of external cells (Barlow et al., 1972). Most of the descendants of the external cells become the **trophoblast** (**trophectoderm**) cells. This group of cells produces no embryonic structures. Rather, these cells form the tissue of the chorion, the embryonic portion of the placenta. The chorion enables the fetus to get oxygen and nourishment from the mother. It also secretes hormones so that the mother's uterus will retain the fetus and produces regulators of the immune response so that the mother will not reject the embryo as she would an organ graft. However, trophoblast cells are not able to produce any cell of the embryo itself. They are necessary for implanting the embryo in the uterine wall (Figure 5.23).

The mouse embryo is derived from the descendants of the inner cells of the 16-cell stage, supplemented by cells dividing from the trophoblast during the transition to the 32-cell stage (Pedersen et al., 1986; Fleming, 1987). These cells generate the **inner cell mass** that will give rise to the embryo and its associated yolk sac, allantois, and amnion. These cells not only look different from the trophoblast cells, but also synthesize different proteins at this early developmental stage. By the 64-cell stage, the inner cell mass (approximately 13 cells) and the trophoblast cells have become separate cell layers, neither contributing cells to the other group (Dyce et al., 1987; Fleming 1987). Thus, the distinction between trophoblast and inner cell mass blastomeres represents the first differentiation event in mammalian development.

Initially, the morula does not have an internal cavity. However, during a process called **cavitation,** the trophoblast cells secrete fluid into the morula to create a blastocoel. The inner cell mass is positioned on one side of the ring of trophoblast cells (see Figures 5.20, 5.22, and 5.23). This structure is called the **blastocyst** and is another hallmark of mammalian cleavage.

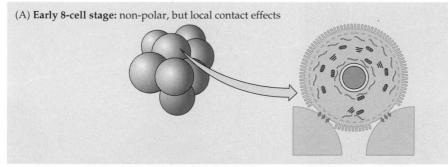

(A) **Early 8-cell stage:** non-polar, but local contact effects

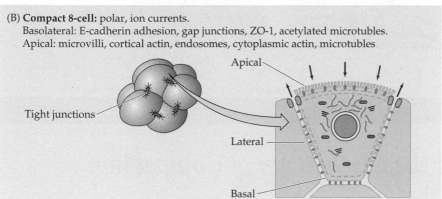

(B) **Compact 8-cell:** polar, ion currents.
  Basolateral: E-cadherin adhesion, gap junctions, ZO-1, acetylated microtubules.
  Apical: microvilli, cortical actin, endosomes, cytoplasmic actin, microtubles

Apical

Tight junctions

Lateral

Basal

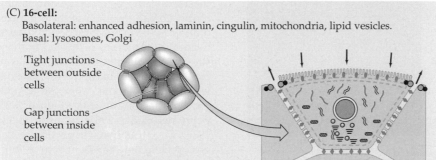

(C) **16-cell:**
  Basolateral: enhanced adhesion, laminin, cingulin, mitochondria, lipid vesicles.
  Basal: lysosomes, Golgi

Tight junctions
between outside
cells

Gap junctions
between inside
cells

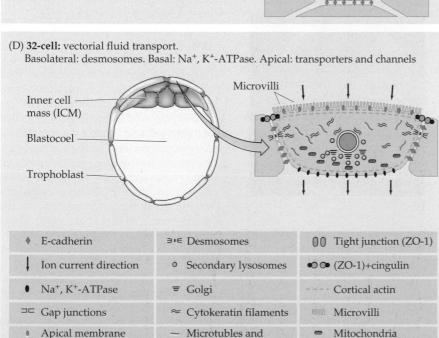

(D) **32-cell:** vectorial fluid transport.
  Basolateral: desmosomes. Basal: Na$^+$, K$^+$-ATPase. Apical: transporters and channels

Microvilli

Inner cell
mass (ICM)

Blastocoel

Trophoblast

◆ E-cadherin	϶ι∈ Desmosomes	〖〗 Tight junction (ZO-1)
↓ Ion current direction	○ Secondary lysosomes	●○●● (ZO-1)+cingulin
● Na$^+$, K$^+$-ATPase	≡ Golgi	--- Cortical actin
⊐⊏ Gap junctions	≈ Cytokeratin filaments	⅏ Microvilli
◊ Apical membrane proteins	— Microtubules and cytoplasmic actin	⬡ Mitochondria

Figure 5.22
Compaction and the formation of the mouse blastocyst. (A,B) 8-cell embryo. (C) 16-cell morula. (D) 32-cell blastocyst. The left side represents the entire organism or its cross section. The right side details the changes associated with the maturation of the trophoblast. (Right-hand figures after Fleming, 1992.)

Figure 5.23
Implantation of mammalian blastocysts into the uterus. (A) Mouse blastocysts entering uterus. (B) Initial implantation of the blastocyst onto the uterus in a rhesus monkey. (A from Rugh, 1967; B courtesy of the Carnegie Institution of Washington, Chester Reather, photographer.)

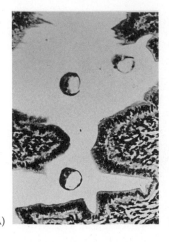

(A)

(B)

Sidelights & Speculations

The Cell Surface and the Mechanism of Compaction

COMPACTION CREATES the circumstances that bring about the first differentiation in mammalian development: the separation of trophoblast from inner cell mass. How is this done? There is growing evidence that compaction is mediated by events occurring at the cell surfaces of adjacent blastomeres. In the first stage of compaction, each of the eight blastomeres interacts with its neighbors to undergo **membrane polarization.** Different components of the cell surface migrate to different regions of the cell (see Figure 5.22; Ziomek and Johnson, 1980). This can be seen by tagging certain cell-surface molecules with fluorescent dyes. One such tag, which recognizes a class of glycoproteins, shows that at the 4-cell stage these glycoproteins are randomly distributed throughout the membrane (Figure 5.24A). However, at the mid-8-cell stage, these molecules are found predominantly at the poles farthest away from the center of the aggregate (Figure 5.24B). Membrane polarization is influenced by cell-cell interactions because it takes place only when the cells are in contact with at least one other blastomere. If a blastomere is separated from the rest of the embryo, it loses its polarization.

Specific cell surface proteins play a role in compaction. One such molecule, **E-cadherin** (also known as uvomorulin),

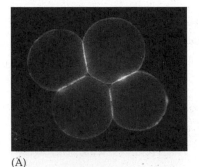

(A)

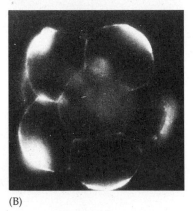

(B)

Figure 5.24 Polarization of membrane components in 8-cell mouse blastomeres. (A) Homogeneous, nonpolar distribution of membrane components labeled with fluorescent concanavalin A at the 4-cell stage. (B) Heterogeneous, polar distribution of these components at the 8-cell stage. (A from Fleming et al., 1986; B from Levy et al., 1986. Photographs courtesy of the authors.)

a 120-kDa adhesive glycoprotein, is synthesized at the 2-cell stage and is uniformly spread throughout the cell membrane. However, as compaction occurs, E-cadherin becomes restricted to those sites on cell membranes that are in contact with adjacent blastomeres. Antibodies to this molecule cause the decompaction of the morula (Figure 5.25; Peyrieras et al., 1983; Johnson et al., 1986). The carbohydrate portion of this glycoprotein may be essential to its function, as tunicamycin (a drug that inhibits the glycosylation of proteins) also prevents compaction.

Recent experiments have shown that the phosphatidylinositol pathway may also be important for initiating compaction. If 4-cell mouse embryos are placed into media containing drugs that activate protein kinase C, premature compaction occurs. Similarly, diacylglycerides can transiently cause these 4-cell embryos to undergo compaction. When this occurs, the E-cadherin accumulates specifically at the junctions between the blastomeres (Winkel et al., 1990). These results suggest that the activation of protein kinase C may initiate compaction by shifting the localization of E-cadherin.

Finally, the cell membrane may also be modified during compaction by cytoskeletal reorganization. Microvilli, extended by actin microfilaments, appear on adjacent cell surfaces and attach one cell to the other. These microvilli may be the sites where E-cadherin is functioning to mediate intercellular adhesion. The flattening of the blastomeres against one another may therefore be brought about by the shortening of the microvilli through actin depolymerization (Pratt et al., 1982; Sutherland and Calarco-Gillam, 1983).

Thus, there is growing evidence that compaction is caused by changes in the architecture of the blastomere cell surface. It is not certain, though, how these events relate to one another or how they are coordinated into the integrated network of events that causes compaction. ■

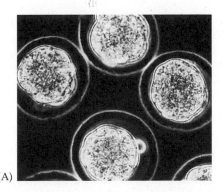 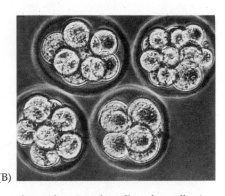

(A) (B)

Figure 5.25 Prevention of compaction by antiserum directed against the cell-surface adhesion glycoprotein E-cadherin. (A) Normal compaction occurring in the absence of antiserum. (B) Proliferation without compaction occurring in the presence of antibodies to E-cadherin. (Photographs courtesy of C. Ziomek.)

Formation of the Inner Cell Mass

The creation of an inner cell mass distinct from the trophoblast is *the* crucial process of early mammalian development. How is a cell directed into one or the other of these paths? How is a cell informed that it is either to give rise to a portion of the adult mammal or to give rise to a rather remarkable supporting tissue that will be discarded at birth? Observations of living embryos suggest that this momentous decision is merely a matter of a cell's being in the right place at the right time. Up through the 8-cell stage, there are no obvious differences in the biochemistry, morphology, or potency of any of the blastomeres. However, compaction forms inner and outer cells with vastly different properties. By labeling the various blastomeres, numerous investigators have found that the cells that happen to be on the outside will form the trophoblast, whereas the cells that happen to be inside will generate the embryo (Tarkowski and Wróblewska, 1967; Sutherland et al., 1990).* Hillman and co-workers (1972) have shown that when each blastomere of a 4-cell mouse embryo is placed on the outside surface of a mass of aggregated blastomeres, the external, transplanted cells only give rise to trophoblast tissue. Therefore, it seems that whether a cell becomes trophoblast or embryo depends on whether it was an external or an internal cell after compaction.

Escape from the Zona Pellucida

While the embryo is moving through the oviduct en route to the uterus, the blastocyst expands within the zona pellucida (the extracellular matrix of the egg that had been essential for sperm binding during fertilization). The cell membranes of the trophectoderm cells contain a sodium pump (a Na^+/K^+-ATPase) facing the blastocoel, and the proteins pump sodium ions into the central cavity. This accumulation of sodium ions draws in water osmotically, thus enlarging the blastocoel (see Figure 5.22; Borland, 1977; Wiley, 1984). During this time, it is essential that the zona pellucida prevent the blastocyst from adhering to the oviduct walls. When such adherence does take place in

---

*The inner cells have been found to come most frequently from the first cell to divide at the 2-cell stage. This cell usually produces the first pair of blastomeres to reach the 8-cell stage, and these cells usually divide in such a way that they are inside the loosely aggregated cluster of blastomeres (Graham and Kelly, 1977).

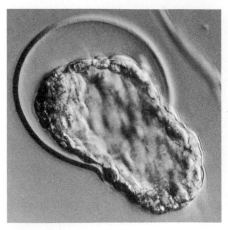

Figure 5.26
Mouse blastocyst hatching from the zona pellucida. (Photograph from Mark et al., 1985, courtesy of E. Lacy.)

humans, it is called an **ectopic** or **tubal pregnancy.** This is a dangerous condition because the implantation of the embryo into the oviduct can cause life-threatening hemorrhage. When the embryo reaches the uterus, however, it must "hatch" from the zona so that it can adhere to the uterine wall.

The mouse blastocyst hatches from the zona by lysing a small hole in it and squeezing through that hole as the blastocyst expands (Figure 5.26). A trypsin-like protease, **strypsin,** is located on the cell membrane and lyses a hole in the fibrillar matrix of the zona (Perona and Wassarman, 1986; Yamazaki and Kato, 1989). Once out, the blastocyst can make direct contact with the uterus. The uterine epithelium "catches" the blastocyst on an extracellular matrix containing collagen, laminin, fibronectin, hyaluronic acid, and heparan sulfate receptors. The trophoblast cells contain the integrins that will bind to the uterine collagen, fibronectin, and laminin, and they synthesize the heparan sulfate proteoglycan dramatically at the time just prior to implantation (see Carson et al., 1993). Once on the uterine epithelial cells, the trophoblast secretes another set of proteases, including **collagenase, stromelysin,** and **plasminogen activator.** These protein-digesting enzymes digest the extracellular matrix of the uterine tissue, enabling the blastocyst to bury itself within the uterine wall (Strickland et al., 1976; Brenner et al., 1989).

Sidelights & Speculations

Twins and Embryonic Stem Cells

THE EARLY CELLS of the embryo can replace each other and compensate for a missing cell. This was first shown in 1952, when Seidel destroyed one cell of a 2-cell rabbit embryo and showed that the remaining cell could produce the entire embryo. Once the inner cell mass (ICM) has become separate from the trophoblast, the ICM cells consititute an "**equivalence group.**" Here, each cell of the ICM has the same potency (in this case, each cell can give rise to all the cell types of the embryo, but not to the trophoblast), and their respective fates will be determined by interactions between their descendants. Gardiner and Rossant (1976) have also shown that if cells of the inner cell mass (but not trophoblast cells) are injected into blastocysts, they contribute to the new embryo. Since its blastomeres can generate any cell type in the body, the inner cell mass has sometimes been referred to as the **pluriblast** (Johnson and Selwood, 1996).

Human twins are classified into two major groups: **monozygotic** (one-egg, or identical) twins and **dizygotic** (two-egg, or fraternal) twins. Fraternal twins are the result of two separate fertilization events, whereas identical twins are formed from a single embryo whose cells somehow dissociated from one another. Identical twins are probably produced by the separation of early blastomeres or even by the separation of the inner cell mass into two regions within the same blastocyst. [cleave3.html]

Identical twins occur in roughly 0.25 percent of human births. About 33 percent of identical twins have two complete and separate chorions, indicating that separation occurred before the formation of the trophoblast tissue at day 5 (Figure 5.27A). The remaining identical twins share a common chorion, suggesting that the split occurred within the inner cell mass after the trophoblast formed. By day 9, the human embryo has completed the construction of anoth-

er extraembryonic layer, the amnion. This tissue forms the amniotic sac (or water sac), which surrounds the embryo with amniotic fluid and protects it from desiccation and abrupt movement (see Chapter 6). If the separation of the embryo were to come after the formation of the chorion on day 5 but before the formation of the amnion on day 9, then the resulting embryos should have one chorion and two amnions (Figure 5.27B). This happens in about two-thirds of human identical twins. A small percentage of identical twins are born within a single chorion and amnion (Figure 5.27C). This means that the division of the embryo came after day 9. Such newborns are at risk of being conjoined ("Siamese") twins. [cleave4.html]

The ability to produce an entire embryo from cells that normally would have contributed to only a portion of the embryo is called *regulation* and is discussed in Chapter 15. Regulation is also seen in the ability of two or more early

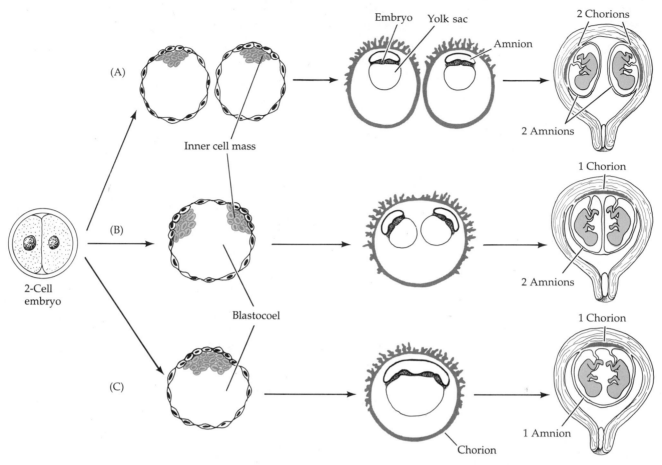

Figure 5.27 Diagram showing the timing of human monozygotic twinning with relation to extraembryonic membranes. (A) Splitting occurs before the formation of the trophectoderm, so each twin has its own chorion and amnion. (B) Splitting occurs after trophectoderm formation but before amnion formation, resulting in twins having individual amniotic sacs but sharing one chorion. (C) Splitting after amnion formation leads to twins in one amniotic sac and a single chorion. (After Langman, 1981).

embryos to form one **chimeric mouse** rather than twins, triplets, or a multi-headed monster. Chimeric mice are the result of two or more early-cleavage (usually 4- or 8-cell) embryos that have been artificially aggregated to form a composite embryo. As shown in Figure 5.28A, the zonae pellucidae of two genetically different embryos are removed and the embryos brought together to form a common blastocyst. These prepared blastocysts are implanted into the uterus of the foster mother. When they are born, the chimeric offspring have some cells from each embryo. This is readily seen when the aggregated blastomeres come from mouse strains that differ in their coat colors. When blastomeres from white and black strains are aggregated, the result is commonly a mouse with black and white bands (Figure 5.28B). There is even evidence (de la Chappelle et al., 1974; Mayr et al., 1979) that human embryos can form chimeras. These individuals have two genetically different cell types (XX and XY) within the same body, each with its own set of genetically defined characteristics. The simplest explanation for such a phenomenon is that these individuals resulted from the aggregation of two embryos, one male and one female, that were developing at the same time. If this explanation were correct, then two fraternal twins fused to create a single composite individual.

Markert and Petters (1978) have shown that three early 8-cell embryos can unite to form a common compacted morula (Figure 5.29) and that the resulting mouse can have the coat colors of the three different strains (Plate 21). Moreover, they showed that each of the three embryos gave rise to precursors of the gametes. When a chimeric (black/brown/white) female mouse was mated to a white-furred (recessive) male, the offspring were each of the three colors.

According to our observations of twin formation and chimeric mice, each blastomere of the inner cell mass should be able to produce any cell of the body. This hypothesis has been confirmed, and it will have very important consequences

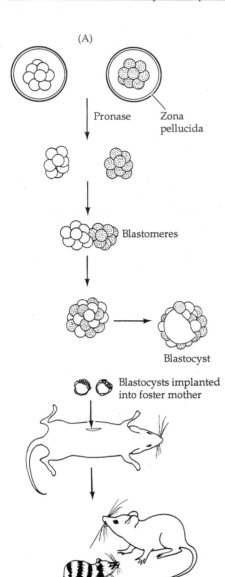

(A)

Pronase Zona
pellucida

Blastomeres

Blastocyst

Blastocysts implanted
into foster mother

(B)

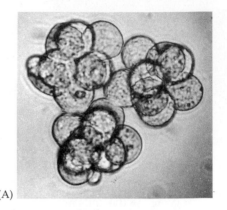

(A)

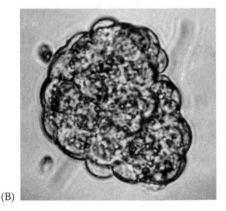

(B)

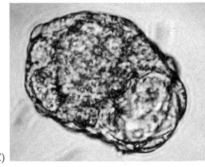

(C)

Figure 5.28 Production of chimeric mice.
(A) The experimental procedures used to
produce chimeric mice. Early 8-cell embryos
of genetically distinct mice (here, those with
coat-color differences) are isolated from
mouse oviducts and brought together after
their zonae are removed by proteolytic
enzymes. The cells form a composite blastocyst, which is implanted into the uterus of a
foster mother. (B) An adult chimeric mouse
showing contributions from the pigmented
(black) and unpigmented (white) embryos.
(Photograph courtesy of B. Mintz.)

Figure 5.29 Aggregation and compaction of
three 8-cell mouse embryos to form a single
compacted morula. Cells from three different
embryos (A) are aggregated together to form
a morula (B), which undergoes compaction
to form a single blastocyst (C). The resulting
chimeric mouse is shown in Plate 21. (From
Markert and Petters, 1978 courtesy of C.
Markert.)

in studying mammalian development.
When inner mass cells are isolated and
grown under certain conditions, they
remain undifferentiated and continue to
divide in culture (Evans and Kaufman,
1981; Martin, 1981). These cells are called
embryonic stem cells (**ES cells**). As
shown in Chapter 2, these cells can be
altered in the petri dish. Cloned genes
can be inserted into their nuclei, or the
existing genes can be mutated. When
these ES cells are injected into blastocysts
of another mouse embryo, the ES cells
can integrate into the host inner cell
mass. The resulting embryo has cells
coming from both the host and the donor
tissue. This technique has become
extremely important in determining the
function of genes during mammalian
development. ■

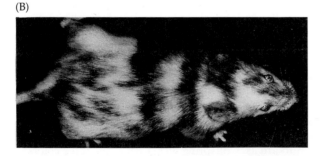

Meroblastic cleavage

As mentioned earlier, yolk concentration plays an important role in cell
cleavage. Nowhere is this more apparent than in the meroblastic cleavage
types. Here, the large concentrations of yolk prohibit cleavage in all but a
small portion of the egg cytoplasm. In **discoidal cleavage,** cell division is

limited to a small disc of yolk-free cytoplasm atop a mound of yolk; in **superficial cleavage,** the centrally located yolk permits cleavage only along the peripheral rim of the egg.

Discoidal Cleavage

Discoidal cleavage is characteristic of birds, fishes, and reptiles.

BIRDS. Figure 5.30 shows the cleavage of an avian egg. The bulk of the oocyte is taken over by the yolk, allowing cleavage to occur only in the **blastodisc,** a region of active cytoplasm about 2–3 mm in diameter at the animal pole of the egg. Because these cleavages do not extend into the yolky cytoplasm, the early-cleavage cells are actually continuous at their bases. The first cleavage furrow appears centrally in the blastodisc, and other cleavages follow to create a single-layered blastoderm. At first, this cellular layer is incomplete, as the cells are still continuous with underlying yolk. Thereafter, equatorial and vertical cleavages divide the blastoderm into a tissue five to six cell layers thick. These cells become linked together with tight junctions (Bellairs et al., 1975; Eyal-Giladi, 1991). Between the blastoderm and the yolk is a space called the **subgerminal cavity.** This space is created when the blastoderm cells absorb fluid from the albumin ("egg white") and secrete it between themselves and the yolk (New, 1956). At this stage, the deep cells in the center of the blastoderm are shed to create a one-cell-thick **area pellucida.** (The shed cells appear to die.) The peripheral ring of blastoderm cells that are not shed constitute the **area opaca.**

By the time a hen has laid an egg, the blastoderm contains some 60,000 cells. Some of these cells are delaminated into the subgerminal cavity to form a second layer (Figure 5.31). Thus, soon after laying, the chick egg contains two layers of cells: the upper **epiblast** and the lower **hypoblast.** Between them lies the blastocoel. We will detail the formation of the hypoblast in the next chapter.

FISHES. In recent years, the zebrafish, *Danio rerio*, has become a favorite organism of those who wish to study vertebrate development. These fish have large broods, breed all year, are easily maintained, have transparent embryos that develop outside the mother (an important feature for microscopy), and can be raised so that mutants can be readily screened and propagated. In addition, they develop rapidly, so that at 24 hours after fertilization, the embryo has formed most of its tissue and organ primordia and displays the characteristic tadpole-like form (see Granato and Nüsslein-Volhard, 1996; Langeland and Kimmel, 1997).

The yolky eggs of fishes develop similarly to those of birds, with cell division occurring only in the animal pole blastodisc. Scanning electron micrographs of fish egg cleavage show beautifully the incomplete nature of dis-

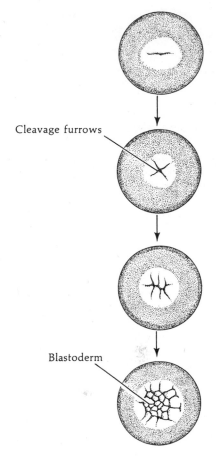

Cleavage furrows

Blastoderm

Figure 5.30
Discoidal cleavage in a chick egg, viewed from the animal pole. The cleavage furrows do not penetrate the yolk, and a blastoderm consisting of a single layer of cells is produced.

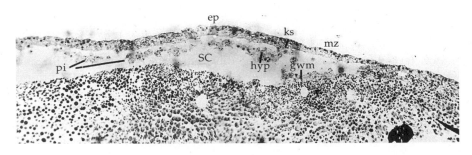

Figure 5.31
Formation of the two-layered chick embryo. This sagittal section of the embryo near the posterior margin shows an upper layer consisting of a central epiblast that trails into the cells of Koller's sickle (ks) and the posterior marginal zone (mz). Certain cells from the epiblast fall (delaminate) from the upper layer to form polyinvagination islands (pi) of 5 to 20 cells each. These cells will be joined by those hypoblast cells (hyp) migrating anteriorly from Koller's sickle to form the lower (hypoblastic) layer. (Sc is subgerminal cavity; gwm is germ wall margin.) (From Eyal-Giladi et al., 1992, courtesy of H. Eyal-Giladi.)

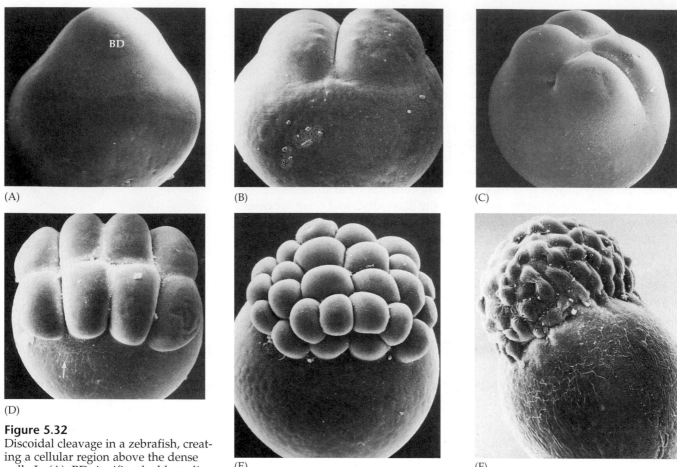

(A)

(B)

(C)

(D)

(E)

(F)

Figure 5.32
Discoidal cleavage in a zebrafish, creating a cellular region above the dense yolk. In (A), BD signifies the blastodisc region. (From Beams and Kessel, 1976, courtesy of the authors.)

coidal cleavage (Figure 5.32). As in amphibian and sea urchin embryos, early-cleavage divisions follow a highly reproducible pattern of meridional and equatorial cleavages. These divisions are rapid, with a periodicity of about 15 minutes each. The first 12 divisions occur synchronously, forming a mound of cells that sits at the animal pole of a large **yolk cell.** Initially, all the cells maintain some open connection with each other and with the underlying yolk cell so that moderately sized (17-kDa) molecules pass freely from one blastomere to the next (Kimmel and Law, 1985). Beginning about the tenth cell division, the onset of the midblastula transition can be detected: zygotic gene transcription begins, cell divisions slow, and cell movement is evident (Kane and Kimmel, 1993).

At this time, two distinct cell populations can be distinguished. The first of these is the **yolk syncytial layer (YSL).** The YSL is formed at the ninth or tenth cell cycle, when the cells at the vegetal edge of the blastoderm fuse with the underlying yolk cell. This produces a ring of nuclei within that part of the yolk cell cytoplasm that sits just beneath the blastoderm. As the blastoderm expands vegetally to surround the yolk cell, some of the yolk syncytial nuclei will move under the blastoderm, to form the **internal YSL**, and some of the nuclei will move vegetally, staying ahead of the blastoderm margin, to form **external YSL** (Figure 5.33A,B). No clear function is yet known for the YSL.

The second cell population distinguished at the midblastula transition is the **enveloping layer (EVL;** see Figure 5.33A). These are the most superficial

(A)

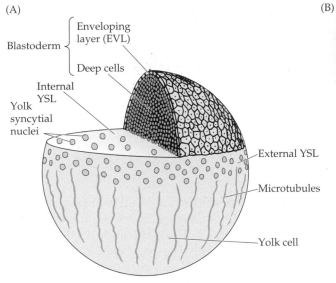

Blastoderm
Enveloping layer (EVL)
Deep cells
Internal YSL
Yolk syncytial nuclei
External YSL
Microtubules
Yolk cell

(C)

Animal pole

Nose, eye
Epidermis
Brain
Neural crest
Spinal cord
Ectoderm
Somite muscle
Head
Mesoderm
Ventral
Pronephros
Noto-chord
Dorsal
Blood Fins Heart Muscle
Intestine Liver
Pharynx
Endoderm
Blastoderm margin
Yolk cell

Vegetal pole

(B)

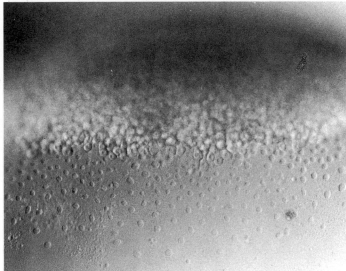

Figure 5.33
The zebrafish blastula. (A) Prior to gastrulation, deep cells are surrounded by the EVL. The animal surface of the yolk cell is flat and contains the nuclei of the YSL. Microtubules extend through the yolky cytoplasm and through the external region of the YSL. (B) Late blastula stage *Fundulus* embryo, showing the external YSL. The nuclei of these cells were derived from cells at the margin of the blastoderm, which released their nuclei into the yolky cytoplasm. (C) Fate map of the deep cells after cell mixing has stopped. The lateral view is shown, and not all organ fates are labeled (for the sake of clarity). The map is generated by injecting cells with a high-molecular-weight dye and then determining which organs the dye-laden cells have generated. (A and C after Langeland and Kimmel, 1996; B from Trinkaus, 1993, courtesy of the author.)

cells of the blastoderm, and the EVL is an epithelial sheet a single cell layer thick. The EVL eventually forms the periderm, an extraembryonic protective covering that is thought to be sloughed off during later development.

Between the outer EVL and the inner YSL are the **deep cells.** These are the cells that give rise to the embryo proper. The fates of the early blastoderm cells are not determined, and cell lineage studies (where a nondiffusible fluorescent dye is injected into one of the cells and the descendants of that cell can be followed) show that there is much cell mixing during cleavage. Moreover, any one cell can give rise to an unpredictable variety of tissue descendants (Kimmel and Warga, 1987; Helde et al., 1994). The fate of the blastoderm cells appears to be fixed shortly before gastrulation begins. At this time, cells in specific regions of the embryo give rise to certain tissues in a highly predictable manner, allowing a fate map to be made (Figure 5.33C; Kimmel et al, 1990).

The processes by which cells contribute to tissues appear to involve a progressive narrowing of possible developmental fates for any given cell. This can be seen in some of the earliest cells to have their fates established—the heart precursor cells (Stainier et al., 1993; Lee et al., 1994). An individual

Figure 5.34
Fate mapping of the deep cells of the zebrafish blastula. Injection of a single blastomere at the early blastula (256- to 512-cell) stage with rhodamine dextran. If the cell were near the margin halfway between the dorsal and ventral poles, the cell's progeny is restricted to forming part of the heart. At this stage, the labeled cells form descendants that can populate both the atrium and the ventricle. If the injection to such a cell were done at a later blastula stage, its descendants populate only one chamber. (After Stainier et al., 1993.)

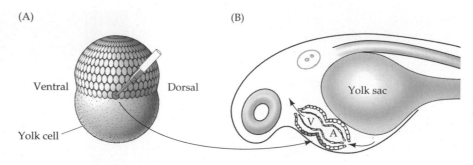

cell at the margin midway between the future dorsal and ventral surfaces of the midcleavage-stage embryo can give rise to cells that populate both the endocardium and the myocardium (Figure 5.34A,B). Slightly later, the descendants of any one cell can populate only the myocardium or the endocardium. Later still, the descendants of a cell will only be able to populate specific subcompartments of the tissue. For example, midblastula cells might contribute progeny to both the atrium and ventricle of the heart.

Superficial Cleavage

Most insect eggs undergo superficial cleavage, whereby a large mass of centrally located yolk confines cleavage to the cytoplasmic rim of the egg. One of the fascinating details of this cleavage type is that the cells do not form until after the nuclei have divided. Cleavage of an insect egg is shown in Figure 5.35. The zygote nucleus undergoes several mitotic divisions within the central portion of the egg. In *Drosophila*, 256 nuclei are produced by a series of nuclear divisions averaging 8 minutes each. The nuclei then migrate to the periphery of the egg, where the mitoses continue, albeit at a progressively slower rate. The embryo is now called a **syncytial blastoderm,** meaning that

Figure 5.35
Superficial cleavage in a *Drosophila* embryo. The numeral above each embryo corresponds to the number of minutes after deposition of the egg; the numeral at the bottom indicates the number of nuclei present. Pole cells (which will form the germ cells) are seen at the 512-nuclei stage even though the cellular blastoderm does not form until nearly 3 hours later. The times are representative, as the duration of each division cycle depends, in part, on the temperature at which the egg is incubated.

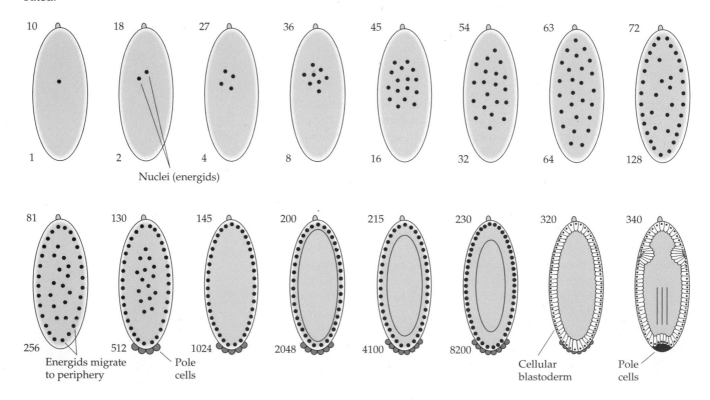

all the cleavage nuclei are contained within a common cytoplasm. No cell membranes exist other than that of the egg itself. Those nuclei migrating to the posterior pole of the egg soon become enveloped by new cell membranes to form the **pole cells** of the embryo. These cells give rise to the germ cells of the adult. Thus, one of the first events of insect development is the separation of the future germ cells from the rest of the embryo.

After the pole cells have been formed, the oocyte membrane folds inward between the nuclei, eventually partitioning off each somatic nucleus into a single cell (Figure 5.36). This creates the **cellular blastoderm,** with all the cells arranged in a single-layered jacket around the yolky core of the egg. Like any other cell formation, the formation of the cellular blastoderm involves a delicate interplay between microtubules and microfilaments. The first phase of blastoderm cellularization is characterized by the invagination of cell membranes and their underlying actin network in the regions between the nuclei. This process is inhibited by drugs that block microtubules. After the membranes and their actin have passed the level of the nuclei, the second phase of cellularization occurs. Here, the rate of invagination increases, and the actin-membrane complex begins to constrict at what will be the basal end of the cell (Schejter and Wieschaus, 1993; Foe et al., 1994). In *Drosophila,* this layer consists of approximately 6000 cells and is formed within 4 hours of fertilization. [cleave5.html]

Although the nuclei originally divide within a common cytoplasm, this does not mean that the cytoplasm is itself uniform. Karr and Alberts (1986) have shown that each nucleus within the syncytial blastoderm is contained within its own little territory of cytoskeletal proteins. When the nuclei reach the periphery during the tenth cleavage cycle, each nucleus becomes sur-

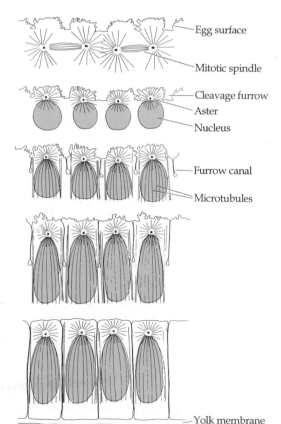

Egg surface

Mitotic spindle

Cleavage furrow

Aster

Nucleus

Furrow canal

Microtubules

Yolk membrane

Figure 5.36
Nuclear elongation and cellularization in *Drosophila* blastoderm. (After Fullilove and Jacobson, 1971.)

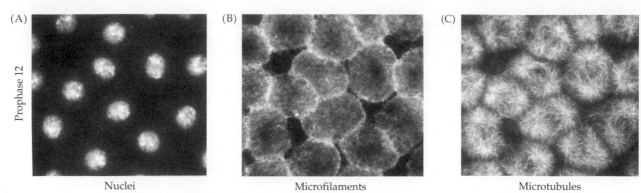

Prophase 12

(A) Nuclei   (B) Microfilaments   (C) Microtubules

Figure 5.37
Localization of the cytoskeleton around nuclei in the syncytial blastoderm of *Drosophila*. A *Drosophila* embryo entering the mitotic prophase of its twelfth division was sectioned and triple-stained. (A) The nuclei were localized by a dye that binds to DNA. (B) Microfilaments were identified using a fluorescent antibody to actin. (C) Microtubules were recognized by a fluorescent antibody to tubulin. Cytoskeletal domains can be seen surrounding each nucleus. (From Karr and Alberts, 1986, courtesy of T. L. Karr.)

rounded by microtubules and microfilaments. The nuclei and their associated cytoplasmic islands are called **energids.** Figure 5.37 shows the nuclei and their essential microfilament and microtubule domains in prophase of the twelfth mitotic division.

After the nuclei reach the periphery, the time required to complete each of the next four divisions becomes gradually longer. While cycles 1–10 are each 8 minutes long, cycle 13, the last cycle in the syncytial blastoderm, takes 25 minutes to complete. The *Drosophila* embryo forms cells in cycle 14 (i.e., after 13 divisions), and cycle 14 is asynchronous. Some groups of cells complete this cycle in 75 minutes, whereas other groups of cells take 175 minutes (Foe, 1989). Transcription from these nuclei (which begins around the eleventh cycle) is greatly enhanced. The slowdown of *Drosophila* nuclear division and the concomitant increase in RNA transcription is often referred to as the **midblastula transition.** Such a transition is also seen in the embryos of numerous vertebrate and invertebrate phyla. The control of this mitotic slowdown (in *Xenopus*, sea urchin, starfish, and *Drosophila* embryos) appears to be effected by the ratio of chromatin to cytoplasm (Newport and Kirschner, 1982a; Edgar et al., 1986). Edgar and his colleagues compared the early development of wild-type *Drosophila* embryos with those of a haploid mutant. The haploid *Drosophila* embryos have half the wild-type quantity of chromatin at each cell division. Hence a haploid embryo at the eighth cell cycle has the same amount of chromatin as a wild-type embryo has at cell cycle 7. These investigators found that whereas wild-type embryos formed their cell layer immediately after the thirteenth division, the haploid embryos underwent an extra, fourteenth, division before cellularization. Moreover, the lengths of cycles 11–14 in wild-type embryos corresponded to cycles 12–15 in the haploid embryos. Thus, the haploid embryos follow a pattern similar to that of the wild-type embryos—only they lag by one cell division.

If their lagging were due to the haploid mutants' having a chromatin-to-cytoplasm ratio of half the wild-type ratio at any given cycle, then one should be able to accelerate cellularization by tying off (ligating) some of the cytoplasm such that the nuclei divide in a smaller volume. When this ligation was performed, the mitotic pattern of the embryos was accelerated. The final blastoderm division, signaling the end of the cleavage period, is reached when there is one nucleus for each 61 μm^3 of cytoplasm. In *Xenopus,* a similar slowdown in mitotic rate is observed after the twelfth cell division. Here, too, the divisions thereafter become asynchronous. Ligation experiments suggest that the timing of this midblastula transition is also due to the chromatin-to-cytoplasm volume ratio (Newport and Kirschner, 1982a,b).

In both *Drosophila* and *Xenopus,* the initiation of transcription can be induced prematurely by artificially lengthening the cell cycle. When cycloheximide (an inhibitor of protein synthesis) delays cell division, the midblastula transition is induced early in *Xenopus,* and a burst of transcription occurs in *Drosophila* (Edgar et al., 1986; Kimelman et al., 1987).

Sidelights & Speculations

Exceptions, Generalizations, and Parasitic Wasp Cleavage

WHAT WE CONSIDER "normal" and what we marginalize as "exceptions" often reflect which animals are most readily accessible to study and most easily domesticated for laboratories. Needless to say, this does not necessarily reflect the condition of the natural world. Rather, our discussions of animal development are often bottlenecked through particular organisms. The development of amphibians is generally represented by *Xenopus laevis,* and the mouse and human are the only mammals whose development is usually studied. Similarly, although there are over 800,000 known species of insects, most developmental biologists know only the development of one species:

Drosophila melanogaster. Drosophila gained preeminence only after it was thought necessary to relate embryological phenomena to particular genes. In 1941, the major compendium of insect development (Johannsen and Butt's *Embryology of Insects and Myriapods*) didn't even mention this species in its index.

Insects are an exceptionally successful and widespread subphylum, however, so it is not surprising to find an enormous amount of variability in their development. The development of the parasitic wasp *Copidosomopsis tanytmemus* differs remarkably from that of the canonical *Drosophila.* Like several other parasitic species, the female *C. tanytmemus* deposits her egg inside the

egg of another species. As the host egg (usually that of a moth) is developing, so is the parasite's egg. However, while the host egg begins development in the usual superficial pattern, the wasp egg divides *holoblastically.* Moreover, instead of differentiating a body axis, the cells of the parasitic embryo divide repeatedly to become a mass of undifferentiated cells called a **polygerm.** By two weeks, the growing polygerm is suspended in the host, remaining loosely attached to the larval brain and trachea (Figure 5.38A; Cruz, 1986a).

Figure 5.38 Development of parasitic wasps (Encyrtidae). (A) Holoblastic cleavage of the *Copidosomopsis tanytmemus* egg produces a polygerm of undifferentiated cells. (B) Precocious larvae of a related genus, *Pentalitomastix,* attacks a larva of *Trathala* inside the same host. The photograph is of a freshly opened host. (A after Cruz, 1986a; B from Cruz, 1981, courtesy of Y. Cruz.)

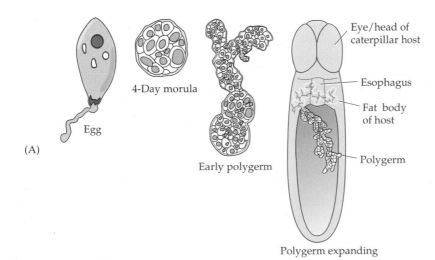

(A)
Egg
4-Day morula
Early polygerm
Eye/head of caterpillar host
Esophagus
Fat body of host
Polygerm
Polygerm expanding

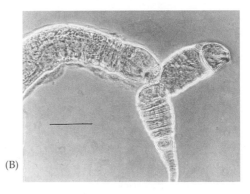

(B)

As the polygerm grows, it splits into dozens (sometimes thousands, depending on the species) of discrete groups of cells. Each of these groups of cells becomes an embryo! The polyembryonic wasp *Copidosoma floridanum* produces up to 2000 individuals from a single fertilized egg (Grbic et al., 1996). This ability of an egg to develop into a mass of cells that routinely forms numerous embryos is called **polyembryony.** (Polyembryony is characteristic of certain insect groups and certain mammalian species, such as the nine-banded armadillo, whose eggs routinely form identical quadruplets.) Most of these parasitic wasp embryos develop into normal wasp larvae that take about 30 days to develop. A smaller group, about 10 percent of the total number of embryos, become **precocious larvae** (Figure 5.38B), which develop within a week. Not only are they formed earlier, but precocious larvae have very little structure and do not undergo metamorphosis. They are essentially a mobile set of jaws. These larvae do not reproduce,

and they die by the time the normal larvae are formed. While they live, however, they go through the host embryo killing the parasitic larvae of other individuals (of different species and of other clones of the same species). In other words, the precocious larvae are predatory forms that kill possible competitors (Cruz, 1981, 1986b; Grbic and Strand, 1992).

As the precocious larvae (and their prey) die, the normal larvae emerge from their first molt, and they begin feeding voraciously on the host's larval organs. By 40 days, the parasitic brood has finished eating its host's muscles, fat bodies, gonads, silk glands, gut, nerve cord, and hemolymph, and the host is little more than a sac of skin holding about 70 pupating wasp larvae. After another 5 or 6 days, the new adults gnaw holes in the host's integument, and in a scene repeated in the movie *Alien,* chew their way out of the host's body. These adults then copulate (often on the body of their dead host), find another host in which to

deposit an egg, and die shortly thereafter.

Such a life cycle discomforted Charles Darwin and made him question the concept of a benign and all-knowing deity. In 1860, he wrote to the American biologist Asa Gray, "I cannot persuade myself that a benevolent and omnipotent God would have designedly created the Ichneumonidae with the express intention of their feeding within the living bodies of Caterpillars." However, in addition to their usefulness in provoking disquieting notions concerning natural order and the nature of "individuality," parasitic wasps may have important economic consequences. *Macrocentrus grandii* is a polyembryonic wasp that parasitizes the European corn borer. The ability of an insect to form from a holoblastically cleaving embryo should also encourage us to appreciate some of the plasticity of nature and discourage us from making sweeping generalizations about an entire subphylum of organisms. ■

■ MECHANISMS OF CLEAVAGE

Regulating the cleavage cycle

The cell cycle of somatic cells is functionally divided into four stages (Figure 5.39A). After mitosis (M), there is a prereplication gap (G_1), after which time the synthesis of DNA takes place (S). After the synthetic period, there is a premitotic gap (G_2), which is followed by mitosis. The progression of these phases is regulated by growth factors. In early-cleavage blastomeres, however, cell division can be much simpler. Early sea urchin blastomeres lack G_1, replicating their DNA during the last portion (telophase) of the previous mitosis (Hinegardner et al., 1964). *Xenopus* and *Drosophila* nuclei have eliminated both the G_1 and the G_2 phases during early cleavage. (*Xenopus* embryos add those phases to the cell cycle sometime after the twelfth cleavage. *Drosophila* adds G_2 during cycle 14 and G_1 during cycle 17.) For the first 12 divisions, *Xenopus* cells divide synchronously in a biphasic cell cycle: S to M and M to S (Figure 5.39B; Laskey et al., 1977; Newport and Kirschner, 1982a).

The factors regulating this biphasic cycle are located in the cytoplasm. Normal enlarging *Xenopus* oocytes are arrested in first meiotic prophase. They are unable to divide. If nuclei from dividing cells are transplanted into these oocytes, they also cease dividing. When normal oocytes are stimulated by progesterone, they resume their meiotic division and stop at the metaphase of the second meiosis. If nuclei from nondividing cells (such as neurons) are placed into the cytoplasm of progesterone-treated oocytes, they, too, initiate division and stop at metaphase (Gurdon, 1968). The cytoplasm

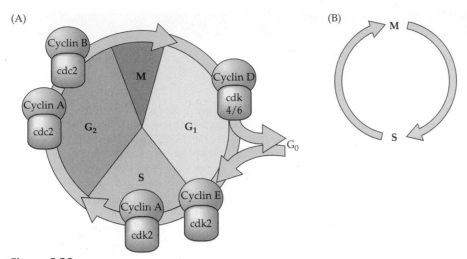

Figure 5.39
Cell cycles of somatic cells and early blastomeres. (A) Cell cycle of a typical somatic cell. Mitosis (M) is followed by an "interphase" condition. This latter period is subdivided into G_1, S (synthesis), and G_2 phases. Cells that are differentiating are usually taken "out" of the cell cycle and are in an extended G_1 phase called G_0. The cyclins and their respective kinases responsible for the progression through the cell cycle are shown at their point of cell cycle regulation. (B) Simpler biphasic cell cycle of the early amphibian blastomeres having only two states, S and M. (A after Nigg, 1995.)

of progesterone-stimulated oocytes still undergoes periodic cortical contractions (characteristic of division), even in the absence of nuclei or centrioles. If cloned DNA fragments are injected into such enucleated embryos, their replication comes under the control of these cycles (Hara et al., 1980; Harland and Laskey, 1980; Karsenti et al., 1984). Thus, the capacity for cell division is regulated by the cytoplasm.

Maturation-Promoting Factor

Some of the factors that govern DNA synthesis and cell division have been identified. The progesterone-induced factor that allows oocyte nuclei to resume their divisions is a two-subunit phosphoprotein called maturation-promoting factor (MPF; now sometimes called *mitosis-promoting factor,* thus keeping the same initials). MPF was first discovered as the major factor responsible for the resumption of meiotic cell divisions in the ovulated frog egg (Smith and Ecker, 1969; Masui and Markert, 1971). This same factor continues to play a role after fertilization, regulating the biphasic cell cycle of early *Xenopus* blastomeres. Gerhart and co-workers (1984) showed that MPF undergoes cyclical changes in level of activity in mitotic cells. The MPF activity of early frog blastomeres is highest during M and undetectable during S. During S phase, MPF exists in an inactive state. This cyclicity is also seen in enucleated blastomeres. Newport and Kirschner (1984) demonstrated that DNA replication (S) and mitosis (M) are driven solely by the gain and loss of MPF activity, even in the absence of protein synthesis. Cleaving cells can be trapped in S phase by incubating them in an inhibitor of protein synthesis. When MPF is microinjected into these cells, they enter M. Their nuclear envelope breaks down and their chromatin condenses into chromosomes. After an hour, MPF is degraded and the chromosomes return to S phase.

MPF and Its Regulators

The Small Subunit of MPF: The cdc2 Kinase (Cyclin-Dependent Kinase, CDK)

MPF has a large and a small subunit. The small subunit of MPF is a protein kinase that, when activated, can phosphorylate a variety of proteins. Thus, MPF functions by adding phosphate groups onto specific proteins. One such target is histone H1, which is bound to DNA. The phosphorylation of this protein may bring about chromosomal condensation. Another target is the nuclear envelope. Within 15 minutes after the addition of MPF, the three major proteins (the lamin proteins) of the nuclear envelope become hyperphosphorylated, and within the next 15 minutes, the envelope has depolymerized and is breaking apart (Miake-Lye and Kirschner, 1985; Arion et al., 1988). The purified MPF kinase subunit has been shown to phosphorylate these nuclear envelope proteins and to bring about their depolymerization in vitro (Peter et al., 1990; Ward and Kirschner, 1990). A third target appears to be RNA polymerase (Cisek and Corden, 1989), and the phosphorylation of RNA polymerase may be responsible for the inhibition of transcription during

mitosis. A fourth target of the kinase appears to be the regulatory subunit of cytoplasmic myosin. When this protein is phosphorylated, it becomes inactive and is unable to function as an ATPase to drive the actin filaments involved in cell division (Satterwhite et al., 1992). The inhibition of this myosin during the early stages of mitosis may prevent the division of the cell until after the chromosomes have separated.

The small subunit of MPF has been remarkably conserved through evolution and is almost identical to a mitosis-inducing phosphoprotein, p34, synthesized by the yeast cdc2 gene (Dunphy et al, 1988; Gautier et al., 1988). In fact, the human gene that encodes the protein corresponding to the small subunit of *Xenopus* MPF can be inserted into the

yeast genome and cause division in cdc2-deficient yeast mutants (Lee and Nurse, 1987). The p34 protein can exist in phosphorylated and unphosphorylated forms. The active form appears to be phosphorylated on threonine-161 (T-161) and unphosphorylated on tyrosine-15 (Y-15). Both these conditions are important for kinase activity (Gould and Nurse, 1989; Solomon, 1993).

The Large Subunit of MPF: Cyclin

How, then, is MPF regulated? Since *Xenopus* cleavage seemed to be regulated by a protein similar to that which regulates yeast cell division, it was thought that whatever regulated the yeast protein might have a counterpart in the animal embryo. One of the most important regulators of the yeast MPF-like protein

Figure 5.40 The development of cell cycle regulation in *Drosophila* embryogenesis. (A) Cyclin and cdc25 (string) protein are both abundant prior to fertilization. Therefore, during the first seven cell cycles, MPF kinase activity is constant and the nuclear divisions proceed as rapidly as the enzymes and substrates function. As cyclin becomes degraded, its synthesis (from mRNAs stored in the cytoplasm) becomes limiting at cycle 8. By cycle 14, the maternal mRNA for cyclin is gone, and it must be synthesized from nuclear genes. Moreover, the degradation of string protein mandates new synthesis from the nucleus. Pre-MPF accumulates but isn't activated until the string phosphatase cleaves the T-14 and Y-15 phosphates from the cdc2 kinase. The mechanisms that relate MPF activity to the completion of DNA synthesis and the initiation of cytokinesis are currently being investigated. (After Edgar et al., 1994.)

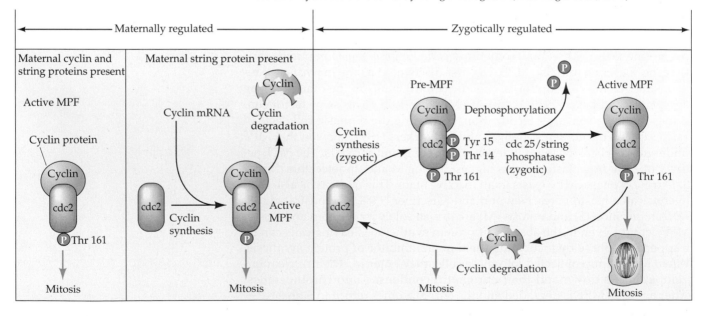

is the product of the *cdc13* gene, a 56-kDa protein called p56^{cdc13}. This gene was cloned, and the sequence of its encoded protein was found to be very similar to that of the cyclin B proteins found in numerous animals (Goebl and Byers, 1988; Solomon et al., 1988). Cyclin B proteins of cleavage-stage cells show a periodic behavior, accumulating during S phase and being degraded during mitosis (Evans et al., 1983; Swenson et al., 1986). Cyclins are often encoded by the mRNAs stored in the oocyte cytoplasm, and if their translation into proteins is selectively inhibited, the cell will not enter mitosis (Minshull et al., 1989). The cyclin B protein combines with the cdc2 kinase of MPF to create the MPF complex. The cyclin enables the cdc2 kinase subunit to become phosphorylated at residues threonine-14 (T-14), tyrosine-15 (Y-15), and threonine-161 (Figure 5.40). The phosphorylation at T-161 is necessary for MPF activity, but phosphorylations at T-14 and Y-15 inhibit it. Thus, when phosphorylated in these positions, the kinase remains inactive but potentially functional. The supply of potentially functional MPF molecules (pre-MPF) accumulates during the late S period.

The cdc25 Phosphatase: Initiator of Mitosis

Mitosis begins with the abrupt dephosphorylation of all these MPF kinase subunits at position 15. This is accomplished by the appearance of **cdc25 phosphatase** (Edgar and O'Farrell, 1989; Gautier et al., 1991; Jessus and Beach, 1992; Lee et al., 1992). In this manner, the gradual accumulation of MPF is converted into the sharp burst of kinase activity that initiates mitosis. This phosphatase (which has been found in numerous organisms) is itself developmentally regulated. In *Drosophila,* the cdc25 phosphatase (the product of the *string* gene) is initially synthesized from stored oocyte mRNA during the first 13 cell cycles. However, during the next cycle, the maternal *string* mRNA is degraded. If the nuclei do not transcribe their own *string* mRNA, then the cells will not divide. Edgar and O'Farrell (1989) have shown that those cells that do divide are synthesizing their own cdc25 phosphatase, while those cells that are not able to join the division cycle are not making it (Figure 5.41). This degradation and need for resynthesis of this protein would explain the switch from cytoplasmic to nuclear control of division seen in cycle 14.

In *Drosophila,* there is a developmental maturation of the regulation of active MPF kinase (see Figure 5.40; Edgar et al., 1994). At ovulation, the pre-MPF complexes stored in the egg are dephosphorylated at T-14 and Y-15 by the newly translated string (cdc25) protein. During the first seven nuclear cycles, active MPF remains at high levels, and the nucleus divides as fast as the DNA synthetic enzymes permit. During cycles 8–13, cyclin begins to be degraded at metaphase, leading to periodic fluctuations of the MPF kinase activity. Cyclin synthesis from stored oocyte mRNA becomes the rate-limiting step for mitosis. The degradation of the oocyte string protein leads to cell cycle arrest at the interphase of cycle 14. Large concentrations of pre-MPF accumulate. The mitoses for divisions 14, 15, and 16 are triggered only when this pre-MPF is dephosphorylated at positions T-14 and Y-15 by string protein. This protein is derived from nuclear transcription at the end of each G2 period. Mitosis has gone from cytoplasmic to nuclear control.

Other Cyclins and Cyclin-Dependent Kinases

MPF is the first discovered member of a family of dimeric proteins having very

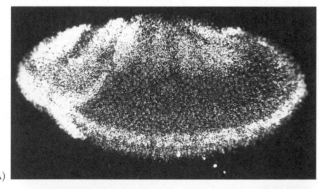

(A)

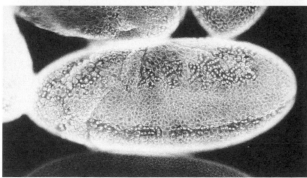

(B)

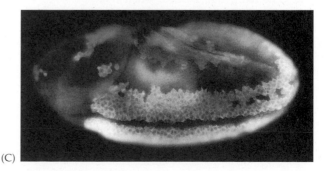

(C)

Figure 5.41 Correlation of *string* gene expression with cell division in *Drosophila* embryos. (A) In this example, a late stage-14 embryo is stained with a radioactive nucleotide sequence that specifically recognizes and binds *string* mRNA (seen here as the white dots in the autoradiograph). (B) A slightly older embryo is stained with fluorescent antibodies to tubulin to show the microtubules of the mitotic spindles. A comparison of the fluorescence photomicrograph and the autoradiograph obtained from the binding of the radioactive probe shows that only those cells capable of dividing synthesize *string* mRNA. (C) Antibodies to the cyclin A protein show that it is degraded after mitosis and is not seen in those regions containing string protein. (From Edgar and O'Farrell, 1989, courtesy of B. A. Edgar.)

similar structures. These proteins each contain a cyclin and a cyclin-dependent kinase. The cyclin-dependent kinase of MPF is called **cdk1.** At least seven other cyclin-dependent kinases are involved in mature vertebrate cells, and over a dozen cyclins have been identified. The roles of some of these cyclin-dependent kinases have been determined, as shown in Figure 5.39). Among these enzymes, one of the most critical is cyclin E/cdk2. While MPF (cyclin B/cdk1) is critical for the entry into mitosis (M), cyclin E/cdk2 is critical for the cell's ability to enter S phase. It allows DNA synthesis to occur. The developmental regulation of this protein is a critical step in *Drosophila* development.

Drosophila embryos add a G2 stage prior to mitosis when the string protein becomes limiting at cycle 14. They add their G1 phase at cycle 17 when **cyclin E** becomes the rate-limiting factor to DNA replication. In early *Drosophila* embryos, cyclin E and cdk2 are always present, their mRNA being supplied by the oocyte and translated throughout the first 15 division cycles. The message for cyclin E is degraded during cycle 16, leading to a deficiency of this protein in cycle 17. Thus, most *Drosophila* cells arrest in G1 of this cycle, not entering the period of DNA synthesis. There they begin differentiating. (The exceptions are the nerve precursor cells, which continue to proliferate, and gut cells, which continue to make DNA in the absence of cell division. In these cases, cyclin E is derived from the zygotic genes.) If cyclin E is induced ectopically, the arrested cells undergo a new round of DNA synthesis (Knoblich et al., 1994). It is thought that cyclin E controls DNA synthesis by phosphorylating certain transcription factors that regulate the transcription of proteins needed for DNA replication (Duronio and O'Farrell, 1994). Indeed, when cells normally exit the cell cycle to begin differentiating, they express the Dacapo protein, an inhibitor of cyclin E/cdk2 (Lane et al., 1996; de Nooij et al., 1996).

The regulation of cyclins is a critical function in development. First, imagine the cartilage cells of our legs undergoing one more cell division; we would be much larger than at present. Worse, imagine that this misregulation occurred in only one leg. Worse still, imagine if the division of the cartilage cells were not coordinated with the division of the skin and blood vessel cells. The regulation of these events is coordinated through hormones and growth factors that eventually regulate the cyclins that control passage through the cell cycle. Second, when a cyclin becomes active without external regulation or when cyclins become stimulated by mutated proteins, cell growth continues without outside control, and a tumor develops. In mature vertebrate cells, the cyclin D/cdk4,6 enzymes play a crucial role in development. In several cell types, they control the dichotomy between cell division and cell differentiation. [cleave6.html]

Checkpoints for Cell Division: DNA and Spindles

The cell cycle demands an exceptionally intricate choreography of cytokinesis, DNA replication, spindle assembly, and cell metabolism. In this ensemble, cyclins and cyclin-dependent kinases are both the targets of regulation and the effectors of regulation. The cyclin-kinase system appears to coordinate these events. For instance, the mitotic spindle fiber cannot form until cyclin B/cdk1 signals the beginning of mitosis; but the proper assembly of the spindle is necessary for the proper functioning of cyclin B (Minshull et al., 1994). If the spindles are incorrectly formed, cyclin B ceases to function, and mitosis ceases. There also appears to be feedback between the replicating chromatin and the cyclin-dependent kinases such that mitosis doesn't begin until the DNA has begun replicating, and so only one round of replication is usually permitted during cell division (Chong et al., 1995; Madine et al., 1995). The molecules that mediate these exchanges are now being studied.

Cytostatic Factor

The synthesis and degradation of MPF leads to the cycling of the cells. However, if cyclin degradation is prevented, MPF remains active, and the cell is frozen in metaphase (Murray et al., 1989). This is apparently what happens during frog oocyte development. The mature frog oocyte stops cell division by producing a protein called cytostatic factor (CSF), which keeps the oocytes arrested in metaphase of the second meiotic division (Figure 5.42). This protein contains the products of the *c-mos* and *cdk-2* genes, and it appears to act by blocking the degradation of cyclin (see Chapter 22). Since cyclin is not degraded, MPF remains active, and the oocyte remains in metaphase. The release of calcium ions during fertilization activates a protease that specifically inactivates CSF (Watanabe et al., 1991). When CSF is degraded, the cyclin can then be degraded, and the cell can return to S phase. Thus, one of the effects of the calcium ions released at fertilization is to initiate the degradation of cyclin and enable the cell to begin DNA replication. After this, the rhythms of cell division are controlled by MPF activity, which is in turn based on the

Figure 5.42 Levels of maturation-promoting factor (MPF) during the early development of the frog *Xenopus laevis*. The normal maturational signal is the hormone progesterone, which stimulates the ovulation of the oocytes and the initiation of meiosis. (After Murray and Kirschner, 1989.)

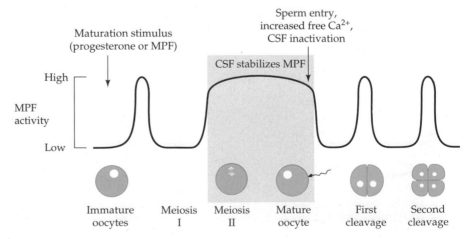

cyclical rhythms of cyclin synthesis and degradation. While the calcium ions are busy turning off mitosis, the fertilization signals that activate protein kinase C are establishing the conditions of interphase: chromatin decondensation and the reformation of the nuclear envelope (Bement and Capco, 1991). ■

The cytoskeletal mechanisms of mitosis

Cleavage is actually the result of two coordinated processes. The first of these cyclical processes is **karyokinesis**—the mitotic division of the nucleus. The mechanical agent of this division is the mitotic spindle, with its microtubules composed of tubulin (the same type of protein that makes the sperm flagellum). The second process is **cytokinesis**—the division of the cell. The mechanical agent of cytokinesis is the **contractile ring** of microfilaments made of actin (the same type of protein that extends the egg microvilli and the sperm acrosomal process). Table 5.2 presents a comparison of these systems of division. The relationship and coordination between these two systems during cleavage is depicted in Figure 5.43A, where a sea urchin egg is shown undergoing first cleavage. The mitotic spindle and contractile ring are perpendicular to each other, and the spindle is internal to the contractile ring. The cleavage furrow eventually bisects the plane of mitosis, thereby creating two genetically equivalent blastomeres.

The actin microfilaments are found in the cortex of the egg rather than in the central cytoplasm. Under the electron microscope, the ring of microfilaments can be seen forming a distinct cortical band 0.1 μm wide (Figure 5.43B). This contractile ring exists only during cleavage and extends 8–10 μm into the center of the egg. It is responsible for exerting the force that splits the zygote into blastomeres; for if it is disrupted, cytokinesis stops. Schroeder (1973) has proposed a model of cleavage wherein the contractile ring splits the egg like an "intercellular purse-string," tightening about the egg as cleavage continues. This tightening of the microfilamentous ring creates the cleavage furrow.

Although karyokinesis and cytokinesis are usually coordinated, they are sometimes separated by natural or experimental conditions. In insect eggs, karyokinesis occurs several times before cytokinesis takes place. Another way to cause this state is to treat embryos with the drug cytochalasin B. This drug inhibits the formation and organization of microfilaments in the contractile ring, thereby stopping cleavage without stopping karyokinesis (Schroeder, 1972). In some instances, nuclei continue to divide and express developmentally regulated proteins even if cleavage is blocked (Lillie, 1902; Whittaker, 1979).

Table 5.2 Karyokinesis and cytokinesis

Process	Mechanical agent	Major protein composition	Location	Major disruptive drug
Karyokinesis	Mitotic spindle	Tubulin microtubules	Central cytoplasm	Colchicine, nocodazole[a]
Cytokinesis	Contractile ring	Actin microfilaments	Cortical cytoplasm	Cytochalasin B

[a]Because colchicine has been found to independently inhibit several membrane functions, including osmoregulation and the transport of ions and nucleosides, nocodazole has become the major drug used to inhibit microtubule-mediated processes (see Hardin, 1987).

Figure 5.43
Role of microtubules and microfilaments in cell division. (A) Diagram of first-cleavage telophase. The chromosomes are being drawn to the centrioles by microtubules while the cytoplasm is pinched in by the contraction of microfilaments. (B) Localization of actin microfilaments in the cleavage furrow. Fluorescent labeling of the actin microfilaments shows the contractile ring in the first-cleavage furrow (arrow) of a telophase sea urchin egg. (C) Fluorescent labeling of tubulin shows the microtubular asters of a sea urchin egg during telophase of first cleavage. (B from Bonder et al., 1988; C from White et al., 1987.)

(A)

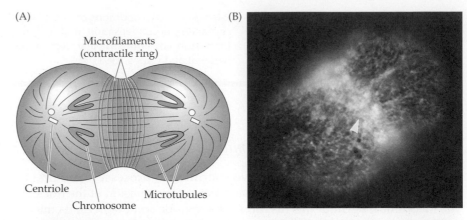

(B)

(C)

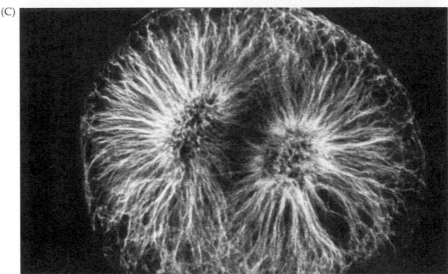

One of the most intriguing unsolved problems of embryonic cleavage is how cytokinesis and karyokinesis are coordinated with each other. In vitro work suggests that DNA replication may control the phosphorylation of the cdc2 kinase subunit of MPF and that this phosphorylation may control the ability of the actin to contract (Smythe and Newport, 1992). However, these observations may not be consistent with observations made in early embryonic cells (Ferrell et al., 1991). Meanwhile, the number and location of the cleavage furrows appear to be controlled by the microtubule asters. These asters (Figure 5.43C) are the microtubular "rays" that extend from the poles of the mitotic spindle to the cell periphery. Normal cleavage occurs only if a pair of asters is present (Wilson, 1901), and polyspermic eggs (obtaining centrioles from each sperm) form multiple cleavage furrows within the same egg (see Figure 4.20). In more recent studies, Raff and Glover (1989) showed that if centrioles migrate into the posterior pole of the *Drosophila* embryo, they can form pole cells even in the absence of nuclei. It thus appears that microtubular asters are the *sine qua non* of cleavage.

The second type of evidence linking asters with cleavage furrow formation comes from experiments in which the direction of cleavage is changed by placing the egg under pressure. Pflüger (1884) discovered that when a frog zygote is gently compressed between two glass plates, the directions of the first three cleavages are all perpendicular to the plane of the plates. Both Driesch and Morgan (reviewed in Morgan, 1927) made similar observations

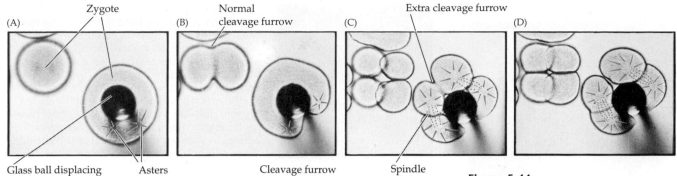

(A) Zygote — Glass ball displacing mitotic apparatus — Asters

(B) Normal cleavage furrow — Cleavage furrow interrupted

(C) Extra cleavage furrow — Spindle

(D)

Figure 5.44
Creation of a new cleavage furrow by displacement of the asters. (A,B) By interrupting the cleavage furrow with a glass ball, a horseshoe-shaped cell is created. At the next division (C,D), a new cleavage furrow is created even though there is no mitotic spindle across it. (From Rappaport, 1961, courtesy of R. Rappaport.)

with sea urchin embryos. In both cases, the plane of the third cleavage (which is normally parallel to the equator of the oocyte) was displaced by 90°. Thus, by changing the location of the mitotic spindle, one can alter the direction of the cleavage furrow.

Rappaport (1961) has extended this type of experiment by displacing the mitotic spindles to the sides of the cells. In Figure 5.44, a glass ball has been used to displace the asters from the center of the cell toward the periphery. The cleavage furrow that results extends only as far as the ball and does not appear on the other side. Thus, a binucleate, horseshoe-shaped cell is formed. At the next division, two spindle apparatuses form between four asters, but three cleavage furrows are generated! Each arm of the horseshoe has its own mitotic spindle and cleavage furrow as expected, but a third furrow appears between the two asters at the top of the arms (Figure 5.44C). This demonstrates clearly that if two asters are close enough together, their interactions cause the formation of a cleavage furrow even in the absence of a mitotic spindle between them. Here again we see that cell division can occur without nuclear division as long as the asters are present.

The formation of new membranes

Our last consideration of embryonic cleavage involves the formation of new cell membranes. Are these membranes newly synthesized or are they merely extensions of the oocyte cell membrane? The answer is probably that both mechanisms contribute to the internal cell membranes.

Amphibian embryos provide evidence that new membrane components are being synthesized during early cleavage. Figure 5.45A shows the first cleavage furrow of a pigmented frog zygote. Whereas the original membrane has a pigmented cortical region associated with it, the new membrane is white. This new membrane also has electrical conductance properties different from those of the original membrane. Byers and Armstrong (1986) have radiolabeled membrane components of newly fertilized *Xenopus* eggs and followed the redistribution of these molecules through cleavage by autoradiography. During the first cleavage, the membrane of the outer surface of the embryo and the membrane of the leading edge of the cleavage furrow are heavily labeled (showing them to be original membrane regions). Between them is a large region devoid of radioactive label (Figure 5.45B). Thus, the furrow membrane is a mosaic of different parts. The membrane at the leading end of the furrow is derived from the preexisting radiolabeled outer surface of the egg, but most of the membrane of the furrow has been derived from regions that were inaccessible to surface labeling. Byers and Armstrong speculate that the domain of heavily labeled membrane at the leading edge

(A)   New unpigmented membrane

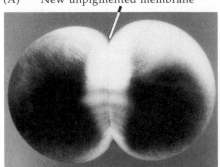

(B)

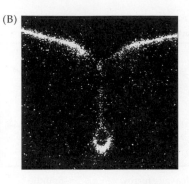

Figure 5.45
Formation of new membranes in the first cleavage of a *Xenopus* egg. (A) Old membrane has pigment granules associated with it. The new membrane appears clear, because it does not have these associated granules. (B) Autoradiograph of membrane proteins in the first-cleavage furrow. The cell surface was radiolabeled prior to division. White spots indicate regions containing proteins found in the zygote surface before division. (A from de Laat and Bluemink, 1974; B from Byers and Armstrong, 1986, courtesy of the authors.)

of the furrow contains the membrane anchors for the underlying ring of cortical microfilaments. The leading edge of the furrow in early cleavage *Xenopus* embryos also contains short, radially arranged, microtubules. It is thought that these microtubules might provide a path for the movement of membrane vesicles toward the place where they are inserted into the membrane (Danilchik and Funk, 1996).

The processes of cleavage divide the cytoplasm of the zygote into numerous cells. Each cell may have the same nuclear genes, but their respective cytoplasms may differ significantly. In the next chapter we will see how these blastomeres move about and interact with each other to lay down the framework of the body.

LITERATURE CITED

Adeslon, D. C. and Humphreys, T. 1988. Sea urchin morphogenesis and cell-hyalin adhesion are perturbed by a monoclonal antibody specific for hyalin. *Development* 104: 391-402.

Arion, D., Meijer, L., Brizuela, L. and Beach, D. 1988. *cdc2* is a component of the M phase-specific histone H1 kinase: Evidence for identity with MPF. *Cell* 55: 371–378.

Balinsky, B. I. 1981. *Introduction to Embryology*, 5th Ed. Saunders, Philadelphia.

Barlow, P., Owen, D. A. J. and Graham, C. 1972. DNA synthesis in the preimplantation mouse embryo. *J. Embryol. Exp. Morphol.* 27: 432–445.

Beams, H. W. and Kessel, R. G. 1976. Cytokinesis: A comparative study of cytoplasmic division in animal cells. *Am. Sci.* 64: 279–290.

Bellairs, R., Breathnach, A. S. and Gross, M. 1975. Freeze–fracture replication of junctional complexes in unincubated and incubated chick embryos. *Cell Tissue Res.* 162: 235–252.

Bement, W. M. and Capco, D. G. 1991. Parallel pathways of cell cycle control during *Xenopus* egg activation. *Proc. Natl. Acad. Sci. USA* 88: 5172–5176.

Bonder, E. M., Fishkind, D. J., Henson, J. H., Cotran, N. M. and Begg, D. A. 1988. Actin in cytokinesis: Formation of the contractile apparatus. *Zool. Sci.* 5: 699–711.

Borland, R. M. 1977. Transport processes in the mammalian blastocyst. *Dev. Mammals* 1: 31–67.

Boycott, A. E., Diver, C., Garstang, S. L. and Turner, F. M. 1930. The inheritance of sinestrality in *Limnaea peregra* (Mollusca: Pulmonata). *Philos. Trans. R. Soc. Lond.* [B] 219: 51–131.

Brenner, C. A., Adler, R. R., Rappolee, D. A., Pedersen, R. A. and Werb, Z. 1989. Genes for extracellular matrix-degrading metalloproteases and their inhibitor, TIMP, are expressed during early mammalian development. *Genes Dev.* 3: 848–859.

Byers, T. J. and Armstrong, P. B. 1986. Membrane protein redistribution during *Xenopus* first cleavage. *J. Cell Biol.* 102: 2176–2184.

Carlson, B. M. 1981. *Patten's Foundations of Embryology.* McGraw-Hill, New York.

Carson, D. D., Tang, J.-P. and Julian, J. 1993. Heparan sulfate proteoglycan (perlecan) expression by mouse embryos during acquisition of attachment competence. *Dev. Biol.* 155: 97–106.

Chong, J. P. J., Mahbubani, H. M., Khoo, C.-Y., and Blow, J. J. 1995. Purification of an MCM-containing complex as a component of the replication licensing system. *Nature* 375: 418–421.

Cisek, L. J. and Corden, J. L. 1989. Phosphorylation of RNA polymerase by the murine homologue of the cell-cycle control protein *cdc2*. *Nature* 339: 679–684

Craig, M. M. and Morrill, J. B. 1986. Cellular arrangements and surface topography during early development in embryos of *Ilyanassa obsoleta*. *Int. J. Invert. Reprod. Dev.* 9: 209–228.

Crampton, H. E. 1894. Reversal of cleavage in a sinistral gastropod. *Ann. N.Y. Acad. Sci.* 8: 167–170.

Cruz, Y. P. 1981. A sterile defender morph in a polyembryonic hymenopteran parasite. *Nature* 294: 446–447.

Cruz, Y. P. 1986a. Development of the polyembryonic parasite *Copidosomopsis tanytmemus* (Hymenoptera: Encyrtidae). *Ann. Entomol. Soc. Am.* 79: 121–127.

Cruz, Y. P. 1986b. The defender role of the precocious larvae of *Copidosomopsis tanytmemus* Caltagirone (Encyrtidae, Hymenoptera). *J. Exp. Zool.* 237: 309–318.

Dan, K. 1960. Cytoembryology of echinoderms and amphibia. *Int. Rev. Cytol.* 9: 321–367.

Danilchik, M. and Funk, C. 1996. Abstracts of the Sixth Internatl. *Xenopus* Conference. Wind Rivers Lodge, Estes Park, CO.

Dan-Sohkawa, M. and Fujisawa, H. 1980. Cell dynamics of the blastulation process in the starfish, *Asterina pectinifera*. *Dev. Biol.* 77: 328–339.

Darwin, C. 1860. Letter to Asa Gray, May 22, 1860. *In* F. Darwin (ed.), *The Life and Letters of Charles Darwin*, Vol. 2. Appleton, New York. [p.105] de Laat, S. W. and Bluemink, J. G. 1974. New membrane formation during cytokinesis in normal and cytochalasin B-treated eggs of *Xenopus laevis*. II. Electrophysical observations. *J. Cell Biol.* 60: 529–540.

de Laat, S. W., Tertoelen, L. G. J., Dorresteijn, A. W. C and van der Biggelaar, J. A. M. 1980. Intercellular communication patterns are involved in cell determination in early muscular development. *Nature* 287: 546–548.

de la Chappelle, A., Schroder, J., Rantanen, P., Thomasson, B., Niemi, M., Tilikainen, A., Sanger, R. and Robson, E. E. 1974. Early fusion of two human embryos? *Ann. Hum. Genet.* 38: 63–75.

de Nooij, J.C., Letendre, M.A. and Hariharan, I.K. 1996. A cyclin-dependent kinase inhibitor, Dacapo, is necessary for timely exit from the cell cycle during *Drosophila* embryogenesis. *Cell* 87: 1237–1247.

Dunphy, W. G., Brizuela, L., Beach, D. and Newport, J. 1988. The *Xenopus cdc2* protein is a component of MPF, a cytoplasmic regulator of mitosis. *Cell* 54: 423–431.

Duronio, R. J. and O'Farrell, P. 1994. Developmental control of a G1-S transcription program in *Drosophila*. *Development* 120: 1503–1515.

Dyce, J., George, M., Goodall, H. and Fleming, T. P. 1987. Do trophectoderm and inner cell mass cells in the mouse blastocyst maintain discrete lineages? *Development* 100: 685–698.

Edgar, B. A. and O'Farrell, P. H. 1989. Genetic control of cell division patterns in the *Drosophila* embryo. *Cell* 57: 177–187.

Edgar, B. A., Kiehle, C. P. and Schubiger, G. 1986. Cell cycle control by the nucleo-cytoplasmic ratio in early *Drosophila* development. *Cell* 44: 365–372.

Edgar, B., Sprenger, F., Duronio, R. J., Leopold, P. and O'Farrell, P. 1994. MPF regulation during the embryonic cell cycles of *Drosophila*. *Genes Dev.* 8: 440–453.

Ettensohn, C. A. and Ingersoll, E. P. 1992. Morphogenesis of the sea urchin embryo. *In* E. F. Rossomondo and S. Alexander (eds.), *Morphogenesis*. Marcel Dekker, New York, pp. 189–262.

Evans, M.J. and Kaufman, M. H. 1981. Establishment in culture of pluripotent cells from mouse embryos. *Nature* 292: 154–156.

Evans, T., Rosenthal, E., Youngblom, J., Distel, D. and Hunt, T. 1983. Cyclin: A protein specified by maternal mRNA in sea urchin eggs that is destroyed at each cleavage division. *Cell* 33: 389–396.

Eyal-Giladi, H. 1991. The early embryonic development of the chick, an epigenetic process. *Crit. Rev. Poultry Biol.* 3: 143–166.

Eyal-Giladi, H., Debby, A. and Harel, N. 1992. The posterior section of the chick's area pellucida and its involvement in hypoblast and primitive streak formation. *Development* 116: 819–830.

Ferrell, J. E., Wu, M., Gergart, J. C. and Martin, G. S. 1991. Cell cycle tyrosine phosphorylation of p34*cdc2* and a microtubule associated protein kinase homologue in *Xenopus* oocytes and eggs. *Mol. Cell. Biol.* 11: 1965–1971.

Fleming, T. P. 1987. Quantitative analysis of cell allocation to trophectoderm and inner cell mass in the mouse embryo. *Dev. Biol.* 119: 520–531.

Fleming, T. P. 1992. Trophectoderm biogenesis in the preimplantation mouse embryo. *In* T. P. Fleming, (ed.) *Epithelial Organization and Development*. Chapman and Hall, London, pp. 111–134.

Fleming, T. P., Pickering, S. J., Qasim, F. and Maro, B. 1986. The generation of cell surface polarity in mouse 8-cell blastomeres: The role of cortical microfilaments analyzed using cytochalasin D. *J. Embryol. Exp. Morphol.* 95: 169–191.

Foe, V. 1989. Mitotic domains reveal early committment of cells in *Drosophila* embryos. *Development* 107: 1–25.

Foe, V. E., Odell, G. M., and Edgar, B. A. 1994. Mitosis and morphogenesis in the *Drosophila* embryo: point and counterpoint. *In The Development of Drosophila melanogaster*, B. M. Bate, (ed.). Cold Spring Harbor Press, Cold Spring Harbor.

Freeman, G. and Lundelius, J. W. 1982. The developmental genetics of dextrality and sinistrality in the gastropod *Limnea peregra*. *Wilhelm Roux Arch. Dev. Biol.* 191: 69–83.

Fullilove, S. L. and Jacobson, A. G. 1971. Nuclear elongation and cytokinesis in *Drosophila montana*. *Dev. Biol.* 26: 560–577.

Gardiner, R. C. and Rossant, J. 1976. Determination during embryogenesis in mammals. *Ciba Found. Symp.* 40: 5–18.

Gautier, C., Norbury, C., Lohka, M., Nurse, P. and Maller, J. 1988. Purified maturation-promoting factor contains the product of *Xenopus* homolog of the fission yeast cell cycle control gene *cdc2*. *Cell* 54: 433–439.

Gautier, J., Solomon, M. J., Booher, R. N., Bazan, J. F. and Kirschner, M. W. 1991. cdc25 is a specific tyrosine phosphatase that directly activates p34^{cdc2}. *Cell* 67: 197–211.

Gerhart, J. C., Wu, M. and Kirschner, M. 1984. Cell dynamics of an M-phase-specific cytoplasmic factor in *Xenopus laevis* oocytes and eggs. *J. Cell Biol.* 98: 1247–1255.

Giudice, A. 1973. *Developmental Biology of the Sea Urchin Embryo*. Academic Press, New York.

Goebl, M. and Byers, B. 1988. Cyclin in fission yeast. *Cell* 54: 739–740.

Gould, K. and Nurse, P. 1989. Tyrosine phosphorylation of the fission yeast cdc2 protein kinase regulates entry into mitosis. *Nature* 342: 39–45.

Graham, C. F. and Kelly, S. J. 1977. Interactions between embryonic cells during early development of the mouse. *In* M. Karkinen-Jaaskelainen, L. Saxén and L. Weiss (eds.), *Cell Interactions in Differentiation*. Academic Press, New York, pp. 45–57.

Granato, M. and Nüsslein-Volhard, C. 1996. Fishing for genes controlling development. *Curr. Opin. Gen. Dev.* 6: 461–468.

Grbic, M. and Strand, M. R. 1992. Sibling rivalry and brood sex ratios in polyembryonic wasps. *Nature* 360: 254–256.

Grbic, M., Nagy, L. M., Carroll, S. B. and Strand, M. 1996. Polyembryonic development: insect pattern formation in a cellularized environment. *Development* 122: 795–804.

Gulyas, B. J. 1975. A reexamination of the cleavage patterns in eutherian mammalian eggs: Rotation of the blastomere pairs during second cleavage in the rabbit. *J. Exp. Zool.* 193: 235–248.

Gurdon, J. B. 1968. Changes in somatic nuclei inserted into growing and maturing amphibian oocytes. *J. Embryol. Exp. Morphol.* 20: 401–414.

Hara, K. 1977. The cleavage pattern of the axolotl egg studied by cinematography and cell counting. *Wilhelm Roux Arch. Entwicklungsmech. Org.* 181: 73–87.

Hara, K., Tydeman, P. and Kirschner, M. W. 1980. A cytoplasmic clock with the same period as the division cycle in *Xenopus* eggs. *Proc. Natl. Acad. Sci. USA* 77: 462–466.

Hardin, J. D. 1987. Archenteron elongation in the sea urchin embryo is a microtubule independent process. *Dev. Biol.* 121: 253–262.

Harland, R. M. and Laskey, R. A. 1980. Regulated replication of DNA microinjected into eggs of *X. laevis*. *Cell* 21: 761–771.

Heasman, J., Ginsberg, D., Goldstone, K., Pratt, T., Yoshidanaro, C. and Wylie, C. 1994. A functional test for maternally inherited cadherin in *Xenopus* shows its importance in cell adhesion at the blastula stage. *Development* 120: 49–57.

Helde, K. A., Wilson, E. T., Cretekos, C. J. and Grunwald, D. J. 1994. Contribution of early cells to the fate map of the zebrafish gastrula. *Science* 265: 517–520.

Hillman, N., Sherman, H. I. and Graham, C. F. 1972. The effects of spatial arrangement of cell determination during mouse development. *J. Embryol. Exp. Morphol.* 28: 263–278.

Hinegardner, R. T., Rao, B. and Feldman, D. E. 1964. The DNA synthetic period during early development of the sea urchin egg. *Exp. Cell Res.* 36: 53–61.

Hörstadius, S. 1939. The mechanics of sea urchin development, studied by operative methods. *Biol. Rev.* 14: 132–179.

Hörstadius, S. 1973. *Experimental Embryology of Echinoderms.* Clarendon Press, Oxford.

Inoué, S. 1982. The role of self-assembly in the generation of biologic form. *In* S. Subtelny and B. P. Green (eds.), *Developmental Order: Its Origin and Regulation,* Alan R. Liss, New York, pp. 35–76.

Jessus, C. and Beach, D. 1992. Oscillation of MPF is accomplished by periodic association between cdc25 and cdc2-cyclin B. *Cell* 68: 323–332.

Johannsen, O.A. and Butt, F.H. 1941. *Embryology of Insects and Myriapods.* McGraw-Hill, NY.

Johnson, M. H. and Selwood, L. 1996. The nomenclature of early development in mammals. *Reprod. Fert. Dev.* 8: 759–764.

Johnson, M. H., Chisholm, J. C., Fleming, T. P. and Houliston, E. 1986. A role for cytoplasmic determinants in the development of the early mouse embryo. *J. Embryol. Exp. Morphol.* [Suppl.]: 97–117.

Kalt, M. R. 1971. The relationship between cleavage and blastocoel formation in *Xenopus laevis.* I. Light microscopic observations. *J. Embryol. Exp. Morphol.* 26: 37–49.

Kane, D. and Kimmel, C. B. 1993. The midblastula transition in zebrafish. *Development* 119: 447–456.

Karr, T. L. and Alberts, B. M. 1986. Organization of the cytoskeleton in early *Drosophila* embryos. *J. Cell Biol.* 102: 1494–1509.

Karsenti, E., Newport, J., Hubble, R. and Kirschner, M. 1984. The interconversion of metaphase and interphase microtubule arrays, as studied by the injection of centrosomes and nuclei into *Xenopus* eggs. *J. Cell Biol.* 98: 1730–1745.

Kimelman, D., Kirschner, M. and Scherson, T. 1987. The events of the midblastula transition in *Xenopus* are regulated by changes in the cell cycle. *Cell* 48: 399–407.

Kimmel, C.B. and Law, R.D. 1985. Cell lineage of zebrafish blastomeres. II. Formation of the yolk syncytial layer. *Dev. Biol.* 108: 86–93.

Kimmel, C. B. and Warga, R. M. 1987. Indeterminate cell lineage of the zebrafish embryo. *Dev. Biol.* 124: 269–280.

Kimmel, C. B., Warga, R. M. and Schilling, T. F. 1990. Origin and organization of the zebrafish fate map. *Development* 108: 581–594.

Knoblich, J. A., Sauer, K., Jones, L., Richardson, H., Saint, R. and Lehner, C. F. 1994. Cyclin E controls S phase progression and its down-regulation during *Drosophila* embryogenesis is required for the arrest of cell proliferation. *Cell* 77: 107–120.

Langeland, J. and Kimmel, C. 1997. The embryology of fish. *In* S. F. Gilbert and A. M. Raunio (eds.), *Embryology: Constructing the Organism.* Sinauer Associates, Sunderland, MA.

Lane, M.E., Sauer, K., Wallace, K., Jan, Y. N., Lehner, C.F. and Vaessin, H. 1996. Dacapo, a cyclin-dependent kinase inhibitor, stops cell proliferation during *Drosophila* development. *Cell* 87: 1225–1235.

Langman, J. 1981. *Medical Embryology,* 4th Ed. Williams & Wilkins, Baltimore.

Laskey, R. A., Mills, A. D. and Morris, N. R. 1977. Assembly of SV40 chromatin in a cell-free system from *Xenopus* eggs. *Cell* 10: 237–243.

Lee, M. and Nurse, P. 1987. Complementation used to clone a human homologue of the fission yeast cell cycle control gene *cdc2+. Nature* 335: 251–254.

Lee, M. S., Ogg, S., Xu, M., Parker, M., Donoghue, D. J., Maller, J. L. and Piwnica-Worms, H. 1992. cdc25 encodes a protein phosphatase the dephosphorylates p34^{cdc2}. *Mol. Biol. Cell* 3: 73–84.

Lee, R. K., Stainier, D. Y. R., Weinstein, B. M. and Fishman, M. C. 1994. Cardiovascular development in the zebrafish. II. Endocardial progenitors are sequestered within the heart field. *Development* 120: 3361–3366.

Lepage, T., Sardet, C. and Gache, C. 1992. Spatial expression of the hatching enzyme gene in the sea urchin embryo. *Dev. Biol.* 150: 23–32.

Levy, J. B., Johnson, M. H., Goodall, H. and Maro, B. 1986. The timing of compaction: Control of a major developmental transition in mouse early embryogenesis. *J. Embryol. Exp. Morphol.* 95: 213–237.

Lillie, F. R. 1898. Adaptation in cleavage. In *Biological Lectures of the Marine Biological Laboratory of Woods Hole.* Ginn, Boston, pp. 43–67.

Lillie, F. R. 1902. Differentiation without cleavage in the egg of the annelid *Chaetopterus pergamentaceous. Wilhelm Roux Arch. Entwicklungsmech. Org.* 14: 477–499.

Lutz, B. 1947. Trends towards non-aquatic and direct development in frogs. *Copeia* 4: 242–252.

Madine, M. A., Khoo, C.-Y., Mills, A. D. and Laskey, R. A. 1995. MCM3 complex required for cell cycle regulation of DNA replication in vertebrate cells. *Nature* 375: 421–424.

Mark, W. H., Signorelli, K. and Lacy, E. 1985. An inserted mutation in a transgenic mouse line results in developmental arrest at day 5 of gestation. *Cold Spring Harbor Symp. Quant. Biol.* 50: 453–463.

Markert, C. L. and Petters, R. M. 1978. Manufactured hexaparental mice show that adults are derived from three embryonic cells. *Science* 202: 56–58.

Martin, G. R. 1981. Isolation of a pluripotent cell line from early mouse embryos cultured in medium conditioned by teratocarcinoma stem cells. *Proc. Natl. Acad. Sci. USA* 78: 7634–7638.

Masui, Y. and Markert, C. L. 1971. Cytoplasmic control of nuclear behavior during meiotic maturation of frog oocytes. *J. Exp. Zool.* 177: 129–146.

Mayr, W. R., Pausch, V. and Schnedl, W. 1979. Human chimaera detectable only by investigation of her progeny. *Nature* 277: 210–211.

Miake-Lye, R. and Kirschner, M. W. 1985. Induction of early mitotic events in a cell-free system. *Cell* 41: 165–175.

Minshull, J., Blow, J. J. and Hunt, T. 1989. Translation of cyclin mRNA is necessary for extracts of activated *Xenopus* eggs to enter mitosis. *Cell* 56: 947–956.

Minshull, J., Sun, H., Tonks, N. K. and Murray, A. W. 1994. A MAP kinase-dependent spindle assembly checkpoint in *Xenopus* egg extracts. *Cell* 79: 475–486.

Mintz, B. 1970. Clonal expression in allophenic mice. *Symp. Int. Soc. Cell Biol.* 9: 15.

Morgan, T. H. 1927. *Experimental Embryology.* Columbia University Press, New York.

Mulnard, J. G. 1967. Analyse microcinematographique du developpement de l'oeuf de souris du stade II au blastocyste. *Arch. Biol.* (Liege) 78: 107–138.

Murray, A. W. and Kirschner, M. W. 1989. Cyclin synthesis drives the early embryonic cell cycle. *Nature* 339: 275–280.

Murray, A. W., Solomon, M. J. and Kirschner, M. W. 1989. The role of cyclin synthesis and degradation in the control of maturation promoting factor activity. *Nature* 339: 280–286.

New, D. A. T. 1956. The formation of sub-blastodermic fluid in hens' eggs. *J. Embryol. Exp. Morphol.* 43: 221–227.

Newport, J. W. and Kirschner, M. W. 1982a. A major developmental transition in early *Xenopus* embryos: I. Characterization and timing of cellular changes at midblastula stage. *Cell* 30: 675–686.

Newport, J. W. and Kirschner, M. W. 1982b. A major developmental transition in early *Xenopus* embryos: II. Control of the onset of transcription. *Cell* 30: 687–696.

Newport, J. W. and Kirschner, M. W. 1984. Regulation of the cell cycle during *Xenopus laevis* development. *Cell* 37: 731–742.

Nieuwkoop, P. D. 1973. The "organization center" of the amphibian embryo: Its origin, spatial organization, and morphogenetic action. *Adv. Morphogenet.* 10: 1–39.

Nigg, E. A. 1995. Cyclin-dependent protein kinases: key regulators of the eukaryotic cell cycle. *BioEssays* 17: 471–480.

Pedersen, R. A., Wu, K. and Batakier, H. 1986. Origin of the inner cell mass in mouse embryos: Cell lineage analysis by microinjection. *Dev. Biol.* 117: 581–595.

Perona, R. M. and Wassarman, P. M. 1986. Mouse blastocysts hatch *in vitro* by using a trypsin-like proteinase associated with cells of mural trophectoderm. *Dev. Biol.* 114: 42–52.

Peter, M., Nakagawa, J., Dorée, M., Labbé, J. C. and Nigg, E. A. 1990. In vitro disassembly of the nuclear lamina and M phase-specific phosphorylation of lamins by cdc2 kinase. *Cell* 61: 591–602.

Peyrieras, N., Hyafil, F., Louvard, D., Ploegh, H. L. and Jacob, F. 1983. Uvomorulin: A non-integral membrane protein of early mouse embryo. *Proc. Natl. Acad. Sci. USA* 80: 6274–6277.

Piko, L. and Clegg, K. B. 1982. Quantitative changes in total RNA, total poly(A), and ribosomes in early mouse embryos. *Dev. Biol.* 89: 362–378.

Pflüger, E. 1884. Uber die Einwirkung der Schwerkraft und anderer Bedingungen auf die Richtung der Zeiltheilung. *Arch. Ges. Physiol.* 3: 4.

Prather, R. S. 1989. Nuclear transfer in mammals and amphibia. *In* H. Schatten and G. Schatten (eds.), *The Molecular Biology of Fertilization.* Academic Press, New York, pp. 323–340.

Pratt, H. P. M., Ziomek, Z. A., Reeve, W. J. D. and Johnson, M. H. 1982. Compaction of the mouse embryo: An analysis of its components. *J. Embryol. Exp. Morphol.* 70: 113–132.

Raff, J. W. and Glover, D. M. 1989. Centrosomes, not nuclei, initiate pole cell formation in *Drosophila* embryos. *Cell* 57: 611–619.

Raff, R. A. and Kaufman, T. C. 1983. *Embryos, Genes, and Evolution: The Developmental-Genetic Basis of Evolutionary Change.* Macmillan, New York.

Rappaport, R. 1961. Experiments concerning cleavage stimulus in sand dollar eggs. *J. Exp. Zool.* 148: 81–89.

Rugh, R. 1967. *The Mouse.* Burgess, Minneapolis.

Satterwhite, L. L., Lohka, M. J., Wilson, K. L., Scherson, T. Y., Cisek, L. J. and Pollard, T. D. 1992. Phosphorylation of myosin-II light chain by cyclin-p34^{cdc2}: A mechanism for the timing of cytokinesis. *J. Cell Biol.* 118: 595–605.

Saunders, J. W., Jr. 1982. *Developmental Biology.* Macmillan, New York.

Schejter, E. D. and Wieschaus, E. 1993. *Bottleneck* acts as a regulator of the microfilament network governing cellularization of the *Drosophila* embryo. *Cell* 75: 373–385.

Schroeder, T. E. 1972. The contractile ring. II. Determining its brief existence, volumetric changes, and vital role in cleaving *Arbacia* eggs. *J. Cell Biol.* 53: 419–434.

Schroeder, T. E. 1973. Cell constriction: Contractile role of microfilaments in division and development. *Am. Zool.* 13: 687–696.

Seidel, F. 1952. Die Entwicklungspotenzen einer isolierten Blastomere des Zweizellenstadiums im Säugetierei. *Naturwissenschaften* 39: 355–356.

Smith, L. D. and Ecker, R. E. 1969. Role of the oocyte nucleus in the physiological maturation in *Rana pipiens. Dev. Biol.* 19: 281–309.

Smythe, C. and Newport, J. W. 1992. Coupling of mitosis to the completion of S phase in *Xenopus* occurs via modulation of the tyrosine kinase that phosphorylates p34^{cdc2}. *Cell* 68: 787–797.

Solomon, M. J. 1993. Activation of the various cyclin/cdc2 protein kinases. *Curr. Opin. Cell Biol.* 5: 180–186.

Solomon, M., Booher, R., Kirschner, M. and Beach, D. 1988. Cyclin in fission yeast. *Cell* 54: 738–739.

Stainier, D. Y. R., Lee, R. K. and Fishman, M. C. 1993. Cardiovascular development in the zebrafish. I. Myocardial fate map and heart tube formation. *Development* 119: 31–40.

Strickland, S., Reich, E. and Sherman, M. I. 1976. Plasminogen activator in early embryogenesis: Enzyme production by trophoblast and parietal endoderm. *Cell* 9: 231–240.

Sturtevant, M. H. 1923. Inheritance of direction of coiling in *Limnaea. Science* 58: 269–270.

Summers, R. G., Morrill, J. B., Leith, A., Marko, M., Piston, D. W. and Stonebraker, A. T. 1993. A stereometric analysis of karyogenesis, cytokinesis, and cell arrangements during and following fourth cleavage period in the sea urchin, *Lytechinus variegatus. Dev. Growth Diff.* 35: 41–57.

Sutherland, A. E. and Calarco-Gillam, P. G. 1983. Analysis of compaction in the preimplantation mouse embryo. *Dev. Biol.* 100: 327–338.

Sutherland, A. E., Speed, T. P. and Calarco, P. G. 1990. Inner cell allocation in the mouse morula: The role of oriented division during fourth cleavage. *Dev. Biol.* 137: 13–25.

Swenson, K. L., Farrell, K. M. and Ruderman, J. V. 1986. The clam embryo protein cyclin A induces entry into M phase and the resumption of meiosis in *Xenopus* oocytes. *Cell* 47: 861–870.

Sze, L. C. 1953. Changes in the amount of deoxyribonucleic acid in the development of *Rana pipiens. J. Exp. Zool.* 122: 577–601.

Tarkowski, A. K. and Wróblewska, J. 1967. Development of blastomeres of mouse eggs isolated at the 4- and 8-cell stage. *J. Embryol. Exp. Morphol.* 18: 155–180.

Trinkaus, J. P. 1993. The yolk syncytial layer of Fundulus: Its origin and history and its significance for early embryogenesis. *J. Exp. Zool.* 265: 258–284.

Tuchmann-Duplessis, H., David, G. and Haegel, P. 1972. *Illustrated Human Embryology,* Vol. 1. Springer-Verlag, New York.

van den Biggelaar, J. A. M. and Guerrier, P. 1979. Dorsoventral polarity and mesentoblast determination as concomitant results of cellular interactions in the mollusc *Patella vulgata. Dev. Biol.* 68: 462–471.

Ward, G. E. and Kirschner, M. W. 1990. Identification of cell cycle-regulated phosphorylation cites on nuclear lamin C. *Cell* 61: 561–577.

Watanabe, N., Hunt, T., Ikawa, Y. and Sagata, N. 1991. Independent inactivation of MPF and cytostatic factor (Mos) upon fertilization of *Xenopus* eggs. *Nature* 352: 247–249.

Welsh, J. H. 1969. Mussels on the move. *Nat. Hist.* 78: 56–59.

White, J. C., Amos, W. B. and Fordham, M. 1987. An evaluation of confocal versus conventional imaging of biological structures by fluorescence light microscopy. *J. Cell Biol.* 105: 41–48.

Whittaker, J. R. 1979. Cytoplasmic determinants of tissue differentiation in the ascidian egg. *In* S. Subtelny and I. R. Konigsberg (eds.), *Determinants of Spatial Organization.* Academic Press, New York, pp. 29–51.

Wiley, L. M. 1984. Cavitation in the mouse preimplantation embryo: Na/K ATPase and the origin of nascent blastocoel fluid. *Dev. Biol.* 105: 330–342.

Wilson, E. B. 1898. Cell lineage and ancestral reminiscences. In *Biological Lectures of the Marine Biological Laboratory of Woods Hole.* Ginn, Boston, pp. 21–42.

Wilson, E. B. 1901. Experiments in cytology. II. Some phenomena of fertilization and cell division in etherized eggs. III. The effect on cleavage of artificial obliteration of the first cleavage furrow. *Wilhelm Roux Arch. Entwicklungsmech. Org.* 13: 353–395.

Winkel, G. K., Ferguson, J. E., Takeichi, M. and Nuccitelli, R. 1990. Activation of protein kinase C triggers premature compaction in the 4-cell stage mouse embryo. *Dev. Biol.* 138: 1–15.

Wolpert, L. and Gustafson, T. 1961. Studies in the cellular basis of morphogenesis of the sea urchin embryo: The formation of the blastula. *Exp. Cell Res.* 25: 374–382.

Wolpert, L. and Mercer, E.H. 1963. An electron microscope study of the development of the blastula of the sea urchin embryo and its radial polarity. *Exp. Cell Res.* 30: 280–300.

Yamazaki, K. and Kato, Y. 1989. Sites of zona pellucida shedding by mouse embryo other than mural trophectoderm. *J. Exp. Zool.* 249: 347–349.

Ziomek, C. A. and Johnson, M. H. 1980. Cell surface interactions induce polarization of mouse 8-cell blastomeres at compaction. *Cell* 21: 935–942.

Gastrulation:
Reorganizing the embryonic cells

*My dear fellow... life is infinitely
stranger than anything which the mind of
man could invent. We would not dare to
conceive the things which are really mere
commonplaces of existence.*
A. CONAN DOYLE (1891)

*It is not birth, marriage, or death, but
gastrulation, which is truly the most
important time in your life.*
LEWIS WOLPERT (1986)

GASTRULATION is the process of highly integrated cell and tissue movements whereby the cells of the blastula are dramatically rearranged. The blastula consists of numerous cells, the positions of which were established during cleavage. During gastrulation, these cells are given new positions and new neighbors, and the multilayered body plan of the organism is established. The cells that will form the endodermal and mesodermal organs are brought inside the embryo, while the cells that will form the skin and nervous system are spread over the outside surface. Thus, the three germ layers—outer ectoderm, inner endoderm, and interstitial mesoderm—are first produced during gastrulation. In addition, the stage is set for the interactions of these newly positioned tissues.

The movements of gastrulation involve the entire embryo, and cell migrations in one part of the gastrulating organism must be intimately coordinated with other movements occurring simultaneously. Although the patterns of gastrulation vary enormously throughout the animal kingdom, there are relatively few mechanisms involved. Gastrulation usually involves combinations of the following types of movements:

- *Epiboly.* The movement of epithelial sheets (usually of ectodermal cells), which spread as a unit rather than individually, to enclose the deeper layers of the embryo.
- *Invagination.* The infolding of a region of cells, much like the indenting of a soft rubber ball when poked.
- *Involution.* The inturning or inward movement of an expanding outer layer so that it spreads over the internal surface of the remaining external cells.
- *Ingression.* The migration of individual cells from the surface layer into the interior of the embryo.
- *Delamination.* The splitting of one cellular sheet into two more or less parallel sheets.

As we look at gastrulation in different types of embryos, we should keep in mind the following questions (Trinkaus, 1984a):

- *What is the unit of migration activity?* Is migration dependent on the movement of individual cells, or are the cells part of a migrating sheet? Remarkable as it first seems, regional migrational properties may be totally controlled by cytoplasmic factors that are independent of cellularization. F. R. Lillie (1902) was able to parthenogeneti-cally activate eggs of the annelid *Chaetopterus* and suppress their cleavage. Many events of early development occurred even in the absence of cells. The cytoplasm of the zygote separated into defined regions, and cilia differentiated in the appropriate parts of the egg. Moreover, the outermost clear cytoplasm migrated down over the vegetal regions in a manner specifically reminiscent of the epiboly of animal hemisphere cells during normal development. This occurred at precisely the time that epiboly would have taken place during gastrulation. Thus, epiboly may be (at least in some respects) independent of the cells that form the migrating region.

- *Is the spreading or folding of a cell sheet due to intrinsic factors within the sheet or to extrinsic forces stretching or distorting it?* It is essential to know the answer to this question if we are to understand how the various cell movements of gastrulation are integrated. For instance, do involuting cells pull the epibolizing cells down toward them, or are the two movements independent?
- *Is there active spreading of the whole tissue, or does the leading edge expand and drag the rest of a cell sheet passively along?*
- *Are changes in cell shape and motility during gastrulation the consequence of changes in cell-surface properties, such as adhesiveness to the substrate or to other cells?*

Keeping these questions in mind, we will observe the various patterns of gastrulation found in echinoderms, amphibians, fish, birds, and mammals.*

Sea urchin gastrulation

The sea urchin blastula consists of a single layer of around 1000 cells. These cells, derived from different regions of the zygote, have different sizes and properties. Figures 6.1 and 6.2 show the fates of the various regions of the zygote as it develops through cleavage and gastrulation to the pluteus larva characteristic of sea urchins. The fate of each cell layer can be seen through the layers' movements during gastrulation.

Ingression of Primary Mesenchyme

FUNCTION OF PRIMARY MESENCHYME CELLS.   Shortly after the blastula hatches from its fertilization membrane, the vegetal side of the spherical blastula begins to thicken and flatten (Figure 6.2, 9 hours). At the center of this flat **vegetal plate,** a cluster of small cells begins to change. These cells show pulsating movements on their inner surfaces, extending and contracting long, thin (30×5 μm) processes called **filopodia.** The cells then dissociate from the epithelial monolayer and ingress into the blastocoel (Figure 6.2, 9–10 hours). These cells are called the **primary mesenchyme** and are derived from the micromeres. The 64 or so primary mesenchyme cells of the sea urchin embryo are the descendants of the four blastomeres that formed by the asymmetrical fourth cleavage.

Gustafson and Wolpert (1961) have used time-lapse films to follow the microscopic movements of these mesenchyme cells within the blastocoel. At

---

*Discussion of *Drosophila* gastrulation will be postponed until Chapter 14, where it occurs in the context of axis formation. Keep in mind the warning of gastrulation researcher Ray Keller (personal communication): "Students should NOT read this material quickly, but too typical a scene is some poor bastard hunkered over this text at 2:30 A.M. with a cup of coffee, frantically scanning the figures to see if he or she can figure out what is happening." Gastrulation is (as Wolpert says in the quotation at the beginning of this chapter) the most important time in your life. It is worth the time to examine it critically and to appreciate it at your leisure.

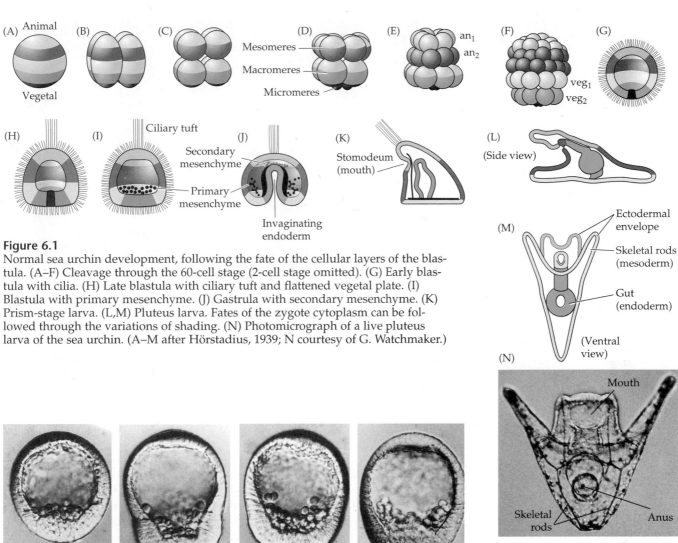

Figure 6.1
Normal sea urchin development, following the fate of the cellular layers of the blastula. (A–F) Cleavage through the 60-cell stage (2-cell stage omitted). (G) Early blastula with cilia. (H) Late blastula with ciliary tuft and flattened vegetal plate. (I) Blastula with primary mesenchyme. (J) Gastrula with secondary mesenchyme. (K) Prism-stage larva. (L,M) Pluteus larva. Fates of the zygote cytoplasm can be followed through the variations of shading. (N) Photomicrograph of a live pluteus larva of the sea urchin. (A–M after Hörstadius, 1939; N courtesy of G. Watchmaker.)

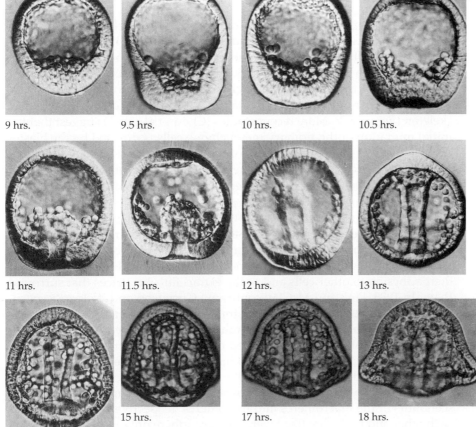

9 hrs.　9.5 hrs.　10 hrs.　10.5 hrs.

11 hrs.　11.5 hrs.　12 hrs.　13 hrs.

13.5 hrs.　15 hrs.　17 hrs.　18 hrs.

Figure 6.2
Entire sequence of gastrulation in *Lytechinus variegatus*. The time shows the length of development at 25°C. (Courtesy of J. Morrill.)

(A)

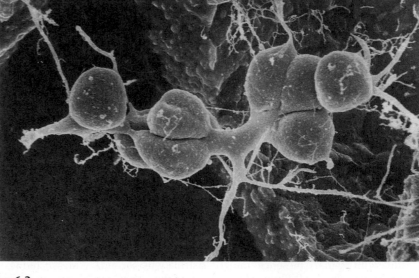

(B)

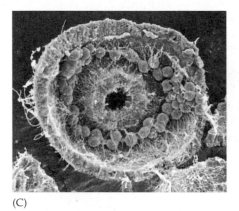

(C)

Figure 6.3

Formation of syncytial cables by mesenchyme cells of the sea urchin. (A) Primary mesenchyme cells in the early gastrula align and fuse to lay down the matrix of the calcium carbonate spicule. (B) Scanning electron micrograph of spicules formed by the fusing of primary mesenchyme cells into syncytial cables. (C) Ring of mesenchyme cells around the archenteron (primitive gut). The animal half and the entire archenteron have been removed. (D) Placement of primary mesenchyme cells in the early larva of *Lytechinus variegatus*. (A and D from Ettensohn, 1990; B and C from Morrill and Santos, 1985; all photographs courtesy of the authors.)

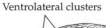

Ventrolateral clusters

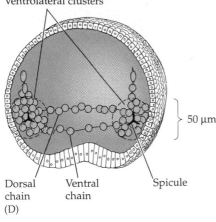

50 μm

Dorsal chain Ventral chain Spicule

(D)

first the cells appear to move randomly along the inner blastocoel surface, actively making and breaking filopodial connections to the wall of the blastocoel. Eventually, these cells become localized within the prospective ventrolateral region of the blastocoel, where their adherence is believed to be strongest. Here the primary mesenchyme cells fuse into **syncytial cables,** which will form the axis of the calcium carbonate spicules of the larval skeleton (Figure 6.3). More recent studies (Cherr et al., 1992) suggest that the initial migration of the primary mesenchyme cells is directed by the blastocoel wall and by the parallel fibrils of the extracellular matrix material that pervades the blastocoel. The micromeres appear to migrate along the blastocoel surface and are enmeshed by these fibrils (Figure 6.4).

IMPORTANCE OF EXTRACELLULAR LAMINA INSIDE THE BLASTOCOEL. Both cytoplasmic and cell-surface events are crucial to the ingression and migration of the primary mesenchyme cells. Gustafson and Wolpert (1967) proposed a model in which the ingression of the micromeres comes about through changes in their adhesion to other cells and to the extracellular matrices that surround them. In 1985, Fink and McClay confirmed Gustafson and Wolpert's speculations by measuring the adhesive strengths of the sea urchin blastomeres for the hyaline layer, for the basal lamina, and for other cells. Originally, all the cells of the blastula are connected on their outer surface to the hyaline layer and on their inner surface by a basal lamina secreted by the cells (see Chapter 3). On their lateral sides, each cell has another cell for a neighbor. Fink and McClay found that the prospective ectoderm and endoderm (descendants of the macromeres and mesomeres, respectively) bind tightly to each other and to the hyaline layer, but they adhere only loosely to the basal lamina (Table 6.1). The micromeres of the blastula originally display a simi-

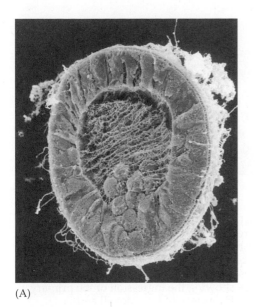

(A)

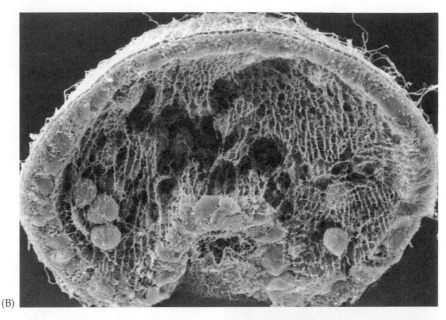

(B)

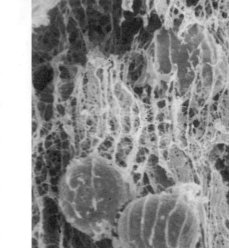

(C)

Figure 6.4
Scanning election microscope stereo photographs of primary mesenchyme cells within the extracellular matrix fibrils of the blastocoel. (A) Primary mesenchyme cells (PMC) enmeshed in the extracellular matrix of early *Strongylocentrotus* gastrula. (B,C) Gastrula-stage mesenchyme cell migration. The extracellular matrix fibrils of the blastocoel have become parallel to the animal-vegetal axis and are intimately associated with the primary mesenchyme cells. (From Cherr et al., 1992; courtesy of G. Cherr.)

lar pattern of binding. However, the micromere pattern changes at gastrulation. Whereas the other cells retain their tight binding to the hyaline layer and to their neighbors, the primary mesenchyme precursors lose their affinity to these structures (to about 2 percent of its original value) while their affinity to components of the basal lamina and extracellular matrix *increases* 100-fold. This change in affinity causes the micromeres to release their attachments to the external hyaline layer and neighboring cells and, drawn in by the basal lamina, migrate up into the blastocoel (Figure 6.5). The changes

Table 6.1 Affinities of mesenchymal and nonmesenchymal cells to cellular and extracellular components[a]

	Dislodgment force (in dynes)		
Cell type	Hyaline	Gastrula cell monolayers	Basal lamina
16-cell-stage micromeres	5.8×10^{-5}	6.8×10^{-5}	4.8×10^{-7}
Migratory-stage mesenchyme cells	1.2×10^{-7}	1.2×10^{-7}	1.5×10^{-5}
Gastrula ectoderm and endoderm	5.0×10^{-5}	5.0×10^{-5}	5.0×10^{-7}

Source: After Fink and McClay, 1985.

[a]Tested cells were allowed to adhere to plates containing hyaline, extracellular basal lamina, or cell monolayers. The plates were inverted and centrifuged at various strengths to dislodge the cells. The dislodgement force is calculated from the centrifugal force needed to remove the test cells from the substrate.

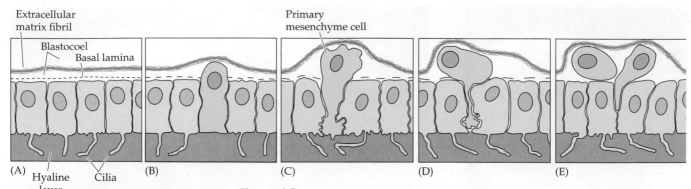

(A) Extracellular matrix fibril
Blastocoel
Basal lamina
Hyaline layer   Cilia
(B)
(C)
Primary mesenchyme cell
(D)
(E)

(F)

Figure 6.5
Ingression of primary mesenchyme cells. (A–E) Interpretative diagrams depicting changes in adhesive interactions in the presumptive primary mesenchyme cells (color). These cells lose their affinities for hyaline and for their neighboring blastomeres while gaining an affinity for the basal lamina. Nonmesenchymal blastomeres retain their original high affinities for hyaline and neighboring cells. (F) Scanning electron micrograph montage showing the ingression of the primary mesenchyme cells of *Lytechinus variegatus*. (F courtesy of J. B. Morrill and D. Flaherty.)

in cell affinity have been correlated with changes in cell-surface molecules that occur during this time (Wessel and McClay, 1985).

As shown in Figure 6.4, there is a heavy concentration of extracellular lamina material around the ingressing primary mesenchyme cells (Galileo and Morrill, 1985; Cherr et al., 1992). Moreover, once inside the blastocoel, the primary mesenchyme cells appear to migrate along the extracellular matrix of the blastocoel wall, extending their filopodia in front of them (Galileo and Morrill, 1985; Karp and Solursh, 1985). The orientation of the fibrils along the animal-vegetal axis may guide the cells in their migration. Three proteins appear to be important in this migration. One is fibronectin, the large (400-kDa) glycoprotein that is a common component of basal laminae, including that of the sea urchin blastocoel (Wessel et al., 1984). Fink and Mc-Clay (1985) showed that during gastrulation, the affinity of the micromeres for this particular molecule increases dramatically. The second set of molecules consists of sulfated proteoglycans found on the cell surface of the ingressing mesenchyme cells (see Chapter 3; Sugiyama, 1972; Lane and Solursh, 1991). If the synthesis (or sulfation) of these proteoglycans is inhibited, the mesenchyme cells enter the blastocoel but do not migrate further* (Figure 6.6; Karp and Solursh, 1974; Anstrom et al., 1987). The third protein, ECM18, is found in the extracellular matrices of the blastocoel cells and is expressed only during gastrulation. Blocking ECM18 with antibodies prevents both primary mesenchyme migration and secondary endoderm invagination (Berg et al., 1996).

But these guidance cues are not sufficient, since the micromeres "know" when to stop their movement and form spicules near the equator of the blastocoel. Primary mesenchyme cells arrange themselves in a ring at a specific position along the animal-vegetal axis. At two sites near the future ventral side of the larva, many of these primary mesenchyme cells cluster together and initiate spicule formation (Figure 6.3). If a labeled micromere from another embryo is injected into the blastocoel of a gastrulating sea urchin em-

---

*In one of the first experiments in chemical embryology, Curt Herbst (1904) found that sea urchin embryos wouldn't gastrulate properly when placed into seawater that lacked sulfate ions. At the time, he couldn't figure out why.

(A) 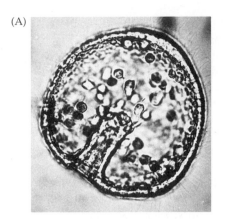 (B)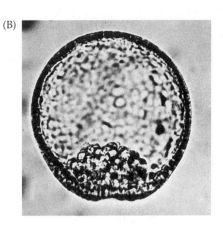

Figure 6.6
Effect of sulfate deprivation on primary mesenchyme movement in the sea urchin *Lytechinus*. (A) Normal gastrula. (B) Abnormal gastrula formed when embryos are grown in sulfate-free seawater. (From Karp and Solursh, 1974, courtesy of M. Solursh.)

bryo, it migrates to the correct location and contributes to the formation of embryonic spicules (Plate 35). If primary mesenchyme cells from older embryos are injected into younger gastrulae, they will delay their differentiation, migrate like younger cells, and incorporate normally into the host's mesenchyme. Moreover, if all the host's primary mesenchyme cells are removed before the older mesenchyme cells are injected, the older mesenchyme cells will repeat the earlier stages of their migration and form a normal mesenchymal ring and skeleton (Ettensohn, 1990). It is thought that this positional information is provided by the prospective ectodermal cells and their basal laminae (von Übisch, 1939; Harkey and Whiteley, 1980). Only the primary mesenchyme cells (and not other cell types or latex beads) are capable of responding to these patterning cues (Ettensohn and McClay, 1986). Miller and colleagues (1995) have reported the existence of extremely fine (0.3-μm-diameter) filopodia on the skeletonogenic mesenchyme; these appear to explore and sense the blastocoel wall (Figure 6.7). These filopodia contain actin and are not thought to be used for locomotion. Rather, like the filopodia on the tips of axonal growth cones, they are thought to sense the environment. These thin extensions may be responsible for picking up dorsoventral and animal-vegetal patterning cues from the ectoderm (Malinda et al., 1995).

First Stage of Archenteron Invagination

As the ring of primary mesenchyme cells forms in the vegetal region of the blastocoel, important changes are occurring in the cells that remain at the

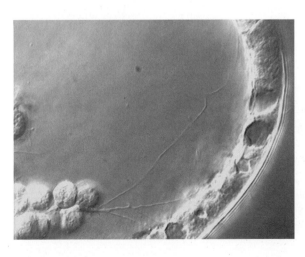

Figure 6.7
Nomarski videomicrograph showing a long, thin filopodium from a primary mesenchyme cell extending to the ectodermal wall of the gastrula, as well as a shorter filopodium extending inward from the ectoderm. The mesenchymal filopodia extend through the extracellular matrix and directly contact the cell membrane of the ectodermal cells. (From Miller et al., 1995; photograph courtesy of D. McClay.)

vegetal plate. These cells remain bound to one another and to the hyaline layer of the egg and move to fill the gaps caused by the ingression of the mesenchyme; therefore, the vegetal plate flattens further. One next sees that the vegetal plate bends inward and extends about one-quarter to one-half the way into the blastocoel (see Figure 6.2, 10.5–11.5 hours; Figure 6.8A). Then invagination suddenly ceases. The invaginated region is called the **archenteron** (primitive gut), and the opening of the archenteron at the vegetal region is called the **blastopore.**

What forces work to invaginate these cells? Lane and co-workers (1993) have provided evidence that the buckling is similar to that produced by heating a bimetallic strip. The hyaline layer is actually made of two laminae, an outer lamina made primarily of hyalin protein and an inner lamina composed of fibropellin proteins* (Hall and Vacquier, 1982; Bisgrove et al., 1991). The vegetal plate cells (and only these cells) secrete a chondroitin sulfate proteoglycan into the inner lamina of the hyaline layer directly beneath them. This hygroscopic (water-absorbing) molecule swells the inner lamina, but does not swell the outer lamina. This causes the vegetal region of the hyaline layer to buckle (Figure 6.8B,C). Slightly later, a second force arising

*Fibropellins are stored in secretory granules within the oocyte. They are secreted from these granules after cortical granule exocytosis releases the hyalin protein. By blastula stage, the fibropellins have formed a meshlike network over the embryo surface.

(A)

Figure 6.8
Invagination of the vegetal plate. (A) Vegetal plate invagination in *Lytechinus variegatus* seen by scanning electron micrography of the external surface of the early gastrula. The blastopore is clearly visible. (B) The hyaline layer consists of inner and outer laminae. Microvilli from the vegetal plate cells extend through the hyaline layer, and their cytoplasm contains secretory vesicles that store a chondroitin sulfate proteoglycan (CSPG). (C) The storage granules secrete the chondroitin sulfate proteoglycan into the inner lamina of the hyaline layer. The proteoglycan absorbs water and swells the inner lamina, while the outer lamina, to which it is attached, does not swell. This causes the bending of the hyaline envelope and its attached epithelium inward. (A from Morrill and Santos, 1985, courtesy of J. B. Morrill; B and C after Lane et al., 1993.)

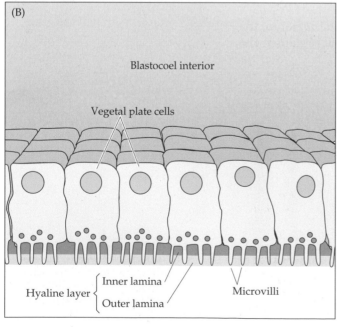

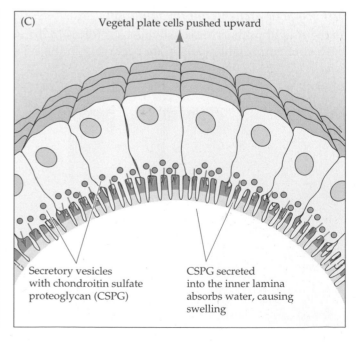

Early gastrulation   Later gastrulation

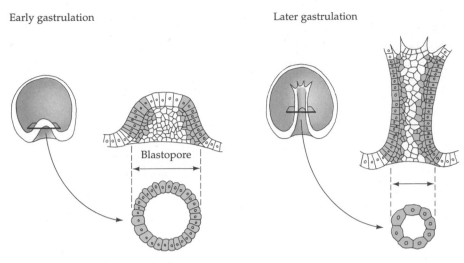

Blastopore

Figure 6.9
Cell rearrangement during the extension of the archenteron in sea urchin embryos. In this species, the early archenteron has 20 to 30 cells around its circumference. Later in gastrulation, the archenteron has a circumference made by only 6 to 8 cells. Fluorescently labeled clones can be seen to stretch apically. (After Hardin, 1990.)

from movements of the epithelial cells adjacent to the vegetal plate may facilitate this invagination by drawing the buckled layer inward (Burke et al., 1991).

Second and Third Stages of Archenteron Invagination

The invagination of the vegetal cells occurs in three discrete stages. After a brief pause, the second phase of archenteron formation begins. During this time, the archenteron extends dramatically, sometimes tripling its length. In the process of extension, the wide, short gut rudiment is transformed into a long, thin tube; yet no new cells are formed (see Figure 6.2, 12 hours; Figure 6.9). To accomplish this extension, the cells of the archenteron rearrange themselves by migrating over one another and by flattening themselves (Ettensohn, 1985; Hardin and Cheng, 1986). This phenomenon, where cells intercalate to narrow the tissue and at the same time move it forward, is called **convergent extension.**

In at least some species of sea urchins, a third stage of archenteron elongation occurs. This last phase of archenteron elongation is initiated by the tension provided by secondary mesenchyme cells, which form at the tip of the archenteron and remain there (see Figure 6.2, 13 hours; Figure 6.10). Filopodia are extended from these cells through the blastocoel fluid to con-

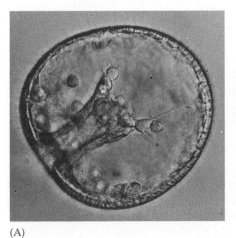

(A)

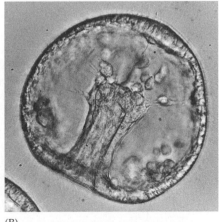

(B)

Figure 6.10
Midgastrula stage of the sea urchin *Lytechinus pictus,* showing filopodial extensions of secondary mesenchyme extending from the archenteron tip to the blastocoel wall. (A) Mesenchyme cells extending filopodia from the tip of the archenteron. (B) Filopodial cables connecting the blastocoel wall to the archenteron tip. The tension of the cables can be seen as they pull on the blastocoel wall at the point of attachment. (Photographs courtesy of C. Ettensohn.)

tact the inner surface of the blastocoel wall (Dan and Okazaki, 1956; Schroeder, 1981). The filopodia attach to the wall at the junctions between the blastoderm cells and then shorten, pulling up the archenteron. Hardin (1988) ablated the secondary mesenchyme cells with a laser, with the result that the archenteron could only elongate about two-thirds of the full length. If a few secondary mesenchyme cells were left, elongation continued, although at a slower rate than in controls. The secondary mesenchyme cells, then, play an essential role in pulling the archenteron up to the blastocoel wall during the last phase of invagination.

But can the secondary mesenchyme filopodia attach to any part of the blastocoel wall, or is there a specific target in the animal hemisphere that must be present for attachment to occur? Is there a region of the blastocoel wall that is already committed to becoming the ventral side of the larva? Studies by Hardin and McClay (1990) show that there is a specific "target" site for the filopodia that differs from other regions of the animal hemisphere. The filopodia of the secondary mesenchyme cells extend, touch the blastocoel wall at random sites, and then retract. However, when the filopodia contact a particular region of the wall, they remain attached there, flatten out against this region, and pull the archenteron toward it. When Hardin and McClay poked in the other side of the blastocoel wall so that the contacts were made most readily with that region, the filopodia continued to extend and retract after touching it. Only when the filopodia found the "target" did they cease these movements. If the gastrula was constricted so that filopodia never reached the target area, the secondary mesenchyme cells continued to explore until they eventually moved off the archenteron and found the target tissue as freely migrating cells. There appears, then, to be a target region on what is to become the ventral side of the larva that is recognized by the secondary mesenchyme cells and that positions the archenteron in the region where the mouth will form.

As the top of the archenteron meets the blastocoel wall in this region, the secondary mesenchyme cells disperse into the blastocoel, where they proliferate to form the mesodermal organs (see Figure 6.2, 13.5 hours). Where the archenteron contacts the wall, a mouth is eventually formed. The mouth fuses with the archenteron to create a continuous digestive tube. Thus, as is characteristic for deuterostomes, the blastopore marks the position of the anus.

Gastrulation in fish

The Midblastula Transition and the Acquisition of Cell Motility

During the tenth cycle of zebrafish cleavage, the cell divisions lose their synchrony, new genes become expressed, and the cells become motile. This midblastula transition (MBT) is also seen in frogs and *Drosophila*. As discussed in Chapter 5, the MBT appears to be regulated by the ratio of chromatin to cytoplasm. Haploid fish enter the MBT one cycle late; tetraploid fish enter the MBT one cycle early (Kane and Kimmel, 1993). It is thought that something in the chromatin is "titrating out" some (as yet unknown) substance from the cytoplasm.

The first cell movement is the *epiboly* of the blastoderm cells over the yolk. In the initial phase, the inner blastoderm cells move outward to intercalate with the more superficial cells (Warga and Kimmel, 1990). Later, the cells move over the surface of the yolk to envelop it completely (Figure 6.11). This movement is not due to the active crawling of the blastomeres, however. Rather, the movement is provided by the autonomously expanding

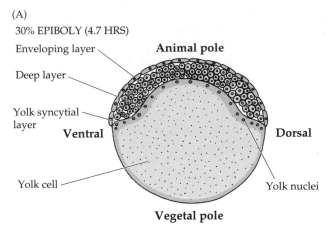

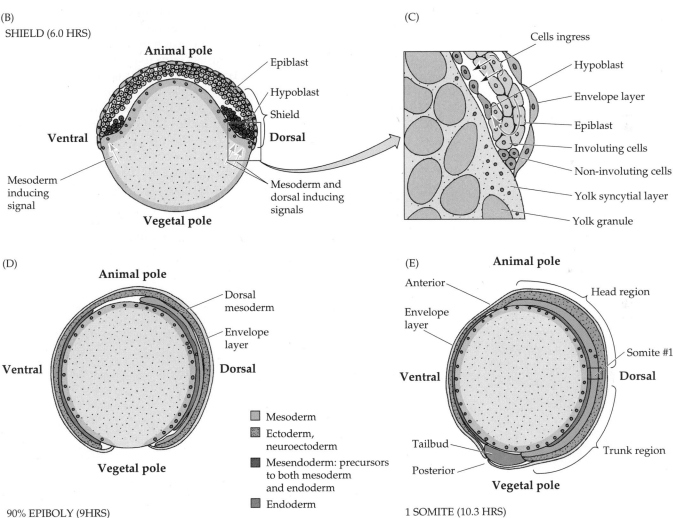

Figure 6.11
Cell movements during gastrulation of the teleost *Danio rerio*. (A) The blastoderm at 30 percent completion of epiboly. (B) Formation of the hypoblast, either by involution of cells at the margin of the epibolizing blastoderm or by delamination of cells from the epiblast. (C) Close-up of the marginal region. (D) At 90 percent epiboly, mesoderm can be seen surrounding yolk, between the endoderm and ectoderm. (E) Completion of gastrulation. (After Driever, 1995, and Langeland and Kimmel, 1997.)

yolk syncytial layer (YSL) "within" the animal cap cytoplasm. The enveloping layer (EVL) is tightly joined to the YSL and is dragged along with it. The deep cells of the blastoderm then fill in the space between the YSL and the EVL as epiboly proceeds. This can be demonstrated by severing the attachments between the YSL and the EVL. When this is done, the blastoderm cells spring back to the top of the yolk, while the YSL continues its expansion around the yolk cell (Trinkaus, 1984b, 1992). The expansion of the YSL is predicated on a network of microtubules in the YSL, and radiation or drugs that block the polymerization of tubulin inhibit epiboly (Strahle and Jesuthasan, 1993; Solnica-Krezel and Driever, 1994).

During migration, one side of the blastoderm becomes noticeably thicker than the other. Cell-labeling experiments indicate that the thinner side marks the site of the future dorsal surface of the embryo (Schmidt and Campos-Ortega, 1995).

Formation of Germ Layers

After the blastoderm cells have covered about half the zebrafish yolk cell (and earlier in fish eggs with larger yolks), a thickening occurs throughout the entire margin. This thickening is called the **germ ring,** and it is composed of a superficial layer, the **epiblast,** and an inner layer, the **hypoblast.** We do not understand how the hypoblast is made. Some laboratories claim that the hypoblast is formed by the *involution* of superficial cells under the margin followed by their migration toward the animal pole (see Figure 6.11). The involution begins at the future dorsal portion of the embryo, but occurs all around the margin. Other laboratories claim that these hypoblast cells *ingress* to form the hypoblast (see Trinkaus, 1996). (It is possible that both mechanisms are at work, with different modes of hypoblast formation predominating in different species.) Once formed, however, the deep cells of both the epiblast and hypoblast intercalate on the future dorsal side of the embryo to form a localized thickening, the **embryonic shield** (Figure 6.12). This shield is functionally equivalent to the dorsal blastopore lip of amphibia, since it can organize a secondary embryonic axis when transplanted to a host embryo (Oppenheimer, 1936; Ho, 1992; see the discussion of amphibian gastrulation).

Thus, as the cells undergo epiboly around the yolk, they are also involuting at the margins and converging anteriorly and dorsally toward the embry-

(A) **Animal pole**

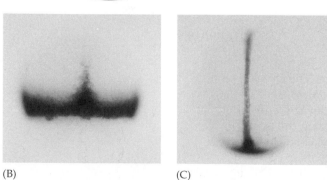

- Extension
- Embryonic shield
- Convergence
- Involution
- Epiboly
- Yolk cell

Figure 6.12
Convergence and extension in the zebrafish gastrula. (A) Dorsal view of the convergence and extension movements during zebrafish gastrulation. Epiboly spreads the blastoderm over the yolk; involution or ingression generates the hypoblast; convergence and extension bring the hypoblast and epiblast cells to the dorsal side to form the embryonic shield. Within the shield, intercalation extends the chordamesoderm toward the animal pole. (B,C). Convergent extension of chordamesoderm is shown by those cells expressing the gene *no tail*, a gene that is expressed by notochord cells. (D,E) Convergent extension of adaxial mesodermal cells (marked by their expression of the *snail* gene) to flank the notochord. (From Langeland and Kimmel, 1997.)

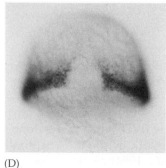

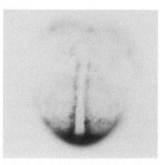

(B) (C) (D) (E)

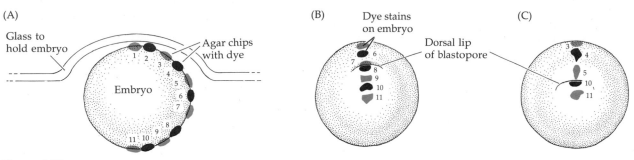

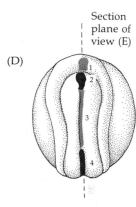

Figure 6.13
Vital dye staining of amphibian embryos. (A) Vogt's method for marking specific cells of the embryo surface with vital dyes. (B–D) Surface views of stain on successively later embryos. (E) Newt embryo dissected in medial plane to show stained cells in the interior. (After Vogt, 1929.)

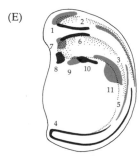

onic shield (Trinkaus, 1992). The hypoblast cells of the embryonic shield converge and extend anteriorly, eventually narrowing along the dorsal midline of the hypoblast. This is the **chordamesoderm,** the primordium of the notochord (Figure 6.12B,C). The cells adjacent to the chordamesoderm, the **adaxial cells,** are the precursors of the mesodermal somites (Figure 6.12D,E). The convergence and extension in the epiblast brings the presumptive brain cells from all over the epiblast into the dorsal midline, where they form the **neural keel.** The rest of the epiblast becomes the skin of the fish. The zebrafish fate map, then, is not much different from that of the frog or other vertebrates (as we will soon see). If one conceptually opens a *Xenopus* blastula at the vegetal pole and stretches the opening into a marginal ring, the resulting fate map closely resembles that of the zebrafish embryo when half of the yolk has been covered by the blastoderm (Langeland and Kimmel, 1997).

Amphibian gastrulation

The study of amphibian gastrulation is both one of the oldest and one of the newest areas of experimental embryology; even though amphibian gastrulation has been extensively studied for the past century, most of our theories concerning the mechanisms of these developmental movements have been revised over the past decade. The study of amphibian gastrulation has been complicated by the fact that there is no single way that all amphibians gastrulate. Different species employ different means toward the same goal (Smith and Malacinski, 1983; Lundmark, 1986). In recent years, the most intensive investigations have focused on *Xenopus,* so we will concentrate on its mode of gastrulation.

Cell Movements During Amphibian Gastrulation

Amphibian blastulae are faced with the same tasks as their echinoderm and piscine counterparts, namely, to bring inside those areas destined to form the endodermal organs, to surround the embryo with cells capable of forming the ectoderm, and to place the mesodermal cells in the proper places between them. The movements whereby this is accomplished can be visualized by the technique of vital dye staining. Vogt (1929) saturated agar chips with dyes, such as neutral red or Nile blue sulfate, that would stain but not damage the embryonic cells. These stained agar chips were pressed to the surface of the blastula, and some of the dye was transferred onto the contacted cells (Figure 6.13). The movements of each group of stained cells were followed throughout gastrulation, and the results were summarized in fate

Figure 6.14
Fate maps of the blastula of the frog *Xenopus laevis*. Fate maps of (A) the exterior and (B) the interior cells indicate that most of the mesodermal derivatives are formed from the interior cells. In this lateral view, the point at which the dorsal blastopore lip forms is indicated by an arrow. (After Keller, 1976.)

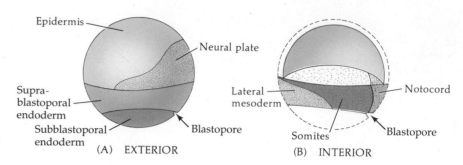

maps. These maps have recently been confirmed and extended by scanning electron microscopy and dye injection techniques (Smith and Malacinski, 1983; Lundmark, 1986).

Vital dye studies by Løvtrup (1975; Landstrom and Løvtrup, 1979) and by Keller (1975, 1976) have shown that cells of the *Xenopus* blastula have different fates depending on whether they are in the deep or the superficial layers of the embryo (Figure 6.14). In *Xenopus*, the mesodermal precursors exist *solely* in the deep layer of cells, while the ectoderm and endoderm arise from the superficial layer on the surface of the embryo. The precursors for the notochord and other mesodermal tissues are located beneath the surface in the equatorial (marginal) region of the embryo. In urodeles (salamanders such as *Triturus* and *Ambystoma*) and in some frogs other than *Xenopus*, the notochord and mesoderm precursors are found in *both* the surface cells and the deep marginal cells (Purcell and Keller, 1993).

Gastrulation in frog embryos is initiated at the future dorsal side of the embryo, just below the equator in the region of the gray crescent (Figure 6.15). Here the prospective local endodermal cells invaginate to form a slit-like blastopore. These cells change their shape dramatically. The main body of each cell is displaced toward the inside of the embryo while it maintains contact with the outside surface by way of a slender neck (Figure 6.16). These **bottle cells** line the initial archenteron. Thus, as in the gastrulating sea urchin, an invagination of cells initiates archenteron formation. However, unlike gastrulation in sea urchins, gastrulation in the frog begins not at the most vegetal region, but in the **marginal zone** near the equator of the blastula, where the animal and vegetal hemispheres meet. Here the endodermal cells are not as large or as yolky as the most vegetal blastomeres.

The next phase of gastrulation involves the involution of the marginal zone cells while the animal cells undergo epiboly and converge at the blastopore. When the migrating marginal cells reach the **dorsal lip of the blastopore,** they turn inward and travel along the inner surface of the outer cell sheets. Thus, the cells constituting the lip of the blastopore are constantly changing. The first cells to compose the dorsal lip are the bottle cells that invaginated to form the leading edge of the archenteron. These cells later become the pharyngeal cells of the foregut. As these first cells pass into the interior of the embryo, the blastopore lip becomes composed of cells that involute into the embryo to become the precursors of the head mesoderm. The next cells involuting into the embryo through the dorsal lip of the blastopore are called the chordamesoderm cells. These cells will form the **notochord,** a transient mesodermal "backbone" that is essential for initiating the differentiation of the nervous system.

As the new cells enter the embryo, the blastocoel is displaced to the side opposite the dorsal blastopore lip. Meanwhile, the blastopore lip expands laterally and ventrally as the processes of bottle cell formation and involu-

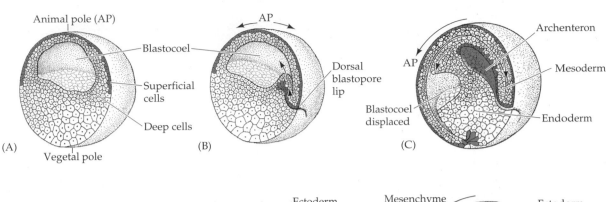

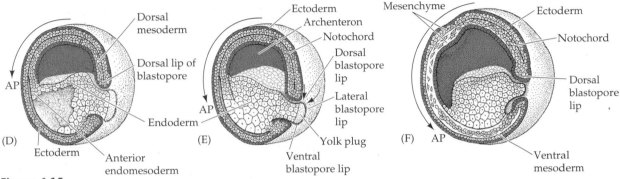

Figure 6.15
Cell movements during frog gastrulation. The sections are cut through the middle of the embryo and are positioned so that the vegetal pole is tilted toward the observer and slightly to the left. The major cell movements are indicated by arrows, and the superficial animal hemisphere cells are colored so that their movements can be followed. (A,B) Early gastrulation. Bottle cells of the margin move inward to form the dorsal lip of the blastopore, and mesodermal precursors involute under the roof of the blastocoel. The AP marks the position of the animal pole, and this will change as gastrulation continues. (C,D) Midgastrulation. The archenteron forms and displaces the blastocoel, and cells migrate from the lateral and ventral blastopore lips into the embryo. The cells of the animal hemisphere migrate down toward the vegetal region, moving the blastopore to the region near the vegetal pole. (E,F) Toward the end of gastrulation, the blastocoel is obliterated, the embryo becomes surrounded by ectoderm, the endoderm has been internalized, and the mesodermal cells have been positioned between the ectoderm and endoderm. (After Keller, 1986.)

Figure 6.16
Structure of the blastopore lip. (A) Diagram of cells seen in a section of gastrulating salamander embryo, showing the extension of the bottle cells from the blastopore. (B) Surface view of an early dorsal blastopore lip of *Xenopus*. The size difference between the animal and vegetal blastomeres is readily apparent. (C) Close-up of the region where the animal hemisphere cells are involuting through the blastopore lip. (A after Holtfreter, 1943; B and C, scanning electron micrographs courtesy of C. Phillips.)

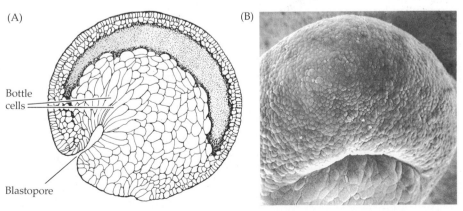

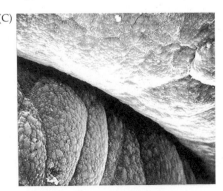

(A)

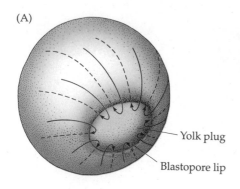

Yolk plug

Blastopore lip

(B)

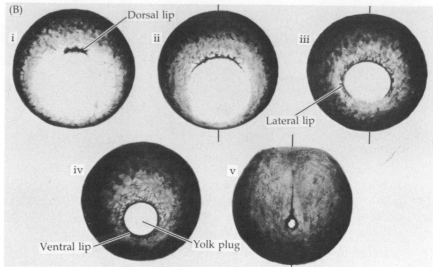

Figure 6.17
Epiboly of the ectoderm. (A) Morphogenetic movements of the cells migrating into the blastopore and then under the surface. (B) Changes in the region around the blastopore as the dorsal, lateral, and ventral lips are formed in succession. When the ventral lip completes the circle, the endoderm becomes progressively internalized. Numbers ii–v correspond to Figures 6.15B–E, respectively. (B from Balinsky, 1975, courtesy of B. I. Balinsky.)

tion continue about the blastopore. The widening blastopore "crescent" develops lateral lips and finally a ventral lip over which additional mesodermal and endodermal precursor cells pass. With the formation of the ventral lip, the blastopore has formed a ring around the large endodermal cells that remain exposed on the vegetal surface. This remaining patch of endoderm is called the **yolk plug;** it, too, is eventually internalized (Figure 6.17). At that point, all the endodermal precursors have been brought into the interior of the embryo, the ectoderm has encircled the surface, and the mesoderm has been brought between them.

Positioning the Blastopore

Having seen the general features of amphibian gastrulation, we can now look at each step in detail. Gastrulation does not exist as an independent process in the life of an animal. In fact, the preparation for gastrulation can be traced back to the literal moment of sperm-egg fusion. The unfertilized egg has a polarity along the animal-vegetal axis. The general fate of these regions can be predicted before fertilization. The surface of the animal hemisphere will become the cells of the ectoderm (skin and nerves), the vegetal hemisphere surface will form the cells of the gut and associated organs (endoderm), and the mesodermal cells will form from the internal cytoplasm around the equator. Thus, the germ layers can be mapped onto the unfertilized ovum; but this tells us nothing about which part of the egg will form the belly and which the back. The dorsal-ventral (back-front), anterior-posterior, and left-right axes have not yet been determined.

The dorsal-ventral and anterior-posterior axes are specified by displacements of the zygote cytoplasm during fertilization. In Chapter 4, we discussed the rotation of the cortical cytoplasm relative to the internal cytoplasm of the frog egg. The internal cytoplasm remains oriented with respect to gravity because of its dense yolk accumulation, while the cortical cytoplasm actively rotates 30° animally ("upward") toward the point of sperm entry (see Figure 4.34). This rotation causes the animal-vegetal axis of the egg surface to be offset 30° relative to the animal-vegetal axis of the internal cytoplasm. In this way, a new state of symmetry is acquired. Whereas the unfertilized egg had been radially symmetrical about the animal-vegetal axis, the fertilized egg now has a dorsoventral axis. It has become bilaterally symmetrical (having right and left sides). The inner cytoplasm moves as well, and fluorescence microscopy of early embryos has shown that the cytoplasmic patterns of presumptive dorsal cells differ from those of the presumptive ventral cells (Plate 7).

These cytoplasmic movements activate the cytoplasm opposite the point of sperm entry to initiate gastrulation (Figure 6.18). The side where the sperm enters marks the future ventral surface of the embryo; the opposite side, where gastrulation is initiated, marks the future dorsum (back) of the embryo (Gerhart et al., 1981, 1986; Vincent et al., 1986). Although the sperm is not needed to induce these movements in the egg cytoplasm, it is important in determining the direction of this rotation. If an artificially stimulated egg is enucleated, the cortical rotation still takes place at the correct time. However, the direction of this movement is unpredictable. (In fact, in dispermic eggs, there is only one direction of rotation.) The sperm appears to provide a spatial cue that orients the autonomous rotation of the cytoplasm, but it is the cytoplasmic rotation that is essential for further development. Moreover, if this cortical rotation is blocked, there is no dorsal development, and the embryo dies as a mass of ventral (primarily gut) cells (Vincent and Gerhart, 1987). The direction of cytoplasmic movement determines which side is to be dorsal and which side is to be ventral.

The directional bias provided by the point of sperm entry can be overridden by mechanically redirecting the spatial relationship between the cortical and subcortical cytoplasms. When the egg is prevented from rotating (by immersing it in a polysaccharide that collapses the perivitelline space between the egg and the fertilization envelope), one can turn the egg on its side 90° so that the animal-vegetal axis is horizontal rather than vertical and the point of sperm entry faces upward (Gerhart et al., 1981; Kirschner and Gerhart, 1981; Cooke, 1986). When fertilized eggs are inclined this way for 30 minutes starting halfway through the first cleavage cycle, the cytoplasm rotates such that almost all the embryos initiate gastrulation on the *same* side as sperm entry (see Figure 6.18).

The preceding discussion suggests that one might be able to get two gastrulation initiation sites if one were to combine the sperm-oriented rotation with an artificially induced rotation of the egg. Black and Gerhart (1985) let the initial sperm-directed rotation occur, but they then immobilized eggs in gelatin and gently centrifuged them so that the internal cytoplasm would go toward the point of sperm entry. When the centrifuged eggs were then allowed to develop in normal water, two sites of gastrulation emerged, leading to conjoined twin larvae (Figure 6.19). Black and Gerhart hypothesize (1986) that such twinning is caused by the formation of two areas of interaction: one axis forms where the normal cortical rotation caused the cytoplasmic interactions in the vegetal region of the cell, the other axis forms where the centrifugation-driven cytoplasm interacts with the vegetal components. Twins can also be produced at normal gravity by placing the sperm entry side of the egg uppermost after removing the egg from its fertilization envelope (Gerhart et al., 1981).

The ability to obtain two functional blastopore lips also suggests that there is nothing unique about the gray crescent, where gastrulation is first seen to begin. Rather, the gastrulation-inducing factors appear to be created

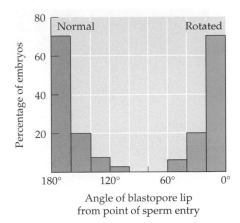

Figure 6.18
Relationship between the point of sperm entry and the dorsal blastopore lip in normal and rotated frog eggs. *Xenopus* eggs were fertilized, dejellied, and placed in Ficoll to dehydrate the perivitelline space. The sperm entry point was marked with dye. Rotated eggs were inclined 90°, with the sperm entry point facing upward, 50–80 minutes after fertilization. (After Gerhart et al., 1981.)

(A)

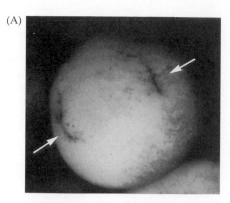

(B)

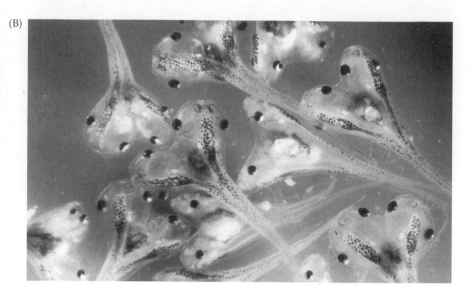

Figure 6.19
Twin blastopores produced by rotating dejellied *Xenopus* eggs ventral side (sperm entry point) up at the time of first cleavage. (A) Two blastopores are instructed to form: the original one (opposite the point of sperm entry) and the new one created by the displacement of cytoplasmic material. (B) These eggs develop two complete axes, which form twin tadpoles, joined ventrally. (Courtesy of J. Gerhart.)

by the interactions of animal and vegetal cytoplasm, interactions that probably activate some component in the vegetal cytoplasm. Gimlich and Gerhart (1984) performed a series of transplantation experiments that confirmed the hypothesis that the factors that initiate gastrulation originally lie in the deep cytoplasm of the dorsal vegetal cells rather than in the gray crescent. They demonstrated that the three most dorsal vegetal blastomeres of 64-cell *Xenopus* embryos are able to induce the formation of the dorsal lip of the blastopore and of a complete dorsal axis in UV-irradiated recipients (which otherwise would have failed to properly initiate gastrulation; Figure 6.20A). Moreover, these three blastomeres, which underlie the prospective dorsal lip region, can also induce a secondary invagination and axis when transplanted into the ventral side of a normal, unirradiated 64-cell embryo (Figure 6.20B). This small cluster of vegetal blastomeres enables its adjacent marginal cells to invaginate and form the dorsal mesodermal axis of the embryo. Holowacz and Elinson (1993) found that cortical cytoplasm from the dorsal vegetal cells of the 16-cell *Xenopus* embryo was able to induce the formation of secondary axes when injected into vegetal ventral cells. Neither cortical cytoplasm from animal cells nor the deep cytoplasm from ventral cells could induce such axes.

It appears, then, that internal rearrangements of the cytoplasm, probably those normally oriented by the entry of the sperm, are responsible for causing the asymmetrical distribution of subcellular factors. This asymmetry creates in the egg a dorsal-ventral distinction that ultimately directs the position of the blastopore above a set of vegetal blastomeres opposite the point of sperm entry. The molecules that may be involved in the formation of the vegetal gastrulation initiation site (the "Nieuwkoop center") will be discussed in Chapter 15.

Cell Movements and the Construction of the Archenteron

THE INITIATION OF GASTRULATION. Amphibian gastrulation is initiated when a group of marginal endoderm cells on the dorsal surface of the blastula sinks into the embryo. The outer (apical) surfaces of these cells contract dramatically, while their inner (basal) ends expand. The apical-basal length of these cells greatly increases to yield the characteristic "bottle" shape. In salamanders, these cells appear to have an active role in the early movements of gas-

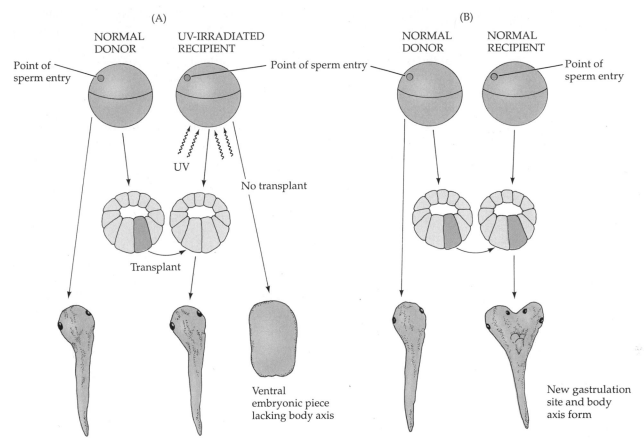

Figure 6.20
Transplantation experiments demonstrating that the vegetal cells underlying the prospective dorsal blastopore lip regions are responsible for causing the initiation of gastrulation. (A) Rescue of irradiated embryos by transplanting the vegetal blastomeres of the most dorsal segment (color) of a 64-cell embryo into a cavity made by the removal of a similar number of vegetal cells. An irradiated zygote without this transplant fails to undergo normal gastrulation. (B) Formation of new gastrulation site and body axis by the transplantation of the most dorsal vegetal cells of a 64-cell embryo into the ventralmost vegetal region of another 64-cell embryo. (After Gimlich and Gerhart, 1984.)

trulation. Johannes Holtfreter (1943, 1944) found that bottle cells from early salamander gastrulae could attach to glass coverslips and lead the movement of those cells attached to them. Even more convincing were Holtfreter's recombination experiments in which dorsal marginal zone cells (which would give rise to the dorsal blastopore lip) were combined with inner endoderm tissue. When the dorsal marginal zone cells were excised and placed on inner prospective endoderm tissue, the blastopore cell precursors formed bottle cells and sank below the surface of the inner endoderm (Figure 6.21). Moreover, as they sank, they created a depression reminiscent of the early blastopore. Thus, Holtfreter claimed that the ability to invaginate into the deep endoderm is an innate property of the dorsal marginal zone cells.

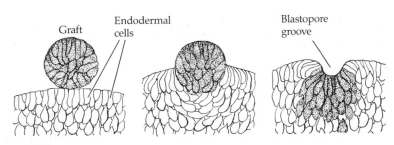

Figure 6.21
A graft of amphibian cells from the dorsal blastopore lip region sinks into a layer of endodermal cells and forms a blastopore groove. (After Holtfreter, 1944.)

The situation in the frog embryo is somewhat different. R. E. Keller and his students (Keller, 1981; Hardin and Keller, 1988) have shown that although the bottle cells of *Xenopus* may play a role in *initiating* the involution of the marginal zone as they become bottle-shaped, they are not essential for gastrulation to continue. The peculiar bottle shape of these cells is needed to initiate gastrulation; it is the constriction of the cells that pulls the marginal zone vegetally while pushing the vegetal cells inward (Figure 6.22A,B). The pulling of the marginal zone vegetally enables the ectoderm to expand vegetally and encircle the embryo, while the pushing of the vegetal cells enables these anterior mesodermal precursors to contact the underside of the poste-

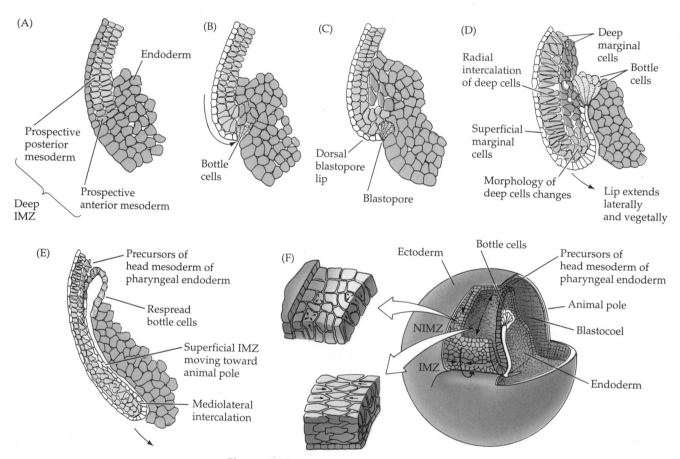

Figure 6.22

Integrative model of cell movements during early *Xenopus* gastrulation. (A) Structure of the involuting marginal zone (IMZ) prior to gastrulation. The deep IMZ consists of the prospective anterior mesoderm and the prospective posterior mesoderm. (B) Constriction of the bottle cells pushes up the prospective anterior mesoderm and rotates the IMZ outward. (C) The anterior mesodermal precursors lead the movement of the mesoderm into the blastocoel. (D) Radial intercalation (interdigitation) of the deep IMZ cells occurs. The mesoderm moves toward the animal pole, pulling along the superficial cells and the bottle cells by involution. (E) As gastrulation continues, the deep marginal cells flatten, and the formerly superficial cells form the wall of the archenteron. (F) Radial intercalation as in (D), looking down at the dorsal blastopore lip from the dorsal surface. In the NIMZ and upper portion of the IMZ, deep (mesodermal) cells are intercalating radially to make a thin band of flattened cells. This thinning of several layers into a few causes extension toward the blastopore lip. Just above the lip, mediolateral intercalation of the cells produces stresses that pull the IMZ over the lip. After involuting over the lip, mediolateral intercalation continues, elongating and narrowing the axial mesoderm. (After Hardin and Keller, 1988; Wilson and Keller, 1991.)

rior mesodermal precursors and begin migrating on the roof of the blasto-coel (Hardin and Keller, 1988).

However, after starting these movements, the *Xenopus* bottle cells are no longer needed. When bottle cells are removed after their formation, involution and blastopore formation and closure continue. The major factor in the movement of cells into the embryo appears to be the involution of the *subsurface* marginal cells rather than the superficial ones. It appears that these subsurface, **deep involuting marginal zone cells** turn inward and migrate toward the animal pole along the inside surfaces of the remaining deep cells (Figure 6.22C–E) and that the superficial layer forms the lining of the archenteron merely because it is attached to the actively migrating deep cells. The movement of the bottle cells deeper into the embryo depends on their attachment to the underlying deep cells. While removal of the *bottle* cells does not affect the involution of the deep or superficial marginal zone cells into the embryo, the removal of the dorsal marginal zone *deep* cells and their replacement with animal region cells (which do not normally undergo involution) stops archenteron formation.

THE FORMATION OF MESODERM DURING *XENOPUS* GASTRULATION. Figures 6.22D–F depict the behavior of these involuting marginal zone (IMZ) cells at successive stages of *Xenopus* gastrulation. (Keller and Schoenwolf, 1977; Keller, 1980, 1981; Hardin and Keller, 1988). Shortly before their involution through the blastopore lip, the several layers of deep IMZ cells intercalate radially to form one thin, broad layer. This intercalation further extends the IMZ vegetally. At the same time, the superficial cells spread out by dividing and flattening. When the deep cells reach the blastopore lip, they involute into the embryo and initiate a second type of intercalation. This intercalation causes a convergent extension along the mediolateral axis (Figure 6.22F) that integrates several mesodermal streams to form a long narrow band. The anterior part of this band migrates toward the animal cap. Thus, the mesodermal stream continues to migrate toward the animal pole, and the overlying layer of superficial cells (including the bottle cells) is passively pulled toward the animal pole, thereby forming the endodermal roof of the archenteron (see Figures 6.15 and 6.22E). Therefore, although the bottle cells may be responsible for creating the initial groove, the motivating force for this involution appears to come from the deep layer of the marginal cells. Furthermore, the radial and mediolateral intercalations of the deep layer of cells appear to be responsible for the continued movement of mesoderm into the embryo.

Migration of the Involuting Mesoderm

As mesodermal movement progresses, convergent extension continues to narrow and lengthen the involuting marginal zone. The IMZ contains the prospective endodermal roof of the archenteron in its superficial layer (IMZ$_S$) and the prospective mesodermal cells, including those of the notochord, in its deep region (IMZ$_D$). During the middle third of gastrulation, the expanding sheet of mesoderm converges toward the midline of the embryo. This process is driven by the continued mediolateral intercalation of cells along the anterior-posterior axis, thereby further narrowing the band. Toward the end of gastrulation, the centrally located notochord separates from the somitic mesoderm on either side of it, and the cells elongate separately (Wilson and Keller, 1991). This convergent extension of the mesoderm appears to be autonomous, because the movements of these cells occur even if this region of the embryo is isolated from the rest of the embryo (Keller, 1986).

During gastrulation, the **animal cap** and **noninvoluting marginal zone (NIMZ) cells** expand by epiboly to cover the entire embryo. The dorsal portion of the noninvoluting marginal cells expands more rapidly than the ventral portion, thus causing the blastopore lips to move toward the ventral side. While those mesodermal cells entering through the dorsal blastopore lip give rise to the dorsal axial mesoderm, the remainder of the body mesoderm (which forms the heart, kidneys, blood, bones, and parts of several other organs) enters through the ventral and lateral blastopore lips to create the **mesodermal mantle.** The endoderm is derived from the IMZ_S cells that form the lining of the archenteron roof and from the subblastoporal vegetal cells that become the archenteron floor (Keller, 1986).

Sidelights & Speculations

Molecular Regulators of Development: Fibronectin and the Pathways for Mesodermal Migration

HOW ARE THE INVOLUTING CELLS informed where to go once they enter the inside of the embryo? In salamanders, it appears that the involuting mesodermal precursors migrate toward the animal pole on a fibronectin lattice secreted by the cells of the blastocoel roof. Shortly before gastrulation, the presumptive ectoderm of

Figure 6.23 Fibronectin and amphibian gastrulation. (A) Immunofluorescence reveals a fibrillar network of fibronectin on the basal surface of the prospective ectodermal cells lining the blastocoel roof in the salamander embryo. (B–E) Scanning electron micrographs of normal (B,C) and abnormal (D,E) salamander gastrulation. The blastocoel in (D) and (E) was injected with the cell-binding fragment of fibronectin, while the normally gastrulating blastula was injected with a control solution. (B) Section during midgastrulation. (C) The yolk plug toward the end of gastrulation. (D,E) The finishing stages of the arrested gastrulation, wherein the mesodermal precursors, having bound the synthetic fibronectin, cannot recognize the normal fibronectin-lined migration route. The archenteron fails to form, and the noninvoluted mesodermal precursors remain on the surface. (ar, archenteron; bc, blastocoel; bl, blastopore; ec, ectoderm; en, endoderm; mes, mesoderm; yp, yolk plug.) (A from Boucaut et al., 1985; B–E from Boucaut et al., 1984, courtesy of J.-C. Boucaut and J.-P. Thiery.)

(A)

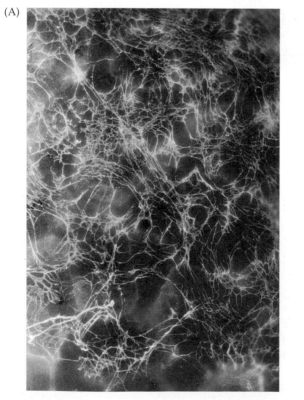

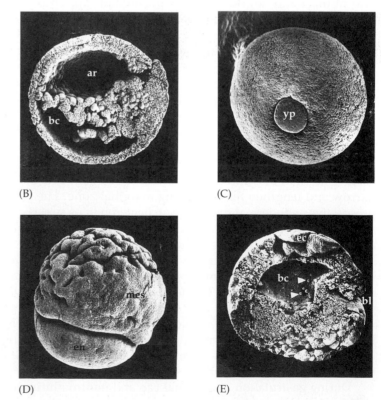

(B)

(C)

(D)

(E)

the blastocoel roof secretes an extracellular matrix that contains fibrils of fibronectin (Figure 6.23A; Boucaut et al., 1984; Nakatsuji et al., 1985). The involuting mesoderm appears to travel on these fibronectin fibers. Confirmation of this was obtained by chemically synthesizing a "phony" fibronectin that can compete with the genuine fibronectin of the extracellular matrix. Cells bind to a certain region of the fibronectin protein that contains a three-amino-acid sequence (Arg-Gly-Asp; RGD). Boucaut and co-workers injected large amounts of a small peptide containing this sequence into the blastocoels of salamander embryos shortly before gastrulation began. If fibronectin were essential for cell migration, then cells binding this soluble peptide fragment instead of the real cell-bound fibronectin should stop. Unable to find their "road," the mesodermal cells should cease their involution. This is precisely what occurred (Figure 6.23B–E). No migrating cells were seen along the underside of the ectoderm. Instead, the mesodermal precursors remained outside the embryo, forming a convoluted cell mass. Other small synthetic peptides (including other fragments of the fibronectin molecule) did not impede migration.

The mesodermal cells are thought to adhere to this fibronectin through the $\alpha v \beta 1$ integrin protein (Alfandari et al., 1995). Mesodermal migration can also be arrested by the microinjection of antibodies against either fibronectin or the $\beta 1$ integrin subunit that serves as part of the fibronectin receptor (D'Arribère et al., 1988, 1990). Alfandari and colleagues (1995) have shown that shortly after fertilization, the α_v subunit of integrin is progressively lost from the blastomere cell membranes. However, just before and during gastrulation, the α_v subunit is expressed on the surface of the migrating mesoderm cells. It appears, then, that the synthesis of this fibronectin receptor may signal the time for the mesoderm to begin and continue migration.

In addition to permitting the attachment of mesodermal cells to the fibronectin lattice, this fibronectin-containing extracellular matrix appears to provide the cues for the *direction* of cell migration. Shi and colleagues (1989) ex-

cised the blastocoel roofs from early salamander gastrulae and deposited them on plastic dishes with their extracellular matrices touching the plastic (Figure 6.24A). The axis from the blastopore to the animal pole was marked, and after 2 hours, the explant was removed, leaving behind its extracellular matrix. A smaller explant from the dorsal marginal zone (DMZ) was then taken from another early gastrula and placed on the matrix with its own blastopore–animal pole axis perpendicular to that of the matrix. Would the cells of this explant migrate on the matrix, and would they migrate in a particular direction? The cells were found to migrate, and the migration could be inhibited by antibodies that block the cells' ability to recognize fi-

bronectin. Moreover, rather than migrating randomly, the DMZ cells migrated to the animal pole of the extracellular matrix that had been absorbed onto the plastic (Figure 6.24B).

In *Xenopus,* convergent extension pushes the migrating cells upward toward the animal pole. However, fibronectin appears to delineate the boundaries within which this movement can occur. The fibronectin of *Xenopus* gastrulae does not form complex lattices. Rather, it is organized into small fibrillar clusters. If fibronectin is synthesized but not organized into these fibrils, the dorsal mesodermal cells will adhere to the basal surface of the presumptive ectoderm but will not migrate (Winklbauer and Nagel, 1991). The fibronectin fibrils are necessary for the head mesodermal cells to flatten and to extend broad (lamelliform) processes in the direction of migration (Winklbauer et al., 1991; Winklbauer and Keller, 1996). The importance of these fibronectin fibrils is also seen in interspecific hybrids that arrest at gastrulation. Delarue and colleagues (1985) have shown that certain inviable hybrids between two species of toads arrest during gastrulation because they do not secrete these fibronectin fibrils. It appears, then, that the extracellular matrix of the blastocoel roof, and particularly its fibronectin component, is important in the migration of the mesodermal cells during amphibian gastrulation. ■

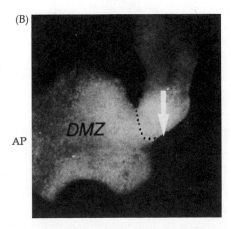

Figure 6.24 Direction of salamander dorsal marginal zone (DMZ) cell migration depends on orientation of the extracellular matrix of the blastocoel roof. (A) Explants of blastocoel roof from the blastopore (BP) to the animal pole (AP) were dissected from early-gastrula-stage salamander embryos and placed on plastic dishes. The extracellular matrix adhered to the dish, and the tissue was then removed. A smaller explant from an early gastrula, containing DMZ cells, was placed on this matrix with its own axis perpendicular to that of the matrix. (B) DMZ cells from the explant migrate toward the animal pole of the matrix. The dotted line indicates the original boundary of the explant, and the white arrow represents its blastopore–animal pole axis. (From Shi et al., 1989, photograph courtesy of the authors.)

Epiboly of the Ectoderm

While involution is occurring at the blastopore lips, the ectodermal precursors are expanding over the entire embryo. Keller (1980) and Keller and Schoenwolf (1977) have used scanning electron microscopy to observe the changes in both the superficial cells and the deep cells of the animal and marginal regions. The major mechanism of epiboly in *Xenopus* gastrulation appears to be an increase in cell number (through division) coupled with a concurrent integration of several deep layers into one (Figure 6.25). During early gastrulation, three rounds of cell division increase the number of the deep cell layers in the animal hemisphere. At the same time, complete integration of the numerous deep cells into one layer occurs. The most superficial layer expands by cell division and flattening. The spreading of cells in the dorsal and ventral marginal zones appears to proceed by the same mechanism, although changes in cell shape appear to play a greater role than in the animal region. The result of these expansions is the epiboly of the superficial and deep cells of the animal cap and noninvoluting marginal regions over the surface of the embryo (Keller and Danilchik, 1988). Most of the marginal region cells, as previously mentioned, involute to join the mesodermal cell stream within the embryo.

Gastrulation in *Xenopus* is the orchestration of several distinct events. The first indication of gastrulation involves the local *invagination* of endodermal bottle cells in the marginal zone at a precisely defined time and place. Next, the *involution* of marginal cells through the blastopore lip begins the formation of the archenteron. These involuting cells at the leading edge of the mesodermal mantle *migrate* along the inner surface of the blastopore roof, and the prospective chordamesoderm behind them narrows and lengthens posteriorly by convergent extension in the dorsal portion of the embryo. At the same time, the ectodermal precursor cells *epibolize* vegetally by cell division and by the integration of previously independent cell layers. The result of these cell movements is the proper positioning of the three germ layers in preparation for their differentiation into the body organs. Molecular studies (to be discussed in Chapter 15) are beginning to give us clues concerning the mechanisms by which the cells are told how to begin and end these migrations.

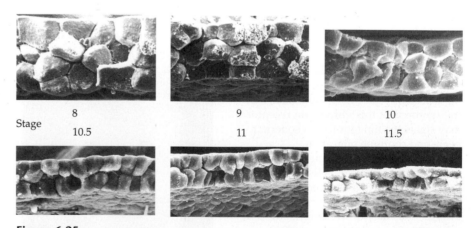

Figure 6.25
Scanning electron micrographs of the *Xenopus* blastocoel roof, showing the changes in cell shape and arrangement. Stages 8 and 9 are blastulae; stages 10–11.5 represent progressively later gastrulae. (From Keller, 1980, courtesy of R. E. Keller.)

Gastrulation in birds

Overview of Avian Gastrulation

Cleavage in avian embryos creates a blastodisc above an enormous volume of yolk. This inert, underlying yolk mass imposes severe constraints on cell movements, and avian gastrulation appears at first glance to be very different from that of a sea urchin or a frog. We will soon see, though, that there are numerous similarities between avian gastrulation and those gastrulations we have already studied. Moreover, we will see that mammalian embryos—which do not have yolk—retain gastrulation movements very similar to those of bird and reptile embryos.

FORMATION OF THE HYPOBLAST AND EPIBLAST. The central cells of the avian blastodisc are separated from the yolk by a subgerminal cavity and appear clear—hence, the center of the blastodisc is called the **area pellucida.** In contrast, the cells at the margin of the area pellucida appear opaque because of their contact with the yolk; they form the **area opaca** (Figure 6.26). Whereas most of the cells remain at the surface, forming the **epiblast,** certain cells migrate individually into the subgerminal cavity to form the **polyinvagination islands (primary hypoblast)**, an archipelago of disconnected clusters containing 5–20 cells each (see Figures 6.26 and 5.33). Shortly thereafter, a sheet of cells from the posterior margin of the blastoderm (Koller's sickle and the marginal zone behind it) migrates anteriorly to join the polyinvagination islands, thereby forming the **secondary hypoblast** (Eyal-Giladi et al., 1992). The two-layered blastoderm (epiblast and hypoblast) is joined together at the margin of the area opaca, and the space between the layers is a blasto-

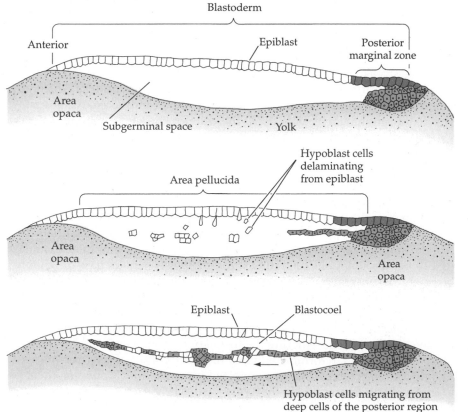

Figure 6.26
Formation of the two-layered blastoderm of the chick embryo. The first hypoblast cells delaminate individually to form islands of cells beneath the epiblast. Cells from the posterior margin (Koller's sickle and the posterior marginal cells behind it) produce a population of cells that migrates beneath the blastodisc and incorporates the polyinvagination islands. This bottom layer becomes the hypoblast. The upper layer is the epiblast. As the hypoblast moves anteriorly, epiblast cells collect at the region anterior to Koller's sickle to form the primitive streak.

coel. Thus, the structure of the avian blastodisc is not unlike that of the amphibian or echinoderm blastula.

The fate map for the avian embryo is restricted to the epiblast. That is, the hypoblast does not contribute any cells to the developing embryo (Rosenquist, 1966, 1972). Rather, the hypoblast cells form portions of the external membranes, especially the yolk sac and the stalk that links the yolk mass to the endodermal digestive tube. All three germ layers of the embryo proper (plus a considerable amount of extraembryonic membrane) are formed from the epiblastic cells. Fate maps of the chick epiblast are shown in Figure 6.27. These maps integrate several types of mapping. Vital dyes and transplants of radioactive cells were useful in mapping out major trends, since they tended to mark groups of cells and diffused as development proceeded. Transplanting genetically marked cells (such as quail cells placed into chick embryos) got around the problem of diffusion but still marked relatively large clusters of cells. Recently, the use of viruses or fluorescent dyes has enabled researchers to follow individual cells throughout development (Schoenwolf, 1991). As can be seen in these figures, there is significant convergent extension as the primitive streak progresses anteriorly. Although cells in a particular region of the gastrula tend to become specific types of cells, they can still form different cell types if transplanted to another region of the embryo.

FORMATION OF THE PRIMITIVE STREAK. The major structure characteristic of avian, reptilian, and mammalian gastrulation is the **primitive streak.** This streak is first visible as a thickening of the epiblast cell layer at the posterior region of the embryo, just anterior to Koller's sickle (Figure 6.27A). This thickening is caused by the ingression of mesodermal cells from the epiblast into the blastocoel and by the migration of cells from the lateral region of the posterior epiblast toward the center (Figure 6.27B; Vakaet, 1984; Bellairs, 1986; Eyal–Giladi et al., 1992). As the thickening narrows, it moves anteriorly and constricts to form the definitive primitive streak. This streak extends 60–75 percent of the length of the area pellucida and marks the anterior-posterior axis of the embryo (Figure 6.27C–E). As the cells converge to form the primitive streak, a depression forms within the streak. This depression is called the **primitive groove,** and it serves as a blastopore through which the migrating cells pass into the blastocoel. Thus, the primitive groove is analogous to the amphibian blastopore. At the anterior end of the primitive streak is a regional thickening of cells called the **primitive knot** or **Hensen's node.** The center of this node contains a funnel-shaped depression (sometimes called the primitive pit) through which cells can pass into the blastocoel. Hensen's node is the functional equivalent of the dorsal lip of the amphibian blastopore. [gast1.html]

As soon as the primitive streak is formed, epiblast cells begin to migrate over the lips of the primitive streak and into the blastocoel (Figure 6.28). Like the amphibian blastopore, the primitive streak has a continually changing cell population. Cells migrating through Hensen's node pass down into the blastocoel and migrate anteriorly, forming foregut, head mesoderm, and notochord; cells passing through the lateral portions of the primitive streak give rise to the majority of endodermal and mesodermal tissues (Schoenwolf et al., 1992). Unlike the *Xenopus* mesoderm, which migrates as sheets of cells into the blastocoel, cells entering the inside of the avian embryo do so as individuals. Rather than forming a tightly organized sheet of cells, the ingressing population creates a loosely connected mesenchyme. Moreover, there is no true archenteron formed in the avian gastrula.

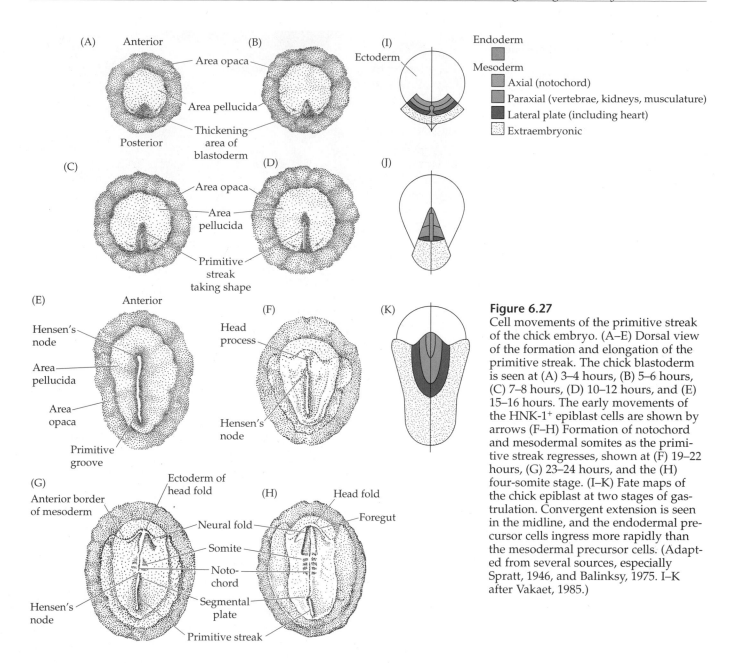

Figure 6.27
Cell movements of the primitive streak of the chick embryo. (A–E) Dorsal view of the formation and elongation of the primitive streak. The chick blastoderm is seen at (A) 3–4 hours, (B) 5–6 hours, (C) 7–8 hours, (D) 10–12 hours, and (E) 15–16 hours. The early movements of the HNK-1$^+$ epiblast cells are shown by arrows (F–H) Formation of notochord and mesodermal somites as the primitive streak regresses, shown at (F) 19–22 hours, (G) 23–24 hours, and the (H) four-somite stage. (I–K) Fate maps of the chick epiblast at two stages of gastrulation. Convergent extension is seen in the midline, and the endodermal precursor cells ingress more rapidly than the mesodermal precursor cells. (Adapted from several sources, especially Spratt, 1946, and Balinksy, 1975. I–K after Vakaet, 1985.)

As the cells enter the primitive streak, the streak elongates toward the future head region. At the same time, the secondary hypoblast cells are continuing to migrate anteriorly from the posterior margin of the blastoderm. The elongation of the primitive streak appears to be coextensive with the anterior migration of these secondary hypoblast cells.

MIGRATION THROUGH THE PRIMITIVE STREAK: FORMATION OF ENDODERM AND MESODERM. The first cells to migrate through the primitive streak are those destined to become the foregut. This situation is again similar to that seen in amphibians. Once inside the blastocoel, these cells migrate anteriorly and eventually displace the hypoblast cells in the anterior portion of the embryo. The hypoblast cells are confined to a region in the anterior portion of the area pellucida. This region, the **germinal crescent,** does not form any embry-

Figure 6.28
Migration of endodermal and mesodermal cells through the primitive streak. (A) Scanning electron micrograph shows epiblast cells passing into the blastocoel and extending their apical ends to become bottle cells. (B) Stereogram of a gastrulating chick embryo, showing the relationship of the primitive streak, the migrating cells, and the two original layers of the blastoderm. The lower layer becomes a mosaic of hypoblast and endodermal cells; but the hypoblast cells eventually sort out to form a layer beneath that of the endoderm and contribute to the yolk sac. (A from Solursh and Revel, 1978, courtesy of M. Solursh; B after Balinsky, 1975.)

(A)

(B)

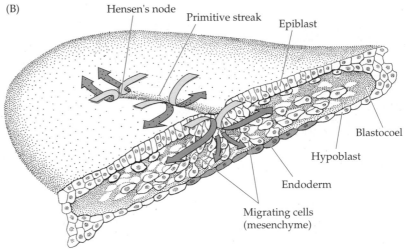

onic structures, but it does contain the precursors of germ cells, which later migrate through the blood vessels to the gonads. The next cells entering the blastocoel through Hensen's node (and the most anterior quarter of the primitive streak) also move anteriorly, but they do not move as far ventrally as the presumptive foregut endodermal cells. These cells remain between the endoderm and the epiblast to form the head mesoderm and the chordamesoderm (notochordal) cells (see Psychoyos and Stern, 1996). These early-ingressing cells have all moved anteriorly, pushing up the anterior midline region of the epiblast to form the **head process** (Figure 6.29). Meanwhile, cells continue migrating inwardly through the lateral portion of the primitive streak. As they enter the blastocoel, these cells separate into two streams. One stream moves deeper and joins the hypoblast along its midline, displacing the hypoblast cells to the sides. These deep-moving cells give rise to all the endodermal organs of the embryo as well as to most of the extraembryonic membranes (the hypoblast forms the rest). The second migrating stream spreads throughout the blastocoel as a loose sheet, roughly midway between the hypoblast and the epiblast. This sheet generates the mesodermal portions of the embryo and extraembryonic membranes. By 22 hours of incubation, most of the presumptive endodermal cells are in the interior of the embryo, although presumptive mesodermal cells continue to migrate inward for a longer time.

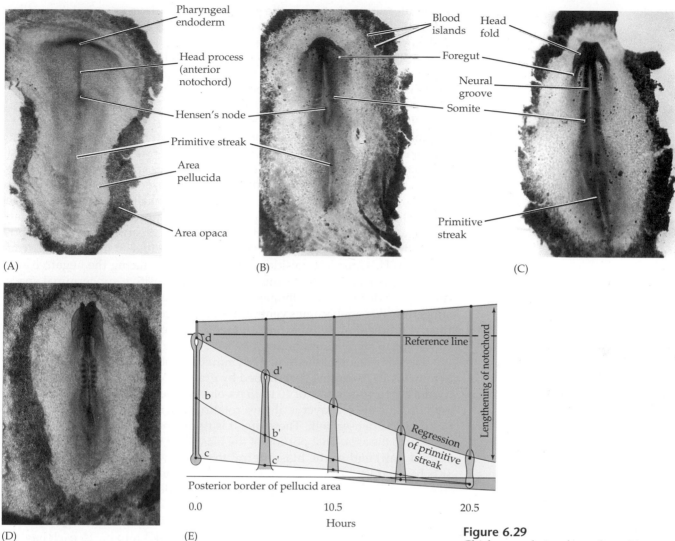

(A)

Pharyngeal endoderm

Head process (anterior notochord)

Hensen's node

Primitive streak

Area pellucida

Area opaca

(B)

Blood islands

Head fold

Foregut

Neural groove

Somite

Primitive streak

(C)

(D)

(E)

d

d'

b

b'

c

c'

Reference line

Regression of primitive streak

Lengthening of notochord

Posterior border of pellucid area

0.0   10.5   20.5
Hours

Figure 6.29
Chick gastrulation from about 24 to about 28 hours. (A) The primitive streak at full extension (24 hours). The head process (anterior notochord) can be seen extending from Hensen's node. (B) Two-somite stage (25 hours). Pharyngeal endoderm is seen anteriorly, while the anterior notochord pushes up the head process beneath it. The primitive streak is regressing. (C) Four-somite stage (27 hours). (D) At 28 hours, the primitive streak has regressed to the caudal portion of the embryo. (E) Regression of the primitive streak, leaving the notochord in its wake. Various points of the streak were followed after it achieved its maximum length. Time represents hours after achieving maximum length at about 18 hours. (Photographs courtesy of K. Linask; E after Spratt, 1947.)

Now a new phase of gastrulation begins. While the mesodermal ingression continues, the primitive streak starts to regress, moving Hensen's node from near the center of the area pellucida to a more posterior position (see Figure 6.29). It leaves in its wake the dorsal axis of the embryo and the head process. As the node moves further posteriorly, the remaining (posterior) portion of the notochord is laid down. Finally, the node regresses to its most posterior position, eventually forming the anal region. By this point, the epiblast is composed entirely of presumptive ectodermal cells.

As a consequence of this two-step gastrulation process, avian (and mammalian) embryos exhibit a distinct anterior-to-posterior gradient of developmental maturity. While cells of the posterior portions of the embryo are undergoing gastrulation, cells at the anterior end are already starting to form organs. For the next several days, the anterior end of the embryo is more advanced in its development (having had a "head start," if you will) than the posterior end.

While the presumptive mesodermal and endodermal cells were moving inward, the ectodermal precursors proliferated to become the only cell population remaining in the upper layer. Moreover, ectodermal cells migrated off the blastodisc to surround the yolk by epiboly. The enclosure of the yolk by the ectoderm (again reminiscent of the epiboly of amphibian ectoderm) is a Herculean task that takes the greater part of 4 days to complete and involves the continuous production of new cellular material and the migration of the presumptive ectodermal cells along the underside of the vitelline envelope. Thus, as avian gastrulation draws to a close, the ectoderm has surrounded the yolk, the endoderm has replaced the hypoblast, and the mesoderm has positioned itself between these two regions.

Mechanisms of Avian Gastrulation

THE ROLE OF THE HYPOBLAST AND THE FORMATION OF THE EMBRYONIC AXES.   The dorsoventral (back-belly) axis is critical to the formation of the hypoblast and to the further development of the embryo. This axis is established when the cleaving cells of the blastoderm establish a barrier between the basic (pH 9.5) albumen above the blastodisc and the acidic (pH 6.5) subgerminal space below the disc. Water and sodium ions are transported from the albumen through the cells and into the subgerminal space, creating a membrane potential difference of 25 mV across the cell layer (positive at the ventral side of the cells). This creates two sides to the cells: a side facing the negative and basic albumen and a side facing the acidic and positive subgerminal space fluid. The side facing the albumen becomes dorsal, while the side facing the subgerminal space becomes ventral. This can be reversed either by reversing the pH gradient or by inverting a potential difference across the cell layer (reviewed in Stern and Canning, 1988).

The conversion of the radially symmetrical blastoderm into a bilaterally symmetrical structure is determined by gravity. As the ovum rolls down the oviduct, it turns at a rate of 10 to 15 revolutions per hour. The cytoplasm that is to become the cell layer is always rotating downward but is displaced upward by the denser yolk. Therefore, it is not at the top of the yolk, but is off slightly to the side. That portion of the blastodisc that is highest becomes the caudal (tail) end of the blastoderm, that part where gastrulation begins (Kochav and Eyal-Giladi, 1971). Thus, the anterior-posterior and dorsolateral axes are determined prior to gastrulation, while the egg is slowly rolling down the oviduct.

The blastoderm of the chick embryo acts as a single integrating system to form a single embryo; if the blastoderm is separated into parts, each having its own marginal zone, each part will form its own embryo (Spratt and Haas, 1960). Control of this field seems to reside in the **posterior marginal zone,** the region where hypoblast formation begins. These posterior marginal cells not only contribute the inducing cells of the hypoblast, but also prevent other regions of the margin from inducing their own hypoblasts. Khaner and Eyal-Giladi (1989) found that if the posterior marginal zone is transposed to a lateral margin area (Figure 6.30A), the posterior tear heals and two primitive streaks emerge. Similarly, if a posterior region is reciprocally transposed with a lateral region (Figure 6.30B), only one primitive streak forms, and it arises from the original posterior region. However, if a posterior marginal zone is placed into an embryo that retains its original posterior margin (Figure 6.30C), only the host's original posterior margin forms the hypoblast that underlies the primitive streak. Khaner and Eyal-Giladi suggest that the marginal zone cells form a gradient of activity whose peak is at the posterior end. These posterior cells will form the hypoblast and at the same time prevent any cells with less activity from forming hypoblasts of their own.*

---

*Earlier investigators (Waddington, 1932; Azar and Eyal-Giladi, 1981) had thought that the hypoblast induced the formation of the primitive streak and provided it with anterior-posterior polarity. However, Khaner (1995) rotated the epiblast with respect to the hypoblast at different stages of chick development and showed that the epiblast initiates primitive streak formation and maintains its polarity independently of the orientation of the hypoblast.

EXPERIMENT   RESULTS   INTERPRETATION

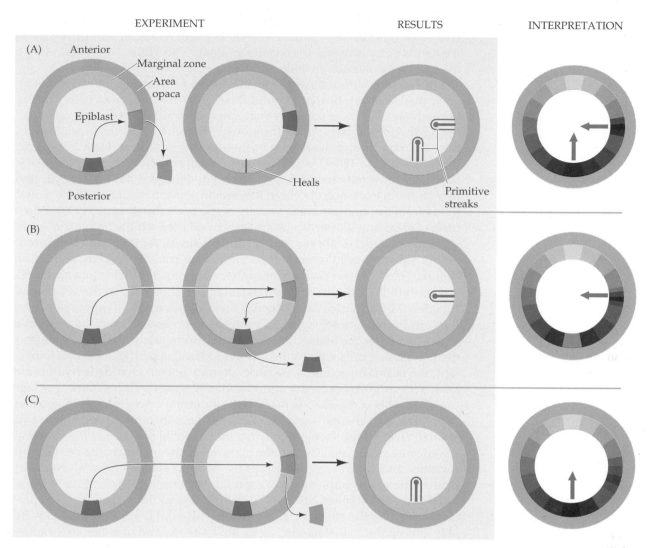

Figure 6.30
Experiments of Khaner and Eyal-Giladi, demonstrating that the posterior part of the marginal zone (PMZ) contributes to the primitive-streak-inducing cells of the hypoblast and prevents other marginal regions from creating their own hypoblasts. (After Khaner and Eyal-Giladi, 1989.)

CELL ACCUMULATION IN THE PRIMITIVE STREAK.   Evidence from the studies of Stern and Canning (1990) suggests that the epiblast is not the homogeneous, undifferentiated tissue that we have long assumed it to be. Rather, there appears to be differentiation in the epiblast cells even before primitive streak formation begins. These studies show that certain cells that are randomly scattered throughout the epiblast can be distinguished by a particular molecule (HNK-1, a sulfated form of glucuronic acid) on their cell surfaces. The cells expressing HNK-1 ingress individually into the blastocoel and migrate to the posterior margin. It is likely that the posterior marginal tissue secretes a chemical that attracts the cells expressing HNK-1, while the anterior marginal tissue secretes a repellent molecule (Jephcott and Stern, quoted in Stern, 1991). The HNK-1-expressing cells that collect at the posterior margin will give rise to the endoderm and mesoderm, and no cell expressing HNK-1 forms ectodermal derivatives. If the HNK-1 cells are selectively destroyed (by antibodies) while they are still on the epiblast, the embryo will not form any mesoderm or endoderm. These HNK-1-positive cells interact with the epiblast cells above them to form the initial rudiment of the primitive streak. The streak rudiment then undergoes a convergent extension process that narrows and extends the streak. When the streak has reached nearly its full extension, the HNK-1-positive cells dissolve the basal lamina of the central epiblast to form a groove through the primitive streak. This allows epiblast cells (which never had expressed HNK-1) to get recruited into the streak as it extends anteriorly and to contribute (along with HNK-1-positive cells) to the embryonic mesoderm and endoderm.

Movement within the amniote blastocoel is by individual cells, not by an epithelial sheet. But, as in amphibian gastrulation, avian cells passing through the blastopore constrict their apical ends to become bottle cells (Figure 6.28). At the anterior tip of the groove, Hensen's node, the breakdown of the basal lamina and the release of these cells from the epiblast may be accomplished by a 190-kDa protein called **scatter factor** (Stern et al., 1990). Scatter factor is only secreted in Hensen's node, and it has been implicated in both the dissociation of cells in this region and the induction of the neural tissue from the epiblast in the vicinity of the node (Streit et al., 1995). When resins containing scatter factor are implanted beneath the epiblast of early gastrulating chick embryos, new primitive streak regions can be induced. Scatter factor binds to receptor tyrosine kinases on adjacent cells, and acting through the G protein cascade, phosphorylates the β-catenins that anchor E-cadherin to the cell membrane (Hartmann et al., 1994). In the absence of functional E-cadherin, the epithelial sheet comes apart in that region, and the cells become mesenchyme. Once cells are released from the primitive streak and enter the blastocoel, they flatten and enter a stream of independent, migrating cells. Extracellular polysaccharides may also play an important role in this migration. One such complex polysaccharide is **hyaluronic acid,** a linear polymer of glucuronic acid and *N*-acetylglucosamine (see Figure 3.35). This compound is made by ectodermal cells and accumulates in the blastocoel, where it coats the surfaces of the incoming cells. Fisher and Solursh (1977) have shown that when this material is digested (by injecting the enzyme hyaluronidase into the blastocoel), the mesenchyme cells clump together and fail to migrate properly. Several studies have shown that hyaluronic acid is important in keeping migrating mesenchyme cells separated from one another. Moreover, hyaluronic acid first accumulates at precisely the time the first cells enter the blastocoel. Hyaluronic acid may be able to keep the cells separated by its ability to expand in water. In an aqueous environment, this polymer can expand to over 1000 times its original volume. Therefore, hyaluronic acid might be an important factor in keeping the mesenchyme cells dispersed during their migration and thereby ensuring that this migration continues.

Hyaluronic acid and other polysaccharides facilitate individual cell migration (see Chapter 3), but they do not appear to direct the movement of these cells (Fisher and Solursh, 1979). Rather, cellular movement of these cells is correlated, once again, with the presence of a fibronectin meshwork in the extracellular basal lamina of the epiblast cells. This fibronectin-rich layer appears on the undersurface of the upper layer shortly before the formation of the primitive streak and disappears in the region of the streak. Within the streak, the cells separate and are then seen to migrate laterally along the fibronectin-rich basement membrane of the epiblast (Duband and Thiery, 1982). However, there is no clear evidence that this fibronectin is essential for the directed cell movements of the cells laterally away from the primitive streak.

SECONDARY FORMATION OF THE NOTOCHORD. While the anterior portion of the notochord is formed by the ingression of cells through Hensen's node and their subsequent anterior migration, the posterior notochord is formed differently. After somite 17 (of the chick), the notochord forms from the condensation of mesodermal tissue that had ingressed through the primitive streak (i.e., not through Hensen's node). This extends posteriorly to form the "tail-bud" of the embryo (including somites 28–50) (Le Douarin et al., 1996).

EPIBOLY OF THE ECTODERM.   During gastrulation, ectodermal precursor cells expand outwardly to encircle the yolk. These cells are joined to each other by tight junctions and travel as a unit rather than as individual cells. In birds, the upper surface of the area opaca adheres tightly to the lower surface of the vitelline envelope and spreads along this inner surface. The same behavior is seen in culture. New (1959) demonstrated that isolated blastoderm will spread normally on isolated vitelline envelope, and Spratt (1963) demonstrated that this spreading will not occur on other substrates. These observations suggest that the vitelline envelope is essential for the spreading of the cell sheet. Interestingly, only the marginal cells (i.e., the cells of the area opaca) attach firmly to the vitelline surface. Most of the blastoderm cells adhere loosely, if at all. These marginal cells are inherently different from the other blastoderm cells, as they can extend enormous (500-μm) cytoplasmic processes onto the vitelline envelope. These elongated filopodia are believed to be the locomotor apparatus of the marginal cells.

There are several lines of evidence indicating that the marginal area opaca cells are the agents of ectodermal epiboly. First, the blastoderm spreads only when the margins are expanding. If the marginal cells are removed, the epiboly of the ectoderm stops. Second, when the marginal cells are cut away from the rest of the blastoderm, they continue to migrate alone. Thus, it appears that the ectodermal precursor cells are carried along by the actively migrating cells of the area opaca (Schlesinger, 1958). There is also a specific relationship between the cell membranes of the marginal cells and the lower surface of the vitelline membrane. New (1959) showed that when the blastoderm is placed on a vitelline envelope upside down (deep layers in contact with the vitelline envelope), the edges of the blastoderm curl inward so that the marginal cells of the upper layer are once again contacting the vitelline surface (Figure 6.31). Lash and his co-workers (1990) expanded these results by showing that fibronectin is present on the inner surface of the vitelline envelope. Then, as in the experiment discussed earlier in which Boucaut and co-workers injected the synthetic fibronectin-attachment-site sequence (RGD) into the salamander blastocoel, Lash and colleagues applied the RGD sequence to the vitelline envelope as the cells were migrating on it. This treatment specifically broke the contact between the marginal cells and the vitelline envelope, caused the retraction of the marginal cell filopodia, and stopped the migration of the blastoderm.

We have identified many of the processes involved in avian gastrulation, but we are ignorant as to how these processes are carried out. We do not yet know how the subgerminal cavity forms, how certain cells are specified to become hypoblast cells, how certain cells express HNK-1 while their neigh-

Figure 6.31
Migratory properties of chick ectodermal precursors. (A) When chick blastoderm is placed on the vitelline envelope with the ectodermal precursors in contact with the vitelline surface, the marginal cells migrate and cover the vitelline envelope with ectoderm. (B) When the deep layers are placed in contact with the vitelline envelope, the blastoderm layer curls to allow the cells of the superficial layer to adhere to and migrate over the vitelline layer. The result is a closed vesicle. (After New, 1959.)

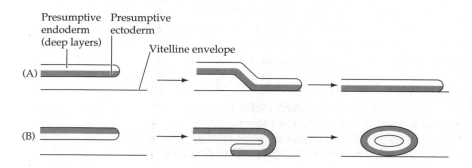

Presumptive endoderm (deep layers)   Presumptive ectoderm

Vitelline envelope

(A)

(B)

bors do not, how HNK-1-expressing cells migrate to the posterior margin and how they interact with the epiblast cells there, how the primitive streak is extended and how it retracts, or how the cells are assigned their respective fates. As Gary Schoenwolf (1991) recently remarked, "Despite all that has been written, it is safe to say that what we know about avian gastrulation and neurulation is considerably less than what still remains to be learned."

Gastrulation in mammals

Birds and mammals are both descendants of reptilian species. Therefore, it is not surprising that mammalian development parallels that of reptiles and birds. What is surprising is that the gastrulation movements of reptilian and avian embryos, which evolved as an adaptation to yolky eggs, are retained even in the absence of large amounts of yolk in the mammalian embryo. The mammalian inner cell mass can be envisioned as sitting atop an imaginary ball of yolk, following instructions that seem more appropriate to its ancestors.

Modifications for Development Within Another Organism

Instead of developing in isolation within an egg, most mammals have evolved the remarkable strategy of developing within the mother herself. The mammalian embryo obtains nutrients directly from its mother and does not rely on stored yolk. This evolution has entailed a dramatic restructuring of the maternal anatomy (such as expansion of the oviduct to form the uterus) as well as the development of a fetal organ capable of absorbing maternal nutrients. This fetal organ—the placenta—is derived primarily from embryonic trophoblast cells, supplemented with mesodermal cells derived from the inner cell mass.

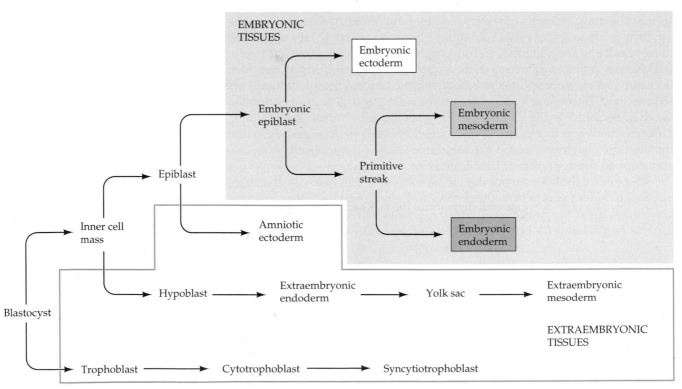

Figure 6.32
Schematic diagram showing the derivation of tissues in human and rhesus monkey embryos. (After Luckett, 1978, and Bianchi et al., 1993.)

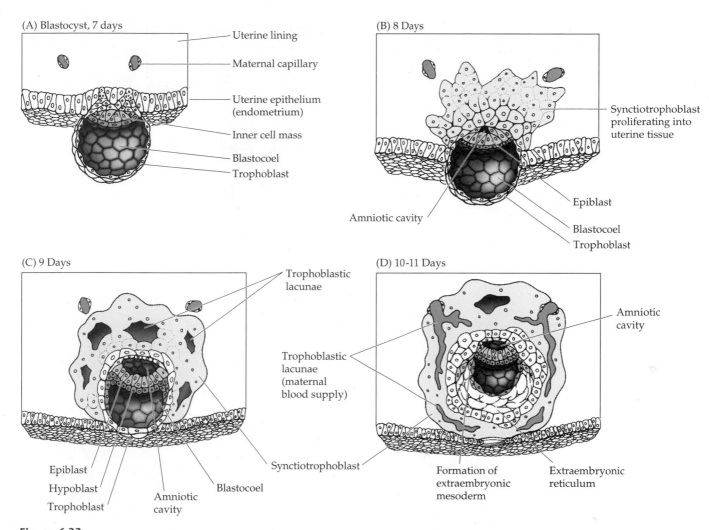

Figure 6.33
Tissue formation in the human embryo between days 7 and 12. (A,B) Human blastocyst immediately prior to gastrulation. The inner cell mass delaminates hypoblast cells that line the trophoblast, thereby forming the primitive yolk sac and a two-layered (epiblast and hypoblast) blastodisc similar to that seen in avian embryos. The trophoblast in some mammals can be divided into the polar trophoblast, which covers the inner cell mass, and the mural trophoblast, which does not. The trophoblast divides into the cytotrophoblast, which will form the villi, and the syncytiotrophoblast, which will ingress into the uterine tissue. (C) Meanwhile, the epiblast has split into the amniotic ectoderm (which encircles the amniotic cavity) and the embryonic epiblast. The adult mammal forms from the cells of the embryonic epiblast. (D) The extraembryonic endoderm forms the yolk sac. (After Gilbert, 1989, and Larsen, 1993.)

The origins of early mammalian tissues are summarized in Figure 6.32. The first segregation of cells within the inner cell mass involves the formation of the hypoblast (sometimes called the primitive endoderm) layer (Figure 6.33). These cells separate from the inner cell mass to line the blastocoel cavity, where they give rise to the **yolk sac endoderm.** As in avian embryos, these cells do not produce any part of the newborn organism. The remaining inner cell mass tissue above the hypoblast is now referred to as the epiblast. The epiblast cells are split by small clefts that eventually coalesce to separate the embryonic epiblast from the other epiblast cells, which form the lining of the **amnion** (Figures 6.33C and 6.34). Once the lining of the amnion is com-

Figure 6.34
Amnion structure and cell movements
during human gastrulation. (A) Human
embryo and uterine connections at 15
days' gestation. In the upper view, the
embryo is cut sagittally through the
midline; the lower view looks down
upon the dorsal surface of the embryo.
(B) The movements of the epiblast cells
into the primitive streak and Hensen's
node and underneath the epiblast are
superimposed on the dorsal surface
view. At days 14 and 15, the ingressing
epiblast cells are thought to replace the
hypoblast cells (which contribute to the
yolk sac lining), while at day 16, the
ingressing cells fan out to form the
mesodermal layer. (After Larsen, 1993.)

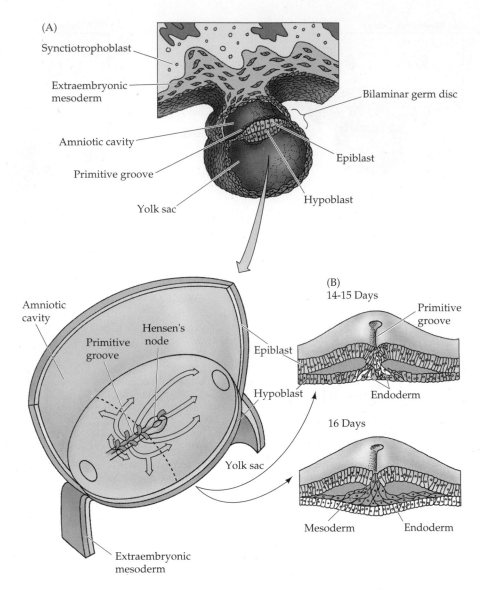

pleted, it fills with a secretion called **amniotic fluid,** which serves as a shock
absorber to the developing embryo while preventing its desiccation.

The embryonic epiblast is believed to contain all the cells that will generate the actual embryo, and it is similar in many ways to the avian epiblast.
By labeling individual cells of the epiblast with horseradish peroxidase,
Kirstie Lawson and her colleagues (1991) were able to construct a detailed
fate map of the mouse epiblast (Figure 6.35). Like the chick epiblast cells, the
mammalian mesoderm and endoderm migrate through a primitive streak.
As they enter the streak, the epiblast cells cease to express the E-cadherin
that holds cells together, and they migrate as individual cells (Bursdal et al.,
1993). Those cells migrating through Hensen's node give rise to the notochord. However, unlike notochord formation in the chick, the cells that form
the mouse notochord are thought to become integrated into the endoderm of
the primitive gut (Jurand, 1974; Sulik et al., 1994). These cells can be seen as
a band of small, ciliated cells extending rostrally from Hensen's node (Figure
6.36). They form the notochord by converging medially and folding off in a
dorsal direction from the roof of the gut.

Figure 6.35
Fate map of the mouse embryo. (A) The egg cylinder stage, 6 days after fertilization. Note that unlike chick and human epiblasts, the mouse epiblast is tightly curved. The parietal and visceral endoderm are derived from hypoblast cells, not from the trophectoderm. (B) Fate map of the 7-day mouse epiblast at early gastrula stage. (The mouse fate map has been flattened and should be seen as rolled up with the primitive streak at the edges.) (After Lawson et al., 1991.)

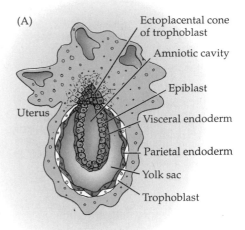

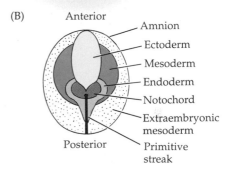

The ectodermal precursors are located anterior to the fully extended primitive streak, similar to their position in the chick epiblast; but whereas the mesoderm of the chick forms from cells posterior to the farthest extent of the streak, the mouse mesoderm forms from cells anterior to the primitive streak. In some instances, clones of cells give rise to descendants in more than one embryonic layer or to both embryonic and extraembryonic derivatives. Thus, at the epiblast stage, these lineages have not become separate from one another. As in avian embryos, the cells migrating between the hypoblast and epiblast layers appear to be coated with hyaluronic acid, which is first synthesized at the time of primitive streak formation (Solursh and Morriss, 1977). It is thought (Larsen, 1993) that the replacement of the hypoblast cells by endoderm precursors occurs on days 14–15 of gestation, while the migration of cells forming the mesoderm doesn't start until day 16.

Formation of Extraembryonic Membranes

While the embryonic epiblast is undergoing cell movements reminiscent of those seen in reptilian or avian gastrulation, the extraembryonic cells are making the distinctly mammalian tissues that enable the fetus to survive within the maternal uterus. Although the initial trophoblast cells of mice and humans appear normal, they give rise to a population of cells wherein nuclear division occurs in the absence of cytokinesis. The initial type of trophoblast cell constitutes a layer called the **cytotrophoblast,** whereas the multinucleated type of cell forms the **syncytiotrophoblast.** The cytotrophoblast adheres to the uterine wall (endometrium) through a series of adhesion molecules that were discussed in Chapter 3. The human cytotrophoblast cells also contain proteolytic enzymes that enable them to enter the uterus and remodel the uterine blood vessels so that the maternal blood bathes fetal blood vessels. The syncytiotrophoblast tissue is thought to further the progression of

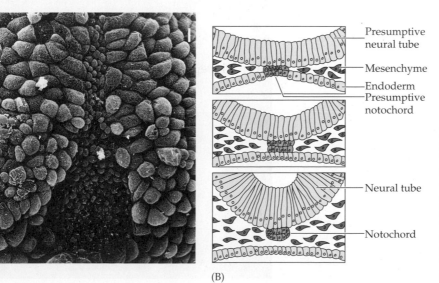

Figure 6.36
Formation of the notochord in the mouse. (A) The ventral surface of the 7.5-day mouse embryo, seen by scanning electron microscopy. The presumptive notochord cells are the small ciliated cells in the midline that are flanked by the larger endodermal cells of the primitive gut. (B) The formation of the notochord by the dorsal infolding of the small, ciliated cells. (From Sulik et al., 1994, courtesy of K. Sulik and G. C. Schoenwolf.)

the embryo into the uterus. This proteolytic activity ceases after the twelfth week of pregnancy (Fisher et al., 1989). The uterus, in turn, sends blood vessels into this area, where they eventually contact the syncytiotrophoblast. Shortly thereafter, mesodermal tissue extends outward from the gastrulating embryo (see Figure 6.33). Recent studies of human and rhesus monkey embryos have suggested that the yolk sac (and hence the hypoblast) is the source of this extraembryonic mesoderm (Bianchi et al., 1993). The extraembryonic mesoderm joins the trophoblastic extensions and gives rise to the blood vessels that carry nutrients from the mother to the embryo. The narrow **connecting stalk** of extraembryonic mesoderm that links the embryo to the trophoblast eventually forms the vessels of the umbilical cord. The fully developed organ, consisting of trophoblast tissue and blood-vessel-containing mesoderm, is called the **chorion,** and the chorion fuses with the uterine wall to create the placenta. Thus, the placenta has both a maternal portion (the uterine endometrium that is modified during pregnancy) and a fetal component (the chorion). The chorion may be very closely apposed to maternal tissues while still being readily separable (as in the *contact placenta* of the pig), or it may be so intimately integrated that the two tissues cannot be separated without damage to both the mother and the developing fetus (as in the *deciduous placenta* of most mammals, including humans).* [gast2.html]

Figure 6.37 shows the relationships between the embryonic and extraembryonic tissues of a 6-week human embryo. The embryo is seen encased in the amnion and is further shielded by the chorion. The blood vessels extending to and from the chorion are readily observable, as are the villi that project from the outer surface of the chorion. These villi contain the

---

*There are numerous types of placentas, and the extraembryonic membranes form differently in different orders of mammals (see Cruz and Pedersen, 1991). Although mice and humans gastrulate and implant in a similar fashion, their extraembryonic structures are distinctive. It is very risky to extrapolate developmental phenomena from one group of mammals to another. Even Leonardo da Vinci got caught (Renfree, 1982). His remarkable drawing of the human fetus inside the placenta is stunning art, but poor science: the placenta is that of a cow.

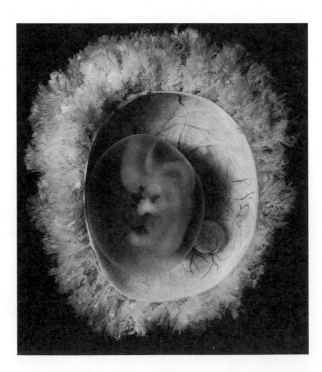

Figure 6.37
Human embryo and placenta after 40 days of gestation. The embryo lies within the amnion, and its blood vessels can be seen extending into the chorionic villi. The sphere to the right of the embryo is the yolk sac. (The Carnegie Institution of Washington, courtesy of C. F. Reather.)

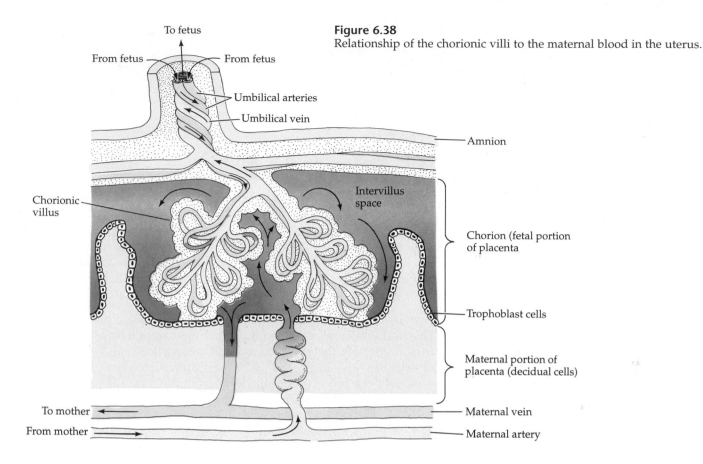

Figure 6.38
Relationship of the chorionic villi to the maternal blood in the uterus.

blood vessels and allow the chorion to have a large area exposed to the maternal blood. Thus, although fetal and maternal circulatory systems normally never merge, diffusion of soluble substances can occur through the villi (Figure 6.38). In this manner, the mother provides the fetus with nutrients and oxygen, and the fetus sends its waste products (mainly carbon dioxide and urea) into the maternal circulation. The blood vessels of the chorionic villi form from extraembryonic mesoderm that enters the mounds of cytotrophoblast tissue called **primary villi** (Figure 6.39). The resulting structures, the **secondary villi,** form during the second week of pregnancy. By the end of the third week, some of this extraembryonic mesoderm has produced blood vessels, and these **tertiary villi** are able to bring nutrients and oxygen from the mother into the embryo. [other.html#gast4]

The trophoblast, therefore, is necessary for adhering to and entering the uterine tissues, and the chorion enables the exchange of gases and nutrients between mother and fetus. But the chorion has even further importance; it is also an endocrine organ. The syncytiotrophoblast portion of the chorion produces three hormones that are essential for mammalian development. First, it produces chorionic gonadotropin, a peptide hormone that is capable of causing other cells in the placenta (and in the maternal ovary) to produce progesterone. Progesterone is the steroid hormone that keeps the uterine wall thick and full of blood vessels. In primates, the ovaries can be removed after the first third of pregnancy without harm to the developing fetus because the chorion itself is able to produce the steroid needed to maintain pregnancy (Zander and von Münstermann, 1956). Placental progesterone is also used by the fetal adrenal gland as a substrate for the production of biologically important corticosteroid hormones. The third hormone produced

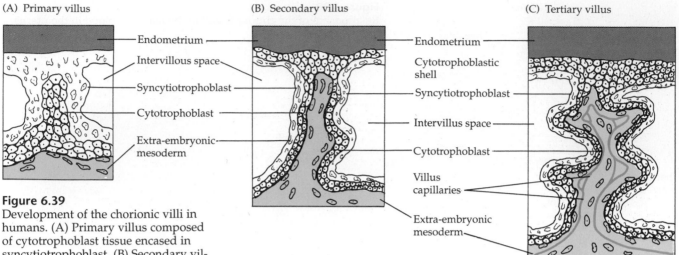

(A) Primary villus
(B) Secondary villus
(C) Tertiary villus

Endometrium
Intervillous space
Syncytiotrophoblast
Cytotrophoblast
Extra-embryonic mesoderm

Endometrium
Cytotrophoblastic shell
Syncytiotrophoblast
Intervillus space
Cytotrophoblast
Villus capillaries
Extra-embryonic mesoderm

Figure 6.39
Development of the chorionic villi in humans. (A) Primary villus composed of cytotrophoblast tissue encased in syncytiotrophoblast. (B) Secondary villus formed when underlying extraembryonic mesoderm penetrates the primary villus. Such secondary villi join adjacent villi to form the cytotrophoblastic shell that will anchor the villi to the endometrium. (C) Within the extraembryonic mesoderm, capillaries form and will connect with branches of the umbilical artery and vein. (After Gilbert, 1989.)

by the chorion is chorionic somatomammotropin (often called placental lactogen). This hormone is responsible for promoting maternal breast development during pregnancy, thereby enabling milk production later.

Recent studies have indicated that the chorion may have yet another function, namely, the protection of the fetus from the immune response of the mother. A person with a normal immune system recognizes and rejects foreign cells within their body; this fact is demonstrated by the rejection of skin and organ grafts from genetically different individuals. The glycoproteins responsible for this rejection are called the major histocompatibility antigens, and they are likely to differ from individual to individual. A human child expresses major histocompatibility antigens from both the mother and the father, and a mother's body will reject her offspring's skin or organs because they contain paternally derived antigens. How, then, can a human fetus remain 9 months within the body of its mother? Why doesn't the mother immunologically reject her fetus as she would an organ from that child? It appears that the chorion has evolved several mechanisms by which it can inhibit the immune response against the fetus (Chaouat, 1990). It can secrete soluble proteins that block the production of antibodies, and it can promote the production of certain types of lymphocytes that suppress the normal immune response within the uterus. The cytotrophoblast cells also contain a placenta-specific form of the major histocompatability antigens which appears to protect the embryo from being recognized by the mother's immune system (Carosella et al., 1996; Pazmany et al., 1996). Thus, the functions of the placenta include not only physical support and nutrient exchange, but also regulation of the endocrine and immunological relationships between the mother and the fetus. [gast3.html]

In gastrulation, we see an incredibly well coordinated series of cell movements whereby the cleavage-stage blastomeres are rearranged and begin to interact with new neighbors. Moreover, although there are differences between the gastrulation movements of sea urchin, amphibian, avian, and mammalian embryos, certain mechanisms are common to them all. Each group has the problem of bringing the mesodermal and endodermal precursor cells inside the body while surrounding the embryo with ectodermal precursors. Given the different amounts and distributions of yolk, as well as other environmental considerations, each type of organism has evolved a way of accomplishing this goal. The stage is now set for the formation of the first organs.

LITERATURE CITED

Alfandari, D., Whittaker, C.A., DeSimone, D. W. and Darribère, T. 1995. Integrin av subunit is expressed on mesodermal cell surfaces during amphibian gastrulation. *Dev. Biol.* 170: 249–261.

Anstrom, J. A., Chin, J. E., Leaf, D. S., Parks, A. L. and Raff, R. A. 1987. Localization and expression of msp130, a primary mesenchyme lineage-specific cell surface protein of the sea urchin embryo. *Development* 101: 255–265.

Azar, Y. and Eyal-Giladi, H. 1981. Interaction of epiblast and hypoblast in the formation of the primitive streak and the embryonic axis in chick, as revealed by hypoblast rotation experiments. *J. Embryol. Exp. Morphol.* 61: 133–141.

Balinsky, B. I. 1975. *Introduction to Embryology,* 4th Ed. Saunders, Philadelphia.

Bellairs, R. 1986. The primitive streak. *Anat. Embryol.* 174: 1–14.

Berg, L. K., Chen, S. W. and Wessel, G. M. 1996. An extracellular matrix molecule that is selectively expressed during development is important for gastrulation in the sea urchin embryo. *Development* 122: 703–713.

Bianchi, D. W., Wilkins-Haug, L. E., Enders, A. C. and Hay, E. D. 1993. Origin of extraembryonic mesoderm in experimental animals: Relevance to chorionic mosaicism in humans. *Am. J. Med. Genet.* 46: 542–550.

Bisgrove, B. W., Andrews, M. E. and Raff, R. A. 1991. Fibropellins, products of an EGF repeat-containing gene, form a unique extracellular matrix structure that surrounds the sea urchin embryo. *Dev. Biol.* 146: 89–99.

Black, S. D. and Gerhart, J. 1985. Experimental control of the site of embryonic axis formation in *Xenopus laevis* eggs centrifuged before first cleavage. *Dev. Biol.* 108: 310–324.

Black, S. D. and Gerhart, J. 1986. High frequency twinning of *Xenopus laevis* embryos from eggs centrifuged before first cleavage. *Dev. Biol.* 116: 228–240.

Boucaut, J.-C., D'Arribère, T., Poole, T. J., Aoyama, H., Yamada, K. M. and Thiery, J. P. 1984. Biologically active synthetic peptides as probes of embryonic development: A competitive peptide inhibition of fibronectin function inhibits gastrulation in amphibian embryos and neural crest cell migration in avian embryos. *J. Cell Biol.* 99: 1822–1830.

Boucaut, J.-C., D'Arribère, T., Li, S. D., Boulekbache, H., Yamada, K. M. and Thiery, J. P. 1985. Evidence for the role of fibronectin in amphibian gastrulation. *J. Embryol. Exper. Morphol.* 89 [Suppl.]: 211–227.

Burke, R. D., Myers, R. L., Sexton, T. L. and Jackson, C. 1991 Cell movements during the initial phase of gastrulation in the sea urchin embryo. *Dev. Biol.* 146: 542–557.

Bursdal, C. A., Damsky, C. H. and Pedersen, R. A. 1993. The role of E-cadherin and integrins in mesoderm differentiation and migration at the mammalian primitive streak. *Development* 118: 829–844.

Carosella, E.D., Dausset, J. and Kirszenbaum, M. 1996. HLA-G revisited. *Immunol. Today* 17: 407–409.

Chaouat, G. 1990. *The Immunology of the Fetus.* CRC Press, Boca Raton, FL.

Cherr, G. N., Summers, R. G., Baldwin, J. D. and Morrill, J. B. 1992. Preservation and visualization of the sea urchin blastocoelic extracellular matrix. *Microsc. Res. Tech.* 22: 11–22.

Cooke, J. 1986. Permanent distortion of positional system of *Xenopus* embryo by brief early perturbation in gravity. *Nature* 319: 60–63.

Cruz, Y. P. and Pedersen, R. A. 1991. Origin of embryonic and extraembryonic cell lineages in mammalian embryos. In *Animal Applications of Research in Mammalian Development.* Cold Spring Harbor Press, Cold Spring Harbor, pp. 147–204.

Dan, K. and Okazaki, K. 1956. Cyto-embryological studies of sea urchins. III. Role of secondary mesenchyme cells in the formation of the primitive gut in sea urchin larvae. *Biol. Bull.* 110: 29–42.

D'Arribère, T., Yamada, K. M., Johnson, K. E. and Boucaut, J.-C. 1988. The 140-kD fibronectin receptor complex is required for mesodermal cell adhesion during gastrulation in the amphibian *Pleurodeles waltii.* *Dev. Biol.* 126: 182–194.

D'Arribère, T., Guida, K., Larjava, H., Johnson, K. E., Yamada, K. M., Thiery, J.-P. and Boucaut, J.-C. 1990. In vivo analysis of integrin β 1 subunit function in fibronectin matrix assembly. *J. Cell Biol.* 110: 1813–1823.

Delarue, M., D'Arribère, T., Aimar, C. and Boucaut, J.-C. 1985. Bufonid nucleocytoplasmic hybrids arrested at the early gastrula stage lack a fibronectin-containing fibrillar extracellular matrix. *Wilhelm Roux Arch. Dev. Biol.* 194: 275–280.

Driever, W. 1995. Axis formation in zebrafish. *Curr. Opin. Genet. Dev.* 5: 610–618.

Duband, J. L. and Thiery, J. P. 1982. Appearance and distribution of fibronectin during chick embryo gastrulation and neurulation. *Dev. Biol.* 94: 337–350.

Ettensohn, C. A. 1985. Gastrulation in the sea urchin embryo is accompanied by the rearrangement of invaginating epithelial cells. *Dev. Biol.* 112: 383–390.

Ettensohn, C. A. 1990. The regulation of primary mesenchyme cell patterning. *Dev. Biol.* 140: 261–271.

Ettensohn, C. A. and McClay, D. R. 1986. The regulation of primary mesenchyme cell migration in the sea urchin embryo: Transplantations of cells and latex beads. *Dev. Biol.* 117: 380–391.

Eyal-Giladi, H., Debby, A. and Harel, N. 1992. The posterior section of the chick's area pellucida and its involvement in hypoblast and primitive streak formation. *Development* 116: 819–830.

Fink, R. D. and McClay, D. R. 1985. Three cell recognition changes accompany the ingression of sea urchin primary mesenchyme cells. *Dev. Biol.* 107: 66–74.

Fisher, M. and Solursh, M. 1977. Glycosaminoglycan localization and role in maintenance of tissue spaces in the early chick embryo. *J. Embryol. Exp. Morphol.* 42: 195–207.

Fisher, M. and Solursh, M. 1979. Influence of local environment on the organization of mesenchymel cells. *J. Embryol. Exp. Morphol.* 49: 295–306.

Fisher, S. J., Cui, T.-Y., Zhang, L., Grahl, K., Guo-Yang, Z., Tarpey, J. and Damsky, C. H. 1989. Adhesive and degradative properties of the human placental cytotrophoblast cells in vitro. *J. Cell Biol.* 109: 891–902.

Galileo, D. S. and Morrill, J. B. 1985. Patterns of cells and extracellular material of the sea urchin *Lytechinus variegatus* (Echinodermata; Echinoidea) embryo, from hatched blastula to late gastrula. *J. Morphol.* 185: 387–402.

Gerhart, J., Ubbels, G., Black, S., Hara, K. and Kirschner, M. 1981. A reinvestigation of the role of the grey crescent in axis formation in *Xenopus laevis.* *Nature* 292: 511–516.

Gerhart, J. and seven others. 1986. Amphibian early development. *BioScience* 36: 541–549.

Gilbert, S. G. 1989. *Pictorial Human Embryology.* University of Washington Press, Seattle.

Gimlich, R. L. and Gerhart, J. C. 1984. Early cellular interactions promote embryonic axis formation in *Xenopus laevis.* *Dev. Biol.* 104: 117–130.

Gustafson, T. and Wolpert, L. 1961. Studies on the cellular basis of morphogenesis in sea urchin embryos: Directed movements of primary mesenchyme cells in normal and vegetalized larvae. *Exp. Cell Res.* 24: 64–79.

Gustafson, T. and Wolpert, L. 1967. Cellular movement and contact in sea urchin morphogenesis. *Biol. Rev.* 42: 442–498.

Hall, H. G. and Vacquier, V. D. 1982. The apical lamina of the sea urchin embryo: Major glycoprotein associated with the hyaline layer. *Dev. Biol.* 89: 168–178.

Hardin, J. 1988. The role of secondary mesenchyme cells during sea urchin gastrulation studied by laser ablation. *Development* 103: 317–324.

Hardin, J. 1990. Context-dependent cell behaviors during gastrulation. *Semin. Dev. Biol.* 1: 335–345.

Hardin, J. D. and Cheng, L. Y. 1986. The mechanisms and mechanics of archenteron elongation during sea urchin gastrulation. *Dev. Biol.* 115: 490–501.

Hardin, J. D. and Keller, R. 1988. The behaviour and function of bottle cells during gastrulation of *Xenopus laevis*. *Development* 103: 211–230.

Hardin, J. and McClay, D. R. 1990. Target recognition by the archenteron during sea urchin gastrulation. *Dev. Biol.* 142: 87–105.

Harkey, M. A. and Whiteley, A. M. 1980. Isolation, culture and differentiation of echinoid primary mesenchyme cells. *Wilhelm Roux Arch. Dev. Biol.* 189: 111–122.

Hartmann, G., Weidner, K. M., Scharz, H. and Birchmeier, W. 1994. The motility signal of scatter factor/hepatocyte growth factor mediated through the receptor tyrosine kinase Met requires intracellular action of Ras. *J. Biol. Chem.* 269: 21936–21939.

Herbst, C. 1904. Über die zur Entwicklung des Seeigellarven notwendigen anorganischen Stoffe, ihre Rolle und Vertretbarkeit. II. *Wilhelm Roux Arch. Entwicklungsmech. Org.* 17: 306–520.

Ho, R. K. 1992. Cell movements and cell fate during zebrafish gastrulation. *Development* Suppl. 1992: 65–73.

Holowacz, T. and Elinson, R. P. 1993. Cortical cytoplasm, which induces dorsal axis formation in *Xenopus*, is inactivated by UV irradiation of the oocyte. *Development* 119: 277–285.

Holtfreter, J. 1943. A study of the mechanics of gastrulation: Part I. *J. Exp. Zool.* 94: 261–318.

Holtfreter, J. 1944. A study of the mechanics of gastrulation: Part II. *J. Exp. Zool.* 95: 171–212.

Hörstadius, S. 1939. The mechanics of sea urchin development, studied by operative methods. *Biol. Rev.* 14: 132–179.

Jurand, A. 1974. Some aspects of the development of the notochord in mouse embryos. *J. Embryol. Exp. Morphol.* 32: 1–33.

Kane, D. A. and Kimmel, C. B. 1993. The zebrafish midblastula transition. *Development* 119: 447–456.

Karp, G. C. and Solursh, M. 1974. Acid mucopolysaccharide metabolism, the cell surface, and primary mesenchyme cell activity in the sea urchin embryo. *Dev. Biol.* 41: 110–123.

Karp, G. C. and Solursh, M. 1985. Dynamic activity of the filopodia of sea urchin embryonic cells and their role in directed mi

gration of the primary mesenchyme in vitro. *Dev. Biol.* 112: 276–283.

Keller, R. E. 1975. Vital dye mapping of the gastrula and neurola of *Xenopus laevis*. I. Prospective areas and morphogenetic movements of the superficial layer. *Dev. Biol.* 42: 222–241.

Keller, R. E. 1976. Vital dye mapping of the gastrula and neurula of *Xenopus laevis*. II. Prospective areas and morphogenetic movements in the deep layer. *Dev. Biol.* 51: 118–137.

Keller, R. E. 1980. The cellular basis of epiboly: An SEM study of deep cell rearrangement during gastrulation of *Xenopus laevis*. *J. Embryol. Exp. Morphol.* 60: 201–243.

Keller, R. E. 1981. An experimental analysis of the role of bottle cells and the deep marginal zone in the gastrulation of *Xenopus laevis*. *J. Exp. Zool.* 216: 81–101.

Keller, R. E. 1986. The cellular basis of amphibian gastrulation. *In* L. Browder (ed.), *Developmental Biology: A Comprehensive Synthesis*, Vol. 2. Plenum, New York, pp. 241–327.

Keller, R. and Danilchik, M. 1988. Regional expression, pattern and timing of convergence and extension during gastrulation of *Xenopus laevis*. *Development* 103: 193–209.

Keller, R. E. and Schoenwolf, G. C. 1977. An SEM study of cellular morphology, contact, and arrangement as related to gastrulation in *Xenopus laevis*. *Wilhelm Roux Arch. Dev. Biol.* 182: 165–186.

Khaner, O. 1995. The rotated hypoblast of the chicken embryo does not initiate an ectopic axis in the epiblast. *Proc. Natl. Acad. Sci. USA* 92: 10733–10737.

Khaner, O. and Eyal-Giladi, H. 1989. The chick's marginal zone and primitive streak formation. I. Coordinative effect of induction and inhibition. *Dev. Biol.* 134: 206–214.

Kirschner, M. W. and Gerhart, J. C. 1981. Spatial and temporal changes in the amphibian egg. *BioScience* 31: 381–388.

Kochav, S. M. and Eyal-Giladi, H. 1971. Bilateral symmetry in the chick embryo determination by gravity. *Science* 171: 1027–1029.

Lane, M. C. and Solursh, M. 1991. Primary mesenchyme cell migration requires a chondroitin sulfate/dermatan sulfate proteoglycan. *Dev. Biol.* 143: 389–397.

Lane, M. C. Koehl, M. A. R., Wilt, F. and Keller, R. 1993. A role for regulated secretion of apical matrix during epithelial invagination in the sea urchin. *Development* 117: 1049–1060.

Landstrom, U. and Løvtrup, S. 1979. Fate maps and cell differentiation in the amphibian embryo: An experimental study. *J. Embryol. Exp. Morphol.* 54: 113–130.

Langeland, J. and Kimmel, C. B. 1997. The embryology of fish. *In* Gilbert, S. F. and Raunio, A. M. (eds.), *Embryology: Construct-

ing the Organism*. Sinauer Associates, Sunderland, MA.

Larsen, W. J. 1993. *Human Embryology*. Churchill Livingston, New York.

Lash, J. W., Gosfield, E. III, Ostrovsky, D. and Bellairs, R. 1990. Migration of chick blastoderm under the vitelline membrane: The role of fibronectin. *Dev. Biol.* 139: 407–416.

Lawson, K. A., Meneses, J. J. and Pedersen, R. A. 1991. Clonal analysis of epiblast fate during germ layer formation in the mouse embryo. *Development* 113: 891–911.

Le Douarin, N., Grapin-Botton, A. and Catala, M. 1996. Patterning of the neural primordium in the avian embryo. *Semin. Dev. Biol.* 7: 157–167.

Lillie, F. R. 1902. Differentiation without cleavage in the egg of the annelid *Chaetopterus pergamentaceus*. *Wilhelm Roux Arch. Entwicklungsmech. Org.* 14: 477–499.

Løvtrup, S. 1975. Fate maps and gastrulation in amphibia: A critique of current views. *Can. J. Zool.* 53: 473–479.

Luckett, W. P. 1978. Origin and differentiation of the yolk sac and extraembryonic mesoderm in presomite human and rhesus monkey embryos. *Am. J. Anat.* 152: 59–98.

Lundmark, C. 1986. Roles of bilateral zones of ingressing superficial cells during gastrulation of *Ambystoma mexicanum*. *J. Embryol. Exp. Morphol.* 97: 47–62.

Malinda, K. M., Fisher, G.W., and Ettensohn, C. A. 1995. Four-dimensional microscopic analysis of the filopodial behavior of primary mesenchyme cells during gastrulation in the sea urchin embryo. *Dev. Biol.* 172: 552–566.

Miller, J. R., Fraser, S. E. and McClay, D. R. 1995. Dynamics of thin filopodia during sea urchin gastrulation. *Development* 121: 2505–2511

Morrill, J. B. and Santos, L. L. 1985. A scanning electron micrographical overview of cellular and extracellular patterns during blastulation and gastrulation in the sea urchin, *Lytechinus variegatus*. *In* R. H. Sawyer and R. M. Showman (eds.), *The Cellular and Molecular Biology of Invertebrate Development*. University of South Carolina Press, pp. 3–33.

Nakatsuji, N., Smolira, M. A. and Wylie, C. C. 1985. Fibronectin visualized by scanning electron microscope immunocytochemistry on the substratum for cell migration in *Xenopus laevis* gastrulae. *Dev. Biol.* 107: 264–268.

New, D. A. T. 1959. Adhesive properties and expansion of the chick blastoderm. *J. Embryol. Exp. Morphol.* 7: 146–164.

Oppenheimer, J. M. 1936. Transplantation experiments on developing teleosts (*Fundulus* and *Perca*). *J. Exp. Zool.* 72: 409–437.

Pazmany, L., Mandelboim, O., Valés-Gómez, M., Davis, D.M., Reyburn, H.T. and Strominger, J.L. 1996. Protection from natural killer cell-mediated lysis by HLA-G expression on target cells. *Science* 274: 792-795.

Psychoyos, D. and Stern, C. D. 1996. Fates and migratory routes of primitive streak cells in the chick embryo. *Development* 122: 1523–1534.

Purcell, S. M. and Keller, R. 1993. A different type of amphibian mesoderm morphogenesis in *Ceratophrys ornata*. *Development* 117: 307–317.

Renfree, M. B. 1982. Implantation and placentation. *In* C. R. Austin and R. V. Short (eds.), *Embryonic and Fetal Development*. Cambridge University Press, Cambridge, pp. 26–69.

Rosenquist, G. C. 1966. A radioautographic study of labeled grafts in the chick blastoderm. Development from primitive-streak stages to stage 12. *Carnegie Inst. Wash. Contrib. Embryol.* 38: 31–110.

Rosenquist, G. C. 1972. Endoderm movements in the chick embryo between the short streak and head process stages. *J. Exp. Zool.* 180: 95–104.

Schlesinger, A. B. 1958. The structural significance of the avian yolk in embryogenesis. *J. Exp. Zool.* 138: 223–258.

Schoenwolf, G. C. 1991. Cell movements in the epiblast during gastrulation and neurulation in avian embryos. *In* R. Keller, W. H. Clark, Jr. and F. Griffin (eds.), *Gastrulation: Movements, Patterns, and Molecules*. Plenum, New York, pp. 1–28.

Schoenwolf, G. C., Garcia-Martinez, V. and Diaz, M. S. 1992. Mesoderm movement and fate during amphibian gastrulation and neurulation. *Dev. Dynam.* 193: 235–248.

Schroeder, T. 1981. Development of a "primitive" sea urchin (*Eucidaris tribuloides*): Irregularities in the hyaline layer, micromeres, and primary mesenchyme. *Biol. Bull.* 161: 141–151.

Shi, D.-L., D'Arribère, T., Johnson, K. E. and Boucaut, J.-C. 1989. Initiation of mesodermal cell migration and spreading relative to gastrulation in the urodele amphibian *Pleurodeles walti*. *Development* 105: 351–363.

Schmidt, B. and Campos-Ortega, J. 1994. Dorsoventral polarity of the zebrafish embryo is distinguishable prior to the onset of gastrulation. *Wilhelm Roux Arch. Dev. Biol.* 203: 374–380.

Smith, J. C. and Malacinski, G. M. 1983. The origin of the mesoderm in an anuran, *Xenopus laevis*, and a urodele, *Ambystoma mexicanum*. *Dev. Biol.* 98: 250–254.

Solnica-Krezel, L. and Driever, W. 1994. Microtubule arrays of the zebrafish yolk cell: organizationand function during epiboly. *Development* 120: 2443–2455.

Solursh, M. and Morriss, G. M. 1977. Glycosaminoglycan synthesis in rat embryos during the formation of the primary mesenchyme and neural folds. *Dev. Biol.* 57: 75–86.

Solursh, M. and Revel, J. P. 1978. A scanning electron microscope study of cell shape and cell appendages in the primitive streak region of the rat and chick embryo. *Differentiation* 11: 185–190.

Spratt, N. T., Jr. 1946. Formation of the primitive streak in the explanted chick blastoderm marked with carbon particles. *J. Exp. Zool.* 103: 259–304.

Spratt, N. T., Jr. 1947. Regression and shortening of the primitive streak in the explanted chick blastoderm. *J. Exp. Zool.* 104: 69–100.

Spratt, N. T., Jr. 1963. Role of the substratum, supracellular continuity, and differential growth in morphogenetic cell movements. *Dev. Biol.* 7: 51–63.

Spratt, N. T. Jr. and Haas, H. 1960. Integrative mechanisms in development of early chick blastoderm. I. Regulated potentiality of separate parts. *J. Exp. Zool.* 145: 97–138.

Stern, C. D. 1991. Mesoderm formation in the chick embryo revisited. *In* R. Keller, W. H. Clark, Jr. and F. Griffin (eds.), *Gastrulation: Movements, Patterns, and Molecules*. Plenum, New York, pp. 29–41.

Stern, C. D. and Canning, D. R. 1988. Gastrulation in birds: A model system for the study of animal morphogenesis. *Experientia* 44: 651–657.

Stern, C. D. and Canning, D. R. 1990. Origin of cells giving rise to mesoderm and endoderm in chick embryo. *Nature* 343: 273–275.

Stern, C. D., Ireland, G. W., Herrick, S. E., Gherardi, E., Gray, J., Perryman, M. and Stoker, M. 1990. Epithelial scatter factor and development of the chick embryonic axis. *Development* 110: 1271–1284.

Strahle, U. and Jesuthasan, S. 1993. Ultraviolet irradiation impairs epiboly in zebrafish embryos: evidence for a microtubule-dependent mechanism of epiboly. *Development* 119: 451–453.

Streit, A., Stern, C. D., Thery, C., Ireland, G. W., Aparicio, S., Sharpe, M. J. and Gherardi, E. 1995. A role for HGF/SF in neural induction and its expression in Hensen's node during gastrulation. *Development* 121: 813–824.

Sugiyama, K. 1972. Occurrence of mucopolysaccharides in the early development of the sea urchin embryo and its role in gastrulation. *Dev. Growth Differ.* 14: 62–73.

Sulik, K., Dehart, D. B., Carson, J. L., Vrablic, T., Gesteland, K. and Schoenwolf, G. C. 1994. Morphogenesis of the murine node and notochordal plate. *Dev. Dynam.* 201: 260–278.

Trinkaus, J. P. 1984a. *Cells into Organs: The Forces that Shape the Embryo*, 2nd Ed. Prentice-Hall, Englewood Cliffs, NJ.

Trinkaus, 1984b. Mechanisms of *Fundulus* epiboly—a current view. *Am. Zool.* 24: 673–688.

Trinkaus, J. P. 1992. The midblastula transition, the YSL transition, and the onset of gastrulation in *Fundulus*. *Development* Suppl., 1992: 75–80.

Trinkaus, J.P. 1996. Ingression during early gastrulation of *Fundulus*. *Dev. Biol.* 177: 356-370.

Vakaet, L. 1984. The initiation of gastrula ingression in the chick blastoderm. *Am. Zool.* 24: 555–562.

Vakaet, L. 1985. Morphogenetic movements and fate maps in the avian blastoderm. *In* G. M. Edelman (ed.), *Molecular Determinants of Animal Form*. Alan R. Liss, New York, pp. 99–109.

Vincent, J. P. and Gerhart, J. C. 1987. Subcortical rotation in *Xenopus* eggs: An early step in embryonic axis formation. *Dev. Biol.* 123: 526–539.

Vincent, J. P., Oster, G. F. and Gerhart, J. C. 1986. Kinematics of gray crescent formation in *Xenopus* eggs. Displacement of subcortical cytoplasm relative to the egg surface. *Dev. Biol.* 113: 484–500.

Vogt, W. 1929. Gestaltungsanalyse am Amphibienkeim mit ortlicher Vitalfarbung. II. Teil Gastrulation und Mesodermbildung bei Urodelen und Anuren. *Wilhelm Roux Arch. Entwicklungsmech. Org.* 120: 384–706.

von Übisch, L. 1939. Keimblattchimarenforschung an Seeigellarven. *Biol. Rev. Cambr. Philos. Soc.* 14: 88–103.

Waddington, C. H. 1932. Experiments in the development of chick and duck embryos cultivated in vitro. *Philos. Trans. R. Soc. Lond.* [B] 13: 221.

Warga, R. M. and Kimmel, C. B. 1990. Cell movements during epiboly and gastrulation in zebrafish. *Development* 108: 569–580.

Wessel, G. M. and McClay, D. R. 1985. Sequential expression of germ layer specific molecules in the sea urchin embryo. *Dev. Biol.* 111: 451–463.

Wessel, G. M., Marchase, R. B. and McClay, D. R. 1984. Ontogeny of the basal lamina in the sea urchin embryo. *Dev. Biol.* 103: 235–245.

Wilson, P. and Keller, R. 1991. Cell rearrangement during gastrulation of *Xenopus*: Direct observation of cultured explants. *Development* 112: 289–300.

Winklbauer, R. and Keller, R. E. 1996. Fibronectin, mesoderm migration, and gastrulation in *Xenopus*. *Dev. Biol.* 177: 413–426.

Winklbauer, R. and Nagel, M. 1991. Directional mesoderm cell migration in the *Xenopus* gastrula. *Dev. Biol.* 148: 573–589.

Winklbauer, R., Selchow, A., Nagel, M., Stoltz, C. and Angres, B. 1991. Mesodermal cell migration in the *Xenopus* gastrula. *In* R. Keller, W. H. Clark, Jr. and F. Griffin (eds.), *Gastrulation: Movements, Patterns, and Molecules.* Plenum, New York, pp. 147–168.

Zander, J. and von Münstermann, A. M. 1956. Progesteron in menschlichem Blut und Geweben. III. *Klin. Wochenschr.* 34: 944–953.

Early vertebrate development: Neurulation and the ectoderm

7

For the real amazement, if you wish to be amazed, is this process. You start out as a single cell derived from the coupling of a sperm and an egg; this divides in two, then four, then eight, and so on, and at a certain stage there emerges a single cell which has as all its progeny the human brain. The mere existence of such a cell should be one of the great astonishments of the earth. People ought to be walking around all day, all through their waking hours calling to each other in endless wonderment, talking of nothing except that cell.
LEWIS THOMAS (1979)

Even more appealing than virgin forest was the jungle lying before me at that moment: the central nervous system.
RITA LEVI-MONTALCINI (1988)

IN 1828, KARL ERNST VON BAER, the foremost embryologist of his day,* proclaimed, "I have two small embryos preserved in alcohol, that I forgot to label. At present I am unable to determine the genus to which they belong. They may be lizards, small birds, or even mammals." Figure 7.1 allows us to appreciate his quandary, and it illustrates von Baer's four general laws of embryology. From his detailed study of chick development and his comparison of those embryos with the embryos of other vertebrates, von Baer derived four generalizations, stated here with some vertebrate examples:

1. *The general features of a large group of animals appear earlier in the embryo than do the specialized features.* All developing vertebrates (fishes, reptiles, amphibians, birds, and mammals) appear very similar shortly after gastrulation. It is only later in development that the special features of class, order, and finally species emerge (see Figure 7.1). All vertebrate embryos have gill arches, notochords, spinal cords, and pronephric kidneys.
2. *Less general characters are developed from the more general, until finally the most specialized appear.* All vertebrates initially have the same type of skin. Only later does the skin develop fish scales, reptilian scales, bird feathers, or the hair, claws, and nails of mammals. Similarly, the early development of the limb is essentially the same in all vertebrates. Only later do the differences between legs, wings, and arms become apparent.
3. *Each embryo of a given species, instead of passing through the adult stages of other animals, departs more and more from them.* The visceral clefts of embryonic birds and mammals do not resemble the gill slits of adult fishes in detail. Rather, they resemble those visceral clefts of *embryonic* fishes and other *embryonic* vertebrates. Whereas fishes preserve and elaborate these clefts into true gill slits, mammals convert them into structures such as the eustachian tubes (between the ear and mouth).
4. *Therefore, the early embryo of a higher animal is never like a lower animal, but only like its early embryo.* Von Baer saw that different groups of animals share certain common features during early embryonic devel-

*K. E. von Baer discovered the notochord, the mammalian egg, and the human egg, as well as making the conceptual advances described here. His work will be discussed further in Chapter 23.

Figure 7.1
Illustration of von Baer's laws. Early vertebrate embryos exhibit features common to the entire subphylum. As development progresses, embryos become recognizable as members of their class, their order, their family, and finally their species. (After Romanes, 1901.)

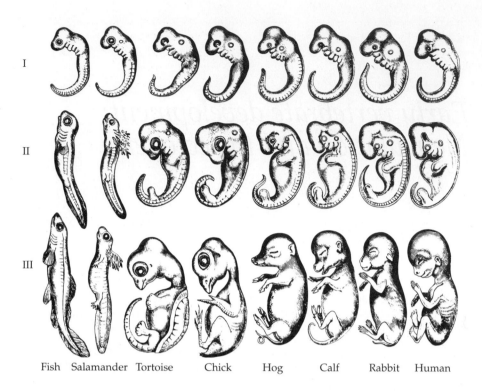

I

II

III

Fish   Salamander   Tortoise   Chick   Hog   Calf   Rabbit   Human

opment and that their features become more and more characteristic of the species as development proceeds. Human embryos never pass through a stage equivalent to an adult fish or bird; rather, human embryos initially share characteristics in common with fish and avian embryos. Later, the mammalian and other embryos diverge, none of them passing through the stages of the other.

Von Baer also recognized that there is a common *pattern* to all vertebrate development: the three germ layers give rise to different organs, and this derivation of the organs is constant whether the organism is a fish, a frog, or a chick. **Ectoderm** forms skin and nerves; **endoderm** forms respiratory and digestive tubes; and **mesoderm** forms connective tissue, blood cells, the heart, the urogenital system, and parts of most internal organs. In this chapter, we follow the early development of ectoderm; this and the following chapter focus on the formation of the nervous system in vertebrates. Chapter 9 will follow the early development of endodermal and mesodermal organs.

■ FORMATION OF THE CENTRAL NERVOUS SYSTEM

Neurulation: An overview

"What is perhaps the most intriguing question of all is whether the brain is powerful enough to solve the problem of its own creation." So Gregor Eichele (1992) ended a recent review of research on mammalian brain development. The construction of an organ that perceives, thinks, loves, hates, remembers, changes, fools itself, and coordinates our conscious and unconscious bodily processes is undoubtedly the most challenging of all developmental enigmas. A combination of genetic, cellular, and organismal approaches is giving us a preliminary understanding of how the basic anatomy of the brain becomes ordered.

The action by which an embryo forms a **neural tube,** the rudiment of the central nervous system, is called **neurulation,** and an embryo undergoing such changes is called a **neurula.** There are two major ways to form a neural tube. In **primary neurulation,** the chordamesoderm directs the ectoderm overlying it to proliferate, invaginate, and pinch off from the surface to form a hollow tube. In **secondary neurulation,** the neural tube arises from a solid cord of cells that sinks into the embryo and subsequently hollows out (cavitates) to form a neural tube (see Schoenwolf, 1991b). The extent to which these modes of construction are used depends on the class of vertebrate. Neurulation in fishes is exclusively secondary. In birds, the anterior portions of the neural tube are constructed by primary neurulation, while the neural tube caudal to the twenty-seventh somite pair (i.e., everything posterior to the hindlimbs) is made by secondary neurulation (Pasteels, 1937; Catala et al., 1996). In amphibians, such as *Xenopus*, most of the tadpole neural tube is made by primary neurulation, but the tail neural tube is derived from secondary neurulation (Gont et al., 1993). In mice (and probably humans, too), secondary neurulation begins at or around the level of somite 35 (Schoenwolf, 1984; Nievelstein et al., 1993).

Primary neurulation

In vertebrates, gastrulation creates an embryo with an internal endodermal layer, an intermediate mesodermal layer, and an external ectoderm. The interaction between the dorsal mesoderm and its overlying ectoderm is one of the most important interactions in all tetrapod development, for it initiates **organogenesis,** the creation of specific tissues and organs. In this interaction, the chordamesoderm directs the ectoderm above it to form the hollow neural tube, which will differentiate into the brain and spinal cord. The events of primary neurulation are diagrammed in Figure 7.2. During primary neurulation, the original ectoderm is divided into three sets of cells: (1)

(A)
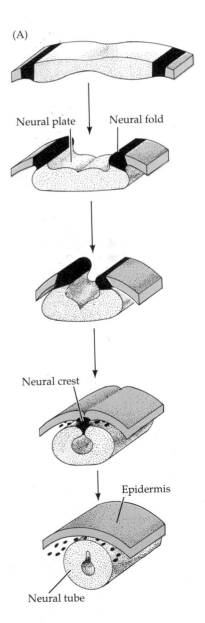

(B)
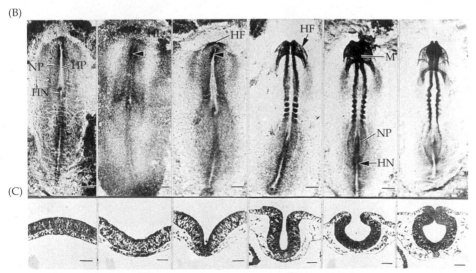

(C)

Figure 7.2
Neurulation in amphibians and amniotes. (A) Diagrammatic representation of neural tube formation. The ectodermal cells are represented either as precursors of the neural crest (black) or as precursors of the epidermis (color). The ectoderm folds in at the most dorsal point, forming an outer epidermis and an inner neural tube connected by neural crest cells. (B) Photomicrographs of neurulation in a 2-day chick embryo. (C) Neural tube formation seen in transverse cross sections of the chick embryo at the region of the future midbrain (arrowheads in B). Each photograph in (C) corresponds to the one above it. (HF, head fold; HP, head process; HN, Hensen's node; M, midbrain; NP, neural plate.) (Photomicrographs courtesy of R. Nagele.)

Figure 7.3
Four views of neurulation in an amphibian embryo, showing early (left), middle (center), and late (right) neurulae in each case. (A) Transverse section through the center of the embryo. (B) The same sequence looking down on the dorsal surface of the whole embryo. (C) Sagittal section through the median plane of the embryo. (D) Computer simulation of the three-dimensional constriction, extension, and rising of the neural plate. (A–C after Balinsky, 1975; D After Jacobson and Gordon, 1976.)

the internally positioned neural tube, which will form the brain and spinal cord, (2) the externally positioned epidermis of the skin, and (3) the neural crest cells, which migrate from the region that had connected the neural tube and epidermis and which will generate the peripheral neurons and glia, the pigment cells of the skin, and several other cell types. The phenomenon of embryonic induction, which initiates neurulation in the dorsal region of the embryo, will be detailed in Chapter 15. In this chapter, we are concerned with the response of the various ectodermal tissues.

The process of primary neurulation in frog embryos is depicted in Figure 7.3 and appears to be similar in amphibians, reptiles, birds, and mam-

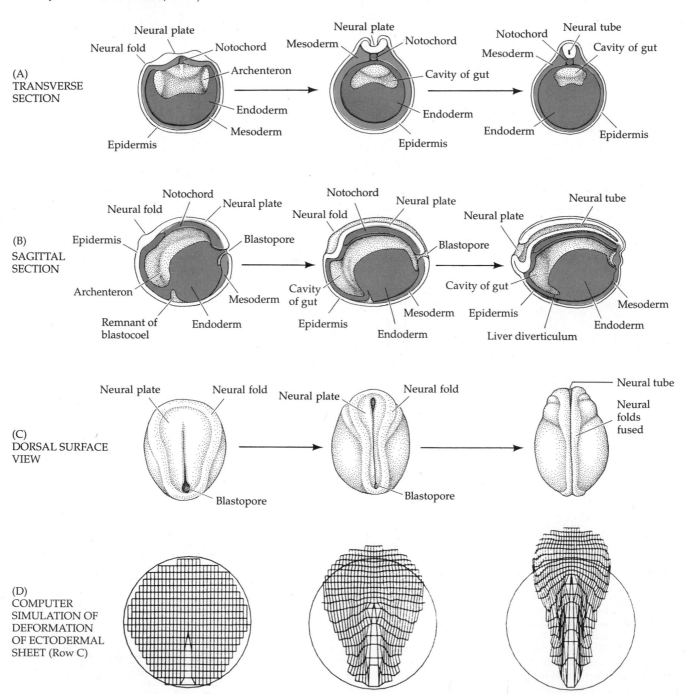

(A) TRANSVERSE SECTION

(B) SAGITTAL SECTION

(C) DORSAL SURFACE VIEW

(D) COMPUTER SIMULATION OF DEFORMATION OF ECTODERMAL SHEET (Row C)

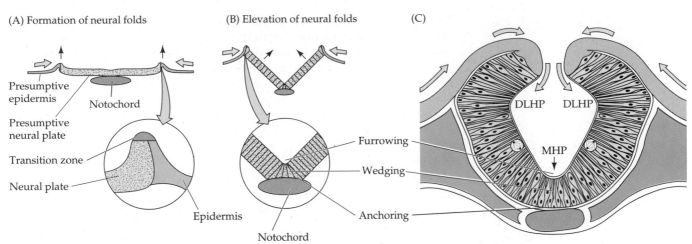

(A) Formation of neural folds

Presumptive epidermis
Notochord
Presumptive neural plate
Transition zone
Neural plate

(B) Elevation of neural folds

Epidermis
Notochord

Furrowing
Wedging
Anchoring

(C)

DLHP   DLHP
MHP

Figure 7.4
Schematic representation of the epithelial bending during chick neurulation. (A) Formation of neural folds occurs when presumptive epidermal cells move medially toward the center of the embryo. This presumptive epidermis pushes the neural plate under it as it moves. (B) As the midline cells of the neural plate (the floor plate cells) are anchored to the notochord, the neural folds are elevated. These movements appear to be continued as the medially moving epidermis drags the neural plate with it, resulting in the apposition of the neural folds. (C) In the three hinge regions (the medial hinge point, MHP, and the two dorsolateral hinge positions, the DLHP), the neural plate cells change their length and constrict at their apices. (After Moury and Schoenwolf, 1995.)

mals (Gallera, 1971). The first indication that a region of ectoderm is destined to become neural tissue is a change in cell shape (Figure 7.4). Midline ectodermal cells become elongated, while the cells destined to form the epidermis become more flattened. The elongation of dorsal ectodermal cells causes these prospective neural regions to rise above the surrounding ectoderm, thereby creating the **neural plate.** As much as 50 percent of the ectoderm is included in this plate. Shortly thereafter, the edges of the neural plate thicken and move upward to form the **neural folds,** while a U-shaped **neural groove** appears in the center of the plate, dividing the future right and left sides of the embryo (see Figures 7.3 and 7.4). The neural folds migrate toward the midline of the embryo, eventually fusing to form the neural tube beneath the overlying ectoderm. The cells at the dorsalmost portion of the neural tube become the **neural crest cells.**

The Mechanics of Primary Neurulation

Neurulation occurs in somewhat different ways in different regions of the body. The head, trunk, and tail each form their region of the neural tube in ways that reflect the relationship of the notochord to its overlying ectoderm. The head and trunk regions both undergo variants of primary neurulation, and this process can be divided into five distinct but spatially and temporally overlapping stages (Schoenwolf, 1991a; Catala et al., 1996): (1) the formation of the neural plate, (2) the formation of the neural floor plate, (3) the shaping of the neural plate, (4) the bending of the neural plate to form the neural groove, and (5) the closure of the neural groove to form the neural tube.

The Formation of the Neural Plate

The *formation* of the neural plate as a region distinct from the other ectodermal cells will be discussed in detail in Chapters 8 and 15. It is generally believed that the underlying dorsal mesoderm (in collaboration with other regions of the embryo) signals the ectodermal cells above it to develop into the columnar neural plate cells (Smith and Schoenwolf, 1989; Keller et al., 1992; discussion later in this chapter). As a result of this **neural induction,** the cells of the prospective neural plate are distinguished from the surrounding ectoderm, which is destined to become the epidermis. The cells of the neural plate and the cells of the epidermis have their own intrinsic movements (Moury and Schoenwolf, 1995). If the epidermis around the neural plate is isolated, its cells move centrally (i.e., toward the area where the neural plate was). If the neural plate is isolated, its cells converge and extend to make a thinner plate, but fail to roll up into a neural tube. These movements of the neural plate and epidermis result in the formation of the neural folds. The ectoderm is first "kinked," and then the presumptive epidermis begins to overlie the neural plate. (Indeed, if the "transition region" containing these two tissues is isolated, it will form small neural folds in culture.) These coordinated movements will eventually cause the elevation and folding of the neural tube (see Figure 7.4; Jacobson and Moury, 1995; Moury and Schoenwolf, 1995).

Formation of the Neural Floor Plate

Previously, it had been thought that only the midline cells of the neural plate formed the floor plate of the neural tube. That is, when the neural plate closed up to form a tube, the most centrally located cells of the neural plate would end up at the bottom of the tube. The most peripheral regions, the neural folds, would become the dorsalmost portion of the tube. This probably is how the head region forms. However, recent evidence suggests that the floor plate of the trunk neural tube has a separate origin—that it arises in part from Hensen's node and is "inserted" into the center of the neural plate.

This model has been proposed by Catala and co-workers (1995) based on their data and earlier studies from several laboratories. To follow the individual embryonic cells from the node, they used the chick-quail chimera system. Quail and chick embryos develop in very similar ways (especially during early development), and when portions of the quail embryo are grafted into a similar region of the chick embryo, the cells become integrated into the embryo and participate in the construction of the appropriate organs. This grafting can be done while the embryo is still inside the egg, and the chick that hatches is a "chimera," having a portion of its body composed of quail cells (Figure 7.5; Le Douarin, 1969; Le Douarin and Teillet, 1973). The chick and quail cells differ, however, in two critical ways. First, the quail heterochromatin in the nucleus is concentrated around the nucleoli. This creates a large, deeply staining mass that is easily distinguishable from the diffuse heterochromatin of chick cells. Second, there are some antigens that are quail-specific and are not seen on chicken cells. Both of these phenomena allow one to readily distinguish individual quail cells, even when the majority of the cell population is chick.

(A)

Figure 7.5
A chick-quail chimera. (A) Two chick-quail chimeras and a control chick 4 days after hatching. In the chimeras, a quail dorsal anterior neural tube replaced a corresponding region of the 12-somite chick embryo. Quail melanocytes of neural crest origin migrate into the head feathers at the level of the graft. (B) A region of the embryo containing both quail cells (with their highly condensed chromatin) and chick cells (with their more diffuse chromatin). (From Le Douarin et al., 1996; photographs courtesy of N. M. Le Douarin.)

(B)

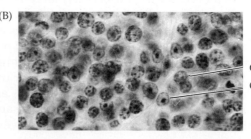

Chick cell
Quail cell

(A)

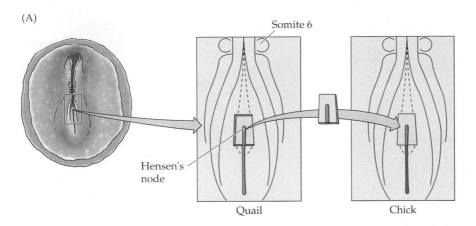

(B)

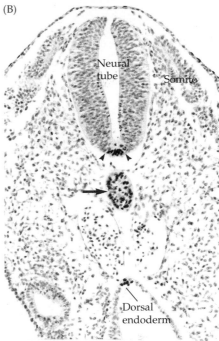

These investigators removed Hensen's node and the caudal end of the elongating notochord from 6-somite (1.5-day) chick embryos and replaced them with their quail counterparts. From that level down to the tail, both the notochord and the floor plate were made of *quail* cells. The walls of the neural tube were made from the chick neural plate (Figure 7.6). Interestingly (as predicted by the regression of the node, discussed in Chapter 6), the floor plate and notochord cells associated with neural plate located more caudally than the node itself. Thus, Hensen's node contains the cells needed to form the caudal floor plate and the notochord.

The floor plate cells become inserted into the center of the dorsal ectoderm. Only later does the notochord separate from the floor plate by the formation of a basement membrane between them (Figure 7.7).* The neural tube, then, has two distinct sources—one ectodermal and one from Hensen's node.

The Shaping and Bending of the Neural Plate

The *shaping* of the neural plate involves forces intrinsic to the neural plate cells. As the cells become more columnar, they cause a narrowing of the neural plate, but the most important shaping of the neural plate is performed by the cells at the midline of the neural plate which lie directly above the notochord. In birds and mammals, these midline neural plate cells are called the **median hinge point** (**MHP**) **cells** and are derived from the neural plate just anterior to Hensen's node and from the anterior midline of Hensen's node (Schoenwolf, 1991a,b; Catala et al., 1996). In both amphibians and amniotes, the neural plate cells undergo convergent extension by intercalating several layers of cells into a few layers (Jacobson and Sater, 1988; Schoenwolf and Alvarez, 1989). In so doing, they lengthen and narrow the neural plate (see Figure 7.3C).

The *bending* of the neural plate is accomplished by forces both intrinsic to and extrinsic to the neural plate cells. In chicks, the neural plate begins bending even as it is being shaped. The MHP cells become anchored to the notochord beneath them and form a hinge that forms a furrow in the dorsal midline. The notochord induces these cells to decrease their height and to become wedge-shaped (van Straaten et al., 1988; Smith and Schoenwolf, 1989). The

Figure 7.6
The floor plate of the trunk neural tube of the chick is derived from Hensen's node. (A) Schema of the operation whereby the node of a 6-somite chick embryo is replaced by its quail counterpart. (B) Analysis of the chimeric axis (stained for the quail-specific antigen), showing quail cells in the notochord (arrow) and floor plate (arrowheads). (B from Catala et al., 1996; photograph courtesy of N. M. Le Douarin.)

---

*While this idea that the notochord and floor plate are derived from the same population of cells has just recently been appreciated, this phenomenon had been documented in one of the famous books of embryology. Hans Spemann's 1938 *Embryonic Induction and Development* has an illustration of the famous Spemann and Mangold grafting experiment. On pages 144 and 146 of that book (and reproduced here in Figure 15.12), the graft of dorsal blastopore lip is shown as giving rise to dorsal mesoderm (notochord and somites) and the floor plate of the neural tube.

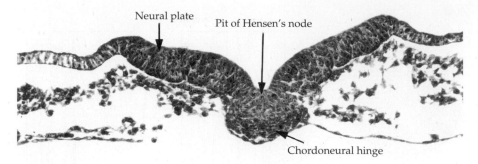

Figure 7.7
Hensen's node contributes to both notochord and neural floor plate. Section through Hensen's node at the 6-somite stage, showing that it contributes to the upper layer of embryonic cells. (From Catala et al., 1996; photographs courtesy of N. M. Le Douarin.)

cells lateral to the MHP do not undergo such a change (Figures 7.4 and 7.8). Shortly thereafter, two other hinge regions form furrows near the connection of the neural plate to the remainder of the ectoderm. These regions are called the **dorsolateral hinge points** (**DLHPs**), and they are anchored to the surface ectoderm of the neural fold. These cells *increase* their height and become wedge-shaped. Cell wedging is intimately linked to changes in cell shape. In the dorsolateral hinge points, microtubules and microfilaments are both involved in these changes. Colchicine, an inhibitor of microtubule polymerization, inhibits the elongation of these cells, while cytochalasin B, an inhibitor of microfilament formation, prevents the apical constriction of these cells, thereby inhibiting wedge formation (Burnside, 1971, 1973; Karfunkel, 1972; Nagele and Lee, 1980, 1987). After the initial furrowing of the neural plate, the plate bends around these hinge regions. Each hinge acts as a pivot that directs the rotation of the cells around it (Smith and Schoenwolf, 1991).

Meanwhile, extrinsic forces are also at work. The surface ectoderm of the chick embryo pushes toward the center of the embryo, providing another motive force for the bending of the neural plate (see Figure 7.4B,C; Alvarez and Schoenwolf, 1992). This movement of the presumptive epidermis and the anchoring of the neural plate to the underlying mesoderm may also be important for ensuring that the neural tube folds *into* the embryo and not outwardly. If small pieces of neural plates are isolated from the rest of the embryo (including the mesoderm), they tend to roll inside out (Schoenwolf, 1991a).

Closure of the Neural Tube

The neural tube closes as the paired neural folds are brought together at the dorsal midline. The folds adhere to each other, and the cells from the two folds merge. In some species, the cells at this junction form the neural crest cells. In birds, the neural crest cells do not migrate from the dorsal region until after the neural tube has been closed at that site. However, in mammals, the cranial neural crest cells (which form facial and neck structures) migrate while the neural folds are elevating, (i.e., prior to neural tube closure), whereas in the spinal cord region, the crest cells wait until closure has occurred (Nichols, 1981; Erickson and Weston, 1983).

The formation of the neural tube does not occur simultaneously throughout the ectoderm. This is best seen in those vertebrates (such as birds and mammals) whose body axis is elongated prior to neurulation. Figure 7.9

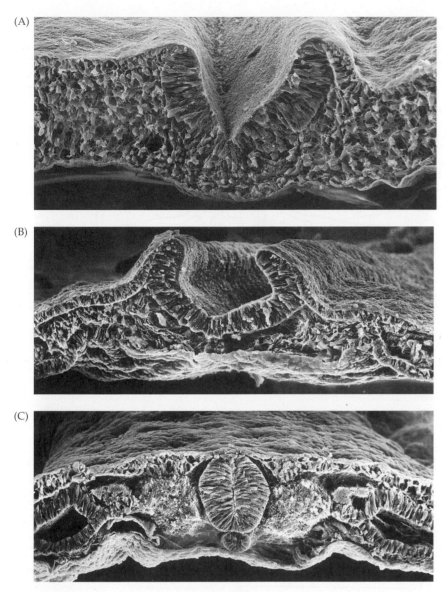

Figure 7.8
Scanning electron micrograph of neural tube formation in the chick embryo. (A) Neural groove surrounded by mesenchymal cells. (B) Elongated neural epithelial cells form a tube as the flattened epidermal cells are brought to the midline of the embryo. The MHP cells form a hinge at the bottom of the tube, while the neural plate cells attached to the basal area of the surface ectoderm form the dorsolateral hinge regions. These three hinges can be seen as furrows. (C) Neural tube formation is completed. The cells that had been the neural plate are now inside the embryo. The presumptive epidermis lies atop the tube, and the neural tube is flanked on its sides by the mesodermal somites and on the bottom by the notochord. (Photographs courtesy of K. W. Tosney.)

depicts neurulation in a 24-hour chick embryo. Neurulation in the cephalic (head) region is well advanced while the caudal (tail) region of the embryo is still undergoing gastrulation. Regionalization of the neural tube also occurs as a result of changes in the shape of the tube. In the cephalic end (where the brain will form), the wall of the tube is broad and thick. Here, a series of swellings and constrictions define the various brain compartments. Caudal to the head region, however, the neural tube remains a simple tube

Figure 7.9
Stereogram of a 24-hour chick embryo. Cephalic portions are finishing neurulation while the caudal portions are still undergoing gastrulation. (From Patten, 1971; after Huettner, 1949.)

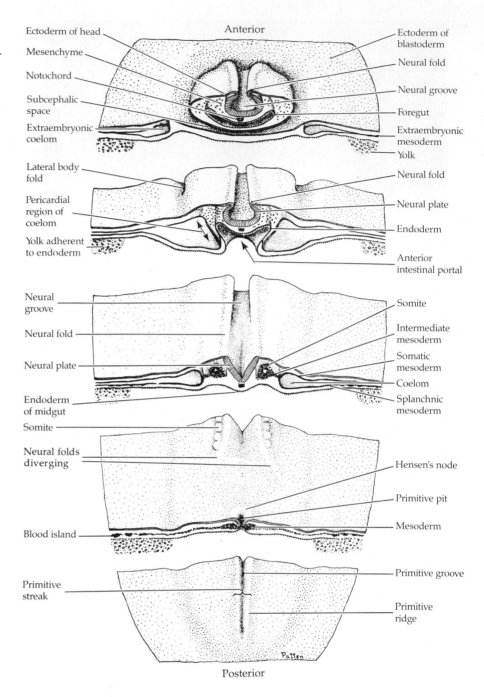

Ectoderm of head
Mesenchyme
Notochord
Subcephalic space
Extraembryonic coelom

Anterior

Ectoderm of blastoderm
Neural fold
Neural groove
Foregut
Extraembryonic mesoderm
Yolk

Lateral body fold
Pericardial region of coelom
Yolk adherent to endoderm

Neural fold
Neural plate
Endoderm
Anterior intestinal portal

Neural groove
Neural fold
Neural plate
Endoderm of midgut
Somite

Somite
Intermediate mesoderm
Somatic mesoderm
Coelom
Splanchnic mesoderm

Neural folds diverging

Blood island

Hensen's node
Primitive pit
Mesoderm

Primitive streak

Primitive groove
Primitive ridge

Patten

Posterior

that tapers off toward the tail. The two open ends of the neural tube are called the **anterior neuropore** and the **posterior neuropore**.

Unlike neurulation in chicks, neural tube closure in mammals is initiated at several places along the anterior-posterior axis (Golden and Chernoff, 1993; Van Allen et al., 1993). Different neural tube defects are caused when various parts of the neural tube fail to close (Figure 7.10). Failure to close the human *posterior* neural tube regions at day 27 (or the subsequent rupture of the posterior neuropore shortly thereafter) results in a condition called **spina bifida,** the severity of which depends on how much of the spinal cord remains open. Failure to close the *anterior* neural tube regions results in a lethal condition, **anencephaly.** Here, the forebrain remains in contact with the amniotic fluid and subsequently degenerates. Fetal forebrain

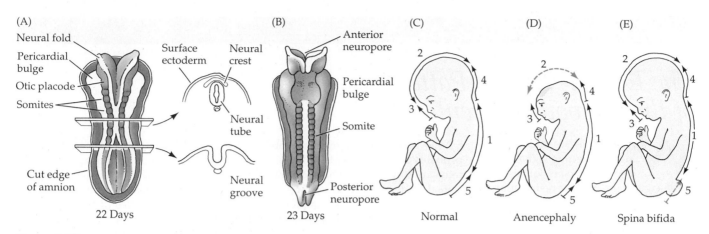

Figure 7.10
Neurulation in human embryos. (A) Dorsal and transverse sections of a 22-day human embryo initiating neurulation. Both anterior and posterior neuropores are open to the amniotic fluid. (B) Dorsal view of a neurulating human embryo a day later. The anterior neuropore region is closing while the posterior neuropore remains open. (C) Regions of neural tube closure postulated by genetic evidence (superimposed on newborn body). (D) Anencephaly due to failure of neural plate fusion in region 2. (E) Spina bifida due to failure of region 5 to fuse (or for the posteriormost neuropore to close). (C–E after Van Allen et al., 1993.)

development ceases, and the vault of the skull fails to form. These abnormalities are not that rare in humans, as they are seen in about 1 in every 500 live births. Neural tube closure defects can often be detected during pregnancy by various physical and chemical tests.

Human neural tube closure has been seen to involve a complex interplay between genetic and environmental factors. Certain genes, such as *Pax3, sonic hedgehog* and *openbrain*, are essential for the mammalian neural tube to form, but dietary factors such as cholesterol and folic acid also appear to be critical.* It has been estimated that around 50 percent of neural tube defects can be prevented by a pregnant woman's taking supplememental folic acid (Vitamin B_{12}), and the U. S. Public Health Service recommends that all women of childbearing age take 0.4 mg of folate daily to reduce the risk of neural tube defects during pregnancy (Milunsky et al., 1989; Czeizel and Dudas, 1992; CDC, 1992). [ecto1.html]

The neural tube eventually forms a closed cylinder that separates from the surface ectoderm. This separation is thought to be mediated by the expression of different cell adhesion molecules. Although the cells that will become the neural tube originally express E-cadherin, they stop producing this protein as the neural tube forms and instead synthesize N-cadherin and N-CAM (see Figure 3.17). As a result, the two tissues no longer adhere to one another. If the surface ectoderm is made to express N-cadherin (by injecting N-cadherin mRNA into one cell of a 2-cell *Xenopus* embryo), the separation of the neural tube from the presumptive epidermis is dramatically impeded (Detrick et al., 1990; Fujimori et al., 1990).

*Cholesterol appears to be necessary for the self-cleavage of the Sonic hedgehog protein. Mutations of Sonic hedgehog can prevent neural tube closure in mice and humans (Chiang et al., 1996; Roessler et al., 1996), and the active portion of Sonic hedgehog is its N-terminal region. This region is cleaved from the precursor molecule in a reaction that requires cholesterol as a cofactor (Porter et al., 1996). In humans, certain syndromes involving failure of neural tube closure have recently been associated with mutations in cholesterol synthesis (Kelley et al., 1996).

Sidelights & Speculations

The Dorsal-Ventral Patterning of the Nervous System

WHILE THE NEURAL TUBE is forming, it is getting signals from two other sets of tissues. These signals instruct the neural tube to have a particular dorsal-ventral polarity. The ventral neural tube appears to be patterned by the **Sonic hedgehog** protein coming from the notochord and floor plate cells (see Figure 7.7B; Ericson et al., 1996). This protein induces certain cells at the ventrolateral side of the neural tube to express the genes that make them into motor neurons (Figure 7.11A; Plate 32). Sonic hedgehog also works by repressing the expression of genes such as *dorsalin*, *Pax3*, and *msx1* from the ventral portion of the embryo. These genes would otherwise be expressed throughout the neural tube, but are inhibited by the ventrally produced the Sonic hedgehog signal. The *dorsal* fates of the neural tube are established by the bone morphogenetic proteins, probably BMP4 and BMP7. These proteins are expressed in the presumptive dorsal epidermis (Figure 7.11B,C), and they have been shown to counteract the effect of Sonic hedgehog (permitting genes such as *msx1* and *Pax3* to be expressed in the dorsal portion of the neural tube) and to promote the expression of other dorsally specific genes (Figure 7.11D). The roles of these proteins were confirmed by in vitro experiments in which isolated neural tubes were exposed to sources of these signals (Yamada et al., 1993; Liem et al., 1995). ■

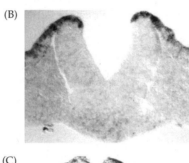
(B)

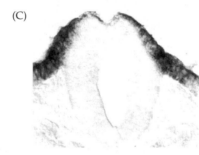
(C)

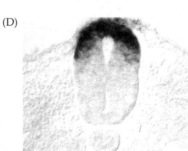
(D)

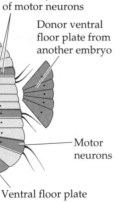

(A)

Secondary set of motor neurons

Donor ventral floor plate from another embryo

Motor neurons

Ventral floor plate

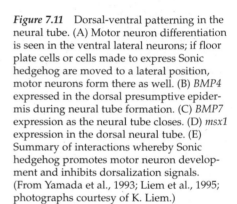
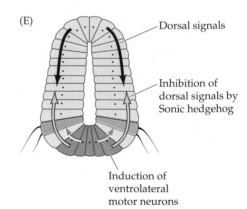

(E)

Dorsal signals

Inhibition of dorsal signals by Sonic hedgehog

Induction of ventrolateral motor neurons

Figure 7.11 Dorsal-ventral patterning in the neural tube. (A) Motor neuron differentiation is seen in the ventral lateral neurons; if floor plate cells or cells made to express Sonic hedgehog are moved to a lateral position, motor neurons form there as well. (B) *BMP4* expressed in the dorsal presumptive epidermis during neural tube formation. (C) *BMP7* expression as the neural tube closes. (D) *msx1* expression in the dorsal neural tube. (E) Summary of interactions whereby Sonic hedgehog promotes motor neuron development and inhibits dorsalization signals. (From Yamada et al., 1993; Liem et al., 1995; photographs courtesy of K. Liem.)

Secondary neurulation

Secondary neurulation involves the making of a **medullary cord** and its subsequent hollowing into a neural tube. In frogs and chicks, this type of neurulation is usually seen in the production of the lumbar and tail vertebrae. In both cases, secondary neurulation can be seen as a continuation of gastrulation. However, instead of involuting into the embryo, the cells of the dorsal

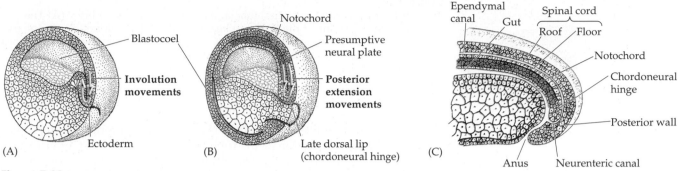

Figure 7.12
Movements of cells during secondary neurulation in *Xenopus*. (A) Involution of the mesoderm at midgastrula stage. (B) Movements of the dorsal blastopore lip at late-gastrula/early-neurula stage. Involution has ceased, and both the ectoderm and the mesoderm of the late blastopore lip move posteriorly. (C) Early tadpole stage, wherein the cells lining the blastopore form the neurenteric canal, part of which becomes the lumen of the secondary neural tube. (From Gont et al., 1993.)

blastopore lip keep growing ventrally (Figure 7.12A,B). The growing region at the tip of the lip is called the **chordoneural hinge** (Pasteels, 1937), and it contains precursors for both the posteriormost portion of the neural plate and the posterior portion of the notochord. The growth of this region converts the roughly spherical gastrula, 1.2 mm in diameter, into a linear tadpole, some 9 mm long. The tip of the tail is the direct descendant of the dorsal blastopore lip, and the cells lining the blastopore form the **neurenteric canal**. The proximal part of the neurenteric canal fuses with the anus, while the distal portion becomes the ependymal canal (i.e., the lumen of the neural tube) (Figure 7.12C; Gont et al., 1993).

In the chick, the tissues located posteriorly to the newly closed neuropore are called the tailbud. Like the tailbud of the frog, this structure is not a mass of undifferentiated cells. By grafting small regions of the quail tailbud into the chick tailbud, Catala and co-workers (1995) have shown that the early tailbud already has cells that have been assigned definite fates. Just as in *Xenopus*, there is a chordoneural hinge, and this region contains the cells that divide to form both the notochord and a medullary cord. Like the situation in the frog, these cells move posteriorly. The neural tube forms as the medullary cord forms small cavities that fuse with each other (Figure 7.13).

Differentiation of the neural tube

The differentiation of the neural tube into the various regions of the central nervous system occurs simultaneously in three different ways. On the gross anatomical level, the neural tube and its lumen bulge and constrict to form the chambers of the brain and the spinal cord. At the tissue level, the cell populations within the wall of the neural tube rearrange themselves to form the different functional regions of the brain and the spinal cord. Finally, on the cellular level, neuroepithelial cells themselves differentiate into the numerous types of neurons and supportive (glial) cells present in the body.

Formation of Brain Regions

The early development of most vertebrate brains is similar, but because the human brain is probably the most organized piece of matter in the solar system and the most interesting organ in the animal kingdom, we will concentrate on the development that is supposed to make *Homo* sapient.

(A)

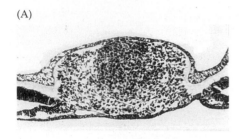

Figure 7.13
Formation of the secondary neural tube in the 25-somite chick embryo. (A) The medullary cord forming at the most caudal end of the chick tailbud. (B) Medullary cord slightly more anterior in the tailbud. (C) Cavitating neural tube and notochord forming. (D) Lumens coalesce to form the central canal of the neural tube. (From Catala et al., 1995; photographs courtesy of N. M. Le Douarin.)

(B)

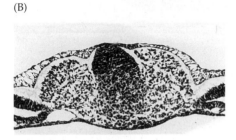

The early mammalian neural tube is a straight structure. However, even before the posterior portion of the tube has formed, the most anterior portion of the tube is undergoing drastic changes. In this anterior region, the neural tube balloons into three primary vesicles (Figure 7.14): forebrain (**prosencephalon**), midbrain (**mesencephalon**), and hindbrain (**rhombencephalon**). By the time the posterior end of the neural tube closes, secondary bulges—the **optic vesicles**—have extended laterally from each side of the developing forebrain.

The forebrain becomes subdivided into the anterior **telencephalon** and the more caudal **diencephalon.** The telencephalon will eventually form the **cerebral hemispheres,** and the diencephalon will form the thalamic and hypothalamic brain regions as well as the region that receives neural input from the retina. Indeed, the retina itself is a derivative of the diencephalon. The mesencephalon (midbrain) does not become subdivided, and its lumen eventually becomes the cerebral aqueduct. The rhombencephalon becomes subdivided into a posterior **myelencephalon** and a more anterior **metencephalon.** The myelencephalon eventually becomes the **medulla oblongata,** whose neurons generate the nerves that regulate respiratory, gastrointestinal, and cardiovascular movements. The metencephalon gives rise to the **cerebellum,** the part of the brain responsible for coordinating movements,

(C)

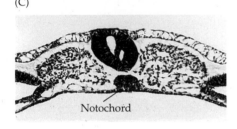

Notochord

(D)

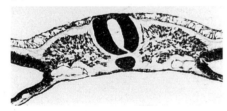

Figure 7.14
Early human brain development. The three primary brain vesicles are subdivided as development continues. To the right are the adult derivatives formed by the walls and cavities of the brain. (After Moore and Persaud, 1993.)

Adult derivatives

3 Primary vesicles

Wall   Cavity

Forebrain (Prosencephalon)

Midbrain (Mesencephalon)

Hindbrain (Rhombencephalon)

5 Secondary vesicles

Telencephalon

Diencephalon

Mesencephalon

Metencephalon

Myelencephalon

Spinal cord

Olfactory lobes	– Smell
Hippocampus	– Memory storage
Cerebrum	– Association ("intelligence")
Retina	– Vision
Epithalamus	– Pineal gland
Thalamus	– Relay center for optic and auditory neurons
Hypothalamus	– Temperature, sleep, and breathing regulation
Midbrain	– Fiber tracts between anterior and posterior brain, optic lobes, and tectum
Cerebellum	– Coordination of complex muscular movements
Pons	– Fiber tracts between cerebrum and cerebellum (mammals only)
Medulla	– Reflex center of involuntary activities

posture, and balance. The hindbrain develops a segmental pattern that specifies the places where certain nerves originate. Periodic swellings called **rhombomeres** divide the rhombencephalon into smaller compartments. These rhombomeres represent separate developmental "territories" in that cells within each rhombomere can mix freely within the rhombomere, but not with cells from adjacent rhombomeres (Guthrie and Lumsden, 1991). Moreover, each rhombomere has a different developmental fate. This has been most extensively studied in the chick, where the first neurons appear in the even-numbered rhombomeres, r2, r4, and r6 (Figure 7.15; Lumsden and Keynes, 1989). Neurons from r2 ganglia form the fifth (trigeminal) cranial nerve; those from r4 form the seventh (facial) and eighth (vestibuloacoustic) cranial nerves; and the ninth (glossopharyngeal) cranial nerve exits from r6. [ecto2.html]

The ballooning of the early embryonic brain is remarkable in its rate, in its extent, and in its being the result primarily of an increase in cavity size, not tissue growth. In chick embryos, the brain volume expands 30-fold between days 3 and 5 of development. This rapid expansion is thought to be caused by positive fluid pressure pressing against the walls of the neural tube by the fluid within it. It might be expected that this fluid pressure might have been dissipated by the spinal cord, but this does not appear to happen. Rather, as the neural folds close in the region between the presumptive brain and spinal cord, the surrounding dorsal tissues push in to constrict the tube at the base of the brain (Figure 7.16; Schoenwolf and Desmond, 1984; Desmond and Schoenwolf, 1986; Desmond and Field, 1992). This occlusion (which also occurs in the human embryo) effectively separates the presumptive brain region from the future spinal cord (Desmond, 1982). If one removes the fluid pressure in the anterior portion of such an occluded neural tube, the chick brain enlarges at a much slower rate and contains many fewer cells than are found in normal, control embryos. The occluded region of the neural tube reopens after the initial rapid enlargement of the brain ventricles.

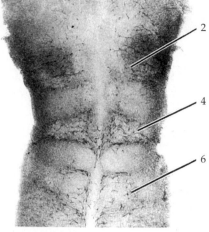

Figure 7.15
A 2-day embryonic chick hindbrain splayed to show lateral walls. Neurons were visualized with an antibody staining neurofilament proteins. Rhombomeres 2, 4, and 6 are distinguished by the high density of axons at this early developmental stage. (From Lumsden and Keynes, 1989, courtesy of A. Keynes.)

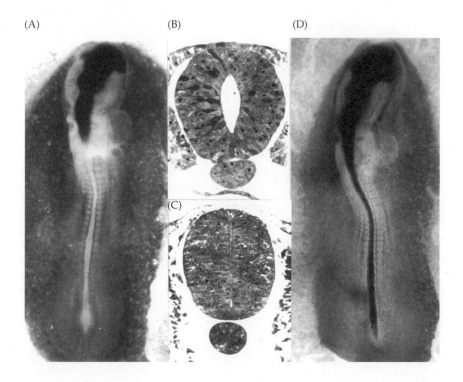

Figure 7.16
Occlusion of the neural tube to allow expansion of the future brain region. (A) Dye injected into the anterior portion of a 3-day chick neural tube fills the brain region but does not pass into the spinal region. (B,C) Section of the chick neural tube at the base of the brain (B) before occlusion and (C) during occlusion. (D) Reopening of occlusion after initial brain enlargement allows dye to pass from the brain region into the spinal cord region. (Courtesy of M. Desmond.)

Determining the Forebrain and Midbrain Regions

THE ANTERIOR-POSTERIOR IDENTITY of each of the mammalian brain vesicles is specified during gastrulation by the prechordal mesoderm and notochord. This specification appears to be stabilized at the neural plate stage by interactions within the plane of the ectoderm. Only the major molecules involved in forebrain and midbrain specification will be discussed here, and the details of hindbrain and spinal cord specification by the *Hox* genes will be discussed in Chapter 16.

The forebrain and midbrain regions are defined by the underlying prechordal mesoderm and anterior notochord. Two genes that are expressed in these anterior mesodermal tissues are *Lim1* and *Otx2*. If either one is missing, the embryo does not form a forebrain or midbrain. Caudal to rhombomere 2, the embryos appear to be normal (Figure 7.17; Acampora et al., 1995; Shawlot and Behringer, 1995).

Rubenstein and Puelles (1994) have proposed that the forebrain is composed of six neuromeric regions called **prosomeres**. Prosomeres p1–p3 comprise the diencephalon, whereas prosomeres p4–p6 comprise the hypothalamus (ventrally) and the telencephalon (dorsally). The prosomeric boundaries coincide with the expression boundaries of several genes that are thought to be important in neural specification. They are also seen to be the boundaries that limit the responses to certain external stimuli. The p2/p3 boundary may be critical in patterning the forebrain region. This boundary corresponds to the *zona limitans*. It is also a source of Sonic hedgehog, a diffusible protein known to induce patterning during gastrulation and limb formation (Figure 7.18; Rubenstein and Puelles, 1994).

One of the critical regions for midbrain development is the metencephalon/mesencephalon border that will normally give rise to the tissues of the isthmus. No morphological boundary can be seen here, but it is marked by the most posterior portion of *Otx2* gene expression. When mid-to-anterior mesencephalon tissue is transplanted to the diencephalon or rhombencephalon, it induces the cells surrounding it to develop mesencephalonic fates (in the diencephalon) or cerebellar fates (in the rhombencephalon) (Figure 7.19A; Bally-Cuif and Wassef, 1994; Marin and Puelles, 1994). When rotated, a "triplication" can ensue, since tissues on both

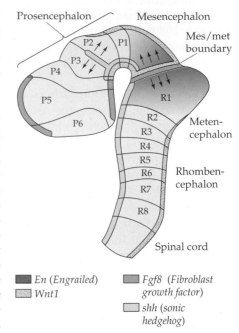

Figure 7.18 Neuromeric structure of the brain with the hypothetical inductive events superpositioned on them. The mesencephalon/metencephalon boundary is positive for both *Fgf8* and *Wnt1* gene expression. The p2/p3 border is thought to be the source of sonic hedgehog protein. (After Bally-Cuif and Wassef, 1995.)

sides of the graft are induced (Figure 7.19B).

This mes/met-inducing region appears to be controlled by **fibroblast growth factor 8** (FGF8). Crossley and colleagues (1996) found that this isthmus-forming tissue secreted FGF8. Moreover, when they transplanted FGF8-containing beads into the diencephalon or rhombencephalon, they obtained the same duplicated midbrain structures. Control beads soaked in saline did not show any such duplications. The FGF8 beads also induced the expression of three genes in the surrounding tissues—*Wnt1*, *Engrailed-2*,

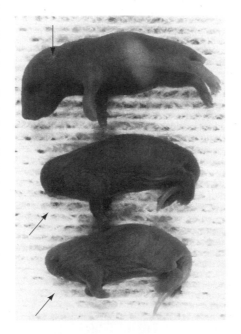

Figure 7.17 Headless phenotype of the *Lim1*-deficient mouse. Two *Lim1*-knockout mice are on the bottom; a wild-type pup is shown on the top. Most *Lim1* mutants die before birth. The ear pinnae (arrows) are the most anterior structures of these mutants. (From Shawlot and Behringer, 1995; courtesy of the authors.)

and *Fgf8* itself. These three genes are normally expressed in the isthmus region. Wnt1 and Engrailed are known to be important in the formation of the cerebellum. Even though the cerebellum does not express *Wnt1* genes, mice deficient in Wnt1 lack their midbrain regions as well as the cerebellum (McMahon and Bradley, 1990; Thomas and Cappecchi, 1990). Wnt1 appears to maintain *Engrailed* gene expression in the cerebellar precursor cells, enabling the cells to proliferate (Dickinson et al., 1994; Danielian and McMahon, 1996). [ecto3.html] ∎

Figure 7.19 The mesencephalon/metencephalon ("mes/met") junction region can act as an inducer of midbrain development and engrailed expression when rotated or transplanted to other regions of the brain. (A) Transplantation of the mes/met junction results in the induction of engrailed gene expression and midbrain and cerebellar structures in ectopic positions. (B) Rotation of the mes/met junction causes "triplications" of certain structures, such as the optic tectum. Abbreviations: gt, griseum tectale; TS, torus semicircularis; P1, pretectal segment; P2, dorsal thalamic segment; cb, cerebellum; ot, optic tectum; ist, isthmus; III, third cranial, or oculomotor, nerve; IV, fourth cranial, or trochlear, nerve. The postulated polarity is represented by arrows. (B after Rubenstein and Puelles, 1994.)

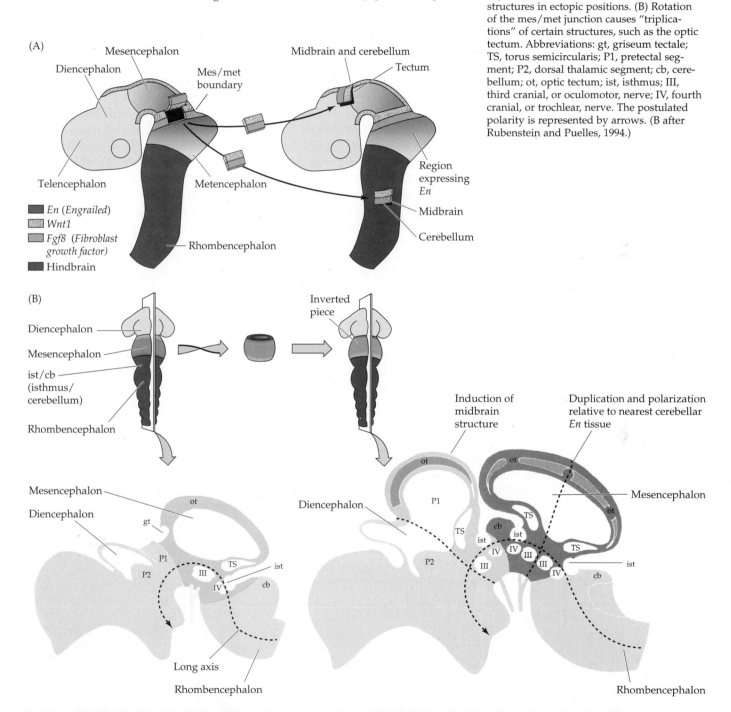

Tissue Architecture of the Central Nervous System

The neurons of the brain cortex are organized into layers, each having different functions and connections. The original neural tube is composed of a germinal neuroepithelium that is one cell layer thick. This is a rapidly dividing cell population. Sauer (1935) and others have shown that all these cells are continuous from the luminal edge of the neural tube to the outside edge but that the nuclei of these cells are at different heights, thereby giving the superficial impression that the wall of the neural tube has numerous cell layers. DNA synthesis (S phase) occurs while the nucleus is at the outside edge of the tube, and the nucleus migrates luminally as mitosis proceeds (Figure 7.20). Mitosis occurs on the luminal side of the cell layer. During early mammalian development, 100 percent of the neural tube cells incorporate radioactive thymidine into DNA (Fujita, 1964). Shortly thereafter, certain cells stop incorporating these DNA precursors, thereby indicating that they are no longer participating in DNA synthesis and mitosis. These neuronal and glial cells can then differentiate at the periphery of the neural tube (Fujita, 1966; Jacobson, 1968).

If dividing cells are labeled by radioactive thymidine at a single point in their development and their progeny are seen in the outer cortex in the adult brain, then the neurons had to migrate to their cortical positions from the germinal neuroepithelium. This happens when the cell divides "vertically" instead of "horizontally." In these cases, the cell adjacent to the lumen keeps connected to the ventricular surface, while the other daughter cell migrates away (Chenn and McConnell, 1995). The time of this horizontal division is the last time that neuron divides and is called the neuron's "birthday." Different types of neuron and glial cells have their birthdays at different times. Labeling at different times during development shows that cells with the earliest birthdays migrate the shortest distances. The cells with later birthdays migrate through these layers to form the more superficial regions of the cortex. Subsequent differentiation depends on the position these neurons occupy once outside the region of dividing cells (Letourneau, 1977; Jacobson, 1991).

Figure 7.20
Schematic section of a chick embryo neural tube, showing the position of the nucleus in a neuroepithelial cell as a function of the cell cycle. Mitotic cells are found near the center of the neural tube, adjacent to the lumen. (B) Scanning electron micrograph of a newly formed chick neural tube, showing cells at different stages of their cell cycles. (A after Sauer, 1935; B courtesy of K. Tosney.)

(A)

Stage of cell cycle

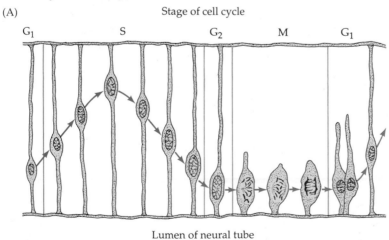

G_1 S G_2 M G_1

Lumen of neural tube

(B)

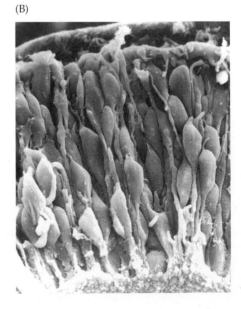

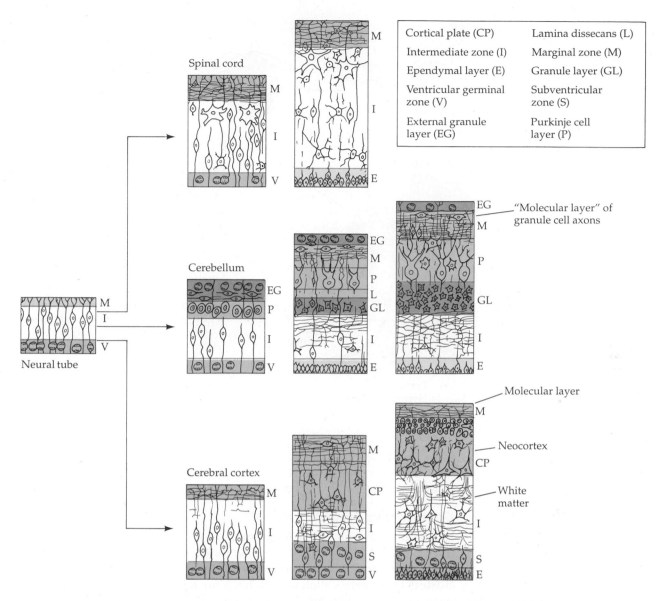

Cortical plate (CP) · Lamina dissecans (L)
Intermediate zone (I) · Marginal zone (M)
Ependymal layer (E) · Granule layer (GL)
Ventricular germinal zone (V) · Subventricular zone (S)
External granule layer (EG) · Purkinje cell layer (P)

Figure 7.21
Differentiation of the walls of the neural tube. Section of a 5-week human neural tube containing three zones: ependymal, mantle, and marginal. In the spinal cord and medulla (top row), the ependyma remains the sole source of neurons and glial cells. In the cerebellum (middle row), a second mitotic layer, the external granular layer, forms at the region furthest removed from the ependyma. Neuroblasts from this layer migrate back into the intermediate zone to form the granule cells. In the cerebral cortex (bottom row), the migrating neuroblasts and glioblasts form a cortical plate containing six layers. (After Jacobson, 1991.)

As the cells adjacent to the lumen continue to divide, the migrating cells form a second layer around the original neural tube. This layer becomes progressively thicker as more cells are added to it from the germinal neuroepithelium. This new layer is called the **mantle** (or **intermediate**) **zone,** and the germinal epithelium is now called the **ventricular zone** (and, later, the **ependyma**) (Figure 7.21). The mantle zone cells differentiate into both neurons and glia. The neurons make connections among themselves and send forth axons away from the lumen, thereby creating a cell-poor **marginal zone.** Eventually, glial cells cover many of these marginal zone axons in myelin sheaths, giving them a whitish appearance. Hence, the mantle zone, containing the cell bodies, is often referred to as the **gray matter;** the axonal, marginal layer is often called the **white matter.**

In the spinal cord and medulla, this basic three-zone pattern of ependymal, mantle, and marginal layers is retained throughout development. The gray matter (mantle) gradually becomes a butterfly-shaped structure surrounded by white matter; and both become encased in connective tissue. As the neural tube matures, a longitudinal groove—the **sulcus limitans**—ap-

(A)

Presumptive
basal
region

Presumptive
alar
region

(B)

(C)

(D)

Epidermis

Sulcus
limitans

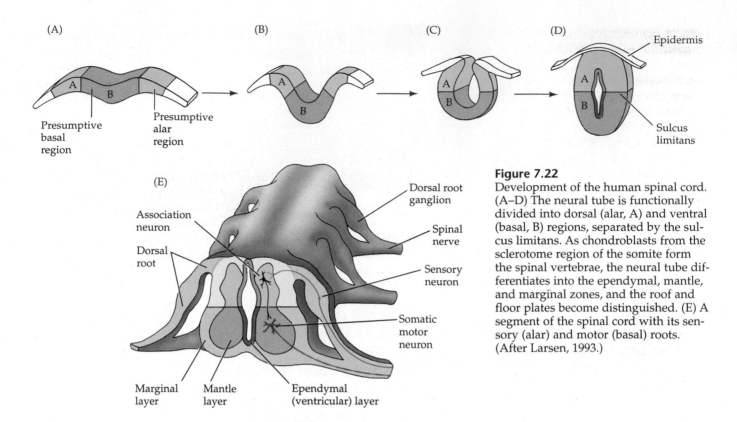

(E)

Association
neuron

Dorsal
root

Dorsal root
ganglion

Spinal
nerve

Sensory
neuron

Somatic
motor
neuron

Marginal
layer

Mantle
layer

Ependymal
(ventricular) layer

Figure 7.22
Development of the human spinal cord.
(A–D) The neural tube is functionally
divided into dorsal (alar, A) and ventral
(basal, B) regions, separated by the sulcus limitans. As chondroblasts from the
sclerotome region of the somite form
the spinal vertebrae, the neural tube differentiates into the ependymal, mantle,
and marginal zones, and the roof and
floor plates become distinguished. (E) A
segment of the spinal cord with its sensory (alar) and motor (basal) roots.
(After Larsen, 1993.)

pears to divide it into dorsal and ventral halves. The dorsal portion receives
input from sensory neurons, whereas the ventral portion is involved in effecting various motor functions (Figure 7.22).

Cerebellar Organization

In the brain, cell migration, differential growth, and selective cell death produce modifications of the three-zone pattern, especially in the cerebellum
and cerebrum. Some neurons enter the white matter to differentiate into
clusters of neurons called **nuclei.** Each nucleus works as a functional unit,
serving as a relay station between the outer layers of the cerebellum and
other parts of the brain. In addition, dividing neuronal precursor cells, **neuroblasts,** migrate to the outer surface of the developing cerebellum, forming
a new germinal zone, the **external germinal layer,** near the outer boundary
of the neural tube. At the outer boundary of the external germinal layer (one
to two cells thick), neuroblasts proliferate. The inner compartment of the external germinal layer contains postmitotic neuroblasts that are the precursors of the major neurons of the cerebellar cortex, the **granular cells.** These
pregranular cell neurons migrate back into the developing cerebellar white
matter to produce granular cell neurons in a region called the **internal granular layer.** Meanwhile, the original ependymal layer of the cerebellum generates a wide variety of neurons and glial cells, including the distinctive and
large **Purkinje neurons.** Each Purkinje neuron has an enormous **dendritic
apparatus,** which spreads like a fan above a bulblike cell body. A typical
Purkinje cell may form as many as 100,000 synapses with other neurons,
more than any other neuron studied. Each Purkinje neuron also emits a slender axon, which connects to other cells in the deep cerebellar nuclei.

The development of spatial organization is critical for the proper functioning of the cerebellum. All impulses eventually regulate the activity of the

Purkinje cells, which are the only output neurons of the cerebellar cortex. For this to happen, the proper cells must differentiate at the appropriate place and time. How is this accomplished?

One mechanism thought to be important for positioning young neurons within the developing mammalian brain is **glial guidance** (Rakic, 1972; Hatten, 1990). Throughout the cortex, neurons are seen to ride "the glial monorail" to their respective destinations. In the cerebellum, the granule cell precursors travel on the long processes of the **Bergmann glia** (Figure 7.23; Rakic and Sidman, 1973; Rakic, 1975). The neural-glial interaction is a complex and fascinating series of events, involving reciprocal recognition between glia and neuroblast (Hatten, 1990; Komuro and Rakic, 1992). The neuron maintains its adhesion to the glial cell through a number of proteins, the most important being an adhesion protein called **astrotactin.** If the astrotactin on the nerve cells is masked by antibodies to that protein, the nerve cell will fail to adhere to the glial processes (Edmondson et al., 1988; Fishell and Hatten, 1991).

Some insight into the mechanisms of spatial ordering may soon come from the analysis of neurological mutations in mice. Over 30 mutations are known to affect the arrangement of cerebellar neurons. Many of the cerebellar mutants have been found because the phenotype of such mutants—namely, the inability to keep balance while walking—can be easily recognized. For obvious reasons, these mutations are given names such as *weaver*, *reeler*, *staggerer*, and *waltzer*. [ecto4.html], [ecto5.html]

Figure 7.23
Neuronal migration on glial cell processes. (A) Diagram of a cortical neuron migrating on a glial cell process. (B) Electron micrograph of the region where the neuron soma adheres to the glial process. (C) Sequential photographs of a neuron migrating on a cerebellar glial process. The leading process has several filopodial extensions. It reaches speeds up to 60 μm/hr as it progresses on the glial process. (A after Rakic, 1975; B from Gregory et al., 1988; C from Hatten, 1990, courtesy of M. Hatten.)

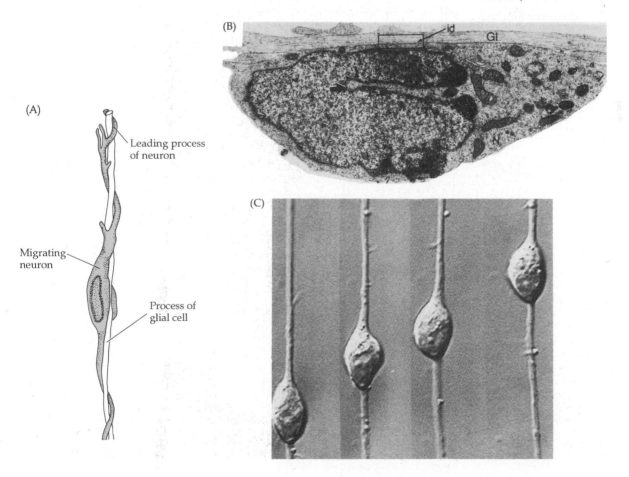

(B)

(A)

Leading process
of neuron

Migrating
neuron

Process of
glial cell

(C)

Cerebral Organization

The three-zone arrangement is also modified in the cerebrum. The cerebrum is organized in two distinct ways. First, like the cerebellum, it is organized vertically into layers that interact with each other (see Figure 7.21). Certain neuroblasts from the mantle zone migrate on the glia through the white matter to generate a second zone of neurons. This new mantle zone is called the **neopallial cortex**. This cortex eventually stratifies into six layers of cell bodies, and the adult forms of these neopallial neurons are not completed until the middle of childhood. Each layer of the cerebral cortex differs from the others in its functional properties, the types of neurons found there, and the sets of connections that emerge. For instance, neurons from layer 4 receive their major input from the thalamus (a region that forms from the diencephalon), while neurons in layer 6 send their major output back to the thalamus. Second, the cerebral cortex is organized horizontally into over 40 regions that regulate anatomically and functionally distinct processes. For instance, neurons to cortical layer 6 of the visual cortex project axons to the lateral geniculate nucleus of the thalamus, while layer 6 neurons of the auditory cortex (located more anteriorly than the visual cortex) project axons to the medial geniculate nucleus of the thalamus.

Neither the vertical nor the horizontal organization is clonally specified. Rather, there is a lot of cell movement that mixes the progenies of various precursor cells. After their final mitosis, most of the newly generated neurons migrate radially along glial processes out from the ventricular (ependymal) zone and form the cortical plate below the pia mater of the brain. Like the cerebellar cortex, those neurons with the earliest "birthdays" form the layer closest to the ventricle. Subsequent neurons travel greater distances to form the more superficial layers of the cortex. This forms an "inside-out" gradient of development (Figure 7.24; Rakic, 1974). A single stem cell in the ventricular layer can produce neurons (and glial cells) in any of the cortical layers (Walsh and Cepko, 1988). But how do the cells know which layer to

Figure 7.24
"Inside-out" gradient of cerebral cortex formation in the rhesus monkey. The "birthdays" of the cortical neurons were determined by injecting pregnant animals intravenously with [³H]-thymidine at certain times of gestation. The animals were born, and those cells undergoing the S phase of their final cell division cycle were found to be heavily labeled. These cells migrated to various regions and were detected by autoradiography of microscopic slices. The figure represents the position of those neurons in the visual cortex indicated. Full gestation in rhesus monkeys is 165 days. The youngest neurons are at the periphery of the neural tube. (After Rakic, 1974).

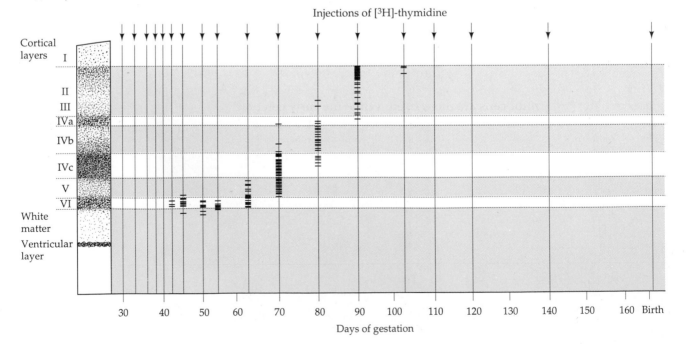

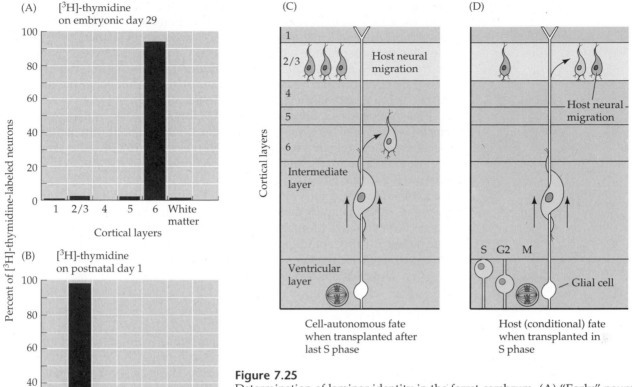

Figure 7.25
Determination of laminar identity in the ferret cerebrum. (A) "Early" neuronal precursors (birthdays on embryonic day 29) migrate to layer 6. (B) "Late" neuronal precursors (birthdays on postnatal day 1) migrate farther, into layers 2 and 3. (C) When early precursors (red) are transplanted into older ventricular zones *after* their last mitotic S phase, the neurons they form migrate into layer 6. (D) If these precursors are transplanted *before* or *during* their last S phase, their neurons migrate (with the host neurons) to layer 2. (After McConnell and Kaznowski, 1991.)

enter? McConnell and Kaznowski (1991) have shown that the determination of laminar identity (i.e., which layer a cell migrates to) is made during the final cell division. Cells transplanted from young brains (where they would form layer 6) into older brains whose migratory neurons are forming layer 2 after their last division are committed to their fate and migrate only to layer 6. However, if the cells are transplanted prior to their final division (during mid S phase), they are uncommitted and can migrate to layer 2 (Figure 7.25). The fates of older progenitor cells are more fixed. While the early cerebral cortical progenitor cells have the potential to become any neuron (at layers 2 or 6, for instance), late cortical progenitor cells only give rise to upper-level (layer 2) neurons (Frantz and McConnell, 1996). We still do not know the nature of the information given to the cell as it becomes committed.

Not all neurons, however, migrate radially. O'Rourke and her colleagues (1992) labeled young neurons with fluorescent dye and followed their migration through the brain. While about 80 percent of the young neurons migrated radially on glial processes from the ventricular zone into the cortical plate, about 12 percent of the new neurons migrated laterally from one functional region of the cortex into another. These observations meshed well with those of Walsh and Cepko (1992), who infected ventricular cells with a retrovirus and were able to stain these cells and their progeny after birth. They found that the neural descendants of a single ventricular cell were dispersed across the functional regions of the cortex. When forebrain neurons from the cortex were transplanted into the region that would form the striatum, these cells took on the morphology of the striatum (Fishell, 1995). Thus, the specification of these cortical areas into specific functions occurs after neurogenesis. Once the cells arrive at their final destination, it is thought that they elaborate particular adhesion molecules that organize them together as brain nuclei (Matsunami and Takeichi, 1995).

The cerebrum is quite plastic. The development of the human neopallial cortex is particularly striking in this regard. The human brain continues to develop at fetal rates even after birth (Holt et al., 1975). Based on morphological and behavioral criteria, Portmann (1941, 1945) suggested that compared with other primates, human gestation should really last 21 months instead of 9. However, no woman could deliver a 21-month-old fetus because the head would not pass through the birth canal; thus, humans give birth at the end of 9 months. Montagu (1962) and Gould (1977) have suggested that during our first year of life, we are essentially extrauterine fetuses, and they speculate that much of human intelligence comes from the stimulation of the nervous system as it is forming during that first year.*

Neuronal types

The human brain consists of over 10^{11} nerve cells (neurons) associated with over 10^{12} glial cells. Those cells that remain integral components of the neural tube lining become **ependymal cells.** These cells can give rise to the precursors of neurons and glial cells. It is thought that the differentiation of these precursor cells is largely determined by the environment that they enter (Rakic and Goldman, 1982) and that, at least in some cases, a given precursor cell can form both neurons and glial cells (Turner and Cepko, 1987). There is a wide variety of neuronal and glial types (as is evident from a comparison of the relatively small granule cell with the enormous Purkinje neuron). The fine extensions of the cell that are used to pick up electrical impulses are called **dendrites** (Figure 7.26). Some neurons develop only a few dendrites, whereas other cells (such as the Purkinje neurons) develop extensive areas for cellular interactions. Very few dendrites can be found on cortical neurons at birth, but one of the amazing things about the first year of human life is the increase in the number of such receptive regions in the cortical neurons. During this year, each cortical neuron develops enough dendritic surface to accommodate as many as 100,000 connections with other neurons. The average cortical neuron connects with 10,000 other neural cells. This pattern of neural connections (synapses) enables the human cortex to function as the center for learning, reasoning, and memory, to develop the capacity for symbolic expression, and to produce voluntary responses to interpreted stimuli.

Another important feature of a developing neuron is its **axon** (sometimes called a neurite). Whereas dendrites are often numerous and do not extend far from the nerve cell body, or **soma,** axons may extend for several feet. The pain receptors on your big toe, for example, must transmit their messages all the way to the spinal cord. One of the fundamental concepts of

---

*Contrary to the claims of a widely circulated anti-abortion film, the human cerebral cortex has no neuronal connections at 12 weeks' gestation (and therefore cannot move in response to a thought, nor experience consciousness or fear). Measurable electrical activity characteristic of neural cells (the electroencephalogram, or EEG, pattern) is first seen at 7 months' gestation. Morowitz and Trefil (1992) put forth the provocative opinion that since society in the United States has defined death as the *loss* of the EEG pattern, perhaps it should accept the *acquisition* of the EEG pattern as the start of human life.

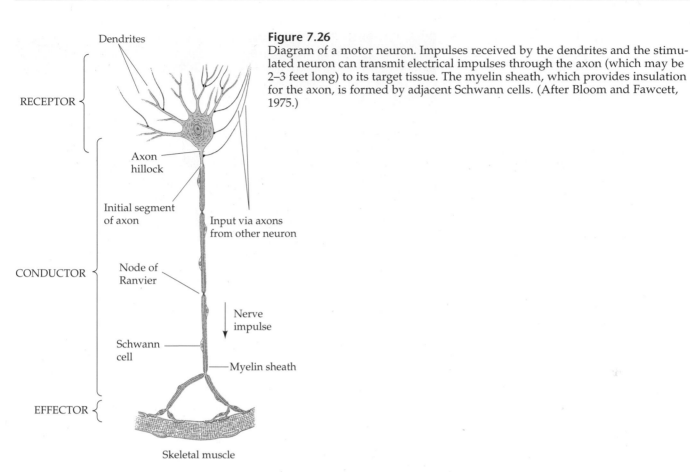

RECEPTOR

Dendrites

Axon
hillock

Initial segment
of axon

Input via axons
from other neuron

CONDUCTOR

Node of
Ranvier

Nerve
impulse

Schwann
cell

Myelin sheath

EFFECTOR

Skeletal muscle

Figure 7.26
Diagram of a motor neuron. Impulses received by the dendrites and the stimulated neuron can transmit electrical impulses through the axon (which may be 2–3 feet long) to its target tissue. The myelin sheath, which provides insulation for the axon, is formed by adjacent Schwann cells. (After Bloom and Fawcett, 1975.)

neurobiology is that the axon is a continuous extension of the nerve cell body. At the turn of the twentieth century, there were many competing theories of axon formation. Schwann, one of the founders of the cell theory, believed that numerous neural cells linked themselves together in a chain to form an axon. Hensen, the discoverer of the embryonic node, thought that the axon formed around preexisting cytoplasmic threads between the cells. Wilhelm His (1886) and Santiago Ramón y Cajal (1890) postulated that the axon was indeed an outgrowth (albeit an extremely large one) of the nerve soma. In 1907, Ross Harrison demonstrated the validity of the outgrowth theory in an elegant experiment that founded both the science of developmental neurobiology and the technique of tissue culture. Harrison isolated a portion of neural tube from a 3-mm frog tadpole. At this stage, shortly after the closure of the neural tube, there is no visible differentiation of axons. He placed these neuroblasts in a drop of frog lymph on a coverslip and inverted the coverslip over a depression slide so he could watch what was happening within this "hanging drop." What Harrison saw was the emergence of the axons as outgrowths from the neuroblasts, elongating at about 56 μm/hr.

Such nerve outgrowth is led by the tip of the axon, called the **growth cone** (Figure 7.27). This cone does not proceed in a straight line but rather "feels" its way along the substrate. The growth cone moves by the elongation and contraction of pointed filopodia called **microspikes.** These microspike filopodia contain microfilaments, which are oriented parallel to the long axis of the axon. (This is similar to the situation seen in the filopodial microfilaments of secondary mesenchyme cells in echinoderms.) Treating neurons with cytochalasin B destroys the actin microspikes, inhibiting their further advance (Yamada et al., 1971; Forscher and Smith, 1988). Within the

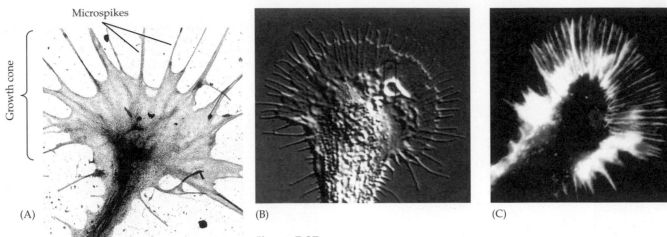

Figure 7.27
Actin microspikes in axon growth cones as seen by (A) transmission electron microscopy, (B) differential interface contrast microscopy, and (C) fluorescence microscopy using fluorescent antibodies to actin. (A from Letourneau, 1979; B and C from Forscher and Smith, 1988. All photographs courtesy of the authors.)

axon itself, structural support is provided by microtubules, and the axon will retract if placed in a solution of colchicine. Thus, the developing neuron retains the same features that we have already noted in the dorsolateral neural tube hinge regions, namely, elongation by microtubules and apical shape changes by microfilaments. As in most migrating cells, moreover, the exploratory filopodia of the axon attach to the substrate and exert a force that pulls the rest of the cell forward. Axons will not grow if the growth cone fails to advance (Lamoureux et al., 1989). In addition to this structural role in axonal migration, filopodia also have a sensory function. Fanning out in front of the growth cone, each filopodium samples the microenvironments and sends signals back to the cell body (Davenport et al., 1993). As we will see in Chapter 8, the filopodia are the fundamental organelles involved in neuronal pathfinding.

Neurons transmit electrical impulses from one region to another. These impulses usually go from the dendrites into the nerve soma, where they are focused into the axon. To prevent dispersion of the electrical signal and to facilitate its conduction, the axon in the central nervous system is insulated at intervals by processes that originate from a type of glial cell called an **oligodendrocyte.** An oligodendrocyte wraps itself around the developing axon. It then produces a specialized cell membrane that is rich in myelin basic protein and that spirals around the central axon (Figure 7.28). This specialized membrane is called a **myelin sheath.** (In the peripheral nervous system, a glial cell called the **Schwann cell** accomplishes this myelination.) The myelin sheath is essential for proper neural function, and demyelination of nerve fibers is associated with convulsions, paralysis, and several severely debilitating or lethal afflictions. In the mouse mutant *trembler,* the Schwann cells are unable to produce a particular protein component of the myelin sheath, so myelination is deficient in the peripheral nervous system but normal in the central nervous system. Conversely, in another mouse mutant, *jimpy,* the central nervous system is deficient in myelin, while the peripheral nerves are unaffected (Sidman et al., 1964; Henry and Sidman, 1988).

The axon must also be specialized for secreting a specific neurotransmitter across the small gaps (synaptic clefts) that separate the axon of one cell

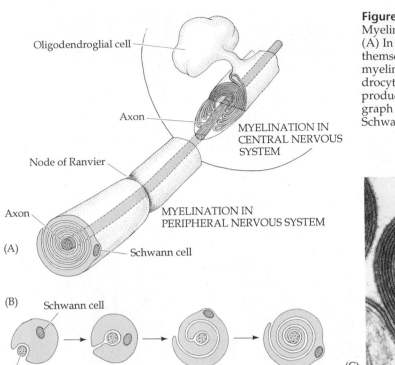

Oligodendroglial cell

Axon

MYELINATION IN
CENTRAL NERVOUS
SYSTEM

Node of Ranvier

Axon

MYELINATION IN
PERIPHERAL NERVOUS SYSTEM

(A)

Schwann cell

(B)   Schwann cell

Axon

Figure 7.28
Myelination in the central and peripheral nervous systems.
(A) In the peripheral nervous system, Schwann cells wrap
themselves around the axon; in the central nervous system,
myelination is accomplished by the processes of oligoden-
drocytes. (B) The mechanism of this wrapping entails the
production of an enormous membrane complex. (C) Micrograph of an axon enveloped by the myelin membrane of a
Schwann cell. (Photograph courtesy of C. S. Raine.)

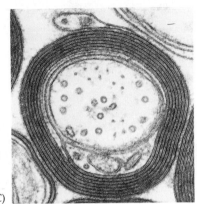

(C)

from the surface of its target cell (the soma, dendrites, or axon of a receiving
neuron or a receptor site on a peripheral organ). Some neurons become able
to synthesize and secrete acetylcholine, while other neurons develop the enzymatic pathways for making and secreting epinephrine, norepinephrine,
octopamine, serotonin, γ-aminobutyric acid, dopamine, or some other neurotransmitter. Each neuron must activate those genes responsible for making
the enzymes that can synthesize its neurotransmitter. Thus, neuronal development involves both structural and molecular differentiation.

Development of the vertebrate eye

An individual gains knowledge of its environment through its sensory organs. The major sensory organs of the head develop from the interactions of
the neural tube with a series of epidermal thickenings called the **cranial ectodermal placodes.** The most anterior placodes are the two olfactory placodes that form the ganglia for the olfactory nerves, which are responsible
for the sense of smell. The auditory placodes similarly invaginate to form
the inner ear labyrinth whose neurons form the acoustic ganglion, which enables us to hear. In this section, we will focus on the eye, because this organ,
perhaps more than any other in the body, must develop with precise coordination.

Dynamics of Optic Development

The story of optic development begins at gastrulation, when the involuting
endoderm and mesoderm interact with the adjacent prospective head ectoderm. This interaction gives a lens-forming bias to the head ectoderm (Saha
et al., 1989).* But not all parts of the head ectoderm eventually form lenses,
and the lens has to be in a precise relationship with the retina. The activation

*The inductions forming the eye will be detailed in Chapter 17.

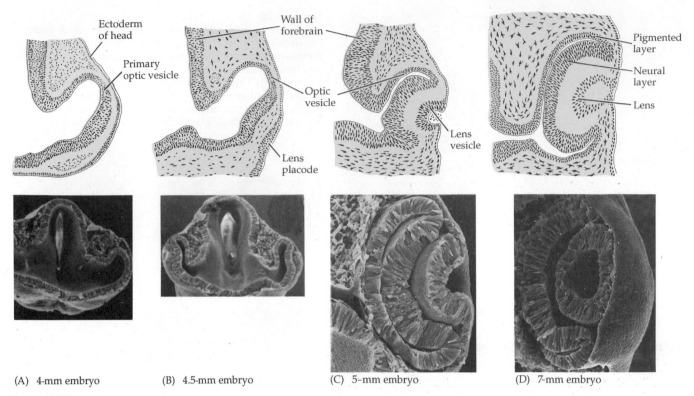

Ectoderm of head

Primary optic vesicle

Wall of forebrain

Optic vesicle

Lens placode

Lens vesicle

Pigmented layer

Neural layer

Lens

(A) 4-mm embryo   (B) 4.5-mm embryo   (C) 5-mm embryo   (D) 7-mm embryo

Figure 7.29
Development of the vertebrate eye. (A) The optic vesicle evaginates from the brain and contacts the overlying ectoderm. (B,C) Overlying ectoderm differentiates into lens cells as the optic vesicle folds in on itself and the lens placode becomes the lens vesicle. (C) The optic vesicle becomes the neural and pigmented retina as the lens is internalized. (D) The lens vesicle induces the overlying ectoderm to become the cornea. (Upper illustrations after Mann, 1964; micrographs A–C from Hilfer and Yang, 1980, courtesy of S. R. Hilfer; D Courtesy of K. Tosney.)

of this latent lens-forming ability and the positioning of the lens in relation to the retina is accomplished by the **optic vesicle.** In humans, the optic vesicles begin as two bulges from the lateral walls of the 22-day embryonic diencephalon (Figure 7.29). These bulges continue to grow laterally from the neural tube and are connected to the diencephalon by the **optic stalks.** Subsequently, when these vesicles contact the head ectoderm, the ectoderm thickens into the **lens placodes.** The necessity for close contact between the optic vesicles and the surface ectoderm is seen in both experimental cases and in certain mutants. For example, in the mouse mutant *eyeless,* the optic vesicles fail to contact the surface, and eye formation ceases (Webster et al., 1984).

Once formed, the lens placode reciprocates and causes changes in the optic vesicle. The vesicle invaginates to form a double-walled **optic cup** (see Figure 7.29C). As the invagination continues, the connection between the optic cup and the brain is reduced to a narrow slit. At the same time, the two layers of the optic cup begin to differentiate in different directions. The cells of the outer layer produce pigment and ultimately become the **pigmented retina** (one of the few tissues other than the neural crest cells that can synthesize its own melanin). The cells of the inner layer proliferate rapidly and generate a variety of glia, ganglion neurons, interneurons, and light-sensitive photoreceptor neurons. Collectively, these cells constitute the **neural retina.** The axons from the ganglion cells of the neural retina meet at the base of the eye and travel down the optic stalk. This stalk is then called the **optic nerve**.

Neural Retina Differentiation

Like the cerebral and cerebellar cortices, the neural retina develops into a layered array of different neuronal types (Figure 7.30). These layers include the light- and color-sensitive photoreceptor (rod and cone) cells, the cell bodies of the ganglion cells, and the bipolar interneurons that transmit the elec-

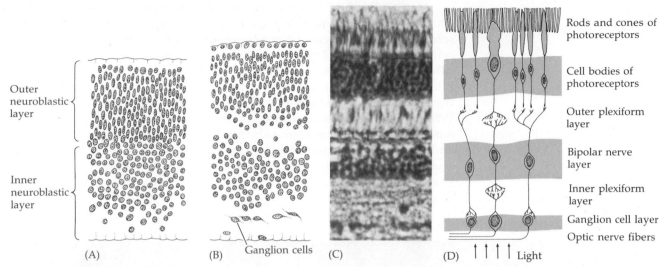

Outer neuroblastic layer

Inner neuroblastic layer

(A) (B) Ganglion cells (C) (D) ↑ ↑ ↑ ↑ Light

Rods and cones of photoreceptors

Cell bodies of photoreceptors

Outer plexiform layer

Bipolar nerve layer

Inner plexiform layer

Ganglion cell layer

Optic nerve fibers

Figure 7.30
Development of the human retina. Retinal neurons sort out into functional layers during development. (A,B) Initial separation of neuroblasts within the retina. (C) The three layers of neurons in the adult retina and the synaptic layers between them. (D) A functional depiction of the major neuronal pathway in the retina. Light traverses the layers until it is received by the photoreceptors. The axons of the photoreceptors synapse with bipolar neurons that transmit the depolarization to the ganglion neurons. The axons of the ganglion cells join to form the optic nerve that enters the brain. (A and B after Mann, 1964; photograph courtesy of G. Grunwald.)

trical stimulus from the rods and cones to the ganglion cells. In addition, there are numerous Müller glial cells that maintain the integrity of the retina, as well as amacrine neurons (which lack large axons) and horizontal neurons that transmit electrical impulses in the plane of the retina.

In the early stages of retinal development, cell division from a germinal layer and the migration and differential death of the resulting cells form the striated, laminar pattern of the neural retina. The formation of this highly structured tissue is one of the most intensely studied problems of developmental neurobiology. It has been shown (Turner and Cepko, 1987) that a single retinal neuroblast precursor cell can give rise to at least three types of neurons or to two types of neurons and a glial cell. This analysis was performed using an ingenious technique to label the cells generated by one particular precursor cell. Newborn rats (whose retinas are still developing) were injected in the back of their eyes with a virus that can integrate into their DNA. This virus contained a β-galactosidase gene (not present in rat retina) that would be expressed only in the infected cells. A month after the rats' eyes were infected, the retinas were removed and stained for the presence of β-galactosidase. Only the progeny of the infected cells should have stained blue. Figure 7.31 shows one of the strips of cells derived from an infected precursor cell. The stain can be seen in five rods, a bipolar neuron, and a retinal (Müller) glial cell.

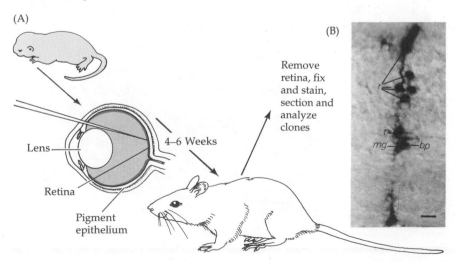

(A)

Lens

Retina

Pigment epithelium

4–6 Weeks

Remove retina, fix and stain, section and analyze clones

(B)

mg — bp

Figure 7.31
Determination of the lineage of a precursor cell in the rat retina. (A) Technique whereby a virus containing a functional β-galactosidase gene is injected into the back of the eye to infect some of the retinal precursor cells. After a month to 6 weeks, the eye is removed and the retina is stained for the presence of β-galactosidase. (B) Stained cells forming a strip across the neural retina, including five rods (r), a bipolar neuron (bp), a rod terminal (t), and a Müller glial cell (mg). The identities of these cells were confirmed by Nomarski-phase contrast microscopy. (Scale bar, 20 µm.) (From Turner and Cepko, 1987, photograph courtesy of D. Turner.)

Many cells of the forebrain express a transcription factor called **Pax6.** This protein appears to be especially important in the development of the retina. Indeed, it appears to be a common denominator for photoreceptive cells in all phyla. The *Pax6* gene probably plays many roles, and one of them is to determine tissues to becoming eyes. If the mouse *Pax6* gene is inserted into the *Drosophila* genome and activated randomly, *Drosophila* eyes form in those cells where the mouse *Pax6* is being expressed (Halder et al., 1995)! While this gene is also expressed in the murine forebrain, hindbrain, and nasal placodes, the eyes seem to be most sensitive to its absence. In humans and mice, *Pax6* heterozygotes have small eyes, while homozygotes in mice and humans (and *Drosophila*) lack eyes altogether (Jordan et al., 1992; Glaser et al., 1994; Quiring et al., 1994). This gene is discussed more in Chapter 23.

Sidelights & Speculations

Why Babies Don't See Well

HUMAN NEWBORNS see poorly. There may be several reasons for this, but one of the most striking is the immaturity of the retinal photoreceptors. Anatomical studies by Yuodelis and Hendrickson (1986) and physical studies by Banks and Bennett (1988) have shown that the cone photoreceptors of the newborn central retina are over 7.5 µm in diameter, shrinking to the normal adult width of 2 µm in about 3 years. During this time, the cone density in this region increases from 18 photoreceptors per 100 µm to 42 photoreceptors per 100 µm, and the photoreceptors develop both their outer segments (that

catch the light) and their basal axonal processes. Figure 7.32 highlights the differences between the photoreceptors of neonatal and adult retinas. The neonatal retina has poorly differentiated photoreceptors, and those photoreceptors that it does have are so wide that not many of them can fit into a given area. Banks and Bennett calculate that this causes the central region of the newborn retina to absorb light about 350 times more poorly than the same region of the adult

retina. This low number of photoreceptors per retinal area also prevents the babies from discriminating between two points at a distance. This may be why a newborn responds to visual stimuli only when they are brought close to the baby's face. The development of the human retinal photoreceptor provides an excellent example of differentiation that begins early in development but which is not complete until years after birth. ▪

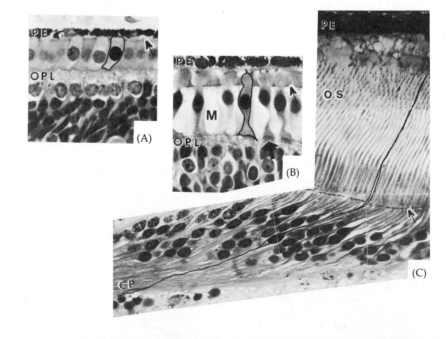

Figure 7.32 Development of the cone photoreceptors in the central region of the human retina. Stained light microscope sections were photographed and one cone in each retina outlined for clarity. The pigment epithelium (PE), outer plexiform layer (OPL), Müller glia (M), and outer segments of the photoreceptor (OS) have been labeled. (A) Fetus of 22 weeks' gestation. (B) Newborn 5 days after birth. (C) 72-year-old. The arrow points to the outer limiting membrane, which served as the original border for the retinal neurons. The axon outlined in (C) is actually shorter than normal, thus allowing the synapse with the bipolar neuron to be shown in the picture. The synapse is formed at the cone synaptic pedicle (CP). (From Yuodelis and Hendrickson, 1986, courtesy of A. Hendrickson.)

Lens and Cornea Differentiation

During its continued development into a lens, the lens placode rounds up and contacts the new overlying ectoderm (see Figure 7.29D). The lens vesicle then induces the ectoderm to form the transparent cornea. Here, physical parameters play an important role in the development of the eye. Intraocular fluid pressure is necessary for the correct curvature of the cornea so that light can be focused on the retina. The importance of such ocular pressure can be demonstrated experimentally; the cornea will not develop its characteristic curve when a small glass tube is inserted through the wall of a developing chick eye to drain away intraocular fluids (Coulombre, 1956, 1965). Intraocular pressure is sustained by a ring of scleral bones (probably derived from the neural crest), which acts as an inelastic restraint.

The differentiation of the lens tissue into a transparent membrane capable of directing light onto the retina involves changes in cell structure and shape as well as synthesis of lens-specific proteins called crystallins (Figure 7.33). Crystallins are synthesized as cell shape changes occur, thereby causing the lens vesicle to become the definitive lens. The cells at the inner portion of the lens vesicle elongate and, under the influence of the neural retina, produce the lens fibers (Piatigorsky, 1981). As these fibers continue to grow, they synthesize crystallins, which eventually fill up the cell and cause the extrusion of the nucleus. The crystallin-synthesizing fibers continue to grow and eventually fill the space between the two layers of the lens vesicle. The anterior cells of the lens vesicle constitute a germinal epithelium, which keeps dividing. These dividing cells move toward the equator of the vesicle, and as they pass through the equatorial region, they, too, begin to elongate (Figure 7.33D). Thus, the lens contains three regions: an anterior zone of dividing epithelial cells, an equatorial zone of cellular elongation, and a posterior and central zone of crystallin-containing fiber cells. This arrangement persists throughout the lifetime of the animal as fibers are continuously being laid down. In the adult chicken, the differentiation from an epithelial cell to a lens fiber takes 2 years (Papaconstantinou, 1967).

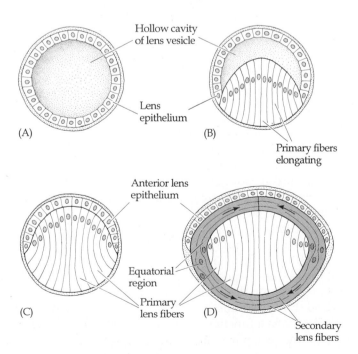

Figure 7.33
Differentiation of the lens cells (A) Lens vesicle as shown in Figure 7.29. (B) Elongation of the interior cells, producing lens fibers. (C) Lens filled with crystallin-synthesizing cells. (D) New lens cells derived from anterior lens epithelium. (E) As the lens grows, new fibers differentiate, and nuclei degenerate. (After Paton and Craig, 1974.)

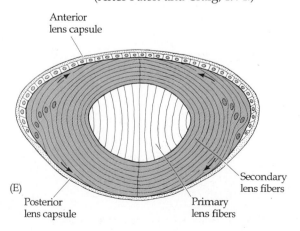

Directly in front of the lens is a pigmented and muscular tissue called the iris. These muscles control the size of the pupil (and give an individual his or her characteristic eye color). Unlike the other muscles of the body (which are derived from the mesoderm), part of the iris is derived from the ectodermal layer. Specifically, this region of the iris develops from a portion of the optic cup that is continuous with the neural retina but does not make photoreceptors.

■ THE NEURAL CREST

The neural crest and its derivatives

Although derived from the ectoderm, the neural crest has sometimes been called the fourth germ layer because of its importance. It has even been said, perhaps hyperbolically, that "the only interesting thing about vertebrates is the neural crest" (quoted in Thorogood, 1989). The neural crest cells originate at the dorsalmost region of the neural tube. Transplantation experiments wherein a quail neural plate is grafted into a chick non-neural ectoderm show that juxtaposing these tissues induces the formation of neural crest cells and that both the prospective neural plate and the prospective epidermis contribute to the neural crest (Selleck and Bronner-Fraser, 1995; Mancilla and Mayor, 1996). These crest cells migrate extensively to generate a bewildering number of differentiated cell types. These include (1) the neurons and glial cells of the sensory, sympathetic, and parasympathetic nervous systems, (2) the epinephrine-producing (medulla) cells of the adrenal gland, (3) the pigment-containing cells of the epidermis, and (4) many of the skeletal and connective tissue components of the head. The fate of the neural crest cells depends, to a large degree, on where the cells migrate and settle. The neural crest can be divided into four main functional (but overlapping) domains:

- *The cephalic (head) neural crest,* whose cells migrate dorsolaterally to produce the craniofacial mesenchyme that differentiates into the cartilage, bone, cranial neurons, glia, and connective tissues of the face. These cells also enter the **pharyngeal pouches** to give rise to thymic cells, odontoblasts of the tooth primordia, and the cartilage of inner ear and jaw.

- *The trunk neural crest,* whose cells take one of two major pathways. Neural crest cells that become the pigment-synthesizing **melanocytes** migrate dorsolaterally into the ectoderm and continue their way toward the ventral midline of the belly. However, most of the trunk neural crest cells pass ventrolaterally through the anterior half of each sclerotome. (**Sclerotomes** are blocks of mesodermal cells that surround the neural tube and differentiate into the vertebral cartilage of the spine.) Those trunk neural crest cells that remain in the sclerotome form the **dorsal root ganglia.** Those cells continuing more ventrally form the **sympathetic ganglia,** the **adrenal medulla,** and the nerve clusters surrounding the aorta.

- *The vagal and sacral neural crest,* whose cells generate the **parasympathetic (enteric) ganglia** of the gut (Le Douarin and Teillet, 1973; Pomeranz et al., 1991). The vagal crest is opposite chick somites 1–7, while the sacral neural crest is posterior to somite 28. Failure of neural crest cell migration to the colon results in the absence of enteric ganglia and thus to the absence of peristaltic movement in this region. This results in a functional obstruction and the dilation and enlargement of the region above it ("megacolon").

Table 7.1 Some derivatives of the neural crest

Derivative	Cell type or structure derived
Peripheral nervous system (PNS)	Neurons, including sensory ganglia, sympathetic and parasympathetic ganglia, and plexuses Neuroglial cells Schwann cells
Endocrine and paraendocrine derivatives	Adrenal medulla Calcitonin-secreting cells Carotid body type I cells
Pigment cells	Epidermal pigment cells
Facial cartilage and bone	Facial and anterior ventral skull cartilage and bones
Connective tissue	Corneal endothelium and stroma Tooth papillae Dermis, smooth muscle, and adipose tissue of skin of head and neck Connective tissue of salivary, lachrymal, thymus, thyroid, and pituitary glands Connective tissue and smooth muscle in arteries of aortic arch origin

Source: After Jacobson, 1991, based on multiple sources.

- *A cardiac neural crest* may be located between the cephalic and trunk crests. There is evidence that such neural crest cells exist from the first to the third somites of chick embryos, overlapping the anterior portion of the vagal neural crest that extends from the first to the seventh somite (Kirby 1987; Kirby and Waldo, 1990). These neural crest cells can develop into melanocytes, neurons, cartilage, and connective tissue (of the third, fourth, and sixth pharyngeal arches). In addition, this region of the neural crest produces the entire musculoconnective tissue wall of the large arteries as they arise from the heart as well as contributing to the septum that separates the pulmonary circulation from the aorta (Le Lièvre and Le Douarin, 1975).

Table 7.1 is a summary of some of the cell types derived from the neural crest.

The trunk neural crest

Migration Pathways of Trunk Neural Crest Cells

As shown in Figure 7.2, the trunk neural crest is a transient structure, its cells dispersing soon after the neural tube closes. There are two major pathways taken by the migrating neural crest cells (Figure 7.34).

THE DORSOLATERAL PATHWAY. The most obvious pathway for trunk neural crest migration is the dorsolateral pathway, by which the melanocyte precursors travel around the periphery of the embryo through the dermal mesoderm underlying the dermis. They enter the ectoderm through minute holes in the basal lamina (which they might make) and colonize the skin and follicles, where they differentiate into melanocytes (Mayer, 1973; Erickson et al., 1992). This pathway was demonstrated in a series of classic experiments by

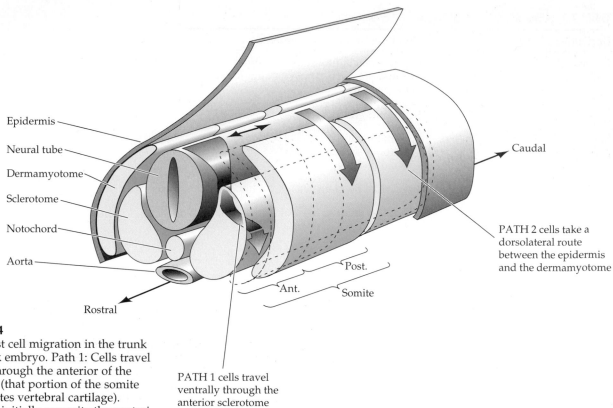

Epidermis

Neural tube

Dermamyotome

Sclerotome

Notochord

Aorta

Rostral

Caudal

PATH 2 cells take a dorsolateral route between the epidermis and the dermamyotome

PATH 1 cells travel ventrally through the anterior sclerotome

Post.

Ant.

Somite

Figure 7.34

Neural crest cell migration in the trunk of the chick embryo. Path 1: Cells travel ventrally through the anterior of the sclerotome (that portion of the somite that generates vertebral cartilage). Those cells initially opposite the posterior portions of the sclerotomes migrate along the neural tube until they come to an opposite anterior region. These cells contribute to the sympathetic and parasympathetic ganglia as well as to the adrenal medullary cells and dorsal root ganglia. Path 2: Somewhat later, cells enter a dorsolateral route beneath the ectoderm. These cells become pigment-producing melanocytes.

Mary Rawles and others (1948), who transplanted the neural tube and crest from a pigmented strain of chicken into the neural tube of an albino chick embryo. The result was a white chicken with a specific region of colored feathers (Figure 7.35A). The neural crest is responsible for the production of all the melanin-containing cells in the organism (with the exception of certain neural derivatives such as the pigmented retina).

THE VENTRAL PATHWAY. By grafting a portion of the chick neural tube and its associated crest from radioactively or genetically marked embryos into other embryos (Weston, 1963; Le Douarin and Teillet, 1974), investigators have been able to trace another major route of trunk neural crest cell migration (Figure 7.35B,C). More recent studies have extended these investigations by using fluorescent antibodies, vital dyes, or virally transformed cells to follow *individual* neural crest cells to their destinations. Those cells leaving by the ventral pathway become sensory (dorsal root) and sympathetic neurons, adrenomedullary cells, and Schwann cells. As can be seen in Figure 7.36 and Plate 19, these trunk neural crest cells migrate ventrally through the *anterior* but not through the *posterior* section of the sclerotomes (Rickmann et al., 1985; Bronner-Fraser, 1986; Loring and Erickson, 1987; Teillet et al., 1987). Teillet and co-workers combined the antibody approach with transplantation of genetically marked quail neural crest cells into chick embryos. The antibody marker recognizes and labels neural crest cells of both species; the genetic marker enables the investigators to distinguish between quail and chick cells. These studies show that neural crest cells initially opposite the *posterior* regions of the somites migrate anteriorly or posteriorly along the neural tube and then enter the anterior region of their own or adjacent somites. These neural crest cells join with the neural crest cells that were ini-

(A)

(B)

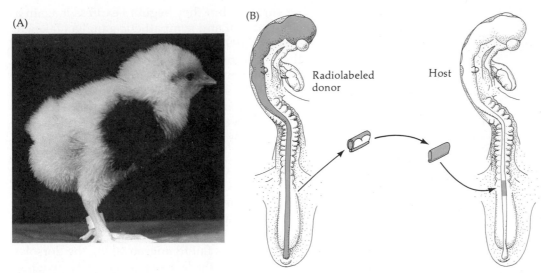

Radiolabeled donor

Host

(C)

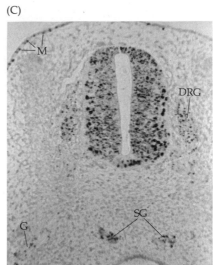

M

DRG

G

SG

Figure 7.35
Neural crest cell migration. (A) Chick resulting from the transplantation of a trunk neural crest region from a pigmented strain of chicken into the trunk neural crest region of an unpigmented strain. The crest cells that gave rise to pigment were able to migrate into the wing skin. (B) Grafting technique for mapping neural crest cells. A piece of the dorsal axis is excised from a donor embryo; the neural tube and its associated crest are isolated and implanted into a host embryo whose neural tube and crest have been excised. When the donor crest cells are radiolabeled (with tritiated thymidine) or genetically labeled (from a different species or strain), their descendants can be traced in the host embryo as development proceeds. (C) Autoradiograph showing locations of neural crest cells that have migrated from transplanted radioactive donor neural crest to form melanoblasts (M), sympathetic ganglia (SG), dorsal root ganglia (DRG), and glial cells (G). (A, original photograph from the archives of B. Willier; B after Weston, 1963; C courtesy of J. Weston.)

tially opposite the anterior portion of the somite, and they form the same structures. Thus, each dorsal root ganglion is composed of three neural crest populations: one from the neural crest opposite the anterior portion of the somite and one from each of the adjacent neural crest regions opposite the posterior portions of the somites. At specific regions of the trunk, crest cells migrating along the same pathway aggregate to form the sympathetic ganglia and the epinephrine-secreting cells of the adrenal medulla. The parasympathetic division of the peripheral nervous system is also formed by neural crest cells migrating by this pathway, but only in the sacral and cervical regions of the embryo.

The Extracellular Matrix and Trunk Neural Crest Migration

Any analysis of migration (be it of birds, butterflies, or neural crest cells) has to ask three questions: How is migration initiated? How do the migratory agents know the path on which to travel? And what signals that the destination has been reached and that migration should end? Moreover, one has to ask when the migrating agent is competent to respond to these cues. Premigratory neural crest cells express the **Slug** protein, a transcription factor. Antisense oligonucleotides against the *slug* mRNA will prevent neural crest cells migration, suggesting that Slug protein may be needed for the nonmotile epithelial cell to become a wanderer (Nieto et al., 1994). Another potential factor in the initiation of neural crest cell migration is the adhesion molecule **N-cadherin.** Originally on the surface of the neural crest cells, this protein is downregulated at the time of cell migration. *Migrating* crest cells

Sclerotoma   Neural
of somite   tube

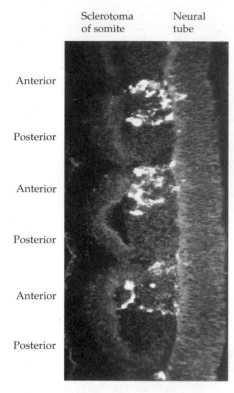

Anterior

Posterior

Anterior

Posterior

Anterior

Posterior

Figure 7.36
Neural crest cell migration. These fluorescence photomicrographs of longitudinal sections of a 2-day chick embryo are stained with antibody HNK-1, which selectively recognizes neural crest cells. Extensive staining is seen in the anterior, but not in the posterior, half of each sclerotome. (From Bronner-Fraser, 1986, courtesy of M. Bronner-Fraser.)

have no N-cadherin on their surfaces, but they begin to express it again as they aggregate to form the dorsal root and sympathetic ganglia (Takeichi, 1988; Akitaya and Bronner-Fraser, 1992). At the same time that the neural crest cells lose their N-cadherin and become able to migrate as individual cells, the extracellular surface surrounding them becomes more adhesive (Perris et al., 1990). There appear to be specific paths for the neural crest cells to follow, and when neural crest cells or their derivatives are placed (either by transplantation or by injection) on their normal migration pathway in a host embryo, they migrate along it. (Bronner-Fraser and Cohen, 1980; Erickson et al., 1980).

The path taken by the neural crest cells is controlled by the extracellular matrix of the embryo (Newgreen and Gooday, 1985; Newgreen et al., 1986). Evidence from salamander development indicates that the direction of neural crest cell migration may be imparted by the extracellular matrices over which they travel. In some axolotl salamanders, there is a mutation in which the neural crest forms but its cells fail to migrate along the dorsolateral pathway. This is most readily seen in the lack of pigment cells anywhere except atop the neural tube of these animals (Figure 7.37), and these cells eventually degenerate. When wild-type neural crests are transplanted into mutant embryos, the crest cells are unable to migrate. However, when crests from mutant embryos are transplanted into wild-type embryos, their cells migrate normally (Spieth and Keller, 1984). Thus, the defect in this mutant is in the environment that the cells encounter, not in the cells themselves. (The road is deficient, not the vehicle.) Löfberg and co-workers (1989) used this information to show that the extracellular matrix contains critical components regulating neural crest cell migration. They adsorbed, onto membrane microcarriers, extracellular matrix from the subepidermal region of the skin (through which the pigment-forming neural crest cells would migrate). They then placed these microcarriers next to the neural crests of mutant and wild-type embryos just before migration would occur. The microcarriers alone did not stimulate migration from either wild-type or mutant neural crests. Microcarriers containing extracellular matrix material from mutant embryos likewise did not stimulate premature neural crest cell migration in either mutant or wild-type embryos. The microcarriers containing wild-type extracellular matrix, however, were able to stimulate neural crest cell migration from both the mutant and wild-type neural crests, demonstrating the importance of the extracellular matrix in neural crest cell migration.

A similar situation exists in chick embryos, since transplanting different regions of the mesoderm into the area adjacent to the neural crest can produce different patterns of migration (Goldstein et al., 1990; Bronner-Fraser and Stern, 1991). The regions allowing neural crest cell migration are determined in the mesoderm before migration occurs.

But what are the molecules that enable or forbid trunk neural crest cell migration? The extracellular matrix apposing the migrating neural crest cells is a rich mix of molecules such as fibronectin, laminin, tenascin, various collagen molecules, and proteoglycans. Experiments undertaken to address this issue must be designed carefully, since neural crest cells may have different migration requirements in different species and even in different parts of the same embryo. One solution is to make antibodies to the regions of the extracellular matrix molecules to which cells bind. When such antibodies are injected into the embryo and block these regions of the matrix, do they perturb neural crest cell migration? The migration of chick *cranial* neural crest cells can be severely altered when antibodies to fibronectin, fibronectin receptors, tenascin, or laminin-heparan sulfate proteoglycan are injected into the devel-

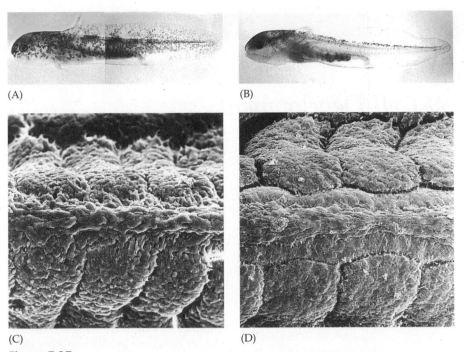

Figure 7.37
Deficiency of neural crest cell migration in the *d/d* mutant of the axolotl. (A) The
larvae of wild-type axolotls are characterized by pigment cells throughout the body
except for the most ventral portions. (B) In the *d/d* mutant, the neural-crest-derived
pigment cells form a stripe along the dorsal midline of the larva. (C,D) Scanning
electron micrographs of the embryonic neural crest show that (C) the neural crest
cells of the wild-type embryos migrate over the neural tube into the somites, while
(D) those crest cells of the mutant remain atop the neural tube. (From Löfberg et al.,
1989, courtesy of the authors.)

oping embryo (Poole and Thiery, 1986; Perris and Bronner-Fraser, 1989).
However, these antibodies did not severely alter the migration of chick *trunk*
neural crest cells.

There are presently two major candidates for the molecules involved in
patterning the trunk neural crest cells. One is the receptor for peanut agglutinin, a compound that binds particular carbohydrate residues of glycoproteins. If chick embryo trunks are treated with peanut agglutinin, the neural
crest cells migrate into both the rostral and caudal halves of the somites at
the same rate (Krull et al., 1995). The other molecule is an Eph-related receptor tyrosine kinase. These molecules are able to guide axons (see Chapter 8)
and their expression correlates with the hindbrain rhombomeres that exclude the neural crest cells (Irving et al., 1996; Weinstein et al., 1996).

In 1963, Weston hypothesized that these specific pathway molecules
were not needed for trunk neural crest migration; rather, neural crest cells
would take advantage of any space open to them, as long as there was no
active inhibition of migration. He rotated the neural crest so that it was now
on the bottom of the neural tube and noted that when the cells emerged
from the ventral region of the neural tube, they migrated in a ventral-to-dor-
sal fashion (the reverse of normal migration) through the anterior half of the
sclerotome. Thus, there seems to be no inherent directionality to the crest
cells' migrations. The cells go where there are openings for them. Both physical and chemical barriers may create the patterns of neural crest cell migration.

Sidelights & Speculations

Analysis of Mutations Affecting Neural Crest Cell Development

THE MIGRATORY PROPERTIES and differentiation of neural crest cells are also being studied at by analyzing mutations that impair one or more of the neural crest lineages. Mutations include the following:

White-spotting. The neural crest cells of these mice lack functional c-kit receptor tyrosine kinase. The homozygous condition is usually lethal, but heterozygotes survive and can be recognized by their spots of pigment-deficient fur. In humans, heterozygotes have a piebald phenotype, where regions of the hair and skin are white, lacking melanocytes (Spritz et al., 1992).

Steel. These mice lack **stem cell factor**, the ligand for the *c-kit* protein kinase. This factor is secreted by tissues along the migration route and is used by the migrating neural crest cells. It stimulates cell division. The homozygous condition is usually lethal, but the heterozygote has a diluted, steel-gray coat color (see Witte, 1990).

Splotch. The mice lack the transcription factor *Pax3*. As mentioned earlier, this protein is expressed on the dorsal region of the neural tube. Mice homozygous for this gene have defects in both neural tube closure and neural crest cell-derived structures, especially cranial ganglia and nerves (Figure 7.38). The heterozygote has regions of pigmentation and regions without pigment. In humans, the heterozygous condition is known as Waardenburg's (type I) syndrome (Tremblay et al., 1995).

Lethal-spotting and *Piebald-lethal.* These are deficiencies of **endothelin-3** and its receptor, the **endothelin-B recep-**

tor, respectively. Endothelin-3 is a growth factor that stimulates cell proliferation in neural crest cells. It is critical for the development of melanocytes and the enteric neurons that cause peristalsis of the digestive tract. The homozygous absence of the endothelin-B receptor genes produces megacolon, the distension of the large intestine due to its inability to evacuate. In humans, this condition is called Hirschsprung's disease. The absence of endothelin-3 also pro-

Figure 7.38 Ventral surface of a mouse heterozygous for the *White* mutation. The mouse has reduced numbers of blood cells, germ cells, and melanocytes. The white spot on the belly is characteristic of heterozygotes, since they do not have enough melanocytes to circumvent the mouse. Viable homozygous *White* animals have no pigment in the trunk. (Courtesy of Jackson Laboratories.)

duces the spotting pattern of melanocytes as well as the lack of ganglia in the bowels (Baynish et al., 1994; Hosoda et al., 1994; Puffenberger et al., 1994; Lahav et al., 1996).

Ret and *GDNF.* Like the endothelin-B receptor, the Ret receptor tyrosine kinase is required for the differentiation of the enteric neurons. Those mice lacking it have no enteric neurons, nor do they have kidneys. (The importance of Ret for kidney development will be discussed in Chapter 17.) In humans, loss of one of the ret genes can produce another form of Hirschsprung's disease—aganglionic megacolon (Edery, 1994; Romeo et al., 1994). The ligand for the Ret protein appears to be glial-derived growth factor (GDNF; Pichel et al., 1996). Mice lacking the GDNF proteins also lack kidneys and enteric neurons.

Microphthalmia. These mice lack a particular transcription factor, leading to deafness and melanocyte deficiencies. The human heterozygous condition yields Waardenburg's (type II) syndrome (Hemesath et al., 1994; Steingrimsson et al., 1994; Tassabehji et al., 1994).

Silky. This mutation in chickens involves the pigmentation pathway. In addition to a phenotype wherein the adult retains the downy feathers of its youth, the internal organs are pigmented by the migration and proliferation of melanocytes into them. In contrast, the feathers remain white. Transplantation studies (Hallet and Ferrand, 1984) show that this defect is not brought about by the melanocyte precursors, but by the environment in which the neural crest cells migrate. [ecto6.html] ∎

The Developmental Potency of Trunk Neural Crest Cells

EVIDENCE FOR PLURIPOTENCY AMONG TRUNK NEURAL CREST CELLS.   One of the most exciting features of neural crest cells is their **pluripotency.** A single neural crest cell can differentiate into several different cell types, depending on its location within the embryo. For example, the parasympathetic neurons formed by the vagal (neck) neural crest cells (opposite somites 1–7) produce acetylcholine as their neurotransmitter; they are therefore **cholinergic** neurons. The sympathetic neurons formed by the thoracic (chest) neural crest cells produce norepinephrine; they are **adrenergic** neurons. But when chick cervical and thoracic neural crests are reciprocally transplanted, the former thoracic crest produces the cholinergic neurons of the parasympathetic ganglia, and the former vagal crest forms adrenergic neurons in the sympathetic ganglia (Le Douarin et al., 1975). Kahn and co-workers (1980) found that premigratory trunk neural crest cells from both the thoracic and the vagal regions have the enzymes for synthesizing both acetylcholine and norepinephrine. Thus, the thoracic crest cells are capable of developing into cholinergic neurons when they are placed into the neck, and the vagal crest cells are capable of becoming adrenergic neurons when they are placed in the trunk.

The pluripotency of some neural crest cells is such that regions of the neural crest that never produce nerves in normal embryos can be made to do so under certain conditions. Mesencephalic neural crest cells normally migrate into the eye and interact with the pigmented retina to become scleral cartilage cells (Noden, 1978). However, if this region of the neural crest is transplanted into the trunk region, it can form sensory ganglia neurons, adrenomedullary cells, glia, and Schwann cells (Schweizer et al., 1983).

The preceding research studied the potential of *populations* of cells. It is still unclear whether most of the cells that leave the neural crest are pluripotent or whether most have already become restricted to certain fates. Bronner-Fraser and Fraser (1988, 1989) provide evidence that some, if not most, of the *individual* neural crest cells are pluripotent as they leave the crest. They injected fluorescent dextran molecules into individual neural crest cells while the cells were still above the neural tube and then looked to see what types of cells the individual crest cells became after migration. The progeny of a single neural crest cell could become sensory neurons, pigment cells, adrenomedullary cells, and glia (Figure 7.39). Moreover, they found that if they labeled individual trunk neural crest cells as they migrated ventrally through the embryo, the label was later found in several cell types, including sensory neurons, sympathetic neurons, and Schwann cells. In mammals, the neural crest cell is similarly seen as a stem cell that can generate further multipotent neural crest cells. However, some of the migrating neural crest cells are pluripotent, while others have more restricted cell fates (Stemple and Anderson, 1992).

EVIDENCE FOR RESTRICTED POTENCY AMONG TRUNK NEURAL CREST CELLS.   Even at the time of emigration, some neural crest cells may be more determined than others. For one thing, potency becomes more restricted as the neural crest ages. Chick trunk crest cells that migrate *earliest* can form a wide variety of derivatives, including pigmented cells, neurons, and adrenergic cells. Late-migrating cells mostly become melanocytes (Serbedzija et al., 1989; Artinger and Bronner-Fraser, 1992). Indeed, the late neural crest emigrés appear to have been already committed to become melanocytes before they entered the dorsolateral pathway (Erickson and Goins, 1995). There is evidence that

Figure 7.39
Pluripotency of trunk neural crest cells. A single neural crest cell is injected with a highly fluorescent dextran molecule. The progeny of this cell will each receive some of these fluorescent molecules. (A) Injection of fluorescent dextran into a neural crest cell shortly before migration of the crest cells is initiated. (B) Two days later, crest-derived tissues contain labeled cells descended from the injected precursor. The figure summarizes data from two different experiments (case 1 and case 2). (After Lumsden, 1988.)

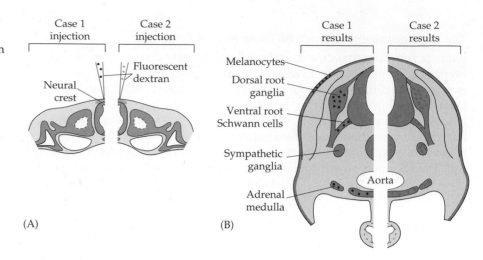

some restriction in potency exists even in some of the early-emigrating cells. Several investigators (see Sieber-Blum and Sieber, 1984; Stocker et al., 1991; Weston, 1991) have found that a significant number of neural crest cells form clones containing relatively few cell types. In addition, transplantation studies by Le Douarin and Smith (1988) suggest that many of the neural-crest-derived cells that form the sensory neurons of the dorsal root ganglia are incapable of forming the autonomic neurons of the sensory ganglia, and vice versa. These studies suggest that some of the neural crest cells have already restricted their potentiality at the time of migration. Two such committed neural crest cells are hypothesized to be a melanocyte-Schwann cell precursor (Nichols and Weston, 1977; Ciment, 1990) and a sympathoadrenal precursor that can give rise to only sympathetic ganglia and adrenomedullary cells (Landis and Patterson, 1981; Anderson and Axel, 1986).

The mechanism for the differential success of these committed cells may involve different requirements for growth factors. The committed dorsal root ganglia neuron precursors appear to need **brain-derived neurotrophic factor** (**BDNF**), a growth factor produced by the neural tube itself. If a thin, nonpermeable membrane is interposed between the neural tube and the prospective dorsal root ganglia region, the ganglia fail to form. Those neural crest cells that have continued their ventral migration to the sympathetic ganglia regions survive. This inhibition of dorsal root ganglia production can be circumvented by coating the barrier with neural tube extract or with BDNF (Kalcheim et al., 1987; Sieber-Blum, 1991). This conclusion is strengthened by the observation that when the *BDNF* gene is knocked out from mouse embryos, the dorsal root ganglia and placodal sensory neurons are missing, but the motor neurons are not changed (Ernfors et al., 1994; Jones et al., 1994). The neural crest cells committed to form the sympathetic ganglia do not need this growth factor for survival. Rather, their differentiation appears to be spurred by basic fibroblast growth factor and glial derived neurotrophic factor (Kalcheim and Neufeld, 1990; Maxwell et al., 1996).

Final Differentiation of the Neural Crest Cells

The final differentiation of the autonomic neural crest cells is determined in large part by the environment in which these cells develop. It does not involve the selective death of those cells already committed to secreting another type of neurotransmitter (Coulombe and Bronner-Fraser, 1987). Heart cells, for example, secrete a protein, leukemia inhibition factor (LIF), that can convert adrenergic sympathetic neurons into cholinergic neurons without changing their survival or growth (Chun and Patterson, 1977; Fukada, 1980;

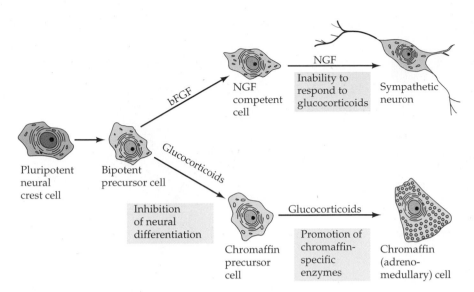

Figure 7.40
Final differentiation of a neural crest cell committed to become either an adreno-medullary (chromaffin) cell or a sympathetic neuron. Glucocorticoids appear to act at two places. First, they inhibit the actions of those factors that promote neuronal differentiation; and second, they induce those enzymes characteristic of the adrenal cells. Those cells exposed sequentially to basic fibroblast growth factor (bFGF) and nerve growth factor (NGF) differentiate into the sympathetic neurons.

Yamamori et al., 1989). Similarly, bone morphogenetic protein 2 (BMP2), a protein secreted by the heart, lung, and dorsal aorta, influences rat neural crest cells to differentiate into cholinergic neurons. Such neurons form the sympathetic ganglia in the region of these organs (Shah et al., 1996). While BMP2 may induce these neural crest cells to become neurons, glial growth factor (GGF; neuregulin) suppresses neuronal differentiation and directs development toward glial fates (Shah et al., 1994). Another paracrine factor, endothelin-3, appears to stimulate the production of melanocytes (Lahav et al., 1996).

Thus, the fate of a committed neural crest cell can be directed by the milieu of the tissue environment in which it settles. The chick trunk neural crest cells that migrate into the region destined to become the *adrenal medulla* can differentiate in two directions. The presence of bone morphogenetic protein 7 (BMP7) may induce these cells to become epinephrine-producing cells (Varley et al., 1995). Such cells usually differentiate into noradrenergic sympathetic neurons. However, if these neural crest cells are given glucocorticoids like those made by the cortical cells of the adrenal gland, they differentiate into adrenomedullary cells (Figure 7.40; Anderson and Axel, 1986; Vogel and Weston, 1990). The type of matrix is important in the differentiation of the salamander neural crest cells, too. If axolotl crest cells are grown on matrices from subepidermal regions (where the pigment cells are found), they become melanocytes. However, if the same crest cells are cultured on matrices from the region of the dorsal root ganglia, they develop a neuronal phenotype (Perris et al., 1988).

The cephalic neural crest

Migratory Pathways of the Cephalic Neural Crest Cells

The "face" is largely the product of the cephalic (cranial) neural crest, and the evolution of the jaws, teeth, and facial cartilage arises from changes in the placement of these cells (see Chapter 23). As mentioned earlier, the hindbrain is segmented along the anterior-posterior axis into rhombomeres. The chick cephalic neural crest cells migrate according to their rhombomeric origin, and there are three major pathways taken by these migrating cells (Figure 7.41; Lumsden and Guthrie, 1991). First, cells from rhombomere 2 mi-

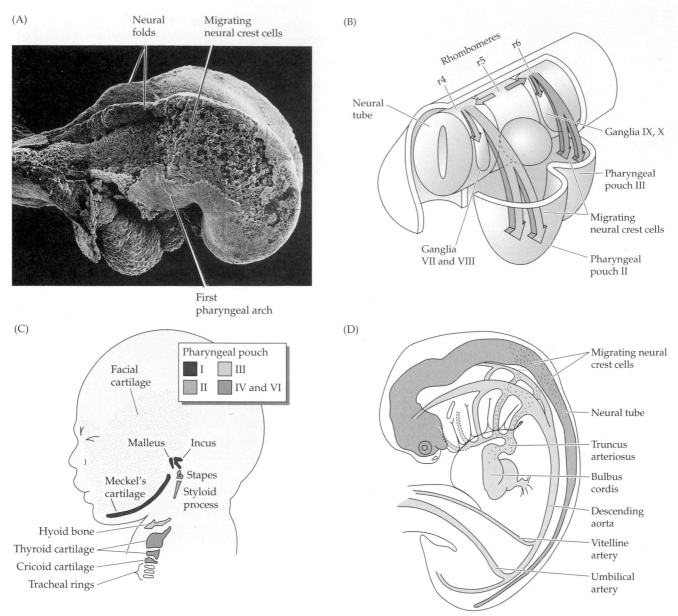

Figure 7.41
Neural crest cell migration in the mammalian head. (A) Scanning electron micrograph of a rat embryo with part of the lateral ectoderm removed from the surface. Neural crest migration can be seen over the midbrain, and the column of neural crest cells migrating into the future first pharyngeal arch is evident. (B) Analysis of cranial neural crest cell migration from rhombomeres 4–6 in the mouse suggests that there is a major migration into the pharyngeal arches and a minor migration to form the cranial nerve ganglia. (C) Structures formed in the human face by the ectomesenchymal cells of the neural crest. The cartilaginous elements of the pharyngeal pouches are indicated by colors, and the stippled region indicates the facial skeleton produced by anterior regions of the cephalic crest. (D) Formation of the truncoconal septa (between the aorta and the pulmonary vein) from the neural crest cells of the cardiac crest. Human hindbrain crest cells migrate to pharyngeal arches 4 and 6 during the fifth week of gestation and enter the truncus arteriosus to generate the septa. (A from Tan and Morriss-Kay, 1985, courtesy of S.-S. Tan; B after Sechrist et al., 1993; D after Kirby and Waldo, 1990.)

grate to the first pharyngeal (mandibular) pouch and also generate the ganglion for the trigeminal nerve. They are also pulled by the expanding epidermis to form the **frontonasal process.** Second, cells from rhombomere 4 populate the second pharyngeal pouch (forming the hyoid cartilage of the neck) and also produce the ganglia for the geniculate and vestibuloacoustic nerves. Third, cells from rhombomere 6 migrate into the third and fourth pharyngeal pouches to form thymus, parathyroid, and thyroid glands as well as the ganglia of the vagus and glossopharyngeal nerves. If the neural crest is removed from those regions including rhombomere 6, the thymus, parathyroid glands, and thyroid fail to form (Bockman and Kirby, 1984). Neural crest cells from rhombomeres 3 and 5 do not migrate through the mesoderm surrounding them, but either undergo apoptotic cell death or enter into the streams of crest cells on either side of them (Graham et al., 1993; Sechrist et al., 1993; Graham et al., 1994).

In mammalian embryos, cranial neural crest cells migrate before the neural tube is closed (Tan and Morriss-Kay, 1985) and give rise to the facial mesenchyme (Johnston et al., 1985). The crest cells originating in the forebrain and midbrain contribute to the nasal process, palate, and mesenchyme of the first pharyngeal pouch. This structure becomes part of the gill apparatus in fishes; in humans, it gives rise to the jawbones and to the incus and malleus bones of the middle ear. The neural crest cells originating in the anterior hindbrain region generate the mesenchyme of the second pharyngeal arch, which generates the human stapes bone as well as much of the facial cartilage (see Figure 7.41C; Table 7.2). The cervical neural crest cells give rise to the mesenchyme of the third, fourth, and sixth pharyngeal arches (the fifth degenerates in humans), which produce the neck bones and muscles.

As we discussed in Chapter 2, a series of genes appears to specify the fates of these cranial neural crest cells and specifies their migration pathways. Chisaka and Capecchi (1991) "knocked out" the *Hoxa-3* gene from inbred mice and found that these mutant mice had severely deficient or absent thymuses, thyroids, and parathyroid glands, shortened neck vertebrae, and malformed major heart vessels. It appears that the *Hoxa-3* genes are responsible for specifying the cranial neural crest cells that give rise to the neck cartilage and pharyngeal arch derivatives. However, this gene does not control the smaller migration of neural crest cells that form the cranial neural ganglia. This migratory pathway is affected when the *Hoxb-1* genes are knocked out. In this mutant, there are defects in the production of the facial nerve*
(Goddard et al., 1996; Studer et al., 1996).

Developmental Potency of the Cephalic Neural Crest Cells

From the preceding discussion, it would appear that all neural crest cells are originally identical in their potencies. This, however, is not the case. Here, again, cranial crest cells differ from trunk crest cells, for only the cells of the cranial neural crest are able to produce the cartilage of the head. Moreover, when transplanted into the trunk region, the cranial neural crest participates in forming trunk cartilage that normally does not arise from neural crest components. In at least some cases, these cranial neural crest cells are instructed quite early as to what tissues they can form. Noden (1983) removed regions of chick neural crest that would normally seed the second pharyngeal arch and replaced them with cells that would migrate into the first pharyngeal arch. These host embryos developed two sets of lower jaw struc-

---

*The phenotype of the *Hoxb-1* mutant mice resembles that of certain human conditions such as Bell's Palsy and the Moebius Syndrome (congenital facial paralysis) and may provide insights into these conditions.

Table 7.2 Some derivatives of the pharyngeal arches

Pharyngeal arch	Skeletal elements (neural crest plus mesoderm)	Arches, arteries (mesoderm)	Muscles (mesoderm)	Cranial nerves (neural tube)
1	Incus and malleus (from neural crest); mandible, maxilla, and temporal bone regions (from crest dermal mesenchyme)	Maxillary branch of the carotid artery (to the ear, nose, and jaw)	Jaw muscles; floor of mouth; muscles of the ear and soft palate	Maxillary and mandibular divisions of trigeminal nerve (V)
2	Stapes bone of the middle ear; styloid process of temporal bone; part of hyoid bone of neck (all from neural crest cartilage)	Arteries to the ear region: cortico-tympanic artery (adult); stapedial artery (embryo)	Muscles of facial expression; jaw and upper neck muscles	Facial nerve (VII)
3	Lower rim and greater horns of hyoid bone (from neural crest)	Common carotid artery; root of internal carotid	Stylopharyngeus (to elevate the pharynx)	Glossopharyngeal nerve (IX)
4	Laryngeal cartilages (from lateral plate mesoderm)	Arch of aorta; right subclavian artery; original spouts of pulmonary arteries	Constrictors of pharynx and vocal cords	Superior laryngeal branch of vagus nerve (X)
6	Laryngeal cartilages (from lateral plate mesoderm)	Ductus arteriosus; roots of definitive pulmonary arteries	Intrinsic muscles of larynx	Recurrent laryngeal branch of vagus nerve (X)

Source: Based on Larsen, 1992.

tures, since the graft-derived cells also produced a mandible. The bases for this instruction will be discussed in Chapter 16.

When chick or quail cephalic neural crest cells are cultured, they turn out to be a heterogeneous population of cells (Baroffio et al., 1991). Some of the cells are pluripotent, and their clones contain cells of many different types. Other crest cells give more restricted derivatives. Interestingly, not all the combinations of cartilage, glia, cholinergic neurons, melanocytes, and adrenergic neurons are seen. Figure 7.42 shows the types of clones that emerge and traces hypothetical routes to the restricted cell types.

The cardiac neural crest

As we will detail in Chapter 9, the heart originally forms in the neck region, directly beneath the pharyngeal arches, and it should not be surprising that it acquires cells from the neural crest. However, the contributions of the neural crest to the heart have only recently been appreciated. The caudal region of the cephalic neural crest is sometimes designated the **cardiac neural crest,** since these neural crest cells (and only these particular neural crest cells) can generate the endothelium of the aortic arch arteries and the septum between the aorta and the pulmonary artery (see Figure 7.41D). In the chick, the cardiac neural crest lies above the neural tube region from rhombomere 7 through the spinal cord apposing the third somite, and these crest

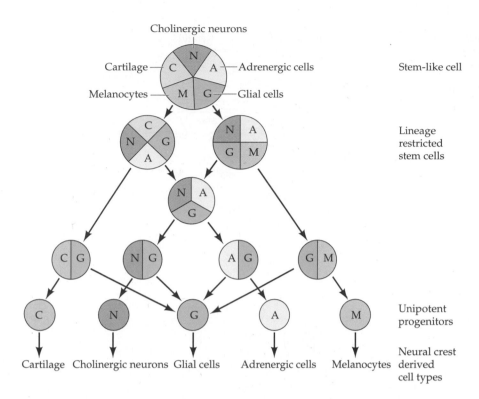

Cholinergic neurons

Cartilage — C N A — Adrenergic cells

Melanocytes — M G — Glial cells

Stem-like cell

Lineage restricted stem cells

Unipotent progenitors

Neural crest derived cell types

Cartilage   Cholinergic neurons   Glial cells   Adrenergic cells   Melanocytes

Figure 7.42
Hypothetical lineage restriction in the cells of the quail cephalic neural crest. A total of 533 clones, each derived from a single cell, were observed for the cell types derived from each cell. The results are consistent with the progressive restriction in cell fate from a pluripotent stem cell through more restricted stem cells to a "unipotential" progenitor cell. (A, adrenergic neuron; C, cartilage; G, glial cells; M, melanocytes; N, cholinergic neurons.) (After Le Douarin et al., 1994.)

cells migrate into pharyngeal arches 3, 4, and 6. The cardiac neural crest is unique in that if it is removed and replaced by anterior cephalic or trunk neural crests, cardiac abnormalities (notably the failure of aortic-pulmonary separation) occur. Thus, the cardiac crest is already determined to generate cardiac cells, and the other regions of the neural crest cannot substitute for it (Kirby, 1989; Kuratani and Kirby, 1991). Congenital heart defects in humans often occur with defects in the parathyroid, thyroid, or thymus glands. It would not be surprising if these were linked to defects in the migration of cells from the neural crest. [ecto7.html]

■ THE EPIDERMIS AND THE ORIGIN OF CUTANEOUS STRUCTURES

The origin of epidermal cells

The cells covering the embryo after neurulation form the presumptive epidermis. Originally, this tissue is one cell layer thick, but in most vertebrates this shortly becomes a two-layered structure. The outer layer gives rise to the **periderm,** a temporary covering that is shed once the bottom layer differentiates to form a true epidermis. The inner layer, called the **basal layer** (or **stratum germinativum**), is a germinal epithelium that gives rise to all the cells of the epidermis (Figure 7.43). The basal layer divides to give rise to another, outer population of cells that constitutes the **spinous layer.** These two epidermal layers are referred to as the **Malpighian layer.** The cells of the Malpighian layer divide to produce the **granular layer** of the epidermis, so called because the cells are characterized by granules of the protein keratin. Unlike the cells remaining in the Malpighian layer, the cells of the granular layer do not divide, but begin to differentiate into epidermal skin cells, or **keratinocytes.** The keratin granules become more prominent as the cells of the granular layer age and migrate outward. Here they form the **cornified layer** (**stratum corneum**), in which the cells have become flattened sacs of

Figure 7.43
Diagram of the layers of the human epidermis. The basal cells are mitotically active, whereas the fully keratinized cells characteristic of external skin are dead and shed off. The keratinocytes obtain this pigment from the transfer of melanosomes from the processes of melanocytes that reside in the basal layer. (After Montagna and Parakkal, 1974.)

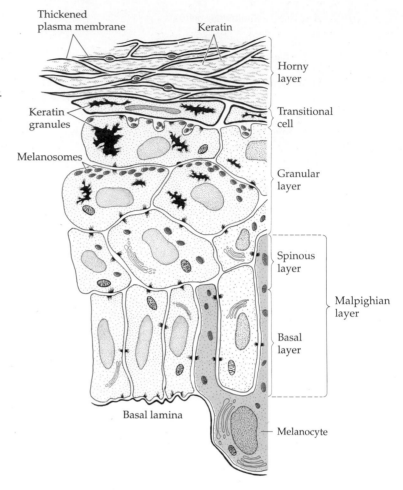

keratin protein. The depth of the cornified layer varies from site to site, but is usually 10 to 30 cells thick. The nuclei of these cells have been pushed to one edge of the cell. Shortly after birth, the cells of the cornified layer are shed and are replaced by new cells coming up from the granular layer. Throughout life, the dead keratinized cells of the cornified layer are being shed (we humans lose about 1.5 grams of these cells each day*) and are replaced by new cells, the source of which is the mitotic cells of the Malpighian layer. The pigment cells from the neural crest also reside in the Malpighian layer, where they transfer their pigment sacs (melanosomes) to the developing keratinocytes.

The epidermal stem cells of the Malpighian layer are bound to the basement membrane by their integrin proteins. However, as they become committed to differentiate, they downregulate their integrins and eventually lose them as the cells migrate into the spinous layer (Jones and Watt, 1993).

There are two main growth factors stimulating the development of the epidermis. The first of these is **transforming growth factor-α** (**TGF-α**). TGF-α is made by the basal cells and stimulates their own division. When a growth factor is made by the same cell that receives it, the factor is called an **autocrine growth factor**. Such factors have to be carefully regulated because if their levels are elevated, more cells are rapidly produced. In adult skin, a cell born in the Malpighian layer takes roughly 8 weeks to reach the stratum

---

*Most of this skin becomes "house dust" on furniture and floors. If you doubt this, burn some of the dust; it will smell just like singed skin.

corneum and remains in the cornified layer about 2 weeks. In individuals with psoriasis, a disease characterized by the exfoliation of enormous amounts of epidermal cells, the time in the cornified layer is only 2 days (Weinstein and van Scott, 1965; Halprin, 1972). This condition has been linked to the overexpression of TGF-α (which occurs secondarily to an immune inflammation) (Elder et al., 1989). Similarly, if the *TGF-α* gene is linked to a promoter for keratin 14 (one of the major skin proteins) and inserted into the mouse pronucleus, the transgenic mice activate the *TGF-α* gene in their skin cells and cannot downregulate it. The result is a mouse with scaly skin, stunted hair growth, and an enormous surplus of keratinized epidermis over its single layer of basal cells (Figure 7.44C; Vassar and Fuchs, 1991).

The other growth factor needed for epidermal production is **keratinocyte growth factor** (**KGF;** also known as fibroblast growth factor 7), a paracrine factor which is produced by the fibroblasts of the underlying (mesodermally derived) dermis. KGF is received by the basal cells above the dermal fibroblasts and is thought to regulate the proliferation of these basal cells. If the *KGF* gene is fused with the keratin 14 promoter and transgenic mice are made, the KGF becomes autocrine. The resulting transgenic mice (Figure 7.44A) have a thickened epidermis, baggy skin, far too many basal cells, and no hair follicles, not even whisker follicles (Guo et al., 1993). These basal cells are "forced" into the epidermal pathway of differentiation. The alternative for the basal cell is to help generate the hair follicle.

Cutaneous appendages

The epidermis and dermis also interact at specific sites to create the sweat glands and the cutaneous appendages: hair, scales, or feathers (depending on the species). The first indication that a hair follicle will form at a particular place is an aggregation of cells in the basal layer of the epidermis. This aggregation is directed by the underlying dermal cells and occurs at differ-

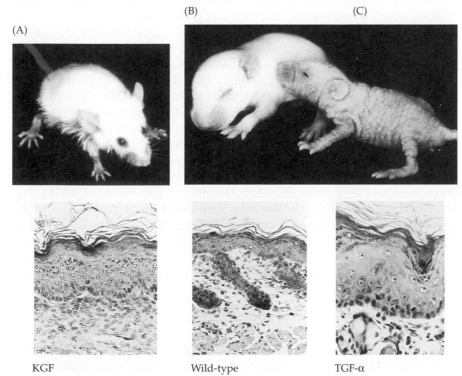

(A)   (B)   (C)

KGF   Wild-type   TGF-α

Figure 7.44
Growth factors and epidermal proliferation. (A) A transgenic mouse expressing low levels of KGF in its keratinocytes. Note the sparsity of hair around the legs, eyes, and nose. (B) A wild-type mouse. (C) A littermate of (B) that is expressing high levels of TGF-α in its keratinocytes. It has scaly skin and very little hair. Below each mouse is a cross section through its skin. The mouse expressing excess KGF has no hair follicles and an increased number of basal epidermal cells. The mouse overexpressing TGF-α has very extensive layers of keratinized epithelia, which it sheds. (From Vassar and Fuchs, 1991, and Guo et al., 1993. Photographs courtesy of E. Fuchs.)

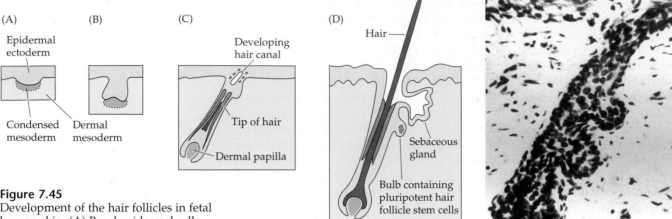

Figure 7.45
Development of the hair follicles in fetal human skin. (A) Basal epidermal cells become columnar and bulge slightly into the dermis. (B) Epidermal cells continue to proliferate, and dermal mesenchyme cells collect at the base of the primary hair germ. (C) Hair shaft differentiation begins in the elongated hair germ. (D) The keratinized hair shaft extends from the hair root, the secondary bud forms the sebaceous gland, and beneath it is a region that may contain the hair stem cells for the next cycle of hair production. (E) Photograph of elongated hair germ. (After Hardy, 1992, and Miller et al., 1993. Photograph courtesy of W. Montagna.)

ent times and different places in the embryo. The basal cells elongate, divide, and sink into the dermis. The dermal cells respond to this ingression of basal epidermal cells by forming a small node (the **dermal papilla**) beneath the epidermal plug. The dermal papilla then pushes up and stimulates the basal stem cells to divide more rapidly and to produce postmitotic cells that will differentiate into the keratinized hair shaft (see Hardy, 1992; Miller et al., 1993). Melanoblasts, which were present among the epidermal cells as they ingressed, differentiate into melanocytes and transfer their pigment to the shaft (Figure 7.45). As this is occurring, two epithelial swellings begin to grow on the side of the follicle. The cells of the lower swelling may retain a population of stem cells that will regenerate the hair shaft periodically when the shaft is shed (Pinkus and Mehregan, 1981; Cotsarelis et al., 1990). The cells of the upper bulge will form the **sebaceous glands,** which produce an oily secretion, **sebum.** In many mammals, including humans, the sebum mixes with the desquamated peridermal cells to form the whitish vernix caseosa, which surrounds the fetus at birth. [ecto8.html]

The first hairs in the human embryo are of a thin, closely spaced type called **lanugo.** This type of hair is usually shed before birth and is replaced (at least in part, by new follicles) by the short and silky **vellus.** Vellus remains on many parts of the human body usually considered hairless, such as the forehead and eyelids. In other areas of the body, vellus gives way to "terminal" hair. During a person's life, some of the follicles that produced vellus can later form terminal hair and still later revert to vellus production. The armpits of infants, for instance, have follicles that produce vellus until adolescence. At that time, terminal shafts are generated. Conversely, in normal masculine "baldness," the scalp follicles revert back to producing unpigmented and very fine vellus hair (Montagna and Parakkal, 1974). The placement and pattern of hair, feathers, scales, and sweat glands involve the interactions of the dermis and the epidermis, and these will be discussed in more detail in Chapter 17. Just as there is a pluripotent neural stem cell whose offspring become neural and glial cells, so there appears to be a pluripotent epidermal stem cell whose progeny can become epidermis, sebaceous gland, or hair shaft.

Coda

In this chapter, we have followed the differentiation of the embryonic ectoderm into a wide variety of tissues. We have seen that the ectoderm produces three sets of cells during neurulation: (1) the neural tube, which gives rise to the neurons, glia, and ependymal cells of the central nervous system; (2) the neural crest cells, which give rise to the peripheral nervous system, pigment cells, adrenal medulla, and certain areas of head cartilage; and (3) the epidermis of the skin, which contributes to the formation of cutaneous structures such as hair, feathers, scales, and sweat and sebaceous glands, as well as forming the outer protective covering of our bodies. We also observed how the interactions of epidermal cells are involved in generating the various tissues of the eye.

Later chapters (16 and 17) discuss in more detail the induction of the neural tube and the coordinated development of the eye. In the next chapter, we will discuss the mechanisms by which neurons are directed to travel to specific sites, thus enabling the development of reflexes and behaviors.

LITERATURE CITED

Acampora, D., Mazan, S., Lallemand, Y., Avantaggiato, V., Maury, M., Simeone, A. and Brulet, P. 1995. Forebrain and midbrain regions are deleted in *Otx2* $^{-/-}$ mutants due to a defective anterior neuroectoderm specification during gastrulation. *Development* 121: 3279–3290.

Akitaya, T. and Bronner-Fraser, M. 1992. Expression of cell adhesion molecules during initiation and cessation of neural crest cell migration. *Dev. Dyn.* 194: 12–20.

Alvarez, I. S. and Schoenwolf, G. C. 1992. Expansion of surface epithelium provides the major extrinsic force for bending of the neural plate. *J. Exp. Zool.* 261: 340–348.

Anderson, D. J. and Axel, R. 1986. A bipotential neuroendocrine precursor whose choice of cell fate is determined by NGF and glucocorticoids. *Cell* 47: 1079–1090.

Artinger, K. B. and Bronner-Fraser, M. 1992. Partial restriction in the developmental potential of late emigrating avian neural crest cells. *Dev. Biol.* 149: 149–157.

Balinsky, B. I. 1975. *Introduction to Embryology,* 4th Ed. Saunders, Philadelphia.

Bally-Cuif, L. and Wassef, M. 1994. Ectopic induction and reorganization of Wnt-1 expression in quail/chick chimeras. *Development* 120: 3379–3394.

Bally-Cuif, L. and Wassef, M. 1995. Determination events in the nervous system of the vertebrate embryo. *Curr. Opin. Genet. Dev.* 5: 450–458.

Banks, M. S. and Bennett, P. J. 1988. Optical and photoreceptor immaturities limit the spatial and chromatic vision of human neonates. *J. Opt. Soc. Am.* 5: 2059–2079.

Baroffio, A., Dupin, E. and Le Douarin, N. M. 1991. Common precursors for neural and mesectodermal derivatives in the cephalic neural crest. *Development* 112: 301–305.

Baynish, A. G., Hosoda, K., Giaid, A., Richardson, J. A., Emoto, N., Hammer, R. E. and Yanagisawa, M. 1994. Interaction of endothelin-3 with endothelin-B receptor is essential for development of epidermal melanocytes and enteric neurons. *Cell* 79: 1277–1285.

Bloom, W. and Fawcett. D. W. 1975. *Textbook of Histology,* 10th Ed. Saunders, Philadelphia.

Bockman, D. E. and Kirby, M. L. 1984. Dependence of thymus development on derivatives of the neural crest. *Science* 223: 498–500.

Bronner-Fraser, M. 1986. Analysis of the early stages of trunk neural crest migration in avian embryos using monoclonal antibody HNK-1. *Dev. Biol.* 115: 44–55.

Bronner-Fraser, M. and Cohen, A. M. 1980. Analysis of the neural crest ventral pathway using injected tracer cells. *Dev. Biol.* 77: 130–141.

Bronner-Fraser, M. and Fraser, S. E. 1988. Cell lineage analysis reveals multipotency of some avian neural crest cells. *Nature* 335: 161–164.

Bronner-Fraser, M. and Fraser, S. 1989. Developmental potential of avian trunk neural crest cells in situ. *Neuron* 3: 755–766.

Bronner-Fraser, M. and Stern, C. 1991. Effects of mesodermal tissues on avian neural crest cell migration. *Dev. Biol.* 143: 213–217.

Burnside, B. 1971. Microtubules and microfilaments in newt neurulation. *Dev. Biol.* 26: 416–441.

Burnside, B. 1973. Microtubules and microfilaments in amphibian neurulation. *Am. Zool.* 13: 989–1006.

Catala, M., Teillet, M.-A. and Le Douarin, N. M. 1995. Organization and development of the tail bud analyzed with the quail-chick chimaera system. *Mech. Dev.* 51: 51–65.

Catala, M., Teillet, M.-A., De Robertis, E. M. and Le Douarin, N. M. 1996. A spinal cord fate map in the avian embryo: while regressing, Hensen's node lays down the notochord and floor plate thus joining the spinal cord lateral walls. *Development* 122: 2599–2610.

Centers for Disease Control. 1992. Recommendations for the use of folic acid to reduce the number of cases of spina bifida and other neural tube defects. *Morb. Mortal. Wkly. Rep.* 41: 1–7.

Chenn, A. and McConnell, S. K. 1995. Cleavage orientation and the asymmetric inheritance of Notch1 immunoreactivity in mammalian neurogenesis. *Cell* 82: 631–641.

Chiang, C., Litingung, Y., Lee, E., Young, K. K., Corden, J. E., Westphal, H. and Beachy, P. A. 1996. Cyclopia and defective axial patterning in mice lacking *Sonic hedgehog* gene function. *Nature* 383: 407–413.

Chisaka, O. and Capecchi, M. 1991. Regionally restricted developmental defects resulting from targeted disruption of the mouse homeobox gene *Hox1.5. Nature* 350: 473–479.

Chun, L. L. Y. and Patterson, P. H. 1977. Role of nerve growth factor in development of rat sympathetic neurons in vitro. Survival, growth, and differentiation of catecholamine production. *J. Cell Biol.* 75: 694–704.

Ciment, G. 1990. The melanocyte/Schwann cell progenitor: A bipotent intermediate in the neural crest lineage. *Commun. Dev. Neurobiol.* 1: 207–223.

Cotsarelis, G., Sun, T.-T. and Lavker, R. M. 1990. Label-retaining cells reside in the bulge area of pilosebaceous unit: Implications for follicular stem cells, hair cycle and skin carcinogenesis. *Cell* 61: 1329–1337.

Coulombe, J. N. and Bronner-Fraser, M. 1987. Cholinergic neurones acquire adrenergic neurotransmitters when transplanted into an embryo. *Nature* 324: 569–572.

Coulombre, A. J. 1956. The role of intraocular pressure in the development of the chick eye. I. Control of eye size. *J. Exp. Zool.* 133: 211–225.

Coulombre, A. J. 1965. The eye. *In* R. De-Haan and H. Ursprung (eds.), *Organogenesis.* Holt, Rinehart & Winston, New York, pp. 217–251.

Crossin, K. L., Chuong, C.-M. and Edelman, G. M. 1985. Expression sequences of cell adhesion molecules. *Proc. Natl. Acad. Sci. USA* 82: 6942–6946.

Crossley, P. H., Martinez, S. and Martin, G. R. 1996. Midbrain development induced by FGF8 in the chick embryo. *Nature* 380: 66–68.

Czeizel, A. and Dudas, I. 1992. Prevention of first occurence of neural tube defects by periconceptional vitamin supplementation. *N. Engl. J. Med.* 327: 1832–1835.

Danielian, P. S. and McMahon, A. P. 1996. *Engrailed-1* as a target of the Wnt-1 signaling pathway in vertebrate midbrain development. *Nature* 383: 332–334.

Davenport, R. W., Doù, P., Rehder, V. and Kater, S. B. 1993. A sensory role for neuronal growth cone filopodia. *Nature* 361: 721–724.

Desmond, M. E. 1982. A description of the occlusion of the lumen of the spinal cord in early human embryos. *Anat. Rec.* 204: 89–93.

Desmond, M. E. and Field, M. C. 1992. Evaluation of neural fold fusion and coincident initiation of spinal cord occlusion in the chick embryo. *J. Comp. Neurol.* 319: 246–260.

Desmond, M. E. and Schoenwolf, G. C. 1986. Evaluation of the roles of intrinsic and extrinsic factors in occlusion of the spinal neurocoel during rapid brain enlargement in the chick embryo. *J. Embryol. Exp. Morphol.* 97: 25–46.

Detrick, R. J., Dickey, D. and Kintner, C. R. 1990. The effects of N-cadherin misexpression on morphogenesis in *Xenopus* embryos. *Neuron* 4: 493–506.

Dickinson, M. E., Krumlauf, R. and McMahon, A. P. 1994. Evidence for a mitogenic effect of Wnt-1 in the developing mammalian central nervous system. *Development* 120: 1453–1471.

Edery, P. and nine others. 1994. Mutations of the RET proto-oncogene in Hirschsprung's disease. *Nature* 367: 378–380.

Edmondson, J. C. Liem, R. K. H., Kuster, J. C. and Hatten, M. E. 1988. Astrotactin: A novel neuronal cell surface antigen that mediates neuronal-astroglial interactions in cerebellar microcultures. *J. Cell Biol.* 106: 505–517.

Eichele, G. 1992. Budding thoughts. *The Sciences* (Jan., 1992): 30–36.

Elder, J. T. and eight others. 1989. Overexpression of transforming growth factor in psoriatic epidermis. *Science* 243: 811–814.

Ericson, J., Morton, S., Kawakami, A., Roelinck, H. and Jessell, T. M. 1996. Two critical periods of Sonic hedgehog signaling required for the specification of motor neuron identity. *Cell* 87: 661–673.

Erickson, C. A. and Weston, J. A. 1983. An SEM analysis of neural crest migration in the mouse. *J. Embryol. Exp. Morphol.* 74: 97–118.

Erickson, C. A. and Goins, T. L. 1995. Avian neural crest cells can migrate in the dorsolateral path only if they are specified as melanocytes. *Development* 121: 915–924.

Erickson, C. A., Tosney, K. W. and Weston, J. A. 1980. Analysis of migrating behaviour of neural crest and fibroblastic cells in embryonic tissues. *Dev. Biol.* 77: 142–156.

Erickson, C. A., Duong, T. D. and Tosney, K. W. 1992. Descriptive and experimental analysis of the dispersion of neural crest cells along the dorsolateral pathway and their entry into ectoderm in the chick embryo. *Dev. Biol.* 151: 251–272.

Ernfors, P., Lee, K. F. and Jaenisch, R. 1994. Mice lacking brain-derived neurotrophic factor develop with sensory deficits. *Nature* 368: 147–150.

Fishell, G. 1995. Striatal precursors adopt cortical identities in response to local cues. *Development* 121: 803–812.

Fishell, G. and Hatten, M. E. 1991. Astrotactin provides a receptor system for glia-guided neuronal migration. *Development* 113: 755–765.

Forscher, P. and Smith, S. J. 1988. Actions of cytochalasins on the organization of actin filaments and microtubules in a neural growth cone. *J. Cell Biol.* 107: 1505–1516.

Frantz, G. D. and McConnell, S. K. 1996. Restriction of late cerebral cortical progenitors to an upper-layer fate. *Neuron* 17: 55–61.

Fujimori, T., Miyatani, S. and Takeichi, M. 1990. Ectopic expression of N-cadherin perturbs histogenesis in *Xenopus* embryos. *Development* 110: 97–104.

Fujita, S. 1964. Analysis of neuron differentiation in the central nervous system by tritiated thymidine autoradiography. *J. Comp. Neurol.* 122: 311–328.

Fujita, S. 1966. Application of light and electron microscopy to the study of the cytogenesis of the forebrain. *In* R. Hassler and H. Stephen (eds.), *Evolution of the Forebrain.* Plenum, New York, pp. 180–196.

Fukada, K. 1980. Hormonal control of neurotransmitter choice in sympathetic neuron cultures. *Nature* 287: 553–555.

Gallera, J. 1971. Primary induction in birds. *Adv. Morphogenet.* 9: 149–180.

Glaser, T., Jepeal, L., Edwards, J. G., Young, S. R., Favor, J. and Maas, R. L. 1994. PAX6 gene dosage effect in a family with congenital cataracts, aniridia, anophthalmia and central nervous system defects. *Nat. Genet.* 7: 463–471.

Goddard, J. M., Rossel, M., Manley, N. R. and Capecchi, M. R. 1996. Mice with targeted disruption of *Hoxb-1* fail to form the motor nucleus of the VIIth nerve. *Development* 122: 3217–3228.

Golden, J. A. and Chernoff, G. F. 1993. Intermittent pattern of neural tube closure in two strains of mice. *Teratology* 47: 73–80.

Goldstein, R. S., Teillet, M.-A. and Kalcheim, C. 1990. The microenvironment created by grafting rostral half-somites is mitogenic for neural crest cells. *Proc. Natl. Acad. Sci. USA* 87: 4476–4480.

Gont, L. K., Steinbeisser, H., Blumberg, B. and De Robertis, E. M. 1993. Tail formation as a continuation of gastrulation: The multiple cell populations of the *Xenopus* tailbud derive from the late blastopore lip. *Development* 119: 991–1004.

Gould, S. J. 1977. *Ontogeny and Phylogeny.* Harvard University Press, Cambridge, MA.

Graham, A., Heyman, I. and Lumsden, A. 1993. Even-numbered rhombomeres control the apoptotic elimination of neural crest cells from odd-numbered rhombomeres of the chick hindbrain. *Development* 119: 233–245.

Graham, A, Francis-West, P., Brickell, P. and Lumsden, A. 1994. The signalling molecule BMP-4 mediates apoptosis in the rhombencephalic neural crest. *Nature* 372: 684–686.

Guo, L., Yu, Q.-C. and Fuchs, E. 1993. Targetting expression of keratinocyte growth factor to keratinocytes elicits striking changes in epithelial differeentiation in transgenic mice. *EMBO J.* 12: 973–986.

Guthrie, S. and Lumsden, A. 1991. Formation and regeneration of rhombomere boundaries in the developing chick hindbrain. *Development* 112: 221–229.

Gregory, W. A., Edmondson, J. C., Hatten, M. E. and Mason, C. A. 1988. Cytology and neural-glial apposition of migrating cerebellar granule cells in vitro. *J. Neurosci.* 8: 1728–1738.

Halder, G., Callaerts, P. and Gehring, W. J. 1995. Induction of ectopic eyes by targeted expression of the *eyeless* gene in *Drosophila.* *Science* 267: 1788–1792.

Hallet, M. M. and Ferrand, R. 1984. Quail melanoblast migration in two breeds of fowl and their hybrids: Evidence for a dominant genic control of the mesodermal pigment pattern through tissue environment. *J. Exp. Zool.* 230: 229–238.

Halprin, K. M. 1972. Epidermal "turnover time"—a reexamination. *J. Invest. Dermatol.* 86: 14–19.

Hardy, M. H. 1992. The secret life of the hair follicle. *Trends Genet.* 8: 55–61.

Harrison, R. G. 1907. Observations on the living developing nerve fiber. *Anat. Rec.* 1: 116–118.

Hatten, M. E. 1990. Riding the glial monorail: A common mechanism for glial-guided neuronal migration in different regions of the mammalian brain. *Trends Neurosci.* 13: 179–184.

Hemesath, T. J. and nine others. 1994. *microphthalmia*, a critical factor in melanocyte development defines a discrete transcription factor family. *Genes Dev.* 8: 2770–2780.

Henry, E. W. and Sidman, R. L. 1988. Long lives for homozygous *trembler* mutant mice despite virtual absence of peripheral nerve myelin. *Science* 241: 344–346.

Hilfer, S. R. and Yang, J.-J. W. 1980. Accumulation of CPC-precipitable material at apical cell surfaces during formation of the optic cup. *Anat. Rec.* 197: 423–433.

His, W. 1886. Zur Geschichte des menschlichen Rückenmarks und der Nervenwurzeln. *Ges. Wiss.* BD 13, S. 477.

Holt, A. B., Cheek, D. B., Mellitz, E. D. and Hill, D. E. 1975. Brain size and the relation of the primate to the non-primate. *In* D. B. Cheek (ed.), *Fetal and Postnatal Cellular Growth: Hormones and Nutrition.* Wiley, New York, pp. 23–44.

Hosoda, K., Hammer, R. E., Richardson, J. A., Baynish, A. G., Cheung, J. C., Giaid, A. and Yanagisawa, M. 1994. Targeted and natural (*Piebald-lethal*) mutations of endothelin-B receptor gene produce megacolon associate with spotted coat color in mice. *Cell* 79: 1267–1276.

Huettner, A. F. 1949. *Fundamentals of Comparative Embryology of the Vertebrates*, 2nd Ed. Macmillan, New York.

Irving, C., Flenniken, A., Alldus, G. and Wilkinson, D. G. 1996. Cell-cell interactions and segmentation in the developing vertebrate hindbrain. *Biochem. Soc. Symp.* 1996: 85–95.

Jacobson, A. G. 1981. Morphogenesis of the neural plate and tube. *In* I. G. Connely et al. (eds.), *Morphogenesis and Pattern Formation.* Raven Press, New York, pp. 233–263.

Jacobson, A. G. and Moury, J. G. 1995. Tissue boundaries and cell behavior during neurulation. *Dev. Biol.* 171: 98–110.

Jacobson, A. G. and Sater, A. K. 1988. Features of embryonic induction. *Development* 104: 341–359.

Jacobson, M. 1968. Cessation of DNA synthesis in retinal ganglion cells correlated with the time of specification of their central connections. *Dev. Biol.* 17: 219–232.

Jacobson, M. 1991. *Developmental Neurobiology*, 2nd Ed. Plenum, New York.

Johnston, M. C., Sulik, K. K., Webster, W. S. and Jarvis, B. L. 1985. Isotretinoin embryopathy in a mouse model: Cranial neural crest involvement. *Teratology* 31: 26A.

Jones, K. R., Farinas, I., Backus, C. and Reichart, L. F. 1994. Targeted disruption of the BDNF gene perturbs brain and sensory neuron development, but not motor neuron development. *Cell* 76: 989–999.

Jones, P. H. and Watt, F. M. 1993. Separation of human epidermal stem cells from transit amplifying cells of the basis of differences in integrin function and expression. *Cell* 73: 713–724.

Jordan, T. and seven others. 1992. The human *PAX6* gene is mutated in two patients with aniridia. *Nat. Genet.* 1: 328–332.

Kahn, C. R., Coyle, J. T. and Cohen, A. M. 1980. Head and trunk neural crest *in vitro*: Autonomic neuron differentiation. *Dev. Biol.* 77: 340–348.

Kalcheim, C. R. and Neufeld, G. 1990. Expression of basic fibroblast growth factor in the nervous system of early avian embryos. *Development* 109: 203–215.

Kalcheim, C., Barde, Y.-A., Thoenen, H. and Le Douarin, N. M. 1987. In vivo effect of brain-derived neurotrophic factor on the survival of neural crest precursor cells of the dorsal root ganglia. *EMBO J.* 6: 2871–2873.

Karfunkel, P. 1972. The activity of microtubules and microfilaments in neurulation in the chick. *J. Exp. Zool.* 181: 289–302.

Keller, R., Shih, J., Sater, A. K. and Moreno, C. 1992. Planar induction of convergence and extension of the neural plate by the organizer of *Xenopus. Dev. Dyn.* 193: 218–234.

Kelley, R. I. and seven others. 1996 Holoprosencephaly in RSH/Smith-Lemli-Opitz syndrome: Does abnormal cholesterol metabolism affect the function of Sonic hedgehog? *Am. J. Med. Genet.* 66: 478–484.

Kirby, M. L. 1987. Cardiac morphogenesis: Recent research advances. *Pediatr. Res.* 21: 219–224.

Kirby, M. L. 1989. Plasticity and predetermination of mesencephalic and trunk neural crest transplanted into the region of the cardiac neural crest. *Dev. Biol.* 134: 401–412.

Kirby, M. L. and Waldo, K. L. 1990. Role of neural crest in congenital heart disease. *Circulation* 82: 332–340.

Komuro, H. and Rakic, P. 1992. Selective role of N-type calcium channels in neuronal migration. *Science* 157: 806–809.

Krull, C. E, Collazo, A., Fraser, S. E. and Bronner-Fraser, M. 1995. Segmental migration of trunk neural crest. Time lapse analysis reveals a role for PNA binding molecules. *Development* 121: 3733–3743.

Kuratani, S. C. and Kirby, M. L. 1991. Initial migration and distribution of the cardiac neural crest in the avian embryo: An introduction to the concept of the circumpharyngeal crest. *Am. J. Anat.* 191: 215–227.

Lamoureux, P., Buxbaum, R. E. and Heidemann, S. R. 1989. Direct evidence that growth cones pull. *Nature* 340: 159–162.

Lahav, R., Ziller, C., Dupin, E. and Le Douarin, N. M. 1996. Endothelin 3 promotes neural crest cell proliferation and mediates a vast increase in melanocyte number in culture. *Proc. Natl. Acad. Sci. USA.* 93: 3892-3897.

Landis, S. C. and Patterson, P. H. 1981. Neural crest cell lineages. *Trends Neurosci.* 4: 172–175.

Larsen, W. J. 1993. *Human Embryology.* Churchill Livingstone, New York.

Le Douarin, N. and Smith, J. 1988. Development of the peripheral nervous system from the neural crest. *Annu. Rev. Cell Biol.* 4: 375–404.

Le Douarin, N. M. 1969. Particularités du noyau interphasique chez la Caille japonaise (*Coturnix coturnix japonica*). Utilisation de ces particularités comme "marquage biologique" dans les recherches sur les interactions tissulaires et les migrations cellulaires au cours de l'ontogenèse. *Bull. Biol. Fr. Belg.* 103: 435–452.

Le Douarin, N. M. and Teillet, M.-A. 1973. The migration of neural crest cells to the wall of the digestive tract in avian embryo. *J. Embryol. Exp. Morphol.* 30: 31–48.

Le Douarin, N. M. and Teillet, M.-A. 1974. Experimental analysis of the migration and differentiation of neuroblasts of the autonomic nervous system and of neuroectodermal mesenchyme derivatives, using a biological cell marking technique. *Dev. Biol.* 41: 162–184.

Le Douarin, N. M., Dupin, E. and Ziller, C. 1994. Genetic and epigenetic control in neural crest development. *Curr. Opin. Genet. Dev.* 4: 685–695.

Le Douarin, N. M., Dieterlen-Lièvre, F. and Teillet, M.-A. 1996. Quail-chick transplantations. *Methods Cell Biol.* 51: 23–61.

Le Douarin, N. M., Renaud, D., Teillet, M.-A. and Le Douarin, G. H. 1975. Cholinergic differentiation of presumptive adrenergic neuroblasts in interspecific chimeras after heterotopic transplantation. *Proc. Natl. Acad. Sci. USA* 72: 728–732.

Le Lièvre, C. S. and Le Douarin, N. M. 1975. Mesenchymal derivatives of the neural crest: Analysis of chimaeric quail and chick embryos. *J. Embryol. Exp. Morphol.* 34: 125–154.

Letourneau, P. C. 1977. Regulation of neuronal morphogenesis by cell-substratum adhesion. *Soc. Neurosci. Symp.* 2: 67–81.

Letourneau, P. C. 1979. Cell substratum adhesion of neurite growth cones, and its role in neurite elongation. *Exp. Cell Res.* 124: 127–138.

Liem, K., Tremmi, G., Roelink, H. and Jessell, T. M. 1995. Dorsal differentiation of neural plate cells induced by BMP-mediated signals from epidermal ectoderm. *Cell* 82: 969–979.

Löfberg, J., Perris, R. and Epperlin, H. H. 1989. Timing in the regulation of neural crest cell migration: Retarded maturation of regional extracellular matrix inhibits pigment cell migration in embryos of the white axolotl mutant. *Dev. Biol* 131: 168–181.

Loring, J. F. and Erickson, C. A. 1987. Neural crest cell migratory pathways in the trunk of the chick embryo. *Dev. Biol.* 121: 220–236.

Lumsden, A. 1988. Multipotent cells in the avian neural crest. *Trends Neurosci.* 12: 81–83.

Lumsden, A. and Guthrie, S. 1991. Alternating patterns of cell surface properties and neural crest cell migration during segmentation of the chick embryo. *Development* [Suppl.] 2: 9–15.

Lumsden, A. and Keynes, R. 1989. Segmental patterns of neuronal development in the chick hindbrain. *Nature* 337: 424–428.

Mancilla, A. and Mayor, R. 1996. Neural crest formation in *Xenopus laevis*: Mechanisms of *Xslug* induction. *Dev. Biol.* 177: 580–589.

Mann, I. 1964. *The Development of the Human Eye.* Grune and Stratton, New York.

Marin, F. and Puelles, L. 1994. Patterning of embryonic avian midbrain after experimental inversion: a polarizing activity for the isthmus. *Dev. Biol.* 163: 19–28.

Matsunami, H. and Takeichi, M. 1995. Fetal brain subdivisions defined by T- and E-cadherins expressions: evidence for the role of cadherin activity in region-specific, cell-cell adhesion. *Dev. Biol.* 172: 466–478.

Mayer, T. C. 1973. The migratory pathway of neural crest cells into the skin of mouse embryos. *Dev. Biol.* 34: 39–46.

Maxwell, G. D., Reid, K., Elefanty, A., Bartlett, P. F. and Murphy, M. 1996. Glial cell line-derived neurotrophic factor promotes the development of adrenergic neurons in mouse neural crest cultures. *Proc. Natl. Acad. Sci. USA* 93: 13274–13279.

McConnell, S. K. and Kaznowski, C. E. 1991. Cell cycle dependence of laminar determination in developing cerebral cortex. *Science* 254: 282–285.

McMahon, A. P. and Bradley, A. 1990. The *Wnt-1* (*int-1*) proto-oncogene is required for the development of a large region of the mouse brain. *Cell* 62: 1073–1085.

Milunsky, A., Jick, H., Jick, S. S., Bruell, C. L., Maclaughlen, D. S., Rothman, K. J. and Willett, W. 1989. Multivitamin folic acid supplementation in early pregnancy reduces the prevalence of neural tube defects. *JAMA* 262: 2847–2852.

Miller, S. J., Lavker, R. M. and Sun, T.-T. 1993. Keratinocyte stem cells of corneal, skin, and hair follicle. *Semin. Dev. Biol.* 4: 217–240.

Montagna, W. and Parakkal, P. F. 1974. The piliary apparatus. *In* W. Montagna (ed.), *The Structure and Formation of Skin.* Academic Press, New York, pp. 172–258.

Montagu, M. F. A. 1962. Time, morphology, and neoteny in the evolution of man. *In* M. F. A. Montagu (ed.), *Culture and Evolution of Man.* Oxford University Press, New York.

Moore K. L. and Persaud, T. V. N. 1993. *Before We Are Born: Essentials of Embryology and Birth Defects.* W. B. Saunders, Philadelphia.

Morowitz, H. J. and Trefil, J. S. 1992. *The Facts of Life: Science and the Abortion Controversy.* Oxford University Press, New York.

Moury, J. D. and Schoenwolf, G. C. 1995. Cooperative model of epithelial shaping and bending during avian neurulation: Autonomous movements of the neural plate, autonomous movements of the epidermis, and interactions in the neural plate/epidermis transition zone. *Dev. Dyn.* 204: 323–337.

Nagele, R. G. and Lee, H. Y. 1980. Studies on the mechanism of neurulation in the chick: Microfilament-mediated changes in cell shape during uplifting of neural folds. *J. Exp. Zool.* 213: 391–398.

Nagele, R. G. and Lee, H. Y. 1987. Studies in the mechanism of neurulation in the chick. Morphometric analysis of the relationship between regional variations in cell shape and sites of motive force generation. *J. Exp. Biol.* 24: 197–205.

Newgreen, D. F. and Gooday, D. 1985. Control of onset of migration of neural crest cells in avian embryos: Role of Ca^{++}-dependent cell adhesions. *Cell Tiss. Res.* 239: 329–336.

Newgreen, D. F., Scheel, M. and Kaster, V. 1986. Morphogenesis of sclerotome and neural crest cells in avian embryos. *In vivo* and *in vitro* studies on the role of notochordal extracellular material. *Cell Tiss. Res.* 244: 299–313.

Nichols, D. H. 1981. Neural crest formation in the head of the mouse embryo as observed using a new histological technique. *J. Embryol. Exp. Morphol.* 64: 105–120.

Nichols, D. H. and Weston, J. A, 1977. Melanogenesis in cultures of peripheral nervous tissue. I. The origin and prospective fates of cells giving rise to melanocytes. *Dev. Biol.* 60: 217–225.

Nieto, M. A., Sargent, M. G., Wilkinson, D. G. and Cooke, J. 1994. Control of cell behavior during vertebrate development by *slug*, a zinc finger gene. *Science* 264: 835–839.

Nievelstein, R. A. J., Hartwig, N. G., Vermeij-Keers, C. and Valk, J. 1993. Embryonic development of the mammalian caudal neural tube. *Teratology* 48: 21–31.

Noden, D. M. 1978. The control of avian cephalic neural crest cytodifferentiation. I. Skeletal and connective tissue. *Dev. Biol.* 69: 296–312.

Noden, D. M. 1983. The role of the neural crest in patterning of avian cranial skeletal, connective, and muscle tissues. *Dev. Biol.* 96: 144–165.

O'Rourke, N. A., Dailey, M. E., Smith, S. J. and McConnell, S. K. 1992. Diverse migratory pathways in the developing cerebral cortex. *Science* 258: 299–302.

Papaconstantinou, J. 1967. Molecular aspects of lens cell differentiation. *Science* 156: 338–346.

Pasteels, J. 1937. Etudes sur la gastrulation des vertébrés méroblastiques. III. Oiseaux. IV. Conclusions générales. *Arch. Biol.* 48: 381–488.

Paton, D. and Craig, J. A. 1974. Cataracts: Development, diagnosis, and management. *Ciba Clin. Symp.* 26(3): 2–32.

Patten, B. M. 1971. *Early Embryology of the Chick,* 5th Ed. McGraw-Hill, New York.

Perris, R. and Bronner-Fraser, M. 1989. Recent advances in defining the role of the extracellular matrix in neural crest development. *Commun. Dev. Neurobiol.* 1: 61–83.

Perris, R., von Boxburg, Y. and Löfberg, J. 1988. Local embryonic matrices determine region-specific phenotypes in neural crest cells. *Science* 241: 86–89.

Perris, R., Löfberg, J. Fällström, C., von Boxburg, Y., Olsson, L. and Newgreen, D. F. 1990. Structural and compositional divergencies in the extracellular matrix encountered by neural crest cells in the *white* mutant axolotl embryo. *Development* 109: 533–551.

Piatigorsky, J. 1981. Lens differentiation in vertebrates: A review of cellular and molecular features. *Differentiation* 19: 134–153.

Pichel. J. G. and eleven others. 1996. Defects in enteric innervation and kidney development in mice lacking GDNF. *Nature* 382: 73–76.

Pinkus, H. and Mehregan, A. H. H. 1981. *A Guide to Dermohistopathology.* Appleton Century Crofts, New York.

Pomeranz, H. D., Rothman, T. P. and Gershon, M. D. 1991. Colonization of the post-umbilical bowel by cells derived from the sacral neural crest: Direct tracing of cell migration using an intercalating probe and a replication-deficient retrovirus. *Development* 111: 647–655.

Poole, T. J. and Thiery, J. P. 1986. *In* H. C. Slavkin (ed.), *Progress in Clinical and Biological Research,* Vol 217. Alan R. Liss, New York, pp. 235–238.

Porter, J. A., Young, K. E. and Beachy, P. A. 1996. Cholesterol modification of Hedgehog signaling proteins in animal development. *Science* 274: 255–259.

Portmann, A. 1941. Die Tragzeiten der Primaten und die Dauer der Schwangerschaft beim Menschen: Ein Problem der vergleichen Biologie. *Rev. Suisse Zool.* 48: 511–518.

Portmann, A. 1945. Die Ontogenese des Menschen als Problem der Evolutionsforschung. *Verh. Schweiz. Naturf. Ges.* 125: 44–53.

Puffenberger, E. G., Hosoda, K., Washington, S. S., Nakao, K., deWilt, D., Yanagisawa, M. and Chakvarti, A. 1994. A missense mutation of the endothelin-B receptor gene in multigenic Hirschsprung's disease. *Cell* 79: 1257–1266.

Quiring, R., Walldorf, U., Kloter, U., and Gehring, W. J. 1994. Homology of the eyeless gene of *Drosophila* to the *Small eye* gene in mice and *Aniridia* in humans. *Science* 265: 785–789.

Rakic, P. 1972. Mode of cell migration to superficial layers of fetal monkey neocortex. *J. Comp. Neurol.* 145: 61–84.

Rakic, P. 1974. Neurons in rhesus visual cortex: Systematic relation between time of origin and eventual disposition. *Science* 183: 425–427.

Rakic, P. 1975. Cell migration and neuronal ectopias in the brain. *In* D. Bergsma (ed.), *Morphogenesis and Malformations of Face and Brain.* Birth Defects Original Article Series II (7): 95–129.

Rakic P. and Goldman, P. S. 1982. Development and modifiability of the cerebral cortex. *Neurosci. Rev.* 20: 429–611.

Rakic, P. and Sidman, R. L. 1973. Organization of cerebellar cortex secondary to deficit of granule cells in *weaver* mutant mice. *J. Comp. Neurol.* 152: 133–162.

Ramón y Cajal, S. 1890. Sur l'origene et les ramifications des fibres neuveuses de la moelle embryonnaire. *Anat. Anz.* 5: 111-119.

Rawles, M. E. 1948. Origin of melanophores and their role in development of color patterns in vertebrates. *Physiol. Rev.* 28: 383–408.

Rickmann, M., Fawcett, J. W. and Keynes, R. J. 1985. The migration of neural crest cells and the growth of motor neurons through the rostral half of the chick somite. *J. Embryol. Exp. Morphol.* 90: 437–455.

Roessler, E. and seven others. 1996. Mutations in the human *Sonic hedgehog* gene cause holoprosencephaly. *Nat. Genet.* 14: 357–360.

Romanes, G. J. 1901. *Darwin and After Darwin.* Open Court Publishing, London.

Romeo, G. and ten others. 1994. Point mutations affecting the tyrosine kinase domain of the RET proto-oncogene in Hirschsprung's disease. *Nature* 367: 377–378.

Rubenstein, J. L. R. and Puelles, L. 1994. Homeobox gene expression during development of the vertebrate brain. *Curr. Top. Dev. Biol.* 29: 1–63.

Saha, M., Spann, C. L. and Grainger, R. M. 1989. Embryonic lens induction: More than meets the optic vesicle. *Cell Differ. Dev.* 28: 153–172.

Sauer, F. C. 1935. Mitosis in the neural tube. *J. Comp. Neurol.* 62: 377–405.

Schoenwolf, G. C. 1984. Histological and ultrastructural studies of secondary neurulation in mouse embryos. *Am. J. Anat.* 169: 361–374.

Schoenwolf, G. C. 1991a. Cell movements driving neurulation in avian embryos. *Development* 2 [Suppl.]: 157–168.

Schoenwolf, G. C. 1991b. Cell movements in the epiblast during gastrulation and neurulation in avian embryos. *In* R. Keller et al. (eds)., *Gastrulation.* Plenum, New York, pp. 1–28.

Schoenwolf, G. C. and Alvarez, I. S. 1989. Roles of neuroepithelial cell rearrangement and division in shaping of the avian neural plate. *Development* 106: 427–439.

Schoenwolf, G. C. and Desmond, N. E. 1984. Descriptive studies of the occlusion and reopening of the spinal canal of the early chick embryo. *Anat. Rec.* 209: 251–263.

Schweizer, G., Ayer-LeLièvre, C. and Le Douarin, N. M. 1983. Restrictions in developmental capacities in the dorsal root ganglia during the course of development. *Cell Differ.* 13: 191–200.

Sechrist, J., Serbedzija, G. N., Scherson, T., Fraser, S. E. and Bronner-Fraser, M. 1993. Segmental migration of the hindbrain neural crest does not arise from its segmental origin. *Development* 118: 691–703.

Selleck, M. A. and Bronner-Fraser, M. 1995. Origins of the avian neural crest: the role of neural plate-epidermal interactions. *Development* 121: 525–538.

Serbedzija, G. N., Bronner-Fraser, M. and Fraser, S. E. 1989. A vital dye analysis of the timing and pathways of avian trunk neural crest cell migration. *Development* 106: 809–816.

Shah, N. M. Groves, A. K. and Anderson, D. J. 1996. Alternative neural crest cell fates are instructively promoted by TGF-β superfamily memebers. *Cell* 85: 331–343.

Shah, N. M., Marchionni, M. A., Isaacs, I., Stroobant, P. and Anderson, D. J. 1994. Glial growth factor restricts mammalian neural crest stem cells to a glial fate. *Cell* 77: 349–360.

Shawlot, W. and Behringer, R. R. 1995. Requirement for *Lim1* in head-organizer function. *Nature* 374: 425–430.

Sidman, R. L., Dickie, M. M. and Appel, S. H. 1964. Mutant mice (*quaking* and *jimpy*) with deficient myelination in the central nervous system. *Science* 144: 309–311.

Sieber-Blum, M. 1991. Role of neurotrophic factors BDNF and NGF in the commitment of pluripotent neural crest cells. *Neuron* 6: 949–955.

Sieber-Blum, M. and Sieber, F. 1984. Heterogeneity among early quail neural crest cells. *Dev. Brain Res.* 14: 241–246.

Smith, J. L. and Schoenwolf, G. C. 1989. Notochordal induction of cell wedging in the chick neural plate and its role in neural tube formation. *J. Exp. Zool.* 250: 49–62.

Smith, J. L. and Schoenwolf, G. C. 1991. Further evidence of extrinsic forces in bending of the neural plate. *J. Comp. Neurol.* 307: 225–236.

Spemann, H. 1938. *Embryonic Development and Induction.* Yale University Press, New Haven.

Spieth, J. and Keller, R. E. 1984. Neural crest cell behavior in white and dark larvae of *Ambystoma mexicanum:* Differences in cell morphology, arrangement, and extracellular matrix as related to migration. *J. Exp. Zool.* 229: 91–107.

Spritz, R. A., Gielbel, L. B. and Holmes, S. A. 1992. Dominant negative and loss of function mutations of the *c-kit* (mast/Stem cell growth factor) proto-oncogene in human piebaldism. *Am. J. Hum. Genet.* 50: 261–269.

Stemple, D. L. and Anderson, D. J. 1992. Isolation of a stem cell for neurons and glia from the mammalian neural crest. *Cell* 71: 973–985.

Steingrimsson, E. and ten others. 1994. Molecular basis of mouse microphthalmia (*mi*) mutations helps explain their developmental and phenotypic consequences. *Nat. Genet.* 8: 256–261.

Stocker, K. M., Sherman, L., Rees, S. and Ciment, G. 1991. Basic FGF and TGF-β1 influence commitment to melanogenesis in neural crest-derived cells of avian embryos. *Development* 111: 635–645.

Studer, M., Lumsden, A., Ariza-McNaughton, L., Bradley, A. and Krumlauf, R. 1996. Altered segmental identity and abnormal migration of motor neurons in mice lacking *Hoxb-1*. *Nature* 384: 630–634.

Takeichi, M. 1988. The cadherins: Cell-cell adhesion molecules controlling animal morphogenesis. *Development* 102: 639–656.

Tan, S.-S and Morriss-Kay, G. 1985. The development and distribution of the cranial neural crest in the rat embryo. *Cell Tiss. Res.* 240: 403–416.

Tassabehji, M., Newton, V. E. and Read, A. P. 1994. Waardenburg syndrome type 2 caused by mutations in the human microphthalmia (*MITF*) gene. *Nat. Genet.* 8: 251–255.

Teillet, M.-A., Kalcheim, C. and Le Douarin, N.M. 1987. Formation of the dorsal root ganglia in the avian embryo: Segmental origin and migratory behavior of neural crest progenitor cells. *Dev. Biol.* 120: 329–347.

Thomas, K. R. and Cappecchi, M. R. 1990. Targeted disruption of the murine *int-1* proto-oncogene resulting in severe abnormalities in midbrain and cerebellar development. *Nature* 346: 847–850.

Thorogood, P. 1989. Review of *Developmental and Evolutionary Aspects of the Neural Crest. Trends Neurosci.* 12: 38–39.

Tremblay, P., Kessel, M. and Gruss, P. 1995. A transgenic neuroantomical marker identifies cranial neural crest deficiencies associated with the *Pax3* mutant *Splotch. Dev. Biol.* 171: 317–329.

Turner, D. L. and Cepko, C. L. 1987. A common progenitor for neurons and glia persists in rat retina late in development. *Nature* 328: 131–136.

Vogel, K. S. and Weston, J. A. 1990. The sympathoadrenal lineage in avian embryos. II. Effects of glucocorticoids on cultured neural crest cells. *Dev. Biol.* 139: 13–23.

Van Allen, M. I. and fifteen others. 1993. Evidence for multi-site closure of the neural tube in humans. *Am. J. Med. Genet.* 47: 723–743.

van Straaten, H. W. M., Hekking, J. W. M., Wiertz-Hoessels, E. J. L. M., Thors, F. and Drukker, J. 1988. Effect of the notochord on the differentiation of a floor plate area in the neural tube of the chick embryo. *Anat. Embryol.* 177: 317–324.

Varley, J. E., Wehby, R. G., Rueger, D. C. and Maxwell, G. D. 1995. Number of adrenergic and islet-1 immunoreactive cells is increased in avian trunk neural crest cultures in the presence of human recombinant osteogenic protein-1. *Dev. Dyn.* 203: 434–447.

Vassar, R. and Fuchs, E. 1991. Transgenic mice provide new insights into the role of TGF-a during epidermal development and differentiation. *Genes Dev.* 5: 714–727.

von Baer, K. E. 1828. *Entwicklungsgeschichte der Thiere: Beobachtung und Reflexion.* Bornträger, Konigsberg.

Walsh, C. and Cepko, C. L. 1988. Clonally related cortical cells show several migration patterns. *Science* 241: 1342–1345.

Walsh, C. and Cepko, C. L. 1992. Widespread dispersion of neuronal clones across functional regions of the cerebral cortex. *Science* 255: 434–440.

Webster, E. H., Silver, A. F. and Gonsalves, N. I. 1984. The extracellular matrix between the optic vesicle and the presumptive lens during lens morphogenesis in an anophthalmic strain of mice. *Dev. Biol.* 103: 142–150.

Weinstein, D. C., Rahman, S. M., Ruiz, J. C. and Hemmati-Brivanlou, A. 1996. Embryonic expression of EPH signaling factors in *Xenopus. Mech. Dev.* 57: 133–144.

Weinstein, G. D. and van Scott, E. J. 1965. Turnover times of normal and psoriatic epidermis. *J. Invest. Dermatol.* 45: 257–262.

Weston, J. 1963. A radiographic analysis of the migration and localization of trunk neural crest cells in the chick. *Dev. Biol.* 6: 274–310.

Weston, J. A. 1991. Sequential segregation and fate of developmentally restricted intermediate cell populations in the neural crest lineage. *Curr. Top. Dev. Biol.* 25: 133–153.

Witte, O. N. 1990. *Steel* locus defines new multipotent growth factor. *Cell* 63: 5–6.

Yamada, K. M., Spooner, B. S. and Wessells, N. K. 1971. Ultrastructure and function of growth cones and axons of cultured nerve cells. *J. Cell Biol.* 49: 614–635.

Yamada, T., Pfaff, S. L., Edlund, T. and Jessell, T. M. 1993. Control of cell pattern in the neural tube: Motor neuron induction by diffusible factors from notochord and floor plate. *Cell* 73: 673–686.

Yamamori, T., Fukada, K., Aebersold, R., Korsching, S., Fann, M.-J. and Patterson, P. H. 1989. The cholinergic neuronal differentiation factor from heart cells is identical to leukemia inhibitory factor. *Science* 246: 1412–1416.

Yuodelis, C. and Hendrickson, A. 1986. A qualitative and quantitative analysis of the human fovea during development. *Vision Res.* 26: 847–855.

Axonal specificity

8

Thus, beyond all questions of quantity there lie questions of pattern which are essential for understanding Nature.
ALFRED NORTH WHITEHEAD (1934)

Like the entomologist in search of brightly colored butterflies, my attention hunted, in the garden of the gray matter, cells with delicate and elegant forms, the mysterious butterflies of the soul.
S. RAMÓN Y CAJAL (1937)

NOT ONLY DO NEURONAL PRECURSOR CELLS MIGRATE to their place of function, but so do their axons. Unlike most cells, whose parts all stay in the same place, the nerve cell is able to elongate axons that may extend meters. The axon has its own locomotory apparatus that resides in the growth cone, and the growth cone can respond to the same types of signals that migrating cells can sense. Thus, axon movement can be directed by chemotaxis, galvanotaxis, and contact guidance, just like migrating cells. The cues for axonal migration, moreover, may be even more specific than those used to get certain cell types to particular areas. The human brain, for instance, is the most ordered piece of matter known. Each of the 10^{11} neurons has the potential to specifically interact with thousands of other cells, and a large neuron (such as a Purkinje cell or motor neuron) can receive input from over 10^5 other cells (Figure 8.1; Gershon et al., 1985). Understanding this generation of ordered complexity is one of the greatest challenges to modern science.

Goodman and Doe (1993) list eight stages of neurogenesis: (1) induction and patterning of a neuron-forming (neurogenic) region; (2) the birth and migration of neurons and glia; (3) the generation of specific cell fates; (4) guidance of the growth cones to specific targets; (5) formation of synaptic connections; (6) binding of trophic factors for survival and differentiation; (7) competitive rearrangement of functional synapses; and (8) continued synaptic plasticity during the organism's life. The first two of these processes were the topics of the previous chapter. Here, we continue investigating the processes of neural development. [axon1.html]

The generation of neuronal diversity

Neurons are fashioned in a hierarchical manner. The first decision is whether a given cell is to be a neuron or anything else. If the cell is to become a neuron, a further decision informs the neuron as to its type. Is it to become a motor neuron, a sensory neuron, a commissural neuron, or some other type? After this fate is determined, still another decision is made that gives the neuron a specific target. To illustrate this progressive specification, we will focus on the motor neurons of vertebrates and *Drosophila*.

Figure 8.1
Connections of axons to a cultured hippocampal neuron. The neuron has been outlined by the synaptic protein synaptotagmin, which is present in the terminals of axons that contact the neuron. (Courtesy of M. Matteoli and P. De Camilli.)

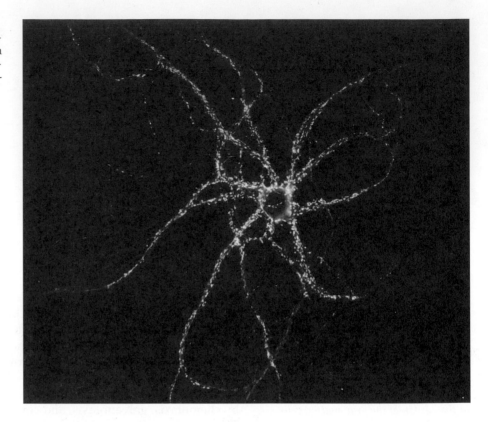

Vertebrate Motor Neuron Specification

Vertebrates form a *dorsal* neural tube, while invertebrates, such as *Drosophila*, form a *ventral* neural tube. Yet, the specification of the neural ectoderm is mediated by the binding of similar proteins. In *Xenopus* (and probably other vertebrates), the notochord secretes **Chordin** and **Noggin** proteins. These proteins bind **BMP4,** and the ectoderm in the vicinity of the notochord develops the capacity to form neurons. If the ectoderm is exposed to BMP4, it becomes epidermal (Sasai et al., 1995; Piccolo et al., 1996; see Chapter 16). In the absence of BMP4 stimulation, the vertebrate ectodermal cells are thought to synthesize a transcription factor (or set of transcription factors) that commits the cells to a neural lineage. Later, the cells will synthesize other proteins (such as NeuroD) that cause them to express their neuronal phenotype (Turner and Weintraub, 1994; Lee et al., 1995).

The decision as to the *type* of neuron appears to be controlled by the position of the neuronal precursor within the neural tube and when it undergoes its last cell division. As described in the previous chapter, neurons at the ventral-lateral margin become the motor neurons, while the sensory neurons are derived from cells at the dorsal region of the tube. Since the grafting of ectopic floor plate cells or notochord (which secrete Sonic hedgehog protein) to lateral areas can respecify dorsolateral cells as motor neurons, this decision as to type of neuron is probably a function of its position relative to the floor plate. Ericson and colleagues (1996) have shown that two periods of Sonic hedgehog signaling are needed to specify the motor neurons: an early period wherein the cells are instructed to become ventral neurons and a later period (which includes the S phase of its birthday division) that specifies that the ventral neuron is to become a motor neuron (rather than an interneuron). The first phase is probably regulated by the secretion of Sonic hedgehog from the notochord, while the latter stage is more likely regulated by the floor plate cells. Sonic hedgehog appears to specify motor neurons by inducing the transcription factor Islet-1. This protein is found in all motor neurons, but in no other neuronal type (Ericson et al., 1992; see Plate 32). Another factor affecting neuronal type is the cell's age when it last divides. As discussed in the previous chapter, a cell's birthday determines which layer of the cortex it will enter.

The next decision involves target specificity. If a cell is to become a neuron and, specifically, a motor neuron, will that motor neuron be one that innervates the thigh, the forelimb, or the tongue? The determination of specificity appears to be regulated by the position of the motor neuron along the anterior-posterior and medial-lateral axes of the neural tube. The neural tube has a distinct anterior-posterior polarity from forebrain through spinal cord. In Chapter 16, we will discuss the *Hox* gene complex that determines this polarity within the spinal cord and which gives target specificity to the respective motor neurons. These genes work in combination to define the positional identity of each region of the embryo. Landmesser (1978) and Hollyday (1980b) showed that motor neurons having the same specificity are grouped together. The cell bodies of motor neurons projecting to a single muscle are clustered in a longitudinal column called a "pool." Pools are grouped together into larger columns according to their target. Motor neurons in the Column of Terni (CT) project ventrally into the sympathetic ganglia. Motor pools of the lateral motor column (LMC) extend to the limb musculature, while those motor neurons in the medial motor column (MMC) project into the axial muscles. The limb and axial columns are subdivided along the medial-lateral axis in a manner that correlates with the dorsal-ventral position of their respective targets (Figure 8.2; Tosney et al., 1995). This arrangement of motor neurons is constant throughout the vertebrates. [axon2.html]

The target specificities of these motor neurons are specified before their axons extend into the periphery. This was shown by Lance-Jones and Landmesser (1980), who reversed segments of the chick spinal cord so that the motor neurons were in new locations. The axons went to their *original* targets, not to the ones expected from their new positions (Figure 8.3). The molecular basis for this specificity may reside in members of the LIM family of proteins (see Figure 8.2; Tsushida et al., 1994). The LIM family includes Islet-1, Islet-2, LIM-1, LIM-2, and LIM-3, and each of these proteins is a tran-

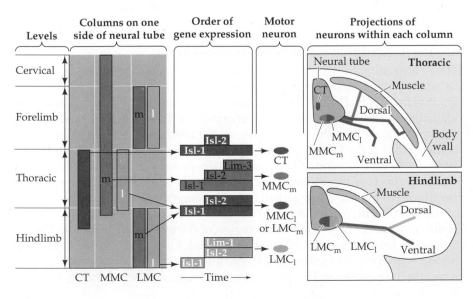

Figure 8.2
Motor neuron organization and LIM specification. On the left is half the spinal cord. Neurons in these columns display specific sets of LIM genes, and neurons within each column make similar pathfinding decisions. CT motor neurons project ventrally into the sympathetic ganglia. The MMC column projects into the axial muscles, and the LMC sends axons to the limb musculature. Where these columns are subdivided, medial (m) subdivisions project into ventral positions and lateral (l) subdivisions send axons to the dorsal regions of the target tissues. (After Tsushida et al., 1994; Tosney et al., 1995.)

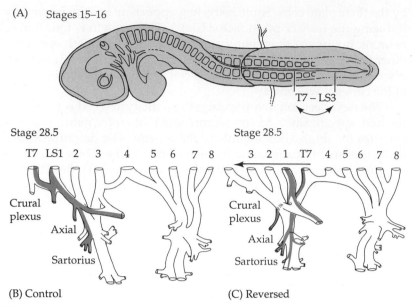

Figure 8.3
Compensation for small dislocations of axonal initiation position in the chick embryo. (A) A length of spinal cord comprising several segments T7–S3 (seventh thoracic to third lumbosacral segments) is reversed in the 2.5-day embryo. (B) Normal pattern of axon projection into the different muscles at 6 days. (C) Projection of axons in the reversed segment. The ectopically placed neurons eventually found their proper neural pathways and innervated the appropriate muscles. (From Lance-Jones and Landmesser, 1980.)

scription factor. LIM proteins have been implicated in specifying cell fate in nematodes (where the LIM gene *mec-3* specifices a touch receptor neuron; Way and Chalfie, 1988) and are important in brain development in mice (Shawlot and Behringer, 1995). For instance, all motor neurons express *Islet-1* and (slightly later) *Islet-2*. If no other of these LIM genes is expressed, the neurons project to the ventral body wall muscles. Those neurons in the medial column of the MMC also express *LIM-3*, distinguishing them from the other motor neurons. The lateral pools of the LMC columns are distinguished by their short expression of *LIM-1*, while the CT motor neurons cease to express *Islet-2*. Thus, each projection is characterized by a particular constellation of *LIM* transcription factors.

Motor Neuron Specification in Drosophila

The specification of neural ectoderm in vertebrates and arthropods appears to be conducted in surprisingly similar ways. The specification of neurogenic ectoderm in *Drosophila* involves the secretion of the *Drosophila* homologue of Chordin, the Short-gastrulation protein. This protein is made by the ventrolateral cells of the blastoderm, and it binds to the *Drosophila* homologue of BMP4, the Decapentaplegic protein (see Figure 15.23 ; Holley et al., 1995).Those cells secreting Short-gastrulation protein are spared from the lateralizing effects of Decapentaplegic, and they become capable of forming the ventral nerve cord. During gastrulation, the vegetalmost cells, the precursors of the mesoderm, invaginate into the yolky blastocoel, causing the neurogenic ectoderm to be in the ventral region of the embryo (Figure 8.4).The ectoderm delaminates about 60 cells (30 per side) into the embryo, and these (along with the cells in the ventral midline) are the precursors of the neurons, the **neuroblasts.** The commitment to become ectoderm is a conse-

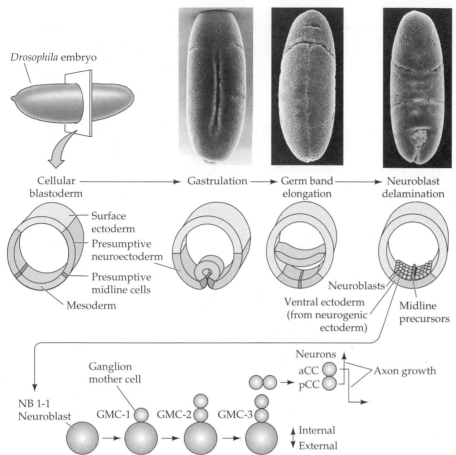

Drosophila embryo

Cellular → Gastrulation → Germ band → Neuroblast
blastoderm   elongation   delamination

Surface
ectoderm
Presumptive
neuroectoderm
Presumptive
midline cells
Mesoderm

Neuroblasts

Ventral ectoderm
(from neurogenic
ectoderm)
Midline
precursors

Neurons
aCC
pCC → Axon growth

Ganglion
mother cell

NB 1-1
Neuroblast
GMC-1   GMC-2   GMC-3

Internal
External

Figure 8.4
Development of the neurogenic region
of insects. In the blastoderm, the presumptive neuroectoderm is located on
either side of the mesodermal precursors. During gastrulation and germ
band extension, the mesoderm invaginates from the surface into the embryo.
The midline neural precursor cells are
now the most ventral cells of the embryo. The ectoderm delaminates neuroblasts into the embryo (along with the
ventral midline cells) to form the central
nervous system. The neuroblasts generate a series of ganglion mother cells,
each of which generates two neurons. In
this case, the 1-1 neuroblast is shown.
(After Goodman and Doe, 1993.)

quence of the position along the dorso-ventral axis of the embryo and will
be discussed in subsequent chapters. The commitment to become a neuroblast instead of an epidermal cell is made by a group of genes called the
proneural genes (Figure 8.5). These are a set of transcription factors that are
found in clusters of about four to six cells in the ectodermal region.* Each
cluster forms a zone of interaction wherein one (and only one) of the cells
becomes a neuroblast. Once a cell has been committed to form a neuroblast,
it inhibits the other cells of its cluster from becoming neuroblasts. This is accomplished through the interaction of a group of genes called the **neurogenic genes.** The proteins Notch and Delta are critical in these reactions.
These proteins interact on the cell membrane. Their interactions suggest that
the cell that is destined to become a neuroblast upregulates its Notch protein, which causes its neighbors to downregulate their Notch proteins. This
decision is communicated through the Delta proteins. In this way, the neuroblast **laterally inhibits** other cells of the cluster from becoming neuroblasts (see Chapter 17). [axon3.html]

Like the situation in vertebrates, neuroblast specification in *Drosophila* is
accomplished by the combinatorial expression of different genes. (Interestingly, these genes had been used earlier to specify each region of *Drosophila*
blastoderm.) If any of these genes are not able to function, the neuroblasts

---

*These transcription factors are members of the achaete-scute family. Interestingly, some of
the transcription factors involved in vertebrate neural determination are also members of this
family (Turner and Weintraub, 1994).

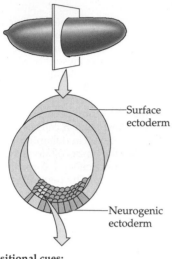

Figure 8.5
Sequential specification of the neuroblast lineage. (A) The neurogenic ectoderm is specified by positional cues along the dorsal-ventral and anterior-posterior axes. (B, C) Clusters of potential neuroblasts are specified by the proneural genes such as *achaete* (shown in F). (D) Interaction between the potential neuroblasts selects one cell from the cluster to be the neuroblast, and this cell inhibits the other cells of the cluster from becoming neuroblasts. (E) The neuroblast buds off ganglion mother cells (in a manner that will be discussed in Chapter 13), each of which forms two neurons. (F) *Drosophila* embryo stained for the *achaete* transcript. The neurogenic clusters express this gene. The bracket indicates a domain of neurogenic activity. (After Goodman and Doe, 1993; photograph from Skeath and Carroll, 1992; courtesy of J. Skeath.)

(A) Positional cues:
segmentation genes (A/P), dorsal/ventral genes

(B) Neuroblast specification:
neuroblast identity genes

(C) Neuroblast formation:
proneural genes

(D) Lateral inhibition:
neurogenic genes

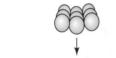

(E) Neuroblast cell lineage:
ganglion mother cells and neural identity genes

Ganglion mother cells — Neurons — Neuroblast

(F)

behave as if they were other types of neurons, often forming nerves that send their axons to the wrong targets (Chu-LaGraff et al., 1995). [axon4.html]

Pattern formation in the nervous system

The functioning of the vertebrate brain depends not only on the differentiation and positioning of the neural cells, but also on the specific connections these cells make among themselves and their peripheral targets. In some manner, nerves from a sensory organ such as the eye must connect to specific neurons in the brain that can interpret visual stimuli, and axons from the nervous system must cross large expanses of tissue before innervating the appropriate target tissue. How does the nerve axon "know" to traverse numerous other potential target cells to make its specific connection? Harrison (1910) suggested that the specificity of axonal growth is due to pioneer nerve fibers, which go ahead of other axons and serve as guides for them.* This observation simplifies, but does not solve, the problem of how neurons form the appropriate patterns of interconnections. Harrison also noted, however, that axons must grow on a solid substrate, and he speculated that differences in the embryonic surfaces might allow axons to travel in certain specified directions. The final connections would occur by complementary interactions on the cell surface:

> That it must be a sort of a surface reaction between each kind of nerve fiber and the particular structure to be innervated seems clear from the fact that sensory and motor fibers, though running close together in the same bundle, nevertheless form proper peripheral connections, the one with the epidermis and the other with the muscle...The foregoing facts suggest that there may be a certain analogy here with the union of egg and sperm cell.

Research on the specificity of neuronal connections has focused on two major types of systems: motor neurons, whose axons travel from the nerve to a specific muscle, and the optic system, wherein axons originating in the retina find their way back into the brain. In both cases, the specificity of axonal connections is seen to unfold in three steps (Goodman and Shatz, 1993):

- *Pathway selection*, wherein the axons travel along a route that leads them to a particular region of the embryo.

*The growth cones of pioneer neurons migrate to their target tissue while embryonic distances are still short and the intervening embryonic tissue is still relatively uncomplicated. Later in development, the other neurons that innervate the target tissue bind (fasciculate) to the pioneer neuron and thereby enter the target tissue. Klose and Bentley (1989) have shown that in some cases, the pioneer neurons die after the other neurons reach their destination. Yet, were that pioneer neuron prevented from differentiating, the other axons would not have reached their target tissue.

- *Target selection*, wherein the axons, once they reach the correct area, recognize and bind to a set of cells with which they may form stable connections.
- *Address selection*, wherein the initial patterns are refined such that each axon binds to a small subset (sometimes only one) of its possible targets.

The first two processes are independent of neuronal activity. The third process involves the interactions between several active neurons and converts the overlapping projections into a fine-tuned pattern of connections. It has been known since the 1930s that motor axons can find their appropriate muscles even if the neural activity of the axons is blocked. Twitty (who had been Harrison's student) and his colleagues found that the embryos of the newt *Taricha torosa* secreted a toxin, tetrodotoxin, that blocked neural transmission in other species. By grafting pieces of *T. torosa* onto other salamander embryos, they were able to paralyze the host embryos for days while development occurred. About the time the tadpoles were to feed, the toxin wore off, and the salamanders swam and fed normally (Twitty and Johnson, 1934; Twitty, 1937). More recent experiments using zebrafish mutants having nonfunctional neurotransmitter receptors similarly demonstrated that motor neurons establish their normal patterns of innervation in the absence of neuronal activity (Westerfield et al., 1990).

But the question remains, How are the axons instructed where to go? As mentioned in Chapter 3, migratory cells get their cues from diffusible substances, ions, or the extracellular matrix over which they travel. The growth cone is able to respond to the same types of cues and leads the axon from the neural cell soma to its target tissue. It should be remembered from the preceding chapter that the growth cone pulls the axon forward. The axon is not extended by pushes coming from the cell body.

Pathway selection: Guidance by the extracellular matrix

The extracellular matrix can provide information for navigation in many forms—some more specific than others. Channels and folds in the extracellular matrix can restrict the path of axon growth to a certain region; this is a very crude type of guidance. In addition, certain basal lamina proteins may be more adhesive than others and bias axon movement along the basement membrane. Finally, molecules in the extracellular matrix may actively repel specific axons, causing the collapse of the growth cone. As we will see, all of these mechanisms appear to be working in the embryo.

Guidance by the Physical Terrain: Contact Guidance

One of the first hypotheses to account for the specificity of axonal growth involves **contact guidance,** or **stereotropism.** Here, physical cues in the substratum direct neural growth. Harrison developed a technique of growing axons on clotted blood, and using this technique, Weiss (1955) noted that the growing axons not only needed a solid substrate on which to migrate but also that the migration tended to follow discontinuities in the clot. When the fibers of the blood clot were randomly oriented, the axons followed this random pattern. But when the clot fibers were made parallel by applying tension to the clot, the nerve axons traveled along these fibers, not veering from the straight and narrow (see Figure 3.31). Singer and his co-workers (1979) found evidence that such physical factors operate in vivo to guide the growth cones. They detected large channels between ependymal cells of the developing newt spinal cord through which growing axons migrated. They hypothesized that these channels provide cues for guiding the axons toward the appropriate regions of the brain. Cellular channels have also been detected in the mouse retina (Silver and Sidman, 1980), and these appear to guide the retinal ganglion cell growth cones into the optic stalk as they develop.

The presence of preexisting channels is probably not critical for the growth of most axons. The growth cone appears capable of digesting its own channels through an extracellular matrix by secreting proteolytic enzymes into its immediate environment (Pittman, 1985).

Guidance by Adhesive Gradients: Haptotaxis

A growing axon's growth cone encounters numerous microenvironments, and some sites may contain molecules that are more adhesive than molecules found in other sites. The ability of a growth cone (or a cell) to migrate up a gradient of adhesivity is called **haptotaxis.** The growth cone has receptors that recognize proteins found in certain basal laminae, and the growth cone leads the axon along the paths coated by these proteins. The **differential adhesive specificity hypothesis** postulates that the growth cone will encounter a patchy environment and recognize its path by having particular receptors for certain molecules in the environment. This can be seen in vitro. When placed into culture, a piece of neural retina tissue will not readily send forth axons onto the plastic dish. However, when the plastic is coated with fibronectin or laminin, long axonal outgrowths are observed (Figure 8.6).

(A)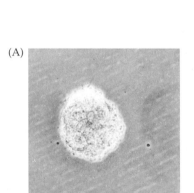

(B)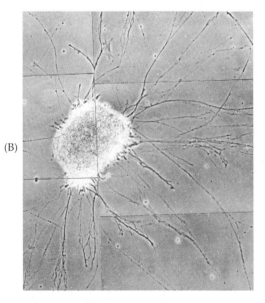

Figure 8.6
Effects of substrate factors on neural outgrowth. (A,B) Effects of fibronectin on neural outgrowth from neural retina aggregates. The aggregate in (A) was cultured 36 hours on untreated tissue culture plastic. The aggregate in (B) was cultured on plastic treated with 50 μg fibronectin per milliliter. (C) Outgrowth of sensory neurons placed on patterned substrate consisting of parallel stripes of laminin applied to a background of type IV collagen. (A and B from Akers et al., 1981, courtesy of J. Lilien; C from Gundersen, 1987, courtesy of R. W. Gundersen.)

(C)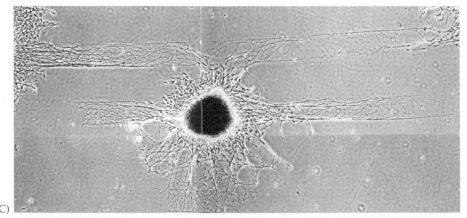

Conversely, glycosaminoglycans, another set of molecules associated with extracellular matrices, appear to impede these neural outgrowths (Tosney and Landmesser, 1985).

The presence of such molecules delineates pathways through the embryo (Akers et al., 1981; Gundersen, 1987), and many of the roads on which axons travel appear to be paved with laminin. Letourneau and co-workers (1988) have shown that the axons of certain spinal neurons travel through the neuroepithelium over a transient laminin-coated surface that precisely marks the path of these axons. Similarly, there is a very good correlation between the elongation of retinal axons and the presence of laminin on the neuroepithelial cells and astrocytes in the embryonic mouse brain (Cohen et al., 1986, 1987; Liesi and Silver, 1988). Punctate laminin deposits are seen on the glial cell surfaces along the pathway leading from the retina to the optic tectum, whereas adjacent areas where the optic nerve fails to grow lack these laminin deposits. After the retinal axons have reached the tectum, the glial cells differentiate and lose their laminin. At this point, the retinal ganglial neurons that have formed the optic nerve lose their integrin receptor for laminin. Laminin deposits may also be necessary for the regeneration of neural tissue. Astroglial cells bearing punctate laminin on their surfaces can induce regeneration when placed into embryos where the normal neuronal pathways of the corpus callosum have been severed.

There are at least four regions of the laminin glycoprotein that can support axonal migration and growth (Figure 8.7). First, integrins in the growth cone can bind to the RGD sequence of laminin protein. Second, another receptor in the growth cone can recognize the amino acid sequence YIGSR in laminin, while a isoleucine-rich 10-amino acid region of the B2 peptide is critical for neurite outgrowth of some neurons (Matsuzawa et al., 1996). The fourth growth cone receptor for laminin is a glycosyltransferase that recognizes particular carbohydrate side chains on the laminin molecule (Begovac and Shur, 1990; Thomas et al., 1990). These carbohydrates may reside in the "neurite outgrowth" domain of the laminin A chain.

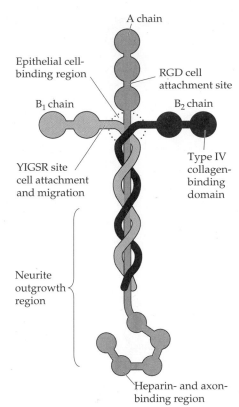

Figure 8.7
Structure of laminin and proposed binding regions.

Guidance by Axon-Specific Migratory Cues: The Labeled Pathways Hypothesis

Because extracellular matrix molecules such as laminin and N-CAM are found in several places throughout the embryo, they can usually provide only general cues for growth cone movement. It would be difficult for such general molecules to direct the growth cones of several different types of neurons in several different directions. Yet in *Drosophila*, grasshoppers, and *Caenorhabditis* (and probably most invertebrates), the patterning of axonal movement is an astoundingly precise process, and adjacent axons are given different migratory instructions by their environment. For example, within each segment of the grasshopper, 61 neuroblasts emerge (30 on each side and 1 in the center). One of them, the 7-4 neuroblast, is a stem cell and gives rise to a family of six neurons, termed C, G, Q1, Q2, Q5, and Q6. This family of neurons is shown in Figure 8.8 and as the yellow neurons in Plate 20. The axonal growth cones of these neurons reach their targets by following specific pathways formed by other earlier neurons. Q1 and Q2 follow a straight path together, traversing numerous other cells, until they meet the axon from the dorsal midline precursor neuron (dMP2), which they then follow posteriorly. The other four neurons of the 7-4 family migrate across the dMP2 axon as if it did not exist. Axons from the C and G neurons progress a long way together, but ultimately, C follows nerves X1 and X2 toward the posterior of the segment, while G adheres to the P1 and P2 axons (which go posteriorly) and moves anteriorly on their surfaces (Goodman et al., 1984; Taghert et al., 1984).

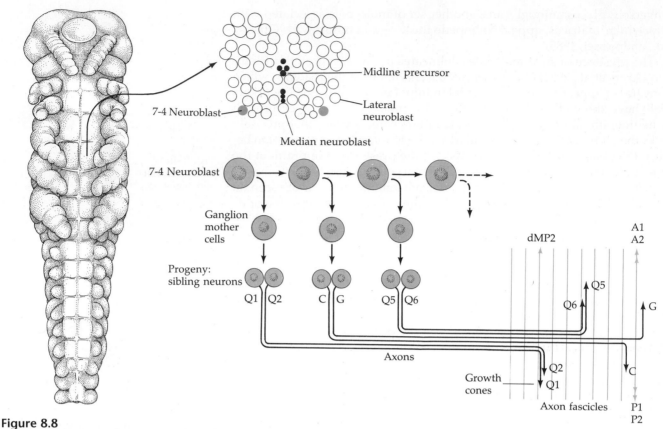

Figure 8.8
Each of the 17 segments of the early grasshopper embryo has the same pattern of neuroblasts. There are 30 lateral neuroblasts on each side, 1 median neuroblast, and 7 midline precursors. The midline neuroblasts divide once, while the lateral neuroblasts are stem cells that divide repeatedly to form "ganglion mother cells." Each of the cells divides once to yield two sibling neurons. Neuroblast 7-4 has nearly 100 neuronal progeny, of which the first 6 are shown here. (After Goodman and Bastiani, 1984.)

The G growth cone will have encountered over 100 different surfaces to which it could have adhered, but it is specific for the P neurons. If the P neurons are ablated by laser, the G growth cone acts abnormally, its filopodia searching randomly for its proper migratory surface. If any of the other hundred or so axons are destroyed, the G growth cone behaves normally.

This formulation of axonal pathfinding in insects has been called the **labeled pathways hypothesis** because it implies that a given neuron can specifically recognize the surface of another neuron that has grown out before it. Evidence for this specificity comes from studies using monoclonal antibodies (Bastiani et al., 1987). Neurons aCC and pCC are sibling neurons in the grasshopper (both are derived from neuroblast 1-1) that have very different fates. Moreover, different sets of axons adhere to each of them, creating independent bundles of axons, called **fascicles**. The specificity of this fasciculation depends on the presence of the protein **fasciclin I**. Fasciclin I is found on the two aCC neurons of each segment in the 10-hour embryo. This protein is not present on the pCC neurons. By hour 11, however, other neurons (but not pCC) are seen to express this cell-surface molecule. These neurons are precisely those (RP1, RP2, U1, U2) whose axons fasciculate with aCC. There are at least four fasciclin molecules expressed on different subsets of neurons, and each of these molecules allows the growth cones of certain neurons to recognize specifically those axons with which they will fasciculate (Harrelson and Goodman, 1988; Zinn et al., 1988).

In other animals with relatively simple nervous systems, such as the leech, there is evidence that each neuron might have qualitatively different cell-surface molecules and that these molecules might be important in synaptic specificity. The nervous system of the leech consists of 34 paired

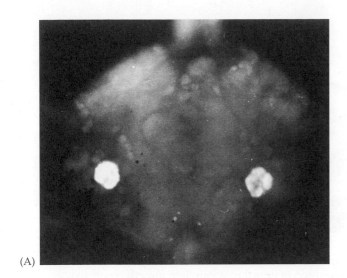

(A)

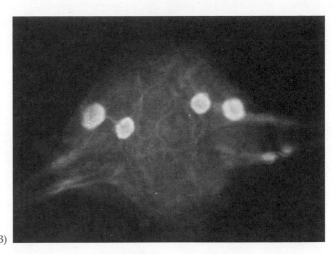

(B)

Figure 8.9
Specific functional neurons stained by monoclonal antibodies to cell-surface components. (A) Lan 3-1 antibodies recognize a single pair of neurons in a particular ganglion. These neurons function in penile eversion. (B) A set of neurons recognized by Lan 3-2 antibodies; these neurons respond to noxious stimulation of the leech's skin. (From Zipser and McKay, 1981, courtesy of B. Zipser.)

ganglia containing about 400 neurons each. Individual neurons have been identified, and the functions of many of these neurons are known. Zipser and McKay (1981) injected the leech nervous system into mice and obtained hundreds of monoclonal antibodies that bound to various regions of the nervous system. In some cases, such differences could be correlated with function. Monoclonal antibody Lan 3-1 bound specifically to a single pair of neurons in each of the midbody ganglia (Figure 8.9). These pairs of neurons are known to control the process of penile eversion in mating leeches. Another monoclonal antibody, Lan 3-2, recognized all four neurons in each ganglion that respond to noxious mechanical stimuli. "The situation," according to Zipser and McKay, "seems quite analogous to colour-coded electrical cable containing many wires, where each wire has its own molecule (dye) to facilitate proper recognition and connection at terminals."

Studies on specifically labeled pathways in vertebrates lag far behind those on invertebrates, but recent studies on zebrafish motor neurons suggest that labeled pathways may function here as well. Zebrafish may become the organism of choice in vertebrate developmental neurobiology because their development is very rapid, numerous individuals can be compared, and the embryos are clear, enabling neurobiologists to observe the growth of axons in living embryos. Neurons can be identified by injecting fluorescently labeled substances into the neuronal precursors (Kimmel and Law, 1985), and axon growth can be followed by eye or with a video recorder. Eisen and her colleagues (1986) watched the axonal elongation of three pioneer motor neurons in these embryos. After the axons left the spinal cord, all three followed the same path along a muscle until they reached a particular site in the embryo. At this point, they diverged into three neuron-specific pathways that led to the appropriate muscles. The labeled pathways hypothesis has been extremely important both as a model for generating research and as a context in which to place existing data about neuronal specificity.

Guidance by Specific Growth Cone Repulsion

In addition to specific adhesion, there is also the possibility of *specific repulsion* by the extracellular matrix. Axons from dorsal root ganglion neurons in the trunk pass through only the anterior portion of each somite area (just as the neural crest cells will migrate through only these regions and not the posterior regions) (Figure 8.10A). The cell surface of the posterior portion of

(A)

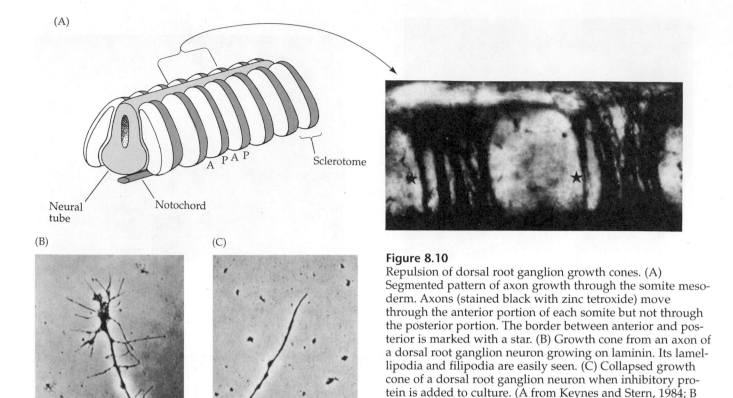

Neural tube

Notochord

Sclerotome

(B)

(C)

Figure 8.10
Repulsion of dorsal root ganglion growth cones. (A) Segmented pattern of axon growth through the somite mesoderm. Axons (stained black with zinc tetroxide) move through the anterior portion of each somite but not through the posterior portion. The border between anterior and posterior is marked with a star. (B) Growth cone from an axon of a dorsal root ganglion neuron growing on laminin. Its lamellipodia and filipodia are easily seen. (C) Collapsed growth cone of a dorsal root ganglion neuron when inhibitory protein is added to culture. (A from Keynes and Stern, 1984; B and C from Raper and Kapfhammer, 1990. All photographs courtesy of the authors.)

the somite may be inhibiting this migration. Davies and colleagues (1990) have shown that membranes isolated from the posterior portion of the somite cause the collapse of growth cones of the dorsal root ganglion neurons (Figure 8.10B,C). Moreover, they have isolated a glycoprotein fraction from chick somites that causes these growth cones to collapse, and components of this fraction are specifically found on the posterior portion of the somites. In insects, **semaphorin I** (also known as fasciclin IV) is a transmembrane protein that is expressed in a band of epithelial cells in the developing limb. This protein appears to inhibit the growth cones of the Ti1 sensory neurons from moving forward, causing them to turn (Figure 8.11; Kolodkin et al., 1992, 1993).

Figure 8.11
The action of semaphorin I in the developing grasshopper limb. Axons from sensory neuron Ti1 project toward the central nervous system. (The long dark arrows represent sequential steps en route. When they reach the band of semaphorin-I-expressing epithelial cells, they reorient their growth cones and extend ventrally along the distal boundary of the semaphorin-I-expressing cells. When their filopodia connect to the Cx1 pair of cells, they cross the boundary and project into the CNS. When semaphorin I is blocked by antibodies, the growth cones search randomly for the Cx1 cells. (After Kolodkin et al., 1993.)

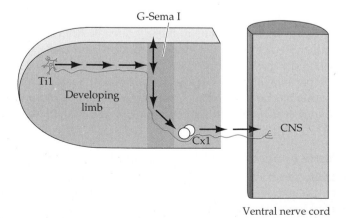

G-Sema I

Ti1

Developing limb

Cx1

CNS

Ventral nerve cord

Sex, Smell, and Specific Adhesion

IN THE LATE NINETEENTH CENTURY, the Johns Hopkins University professor John Mackenzie (1898), the German psychiatrist Wilhelm Fliess (1897), and the Viennese sexologist Richard von Krafft-Ebing (1886) all shared the mistaken view that there were similarities in the development of the penis and the nose. All three of these investigators used the same case study as evidence: the report of a man who had no sense of smell—no nasal or olfactory nerves— and whose genitalia were much smaller than normal.

Such people are now known to have **Kallmann syndrome,** an X-linked disease characterized by anosmia (no sense of smell), small genitalia, and sterile gonads. The anosmia is due to the lack of neurons in the brain that receive input from the axons coming from nasal neurons. The small gonads and genitalia are the result of a lack of gonadotropin-releasing hormone (GnRH). GnRH is a peptide hormone secreted by the *hypothalamus* that instructs the anterior pituitary to secrete luteinizing hormone, the hormone required for gonadal development and genital maturation. What links these two problems? In 1989, two laboratories (Schwanzel-Fukada and Pfaff, 1989; Wray et al., 1989) made the sur-

prising discovery that the GnRH-secreting neurons do not originate in the hypothalamus. Rather, they originate in the olfactory epithelium (the vomeronasal organ) in the nose rudiment and *migrate* into the hypothalamic region of the brain during fetal development (Figure 8.12). The olfactory receptor neurons of the nose originate from the same place. The axons from the olfactory receptor neurons enter the brain to synapse with the olfactory bulb, while the cell bodies of these neurons remain in the developing nose. Patients with Kallmann syndrome have no olfactory bulb in the brain, as the development of this bulb requires innervation from the olfactory receptor neurons (Stout and Gradziadi, 1980).

The defect in Kallman syndrome can be traced to the failure of the GnRH-secreting neurons and the olfactory neuron growth cones to migrate into the brain from the olfactory placode (Schwanzel-Fukada et al., 1989). It is thought that the olfactory axons migrate first and that the GnRH-secreting neurons follow the ol-

factory nerve fascicles into the brain (Livne et al., 1993). The gene whose absence or abnormality causes the syndrome has been cloned, and its cDNA sequence predicts a cell adhesion protein of the immunoglobulin superfamily (Franco et al., 1991; Legouis et al., 1991). Members of this class of proteins are known to mediate cell-cell or axon-axon adhesion (Grumet, 1991), and they include N-CAM, L1, LFA-1, CD4, fasciclin II, contactin, and neuroglian. The Kallmann syndrome protein also contains regions that resemble the fibronectin molecule, a molecule of the extracellular matrix that is critically important in numerous cell migrations during development. However, the tests to see if this protein is on the tracts followed by the migrating cells and elongating axons from the olfactory epithelium have not yet been done in mammals, nor has it been determined that the axons or cells from this region actually bind to this protein. ■

Figure 8.12 Model for the etiology of Kallman syndrome. In the left illustration, sensory neurons from the olfactory epithelium extend axons into the olfactory bulb of the brain. In Kallmann syndrome, the olfactory bulb has degenerated, and this loss is thought to be secondary to the lack of axons from the sensory neurons. The series of sagittal head sections from embryonic mice shows the migration of GnRH-secreting neurons (color) from the nose anlagen into the hypothalamic portion of the brain. This migration does not occur in Kallmann syndrome. (After Calof, 1992.)

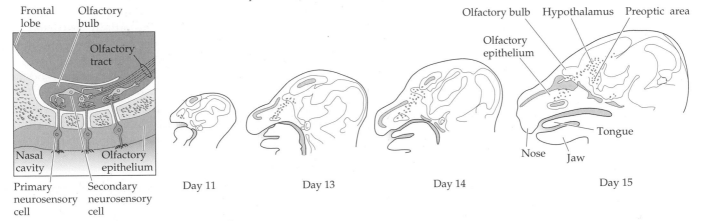

(A)

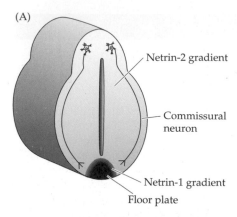

Netrin-2 gradient

Commissural neuron

Netrin-1 gradient

Floor plate

(B)

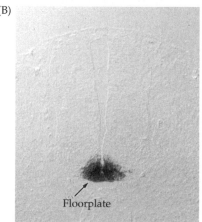

Floorplate

Figure 8.13
Trajectory of the commissural axons in the rat spinal cord. (A) Schematic drawing of a model wherein commissural neurons first experience a gradient of netrin-2 and then a steeper gradient of netrin-1. The commissural axons are chemotactically guided ventrally down the lateral margin of the spinal cord toward the floor plate. Upon reaching the floor plate, the commissural axons change their direction owing to contact guidance from the floor plate cells. (B) Autoradiographic localization of netrin-1 mRNA by in situ hybridization to the hindbrain of a stage 16 rat embryo using antisense RNA. Hybridization gives an intense signal from the floor plate neurons. (B from Kennedy et al., 1994; photograph courtesy of M. Tessier-Lavigne.)

Pathway selection: Guidance by diffusible molecules

The idea that chemotactic cues guide axons in the developing nervous system was first proposed by Ramón y Cajal (1892). He suggested that the commissural neurons of the spinal cord might travel from their dorsal positions to the ventral floor plate by such diffusible factors. The axons of these neurons begin growing ventrally down the side of the neural tube. However, about two-thirds of the way down, their direction changes, and they project through the ventrolateral (motor) neuron area of the neural tube toward the floor plate cells (Figure 8.13). [axon1.html], [axon5.html]

In 1994, Serafini and colleagues developed an assay that would allow them to screen for such a diffusible molecule. When dorsal spinal cord explants were plated onto collagen gels, the presence of floor plate cells near them would promote the outgrowth of commissural axons from these explants. Serafini and his co-workers took fractions of embryonic chick brain homogenate and tested them to see if any of the proteins mimicked this activity. This resulted in the identification of two proteins, **netrin-1** and **netrin-2.** Netrin-1 is made by and secreted from the floor plate cells, whereas netrin-2 is synthesized in the lower region of the spinal cord, but not in the floor plate (see Figure 8.13). The chemotactic effects of these netrins were shown by tranforming COS cells (which usually do not make these proteins) with a vector containing an active *netrin* gene (Kennedy et al., 1994). Aggregates of the netrin-secreting COS cells elicited commissural axon outgrowth from dorsal rat spinal cord explants, while those COS cells that were given the vector without the active *netrin* gene did not elicit such activity (Figure 8.14). Both netrins become associated with the extracellular matrix. * It is possible that the commissural neurons first encounter a gradient of netrin-2, which brings them into the domain of the steeper netrin-1 gradient (see Figure 8.13).

The netrins have numerous regions of homology with **UNC-6**, a protein implicated in directing the circumferential migration of axons around the body of *Caenorhabditis elegans*. In the wild-type nematode, UNC-6 induces axons from certain centrally located positions to move ventrally, and it induces some ventrally placed cell bodies to extend axons dorsally (Figure 8.15). In loss-of-function mutations of *unc-6*, neither of these axonal movements occurs (Hedgecock et al., 1990; Ishii et al., 1992; Hamelin et al., 1993). Mutations of the *unc-40* gene disrupt the ventral (but not the dorsal) axonal migration, while mutations of the *unc-5* gene prevent only the dorsal migration. Culotti (1994) has proposed that the UNC-6 protein can attract the set of axons synthesizing UNC-40 and repel the axons making UNC-5. Recent studies (Wadsworth et al., 1996) show that UNC-6 is spatially restricted to the ventralmost cells of the hypodermis (skin) and nervous system and that the attractive and repulsive properties of this molecule are mediated by different regions of the protein. Moreover, the netrin cues also guide mesodermal cells as well as axons.[†]

If UNC-6 is attractive to some neurons and repulsive to others, one might think that this dual role might also be ascribed to netrins. Colamarino and Tessier-Lavigne (1995) have shown this to be the case by looking at the

---

*The binding of a soluble factor to the extracellular matrix makes for an interesting ambiguity between chemotaxis, haptotaxis, and labeled pathways. Nature doesn't necessarily conform to our categories.

[†]Not only is UNC-6 homologous to netrin, but UNC-40 is homologous to the netrin receptor of mammals and *Drosophila* (Chan et al., 1996; Keino-Masu et al., 1996; Kolodziej et al., 1996). In all types of organisms, the netrin molecule appears to provide guidance for the migration of the cells bearing its receptor.

(A)

(B)

(C)

(D)

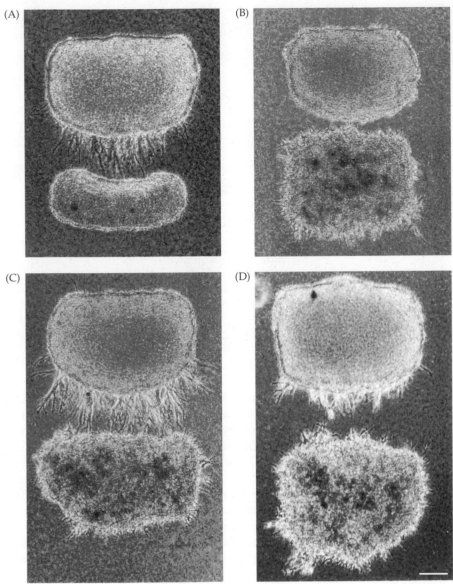

Figure 8.14
Aggregated COS cells secreting netrins elicit commissural axon outgrowth from 11-day embryonic rat dorsal spinal cord explants. (A) Outgrowth of commissural neurons is seen when the rat dorsal spinal cord explant (upper tissue) encounters a floor plate explant. (B) No outgrowth is seen when the dorsal explant is exposed to aggregated COS cells that have been transfected with the cloning vector alone (no *netrin* gene). (C,D) Commissural neuron outgrowths from aggregated COS cells that were expressing the gene for netrin-1 (C) and netrin-2 (D). Their identity as commissural neurons was confirmed by immunohistology showing commissural-specific antigens on these axons. (Scale bar, 100 μm.) (From Kennedy et al., 1994; photographs courtesy of M. Tessier-Lavigne.)

trajectory of the trochlear (fourth cranial) nerve. On its way to innervate an eye muscle, the axons of the trochlear nerve originate near the floor plate of the brainstem and migrate dorsally away from the floor plate region. This pathway is maintained when the brainstem regions are explanted into collagen gels. The dorsal outgrowth of the trochlear neurons could be prevented by placing floor plate cells or netrin-1-secreting COS cells within 450 μm of the dorsal portion of the explant. This dorsal outgrowth was not prevented by dorsal explants of the neural tube or by COS cells that did not contain the active *netrin-1* gene (Figure 8.16). Therefore, netrins and UNC-6 appear to be chemotactic to some neurons and chemorepulsive to others.

The **semaphorin** family comprises another set of chemorepulsive molecules (no attractive members of this family have yet been found). Semaphorin I is found in insects and is a membrane-bound protein that inhibits the branching of axons when they encounter it in the limb (Kolodkin et al., 1992). Semaphorin II is secreted in *Drosophila* by a single large thoracic muscle. In this way, the thoracic muscle prevents itself from being innervated by

Figure 8.15
UNC expression and function in axonal guidance. (A) In the body of the wild-type *C. elegans* embryo, sensory neurons project ventrally and motor neurons project dorsally. The ventral body wall epidermoblasts expressing *unc-6* are filled in. (B) In the *unc-6* mutant embryos, neither of these migrations occur. (C) The *unc-5* loss-of-function mutations affect only the dorsal movements of the motor neurons. (D) The *unc-40* loss-of-function mutations affect only the ventral migration of the sensory growth cones. (After Goodman, 1994.)

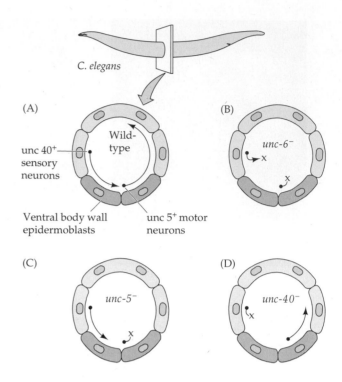

inappropriate axons (Matthes et al., 1995). Semaphorin III, found in mammals and birds, is also known as **collapsin** (Luo et al., 1993). This secreted protein was found to collapse the growth cones of axons originating in the dorsal root ganglia (see Figure 8.10C). There are several types of axons in the dorsal root ganglia that enter into the dorsal spinal cord. Most of these axons are prevented from travelling further and from entering the ventral spinal cord. However, a subset of these neurons do travel ventrally through the neural cells (Figure 8.17). These particular (NT-3-responsive) neurons are not inhibited by semaphorin III, while the other neurons are (Messersmith et al., 1995). This suggests that semaphorin/collapsin patterns sensory projections from the dorsal root ganglia by selectively repelling axons that so that they terminate dorsally.

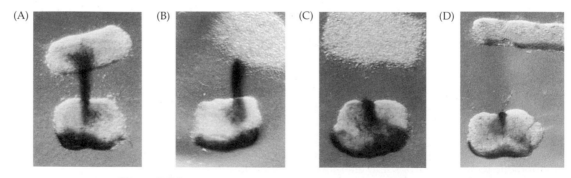

Figure 8.16
Netrin inhibits the outgrowth of trochlear axons from the dorsal spinal cord. Trochlear axons, stained for trochlear-axon-specific antigen, emerge dorsally and are not inhibited by a dorsal spinal cord explant (A) or by COS cells (B). They are inhibited by COS cells secreting netrin-1 (C) or by the floor plate of the spinal cord (D). (After Colamarino and Tessier-Lavigne, 1995; photographs courtesy of M. Tessier-Lavigne.)

(A)

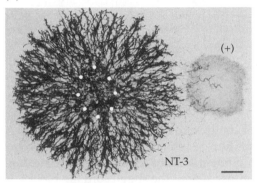

- ● Ia afferent neurons (responsive to NT-3)
- ● Afferent to low-threshold mechanoreceptors
- ● Temperature and pain receptors

(B)

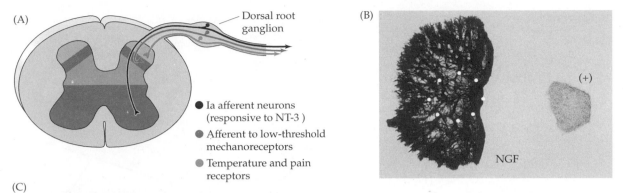

NGF

(C)

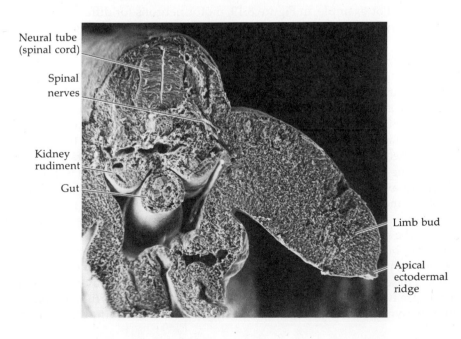

NT-3

Figure 8.17
Semaphorin III as a selective inhibitor of axonal projections into the ventral spinal cord. (A) Trajectory of axons in relation to semaphorin III expression in the 14-day embryonic rat spinal cord. The neurotrophin-3-responsive neurons can travel to the ventral region of the spinal cord, but the afferent neurites for the mechanoreceptors and temperature and pain receptor neurons terminate dorsally. (B) Semaphorin-III-secreting COS cells inhibit outgrowths of mechanoreceptor axons (here shown growing in medium treated with NGF, but inhibited from growing toward the source of semaphorin III). (C) The neurons that are responsive to NT-3 for growth are not inhibited from extending toward the source of semaphorin III. (A after Marx, 1995; B and C from Messersmith et al., 1995; photographs courtesy of A. Kolodkin.)

Multiple guidance cues

Vertebrate Motor Neurons

One of the most important research programs in vertebrate developmental neurobiology concerns the innervation of the limb muscles. Axonal outgrowth of motor neurons occurs very early in development, before the soma of the motor neurons have migrated to their definitive positions in the spinal cord and before the muscles have condensed out of mesenchyme (Landmesser, 1978; Hollyday, 1980a). This stage can be seen in Figure 8.18. To in-

Neural tube (spinal cord)

Spinal nerves

Kidney rudiment

Gut

Limb bud

Apical ectodermal ridge

Figure 8.18
Scanning electron micrograph of a cross section of a 4-day chick embryo, showing the emergence of spinal nerves into the developing limb bud. (From Tosney and Landmesser, 1985, courtesy of K. Tosney.)

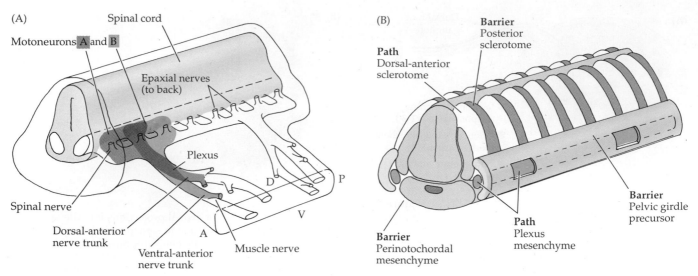

Figure 8.19
Motor axon pathways in the hindlimb region of the chick embryo. (A) Neural pattern of the hindlimb. Motor neuron axons merge in the plexus and then separate into dorsal and ventral nerve trunks. One plexus is anterior; the other, posterior. (B) The environmental components that create the neural patterning. The segmentation of the spinal nerves is created by the sclerotome. Anterior dorsal sclerotome is permissive to migration, whereas posterior dorsal and all ventral sclerotome (the perinotochordal mesenchyme) is a barrier to motor nerve axons. The plexus mesenchyme is permissive, but the pelvic girdle forms a barrier. The two holes in this barrier allow the nerve trunks to extend through them. (After Tosney, 1991.)

nervate the limb musculature, the axon extends over hundreds of cells in a complex and changing environment. Recent research has discovered several paths and several barriers that help guide the axons to their appropriate destinations. As mentioned earlier, on either side of the spinal cord are blocks of mesodermal tissue called somites. Just before the axons begin their extension, the somite splits into two types of tissue. The dorsal portion becomes the dermamyotome (which produces the dermis and musculature of the back), while the ventral portion of the somite becomes the sclerotome (which produces the vertebral cartilage). Lateral to the somites, at the base of the limb bud, lies the plexus mesenchyme and the prospective shoulder girdle cells. The cell bodies of the motor neurons lie in the ventrolateral regions of the neural tube (Figure 8.19A). Axons from the motor neurons that will innervate the limb muscles are mixed together when they exit from the spinal cord. Axon populations from several segmental levels of spinal cord can form into a common spinal nerve. These spinal nerves meet at a **plexus.** At these plexuses, however, the axons from different regions go along different pathways. For example, in Figure 8.19, motor neurons for different muscles diverge into the appropriate nerve trunks and eventually project into single muscles.

By various surgical manipulations of the early chick embryo, some of the environmental cues directing this migration have been discovered. The ventral part of the sclerotome that encircles the notochord forms a barrier against motor axon elongation. Even though cells in this region appear loose and easily circumvented, they repel axons in their vicinity. When the neural tube is rotated such that the motor axons emerge ventrally into this region, they immediately turn to avoid it and travel only through the dorsal anterior sclerotome (Figure 8.19B). Thus, the perinotochordal sclerotome is a barrier to motor axon growth, while the dorsal anterior sclerotome is a pathway (Tosney and Oakley, 1990; Tosney, 1991). Once the axons have progressed through the anterior dorsal sclerotome (along with the neural crest cells that follow the same route), they come to the plexus mesenchyme at the base of the limb bud. This, too, is a permissive environment for axonal growth. However, just past the plexus mesenchyme lie the pelvic girdle precursor cells. These cells inhibit axon growth, and axons turn away from them. There are two holes in the pelvic girdle precursor tissue that fill with plexus mesenchyme. Axons extend through these holes to form the anterior and posterior nerve trunks that enter the limb. If other holes are made experimentally in the pelvic girdle precursor tissue, the axons readily go through them (Tosney and Landmesser, 1984, 1985).

The nerves can even reach their respective destinations if the muscle-forming cells have been removed (Phelan and Hollyday, 1990), and it is probable that the other mesenchymal cells in the limb bud (such as those that form the dermis or cartilage) might be providing the directional cues. These paths to the limb muscle regions appear to be very well defined. First, if one redirects axons from a different source (such as a different ganglion) into the limb, they branch like the axons that originally innervated the limb. In other words, the limb is able to dictate the pattern of innervation to a set of axons that normally would not enter the limb (Hamburger, 1939; Hollyday et al., 1977). Moreover, if segments of the chick spinal cord are reversed, so that the motor neurons are in new locations, their axons will find their original targets (see Figure 8.3; Lance-Jones and Landmesser, 1980). Yet, when limbs develop with duplicated areas (such as two thighs), the neurons innervating the second thigh are not "thigh"-specific neurons, but are neurons that usually innervate the calf (Whitelaw and Hollyday, 1983). These experiments present a paradox that has yet to be resolved: "Particular axons are biased to grow to specific places, yet axons from different motor neuron pools can substitute for one another in the establishment of normal nerve patterns" (Purves and Lichtman, 1985). The most plausible explanation is that several mechanisms act simultaneously to ensure that the axons get to their appropriate places. One of these mechanisms appears to be the guidance from the cell surface of non-muscle-forming mesenchyme cells in the limb bud, while another mechanism probably involves chemotaxis from the limb bud myoblasts (Goodman and Shatz, 1993).

Retinal Axons

Multiple guidance cues have also been postulated to explain how individual retinal neurons are able to send axons to the appropriate area of the brain, even when transplanted away from the optic nerve (Harris, 1986). This ability to guide the axons of translocated neurons to their appropriate sites implies that the guidance cues are not distributed solely along the normal pathway but exist throughout the embryonic brain. Guiding an axon from the nerve cell body to its destination across the embryo is a complex phenomenon, and several different types of cues might be used simultaneously to ensure that the correct connections get established.

The first steps in getting the retinal axons to their specific regions of the optic tectum take place within the retina (Figure 8.20A). As the retinal ganglion cells differentiate, their position in the inner margin of the retina is determined by cadherin molecules (N-cadherin as well as retina-specific R-cadherin) on their cell membranes (Matsunaga et al., 1988; Inuzuka et al., 1991). The axons from these cells grow along the inner surface of the retina toward the optic nerve head (Figure 8.20B). The adhesion and growth of the retinal cell axons along the inner surface of the retina may be governed by the laminin-containing basal lamina. However, the attachment to laminin cannot explain the directionality of the growth. It is possible that a gradient of the inhibitory extracellular matrix molecule chondroitin sulfate proteoglycan plays a role in specifying which way to grow (Hynes and Lander, 1992).

When the axons enter the optic nerve, they grow on glial cells toward the brain. In vitro studies suggest that numerous cell adhesion molecules—N-CAM, cadherins, and integrins—play roles in orienting the axon toward the optic tectum (Neugebauer et al., 1988). N-CAM appears to be especially important here, since the directional migration of the retinal ganglion

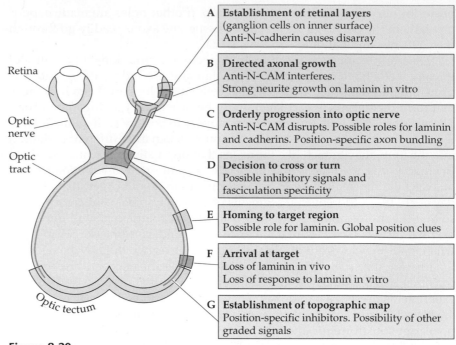

A	**Establishment of retinal layers** (ganglion cells on inner surface) Anti-N-cadherin causes disarray
B	**Directed axonal growth** Anti-N-CAM interferes. Strong neurite growth on laminin in vitro
C	**Orderly progression into optic nerve** Anti-N-CAM disrupts. Possible roles for laminin and cadherins. Position-specific axon bundling
D	**Decision to cross or turn** Possible inhibitory signals and fasciculation specificity
E	**Homing to target region** Possible role for laminin. Global position clues
F	**Arrival at target** Loss of laminin in vivo Loss of response to laminin in vitro
G	**Establishment of topographic map** Position-specific inhibitors. Possibility of other graded signals

Figure 8.20
Multiple guidance cues direct the movement of retinal ganglia axons to the optic tectum. (After Hynes and Lander, 1992.)

growth cones depends on the N-CAM-expressing glial endfeet at the inner retinal surface (Figure 8.20C; Stier and Schlosshauer, 1995). Upon their arrival at the optic nerve, the axons fasciculate with axons that are already present. N-CAM is also critical to this fasciculation, and antibodies against N-CAM (or removal of its polysialic acid component) cause the axons to enter the optic nerve in a disorderly fashion, which in turn causes them to emerge at the wrong positions in the tectum (Thanos et al., 1984; Yin et al., 1995).

Upon entering the brain, mammalian retinal axons reach the **optic chiasm,** where they have to "decide" if they are to continue straight or if they are to turn 90° and enter the other side of the brain (Figure 8.20D). It appears that those axons that are not destined to cross to the other side of the brain are repulsed from doing so when they enter the chasm (Godement et al., 1990), but the molecular basis of this repulsion is not known. On their way to the optic tectum, the axons travel on a pathway (the **optic tract**) over glial cells whose surfaces are coated with laminin (Figure 8.20E). Very few areas of the brain have laminin, and the laminin in this pathway exists only when the optic nerve fibers are growing on it (Cohen et al., 1987).

The axon migrating from the retina to the tectum encounters numerous other cells and potential targets for innervation. However, a combination of several guidance cues, probably involving both attraction and repulsion, guides the axon along its way. At this point, the retinal axons have reached the optic region of the brain (Figure 8.20F), and target selection begins.

Target selection

When the axons come to the end of this laminin-lined tract, they spread out and find their specific targets. Studies on frogs and fish (where retinal neurons from each eye project to the opposite side of the brain) have indicated that each retinal axon sends its impulse to one specific site (a cell or small

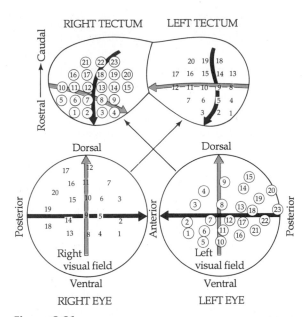

Figure 8.21
Map of the normal retinotectal projection in the adult *Xenopus*. The right eye innervates the left tectum, and the left eye innervates the right tectum. The numbers on the visual fields (retina) and tecta show regions of correspondence; that is, stimulation of spot 15 on the right retina sends electrical impulses to left tectal region 15. The black and colored arrows summarize the pattern of retinotectal connections. (From Jacobson, 1967.)

group of cells) within the tectum (Sperry, 1951). As shown in Figure 8.21, there are two optic tecta in the frog brain. The axons from the right eye enter the left optic tectum, while those from the left eye form synapses with the cells of the right optic tectum. The growth of the neurons in the *Xenopus* optic tract appears to be mediated by fibroblast growth factors secreted by the cells lining the tract. The retinal ganglion axons express FGF receptors in their growth cones. However, as the ganglion cells reach the tectum, the amount of FGF rapidly diminishes, perhaps slowing down the axons and allowing them to find their targets (McFarlane et al., 1995).

The map of retinal connections to the frog optic tectum (the retinotectal projection) was detailed by Marcus Jacobson (1967). Jacobson defined this map by shining a narrow beam of light on a small, limited region of the retina and noting, by means of a recording electrode in the tectum, which tectal cells were being stimulated. The retinotectal projection of *Xenopus laevis* is shown in Figure 8.21. Light illuminating the ventral part of the retina stimulates cells on the lateral surface of the tectum. Similarly, light focused on the posterior part of the retina stimulates cells in the caudal portion of the tectum. These studies demonstrated a point-for-point correspondence between the cells of the retina and the cells of the tectum. When a group of retinal cells is activated, a very small and specific group of tectal cells is stimulated. We also can observe that the points form a continuum; in other words, adjacent points on the retina project onto adjacent points on the tectum. This arrangement enables the frog to see an unbroken image. This intricate specificity caused Sperry (1965) to hypothesize the **chemoaffinity hypothesis:**

> *The complicated nerve fiber circuits of the brain grow, assemble, and organize themselves through the use of intricate chemical codes under genetic control. Early in development, the nerve cells, numbering in the millions,*

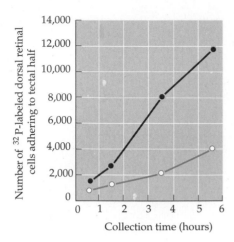

Figure 8.22
Differential adhesion of dorsal radioactive chick retinal cells to dorsal and ventral tectal halves. Radioactive cells from the dorsal half of 7-day chick retinas were added to dorsal (color) and ventral (black) halves of 12-day chick optic tecta. The data show the selective adhesion of the dorsal retinal cells to the ventral tectal tissue. (After Roth and Marchase, 1976.)

acquire and retain thereafter, individual identification tags, chemical in nature, by which they can be distinguished and recognized from one another.

Current theories do not propose a point-for-point specificity between each axon and the nerve that it contacts. Rather, evidence now demonstrates that gradients of adhesivity (especially those involving repulsion) play a role in defining the territories that the axons enter and that activity-driven competition between these neurons determines the final connection of each axon.*

Adhesive Specificities in Different Regions of the Tectum

There is good evidence that retinal ganglion cells can distinguish between regions of the tectum. Cells prepared from the ventral half of chick neural retina preferentially adhere to dorsal halves of the tectum (Figure 8.22; Roth and Marchase, 1976). Gottlieb and co-workers (1976) found that neurons taken from the dorsalmost part of the chick retina adhere preferentially to the ventralmost portion of the tectum and that the extreme ventral neurons of the retina preferentially adhere to the dorsalmost extremes of the tectum. These results were confirmed in other experimental conditions using axonal tips rather than entire neurons (Halfter et al., 1981).

One gradient that has been identified functionally is a gradient of repulsion that is highest in the posterior tectum and weakest in the anterior tectum. Bonhoeffer and colleagues (Walter et al., 1987) prepared a "carpet" of membranes having alternating stripes derived from the posterior and anterior tecta. They then let cells from the nasal (anterior) or temporal (posterior) regions of the retina extend axons into this carpet. The ganglion cells from the *nasal* portion of the retina extended axons equally well on both the anterior and posterior tectal membranes. The neurons from the *temporal* side of the retina, however, extended axons only on the anterior tectal membranes (Figure 8.23). The basis for this specificity appears to be a repulsive factor on the posterior tectal cell membranes. When the growth cone of a temporal retinal axon contacts a posterior tectal cell membrane, the filopodia of the growth cone withdraw, and the growth cone collapses and retracts (Cox et al., 1990). Baier and Bonhoeffer (1992) demonstrated that a gradient of inhibitory substance isolated from the posterior portion of the tectum is capable of guiding the temporal axons from the retina.

Two of these repulsive molecules have been identified in chick embryos. Called **RAGS** (repulsive axon guidance signal) and **ELF-1** (Eph ligand family 1), they are present in a caudal-to-rostral gradient across the tectum, and the cloned protein has the ability to repulse axons (Figure 8.24; Drescher et al., 1995). RAGS and ELF-1 turn out to be ligands for a family of receptor tyrosine kinases called the **Eph receptor kinases.** Eph receptor kinases have been found on the chick retinal ganglion cells, and they are expressed in a temporal-to-nasal gradient along the retinal axons (Cheng et al., 1995). There appear to be several Eph receptors in the retina and ligands in the tectum, and they may play push-and-pull roles in guiding the temporal retinal axons to the anterior tectum and allowing the nasal retinal axons to project to the posterior portion of the tectum.

---

*In recent years, researchers have discovered dozens of mutants in zebrafish that affect the migration of the retinal axons to the tectum or the specificity of the retinotectal connections. These mutants are only now being analyzed, but they promise to provide major insights into the mechanisms by which our sensory input enters the brain. The December, 1996 (Volume 123) issue of *Development* contains several articles mapping the genes involved in the migration of the axon from the retina to the optic cortex. Over 30 mutant genes have been found that effect either the ability of zebrafish retinal axons to find the optic tectum or the ability of the axons to find their appropriate connections within the tectum (Karlstrom et al., 1997).

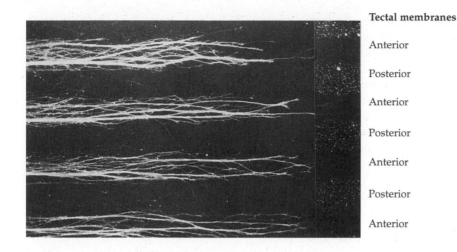

Tectal membranes

Anterior

Posterior

Anterior

Posterior

Anterior

Posterior

Anterior

Figure 8.23
Differential repulsion of temporal retinal axons on tectal membranes. Alternating stripes of anterior and posterior tectal membranes were absorbed onto filter paper. When axons from temporal (posterior) retinal ganglion cells grew on such alternating carpets, they preferentially extended axons on the anterior tectal membranes. (From Walter et al., 1987.)

The possible importance of ELF-1 in the optic tectum was demonstrated by Nakamoto and colleagues (1996). When they infected chick hindbrain regions with an ELF-expressing virus, patches of ELF-1 became expressed in regions of the tectum which normally did not express much of this molecule. Axons from the temporal (but not those from the nasal) regions of the retina

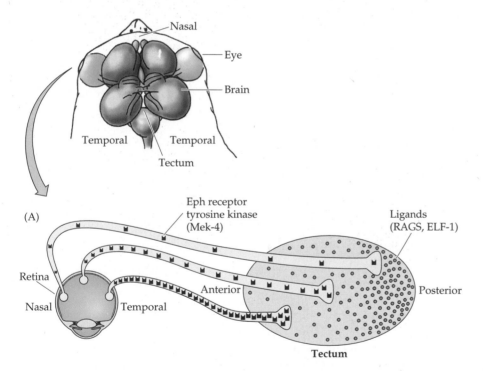

Nasal

Eye

Brain

Temporal   Temporal

Tectum

(A)

Eph receptor
tyrosine kinase
(Mek-4)

Ligands
(RAGS, ELF-1)

Retina

Nasal   Temporal

Anterior   Posterior

Tectum

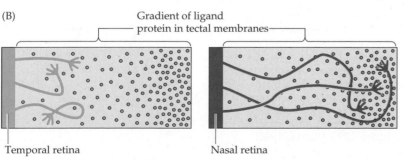

(B)

Gradient of ligand
protein in tectal membranes

Temporal retina   Nasal retina

Figure 8.24
Differential retinotectal adhesion by gradients of Eph receptor tyrosine kinases and their ligands. (A) Representation of the dual gradients of Eph receptor tyrosine kinase (Mek-4) in the retina and its ligand (RAGS, ELF-1) in the tectum. (B) Experiment showing that temporal, but not nasal, retinal axons respond to a gradient of posterior tectal membranes by turning away or slowing down. (After Barinaga, 1995.)

avoided the ELF-1 expressing regions. Thus, ELF-1 could provide negative cues to the temporal retinal regions.

The appearance of RAGS and ELF-1 is regulated by the expression of the Engrailed protein. The Engrailed protein is expressed on day 2 of chick development in a band that includes the caudal (posterior) portion of the future optic tectum (see Figure 7.18). If the Engrailed protein is experimentally induced in the rostral portion of the tectum as well, the rostral portion of the tectum adopts a caudal phenotype. When this occurs, RAGS and ELF-1 are expressed throughout the tectum, and the temporal axons are repelled from

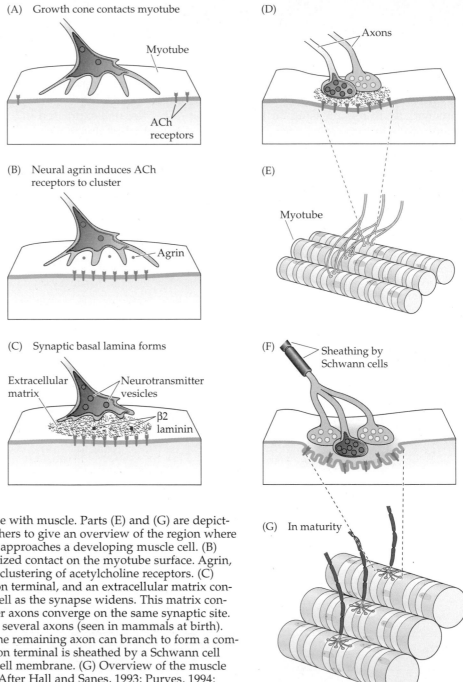

(A) Growth cone contacts myotube

Myotube

ACh receptors

(B) Neural agrin induces ACh receptors to cluster

Agrin

(C) Synaptic basal lamina forms

Extracellular matrix

Neurotransmitter vesicles

β2 laminin

(D)

Axons

(E)

Myotube

(F)

Sheathing by Schwann cells

(G) In maturity

Figure 8.25
Differentiation of motor neuron synapse with muscle. Parts (E) and (G) are depicted at a lower magnification than the others to give an overview of the region where axon meets muscle. (A) A growth cone approaches a developing muscle cell. (B) The axon stops and forms an unspecialized contact on the myotube surface. Agrin, released by the neural tube, causes the clustering of acetylcholine receptors. (C) Neurotransmitter vesicles enter the axon terminal, and an extracellular matrix connects the axon terminal to the muscle cell as the synapse widens. This matrix contains a nerve-specific laminin. (D) Other axons converge on the same synaptic site. (E) Overview of muscle innervation by several axons (seen in mammals at birth). (F) All axons but one are eliminated. The remaining axon can branch to form a complex junction with the muscle. Each axon terminal is sheathed by a Schwann cell process, and folds form in the muscle cell membrane. (G) Overview of the muscle innervation several weeks after birth. (After Hall and Sanes, 1993; Purves, 1994; Hall, 1995.)

both halves (Logan et al., 1996). Thus, the early expression of Engrailed appears to induce the expression of RAGS and ELF-1, and these two proteins mediate the exclusion of temporal retinal axons from the caudal (posterior) portion of the tectum.

Address selection: Activity-dependent development

When an axon contacts its "target" (usually either a muscle or another neuron), it forms a specialized junction called a **synapse.** Neurotransmitters from the axon terminal are released at these synapses to depolarize or hyperpolarize the membrane of the cell across the synaptic cleft. The construction of the synapse involves several steps (Figure 8.25). When motor neurons in the spinal cord extend axons to muscles, growth cones that contact the newly formed muscle cells migrate over their surfaces. When the growth cone first adheres to the muscle cell membrane, no specializations can be seen in either membrane. However, the axon terminals soon begin to accumulate neurotransmitter-containing synaptic vesicles, the membranes of both cells thicken at the region of contact, and the cleft between the cells fills with extracellular matrix that includes a specific form of laminin. This muscle-derived laminin specifically binds the growth cones of motor neurons and may act as a "stop signal" for axonal growth (Martin et al., 1995; Noakes et al., 1995). After this first contact is made, growth cones from other axons converge at this site to form additional synapses. During development, all mammalian muscles studied are seen to be innervated by at least two axons. However, this polyneuronal innervation is transient. During early postnatal life, all but one of these axon branches are retracted. This rearrangement is based on "competition" between the axons (Purves and Lichtman, 1980; Thompson, 1983). When one of the motor neurons is active, it suppresses the synapses of the other neurons, possibly through a nitric oxide-dependent mechanism (Dan and Poo, 1992; Wang et al., 1995). Eventually, the less active synapses are eliminated. The remaining axon terminal expands and is ensheathed by a Schwann cell.

Activity-dependent synapse formation also appears to be involved in the final stages of the retinal projection to the brain. In frog, bird, and rodent embryos treated with tetrodotoxin, axons will grow normally to their respective territories and will make synapses with the tectal neurons. However, the retinotectal map is a coarse one, lacking fine resolution. Just as in the final specification of the motor neuron synapse, neural activity is needed for the point-for-point retinal projection onto the tectal neurons (Harris, 1984; Fawcett and O'Leary, 1985; Kobayashi et al., 1990). This elimination of transient retinal contacts by the tectum may also involve nitric oxide expression by the target tectal cells (Wu et al., 1994).

Differential survival after innervation: Neurotrophic factors

Reflecting on his life as an embryo, Lewis Thomas (1992) writes,

> By the time I was born, more of me had died than survived. It is no wonder I cannot remember; during that time I went through brain after brain for nine months, finally contriving the one model that could be human, equipped for language.

Indeed, one of the most puzzling phenomena in the development of the nervous system is neuronal cell death. In many parts of the vertebrate central and peripheral nervous systems, over half the neurons die during the normal course of development. Moreover, there do not seem to be similarities across species. For instance, in the cat retina, around 80 percent of the retinal ganglion cells die, while in the chick retina, this figure is only 40 percent. In the retinas of fishes and amphibians, no retinal ganglion cells appear to die (Patterson, 1992).

The demise of a neuron is not caused by any obvious defect. Indeed, these neurons have differentiated and successfully extended axons to their targets. Rather, it appears that the target tissue regulates the number of axons innervating it by limiting a supply of some critical survival factor. There seems to be competition for this limited factor. For instance, if more target tissue is grafted into the original target, more of the neurons survive, and if the target tissue is removed before the axons reach it, nearly all the neurons die. These **neurotrophic factors** have been isolated and shown to regulate the survival of different subsets of neurons.

The best-characterized neurotrophic factor is **nerve growth factor** (**NGF**), a glycoprotein composed of two identical 13-kDa subunits. NGF is necessary for the survival of sympathetic and sensory neurons. Treating mouse embryos with anti-NGF antibodies reduces the number of trigeminal sympathetic and dorsal root ganglion neurons to 20 percent of their control amounts (Levi-Montalcini and Booker, 1960; Pearson et al., 1983). NGF appears to function after innervation has taken place, as NGF is not secreted by the target tissues until after innervation, and growing axons lack the receptors for NGF, so they couldn't respond any earlier (Davies et al., 1987). Removal of these target tissues causes the death of the neurons that would have innervated them, and there is a good correlation between the amount of NGF secreted and the survival of neurons that innervate these tissues (Korsching and Thoenen, 1983; Harper and Davies, 1990). [axon1.html]

Other neurotrophic proteins have been characterized. Two of these proteins—**brain-derived neurotrophic factor** (**BDNF**) and **neurotrophin 3** (**NT-3**)—share the same basic structure as NGF. However, they favor the survival of somewhat different groups of neurons. While some neurons respond to all three factors, other neurons respond to only one or two (Figure 8.26; Oppenheim et al., 1992). NGF supports the growth and differentiation of sympathetic ganglion cells and certain sensory neurons, but it does not appear to influence motor neuron survival. BDNF, however, can rescue fetal motor neurons in vivo from normally occurring cell death and from induced cell death following the removal of its target tissue. The results of these in vitro studies have been corroborated by gene knockout experiments, wherein the deletion of particular neurotrophic factors causes the loss of only particular subsets of neurons (Crowley et al., 1994; Jones et al., 1994). NT-3 made by target tissues can support the survival of visceral neurons that are not responsive to NGF (Hohn et al., 1990; Maisonpierre et al., 1990). BDNF, NT-3, and two other neurotrophic molecules—neurotrophin 4/5 (NT-4/5) and fibroblast growth factor 5 (FGF5)—are each synthesized in embryonic rat limb muscle cells when motor neuron axons are growing into the muscle and competing for such survival factors. Moreover, BDNF, NT-3, NT-4/5, and several FGFs prevent the death of the cultured rat embryonic motor neurons. (Henderson et al., 1993; Hughes et al., 1993). Another newly discovered neurotrophin, glial cell line–derived neurotrophic factor (GDNF), enhances the survival of another group of neurons: the midbrain dopaminergic neurons whose destruction characterizes Parkinson's disease (Lin et al., 1993). This factor can prevent the death of these neurons in adult brains (see Lindsay, 1995). Yet another neurotrophin, ciliary neurotrophic factor (CNTF), seems to supports the survival of embryonic motor neurons; CNTF is able to

(A) Sympathetic (B) Dorsal root (C) Nodose

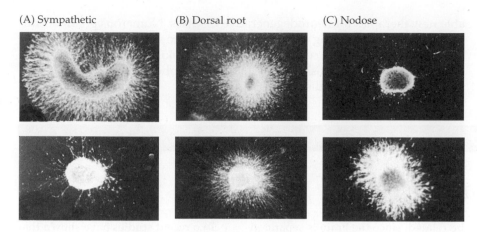

Figure 8.26
Effects of NGF (top row) and BDNF (bottom) on neurite outgrowths from (A) sympathetic ganglia, (B) dorsal root ganglia, and (C) nodose ganglia. While both NGF and BDNF had a mild stimulatory effect on dorsal root ganglion axonal outgrowth, the sympathetic ganglia responded to NGF and hardly at all to BDNF, while the converse was true with the nodose ganglia. (From Ibáñez et al., 1991.)

prevent the degeneration of motor neurons in a mouse mutant characterized by a progressive loss of motor neurons after birth (Sendtner et al., 1992).

The actual survival of any given neuron in the embryo may depend on a combination of agents. Schmidt and Kater (1993) have shown that neurotrophic factors, depolarization, and interactions with the substrate all combine synergistically to determine neuronal survival. For instance, the survival of chick ciliary ganglion neurons in culture was promoted by FGF, laminin, or depolarization. However, FGF did not promote survival when the laminin was absent, and the combined effects of laminin, FGF, and depolarization were greater than the added effects of each of them (Figure 8.27). The neurotrophic factors and the other environmental agents appear to function by suppressing a "suicide program" that would be constitutively expressed unless repressed by these factors (see Chapter 13; Raff et al., 1993). The survival of retinal ganglion cells in culture is predicated on neurotrophic factors, but these cells can respond to these factors only if they have been depolarized (Meyer-Franke et al., 1995). Moreover, since neural activity stimulates the production of neurotrophic factors by the active nerves, it is likely that neurons receiving a signal would produce more neurotrophic factor (Thoenen, 1995). This factor could have an effect on nearby synapses that are active (i.e., capable of responding to that factor), thereby stabilizing a set of active synapses to the exclusion of inactive ones.

The discovery and purification of these neurotrophic proteins and the analysis of their interactions with substrate and electrical conditions may en-

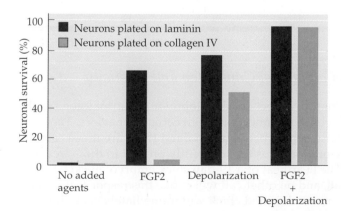

Figure 8.27
Interactions between substrate, depolarization, and neurotrophic factor basic FGF (FGF2) in the survival of ciliary ganglion neurons. Neurons were plated either on laminin (a survival-enhancing substrate) or collagen IV (which does not enhance neuron survival) and observed after 24 hours of culture in the presence or absence of depolarization or FGF2. When cells were depolarized and grown in the presence of FGF2, it did not matter on what substrate they grew. However, when FGF2 was present without depolarization, the substrate made a large difference. (From Schmidt and Kater, 1993.)

able new therapies for neurodegenerative diseases. Numerous pharmaceutical companies are starting clinical trials of neurotrophic factors for the possible alleviation of spinal cord injuries (NGF), Parkinson's disease (GDNF), amotrophic lateral sclerosis (BDNF, CNTF), peripheral neuropathies (NGF, NT-3), and Alzheimer's disease (NGF, GDNF).

Sidelights & Speculations

Fetal Neurons in Adult Hosts

I N 1976, Lund and Hauschka implanted fetal rat brain tissue into the brain of a newborn rat. The fetal neurons made the appropriate connections within the host brain. This study offered the possibility that transplants of fetal neurons might be able to repair damaged regions in human brains. There are many neural degenerative diseases, and Parkinson's disease is one of the most prevalent, afflicting about a million Americans. In Parkinson's disease, dopamine-producing neurons of the substantia nigra (a cluster of cells in the brainstem) are destroyed, and their axon terminals in the caudate nucleus and putamen (two brain nuclei) degenerate. This leads to muscle tremors, difficulty in initiating voluntary movements, and problems in cognition. The injection of L-dopa (which the body metabolizes into dopamine) relieves the symptoms temporarily, but L-dopa loses its effect with prolonged use, and it sometimes has adverse side effects.

In 1990, Lindvall and colleagues implanted human neural cells from the substantia nigra of 8 to 9 week fetuses into a patient with Parkinson's disease. The donor and recipient did not have to be related, since the brain is separated from the immune system by the blood-brain barrier, which shelters transplants of tissue implanted into the brain from rejection by the immune system. Within 5 months, the transplant had restored much of the dopamine normally made by the substantia nigra, as well as the patient's capacity for voluntary movement. Two other laboratories have reported similar restoration of function following the transplantation of fetal neurons into such patients (Freed et al., 1992; Spencer et al., 1992). According to Björkland (1987), the optimal donor tissue is that containing presumptive dopamine-secreting neurons that have undergone their last cell division but that have not yet formed extensive synaptic connections. In 1992, Widner and colleagues showed that grafts from fetal mesencephalons were able to restore motor functions in two patients who had destroyed their substantiae nigrae by injecting themselves with a synthetic heroin contaminated with the byproduct MPTP. This compound had created a condition that resembled severe Parkinson's disease.

Two recent studies have shown that fetal human cell transplants are not the only way to restore the functional anatomy of the substantia nigra in Parkinson patients. First, studies of Isacson and colleagues (1995) suggest that the donor embryonic cells need not be human. Embryonic pig mesencephalon cells reconstructed the normal neuronal connections when injected into the striata of adult rat with a Parkinson-like disease. Second, when Gash and colleagues (1996) injected glia-derived growth factor into the cerebra of monkeys induced to have Parkinson-like syndromes due to MPTP, the injected monkeys showed a functional recovery from their symptoms. Moreover, they had substantially more dopamine and dopamine-producing neurons. Since Parkinson's disease is progressive, it is not known if the grafted or newly divided neurons will fall to the same process that destroyed the endogenous neurons. However, it appears likely that fetal grafts and new neurons are able to reestablish the synaptic connections that the destroyed neurons once made. ∎

The development of behaviors: Constancy and plasticity

One of the most fascinating aspects of developmental neurobiology is the correlation of certain neuronal connections with certain behaviors. There are two remarkable aspects of this phenomenon. First there are those cases in which complex behavioral patterns are inherently present in the "circuitry" of the brain at birth. The heartbeat of a 19-day chick embryo quickens when it hears the distress call, and no other call will evoke this response (Gottlieb, 1965). Furthermore, a newly hatched chick will immediately seek shelter if presented with the shadow of a hawk. The actual hawk is not needed; the shadow cast by a paper silhouette will suffice, and the shadow of no other bird will cause this response (Tinbergen, 1951). There appear, then, to be certain neuronal connections that lead to inherent behaviors in vertebrates.

Equally remarkable are those instances in which the nervous system is so plastic that new experiences can modify the original set of neuronal connections, causing the creation of new neurons or the formation of new synapses between existing neurons. We will discuss neural plasticity at greater length in Chapter 21, but suffice it to say here that the brain does not stop developing at birth. The Nobel Prize–winning research of Hubel and Wiesel (1962, 1963) demonstrated that there was competition between the retinal neurons of each eye for targets in the cortex, and that their connections had to be strengthened by experience. In songbirds, moreover, new neurons are generated and new synapses formed when the birds learn their songs (Alvarez-Buylla et al., 1990), and when adult rats learn new tricks, their cortical neurons develop new synapses (Black et al., 1990). Thus, the nervous system continues to develop in adult life, and the pattern of neuronal connections is a product of inherited patterning and patterning produced by experience. [axon6.html]

As one investigator (Purves, 1994) recently concluded his analysis of brain development:

> *Although a vast majority of this construction must arise from developmental programs laid down during the evolution of each species, neural activity can modulate and instruct this process, thus storing the wealth of idiosyncratic information that each of us acquires through individual experience and practice.*

LITERATURE CITED

Akers, R. M., Mosher, D. F. and Lilien, J. E. 1981. Promotion of retinal neurite outgrowth by substratum-bound fibronectin. *Dev. Biol.* 86: 179–188.

Alvarez-Buylla, A., Kirn, J. R. and Nottebohm, F. 1990. Birth of projection neurons in adult avian brain may be related to perceptual or motor learning. *Science* 249: 1444–1446.

Baier, H. and Bonhoeffer, F. 1992. Axon guidance by gradients of a target-derived component. *Science* 255: 472–475.

Barinaga, M. 1995. Receptors find work as guides. *Science.* 269: 1668–1670.

Bastiani, M. J., Harrelson, A. L., Snow, P. M. and Goodman, C. S. 1987. Expression of fasciclin I and II glycoproteins on subsets of axon pathways during neuronal development in the grasshopper. *Cell* 48: 745–755.

Begovac, P. C. and Shur, B. D. 1990. Cell surface galactosyltransferase mediates the initiation of neurite outgrowth from PC12 cells on laminin. *J. Cell Biol.* 110: 461–470.

Björkland, A. 1987. Brain implants, transplants. *In* G. Adelman (ed.), *Encyclopedia of Neuroscience*, Vol. 1. Birkhauser, Boston, pp. 165–167.

Black, J. E., Issacs, K. R., Anderson, B. J. Alcantara, A. A. and Greenough, W. T. 1990. Learning causes synaptogenesis, whereas motor activity causes angiogenesis, in cerebellar cortex of adult rats. *Proc. Natl. Acad. Sci. USA* 87: 5568–5572.

Calof, A. L. 1992. Sex, nose, and genotype. *Curr. Biol.* 2: 103–105.

Chan, S. S.-Y., Zheng, H., Su, M.-W., Wilk, R., Killeen, M. T., Hedgecock, E. M. and Culotti, J. G. 1996. UNC-40, a C. elegans homolog of DCC (Deleted in Colonorectal Cancer), is required in motile cells responding to UNC-6 netrin cues. *Cell* 87: 186–195.

Cheng, H.-J., Nakamoto, M., Bergemann, A. D. and Flanagan, J. G. 1995. Complementary gradients in expression and binding of ELF-1 and Mek4 in development of the topographic retinotectal projection map. *Cell* 82: 371–381.

Chu-LaGraff, Q., Schmid, A., Leidel, J., Brönner, G., Jäckle, H. and Doe, C. 1995. *huckebein* specifies aspects of CNS precursor identity required for motoneuron axon pathfinding. *Neuron* 15: 1041–1051.

Cohan, C. S. and Kater, S. B. 1986. Suppression of neurite elongation and growth cone motility by electrical activity. *Science* 232: 1638–1640.

Cohen, J., Burne, J. F., Winter, J. and Bartlett, P. 1986. Retinal ganglial cells lose responsiveness to lamina with maturation. *Nature* 322: 465–467.

Cohen, J., Burne, J. F., McKinlay, C. and Winter, J. 1987. The role of laminin and the laminin/fibronectin receptor complex in the outgrowth of retinal ganglial cell axons. *Dev. Biol.* 122: 407–418.

Colamarino, S. A. and Tessier-Lavigne, M. 1995. The axonal chemoattractant netrin-1 is also a chemorepellent for trochlear motor axons. *Cell* 81: 621–629.

Cox, E. C., Müller, B. and Bonhoeffer, F. 1990. Axonal guidance in chick visual system: Posterior tectal membranes induce collapse of growth cones from temporal retina. *Neuron* 2: 31–37.

Crowley, C. and ten others. 1994. Mice lacking nerve growth factor display perinatal loss of sensory and sympathetic neurons yet develop basal forebrain cholinergic neurons. *Cell* 76: 1001–1011.

Culotti, J. G. 1994. Axon guidance mechanisms in *Caenorhabditis elegans*. *Curr. Opin. Genet. Dev.* 4: 587–595.

Dan, Y. and Poo, M.-M. 1992. Hebbian depression of isolated neuromuscular synapses in vitro. *Science* 256: 1570–1573.

Davies, A. M., Brandtlow, C., Heumann, R., Korsching, S., Rohrer, H. and Thoenen, H. 1987. Timing and site of nerve growth factor synthesis in developing skin in relation to innervation and expression of the receptor. *Nature* 326: 353–358.

Davies, J. A., Cook, G. W. M., Stern, C. D. and Keynes, R. J. 1990. Isolation from chick somites of a glycoprotein fraction that causes collapse of dorsal root ganglion growth cones. *Neuron* 2: 11–20.

Drescher, U., Kremoser, C., Handwerker, C., Löschinger, J., Noda, M. and Bonhoeffer, F. 1995. In vitro guidance of retinal ganglion cell axons by RAGS, a 25 kDa protein related to ligands for Eph receptor tyrosine kinases. *Cell* 82: 359–370.

Eisen, J., Meyers, P. Z. and Westerfield, M. 1986. Pathway selection by growth cones of identified motoneurones in live zebra fish embryos. *Nature* 320: 269–271.

Ericson, J., Thor, S., Edlund, T., Jessell, T. J. and Yamada, T. 1992. Early stages of motor neuron differentiation revealed by expression of homeobox gene *islet-1*. *Science* 256: 1555–1560.

Ericson, J., Morton, S., Kawakami, A., Roelink, H. and Jessell, T. M. 1996. Two critical periods of sonic hedgehog signaling required for the specification of motor neuron identity. *Cell* 87: 661–673.

Fawcett, J. W. and O'Leary, D. D. M. 1985. The role of electrical activity in the formation of topographic maps in the nervous system. *Trends Neurosci.* 8: 201–206.

Fliess, W. 1897. Quoted in J. Geller, 1992. The cultural construction of the other. *In* H. Eilberg-Schwartz (ed.), *People of the Body*. State University of New York Press, Albany, pp. 243–282.

Flor, H. and seven others. 1995. Phantom-limb pain as a perceptual correlate of cortical reorganization following arm amputation. *Nature* 375: 482–484.

Franco, B. and fifteen others. 1991. A gene deleted in Kallmann's syndrome shares homology with neural cell adhesion and axonal path-finding molecules. *Nature* 353: 529–536.

Freed, C. R. and eighteen others. 1992. Survival of implanted dopamine cells and neurological iprovement 12 to 46 months after transplantation for Parkinson's disease. *N. Engl. J. Med.* 327: 1549–1555.

Gash, D. M. and eleven others. 1996. Functional recovery in parkinsonnian monkeys treated with GDNF. *Nature* 380: 252–255.

Gershon, M. D., Schwartz, J. H. and Kandel, E. R. 1985. Morphology of chemical synapses and pattern of interconnections. *In* E. R. Kandel and J. H. Schwartz (eds.), *Principles of Neural Science*, 2nd Ed. Elsevier, New York, pp. 132–147.

Godement, P., Salaun, J. and Mason, C. A. 1990. Retinal axon pathfinding in the optic chiasm: Divergence of crossed and uncrossed fibers. *Neuron* 5: 173–186.

Goodman, C. S. 1994. The likeness of being: Phylogenetically conserved molecular mechanisms of growth cone guidance. *Cell* 78: 353–356.

Goodman, C. S. and Bastiani, M. J. 1984. How embryonic nerve cells recognize one another. *Sci. Am.* 251(6): 58–66.

Goodman, C. S. and Doe, C. Q. 1993. Embryonic development of the *Drosophila* central nervous system. *In* Bate, M. and Martinez Arias, A., *The Development of Drosophila melanogaster*. Cold Spring Harbor Press, Cold Springs Harbor, NY, pp. 1131–1206.

Goodman, C. S. and Shatz, C. J. 1993. Developmental mechanisms that generate precise patterns of neuronal connectivity. *Neuron* 10 [Suppl.]: 77–98.

Goodman, C. S., Bastiani, M. J., Doe, C. Q., du Lac, S., Helfand, S. L., Kuwada, J. Y. and Thomas, J. B. 1984. Cell recognition during neuronal development. *Science* 225: 1271–1287.

Gottlieb, D. I., Rock, K. and Glaser, L. 1976. A gradient of adhesive specificity in developing avian retina. *Proc. Natl. Acad. Sci. USA* 73: 410–414.

Gottlieb, G. 1965. Prenatal auditory sensitivity in chickens and ducks. *Science* 147: 1596–1598.

Grumet, M. 1991. Cell adhesion molecules and their subgroups in the nervous system. *Curr. Opin. Neurobiol.* 1: 370–376.

Gundersen, R. W. 1987. Response of sensory neurites and growth cones to patterned substrata of laminin and fibronectin in vitro. *Dev. Biol.* 121: 423–431.

Halfter, W., Claviez, M. and Schwarz, U. 1981. Preferential adhesion of tectal membranes to anterior embryonic chick retina neurites. *Nature* 292: 67–70.

Hall, Z. W. 1995. Laminin β2 (S-laminin): A new player at the synapse. *Science* 269: 362–363.

Hall, Z. W. and Sanes, J. R. 1993. Synaptic structure and development: The neuromuscular junction. *Neuron* 10 [Suppl.]: 99–121.

Hamburger, V. 1939. The development and innervation of transplanted limb primordia of chick embryos. *J. Exp. Zool.* 80: 347–389.

Hamelin, M., Zhou, Y., Su, M.-W., Scott, I. M. and Culotti, J. G. 1993. Expression of the unc-5 guidance receptor in the touch neurons of *C. elegans* steers their axons dorsally. *Nature* 364: 327–330.

Harper, S. and Davies, A. M. 1990. NGF mRNA expression in developing cutaneous epithelium related to innervation density. *Development* 110: 515–519.

Harrelson, A. L. and Goodman, C. S. 1988. Growth cone guidance in insects: Fasciclin II is a member of the immunoglobulin superfamily. *Science* 242: 700–708.

Harris, W. A. 1984. Axonal pathfinding in the absence of normal pathways and impulse activity. *J. Neurosci.* 4: 1153–1162.

Harris, W. A. 1986. Homing behavior of axons in the embryonic vertebrate brain. *Nature* 320: 266–269.

Harrison, R. G. 1910. The outgrowth of the nerve fiber as a mode of protoplasmic movement. *J. Exp. Zool.* 9: 787–848.

Hedgecock, E. M., Culotti, J. G. and Hall, D. H. 1990. The unc-5, unc-6, and unc-40 genes guide circumferential migrations of pioneer axons and mesodermal cells on the epidermis in *C. elegans*. *Neuron* 2: 61–85.

Henderson, C. E. and twelve others. 1993. Neurotrophins promote motor neuron survival and are present in embryonic limb bud. *Nature* 363: 266–270.

Hohn, A., Leibrock, J., Bailey, K. and Barde, Y.-A. 1990. Identification and characterization of a novel member of the nerve growth factor/brain-derived neurotrophic factor family. *Nature* 344: 339–341.

Holley S. A., Jackson, P. D., Sasai, Y., Lu, B., De Robertis, E. M., Hoffmann, F. M. and Ferguson, E. L. 1995. A conserved system for dorsal-ventral patterning in insects and vertebrates involving *sog* and *chordin*. *Nature* 376: 249–253.

Hollyday, M. 1980a. Motoneuron histogenesis and the development of limb innervation. *Curr. Top. Dev. Biol.* 15: 181–215.

Hollyday, M. 1980b. Organization of motor pools in the chick lumbar lateral motor column. *J. Comp. Neurol.* 194: 143–170.

Hollyday, M., Hamburger, V. and Farris, J. M. G. 1977. Localization of motor neuron pools supplying identified muscles in normal and supernumerary legs of chick embryos. *Proc. Natl. Acad. Sci. USA* 74: 3582–3586.

Hubel, D. H. and Wiesel, T. N. 1962. Receptive fields, binocular interaction and functional architecture in the cat's visual cortex. *J. Physiol.* 160: 106–154.

Hubel, D. H. and Wiesel, T. N. 1963. Receptive fields of cells in striate cortex of very young, visually inexperienced kittens. *J. Neurophysiol.* 26: 944–1002.

Hughes, R. A., Sendtner, M., Goldfarb, M., Linholm, D. and Thoenen, H. 1993. Evidence that fibroblast growth factor 5 is a major muscle-derived survival factor for cultured spinal motoneurons. *Neuron* 10: 369–377.

Hynes, R. O. and Lander, A. D. 1992. Contact and adhesive specificities in the associations, migrations, and targeting of cells and axons. *Cell* 68: 303–322.

Ibáñez, C. F., Ebendal, T. and Persson, H. 1991. Chimeric molecules with multiple neurotrophic activities reveal structural elements determining the specificities of NGF and BDNF. *EMBO J.* 10: 2105–2110.

Inuzuka, H., Miyatani, S. and Takeichi, M. 1991. R-cadherin: A novel Ca^{2+}-dependent cell-cell adhesion molecule expressed in the retina. *Neuron* 7: 69–79.

Ishii, N., Wadsworth, W. G., Stern, B. D., Culotti, J. G. and Hedgecock, E. M. 1992. UNC-6, a laminin-related protein, guides pioneer axon migrations in *C. elegans*. *Neuron* 9: 873–881.

Isacson, O., Deacon, T., Pakzaban, P., Galpern, W. R., Dinsmore, J. and Burns, L. H. 1995. Transplanted xenogeneic neural cells in neurodegenerative disease models exhibit remarkable axonal target specificity and distinct growth patterns of glial and axonal fibres. *Nat. Med.* 1: 1189–1194.

Jacobson, M. 1967. Retinal ganglion cells: Specification of central connections in larval *Xenopus laevis*. *Science* 155: 1106–1108.

Jones, K. R., Farinas, I., Backus, C. and Reichardt, L. F. 1994. Targeted disruption of the BDNF gene purturbs brain and sensory neuron development but not motor neuron development. *Cell* 76: 989–999.

Karlstrom, R. O., Trowe, T. and Bonhoeffer, F. 1997. Genetic analysis of axon guidance and mapping in the zebrafish. *Trends Neurosci.* 20: 3–8.

Keino-Masu, K. Masu, M., Hinck, L. Leonardo, E. D. Chan, S. S.-Y., Culotti, J. G. and Tessier-Lavigne, M. 1996. Deleted in Colonorectal Cancer (DCC) encodes a netrin receptor. *Cell* 87: 175–185.

Kennedy, T. E., Serafini, T., de la Torre, J. R. and Tessier-Lavigne, M. 1994. Netrins are diffusible chemotropic factors for commissural axons in the embryonic spinal cord. *Cell* 78: 425–435.

Keynes, R. J. and Stern, C. D. 1984. Segmentation in the vertebrate nervous system. *Nature* 310: 786–787.

Kimmel, C. B. and Law, R. D. 1985. Cell lineage of zebrafish blastomeres I. Cleavage pattern and cytoplasmic bridges between cells. *Dev. Biol.* 108: 78–101.

Klose, M. and Bentley, D. 1989. Transient pioneer neurons are essential for formation of an embryonic peripheral nerve. *Science* 245: 982–984.

Kobayashi, T. Nakamura, H. and Yasuda, M. 1990. Disturbance of refinement of retintectal projection in chick embryos by tetrodotoxin and grayanotoxin. *Dev. Brain Res.* 57: 29–35.

Kolodkin, A. L., Matthes, D. J. and Goodman, C. S. 1993. The semaphorin genes encode a family of transmembrane and secreted growth cone guidance molecules. *Cell* 75: 1389–1399.

Kolodkin, A. L., Matthes, D. J., O'Connor, T. P., Patel, N. H., Bentley, D. and Goodman, C. S. 1992. Fasciclin IV: Sequence, expression, and function during growth cone guidance in the grasshopper embryo. *Neuron* 9: 831–845.

Kolodziej, P. A., Timpe, L. C., Mitchell, K. J., Fried, S. R., Goodman, C. S., Jan, L. Y. and Jan, Y. N. 1996. *Frazzled* encodes a Drosophila member of the DCC immunoglobulin subfamily and is required for CNS and motor axon guidance. *Cell* 87: 197–204.

Korsching, S. and Thoenen, H. 1983. Nerve growth factor in sympathetic ganglia and corresponding target organs of the rat: correlation with density of sympathetic innervation. *Proc. Natl. Acad. Sci. USA* 80: 3513–3516.

Krafft-Ebing, R. von. 1886. *Psychopathia Sexualis*. Enke, Stuttgart.

Lance-Jones, C. and Landmesser, L. 1980. Motor neuron projection patterns in chick hindlimb following partial reversals of the spinal cord. *J. Physiol.* 302: 581-602.

Landmesser, L. 1978. The development of motor projection patterns in the chick hindlimb. *J. Physiol.* 284: 391–414.

Lee, J. E., Hollenberg, S. M., Snider, L., Turner, D. L., Lipnick, N. and Weintraub, H. 1995. Conversion of *Xenopus* ectoderm into neurons by NeuroD, a basic helix-loop-helix protein. *Science* 268: 836–844.

Legouis, R. and fourteen others. 1991. The candidate gene for X-linked Kallmann syndrome encodes a protein related to adhesion molecules. *Cell* 67: 423–435.

Letourneau, P., Madsen, A. M., Palm, S. M. and Furcht, L. T. 1988. Immunoreactivity for laminin in the developing ventral longitudinal pathway of the brain. *Dev. Biol.* 125: 135–144.

Levi-Montalcini, R. and Booker, B. 1960. Destruction of the sympathetic ganglia in mammals by an antiserum to the nerve growth factor protein. *Proc. Natl. Acad. Sci. USA* 46: 384–390.

Liesi, P. and Silver, J. 1988. Is astrocyte laminin involved in axon guidance in the mammalian CNS? *Dev. Biol.* 130: 774–785.

Lin, L.-F. H., Doherty, D. H., Lile, J. D., Bektesh, S. and Collins, F. 1993. GDNF: A glial cell-line derived neurotrophic factor for midbrain dopaminergic neurons. *Science* 260: 1130–1132.

Lindsay, R. M. 1995. Neuron saving schemes. *Nature* 373: 289–290.

Lindvall, O., Brundin, P. and Widner, H. 1990. Grafts of fetal dopamine neurons survive and improve motor functions in Parkinson's disease. *Science* 247: 574–577.

Livne, I., Gibson, M. J. and Silverman, A. J. 1993. Biochemical differentiation and intercellular interactions of migratory GnRH cells in the mouse. *Dev. Biol.* 159: 643–666.

Logan, C., Wizenmann, A., Drescher, U., Monschau, B., Bonhoeffer, F. and Lumsden, A. 1996. Rostral optic tectum acquires caudal characteristics following ectopic *Engrailed* expression. *Curr. Biol.* 6: 1006–1014.

Lund, R. D. and Hauschka, S. D. 1976. Transplanted neural tissue develops connection with host rat brain. *Science* 193: 582–584.

Luo, Y., Raibile, D. and Raper, J. A. 1993. Collapsin: A protein in brain that induces the collapse and paralysis of neuronal growth cones. *Cell* 75: 217–227.

Mackenzie, J. L. 1898. The physiological and pathological relations between the nose and the sexual apparatus of man. *Johns Hopkins Hosp. Bull.* 82: 10–17.

Maisonpierre, P. C., Belluscio, L., Squinto, S., Ip, N. Y., Furth, M. E., Lindsay, R. M. and Yancopoulos, G. D. 1990. Neurotrophin-3: A neurotrophic factor related to NGF and BDNF. *Science* 247: 1446–1451.

Martin, P. T., Ettinger, A. J. and Sanes, J. R. 1995. Synaptic localization domain in the synaptic cleft protein laminin β2 (s-laminin). *Science* 269: 413–416.

Marx, J. 1995. Helping neurons find their way. *Science* 268: 971–973.

Matthes, D. J., Sink, H., Kolodkin, A. L. and Goodman, C. S. 1995. Semaphorin II can function as a selective inhibitor of specific synaptic arborizations. *Cell* 81: 631–639.

Matsunaga, M., Hatta, K. and Takeichi, M. 1988. Role of N-cadherin cell adhesion molecules in the histogenesis of neural retina. *Neuron* 1: 289–295.

Matsuzawa, M., Weight, F. F., Potember, R. S. and Liesi, P. 1996. Directional neurite outgrowth and axonal differentiation of embryonic hippocampal neurons is promoted by a neurite outgrowth domain of the B2-chain of laminin. *Inter. J. Dev. Neurosci.* 14: 283–295.

McFarlane, S., McNeill, L. and Holt, C. E. 1995. FGF signaling and target recognition in the developing Xenopus visual system. *Neuron* 15: 1017–1028.

Messersmith, E. K., Leonardo, E. D., Shatz, C. J., Tessier-Lavigne, M., Goodman, C. S. and Kolodkin, A. 1995. Semaphorin III can function as a selective chemorepellent to pattern sensory projections in the spinal cord. *Neuron* 14: 949–959.

Meyer-Franke, A., Kaplan, M. R., Pfrieger, F. W. and Barres, B. A. 1995. Characterization of the signaling interactions that promote the survival and growth of developing retinal ganglion cells in culture. *Neuron* 15: 805–819.

Nakamoto, M. and several others. 1996. Topographically specific effects of ELF-1 on retinal axon guidance in vitro and retinal axon mapping in vivo. *Cell* 86: 755–766.

Neugebauer, K. M., Tomaselli, K. J., Lilien, J. and Reichardt, L. F. 1988. N-cadherin, N-CAM, and integrins promote retinal neurite outgrowth on astrocytes in vitro. *J. Cell Biol.* 107: 1177–1187.

Noakes, P. G., Gautam, M., Mudd, J., Sanes, J. R. and Merlie, J. P. 1995. Aberrant differentiations of neuromuscular junctions in mice lacking s-laminin/laminin β2. *Nature* 374: 258–262.

Oppenheim, R. W., Qin-Wei, Y., Prevette, D. and Yan, Q. 1992. Brain-derived neurotrophic growth factor rescues developing avian motoneurons from cell death. *Nature* 360: 755–757.

Patterson, P. H. 1992. Neuron-target interactions. *In* Z. Hall, (ed.), *An Introduction to Molecular Neurobiology*. Sinauer Associates, Sunderland, MA, pp. 428–459.

Pearson, J., Johnson, E. M., Jr. and Brandeis, L. 1983. Effects of antibodies to nerve growth factor on intrauterine development of derivatives of cranial neural crest and placode in the guinea pig. *Dev. Biol.* 96: 32–36.

Phelan, K. A. and Hollyday, M. 1990. Axon guidance in muscleless chick wings: The role of muscle cells in motoneural pathway selection and muscle nerve formation. *J. Neurosci.* 10: 2699–2716.

Piccolo, S., Sasai, Y., Lu, B. and De Robertis, E. M. 1996. Dorsoventral patterning in *Xenopus*: Inhibition of ventral signals by direct binding of chordin to BMP-4. *Cell* 86: 589–598.

Pittman, R. N. 1985. Release of plasminogen activator and a calcium-dependent metalloprotease from cultured sympathetic and sensory neurons. *Dev. Biol.* 110: 91–101.

Purves, D. 1994. *Neural Activity and the Growth of the Brain*. Cambridge University Press, New York.

Purves, D. and Lichtman, J. W. 1980. Elimination of synapses in the developing nervous system. *Science* 210: 153–157.

Purves, D. and Lichtman, J. W. 1985. *Principles of Neural Development*. Sinauer Associates, Sunderland, MA.

Raff, M. C., Barres, B. A., Burne, J. F., Coles, H. S., Ishizaki, Y. and Jacobson, M. D. 1993. Programmed cell death and the control of cell survival: Lessons from the nervous system. *Science* 262: 695–700.

Ramón y Cajal, S. 1892. Le Rétine de Vertébrés. *La Cellule* 9: 119–258.

Raper, J. A. and Kapfhammer, J. P. 1990. The enrichment of a neuronal growth cone collapsing activity from embryonic chick brain. *Neuron* 4: 21–29.

Roth, S. and Marchase, R. B. 1976. An in vitro assay for retinotectal specificity. *In* S. H. Barondes (ed.), *Neuronal Recognition*. Plenum, New York, pp. 227–248.

Sasai, Y., Lu, B., Steinbeisser, H. and De Robertis, E. M. 1995. Regulation of neural induction by the Chd and Bmp-4 antagonistic patterning signals in Xenopus. *Nature* 376: 333–336.

Schmidt, M. and Kater, S. B. 1993. Fibroblast growth factors, depolarization, and substrate interact in a combinatorial way to promote neuronal survival. *Dev. Biol.* 158: 228–237.

Schwanzel-Fukada, M. and Pfaff, D. W. 1989. Origin of luteinizing hormone-releasing hormone neurons. *Nature* 338: 161–164.

Schwanzel-Fukada, M., Bick, D. and Pfaff, D. W. 1989. Luteinizing hormone-releasing hormone (LHRH)-expressing cells do not migrate normally in an inherited hypogonadal (Kallman) syndrome. *Mol. Brain Res.* 6: 311–326.

Sendtner, M., Schmalbruch, H., Stöckli, K. A., Carroll, P., Kreutzberg, G. W. and Thoenen, H. 1992. Ciliary neurotrophic factor prevents degeneration of motor neurons in mouse mutant progressive motor neuropathy. *Nature* 358: 502–504.

Serafini, T., Kennedy, T. E., Galko, M. J., Mirayan, C., Jessell, T. M. and Tessier-Lavigne, M. 1994. The netrins define a family of axon outgrowth–promoting proteins homologous to *C. elegans* UNC-6. *Cell* 78: 409–424.

Shawlot, W. and Behringer, R. R. 1995. Requirement for *Lim1* in head organizer function. *Nature* 374: 425–430.

Silver, J. and Sidman, R.L. 1980. A mechanism for guidance and topologic patterning of retinal ganglion cell axons. *J. Comp. Neurol.* 189: 101–111.

Singer, M., Norlander, R. and Egar, M. 1979. Axonal guidance during embryogenesis and regeneration in the spinal cord of the newt: The blueprint hypothesis of neuronal pathway patterning. *J. Comp. Neurol.* 185: 1–22.

Spencer, D. D. and fifteen others. 1992. Unilateral transplantation of human fetal mesencephalic tissue into the caudate nucleus of patients with Parkinson disease. *N. Engl. J. Med.* 327: 1541–1548.

Sperry, R. W. 1951. Mechanisms of neural maturation. *In* S. S. Stevens (ed.), *Handbook of Experimental Psychology*. Wiley, New York, pp. 236–280.

Sperry, R. W. 1965. Embryogenesis of behavioral nerve nets. *In* R. L. DeHaan and H. Ursprung (eds.), *Organogenesis*. Holt, Rinehart & Winston, New York, pp. 161–186.

Stier, H. and Schlosshauer, B. 1995. Axonal guidance in the chicken retina. *Development* 121: 1443–1454.

Stout, R. P. and Gradziadi, P. P. C. 1980. Influence of the olfactory placode on the development of the brain in *Xenopus laevis* (Daudin). *Neuroscience* 5: 2175–2186.

Taghert, P. H., Doe, C. Q. and Goodman, C. S. 1984. Cell determination and regulation during development of neuroblasts and neurones in grasshopper embryo. *Nature* 307: 163–165.

Thanos, S., Bonhoeffer, F. and Rutischauser, U. 1984. Fiber-fiber interaction and tectal cues influence the development of the chicken retinotectal projection. *Proc. Natl. Acad. Sci. USA* 81: 1906–1910.

Thoenen, H. 1995. Neurotrophins and neuronal plasticity. *Science* 270: 593–598.

Thomas, L. 1992. *The Fragile Species*. Macmillan, New York.

Thomas, W. A., Schaeffer, A. W. and Treadway, R. M. Jr. 1990. Galactosyl transferase-dependence of neurite outgrowth on substratum-bound laminin. *Development* 110: 1101–1114.

Thompson, W. J. 1983. Synapse elimination in neonatal rat muscle is sensitive to pattern of muscle use. *Nature* 302: 614–616.

Tinbergen, N. 1951. *The Study of Instinct*. Clarendon Press, Oxford.

Tosney, K. W. 1991. Cells and cell interactions that guide motor axons in the developing chick embryo. *BioEssays* 13: 17–23.

Tosney, K. W. and Landmesser, L. T. 1984. Pattern and specificity of axonal outgrowth following varying degrees of limb ablation. *J. Neurosci.* 4: 2158–2527.

Tosney, K. W. and Landmesser, L. T. 1985. Development of the major pathways for neurite outgrowth in the chick hindlimb. *Dev. Biol.* 109: 193–214.

Tosney, K. W. and Oakley, R. A. 1990. Perinotochordal mesenchyme acts as a barrier to axon advance in the chick embryo: Implications for a general mechanism of axon guidance. *Exp. Neurol.* 109: 75–89.

Tosney, K. W., Hotary, K. B. and Lance-Jones, C. 1995. Specificity of motoneurons. *BioEssays* 17: 379–382.

Tsushida, T., Ensini, M., Morton, S. B., Baldassare, M., Edlund, T., Jessell, T. M. and Pfaff, S. L. 1994. Topographic organization of embryonic motor neurons defined by expression of LIM homeobox genes. *Cell* 79: 957–970.

Turner, D. L. and Weintraub, H. 1994. Expression of achaete-scute homolog 3 in Xenopus embryos converts ectodermal cells to a neural fate. *Genes Dev.* 8: 1434–1447.

Twitty, V. C. 1937. Experiments on the phenomenon of paralysis produced by a toxin occuring in *Triturus* embryos. *J. Exp. Zool.* 76: 67–104.

Twitty, V. C. and Johnson, H. H. 1934. Motor inhibition in *Amblystoma* produced by *Triturus* transplants. *Science* 80: 78–79.

Wadsworth, W. G., Bhatt, H. and Hedgecock, E. M. 1996. Neuroglia and pioneer neurons express UNC-6 to provide global and local netrin cues for guiding migrations in *C. elegans*. *Neuron* 16: 35–46.

Walter, J., Henke-Fahle, S. and Bonhoeffer, F. 1987. Avoidance of posterior tectal membranes by temporal retinal axons. *Development* 101: 909–913.

Wang, T., Xie, Z. and Lu, B. 1995. Nitric oxide mediates activity-dependent synaptic suppression at developing neuromuscular synapses. *Nature* 374: 262–266.

Way, J. C. and Chalfie, M. 1988. *mec-3*, a homeobox-containing gene that specifies differentiation of the touch receptor neurons in *C. elegans*. *Cell* 54: 5–16.

Weiss, P. A. 1955. Nervous system. *In* B. H. Willier, P. A. Weiss and V. Hamburger (eds.), *Analysis of Development*. Saunders, Philadelphia, pp. 346–402.

Westerfield, M., Liu, D. W., Kimmel, C. B. and Walker, C. 1990. Pathfinding and synapse formation in a zebrafish mutant lacking functional acetylcholine receptors. *Neuron* 4: 867–874.

Whitelaw, V. and Hollyday, M. 1983. Position-dependent motor innervation of the chick hindlimb following serial and parallel duplications of limb segments. *J. Neurosci.* 3: 1216–1225.

Widner, H. and eight others. 1992. Bilateral fetal mesencephalic grafts in two patients with parkinsonism induced by 1-methyl-4-phenyl-1,2,3,6 tetrahydropryridine (MPTP). *N. Engl. J. Med.* 327: 1556–1563.

Wray, S., Grant, P. and Gainer, H. 1989. Evidence that cells expressing luteinizing hormone-releasing hormone mRNA in the mouse are derived from progenitor cells on the olfactory placode. *Proc. Natl. Acad. Sci. USA* 86: 8132–8136.

Wu, H. H., Williams, C. V. and McLoon, S. C. 1994. Involvement of nitic oxide in the elimination of a transient retinotectal projection in development. *Science* 265: 1593–1596.

Yin, X., Watanabe, M. and Rutishauser, U. 1995. Effects of polysialic acid on the behavior of retinal ganglion cell axons during growth into the optic tract and tectum. *Development* 121: 3439–3446.

Zinn, K., McAllister, L. and Goodman, C. S. 1988. Sequence analysis and neuronal expression of fasciclin I in grasshopper and *Drosophila*. *Cell* 53: 577–587.

Zipser, B. and McKay, R. 1981. Monoclonal antibodies distinguish identifiable neurones in the leech. *Nature* 289: 549–554.

Early vertebrate development: Mesoderm and endoderm

9

Of physiology from top to toe I sing,
Not physiognomy alone or brain alone is
worthy for the Muse,
I say the form complete is worthier far,
The Female equally with the Male I sing.
WALT WHITMAN (1867)

Theories come and theories go. The frog
remains.
JEAN ROSTAND (1960)

IN CHAPTERS 7 AND 8, we followed the various tissues formed by the developing ectoderm. In this chapter, we will follow the early development of the mesodermal and endodermal germ layers. We will see that endoderm forms the lining of the digestive and respiratory tubes, with their associated organs; mesoderm will be seen to generate all the organs between the ectodermal wall and the endodermal tissues.

■ MESODERM

The mesoderm of a neurula-stage embryo can be divided into five regions (Figure 9.1). The first region is the **chordamesoderm.** This tissue forms the notochord, a transient organ whose major functions include inducing the formation of the neural tube and establishing the anterior-posterior body axis. As we saw in Chapter 6, the chordamesoderm forms in the center of the embryo on the future dorsal side. The second region is the **somitic dorsal mesoderm.** The term *dorsal* refers to the observation that the tissues developing from this region will be in the *back* of the embryo, along the spine. The cells in this region will form *somites*, blocks of mesodermal cells on both sides of the neural tube that will produce many of the connective tissues of the back (bone, muscle, cartilage, and dermis). The **intermediate mesoderm** forms the urinary system and genital ducts; we will discuss this region in detail in later chapters. Further away from the notochord, the **lateral plate mesoderm** gives rise to the heart, blood vessels, and blood cells of the circulatory system, as well as to the lining of the body cavities and to all the mesodermal components of the limbs except the muscles. It will also form a series of extraembryonic membranes that are important for transporting nutrients to the embryo. Lastly, the **head mesenchyme** will contribute to the connective tissues and musculature of the face.

Dorsal mesoderm: The notochord and the differentiation of somites

Paraxial Mesoderm

One of the major tasks of gastrulation is to create a mesodermal layer between the endoderm and the ectoderm. As shown in Figure 9.2, the forma-

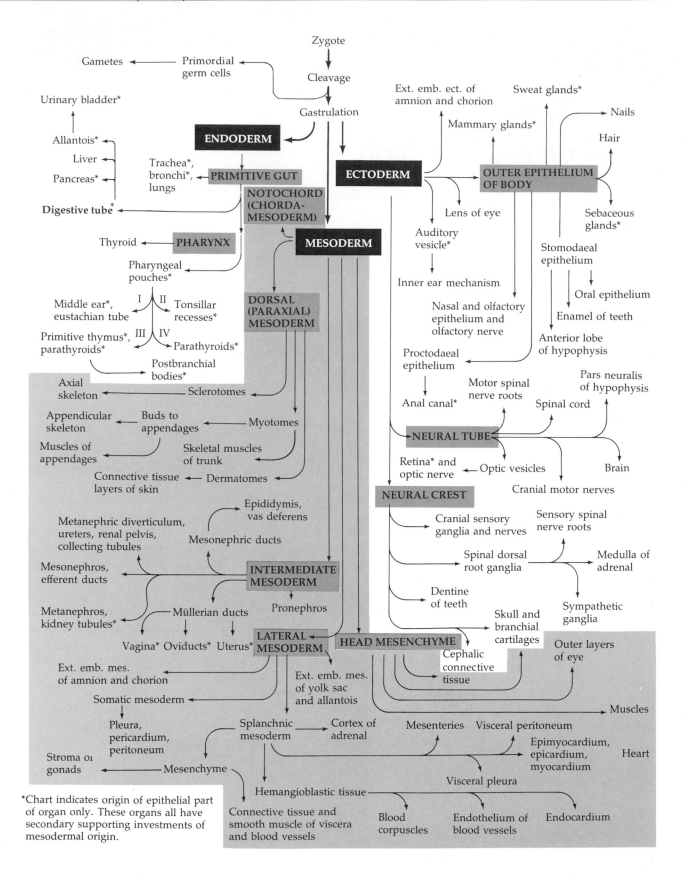

*Chart indicates origin of epithelial part of organ only. These organs all have secondary supporting investments of mesodermal origin.

◀ **Figure 9.1**
Chart depicting the lineage of the specialized parts of the body through the three primary germ layers. The germ cells are represented as a line of cells separate from those of the three somatic germ layers because, although the germ cell precursors are located in the presumptive endoderm or mesoderm, they are probably a unique cell type. (After Carlson, 1981.)

tion of mesodermal and endodermal organs is not subsequent to neural tube formation, but occurs synchronously. The formation of the notochord was discussed in Chapter 6. This epithelial rod extends from the base of the head into the tail. On either side of the notochord are thick bands of mesodermal cells. These bands of **paraxial mesoderm** are referred to as the *segmental plate* (in birds) and the *unsegmented mesoderm* (in mammals). As the primitive streak regresses and the neural folds begin to gather at the center of the embryo, the paraxial mesoderm separates into blocks of cells called **somites.** Although somites are transient structures, they are extremely important in organizing the segmental pattern of vertebrate embryos. As we saw in the preceding chapter, the somites determine the migration paths of neural crest cells and spinal nerve axons. Somites give rise to the cells that form (1) the vertebrae and ribs, (2) the dermis of the dorsal skin, (3) the skeletal muscles of the back, and (4) the skeletal muscle of the body wall and limbs.

Somitomeres and the Initiation of Somite Formation

The first somites appear in the anterior portion of the embryo, and new somites "bud off" from the rostral end of the paraxial mesoderm at regular intervals (Figures 9.2C,D and 9.3). Because embryos can develop at slightly different rates (as when chicken embryos are incubated at slightly different temperatures), the number of somites present is usually the best indicator of

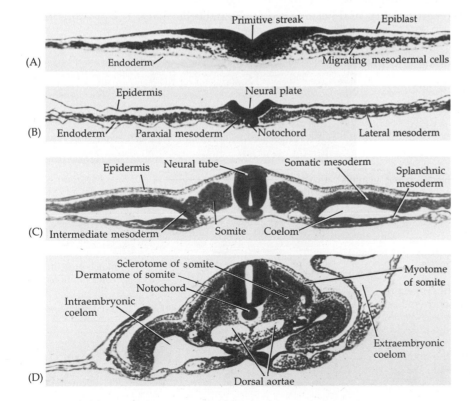

Figure 9.2
The progressive development of the chick embryo, focusing on the mesodermal component. (A) Primitive streak region, showing migrating mesodermal and endodermal precursors. (B) Formation of the notochord and paraxial mesoderm. (C,D) Differentiation of the somites, coelom, and the two aortae (which will eventually fuse). A–C, 24-hour embryos; D, 48-hour embryo.

Figure 9.3
Neural tube and somites. Scanning electron micrograph showing well-formed somites and paraxial mesoderm (bottom right) that has not yet separated into distinct somites. A rounding of the paraxial mesoderm into a somitomere can be seen at the lower left, and neural crest cells can be seen migrating ventrally from the roof of the neural tube. (Courtesy of K. W. Tosney.)

how far development has proceeded. The total number of somites formed is characteristic of a species.

The mechanism for somite formation is not well established, but several studies in chicks have shown that the cells of the segmental plate are organized into whorls of cells called **somitomeres** (Meier, 1979; Packard and Meier, 1983). Conversion from somitomere to somite is seen as the cells of the most anterior somitomere become compacted. This transition from a loosely compacted somitomere to an epithelial somite is correlated with the synthesis of two extracellular matrix proteins, fibronectin and N-cadherin (Figure 9.4A; Ostrosky et al., 1984; Lash and Yamada, 1986; Hatta et al., 1987). These proteins, in turn, may be regulated by the expression of *Notch1* and *Paraxis*. The *Notch1* gene encodes a transcription factor that is active in the most anterior region of the unsegmented dorsal mesoderm, and mice lacking this factor develop misaligned somites of various sizes (Figure 9.4B,C; Conlon et al., 1995). *Paraxis*, a gene encoding another transcription factor, is expressed at the rostral (anterior) end of the unsegmented mesoderm of the mouse and chick embryos. Injection of antisense oligonucleotides complementary to *Paraxis* produces somitic segmentation defects (Burgess et al., 1995; Barnes et al., 1997). Cells of the normal, newly formed somite are organized randomly but soon become organized into a ball of columnar epithelial cells that surround a small cavity filled with loosely connected cells. The epithelial cells become attached to each other by tight junctions. Paraxis is an essential part of this conversion from mesenchyme to epithelium (Burgess et al., 1996).

Generation of the Somitic Cell Types

When the somite is first formed, any of its cells can become any of the somite-derived structures. However, as the somite matures, the various regions of the somite become committed to forming only certain cell types. The *ventral medial* cells of the somite (those cells located farthest from the back but close to the neural tube) undergo mitosis, lose their round epithelial characteristics, and become mesenchymal cells again. The portion of the

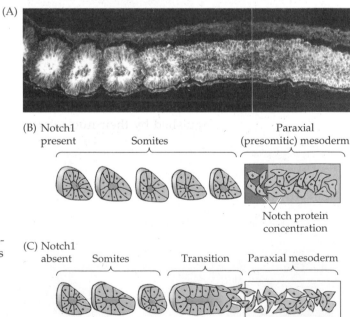

Figure 9.4
Transition from somitomere to somite. (A) N-cadherin expression correlates with the conversion of loose mesenchyme cells into the epithelial somite. (B) In wild-type embryos, *Notch1* expression is seen at the anteriormost region of the unsegmented paraxial mesoderm (i.e., the portion being organized into a somite). (C) In *Notch1*-deficient embryos, the organization of the somites is deranged. (A from Hatta et al., 1987, courtesy of M. Takeichi; B and C after Conlon et al., 1995.)

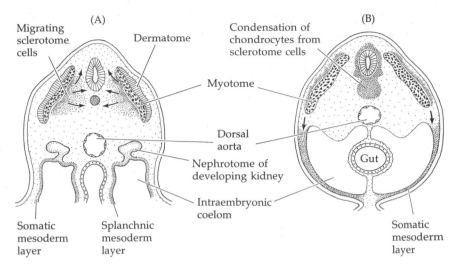

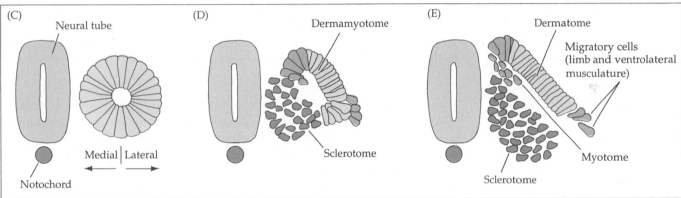

Figure 9.5
Diagram of a transverse section through the trunk of (A) an early 4-week and (B) a late 4-week human embryo, showing formation of somite structures. (A) The sclerotome cells are beginning to migrate away from the myotome and dermatome. (B) By the end of the fourth week, the sclerotome cells are condensing to form cartilaginous vertebrae, the dermatome is starting to form the dermis, and the myotome cells are extending ventrally down the walls of the embryo. (C–E) The changing structure of the chick somite as cell migrations occur. (A and B after Langman, 1981; C–E after Ordahl, 1993.)

somite that gives rise to these cells is called the **sclerotome,** and these mesenchymal cells ultimately become the vertebral **chondrocytes** (Figures 9.2 and 9.5). Chondrocytes are responsible for secreting the special types of collagens and GAGs (such as chondroitin sulfate) characteristic of cartilage. These particular chondrocytes will be responsible for constructing the axial skeleton (vertebrae, ribs, cartilage, and ligaments). The cells of the *lateral* portion of the somite (that region farthest from the neural tube) also disperse. These cells form the precursors of the muscles of the limb and body wall. Ordahl and Le Douarin (1992) have followed these cells by transplanting portions of quail somites into somites of chicken embryos. The chick and quail cells can be distinguished by their nucleolar morphology. The researchers find that whatever somite cells are farthest from the neural tube migrate to form body wall and limb musculature, even if those donor cells were originally from the medial portion of a somite.

Once the cells of the sclerotome and the body wall and limb muscle cell precursors have migrated away from the somite, the somitic cells closest to the neural tube migrate ventrally down the remaining epithelial portion of the somite to form a solid double-layered epithelium called the **dermamyotome** (see Figures 9.2 and 9.5). The dorsal layer of this structure is called the **dermatome,** and it generates the mesenchymal connective tissue of the dorsal skin: the dermis. (The dermis of other areas of the body forms from other mesenchymal cells, not from the somites.) The inner layer of cells is called the **myotome,** and these cells give rise to the vertebral muscles that span the vertebrae and allow the back to bend (Chevallier et al., 1977; Christ

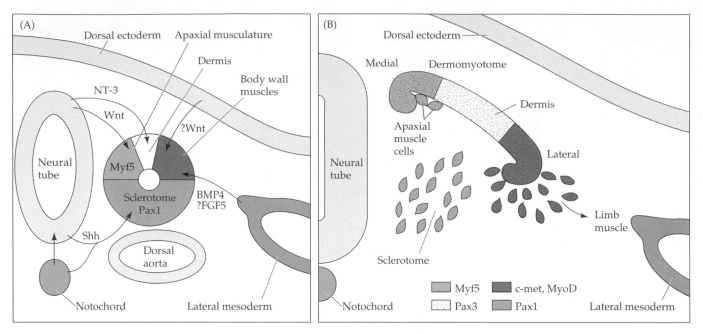

Figure 9.6
Model of major postulated interactions in the patterning of the somite. (A) Sonic hedgehog from the notochord and floor plate cells induces sclerotome formation; Wnt from the neural tube induces the region of the myotome that forms apaxial musculature, and the combination of Wnt protein from the epidermis and BMP4 (and perhaps FGF5) from the lateral plate mesoderm induces the portion of the myotome that gives rise to the body wall muscles. Neurotrophin 3 from the neural tube may cause the differentiation of the dermamyotome cells. (B) Different transcription factors in the different regions of the somite presage cell fate. The cells of the sclerotome express Pax1, while the medial dermamyotome cells express myogenic protein Myf5. The lateral dermamyotome cells express myogenic transcription factor MyoD as well as the c-met receptor for scatter factor. The central portion of the dermamyotome becomes the dermis and expresses Pax3. (After Cossu et al., 1996b.)

et al., 1977). Thus, the somites are critical for forming the back of our bodies: the vertebrae that surround the spinal cord, the muscles and connective tissue that hold the vertebrae together, the dermal underlayer of the skin of the back, and the back musculature. And what happens to the notochord, that central mesodermal structure? After it has provided the axial integrity of the early embryo and has induced the formation of the dorsal neural tube, most of it degenerates. Wherever the sclerotome cells have formed a vertebral body, the notochordal cells die. However, in between the vertebrae, the notochordal cells form the tissue of the intervertebral discs, the **nuclei pulposi.** These are the discs that "slip" in certain back injuries.

The specification of the somite is accomplished by the interaction of several tissues that form its environment. The ventral-medial portion of the somite is induced to become the sclerotome by factors, especially **Sonic hedgehog** protein, secreted from the notochord and neural tube floor plate (Fan and Tessier-Lavigne, 1994; Johnson et al., 1994). If portions of the notochord (or another source of Sonic hedgehog) are transplanted next to other regions of the somite, those regions, too, will become sclerotome cells. These sclerotome cells express a new transcription factor, Pax1, which activates the cartilage-specific genes and whose presence is necessary for the formation of the vertebrae (Figure 9.6; Smith and Tuan, 1996). They also express I-mf, an inhibitor of the MyoD family of transcription factors that initiate muscle formation (Chen et al., 1996). In similar ways, the myotome is seen to be induced by two distinct signals. The epaxial muscle cells (which surround the body axis) come from the medial portion of the somite and are induced by factors from the dorsal neural tube, probably members of the Wnt family (Münsterberg et al., 1995; Stern et al., 1995). The hypaxial muscles (which are formed from the medial portion of the somite and form the musculature of the limbs and body wall) are probably induced through the combination of Wnt proteins from the epidermis and bone morphogenetic protein-4 (BMP4) protein from the lateral plate mesoderm (Cossu et al., 1996a; Pourquié et al., 1996). These factors cause the myotome cells to express particular transcription factors (**MyoD** and **Myf5**) that activate the muscle-specific genes. The dermatome differentiates in response to another factor secreted by the neural tube, neurotrophin 3 (NT-3). Antibodies that block the activities of NT-3 prevent the conversion of the epithelial dermatome into the loose dermal mesenchyme that migrates beneath the epidermis (Brill et al., 1995). In addition to these positive signals, there are at least two other sets of proteins necessary for the patterning of the somite into its particular regions. One of these factors prevents the activation of one group of cells by inappropriate proteins. For example, the BMP4 signal from the lateral plate mesoderm is countered by a factor from the neural tube that prevents low levels of BMP4 from acting on the more medial cells. The other set of proteins is necessary for maintaining the pattern of gene expression initiated by the original signal (Pownall et al., 1996). [mesend1.html]

Myogenesis: Differentiation of Skeletal Muscle

A skeletal muscle cell is an extremely large, elongated cell that contains many nuclei. In the mid-1960s, developmental biologists debated whether each of these cells (often called **myotubes**) was derived from the fusion of several mononucleate muscle precursor cells (**myoblasts**) or from a single myoblast that undergoes nuclear division without cytokinesis. Evidence for the fusion of skeletal myoblasts to form multinucleated myotubes came from two independent sources. The critical evidence for myoblast fusion came from chimeric mice. These mice can be formed from the fusion of two early embryos, which regulate to produce a single mouse having two distinct cell populations (see Figure 5.28). Mintz and Baker (1967) fused mouse embryos that produced different types of the enzyme isocitrate dehydrogenase. This enzyme, found in all cells, is composed of two identical subunits. Thus, if myotubes are formed from one cell whose nuclei divide without cytokinesis, one would expect to find two distinct forms of the enzyme, that is, the two parental forms, in the allophenic mouse (Figure 9.7). But if myotubes are formed by fusion between cells, one would expect to find muscle cells expressing not only the two parental types of enzymes (AA and BB), but also a third class composed of a subunit from each of the parental types (AB). The different forms of isocitrate dehydrogenase can be separated and identified by their electrophoretic mobility. The results clearly demonstrated that although only the two parental types of enzyme were present in all the other tissues of the allophenic mice, the hybrid (AB) enzyme was present in extracts of skeletal muscle tissue. Thus, the myotubes must have been formed from the fusion of numerous myoblasts.

This evidence was important in showing that myoblast fusion actually occurred within the embryo. The analysis of *how* this fusion takes place was based on such fusion events occurring in culture. Konigsberg (1963) found that myoblasts isolated from chick embryos would proliferate in collagen-coated petri dishes. After about 2 days, however, these myoblasts stopped dividing and began to fuse with their neighbors to produce extended myotubes synthesizing muscle-specific proteins. DNA synthesis and nuclear division were not seen in the multinucleated myotubes. This fusion process is a complex orchestration of biochemical events at the myoblast cell surface. The first step appears to be the withdrawal of the cells from the division cycle. As long as there are particular growth factors in the medium (particularly fibroblast growth factors), the myoblasts will proliferate without differentiating. When these factors are depleted, the myoblasts cease dividing, secrete fibronectin onto their extracellular matrix, and bind to it through their $\alpha 5\beta 1$ integrin, the major fibronectin receptor (Menko and Boettiger, 1987; Boettiger et al., 1995). If this adhesion is blocked, no further muscle development ensues, and it appears that the signal from the integrin-fibronectin attachment is critical for initiating the myoblast to differentiate into muscle

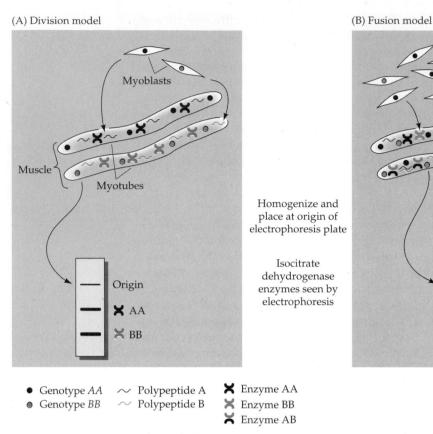

(A) Division model   (B) Fusion model

Myoblasts

Muscle

Myotubes

Muscle

Homogenize and place at origin of electrophoresis plate

Isocitrate dehydrogenase enzymes seen by electrophoresis

Origin

✘ AA

✘ BB

Hybrid enzyme formed

Origin

✘ AA
✘ AB
✘ BB

● Genotype *AA*   ∼ Polypeptide A   ✘ Enzyme AA
● Genotype *BB*   ∼ Polypeptide B   ✘ Enzyme BB
                                      ✘ Enzyme AB

Figure 9.7
The two possible mechanisms of skeletal muscle formation and how to distinguish between them. Chimeric mice are made from the fusion of mouse embryos from two different strains, each strain making a different form of the enzyme isocitrate dehydrogenase. This enzyme is composed of two subunits; one strain of mouse makes AA isocitrate dehydrogenase (indicated in black) and the other makes BB (color). (A) If the enzymes are made in a single cell or in multinucleate cells arising from the nuclear divisions within a single cell, the enzyme will be purely AA or BB. (B) If there are two different nuclei in the same cell, however, one might code for B subunits while the other might code for A, with the result that some molecules of the enzyme will be hybrid (AB). Electrophoresis can separate these three types of molecules. The presence of the AB molecule in skeletal muscle cells (but not in other cell types) confirms the fusion model. (After Mintz and Baker, 1967.)

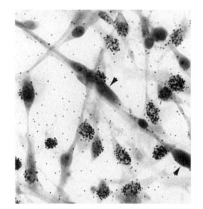

Figure 9.8
Autoradiograph showing DNA synthesis in myoblasts and the exit of fusing cells from the cell cycle. Phospholipase C can "freeze" myoblasts after they have aligned with other myoblasts but before their membranes fuse. Such cultured myoblasts were treated with phospholipase C and then exposed to radioactive thymidine. Unattached myoblasts still divide and incorporate the radioactive thymidine into their DNA. Lined-up (but not yet fused) cells (arrows) do not incorporate the label. (From Nameroff and Munar, 1976, courtesy of M. Nameroff.)

cells (Figure 9.8). The second step is the alignment of the myoblasts together into chains. This step is mediated by cell membrane glycoproteins, including several cadherins and CAMs (Knudsen, 1985: Knudsen et al., 1990). Recognition and alignment between cells only takes place if the two cells are myoblasts. Fusion can even occur between chick and rat myoblasts (Yaffe and Feldman, 1965); the identity of the species is not critical in culture.

The third step is the cell fusion event itself. Like most membrane fusions, calcium ions are critical, and fusion can be activated by calcium ionophores, such as A23187, which carry calcium ions across cell membranes (Shainberg et al., 1969; David et al., 1981). Fusion appears to be mediated by a set of metalloproteinases called **meltrins.** These proteins were discovered during a search for myoblast proteins that would be homologous to fertilin, a protein implicated in sperm-egg membrane fusion. Yagami-Hiromasa and colleagues (1995) found that one of these meltrins (meltrin-α) is expressed in myoblasts at about the same time that fusion begins and that antisense RNA to the *meltrin-α* message inhibited fusion when added to myoblasts.

Muscle Building and the MyoD Family of Transcriptional Regulators

OW IS an embryonic mesenchymal cell instructed to form a muscle cell instead of a cartilage cell, a fibroblast, or an adipose cell? What molecules commit its fate to one lineage and not another? In 1986, Lassar and co-workers took DNA from myoblast cells and transfected it into a certain embryonic mouse cell type, the C3H10T$\frac{1}{2}$ cell. This cell has a fibroblast-like appearance, but it resembles primitive mesenchyme, since it can become an adipose cell, a muscle cell, or cartilage. When the muscle DNA was added to these cells, the C3H10T$\frac{1}{2}$ cells were transformed into muscle cells. DNA isolated from fibroblasts or other cell types cannot accomplish this conversion. By subtraction cloning (see Chapter 2), a myoblast-specific mRNA was found that could also effect this change in differentiated phenotype. The myoblast mRNA encoded a protein called myoblast determination 1 protein, or, more commonly, **MyoD** (Davis et al., 1987). The *MyoD* gene is expressed only in cells of the muscle lineages. It appears to be a "master switch" gene in that it can convert other cell types into muscles if this gene is active in them. This hypothesis was tested by cloning the *MyoD* gene into a viral vector so that it was under the control of a constitutively active viral promoter (it was always "on"). When this *MyoD* fusion gene was transfected into various cells, pigment cells, nerve cells, fat cells, fibroblasts, and liver cells were converted into muscle-like cells (Figure 9.9; Weintraub et al., 1989). Thus, MyoD appears to be sufficient to activate the muscle-specific genes that make up the muscle phenotype.

MyoD encodes a nuclear DNA-binding protein that can bind to regions of the DNA adjacent to muscle-specific genes and activate these genes. For instance, the MyoD protein appears to directly activate the muscle-specific creatine phosphokinase gene by binding to the DNA immediately upstream from it (Lassar et al., 1989). Similarly, there are

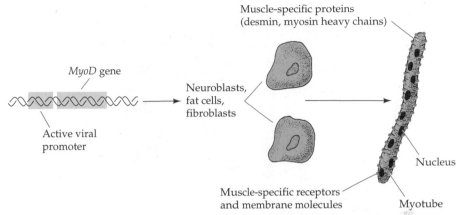

Figure 9.9 Summary of several experiments in which the *MyoD* gene was activated by a viral promoter and transfected into nonmuscle cells. The MyoD protein appears to override the original regulators of the cell phenotype and convert the cells into muscles.

two MyoD-binding sites on the DNA adjacent to a subunit gene of the chicken muscle acetylcholine receptor (Piette et al., 1990). It also directly activates itself. Once the *MyoD* gene is on, its protein product binds to the DNA immediately upstream of the *MyoD* gene and keeps this gene from being turned off (Thayer et al., 1989). In other cases, the effects of MyoD may be indirect. Not all genes involved in producing the muscle phenotype may be directly activated by MyoD protein. MyoD probably acts indirectly by turning on other regulatory genes, which then activate the structural muscle-specific genes.

MyoD is not the only muscle switch gene. There is a family of MyoD-like proteins that have very similar structures and appear to be able to substitute for each other to a large degree. This family of proteins (sometimes referred to as the "MyoD family" or the "myogenic bHLH proteins") includes **myogenin, Myf5,** and **MRF4**, and these proteins appear to bind to similar sites on the DNA (to be discussed in Chapter 10). Transfection of any of these myogenic genes into a wide range of cultured cells also converts these cells into muscles. *MyoD* expression leads to myogenin expression, and

the transfection of myogenin genes activates *MyoD* expression. Thus, there is a reciprocal positive feedback loop such that when either *myogenin* or *MyoD* is activated, so is the other gene (Thayer et al., 1989). In the chick, *MyoD* is activated in somitic cells that generate the abdominal and limb musculature, while *myf5* is activated in cells producing the back muscles. In both cases, this activation commits the somitic cell to the myogenic lineage. Both groups of cells then express *myogenin* and *MRF4* to produce their myotubes and myofibers (Figure 9.10; Lyons and Buckingham, 1992; Pownall and Emerson, 1992a,b; Braun and Arnold, 1996; Cossu et al., 1996a).

In some instances, these myogenic transcription factors can compensate for the loss of one or the other. Using a gene-targeting technique (see Chapter 2), Rudnicki and colleagues (1992) showed that Myf5 and MyoD can accomplish the same functions. When mice lack both *MyoD* genes, the expression of the myf5 gene takes over. The resulting mice have normal muscle development. When the mice lack their *myf5* genes, they also have normal muscle development. However, the absence of Myf5 protein causes the delay of myotome formation

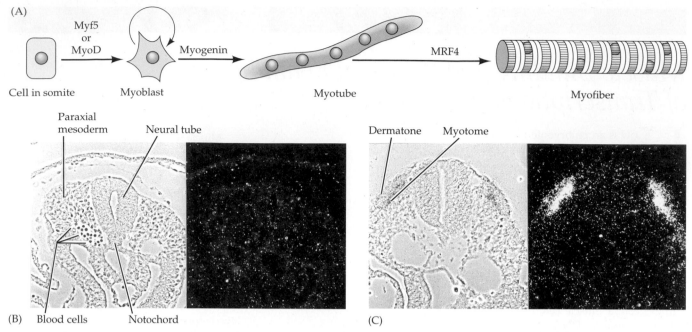

(B)   Blood cells   Notochord   (C)

Figure 9.10   Muscle commitment and differentiation mediated by the MyoD family of transcription factors. (A) Postulated roles of myogenic proteins during skeletal muscle formation in the mouse. (B) In situ hybridization indicating the absence of *myf5* mRNA in the unsegmented paraxial mesoderm of the mouse embryo. Left-hand side shows light microscope photograph of the area. (C) In situ hybridization showing the presence of *myf5* mRNA in the myotome of the mouse embryonic somite. (A after Rudnicki et al., 1993; photographs courtesy of G. Lyons.)

by several days, and this causes a failure of the lateral portion of the sclerotome to form properly. Although these mice have normal muscles, their rib cages are distorted and they are unable to breathe (Braun et al., 1992). Recent experiments in Rudolf Jaenisch's laboratory (Rudnicki et al., 1993) show that when the *myf5* and *MyoD* genes are *both* absent from the embryo, no muscles or ribs form.* While MyoD and Myf5 can substitute for one another, there does not appear to be redundancy in the functions of myogenin. Mice homozygous for a targeted mutation in the *myogenin* gene die shortly after birth because of defects in their muscle cell formation (Hasty et al., 1993; Nabeshima et al., 1993). The somites formed normally and were compartmentalized into myotome, sclerotome, and dermatome, but the myoblasts failed to differentiate into myofibers (Venuti et al., 1995).

MyoD and its relatives appear to be critical in removing the myoblast from the cell cycle. As mentioned earlier, dividing myoblasts do not differentiate. This distinction between division and differentiation is characteristic of several cell types that are derived from stem cell populations (Bischoff and Holtzer, 1969; Holtzer et al., 1975). There appear to be two ways that the myoblast removes itself from the cell cycle. The first mechanism is to inhibit the cell division pathway. To do this, the MyoD protein induces the expression of p21, an inhibitor of cyclin-dependent kinases (Figure 9.11; Halevy et al., 1995). The second mechanism of withdrawing from the cell cycle involves the myoblast's down regulation of its growth factor receptors. One of the major growth factors promoting myoblast cell division is basic fibroblast growth factor. FGF2 promotes myoblast cell division, while at the same time inhibiting myoblast differentiation by suppressing the transcription of *MyoD* and *myogenin* (Vaidya et al., 1989; Brunetti and Goldfine, 1990). The FGF receptors are lost when the myoblast differentiates into a muscle cell (Olwin and Hauschka, 1988; Moore et al., 1991).

How are the MyoD family proteins turned on? New experiments have provided the bases for some fascinating speculations. George-Weinstein and her colleagues (1996) have demonstrated that when chick epiblasts are isolated from the rest of the gastrula and separated into their individual cells, these epiblast cells become muscle. Moreover, the researchers find that *MyoD* mRNA (and perhaps protein) is present in these cells. It appears that epiblast cells have the "preferred" ability to become committed to myoblasts, and it is only their interactions with other cell types that prevent their becoming muscles. In this case, the factors that promote myogenesis (such as the Wnt proteins) may do so by repressing the inhibitors. One of these inhibitors may be the **Twist** protein. The twist protein is a DNA-binding protein

*This means that there is some redundancy in the development of skeletal muscles. Such redundancy has long been known to embryologists (Spemann, 1938), but geneticists are rediscovering it (to their dismay, as it muddies up the interpretation of such experiments). Gould (1990) considers developmental redundancy essential for evolution to occur, since one of the redundant partners is free to get a new function while the other partner maintains the original function.

(A)

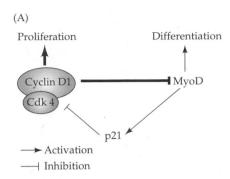

(B)

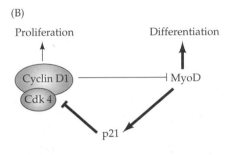

Figure 9.11 Switch between proliferation and differentiation. (A) Conditions favoring proliferation (as when there are abundant fibroblast growth factors in the medium) enable the continued expression of cyclin-dependent kinase 4. This kinase is able to repress the expression of MyoD. (B) Conversely, once MyoD is formed, it can suppress cdk4 through its activation of the p21 protein. In this manner, the dividing cells will not differentiate and the differentiated cells will not divide. (After Halevy et al., 1995.)

that looks very much like MyoD. However, it appears to inhibit MyoD and other such proteins from binding to the promoters of their target muscle-specific genes. The *twist* gene is originally present throughout the early somite but then becomes absent specifically in the myotome (Spicer et al., 1996). It is possible that MyoD and other myogenic bHLH proteins are already present in the epiblast cells but are prohibited from functioning until the twist protein is downregulated. This downregulation could possibly come as a result of the secretion of Wnt proteins (by the epidermis or neural tube), which could abrogate an inhibitory effect mediated by Notch1.

In addition to the myogenic bHLH proteins, another transcription factor, **MEF2A,** appears to be important for skeletal muscle development. MEF2A also induces fibroblasts to become muscles, and it appears to cooperate with MyoD on the enhancers of muscle-specific genes. Kaushal and colleagues (1994) speculate that MEF2A provides additional specifity to MyoD binding such that MyoD doesn't inadvertently activate non-muscle genes that have a regulating sequence capable of binding bHLH proteins. ■

Osteogenesis: The Development of Bones

Some of the most obvious structures derived from the somitic mesoderm are the bones. In this chapter, we can only begin to outline the mechanisms of bone formation, and students wishing further details are invited to consult histology textbooks that devote entire chapters to this topic. There are three distinct lineages that generate the skeleton. The sclerotome generates the axial skeleton, the lateral plate mesoderm generates the limb skeleton, and the cranial neural crest gives rise to the branchial arch and craniofacial bones and cartilage.* There are two major modes of bone formation, or **osteogenesis,** and both involve the transformation of a preexisting mesenchymal tissue into bone tissue. The direct conversion of mesenchymal tissue into bone is called **intramembranous ossification.** This occurs primarily in the bones of the skull. In other cases, the mesenchymal cells differentiate into cartilage, and this cartilage is later replaced by bone. This process by which a cartilage intermediate is formed and replaced by bone cells is called **endochondral ossification.**

INTRAMEMBRANOUS OSSIFICATION. Intramembranous ossification is the characteristic way in which the flat bones of the skull are formed. Mesenchymal cells derived from the neural crest interact with the extracellular matrix of the head epithelial cells to form bone. If the mesenchymal cells do not contact this matrix, no bone will be formed (Tyler and Hall, 1977; Hall, 1988). This was shown in vitro by Hall and colleagues (1983), who isolated head

---

*Craniofacial cartilage development was discussed in Chapter 7 and will be revisited in Chapter 23; the development of the limb will be detailed in Chapter 18.

Figure 9.12
Schematic diagram of membranous ossification. (A) Mesenchymal cells, probably derived from the neural crest, condense to produce osteoblasts, which deposit osteoid matrix. These osteoblasts become arrayed along the calcified region of the matrix. Osteoblasts that are trapped within the bone matrix become osteocytes. (B) Spread of bone spicules from the primary ossification site in the flat skull bones of a 3-month-old human embryo. The bones shown in black are formed by endochondral ossification. (After Langman, 1981.)

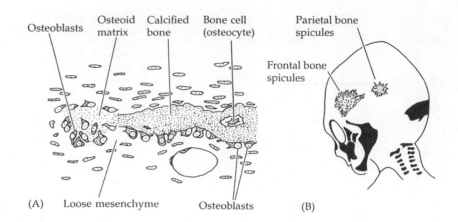

mesenchymal cells and plated them into culture dishes. If no extracellular matrix was present on the surface of these dishes, the cells remained mesenchymal. However, if head epithelial cells had first secreted an extracellular matrix upon the surface, the cells differentiated into bone cells.

The mechanisms responsible for this conversion of mesenchymal cells into bone are still unknown, but recent evidence points to a particular group of molecules at the epithelial-mesenchymal junction. **Bone morphogenetic proteins** can be isolated from adult bone and injected into embryonic muscle or connective tissues. When this is done, cartilage develops from cells within these tissues, and this cartilage is then replaced by bone cells (Syftestad and Caplan, 1984; Urist et al., 1984; see Chapter 17).

During intramembranous ossification, the mesenchymal cells proliferate and condense into compact nodes. Some of these cells develop into capillaries; others change their shape to become **osteoblasts,** cells capable of secreting the bone matrix. The secreted collagen-proteoglycan matrix is able to bind calcium salts, which are brought to the region through capillaries. In this way, the matrix becomes calcified. In most cases, osteoblasts are separated from the region of calcification by a layer of the prebone (osteoid) matrix they secrete. Occasionally, though, osteoblasts become trapped in the calcified matrix and become **osteocytes**—bone cells. As calcification proceeds, the bony spicules radiate out from the center, where ossification began (Figure 9.12). Furthermore, the entire region of calcified spicules becomes surrounded by compact mesenchymal cells that form the **periosteum.** The cells on the inner surface of the periosteum also become osteoblasts and deposit bone matrix parallel to that of the existing spicules. In this manner, many layers of bone are formed.

ENDOCHONDRAL OSSIFICATION. Endochondral ossification involves the formation of cartilage tissue from aggregated mesenchymal cells and the subsequent replacement of this cartilage tissue by bone (Horton, 1990). The cartilage tissue is a model for the bone that follows. The skeletal components of the vertebral column, the pelvis, and the extremities are first formed of cartilage and then later changed into bone. This remarkable process coordinates **chondrogenesis** (cartilage production) with osteogenesis (bone growth); the skeletal elements are simultaneously bearing a load, growing in width, and responding to local stresses. The cells that form cartilage tissues expres **Scleraxis**, a transcription factor that is thought to activate the cartilage-specific genes (see Figure 9.13; Titlepage; Cserjesi et al., 1995). Thus, Scleraxis is expressed in the sclerotome, in the facial mesenchyme that forms cartiligenous

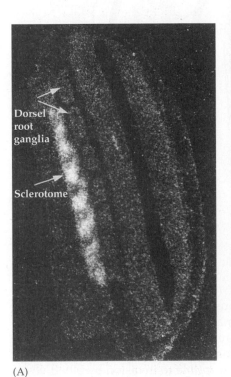

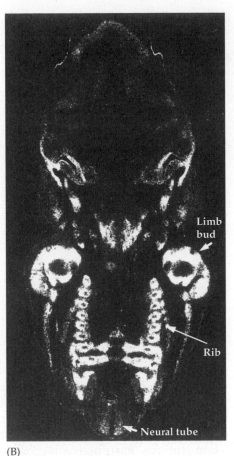

(A)   (B)

Figure 9.13
Localization of the *scleraxis* message at the sites of chondrocyte formation. (A) Expression of *scleraxis* in the somites of a 12.5-day mouse embryo. This section was cut tangentially, and the neural tube runs along the anterior-posterior axis. (B) Section through a 11.5-day mouse embryo where scleraxis transcripts are seen in the condensing cartilage of the nose and face and in the precursors of the limbs and ribs. (After Cserjesi et al., 1995; photographs courtesy of Dr. E. Olson.)

precursors to bone, and in the limb mesenchyme. This protein remains active until the cartilage begins to be replaced by bony tissue. [mesend2.html]

Cartilage formation can be divided into three phases: mesenchyme proliferation, condensation of the precartiligenous mesenchyme, and differentiation of the chondrocyte. Chondrogenesis is initiated when the dividing precartilage mesenchyme cells begin expressing extracellular matrix proteins that cause them to condense into nodules. N-cadherin appears to be important in the initiation of these condensations, and N-CAM seems to be critical for maintaining them (Oberlender and Tuan, 1994; Hall and Miyake, 1995). Once condensed, the cells become **chondrocytes** and begin secreting the chondrocyte-specific extracellular proteoglycans and collagens.*

In humans, the "long bones" of the embryonic limb buds form from mesenchymal cells that form nodules in those regions that will become bone. These cells become chondrocytes, and they secrete the cartilage extracellular matrix. The mesenchymal cells around them become the periosteum

---

*Mutations that affect the formation of nodules often cause limb abnormalities. In chicks, the *talpid* mutations are characterized by duplicated and fused limbs. This, in turn, has been found to be caused by abnormally large prechondrogenic condensations. These large nodules are caused by the excessive stickiness of the mesenchyme cells in these condensates, and this has been linked to an overexpression of N-CAM (Ede 1983; Chuong et al., 1993). In humans, the *SOX9* gene is expressed in the precartilaginous condensations, and this encodes a DNA-binding protein. Mutations of the *SOX9* gene cause camptomelic dysplasia, a rare disorder of skeletal development, causing deformities of most of the bones of the body. Most of the affected babies die from respiratory failure due to poorly formed tracheobronchial and rib cartilage (Wright et al., 1995).

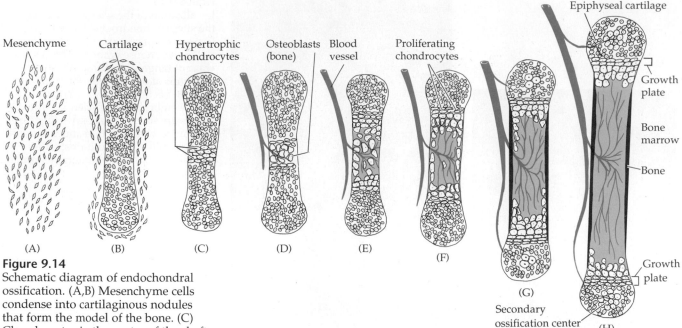

Mesenchyme   Cartilage   Hypertrophic chondrocytes   Osteoblasts (bone)   Blood vessel   Proliferating chondrocytes   Epiphyseal cartilage   Growth plate   Bone marrow   Bone   Growth plate   Secondary ossification center

(A)   (B)   (C)   (D)   (E)   (F)   (G)   (H)

Figure 9.14
Schematic diagram of endochondral ossification. (A,B) Mesenchyme cells condense into cartilaginous nodules that form the model of the bone. (C) Chondrocytes in the center of the shaft undergo hypertrophy and change their extracellular matrix, allowing blood vessels to enter. (D,E) Blood vessels bring in osteoblasts, which bind to the degenerating cartilaginous matrix and deposit bone matrix. (F–H) Formation of the epiphyseal growth plates by chondrocytes, which proliferate before undergoing hypertrophy. Secondary ossification centers also form as blood vessels enter near the tips of the bone. (After Horton, 1990.)

(Figure 9.14). Soon after the cartilaginous "model" is formed, the cells in the central part of the model become dramatically larger and begin secreting a different type of matrix, one that contains different types of collagen, more fibronectin, and less protease inhibitor. These cells are the **hypertrophic chondrocytes**. Their matrix is more susceptible to invasion by blood vessel cells from the periosteum. A capillary from the periosteum then invades the center of the previously avascular cartilage shaft. As the cartilage matrix is degraded, the hypertrophic cartilage cells die, and osteoblasts (bone-forming cells) carried by the blood vessels begin to secrete bone matrix on the partially degraded cartilage (Hattori et al., 1995). Eventually, all the cartilage is replaced by bone.

As the center of the cartilage model is converted into bone, an ossification front is formed between the newly synthesized bone and the remaining cartilage. The cartilage side of this front contains the hypertrophic cartilage that prepares the shaft for invasion by the blood vessels, and the bone side contains the osteoblast cells laying down the bone matrix. This front spreads outward in both directions from the center as more cartilage is turned to bone. If this were all, however, there would be no growth, and our bones would just be as large as the original cartilaginous model. However, as the ossification front nears the ends of the cartilage model, the chondrocytes near the ossification front proliferate prior to undergoing hypertrophy. This pushes out the cartilaginous ends of the bone, providing a source of new cartilage. These cartilaginous regions at the end of the long bones are called **epiphyseal growth plates.** These plates contain three regions: a region of chondrocyte proliferation, a region of mature chondrocytes, and a region of hypertrophying chondrocytes (Figure 9.15; Chen et al., 1995). As this cartilage hypertrophies and the ossification front extends further outward, the remaining cartilage in the epiphyseal plate proliferates. This cartilage forms the growth area of the bone. Thus, the bone keeps growing because of the production of new cartilage cells that undergo hypertrophy, enable the blood vessels to enter, and die as the bony matrix is deposited. As long as

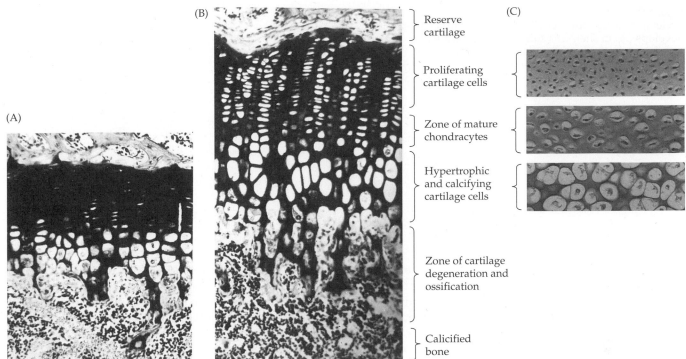

(B)

- Reserve cartilage
- Proliferating cartilage cells
- Zone of mature chondracytes
- Hypertrophic and calcifying cartilage cells
- Zone of cartilage degeneration and ossification
- Calicified bone

(A)

(C)

Figure 9.15
Proliferation of cells in epiphyseal plate in response to growth hormone. (A) Cartilaginous region in a young rat that was made growth hormone–deficient by removal of its pituitary. (B) Same region in the rat after injection of growth hormone. (C) Stained cartilage in particular regions of the growth plate. (I. Gersh's photographs from Bloom and Fawcett, 1975; C from Chen et al., 1995, courtesy of P. Goetinck.)

the epiphyseal growth plates are able to produce chondrocytes, the bone continues to grow.

The epiphyseal growth plate cells are very responsive to hormones, and their proliferation is stimulated by growth hormone and insulin-like growth factors. Nilsson and colleagues (1986) have recently shown that growth hormone stimulates the production of **insulin-like growth factor I (IGF-I)** in these chondrocytes and that these chondrocytes respond to it by proliferating. When they added growth hormone to the tibial growth plate of young mice (who could not manufacture their own growth hormone because their pituitaries had been removed), the growth hormone stimulated the formation of IGF-I from the chondrocytes in the proliferative zone (see Figure 9.15). The combination of growth hormone and IGF-I appears to provide an extremely strong mitotic signal. The pygmies of the Ituri Forest of Zaire have normal growth hormone and IGF-I levels until puberty. However, at puberty, the pygmies' IGF-I levels fall to about one-third that of other adolescents. It appears that IGF-I is essential for the normal growth spurt at puberty (Merimee et al., 1987). Hormones are also responsible for the cessation of growth. At the end of puberty, high levels of estrogen or testosterone cause the remaining epiphyseal plate cartilage to undergo hypertrophy. These cartilage cells grow, die, and are replaced by bone. Without any further cartilage, growth of these bones ceases.

The replacement of chondrocytes by osteoblasts appears to depend on the mineralization of the extracellular matrix. In chick embryos, the source of calcium is the calcium carbonate of the eggshell, and during its development, the circulatory system of the chick translocates about 120 mg of calcium from the shell to the skeleton (Tuan, 1987). When chick embryos are removed from their shells at day 3 and grown in shell-less culture (in plastic wrap) for the duration of their development, much of the calcium-deficient cartilaginous skeleton fails to mature into bony tissue (Figure 9.16; Tuan and Lynch, 1983). In mammals, calcium is transferred across the placenta and is

Figure 9.16
Skeletal mineralization in 17-day chick embryos that develops (A) in shell-less culture and (B) inside the egg during normal incubation. The embryos were fixed and stained with Alizarin red to show the calcified matrix. (From Tuan and Lynch, 1983, courtesy of R. Tuan.)

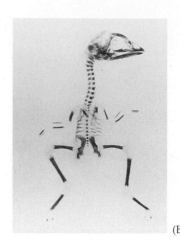

(A)

(B)

deposited onto the matrix by the chondrocytes. It has been shown that hypertrophic chondrocytes switch from aerobic to anaerobic respiration (Brighton and Hunt, 1974; Brighton, 1984), causing a decrease in cellular ATP and the employment of a phosphocreatine-mediated energy pathway, such as used in oxygen-depleted muscle (Shapiro et al., 1992). By some as yet unknown mechanisms, these metabolic changes are thought to result in the deposition of calcium in the extracellular matrix, within tiny membrane-bound structures known as **matrix vesicles** (Wuthier, 1982). This initiates the process of calcification and allows osteoblasts to bind and initiate bone formation (Figure 9.17).

As new bone material is added peripherally from the internal surface of the periosteum, there is a hollowing out of the internal region to form the bone marrow cavity. This destruction of bone tissue is due to **osteoclasts,** multinucleated cells that enter the bone through the blood vessels (Kahn and Simmons, 1975; Manolagas and Jilka, 1995). Osteoclasts are probably derived from the same precursors as blood cells, and they dissolve both the inorganic and the protein portions of the bone matrix (Ash et al., 1980; Blair et al., 1986). The osteoclast extends numerous cellular processes into the matrix and pumps out hydrogen ions from the osteoclast onto the surrounding material, thereby acidifying and solubilizing it (Figure 9.18; Baron et al., 1985, 1986). The blood vessels also import the blood-forming cells, which will reside in the marrow for the duration of the organism's life.

Chondrocytes

Calcium on
extracellular matrix

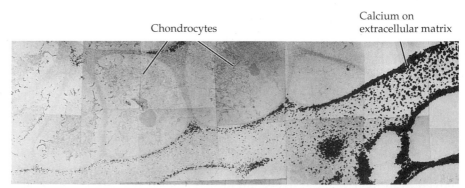

Figure 9.17
Deposition of calcium by chondrocytes in the distal region of the hypertrophic zone. Calcium (stained darkly in this electron micrograph montage) is placed onto the matrix by the enlarging cells. (From Brighton and Hunt, 1974, courtesy of C. T. Brighton.)

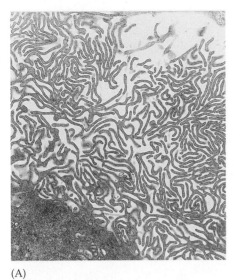

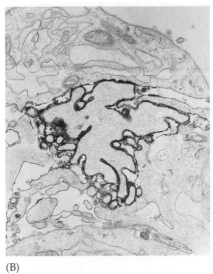

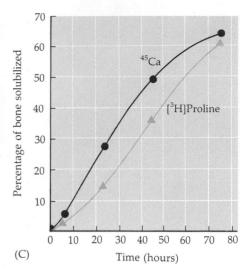

(A)

(B)

(C)

Figure 9.18
Osteoclast activity on the bone matrix. (A) Electron micrograph of the ruffled membrane of a chick osteoclast cultured on reconstituted bone matrix. (B) Section of ruffled membrane stained for the presence of an ATPase capable of transporting hydrogen ions from the cell. The ATPase is restricted to the membrane of the cell process. (C) Solubilization of inorganic and collagenous matrix components (as measured by the release of [^{45}Ca] and [^{3}H] proline, respectively) by 10,000 osteoclasts incubated on labeled bone fragments. (A and C from Blair et al., 1986; B from Baron et al., 1986, courtesy of the authors.)

Sidelights & Speculations

Control of Chondrogenesis at the Growth Plate

RECENT DISCOVERIES of human and murine mutations of skeletal development have provided remarkable insights into how the differentiation, proliferation, and patterning of chondrocytes are regulated.

Fibroblast Growth Factor Receptors
The proliferation of the epiphyseal growth plate cells and facial cartilage can be halted by the presence of **fibroblast growth factors** (Deng et al., 1996; Webster and Donoghue, 1996). These factors appear to instruct the cartilage precursors to differentiate rather than to divide. In humans, mutations of the receptors for fibroblast growth factors can cause these receptors to become activated prematurely. This gives rise to the major types of human dwarfisms. **Achondroplasia** is a dominant mutation caused by mutations in the transmem-

brane region of fibroblast growth factor receptor 3 (FGFR3). Roughly 95 percent of the achondroplastic dwarfs have the same mutation of FGFR3, a base pair substitution that converts glycine to arginine at position 380 in the transmembrane region of the protein. In addition, mutations in the extracellular portion of the FGFR3 protein or in the tyrosine kinase intracellular domain have resulted in **thanatophoric dysplasia,** a lethal form of dwarfism that resembles homozygous achondroplasia (Figure 9.19; Bellus et al., 1995; Tavormina et al., 1995). Mutations in FGFR1 can cause **Pfeiffer syndrome,** characterized by limb defects and premature fusion of the cranial sutures (craniosynostosis), resulting in abnormal skull and facial shape. Different mutations in FGFR2 can give rise to various abnormalities of the limbs and/or face (Park et al., 1995; Wilkie et al., 1995). [cell7.html]

The extracellular matrix of cartilage is also critical for the proper differentiation and organization of growth plate chondrocytes. Mutations that affect type XI collagen or the sulfation of cartilage proteoglycans can cause severe skeletal abnormalities. Mice with deficiencies of type XI collagen die at birth with abnormalities of limb, mandible, rib, and tracheal cartilage (Li et al., 1995). Failure to add sulfate groups to cartilage proteoglycans causes diastrophic dysplasia, a human dwarfism characterized by severe curvature of the spine, clubfoot, and deformed earlobes (Hästbacka et al., 1994). [cell6.html]

Estrogen Receptors
Hormones have also been known to have a marked effect on human epiphyses. The pubertal growth spurt and subsequent maturation of the epiphyseal plate (i.e., the conversion of proliferating

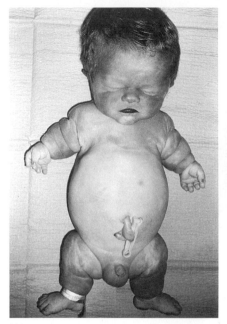

(A)

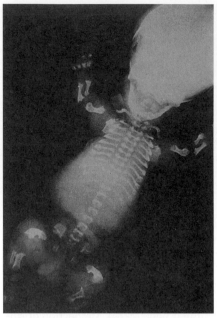

(B)

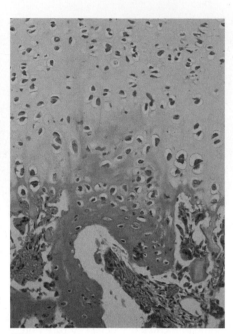

(C)

Figure 9.19   Human bone dysplasia caused by dominant activating mutations of fibroblast growth factor receptor 3. (A) Thanatophoric dysplasia, a fatal condition characterized by severe shortening of the ribs and limbs due to the epiphyses being covered by bony tissue. Death is due to breathing problems. (B) X-ray of an infant born with thanatophoric dysplasia. (C) Microscopic section showing the disorganization of an epiphysis in thanatophoric dysplasia. Note the absence of dividing chondrocytes. (From Gilbert-Barness and Opitz, 1996.)

cells to mature cartilage and bone) are induced by sex hormones (Kaplan and Grumbach, 1990). In conditions of precocious puberty, there is an initial growth spurt (making the individual taller than his or her peers), followed by the cessation of epiphyseal cell division (allowing that person's peers to catch up and surpass his or her height). In males, it was not thought that estrogen played any role in these events. However, in 1994, Smith and colleagues published the case history of a man whose growth was still linear despite undergoing a normal puberty. His epiphyseal plates had not matured, and he still had proliferating chondrocytes at 28 years of age. His "bone age"—the amount of ephiphyseal cartilage he retained—was roughly half his chronological age. This person was found to lack any functional estrogen receptor. Therefore, estrogen plays a role in epiphyseal maturation in males as well as in females. Thyroid hormone and parathyroid-related hormone are also important in regulating the maturation and hypertrophy program of the epiphyseal growth plate (Ballock and Reddi, 1994). Thus, children with hypothyroidism are prone to developing growth plate disorders. [limb3.html] ■

Lateral plate mesoderm

Not all of the mesodermal mantle is organized into somites. Adjacent to somitic mesoderm is the **intermediate mesodermal** region. This cord of mesodermal cells develops into the pronephric tubule, which is the precursor of kidney and genital ducts. The development of these organ systems will be discussed in detail in Chapters 17 and 19, respectively. Further laterally on each side we come to the **lateral plate mesoderm.** These plates split horizontally into the dorsal **somatic (parietal) mesoderm,** which underlies the ectoderm, and a ventral **splanchnic (visceral) mesoderm,** which overlies the endoderm (see Figure 9.2C). Between these layers is the body cavity—the **coelom**—which stretches from the future neck region to the posterior of the body. During later development, the right- and left-side coeloms fuse, and folds extend from the somatic mesoderm, dividing the coelom into separate cavities. In mammals, the coelom is subdivided into the pleural, peri-

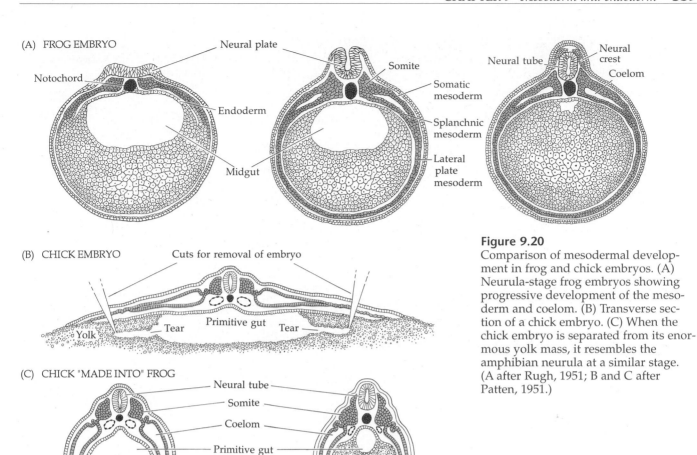

(A) FROG EMBRYO

Neural plate

Notochord

Endoderm

Midgut

Somite

Somatic mesoderm

Splanchnic mesoderm

Lateral plate mesoderm

Neural tube

Neural crest

Coelom

(B) CHICK EMBRYO

Cuts for removal of embryo

Yolk

Tear

Primitive gut

Tear

(C) CHICK "MADE INTO" FROG

Neural tube

Somite

Coelom

Primitive gut

Yolk

CHICK EMBRYO
(removed from yolk;
edges pulled together)

FROG EMBRYO

Figure 9.20
Comparison of mesodermal development in frog and chick embryos. (A) Neurula-stage frog embryos showing progressive development of the mesoderm and coelom. (B) Transverse section of a chick embryo. (C) When the chick embryo is separated from its enormous yolk mass, it resembles the amphibian neurula at a similar stage. (A after Rugh, 1951; B and C after Patten, 1951.)

cardial, and peritoneal spaces, enveloping the thorax, heart, and abdomen, respectively. The mechanism for creating mesodermal somites and body linings has changed little throughout vertebrate evolution, and the development of the chick mesoderm can be compared with similar stages of frog embryos (Figure 9.20).

Formation of Extraembryonic Membranes

In reptiles, birds, and mammals, embryonic development has taken a new direction. Reptiles evolved a mechanism for laying eggs on dry land, thus freeing them to explore niches that were not close to ponds. To accomplish this, the embryo produced four sets of **extraembryonic membranes** to mediate between it and the environment, and even though most mammals have evolved placentas instead of shells, the basic pattern of extraembryonic membranes remains the same. In developing reptiles, birds, and mammals, there initially is no distinction between embryonic and extraembryonic domains. However, as the body of the embryo takes shape, the border epithelia divide unequally to create the body **folds,** isolating the embryo from the yolk and delineating which areas are to be embryonic and which extraembryonic (Miller et al., 1994).

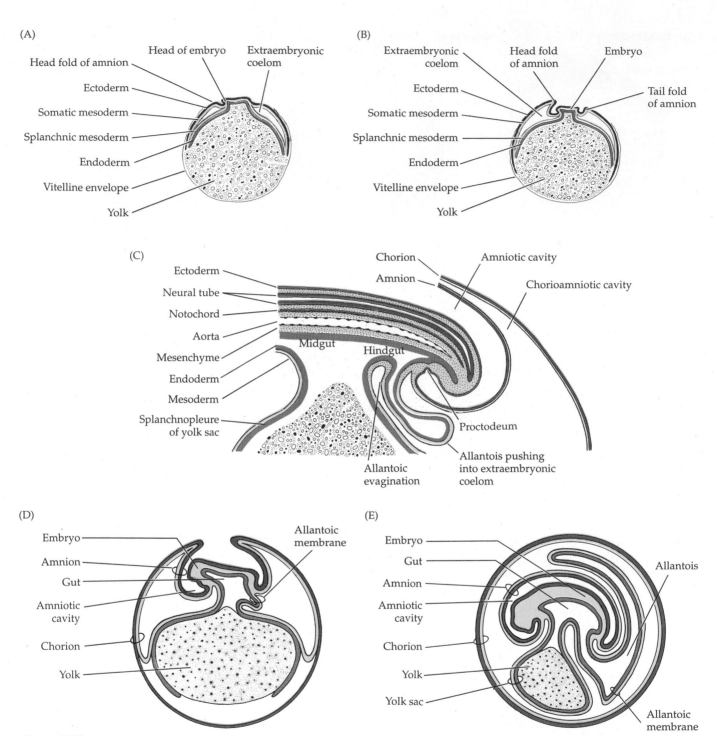

(A)
- Head fold of amnion
- Ectoderm
- Somatic mesoderm
- Splanchnic mesoderm
- Endoderm
- Vitelline envelope
- Yolk
- Head of embryo
- Extraembryonic coelom

(B)
- Extraembryonic coelom
- Ectoderm
- Somatic mesoderm
- Splanchnic mesoderm
- Endoderm
- Vitelline envelope
- Yolk
- Head fold of amnion
- Embryo
- Tail fold of amnion

(C)
- Ectoderm
- Neural tube
- Notochord
- Aorta
- Mesenchyme
- Endoderm
- Mesoderm
- Splanchnopleure of yolk sac
- Midgut
- Hindgut
- Chorion
- Amnion
- Amniotic cavity
- Chorioamniotic cavity
- Proctodeum
- Allantois pushing into extraembryonic coelom
- Allantoic evagination

(D)
- Embryo
- Amnion
- Gut
- Amniotic cavity
- Chorion
- Yolk
- Allantoic membrane

(E)
- Embryo
- Gut
- Amnion
- Amniotic cavity
- Chorion
- Yolk
- Yolk sac
- Allantois
- Allantoic membrane

Figure 9.21
Schematic drawings of the extraembryonic membranes of the chick. The embryo is cut longitudinally, and the albumen and shell coatings are not shown. (A) A 2-day embryo. (B) A 3-day embryo. (C) Detailed schematic diagram of the caudal (hind) region of the chick embryo, showing the formation of the allantois. (D) A 5-day embryo. (E) A 9-day embryo. (After Carlson, 1981.)

The membranous folds are formed by the extension of ectodermal and endodermal epithelium underlaid with mesoderm. The combination of ectoderm and mesoderm, often referred to as the **somatopleure,** forms the amnion and chorion membranes; the combination of endoderm and mesoderm—the **splanchnopleure**—forms the yolk sac and allantois. The endodermal or ectodermal tissue acts as the functioning epithelial cells; and the mesoderm generates the essential blood supply to and from this epithelium. The formation of these folds can be followed in Figure 9.21.

The first problem of a land-dwelling egg is desiccation. Embryonic cells would quickly dry out if they were not in an aqueous environment. This environment is supplied by the **amnion.** The cells of this membrane secrete amniotic fluid; thus, embryogenesis still occurs in water. This evolutionary advance is so significant and characteristic that reptiles, birds, and mammals are grouped together as the **amniote vertebrates.**

The second problem of a land-dwelling egg is gas exchange. This exchange is provided for by the **chorion,** the outermost extraembryonic membrane. In birds and reptiles, this membrane adheres to the shell, allowing the exchange of gasses between the egg and the environment. In mammals, as we have seen, the chorion has evolved into the placenta, which has many other functions besides respiration.

The **allantois** stores urinary wastes and mediates gas exchange. In reptiles and birds, the allantois becomes a large sac, as there is no other way to keep toxic by-products of metabolism from the developing embryo. The mesodermal layer of allantoic membrane often reaches and fuses with the mesodermal layer of the chorion to create the **chorioallantoic membrane.** This extremely vascular envelope is crucial for chick development and is responsible for transporting calcium from the eggshell into the embryo for bone production (Tuan, 1987). In mammals, the size of the allantois depends on how well nitrogenous wastes can be removed by the chorionic placenta. In humans, the allantois is a vestigial sac, whereas in pigs it is a large and important organ.

The **yolk sac** is the first extraembryonic membrane to be formed, as it mediates nutrition in developing birds and reptiles. It is derived from endodermal cells that grow over the yolk to enclose it. The yolk sac is connected to the midgut by an open tube, the yolk duct, so that the walls of the yolk sac and the walls of the gut are continuous. The blood vessels within the mesoderm of the splanchnopleure transport nutrients from the yolk into the body, for yolk is not taken directly into the body through the yolk duct. Rather, endodermal cells digest the protein into soluble amino acids that can then be passed on to the blood vessels surrounding the yolk sac. Other nutrients, including vitamins, ions, and fatty acids, are stored in the yolk sac and transported by the yolk sac into the embryonic circulation. In these ways, the four extraembryonic membranes enable the embryo to develop on land.

The Heart

The circulatory system is one of the great achievements of the lateral plate mesoderm. Consisting of a heart, blood cells, and an intricate system of blood vessels, the circulatory system provides nourishment to the developing vertebrate embryo. The circulatory system is the first functional unit in the developing embryo, and the heart is the first functional organ. The vertebrate heart arises from two regions of splanchnic mesoderm that have interacted with adjacent tissue to become specified for heart development. These cardiogenic cells migrate to a ventral midline position and fuse to become a simple tube of contracting muscle cells. This tubular heart then contorts to form an S-shaped structure with a single atrium and a single ventricle. As development continues, the ventricle forms its layers and proliferates more rapidly than the atrium, the septa separate the chambers of the heart, and the valves develop.

FUSION OF THE HEART RUDIMENTS. In amphibians, the two presumptive heart-forming regions are initially found at the most anterior position of the mesodermal mantle. While the embryo is undergoing neurulation, these two re-

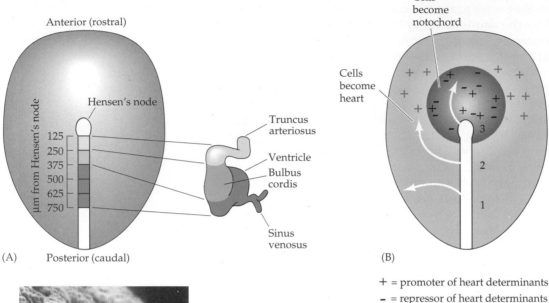

+ = promoter of heart determinants

− = repressor of heart determinants

Figure 9.22
Heart-forming cells of the ckick embryo. (A) Origin of heart cells in the early (stage 3b) chick embryo. The general anterior-posterior pattern in the primitive streak is seen in the endocardium and epimyocardium of the heart. (B) Model for the specification of cardiac mesoderm. The routes of mesodermal migration at various regions of the primitive streak are represented by arrows. Signals that induce cardiac myogenesis are represented by +, and inhibitors of cardiac induction are represented as −. Migrating mesoderm in region 1 encounters neither inducers nor repressors. Cells migrating from region 3 encounter both. Only those cells migrating from region 2 encounter the inducer without the inhibitor. (C) Scanning electron micrograph of the heart-forming mesoderm in the 24-hour chick embryo. The mesoderm is readily separated from the ectoderm but remains in close association with the endoderm. (A after Garcia-Martinez and Schoenwolf, 1993; B after Schultheiss et al., 1995; C from Linask and Lash, 1986, courtesy of K. Linask.)

gions come together in the ventral region of the embryo to form a common **pericardial cavity.** In birds and mammals, the heart also develops by fusion of paired primordia, but the fusion of these two rudiments does not occur until much later in development. In such amniote vertebrates, the embryo is a flattened disc, and the lateral plate mesoderm does not completely encircle the yolk sac. The presumptive heart cells originate in the early primitive streak, just posterior to Hensen's node and extending about halfway down its length (Figure 9.22A). These cells migrate through the streak and form two groups of mesodermal cells lateral to (and at the same level as) Hensen's node (Figure 9.22B; Garcia-Martinez and Schoenwolf, 1993). When the chick embryo is only 18–20 hours old, these presumptive heart cells move anteriorly between the ectoderm and endoderm toward the middle of the embryo, remaining in close contact with the endodermal surface (Figure 9.22C; Linask and Lash, 1986). When the cells reach the area where the gut has extended into the anterior region of the embryo, migration ceases. The directionality for this migration appears to be provided by the endoderm. If the cardiac region endoderm is rotated with respect to the rest of the embryo, migration of the precardiac mesoderm cells is reversed. It is thought that the endodermal component responsible for this movement is an anterior-to-posterior concentration gradient of fibronectin. Antibodies against fibronectin stop the migration, while antibodies against other extracellular matrix components do not (Linask and Lash, 1988a,b).

The endoderm also causes the precardiac cells to begin their development as heart muscles. The anterior endoderm can cause non-cardiac mesoderm cells to express heart-specific proteins in both chicks and amphibians (Jacobson, 1961; Sugi and Lough, 1994; Nascone and Mercola, 1995; Schultheiss et al., 1995). This differentiation occurs independently in the two heart-forming primordia migrating toward each other. The presumptive heart cells of birds and mammals form a double-walled tube consisting of an inner **endocardium** and an outer **epimyocardium.** The endocardium will form the inner lining of the heart, and the outer layer will form the layer of heart muscles that will pump for the lifetime of the organism.

As neurulation proceeds, the foregut is closed by the inward folding of the splanchnic mesoderm (Figure 9.23). This movement brings the two tubes together, eventually uniting the epimyocardium into a single tube. The two endocardia lie within the common chamber for a short while, but these will also fuse. At this time, the originally paired coelomic chambers unite to form the body cavity in which the heart resides. The bilateral origin of the heart can be demonstrated by surgically preventing the merger of the lateral plate mesoderm (Gräper, 1907; DeHaan, 1959). This results in a condition called **cardia bifida,** in which a separate heart forms on each side of the body (Figure 9.24). The next step in heart formation is the fusion of the endocardial tubes to form a single pumping chamber (see Figure 9.23C,D). This fusion occurs at about 29 hours in chick development and at 3 weeks in human gestation. The unfused posterior portions of the endocardium become the openings of the **vitelline veins** into the heart (Figure 9.25). These veins will carry nutrients from the yolk sac into the **sinus venosus.** The blood then passes through a valvelike flap into the atrial region of the heart. Contractions of the **truncus arteriosus** speed the blood into the **aorta.**

Pulsations of the heart begin while the paired primordia are still fusing. The pacemaker of this contraction is the sinus venosus. Contractions begin here, and a wave of muscle contraction is then propagated up the tubular heart. In this way, the heart can pump blood even before its intricate system of valves has been completed. Heart muscle cells have their own inherent ability to contract, and isolated heart cells from 7-day rat or chick embryos will continue to beat in petri dishes (Harary and Farley, 1963; DeHaan, 1967). In the embryo, these contractions become regulated by electrical stimuli from the medulla oblongata via the vagus nerve, and by 4 days, the electrocardiogram of a chick embryo approximates that of an adult.

FORMATION OF THE HEART CHAMBERS.   In 3-day chick embryos and 5-week human embryos, the heart is a two-chambered tube, with one atrium and one ventricle. In the chick embryo, the unaided eye can see the remarkable cycle of blood entering the lower chamber and being pumped out through the aorta. The partitioning of this tube into a distinctive atrium and ventricle is accomplished when cells from the myocardium produce a factor (probably transforming growth factor β3) that causes cells from the adjacent endocardium to detach and enter the hyaluronate-rich "cardiac jelly" between the two layers (Markwald et al., 1977; Potts et al., 1991). In humans, these cells cause the formation of an **endocardial cushion** that divides the tube into right and left atrioventricular channels (Figure 9.26). Meanwhile, the primitive atrium is partitioned by the growth of two septa that grow ventrally toward the endocardial cushions. The septa, however, have holes in them, so blood can still cross from one side into the other. This crossing of blood is

Figure 9.23
Formation of the heart. Transverse sections through the heart-forming region of the chick embryo at (A) 25 hours, (B) 26 hours, (C) 28 hours, and (D) 29 hours. (After Carlson, 1981.)

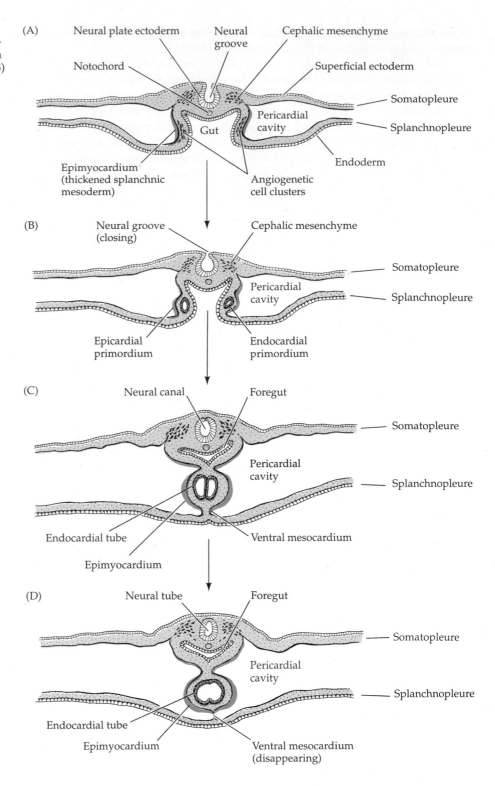

needed for survival of the fetus before the circulation to functional lungs has been established. Upon the first breath, however, these holes close, and the left and right circulatory loops become established (see Sidelights & Speculations, page 372). The partitioning of the ventricles is accomplished by the growth of the ventricular septum toward the endocardial cushion. With this separation (which usually occurs in the seventh week of human develop-

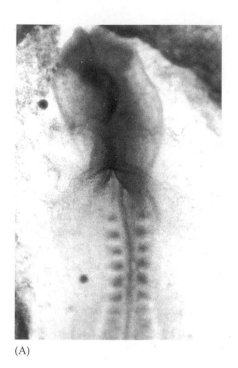

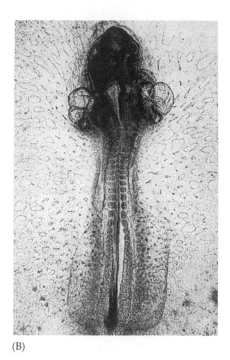

(A)

(B)

Figure 9.24
Fusion of the right and left heart rudiments to form a single heart tube. (A) Chick embryo (±30 hours) showing the paired heart primordia meeting at the ventral midlines. (B) Cardia bifida in chick embryo caused by preventing the two heart primordia from fusing. (A courtesy of K. Linask; B courtesy of R. L. DeHaan.)

ment), the heart is a four-chambered structure with the pulmonary trunk connected to the right ventricle and the aorta connected to the left.

One question that arises in these studies is, How does the left-right polarity emerge in the heart when the sides start off equally? Why should the left side of the heart become different from the right side? Studies of fetuses

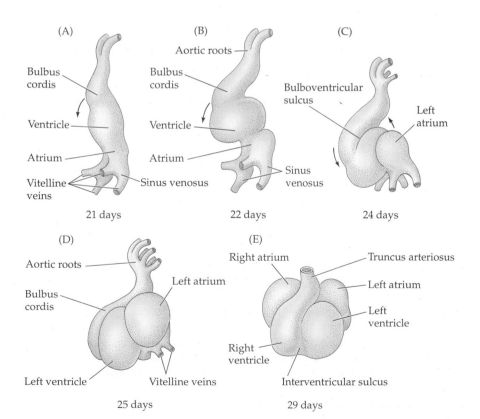

(A)

Bulbus cordis

Ventricle

Atrium

Vitelline veins

Sinus venosus

21 days

(B)

Aortic roots

Bulbus cordis

Ventricle

Atrium

Sinus venosus

22 days

(C)

Bulboventricular sulcus

Left atrium

24 days

(D)

Aortic roots

Left atrium

Bulbus cordis

Left ventricle

Vitelline veins

25 days

(E)

Right atrium

Truncus arteriosus

Left atrium

Left ventricle

Right ventricle

Interventricular sulcus

29 days

Figure 9.25
Heart chamber formation during the third week of human development, showing formation of the heart chambers from a simple tube. Views A–D show the developing heart from the left side; E is a frontal view. Although the atria are distinguished externally, they are not separated inside the heart. Note that there are two aortic roots and that these branch out to form the aortic arches (see Figure 9.27). (After Langman, 1981, and Larsen, 1993.)

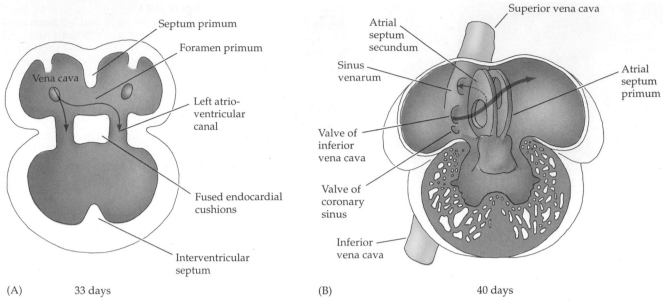

Figure 9.26
Formation of the chambers of the heart. (A) Diagrammatic cross section of the human heart at 4.5 weeks. The atrial and vetricular septa are growing toward the endocardial cushion. (B) Cross section of the human heart just prior to birth. Blood can cross from the right side of the heart to the left side through the openings in the primary and secondary atrial septa. (After Larsen, 1993.)

with malformed hearts having either two left or two right sides show a correlation between the presence of the spleen and the left side of the heart. Polysplenia (a spleen in both the left and right sides of the body) is associated with hearts that have two left sides, while asplenia (absence of the spleen) is associated with hearts having two right sides (Anderson et al., 1990; Ho et al., 1991). The mechanism for the left-right assymetry is not understood, but Tsuda and colleagues (1996) have shown an earlier asymmetrical deposition of the extracellular matrix protein flectin, which may predispose one side of the heart to develop differently than the other (Plate 33).*

Formation of Blood Vessels

CONSTRAINTS ON HOW BLOOD VESSELS MAY BE CONSTRUCTED. There are three major constraints on the construction of blood vessels. The first constraint is *physiological*. Unlike new machines, which do not need to function until they have left the assembly line, new organisms have to function even as they develop. The embryonic cells must obtain nourishment before there is an intestine, use oxygen before there are lungs, and excrete wastes before there are kidneys. Therefore, the circulatory physiology of the developing embryo differs from that of the adult organism, and its circulatory system reflects those differences. Food is absorbed not through the intestine, but from either the yolk or the placenta, and respiration is not conducted through the gills or lungs, but through the chorionic or allantoic membranes. The major embryonic blood vessels must be constructed to serve these extraembryonic structures.

The second constraint is *evolutionary*. The mammalian embryo will extend blood vessels to the yolk sac even though there is no yolk inside. Moreover, the blood leaving the heart loops over the foregut to form the dorsally located aorta. These six pairs of aortic arches loop over the pharynx (Figure 9.27). In primitive fishes, these arches persist and enable the gills to oxygenate the blood. In the adult bird or mammal, where lungs oxygenate the blood, such a system makes little sense, but all six pairs of aortic arches are formed in mammalian and avian embryos before the system eventually be-

*We will discuss right-left polarity in Chapter 16.

Figure 9.27
The aortic arches of the human embryo. (A) Originally, the truncus arteriosus pumps blood into the aorta, which branches on either side of the foregut. The six aortic arches take blood from the ventral aorta and allow it to flow into the dorsal aorta. (B) The arches begin to disintegrate or become modified; the dotted lines indicate degenerating structures. (C) Eventually, the remaining arches are modified and the adult arterial system is formed. (After Langman, 1981.)

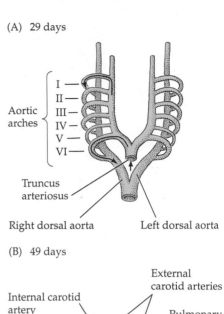

(A) 29 days

Aortic arches { I — II — III — IV — V — VI — }

Truncus arteriosus

Right dorsal aorta   Left dorsal aorta

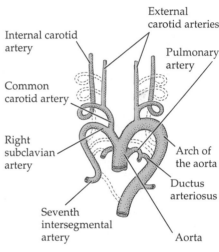

(B) 49 days

Internal carotid artery

External carotid arteries

Common carotid artery

Pulmonary artery

Right subclavian artery

Arch of the aorta

Ductus arteriosus

Seventh intersegmental artery

Aorta

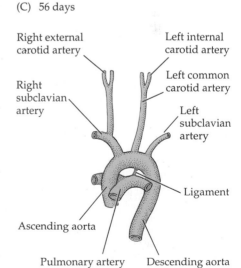

(C) 56 days

Right external carotid artery

Left internal carotid artery

Right subclavian artery

Left common carotid artery

Left subclavian artery

Ligament

Ascending aorta

Pulmonary artery   Descending aorta

comes simplified into a single aortic arch. Thus, even though our physiology does not require such a structure, our embryonic condition reflects our evolutionary history.

The third set of constraints is *physical*. According to the laws of fluid movement, the most effective transport of fluids is performed by large tubes. As the radius of the blood vessel gets smaller, resistance to flow increases as r^{-4} (Poiseuille's law). A blood vessel that is half as wide as another has a resistance to flow 16 times greater. However, diffusion of nutrients can take place only when blood flows slowly and has access to the membrane. So here is a paradox: The constraints of diffusion mandate that the vessels be small, while the laws of hydraulics mandate that the vessels be large. Living organisms have solved this paradox by evolving circulatory systems with a hierarchy of vessel sizes (LaBarbera, 1990). This hierarchy is formed very early in development, as can be seen in the 3-day chick embryo. In dogs, blood in the large vessels (aorta and vena cava) flows over 100 times faster than it does in the capillaries. By having large vessels specialized for transport and small vessels specialized for diffusion (where the blood spends most of its time), nutrients and oxygen can reach the individual cells of the growing organism. But this is not the entire story. If fluid under constant pressure moves directly from a large-diameter pipe into a small-diameter pipe (as in a hose nozzle), the fluid velocity increases. The evolutionary solution to this problem was the emergence of many smaller vessels branching out from a larger one, making the collective cross-sectional area of all the smaller vessels greater than that of the larger vessel. This relationship (known as Murray's law) is that the cube of the radius of the parent vessel approximates the sum of the cubes of the radii of the smaller vessels. The construction of any circulation system must negotiate among these physical, physiological, and evolutionary constraints.

VASCULOGENESIS: FORMATION OF BLOOD VESSELS FROM BLOOD ISLANDS. The creation of blood vessels de novo from the mesoderm is called **vasculogenesis** (Pardanaud et al., 1989). In the gut, lung, and aorta and in the splanchnic mesoderm lining the yolk sac, capillary networks arise independently within the tissues themselves (Auerbach et al., 1985; Pardanaud et al., 1989). In such cases, the capillaries do not arise as smaller and smaller extensions of the major blood vessels growing from the heart. Rather, the mesoderm of each of these organs contains cells called **angioblasts** that arrange themselves into capillary vessels. These organ-specific capillary networks eventually become linked to the extensions of the major blood vessels.

In the chick, there are two sources of angioblasts (Figure 9.28; Pardanaud et al., 1996). The first source is paraxial mesoderm. The cephalic paraxial mesoderm provides angioblasts for the blood vessels of the head (Couly et al., 1995), while the somitic paraxial mesoderm of the trunk contains angioblasts that migrate to form the vessels of the body wall, limb, kidney, and dorsal portions of the aorta. The second source of angioblasts is the splanchnopleural mesoderm. These angioblasts colonize the visceral organs, gut, and the floor of the aorta. These angioblasts are actually heman-

Figure 9.28
Two sources of angioblasts in the chick embryo form the endothelia of separate regions. The angioblasts from the somites migrate through the intermediate mesoderm (kidney), somatopleure, and the roof and lateral regions of the aorta. The angioblasts from the splanchnopleure form the vessels of the gut and visceral organs as well as the floor of the aorta. The angioblasts at the floor of the aorta also produce blood cells. (After Pardanaud et al., 1996.)

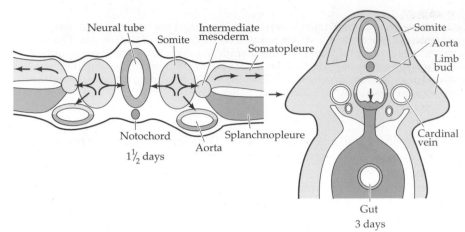

gioblasts, because they not only generate endothelial lining, but also provide the blood cell precursors (Pardanaud et al., 1996).

The aggregation of the splanchnic mesoderm cells is a critical step in amniote development, for these **angiogenetic clusters** (sometimes called **blood islands**) that line the yolk sac produce the **vitelline (omphalomesenteric) veins** that bring nutrients to the body and transport gasses to and from the sites of respiratory exchange (Figure 9.29). These cells are first seen in the area opaca at the headfold stage of chick embryogenesis, when the primitive streak is at its fullest extent (Pardanaud et al., 1987). These cords of cells soon hollow out into double-walled tubes analogous to the double tube of the heart. The inner wall becomes the flat endothelial cell lining of the vessel, and the outer cells become the smooth muscle. Between these layers is a basal lamina containing a type of collagen specific for blood vessels. It is thought that this basal lamina initiates the differentiation of the cell types in the vessel (Murphy and Carlson, 1978; Kubota et al., 1988). The central cells of the blood islands differentiate into the embryonic blood cells. As the blood islands grow, they eventually merge to form the capillary network draining into the two vitelline veins, which bring the food and blood cells to the newly formed heart.

Three growth factors may be responsible for initiating vasculogenesis. One of these, **basic fibroblast growth factor (FGF2)** is required for the generation of angioblasts from the mesoderm. When cells from quail blastodiscs are dissociated in culture, they will not form blood islands or endothelial cells. However, when these cells are cultured in FGF2, blood islands emerge in culture, and these form endothelial cells (Flamme and Risau, 1992). FGF2 is synthesized in the chick embryo chorioallantoic membrane and is responsible for the vascularization of this tissue (Ribatti et al., 1995). The second

Figure 9.29
Vasculogenesis. Blood vessel formation is first seen in the wall of the yolk sac, where (A) undifferentiated mesenchyme condenses to form (B) angiogenetic cell clusters. (C) The centers of these clusters form the blood cells, and the outside of the clusters develop into blood vessel endothelial cells. (After Langman, 1981.)

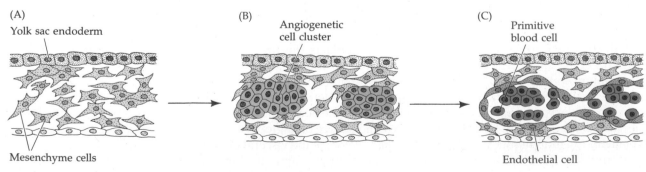

Condensation of
chondrocytes from
sclerotome cells

Nephrotome of
developing kidney

Intraembryonic
coelom

Somatic
mesoderm
layer

Splanchnic
mesoderm
layer

Somatic
mesoderm
layer

Migratory cells
(limb and ventrolateral
musculature)

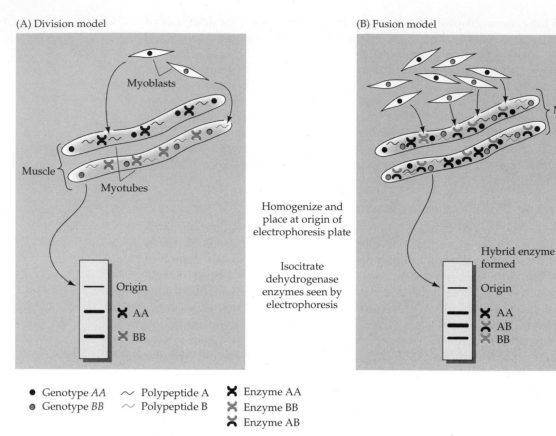

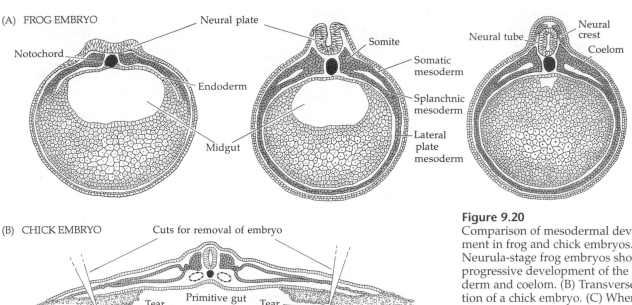

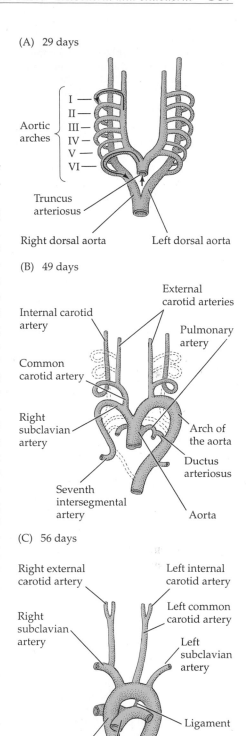

(A)

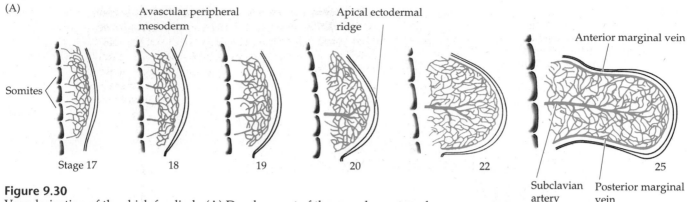

Avascular peripheral mesoderm

Apical ectodermal ridge

Anterior marginal vein

Somites

Stage 17   18   19   20   22   25

Subclavian artery   Posterior marginal vein

Figure 9.30
Vascularization of the chick forelimb. (A) Development of the vascular system during the early devlopment of the chick wing bud. The periphery of the bud is avascular, and the more avascular regions will form at the regions where chondrocytes will condense to form the cartilaginous precursors to the bones. (B) Dorsal view of a stage 22 wing bud injected with India ink. (A after Feinberg, 1991; B from Feinberg and Cafasso, 1995, photograph courtesy of Dr. R. N. Feinberg.)

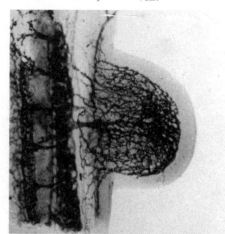

(B)

protein is **vascular endothelial growth factor** (**VEGF**). The latter protein appears to be specific for enabling the differentiation of the angioblasts and their multiplication to form the endothelial tubes. Moreover, the *receptors* for VEGF are found in the blood islands and in other places where VEGF is thought to be active (Millauer et al., 1993). If mouse embryos lack the genes encoding the major receptor for VEGF (the Flk1 receptor tyrosine kinase), yolk sac blood islands fail to appear, and vasculogenesis fails to take place. Mice lacking genes for the second receptor for VEGF (the Flt1 receptor tyrosine kinase) have differentiated endothelial cells and blood islands, but these cells are not organized into blood vessels (Fong et al., 1995; Shalaby et al., 1995). A third factor, **angiopoietin-1,** mediates the interaction between the endothelial cells and the smooth muscles they recruit to cover them. Mutations of either angiopoietin or its receptor lead to malformed blood vessels, deficient in the smooth muscles that usually surround them (Davis et al., 1996; Suri et al., 1996; Vikkula et al., 1996).

ANGIOGENESIS: SPROUTING OF BLOOD VESSELS.   Vasculogenesis is not the only way to make blood vessels. In other organs (notably the limb buds, kidney, and brain), existing blood vessels sprout and send endothelial cells into the developing organ (Wilson, 1983; Sariola, 1985). This type of blood vessel formation, in which new vessels emerge from the proliferation of preexisting blood vessels, is called **angiogenesis.** In the forelimb bud, for instance, the capillary network is derived by the sprouting of cells from the aorta (Evans, 1909; Feinberg, 1991). Within this capillary network, a central artery (which becomes the subclavian) forms as the major feeding vessel. Blood returns to the body through a marginal vein that forms from the anterior and posterior capillaries (Figure 9.30). The organ-forming regions are thought to secrete **angiogenesis factors** that promote the mitosis and migration of endothelial cells into that area. VEGF (mentioned earlier as a vasculogenesis factor) also promotes the migration of endothelial cells into these organs from preexisting blood vessels on the organs' surface. The amount of limb vascularization is correlated with levels of VEGF in the limb bud, and the spatiotemporal patterns of VEGF expression correlate well with the times and places where blood vessels enter into the kidney and brain (Figure 9.31; Breier et al., 1992; Millauer et al., 1993; Flamme et al., 1995).

Figure 9.31
Angiogenesis factor production by fetal mouse tissue. In situ hybridization shows that mRNA for secreted VEGF is synthesized by the glomeruli of the 15-day fetal mouse kidney. The bright-field micrograph on the left corresponds to the dark-field autoradiograph on the right. (From Breier et al., 1992, courtesy of W. Risau.)

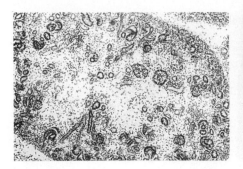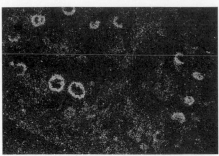

Some organs appear to make their own angiogenesis factors. The placenta is one organ whose function depends on redirecting existing blood vessels into it. When the placenta is first being formed, it induces angiogenesis by secreting **proliferin** (**PLF**), a factor that resembles growth hormone. When the placental blood vessels have become established (after day 12 in the mouse), the placenta secretes proliferin-related protein (PRP), a peptide that acts as an inhibitor of angiogenesis (Jackson et al., 1994). The developing bone is another organ that redirects blood vessels into it while it is forming. As mentioned earlier, cartilage is usually an avascular tissue, except when capillaries invade the growth plate to convert the cartilage into bone. Hypertrophic cartilage (but not mature or dividing cartilage) secretes a 120-kDa angiogenesis factor (Alini et al., 1996). It is interesting that this factor is made only when the early hypertrophic chondrocytes have been exposed to vitamin D. This would help explain the bone deformities seen in patients with rickets.

Angiogenesis is critical in the growth of any tissue, including tumors. Tumors are "successful" only when they are able to direct blood vessels into them. Therefore, tumors secrete angiogenesis factors. The ability to inhibit such factors may become an extremely important way to prevent tumor growth and metastasis (Fidler and Ellis, 1994). [mesend3.html]

EMBRYONIC CIRCULATION. The embryonic circulatory system to and from the chick embryo and yolk sac is shown in Figure 9.32. Blood pumped through the dorsal aorta passes over the aortic arches and down into the embryo. Some of this blood leaves the embryo through the **vitelline arteries** and enters the yolk sac. Nutrients and oxygen are absorbed, and the blood returns through the vitelline veins back into the heart through the sinus venosus. In mammalian embryos, food and oxygen are obtained from the placenta. Thus, although the mammalian embryo has vessels analogous to the vitelline veins, the main supply of food and oxygen comes from the **umbilical vein,** which unites the embryo with the placenta (Figure 9.33). This vein, which takes the oxygenated and food-laden blood back into the embryo, is derived from what would be the right vitelline vein in birds. The **umbilical artery,** carrying wastes to the placenta, is derived from what would have become the allantoic artery of the chick. It extends from the caudal portion of the aorta and proceeds along the allantois and then out to the placenta.

After entering the embryonic mammalian heart, the blood is pumped into a series of aortic arches that encircle the pharynx to bring the blood dorsally. In mammals, the left member of the fourth pair of aortic arches is the only one surviving to reach the aorta. The right member of this pair has become the root of the subclavian artery. The third aortic arches have been modified to form the common carotid arteries, which supply blood to the brain and head. The sixth arch is modified to form the pulmonary artery;

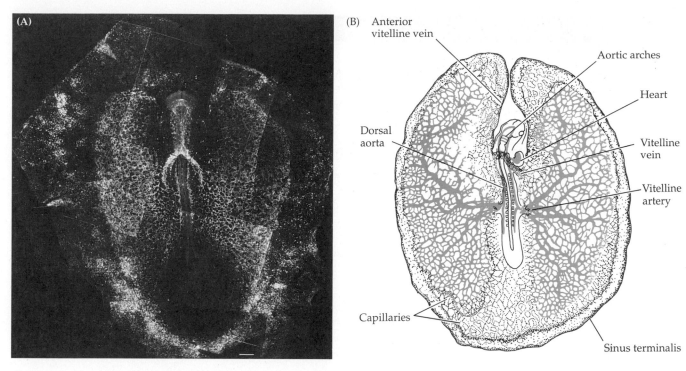

(A)

(B) Anterior vitelline vein

Aortic arches

Heart

Dorsal aorta

Vitelline vein

Vitelline artery

Capillaries

Sinus terminalis

Figure 9.32
Circulatory system of the early avian embryo. (A) Construction of the vasculatur in a 7-somite quail embryo, stained with a fluorescent antibody that recognizes endothelial cells. The blood islands can be seen at the edges. (B) Circulatory system of a 44-hour chick embryo. This view shows arteries in color; the veins are stippled. The sinus terminalis is the outer limit of the circulatory system and the site of blood cell generation. (Photographic montage from Pardanaud et al., 1987; courtesy of Dr. F. Dieterlen-Lièvre; B after Carlson, 1981.)

and the first, second, and fifth arches degenerate. The aorta and pulmonary artery, therefore, have a common opening to the heart for much of their development. Eventually, partitions form within the truncus arteriosus to create two different vessels. Only when the first breath of the newborn animal indicates that the lungs are ready to handle the oxygenation of the blood does the heart become modified to pump blood separately to the pulmonary artery.

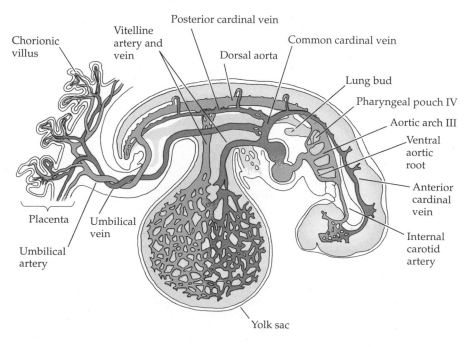

Chorionic villus

Vitelline artery and vein

Posterior cardinal vein

Dorsal aorta

Common cardinal vein

Lung bud

Pharyngeal pouch IV

Aortic arch III

Ventral aortic root

Anterior cardinal vein

Internal carotid artery

Placenta

Umbilical vein

Umbilical artery

Yolk sac

Figure 9.33
Circulatory system of a 4-week human embryo. Although at this stage all the major blood vessels are paired left and right, only the right vessels are shown. Arteries are shown in color. (From Carlson, 1981.)

Redirecting Blood Flow in the Newborn Mammal

ALTHOUGH THE DEVELOPING FETUS shares with the adult the need to get oxygen and nutrients to its tissues, the physiology of the mammalian fetus differs drastically from that of the adult. Chief among these differences is the lack of functional lungs and intestines. All oxygen and nutrients must come from the placenta. This raises two questions. First, how does the fetus obtain oxygen from maternal blood? And second, how is blood circulation redirected to the lungs once the umbilical cord is cut and breathing is made necessary?

The solution to the fetus's problems in getting oxygen from its mother's blood involves the development of a fetal hemoglobin. The hemoglobin in fetal red blood cells differs slightly from that in adult corpuscles. Two of the four peptides of the fetal and adult hemoglobin chains are identical—the alpha (α) chains—but adult hemoglobin has two beta (β) chains where the fetus has two gamma (γ) chains (Figure 9.34A). Normal β chains bind the natural regulator diphosphoglycerate, which assists in the unloading of oxygen. The γ-chain isoforms do not bind diphosphoglycerate as well and therefore have a higher affinity for oxygen. In the low-oxygen environment of the placenta, oxygen is released from adult hemoglobin. In this same environment, fetal hemoglobin does not give away oxygen, but binds it. This small difference in oxygen affinity mediates the transfer of oxygen from the mother to the fetus (Figure 9.34B). Within the fetus, the myoglobin of the fetal muscles has an even higher affinity for oxygen, so oxygen molecules pass from fetal hemoglobin for storage and use in the fetal muscles. Fetal hemoglobin is not deleterious to the newborn, and in humans, the replacement of fetal-hemoglobin-containing blood cells by adult-hemoglobin-containing blood cells is not complete until about 6 months after birth.*

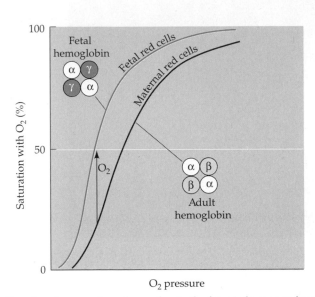

Figure 9.34 Transfer of oxygen from the mother to the fetus in human embryos. Adult and fetal hemoglobin molecules differ in their protein subunits. The fetal γ chain binds diphosphoglycerate less avidly than does the adult β chain. Consequently, fetal hemoglobin can bind oxygen more efficiently than can adult hemoglobin. In the placenta, there is a net flow (arrow) of oxygen from the mother's blood (which gives oxygen up to the tissue at the lower oxygen pressure) to the fetal blood, which is still picking it up.

But once the fetus is not getting its oxygen from the mother, how does it restructure its circulation to get oxygen from its own lungs? During fetal development, an opening—the **ductus arteriosus**—diverts the passage of blood from the pulmonary artery into the aorta (and thus to the placenta). Because blood does not return from the pulmonary vein in the fetus, the developing mammal has to have some other way of getting blood into the left ventricle to be pumped. This is accomplished by the **foramen ovale,** a hole in the septum separating the right and left atria. Blood can enter the right atrium, pass through the foramen to the left atrium, and then enter the left ventricle (Figure 9.35).

*The molecular basis for this switch in globins will be discussed in Chapter 11.

When the first breath is drawn, the oxygen in the blood causes the muscles surrounding the ductus arteriosus to close the opening. As the blood pressure in the left side of the heart increases, it causes the septa over the foramen ovale to close, thereby separating the pulmonary and systemic circulation.† Thus, when breathing begins, the respiratory circulation is shunted from the placenta to the lungs. [other.html#4] ∎

†In some infants, the septa fail to close, and the foramen ovale is left open. Usually the opening is so small that such children have no physical symptoms, and the foramen eventually closes. However, if the septum secundum fails to form, the atrial septal opening may cause enlargement of the right side of the heart, which can lead to heart failure during early adulthood.

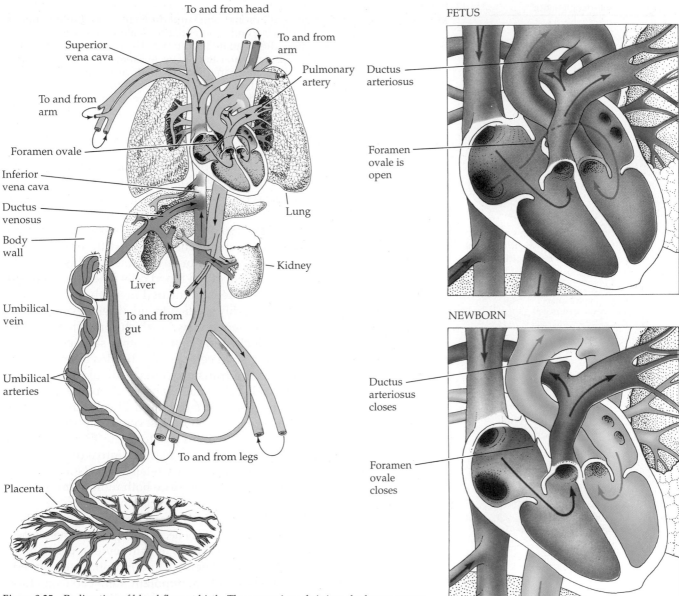

Figure 9.35 Redirection of blood flow at birth. The expansion of air into the lungs causes pressure changes that redirect the flow of blood in the newborn infant. The ductus arteriosus squeezes shut, breaking off the connection between the aorta and the pulmonary artery, and the foramen ovale, a passageway between the left and right atria, also closes. In this way, pulmonary circulation is separated from systemic circulation.

The development of blood cells

The Stem Cell Concept

While many of the cells we have now are the same cells we acquired as embryos, there are several populations of cells that are constantly regenerating. We lose and replace about 10^{11} red blood cells and small intestinal cells each day. Where are replacement cells coming from? They come from populations of **stem cells.** A stem cell is capable of extensive proliferation, creating more stem cells (self-renewal) as well as more differentiated cellular progeny.

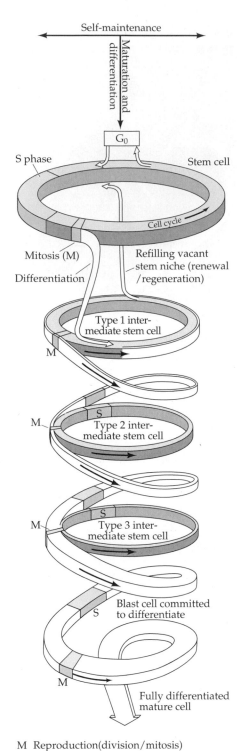

Self-maintenance

Maturation and differentiation

G_0

S phase

Stem cell

Cell cycle

Mitosis (M)

Refilling vacant stem niche (renewal /regeneration)

Differentiation

Type 1 intermediate stem cell

M

Type 2 intermediate stem cell

M   S

Type 3 intermediate stem cell

M   S

Blast cell committed to differentiate

S

M

Fully differentiated mature cell

M Reproduction(division/mitosis)
▨ Self–reproduction/replication
☐ Reproduction/replication

Figure 9.36
Model of the dynamics of stem cell proliferation and differentiation. Proliferation is represented by the horizontal circles, and differentiation is along the vertical axis progressing downward to more differentiated cell types. The initial stem cells (S) can remain quiescent (in G_0 phase) or enter the cell cycle. Stem cells that produce more stem cells remain at one level, but can divide to produce a transition cell type that "falls" to the next level. At each lower level, the probability of falling still lower at the next division is increased. Eventually, a mature differentiated cell is generated. (After Potten and Loeffler, 1990.)

Stem cells are, in effect, an embryonic population of cells, continuously producing cells that can undergo further development within an adult organism. Our blood cells, intestinal crypt cells, epidermis, and (in males) spermatocytes are populations in a steady-state equilibrium in which cell production balances cell loss (Hay, 1966). In most cases, stem cells can regulate the production of either more stem cells or more differentiated cells when the equilibrium is stressed by injury or environment. (This is seen by the production of enormous numbers of red blood cells when the body suffers from anoxia.) Stem cells have been identified in all the tissues mentioned earlier, but they are most readily studied in blood cell development.

Potten and Loefffler (1990) present a view in which some stem cells are noncycling *potential* stem cells locked in G_0, while other stem cells are actively in the cell cycle. A cycling stem cell usually divides to create more stem cells, but it can also generate a transitory intermediate stem cell type (T1). A T1 cell can regenerate itself, but usually it proceeds to make a second type of transitory cell, T2. (Under some conditions, a T1 cell can regenerate the original stem cell if the stem cell population is severely depleted.) The T2 cell can also maintain itself, but usually it divides to create T3 cells. Eventually, a transitory cell type is made that *always* matures into a differentiated cell type (Figure 9.36). Thus, the vertebrate body retains populations of stem cells, and these stem cell populations can produce both more stem cells and a population of cells that can undergo further development.

The path of development that a stem cell descendant enters depends on the molecular milieu in which it resides. This became apparent when experimental evidence showed that red blood cells (erythrocytes), white blood cells (granulocytes, neutrophils, and platelets), and lymphocytes shared a common precursor—the **pluripotential hematopoietic stem cell** (sometimes called the long-term repopulating hematopoietic stem cell).

Pluripotential Stem Cells and Hematopoietic Microenvironments

THE CFU-S.   The pluripotential hematopoietic stem cell is one of our body's most impressive cells. From it will emerge erythrocytes, neutrophils, basophils, eosinophils, platelets, mast cells, monocytes, tissue macrophages, osteoclasts, and the T and B lymphocytes. The existence of a pluripotential hematopoietic stem cell was shown by Till and McCulloch (1961), who injected bone marrow cells into lethally irradiated mice of the same genetic strain as the marrow donors. (Irradiation kills the hematopoietic cells of the host, enabling one to see the new colonies from the donor mouse.) Some of these donor cells produced discrete nodules on the spleens of the host animals (Figure 9.37). Microscopic studies showed these nodules to be composed of erythrocyte, granulocyte, and platelet precursors. Thus, a single cell from the bone marrow was capable of forming many of the different blood cell types. The cell responsible was called the **CFU-S,** the *c*olony-*f*orming *u*nit of the *s*pleen. Further studies used chromosomal markers to prove that the different types of cells within the colony were formed from the same

CFU-S. Here, marrow cells were irradiated so that very few survived. Many of those that did survive had abnormal chromosomes that could be detected microscopically. When such irradiated CFU-S cells were injected into a mouse whose own blood-forming stem cells had been destroyed, each cell of a spleen colony, be it granulocyte or erythrocyte precursor, had the same chromosomal anomaly (Becker et al., 1963). An important part of the stem cell concept is the requirement that the stem cell be able to form more stem cells in addition to its differentiated cell types. This has indeed been found to be the case. When spleen colonies derived from a single CFU-S are resuspended and injected into other mice, many spleen colonies emerge (Jurśšková and Tkadleček, 1965; Humphries et al., 1979). Thus, we see that a single marrow cell can form numerous different cell types and can also undergo self-renewal; in other words, the CFU-S is a pluripotential hematopoietic stem cell.

The preceding data indicate that although the CFU-S can generate many of the blood cell types, it is not capable of generating lymphocytes. This conclusion is supported by the experiments of Abramson and her colleagues (1977), who have shown that both the CFU-S and lymphocytes are derived from yet another pluripotential hematopoietic stem cell, sometimes called the *c*olony-*f*orming *u*nit of the *m*yeloid and *l*ymphoid cells, or **CFU-M,L.** When they injected irradiated bone marrow cells into mice having a hereditary deficiency of blood-forming cells, the researchers found the same chromosomal abnormalities in both the spleen colonies and the circulating lymphocytes. This work has been confirmed by studies in which marrow cells are injected with certain viruses that become incorporated into cellular DNA at various random places. The same virally derived genes are seen in the same region of the genome in lymphocytes and blood cells (Keller et al., 1985; Lemischka et al., 1986). In 1995, Berardi and colleagues isolated a fraction of cells that may be the human CFU-M,L. By killing all the cells that divided when exposed to cytokines that would activate more committed stem cells, they were left with about one nucleated cell for every 10,000 originally in the bone marrow. These cells could generate both the blood and lymphoid lineages.

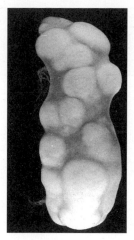

Figure 9.37
Isolated blood-forming colonies. When bone marrow containing hematopoietic stem cells is injected into an irradiated mouse, discrete colonies of blood cells are seen on the surface of the spleen of that mouse. (From Till, 1981, courtesy of J. E. Till.)

BLOOD AND LYMPHOCYTE LINEAGES. Figure 9.38 summarizes several studies. The first pluripotential hematopoietic stem cell is the CFU-M,L. The development of this CFU-M,L appears to be dependent on the transcription factor SCL. Mice lacking this protein die from the absence of all blood and lymphocyte lineages. SCL may specify the ventral mesoderm to a blood cell fate, or it may involve the formation or maintenance of the CFU-M,L cells (Porcher et al., 1996; Robb et al., 1996). This cell gives rise to the CFU-S (blood cells) and the CFU-L (lymphocytes). The CFU-S and the CFU-L also are pluripotential stem cells because their progeny can differentiate into numerous cell types. The immediate progeny of the CFU-S, however, are *lineage-restricted* stem cells. Each can produce only one type of cell in addition to renewing itself. The BFU-E (burst-forming unit, erythroid), for instance, is formed from the CFU-S, and it can form only one cell type in addition to itself. This new cell is the CFU-E (colony-forming unit, erythroid), which is capable of responding to the hormone **erythropoietin** to produce the first recognizable differentiated member of the erythrocyte lineage, the **proerythroblast.** Erythropoietin is a glycoprotein that rapidly induces the synthesis of the mRNA for globin (Krantz and Goldwasser, 1965). It is produced predominantly in the kidney, and its synthesis is responsive to environmental conditions. If the level of blood oxygen falls, erythropoietin production is increased, an event leading to the production of more red blood cells. As a red blood cell

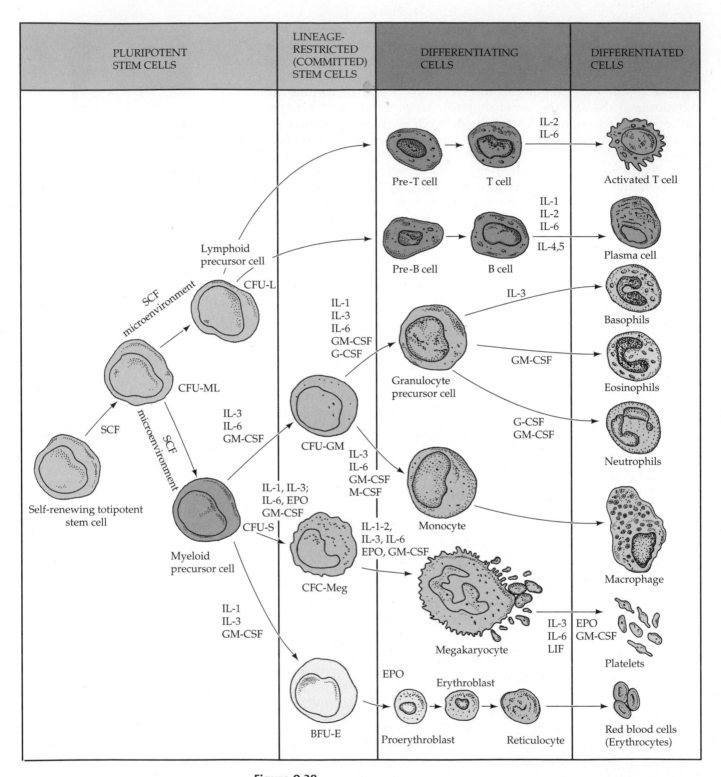

Figure 9.38
A model for the origin of mammalian blood and lymphoid cells. (Other models are consistent with the data, and this one summarizes features from several models.) EPO, erythropoietin; G-CSF, granulocyte colony stimulating factor; GM-CSF, granulocyte-macrophage colony stimulating factor; IL, interleukin; LIF, leukemia inhibiting factor; M-CSF, macrophage colony stimulating factor; SCF, stem cell factor. (After Nakauchi and Gachelin, 1993.)

matures, it becomes an **erythroblast,** synthesizing enormous amounts of hemoglobin. Eventually, the mammalian erythroblast expels its nucleus, becoming a **reticulocyte.** Reticulocytes can no longer synthesize globin mRNA but can still translate existing messages into globin. The final stage of differentiation is the **erythrocyte** stage. Here, no division, RNA synthesis, or protein synthesis takes place. The cell leaves the bone marrow to undertake its role of delivering oxygen to the bodily tissues. Similarly, there are lineage-restricted stem cells for platelets and for granulocytes (neutrophils, basophils, and eosinophils) and macrophages.

Some hematopoietic growth factors (such as IL-3) stimulate the division and maturation of the more primitive stem cells, thus increasing the numbers of all blood cell types. Other factors (such as erythropoietin) are specific for certain cell lineages only. A cell's ability to respond to these factors is dependent upon the presence of receptors for these factors on its surface. The number of these receptors is quite low. There are only about 700 receptors for erythropoietin on a CFU-E, and most other progenitor cells have similar low numbers of growth factor receptors. The exception is the receptor for macrophage colony-stimulating factor—M-CSF, also known as CSF-1—which can number up to 73,000 per cell on certain progenitor cells.

HEMATOPOIETIC INDUCTIVE MICROENVIRONMENTS.   Some hematopoietic growth factors are made by the stromal cells (fibroblasts and other connective tissue elements) of the bone marrow itself. Other growth factors travel through the blood and are retained by the extracellular matrix of the stromal cells. In the spleen, stem cells are predominantly committed to erythroid development. In the bone marrow, granulocyte development predominates. The developmental path taken by the descendants of a pluripotential stem cell depends on which growth factors it meets, and this is determined by the stromal cells of the bone marrow. Wolf and Trentin (1968) demonstrated that short-range interactions between stromal cells and the stem cells determine the developmental fates of the stem cells' progeny. These investigators placed plugs of bone marrow into the spleen and then injected stem cells. Those colonies in the spleen were predominantly erythroid, whereas those forming in the marrow plugs were predominantly granulocytic. In fact, those colonies that straddled the borders were predominantly erythroid in the spleen and granulocytic in the marrow. The regions of determination are referred to as **hematopoietic inductive microenvironments (HIMs).**

The stromal cells of the bone marrow create HIMs by their ability to bind the hematopoietic growth factors (Hunt et al., 1987; Whitlock et al., 1987). GM-CSF and the multilineage growth factor IL-3 both bind to the heparan sulfate glycosaminoglycan of the bone marrow stroma (Gordon et al., 1987; Roberts et al., 1988). Moreover, they remain active when bound. In this way, the growth factors may be concentrated and compartmentalized, stimulating stem cells in one area to differentiate into one cell type while allowing the same type of stem cells in another area to differentiate into another cell type. Without these growth factors, the stem cells die.

Osteoclast Development

As we have seen, stem cells are influenced by numerous hematopoietic growth factors. Moreover, these factors are themselves influenced by the hormonal milieu of the body. This fact may be extremely important in postmenopausal osteoporosis. Loss of ovarian function in many female mammals causes a loss of bone mass that can often be prevented by giving the individual estrogen, and such bone loss has been associated with the increased production of osteoclasts. It is thought that the osteoclast (the cell responsible for hollowing bones, as described earlier) comes from the same stem cell as macrophages and granulocytes, the CFU-GM (Kurihara et al., 1990; Hattersley et al., 1991). The growth factor interleukin 6 (IL-6) stimulates the production of osteoclasts. However, the production of IL-6 is inhibited by estrogen, and when estrogen is added to cultured mouse marrow cells, both IL-6 and osteoclast production are inhibited (Girasole et al, 1992). Jilka and colleagues (1992) showed that the removal of mouse ovaries causes an increase in the number of CFU-GMs, enhanced osteoclast development, and an increase in the number of osteoclasts found in bone. These changes could be prevented by injecting these mice with either estrogen or an antibody to IL-6. This suggests that estrogen usually suppresses IL-6 production and osteoclast formation in female mammals and that postmenopausal loss of bone mass may be due to the production of new osteoclasts by IL-6.* [mesend4.html]

Sites of Hematopoiesis

In avian and amphibian species, the first blood cells derive from the yolk or yolk sac. This cell population, however, is transitory; the hematopoietic stem cells that last the lifetime of the organism are derived from the mesodermal area surrounding the aorta. This was shown in the chick by a series of elegant experiments by Dieterlen-Lièvre, who grafted the blastoderm of chickens onto the yolk of the Japanese quail (Figure 9.39). Chick cells are readily distinguishable from quail cells because the quail cell nucleus stains much more darkly (owing to its dense nucleoli), thus providing a permanent marker for distinguishing the two cell types. Using these "yolk sac chimeras," Dieterlen-Lièvre and Martin (1981) showed that the yolk sac stem

* So how come males—who don't have ovaries or as much estrogen—do not usually suffer osteoporotic bone loss? It seems that testosterone also suppresses osteoclast development (Bellido et al., 1995). In human males, testosterone production is usually maintained with age. Given the physiology of the osteoclast, we can now appreciate H. L. Menken's (1919) prescient intuition: "Life is a struggle, not against sin, not against the Money Power, not against malicious animal magnetism, but against hydrogen ions."

Figure 9.39
Blood cell mapping by chick-quail chimeras. (A) Photograph of a "yolk sac chimera," wherein the blastoderm of a quail was transplanted onto the yolk sac of a chick. (B) Photograph of chick and quail cells in the thymus of a chimeric animal, showing the difference in the nuclear staining. The lymphoid cells are all chick, whereas the structural cells of the thymus are of quail origin. (C) Section through the aorta of a 3-day chick embryo, showing the cells (arrows) that give rise to the hematopoietic stem cells. If cells from this region are taken from quail embryos and placed into chick embryos, the chick embryos have quail blood. (From Martin et al., 1978, and Dieterlen-Lièvre and Martin, 1981, photographs courtesy of F. Dieterlen-Lièvre.)

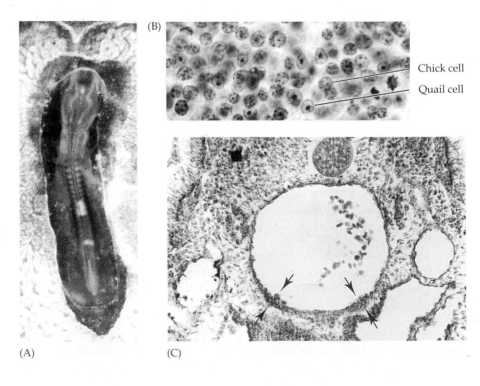

(B)

Chick cell

Quail cell

(A)

(C)

cells do not contribute cells to the adult animal but that the true stem cells are formed within nodes of mesoderm that line the mesentery and the major blood vessels. These are the hemangioblasts that are derived from the splanchnopleure (see Figure 9.28; Pardanaud et al., 1996). In the 4-day chick embryo, the aortic wall appears to be the most important source of new blood cells, and it has been found to contain numerous hematopoietic stem cells (Cormier and Dieterlen-Lièvre, 1988).

In mammals, the situation is more controversial; but the situation is beginning to look very similar to that of the chick. The first blood islands in the mouse embryo appear in the extraembryonic and yolk sac mesoderm. These blood cells appear to have CFU-C activity. This yolk sac-derived population is probably transient or may provide only for the respiratory needs of the embryo (producing nucleated red blood cells). By day 11, pluripotential hematopoietic stem cells and CFU-S cells can be found in the embryonic mouse mesodermal region that includes the aorta, gonads and mesonephros (the **AGM region;** Kubai and Auerbach, 1983; Godlin et al., 1993; Medvinsky et al., 1993). These are the blood cell precursors that will colonize the liver and constitute the fetal and adult circulatory system (Medvinsky and Dzierak, 1996). Müller and colleagues (1994) have proposed that two waves of cells colonize the fetal liver. The minor population of these cells would come from the yolk sac and would be predominantly CFU-C cells. The majority population would come from the AGM sites and constitute both CFU-S and pluripotential hematopoietic stem cells (Figure 9.40). This proposal was strengthened by the finding that mice deficient in the transcription factor AML1 have normal yolk sac hematopoiesis, but no definitive (AGM) hematopoiesis (Okuda et al., 1996). These mutant mice die at embryonic day 12.5. Their liver contains a small number of primitive nucleated red blood

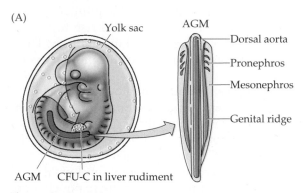

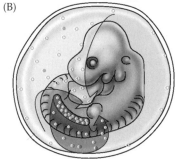

(A)

(B)

Figure 9.40

Colonization of the mouse liver by two waves of hematopoietic stem cells. The two main sources of hematopoietic progenitor cells are the yolk sac and the AGM region. (A) At day 9 the yolk sac contributes an early line of CFU-C cells that probably does not last long after birth and which makes a population of predominantly red blood cells. This is thought to be the major source of the first wave of hematopoiesis in the liver. (B) At 10 days, the AGM-derived cells provide CFU-S cells and pluripotential hematopoietic stem cells. These constitute the major cells of the second wave. (After Dzierzak and Medvinsky, 1995.)

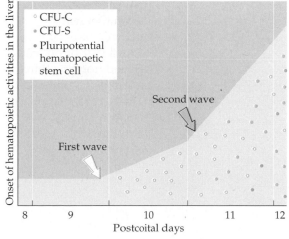

cells, whereas control livers are full of blood cells derived from the AGM. The AML protein is critical for activating the genes involved in definitive hematopoiesis. Around the time of birth, stem cells from the liver populate the bone marrow, which then becomes the major site of blood formation throughout adult life.

■ ENDODERM

Pharynx

The function of embryonic endoderm is to construct the linings of two tubes within the body. The first tube, extending throughout the length of the body, is the digestive tube. Buds from this tube form the liver, gallbladder, and pancreas. The second tube, the respiratory tube, forms as an outgrowth of the digestive tube, and it eventually bifurcates into two lungs. The digestive and respiratory tubes share a common chamber in the anterior region of the embryo; this region is called the **pharynx.** Epithelial outpockets of the pharynx give rise to the tonsils, thyroid, thymus, and parathyroid glands.

The respiratory and digestive tubes are both derived from the primitive gut (Figure 9.41). As the endoderm pinches in toward the center of the embryo, the foregut and hindgut regions are formed. At first, the oral end is blocked by a region of ectoderm called the **oral plate,** or **stomodeum.** Eventually (about 22 days in human embryos), the stomodeum breaks, thereby creating the oral opening of the digestive tube. The opening itself is lined by ectodermal cells. This arrangement creates an interesting situation, because the oral plate ectoderm is in contact with the brain ectoderm, which has curved around toward the ventral portion of the embryo. The two ectodermal regions mutually interact with each other. The roof of the oral region forms Rathke's pouch and becomes the *glandular* part of the pituitary gland. The neural tissue on the floor of the diencephalon gives rise to the infundibular process, which becomes the *neural* portion of the pituitary. Thus, the pituitary gland has a dual origin; this dual nature of the pituitary is reflected in its adult functions.

The endodermal portion of the digestive and respiratory tubes begins in the pharynx. Here, the mammalian embryo produces four pairs of **pharyngeal pouches** (Figure 9.42). In aquatic vertebrates, these structures produce the gills, but mammalian pharyngeal pouches have been modified for the terrestrial environment. As discussed in Chapter 7, cranial neural crest cells migrate into these pouches to form the cartilaginous or mesenchymal component of these endodermally lined structures. Between these pouches are the **pharyngeal arches.** The first pair of pharyngeal pouches becomes the auditory cavities of the middle ear and the associated eustachian tubes. The second pair of pouches gives rise to the walls of the tonsils. The thymus is derived from the third pair of pharyngeal pouches; it will direct the differentiation of T lymphocytes during later stages of development. One pair of parathyroid glands is also derived from the third pair of pharyngeal pouches, and the other pair is derived from the fourth. In addition to these paired pouches, a small, central diverticulum is formed between the second pharyngeal pouches on the floor of the pharynx. This pocket of endoderm and mesenchyme will bud off from the pharynx and migrate down the neck to become the thyroid gland.

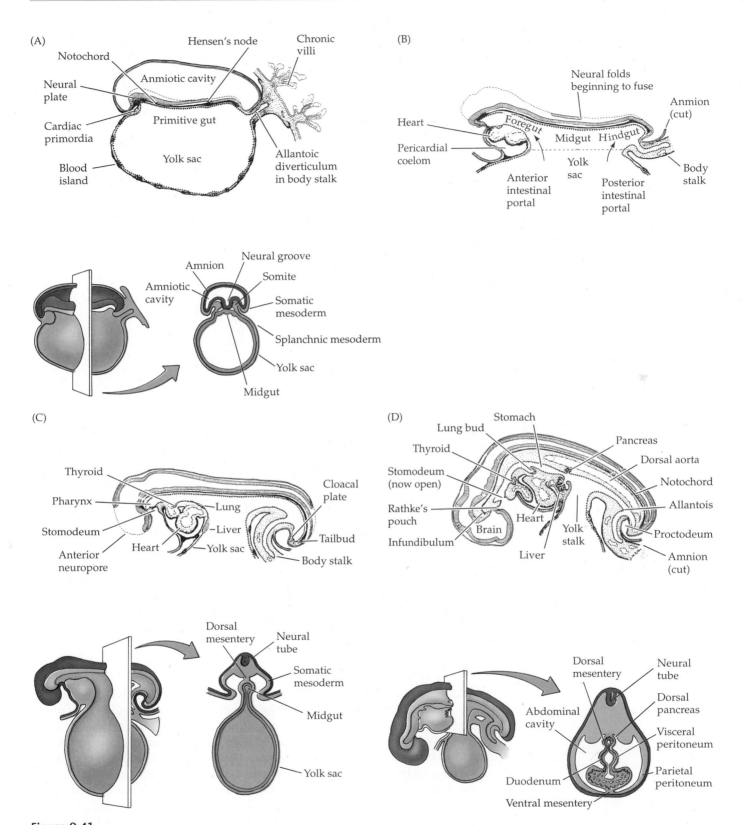

Figure 9.41
Formation of the human digestive system, depicted at about (A) 16 days, (B) 18 days, (C) 22 days, and (D) 28 days. (After Crelin, 1961.)

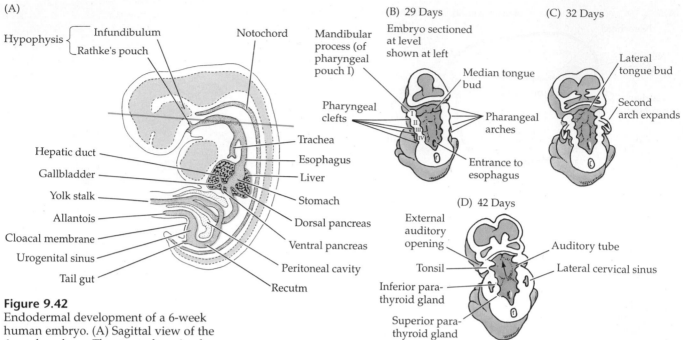

Figure 9.42
Endodermal development of a 6-week human embryo. (A) Sagittal view of the 6-week embryo. The stomach region has begun to dilate, and the pancreas is represented by two buds that will eventually fuse. (B–D) Sections through the 6-week embryo at the planes in (A), showing fates of the pharyngeal clefts. The first cleft forms the external auditory passages, while the second cleft expands, eventually covering clefts 2, 3, and 4. (After Larsen, 1993.)

The digestive tube and its derivatives

Posterior to the pharynx, the digestive tube constricts to form the esophagus, which is followed in sequence by the stomach, small intestine, and large intestine. The endodermal cells generate only the lining of the digestive tube and its glands, for mesodermal mesenchyme cells will surround this tube to provide the muscles for peristalsis.

Figure 9.42 shows that the stomach develops as a dilated region close to the pharynx. More caudally, the intestines develop, and the connection between the intestine and yolk sac is eventually severed. At the caudal end of the intestine, a depression forms where the endoderm meets the overlying ectoderm. Here, a thin **cloacal membrane** separates the two tissues. It eventually ruptures, forming the opening that will become the anus. The development of the distinct regions of the digestive tube will be detailed in Chapter 17.

Liver, Pancreas, and Gallbladder

Endoderm also forms the lining of three accessory organs that develop immediately caudal to the stomach. The **hepatic diverticulum** is the tube of endoderm that extends out from the foregut into the surrounding mesenchyme. The mesenchyme induces the endoderm to proliferate, to branch, and to form the glandular epithelium of the liver. A portion of the hepatic diverticulum (that region closest to the digestive tube) continues to function as the drainage duct of the liver, and a branch from this duct produces the gallbladder (Figure 9.43).

The pancreas develops from the fusion of distinct dorsal and ventral diverticula. Both of these primordia arise from the endoderm immediately caudal to the stomach, and as they grow, they come closer together and eventually fuse. In humans, only the ventral duct survives to carry digestive enzymes into the intestine. In other species (such as the dog), both the dorsal and ventral ducts empty into the intestine. As with other endoderm or-

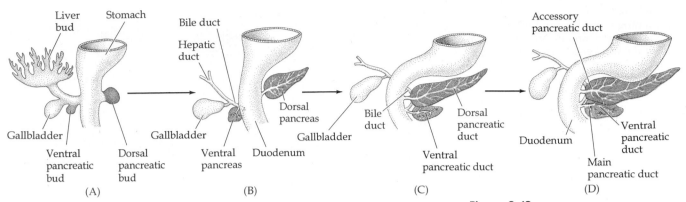

Figure 9.43
Pancreatic development in humans. (A) At 30 days, the ventral pancreatic bud is close to the liver primordium. (B) By 35 days it begins migrating posteriorly and (C) comes into contact with the dorsal pancreatic bud during the sixth week of development. (D) In most individuals, the dorsal pancreatic bud loses its duct into the duodenum; however, in about 10 percent of the population, the dual duct system persists. (After Langman, 1981.)

gans, the pancreas develops through interactions between the epithelium and its associated mesenchyme. Both these tissues have specificities provided by their position along the anterior-posterior axis (to be discussed in Chapters 16 and 17). If pancreatic epithelium is cultured in a permissive environment in the absence of mesenchyme, it differentiates almost entirely into insulin- and glucagon-secreting islet cells. No acinar (chymotrypsin- or amylase-secreting) or ductal structures are made (Gittes et al., 1996). This suggests that the "default" state of the pancreatic epithelium is to produce endocrine hormones and that the secretory cells and ducts characteristic of its digestive (exocrine) function are the result of its interactions with the mesenchyme. The *pdx-1* gene appears to give the pancreatic epithelium the ability to respond to its mesenchyme. Mice without this gene lack any pancreas, although their epithelium does differentiate into pre-islet cells that synthesize small amounts of glucagon and insulin (Jonnson et al., 1994; Ahlgren et al., 1996; Offield et al., 1996). The pancreatic epithelium, therefore, may have autonomous endocrine capacity, but it needs to interact with the mesenchyme to form the exocrine cells and the ducts that take their secretions into the duodenum.

The Respiratory Tube

The lungs also are a derivative of the digestive tube, even though they serve no role in digestion. In the center of the pharyngeal floor, between the fourth pair of pharyngeal pouches, the **laryngotracheal groove** extends ventrally (Figure 9.44). This groove then bifurcates into the two branches that form the pair of bronchi and lungs. The laryngotracheal endoderm becomes the lining of the trachea, the two bronchi, and the air sacs (alveoli) of the lungs. As we will see in a later chapter, the branching of this endodermal tube depends on interactions with the different types of mesodermal cells in its path.

The lungs are an evolutionary novelty, and they are among the last of the mammalian organs to fully differentiate. The lungs must be able to draw in oxygen at the baby's first breath. To accomplish this, alveolar cells secrete a *surfactant* into the fluid bathing the lungs. This surfactant, consisting of phospholipids such as sphingomyelin and lecithin, is secreted very late in gestation, and it usually reaches physiologically useful levels around week 34 of human gestation. These compounds enable the alveolar cells to touch one another without sticking together. Thus, infants born prematurely often have difficulty breathing and have to be placed in respirators until their surfactant-producing cells mature.

This concludes our survey of the early features of animal development. We now attend to the *mechanisms* that enable this development to take place.

Figure 9.44
Partitioning of the foregut into the esophagus and respiratory diverticulum during the third and fourth weeks of human gestation. (A) Lateral view, end of week 3. (B,C) Ventral views, week 4. (After Langman, 1981.)

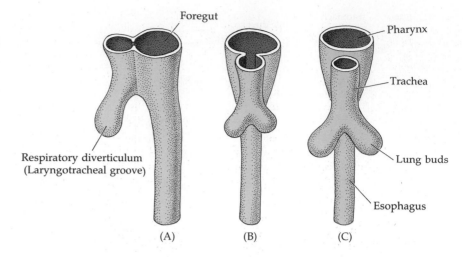

In Part III, we focus on the molecular events that direct cell differentiation. In Part IV, we see the roles these molecules play in forming the embryonic body axes. Part V will discuss the genetic, cellular, and environmental forces that interact during organ formation.

LITERATURE CITED

Abramson, S., Miller, R. G. and Phillips, R. A. 1977. The identification in adult bone marrow of pluripotent and restricted stem cells of the myeloid and lymphoid systems. *J. Exp. Med.* 145: 1567–1579.

Ahlgren, U., Jonnson, J. and Edlund, H. 1996. The morphogenesis of the pancreatic mesenchyme is uncoupled from that of the pancreatic epithelium in IPF/PDX1-deficient mice. *Development* 122: 1409–1416.

Alini, M., Marriott, A., Chen, T., Abe, S. and Poole, A. R. 1996. A novel angiogenic molecule produced at the time of chondrocyte hypertrophy during endochondral bone formation. *Dev. Biol.* 176: 124–132.

Anderson, C., Devine, W. A., Anderson, R. H., Debich, D. E. and Zuberbuhler, J. R. 1990. Abnormalities of the spleen in relation to congenital malformation of the heart: A survey of necropsy findings in children. *Br. Heart J.* 63: 122–128.

Ash, P. J., Loutit, J. F. and Townsend, K. M. S. 1980. Osteoclasts derived from haematopoietic stem cells. *Nature* 283: 669–670.

Auerbach, R., Alby, L., Morrissey, L., Tu, M. and Joseph, J. 1985. Expression of organ-specific antigens on capillary endothelial cells. *Microvasc. Res.* 29: 401–411.

Ballock, R. T. and Reddi, A. H. 1994. Thyroxine is the serum factor that regulates morphogenesis of columnar cartilage from isolated chondrocytes in chemically defined medium. *J. Cell Biol.* 126: 1311–1318.

Barnes, G. L., Hsu, C. W., Mariani, B. D. and Tuan, R. S. 1997. Cloning and characterization of chicken paraxis: A regulator of somite segmentation. *Dev. Biol.* In press.

Baron, R., Neff, L., Louvard, D. and Courtoy, P. J. 1985. Cell mediated extracellular acidification and bone resorption: Evidence for a low pH in resorbing lacuna and localization of a 100-kD lysosomal membrane protein at the osteoclast ruffled border. *J. Cell Biol.* 101: 2210–2222.

Baron, R., Neff, L., Roy, C., Boisvert, A. and Caplan, M. 1986. Evidence for a high and specific concentration of (Na+,K+) ATPase in the plasma membrane of the osteoclast. *Cell* 46: 311–320.

Becker, A. J., McCulloch, E. A. and Till, J. E. 1963. Cytological demonstration of the clonal nature of spleen cells derived from transplanted mouse marrow cells. *Nature* 197: 452–454.

Bellido, T. and seven others. 1995. Regulation of interleukin-6, osteoclastogenesis, and bone mass by androgens: The role of the androgen receptor. *J. Clin. Invest.* 95: 2886–2895.

Bellus, G. A. and eight others. 1995. A recurrent mutation in the tyrosine kinase domain of fibroblast growth factor receptor 3 causes hypochondroplasia. *Nat. Genet.* 10:357–359.

Berardi, A. C., Wang, A., Levine, J. D., Lopez, P. and Scadden, D. T. 1995. Functional characterization of human hematopoietic stem cells. *Science* 267: 104–108.

Bischoff, R. and Holtzer, H. 1969. Mitosis and processes of differentiation of myogenic cells in vitro. *J. Cell Biol.* 41: 188–200.

Blair, H. C., Kahn, A. J., Crouch, E. C., Jeffrey, J. J. and Teitelbaum, S. L. 1986. Isolated osteoclasts resorb the organic and inorganic components of bone. *J. Cell Biol.* 102: 1164–1172.

Bloom, W. and Fawcett, D. W. 1975. *Textbook of Histology*, 10th Ed. Saunders, Philadelphia.

Boettiger, D., Enomoto-Iwamoto, M., Yoon, H. Y., Hofer, U., Menko, A. S. and Chiquet-Ehrismann, R. 1995. Regulation of integrin α5β1 affinity during myogenic differentiation. *Dev. Biol.* 169: 261–272.

Braun, T. and Arnold, H.-H., 1996. Myf-5 and myoD genes are activated in distinct mesenchymal stem cells and determine different skeletal muscle lineages. *EMBO J.* 15: 310–318.

Braun, T., Rudnicki, M. A., Arnold, H.-H. and Jaenisch, R. 1992. Targeted inactivation of the muscle regulatory gene *Myf-5* results in abnormal rib development and perinatal death. *Cell* 71: 369–382.

Breier, G., Albrecht, U., Sterrer, S. and Risau, W. 1992. Expression of vascular endothelial growth factor during embryonic angiogenesis and endothelial cell differentiation. *Development* 114: 521–532.

Brighton, C. T. 1984. The growth plate. *Orthop. Clin. North Am.* 15: 571–594.

Brighton, C. T. and Hunt, R. M. 1974. Mitochondrial calcium and its role in calcification. *Clin. Orthop.* 100: 406–416.

Brill, G., Kahane, N., Carmeli, C., von Schack, D., Barde, Y.-A. and Kalcheim, C. 1995. Epithelial-mesenchymal conversion of dermatome progenitors requires neural tube-derived signals: Characterization of the role of neurotrophin-3. *Development* 121: 2583–2594.

Brunetti, A. and Goldfine, I. D. 1990. Role of myogenin in myoblast differentiation and its regulation by fibroblast growth factor. *J. Biol. Chem.* 265: 5960–5963.

Burgess, R., Cserjesi, P., Ligon, K. L. and Olson, E. N. 1995. Paraxis: A basic helix-lop-helix protein expressed in paraxial mesoderm and developing somites. *Dev. Biol.* 168: 296–306.

Burgess, R., Rawls., Brown, D., Bradley, A. and Olson, E. N. 1996. Requirement of the *praxis* gene for somite formation and musculoskeletal patterning. *Nature* 384: 570–573.

Carlson, B. M. 1981. *Patten's Foundations of Embryology*. McGraw-Hill, New York.

Chen, C.-M., Kraut, N., Groudine, M. and Weintraub, H. 1996. I-mf, a novel myogenic repressor, interacts with members of the MyoD family. *Cell* 86: 731–741.

Chen, Q., Johnson, D. M., Haudenschild, D. R. and Goetinck, P. F. 1995. Progression and recpitulation of the chondrocyte differentiation program: Cartilage matrix protein is a marker for cartilage maturation. *Dev. Biol.* 172: 293–306.

Chevallier, A., Kieny, M., Mauger, A. and Sengel, P. 1977. Developmental fate of the somitic mesoderm in the chick embryo. *In* D. A. Ede, J. R. Hinchliffe and M. Balls (eds.), *Vertebrate Limb and Somite Morphogenesis.* Cambridge University Press, Cambridge, pp. 421–432.

Christ, B., Jacob, H. J. and Jacob, M. 1977. Experimental analysis of the origin of the wing musculature in avian embryos. *Anat. Embryol.* 150: 171–186.

Chuong, C.-M., Widelitz, R. B., Jiang, T.-X., Abbott, U. K., Lee, Y.-S. and Chen, H.-M. 1993. Roles of adhesion molecules NCAM and tenascin in limb skeletogenesis: Analysis with antibody pertubations, exogenous gene expression, talpid2 mutants and activin stimulation. *In* J. F. Fallon, (ed.) *Limb Development and Regeneration.* Wiley-Liss, New York, pp. 465–474.

Conlon, R. A., Reaume, A. G. and Rossant, J. 1995. Notch1 is required for the coordinate segmentation of somites. *Development* 121: 1533–1545.

Cormier, F. and Dieterlen-Lièvre, F. 1988. The wall of the chick aorta harbours M-CFC, G-CFC, GM-CFC and BFU-E. *Development* 102: 279–285.

Cossu, G., Kelly, R., Tajbakhsh, S., Di Donna, S., Vivarelli, E. and Buckingham, M. 1996a. Activation of different myogenic pathways: myf-5 is induced by the neural tube and MyoD by the dorsal ectoderm in mouse paraxial mesoderm. *Development* 122: 429–437.

Cossu, G., Tajbakhsh, S. and Buckingham, M. 1996b. How is myogenesis initiated in the embryo? *Trends Genet.* 12: 218–223.

Couly, G., Coltey, P., Eichmann, A. and LeDouarin, N. M. 1995. The angiogenic potentials of the cephalic mesoderm and the origin of brain and head blood vessels. *Mech. Dev.* 53: 97–112.

Crelin, E. S. 1961. Development of the gastrointestinal tract. *Clin. Symp.* 13: 68–82.

Cserjesi, P. and seven others. 1995. A basic helix-loop-helix protein that prefigures skeletal formation during mouse embryogenesis. *Development* 121: 1099–1110.

David, J. D., See, W. M. and Higginbotham, C. A. 1981. Fusion of chick embryo skeletal myoblasts: Role of calcium influx preceding membrane union. *Dev. Biol.* 82: 297–307.

Davis, R. L., Weintraub, H. and Lassar, A. B. 1987. Expression of a single transfected cDNA converts fibroblasts into myoblasts. *Cell* 51: 987–1000.

Davis, S. and ten others. 1996. Isolation of angiopoietin-1, a ligand for the TIE2 receptor, by secretion trap expression cloning. *Cell* 87: 1161–1169.

DeHaan, R. 1959. *Cardia bifida* and the development of pacemaker function in the early chicken heart. *Dev. Biol.* 1: 586–602.

DeHaan, R. L. 1967. Regulation of spontaneous activity and growth of embryonic chick heart cells in tissue culture. *Dev. Biol.* 16: 216–249.

Deng, C., Wynshaw-Boris, A., Zhou, F., Kuo, A. and Leder, P. 1996. Fibroblast growth factor receptor-3 is a negative regulator of bone growth. *Cell* 84: 911–921.

Dieterlen-Lièvre, F. and Martin, C. 1981. Diffuse intraembryonic hemopoiesis in normal and chimeric avian development. *Dev. Biol.* 88: 180–191.

Dzierzak, E. and Medvinsky, A. 1995. Mouse embyonic hematopoiesis. *Trends Genet.* 11: 359–366.

Ede, D. A. 1983. Cellular condensations and chondrogenesis. *In* B. K. Hall (ed.), *Cartilage, Volume 2: Development, Differentiation, and Growth.* Academic Press, New York, pp. 143–185.

Evans, H. M. 1909. On the earliest blood vessels in the anterior limbs of birds and their relation to the primary subclavian artery. *Am. J. Anat.* 9: 281–319.

Fan, C. M. and Tessier-Lavigne, M. 1994. Patterning of mammalian somites by surface ectoderm and notochord: Evidence for sclerotome induction by a hedgehog homolog. *Cell* 79: 1175–1186.

Feinberg, R. N. 1991. Vascular development in the embryonic limb bud. *In* R. N. Feinberg, G. K. Sherer, and R. Auerbach (eds.), *The Development of the Vascular System. Issues in Biomedicine 14.* Karger, Basel, pp. 136–148.

Feinberg, R. N. and Cafasso, E. 1995. Macromolecular permeability of chick wing microvessels: An intravital study. *Anat. Embryol.* 191: 337–342.

Fidler, I. J. and Ellis, L. M. 1994. The implications of angiogenesis for the biology and therapy of cancer metastsis. *Cell* 70: 185–188.

Flamme, I. and Risau, W. 1992. Induction of vasculogenesis and hematogenesis in vitro. *Development* 116: 435–439.

Flamme, I., von Reutern, M., Dexter, H. C. A., Syedali, S. and Risau, W. 1995. Overexpression of vascular endothelial growth factor in avian embryos induces hypervascularization and increased vascular permeability without alterations of embryonic pattern formation. *Dev. Biol.* 171: 399–414.

Fong, G.-H., Rossant, J., Gertenstein, M. and Breitman, M. L. 1995. Role of the Flt-1 receptor tyrosine kinase in regulating the assembly of vascular endothelium. *Nature* 376: 66–70.

Garcia-Martinez, V. and Schoenwolf, G. C. 1993. Primitive-streak origin of the cardiovascular system in avian embryos. *Dev. Biol.* 159: 706–719.

George-Weinstein, M. and nine others. 1996. Skeletal myogenesis: The preferred pathway of chick embryo epiblast cells *in vitro. Dev. Biol.* 173: 279–291.

Gilbert-Barness, E. and Opitz, J. M. 1996. Abnormal bone development: Histopathology and skeletal dysplasias. *In* M. E. Martini-Neri, G. Neri, and J. M. Opitz, (eds.), *Gene Regulation and Fetal Development: March of Dimes Birth Defects Foundation Original Article Series 30 (1),* Wiley-Liss, NY. pp. 103–156.

Girasole, G., Jilka, R. L., Passeri, G., Boswell, S., Boder, G., Williams, D. C. and Manolanas, S. C. 1992. 17-Estradiol inhibits interleukin-6 production by bone marrow derived stromal cells and osteoblasts in vitro: A potential mechanism for the antiosteoporotic effect of estrogens. *J. Clin. Invest.* 89: 883–891.

Gittes, G., Galante, P. E., Hanahan, D., Rutter, W. J. and Debas, H. T. 1996. Lineage-specific morphogenesis in the developing pancreas: Role of mesenchymal factors. *Development* 122: 439–447.

Godlin, I. E., Garcia-Porrero, J. A., Coutinho, A., Dieterlen-Lièvre, F. and Marcos, M. A. R. 1993. Para-aortic splanchnopleura from early mouse embryos contain B1a cell progenitors. *Nature* 364: 67–70.

Gordon, M. Y., Riley, G. P., Watt, S. M. and Greaves, M. F. 1987. Compartmentalization of a haematopoietic growth factor (GM-CSF) by glycosaminoglycans in the bone marrow microenvironment. *Nature* 326: 403–405.

Gould, S. J. 1990. An earful of jaw. *Nat. Hist.* 1990(3): 12–23.

Gräper, L. 1907. Untersuchungen über die Herzbildung der Vögel. *Wilhelm Roux Arch. Entwicklungsmech. Org.* 24: 375–410.

Halevy, O. and seven others. 1995. Correlation of terminal cell cycle arrest of skeletal muscle with induction of p21 by MyoD. *Science* 267: 1018–1021.

Hall, B. K. 1988. The embryonic development of bone. *Am. Sci.* 76: 174–181.

Hall, B. K. and Miyake, T. 1995. Divide, accumulate, differentiate: Cell condensations in skeletal development revisited. *Int. J. Dev. Biol.* 39: 881–893.

Hall, B. K., van Exan, R. J. and Brunt, S. L. 1983. Retention of epithelial basal lamina allows isolated mandibular mesenchyme to form bone. *J. Craniofac. Genet. Dev. Biol.* 3: 253–267.

Harary, I. and Farley, B. 1963. *In vitro* studies on single beating rat heart cells. II. Intercellular communication. *Exp. Cell Res.* 29: 466–474.

Hästbacka, J., de la Chapelle, A. and Mahtani, M. M. 1994. The diastrophic dysplasia gene encodes a novel sulfate transporter: Positional cloning by fine-structure linkage disequilibrium mapping. *Cell* 78: 1073–1087.

Hasty, P., Bradley, A., Morris, J. H., Edmondson, D. G., Venuti, J. M., Olson, E. and Klein, W. H. 1993. Muscle deficiency and neonatal death in mice with a targeted mutation in the *myogenin* gene. *Nature* 364: 501–506.

Hatta, K., Takagi, S., Fujisawa, H. and Takeichi, M. 1987. Spatial and temporal expression of N-cadherin cell adhesion molecule correlated with morphogenetic processes of chicken embryo. *Dev. Biol.* 120: 218–227.

Hattersley, G., Kirby, J. A. and Chambers, T. J. 1991. Identification of osteoclast precursors in multilineage hematopoietic colonies. *Endocrinology* 128: 259–262.

Hattori, M., Klatte, K. J., Teixeira, C. C. and Shapiro, I. M. 1995. End labeling studies of fragmented DNA in avian growth plate: Evidence for apoptosis in terminally differentiating chondrocytes. *J. Bone Miner. Res.* 10: 1960–1968.

Hay, E. 1966. *Regeneration*. Holt, Rinehart & Winston, New York.

Ho, S. Y., Cook, A., Anderson, R. H., Allan, L. D. and Fagg, N. 1991. Isomerism of the atrial appendages in the fetus. *Pediatr. Pathol.* 11: 589–608.

Holtzer, H., Rubinstein, N., Fellini, S., Yeoh, G., Chi, J., Birbaum, J. and Okayama, M. 1975. Lineages, quantal cell cycles, and the generation of cell diversities. *Q. Rev. Biophys.* 8: 523–557.

Horton, W. A. 1990. The biology of bone growth. *Growth Genet. Horm.* 6(2): 1–3.

Humphries, R. K., Jacky, P. B., Dill, F. J., Eaves, A. C. and Eaves, C. J. 1979. CFUs in individual erythroid colonies derived in vitro from adult mature mouse marrow. *Nature* 279: 718–720.

Hunt, P., Robertson, D., Weiss, D., Rennick, D., Lee, F. and Witte, O. N. 1987. A single bone marrow stromal cell type supports the in vitro growth of early lymphoid and myeloid cells. *Cell* 48: 997–1007.

Jacobson, A. G. 1961. Heart determination in the newt. *J. Exp. Zool.* 146: 139–152.

Jackson, D., Volpert, O. V., Bouk, N. and Linzer, D. I. H. 1994. Stimulation and inhibition of angiogenesis by placental proliferin and proliferin-related protein. *Science* 266: 1581–1584.

Jilka, R. L. and eight others. 1992. Increased osteoclast development after estrogen loss: Mediation by interleukin-6. *Science* 257: 88–91.

Jonnson, J., Carlsson, L., Edlund, T., and Edlund, H. 1994. Insulin-promote-factor 1 is required for pancreas development in mice. *Nature* 371: 606–608.

Johnson, R. L., Laufer, E., Riddle, R. D. and Tabin, C. 1994. Ectopic expression of *Sonic hedgehog* alters dorsal-ventral patterning of somites. *Cell* 79: 1165–1173.

Jurśśková V. and Tkadleček, L. 1965. Character of primary and secondary colonies of haematopoiesis in the spleen of irradiated mice. *Nature* 206: 951–952.

Kahn, A. J. and Simmons, D. J. 1975. Investigation of cell lineage in bone using a chimaera of chick and quail embryonic tissue. *Nature* 258: 325–327.

Kaplan, S. L. and Grumbach, M. M. 1990. Pathophysiology and treatment of sexual precocity. *J. Clin. Endocrinol. Metab.* 71: 785–789.

Kaushal, S., Schneider, J. W., Nadel-Ginard, B. and Mahdavi, V. 1994. Activation of the myogenic lineage by MEF2A, a factor that induces and cooperates with MyoD. *Science* 266: 1236–1240.

Keller, G., Paige, C., Gilboa, E. and Wagner, E. F. 1985. Expression of a foreign gene in myeloid and lymphoid cells derived from multipotent hematopoietic precursors. *Nature* 318: 149–154.

Knudsen, K. A. 1985. The calcium-dependent myoblast adhesion that precedes cell fusion is mediated by glycoproteins. *J. Cell Biol.* 101: 891–897.

Knudsen, K. A., McElwee, S. A. and Myers, L. 1990. A role for the neural cell adhesion

molecule, N-CAM, in myoblast interaction during myogenesis. *Dev. Biol.* 138: 159–168.

Konigsberg, I. R. 1963. Clonal analysis of myogenesis. *Science* 140: 1273–1284.

Krantz, S. B. and Goldwasser, E. 1965. On the mechanism of erythropoietin induced differentiation. II. The effect on RNA synthesis. *Biochim. Biophys. Acta* 103: 325–332.

Kubai, L. and Auerbach, R. 1983. A new source of embryonic lymphocytes in the mouse. *Nature* 301: 154–156.

Kubota, Y., Kleinman, H. K., Martin, G. R. and Lawley, T. J. 1988. Role of laminin and basement membrane in the morphological differentiation of human endothelial cells into capillary-like structures. *J. Cell Biol.* 107: 1589–1598.

Kurihara, N., Chenu, C., Miller, M., Civin, C. and Roodman, G. D. 1990. Identification of committed mononuclear precursors for osteoclast-like cells in long term human marrow cultures. *Endocrinology* 126: 2733–2741.

LaBarbera, M. 1990. Principles of design of fluid transport systems in zoology. *Science* 249: 992–1000.

Langman, J. 1981. *Medical Embryology*, 4th Ed. Williams & Wilkins, Baltimore.

Larsen, W. J. 1993. *Human Embryology*. Churchill-Livingstone, New York.

Lash, J. W. and Yamada, K. M. 1986. The adhesion recognition signal of fibronectin: A possible trigger mechanism for compaction during somitogenesis. *In* R. Bellairs, D. H. Ede and J. W. Lash (eds.), *Somites in Developing Embryos*. Plenum, New York, pp. 201–208.

Lassar, A. B., Paterson, B. M. and Weintraub, H. 1986. Transfection of a DNA locus that mediates the conversion of 10T1/2 fibroblasts into myoblasts. *Cell* 47: 649–656.

Lassar, A. B., Buskin, J. N., Lockshon, D., Davis, R. L., Apone, S., Hauschka, S. D. and Weintraub. H. 1989. MyoD is a sequence-specific DNA binding protein requiring a region of *myc* homology to bind to the muscle creatine kinase enhancer. *Cell* 58: 823–831.

Lemischka, I. R., Raulet, D. H. and Mulligan, R. C. 1986. Developmental potential and dynamic behavior of hematopoietic stem cells. *Cell* 45: 917–927.

Li, Y. and sixteen others. 1995. A fibrillar collagen gene, Col11a1, is essential for skeletal morphogenesis. *Cell* 80: 423–430.

Linask, K. K. and Lash, J. W. 1986. Precardiac cell migration: Fibronectin localization at mesoderm-endoderm interface during directional movement. *Dev. Biol.* 114: 87–101.

Linask, K. K. and Lash, J. W. 1988a. A role for fibronectin in the migration of avian precardiac cells I. Dose-dependent effects of fibronectin antibody. *Dev. Biol.* 129: 315–323.

Linask, K. K. and Lash, J. W. 1988b. A role for fibronectin in the migration of avian precardiac cells II. Rotation of the heart-forming region during different stages and their effects. *Dev. Biol.* 129: 324–329.

Lyons, G. E. and Buckingham, M. E. 1992. Developmental regulation of myogenesis in the mouse. *Semin. Dev. Biol.* 3: 243–253.

Manolagas, S. and Jilka, R. L. 1995. Bone marrow, cytokines, and bone remodeling. *N. Engl. J. Med.* 332: 305–310.

Markwald, R. R., Fitzharris, T. P. and Manasek, J. J. 1977. Structural development of endocardial cushions. *Am. J. Anat.* 148: 85–120.

Martin, C., Beaupain, D. and Dieterlen-Lièvre, F. 1978. Developmental relationships between vitelline and intraembryonic haemopoiesis studied in avian yolk sac chimeras. *Cell Differ.* 7: 115–130.

Medvinsky, A. and Dzierak,E. 1996. Definitive hematopoiesis is autonomously initiated by the AGM region. *Cell* 86: 897–906.

Medvinsky, A. L., Samoylina, N. L., Müller, A. M. and Dzierzak, E. A. 1993. An early pre-liver intraembryonic source of CFU-S in the developing mouse. *Nature* 364: 64–67.

Meier, S. 1979. Development of the chick mesoblast: Formation of the embryonic axis and the establishment of the metameric pattern. *Dev. Biol.* 73: 25–45.

Mencken, H. L. 1919. Exeunt omnes. *Smart Set* 60: 138–145.

Menko, A. S. and Boettiger, D. 1987. Occupation of the extracellular matrix integrin is a control point for myogenic differentiation. *Cell* 51: 51–57.

Merimee, T. J., Zapf, J., Hewlett, B. and Cavalli-Sforza, L. L. 1987. Insulin-like growth factors in pygmies. The role of puberty in determining final stature. *N. Engl. J. Med.* 316: 906–911.

Millauer, B., Wizigmann-Voos, Schnürch, H., Martinez, R., Müller, N. P. H., Risau, W. and Ullrich, A. 1993. High-affinity VEGF binding and developmental expression suggest *flk-1* as a major regulator of vasculogenesis and angiogenesis. *Cell* 72: 835–846.

Miller, S. A., Bresee, K. L., Michaelson, C. L. and Tyrell, D. A. 1994. Domains of differential cell proliferation and formation of amnion folds in chick embryo ectoderm. *Anat. Rec.* 238: 225–236.

Mintz, B. and Baker, W. W. 1967. Normal mammalian muscle differentiation and gene control of isocitrate dehydrogenase synthesis. *Proc. Natl. Acad. Sci. USA* 58: 592–598.

Moore, J. W., Dionne, C., Jaye, M. and Swain, J. 1991. The mRNAs encoding acidic FGF, basic FGF, and FGF receptor are coordinately downregulated during myogenic differentiation. *Development* 111: 741–748.

Müller, A. M., Medvinsky, A., Strouboulis, J., Grosveld, F. and Dzierzak, E. 1994. Development of hematopoietic stem cell activity in the mouse embryo. *Immunity* 1: 291–301.

Münsterberg, A. E., Kitajewski, J., Bumcroft, D. A., McMahon, A. P. and Lassar, A. B. 1995. Combinatorial signaling by sonic hedgehog and Wnt family members induce myogenic bHLH gene expression in the somite. *Genes Dev.* 9: 2911–2922.

Murphy, M. E. and Carlson, E. C. 1978. Ultrastructural study of developing extracellular matrix in vitelline blood vessels of the early chick embryo. *Am. J. Anat.* 151: 345–375.

Nabeshima, Y., Hanaoka, K., Hayasaka, M., Esumi, E., Li, S., Nonaka, I. and Nabeshima, Y. 1993. *Myogenin* gene disruption results in perinatal lethality because of severe muscle defect. *Nature* 364: 532–535.

Nakauchi, H. and Gachelin, G. 1993. Les celules souches. *La Recherche* 254: 537–541.

Nameroff, M. and Munar, E. 1976. Inhibition of cellular differentiation by phospholipase C. II. Separation of fusion and recognition among myogenic cells. *Dev. Biol.* 49: 288–293.

Nascone, N. and Mercola, M. 1995. An inductive role for the endoderm in *Xenopus* cardiogenesis. *Development* 121: 515–523.

Nilsson, A., Isgaard, J., Lindahl, A., Dahlström, A., Skottner, A. and Isaksson, O. G. P. 1986. Regulation by growth hormone of number of chondrocytes containing IGF-I in rat growth plate. *Science* 233: 571–574.

Oberlender, S. A. and Tuan, R. S. 1994. Expression and functional involvement of N-cadherin in embryonic limb chondrogenesis. *Development* 120: 177–187.

Offield, M. F. and seven others. 1996. PDX-1 is required for pancreatic outgrowth and differentiation of the rostral duodenum. *Development* 122: 983–995.

Okuda, T., van Deursen, J., Hiebert, S. W., Grosveld, G. and Downing, J. R. 1996. AML, the target of multiple chromosomal translocations in human leukemia, is essential for normal fetal liver hematopoiesis. *Cell* 84: 321–330.

Olwin, B. B. and Hauschka, S. D. 1988. Cell surface fibroblast growth factor and epidermal growth factor receptors are permanently lost during skeletal muscle terminal differentiation in culture. *J. Cell Biol.* 107:761–769.

Ordahl, C. P. 1993. Myogenic lineages within the developing somite. *In* M. Bernfield (ed.), *Molecular Basis of Morphogenesis*. Wiley-Liss, New York, pp. 165–170.

Ordahl, C. P. and Le Douarin, N. 1992. Two myogenic lineages within the developing somite. *Development* 114: 339–353.

Ostrovsky, D., Cheney, C. M., Seitz, A. W. and Lash, J. W. 1984. Fibronectin distribution during somitogenesis in the chick embryo. *Cell Differ.* 13: 217–223.

Packard, D. S., Jr. and Meier, S. 1983. An experimental study of somitomeric organization of the avian vegetal plate. *Dev. Biol.* 97: 191–202.

Pardanaud, L., Altmann, C., Kitos, P., Dieterlen-Lièvre, F. and Buck, C. 1987. Vasculogenesis in the early quail blastodisc as studied with a monoclonal antibody recognizing endothelial cells. *Development* 100: 339–349.

Pardanaud, L., Yassine, F. and Dieterlen-Lièvre, F. 1989. Relationship between vasculogenesis, angiogenesis, and hemopoiesis during avian ontogeny. *Development* 105: 473–485.

Pardanaud, L., Luon, D., Prigent, M., Bourcheix, L.-M., Catala, M., and Dieterlen-Lievre, F. 1996. Two distinct endothelial lineages in ontogeny, one of them related to hemopoiesis. *Development* 122: 1363–1371.

Park, W.-J., Bellus, G. and Jabs, E. W. 1995. Mutations in fibroblast growth factor receptors: Phenotypic consequences during eukaryotic development. *Am. J. Hum. Genet.* 57: 748–754.

Patten, B. M. 1951. *Early Embryology of the Chick*, 4th Ed. McGraw-Hill, New York.

Piette, J., Bessereau, J.-L., Huchet, M. and Changeaux, J.-P. 1990. Two adjacent MyoD1-binding sites regulate expression of the acetylcholine receptor a-subunit gene. *Nature* 345: 353–355.

Porcher, C., Swat, W., Rockwell, K., Fujiwara, Y., Alt, F. W. and Orkin, S. H. 1996. The T cell leukemia oncoprotein SCL/tal-1 is essential for development of all hematopoietic lineages. *Cell* 86: 47–57.

Pourquié, O. and nine others. 1996. Lateral and axial signals involved in somite patterning: A role for BMP4. *Cell* 84: 461–471.

Potten, C. S. and Loeffler, M. 1990. Stem cells: attributes, spirals, pitfalls, and uncertainties. Lessons for and from the Crypt. *Development* 110: 1001–1020.

Potts, J. D., Dagle, J. M., Walder, J. A., Weeks, D. L. and Runyon, R. B. 1991. Epithelial-mesenchymal transformation of embryonic cardiac endothelial cells is inhibited by a modified antisense oligodeoxynucleotide to transforming growth factor b3. *Proc. Natl. Acad. Sci. USA* 88: 1516–1520.

Pownall, M. E. and Emerson, C. E., Jr. 1992a. Sequential activation of three myogenic regulatory genes during somite morphogenesis in quails. *Dev. Biol.* 151: 67–79.

Pownall, M. E. and Emerson, C. E., Jr. 1992b. Molecular and embryological studies of avian myogenesis. *Semin. Dev. Biol.* 3: 229–241.

Pownall, M. E., Strunk, K. E. and Emerson, C. E. Jr. 1996. Notochord signal controls the transcriptional cascade of myogenic bHLH genes in somites of quail embryos. *Development* 122: 1475–1488.

Ribatti, D., Urbinati, C., Nico, B., Rusnati, M., Roncali, L. and Presta, M. 1995. Endogenous basic fibroblast growth factor is implicated in the vascularization of the chick embryo chorioallantoic membrane. *Dev. Biol.* 170: 39–49.

Robb, L., Elwood, N. J., Elefanty, A. G. Köntgen, F., Li, R., Barnett, L. D. and Begley, C. G. 1996. The *scl* gene product is required for the generation of all hematopoietic lineages in the adult mouse. *EMBO J.* 15: 4123–4129.

Roberts, R., Gallagher, J., Spooncer, E., Allen, T. D., Bloomfield, F. and Dexter, T. M. 1988. Heparan sulphate-bound growth factors: A mechanism for stromal cell mediated haemopoiesis. *Nature* 332: 376–378.

Rudnicki, M. A., Braun, T., Hinuma, S. and Jaenisch, R. 1992. Inactivation of *MyoD* in mice leads to up-regulation of the myogenic HLH gene *Myf-5* and results in apparently normal muscle development. *Cell* 71: 383–390.

Rudnicki, M. A., Schnegelsberg, P. N. J., Stead, R. H., Braun, T., Arnold, H.-H. and Jaenisch, R. 1993. MyoD or Myf-5 is required in a functionally redundant manner for the formation of skeletal muscle. *Cell* 75: 1351–1359.

Rugh, R. 1951. *The Frog: Its Reproduction and Development.* Blakiston, Philadelphia.

Sariola, H. 1985. Interspecies chimeras: An experimental approach for studies on embryonic angiogenesis. *Med. Biol.* 6: 43–65.

Schultheiss, T. M., Xydas, S. and Lassar, A. B. 1995. Induction of avian cardiac myogenesis by anterior endoderm. *Development* 121: 4203–4214.

Shainberg, A., Yagil, G. and Yaffe, D. 1969. Control of myogenesis in vitro by Ca2+ concentration in nutritional medium. *Exp. Cell Res.* 58: 163–167.

Shalaby, F., Rossant, J., Yamaguchi, T. P., Gertenstein, M., Wu, X.-F., Breitman, M. L. and Schuh, A. C. 1995. Failure of blood-island formation and vasculogenesis in flk-1-deficient mice. *Nature* 376: 62–66

Shapiro, I., DeBolt, K., Funanage, V., Smith, S. and Tuan, R. 1992. Developmental regulation of creatine kinase activity in cells of the epiphyseal growth plate. *J. Bone Miner. Res.* 7: 493–500.

Smith, C. A. and Tuan, R. S. 1996. Functional involvement of Pax-1 in somite development: Somite dysmorphogenesis in chick embryos treated with pax-1 paired-box antisense oligonucleotide. *Teratology* 52: 333–345.

Smith, E. P. and eight others. 1994. Estrogen resistance caused by a mutation in the estrogen-receptor gene in a man. *N. Engl. J. Med.* 331: 1056–1061.

Spemann, H. 1938. *Embryonic Development and Induction.* Yale University Press, New Haven.

Spicer, D. B., Rhee, J., Cheung, W. L. and Lassar, A. B. 1996. Inhibition of myogenic bHLH and MEF2 transcription factors by the bHLH protein Twist. *Science* 272: 1476–1480.

Stern, H. M., Brown, A. M. C. and Hauschka, S. D. 1995. Myogenesis in paraxial mesoderm: preferential induction by dorsal neural tube and by cells expressing Wnt-1. *Development* 121: 3675–3686.

Sugi, Y. and Lough, J. 1994. Anterior endoderm is a specific effector of terminal cardiac myocyte differentiation of cells from the embryonic heart forming region. *Dev. Dyn.* 200: 155–162.

Suri, C. and seven others. 1996. Requisite role of angiopoietin-1, a ligand for the TIE2 receptor, during embryonic angiogenesis. *Cell* 87: 1171–1180.

Syftestad, G. T. and Caplan, A. I. 1984. A fraction from extracts of demineralized adult bone stimulates conversion of mesenchymal cells into chondrocytes. *Dev. Biol.* 104: 348–356.

Tavormina, P. L. and nine others. 1995. Thanatophoric dysplasia (types I and II) caused by distinct mutations in fibroblast growth factor receptor 3. *Nat. Genet.* 9: 321–328.

Thayer, M. J., Tapscott, S. J., Davis, R. L., Wright, W. E., Lassar, A. B. and Weintraub, H. 1989. Positive autoregulation of the myogenic determination gene *MyoD1. Cell* 58: 241–248.

Till, J. E. 1981. Cellular diversity in the blood-forming system. *Am. Sci.* 69: 522–527.

Till, J. E. and McCulloch, E. A. 1961. A direct measurement of the radiation sensitivity of normal mouse bone marrow cells. *Radiat. Res.* 14: 213–222.

Tsuda, T., Philp, N., Zile, M. H. and Linask, K. K. 1996. Left-right asymmetric localization of flectin in the extracellular matrix during heart looping. *Dev. Biol.* 173: 39–50.

Tuan, R. 1987. Mechanisms and regulation of calcium transport by the chick embryonic chorioallantoic membrane. *J. Exp. Zool.* [Suppl.] 1: 1–13.

Tuan, R. S. and Lynch, M. H. 1983. Effect of experimentally induced calcium deficiency on the developmental expression of collagen types in chick embryonic skeleton. *Dev. Biol.* 100: 374–386.

Tyler, M. S. and Hall, B. K. 1977. Epithelial influence on skeletogenesis in the mandible of the embryonic chick. *Anat. Rec.* 206: 61–70.

Urist, M. R. and eight others. 1984. Purfication of bovine bone morphogenetic protein by hydroxyapatite chromatography. *Proc. Natl. Acad. Sci. USA* 81: 371–375.

Vaidya, T. B., Rhodes, S. J., Taparowsky, E. J. and Konieczny, S. F. 1989. Fibroblast growth factor and transforming growth factor-β repress transcription of the myogenic regulatory gene MyoD1. *Mol. Cell Biol.* 9: 3576-3579.

Venuti, J. M., Morris, J. H., Vivian, J. L., Olson, E. N. and Klein, W. H. 1995. Myogenesis is required for late but not early aspects of myogenesis during mouse development. *J. Cell Biol.* 128: 563–576.

Vikkula, M. and eleven others. 1996. Vascular dysmorphogenesis caused by an activating mutation in the receptor tyrosine kinase TIE2. *Cell* 87: 1181–1190.

Webster, M. K. and Donoghue, D. J. 1996. Constitutive activation of fibroblast growth factor receptor 3 by the transmembrane domain point mutation found in achondroplasia. *EMBO J.* 15: 520–527.

Weintraub, H., Tapscott, S. J., Davis, R. L., Thayer, M. J., Adam, M. A., Lassar, A. B. and Miller, D. 1989. Activation of muscle-specific genes in pigment, nerve, fat, liver, and fibroblast cell lines by forced expression of MyoD. *Proc. Natl. Acad. Sci. USA* 86: 5434–5438.

Whitlock, C. A., Tidmarsh, G. F., Muller-Sieburg, C. and Weissman, I. L. 1987. Bone marrow stromal cell lines with lymphopoietic activity express high levels of a pre-B neoplasia-associated molecule. *Cell* 48: 1009–1021.

Wilkie, A. O. M., Morriss-Kay, G. M., Jones, E. Y. and Heath, J. K. 1995. Functions of fibroblast growth factors and their receptors. *Curr. Biol.* 5: 500–507.

Wilson, D. 1983. The origin of the endothelium in the developing marginal vein of the chick wing bud. *Cell Differ.* 13: 63–67.

Wolf, N. S. and Trentin, J. J. 1968. Hemopoietic colony studies. V. Effect of hemopoietic organ stroma on differentiation of pluripotent stem cells. *J. Exp. Med.* 127: 205–214.

Wright, E. and eight others. 1995. The *Sry*-related gene *Sox9* is expressed during chondrogenesis in mouse embryos. *Nat. Genet.* 9: 15–20.

Wuthier, R. 1982. A review of the primary mechanism of endochondral ossification with special emphasis on the role of cells, mitochondria, and matrix vesicles. *Clin. Orthop. Rel. Res.* 169: 219–242.

Yaffe, D. and Feldman, M. 1965. The formation of hybrid multinucleated muscle fibres from myoblasts of different genetic origin. *Dev. Biol.* 11: 300–317.

Yagami-Hiromasa, T., Sato, T., Kurisaki, T., Kamijo, K., Nabeshima, Y.-I. and Fujisawa-Sehara, A. 1995. A metalloprotease-disintegrin participating in myoblast fusion. *Nature* 377: 652–656.

Mechanisms of
Cellular Differentiation

Transcriptional regulation
of gene expression: Transcription factors and the activation of specific promoters

10

But whatever the immediate operations of
the genes turn out to be, they most certainly belong to the category of developmental processes and thus belong to the
province of embryology. This central
problem of fundamental biology at the
present time is being attacked from many
sides, both by physiologists and bio-
chemists and by geneticists; but it is
essentially an embryological problem.
C. H. WADDINGTON (1956)

We have entered the cell, the mansion of
our birth, and have started the inventory
of our acquired wealth.
ALBERT CLAUDE (1974)

DIFFERENT CELL TYPES make different sets of proteins, even though their genomes are identical. Each human being has roughly 150,000 genes in each nucleus, but each cell uses only a small subset of these genes. Moreover, different cell types use different subsets of these genes. Red blood cells make globins, lens cells make crystallins, nerve cells make neurotransmitters, and endocrine glands make their specific hormones. *Developmental genetics* is a discipline that examines how the genotype is transformed into the phenotype, and the major paradigm of developmental genetics is *differential gene expression from the same nuclear repertoire.* This regulation of gene expression can be accomplished at several levels:

- *Differential gene transcription*, regulating which of the nuclear genes are transcribed into RNA
- *Selective nuclear RNA processing*, regulating which of the transcribed RNAs get into the cytoplasm to become messenger RNAs
- *Selective messenger RNA translation*, regulating which of the mRNAs in the cytoplasm get translated into protein
- *Differential protein modification*, regulating which proteins are allowed to remain or function in the cell

Some genes (such as those coding for the globin proteins of hemoglobin) are regulated at each of these levels. This and the following chapter will discuss the mechanisms of differential gene transcription: how different genes are activated in different types of cells at particular times. We discussed the basic phenomena of differential gene transcription in Chapter 2. The puffs of the polytene chromosomes represent the activation of groups of genes in response to a hormone produced in the larval insect. Similarly, the expression of endoderm-specific genes in sea urchin larvae was controlled at the level of gene transcription. In these chapters, we will be discussing the mechanisms whereby different genes are able to become activated or repressed in specific cells as the cells differentiate.

Exons and introns

When one looks at genes, the first thing that becomes apparent is that most eukaryotic genes are not like most prokaryotic genes. Eukaryotic genes are not co-linear with their peptide products. Rather, the 5′ and 3′ ends of eukaryotic mRNA come from noncontiguous regions on the chromosome. Between the regions of DNA coding for the proteins—**exons**—are intervening sequences—**introns**—that have nothing whatsoever to do with the amino acid sequence of the protein.* The structure of the human β-globin gene is shown in Figure 10.1. This gene consists of the following elements:

1. A **promoter region** responsible for the binding of RNA polymerase and for the subsequent initiation of transcription. This promoter region of the human β-globin gene has three distinct units and extends from 95 to 26 base pairs before ("upstream from") the transcription initiation site (i.e., from −95 to −26).

2. The sequence ACATTTG, where *transcription* is initiated. This is often called the **cap sequence** because it represents the 5′ end of the RNA, which will receive a "cap" of modified nucleotides soon after it is transcribed. The specific cap sequence varies among genes.

3. The ATG codon for the initiation of *translation*. This codon is located 50 base pairs after the initiation point of transcription (although this distance differs greatly in different genes). The intervening sequence of 50 nucleotide pairs between the initiation points of transcription and translation is called the **leader sequence**. The leader sequence can determine the rate at which translation is initiated.

4. The first exon containing 90 base pairs coding for amino acids 1–30 of human β-globin.

5. An intron containing 130 base pairs with no coding sequences for globin. The structure of this intron is important in enabling the RNA to be processed into messenger RNA and exit from the nucleus.

6. An exon containing 222 base pairs coding for amino acids 31–104.

7. A large intron—850 base pairs—having nothing to do with the globin protein structure.

8. An exon containing 126 base pairs coding for amino acids 105–146.

9. A **translation termination codon,** TAA.

10. A **3′ untranslated region** that, although transcribed, is not translated into protein. This region includes the sequence AATAAA, which is needed to place a "tail" of some 200 to 300 adenylate residues on the RNA transcript. This poly(A) tail confers stability and translatability on the mRNA, and it is inserted into the RNA about 20 bases downstream of the AAUAAA sequence. However, transcription continues beyond the AATAAA site for about 1000 nucleotides before being terminated. Within the 3′ transcribed but untranslated sequence (about 600 to 900 base pairs from the AATAAA site) is a DNA sequence that serves as an enhancer. This sequence is necessary for the temporal and tissue-specific expression of the β-globin gene in adult red blood cell precursors (Trudel and Constantini, 1987).

*The term *exon* has taken on two overlapping meanings. In the original sense, it is anatomically defined as a nucleotide sequence whose RNA "exits" the nucleus. It has taken on the functional definition of a protein-encoding nucleotide sequence. For discussion here, we will use the former definition, and we will define the leader sequences and 3′ untranslated sequences as exons that are not translated. Certain eukaryotic genes (such as the genes for histones) lack intervening sequences, and any hypothesis concerning intron function must take these exceptions into account. By convention, upstream, downstream, 5′, and 3′ directions are specified in relation to the RNA. Thus, the promoter is *upstream* of the gene, near its 5′ end.

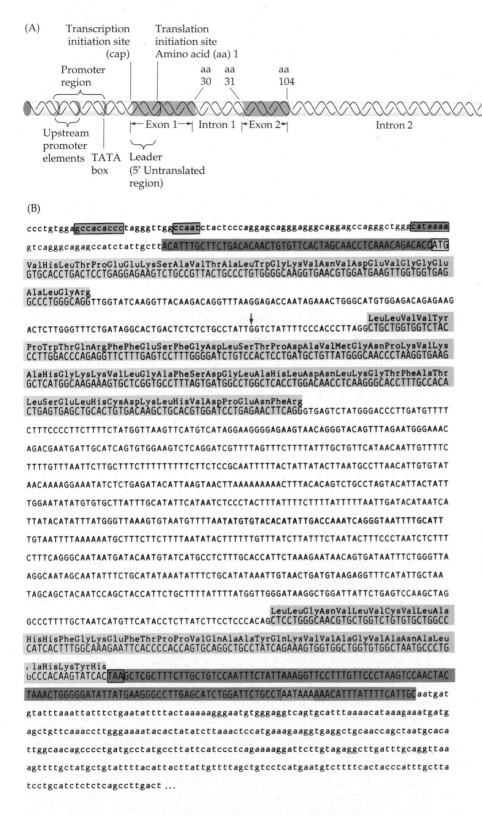

(A)

Figure 10.1
Nucleotide sequence of the human β-globin gene. (A) Schematic representation of the locations of the promoter region, transcription initiation (cap) site, leader sequence, and exons and introns of the β-globin gene. Exons are in color; the numbers flanking them indicate the amino acid positions they encode in β-globin. (B) The nucleotide sequence of the β-globin gene, shown from the 5' end to the 3' end of the RNA. The promoter sequences are boxed, as are the translation initiation and termination codes ATG and TAA. The large capital letters boxed in color correspond to exons, and the amino acids for which they code are abbreviated above them. The small capital letters are the bases of the intervening sequences. The codons represented by capital letters after the translation terminator are in the globin mRNA but are not translated into proteins. Within this group is the sequence thought to be needed for polyadenylation. A G in the first intron (arrow) is mutated to an A in one form of β⁺-thalassemia. (Sequence from Lawn et al., 1980.)

The original nuclear RNA transcript for such a gene contains the capping sequence, the leader sequence, the exons, the introns, and the 3' untranslated region (Figure 10.2). In addition, both its ends become modified. A cap consisting of methylated guanosine is placed on the 5' end of the RNA in opposite polarity to the RNA itself. Thus, whereas all the bases in the message

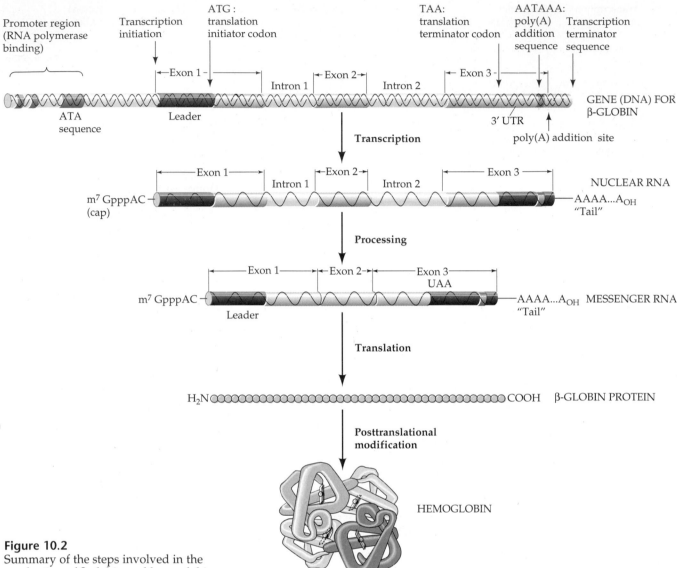

Figure 10.2
Summary of the steps involved in the production of β-globin and hemoglobin.

precursor are linked 5′ to 3′, the cap structure is linked 5′ to 5′. This means that there is no free 5′ phosphate group on the nuclear RNA (Figure 10.3). Messenger RNA molecules are likewise "capped," although it is not certain whether the mRNA cap is the original one it received in the nucleus. The 5′ cap is necessary for the binding of mRNA to the ribosome and for subsequent translation (Shatkin, 1976).

The 3′ terminus is usually modified in the nucleus by having roughly 200 adenylate residues added on as a tail. These adenylic acid residues are put together enzymatically and are added to the transcript. They are not part of the gene sequence. Both the 5′ and 3′ modifications may protect the RNA from exonucleases (Sheiness and Darnell, 1973; Gedamu and Dixon, 1978), thereby stabilizing the message and its precursor.

Promoter structure and function

In addition to the gene structure we have just discussed, there are regulatory sequences that can be on either end of the gene (or even within it). These se-

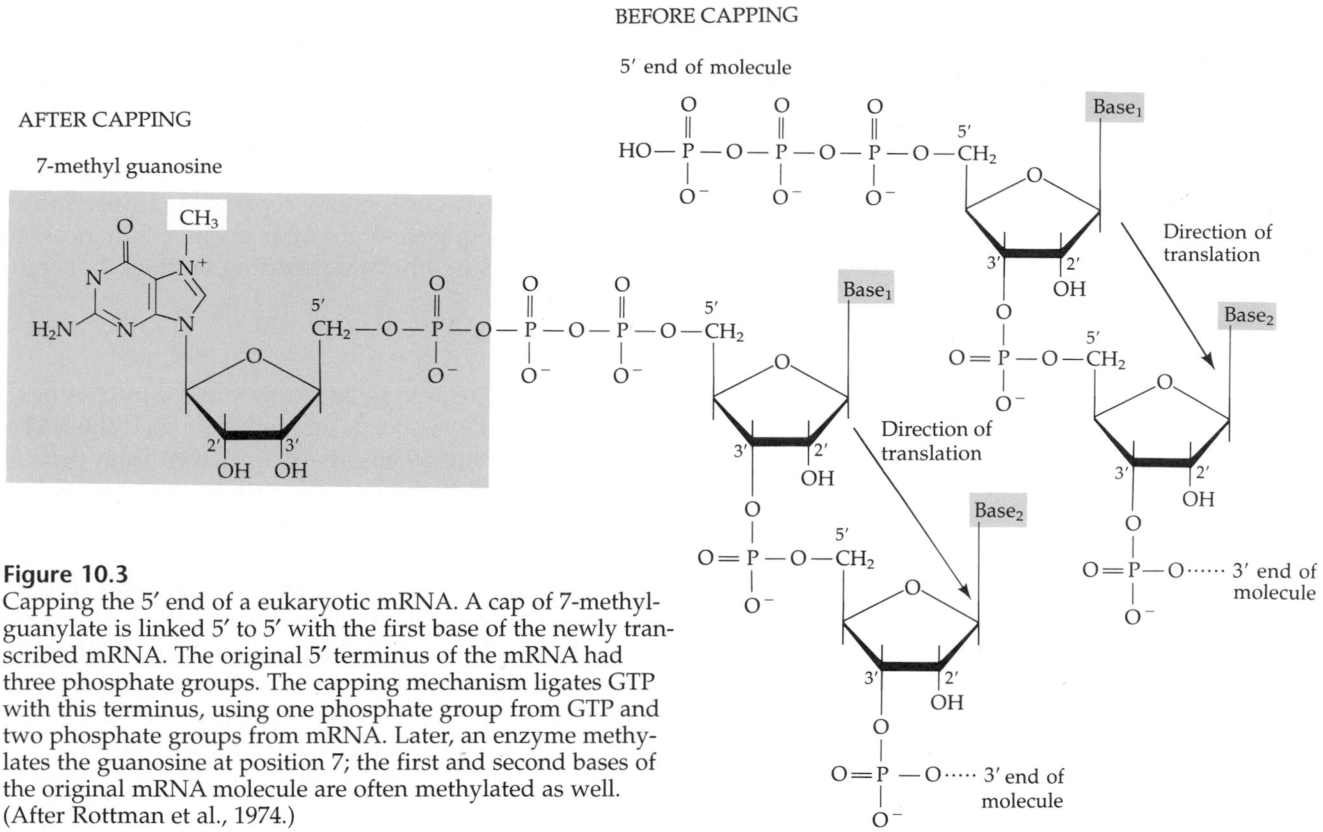

BEFORE CAPPING

5' end of molecule

AFTER CAPPING

7-methyl guanosine

Figure 10.3
Capping the 5' end of a eukaryotic mRNA. A cap of 7-methyl-guanylate is linked 5' to 5' with the first base of the newly transcribed mRNA. The original 5' terminus of the mRNA had three phosphate groups. The capping mechanism ligates GTP with this terminus, using one phosphate group from GTP and two phosphate groups from mRNA. Later, an enzyme methylates the guanosine at position 7; the first and second bases of the original mRNA molecule are often methylated as well. (After Rottman et al., 1974.)

quences, the promoters and enhancers (introduced in Chapter 2), are necessary for controlling where and when a particular gene is transcribed.

Two types of regulatory elements are needed to effect transcription at the proper sites. The first set of regulatory elements are called ***cis*-regulators.** These represent specific DNA sequences on a given chromosome. *Cis*-regulators act only on adjacent genes. The second group of regulatory elements are called ***trans*-regulators.** These are soluble molecules (including proteins and RNAs) that are made by one gene and interact with genes on the same or different chromosomes. If one recalls gene induction in the *lac* operon in *E. coli*, one will remember that a repressor gene makes a repressor protein that interacts with the operator sequence of the *lac* operon genes. In this case, the operator DNA is a *cis*-regulatory element because it controls only the adjacent *lac* operon on its own chromosome. The repressor protein, however, is a *trans*-regulator, because it can be made by one chromosome and bind to the *cis*-regulatory operator on another chromosome.

In eukaryotic genes that encode messenger RNA, two types of *cis*-regulatory DNA sequences have been discovered that influence which genes become transcribed in which cells. These are the promoters and the enhancers. **Promoters** are typically located immediately upstream from the site where transcription begins and are generally hundreds of base pairs long. The promoter site is required for the binding of RNA polymerase II and the accurate initiation of transcription. Eukaryotic RNA polymerases require additional protein factors to bind efficiently to the promoter. The **enhancer** is a DNA sequence that can activate the utilization of the promoter, controlling the efficiency and rate of transcription from that particular promoter. Enhancers can activate only *cis*-linked promoters (i.e., promoters on the same chromosome), but they can do so at great distances (some as great as 50 kilobases away from the promoter). Moreover, enhancers do not need to be at the 5′ (upstream) side of the gene. They can be at the 3′ end, in the introns, or even on the complementary DNA strand (Maniatis et al., 1987). Like the promoter, enhancers function by binding specific *trans*-regulatory proteins called **transcription factors**.

One type of enhancer is a "negative enhancer," also called a **silencer**. When transcription factors bind to silencers, they repress the transcription from *cis*-linked promoters. Some sequences can act as positive enhancers in some cells and negative enhancers in others, depending on the other transcription factors present in the cell.

Promoter Structure

Promoters of genes that transcribe relatively large amounts of mRNA have similar structures. They have a TATA sequence (sometimes called the **TATA box** or **Goldberg-Hogness box**) about 30 base pairs upstream from the site where transcription begins, as well as one or more promoter elements further upstream (Figure 10.4; Grosschedl and Birnstiel, 1980; McKnight and Tjian, 1986). The "functional anatomy" of a promoter region can be analyzed by determining which of its bases are necessary for efficient transcription. Cloned genes can be accurately transcribed when they are placed into the nuclei of frog oocytes or fibroblasts or when they are incubated with RNA polymerase in the presence of nucleotides and nuclear extracts (Wasylyk et al., 1980). After the transcription of a gene is confirmed, one uses restriction enzymes to make specific deletions in the gene or in the regions surrounding it. One can then see whether such a modified gene will still be accurately transcribed. Such studies on β-globin genes (Grosveld et al., 1982; Dierks et al., 1983) showed that the first 109 base pairs preceding the cap site were sufficient for the correct initiation of β-globin gene transcription by RNA polymerase.

Myers and co-workers (1986) refined this analysis by cloning the region of a mouse β-globin gene from 106 base pairs upstream from the start of transcription (−106) through the first 475 base pairs (+475) of the first exon. These clones were subjected to in vitro mutagenesis (wherein specific mutations can be placed into a cloned gene). In this way, 130 different single-base substitutions were introduced into the promoter region of the globin gene. These cloned genes were placed into plasmids containing an enhancer from a gene normally expressed in all tissues. The recombinant plasmids were then introduced by transfection into cultured cells that do not usually produce globin. Would the cells transcribe a truncated globin message (475 bases) from the clones? Figure 10.5 shows the results. In most cases, mutating a base in the 5′ flanking region did not affect the efficiency of globin gene transcription. But there were three clusters of nucleotides in which mutations drastically reduced transcription. One cluster was in the TATA box, another was in the CAAT upstream promoter element, and a third was in the CACCC region, about 95 to 87 base pairs upstream from the cap site.

The CAAT and TATA boxes have been found to be critical elements in numerous eukaryotic promoters (Efstratiadis et al., 1980), but the CACCC sequence is seldom seen except in the β-globin gene promoters in several species. In humans, this sequence appears to be critical. A naturally occurring mutation in this sequence causes a total loss of β-globin gene transcription (Orkin and Kazazian, 1984), and this sequence is recognized by an erythroid-specific transcription factor (Mantovani et al., 1988). Two mutations, at positions −78 and −79, have actually increased transcription to three times the wild-type level. It is thought that these changes facilitate the interaction of the promoter with the *trans*-regulatory proteins.

CCAAT
GGGGCGG
GCCACACCC
ATGCAAAT

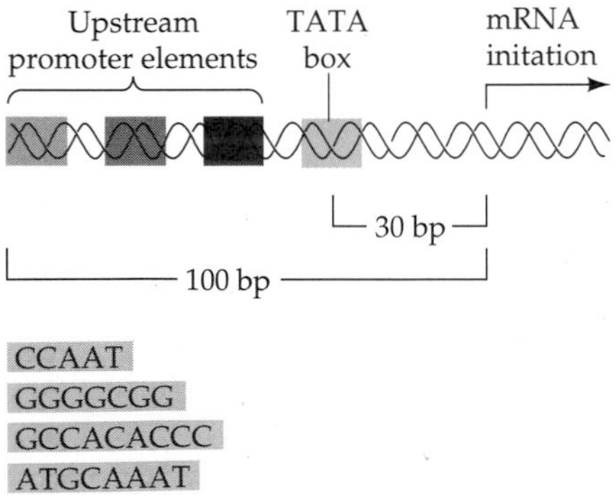

Figure 10.4
Typical promoter region for a protein-coding eukaryotic gene. The gene diagrammed here contains a TATA box and three upstream promoter elements. Examples of such upstream elements are shown beneath the diagram. (After Maniatis et al., 1987.)

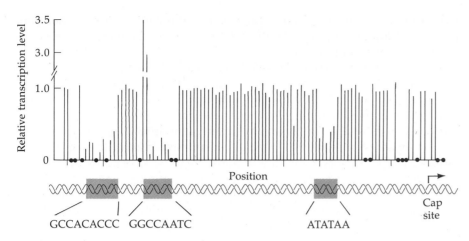

Figure 10.5
The effect of specific point mutations in the mouse β-globin promoter on the ability of that promoter to initiate transcription. Each line represents the transcription level of a mutant promoter relative to the transcription level of a wild-type globin promoter assayed concurrently. The black dots represent nucleotides for which no mutation had been generated. Below the histogram is a diagram showing the position of the TATA box and the two upstream promoter elements of the mouse β-globin gene. (After Myers et al., 1986.)

Promoter Function

Promoters can function not only to bind RNA polymerase, but also to specify the places and times that transcription can occur from that gene. This function of promoters can be vividly demonstrated in certain transgenic animals. Here, a new gene is constructed wherein the normal promoter of a particular gene is replaced with the promoter of some other gene, and the fused gene is placed into a pronucleus of a mammalian zygote. Palmiter and co-workers (1982) isolated the rat growth hormone gene and deleted its 5′ promoter region. Into this space, they substituted the promoter sequence of another gene—*Mt-1* for mouse metallothionein 1, a small protein involved in regulating serum zinc levels. This hybrid gene is shown in Figure 10.6A. The *Mt-1* gene can be induced by the presence of heavy metals such as zinc or cadmium, and the sequences responsible for this induction are in the promoter of this gene. By fusing this metallothionein promoter region to the rat growth hormone gene (*rGH*), the rat growth hormone gene is placed under

Figure 10.6
Promoter function seen in transgenic mice. (A) Recombinant plasmid containing rat growth hormone structural gene, mouse metallothionein regulatory region, and bacterial plasmid *pBR322*. The plasmid, p*MGH*, was injected into the mouse oocytes. The dark boxes on the injected plasmid correspond to the exons of the *GH* gene. The direction of transcription is indicated by an arrow. (B) A mouse derived from the eggs injected with p*MGH* (left) and a normal littermate (right). (From Palmiter et al., 1982, photograph courtesy of R. L. Brinster.)

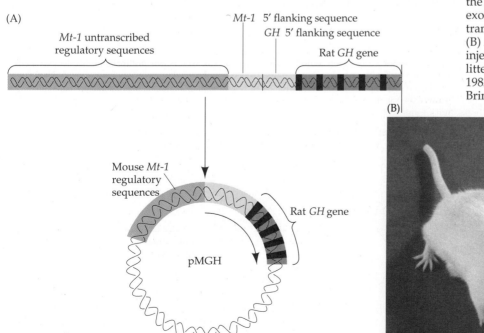

the control of the metallothionein promoter. In this case, rat growth hormone message should be made when the *Mt-1* promoter is activated by the presence of zinc or cadmium.

A plasmid containing this fused gene was grown in bacteria (see Chapter 2), the *Mt-1/rGH* piece was isolated, and about 600 copies of this fragment were injected into pronuclei of recently fertilized mouse eggs. DNA hybridization showed that many of these newborn mice had incorporated numerous copies of the rat growth hormone gene into their chromosomes. These transgenic mice were then fed a diet supplemented with zinc. The zinc induced large amounts of rat growth hormone to be secreted by the livers of these mice. (The liver is where metallothionein is usually made. Growth hormone is usually secreted from the pituitary gland.) The amount of growth hormone secreted correlated with the size of these mice. The transgenic mice became enormous, up to 80 percent larger than their normal littermates (Figure 10.6B).* The metallothionein promoter regulated the synthesis of growth hormone in these transgenic mice.

This strategy is currently being used by pharmaceutical companies to make large quantities of protein products such as peptide hormones, α_1-antitrypsin (used for patients with emphysema), and blood-clotting factors. Pronuclei from cows, sheep, and goats have been injected with recombinant DNA containing the gene sequence for the desired protein fused with the promoters of genes for casein, lactalbumin, or β-lactoglobulin (three major milk proteins). Lactating cows synthesize enormous quantities of milk proteins, and much of this production is regulated by the transcription of new messages. The hope is that when the animal transcribes the genes for casein or lactalbumin (in response to the hormone prolactin), the genes for these therapeutic proteins will be transcribed and synthesized as well. For instance, in one case, a human gene for α_1-antitrypsin protein was fused to a β-lactoglobulin promoter and injected into sheep zygote pronuclei. One of those sheep embryos developed into a female whose milk contained 35 g of human α_1-antitrypsin protein per liter (Figure 10.7; Wright et al., 1991).[†] Promoters, then, play a role in specifiying which gene is transcribed in which cells during development.

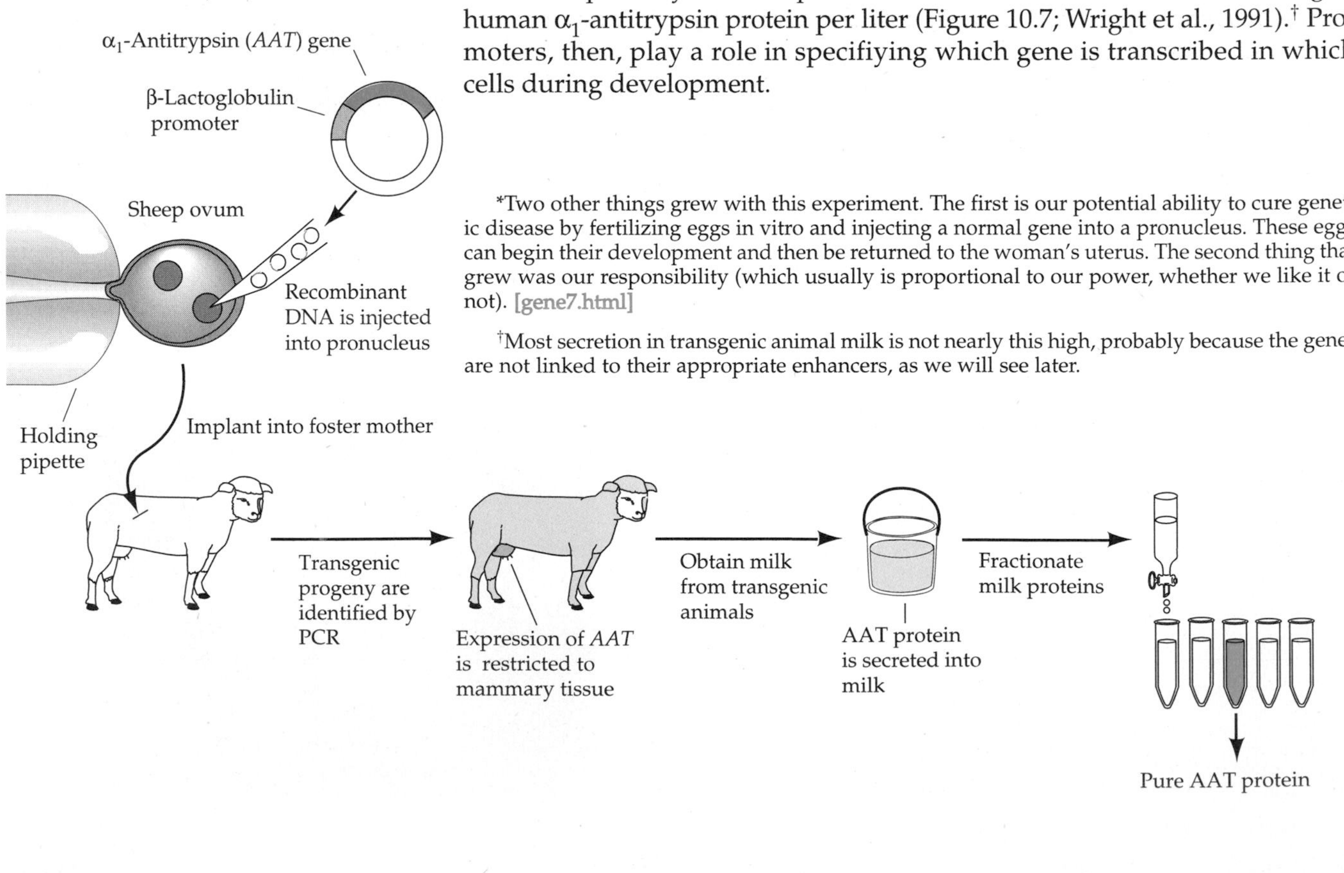

*Two other things grew with this experiment. The first is our potential ability to cure genetic disease by fertilizing eggs in vitro and injecting a normal gene into a pronucleus. These eggs can begin their development and then be returned to the woman's uterus. The second thing that grew was our responsibility (which usually is proportional to our power, whether we like it or not). [gene7.html]

[†]Most secretion in transgenic animal milk is not nearly this high, probably because the genes are not linked to their appropriate enhancers, as we will see later.

Figure 10.7
Promoter function seen in transgenic sheep. The structural gene for a pharmaceutically important protein such as α1-antitrypsin or clotting factor peptides is linked to the promoter for sheep milk β-lactalbumin (or casein). The recombinant gene is injected into a pronucleus of a newly fertilized sheep egg, and the egg is implanted into the uterus of a foster mother. The newborn sheep are screened (by PCR or Southern blotting) for the presence of the transgene. When female transgenic sheep mature, the transgene should be activated in the mammary gland and the protein secreted into the milk. From the milk, one can isolate the pharmaceutically important compound. (After Watson et al., 1992.)

RNA Polymerase and the trans-Regulatory Factors at the Promoter

TRANSCRIPTION REQUIRES the interaction of RNA polymerase with promoter DNA. In eukaryotic cells, there are three different types of RNA polymerases, each having particular functions and properties (Rutter et al., 1976). **RNA polymerase I** is found in the nucleolar region of the nucleus and is responsible for transcribing the large ribosomal RNAs; **RNA polymerase II** transcribes messenger RNA precursors; and **RNA polymerase III** transcribes small RNAs such as transfer RNA, 5S ribosomal RNA, and other small DNA sequences.* None of the eukaryotic RNA polymerases can bind efficiently to DNA. Rather, there are families of DNA-binding proteins that first bind to DNA and, once bound, interact with the RNA polymerase to initiate RNA synthesis.

The TATA element and RNA polymerase II.

The classic diagram of transcription shows that DNA, in the presence of RNA polymerase and ribonucleoside triphosphates, transcribes RNA molecules. But this simple scheme does not account for the difficulties in (1) getting the RNA to start in the correct place and (2) having transcription of a particular gene occur only at particular times and in particular types of cells. There have to be factors that enable RNA polymerase to bind solely to the promoters of particular genes. Here, we will discuss those proteins and DNA sequences that localize the RNA polymerase to the promoter sites. The enzyme responsible for the transcription of messenger RNAs is RNA polymerase II. However, accurate transcription of cloned genes in vitro will not occur if the genes are incubated with purified RNA

polymerase II and nucleoside triphosphates. Nuclear extracts must be added if accurate transcription is to commence. What are these factors that allow transcription to be initiated? At least six nuclear proteins have been shown to be necessary for the proper initiation of transcription by RNA polymerase II (Figure 10.8; Buratowski et al., 1989; Sopta et al., 1989).

TFIID and TFIIA.[†] In the first step of mRNA transcription, the TFIID complex binds to the TATA box. This was shown by DNase protection experiments in which TFIID was added to cloned genes and the DNA was then digested with DNase. The only way to "rescue" the DNA from digestion was for the TFIID to bind to the DNA, thereby preventing the DNase from reaching it. In this way, Sawadogo and Roeder (1984) showed that TFIID specifically bound to the TATA region of the genes. TFIID is a multimeric protein, one of whose components—the **TATA-binding protein (TBP)**—binds directly into the minor groove of the TATA sequence (Lee et al., 1991; Starr and Hawley, 1991). The TFIID complex has several activities; the first is to bind the TATA sequence and to serve as the foundation for the transcriptional complex. Another role of TFIID is to prevent the stabilization of nucleosomes in the promoter region. When promoter-containing DNA is incorporated into nucleosomes, these genes are unable to be transcribed when TFIID, RNA polymerase II, and other factors are added later. However, when TFIID is added to the genes before or during nucleosome formation, the resulting chromatin is transcriptionally active (Workman and Roeder, 1987). In blocking nucleosome production, TFIID appears to act antago-

nistically to histone H1. Histone H1 (as we will see in the next chapter) stabilizes nucleosomes and prevents transcription in the region where it binds. The addition of histone H1 prevents TFIID from finding the TATA sites and prevents transcription; however, this inhibition is overcome if TFIID is added first (Laybourn and Kadonaga, 1991). The binding of TFIID is facilitated and stabilized by the transcription factor TFIIA (Buratowski et al., 1989; Maldonado et al., 1990). Many transcription factors activate transcription function by recruiting TFIID and "activating" it so that TFIID can bind the other members of the transcription complex (Chi and Carey, 1996; Stargell and Struhl, 1996). Thus, the decision to transcribe or not to transcribe a particular gene often depends on the balance between inhibitory factors (such as histones) and TFIID and TFIIA.

TFIIB and RNA polymerase II. The TFIID/TFIIA complex cannot form a stable complex directly with RNA polymerase II. Rather, TFIID binds factor **TFIIB.** The binding of TFIIB to TFIID appears to be the key rate-limiting step in the transcription of numerous genes. This rate can be dramatically increased by the proximity of certain promoter- and enhancer-binding transcription factors. These transcription factors are sequence-specific and can regulate which genes will be transcribed. The *activator domain* of these transcription factors binds directly to TFIIB and facilitates its assembly with TFIID (Lin and Greene, 1991; Lin et al., 1991). Once TFIIB is in place, it can bind RNA polymerase II. Most of the RNA polymerase is positioned by its interaction with TFIIB, but the carboxy-terminal tail of the large subunit of RNA polymerase II interacts directly with TFIID (see Figure 10.8D). In this way, RNA polymerase II is placed at the promoter.

TFIIE/F and TFIIH. Either directly before or during its binding to TFIIB,

---

*In most cells, ribosomal and transfer RNAs are constitutively synthesized. However, animals have evolved remarkable mechanisms for upregulating rRNA synthesis in their oocytes. We will therefore postpone our discussion of RNA polymerases I and III until we detail the events of oogenesis in Chapter 22.

[†]*TF* stands for *transcription factor*; *II* indicates that the factor was first found to be needed for RNA polymerase *II*; and the letter designations refer to fractions from phosphocellulose columns that had the activity.

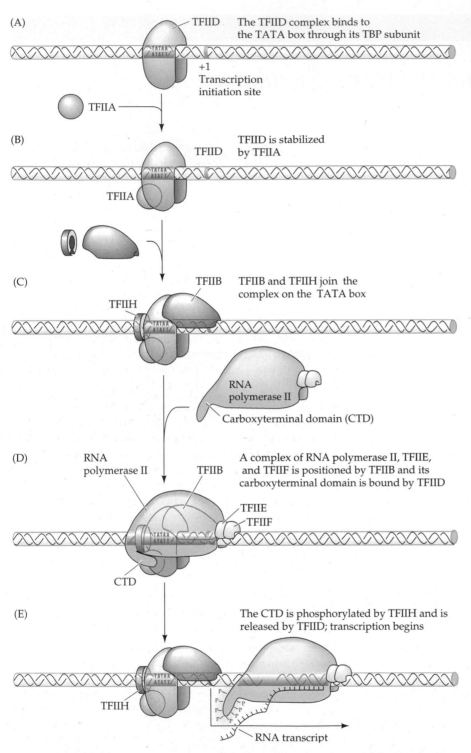

(A)   TFIID   The TFIID complex binds to the TATA box through its TBP subunit

+1 Transcription initiation site

TFIIA

(B)   TFIID   TFIID is stabilized by TFIIA

TFIIA

(C)   TFIIB   TFIIB and TFIIH join the complex on the TATA box

TFIIH

RNA polymerase II

Carboxyterminal domain (CTD)

(D)   RNA polymerase II   TFIIB   A complex of RNA polymerase II, TFIIE, and TFIIF is positioned by TFIIB and its carboxyterminal domain is bound by TFIID

TFIIE
TFIIF

CTD

(E)   The CTD is phosphorylated by TFIIH and is released by TFIID; transcription begins

TFIIH

RNA transcript

Figure 10.8 The formation of the active eukaryotic initiation complex. The diagrams represent the complexes formed on the TATA box by the transcription factors and RNA polymerase II. (A) The TFIID complex binds to the TATA box through its TBP subunit. (B) TFIID is stabilized by TFIIA. (C) TFIIB and TFIIH join the complex on the TATA box while TFIIE and TFIIF associate with RNA polymerase II. (D) RNA polymerase is positioned by TFIIB, and its carboxy-terminal domain (CTD) is bound by TFIID. (E) The CTD is phosphorylated by TFIIH and is released by TFIID. The RNA polymerase II is now competent to transcribe mRNA from the gene.

RNA polymerase II becomes associated with TFIIF and TFIIE (Buratowski et al., 1991; Conaway et al., 1991). **TFIIF** has an enzymatic activity needed to unwind the DNA helix. **TFIIE** is a DNA-dependent ATPase and is probably needed for generating the energy for transcription (Bunick et al., 1982; Sawadogo and Roeder, 1984). But what good is all this if the RNA polymerase remains bound to this complex on the TATA box? For transcription to occur, the RNA polymerase must be released from the promoter region. This releasing activity appears to be the function of **TFIIH**. The RNA polymerase is tightly bound by its carboxy-terminal domain (CTD) to TFIID. However, TFIID will only bind the *unphosphorylated* form of the CTD. In mammals, the CTD contains 52 repeats of the seven-amino-acid sequence YSPTSPS. When the initiation complex is formed, the completed complex activates the serine/threonine protein kinase activity of TFIIH. TFIIH then phosphorylates each of the 52 repeats (see Figure 10.8E; Koleske et al., 1992; Lu et al., 1992; Usheva et al., 1992). TFIID cannot bind this heavily phosphorylated region and releases the RNA polymerase. While the first phosphodiester bond can be made without the phosphorylation of the CTD, this phosphorylation appears to be essential for the further transcription of messenger RNA (Akoulitchev et al., 1995).

TAFs and the activation of basal transcription. TFIID is a multimeric protein, only one of its subunits actually binding to the TATA sequence. Some of the other subunits are called **TATA-binding protein-associated factors** (**TAFs**). Purification of the TAFs from human and *Drosophila* TFIID showed that they were composed of a similar set of proteins (Figure 10.9; Dynlacht et al., 1991). These TAFs are thought to serve two functions: (1) they can determine whether or not the TFIID remains on the promoter, and (2) they can function as co-activators, bridging the enhancer-bound proteins to the transcription complex through protein-protein interactions.

Whether the TATA-binding protein stays on the promoter is of great concern to a gene. If it comes off, the gene will not be transcribed. Verrijzer and col-

(A) A minimal complex of TBP and a TAF fails to activate transcription (Sp1 and NTF cannot associate with TBP)

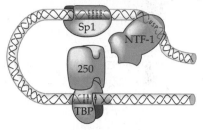

Figure 10.9 Outline of experiments suggesting that different TAFs interacted with different transcription factors to activate transcription. The full complement of TAFs is shown in (D). The "activator" in (D) is a protein bound to a DNA sequence that has been stabilized by the other interactions. (After Chen et al., 1994.)

(B) Addition of the p150 TAF or the p60 TAF allows transcriptional activation by NTF, but not Sp1

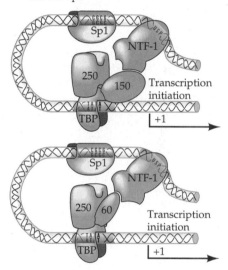

(C) Addition of the p110 TAF and the p150 TAF allows activation by both NTF and Sp1

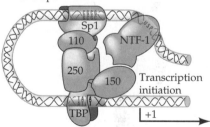

(D) The holo-TFIID supports activation by numerous factors and will allow other activation proteins access to the transcriptional complex

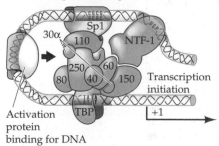

leagues (1995) have shown that the 250- and 150-kDa TAFs are critical in determining whether the TBP remains bound at the TATA box. These TAFs recognize upstream elements of the promoter, which, if present, stabilize or destabilize the TBP at the promoter. This means that some promoters are intrinsically more "difficult" to transcribe from and that certain factors might have to be present to make these promoters transcribable. As we will see later, some promoters (such as the one for human interferon-β), are transcribed only after a great amount of effort has been put into bending them, contorting them, and bracing up the fragile transcription complex.

The association of TBP with different TAFs enables the transcription complex to be activated by proteins bound at the enhancer and upstream promoter sites. Moreover, different TAFs are able to co-activate with different *trans* factors. For instance, one of the most common transcription factors is Sp1. This non-TAF protein binds to the promoter or enhancer GGGCGG sequences through its carboxyl end but regulates transcriptional activity via its amino terminus (Dynan and Tjian, 1985; Kadonaga et al., 1988). This factor is probably found in all cells and so cannot regulate differential gene expression. However, it appears to be involved with the interactions between the promoter region and the enhancer region in ways that do result in the differential transcription of particular genes in particular cells. Sp1 needs to bind to the 110-kDa TAF for its activation of the transcription complex. Thus, this TAF bridges Sp1 to the TBP, forming a loop in the DNA (Hoey et al., 1993; Chen et al., 1994). TAFs would enable the looping of DNA so that Sp1 elements in the enhancer would meet the TFIID protein at the promoter (see Figure 10.9). The Bicoid transcription factor binds to the 110-kDa and 60-kDa TAFs, and mutations in either of these TAFs reduce Bicoid-dependent transcription in *Drosophila* embryos (Sauer et al., 1996). Similarly, the *Drosophila* NTF-1 transcrip-

tion factor binds to both the 60-kDa and 150-kDa TAFs, and in humans, the transcriptional activation by the estrogen receptor is accomplished by its binding to a 30-kDa TAF (Jacq et al., 1994). Indeed, Jacq and colleagues found that not all TBPs had the 30-kDa TAF. It appears that there are some TAFs that are found in all TFIIDs, while other TAFs may be more specific.

Promoters lacking TATA elements.
There are many genes (mostly those encoding general metabolic proteins and not cell-specific proteins) that use RNA polymerase II, but whose promoters lack the TATA sequence. In these cases, some other protein binds to the promoter region. These are usually general promoter-binding proteins such as Sp1. The Sp1 protein on the GC-rich promoter element then binds TFIID either directly or through a TAF. The TFIID is now able to begin the cascade of factors that will form the transcription initiation complex and bind an RNA polymerase II protein to the promoter region (Figure 10.10; Pugh and Tjian, 1991; Rigby, 1993). So even though these promoters lack a TATA box sequence, TFIID is still the deciding factor in regulating whether transcription occurs. ■

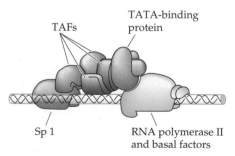

Figure 10.10 Possible configuration for transcription factors mediating RNA polymerase II binding to a TATA-less promoter containing an Sp1-binding site. (After Pugh and Tjian, 1991; Comai et al., 1992.)

Enhancer structure and function

Requirement for Enhancers

In addition to the promoters, enhancers are also important in regulating the transcription of nearby genes. One of the first cellular enhancers found was seen to control the cell specificity of immunoglobulin gene transcription. B cells are the only cells in the body that produce immunoglobulin (antibody) protein. Gillies and his co-workers (1983) transfected a cloned immunoglobulin heavy chain gene into cultured B lymphocyte tumor cells that had lost the ability to make their own heavy chain. These transfected cells were then able to synthesize the heavy chain encoded by the incorporated gene. However, if these researchers added the same gene—but lacking a small region of a particular intron—into these defective B cells, they observed very little transcription of the inserted gene. There was an enhancer region within the intron that was necessary for transcription (Figure 10.11).

Enhancers are also the primary elements responsible for tissue-specific transcription: the cloned immunoglobulin genes are *not* transcribed when inserted into the nuclei of cells other than B cells (Banerji et al., 1983; Gillies et al., 1983). Moreover, when the enhancer region of the immunoglobulin

Figure 10.11
Tissue specificity of enhancer element effect. The immunoglobulin heavy chain gene was isolated and cloned from an IgG-producing myeloma cell line. Some of these clones were kept intact, while in others, various regions of the intron between the VDJ and the C_γ exons (which will be discussed later) were excised by restriction enzymes. DNA from the resulting clones was transfected into myeloma cells that had lost their own immunoglobulin genes. The accumulated mRNA from the transfected cells was isolated and separated by electrophoresis on a polyacrylamide gel, along with mRNA from a normal mouse fibroblast and from the original myeloma cell. The RNA was blotted onto nitrocellulose paper and hybridized with a radioactive restriction enzyme fragment of the C_γ region. If the C_γ region was being transcribed from these cloned genes, the radioactive probe should bind to it. The probe detected C_γ message only from those clones that contained a certain region of the intron (indicated by the colored bar within the intron). When transfected into mouse fibroblast cells, however, even the entire cloned gene would not transcribe C_γ mRNA. (Protocol and data of Gillies et al., 1983.)

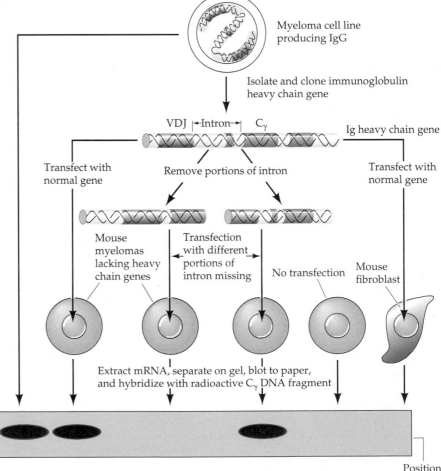

heavy chain is inserted into a cloned gene for β-globin, it stimulates the transcription of that hemoglobin gene only when the gene is inserted into a B cell. Both *cis*-regulatory elements and *trans*-regulatory factors are needed for cell-specific gene transcription.

Enhancer Function: Temporal and Spatial Patterns of Transcription

Enhancers can regulate the temporal and tissue-specific expression of all differentially regulated genes, and genes active in adjacent cell types have different enhancers. In the pancreas, for instance, the exocrine protein genes (for the proteins chymotrypsin, amylase, and trypsin) have enhancers different from that of the gene for the endocrine protein insulin. These enhancers lie in the 5′ flanking sequences of their respective genes. Walker and colleagues (1983) placed these flanking regions onto the gene for bacterial chloramphenicol acetyltransferase (CAT), a gene whose enzyme product is not found in mammalian cells. *CAT* activity is easy to assay in mammalian cells and is used as a "reporter gene" to tell investigators whether a particular enhancer is functioning. They then transfected these hybrid genes into (1) ovary cells (which do not secrete insulin or chymotrypsin), (2) an insulin-secreting cell line, and (3) an exocrine cell line, and measured the activity of the marker enzyme in each of these cells. As shown in Figure 10.12, neither enhancer sequence caused the enzyme to be made in the ovarian cells. In the insulin-secreting cell, however, the 5′ flanking region of the insulin gene enabled the chloramphenicol acetyltransferase gene to be expressed, but the 5′ flanking region of the chymotrypsin gene did not. Conversely, when the clones were placed into the exocrine pancreatic cell line, the chymotrypsin 5′ flanking sequence allowed *CAT* expression, while the insulin enhancer did not. The enhancers for 10 exocrine proteins share a 20-base-pair consensus sequence, suggesting that these similar sequences play a role in activating these genes in the exocrine cells of the pancreas (Boulet et al., 1986). Thus, the expression of genes in exocrine and endocrine cells of the pancreas appears to be controlled by different enhancers.

Figure 10.12
Tissue specificity of pancreatic gene enhancers. The 5′ flanking regions of the insulin gene (I) and the chymotrypsin gene (C) were separately inserted next to the gene for bacterial *CAT*. As a positive control, the enhancer from Rous sarcoma virus (V), which appears to operate in all cell types, was also placed by the *CAT* gene. The three clones were transfected into three types of cells, (A) an ovarian cell line making neither insulin nor chymotrypsin, (B) an insulin-secreting cell line, or (C) a chymotrypsin-secreting cell line. The CAT activity assayed on the cell lysates. The inserts show typical autoradiographs of the CAT assay wherein radioactive chloramphenicol (the substrate of the CAT reaction) can be separated from chloramphenicol monoacetate (the product of the CAT reaction) by chromatography. (After Walker et al., 1983.)

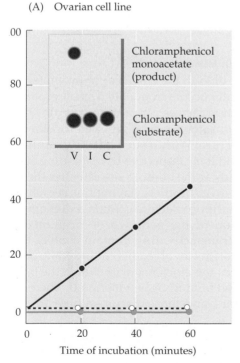

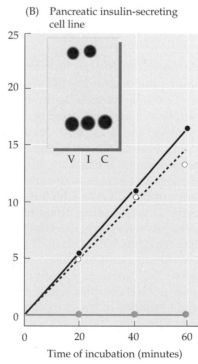

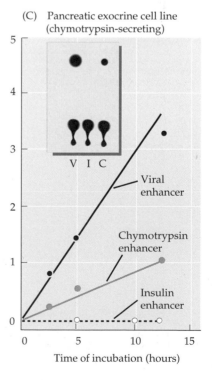

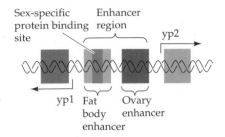

Figure 10.13
Modular structure of the yolk protein enhancer region of *Drosophila*. The two yolk protein genes (*yp1, yp2*) are regulated by an enhancer between them. One region of the enhancer binds transcription factors in the ovary nuclei and enables these genes to be expressed in the ovary. Another region of the enhancer enables the gene to become expressed in the fat bodies. Within the region controlling the expression of the yolk protein genes in the fat bodies are DNA sequences that bind the sex-specific transcription factors.

Enhancers are critical in the regulation of normal development, and over the past decade, five generalizations that emphasize their importance for differential gene expression have been made:

1. Most genes require enhancers for their transcription.
2. Enhancers are the major determinant of differential transcription in space (cell type) and time.
3. Having the enhancer at a relatively large distance from the promoter means that there can be multiple signals to determine whether a given gene is transcribed. A given gene can have several enhancer sites linked to it, and each enhancer can be bound by more than one factor (which may regulate whether it stimulates or inhibits transcription).
4. The interaction between the proteins bound to the enhancer sites with the transcription apparatus assembled at the promoter is thought to regulate transcription. The mechanism of this association is not fully known, nor do we comprehend how the promoter integrates all these signals.
5. Enhancers are modular. There are DNA elements that confer temporal and spatial gene expression, and these can be mixed and matched. For example, the yolk protein enhancer of *Drosophila melanogaster* is so constructed that one DNA element permits the gene to be expressed in the fat bodies, another DNA element enables the gene to be expressed in the ovaries, and a third element binds sex-specific proteins (the Doublesex proteins). The female-specific Doublesex protein stimulates transcription; the male-specific Doublesex protein represses transcription. Thus, the *yolk protein* gene is turned on only in the fat bodies and ovaries of the female fly (Figure 10.13; Garabedian et al., 1985; An and Wensink, 1995). The DNA element for expression in fat bodies is shared with other genes that are expressed in this organ, and the DNA element binding the Doublesex proteins is similarly shared among genes whose expression is sex-specific.

Transcription factors:
The trans-regulators of promoters and enhancers

Transcription factors are proteins that bind to the enhancer or promoter regions and interact such that transcription occurs from only a small group of promoters in any cell. Most transcription factors can bind to specific DNA sequences, and these *trans*-regulatory proteins can be grouped together in families based on similarities in structure. Within such a family, proteins share a common framework structure in their respective DNA-binding sites, and slight differences in the amino acids at the binding site can alter the sequence of the DNA to which it binds. In addition to having this sequence-specific **DNA-binding domain,** transcription factors contain a domain involved in activating the transcription of the gene whose promoter or enhancer it has bound. Usually, this ***trans*-activating domain** enables that transcription factor to interact with proteins involved in binding RNA polymerase. This interaction often enhances the efficiency with which the basal transcriptional complex can be built and bind RNA polymerase II. There are several families of transcription factors, and those discussed here are just some of the main types.

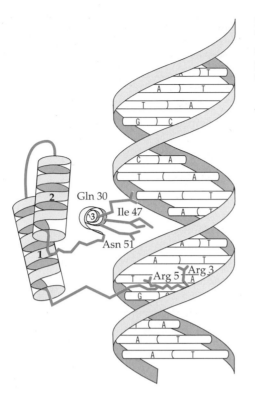

Figure 10.14
The homeodomain of the Engrailed protein binds to a particular site in the DNA. Helix 3 contacts the base pairs in the major groove, while the amino-terminal portion of the homeodomain enters the minor groove. (After Pabo and Sauer, 1992.)

Homeodomain Proteins

One extremely important family of *trans*-regulatory factors is the set of **homeodomain proteins.** These proteins are critical for specifying the anterior-posterior body axes throughout the animal kingdom; they will be detailed further in Chapters 14 and 16. The homeodomain consists of 60 amino acids arranged in a helix-turn-helix, such that the third helix extends into the major groove of the DNA that it recognizes. Amino acids in the amino-terminal portion of the homeodomain also contact the bases in the minor groove (Figure 10.14). This homeodomain was first seen in proteins that specify segment identity in *Drosophila*. Mutations of these proteins caused one body segment to be transformed into another (a transformation known as homeosis, which will be discussed in detail in Chapter 14). Several of the homeodomain proteins of *Drosophila melanogaster* have been cloned, sequenced, and tested for their ability to regulate transcription. Table 10.1 shows nine homeodomain-containing *Drosophila* proteins and the DNA sequences they recognize. The recognition of specific promoters by homeodomain-containing proteins has been shown to be essential for *Drosophila* development. The Bicoid protein, for instance, is a homeodomain transcription factor that binds to the promoters on the *hunchback* gene. This binding activates transcription of the *hunchback* gene; the resulting Hunchback protein is also a transcription factor, and it binds to the enhancers of those genes necessary for *Drosophila* head and thorax formation (Driever and Nüsslein-Volhard, 1989; Struhl et al., 1989). Small changes in the amino acid composition of the DNA-binding site can change the DNA sequence recognized by the protein. Treisman and her colleagues (1989) found that by altering a single amino acid in the homeodomain, one could change the promoters that this protein would activate.

Table 10.1 Major homeodomain proteins of *Drosophila melanogaster* and their DNA-binding sites

Protein	DNA-binding site(s)
Abdominal B	T A A T T G C A T
	T C A A T T A A A T
Antennapedia	T A A T A A T A A T A A T A A
Bicoid	T C C T A A T C C C
Engrailed	T C A A T T A A A T
Even-skipped	T C A A T T A A A T
	T A A T A A T A A T A A T A A
	T C A G C A C C G
Fushi tarazu	T C A A T T A A A T
	T A A T A A T A A T A A T A A
Paired	T C A A T T A A A T
Ultrabithorax	T A A T A A T A A T A A T A A
Zerknült	T C A A T T A A A T

The POU Transcription Factors

Some transcription factors have both a homeodomain and a second DNA-binding region (Figure 10.15). In some instances, this region that comprises the homeodomain and the second DNA-binding region is called the **POU domain** (Herr et al., 1988). The initials are taken from the four proteins first seen to have such domains: *P*it-1 (also called GHF1), a pituitary-specific factor that activates the genes encoding growth hormone, prolactin, and other pituitary proteins; *O*ct1, a ubiquitous protein that recognizes a certain eight-base-pair sequence called the octa box, and *O*ct2, the B-cell-specific protein that recognizes the octa box and activates immunoglobulin genes; and *U*NC-86, a nematode gene product involved in determining neuronal cell fates. The homeodomain of Pit-1 recognizes the sequence ATATTCAT, while the homeodomain of Oct2 recognizes the similar sequence ATTTGCAT. If the DNA element recognized by Pit-1 is altered in two places, it becomes an Oct2-binding site, and the prolactin gene becomes expressed in B lymphocytes (Elsholtz et al., 1990). Thus, a two-base change in the enhancer recognized by a POU protein can convert pituitary-specific transcription into lymphocyte-specific transcription.

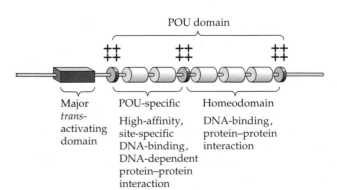

Figure 10.15
The domains of POU family transcription factors.

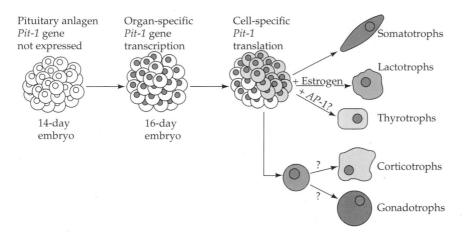

Pituitary anlagen *Pit-1* gene not expressed

Organ-specific *Pit-1* gene transcription

Cell-specific *Pit-1* translation

Somatotrophs

Lactotrophs

+ Estrogen
+ AP-1?

Thyrotrophs

? Corticotrophs

? Gonadotrophs

14-day embryo

16-day embryo

Figure 10.16
Determination of cell type by combinations of *trans*-regulatory proteins. The mRNA for the Pit-1 transcription factor is transcribed in all cell types destined to reside in the anterior pituitary. However, Pit-1 protein is only translated in thyrotroph, somatotroph, and lactotroph cells. Somatotrophs synthesize growth hormone upon the translation of *Pit-1*. Lactotrophs synthesize some prolactin concomitant with *Pit-1* translation but produce significant prolactin only when co-stimulated with estrogen (and its *trans*-regulatory receptor protein). The mechanism distinguishing thryroid-stimulating hormone transcription is thought to involve the silencer region (see Figure 10.17). (After Simmons et al., 1990.)

The **Pit-1** protein is found only in pituitary cells. When growth hormone genes (which are active in the pituitary) are cloned and placed in nuclear extracts of nonpituitary cells, these genes are not transcribed, whereas transcription will take place if the growth hormone genes are placed into nuclear extracts of anterior pituitary cells. Moreover, the addition of Pit-1 to the nonpituitary nuclear extract will enable it to transcribe the growth hormone gene (Bodner and Karin, 1987). It appears, then, that the specific expression of the growth hormone gene in the anterior pituitary is mediated by a tissue-specific *trans*-regulating protein. Conversely, when other genes (such as those for globin) are cloned next to the Pit-1-binding sequences and are used to construct transgenic mice, these genes show pituitary-specific transcription (Behringer et al., 1988).

The Pit-1 enhancer-binding protein is itself developmentally regulated (Dollé et al., 1990; Simmons et al., 1990). Transcription of the mouse *Pit-1* gene is detectable within a day following the histological appearance of Rathke's pouch, the anlage of the anterior pituitary, but translation of this mRNA into protein does not occur for another 2 to 3 days. Growth hormone mRNA is first detected when the *Pit-1* gene is translated. Interestingly, in two types of anterior pituitary cells, the corticotrophs that synthesize corticotropin (also known as adrenocorticotropic hormone, or ACTH) and the gonadotrophs that synthesize gonadotropins, *Pit-1* mRNA is made but remains untranslated (Figure 10.16). Only in the pituitary cells of somatotrophs (growth hormone-producing), lactotrophs (prolactin-producing), and thyrotrophs (which synthesize thyroid-stimulating hormone) is the *Pit-1* message translated into the DNA-binding nuclear protein. The Pit-1 enhancer-binding protein not only mediates the transcription of the differentiated products of these cells, but is necessary for the formation of the anterior pituitary cells in Rathke's pouch. Two *dwarf* mutations in mice are caused by a mutation in Pit-1 protein. These mice lack the thyrotroph, lactotroph, and somatotroph cells of the anterior pituitary (Li et al., 1990).

COMBINATORIAL INTERACTIONS AND LOOPING IN REGULATING PROLACTIN GENE TRANSCRIPTION. The activity of the Pit-1 protein on the gene for prolactin illustrates many of the characteristics of transcription factors. First, the enhancer of the *prolactin* gene binds several different factors whose interactions regulate transcription. The *prolactin* gene is activated during pregnancy to make the pituitary hormone (prolactin) that stimulates milk production; this gene is maximally stimulated by the *combination* of Pit-1 and estrogen. This combination regulates the place (the pituitary gland) and the time (pregnancy and shortly thereafter) of prolactin synthesis. Simmons and colleagues

Figure 10.17
Combinatorial synergism in the *prolactin* enhancer. (A) Synergism between sites of the rat *prolactin* enhancer. The *prolactin* enhancer was fused to a reporter (*luciferinase*) gene and the level of reporter gene activity assayed when the fused gene was added to cultured cells. When either Pit-1 or estrogen was present in these cells, there was only minor increase in the level of transcription. However, the addition of *both* substances caused a 1400-fold increase in the level of transcription. (B) Synergism between enhancer and promoter sites of the mouse *prolactin* gene. Fused genes were made carrying the reporter gene and (1) the entire 5′ *prolactin* enhancer region, flanking sequences, and *prolactin* promoter; (2) just the tissue-specific enhancer and the *prolactin* promoter; (3) just the *prolactin* promoter; (4) the *prolactin* enhancer plus the promoter of a *thymidine kinase* (*TK*) gene; and (5) just the *TK* gene promoter. These constructs were placed into mouse zygotes, and the reporter gene expression was monitored in lactotrophic and thyrotrophic cells of the pituitary. (C) Model for the regulation of *prolactin* gene expression. The promoter and enhancer both have four Pit-1-binding sites. The estrogen receptor binds to the estrogen-reponsive element (ERE) of the enhancer region. The regions that inhibit *prolactin* transcription in thryotrophs and somatotrophs are in black. (A after Simmons et al., 1990; B after Crenshaw et al., 1989.)

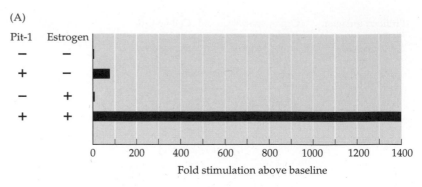

(A)

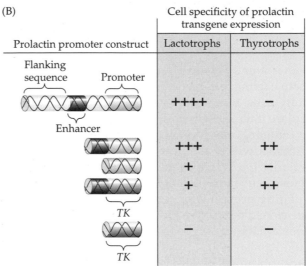

(B)

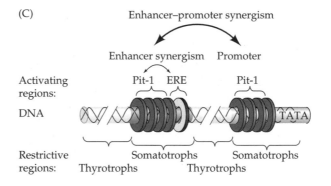

(1990) showed that this synergism occurs in the enhancer region of the gene. The Pit-1 protein binds to one region of the enhancer, while the estrogen binds through its receptor protein to another region of the enhancer. When these factors are present together, transcription is much greater than when either is added separately (Figure 10.17A). Moreover, there appear to be silencer regions flanking the enhancer that are needed for turning off the *prolactin* gene in thyrotrophs (which could otherwise activate the *prolactin* gene) (Figure 10.17B,C; Crenshaw et al, 1989). Thus, Pit-1 acts in a combinatorial way with other transcription factors to regulate its target genes.

Second, there is synergism between the enhancer and the promoter of the *prolactin* gene. The *prolactin* gene enhancer will not stimulate the promoter of another gene nearly as efficiently as it will stimulate its own promoter (Figure 10.17B,C; Crenshaw et al., 1989). This synergism between promoter and enhancer appears to be caused by DNA looping between the enhancer and the promoter sites. In the rat prolactin gene, the enhancer is located over 1300 base pairs upstream from its promoter. Using an assay that would fuse DNA brought together by protein-protein interactions, Cullen and colleagues (1993) showed that the promoter and enhancer regions are brought together only when Pit-1 and estrogen are both present. It appears that the hormone-bound estrogen receptor on the enhancer is able to stabilize the interaction between the enhancer and the promoter regions, thus allowing the interaction between the enhancer-bound proteins (Pit-1 and the estrogen receptor) with the transcription apparatus of the promoter.

Third, the Pit-1 protein positively regulates its own synthesis. One of the targets for the Pit-1 protein is the enhancer of the *Pit-1* gene itself (Rhodes et al., 1993). Once the *Pit-1* gene has been activated (by other transcription factors), Pit-1 protein binds to its own enhancer and maintains *Pit-1* gene transcription. This type of positive autoregulation is important as a mechanism for committing a cell to a given pathway of development. Thus, the *Pit-1* gene, once active, maintains the pituitary phenotype. Such autoregulation is also seen for the MyoD protein (which commits cells to the path of muscle cell development) and for various *Drosophila* proteins that maintain the individual's sex and segment-specific boundaries.

Sidelights & Speculations

Regulation of Transcription From Immunoglobulin Light Chain Genes

THE ABILITY to obtain clonal lines of cells "frozen" at a particular stage in their development enables us to sample the transcription factors present during that particular time. Such clones are obtainable from tumors of certain tissues, and leukemias of the B lymphocytes (B cells) of the immune system have enabled researchers to identify many of the the *cis-* and *trans*-regulatory elements necessary for the development of the B cell lineage.

The Structure of the Immunoglobulin Genes

So far, our model has been that every cell in the body contains the same exact genes. This is probably true for most cell types, but the lymphocytes are different. In each B cell, the genome has been rearranged so that each B cell is able to make only one type of antibody (out of a possible repertoire of over 10 million antibodies).

Antibodies are produced when a for-eign substance—the antigen—comes into contact with B cells, which reside in the lymph nodes and spleen. Even before contact with an antigen, each of the resting B cells makes immunoglobulin proteins but does not secrete them. Rather, the immunoglobulin molecules are inserted into the B cell membranes and are used as antigen receptors. These receptors can signal the cell to divide and further differentiate when they bind antigens. Each B cell makes an antibody that recognizes one and only one antigenic shape. Therefore, an antibody recognizing the protein shell of poliovirus would not be expected to recognize cholera toxin, *E. coli* membranes, or zebra dander.

All antibody proteins on the B cell membrane have a very similar structure. Each consists of two pairs of polypeptide subunits. There are two identical heavy chains and two identical light chains; the chains are linked together by disulfide bonds (Figure 10.18). The *specificity* of the antibody molecule (i.e., whether it will bind to a poliovirus, an *E. coli* cell, or some other molecule) is determined by the amino acid sequence of the *variable region*. This region is made up of the amino-terminal ends of the heavy and light chains. The variable regions of the antibody molecules are attached to *constant regions* that give the antibody its effector properties needed for inactivating the antigen. For decades, immunologists puzzled over how the immune system could possibly generate so many different types of antibodies. Could all the 10^7 different types of antibody proteins be encoded in the genome? This would take up an enormous amount of chromosomal space. Moreover, how could the immune system "know" how to make an antibody to some molecule that isn't even found outside the laboratory? Surprisingly, it was discovered that the production of specific antibody molecules involves the creation of *new genes* during B cell differentiation.

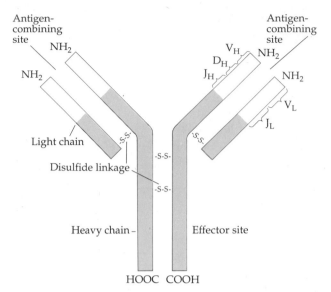

Figure 10.18 Structure of a typical immunoglobulin (antibody) protein. Two identical heavy chains are connected by disulfide linkages. The antigen-combining site is composed of the variable regions (white) of the heavy and light chains, whereas the effector site of the antibody (which controls whether it agglutinates antigens, binds to macrophages, or enters into mucous secretions) is determined by the amino acid sequence of the heavy chain constant (color) region.

Creation of Antibody Light Chain Genes

Antibody heavy and light chain genes are organized in segments. Mammalian light chain genes contain three segments (Figure 10.19). The first gene segment, V, encodes the first 97 amino acids of the light chain variable region. There are about 300 different V sequences linked tandemly on the mouse genome. The second segment, J, consists of four or five possible DNA sequences for the last 15–17 residues of the variable region of the antibody light chain. The third segment is the constant (C) region of the light chain. During B cell differentiation, one of the 300 V segments and one of the five J segments combine to form the variable region of the antibody gene. This is done by moving a V segment sequence

to a J segment sequence, a rearrangement that eliminates the intervening DNA.

This gene rearrangement was first shown by Hozumi and Tonegawa (1976), who isolated DNA from a mouse embryo and from a light chain secreting B cell tumor.* They digested these two DNAs separately with the restriction enzyme BamHI (see Chapter 2), which cleaves DNA wherever it locates the sequence GGATCC. The result was a series of DNA fragments, the size of each fragment determined by the length of the DNA molecule between two cleavage sites. These DNA fragments were placed into a well at one end of a gel and electrophoresed through the gel. As the DNA migrated toward the positive electrode, the smaller fragments moved faster than the larger ones, effectively separating the fragments

by size. The gel, with the DNA spread throughout it, was cut into several pieces, each slice containing pieces of DNA of a certain size. The DNA from each slice of the gel was eluted and denatured into single strands. Part of this DNA was hybridized with radioactive RNA that coded for the entire light chain and had been isolated from the original B cell tumor. The other part was hybridized with radioactive RNA coding only for the C region of the light chain (the 3' half of the RNA message). The DNA from the embryo bound the light chain mRNA in two slices. The DNA of the first slice had a molecular weight (MW) of about 6 million; the DNA of the second band was of MW 3.9 million. When the mouse embryo DNA was hybridized with the light chain C region mRNA, only the 6-million-MW DNA bound the RNA. Thus, in the mouse *embryo*, the C region was coded within DNA fragments having a molecular weight of 6 million (between *Bam*HI sites), while the V region was coded within a region of MW 3.9 million (Figure 10.20).

The lymphocyte tumor DNA, however, gave a very different result. The only lymphocyte DNA that bound the light chain mRNA had a molecular weight of 2.4 million. Moreover, it bound the light chain C region mRNA segment as well. Both the C and the V regions were found to be coded on the same fragment of DNA! The simplest explanation (and one confirmed by numerous other laboratories and methods; see Bernard et al., 1978; Brack et al., 1978) was that two gene fragments, one encoding the light chain C region and one encoding a specific light chain V region, had fused together *to form a new gene* during the development of the lymphocyte. Their model for such gene synthesis is presented in Figure 10.21.

*B cell tumors (myelomas) were used because they produce an enormous amount of one specific immunoglobulin (and of the mRNA for that immunoglobulin).

Figure 10.19 Rearrangement of the light chain genes during B lymphocyte development. While the developing B cell is still maturing in the bone marrow, one of the 300 or more V gene segments combines with one of the five J gene segments and moves closer to the constant (C) gene segment.

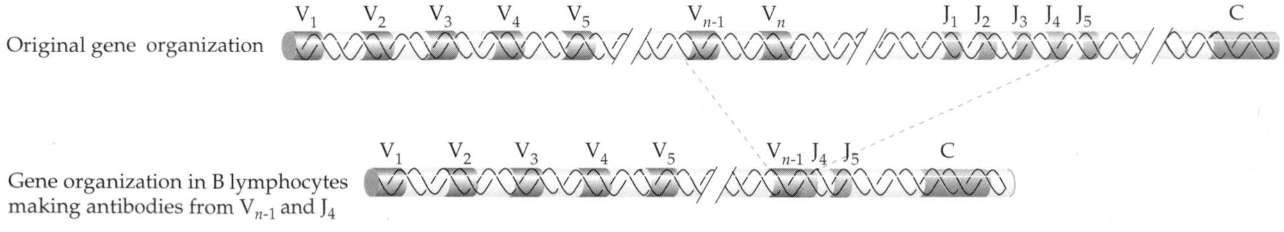

V_1 V_2 V_3 V_4 V_5 V_{n-1} V_n J_1 J_2 J_3 J_4 J_5 C

Original gene organization

V_1 V_2 V_3 V_4 V_5 V_{n-1} J_4 J_5 C

Gene organization in B lymphocytes making antibodies from V_{n-1} and J_4

Figure 10.20 Protocol and results of the Hozumi and Tonegawa experiment. DNAs from mouse embryo cells and B cell tumors (myelomas) were separately digested in *Bam*HI, separated by electrophoresis, and eluted from the gels. After being denatured, each eluted DNA sample was hybridized to either radioactive mRNA coding for the V and C regions of the immunoglobulin light chain (whole) or a fragmented radioactive mRNA encoding only the C region of that light chain protein (3′ half). For the embryonic DNA, the V and C regions of the light chain protein were found on two different pieces of DNA (the V region on a piece having a molecular weight of 3.9×10^6, the C region on a DNA fragment of MW 6×10^6). In the lymphocyte tumor, the V and C regions were found together on a single DNA fragment of MW 2.4×10^6.

Creation of Antibody Heavy Chain Genes

The heavy chain genes of antibodies contain even more segments than the light chain genes. Heavy chain gene segments include a V segment (200 different sequences for the first 97 amino acids), a D segment (10 to 15 different sequences encoding 3 to 14 amino acids), and a J segment (four sequences for the last 15 to 17 amino acids of the V region). The next segment codes for the C region. The heavy chain variable region is formed by adjoining one V segment and one D segment to one J segment (Figure 10.22A,B). This VDJ variable region sequence is now adjacent to the first constant region of the heavy chain genes—the C_μ region specifically for antibodies that can be inserted into the plasma membrane. Thus, an immunoglobulin molecule is formed from two genes created during the antigen-independent stage of B lymphocyte development. About 10^3 light chain genes and about 10^4 heavy chain genes can be formed. Since each is formed independently of the other, about 10^7 types of antibody can be created from the union of the light chain and the heavy chain within a cell. Each cell makes only one of these 10^7 antibody types and places the antibodies in the cell membrane to be used as antigen receptors. The genome of each lymphocyte clone can be markedly different from that of any other cell.

Later in adult development, some of the B cells undergo another rearrangement called **class switching.** Here, the entire VDJ sequence of the heavy chain *gene* is transferred from the C_μ region to another C region. The new C-region will have different properties. In Figure 10.22C and D, the VDJ sequence is transferred to a C_α region. The C_α sequence encodes the constant region that enables the antibodies to be secreted into the mucosa and digestive system, where they protect the body against airborne and food-borne antigens.

The B cell is not the only cell type that alters its genome during differentiation. The other major cell of the immune system, the T lymphocyte, also deletes a portion of its genome in the construction of its antigen receptor (Fujimoto and Yamagishi, 1987). The enzymes responsible for mediating the DNA recombination events appear to be the same in the B and T cell lineages. Called **recombinases** (Schatz et al., 1989; Oettinger et al., 1990), these two proteins recognize the

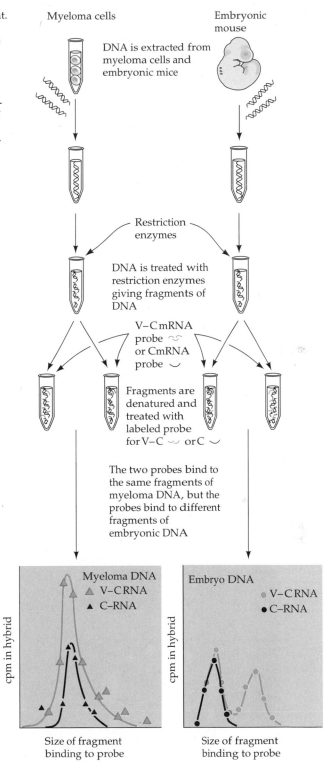

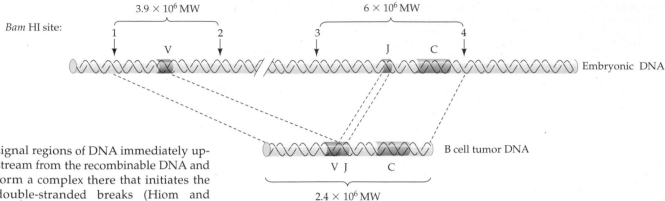

Figure 10.21   Model of the changes in DNA between embryonic cells and the B lymphocyte, according to the data of Hozumi and Tonegawa (1976).

signal regions of DNA immediately up-stream from the recombinable DNA and form a complex there that initiates the double-stranded breaks (Hiom and Gellert, 1997). Moreover, the genes for these enzymes are active only in the pre-B cells and pre-T cells, where the genes are being recombined. These recombinase genes are not active in mature B and T cells, nor are they seen in most other cell types.*

cis-Regulatory Regions of the Immunoglobulin Genes

The discovery of V(D)J joining and class switching solved most of the problems concerning the origin of antibody diversity. However, it created some other problems. Each V segment has its own promoter attached to it. Why isn't every V segment gene expressed? Deletion mapping of cloned genes has shown that the promoter of the immunoglobulin light chain contains several critical regions for transcription. One of these upstream promoter elements is called an **octa box** (Bergman et al., 1984; Parslow et al., 1984) because of its eight base pairs: ATTTGCAT. This octa sequence has been found in every light chain immunoglobulin promoter studied. The heavy chain immunoglobulin gene promoters have an *inverted* octa box: ATGCAAAT. When the octa box is placed up-

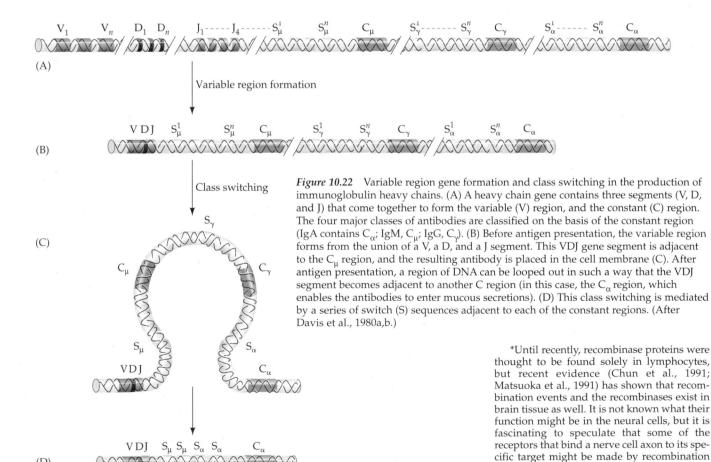

Figure 10.22   Variable region gene formation and class switching in the production of immunoglobulin heavy chains. (A) A heavy chain gene contains three segments (V, D, and J) that come together to form the variable (V) region, and the constant (C) region. The four major classes of antibodies are classified on the basis of the constant region (IgA contains C_α; IgM, C_μ; IgG, C_γ). (B) Before antigen presentation, the variable region forms from the union of a V, a D, and a J segment. This VDJ gene segment is adjacent to the C_μ region, and the resulting antibody is placed in the cell membrane (C). After antigen presentation, a region of DNA can be looped out in such a way that the VDJ segment becomes adjacent to another C region (in this case, the C_α region, which enables the antibodies to enter mucous secretions). (D) This class switching is mediated by a series of switch (S) sequences adjacent to each of the constant regions. (After Davis et al., 1980a,b.)

*Until recently, recombinase proteins were thought to be found solely in lymphocytes, but recent evidence (Chun et al., 1991; Matsuoka et al., 1991) has shown that recombination events and the recombinases exist in brain tissue as well. It is not known what their function might be in the neural cells, but it is fascinating to speculate that some of the receptors that bind a nerve cell axon to its specific target might be made by recombination of several gene regions.

(A)   Germ-line gene: No transcription

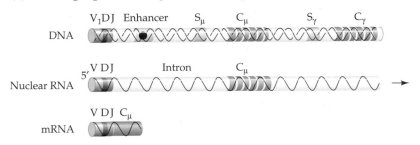

(B)   Rearranged gene: Transcription of immunoglobulin

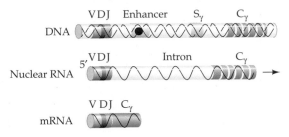

Figure 10.23   Model for immunoglobulin enhancer activity. (A) The immunoglobulin heavy chain gene enhancer region appears to involve sequences between the J gene segments and the switch sequences (S_μ) preceding C_μ. If the enhancer is deleted, transcription is greatly diminished. The 5′ promoter precedes each of the V region gene segments and is originally very distant from the enhancer. (B) The VDJ gene rearrangement brings one promoter near the enhancer and allows transcription to take place. (C) During class switching, the enhancer stays with the VDJ segments as they are placed near a new constant region (C_γ).

(C)   Class switched gene: Transcription of new class of immunoglobulin

This placement occurs during the construction of the immunoglobulin gene.

The gene rearrangements bring a particular promoter into the proximity of the enhancer. A similar event occurs for the heavy chain gene, and during class switching (when a region of DNA is looped out and deleted), the enhancer region remains by the VDJ piece (Figure 10.23).

trans-Regulation of Immunoglobulin Synthesis

Gene rearrangement itself, however, is not sufficient for gene activation, since a rearranged immunoglobulin gene will not actively transcribe when placed in a fibroblast or liver cell. There have to be cell-specific *trans*-regulatory factors. Two promoter-binding factors were identified by Staudt and co-workers in 1986. The researchers incubated a small piece of DNA containing the octa box in nuclear extracts from various cells. They then ran the resulting products on a gel. If the nuclear extract did not have a protein that bound

stream from a globin gene in a cultured B lymphocyte, the globin gene transcription increases 11 to 18 times. This increase is seen only in lymphoid cells and has not been observed in fibroblasts (Wirth et al., 1987).

The *enhancer* sequence of the light chain immunoglobulin gene is located within the first intron between the VJ sequence and the C region (Queen and Baltimore, 1983; Bergman et al., 1984).

When this sequence is translocated onto cloned globin genes, globin gene transcription can also occur specifically in lymphocytes (Picard and Schaffner, 1984). For transcription to occur in the lymphocyte, the promoter must be brought into proximity of the enhancer. All the V segments carry a promoter, but only the V segment brought near the constant region (with its enhancer) will be activated (Mather and Perry, 1982).

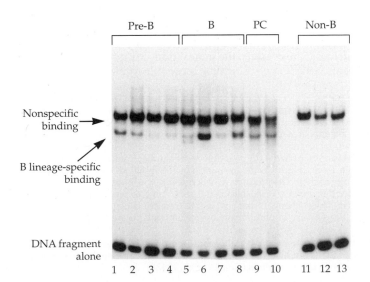

Figure 10.24   Gel mobility shift assay. Nuclear extracts from B cell lineage cells [pre-B, B, and plasma cell (PC)] and non-B cells (cervical, fibroblast, and red blood cell precursor cell lines) were mixed with a small segment of DNA containing the octamer. After incubation, the mixtures were electrophoretically separated on a gel, blotted to nitrocellulose paper, and hybridized with radioactive DNA complementary to the octamer sequence. Without any binding, the octamer-containing fragment rapidly moves to the bottom of the gel. All nuclei contain a protein that nonspecifically binds to the octamer and greatly impedes migration. The nuclei from B cell lineage cells, however, also contain another protein that inhibits migration by binding to the octamer sequence. (From Staudt et al., 1986, courtesy of D. Baltimore.)

Figure 10.25 Regulation of NF-κB by I-κB. I-κB binds to the larger subunit of NF-κB and prevents the complex from entering the nucleus. I-κB can be phosphorylated by several kinases that are activated by replicating viruses, antigens, lipopolysaccharides, or tumor necrosis factor α. The phosphorylation of I-κB releases NF-κB which can then enter the nucleus and bind to those promoter and enhancer sites it recognizes. Such genes include those encoding the kappa immunoglobulin light chain, tumor necrosis factor, interleukin 2, and the receptor for interleukin 2. The human immunodeficiency virus also has sites for NF-κB binding.

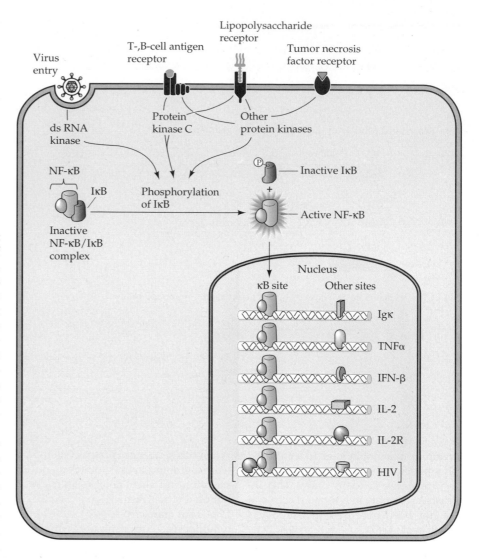

to this DNA, the small DNA fragment should migrate rapidly through the gel. If a protein *did* bind to this DNA, however, the migration should be impeded. This **mobility shift assay** (Figure 10.24) demonstrated that every nucleus had at least one factor capable of binding to the DNA fragment. However, the B cell lineage (pre-B cells, B cells, and plasma cells) contained, in addition, another factor capable of binding specifically to the octa box DNA. These binding proteins were isolated, and the lymphocyte-specific protein was termed Oct2 (NF-A2).

Similar assays were used to find a nuclear protein restricted to B cell lineage cells that binds specifically to the enhancer sequences of light chain immunoglobulin genes (Sen and Baltimore, 1986a; Atchinson and Perry, 1987). One such protein, **NF-κB**, was found only in mature B cells and plasma cells, and it recognized the binding sequence 5′–GGGACTTTCC–3′ in the light chain enhancer. Moreover, when pre-B cells are induced to become B cells, active NF-κB protein appears.*

This brings the problem of differential gene activity to another level: What controls the synthesis of the B-cell-specific *trans*-regulatory factors such as Oct2 or NF-κB? (That is, to explain how immunoglobulins are produced in a cell-specific manner, we now have to explain how these nuclear proteins are generated in a cell-specific manner.) It appears that the NF-κB protein is actually present in numerous cell types, but is bound by a 65-kDa protein called **IκB** (inhibitor of kappa). In its bound state, NF-κB cannot enter the nucleus and bind to DNA (Figure 10.25; Henkel et al., 1992). Thus, NF-κB can recognize its enhancer region only in those cells that either do not synthesize or do not activate their IκB—that is, in

mature lymphocytes (Sen and Baltimore, 1986b; Baeuerle and Baltimore, 1988). The synthesis of immunoglobulin light chains is probably initiated when a signal on the cell surface activates protein kinases that can phosphorylate IκB. This phosphorylation is thought to release NF-κB and allow it to enter the nucleus (Ghosh and Baltimore, 1990; Kerr et al., 1991).

In addition to the positive regulation of immunoglobulin gene transcription seen in B cells, there is also negative regulation of immunoglobulin production in non-B cells. There appear to be several sites flanking the immunoglobulin genes that are bound by proteins that inhibit the genes' transcription (Calame, 1989). Proteins capable of binding to these si-

*B cells are the only cells that synthesize immunoglobulins. The pre-B cell can make the heavy chain but not the light chain of the immunoglobulin protein. Active NF-κB has also been found in activated (but not unactivated) T cells. T cell-specific genes such as those encoding interleukin 2 (T cell growth factor) and its receptor have enhancers that bind NF-κB. This NF-κB responsiveness may be important in the propagation of the human immune deficiency virus (HIV). When HIV infects a T cell, it induces the formation of active NF-κB (Sen and Baltimore, 1986b; Lenardo and Baltimore, 1989). The production of active NF-κB stimulates the T cell genes that have enhancer sites for this protein. Thus, the T cell is stimulated to proliferate. At the same time, HIV also has NF-κB enhancer elements that enable it to transcribe its products rapidly, too. It is obvious that NF-κB plays a very important role in normal and altered lymphocyte development.

One might wonder what the nonspecific octa-box-binding protein is doing in non-B cells. Although Oct1 and Oct2 proteins can bind to the same octa box, they exert their effects by interacting with other proteins at certain promoters (Tanaka et al., 1992).

lencer regions of immunoglobulin genes have been seen in non-B cells. Thus, transcription may be stimulated or inhibited by *trans*-regulatory proteins, depending on a cell's developmental history.

The analysis of cell-specific transcription of immunoglobulin light chain genes has moved, then, to a level beyond looking at the terminal differentiated cell product. We know that the transcription of this immunoglobulin gene depends on the prior activity of two nuclear proteins, Oct2 and NF-κB, whose activity is seen only in the B cell lineage. ∎

Basic Helix-Loop-Helix Transcription Factors

Another prominent arrangement seen in enhancer and promoter DNA-binding proteins is the **basic helix-loop-helix (bHLH)** motif. The muscle-specific transcription factors MyoD and myogenin (discussed in Chapter 9) contain this motif, as do several *Drosophila* proteins that determine the cells of *Drosophila*'s peripheral nervous system: the products of the *daughterless*, *achaete-scute*, and *extramacrochaetae* genes. As we will see in Chapter 20, the genes that determine the sex of *Drosophila* also contain the bHLH pattern. The bHLH proteins bind to DNA through a region of basic amino acids (typically 10 to 13 residues) that precedes the first α-helix (Figure 10.26). The helices contain hydrophobic amino acids at every third or fourth position, so that the helix presents a surface of hydrophobic residues to the environment. This enables the protein to pair by hydrophobic interaction with the same protein or with a related protein that displays such a surface (Jones, 1990).

Recent studies have shown that homodimers (between two identical bHLH proteins) do not bind well to DNA. Rather, the bHLH proteins recognize their promoter sequences according to the following paradigm (Table 10.2). There is a ubiquitous bHLH protein synthesized in most cells that can form a dimer with either of two potential partners. One of these possible partners is the *positive* regulator (which stimulates transcription); the other partner is the negative regulator. When the positive regulator dimerizes with the ubiquitous protein, it forms an activator complex that stimulates transcription from the genes it recognizes. When the negative regulator dimerizes with the ubiquitous bHLH protein, it forms an inhibitory complex that represses transcription from those same genes. For instance, the MyoD family of proteins are active in promoting myogenesis when complexed to either the E12 or E47 proteins—two ubiquitous bHLH proteins (French et al., 1991; Lassar et al., 1991). Muscle development is inhibited, though, when the MyoD, E12, or E47 proteins are bound to the **Id** (*i*nhibitor of *d*ifferentiation) protein. The Id protein contains the HLH motif, but lacks the basic region that binds to the DNA. Dimerization of Id with MyoD, E12, or E47 interferes with the ability of these proteins to bind DNA, and the expression of Id in cells prevents the activity of the MyoD proteins (Benezra et al., 1990). The Id protein is made while the muscle cell precursors are still dividing, and they disappear when the myoblasts leave the cell cycle to begin differentiating into myotubes. If Id is overexpressed in cultured myoblasts, they will not differentiate into myotubes (Jen et al., 1992).

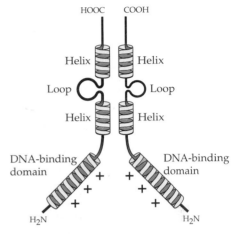

Figure 10.26
Domains of the basic helix-loop-helix transcription factors.

Table 10.2 Developmental bHLH dimers

Dimer	*Drosophila* neurogenesis	Mammalian myogenesis	Mammalian cell division
Ubiquitous protein	daughterless	E12, E47	Max
Positive regulator	achaete-scute	MyoD family	myc
Negative regulator	extramacrochaetae	Id	Mad or Max

Regulating the Myogenic bHLH Proteins: Governing the Switch Between Muscle Cell Proliferation and Differentiation

THERE ARE WAYS to regulate myogenic bHLH proteins other than dimerizing them with Id. It has long been known (Stockdale and Holtzer, 1961; Bischoff and Holter, 1969) that muscle cells do not generally become differentiated until after they have finished proliferating. Proliferating muscle cells do not express the muscle-specific phenotype, while differentiated muscles no longer divide. Growth and differentiation are considered mutually exclusive states in skeletal muscle development, and once the muscle cell has left the cell cycle, it cannot return, even if growth factors are provided (Konigsberg et al., 1960; Nadal-Ginard, 1978). Such mutual exclusivity between differentiation and proliferation is also seen in the development of neurons, adipocytes, blood cells, and skin keratinocytes. The mechanisms for this switch during myogenesis appear to involve the regulation of myogenic bHLH proteins.

The first of these mechanisms is responsible for preventing premature muscle differentiation when the myogenic bHLH proteins first appear. Myogenic bHLH proteins are extremely sensitive to growth factors. As long as growth factors are present to stimulate mitosis, myogenesis does not occur,

even if MyoD or Myf-5 proteins are present in the cells (Olson, 1992). The inactivation of these bHLH proteins is associated with their inability to bind the CAN*N*TG (where N is any base) sequence of the DNA (Brennan et al., 1991). Why can't these myogenic bHLH proteins function? Growth factors such as fibroblast growth factor (FGF) not only stimulate Id transcription but also activate protein kinase C. This kinase induces the phosphorylation of the myogenic bHLH proteins right in their DNA-binding sites (Li et al., 1992). When this site is phosphorylated, the myogenic bHLH will not bind DNA. Thus, as long as the growth factors are present and able to be received, myogenesis will not occur.

The second type of regulation invloves preventing MyoD expression where it is not wanted. MyoD is one of the most powerful transcriptional regulators. As discussed in Chapter 9, if it is expressed within most cells, those cells become muscle. This means that it has to be controlled very tightly. One mechanism is to have cells synthesize a potent inhibitor of MyoD function and to release that control just in certain areas. This appears to be the case in the expression of MyoD in the somite where **Twist**

inhibits MyoD expression in the dermatome and sclerotome (Chapter 9).

MyoD may also be suppressed by signaling from the Notch receptor protein. The activated Notch protein stimulates the transcription of the *hes*-1 gene that encodes another bHLH protein. This protein appears to bind to MyoD and inhibit its ability to function as a muscle-inducing transcription factor (Sasai et al., 1992; Kopan et al., 1994; Jarriault et al., 1995). Since chick epiblast cells express MyoD and become muscles when dissociated in culture, it is possible that some juxtacrine cell contact-mediated signal such as Notch is responsible for inhibiting this expression and retaining the pluripotency of the epiblast (Kopan et al., 1994; George-Weinstein et al., 1996).

MyoD is often called a "master regulatory gene" since its product is capable of converting nearly any cell into muscle. The paradox is that any "master control gene" has to be masterfully regulated. Their products are so powerful that the cell has evolved numerous means—at several levels—to prevent their expression in the wrong cells or at inappropriate times. ■

Basic Leucine Zipper Transcription Factors

The structure of **basic leucine zipper (bZip)** transcription factors is very similar to that of the bHLH proteins. The bZip proteins are dimers, each of whose subunits contains a basic DNA-binding domain at the carboxyl end, followed closely by an α helix containing several leucine residues. These leucines are placed in the helix such that they interact with similarly spaced leucine residues on other bZip proteins to form a "leucine zipper" between them, causing dimers to form. This domain is followed by a regulatory domain that can interact with the promoter to stimulate or repress transcription (Landschulz et al., 1988; Pathak and Sigler, 1992). The C/EBP, AP1, and yeast GCN4 transcription factors are members of the bZip family. Genetic and X-ray crystallographic methods have converged on a model of DNA

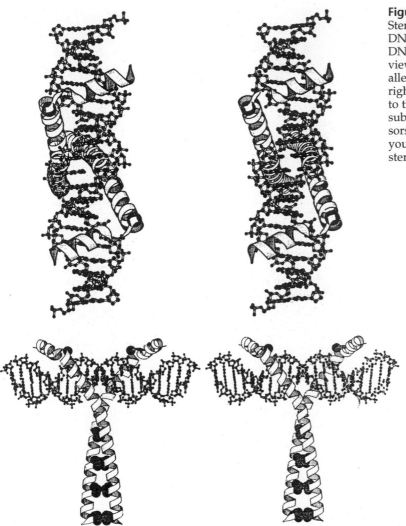

Figure 10.27
Stereoscopic representation of bZip protein C/EBP DNA-binding region interacting with 20 base pairs of DNA containing the CCAAT sequence. (Top) "Dorsal view," looking down at a DNA double helix and parallel to the leucine zipper. (Bottom) "Side view," at right angles to the upper diagram and perpendicular to the DNA axis. Leucine residues connecting the two subunits can be seen at the bottom, as can the "scissors grip" in the DNA. (If you aren't used to crossing your eyes to see the composite stereoimage, borrow a stereopticon.) (From Pathak and Sigler, 1992.)

binding shown in Figure 10.27 (Vinson et al., 1989; Pu and Struhl, 1991). In the figure, the two α helices containing the DNA-binding region are inserted into the major groove of the DNA, each helix finding an identical DNA sequence. The resulting binding looks like that of a scissors or hemostat.

There appear to be several bZIP proteins that can bind to the sequence CCAAT; one of the most important is called the **CCAAT enhancer-binding protein** (**C/EBP**). C/EBP plays a role in adipogenesis similar to that of the myogenic bHLH proteins in myogenesis. Precocious expression of C/EBP in dividing, pre-adipose cells causes the cessation of cell division and the initiation of the adipose phenotype (Umek et al., 1991). (Unlike the myogenic bHLH proteins, which can convert nerve cells and fibroblasts into muscles, C/EBP does not appear to convert other types of cells into the adipocyte lineage.) The C/EBP bZIP protein binds to the enhancers of numerous adipose-specific genes when adipogenesis is initiated in culture (Figure 10.28; Christy et al., 1989; Kaestner et al., 1990). Antisense mRNA against C/EBP suppresses the coordinate expression of adipocyte-specific messages and the differentiation of pre-adipocytes into adipocytes (Samuelsson et al., 1991; Lin and Lane, 1992).

C/EBP is also enriched in liver cells, and it is one of the major regulators of liver-specific gene expression. In mouse hepatocytes, other transcription

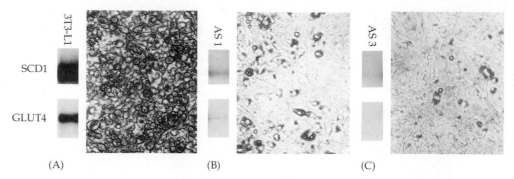

Figure 10.28
Adipogenesis (fat cell formation) mediated through the C/EBP transcription factor. Lipid staining is shown in the box on the right. The left-hand column shows the mRNAs for the SCD1 and GLUT4 proteins that are involved in adipocyte differentiation. (A) Normal adipogenesis of the 3T3-L1 pre-adipocyte cell line in culture. The *SCD1* and *GLUT4* genes are activated, and the cells synthesize and accumulate large amounts of triglycerides. (B,C) Two lines of 3T3-L1 cells that have been transfected with antisense RNA against the C/EBP message. Neither gene is well expressed, and triglyceride levels are 15 and 5 percent of normal. (From Lin and Lane, 1992.)

factors bind to the promoter and enhancer regions of the liver-specific genes during development. However, these genes do not transcribe large amounts of proteins (such as albumin) until C/EBP is expressed in these cells immediately prior to birth (Milos and Zaret, 1992). Another liver-specific gene activated by the C/EBP protein is the gene for blood-clotting factor IX. Mutations in this clotting factor gene cause hemophilia B. In some patients, the cause of this disease has been traced to mutations in the C/EBP-binding site in the promoter region of the factor IX gene. These mutations prevent the binding of C/EBP to this gene (Crossley and Brownlee, 1990).

Sidelights & Speculations

Enhancer Traps: Natural and Experimental

Translocation-Induced Leukemias
One critical transcription factor that is very important in regulating cell division is the **c-Myc** protein. It is a member of a class of DNA-binding proteins that have incorporated a leucine zipper and a basic helix-loop-helix motif. The c-Myc proteins behave similarly to the bHLH and bZIP proteins in that they form heterodimers that bind DNA (see Table 10.2). The c-Myc forms an activator complex when joined to the ubiquitous protein Max. The complex between Max and an inhibitory protein, Mad, creates the Mad-Max repressor protein, which

binds to the same site as the c-Myc/Max complex, CACGTG (Ayer et al., 1993).

The c-*myc* gene is the cellular homologue to the cancer-producing gene, or **oncogene,** v-*myc* of the avian myelocytomatosis virus (Donner et al., 1982). The c-*myc* gene synthesizes very short-lived mRNA and protein products when stimulated by a variety of growth factors (Kelly et al., 1983). These c-*myc* gene products appear suddenly as cells are induced from the G_0 state into G_1, and they are degraded shortly thereafter. The c-Myc proteins signal cell division, and if they are not degraded rapidly, the cell

will continue proliferating, thereby increasing the risk of tumor formation.

Given that enhancers are able to control the expression of reporter genes that are unrelated to their normal targets, it would appear that these sequences are very powerful regulators of the specificity of gene transcription. What would happen if a spontaneous chromosomal translocation brought an enhancer for one protein adjacent to a structural gene for another protein? In most cases, it probably wouldn't matter. Even if the enhancer did stimulate the expression of the gene in the "wrong" cell, the product of only one cell would be abnormally regulated. Perhaps that cell would die and be replaced by another. Perhaps the descendants of that cell would form a clone of cells expressing a protein that other cells in the tissue did not make. However, if the translocation brought the c-*myc* gene next to an enhancer for a very actively transcribed gene, the c-*myc* gene would be activated to transcribe large amounts of message as the cell differentiated. In such cases, the single cell could give rise to a tumor.

Chromosomal translocations involving the c-*myc* gene appear to be responsible for tumors of the immunoglobulin-synthesizing B cells of our immune system (Croce, 1987). Here, the c-*myc* gene on the end of the small arm of human chromosome 8 has been translocated to chromosome 14, 22, or 2. These three chromosomes contain the genes for the immunoglobulin proteins, and the c-*myc* gene has been translocated to the region of the immunoglobulin gene enhancers (Leder et al., 1983; Croce, 1985). The amount of c-*myc* mRNA transcribed from these translocated chromosomes correlates with the activation of the immunoglobulin genes. Thus, when the enhancers of the immunoglobulin genes are activated (during B cell development), they turn on the adjacent gene—which is now c-*myc*. The c-*myc* mRNA is made in enormous amounts and is translated into the c-Myc transcription factor. This factor instructs the cells to divide. In the continuous presence of c-Myc protein, the cells keep dividing, thus forming a tumor called Burkitt's lymphoma (Nishikura et al., 1983; Croce et al., 1984). In these situations, the translocation of the c-*myc* gene to an immunoglobulin enhancer in a single cell can cause the entire leukemia. Indeed, most leukemias result from a single cell.

Several types of leukemias are caused when other transcription factor genes have been translocated to the regions of the immunoglobulin gene enhancers. These types of rearrangements between transcription-factor-encoding regions and lymphocyte-specific regulatory regions may be errors caused by the VDJ recombinases that are specific to lymphocytes and would explain why these translocations are seen in so many leukemias (Rabbitts, 1991).

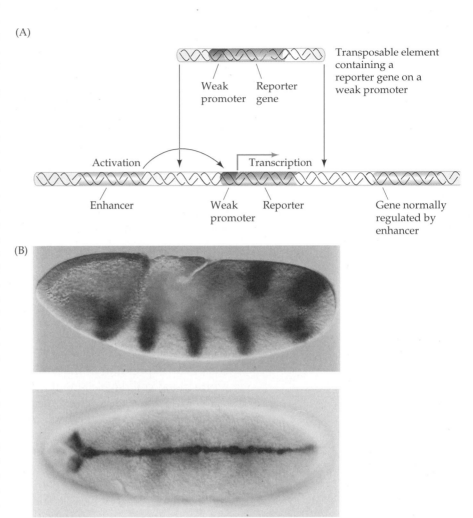

Figure 10.29 Enhancer trap technique. (A) A reporter gene is fused to a weak promoter that cannot direct transcription on its own. The construct is injected into the nucleus of the egg and integrates randomly into the genome. If it integrates near an enhancer, the reporter gene will become expressed when that enhancer is activated, showing the normal expression pattern of a gene normally associated with that enhancer. (B) Reporter gene expression in *Drosophila* injected with an enhancer trap. These enhancers are active in the development of the insect nervous system and were unrecognized before this procedure. (Photographs courtesy of Y. Hiromi.)

Identifying Enhancers by Reporter Genes

This ability of an enhancer to activate other genes has been used by scientists to find new enhancers and the genes regulated by them. To do this, one makes an **enhancer trap.** The trap consists of a reporter gene (such as the gene for *E. coli* β-galactosidase or green fluorescent protein) fused to a relatively weak eukaryotic promoter. This promoter will not initiate the transcription of the reporter gene without the help of an enhancer.

This recombinant enhancer trap is then introduced into the egg or oocyte by various means (see Chapter 2), where it integrates randomly into the genome. If the reporter gene becomes expressed, that means that it has come within the domain of an active enhancer (Figure 10.29). By isolating this region of the genome in wild-type flies or mice, the normal gene activated by this enhancer can be discovered (O'Kane and Gehring, 1987). [transcr1.html] ■

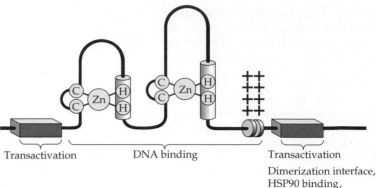

Figure 10.30
Functional domains of zinc finger transcription factors. Cysteine (C) and histidine (H) coodinate a zinc atom, causing the looping out of the "zinc fingers."

The labels in the figure read: Transactivation, DNA binding, Transactivation, Dimerization interface, HSP90 binding, inhibitory function, transactivation

Zinc Finger Transcription Factors

Another type of DNA-binding domain is the **zinc finger** motif. Zinc finger proteins include WT-1 (a important transcription factor critical in the formation of the kidney and gonads); the ubiquitous transcription factor Sp1; *Xenopus* 5S rRNA transcription factor TFIIIA; Krox 20 (a protein that regulates gene expression in the developing hindbrain); Egr-1 (which commits white blood cell development to the macrophage lineage); Krüppel (a protein that specifes abdominal cells in *Drosophila*); and numerous steroid-binding transcription factors. Each of these proteins has two or more "DNA-binding fingers," α-helical domains whose central amino acids tend to be basic. These domains are linked together in tandem and are each stabilized by a centrally located zinc ion coordinated by two cysteines (at the base of the helix) and two internal histidines (Figure 10.30). The crystal structure shows that the zinc fingers bind in the major groove of the DNA.

The WT-1 protein contains four zinc finger regions, and it is usually expressed in the fetal kidney and gonads. People with one mutant *WT1* allele (usually a deletion of the gene or of a zinc finger region) have urogenital malformations and develop Wilm's tumor of the kidney (Haber et al., 1990; Bruening et al., 1992; see Chapter 17). In mice, *both WT1* genes can be deleted by gene targeting, and the resulting mice die in utero, having neither kidneys nor gonads (Kriedberg et al., 1993). The WT1 factor binds to the regulatory regions of several genes that are active during kidney development and also is thought to inhibit the expression of certain growth factors (especially insulin-like growth factor II) in the developing kidney (Drummond et al., 1992).

Nuclear Hormone Receptors and Their Hormone-Responsive Elements

Specific steroid hormones are known to increase the transcription of specific sets of genes. Once the hormone has entered the cell, it binds to its specific receptor protein, converting that receptor into a conformation that is able to enter the nucleus and bind particular DNA sequences (Miesfeld et al., 1986; Green and Chambon, 1988). The family of steroid hormone receptors includes proteins that recognize estrogen, progesterone, testosterone, and cortisone as well as nonsteroid lipids such as retinoic acid, thyroxine, and vitamin D. The DNA sequences capable of binding nuclear hormone receptors are called **hormone-responsive elements,** and they can be in either enhancers or promoters. One set of steroids includes the glucocorticoid hor-

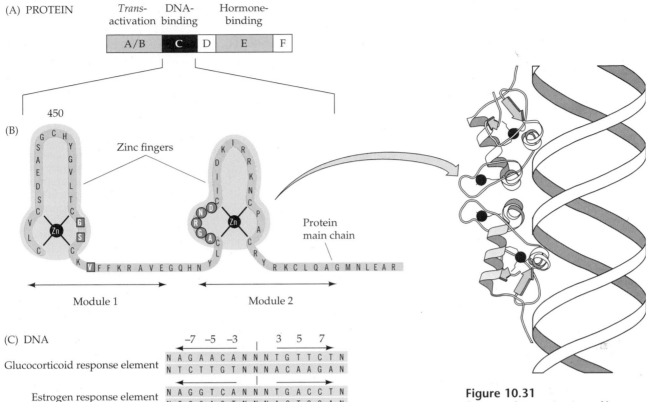

(A) PROTEIN

Trans-activation DNA-binding Hormone-binding

A/B   C   D   E   F

(B)

450

Zinc fingers

Module 1   Module 2

Protein main chain

(C) DNA

−7 −5 −3   3 5 7

Glucocorticoid response element
N A G A A C A N N N T G T T C T N
N T C T T G T N N N A C A A G A N

Estrogen response element
N A G G T C A N N N T G A C C T N
N T C C A G T N N N A C T G G A N

Thyroxin and retinoic acid response element
N N A G G T C A T G A C C T N N
N N T C C A G T A C T G G A N N

Figure 10.31
Structural organization of hormone receptor DNA-binding proteins. (A) Generalized structure of a steroid hormone binding protein. The functions ascribed to each fraction have been determined by analyzing the effects of mutations in each of these regions and by making chimeric protein molecules having regions derived from different receptor proteins. A similar structure is seen in the retinoic acid receptor and the thyroid hormone receptor. (B) Zinc finger DNA-binding region of the glucocorticoid receptor. Boxed residues in module 1 discriminate between estrogen- and glucocorticoid-responsive elements. Circled residues are involved in dimerization. (C) The zinc finger region of the glucocorticoid receptor bound to its responsive element. The DNA sequences for the responsive elements are shown to the left. Note that they are inverted palindromes, so each dimer is exposed to the same site. N, any base; GRE, glucocorticoid (and progesterone)-responsive element; ERE, estrogen-responsive element; TRE, thyroxine- and retinoic-acid-responsive element. The distinction between binding glucocorticoid or progesterone receptors versus binding thyroxine or retinoic acid receptors is determined by the spacing of the responsive elements, by limiting amounts of receptors, and by other interacting *cis* elements. (After Kaptein, 1992.)

mones (cortisone, hydrocortisone, and the synthetic hormone dexamethasone). These bind to the glucocorticoid hormone receptors and enable the bound receptors to bind to the glucocorticoid-responsive elements in the chromosomes (Figure 10.31).

The steroid-hormone-responsive elements are extremely similar to one another and are recognized by closely related proteins. These steroid receptor proteins each contain three functional domains: (1) a **hormone-binding domain,** (2) a **DNA-binding domain** that recognizes the hormone-responsive element, and (3) a *trans*-**activation domain,** which is involved in mediating the signal to initiate transcription. These functions can overlap, and all domains appear to have some role in activating transcription (Beato, 1989). For transcriptional activation to occur, the receptor has to enter the nucleus and dimerize with a similar hormone-binding protein. The binding of hormone to the hormone-binding domain appears necessary for dimerization, translocation into the nucleus, and the ability of the DNA-binding region to recognize the hormone-responsive element (Kumar et al., 1987).

The hormone-responsive elements within the DNA were originally discovered by competitive binding assays (Pfahl, 1982; Karin, 1984) wherein specific restriction fragments of DNA were tested for their ability to bind hormone receptors carrying radioactive hormone. By using various restriction-enzyme-derived fragments of DNA and comparing the sequences of numerous glucocorticoid-responsive elements, it was found that the consensus sequence of the glucocorticoid-response element is AGAACANNNT-GTTCT (where *N* can be any base). These glucocorticoid-binding sequences were shown to act as enhancers: when the glucocorticoid-response element

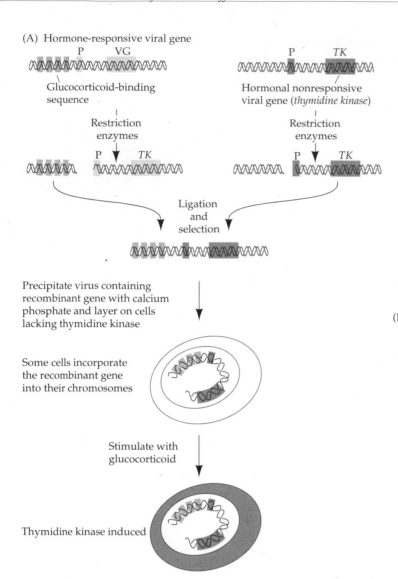

(A) Hormone-responsive viral gene

P   VG

Glucocorticoid-binding sequence

Restriction enzymes

P   TK

Hormonal nonresponsive viral gene (*thymidine kinase*)

Restriction enzymes

P   TK

Ligation and selection

Precipitate virus containing recombinant gene with calcium phosphate and layer on cells lacking thymidine kinase

Some cells incorporate the recombinant gene into their chromosomes

Stimulate with glucocorticoid

Thymidine kinase induced

Figure 10.32
Test for the glucocorticoid enhancer sequence. (A) A recombinant virus containing the glucocorticoid-responsive enhancer of the mouse mammary tumor virus and the *thymidine kinase* gene of Herpes simplex virus can integrate into the genome of a cell lacking *thymidine kinase* gene. Upon treatment with glucocorticoids, the recombinant gene transcribes viral thymidine kinase. P, promoter region; *TK*, *thymidine kinase* gene; VG, viral gene of mouse mammary tumor virus. (B) Electron micrograph of glucocorticoid enhancer elements, showing hormone receptor bound to that region of the gene. (A modified from Chandler et al., 1983; B from Payvar et al., 1983, courtesy of K. R. Yamamoto.)

(B)

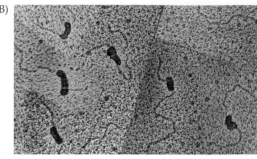

was ligated onto genes that are not usually hormone-dependent, those genes gained responsiveness to the glucocorticoid (Figure 10.32; Chandler et al., 1983).

The binding of the receptor protein to the hormone-responsive enhancer element is accomplished by a zinc finger region in the DNA-binding domain (Green et al., 1988). When chimeric proteins are made wherein the zinc finger domain of the estrogen receptor substitutes for the same region of a glucocorticoid receptor, the protein recognizes the DNA that contains estrogen-responsive elements and causes the gene to be responsive to glucocorticoids. The critical amino acids appear to reside at the "knuckle" of the zinc finger (Danielsen et al., 1989; Umesono and Evans, 1989). Changing as few as two amino acids at the knuckle of the zinc finger region will change the specificity of the binding protein. Thus, while these DNA-binding domains of the hormone receptor proteins are very similar, they can distinguish subtle differences in the enhancer sequences. For example, the (palindromic) sequence 5'–GGTCACTGTGACC–3' is a strong estrogen-responsive enhancer element that will bind the estrogen-containing receptor protein. Two symmetrical mutations in this sequence, making it 5'–GG**A**CACTGTG**T**CC–3', convert this DNA into a glucocorticoid-responsive enhancer (Klock et al., 1987; Martinez et al., 1987). Given the similarities among hormone receptor proteins and the similarities among hormone-responsive elements, it is likely that each steroid hormone mediates its transcriptional activation by the same general mechanism.

DNA-bending proteins

In addition to DNA-binding proteins, there is also a set of transcription factors that function primaily as DNA-bending proteins. Most of these proteins are characterized by a DNA-binding element called the HMG box, a set of approximately 80 amino acids that mediate the binding of these proteins to the minor groove of the DNA. These proteins include the Y chromosome sex-determining factor, SRY (to be discussed in Chapter 20), the lymphocyte enhancer protein LEF-1, and the chromatin proteins HMG-1(Y) and HMG-2. These proteins are not thought to activate transcription by directly interacting with the transcription apparatus. Rather, they are thought to bend the DNA so that the activators and repressors can be brought into contact. For example, the binding of SRY to the DNA causes a 70°–80° bend in the helix, converting the "I" into an "L." Point mutations that prevent this bending also prevent the protein from mediating the formation of testes. It is thought that the SRY protein bends the DNA so that factors that otherwise might be apart on the chromosome are brought into contact (see Figure 20.5; Werner et al., 1995).

The DNA-bending proteins can create specific three-dimensional structures called "enhanceosomes" (Thanos and Maniatis, 1995). A model for one such enhanceosome is shown in Figure 10.33A. Thanos and Maniatis have shown that direct protein-protein interactions between transcription factors are greatly facilitated by the presence of HMG-1, a DNA-bending protein. There are three sites for this protein within the enhancer for the human interferon-β (*IFNβ*) gene, and these sites are essential for the synergistic activation of the complex by the transcription factors. They also showed that the mere presence of these transcription factors is not sufficient for the activation of the *IFNβ* promoter. The transcription factors had to be in the correct order in the enhancer. By shuffling around the binding elements of the DNA, they made different combinations. Only the wild-type combination worked. Thanos and Maniatis also showed that the helical phasing was important. By adding a bit of DNA that caused a half-helical turn, the enhancer was inactivated. By inserting another half-helical turn, the enhancer was made to function again. Therefore, the linear arrangement of the transcription factors and their three-dimensional organization were critical. The HMG-1 protein bound them all together in an enhanceosome. When this enhanceosome was completed, the DNA, which had a natural bend of –20°, now had a bend of +26°. Moreover, an inactive enhancer had been turned into an activator (Figure 10.33B; Falvo et al., 1995).

Context-dependent activation or silencing

The interactions between steroid receptors and other transcriptional regulatory proteins can determine whether the steroid's effect is positive or negative. Diamond and co-workers (1990) demonstrated that the effect of glucocorticoid hormones on the transcription of the mouse *Proliferin* gene can be positive or negative depending on the prior physiological state of the cell. A 25-base-pair sequence upstream from the *Proliferin* gene can bind the glucocorticoid receptor and the bZip dimer of either c-Jun and c-Jun or c-Jun and c-Fos. The binding of c-Jun to this site is necessary for the function of the

(A)

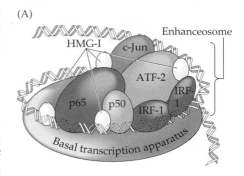

(B)

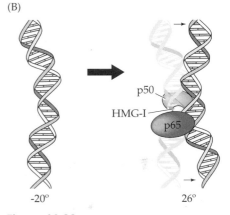

Figure 10.33
Structure of an enhanceosome. (A) The HMG-I(Y) DNA-bending protein wraps a 60-base-pair coil of DNA around the transcriptional activators NF-κB (the p50/p65 complex), IRF1, and ATF2/c-Jun. The HMG-I(Y) is in the minor groove, while the other transcription factors operate in the major groove of the double helix. Once the enhanceosome is assembled, it contacts the basal transcription apparatus at several sites. (B) The DNA prior to enhanceosome formation has a –20° bend in it. After the enhanceosome forms, it bends in the opposite direction, +26°. This latter complex stimulates transcription. (A after Thanos and Maniatis, 1995; B after Falvo et al., 1995.)

Figure 10.34
Alternative enhancing and silencing effects of the glucocorticoid-response element upstream from the mouse *Proliferin* gene. The effect of the glucocorticoid depends on the prior condition of the cell (i.e., whether high concentrations of c-Fos were being synthesized). The arrows represent transcription of the *Proliferin* gene. The large circle represents the hormone-bound glucocorticoid receptor. The c-Jun protein is represented as a small circle, while the c-Fos protein is a small square. (After Diamond et al., 1990.)

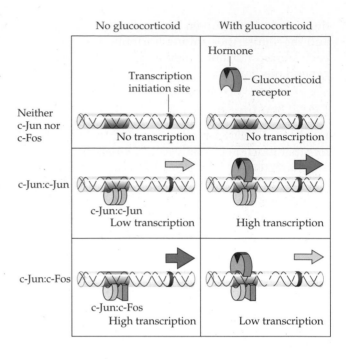

glucocorticoid receptor. If the c-Jun/c-Jun dimer were present at this site (without the glucocorticoid receptor), very little transcription would occur. This transcription is dramatically *enhanced* by the addition of glucocorticoids (Figure 10.34). If the c-Jun/c-Fos dimer were present at this site, however, it could direct extremely efficient transcription of the *Proliferin* gene. This transcription is *inhibited* by the presence of glucocorticoids. Thus, whether the glucocorticoid has a stimulatory or inhibitory effect on *Proliferin* transcription depends on the prior physiological state of the cell. A single DNA sequence binding a particular hormone receptor may be both an enhancer and a silencer for the same protein.

There are other ways for a *cis*-regulatory element to be an activator in some situations and a repressor in others. For example, the *Drosophila* transcription factor Krüppel (a protein whose activity, we will see in Chapter 14, is responsible for forming the fly's thorax and upper abdomen) is an activator at low concentrations and a repressor at high concentrations. At low concentrations, it binds to its *cis*-regulatory element on the DNA, and it interacts with TFIIB to facilitate the assembly of the transcription initiation complex. At high concentrations, however, it binds to itself, and the resulting dimers do not complex with TFIIB (Sauer et al., 1995). Instead, the dimers interact with TFIIE and may block its function. Whether the tumor suppressor protein p53 is an activator or repressor depends on the structure of the specific gene's promoter. If a p53-binding element exists in the promoter, p53 acts as an activator. If there is no p53 element in the promoter, p53 can bind to a TAF in TFIID and prevent transcription. It can also interact with the WT1 transcription factor. WT1 is usually an activator of transcription, but if it is bound by p53, it becomes a repressor* (Figure 10.35; Seto et al., 1992; Maheswaran et al., 1993).

---

*We have good news and bad news. The good news is that by the end of this decade, we may well know most, if not all, of the transcription factors active in many cell types and how they interact to initiate or repress transcription. The bad news is that many of us will have to learn physical chemistry to understand these data.

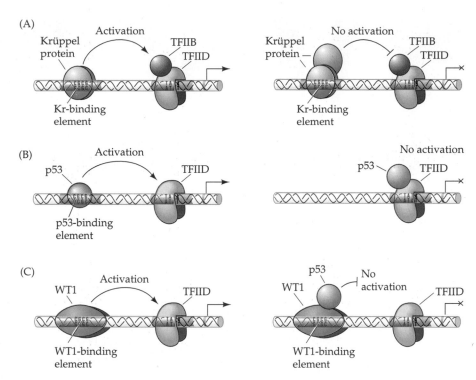

Figure 10.35
Transcription factors can be activators or repressors, depending on the context. (A) Krüppel protein in low concentrations stimulates TFIIB and activates transcription. In high concentrations, it forms dimers that do not bind TFIIB (and which may interfere with TFIIE). (B) The p53 protein is an activator where specific binding sites are present. On some promoters, however, it can bind to TFIID and inactivate it when such sites are absent. (C) The WT1 protein is an activator when p53 is absent. In the presence of high concentrations of p53, the binding of WT1 blocks transcription.

Regulation of transcription factor activity

If transcription factors are proteins that regulate the expression of particular genes, then how are the transcription factors themselves regulated? One obvious way is to regulate the synthesis of transcription factors by other transcription factors. This method is used in *Drosophila* development, in which there is a cascade of transcription factor synthesis (see Chapter 14). In mammals, several transcription factors are similarly regulated by the synthesis of other transcription factors. The activation of the Pit-1 transcription factor in the mammalian pituitary, for instance, is accomplished by the binding of the Prop1 homeodomain-containing transcription factor to the 5' flanking sequence of the *Pit-1* gene during an earlier stage of pituitary development (Sornson et al., 1996).

Moreover, transcription factors often have very complex enhancers and promoters that enable them to be expressed only in certain cells. The mouse *Myogenin* gene, for instance, is expressed in the myotome, pharyngeal arches, and limb buds. There appear to be at least three separable sites in the upstream regulatory region for this gene. The closest site is necessary for the transcription of this gene in the limb buds. If this site is mutated, the *Myogenin* gene is not transcribed there. A second site, further upstream, is needed for expression of *Myogenin* in the limb buds, pharyngeal arches, and central cells of the posterior somites (Figure 10.36; Cheng et al., 1993; Yeo and Rigby, 1993). A third site, still further upstream, is necessary for the increased efficiency of transcription of the gene. These three sites bind different transcription factors. A similar situation appears to exist for *myf-5*, wherein different regions of DNA regulate the different expression pattern elements (Plate 24; Patapoutian et al., 1993).

Another mechanism of regulating transcription factor activity is by phosphorylation. In one set of cases, the transcription factor protein is present, but inactive, and phosphorylation activates the dormant protein. As we have discussed earlier, the phosphorylation of a sequestered transcription

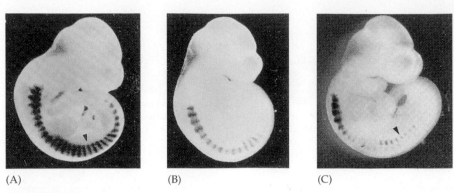

(A)   (B)   (C)

Figure 10.36

Myogenin expression in the 10.5-day mouse embryo. A β-galactosidase reporter gene was linked to the upstream regulatory sequences of the *Myogenin* gene, and this was used to make transgenic mice. The transgenic embryos were stained for the presence of bacterial β-galactosidase at 10.5 days. (A) Wild-type *Myogenin* promoter region, showing all the places where the *Myogenin* gene is usually expressed. (B) Expression from a *Myogenin* promoter with a mutation in a site close to the *Myogenin* gene. There is no transcription from this gene in the limb buds. (C) Expression from a *Myogenin* promoter with a mutation in a site further upstream from the gene. No transcription from this promoter is seen in the pharyngeal arches, limbs, or central posterior myotome cells. (From Cheng et al., 1993.)

factor or its inhibitor (such as IκB) can release the inhibition and enable the transcription factor (in this case, NF-κB) to enter the nucleus and bind its DNA sequence. Phosphorylation can also work more directly. As discussed in Chapter 3, the JAK/STAT transcription factors are present in the cytoplasm but only enter the nucleus when they are phosphorylated in response to signals at the cell membrane. Phosphorylation can also be used to repress transcription factors, as when the binding of DNA by Pit-1, Oct1, or myogenin is inhibited by their being phosphorylated (Hunter and Karin, 1992).

So we see that the activity of transcription factors can be regulated at several levels. Given that each gene is often regulated by several transcription factors, the cell is given many options on how to express certain genes in only certain cell types. Within the past five years, our knowledge of transcription factors has progressed enormously and has given us a new, dynamic view of gene expression. The gene itself is no longer seen as an independent entity controlling the synthesis of proteins. Rather, the gene both *directs* and *is directed by* protein synthesis. Angier (1992) writes:

> *A series of new discoveries suggests that DNA is more like a certain type of politician, surrounded by a flock of protein handlers and advisers that must vigorously massage it, twist it and, on occasion, reinvent it before the grand blueprint of the body can make any sense at all.*

Certainly, the interactions between DNA and its transcription factors are taking the interactive relationship of nucleus and cytoplasm to new and splendidly complex levels. So far, we have been focusing our attention on the relationship of the transcription factors to DNA. But transcription factors do not see mere DNA. Rather, they see a highly structured DNA-protein complex called chromatin. To initiate transcription, one also must deal with the higher-order structures of the cell, and we will continue our discussion of the transcriptional regulation of development into the next chapter.

LITERATURE CITED

Akoulitchev, S., Mäkelä, T. P., Weinberg, R. A. and Reinberg, D. 1995. Requirement for TFIIH kinase activity by RNA polymerase II. *Nature* 377: 557–560.

An, W. and Wensink, P. C. 1995. Three protein binding sites form an enhancer that regulates sex- and fat body-specific transcription of *Drosophila yolk protein* gene. *EMBO J.* 14: 1221–1230.

Angier, N. 1992. A first step in putting genes into action: Bend the DNA. *New York Times*, August 4, 1992, pp. C1, C7.

Atchinson, M. L. and Perry, R. P. 1987. The role of κ-enhancer and its binding factor NF-κB in the developmental regulation of κ gene transcription. *Cell* 48: 121–128.

Ayer, D. E., Kretzner, L. and Eisenman, R. N. 1993. Mad: A heterodimeric partner for Max that antagonizes Myc transcriptional activity. *Cell* 72: 211–222.

Baeuerle, P. A. and Baltimore, D. 1988. IκB: A specific inhibitor of the NF-kB transcription factor. *Science* 242: 540–545.

Banerji, J., Olson, L. and Schaffner, W. 1983. A lymphocyte-specific cellular enhancer is located downstream of the joining region in immunoglobulin heavy chain genes. *Cell* 33: 729–740.

Beato, M. 1989. Gene regulation by steroid hormones. *Cell* 56: 335–344.

Behringer, R. R., Mathews, L. S., Palmiter, R. D. and Brinster, R. L. 1988. Dwarf mice produced by genetic ablation of growth hormone-expressing cells. *Genes Dev.* 2: 453–461.

Benezra, R., Davis, R. L., Lockshon, D., Turner, D. L. and Weintraub, H. 1990. The protein Id: A negative regulator of helix-loop-helix DNA binding proteins. *Cell* 61: 49–59.

Bergman, Y., Rice, D., Grosschedl, R. and Baltimore, D. 1984. Two regulatory elements for immunoglobulin κ light chain gene expression. *Proc. Natl. Acad. Sci. USA* 81: 7041–7045.

Bernard, O., Hozumi, N. and Tonegawa, S. 1978. Sequences of mouse immunoglobulin light chain genes before and after somatic change. *Cell* 15: 1133–1144.

Bischoff, R. and Holtzer, H. 1969. Mitosis and the processes of differentiation of myogenic cells in vitro. *J. Cell Biol.* 41: 188–200.

Bodner, M. and Karin, M. 1987. A pituitary-specific *trans*-acting factor can stimulate transcription from the growth hormone promoter in extracts of non-expressing cells. *Cell* 50: 267–275.

Boulet, A. M., Erwin, C. R. and Rutter, W. J. 1986. Cell-specific enhancers in the rat exocrine pancreas. *Proc. Natl. Acad. Sci. USA* 83: 3599–3603.

Brack, C., Hirama, M., Lenhard-Schuller, R. and Tonegawa, S. 1978. A complete immunoglobulin gene is created by somatic recombination. *Cell* 15: 1–14.

Brennan, T., Edmondson, D. G., Li, L. and Olson, E. N. 1991. Transforming growth factor β represses the actions of myogenin through a mechanism independent of DNA binding. *Proc. Natl. Acad. Sci. USA* 88: 3822–3826.

Bruening, W. and seven others. 1992. Germline intronic and exonic mutations in the Wilms' tumor gene (*WT1*) affecting urogenital development. *Nat. Genet.* 1: 144–148.

Bunick, D., Zandomeni, R., Ackerman, S. and Weinmann, R. 1982. Mechanism of RNA polymerase II-specific initiation of transcription in vitro: ATP requirement and uncapped runoff transcripts. *Cell* 29: 877–886.

Buratowski, S., Hahn, S., Guarente, L. and Sharp, P. A. 1989. Five initiation complexes in transcription initiation by RNA polymerase II. *Cell* 56: 549–561.

Buratowski, S., Sopta, M., Greenblatt, J. and Sharp, P. 1991. RNA polymerase II-associated proteins are required for a DNA conformation change in the transcription initiation complex. *Proc. Natl. Acad. Sci. USA* 88: 7509–7513.

Calame, K. L. 1989. Immunoglobulin gene transcription: Molecular mechanisms. *Trends Genet.* 5: 395–399.

Chandler, V. L., Maier, B. A. and Yamamoto, K.R. 1983. DNA sequences bound specifically by glucocorticoid receptor in vitro render a heterologous promoter hormone responsive in vivo. *Cell* 33: 489–499.

Chen, J. -L., Attardi, L. D., Verrijzer, C. P., Yokomori, K. and Tjian, R. 1994. Assembly of recombinant TFIID reveals differential cofactor requirements for distinct transcriptional activators. *Cell* 79: 93–105.

Chen, P.-L., Scully, P., Shew, J.-Y., Wang, J. Y. J. and Lee, W.-H. 1989. Phosphorylation of the retinoblastoma gene product is modulated during the cell cycle and cellular differentiation. *Cell* 58: 1193–1198.

Cheng, T.-C., Wallace, M., Merlie, J. P. and Olson, E. N. 1993. Separable regulatory elements govern myogenin transcription in mouse embryogenesis. *Science* 261: 215–218.

Chi, T. and Carey, M. 1996. Assembly of the isomerized TFIIA-TFIID-TATA ternary complex is necessary and sufficient for gene activation. *Genes Dev.* 10: 2540–2550.

Christy, R. J. and seven others. 1989. Differentiation-induced gene expression in 3T3–L1 preadipocytes: CCAAT/enhancer binding protein interacts with and activates the promoter of two adipocyte-specific genes. *Genes Dev.* 3: 1323–1335.

Chun, J. J. M., Schatz, D. G., Oettinger, M. A., Jaenisch, R. and Baltimore, D. 1991. The recombination activating gene 1 (*RAG-1*) is present in the murine central nervous system. *Cell* 64: 189–200.

Comai, L., Tanese, N. and Tjian, R. 1992. The TATA-binding protein and associated factors are integral components of the RNA polymerase I transcription factor, SL1. *Cell* 68: 965–976.

Conaway, R. C., Pfeil-Garrett, K., Hanley, J. P. and Conaway, J. W. 1991. Mechanism of promoter selection by RNA polymerase II: Mammalian transcription factors a and bg promote entry of polymerase into the preinitiation complex. *Proc. Natl. Acad. Sci. USA* 88: 6205–6209.

Crenshaw, E. B., Kalla, K., Simmons, D. M., Swanson, L. W. and Rosenfeld, M. G. 1989. Cell-specific expression of the prolactin gene in transgenic mice is controlled by synergistic interactions between promoter and enhancer elements. *Genes Dev.* 3: 959–972.

Croce, C. M. 1985. Chromosomal translocations, oncogenes, and B-cell tumors. *Hosp. Pract.* 20(1): 41–48.

Croce, C. M. 1987. Role of chromosome translocations in human neoplasia. *Cell* 49: 155–156.

Croce, C. M., Thierfelder, W., Erikson, J., Nishikura, K., Finan, J., Lenoir, G. M. and Nowell, P. C. 1984. Transcriptional activation of an unarranged and untranslocated c-*myc* oncogene by translocation of a C_λ locus in Burkitt lymphoma cells. *Proc. Natl. Acad. Sci. USA* 80: 6922–2926.

Crossley, M. and Brownlee, G. G. 1990. Disruption of a C/EBP binding site in the factor IX promoter is associated with haemophilia B. *Nature* 345: 444–446.

Cullen, K. E., Kladde, M. P. and Seyfred, M. A. 1993. Interaction between transcriptional regulatory regions of prolactin chromatin. *Science* 261: 203–206.

Danielsen, M. Hinck, L. and Ringold,G. M. 1989. Two amino acids within the knuckle of the first zinc finger specify DNA response element activation by the glucocorticoid receptor. *Cell* 57: 1131–1138.

Davis, M. M., Calame, K., Early, P. W., Livant, D. L., Joho, R., Weissman, I. L. and Hood, L. 1980a. An immunoglobulin heavy chain gene is formed by at least two recombinational events. *Nature* 283: 733–739.

Davis, M. M., Kim, S. K. and Hood, L. 1980b. Immunoglobulin class switching: Developmentally regulated DNA rearrangements during differentiation. *Cell* 22: 1–2.

Diamond, M. I., Miner, J. N., Yoshinaga, S. K. and Yamamoto, K. R. 1990. Transcription factor interactions: Selectors of positive and negative regulation from a single DNA element. *Science* 249: 1266–1272.

Dierks, P., van Ooyen, A., Chochran, M. D., Dobkin, C., Reiser, J. and Weissman, C. 1983. Three regions upstream from the cap site are required for efficient and accurate transcription of the rabbit β-globin gene in mouse 3T6 cells. *Cell* 32: 695–706.

Dollé, P., Castrillo, J.-L., Theill, L. E., Deerinck, T., Ellisman, M. and Karin, M. 1990. Expression of GHF-1 protein in mouse pituitaries correlates both temporally and spatially with the onset of growth hormone gene activity. *Cell* 60: 809–820.

Donner, P., Greiser-Wilka, I. and Moelling, K. 1982. Nuclear localization and DNA binding of the transforming gene product of avian myelocytomatosis virus. *Nature* 296: 262–266.

Driever, W. and Nüsslein-Volhard, C. 1989. The bicoid protein is a positive regulator of *hunchback* transcription in the early *Drosophila* embryo. *Nature* 337: 138–143.

Drummond, I. A., Madden, S. L., Rohwer, N. P., Bell, G. I., Sukhatme, V. P. and Rauscher, F. III. 1992. Repression of the insulin-like growth factor II gene by the Wilms' tumor suppressor gene *WT1. Science* 257: 674–678.

Dynan, W. S. and Tjian, R. 1985. Control of eukaryotic messenger RNA synthesis by sequence-specific DNA-binding proteins. *Nature* 316: 774–778.

Dynlacht, B. D., Hoey, T. and Tjian, R. 1991. Isolation of cofactors associated with the TATA-binding protein that mediate transcriptional activation. *Cell* 66: 563–576.

Efstratiadis, A. and fourteen others. 1980. The structure and evolution of the human β-globin gene family. *Cell* 21: 653–668.

Elsholtz, H. P., Albert, V. R., Treacy, M. N. and Rosenfeld, M. G. 1990. A two-base change in a POU factor-binding site switches pituitary specific to lymphoid-specific gene expression. *Genes Dev.* 4: 43–51.

Falvo, J. V., Thanos, D. and Maniatis, 1995. Reversal of intrinsic DNA bends in the IFNβ gene enhancer by transcription factors and the architectural protein HMG I(Y). *Cell* 83: 1101–1111.

French, B. A., Chow, K.-L., Olson, E.N. and Schwartz, R. J. 1991. Heterodimers of myogenic regulatory factors and E12 bind a complex element governing myogenic induction of the avian a-actin promoter. *Mol. Cell. Biol.* 11: 2439–2450.

Fujimoto, S. and Yamagishi, H. 1987. Isolation of an excision product of T cell receptor a-chain gene rearrangements. *Nature* 327: 242–243.

Garabedian, M. J., Hung, M.-C. and Wensink, P. C. 1985. Independent control elements that determine yolk protein gene expression in alternative *Drosophila* tissues. *Proc. Natl. Acad. Sci. USA* 82: 1396–1400.

Gedamu, L. and Dixon, G. H. 1978. Effect of enzymatic decapping on protamine messenger RNA translation in wheat-germ S-30. *Biochem. Biophys. Res. Commun.* 85: 114–124.

George-Weinstein, M. and nine others. 1996. Skeletal myogenesis: The preferred pathway of chick epiblast cells in vitro. *Dev. Biol.* 173: 279–291.

Ghosh, S. and Baltimore, D. 1990. Activation in vitro of NF-κB by phosphorylation of its inhibitor, IκB. *Nature* 344: 678–682.

Gillies, S. D., Morrison, S. L., Oi, V. T. and Tonegawa, S. 1983. A tissue-specific transcription enhancer element is located in the major intron of a rearranged immunoglobulin heavy chain gene. *Cell* 33: 717–728.

Green, S. and Chambon, P. 1988. Nuclear receptors enhance our understanding of transcriptional regulation. *Trends Genet.* 4: 309–314.

Green, S., Kumar, V., Thenlaz, I., Wahli, W. and Chambon, P. 1988. The *N*-terminal DNA binding zinc finger of the oestrogen and glucocorticoid receptors determines target gene specificity. *EMBO J.* 7: 3037–3044.

Grosschedl, R. and Birnstiel, M. L. 1980. Spacer DNA upstream from the TATAATA sequence are essential for promotion of H2A histone gene transcription in vivo. *Proc. Natl. Acad. Sci. USA* 77: 7102–7106.

Grosveld, F., de Boer, E., Shewmaker, C. K. and Flavell, R. A. 1982. DNA sequences necessary for the transcription of the rabbit β-globin gene in vivo. *Nature* 295: 120–126.

Haber, D. A and seven others. 1990. An internal deletion within an 11p13 zinc finger gene contributes to the development of Wilms' tumor. *Cell* 61: 1257–1269.

Henkel, T., Zabel, U., van Zee, K., Müller, J. M., Fanning, E. and Baeuerle, P. A. 1992. Intramolecular masking of the nuclear location signal and dimerization domain in the precursor for the p50 NFvκB subunit. *Cell* 68: 1121–1133.

Herr, W. and eleven others. 1988. The POU domain: A large conserved region in the mammalian *pit-1, oct-1, oct-2,* and *Caenorhabditis elegans unc-86* gene products. *Genes Dev.* 2: 1513–1516.

Hiom, K. and Gellert, M. 1997. A stable RAG1-RAG2-DNA complex that is active in V(D)J cleavage. *Cell* 88: 65-72.

Hoey, T., Weinzierl, R. O. J., Gill, G., Chen, J.-L., Dynlacht, B. D. and Tjian, R. 1993. Molecular cloning and functional analysis of *Drosophila* TAF110 reveal properties expected of coactivators. *Cell* 72: 247–260.

Hozumi, N. and Tonegawa, S. 1976. Evidence for somatic rearrangement of immunoglobulin genes coding for variable and constant regions. *Proc. Natl. Acad. Sci. USA* 73: 3628–3632.

Hunter, T. and Karin, M. 1992. The regulation of transcription by phosphorylation. *Cell* 70: 375–387.

Jacq, X., Brou, C., Lutz, Y., Davidson, I., Chambon, P. and Tora, L. 1994. Human TAF_{II}30 is present in a distinct TFIID complex and is required for transcriptional activation by the estrogen receptor. *Cell* 79: 107–117.

Jarriault, S., Brou, C., Logeat, C., Schroeter, E. H., Kopan, R. and Israel, A. 1995. Signalling downstream of activated mammalian Notch. *Nature* 377: 355–358.

Jen, Y., Weintraub and Benezra, R. 1992. Overexpression of Id protein inhibits the muscle differentiation program: In vivo association of Id and E2A proteins *Genes Dev.* 6: 1466–1479.

Jones, N. 1990. Transcriptional regulation by dimerization: Two sides to an incestuous relationship. *Cell* 61: 9–11.

Kadonaga, J. T., Courey, A. J., Ladika, J. and Tjian, R. 1988. Distinct regions of Sp1 modulate DNA binding and transcriptional activation. *Science* 242: 1566–1570.

Kaestner, K. H., Christy, R. J. and Lane, M. D. 1990. Mouse insulin-responsive glucose transporter gene: Characterization of the gene and *trans*-activation by the CCAAT/enhancer binding protein. *Proc. Natl. Acad. Sci. USA* 87: 251–255.

Kaptein, R. 1992. Zinc-finger structures. *Curr. Opin. Struct. Biol.* 2: 109–115.

Karin, M., Haslinger, A., Holtgreve, H., Richards, R. I., Krautner, P., Westphal, H. M. and Beato, M. 1984. Characterization of DNA sequences through which cadmium and glucocorticoid hormones induce human metallothionein-II gene. *Nature* 308: 513–519.

Kelly, K., Cochrane, B. H., Stiles, C. D. and Leder, P. 1983. Cell-specific regulation of the c-*myc* gene by lymphocyte mitogens and platelet-derived growth factor. *Cell* 35: 603–610.

Kerr, L. D., Inoue, J.-I., Davis, N., Link, E., Baeurle, P. A., Bose, H. R. Jr. and Verma, I. M. 1991. The rel-associated pp40 protein prevents DNA binding at rel and NF-κB: Relationship with IκB and regulation by phosphorylation. *Genes Dev.* 5: 1464–1476.

Klock, G., Strähle, U. and Schüutz, G. 1987. Oestrogen and glucocorticoid responsive elements are closely related but distinct. *Nature* 329: 734–736.

Koleske, A. J., Buratowski, S., Nonet, M. and Young, R. A. 1992. A novel transcription factor reveals a functional link between the RNA polymerase II CTD and TFIID. *Cell* 69: 883–894.

Konigsberg, I. R., McElvain, N., Tootle, M. and Herrmann, H. 1960. The dissociability of deoxyribonucleic acid synthesis from the development of multinuclearity of muscle cells in culture. *J. Biophys. Biochem. Cytol.* 8: 333–343.

Kopan, R., Nye, J. S. and Weintraub, H. 1994. The intracellular domain of mouse *Notch*: a constutively actived repressor of myogenesis directed at the basic helix-loop-helix region of MyoD. *Development* 120: 2421–2430.

Kreidberg, J. A., Saviola, H., Loring, J. M., Maeda, M., Pelletier, J., Housman, D. and Jaenisch, R. 1993. WT-1 is required for early kidney development. *Cell* 74: 679–691.

Kumar, V., Green, S., Stack, G., Berry, M., Jin, J.-R. and Chambon, P. 1987. Functional domains of the human estrogen receptor. *Cell* 51: 941–951.

Landschulz, W. H., Johnson, P. F. and McKnight, S. L. 1988. The leucine zipper: A hypothetical structure common to a new class of DNA-binding proteins. *Science* 240: 1759–1764.

Lassar, A. B. and seven others. 1991. Functional activity of myogenic HLH proteins requires hetero-oligomerization with E12/E47-like proteins in vivo. *Cell* 66: 305–315.

Lawn, R. M., Efstratiadis, A., O'Connell, C. and Maniatis, T. 1980. The nucleotide sequence of the human β-globin gene. *Cell* 21: 647–651.

Laybourn, P. J. and Kadonaga, J. T. 1991. Role of nucleosome cores and histone H1 in regulation of transcription by RNA polymerase II. *Science* 254: 238–245.

Leder, P. and seven others. 1983. Translocations among antibody genes in human cancer. *Science* 222: 765–771.

Lee, D. K., Horikoshi, M. and Roeder, R. G. 1991. Interaction of TFIID in the minor groove of the TATA element. *Cell* 67: 1241–1250.

Lenardo, M. J. and Baltimore, D. 1989. NF–κB: A pleiotropic mediator of inducible and tissue-specific gene control. *Cell* 58: 227–229.

Li, L., Zhou, J., Guy, J., Heller-Harrison, R., Czech, M. P. and Olson, E. N. 1992. FGF inactivates myogenic helix-loop-helix proteins through phosphorylation of a conserved protein kinase C site in their DNA-binding domains. *Cell* 71: 1181–1194.

Li, S., Crenshaw, E. B. III, Rawson, E. J., Simmons, D. M., Swanson, L. W. and Rosenfeld, M. G. 1990. *Dwarf* locus mutants lacking three pituitary cell types result from mutations in the POU-domain gene *pit-1*. *Nature* 347: 528–533.

Lin, F.-T. and Lane, M. D. 1992. Antisense CCAAT/enhancer-binding protein RNA suppresses coordinate gene expression and triglyceride accumulation during differen-tiation of 3T3-L1 pre-adipocytes. *Genes Dev.* 6: 533–544.

Lin, Y.-S. and Green, M. R. 1991. Mechanism of action of an acidic transcriptional activator *in vitro*. *Cell* 64: 971–981.

Lin, Y.-S., Ha, I., Maldonado, E., Reinberg, D. and Green, M. R. 1991. Binding of general transcription factor TFIIB to an acidic activating region. *Nature* 353: 569–571.

Lu, H., Zawel, L., Fisher, L., Egly, M. and Reinberg, D. 1992. Human general transcription factor IIH phosphorylates the C-terminal domain of RNA polymerase II. *Nature* 358: 641–645.

Maheswaran, S. , Park, S., Bernard, A., Morris, J., Rauscher, F. J. III, Hill, D. E. and Haber, D. A. 1993. Physical and functional interactions between WT1 and p53 proteins. *Proc. Natl. Acad. Sci* . *USA* 90: 5100–5104

Maldonado, E., Ha, I., Cortes, P., Weis, L. and Reinberg, D. 1990. Factors involved in specific transcription by mammalian RNA polymerase II: Role of transcription factors IIA, IID, and IIB during formation of a transcription competent complex. *Mol. Cell Biol.* 10: 6335–6347.

Maniatis, T., Goodbourn, S. and Fischer, J. A. 1987. Regulation of inducible and tissue-specific gene expression. *Science* 236: 1237–1245.

Mantovani, R. and eight others. 1988. An erythroid-specific nuclear factor binding to the proximal CACCC box of the β-globin gene promoter. *Nucleic Acid Res.* 16: 4299–4313.

Martinez, E., Givel, F. and Wahl, W. 1987. An estrogen-responsive element as an inducible enhancer: DNA sequence requirements and conversion to a glucocorticoid-responsive element. *EMBO J.* 6: 3719–3727.

Mather, E. L. and Perry, R. P. 1982. Transcriptional regulation of the immunoglobulin V genes. *Nucleic Acids Res.* 9: 6855–6867.

Matsuoka, M., Nagawa, F., Okazaji, K., Kingsbury, L., Yoshida, K., Muller, U., Larue, D.T., Winer, J. A. and Sakano, H. 1991. Detection of somatic DNA recombination in the transgenic mouse brain. *Science* 254: 81–86.

McKnight, S. and Tjian, R. 1986. Transcriptional selectivity of viral genes in mammalian cells. *Cell* 46: 795–805.

Miesfeld, R. and seven others. 1986. Genetic complementation of a glucocorticoid receptor deficiency by a cloned receptor cDNA. *Cell* 46: 389–399.

Milos, P. M. and Zaret, K. S. 1992. A ubiquitous factor is required for C/EBP-related proteins to form stable transcription complexes on an ovalbumin promoter segment in vitro. *Genes Dev.* 6: 183–196.

Myers, R. M., Tilly, K. and Maniatis, T. 1986. Fine structure genetic analysis of a β-globin promoter. *Science* 232: 613–618.

Nadal-Ginard, B. 1978. Commitment, fusion, and biochemical differentiation of a myogenic cell line in the absence of DNA synthesis. *Cell* 15: 855–866.

Nishikura, K., Rushdim, A., Erikson, J., Watt, R., Rovera, G. and Croce, C. M. 1983. Differential expression of the normal and of the translocated human c-*myc* oncogenes in B cells. *Proc. Natl. Acad. Sci. USA* 80: 4822–4286.

Oettinger, M. A., Schatz, D. G., Gorka, C. and Baltimore, D. 1990. *RAG-1* and *RAG-2*, adjacent genes that synergistically activate V(D)J recombination. *Science* 248: 1517 –1522.

O'Kane, C. J. and Gehring, W. J. 1987. Detection *in situ* of genomic regulatory elements in *Drosophila*. *Proc. Natl. Acad. Sci. USA* 84: 9123–9127.

Olson, E. N. 1992. Interplay between proliferation and differentiation within myogenic lineage. *Dev. Biol.* 154: 261–272.

Orkin, S. and Kazazian, H. H. 1984. The mutation and polymorphism of the human β-globin gene and its surrounding DNA. *Annu. Rev. Genet.* 18: 131–171.

Pabo, C. O. and Sauer, R. T. 1992. Transcription factors: Structural families and principles of DNA recognition. *Annu. Rev. Biochem.* 61: 1053–1095.

Palmiter, R. D., Brinster, R. L., Hamm, R. E., Trumbauer, M. E., Rosenfeld, M. G., Birnberg, N. C. and Evan, R. M. 1982. Dramatic growth of mice that develop from eggs microinjected with metallothionein growth hormone fusion genes. *Nature* 300: 611–615.

Parslow, T. G., Blair, D. L., Murphy, W. J. and Granner, D. K. 1984. Structure of the 5′ ends of immunoglobulin genes: A novel conserved sequence. *Proc. Natl. Acad. Sci. USA* 81: 2650–2654.

Patapoutian, A., Miner, J. H., Lyons, G. E. and Wold, B. 1993. Isolated sequences from the linked *Myf-5* and *MRF-4* genes drive distinct patterns of muscle-specific expression in transgenic mice. *Development* 118: 61–69.

Pathak, D. and Sigler, P. B. 1992. Updating structure-function relationships in the bZip family of transcription factors. *Curr. Opin. Struct. Biol.* 2: 116–123.

Payvar, F. and seven others. 1983. Sequence-specific binding of glucocorticoid receptor to MTV DNA at sites within and upstream of the transcribed region. *Cell* 35: 381–392.

Pfahl, M. 1982. Specific binding of the glucocorticoid-receptor complex to the mouse mammary tumor proviral promoter region. *Cell* 31: 475–482.

Picard, D. and Schaffner, W. 1984. A lymphocyte-specific enhancer in the mouse immunoglobulin κ gene. *Nature* 307: 80–82.

Pu, W.T. and Struhl, K. 1991. The leucine zipper symmetrically positions the adjacent regions for specific DNA binding. *Proc. Natl. Acad. Sci. USA* 88: 6901–6905.

Pugh, B. F. and Tjian, R. 1991. Transcription from a TATA-less promoter requires a multisubunit TFIID complex. *Genes Dev.* 5: 1935–1944.

Queen, C. and Baltimore, D. 1983. Immunoglobulin gene transcription is activated by downstream sequence elements. *Cell* 33: 741–748.

Rabbitts, T, H, 1991. Translocations, master genes, and the differences between the origins of acute and chronic leukemias. *Cell* 67: 641–644.

Rhodes, S. J. and seven others. 1993. A tissue-specific enhancer confers Pit-1-dependent morphogen inducibility on the *pit-1* gene. *Genes Dev.* 7: 913–932.

Rigby, P. W. J. 1993. Three in one and one in three: It all depends on TBP. *Cell* 72: 7–10.

Rottman, F. A., Shatkin, A. J. and Perry, R. P. 1974. Sequences containing methylated nucleotides at the 5′ termini of messenger RNAs: Possible applications for processing. *Cell* 3: 197–199.

Rutter, W., Jr., Valenzuela, P., Ball, G. E., Holland, M., Hager, G. L., Degennero, L. J. and Bishop, R. J. 1976. The role of DNA-dependent RNA polymerase in transcriptive specificity. *In* E. M. Bradbury and K. Jaucherian (eds.), *The Organization and Expression of the Eukaryotic Genome*. Academic Press, New York, pp. 279–293.

Samuelsson, L., Strömberg, K., Vikman, K., Bjursell, G. and Enerbäck, S. 1991. The CCAAT/enhancer binding protein and its role in adipocyte differentiation: Evidence for direct involvement in terminal adipocyte development. *EMBO J.* 10: 3787–3793.

Sasai, Y., Kageyama, R., Tagawa, Y., Shigemoto, R. and Nakanishi, S. 1992. Two mammalian helix-loop-helix factors structurally related to *Drosophila hairy* and *Enhancer of split*. *Genes Dev.* 6: 2620–2634.

Sauer, F., Wassarman, D. A., Rubin, G. M., and Tjian, R. 1996. TAF$_{II}$s mediate activation of transcription in the *Drosophila* embryo. *Cell* 87: 1271–1284.

Sauer, F., Fondell, J. D., Ohkuma, Y., Roeder, R. G. and Jäckle, H. 1995. Control of transcription by Krüppel through interactions with TFIIB and TFIIEβ. *Nature* 375: 162–164.

Sawadogo, M. and Roeder, R. G. 1984. Energy requirement for specific transcription by the human RNA polymerase II system. *J. Biol. Chem.* 259: 5321–5326.

Schatz, D. G., Oettinger, M. A. and Baltimore, D. 1989. The V(D)J recombination activating gene, *RAG-1*. *Cell* 59: 1035–1048.

Sen, R. and Baltimore, D. 1986a. Multiple nuclear factors interact with the immunoglobulin enhancer sequences. *Cell* 46: 705–716.

Sen, R. and Baltimore, D. 1986b. Inducibility of κ immunoglobulin enhancer-binding protein NF-κB by a posttranslational mechanism. *Cell* 47: 921–928.

Seto, E. and seven others. 1992. Wild-type p53 binds to TATA-binding protein and represses transcription. *Proc. Natl. Acad. Sci. USA* 89: 12028–12032.

Shatkin, A. J. 1976. Capping of eucaryotic mRNAs. *Cell* 9: 645–653.

Sheiness, D. and Darnell, J. E. 1973. Polyadenylic segment in mRNA becomes shorter with age. *Nat. New Biol.* 241: 265–268.

Simmons, D. M., Voss, J. W., Ingraham, H. A., Holloway, J. M., Broide, R. S., Rosenfeld, M. G. and Swanson, L. W. 1990. Pituitary cell phenotypes involve cell-specific *Pit-1* mRNA translation and synergistic interactions with other classes of transcription factors. *Genes Dev.* 4: 695–711.

Sopta, M., Burton, Z. F. and Greenblatt, J. 1989. Structure and associated DNA helicase activity of a general transcription factor that binds to RNA polymerase II. *Nature* 341: 410–414.

Sornson, M. W. and fourteen others. 1996. Pituitary lineage determination by the *Prophet of Pit-1* homeodomain factor defective in Ames dwarfism. *Nature* 384: 327-333.

Stargell, L. A. and Struhl, K. 1996. Mechanism of transcriptional activation *in vivo*: Two steps forward. *Trends Genet.* 12: 311-315.

Starr, D. B. and Hawley, D. K. 1991. TFIID binds in the minor groove of the TATA box. *Cell* 67: 1231–1240.

Staudt, L. M., Singh, H., Sen, R., Wirth, T., Sharp, P. A. and Baltimore, D. 1986. A lymphoid-specific protein binding to the octamer motif of immunoglobulin genes. *Nature* 323: 640–643.

Stockdale, F. E. and Holter, H. 1961. DNA synthesis and myogenesis. *Exp. Cell Res.* 24: 508–520.

Struhl, G., Struhl, K. and Macdonald, P. M. 1989. The gradient morphogen *bicoid* is a concentration-dependent transcriptional activator. *Cell* 57: 1259–1273.

Tanaka, M., Lai, J.-S. and Herr, W. 1992. Promoter-specific activation domains in *Oct-1* and *Oct-2* direct differential activation of an snRNA and mRNA promoter. *Cell* 68: 755–767.

Thanos, D. and Maniatis, T. 1995. Virus induction of human IFNb gene expression requires the assembly of an enhanceosome. *Cell* 83: 1091–1100.

Treisman, J., Gönczy, P., Vashishta, M., Harris, E. and Desplan, C. 1989. A single amino acid can determine the DNA binding specificity of homeodomain proteins. *Cell* 59: 553–562.

Trudel, M. and Constantini, F. 1987. A 3′ enhancer contributes to the stage-specific expression of the human β-globin gene. *Genes Dev.* 1: 954–961.

Umek, R. M., Friedman, A. D. and McKnight, S. L. 1991. CCAAT-enhancer binding protein: A component of a differentiation switch. *Science* 251: 288–292.

Umesono, K. and Evans, R. M. 1989. Determinants of target gene specificity for steroid/thyroid hormone receptors. *Cell* 57: 1139–1146.

Usheva, A., Maldonado, E., Goldring, A., LuH., Houbavi, C., Reinberg, D. and Aloni, Y. 1992. Specific interaction between the nonphosphorylated form of RNA polymerase II and the TATA-binding protein. *Cell* 69: 871–881.

Verrijzer, C. P., Chen, J.-L. and Yokomori, K. 1995. Binding of TAFs to core elements directs promoter selectivity by RNA polymerase II. *Cell* 81: 1115–1125.

Vinson, C. R., Sigler, P. B. and McKnight, S. L. 1989. Scissors-grip model for DNA recognition by a family of leucine zipper proteins. *Science* 246: 911–916.

Walker, M. D., Edlund, T., Boulet, A. M. and Rutter, W. J. 1983. Cell-specific expression controlled by the 5′ flanking region of the insulin and chymotrypsin genes. *Nature* 306: 557–561.

Wasylyk, B., Kedinger, C., Corden, J., Brison, D. and Chambon, P. 1980. Specific in vitro initiation of transcription on conalbumin and ovalbumin genes and comparison with adenovirus 2 early and late genes. *Nature* 285: 367–373.

Watson, J. D., Gilman, M., Witkowski, J. and Zoller, M. 1992. *Recombinant DNA*, 2nd Ed. Scientific American Books, New York.

Werner, M. H., Huth, J. R., Gronenborn, A. M. and Clore, G. M. 1995. Molecular basis of human 46 X,Y sex reversal revealed from the three dimensional solution structure of the human SRY-DNA complex. *Cell* 81: 705–714.

Wirth, T., Staudt, L. and Baltimore, D. 1987. An octamer oligonucleotide upstream of a TATA motif is sufficient for lymphoid-specific promoter activity. *Nature* 329: 174–178.

Workman, J. L. and Roeder, R, G. 1987. Binding of transcription factor TFIID to the major late promoter during in vitro nucleosome assembly potentiates subsequent initiation by RNA polymerase II. *Cell* 51: 613–622.

Wright, G. and eight others. 1991. High level expression of active human α1-antitrypsin in the milk of transgenic sheep. *BioTech.* 9: 830–834.

Yeo, S. P. and Rigby, P. W. J. 1993. The regulation of *myogenin* gene expression during embryonic development of the mouse. *Genes Dev.* 7: 1277–1287.

Transcriptional regulation of gene expression: The activation of chromatin

11

While my companion contemplated with a serious and satisfied spirit the magnificent appearances of things, I delighted in investigating their causes....Curiosity, earnest research to learn the hidden laws of nature, gladness akin to rapture, as they unfolded to me, are among the earliest sensations I can remember.
MARY WOLLSTONECRAFT SHELLEY (1817)

Hence, we cannot categorically deny that perhaps we may be able to grind genes in a mortar and cook them in a beaker after all.
H. J. MULLER (1922)

U P TO THIS POINT, we have limited our discussion of messenger RNA transcription to the structure of the gene itself. But genes do not exist in an uncovered state within the nucleus, readily accessible to RNA polymerase or to any enhancer- or promoter-binding protein. Rather, eukaryotic chromosomes contain as much protein (by weight) as nucleic acid, and this DNA-protein complex is called **chromatin.** The most abundant of the chromatin proteins are basic polypeptides called **histones,** which are organized into **nucleosomes.**

In addition to nucleosomes, which are general inhibitors of transcription, there are higher-order chromatin elements that also may be important in regulating gene expression. There are **locus control regions** (**LCRs**) which regulate the expression of a region of the chromosome; there are **matrix-associated regions** (**MARs**) where the DNA is anchored to the nuclear matrix and where DNA-unwinding proteins may be active; and there are **insulators,** sequences that separate regulatory "domains" and thereby prevent positive and negative regulatory elements in one domain from acting on genes in the adjacent domain.

Nucleosomes and the activation of repressed chromatin

The nucleosome is the basic unit of chromatin structure. It is composed of a histone octamer (two molecules each of histones H2A-H2B and histones H3-H4) wrapped about with two loops of approximately 140 base pairs of DNA (Figure 11.1; Kornberg and Thomas, 1974). Chromatin can thus be visualized as a string of nucleosome beads linked by from 10 to 100 base pairs of DNA. While classical geneticists have likened genes to "beads on a string," molecular geneticists liken genes to "string on the beads."

Transcription factors must be able to find the DNA sequences despite the fact that most of the DNA is packaged in nucleosomes. It is currently thought that rendering a gene competent to transcribe RNA involves (1) the binding of transcription factors to the DNA and (2) the exclusion of nucleosomes from the promoter region of the gene. The interactions between specific transcription factors and the DNA they bind cause the phenomenon of tissue-specific and temporally specific gene transcription. Once the RNA

Figure 11.1
Nucleosome and chromatin structure. (A) Model of nucleosome structure as seen by X-ray crystallography at a resolution of 0.33 nm. The central biconcave protein unit (white) is the H3-H4 tetramer. Each of the two dark ovoids flanking the tetramer is an H2A-H2B dimer. (B) Relationship of histone H1 to the core nucleosome (containing two copies each of histones H2A, H2B, H3, and H4). (C) H1 can link the DNA into compact forms and can draw nucleosomes together. About 140 base pairs of DNA encircle the histone octamer, and about 60 base pairs of DNA link the nucleosomes together. (D) Model for the arrangement of nucleosomes into the highly compacted, solenoidal chromatin structure. An alternative model, placing H1 between the nucleosome Octamer and one coil of DNA, has recently been proposed and is currently being tested (Pruss et al., 1996). Here, one end of H1 binds to the necleosome, while the other binds the DNA. (A from Burlingame et al., 1985; B–D after Wolfe, 1993.)

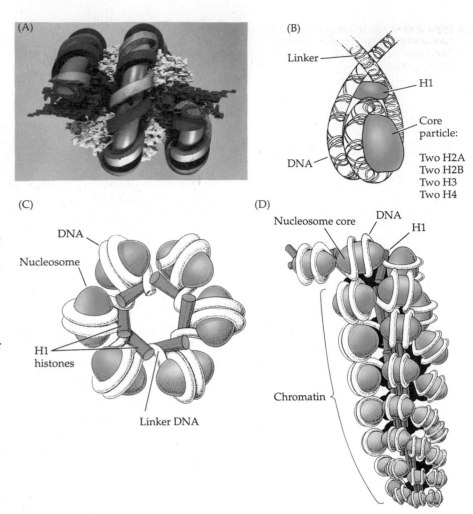

polymerase begins transcribing, it can temporarily displace DNA from the nucleosome and synthesize RNA (Clark and Felsenfeld, 1992; see Lewin, 1994). [chrom1.html]

Nucleosomes aren't the only structural impediment to the binding of transcription factors to their DNA sequences, because the nucleosomes are themselves wound into tight "solenoids" that are stabilized by histone H1. Histone H1 is found in the 60 or so base pairs of "linker" DNA between the nucleosomes (Figures 11.1 and 11.2; Weintraub, 1984). This H1-dependent conformation of nucleosomes inhibits the transcription of genes in somatic cells by packing adjacent nucleosomes together into tight arrays that prohibit the access of transcription factors and RNA polymerases (Thoma et al., 1979; Schlissel and Brown, 1984).

Accessibility to trans-*Regulatory Factors*

It is incredible that DNA can become accessible to *trans*-regulatory factors at all. There is enough DNA in a single human body to extend the diameter of the solar system (Crick, 1966), and this enormous length must be packed up tightly into the nuclei of our cells. Yet our genetic library can be accessed specifically in each cell type. Solution hybridization experiments suggest that there is a minimum of 10,000 tissue-specific genes in the genome of most vertebrates; so it is not surprising that in any given cell type, most of these genes are repressed. It is generally thought, then, that the "default" condition of chromatin is a repressed state and that tissue-specific genes be-

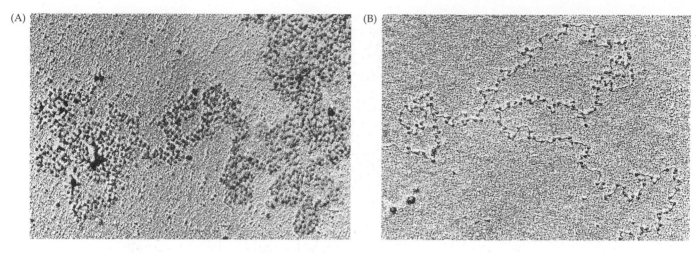

(A)   (B)

Figure 11.2
The role of H1 in compaction of chromatin. (A) Chicken liver chromatin observed in the electron microscope. The beads represent the nucleosomes. (B) The same chromatin after the removal of histone H1 by salt elution. The chromatin has become far less compact. (From Oudet et al., 1975; photographs courtesy of P. Chambon.)

come activated by locally interrupting the repressive factors (Weintraub, 1985). As already mentioned, the major mechanism of general gene repression is probably the compaction of DNA into clusters of nucleosomes, and the initiation of transcription depends on removing nucleosomes from the promoter region of the gene. There are two main ways that this might be accomplished. First, during DNA synthesis (S phase in the cell cycle), nucleosomes are removed from a strand of the DNA and are replaced shortly thereafter. At this time of replacement, there could be competition for the promoter sites between histones and transcription factors such as TATA-binding TFIID. Second, there appear to be certain transcriptional activators (such as the glucocorticoid receptor) that can bind to existing nucleosomes and disrupt them (Rigaud et al., 1991; Adams and Workman, 1993). Once the nucleosomes are dissociated in the promoter region, other transcription factors can bind (Figure 11.3).

The ability of transcription factors to remove nucleosomes from active genes and their promoters can be seen by nuclease experiments. The accessibility of a gene to nuclear proteins can be detected by treating the chromatin of a tissue with small amounts of **DNase I.** This pancreatic DNase digests accessible regions of DNA, but the DNA covered by nucleosomes is protected. After digestion, the DNA of the treated chromatin is extracted and mixed with radioactive cDNA for a particular gene (Figure 11.4). If the cDNA finds sequences to bind to, then the gene has been protected from digestion by chromatin proteins—that is, it was not accessible to the DNase, and it probably would not be accessible to transcription factors or RNA polymerase either. However, if the cDNA probe does not find sequences to bind, then the gene has been exposed to the DNase and probably would be accessible to RNA polymerase and *trans*-regulatory factors.

The DNase I sensitivity of a given gene was found to be dependent on the cell type in which it resides (Table 11.1; Weintraub and Groudine, 1976). When chromatin from developing chick red blood cells was treated with DNase I and its DNA extracted and mixed with radioactive globin cDNA, the globin cDNA found little with which to bind. The globin genes in the chromatin had been digested by the small amounts of DNase I. However, treating brain cell chromatin with the same amounts of DNase I did not destroy the globin genes. Therefore, the globin gene was accessible to outside enzymes in developing red blood cell chromatin but not in brain cell chromatin. Similarly, the ovalbumin (egg white) gene is susceptible to DNase I digestion in oviduct chromatin but not in red blood cell chromatin. When

Figure 11.3
Binding of a transcription factor (TF) to the nucleosome can destabilize the nucleosome, enable the histones to be removed, and open the region for other transcription factors. (After Adams and Workman, 1993.)

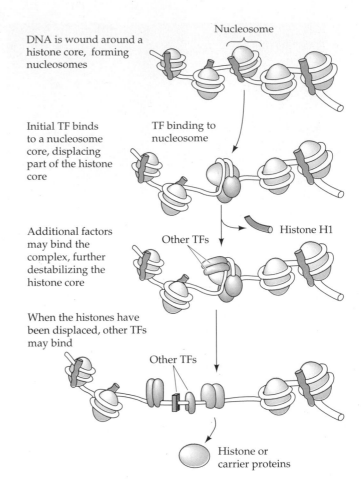

the chromatin is treated with DNase I and the DNA is extracted, ovalbumin cDNA is able to find sequences in the erythrocyte preparation but not in the DNA from the treated oviduct chromatin. Here, then, we have clear correlation of differential gene regulation and chromatin structure.

DNase I-Hypersensitive Sites

There are some regions of the chromatin referred to as **DNase I-hypersensitive sites**. These sites, identified on Southern blots by small radioactive DNA fragments, are destroyed by extremely minute amounts of DNase, in-

Table 11.1 Binding studies with DNase I-treated chromatin

Source (chick)	DNase treatment	Radioactive cDNA probe	Percent maximum binding of radioactive cDNA to DNA extracted from treated chromatin
Red blood cell DNA	–	Globin cDNA	94
Brain cell chromatin	+	Globin cDNA	90–100
Fibroblast chromatin	+	Globin cDNA	90–100
Red blood cell chromatin	+	Globin cDNA	25
Red blood cell chromatin	+	Ovalbumin cDNA	90–100

Source: After Weintraub and Groudine, 1976.

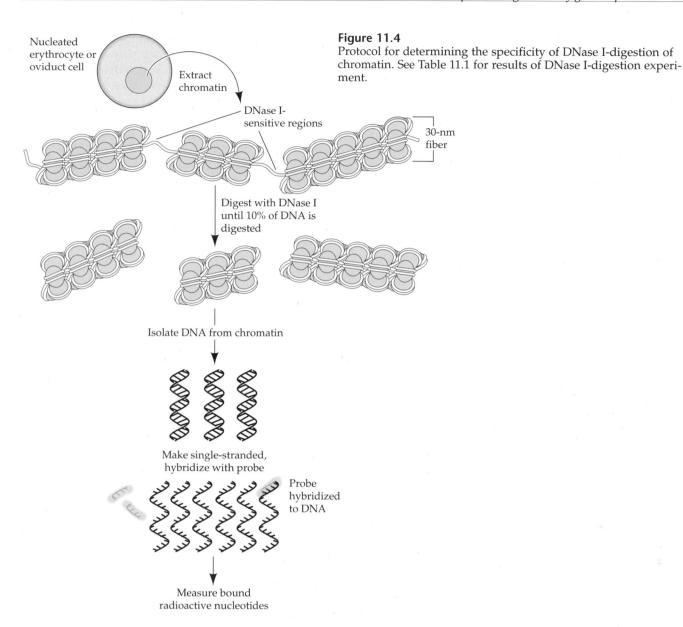

Figure 11.4
Protocol for determining the specificity of DNase I-digestion of chromatin. See Table 11.1 for results of DNase I-digestion experiment.

dicating that they are extremely accessible to outside molecules. This accessibility appears to arise from the nearly total absence of nucleosomes in this region of DNA in the tissues that express them (Elgin, 1988). The DNase I-hypersensitive sites mark regions of the chromatin, such as active promoters and enhancers, where DNA-binding proteins are bound. DNase I-hypersensitive regions are therefore associated with tissue-specific developmentally regulated genes (Elgin, 1981; Conklin and Groudine, 1984). For example, globin genes in red blood cells and their immediate precursors contain DNase I-hypersensitive sites, but globin genes in other cells do not (Stalder et al., 1980; Groudine et al., 1983). The 5′ flanking region of the chick vitellogenin gene contains several hypersensitive sites in the chromatin from the liver of laying hens; but these sites are not present in chromatin from male liver, embryonic liver, brain, or lymphocyte (Burch and Weintraub, 1983).

DNase I-hypersensitive sites often reside within or adjacent to sites that have enhancer functions, and certain *trans*-regulatory factors are able to induce the formation of these hypersensitive sites. Zaret and Yamamoto (1984), studying the glucocorticoid-responsive enhancer of the mouse mammary

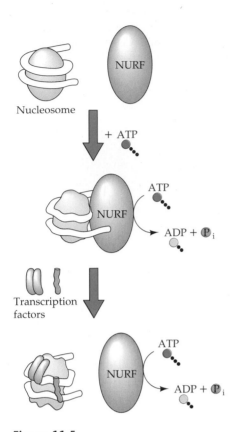

Figure 11.5
Model for the proposed mechanism of NURF on a nucleosome. NURF can hydrolyze ATP and utilize the energy to reconfigure histone-DNA (or histone-histone) interactions. These perturbations are thought to increase the accessibility of transcription factors to the nucleosomal DNA. This can lead to further changes in nucleosome structure. (After Tsukiyama and Wu, 1995.)

tumor virus, demonstrated that before the addition of the hormone to cells containing the virus, this enhancer sequence showed no special DNase I sensitivity. After the hormone was administered, a discrete DNase I-hypersensitive site developed in this region. The formation of the hypersensitive site coincided with the initiation of viral gene transcription; when the hormone was withdrawn, both the hypersensitive site and viral gene transcription disappeared. Zaret and Yamamoto speculated that the interaction between the glucocorticoid receptor complex and the enhancer DNA alters the chromatin configuration to facilitate transcription from the nearby promoter. This would be the case if, as already mentioned, the glucocorticoid receptor could remove nucleosomes from the region of its DNA-binding sequence.

Nucleosome Disruption and Reorganization: The Role of Disruptional Complexes

How can nucleosomes be removed? Recent studies have identified two factors that may be important in this process. The GAGA transcription factor (a constitutively expressed protein in *Drosophila* that is bound on numerous promoters having GA sequences) is one of those transcription factors that can disrupt nucleosomes. When it binds to a nucleosome containing the TATA box of the *hsp70* gene (encoding the 70-kDa heat shock protein), it disrupts the nucleosome and creates a DNase I-hypersensitive site at the TATA box site. It does this most efficiently when the nucleosomes lack histone H1. The GAGA factor is unable to do this alone, but works in concert with a four-peptide protein called the **nucleosome-remodeling factor** (**NURF**). In the absence of transcription factors, NURF can perturb the nucleosome in an ATP-dependent manner. This permits transcription factors such as GAGA to bind to the promoter regions and disrupt the nucleosome further (Figure 11.5; Tsukiyama et al., 1994; Tsukiyama and Wu, 1995). Another protein complex capable of disrupting nucleosomes, the **SW1/SNF complex,** was originally discovered in yeast, but has since been found in *Drosophila* and humans (Peterson and Tamkun, 1995). When the TATA box is incorporated into the nucleosomal DNA, it is not accessible to the TATA-binding protein, and transcription is severely reduced. This inhibition can be abrogated by ATP-dependent modifications of the nucleosome effected by SW1/SNF (Imbalzano et al., 1994). Similarly, SW1/SNF can disrupt the nuclesosomes in the enhancer regions and allow transcription factors to bind (Kwon et al., 1994; Pazin et al., 1994). It appears that the SWI/SNF complex is actually part of the RNA polymerase and is attached to its carboxy-terminal domain (Wilson et al., 1996). This complex may be activated by transcription factors capable of disrupting nucleosomes.

One of the major ways to disrupt nucleosomes is through acetylation. There is a good correlation between **histone acetylation** and the transcriptional activity of a particular region of chromatin. Extremely active transcriptional regions have nuclesosomes that are highly acetylated, while transcriptionally repressed domains have hypoacetylated histones on their nucleosomes (Braunstein et al., 1993; Jeppesen and Turner, 1993; Hebbes et al., 1994). When acetyl groups are placed on the lysines of the histone tails, the entire nucleosome structure changes (Figure 11.6; Lee et al., 1993; Garcia-Ramirez et al., 1995). The tails are moved outward, so that they no longer contact the double helix, thereby severely loosening their hold on the DNA. The DNA becomes much more accessible to transcription factors. A histone acetyltransferase that acetylates nucleosomal histones has been identified in *Tetrahymena*, and it is a homologue of a yeast transcriptional activator (Brownell et al., 1996). Moreover, the 250-kDa TAF subunit of TFIID has recently been found to have the ability to acetylate histones H3 and H4 (Mizzen et al., 1996). This enzymatic activity may play an important role in allowing TFIID to replace nucleosomes.

Nucleosome Disruption and Reorganization: The Role of Histone Competition

Competition between histones and transcription factors was first suggested for the regulation of the 5S rRNA genes of *Xenopus*. The transcription factor TFIIIA was seen to compete with histone H1 for the sites regulating 5S rRNA synthesis. If this gene were incubated with TFIIIA before histone H1, even in the presence of the core histones, the transcriptional complex would form. If H1 were present before TFIIIA, transcription was blocked (Schlissel and Brown, 1984). Prioleau and colleagues (1994) have related competition between histones and the TATA-binding protein to the occurrence of the midblastula transition. Genes activated at the midblastula transition are repressed during early cleavage. When such genes are injected into *Xenopus* nuclei at fertilization or early-cleavage stages, they become wrapped up in chromatin and are repressed. After the midblastula transition, the injected genes become transcribed. This repression during early cleavage can be alleviated by pre-incubating the injected genes with TATA-binding protein. In some systems, then, it is possible that competition between transcription factors and histones can regulate gene expression. The degree of DNA methylation (to be discussed shortly) might be critical to this competition, since histone H1 binds more avidly to methylated DNA than to unmethylated DNA (McArthur and Thomas, 1996).

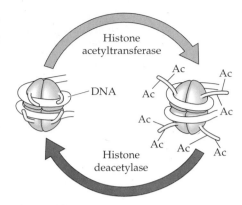

Figure 11.6
Histone acetyltransferase can modify histone tails and change their conformation with the nucleosomal DNA. This allows the loosening of the DNA from the nucleosome core. (After Lee et al., 1993.)

Locus control regions: Globin gene transcription

Locus control regions (LCRs) are sequences of DNA that are essential for establishing an "open" chromatin configuration. That is, these regions are able to inhibit the normal repression of transcription over a relatively large area containing several genes. One of the best-studied LCRs is that regulating the tissue-specific expression of the β-globin family of genes in humans, mice, and chicks.

In many species, including chicks and humans, the embryonic or fetal hemoglobin differs from that found in adult red blood cells. A schematic diagram of human hemoglobin types and the genes that code for them is shown in Figure 11.7. Human **embryonic hemoglobin** consists largely of two zeta (ζ) globin chains, two epsilon (ε) globin chains, and four molecules of heme. During the second month of human gestation, ζ- and ε-globin synthesis abruptly ceases, while alpha (α) and gamma (γ) globin synthesis increases (Figure 11.8). The association of two γ-globin chains with two α-globin chains produces **fetal hemoglobin** ($\alpha_2\gamma_2$). At 3 months gestation, the beta (β) globin and delta (δ) globin genes begin to be active, and their products slowly increase while γ-globin levels gradually decline. This switchover is greatly accelerated after birth, and fetal hemoglobin is replaced by **adult hemoglobin:** $\alpha_2\beta_2$. The normal adult hemoglobin profile is 97 percent $\alpha_2\beta_2$, 2–3 percent $\alpha_2\delta_2$, and 1 percent $\alpha_2\gamma_2$. In humans, the ζ- and α-globin genes are on chromosome 16, but the ε-, γ-, δ-, and β-globin genes are linked together, in order of appearance, on chromosome 11. It appears, then, that there is a mechanism that directs the sequential switching of the chromosome 11 genes from embryonic, to fetal, to adult globins.

In addition to the DNase-hypersensitive sites in the promoters and enhancers near and within each globin gene, there is a locus control region located far upstream from the most 5′ member (ε) of the β-globin gene com-

Figure 11.7
Sequential gene activation in hemoglobin synthesis during development.

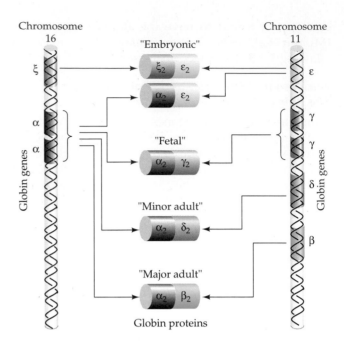

plex. This LCR contains four sites that are DNase I-hypersensitive only in erythroid precursor cells and that appear to be necessary for activating high levels of erythroid cell-specific transcription from the entire β-globin gene family (ε-, γ-, β-, and δ-globins) on human chromosome 11 (Grosveld et al., 1987). Deletion or mutation of the LCR causes the silencing of all these genes. Conversely, if the LCR is placed adjacent to genes that are not usually expressed in red blood cells (such as the T-cell-specific *thy-1* gene) and then transfected into erythroid precursor cells, these new genes are expressed in red blood cells. This effect is specific for red blood cell precursors, since only they would have the appropriate *trans*-regulatory factors to bind to this region (Blom van Assendelft et al., 1989; Fiering et al., 1993).

The LCR is responsible for permitting gene expression in an entire region. Moreover, if the globin genes remain attached to the LCR, they can be expressed in erythroid cells no matter where they reside in the genome. If they are separated from the LCR, the globin genes are repressed, even in the erythroid cells that would be transcribing the globin genes. Ryan and co-

Figure 11.8
Percentages of hemoglobin polypeptide chains as a function of human development. The physiological importance of the γ-globin chain in fetal hemoglobin was examined in Chapter 9. (After Karlsson and Nienhaus, 1985.)

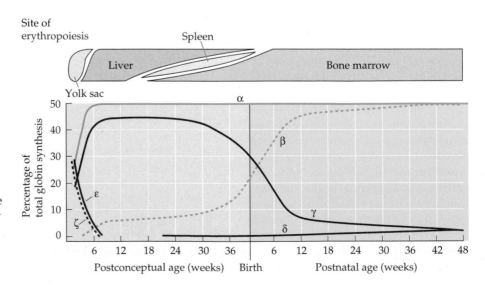

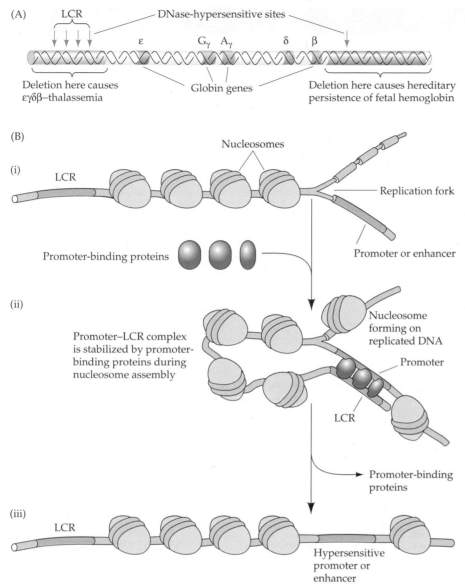

(A)

LCR — DNase-hypersensitive sites

ε   G_γ A_γ   δ   β

Deletion here causes
εγδβ–thalassemia

Globin genes

Deletion here causes hereditary
persistence of fetal hemoglobin

(B)

(i)

LCR

Nucleosomes

Replication fork

Promoter-binding proteins

Promoter or enhancer

(ii)

Promoter–LCR complex
is stabilized by promoter-
binding proteins during
nucleosome assembly

Nucleosome
forming on
replicated DNA

Promoter

LCR

Promoter-binding
proteins

(iii)

LCR

Hypersensitive
promoter or
enhancer

Figure 11.9
Diagram of the human β-globin family of genes on chromosome 11. (A) The erythroid-specific LCR region is located 6 to 22 kilobases upstream of the ε-globin gene. The four DNase I-hypersensitive sites within this region are marked by arrows. A fifth DNase I-hypersensitive site downstream from the β-globin gene is also marked, and a deletion of this region causes the persistence of γ-globin gene transcription. Downstream from the ε-globin (embryonic) gene are two nearly identical γ-globin (fetal) genes. These are followed by the adult δ- and β-globin genes. (B) One possible model for LCR activity. Transcription factors binding to the globin promoters are stabilized at the replication fork by binding to the LCR. The complex would not be dissociated, and the regions associated with the LCR would remain devoid of nucleosomes. (A after Ryan et al., 1989; B after Felsenfeld, 1992.)

workers (1989) constructed transgenic mice containing the human β-globin gene and its immediate promoters and enhancers. These transgenic mice made only small amounts of human β-globin (less than 0.3 percent of the total cell β-globin). However, when the researchers added the LCR, human β-globin accounted for over half the total globin in these mice. This result explains medical observations that patients who lacked this region had deficiencies of ε, γ-, δ-, and β-globins, even though their genes for these proteins were intact and the globin genes on the other chromosome functioned normally (Tuan et al., 1987).

The locus control region is crammed with binding sites for *trans*-regulatory factors. As Gary Felsenfeld (1992) observed, "The domains look as though they were put together by an overenthusiastic student determined to construct a powerful *cis*-acting element." He suggests that one of the functions of the LCR is to loop around to one of the promoter regions during DNA replication and bind to it in a manner that prevents nucleosomes from forming on that globin promoter (Figure 11.9). Indeed, the globin promoters are not DNase I-hypersensitive except in the presence of the LCR.

Globin Gene Switching

ALTHOUGH IT IS OBVIOUS that globin gene transcription switches from embryonic to fetal to adult isoforms during development, we do not know the mechanism by which this takes place. Recent models of globin switching have focused on competition and cooperation between enhancers, promoters, and the locus control region. [chrom2.html]

Human Globin Gene Regulation

The gene expression switching system for human globins is complicated. There are several cis-regulatory elements for β-globin. We already discussed the β-globin promoter region and the locus control region that keeps the entire region of the chromosome ready to be transcribed. In addition, there is a 3′ enhancer (upstream from the β-globin gene) that regulates the temporal expression of the β-globin gene, and there is another, intragenic enhancer that helps regulate the tissue specificity of β-globin gene expression. This latter enhancer is actually located *within* the third exon of the β-globin gene itself (Behringer et al., 1987; Trudel and Constantini, 1987). As shown

in Figure 11.10, these *cis*-regulatory regions have numerous sites for ubiquitous and erythroid-specific transcription factors. One of the most important of the erythroid-specific factors is a zinc finger protein, **GATA-1** (Orkin, 1992). This factor binds to GATA sequences, which are found throughout the LCR as well as in the promoters and enhancers of numerous genes that are expressed in red blood cells (including some of the genes for globin, heme synthesis, and the erythropoietin receptor). Gene-targeting experiments show that mice without the gene for GATA-1 produce no erythroid cell lineage (Pevny et al., 1991). A second critical transcription factor appears to be **NF-E2**. This erythroid-specific bZIP transcription factor binds to areas of the LCR and may mediate the communication between the LCR and the promoter regions (perhaps by binding to GATA-1) (Talbot and Grosveld, 1991; Gong and Dean, 1993). GATA-1 and NF-E2 are also required for the formation of one of the DNase I-hypersensitive sites in the LCR (Stamatoyannopoulos et al., 1995).

Hereditary Persistence of Fetal Hemoglobin

Whereas most persons switch from fetal to adult globins around the time of birth, there are some people who do not. These individuals retain transcription from their γ-globin gene and are said to have

hereditary persistence of fetal hemoglobin (HPFH). It does them no harm.* The mutations that give rise to HPFH cluster in the *cis*-regulatory regions. Deletion mutations that remove the β-globin enhancer or promoter regions suffice to elevate γ-globin levels in adult cells. Point mutations in the promoters for either of the γ-globin genes can also cause HPFH (Martin et al., 1989). One of these mutations creates a new binding site for GATA-1, whereas another creates a strong binding site for the ubiquitous factor Sp1 (Ottolenghi, 1992). Thus, there appears to be competition between the promoters of the γ-globin genes and the promoters of the β-globin genes (Enver et al., 1990). This competition is influenced by the presence of enhancing and

*Individuals with HPFH are phenotypically normal and are identified through the screening of populations for other globin abnormalities (such as thalassemia and sickle-cell anemia). Researchers would love to know how to safely reactivate the γ-globin gene in people suffering from β-thalassemia, sickle-cell anemia, and other diseases of β-globin. If the γ-globin gene were reactivated to even a small extent, it could alleviate many of the symptoms of these diseases. Recent studies suggest that administration of butyrate or the combination of intravenous hydroxyurea and erythropoietin may be efficacious in elevating fetal hemoglobin in newly generated red blood cells (Perrine et al., 1993; Rodgers et al., 1993). Studies are under way to evaluate these procedures.

Figure 11.10 Schematic representation of the human β-globin gene and its regulatory regions. The shaded shapes represent different transcription factors, and the double arrows indicate that more than one factor can bind at that site. (After Ottolenghi, 1992.)

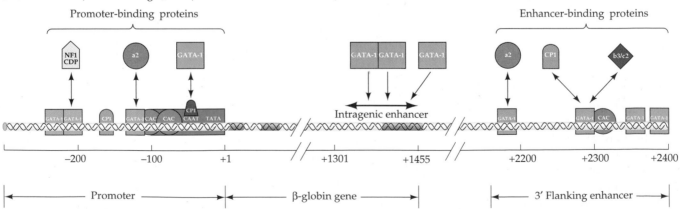

silencing *trans*-regulatory factors. There are also point mutations that cause HPFH by preventing the binding of a negative regulator to the γ-globin promoter in adult cells (Berry et al., 1992). Bacon and colleagues (1995) have evidence that the ratio between GATA-1 and Sp1 transcription factors changes over time and that the type of globin may depend on the relative concentrations of these factors.

The LCR and Globin Switching in Humans

The major competition may not be for the activation by the 3′ enhancer (which may function locally), but for the LCR. While some investigators think that the major effect of the LCR is to maintain the globin loci-containing chromatin in a transcriptionally permissive conformation (see Martin et al., 1996), other researchers envision specific interactions between the different globin gene promoters and the regions of the LCR. Interestingly, the *distance* between the LCR and the globin genes affects their activation (Hanscombe et al., 1991). When linked closely to the LCR, the human β-globin gene becomes expressed in transgenic mouse embryonic cells. Its correct activation (in adult cells only) is restored only when it is placed farther away from the LCR. Similarly, the human γ-globin gene is repressed earlier (like the normal β-globin gene) when it is further separated from the LCR. This suggests that the interaction between the LCR and the globin genes is polarized (see Figure 11.10; Hanscombe et al., 1991): those globin genes closest to the LCR are turned on earliest, while those genes more distal are turned on later. Presumably, there is physical contact between the LCR and the gene-specific promoters and enhancers.

The mechanism by which distance from the LCR might regulate the activation of different promoters at different times remains to be explained. One model (Figure 11.11; Ellis et al., 1996) has recently been proposed on the basis of transgenic mice containing pieces of the LCR in their genome. In this model, the third DNase hypersensitive site is opened up by a *trans*-acting transcrip-

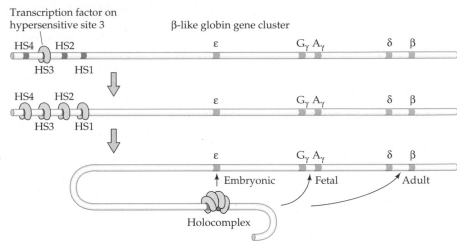

Figure 11.11 Proposed mechanism for the activation of the β-globin family by the LCR. (A) The hypersensitive site 3 of the LCR is activated by a *trans*-acting factor. (B) Once site 3 is open, the other hypersensitive sites open and bind to their transcription factors. Interactions between these proteins form a "holocomplex" of DNA and protein. This can loop over to interact with the promoters of the globin genes. Competition for this interaction by the presence of different concentrations of transcription factors would allow the differential and sequential activation of these genes during erythrocyte development. (After Ellis et al., 1996.)

tion factor. Once this site is opened, the other three tissue-specific sites also open. Protein-protein interactions between these sites bind the sites together to form a "holocomplex" that spreads the alteration in chromatin structure throughout the β-globin region. The holocomplex would then loop out to interact with each of the globin genes and interact sequentially with the promoter regions of each gene. The transcription factors involved in the HPFH syndromes might be those that mediate the interactions between the promoters and the hypersensitive sites of the LCR. Moreover, these sites are not interchangeable (i.e., site 4 cannot substitute for site 3), so each of the hypersensitive sites may play a different role in these interactions (Bungert et al., 1995).

In these models, there is competition between the promoters for the LCR. A newly discovered transcription factor, EKLF (erythroid Krüppel-like factor) may be crucial in regulating this competition by stabilizing the interactions between the LCR and the β-globin promoter. In mice lacking EKLF, but having a functional human β-globin gene system (including the human LCR), the ε- and γ-globins are made normally. How-

ever the switch to β-globin is not made. This suggests that the LCR holocomplex would continue to interact with the γ-globin gene promoters unless stabilized to the β-globin gene promoter by EKLF (Wijgerde et al., 1996). Conversely, transcription factors such as GATA1 and YY1 may intefere with the LCR's binding to one promoter while they enhance the binding of the LCR to another promoter (Raich et al., 1996; Wandersee et al., 1996). The interactions between the LCR sites and the promoters and how they might be regulated by different ratios or types of transcription factors still must be elucidated, but these interactions between the enhancers, promoters, and the LCR should provide a fascinating story of differential gene expression in human cells.* ■

*If you think things are getting complicated, you're right. Harold Weintraub, who was one of the leading figures of chromatin research, noted: "A complex enhancer can contain 10 binding sites, an LCR perhaps as many, a transcription complex may contain 15 proteins, and RNA polymerase maybe 12. What a mess!" (H. Weintraub, personal communication). We don't know why there are so many different factors regulating the transcription of these genes.

DNA methylation and gene activity

It is often assumed that a gene contains exactly the same nucleotides whether it is active or inactive. A β-globin gene in a red blood cell precursor should have the same nucleotides as the β-globin gene in a fibroblast or retinal cell of the same animal. There is, however, a subtle difference in the DNA. In 1948, R. D. Hotchkiss discovered a "fifth base" in DNA, **5-methylcytosine.** In some eukaryotes, this base is made enzymatically after DNA is replicated, and about 5 percent of the cytosines in mammalian DNA are converted to 5-methylcytosine. This conversion can only occur when the cytosine residue is followed by a guanosine (CpG). Recent studies have shown that the degree to which the cytosines of a gene are methylated may also control the gene's transcription. In other words, DNA methylation might change the structure of the gene and, in so doing, regulate its activity. Cytosine methylation appears to be a major mechanism of transcriptional regulation in vertebrates. However, *Drosophila*, nematodes, and perhaps most invertebrates do not methylate their DNA.

There are three areas in which DNA methylation is thought to contribute to differential gene activity. First, the methylation of promoter sequences contributes to the temporal and spatial regulation of genes encoding tissue-specific proteins. Second, DNA methylation is thought to be responsible for distinguishing between certain egg-derived and sperm-derived genes in mammals, thereby allowing only one of them to be expressed during early development. Third, DNA methylation is thought to be responsible for the continued repression of the genes on one of the two X chromosomes in each female mammalian cell.

Correlations Between Promoter Methylation and Gene Inactivity

The first evidence that DNA methylation helps regulate gene activity comes from studies showing a correlation between gene activity and low amounts of cytosine methylation (hypomethylation), especially in the promoter region of the gene. In developing human and chick red blood cells, the DNA involved in globin synthesis is completely (or nearly completely) unmethylated, whereas the same genes are highly methylated in cells that do not produce globin (Figure 11.12). Fetal liver cells that produce hemoglobin in early development have unmethylated genes for fetal hemoglobin. These genes become methylated in the adult tissue (van der Ploeg and Flavell, 1980; Groudine and Weintraub, 1981; Mavilio et al., 1983).

Organ-specific methylation patterns are also seen in the chick ovalbumin gene; the gene is unmethylated in the oviduct cells but is methylated in other chick tissues (Mandel and Chambon, 1979). Demethylation accompanies class switching in immunoglobulin synthesis (Rogers and Wall, 1981) and correlates with the ability of murine lymphocytes to produce the metal-binding protein metallothionein I (Compere and Palmiter, 1981). In mouse somites, demethylation of a *MyoD* enhancer precedes *MyoD* transcription and is essential for specifying these cells to be muscle precursors (Brunk et al., 1996). Thus, the absence of DNA methylation correlates well with the tissue-specific expression of certain genes.

The second type of evidence for DNA methylation as a regulatory process comes from experiments in which the expression of cloned genes is altered by introducing or removing methyl groups from their cytosine residues. When Busslinger and co-workers (1983) added cloned globin genes to cells (by co-precipitation with calcium phosphate), the cells took up the DNA and, in many cases, incorporated the DNA into the nucleus. In such cases, the cloned globin genes were transcribed. By protecting certain re-

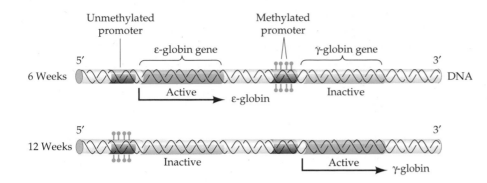

Figure 11.12
Methylation of globin genes in human embryonic blood cells. The activity of the globin genes correlates inversely with the methylation of their promoters. (After Mavilio et al., 1983.)

gions of the cloned globin genes from methylation before adding them to the cells, it is possible to create clones in which the globin genes have identical sequences but different methylation patterns. A completely unmethylated gene is transcribed, whereas a completely methylated gene (methyl groups on every appropriate C residue) is not transcribed. Using partially methylated clones, Busslinger and co-workers showed that methylation in the 5′ region of the globin gene (nucleotides –760 to +100) prevents transcription. Thus, methylation in the 5′ end of a gene appears to play a direct role in the regulation of gene expression. In general, methylation of the promoter region inhibits the transcription of genes.

Methylation and the Maintenance of Transcription Patterns

Methylation differences may be responsible for *maintaining* (as opposed to *initiating*) a pattern of transcriptional activity through several cell generations (Holliday, 1987). During replication, each DNA strand serves as a template for its complementary strand. In regions of methylation, the methyl groups are usually on both strands of the double helix, since a CpG on one side is mirrored by an antiparallel CpG on the other. If the C on one strand is methylated, so is the C on the other strand (Figure 11.13). During replication, one strand of the DNA (the template strand) would have the methylation pattern, while the newly synthesized strand would not. However, the enzyme DNA (cytosine-5)-methyltransferase has a strong preference for DNA that has one methylated strand, and when it sees a methyl-CpG on one side of the DNA, it methylates the new C on the other side (Gruenbaum et al., 1982; Bestor and Ingram, 1983).

It is doubtful that changes in methylation actually *initiate* changes in gene activity, since the DNA methyltransferase has no inherent sequence specificity (aside from a general bias toward CpG-rich areas). Since the methylation pattern should be inherited after each cell division, something else must recognize the genes of differentiated cells in their methylated state and subsequently demethylate them. This has been shown by transfecting a methylated α-actin gene into cultured myoblast cells (which normally transcribe that gene). When transfected into myoblasts, this gene was demethylated and transcribed. However, if transferred into other cell types, this mus-

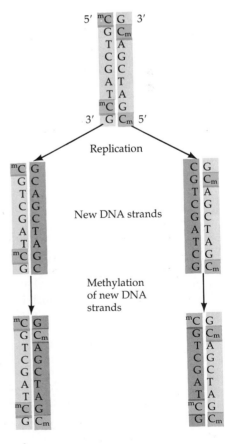

Figure 11.13
Model for the propagation of methylation patterns. When DNA replicates, only one of the two strands (the "old" strand) retains the original methylation pattern. The other strand (the "new" strand) is unmethylated. A CpG-specific methylating enzyme would be able to bind to CpG pairs where a C residue was methylated and then methylate the C residue on the complementary strand. (After Browder, 1984.)

cle-specific gene remained methylated. New DNA synthesis was not required, but certain *cis*-DNA sequences were necessary for the muscle-specific demethylation to occur (Yisraeli et al., 1986; Paroush et al., 1990). A similar situation has been seen in the demethylation of the immunoglobulin and vitellogenin (yolk protein) genes (Frank et al., 1990; Jost, 1993). Therefore, methylation may be needed to stabilize the pattern of gene transcription, but the initial activation of the gene is probably accomplished by tissue-specific transcription factors.

How does methylation prohibit transcription? One possibility is that transcription factors cannot bind to their enhancer or promoter sequences if the DNA is methylated (Iguchi-Ariga and Schaffner, 1989). Another possibility is that certain proteins specifically recognize the methylated DNA and compete against the transcription factors for these sites. Boyse and Bird (1991, 1992) have provided evidence for the second model by showing that methylated promoter sequences are bound by a **methyl-CpG-binding protein.** This protein appears to compete with the binding of transcription factors, thereby reducing transcription from these sites.

DNA methylation may also influence the formation of nucleosomes. Keshet and co-workers (1986) have found that methylation affects chromatin structure and suggest that demethylation creates DNase I-hypersensitive sites. When they transfected *un*methylated globin genes into mouse fibroblast nuclei, the genes became packaged into DNase-sensitive chromatin (regardless of the transcriptional ability of the gene). When the same genes were methylated at every CpG site, the DNase-sensitive regions failed to form, presumably because their methylation caused them to be packaged into an inaccessible form. It is possible that when *trans*-regulating factors remove the nucleosomes from the DNA, these regions become demethylated. The demethylation may be necessary to stabilize these regions of activity. The methyl groups would interact with the histones to enable nucleosomes to form only on methylated DNA and not on unmethylated DNA, leaving the active regions free of nucleosomes (Keshet et al., 1986). Once these regions were established, it would become easier for other *trans*-regulatory elements to find these nucleosome-free regions.

Sidelights & Speculations

Methylation and Gene Imprinting

IN CHAPTER 4, we saw that the genomes of the mammalian sperm and egg are not equivalent. Zygotes do not develop properly if their nuclei are derived from two egg pronuclei or from two sperm pronuclei. This inability to develop is probably due to certain genes that are active only if they are derived from the sperm or from the egg. Unlike most genes where (as predicted by Mendelian genetics), it does not matter whether the gene comes from the mother or the father, there are about a dozen known genes for which it does matter whether the gene comes from the sperm or the egg. In some mutations of mice and humans, a severe or lethal condition arises if the mutant gene is derived from one parent, but that same mutant gene has no deleterious effects if inherited from the other parent. For instance, in mice, the gene for insulin-like growth factor-II (*Igf-2*) on chromosome 7 is active in early embryos only on the chromosome transmitted from the father. Conversely, the gene (*Igf-2r*) for a binding protein of this growth factor on chromosome 17 is active only when transmitted from the mother (Barlow et al., 1991; DeChiara et al., 1991; Bartolomei and Tilghman, 1992). Igf-2r acts to bind and degrade excess Igf-2. A mouse pup that inherits a deletion of the *Igf-2r* gene from its father is normal, but if the same deletion is inherited from the mother, the fetus' growth is enhanced, and it dies late in gestation.* In humans, the loss of a particular segment of the long arm of chromosome 15 results in different phe-

Table 11.2 Evidence that genomic imprinting affects phenotype in human gene disorders at chromosome 15 (locus 11q13)

| Parental origin | | Phenotype |
Mother	Father	
Normal allele	Mutant allele	Prader-Willi syndrome
Mutant allele	Normal allele	Angelman's syndrome
Two copies of allele	Allele absent	Prader-Willi syndrome
Allele absent	Two copies of allele	Angelman's syndrome

Source: After Nicholls et al., 1993.

notypes, depending on whether the loss is in the male-derived or the female-derived chromosome (Table 11.2). If the chromosome with the defective or missing segment comes from the father, the child is born with Prader-Willi syndrome, a disease associated with mild mental retardation, obesity, small gonads, and short stature. If the defective or missing gene comes from the mother, the child has Angelman syndrome, characterized by severe mental retardation, seizures, lack of speech, and inappropriate laughter (Knoll et al., 1989; Nicholls et al., 1989). [chrom3.html]

It is now generally thought that most, if not all, of the differences between the mammalian male and female pronuclear genes involve differences in their DNA methylation patterns. The distribution of the methylated and unmethylated CG doublets can be analyzed by cutting the DNA with two restriction enzymes, *Hpa*II and *Msp*I (McGhee and Ginder, 1979). Both of these enzymes cut at the same site—CCGG—but *Hpa*II will not cut DNA if the central C is methylated, whereas *Msp*I cuts whether the sequence is methylated or not. Therefore, DNA from a particular cell type can be digested separately with *Hpa*II and *Msp*I and the cleaved DNA fragments Southern blotted and hybridized with a radioactive gene-specific probe (see Chapter 2). Differences in band patterns in the autoradiographs of the *Msp*I-cleaved and *Hpa*II-cleaved fragments can then be ascribed to methylation differences.

Figure 11.14 shows the results of an experiment in which DNA from sperm was isolated and treated with *Hpa*II or *Msp*I. The probe was a radioactive DNA from the second exon of the β-globin gene. The autoradiograph of fragments from the *Msp*I digestion shows that this probe binds to DNA fragments that have 1400 base pairs between CCGG sites. The *Hpa*II digestion autoradiograph shows that in the sperm these sites (and probably numerous others) are methylated and that this DNA sequence now resides in a 25,000-base-pair stretch of DNA wherein all CCGG sites are methylated (Groudine and Conklin, 1985).

This technique showed that the primordial germ cell nuclei of both male and female mammals are strikingly hypomethylated (Monk et al., 1987; Driscoll and Migeon, 1990), but both the sperm and egg genes undergo extensive methylation as the gametes mature. It appears that during germ cell formation, previous methylation information is erased, and then, during meiosis, new information is introduced onto the genome. The pattern of methylation on a given gene can differ between egg and sperm, and these gene-specific methylation differences are seen in the chromosomes of embryonic cells (Reik et al., 1987; Sanford et al., 1987; Sapienza et al., 1987; Chaillet et al., 1991; Kafri et al., 1992). Thus, methylation differences between sperm and egg genes may specify whether a gene came from the father or mother. This maternal and paternal imprinting adds additional information to the inherited genomes, information that may regulate spatial and temporal gene activity and chromosome behavior.

Swain and co-workers (1987) were able to follow these events by looking at a specific gene that undergoes differential methylation in the sperm and egg. They constructed a strain of transgenic mice in which a particular gene, c-*myc*, was inserted into a particular region of the mouse genome. When this gene was inherited from the male parent, it was transcribed specifically in the heart and in no other tissue. When it was inherited from the female parent, it was not expressed at all. The pattern of expression correlated with the degree of methylation; this gene is methylated during egg maturation but remains hypomethylated during sperm formation. In animals that inherit the transgene from males, the gene is unmethylated and is expressed in the heart. In animals that acquire this transgene from their mothers, the gene is methylated and silent. In both the males and females, the pattern of methylation is erased in the germ cells (Chaillet et al., 1991; Kafri et al., 1992). In mice, methylation differences from the gametes are also seen in the imprinted genes for Igf-2r and H19 (Ferguson-Smith et al., 1993; Stöger et al., 1993). Moreover, if these

*The 30 percent increase in growth is caused by an excess of Igf-2. The lethality is most likely due to lysosomal defects, since the Igf-2-binding protein also serves to target lysosomal enzymes to that organelle (Wang et al., 1994).

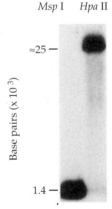

Figure 11.14 Detection of methylation sites on DNA. DNA was isolated from chicken sperm and digested with either *Msp*I (lane 1) or *Hpa*II (lane 2). The fragments were separated by electrophoresis, blotted onto paper, and hybridized to a radioactive DNA probe from the second exon of the β-globin gene. This probe bound to a fragment 1400 bases long in the *Msp*I digest, but to a fragment about 25,000 bases long in the *Hpa*II digest. (From Groudine and Conklin, 1985; photograph courtesy of M. Groudine.)

genes are placed in a mutant mouse strain that lacks the enzyme capable of methylating the CpG sites, transcription of the *H19* gene occurs from the previously silent allele, whereas transcription of *Igf-2r* is lost (Li et al., 1993). Thus, in imprinted genes, methylation can act as either a negative or positive signal for transcription. Such gamete-specific methylation differences provide a plausible explanation for the failure of parthenogenetic mammals to develop and for the need for both male and female pronuclei in the zygote. It also provides a reminder that the organism cannot be explained solely by its genes. One needs knowledge of developmental parameters as well as genetic ones. ■

Mammalian X-chromosome dosage compensation

In animals as diverse as *Drosophila* and humans, females are characterized as having two X chromosomes per cell, while males are characterized as having a single X chromosome per cell. Unlike the Y chromosome, the X chromosome has on it thousands of genes that are essential for cell activity. Yet, despite the female's cells having double the number of X chromosomes as the male's, male and female cells have approximately equal amounts of X chromosome-encoded gene products. This equalization is called **dosage compensation.** The transcription rates of the X chromosomes have been altered so that male and female cells transcribe the same amount of RNAs from their X chromosomes. In *Drosophila*, both X chromosomes in the female are active, but there is increased transcription from the male's X chromosome, so that the single X chromosome of the male cells produces as much product as the two X chromosomes in the female cells (Lucchesi and Manning, 1987). This is accomplished by the binding of particular transcription factors to hundreds of sites along the male X chromosome (Kuroda et al., 1991).

In mammals, X chromosome dosage compensation occurs by inactivating one X chromosome in each female cell. Thus, each mammalian somatic cell, whether male or female, has only one functioning X chromosome. This phenomenon is called **X chromosome inactivation.** The chromatin of the inactive X chromosome is converted into **heterochromatin**—chromatin that remains condensed throughout most of the cell cycle and replicates after most of the chromatin (the **euchromatin**) of the nucleus. This heterochromatin, in a formation called the **Barr body** (Figure 11.15), is often seen on the nuclear envelope of female cells (Barr and Bertram, 1949). X chromosome inactivation must occur early in development. Using a mutated X chromosome that would not inactivate, Tagaki and Abe (1990) showed that the expression of two X chromosomes per cell in mouse embryos leads to ectodermal cell death and the absence of mesoderm formation, eventually causing embryonic death at day 10 of gestation.

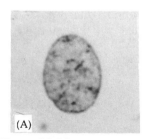 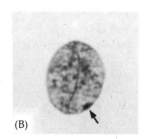 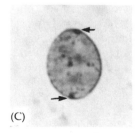

(A)   (B)   (C)

Figure 11.15
Nuclei of human oral epithelial cells stained with Cresyl violet. (A) Cell from a normal XY male, showing no Barr body. (B) Cell from a normal XX female, showing a single Barr body (arrow). (C) Cell from a female with three X chromosomes. Two Barr bodies can be seen, and only one X chromosome per cell is active. (From Moore, 1977.)

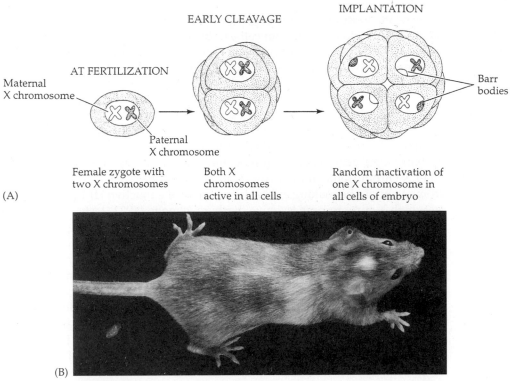

(A)

EARLY CLEAVAGE

IMPLANTATION

AT FERTILIZATION

Maternal
X chromosome

Paternal
X chromosome

Barr
bodies

Female zygote with
two X chromosomes

Both X
chromosomes
active in all cells

Random inactivation of
one X chromosome in
all cells of embryo

(B)

Figure 11.16
X chromosome inactivation in mammals. (A) Schematic diagram illustrating random X chromosome inactivation. The inactivation is
believed to occur at about the time of implantation. (B) A female
mouse heterozygous for the X-linked coat color gene *dappled*. Distinctly different pigmented regions are seen. (Photograph courtesy of
M. F. Lyon.)

The early inactivation of one X chromosome per cell has important phenotypic consequences. One of the earliest analyses of X chromosome inactivation was performed by Mary Lyon (1961), who observed coat color patterns in mice. If a mouse is heterozygous for an *autosomal* gene controlling
hair pigmentation, then the mouse resembles one of the two parents or has a
color intermediate between the two. In either case, the mouse is a single
color. But if a female mouse is heterozygous for a pigmentation gene on the
X chromosome, a different result is seen: patches of one parental color alternate with patches of the other parental color (Figure 11.16). Lyon proposed
the following hypothesis to account for these results:

1. Very early in the development of female mammals, both X chromosomes are active.
2. As development proceeds, one X chromosome is turned off in each
   cell.
3. This inactivation is random. In some cells, the paternally derived X
   chromosome is inactivated; in other cells, the maternally derived X
   chromosome is shut off.
4. This process is irreversible. Once an X chromosome has been inactivated, the same X chromosome is inactivated in all that cell's progeny. (The areas of pigment in these mice are large patches, not a "salt
   and pepper" pattern.) Thus, all tissues in female mammals are mosaics of two cell types.

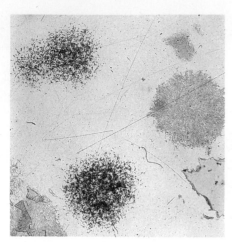

Figure 11.17
Retention of X chromosome inactivation. Approximately 30 cells from a woman heterozygous for HPRT deficiency were placed into a petri dish and allowed to grow. The cells were visualized by autoradiography after incubation in a medium containing radioactive hypoxanthine. Cells with HPRT incorporate the radiolabeled compound into their RNA and darken the photographic emulsion placed over them. The clones of cells without HPRT appear lighter because their cells cannot incorporate the radioactive compound. (From Migeon, 1971, courtesy of B. Migeon.)

Some of the most impressive evidence for this model comes from biochemical studies on clones of human cells. In humans, there is a genetic disease—Lesch-Nyhan syndrome—that is characterized by the lack of the X-linked enzyme hypoxanthine phosphoribosyltransferase (HPRT). Lesch-Nyhan syndrome is transmitted through the X chromosome—that is, males who have this mutation in their one X chromosome suffer (and die) from the disease. In females, however, the presence of the mutant *HPRT* gene can be masked by the other X chromosome, which carries the wild-type allele. A woman who has sons with this disease is said to be a carrier, as she has a mutant HPRT gene on one chromosome and a wild-type *HPRT* gene on the other X chromosome. If the Lyon hypothesis is correct, each cell from such a woman should be making either the active or the inactive HPRT, depending on which X chromosome is active. Barbara Migeon (1971) tested this prediction by taking individual skin cells from a woman heterozygous for the *HPRT* gene and placing them in culture. Each of these cells divided to form a clone of cells. When Migeon stained the clones for the presence of wild-type *HPRT*, approximately half of the clones had the enzyme and the other half did not (Figure 11.17).

The Lyon hypothesis of X chromosome inactivation provides an excellent account of differential gene inactivation at the level of transcription. Some interesting exceptions to the general rules further show its importance. First, X chromosome inactivation holds true only for *somatic* cells, not *germ* cells. In female germ cells, the inactive X chromosome is reactivated shortly before the cells enter meiosis (Gartler et al., 1973, 1980; Migeon and Jelalian, 1977; Kratzer and Chapman, 1981). Thus, in mature oocytes, both X chromosomes are active. In each generation, X chromosome inactivation has to be established anew.

Second, there are some exceptions to the rule of randomness in the inactivation pattern. The first X chromosome inactivation in the mouse is seen in the trophectoderm, where the *paternally* derived X chromosome is specifically inactivated (Tagaki, 1974; West et al., 1977). Third, X chromosome inactivation does not extend to every gene on the human X chromosome. There are several genes on the short arm of the X chromosome (such as that encoding steroid sulfatase) that "escape" this dosage-related inactivation (Mohandas et al., 1980; Brown and Willard, 1990), and even on the long arm, there are some genes that are transcribed from both X chromosomes in each female somatic cell. Thus, in humans, heterochromatization does not extend throughout the entire X chromosome.

The fourth exception really ends up proving the rule. There are a few male mammals with coat color patterns we would not expect to find unless the animals exhibited X chromosome inactivation. Male calico and tortoiseshell cats are among these examples. These spotted coat patterns are normally seen in females and are thought to result from random X chromosome inactivation. But rare males exhibit these coat patterns as well. How can this be? It turns out that these cats are XXY. The Y chromosome makes them male (see Chapter 20), but one X chromosome undergoes inactivation, just as in females, so there is only one active X per cell (Centerwall and Benirschke, 1973). Thus, these cats have cells with a Barr body and random X chromosome inactivation. It is clear, then, that one mechanism for transcription-level control of gene regulation is to make a large number of genes heterochromatic and thus transcriptionally inert.

The Mechanism of X Chromosome Inactivation

THE MECHANISM of X chromosome inactivation is still poorly understood, but new research is giving us some indication of the factors that may be involved in initiating and maintaining a heterochromatic X chromosome.

Initiation of X Chromosome Inactivation: The *Xist* Gene

The first clues for an initiator of X chromosome inactivation came from genetic studies wherein certain rearranged X chromosomes in mice were unable to become inactive (Russell, 1963; Cattanach et al., 1969; Mattei et al., 1981). These chromosomes lacked a particular region thereafter called the X chromosome inactivation center (XIC). In 1991, Brown and her colleagues found an RNA transcript that was made solely from the *inactive* X chromosome of humans. (In those loci that escape X chromosome inactivation in humans, both X chromosomes synthe-

size the transcripts. Here, the transcript was coming only from the "inactive" X chromosome.) This transcript, *XIST*, was being made by a gene within the XIC region. Moreover, this transcript does not appear to encode a protein. It stays within the nucleus and interacts with the inactive X chromatin Barr body (Brown et al., 1992). A similar situation exists in the mouse, where the *Xist* gene of the inactive X chromosome synthesizes a nuclear RNA whose sequence cannot encode a protein* (Borsani et al., 1991; Brockdorrf et al., 1992).

The *Xist* gene is an excellent candidate for the initiator of X inactivation. First, the transcripts from the *Xist* gene are seen in mouse embryos prior to X chromosome inactivation, which would be expected if this gene plays a role in initiating X chromosome inactivation (Kay et al., 1993). Second, knocking out one *Xist* locus in an XX cell prevents X inactivation from occurring on that chro-

mosome (Penny et al., 1996). Third, the transferal of a 450-kilobase segment containing the mouse *Xist* gene into an *autosome* of *male* embryonic stem cells causes the random inactivation of either that autosome or the endogenous X chromosome (Lee et al., 1996). The autosome is "counted" as an X chromosome. *Xist* expression is only needed for the initiation of X chromosome inactivation. Once inactivation occurs, it is dispensable (Brown and Willard, 1994). It is still not known what this *Xist* RNA is doing to inactivate the chromosome.

The *Xist* locus is imprinted in the gametes, and the imprinting is effected by differential methylation in the *Xist* promoter region. During spermatogenesis, three CG sites on the *Xist* promoter are demethylated, while the same sites are completely methylated during oogenesis. In somatic cells, the active *Xist* gene (on the inactive X chromosome) is largely unmethylated, while the inactive *Xist* gene (on the active X chromosome) is completely methylated (Figure 11.18; Norris et al., 1994; Ariel et al., 1995; Zuccotti and Monk, 1995). This pattern of *Xist* expression is maintained in mouse extraembryonic tissues (such that the paternally derived *Xist* is demethylated and active, leading to the inactivation of that chromosome). The cells of the embryonic epiblast, however, lose their parental imprinting patterns and reestablish the methylation differences randomly.

Germ cells:

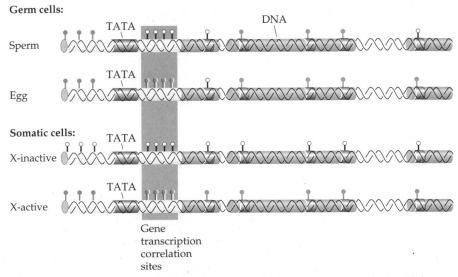

Sperm

Egg

Somatic cells:

X-inactive

X-active

Gene
transcription
correlation
sites

Figure 11.18 Summary of the *Xist* methylation patterns in sperm, egg, and the two X chromosomes in somatic cells. Open squares represent unmethylated CG sites; filled squares represent methylated CG sites. The shaded area indicates the sites that correlate with the gene's transcription. (After Zuccotti and Monk, 1995.)

*The list of RNA types is growing. In addition to the well known mRNA, rRNA, tRNA, and small nuclear RNAs (involved in RNA splicing), there are also H19 RNA and Xist RNA, neither of which encodes proteins. In later chapters, we will discuss translational control (natural antisense) RNAs, and RNAs such as Xlsrt which are used to localize messages to regions of the oocyte cytoplasm. The embryo uses RNAs in many more creative ways than the adult organisms.

Forbidding Transcription: The Unacetylated Nucleosome

What is preventing the DNA of the inactive X chromosome from being transcribed like the active one? A recent study by Jeppeson and Turner (1993) suggests that the inactive and active X chromosomes differ from one another by the acetylation of their respective H4 histones. One of the best ways to release protein from DNA is to add negative charges to the protein. This can be accomplished by adding phosphate or acetate groups to the DNA-binding regions of the protein. Acetylation of histone H4 has been correlated with actively transcribing genes. Nucleosomes from nonmethylated CG promoter regions have highly acetylated H4 proteins, and the acetylation of histone H4 correlates with the activation of certain genes (Chahal et al., 1980; Tazi and Bird, 1990). Moreover, whereas transcription factor TFIIIA cannot bind to the 5S RNA gene if the gene is wrapped up in a nucleosome, that factor can bind to the gene if the nucleosome's histones have been acetylated (Lee et al., 1993). Histone acetylation does not appear to be obligatory for gene transcription, but it may facilitate transcription in several systems (Turner, 1991). Using an antibody that recognizes acetylated histone H4 (but not unacetylated histone H4), Jeppesen and Turner found that the active X chromosomes of humans and mice have about as much acetylated histone H4 as most other chromosomes. However, the inactive X chromosome has hardly any acetylated histone H4 (Figure 11.19). It is not known how *Xist* expression would cause the occurrence of unacetylated H4 in the nucleosomes of the inactive X chromosomes.

Locking in Transcription Patterns: DNA Methylation

The locking in of the transcriptionally inactive state is done by methylation. The first evidence for such *"cis"* differences between active and inactive X-chromosomal DNA came when Liskay and Evans (1980) transfected the X-linked gene for HPRT into HPRT-deficient mouse cells in culture. When the DNA came from a clone of cells in which the gene for HPRT was on the *inactive* X

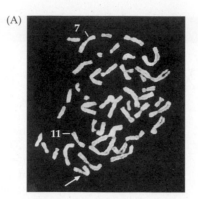 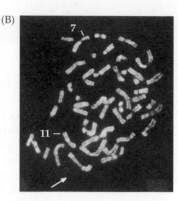

Figure 11.19 The inactive X chromosome of human female cells contains underacetylated histone H4. (A) Metaphase spread of a human female fibroblast cell stained with Hoechst 33258, which stains chromatin. (Chromosomes 7 and 11 are numbered, and an arrow points to the inactive X.) (B) Same preparation stained with fluorescent antibody to acetylated histone H4. While all the other chromosomes are clearly visible, the inactive X is not. (From Jeppesen and Turner, 1993; photographs courtesy of the authors.)

chromosome, the DNA was unable to make HPRT in the HPRT-deficient host cell. However, when the DNA was derived from a clone of cells expressing the *HPRT* gene on its *active* X chromosome, the transfected cells made HPRT from this gene. Shortly thereafter, Mohandas and colleagues (1981) found that 5-azacytidine (a drug that inhibits cytosine methyltransferase) could locally reactivate these genes on the inactive X chromosome. Further investigations, using restriction enzymes and cDNA probes, showed that the CG "islands" at the promoter sites of several genes are methylated in the inactive X chromosome and unmethylated in the active X chromosome (Wolf et al., 1982; 1984; Keith et al., 1986). These methylation patterns are removed during germ cell formation, thereby enabling a new pattern of X

chromosome inactivation to occur in the next generation. During the first trimester of human development, the adult pattern of methylation and X chromosome inactivation is established anew (Migeon et al., 1991).*

We are still ignorant of the mechanisms by which the *Xist* transcript regulates the state of the chromatin and how the spreading of inactivation occurs. We are still ignorant of the ways in which *Xist* transcription, nucleosome modification, and DNA methylation are linked to the heterochromatization of an X chromosome. We do not yet know how the choice between the two X chromosomes is originally made or how the *Xist* RNA is transcribed from a region surrounded by inactivated genes. There is still much to learn about this critical mammalian phenomenon. ■

*As mentioned in earlier chapters, it is difficult to extrapolate from one mammalian group to another. This is certainly the case with X chromosome inactivation. Just because X chromosome inactivation happens this way in the mouse placenta does not mean it happens this way in the placenta of all mammals. In human chorionic villi, some cells contain two active X chromosomes, and those X chromosomes that are inactivated can be reactivated (Migeon et al., 1985, 1986). Also, X chromosome inactivation in the *human* placenta appears to be random; either the paternally or maternally derived X chromosome can be turned off. In marsupials, the paternally derived X chromosome is preferentially inactivated throughout the embryo (Cooper et al., 1971; Sharman, 1971; Samollow et al., 1987). Moreover, in humans, there are obvious regions of the X chromosome that escape inactivation. The somatic differences between humans with XX and XO karyotypes also predict that there would be some X-linked genes that would be needed in two doses for normal female development. In the mouse, X chromosome inactivation appears to extend throughout the entire chromosome (Ashworth et al., 1991). In sex determination (Chapter 20), it is crucial that the genes for X-dosage compensation be linked to the genes responsible for the sexual phenotype. If the dosage is not equalized, the embryo usually dies.

Association of active DNA with the nuclear matrix

Attachment of Active Chromatin to a Nuclear Matrix

Within the nucleus, the replicative enzymes must somehow find their sites for initiating DNA synthesis; the transcription factors and polymerases must find their promoters and enhancers; the RNA-processing factors must find their RNA-splicing sites; and the messenger RNA must efficiently find the pores through which to exit the nucleus. This is a great deal to ask of molecules in solution. We would have to expect the various factors involved in transcription to be floating around in the nuclear sap, bumping randomly into DNA. The RNA so formed would then be spliced and bump around the nucleus until it found a nuclear pore through which to leave.

An alternative model suggests that RNA is transcribed on a solid substrate in which all the enzymes needed for transcription, processing, and transport are collected together. There are precedents for thinking in such terms. The electron transport chain of the mitochondria is such an ordered aggregate, and the DNA-synthetic enzymes of bacteria have long been known to reside on the inner surface of the cell membrane. What one has to ask, then, are the following questions: Is there a nuclear reticulum wherein such enzymes might be found? If such a network exists, are the transcriptionally active genes located there? If such a network exists, are the RNA-synthesizing enzymes found there?

A nuclear matrix can be isolated by dissolving nuclei in lipid detergents and solubilizing most of the DNA with DNases (Berezney and Coffey, 1977; Capco et al., 1982). Transmission electron microscopy of such complexes shows a meshwork of proteins that extends throughout the nucleus and connects to the cytoskeleton at the nuclear envelope (Figure 11.20). This matrix is seen in all eukaryotic nuclei thus far examined (Wilson, 1895; Nelson et al., 1986).

When one isolates such a matrix, DNase has removed about 98 percent of the DNA. Is the DNA that is still bound to this matrix (and presumably protected from the DNase by being so tightly associated with it) enriched for actively transcribing genes? There is evidence that this is so for at least some genes. The ovalbumin gene is preferentially associated with the nuclear matrix in adult chick oviduct cells, but not in chick liver or erythrocyte cells. The globin genes, moreover, are not associated with the nuclear matrix of the oviduct cells (Robinson et al., 1982; Thorburn and Knowland, 1993). Ciejek and co-workers (1983) confirmed and extended these observations, showing that the entire hormone-inducible transcription unit of the ovalbumin gene is bound to the nuclear matrix. No other genes within 100,000 base pairs of this unit are matrix-associated. Moreover, when estrogen was withdrawn from the animals, the specific attachment of these genes to the nuclear matrix was abolished. It appears that these genes are bound to the nuclear matrix only when activated.

In 1985, Hutchinson and Weintraub showed that DNase I-sensitive sites are not found uniformly throughout the nucleus. They treated nuclei with DNase I, and then repaired the nicks with radioactive nucleotides. The labeled DNA should represent just the actively transcribing (i.e., DNase I-sensitive) genes. The results of such treatment showed that the DNase I-sensitive DNA existed at the periphery of the nuclei and along channels or fibers that connected to the nuclear envelope (Figure 11.21). It is possible, then, that active genes are specifically associated with the nuclear envelope or matrix.

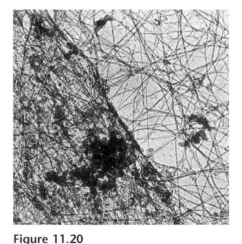

Figure 11.20
Transmission electron micrograph (47,000×) of a portion of nuclear matrix (left) and surrounding cytoplasm. Cytoskeletal filaments are clearly visible. The mouse fibroblast cells were extracted with detergent to remove lipids and further treated with DNase I. In 1895, E. B. Wilson, looking through the light microscope, reported that the nucleus was traversed by fibers that were continuous with those of the cytoplasmic reticulum and that surrounded the chromatin. (From Capco et al., 1982, courtesy of S. Penman.)

(A)

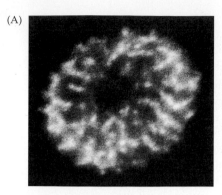

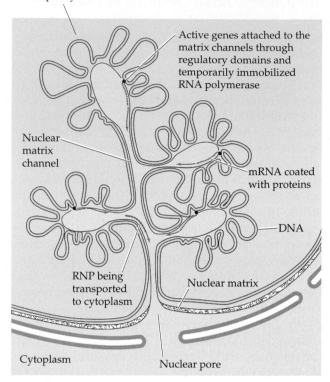

(B) Chromosomal DNA loops attached to the nuclear matrix through replication origins. The packaging into nucleosomes and 30 nm fibers is not shown for simplicity

Figure 11.21
Presence of active chromatin along the nuclear periphery and channels. (A) Erythrocyte nuclei were treated with DNase I, which is thought to nick actively transcribing chromatin regions. This nick was "healed" by nick translation within the nucleus in the presence of nucleotides whose presence could be detected by fluorescence. The labeled nucleotides were found at the periphery of the nucleus and along structures leading inward from the nuclear envelope. (B) Speculative model of interphase chromatin organization, envisioning the nuclear matrix as a series of internal channels. (A from Hutchinson and Weintraub, 1985, courtesy of N. Hutchinson; B after Razin and Gromova, 1995.)

Another type of evidence for nuclear matrix participation in transcription involves the finding that most (some report 95 percent) of the newly synthesized RNA appears to be attached to the nuclear matrix (Herman et al., 1978; Miller et al., 1978; van Eekelen and van Venrooij, 1981; Mariman et al., 1982). This binding appears to be mediated by a set of nuclear matrix proteins. These proteins include lamin B1, a major component of the nuclear envelope (Ludérus et al., 1992), a thymus-specific DNA-binding protein that unwinds the DNA adjacent to its binding site on the DNA (Dickinson et al., 1992), and the transcription factor YY1/NF-E1 that was found to be identical to nuclear matrix protein 1 (NMP-1) (Guo et al., 1995). Given that active genes, RNA polymerase, and nascent transcripts appear to be bound to a nuclear matrix, Jackson and Cook (1985) have proposed that transcription does not occur by a mobile polymerase traveling down the length of a gene. Rather, they envision the RNA polymerase as tethered to the nuclear matrix, with the DNA traveling through it.

There is also some evidence that active DNA may be attached to the nuclear matrix by AT-rich DNA sequences called **matrix-associated regions** (**MARs**), or scaffold-associated regions (Gasser and Laemmli, 1986). Most of these MARs map near or within enhancers or promoters. The importance of these regions was shown by Stief and co-workers (1989), who identified two MARs of the chick lysozyme gene. In this case, the MARs were not in the enhancer and so could be separated from it. When they fused the chick lysozyme enhancer and promoter to the reporter gene *CAT* and transfected the clone into lysozyme-producing cells, it did not produce much CAT protein. They then made a similar gene that contained the promoter, enhancer, and *CAT* sequences and that was flanked by the two MARs. When this clone

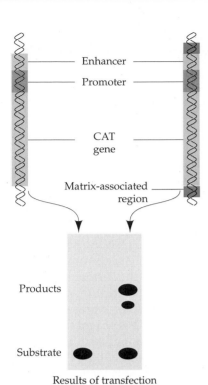

Figure 11.22
Importance of matrix-associated regions in transcription. When clones consisting of lysozyme promoter, enhancer, and the *CAT* gene are transfected into a lysozyme-secreting cell line, very little CAT protein is made, as assayed by CAT enzyme activity. However, if the two MARs are also included in the cloned gene, much more CAT protein can be found in these cells. (After Stief et al., 1989.)

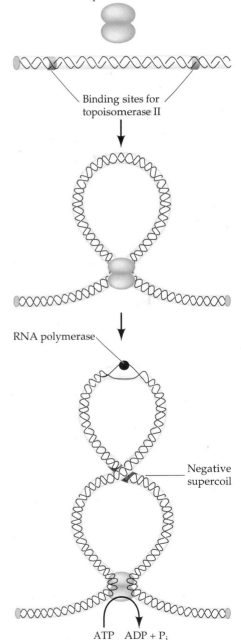

was transfected into the lysozyme-producing cells, CAT synthesis was enormously enhanced (Figure 11.22). Similarly, two MARs flank an enhancer of the mouse immunoglobulin μ heavy chain locus, and the transcription of this gene requires the presence of both the enhancer and its two MARs. The MARs appear to cooperate with the enhancer to extend a region of factor-accessible chromatin to the immunoglobulin gene promoter (Forrester et al., 1994; Jenuwein et al., 1997).

Topoisomerases and Gene Transcription

In several studies, matrix-associated regions have been seen to contain or be adjacent to a DNA sequence that is recognized by an enzyme—**topoisomerase II**—that might be essential for transcription (Cockerill and Garrard, 1986; Adachi et al., 1989; Scheuermann and Chen, 1989). Recent studies have suggested that the unwinding of the DNA helix is very important in facilitating transcription. Transcriptionally active chromatin has to be twisted to enable the strands to unwind (Ryoji and Worcel, 1984), and this twisting is accomplished by supercoiling the DNA helix (Figure 11.23). Villeponteau and co-workers (1984) have shown that DNase I-sensitive sites in active genes are formed only when the genes are under torsional strain. Topoisomerase II is the enzyme responsible for twisting the DNA and allowing the strands to separate. Using antibodies to this protein, Berrios and colleagues

Figure 11.23
Supercoiling of DNA during transcription. Topoisomerase II joins together two regions of DNA and introduces supercoiling by transiently breaking and recombining the DNA strands. As a result of the strain, a portion of the double helix separates into two strands, enabling RNA polymerase (and presumably other *trans*-regulatory factors) to initiate transcription. The topoisomerase-binding sites have been found on matrix-bound DNA (Cockerill and Garrard, 1986). (After Darnell et al., 1986.)

Figure 11.24
One of four regions of the heavy chain immunoglobulin gene enhancer protected by the NF-μNR protein. NF-μNR was added to the enhancer region DNA, and the DNA was digested with DNase. Only those sequences covered by NF-μNR would be preserved. The protected region (gray) includes a matrix-associated region sequence and a topoisomerase II-binding site (color). (After Scheuermann and Chen, 1989.)

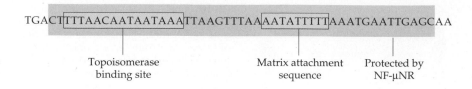

(1985) demonstrated that topoisomerase II is located on the nuclear matrix-nuclear envelope complex. The proximity of the MARs and the topoisomerase II-binding regions suggests that the anchoring of the DNA to the matrix may be necessary to prevent free rotation of the DNA so as to permit the matrix-bound topoisomerases to twist the chromatin (Bode et al., 1992)

If binding to the nuclear matrix is essential for RNA transcription, it is possible that negative regulatory proteins such as silencers may inhibit this association. This possibility was suggested by Scheuermann and Chen (1989), who isolated a protein that inhibits the transcription of the heavy chain immunoglobulin gene. This protein, NF-μNR, is expressed in non-B cells and in early stages of B cell development, but it is absent from the mature B cells that transcribe large amounts of immunoglobulin genes. This protein binds to four places flanking the heavy chain enhancer: two MAR consensus sequences and two topoisomerase II consensus sequences (Figure 11.24). It is possible, then, that when NF-μNR is present in nuclei, it binds to the enhancer-flanking regions and prevents the association of the heavy chain immunoglobulin gene with the nuclear matrix and topoisomerase. When this protein is not present, these associations may not occur, resulting in transcription of the gene. The finding that the nuclear matrix protein NMP-1 is the same as transcription factor YY1 is particularly interesting since YY1 has been implicated in the silencing of the ε-globin gene once the γ-globin genes become expressed (Raich et al., 1995; Wandersee et al., 1996).

Insulators and domains

The eukaryotic genome is not merely parsed into particular genes. Rather, it appears to be divided into relatively independent developmental regions often referred to as **domains.** Evidence for domains came from studies where blocks of DNA were placed near reporter genes that could otherwise be activated by an enhancer. Certain sequences stopped the enhancer from activating the reporter gene, while other sequences did not (Geyer and Corces, 1992). It was proposed that these **insulator** sequences bound proteins that prevented the enhancers from interacting with promoters on the other side of them. Thus, they could establish boundaries: activation could occur on one side of them, but this activation could not cross over it. Some of these boundary sequences have been isolated from *Drosophila* DNA, and some of their binding proteins have been isolated. Kellum and Schedl (1991) showed that the *hsp70* gene (for the *Drosophila* heat shock protein) was bounded by two sequences, *scs* and *scs'*, that prevented the effects of adjacent chromatin from influencing their transcription. Zhao and colleagues (1995) have identified a 32-kDa protein that binds to the *scs'* boundary element and is localized between the bands of numerous genes in *Drosophila* (see Plate 31; Zhao et al., 1995). This can be visualized when these genes puff and the staining for these proteins shows them to be at the puff boundaries. The *scs'* site in the Bithorax complex appears to be after the last gene (*AbdB*), so the entire unit may be regulated as a single genetic locus.

Coda

Differential gene transcription is a major way to regulate development. The *cis*-regulatory regions in the DNA and the *trans*-regulatory proteins that activate and repress transcription are being identified, and their mechanisms of action are being delineated. It appears that certain transcription factors are able to disrupt or prevent nucleosome formation in enhancers and promoter regions, thereby enabling RNA polymerase II to bind to the promoter and transcribe the gene. Certain transcription factors are able to stimulate transcription by interacting with the transcriptional complex and to accelerate its formation. Demethylation and the unwinding of genetic regions on the nuclear matrix are also probably involved in regulating gene expression. As Albert Claude said, we are just beginning to appreciate our acquired wealth.

LITERATURE CITED

Adachi, Y., Käs, E. and Laemmli, U. 1989. Preferential, cooperative binding of DNA topoisomerase II to scaffold-associated regions. *EMBO J.* 8: 3997–4006.

Adams, C. C. and Workman, J. L. 1993. Nucleosome displacement in transcription. *Cell* 72: 305–308.

Ariel, M., Robinson, E., McCarrey, J. R. and Cedar, H. 1995. Gamete-specific methylation correlates with imprinting of the murine *Xist* gene. *Nat. Genet.* 9: 312–315.

Ashworth, A., Rastan, S., Lovell-Badge, R. and Kay, G. F. 1991. X inactivation may explain the difference in viability of XO humans and mice. *Nature* 351: 406–408.

Bacon, E. R., Dalyot, N., Filon, D., Schreiber, L., Rachmilewitz, E. A. and Oppenheim, A. 1995. Hemoglobin switching in humans is accompanied by changes in the ratio of the transcription factors GATA-1 and SP1. *Molec. Med.* 1: 295–305.

Barlow, D. P., Stoger, R., Herrmann, B. G., Saito, K. and Schweifer, N. 1991. The mouse insulin-like growth factor type-2 receptor is imprinted and closely linked to the *Tme* locus. *Nature* 349: 84–87.

Barr, M. L. and Bertram, E. G. 1949. A morphological distinction between neurones of the male and female, and the behavior of the nucleolar satellite during accelerated nucleoprotein synthesis. *Nature* 163: 676.

Bartolomei, M. S. and Tilghman, S. M. 1992. Parental imprinting of mouse chromosome 7. *Semin. Dev. Biol.* 3: 107–117.

Behringer, R., Hammer, R., Brinster, R., Palmiter, R. and Townes, T. 1987. Two 3′ sequences direct adult erythroid-specific expression of human β-globin genes in transgenic mice. *Proc. Natl. Acad. Sci. USA* 84: 7056–7060.

Berezney, R. and Coffey, D. S. 1977. Nuclear matrix: Isolation and characterization of a framework structure from rat liver nuclei. *J. Cell Biol.* 73: 616–637.

Berrios, M., Osheroff, N. and Fisher, P. A. 1985. *In situ* localization of DNA topoisomerase II, a major polypeptide component of the *Drosophila* nuclear matrix fraction. *Proc. Natl. Acad. Sci. USA* 82: 4142–4146.

Berry, M., Grosveld, F. and Dillon, N. 1992. A single point mutation is the cause of the Greek form of hereditary persistence of fetal hemoglobin. *Nature* 358: 499–502.

Bestor, T. H. and Ingram, V. M. 1983. Two DNA methyltransferases from murine erythroleukemia cells: Purification, sequence specificity, and mode of interaction with DNA. *Proc. Natl. Acad. Sci. USA* 82: 2674–2678.

Blom van Assendelft, G., Hanscombe, O., Grosveld, F. and Greaves, D. R. 1989. The β-globin dominant control region activates homologous and heterologous promoters in a tissue-specific manner. *Cell* 56: 969–977.

Bode, J., Kohwi, Y., Dickinson, L., Joh, T., Klehr, D., Mielke, C. and Kohwi-Shigematsu, T. 1992. Biological significance of unwinding capability of nuclear matrix-associated DNA. *Science* 255: 195–197.

Borsani, G. and thirteen others. 1991. Characterization of a murine gene expressed from the inactive X chromosome. *Nature* 351: 325–329.

Boyse, J. and Bird, A. 1991. DNA methylation inhibits transcription indirectly via a methyl-CpG binding protein. *Cell* 64: 1123–1134.

Boyse, J. and Bird, A. 1992. Repression of genes by DNA methylation depends upon CpG density and promoter strength: Evidence for involvement of a methyl-CpG binding protein. *EMBO J.* 11: 327–333.

Braunstein, M., Rose, A. B., Holmes, S. G., Allis, C. D. and Broach, J. R. 1993. Transcriptional silencing in yeast is associated with reduced nucleosome acetylation. *Genes Dev.* 7: 592–604.

Brockendorrf, N. and seven others. 1992. The product of the mouse *Xist* gene is a 15-kb inactive X-specific transcript containing no conserved ORF and located in the nucleus. *Cell* 71: 515–526.

Browder, L. W. 1984. *Developmental Biology*, 2nd Ed. Saunders, Philadelphia.

Brown, C. J. and Willard, H. F. 1990. Localization of a gene that escapes inactivation to the X chromosome proximal short arm: Implications for X inactivation. *Am. J. Hum. Genet.* 46: 273–279.

Brown, C. J. and Willard, H. F. 1994. The human X-inactivation centre is not required for maintenance of X-chromosome inactivation. *Nature* 368: 154–156.

Brown, C. J. and six others. 1991a. A gene from the region of the human X inactivation center is expressed exclusively from the inactive X chromosome. *Nature* 349: 38–44.

Brown, C. J. and nine others. 1991b. Localization of the X chromosome inactivation center on the human X chromosome. *Nature* 349: 82–84.

Brown, C. J., Hendrich, B. D., Rupert, J. L., Lafreniere, R. G., Xing, Y., Lawrence, J. and Willard, H. F. 1992. The human *XIST* gene: Analysis of a 17-kb inactive X-specific RNA that contains conserved repeats and is highly localized within the nucleus. *Cell* 71: 527–542.

Brownell, J. E., Zhou, J., Ranalli, T., Kobayashi, R., Edmonson, D. G., Roth, S. Y. and Allis, C. D. 1996. *Tetrahymena* histone acetyltransferase A: A homolog to yeast GCN5 linking histone acetylation to gene activation. *Cell* 84: 843–851.

Brunk, B. P., Goldhammer, D. J. and Emerson, C. P. Jr. 1996. Regulated demethylation of the *myoD* distal enhancer during skeletal myogenesis. *Dev. Biol.* 177: 490–503.

Bungert, J., Davé, U., Lim, K.-C., Lieuw, K. H., Shavit, J. A., Liu, Q. and Engel, J. D. 1995. Synergistic regulation of human β-globin gene switching by locus control region elements HS3 and HS4. *Genes Dev.* 9: 3083–3096.

Burch, J. B. and Weintraub, H. 1983. Temporal order of chromatin structral changes associated with activation of the major chick vitellogenin gene. *Cell* 33: 64–76.

Burlingame, R. W., Love, W. E., Wang, B.-C., Hamlin, R., Xuong, N.-H. and Moudrianakis, E. N. 1985. Crystallographic structure of the octameric histone core of the nucleosome at a resolution of 3.3 Å. *Science* 228: 246–253.

Busslinger, M., Hurst, J. and Flavell, R. A. 1983. DNA methylation and the regulation of globin gene expression. *Cell* 34: 197–206.

Capco, D. G., Wan, K. M. and Penman, S. 1982. The nuclear matrix: Three-dimensional architecture and protein composition. *Cell* 29: 847–858.

Cattanach, B. M., Pollard, C. E. and Perez, J. N. 1969. Controlling elements in the mouse X chromosome. 1. Interaction with the X-linked genes. *Genet. Res.* 14: 223–235.

Centerwall, W. R. and Benirschke, K. 1973. Male tortoiseshell and calico (T-C) cats. *J. Hered.* 64: 272–278.

Chahal, S. S., Matthews, H. R. and Bradbury, E. M. 1980. Acetylation of histone H4 and its role in chromatin structure and function. *Nature* 287: 76–79.

Chaillet, J. R., Vogt, T. F., Beier, D. R. and Leder, P. 1991. Parental-specific methylation of an imprinted transgene is established during gametogenesis and progressively changes during embryogenesis. *Cell* 66: 77–83.

Ciejek, E. M., Tsai, M.-J. and O'Malley, B. W. 1983. Actively transcribed genes are associated with the nuclear matrix. *Nature* 306: 607–609.

Clark, D. J. and Felsenfeld, G. 1992. A nucleosome core is transferred out of the path of a transcribing polymerase. *Cell* 71: 11–22.

Cockerill, P. N. and Garrard, W. T. 1986. Chromosome loop anchorage of the κ immunoglobulin gene occurs next to the enhancer in a region containing topoisomerase II sites. *Cell* 44: 273–282.

Compere, S. J. and Palmiter, R. D. 1981. DNA methylation controls the inducibility of the mouse metallothionein-I gene in lymphoid cells. *Cell* 25: 233–240.

Conklin, K.F. and Groudine, M. 1984. Chromatin structure and gene expression. *In* A. Razin, H. Cedar and A. D. Riggs (eds.), *DNA Methylation.* Springer-Verlag, New York, pp. 293–351.

Cooper, D. W., Vandeberg, J. L., Sharmen, G. B. and Poole, W. E. 1971. Phosphoglycerate kinase polymorphism in kangaroos provides further evidence for paternal X inactivation. *Nat. New Biol.* 230: 155–157.

Crick, F. H. C. 1966. *Of Molecules and Men.* University of Washington Press, Seattle.

Darnell, J., Lodish, H. and Baltimore, D. 1986. *Molecular Cell Biology.* Scientific American Books, New York.

DeChiara, T. M., Robertson, E. J. and Efstratiadis, A. 1991. Parental imprinting of the mouse insulin-like growth factor II gene. *Cell* 64: 849–859.

Dickinson, L. A., Joh, T., Kohwi, Y. and Kohwi-Shigematsu, T. 1992. A tissue-specific MAR/SAR DNA-binding protein with unusual binding site recognition. *Cell* 70: 631–645.

Driscoll, D. J. and Migeon, B. R. 1990. Sex difference in methylation of single-copy genes in human meiotic germ cells: Implications for X chromosome inactivation, parental imprinting, and the origin of PGC mutations. *Somat. Cell Mol. Genet.* 16: 267–268.

Elgin, S. 1981. DNase I-hypersensitive sites of chromatin. *Cell* 27: 413–415.

Elgin, S. C. R. 1988. The formation and function of DNase-I hypersensitivity sites in the process of gene activation. *J. Biol. Chem.* 263: 9259–9262.

Ellis, J., Tan-Un, K. C., Harper, A., Michaelovich, D., Yannoutsos, N., Philipsen, S. and Grosveld, F. 1996. A dominant chromatin-opening activity in 5' hypersensitive site 3 of the human β-globin locus control region. *EMBO J.* 15: 562–568.

Enver, T., Raich, N., Ebens, A. J., Papayannopoulou, T., Costantini, F. and Stamatoyannopoulos, G. 1990. Developmental regulation of human fetal-to-adult globin gene switching in transgenic mice. *Nature* 344: 309–312.

Felsenfeld, G. 1992. Chromatin as an essential part of the transcriptional mechanism. *Nature* 355: 219–224.

Ferguson-Smith, A. C., Sasaki, H., Cattanach, B. M. and Surani, M. A. 1993. Parental origin-specific epigenetic modification of the mouse H19 gene. *Nature* 362: 751–755.

Fiering, S., Kim, C. G., Epner, E. M. and Groudine, M. 1993. An "in-out" strategy using gene targeting and FLP recombinase for the functional dissection of complex DNA regulatory elements: Analysis of the β-globin locus control region. *Proc. Natl. Acad. Sci. USA* 90: 8469–8473.

Forrester, W. C., Genderen, C. van, Jenuwein, T. and Grosschedl, R. 1994. Dependence of enhancer-mediated transcription of the immunoglobulin μ gene on nuclear matrix attachment regions. *Science* 265: 1221–1225.

Frank, D., Lichtenstein, M., Paroush, Z., Bergmann, Y., Shani, M., Razin, A. and Ceder. H. 1990. Demethylation of genes in animal cells. *Philos. Trans. R. Soc. Lond. [B]* 326: 241–251.

Garcia-Ramirez, M., Rocchini, C. and Ausio, J. 1995. The modulation of chromatin folding by histone acetylation. *J. Biol. Chem.* 270: 17923–17928.

Gartler, S. M., Liskay, R. M. and Grant, N. 1973. Two functional X chromosomes in human fetal oocytes. *Exp. Cell Res.* 82: 464–466.

Gartler, S. M., Rivest, M. and Cole, R. E. 1980. Cytological evidence for an inactive X chromosome in murine oogonia. *Cytogenet. Cell Genet.* 28: 203–207.

Gasser, S. M. and Laemmli, U. K. 1986. Cohabitation of scaffold binding regions with upstream/enhancer elements of three developmentally regulated genes in *D. melanogaster*. *Cell* 46: 521–530.

Geyer, P. K. and Corces, V. G. 1992. DNA position-specific repression of transcription by a *Drosophila* zinc finger protein. *Genes Dev.* 6: 1865–1873.

Gong, Q. and Dean, A. 1993. Enhancer-dependent transcription of the ε-globin promoter requires promoter-bound GATA-1 and enhancer bound AP-/NF-E2. *Mol. Cell Biol.* 13: 911–917.

Grosveld, F., Blom van Assendelft, G., Greaves, D. R. and Kollins, G. 1987. Position-dependent high-level expression of the human β-globin gene in transgenic mice. *Cell* 51: 975–985.

Groudine, M. and Conklin, K. F. 1985. Chromatin structure and *de novo* methylation of sperm DNA: Implications for activation of the paternal genome. *Science* 228: 1061–1068.

Groudine, M. and Weintraub, H. 1981. Activation of globin genes during chick development. *Cell* 24: 393–401.

Groudine, M., Kohwi-Shigematsu, T., Gelinas, R., Stamatoyannopoulos, G. and Papayannopoulo, T. 1983. Human fetal to adult hemoglobin switching: Changes in chromatin structure of the β-globin gene locus. *Proc. Natl. Acad. Sci. USA* 80: 7551–7555.

Gruenbaum, Y., Ceder, H. and Razin, A. 1982. Substrate and sequence specificity of a eukaryotic DNA methylase. *Nature* 295: 620–622.

Guo, B and eight others. 1995. The nuclear matrix protein NMP-1 is the transcritpion factor YY1. *Proc. Natl. Acad. Sci. USA* 92: 10526–10530.

Hanscombe, O., Whyall, D., Fraser, P., Yannoutsos, N., Greaves, D., Dillon, N. and Grosveld, F. 1991. Importance of globin gene order for correct developmental expression. *Genes Dev.* 5: 1387–1394.

Hebbes, T. R., Clayton, A. L., Thorne, A. W. and Crane-Robertson, C. 1994. Core histone acetylation co-maps with generalized DNase I sensitivity in the chick β-globin chromsomal domain. *EMBO J.* 7: 1395–1402.

Herman, R., Weymouth, L. and Penman, S. 1978. Heterogeneous nuclear RNA-protein fibers in chromatin depleted nuclei. *J. Cell Biol.* 78: 663–674.

Holliday, R. 1987. The inheritance of epigenetic defects. *Science* 238: 163–170.

Hotchkiss, R. D. 1948. The quantitative separation of purines, pyrimidines, and nucleosides by paper chromatography. *J. Biol. Chem.* 175: 315–332.

Hutchinson, N. and Weintraub, H. 1985. Localization of DNase I-sensitive sequences to specific regions of interphase nuclei. *Cell* 43: 471–482.

Iguchi-Ariga, S. M. M. and Schaffner, W. 1989. CpG methylation of the cAMP-responsive enhancer/promoter sequence TGACGTCA abolishes specific factor binding as well as transcriptional activation. *Genes Dev.* 3: 612–619.

Imbalzano, A. N., Kwon, H., Green, M. R. and Kingston, R. E. 1994. Facilitated binding of TATA-binding protein to nucleosome DNA. *Nature* 370: 481–485.

Jackson, D. A. and Cook, P. R. 1985. Transcription occurs at a nucleoskeleton. *EMBO J.* 4: 919–925.

Jenuwein, T., Forrester, W. C., Fernandez-Herrero, L. A., Laible, G., Dull, M and Grosschedl, R. 1997. Extension of chromatin accessibility by nuclear matrix attachment regions. *Nature* 385: 269–272.

Jeppesen, P. and Turner, B. M. 1993. The inactive X chromosome in female mammals is distinguished by a lack of histone H4 acetylation, a cytogenetic marker for gene expression. *Cell* 74: 281–289.

Jost, J. P. 1993. Nuclear extracts of chicken embryos promote an active demethylation of DNA by excision repair of 5-methyldeoxycytidine. *Proc. Natl. Acad. Sci. USA* 90: 4684–4688.

Kafri, T. and seven others. 1992. Developmental pattern of gene-specific DNA methylation in the mouse embryo and germ line. *Genes Dev.* 6: 705–714.

Karlsson, S. and Nieuhaus, A. W. 1985. Developmental regulation of human globin genes. *Annu. Rev. Biochem.* 54: 1071–1108.

Kay, G. F., Penny, G. D., Patel, D., Ashworth, A., Brockdorff, N. and Rastan, S. 1993. Expression of *Xist* during mouse development suggests a role in the initiation of X chromosome inactivation. *Cell* 72: 171–182.

Keith, D. H., Singersam, J. and Riggs, A. D. 1986. Active X-chromosome DNA is unmethylated at eight CCGG sites clustered in a guanine-plus-cytosine-rich island at the 5' end of the gene for phosphoglycerate kinase. *Mol. Cell Biol.* 6: 4122–4125.

Kellum, R. and Schedl, P. 1991. A position-effect assay for boundaries of higher order chromatin domains. *Cell* 64: 941–950.

Keshet, I., Lieman-Hurwitz, J. and Cedar, H. 1986. DNA methylation affects the formation of active chromatin. *Cell* 44: 535–543.

Knoll, J. H. M., Nicholls, R. D., Magenis, R. E., Graham, J. M., Jr., Lalande, M. and Latt, S. A. 1989. Angelman and Prader-Willi syndromes share a common chromosome 15 deletion but differ in the parental origin of the deletion. *Am. J. Med. Genet.* 32: 285–290.

Kornberg, R. D. and Thomas, J.D. 1974. Chromatin structure: Oligomers of histones. *Science* 184: 865–868.

Kratzer, P. G. and Chapman, V. M. 1981. X-chromosome reactivation in oocytes of *Mus caroli. Proc. Natl. Acad. Sci. USA* 78: 3093–3097.

Kuroda, M. I., Kernan, M. J., Kreber, R., Ganetzky, B. and Baker, B. S. 1991. The maleless protein associates with the X chromosome to regulate dosage compensation in *Drosophila. Cell* 66: 935–947.

Kwon, H., Imbalzano, A. N., Khavari, P. A., Kingston, R. E.,and Green, M.R. 1994. Nucleosome disruption and enhancement of activator binding by a human SW1/SNF complex. *Nature* 370: 477–481.

Lee, D. Y., Hayes, J. J., Pruss, D. and Wolffe, A. P. 1993. A positive role for histone acetylation on transcription factor access to nucleosomal DNA. *Cell* 72: 73–84.

Lee, J. T., Strauss, W. M., Dausman, J. A. and Jaenisch, R. 1996. A 450 kb transgene displays properties of the mammalian X-inactivation center. *Cell* 86: 83–94.

Lewin, B. 1994. Chromatin and gene expression: Constant questions, but changing answers. *Cell* 79: 397–406.

Li, E., Beard, C. and Jaenisch, R. 1993. The role of DNA methylation in genomic imprinting. *Nature* 36: 362–365.

Liskay, R. M. and Evans, R. 1980. Inactive X chromosome DNA does not function in DNA-mediated cell transformation for the hypoxanthine phosphoribosyltransferase gene. *Proc. Natl. Acad. Sci. USA* 77: 4895–4898.

Lucchesi, J. C. and Manning, J. E. 1987. Gene dosage and compensation in *Drosophila melanogaster. Adv. Genet.* 24: 371–429.

Ludérus, L. A., M. E. E., de Graaf, A., Mattia, E., den Blaauwen, J. L., Grande, M. A., de Jong, L. and van Driel, R. 1992. Binding of matrix attachment regions to lamin B1. *Cell* 70: 949–959.

Lyon, M. F. 1961. Gene action in the X chromosome of the mouse (*Mus musculus* L.) *Nature*: 190: 372–373.

Mandel, J. L. and Chambon, P. 1979. DNA methylation differences: Organ-specific variations in methylation pattern within and around ovalbumin and other chick genes. *Nucleic Acids Res.* 7: 2081–2103.

Mariman, E. C. M., van Eekelen, C. A. G., Reinders, R. J., Berns, A. J. M. and van Venrooji, W. J. 1982. Adenoviral heterogenous nuclear RNA is associated with the host nuclear matrix during splicing. *J. Mol. Biol.* 154: 103–119.

Martin, D. I. K., Tsai, S.-F. and Orkin, S. H. 1989. Increased γ-globin expression in a nondeletion HPFH mediated by an erythroid-specific DNA-binding factor. *Nature* 338: 435–437.

Martin, D. I. K., Fiering, S. and Groudine, M. 1996. Regulation of β-globin gene expression: straightening out the locus. *Curr. Opin. Genet. Dev.* 6: 488–495.

Mattei, M. G., Mattei, J. F., Vidal, I. and Giraud, F. 1981. Structural anomalies of the X chromosome and activation center. *Hum. Genet.* 56: 401–408.

Mavilio, F. and nine others. 1983. Molecular mechanisms for human hemoglobin switching: Selective undermethylation and expression of globin genes in embryonic, fetal, and adult erythroblasts. *Proc. Natl. Acad. Sci. USA* 80: 6907–6911.

McArthur, M. and Thomas, J. O. 1996. A preference of histone H1 for methylated DNA. *EMBO J.* 15: 1705–1714.

McGhee, J. D. and Ginder, G. D. 1979. Specific DNA methylation sites in the vicinity of the chick β-globin genes. *Nature* 280: 419–420.

Migeon, B. R. 1971. Studies of skin fibroblasts from ten families with HGPRT deficiency, with reference to X-chromosomal inactivation. *Am. J. Hum. Genet.* 23: 199–209.

Migeon, B. R. and Jelalian, K. 1977. Evidence for two active X chromosomes in germ cells of female before meiotic entry. *Nature* 269: 242–243.

Migeon, B. R., Wolf, S. F., Axelman, J., Kaslow, D.C. and Schmidt, M. 1985. Incomplete X chromosome dosage compensation in chorionic villi of human placenta. *Proc. Natl. Acad. Sci. USA* 82: 3390–3394.

Migeon, B. R., Schmidt, M., Axelman, J. and Cullen, C. R. 1986. Complete reactivation of X chromosomes from human chorionic villi with a switch to early DNA replication. *Proc. Natl. Acad. Sci. USA* 83: 2182–2186.

Migeon, B. R., Holland, M. M., Driscoll, D. J. and Robinson, J. C. 1991. Programmed demethylation in CpG islands during human fetal development. *Somatic Cell Molec. Genet.* 17: 159–168.

Miller, T. E., Huang, C.-Y. and Pogo, A. O. 1978. Rat liver nuclear skeleton and ribonucleoprotein complexes containing hnRNA. *J. Cell Biol.* 76: 675–691.

Mizzen, C. A. and eleven others. 1996. The TAF$_{II}$250 subunit of TFIID has histone acetyltransferase activity. *Cell* 87: 1261–1267.

Mohandas, T., Sparkes, R. S., Hellkuhl, B., Brzeschik, K. H. and Shapiro, L. J. 1980. Expression of an X-linked gene from an inactive human X chromosome in mouse-human hybrid cells: Further evidence for the non-inactivation of the steroid sulfatase locus in man. *Proc. Natl. Acad. Sci. USA* 77: 6759–6763.

Mohandas, T., Sparkes, R. S. and Shapiro, L. J. 1981. Reactivation of an inactive human X chromosome: Evidence for X inactivation by DNA methylation. *Science* 211: 393–396.

Monk, M., Boubelik, M. and Lehnert, S. 1987. Temporal and regional changes in DNA methylation in the embryonic, extraembryonic, and germ cell lineages during mouse embryo development. *Development* 99: 371–382.

Moore, K. L. 1977. *The Developing Human.* Saunders, Philadelphia.

Nelson, W. G., Pienta, K. J., Barrack, E. R. and Coffey, D. S. 1986. The role of the nuclear matrix in the organization and function of DNA. *Annu. Rev. Biophys. Chem.* 15: 457–475.

Nicholls, R. D., Kroll, J. H. M., Butler, M. G., Karma, S. and Lalande, M. 1989. Genetic imprinting suggested by maternal heterodisomy in non-deletion Prader-Willi syndrome. *Nature* 342: 281–285.

Norris, D. P. Patel, D., Kay, G. F., Penny, G. D., Brockdorff, N., Sheardown, S. A. and Rastan, S. 1994. Evidence that random and imprinted *Xist* expression is controlled by preemptive methylation. *Cell* 77: 41–51.

Orkin S. H. 1992. GATA-binding transcription factor in hematopoietic cells. *Blood* 80: 575–581.

Ottolenghi, S. 1992. Developmental regulation of human globin genes: A model for cell differentiation in the hematopoietic system. *In* V. E. Russo et al., (eds.), *Development: The Molecular Genetic Approach.* Springer-Verlag, New York, pp. 519–536.

Oudet, P., Gross-Bellard, M. and Chambon, P. 1975. Electron microscope and biochemical evidence that chromatin structure is a repeating unit. *Cell* 4: 281–300.

Paroush, Z., Keshet, I., Yisraeli, J. and Cedar, H. 1990. Dynamics of demethylation and activation of the α-actin gene in myoblasts. *Cell* 63: 1229–1337.

Pazin, M. J., Kamakaka, R. T.,and Kadonaga, J. T. 1994. ATP-dependent nucleosome reconfiguration and transcriptional activation from preassembled chromatin templates. *Science* 266: 2007–2011.

Penny, G. D., Kay, G. F., Sheardown, S. A., Rastan, S. and Brockendorff, N. 1996. Requirement for *Xist* in X chromosome inactivation. *Nature* 379: 131–137.

Perrine, S. P. and eight others. 1993. A short-term trial of butyrate to stimulate fetal globin gene expression in the β-globin disorders. *N. Engl. J. Med.* 328: 81–86.

Peterson, C. L. and Tamkun, J. W. 1995. The SW1/SNF complex: A chromatin remodeling machine? *Trends Biochem. Sci.* 20: 143–146.

Pevny, L. and seven others. 1991. Erythroid differentiation in chimeric mice blocked by a targeted mutation in the gene for transcription factor GATA-1. *Nature* 349: 257–261.

Prioleau, M.-N., Huett, J., Sentenac, A. and Méchali, M. 1994. Competition between chromatin and transcription complex assembly regulates gene expression during early development. *Cell* 77: 439–449.

Pruss, D., Bartholomew, B., Persinger, J., Hayes, J., Arents, G., Moudrianakis, E. and Wolffe, A. P. 1996. An asymmetric model for the nucleosome: A binding site for linking histones inside the DNA gyres. *Science* 274: 614–617.

Raich, N., Clegg, C. H., Grofti, J., Roméo, P.-H. and Stamatoyannopoulos, G. 1995. GATA1 and YY1 are developmental repressors of the human ε-globin gene. *EMBO J.* 14: 801–809.

Razin, S. V. and Gromova, I. I. 1995. The channels model of nuclear matrix structure *BioEssays* 17: 443–450.

Reik, W., Collick, A., Norris, M. L., Barton, S. C. and Surani, M. A. 1987. Genomic imprinting determines methylation of paternal alleles in transgenic mice. *Nature* 328: 248–250.

Rigaud, G., Roux, J., Pictet, R. and Grange, T. 1991. In vivo footprinting of rat TAT gene: Dynamic interplay between glucocorticoid receptor and a liver-specific factor. *Cell* 67: 977–986.

Robinson, S. I., Nelkin, B. D. and Vogelstein, B. 1982. The ovalbumin gene is associated with the nuclear matrix of chicken oviduct cells. *Cell* 28: 99–106.

Rodgers, G. P., Dover, G. J., Vyesaka, N., Noguchi, C. T., Schecter, A. N. and Nieuhuis, A. W. 1993. Augmentation by erythropoietin of the fetal hemoglobin response to hydroxyurea in sickle cell disease. *N. Engl. J. Med.* 328: 73–80.

Rogers, J. and Wall, R. 1981. Immunoglobulin heavy-chain genes: Demethylation accompanies class switching. *Proc. Natl. Acad. Sci. USA* 78: 7497–7501.

Russell, L. B. 1963. Mammalian X chromosome action: Inactivation limited in spread and region of origin. *Science* 140: 976–978.

Ryan, T. M., Behringer, R. B., Martin, N. C., Townes, T. M., Palmiter, R. D. and Brinster, R. L. 1989. A single erythroid-specific DNase I super-hypersensitivity site activates high levels of human β-globin gene expression in transgenic mice. *Genes Dev.* 3: 314–323.

Ryoji, M. and Worcel, A. 1984. Chromatin assembly in *Xenopus* oocytes: In vitro studies. *Cell* 37: 21–32.

Samollow, P. B., Ford, A. L. and VandeBerg, J. L. 1987. X-linked gene expression in the Virginia opossum: Differences between the paternally derived *Gpd* and *Pgk-A* loci. *Genetics* 115: 185–195.

Sanford, J. P., Clark, H. J., Chapman, V. M. and Rossant, J. 1987. Differences in DNA methylation during oogenesis and spermatogenesis and their persistence during early embryogenesis in the mouse. *Genes Dev.* 1: 1039–1046.

Sapienza, C., Peterson, A. C., Rossant, J. and Balling, R. 1987. Degree of methylation of transgenes is dependent on gamete of origin. *Nature* 328: 251–254.

Scheuermann, R. H. and Chen, U. 1989. A developmental-specific factor binds to suppressor sites flanking the immunoglobulin heavy-chain enhancer. *Genes Dev.* 3: 1255–1266.

Schlissel, M. S. and Brown, D. D. 1984. The transcriptional regulation of *Xenopus* 5S RNA genes in chromatin: The roles of active stable transcription complex and histone H1. *Cell* 37: 903–913.

Sharman, G. B. 1971. Late DNA replication in the paternally derived X chromosome of female kangaroos. *Nature* 230: 231–232.

Stalder, J., Larsen, A., Engel, J. D., Dolan, M., Groudine, M. and Weintraub, H. 1980. Tissue-specific DNA cleavages in the globin chromatin domain introduced by DNase I. *Cell* 20: 451–460.

Stamatoyannopoulos, J. A., Goodwin, A., Joyce, T. and Lowrey, C. M. 1995. NF-E2 and GATA binding motifs are required for the formation of DNase-I hypersensitive site-4 of the human P-globin locus control region. *EMBO J.* 14: 106–116.

Stief, A., Winter, D. M., Strätling, W. H. and Sippel. A. E. 1989. A nuclear DNA attachment element mediates elevated and position-dependent gene activity. *Nature* 341: 343–345.

Stöger, R., Kublicka, P, Liu, C.-G., Kafri, T., Razin, A., Cedar, H. and Barlow, D. P. 1993. Maternal-specific methylation of the imprinted mouse *Igf2r* locus identifies the expressed locus as carrying the imprinting signal. *Cell* 73: 61–71.

Swain, J. L., Stewart, T. A. and Leder, P. 1987. Parental legacy determines methylation and expression of an autosomal transgene: A molecular mechanism for parental imprinting. *Cell* 50: 719–727.

Tagaki, N. 1974. Differentiation of X chromosomes in early female mouse embryos. *Exp. Cell Res.* 86: 127–135.

Tagaki, N. and Abe, K. 1990. Detrimental effects of two active X chromosomes on early mouse development. *Development* 109: 189–201.

Talbot, D. and Grosveld, F. 1991. The 5′ HS2 of the globin locus control region enhances transcription through the interaction of a multimeric complex binding at two functionally distinct NF-E2 binding sites. *EMBO J.* 10: 1391–1398.

Tazi, J. and Bird, A. P. 1990. Alternative chromatin structure at CpG islands. *Cell* 60: 909–920.

Thoma, F., Koller, T. and Klug, A. 1979. Involvement of histone H1 in the organization of the nucleosome and of the salt-dependent superstructures of chromatin. *J. Cell Biol.* 83: 403–427.

Thorburn, A. and Knowland, J. 1993. Attachment of vitellogenin genes to the nucleoskeleton accompanies their activation. *Biochem. Biophys. Res. Commun.* 191: 308–313.

Trudel, M. and Constantini, F. 1987. A 3′ enhancer contributes to the stage-specific expression of the human β-globin gene. *Genes Dev.* 1: 954–961.

Tsukiyama, T. and Wu, C. 1995. Purification and properties of an ATP-dependent nucleosome remodeling factor. *Cell* 83: 1011–1020.

Tsukiyama, T., Becker, P. B. and Wu, C. 1994. ATP-dependent nucleosome disruption at a heat-shock promoter mediated by binding of GAGA transcripion factor. *Nature* 367: 525–531.

Tuan, D., Abeliovich, A., Lee-Oldham, M. and Lee, D. 1987. Identification of regulatory elements in human β-like globin genes. *In* G. Stamatoyannopoulos and A. W. Nienhuis (eds.), *Developmental Control of Globin Gene Expression.* Alan R. Liss, New York, pp. 211–220.

Turner, B. M. 1991. Histone acetylation and control of gene expression. *J. Cell Sci.* 99: 13–20.

van der Ploeg, L. H. T. and Flavell, R. D. 1980. DNA methylation in the human γ–δ–β globin locus in erythroid and non-erythroid cells. *Cell* 19: 947–958.

van Eekelen, C. A. G. and van Venrooij, W. J. 1981. HnRNA and its attachment to a nuclear protein matrix. *J. Cell Biol.* 88: 554–563.

Villeponteau, B., Lundell, M. and Martinson, H. 1984. Torsional stress promotes the DNase hypersensitivity of active genes. *Cell* 39: 469–478.

Wandersee, N. J., Ferris, R. C., and Ginder, G. D. 1996. Intronic and flanking sequences are required to silence enhancement of an embryonic beta-type globin. *Mol. Cell Biol.* 16: 236–246.

Wang, Z.-Q., Fung, M. R., Barlow, D. P. and Wagner, E. F. 1994. Regulation of embryonic growth and lysosomal targeting by the imprinted Igf2/Mpr gene. *Nature* 372: 464–467.

Weintraub, H. 1984. Histone H1-dependent chromatin superstructures and the suppression of gene activity. *Cell* 38: 17–27.

Weintraub, H. 1985. Assembly and propagation of repressed and derepressed chromosomal states. *Cell* 42: 705–711.

Weintraub, H. and Groudine, M. 1976. Chromosomal subunits in active genes have an altered configuration. *Science* 193: 848–856.

West, J. D., Frels, W. I., Chapman, V. M. and Papaioannou, V. E. 1977. Preferential expression of the maternally derived X chromosome in the mouse yolk sac. *Cell* 12: 873–882.

Wijgerde, M., Gribnau, J., Trimborn, T., Nuez, B., Philipsen, S., Grosvelde, F. and Fraser, P. 1996. The role of EKLF in human β-globin gene competition. *Genes Dev.* 10: 2894–2902.

Wilson, C. J., Chao, D. M., Imbalzano, A. N., Schnitzler, G. R., Kingston, R. E. and Wilson, E. B. 1895. *An Atlas of the Fertilization and Karyogenesis of the Ovum.* Macmillan, New York.

Wolf, S. F., Jolly, D. J., Lunnen, K., Friedmann, T. and Migeon, B. R. 1982. Methylation of the hypoxanthine phosphoribosyltransferase locus on the human X chromosome: Implications for X chromosome inactivation. *Proc. Natl. Acad. Sci. USA* 81: 2806–2810.

Wolf, S. F., Dintgis, S., Toniolo, D., Persico, G., Lunnen, K. D., Axelman, J. and Migeon, B. R. 1984. Complete concordance between glucose-6-phosphate dehydrogenase activity and hypomethylation of 3′ CpG clusters: Implication for X chromosome dosage compensation. *Nucleic Acids Res.* 12: 9333–9348.

Wolfe, S. L. 1993. *Molecular and Cellular Biology.* Wadsworth, Belmont, CA.

Yisraeli, J., Adelstein, R. S., Melloui, D., Nudel, U., Yaffe, D. and Ceder, H. 1986. Muscle-specific activation of a methylated chimeric actin gene. *Cell* 46: 409–416.

Young, R. A. 1996. RNA polymerase II holoenzyme contains SWI/SNF regulators involved in chromatinremodeling. *Cell* 84: 235–244.

Zaret, K. S. and Yamamoto, K. R. 1984. Reversible and persistent changes in chromatin structure accompany activation of a glucocorticoid-dependent enhancer element. *Cell* 38: 29–38.

Zhao, K., Hart, C. M. and Laemmli, U. K. 1995. Visualization of chromosomal domains with boundary element-associated factor BEAF-32. *Cell* 81: 879–889.

Zuccotti, M. and Monk, M. 1995. Methylation of the mouse Xist gene in sperm and eggs correlates with imprinted Xist expression and paternal X-inactivation. *Nat. Genet.* 9: 316–320.

Control of development by differential RNA processing and translation

12

Between the conception
And the creation...
Between the potency
And the existence
Between the essence
And the descent
Falls the Shadow.
T. S. ELIOT (1936)

There is no rest for the messenger til the
message is delivered.
JOSEPH CONRAD (1920)

THE REGULATION OF GENE EXPRESSION is not confined to the differential transcription of the DNA. Even though a particular RNA transcript is synthesized, there is no guarantee that it will create a functional protein in the cell. To become an active protein, the RNA must be (1) processed into a messenger RNA by the removal of introns, (2) translocated from the nucleus to the cytoplasm, and (3) translated by the protein-synthesizing apparatus. In some cases, the protein synthesized is not in its mature form and (4) must be posttranslationally modified to become active. Regulation can occur at any of these steps during development.

■ CONTROL OF DEVELOPMENT BY DIFFERENTIAL RNA PROCESSING

The essence of differentiation is the production of different sets of proteins in different types of cells. In bacteria, differential gene expression can be effected at the levels of transcription, translation, and protein modification. In eukaryotes, however, another possible level of regulation exists, namely, control at the level of RNA processing and transport. This chapter will present two ways in which differential RNA processing can regulate development. The first involves the "censoring" of which nuclear transcripts can be processed into cytoplasmic messages. Here, different cells can select different nuclear transcripts to be processed and placed into the cytoplasm as messenger RNA. The same pool of nuclear transcripts can thereby give rise to different populations of cytoplasmic mRNAs in different cell types. The second mode of differential RNA processing concerns splicing the mRNA precursors into different proteins by using different combinations of potential exons. If an mRNA precursor were to have five potential exons, one cell might use exons 1, 2, 4, and 5; a different cell might utilize exons 1, 2, and 3; and yet another cell type might use a different combination. Thus, one gene may create a family of related proteins.

Control of early development by nuclear RNA selection

In the late 1970s, numerous investigators found that mRNA was not the primary transcript from the genes. Rather, the genes transcribed a *nuclear* RNA (nRNA), sometimes called *heterogeneous nuclear* RNA (hnRNA) because of its wide size range. This nRNA was often many times longer than the message, and it seemed to decay more rapidly. We now know that the nuclear RNA contains introns that get spliced out during the passage from nucleus to cytoplasm. In addition, studies in sea urchins suggested that entire transcripts are being degraded in some cells and processed into mRNA in others. In other words, these studies suggested that different cell types may be transcribing the same type of nuclear RNA, but that different subsets of this population are being processed to mRNA in different types of cells. [RNA1.html]

Kleene and Humphreys (1977, 1985) showed that nuclear RNA from whole pluteus larvae and from blastulae were (within experimental error) identical to each other. Both RNAs bound to the same 30 percent of the genome. When the RNA complexity was analyzed, nRNA from blastula cells was found to bind to 15 percent of this DNA (i.e., to 30 percent of the genomic single-copy DNA). Similarly, pluteus-stage nRNA, even when present in great excess, also bound to 30 percent of the single-copy DNA. Are these two sets of DNA sequences the same, or are they different? This question was approached by mixing blastula and pluteus nuclear RNAs and adding them to the denatured single-copy DNA. If the sequences were totally different, one would expect 30 percent of the DNA to be bound (i.e., 60 percent of the genome would be coding for the combined set of blastula and pluteus messages). If they were identical, one would expect 15 percent of the DNA to be bound. The result is shown in Figure 12.1. The mixture bound to only 15 percent of the DNA. The nRNA sequences of the blastula and pluteus bound to the same DNA. Within experimental error, the nRNA was identical in blastula and pluteus cells. Wold and her colleagues (1978) extended these observations by showing that sequences present in blastula *messenger* RNA (isolated from translating polysomes) but absent in gastrula and adult tissue mRNA were nonetheless present in the *nuclear* RNA of the gastrula and adult tissues. These results were interpreted to indicate that more genes are transcribed in the nucleus than are allowed to become mRNAs in the cytoplasm (Figure 12.2; Table 12.1).

Figure 12.1
Hybridization of nuclear RNA from sea urchin embryos with single-copy [³H]DNA. Radioactive single-copy DNA was mixed with either blastula RNA, pluteus RNA, or a mixture of blastula and pluteus RNA. The mixtures were incubated to allow all complementary sequences to pair. (The RNA C_0t axis is the concentration of the RNA times the time allowed to incubate.) In all three cases, about 15 percent of the DNA hybridized to the RNA. (After Kleene and Humphreys, 1977.)

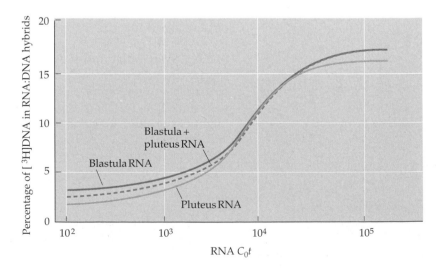

Table 12.1 Intertissue comparisons of structural gene sequences in messenger RNA and nuclear RNA

Reference tracer complementary to	Normalized reaction with parent mRNA		Normalized reaction with other mRNA		Normalized reaction with nRNA	
	mRNA	%	mRNA	%	nRNA	%
SEA URCHIN						
Blastula mRNA (single-copy DNA)	Blastula	100	Intestine	12	Intestine	97
			Coelomocyte	13	Coelomocyte	101
MOUSE						
Brain mRNA (total cDNA)	Brain	100	Kidney	78	Kidney	102
Brain mRNA (cDNA representing rare messages)	Brain	100	Kidney	56	Kidney	100

Source: Davidson and Britten, 1979.

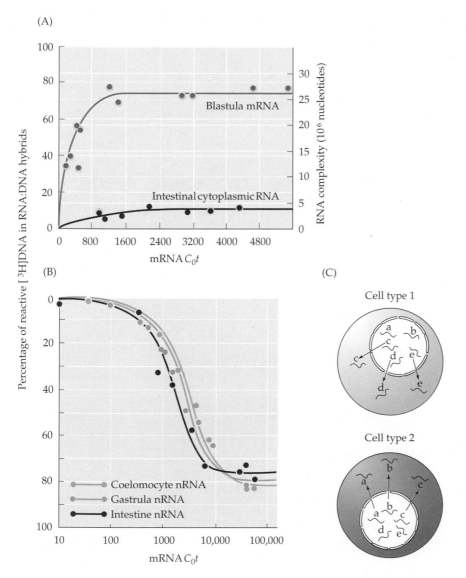

Figure 12.2
Sequences found in the nuclear RNA of several cell types but not found in the mRNA. (A) Specificity of sea urchin blastula messenger cDNA. Hybridization of blastula messenger cDNA (cDNA to blastula mRNA) with blastula mRNA and intestinal cytoplasmic RNA shows that the mRNAs are very different. (B) Hybridization of the blastula messenger cDNA to the nuclear RNAs (nRNAs) of gastrulae and adult coelomocytes and intestinal cells suggests the identity of all nuclear RNAs. (C) Speculative model based on differential RNA processing. In both cell types, the same RNAs (a, b, c, d, e) are transcribed, but in one cell type, sequences c, d, and e are processed to cytoplasmic mRNA, whereas in another cell type, sequences a, b, and c are processed and sent into the cytoplasm. (A and B after Wold et al., 1978.)

Figure 12.3
Assays for detecting the accumulation of a message in the cytoplasm. (A) Ribonuclease protection assay. RNA is isolated and purified from embryonic tissue. A radioactive RNA probe is synthesized that is complementary to a small stretch of the RNA one is assaying. If the specific RNA is present, the radioactive probe will bind to it. RNase is then added, destroying all the RNA except for the double-stranded region that contains the radioactive oligonucleotide. This can be electrophoresed on a gel and autoradiographed. (B) Nuclear run-on assay. Nuclei are isolated from embryonic tissue. Radioactive UTP is added to the nuclei. The mRNA that is being synthesized incorporates radioactive label as it continues being transcribed. The mRNA can be isolated and hybridized to complementary DNA sequences immobilized on paper. If the radioactive transcript binds, it is detected by autoradiography.

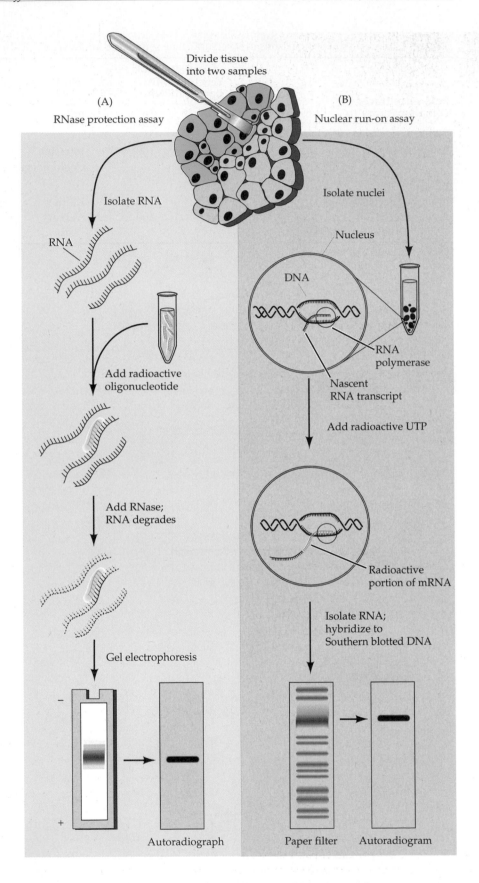

This posttranscriptional control of RNA processing has been confirmed for specific messages by ribonuclease protection assays and nuclear run-on transcription assays. The **ribonuclease protection assay** is a sensitive test for determining the presence (or absence) of a particular sequence in a population of RNAs (Figure 12.3). One makes a relatively small radioactively labeled RNA that is complementary to a specific sequence in the RNA one wishes to detect. This RNA "probe" is mixed with cellular RNA that one is testing for the presence of the particular sequence. If the sequence is present, the probe binds to it. Then ribonuclease (RNase, an enzyme that digests single-stranded RNA) is added to the mixture to cleave away all the RNA that is not hybridized. However, the double-stranded RNA formed by the hybridization of radioactive and cellular RNAs is not cleaved. This double-stranded RNA can be run on a gel and detected by autoradiography. If the RNA sequence is present, a band of a particular size should show up on the autoradiograph. Gagnon and his colleagues (1992) performed such an analysis on the transcripts from the *Spec1* and *CyIIIa* genes of the sea urchin *Strogylocentrotus purpuratus*. These genes encode calcium-binding and actin proteins, respectively, that are expressed only in the aboral ectoderm of the pluteus larva. Using probes that bound to an exon and to an intron, they found that these genes were being transcribed not only in the ectoderm cells, but also in the mesoderm and endoderm. The analysis of the *CyIIIa* gene showed that the concentration of *introns* was the same in both the gastrula ectoderm and in the mesoderm/endoderm samples, suggesting that this gene was being transcribed at the same rate in all cell types (Figure 12.4A). The *exon*, however, accumulated in the ectoderm (which expresses the protein), but not in the mesoderm or endoderm (which do not). Thus, while the genes appear to be transcribed at a similar rate in the ectoderm and in other tissues, the messenger RNA for these proteins (represented by the exons) accumulates only in the ectoderm.

This conclusion was confirmed when nuclei were isolated from the gastrula tissues. Radioactive UTP enabled the researchers to follow whatever RNAs were being transcribed at the time the nuclei were isolated. Both ectoderm and endoderm/mesoderm nuclei were transcribing the *Spec1* gene at the gastrula stage (Figure 12.4B). Thus, the expression of the *Spec1* and *CyIIIa* genes is at the level of RNA processing in the gastrula. Later in development (by the pluteus stage), these genes come under transcriptional control, where transcription of the genes stops in the cells not expressing these proteins. It appears, then, that RNA processing plays a major role in the control of sea urchin gene expression in early embryos.

The mechanisms of RNA splicing: Spliceosomes

The splicing of pre-mRNA is mediated through a 60S nuclear particle called a **spliceosome.** The spliceosome is composed of five **small nuclear (sn) RNAs** (the U1, U2, U4, U5, and U6 snRNAs) and numerous proteins. These proteins often associate with the snRNAs to form small nuclear ribonucleoprotein partcles (**snRNPs**) that are named for their associated snRNAs (such as the U2 snRNP). The spliceosome does not exist as a preformed complex floating in the nucleus, but is assembled on the pre-mRNA in a multistep process. The 5′ splice site is first identified by the U1 snRNA by base complementarity. The 5′ splice site at the beginning of each intron has a consensus sequence that is recognized by the U1 snRNA (Figure 12.5). The 3′ end of the intron is recognized by the U2 snRNP auxiliary factor, **U2AF** (Ruskin et al., 1988; Wu and Maniatis, 1993). This recognition of the splice sites establishes a "commitment complex" wherein other snRNPs will associate with

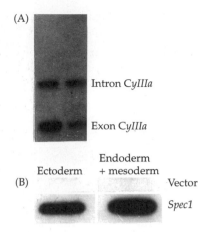

(A)

Intron *CyIIIa*

Exon *CyIIIa*

Ectoderm | Endoderm + mesoderm

(B) | Vector

Spec1

Figure 12.4
Regulation of ectoderm-specific gene expression by RNA processing. (A) Autoradiographs of ribonuclease protection assay. The left-hand column represents RNA isolated from the gastrula ectodermal tissue; the right-hand column represents RNA isolated from the endodermal and mesodermal tissues. The upper band is the RNA protected by a probe that binds to an intron sequence (which should be found only in the nucleus) of *CyIIIa*. The lower band represents the RNA protected by a probe complementary to an exon sequence. (B) Results of a nuclear transcription run-on assay. Radioactive RNA synthesized in vitro by ectodermal nuclei (left) and mesodermal and endodermal nuclei (right) was hybridized to an intron of the *Spec1* gene affixed to a filter. Both ectoderm and endoderm/mesoderm nuclei were transcribing this gene in the sea urchin gastrula, even though the *Spec1* message is seen only in the ectodermal cells. (From Gagnon et al., 1992, courtesy of R. and L. Angerer.)

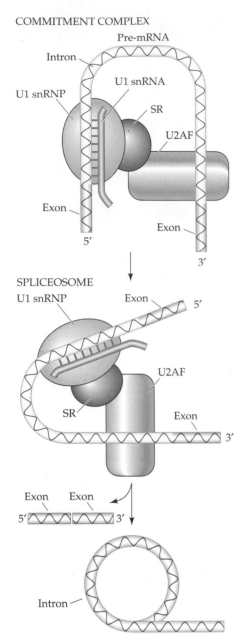

COMMITMENT COMPLEX

SPLICEOSOME

Figure 12.5
Splicing the intron and connecting the adjacent exons. In the "commitment complex" leading to the formation of the spliceosome, the 5' and 3' splicing junctions of the intron have been recognized by the U1 snRNA and the U2AF protein, respectively. The U1 snRNP and the U2AF are stabilized by proteins of the SR family. The U2AF and SR proteins are thought to be replaced by small nuclear riboproteins (snRNPs) that facilitate the cleaving of the intron and the ligation of the two adjacent exons. (After Hodges and Beggs, 1994.)

these proteins and eventually catalyze the removal of the intron (Hodges and Beggs, 1994). [RNA2.html]

The average vertebrate pre-mRNA consists of relatively short exons (averaging about 140 bases) separated by introns that are usually much longer. Any mechanism for the coordination of splicing in a multi-exon RNA must provide an explanation of how the small exons are conserved and separated from the larger introns. Berget (1995) has proposed the notion that splicing is done from one end of the *exon* to another, rather than across the intron. This hypothesis of **exon definition** contends that the small size of the exons allows the U2 snRNA (at the 5' end of the exon) to connect with the U1 snRNA at the other end. Following this "definition" of the exon boundaries, the various exons are brought together.

The processing of mature mRNA also requires the addition of a poly(A) tail to the nuclear mRNA. The 3' end of most eukaryotic mRNAs (histone messages being the sole known exceptions) is formed by cleaving the original transcript and adding segments of adenylate residues. The 3' untranslated region of most mRNA precursors contains the sequence AAUAAA, which is essential for the cleavage of the RNA 10 to 30 bases downstream from this site (Proudfoot and Brownlee, 1976). Mutations of this sequence prevent the 3' end formation of mRNA (Wickens and Stephenson, 1984; Orkin et al., 1985). Another *cis*-acting element is a GU- or U-rich sequence, usually located further downstream (3') from the cleavage. This sequence appears to be critical for the efficient cleavage of the nuclear RNA at the 3' processing site (McDevitt et al., 1984; Christofori and Keller, 1988). [RNA3.html]

Alternative RNA splicing: Creating alternative proteins from the same gene

One Gene, Many Related Proteins

In addition to deciding which RNAs are allowed into the cytoplasm, the regulation of development by RNA processing can also occur by **alternative RNA splicing.** Most mammalian pre-mRNAs contain numerous introns. By the selective recognition of these one can have alternative RNA splicing. This can occur in several ways (see Figure 12.6). Cells could differ in their ability to recognize the 5' splice site or the 3' splice site. Or some cells would not recognize a sequence as an intron at all, retaining it within the message. Whether a cell recognizes the splice sites depends on certain factors in the nucleus that can interact with these sites and compete or cooperate with the proteins that normally recognize them. The 5' splice site is recognized by U1 snRNA, but only with the cooperation of a protein called **splicing factor 2** (**SF2**; alternative splicing factor). In at least some cases, the choice between the alternative 5' splice sites is influenced by the ratio of protein SF2 and another protein, hnRNP-A1. In general, an excess of SF2 results in the utilization of the proximal (closer) 5' splice site, while an excess of hnRNP-A1 results in the spliceosome's utilizing the distal (farther) 5' splice site (Mayeda and Krainer, 1992). The choice of alternative 3' splice sites is often controlled by which splice site can best bind U2AF.

What is an intron in one cell's nucleus may be an exon in another cell's nucleus. Alternative RNA processing has been found to control the alternative forms of expression of over 100 proteins. The deletion of certain potential exons in some cells but not in others enables one gene to create a family of closely related proteins. Instead of one gene-one polypeptide, one can

CONSTITUTIVE SPLICING

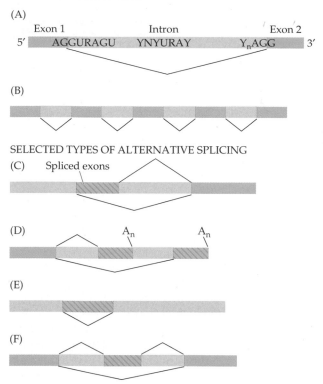

Figure 12.6
Schematic diagram of alternative pre-mRNA splicing. Exons are represented as shaded boxes, alternatively spliced exons are represented by hatched boxes, and introns are represented by broad lines. By convention, the path of splicing is shown by fine V-shaped lines. (A) The exon-intron borders, showing the consensus sequences at the 5' and 3' ends of the intron. R represents either purine, Y either pyrimidine, and N any nucleotide. (B) The splicing of a pre-mRNA that has five exons. (C–F) Alternative splicing by (C) alternative 5' splice sites, (D) alternative 3' splice sites (in some cases, this would provide different termini to the mRNA, and both sites would need a polyadenylation sequence, shown here as A_n), (E) a splice/no splice decision, and (F) exon inclusion/exon skipping. (After Horowitz and Krainer, 1994.)

have one gene–one family of proteins. For instance, the mRNA precursor for the cell adhesion molecule N-CAM can be alternatively processed into over 100 different forms, depending on which exons are included in the mRNA. Although only four major forms of this protein are usually seen in any embryo, some of the minor forms are seen in the brain and heart (Zorn and Krieg, 1992). Similarly, alternative RNA splicing enables the β-tropomyosin gene to encode both the skeletal muscle and fibroblast forms of this protein. The nuclear RNA for β-tropomyosin contains 11 exons. Exons 1–5, 8, and 9 are common to all mRNAs expressed from this gene. Exons 6 and 11 are also used in fibroblasts and smooth muscle cells, while exons 7 and 10 are used in the synthesis of skeletal muscle β-tropomyosin (Figure 12.7). In smooth muscles and fibroblasts, a protein is made that blocks spliceosomes from forming at the the skeletal muscle-specific splice sites (Guo et al., 1991; d'Orval et al., 1991). In the nervous system, K+ channel diversity plays an important role in regulating the membrane excitability. These kinetic differences have been correlated to the alternative splicing of the message precursor from the *shaker* gene (Mottes and Iverson, 1995).

Figure 12.7
Schematic diagram of alternative RNA splicing in the β-tropomyosin mRNA precursor. In the skeletal muscle, possible exons 6 and 11 are skipped (and made into introns), while in fibroblasts and smooth muscle cells, possible exons 7 and 10 are made into introns, and 6 and 11 are used as exons.

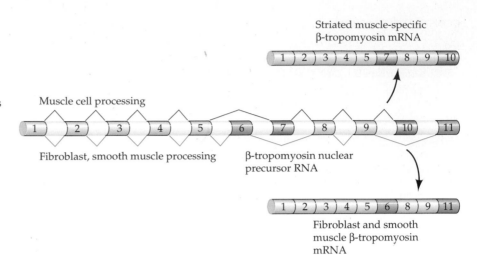

In some cases, the properties of the alternatively spliced proteins may have important consequences during development. The five different human fibronectin proteins are generated from one pair of identical fibronectin genes. The diverse (and in some instances organ-specific) forms of fibronectin come from different mRNAs generated by splicing together different exons of the fibronectin mRNA precursor (Tamkun et al., 1984; Hynes, 1987). Some forms of fibronectin are found on the pathways over which embryonic cells migrate, whereas other forms are not, suggesting that the alternatively spliced forms of fibronectin have different embryonic functions (ffrench-Constant and Hynes, 1989). Certain splicing isoforms of fibroblast growth factor 8 (FGF8) will only interact with receptors generated from particularly spliced RNA. Thus, FGF8b (one of seven splicing variants of FGF8) will bind to FGF receptors 2c and 3c but not to FGF receptors 1b, 1c, 2b, or 3b. In the developing mouse, FGF8b is produced from the apical ectodermal ridge of the limb bud and from the branchial arches and nasal pits of the head. In each of these places, the underlying mesenchyme expresses the FGFR2c receptor (Cover photograph; Crossley and Martin, 1995; MacArthur et al., 1995).

Alternative RNA Processing and Drosophila *Sex Determination*

Alternative RNA splicing can generate families of proteins whose members can have different functions. There is nothing to prevent such alternative RNA splicing from making alternative transcription factor proteins, and this technique has been utilized by *Drosophila* to control its sexual differentiation. Moreover, sex determination in *Drosophila* is regulated by a cascade of RNA-processing events (see Baker et al., 1987; MacDougall et al., 1995).

As we will see in Chapter 20, the development of the sexual phenotype in *Drosophila* is mediated by a series of genes that convert the X chromosome-to-autosome ratio into either a male or a female cell. When the X-to-autosome ratio is 1 (i.e., when there are two X chromosomes per diploid cell), the embryo develops into a female fly. When the ratio is 0.5 (i.e., when the fly is XY with only one X chromosome per diploid cell), the embryo develops into a male (Figure 12.8).

One of the key genes of this pathway is *transformer* (*tra1*). This gene is necessary for the production of females, and its loss results in male flies, irrespective of the chromosomal ratio. Throughout the larval period, the *tra1* gene actively synthesizes a transcript that is processed into a general mRNA (found in both females and males) or into a female-specific mRNA (Figure

12.9). Only females contain the alternatively spliced message. The general mRNA found in both males and females contains a stop codon (UGA) early in the second exon, and the small protein produced by this mRNA is not functional. Therefore, the general nonspecific transcript has no bearing on sex determination (Belote et al., 1989). However, in the female-specific message, this UGA codon is in an *intron* that is spliced out during mRNA formation and does not interfere with the translation of the message. In other words, the female transcript is the only functional transcript of this gene. In fact, when the cDNA of this female-specific transcript is incorporated into the genomes of XY flies, these flies become female. The protein encoded by the female-specific mRNA appears to be an arginine-rich peptide 196 amino acids long (Boggs et al., 1987).

What makes the *transformer* (*tra1*) gene process a female-specific transcript in XX cells and not in XY cells? It appears that sex-specific alternative splicing of *tra1* nRNA involves competition between two possible 3' (acceptor) splice sites in the intron. Sosnowski and her colleagues (1989) have provided evidence that this competition is altered by the presence or absence of a functional *Sex-lethal* gene product. The *Sex-lethal* (*Sxl*) gene is one of the first genes on the sex phenotype pathway, and it acts prior to *transformer*. If the X-to-autosome ratio is 1, functional Sxl protein is made.* This gene does not make a functional protein in XY embryos or larvae. When the *Sex-lethal* gene is functional, the *transformer* gene makes both the general and the female-specific transcript and the fly becomes female. If the *Sex-lethal* gene is deleted or mutated, the *transformer* gene makes only the nonfunctional general transcript and the fly becomes male. It appears that the product of the *Sex-lethal* gene is controlling which of the 3' splice sites is being used.

There are two major ways that the Sex-lethal protein could control which 3' splice site is used. One way is to block the use of the general acceptor site so that only the alternative, female-specific acceptor site can be used. The other model is to activate the female-specific acceptor site in a positive manner. Valcárcel and colleagues (1993) have shown that the Sex-lethal protein inhibits splicing at the general (non-female-specific) acceptor site by specifically binding to its polypyrimidine tract. This blocks the binding of a splicing factor, U2AF, to the general site and causes it to use the lower-affinity female-specific acceptor site. It appears, then, that if the general 3' acceptor splice site of *transformer* is blocked (either by mutation or by the Sex-lethal protein), the alternative, female-specific acceptor site is used (see Figure 12.9). The result is a female fly.

The Transformer-1 protein is, itself, an alternative splicing factor, and it regulates the splicing of the nuclear transcript of the *doublesex* gene (*dsx*). This gene is needed for the production of either sexual phenotype, and mutations of *dsx* can reverse the expected sexual phenotype, causing XX embryos to become males or XY embryos to become females. During pupation, *doublesex* makes a transcript that can be processed in two alternative ways. It can generate a female-specific mRNA or a male-specific mRNA (see Figure 12.9; Nagoshi et al., 1988). In females and males, the first three exons are the same. However, the fourth exons are different. The male-specific RNA deletes a large section of the precursor RNA that includes the female-specific exon.

Tian and Maniatis (1992) have shown that sex-specific processing of *dsx* pre-mRNA involves the activation of the female-specific 3' splice site by the products of the *transformer* and *transformer-2* genes. The polypyrimidine

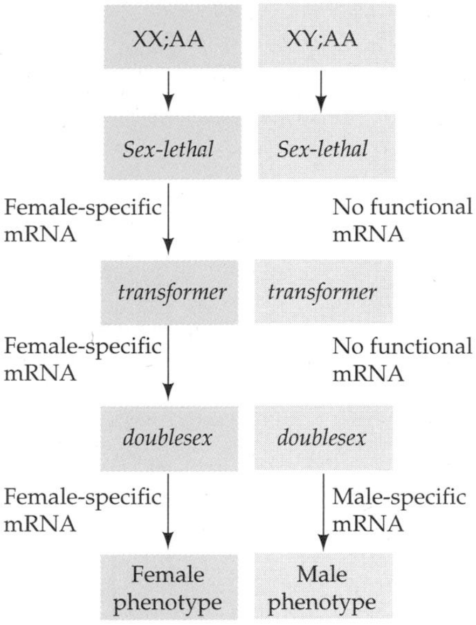

Figure 12.8
Sex determination in *Drosophila*. This simplified scheme shows that the X-to-autosome ratio is monitored by the *Sex-lethal* gene. If this gene is active, it processes the *transformer* pre-mRNA into a functional female-specific message. In the presence of the female-specific Transformer protein, the *doublesex* gene transcript is processed in a female-specific fashion, leading to the production of the female phenotype. If the *transformer* gene does not make a female-specific product (i. e., if the *Sex-lethal* gene is not activated), the *doublesex* transcript is spliced in the male-specific manner, leading to the realization of the male phenotype. (The details of this pathway will be discussed in Chapter 20.)

---

*The Sxl protein is itself a product of a complex type of alternative RNA splicing. More will be said about that in Chapter 20.

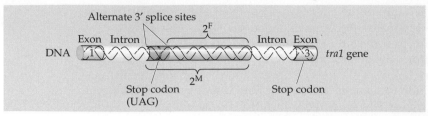

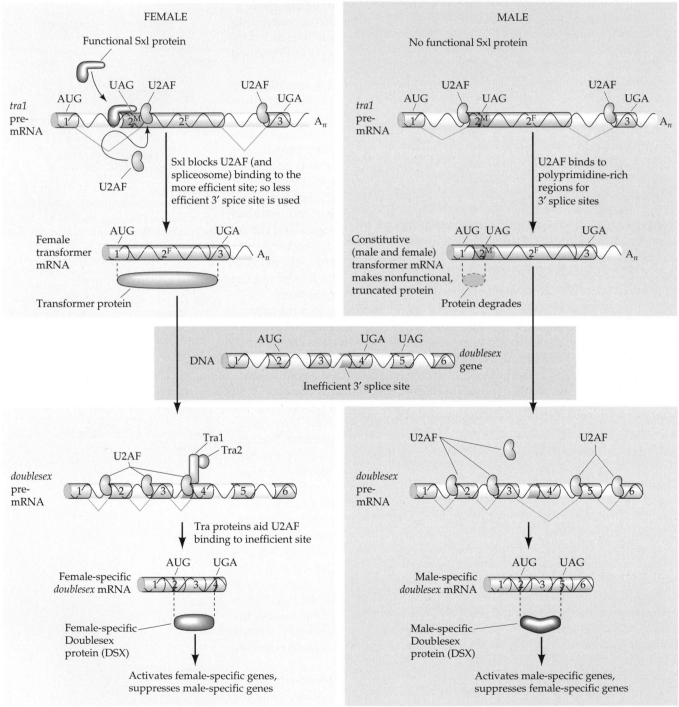

◀ **Figure 12.9**
Schematic representation of alternative splicing events in the sex determination pathway of *Drosophila*. The female pathway is on the left, the male pathway is on the right, and the *transformer-1* and *doublesex* genes are in the center. In the female pathway, active Sxl protein is made when the X-to-autosome ratio is 1. (How that happens will be discussed in Chapter 20.) Active Sxl protein blocks the usual 3' splice site of the first intron of the *tra1* pre-mRNA. This forces the spliceosome to use another 3' splice site. In the male pathway, no Sxl protein is made, and the spliceosome uses the more efficient site. This leads to the incorporation of sequences into the mRNA that encode a stop codon (UAG) early in the message. The truncated Tra1 peptide made from this message does not appear to have a function. It is as if the gene were inactive. In the female pathway, the active Tra1 protein combines with Tra2 protein to stabilize U2AF and spliceosome assembly at the 3' splice site of the third intron of the *doublesex* pre-mRNA. This leads to the formation of an mRNA containing the fourth exon. In males, the absence of Tra1 protein prevents U2AF binding and spliceosome assembly at that site. Instead, exons 5 and 6 are utilized in the male mRNA. (After MacDougall et al., 1995.)

(U/C-rich) tract in front of exon 4 in the *doublesex* mRNA precursor is usually a weak binder of U2AF, because it is broken by a group of purine residues (represented by a serrated line in Figure 12.9). Therefore, it is not usually an efficient 3' splice site. However, in the presence of the Transformer (and Transformer-2) proteins, this site becomes an efficiently utilized site (Tian and Maniatis, 1993). This means that the splice site will be used in females (which have the active Transformer proteins) but not in males (which lack them). The male *doublesex* mRNA will not have exon 4, while the female transcript will. The Doublesex proteins made by these mRNAs are both transcription factors. Moreover, they recognize the same sequence of DNA. However, while the "female" Doublesex protein will activate female-specific enhancers (such as those that make yolk proteins), the "male" Doublesex proteins will inhibit transcription from those same enhancers (Coschigano and Wensink, 1993; Jursnich and Burtis, 1993). Conversely, the "female" Doublesex protein can inhibit transcription from genes that are otherwise activated by the "male" Doublesex protein. Research into *Drosophila* sex determination shows that differential RNA processing plays enormously important roles throughout development.

Widespread Use of RNA Processing to Control Gene Expression

We still know relatively little about the mechanisms of alternative mRNA processing or about the ways in which some cells process transcripts that other cells do not. The mechanism underlying such differential RNA processing may give us insights into the very core of cell differentiation and embryonic determination.

■ TRANSLATIONAL REGULATION OF DEVELOPMENTAL PROCESSES

After a messenger RNA has been transcribed, processed, and exported from the nucleus, it still needs to be translated to form the protein encoded in the genome. In the sections that follow, we will see that regulation at the level of translation is an extremely important mechanism in the control of gene expression. In such cases, the message is already present in the cytoplasm but may or may not be translated, depending on certain cellular conditions. Thus, translational control of gene expression can be used when a burst of protein synthesis is needed immediately (as in the case of newly fertilized eggs), or it can be used as a fine-tuning mechanism to ensure that a very precise amount of protein is made from the available supply of messages (as in hemoglobin synthesis). We will also see that there are several ways to effect translational control and that different cells have evolved different means to do so.

Mechanisms of eukaryotic translation

Translation is the process by which the information contained in the nucleotide sequence of mRNA instructs the synthesis of a particular polypeptide. This process, outlined in Figure 12.10, has been divided into three phases—initiation, elongation, and termination—and it is regulated by soluble proteins called (appropriately) initiation factors, elongation factors, and termination factors (Hershey, 1989; Safer, 1989).

Initiation consists of the reactions wherein the first aminoacyl-transfer RNA and the mRNA are bound to the ribosome. The only transfer RNA (tRNA) capable of initiating translation is a special initiator tRNA (tRNA$_i$), which carries the amino acid methionine. As shown in Figure 12.11, the first reactions involve the formation of an **initiation complex** consisting of methionyl-initiator tRNA bound to a 40S ("small") ribosomal subunit. This reaction is catalyzed by the active form of **eukaryotic initiation factor 2** (eIF2-GTP), which binds the initiator Met-tRNA to the 40S ribosomal subunit. Note that this binding occurs in the absence of mRNA. The mRNA is added next. First, a **cap-binding protein—eIF4E**—binds to the 7-methylguanosine cap at the 5′ end of the message. Without this cap, the binding of mRNA to the ribosomal subunit is often not completed (Shatkin, 1976, 1985), and eIF4E is critical for the translation to proceed. However, there is less eIF4E than the number of messages in the cell, so it is thought that each mRNA has to compete for this cap-binding protein (Thach, 1992). Initiation factor 4A then complexes with eIF4E and positions itself on a helical hairpin loop in the leader sequence of the mRNA. The eIF4A (stimulated by eIF4B and ATP) unwinds the helix. This step can be rate-limiting if the hairpin loop helix is hidden by some other stable secondary structure. The 40S ribosomal subunit then travels down the message until it reaches an AUG codon in the proper context. Kozak (1986) has shown that not just any AUG will do. For the 40S

Figure 12.10
Schematic representation of the events of eukaryotic translation. The initiation steps bring together the 40S and 60S ribosomal subunits, mRNA, and the initiator tRNA, which is complexed to the amino acid methionine (Met). During elongation, amino acids are brought to the polysome, and peptide bonds are formed between the amino acids. The sequence of amino acids in the growing protein is directed by the sequence of nucleic acid codons in the mRNA. After the last peptide bond of the protein has been made, one of the codons UAG, UGA, or UAA signals the termination of translation. The ribosomal subunits and message can be reutilized.

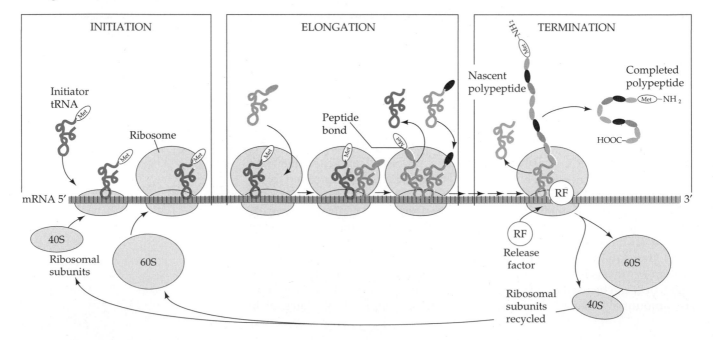

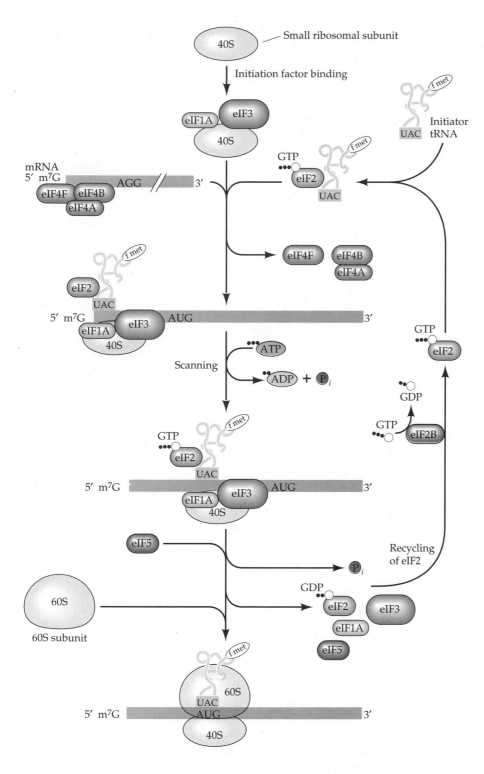

Figure 12.11
Initiation phase of eukaryotic translation. All the initiation factors are represented as circles. The first complex is made by the union of the 40S ribosomal subunit with initiator tRNA. The initiator tRNA has been complexed with the active (GTP) form of initiation factor 2. After this complex is formed, the mRNA is positioned with the aid of cap-binding protein (eIF4E) and other eIF4 subunits. Once the mRNA is in place, eIF5 mediates the joining of the 60S ribosomal subunit and the release of the previous initiation factors. The eIF2, now in its inactive (GDP) form, is reactivated by eIF2B. (After Hershey, 1989; Thach, 1992; Cooper, 1996.)

ribosomal subunit to stop and initiate translation, the nucleotides around AUG are also important. By mutating cloned genes and assaying the translation of their RNAs, Kozak found the "optimum" sequence to be AC-**CA**UG**G**. Mutations in the flanking nucleotides could reduce translation 20-fold. The importance of the flanking sequence also has been seen in vivo. Morle and co-workers (1985) have reported a patient whose α-thalassemia (deficiency of the α-globin subunit of hemoglobin) was due to a change in this sequence from ACCAUGG to CCCAUGG. The binding of the 40S sub-

Figure 12.12
Individual polysome transcribing the giant mRNA from the BR2 puff of *Chironomus tentans*. (A) Electron micrograph of a polysome containing 74 ribosomes. The nascent proteins can be seen extending from the ribosomes and growing as the ribosomes move from the 5′ end of the message to the 3′ end. Near the 3′ end are ribosomes from which the protein has detached. (B) Higher magnification of such a polysome; the polysome has been stretched during the preparation of the specimen. The relationship of the mRNA to the ribosomal subunits and the nascent polypeptide can be seen. (From Francke et al., 1982; photographs courtesy of J. E. Edstrom.)

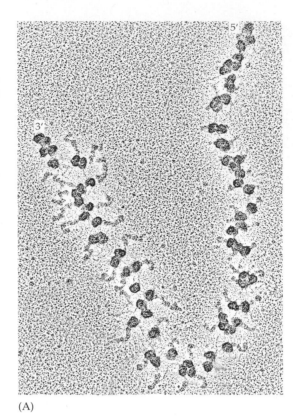

(A)

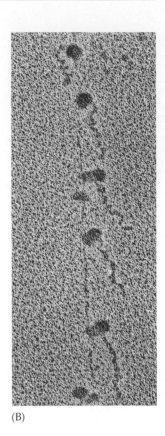

(B)

unit to the AUG of the message positions the initiator tRNA over the AUG codon. Only after the mRNA has been properly positioned on the small ribosomal subunit can the 60S ("large") ribosomal subunit bind. This completes the initiation reaction. During this process, the GTP on eIF2 is hydrolyzed to GDP. For the eIF2 to pick up a new initiator tRNA, it must be regenerated to eIF-GTP by eIF2B.

Elongation involves the sequential binding of aminoacyl-tRNAs to the ribosome and the formation of peptide bonds between the amino acids as they sequentially relinquish their tRNA carriers (see Figure 12.10). As amino acids are joined together, the ribosome travels down the message, thereby exposing new codons for tRNA binding. This allows another ribosome to initiate on the 5′ end of the message and begin its traveling. Thus, any mRNA usually will have several ribosomes attached to it. This structure is then called a polyribosome—or, more commonly, a **polysome** (Figure 12.12). The **termination** of protein synthesis takes place when one of the mRNA codons UAG, UAA, or UGA is exposed on the ribosome. These nucleotide triplets (called *termination codons*) are not recognized by tRNAs and hence do not code for any amino acids. Rather, they are recognized by release factors, which hydrolyze the peptide from the last tRNA, freeing it from the ribosome. The ribosome separates into its two subunits, and the cycle of translation begins anew.

Control of protein synthesis by differential longevity of mRNA

One of the chief ways of regulating gene expression at the translational level involves the selective degradation or stabilization of the mRNA. If the mRNA were degraded rapidly after it entered the cytoplasm, it could generate very few proteins. However, if a message with a relatively short half-life were selectively stabilized in certain cells at certain times, then it would make large amounts of that particular protein only at certain times and places.

Selective Degradation of mRNAs

LONGEVITY OF AN MRNA AND ITS 3′ UNTRANSLATED REGION. Not all mRNAs have the same stability within a cell. A stable message such as β-globin has a half-life of around 17 hours, while the mRNAs for various growth factors have half-lives of less than 30 minutes. Thus, the amount of protein made from a single globin message should be much greater than that from a single growth factor message. Recent data, summarized by Decker and Parker (1994), suggest that RNA sequences within the **3′ untranslated region (3′ UTR)** can promote the rapid deadenylation of the 3′ polyadenylate tail. This leads to the decapping of the 5′ end of the message and its subsequent 5′-to-3′ degradation. Therefore, the major determinants regulating message half-life appear to reside in the 3′ UTR. The more ephemeral RNA species contain one or more AU-rich sequences in this region. Shaw and Kamen (1986) inserted a 51-base-pair AT-rich region from the 3′ UTR of the gene for growth factor GM-CSF into the 3′ UTR of the rabbit β-globin gene (Figure 12.13). The resulting globin message had a half-life of less than 30 minutes. A similar sequence, but containing 14 G and C residues, was inserted into another β-globin gene as a control. Its globin message had the normally long half-life.

The ability to differentially degrade mRNAs is critical for cell function. For example, the *c-fos* gene encodes a transcription factor required for normal fibroblast cell division (Holt et al., 1986). Like the GM-CSF growth factor message, the mRNA for *c-fos* contains large 3′ untranslated regions rich in AU sequences. If these regions are deleted (either experimentally or by natural mutation), the messages gain a longer half-life. Consequently, there is more c-Fos protein being made, and the cell is continuously signaled to divide. The result is a tumor of those cells having the *c-fos* gene that lacks the AU-rich 3′ UTR (Meijlink et al., 1985). Wilson and Treisman (1988) found that this region stimulates the removal of the poly(A) tail when that message is translated. If the AU-rich region was deleted or replaced by some other sequence, the poly(A) tail remained, and the message had a longer half-life. Several proteins have been found that recognize these 3′ UTR AU-rich regions, and these may accelerate the decay of the messages when they bind to them (Chen et al., 1992, 1994). Conversely, the 3′ UTR of the long-lived α-globin mRNA contains three C-rich regions that bind proteins that appear to stabilize the message (Kiledjian et al., 1995).

Differential poly(A) tail shortening plays a decisive role in the life cycle of slime mold *Dictyostelium*. In the slime mold, a new set of messages are transcribed during the switch from vegetative growth (amoeba) to develop-

Figure 12.13
Regulation of mRNA longevity by a sequence in the 3′ untranslated region. (A) The 3′ end of the rabbit β-globin gene was altered by the insertion of a 62-base-pair fragment derived from the 3′ end of the human *GM-CSF* gene or a related sequence in which several of the AT pairs had been replaced by GC pairs (indicated in color). (B) The clones were injected into cultured mouse cells, and the presence of the message after 30 hours was measured by incubating cell extracts with a ^{32}P-labeled DNA complementary to the 5′ end of the message. If the rabbit β-globin message still existed, the radioactive cDNA would bind to it and therefore be resistant to S1 nuclease (which destroys single-stranded nucleic acids only). If the message were not present, the added S1 nuclease would digest the probe to mononucleotides, and no radioactive cDNA would be bound. The resultant solutions were run on a gel and autoradiographed. Lane 1: Extract from cells incorporating the wild-type (WT) cloned rabbit β-globin gene. Lane 2: Extract from cells incorporating the rabbit β-globin gene with the AT-rich 3′ end (showing no mRNA after 30 hours). Lane 3: Extract from cells incorporating the rabbit β-globin gene with the GC-substituted 3′ end (showing stable mRNA after 30 hours). The gene and probe for β$_2$-microglobulin (producing a long-lived mRNA) was used as a control. (After Shaw and Kamen, 1986.)

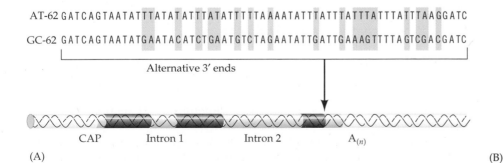

AT-62 GATCAGTAATATTTATATATTTATATTTTTAAAATATTTATTTATTTATTTATTTAAGGATC

GC-62 GATCAGTAATATGAATACATCTGAATGTCTAGAATATTGATTGAAAGTTTTAGTCGACGATC

Alternative 3′ ends

CAP Intron 1 Intron 2 A$_{(n)}$

(A)

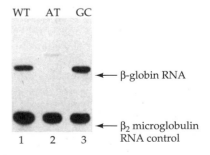

WT AT GC

←β-globin RNA

←β$_2$ microglobulin RNA control

1 2 3

(B)

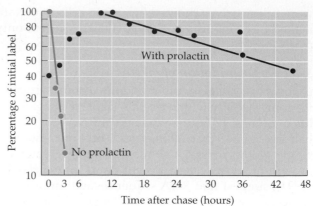

Figure 12.14
Degradation of casein mRNA in the presence and absence of prolactin. Cultured mammary cells were given radioactive RNA precursors (pulse) and after a given time were washed and fed nonradioactive precursors (chase). The casein RNA synthesized during the pulse time was then isolated and counted. In the absence of prolactin, the newly synthesized casein mRNA decayed rapidly, with a half-life of 1.1 hours. When the same experiment was done in medium containing prolactin, the half-life was extended to 28.5 hours. (After Guyette et al., 1979.)

ment (grex). At the same time, the poly(A) tails of the existing vegetative-stage mRNAs are dramatically shortened. As a result, the newly transcribed messages are translated, whereas the preexisting messages are not (Palatnik et al., 1984). This mechanism has also been seen in larval *Drosophila* salivary glands (Restifo and Guild, 1986).

HORMONAL STABILIZATION OF SPECIFIC MESSENGER RNAS. Differentiated gene products are often synthesized in response to hormonal induction. In some cases, hormones do not increase the transcription of certain messages, but act at the level of translation. One such case involves the synthesis of casein by lactating mammals. Casein is the major phosphoprotein of milk and is therefore a differentiated product of the mammary gland. As will be discussed more fully in Chapter 19, the mammary gland is prepared by the sequential actions of several hormones. Prolactin, however, is the hormone responsible for lactation—that is, actual milk production. Prolactin augments the transcription of casein messages only about twofold; its major effect appears to be the stabilization of casein mRNA (Guyette et al., 1979). Prolactin somehow increases the longevity of the casein message so that it exists 25 times longer than most other messages in the cell (Figure 12.14). Consequently, each casein mRNA can be used for more rounds of translation. In this way, a greater than normal number of casein molecules can be synthesized from each casein message. Table 12.2 summarizes these data and shows that other hormones also increase the stability of specific messenger RNAs.

Translational control of oocyte messages

In most animal species, the diploid nucleus is not immediately expressed. Evidence that early development is controlled by factors stored in or made by the oocyte came from several experiments at the end of the last century (reviewed in Davidson, 1976). These experiments clearly demonstrated the dominance of maternal traits during the initial stages of embryogenesis and a switch to paternal or hybrid characteristics only later in development. Such far-reaching maternal effects have already been alluded to in our dis-

Table 12.2 Stabilization of specific messenger RNAs by hormones

mRNA	Cell or tissue	Regulatory effector	Half-life (hours)	
			+ Effector	− Effector
Vitellogenin	*Xenopus* liver	Estrogen	500	16
Albumin	*Xenopus* liver	Estrogen	10	3
Vitellogenin	Avian liver	Estrogen	22	~2.5
Apo VLDL II	Avian liver	Estrogen	26	3
Casein	Rat mammary gland	Prolactin	92	5
Growth hormone	Rat pituitary cell cultures	Dexamethasone and thyroxine	20	2
Insulin	Rat pancreatic islet cells	Glucose	77	29
Ovalbumin	Hen oviduct	Estrogen, progesterone	~24	2–5

Source: After Shapiro et al., 1987.

cussion of the cleavage orientation in snail embryos, in which the oocyte cytoplasm contains a factor that directs the rotations of the cleavage planes in a dextral or sinistral direction. [cleave1.html]

Characterization of Stored Oocyte Messenger RNAs

TYPES OF STORED OOCYTE MRNAS. Davidson and his colleagues have estimated the complexity of the oocyte mRNA in a manner similar to their analysis of nuclear RNA complexity. RNA (in large excess) was hybridized to denatured DNA, and the half C_0t value of the hybridization was found to be proportional to the amount of different RNA sequences present. By this analysis, they estimated that each oocyte (in numerous phyla) had enough different nucleotide sequences to account for roughly 1600 copies each of 20,000 to 50,000 RNA types (Galau et al., 1976; Hough-Evans et al., 1977). This is the greatest message complexity of any known cell type, and it reflects the enormous developmental potential of the oocyte. Relatively few of these messages have been characterized. Moreover, many of these mRNAs are not used in the oocyte but are stored there and translated after fertilization. This was first shown by using protein synthesis inhibitors; but more recently, PCR and RNase protection assays have also shown the existence of mRNAs that are stored in the oocyte and first translated during oocyte maturation (immediately prior to and during ovulation), fertilization, or in early cleavage. Table 12.3 gives a partial list of these stored mRNAs. [RNA4.html]

Some of these mRNAs are for proteins that will be needed during cleavage, when the embryo makes enormous amounts of chromatin, cell membranes, and cytoskeletal components. One of the most remarkable instances is the storage of the information needed to make ribonucleotide reductase for the clam embryo. The large subunit is stored as a protein in the oocyte cytoplasm. The small subunit is stored as an untranslated maternal message. Only after fertilization, when the mRNA for the small subunit is translated, can the newly synthesized small subunit combine with the preformed larger subunit to generate the functional enzyme (Standart et al., 1986).

Some of these stored mRNAs regulate the timing of early cell division. In many species (including sea urchins and *Drosophila*) the rate and pattern of early cell divisions do not require a nucleus. Rather, they require continued protein synthesis from stored maternal mRNAs (Wagenaar and Mazia, 1978). The reason for this dependency on stored messages was shown in 1983, when Evans and colleagues found a class of proteins they called *cyclins*.

Table 12.3 Some mRNAs stored in oocyte cytoplasm and translated at or near fertilization

mRNAs encoding	Function(s)	Organism(s)
Cyclins	Cell division regulation	Sea urchin, clam, starfish, frog
Actin	Cell movement and contraction	Mouse, starfish
Tubulin	Formation mitotic spindles, cilia, flagella	Clam, mouse
Small subunit of ribonucleotide reductase	DNA synthesis	Sea urchin, clam, starfish
Hypoxanthine phosphoribosyl-transferase	Purine synthesis	Mouse
Vg1	Mesodermal determination(?)	Frog
Histones	Chromatin formation	Sea urchin, frog, clam
Cadherins	Blastomere adhesion	Frog
Metalloproteinases	Implantation in uterus	Mouse
Growth factors	Cell growth; uterine cell growth(?)	Mouse
Sex determination factor FEM-3	Sperm formation	*C. elegans*
PAR gene products	Segregate morphogenetic determinants	*C. elegans*
SKN-1 morphogen	Blastomere fate determination	*C. elegans*
Hunchback morphogen	Anterior fate determination	*Drosophila*
Caudal morphogen	Posterior fate determination	*Drosophila*
Bicoid morphogen	Anterior fate determination	*Drosophila*
Nanos morphogen	Posterior fate determination	*Drosophila*
GLP-1 morphogen	Anterior fate determination	*C. elegans*
Germ cell-less protein	Germ cell determination	*Drosophila*
Oskar protein	Germ cell localization	*Drosophila*
Ornithine transcarbamylase	Urea cycle	Frog
Elongation factor 1α	Protein synthesis	Frog
Ribosomal proteins	Protein synthesis	Frog, *Drosophila*

Sources: Compiled from numerous sources, including Raff, 1980; Shiokawa et al., 1983; Rappollee et al., 1988; Brenner et al., 1989; Standart, 1992.

These proteins regulate cell division (as discussed in Chapter 5) and are encoded by maternal mRNAs. What is striking about cyclins is that they are destroyed upon cell division and have to be resynthesized anew from the stored messages after the completion of each cleavage. Cyclin synthesis from stored messages is seen to decline as the embryo nears the end of the blastula stage.

Other stored messages encode proteins that determine the fate of cells. The *bicoid* and *nanos* messages of *Drosophila*, the *vg1* mRNA of *Xenopus*, and the *glp-1* mRNA of *C. elegans* are all critical in determining cell fate. As we will see later in this chapter, not only is the timing of their translation critical, but so is the location of the mRNA when it does become translated.

(A)

An2

RNA: T   A   V

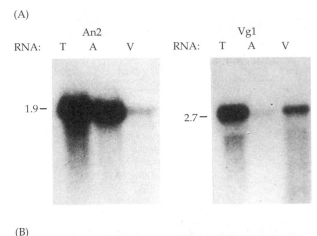

1.9 —

Vg1

RNA: T   A   V

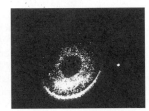

2.7 —

(B)

Figure 12.15
Demonstration of localized messages in the animal and vegetal poles of the *Xenopus* oocyte. RNA was obtained from the entire egg (T), the animal cap (A), or the vegetal cap (V) and electrophoretically separated on gels. The RNA was transferred to paper by the Northern blot procedure, and the paper was incubated with radioactive DNA from clones derived from cDNA complementary to oocyte messages. The radioactive DNA from clone *An*2 hybridizes to a message present in the animal pole but not in the vegetal pole. The reverse distribution is seen for the messages hybridizing to the DNA from clone *vg1*. (B) In situ hybridization showing the *vg1* message at different stages of localization in the *Xenopus* oocyte. In the mature oocyte, it resides solely in the vegetal cortex. (The *vg1* RNA has recently been shown to encode a critical factor for determining the dorsal-ventral axis in vertebrates and will be discussed more fully in Chapter 15.) (A from Rebagliati et al., 1985, courtesy of D. Melton; B from Melton, 1987.)

LOCALIZATION OF STORED MRNAS. Some stored mRNAs are not uniformly distributed throughout the oocyte (Rodgers and Gross, 1978). Rebagliati and co-workers (1985) have shown that whereas most of the maternal messages are found uniformly throughout the unfertilized *Xenopus* egg, some stored mRNAs are localized in the animal or vegetal pole of the cytoplasm. They extracted poly(A)-containing RNA from oocytes and used reverse transcriptase and DNA polymerase to convert the RNAs into a population of double-stranded DNAs. These double-stranded DNAs were then inserted into cloning vectors and grown separately in *E. coli*. About 2 million clones were derived in this manner. The DNA from these clones (the *Xenopus* oocyte library) was then transferred to two pieces of filter paper and denatured under conditions yielding single-stranded DNA. The investigators then cut the animal or vegetal pole from the egg and extracted the poly(A)-containing RNA from these regions. Radioactive cDNAs were made from the RNAs, and one group of the DNA-containing filters was incubated in the cDNAs from the animal messages, while the other group was incubated in the cDNAs from the vegetal messages. When the binding of radioactive cDNAs was measured, most clones bound equal amounts of cDNA from the animal and vegetal poles, indicating that those messages were equally distributed. However, about 1.2 percent of the clones only bound cDNA made from animal pole messages, and about 0.2 percent of the clones only bound to the cDNA derived from vegetal pole mRNA.

The DNA of the clones specific for animal or vegetal messages could then be used to identify the localized mRNAs. RNA was extracted from whole eggs or their animal and vegetal poles and run on gels. The electrophoretically separated RNAs were blotted onto nitrocellulose paper (Northern blots) and probed with radioactive DNA from each of the region-specific clones. Two of these results are shown in Figure 12.15 and Plate 8. The localization of these mRNAs to specific regions of the egg is accomplished through the cytoskeleton and will be detailed in Chapters 13 and 20.

Determining Cell Fate by Localized Oocyte mRNA

MANY OF THE ACTIVITIES of early development occur without the embryonic nuclei being activated. Since the specification of the embryonic axes is usually one of the first processes of embryogenesis, it was long thought that these events might be regulated by maternal mRNAs. In recent years, this has been shown to be the case.

The anterior-posterior (head-tail) axis of *Drosophila* is specified primarily by the proteins encoded by the *bicoid, nanos,* and *hunchback* genes (Figure 12.16). Although we will discuss their activities more fully in subsequent chapters, we will discuss their translational regulation here. First, the *bicoid* and *nanos* mRNAs are transported into the egg from ovarian follicle cells. The *bicoid* mRNA stays in the anteriormost region of the oocyte, while the nanos message goes to the posterior pole. The *bicoid* message appears to be tethered by its 3′ UTR to anterior microtubules by the products of the *swallow* and *staufen* genes (Ferrandon et al., 1994; see Chapter 22). The *bicoid* message placed into the oocyte has a relatively short poly(A) tail, roughly 70 residues. However, within the first division cycle, the message is polyadenylated such that its tail length nearly doubles (Sallés et al., 1994; Lieberfarb et al., 1996). This polyadenylation coincides with the ability of the *bicoid* message to be translated. The Bicoid protein diffuses throughout the anterior portion of the early embryo, and it is responsible for specifying the head and thoracic regions of the *Drosophila* larva. The products of the *cortex* and *grauzone* genes appear to be critical for the polyadenylation of *bicoid* mRNA, since mutations in these genes prevent the polyadenylation and translation of the *bicoid* message.

The *nanos* mRNA is localized to the posterior pole of the egg during oogenesis, but significant amounts remain distributed throughout the cytoplasm. During early development, however, only the *nanos* message bound to the posterior cytoplasm is translated (Gavis and Lehmann, 1994). The correct positioning

of this message is also due to its 3′ UTR, and if other mRNAs (such as that for tubulin) are given the *nanos* 3′ UTR, those messages become localized to the posterior of the oocyte. When translated, the Nanos protein diffuses through the posterior portion of the embryo and specifies those cells retaining it to be abdominal. The *nanos* message appears to be translationally repressed by the binding of the Smaug protein to two sites in its 3′ UTR (Smibert et al., 1996). This repression is abrogated once the *nanos* mRNA becomes localized at the posterior pole. Here, the Oskar protein may be involved in binding the *nanos* message to the cytoskeleton, and the Vasa protein may act as a helicase to unravel the RNA (Liang et al., 1994). The Oskar protein itself is regulated translationally. However, in contrast to the *nanos* message, which is translated after fertilization, the *oskar* mRNA is synthesized and transported into the oocyte early in oogenesis and becomes localized to the posterior pole during the middle stages of oogenesis. Once it is localized, the message is translated into protein that remains at the posterior pole (Kim-Ha et al., 1995). Like the *nanos* and *bicoid* messages, its localization and timing of translation depends on its 3′

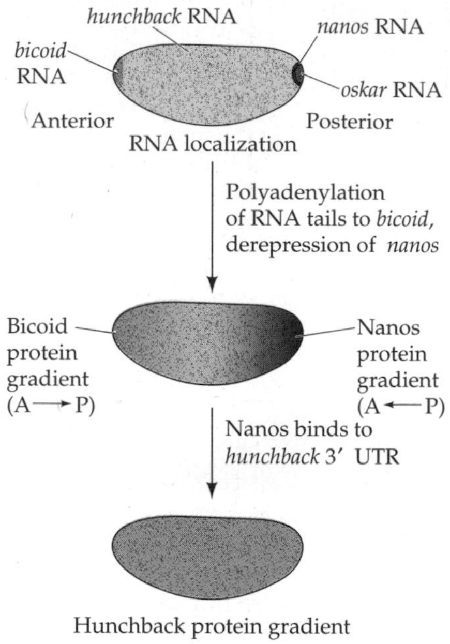

Figure 12.16 Control of anterior-posterior polarity in *Drosophila* embryos by maternal mRNA. (A) The *bicoid, nanos, oskar,* and *hunchback* messages are provided to the egg by the ovarian nurse cells. They are encoded by the mother's genome. The *oskar* mRNA enters earliest, is transported to the posterior pole, and is translated at mid oogenesis, before the other messages arrive. The *bicoid* message is tethered to the anterior of the egg by its 3′ UTR. The *nanos* message is directed by its 3′ UTR to go to the posterior pole, where it interacts with Oskar protein. The *hunchback* mRNA is seen throughout the oocyte. (B) Upon fertilization and egg activation, *bicoid* mRNA is polyadenylated and becomes translationally active, forming an anterior-posterior gradient of Bicoid protein. The *nanos* mRNA at the posterior pole becomes translationally competent and begins to make a posterior-to-anterior gradient of Nanos protein. (C) The Nanos protein binds to the 3′ UTR of the *hunchback* message to prevent translation. The Bicoid protein binds to the enhancer region of the *hunchback* gene to promote transcription of new *hunchback* messages. The result is a steep gradient of Hunchback protein. This protein will activate different genes at different concentrations, thereby specifying the different regions of the embryo.

UTR. The *nanos* and *oskar* mRNAs have relatively short tracts of poly(A) that are not significantly lengthened when the messages become translatable. This suggests that there are at least two mechanisms of activating the maternal messages in *Drosophila* eggs (Salles et al., 1994). The first is position-dependent and does not involve increasing the poly(A) tail (e.g., *oskar* and *nanos*). The second is position-independent and requires poly(A) synthesis (*bicoid,* also *Toll* and *torso* messages).

The Bicoid and Nanos proteins perform their functions by regulating the synthesis of Hunchback protein. Hunchback protein will eventually specify the head and thorax of the fly in a concentration-dependent manner. The Bicoid protein (now active in the anterior) acts as a transcription factor to activate the hunchback gene, thereby producing more *hunchback* message and protein in the anterior of the embryo. The Nanos protein, however, works at the level of *translation* to inhibit the production of Hunchback protein from existing *hunchback* mRNA. This creates a gradient whereby Hunchback protein synthesis is increased in the anterior of the embryo and actively repressed in the posterior (Wharton and Struhl, 1991; Wang et al., 1994). The 3′ UTR of the *hunchback* mRNA contains several sites that bind two factors (Pumilio and a 55-kDa protein) that do not themselves block *hunchback* translation. However, these two proteins appear to form a "landing site" for the Nanos protein. When Nanos binds, the translation of *hunchback* mRNA is inhibited (Murata and Wharton, 1995).

The Bicoid protein also works at the level of translation to block the synthesis of the Caudal protein. Like Nanos, the

(A)

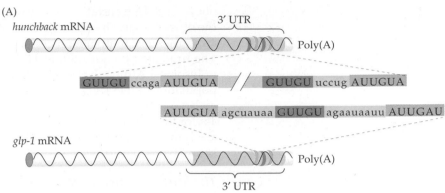

(B)

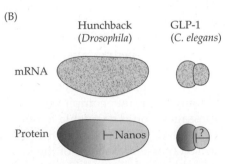

Figure 12.18 Similarities in the regulation of *hunchback* and *glp-1* mRNAs through their 3′ UTRs. (A) The 3′ UTR of *hunchback* message contains two regions considered essential for Nanos binding and the suppression of translation. These "*nanos* response elements" consist of the motifs GUUGU and AUUGUA. The same elements can be seen in the 3′ UTR of the *glp-1* message. (B) Model for the regulation of *hunchback* and *glp-1* translation. Both messages are uniformly distributed throughout the egg and early embryo. In both cases, the message is repressed in the posterior portion of the embryo. The regulator of *hunchback* translation is the posteriorly localized Nanos protein. The regulator of *glp-1* translation is not yet known but may be the PAL-1 proteins. (After Evans et al., 1994.)

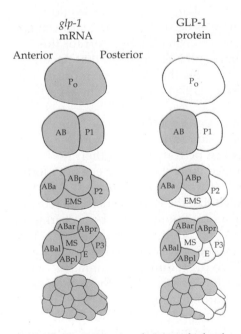

Figure 12.17 Comparison between the localization of the GLP-1 protein and the *glp-1* message. The maternal *glp-1* message is found in every cell of the *C. elegans* embryo throughout early development. The GLP-1 protein, however, is seen only in the progeny of the anterior cell formed at the first division. (After Evans et al., 1994.)

Caudal protein is critical in establishing the posterior segments of the fly, but unlike *nanos* message, the maternal *caudal* mRNA is uniformly distributed throughout the *Drosophila* egg. The Bicoid protein binds to the 3′ UTR of *caudal* mRNA, preventing it from being translated in the anterior of the embryo (Dubnau and Struhl, 1996). The mechanisms by which translational gene regulation determines the anterior-posterior axis of the *Drosophila* embryo will be detailed more fully in Chapter 14.

This may seem to be a workable solution to axis specification when the early-cleavage-stage embryo remains a syncytium that allows such gradients to form. However, recent experiments (Evans et al., 1994) have shown that such translational control of cell specification can also occur in embryos that form cells shortly after fertilization. In the nematode *C. elegans*, much of early embryogenesis depends on the protein GLP-1. The GLP-1 protein is a cell-surface receptor that receives signals from posterior cells

to specify the anterior cell fates (see Chapter 17). It is active between the 4- and 28-cell stages of cleavage. Antibody staining shows that the only cells that have this protein are the descendants of the anterior cell at the 2-cell stage. However, in situ hybridization shows that the maternal message for this protein is found in every cell of the embryo (Figure 12.17). The posterior cell of the 2-cell stage and its progeny appear to have an inhibitor of *glp-1* mRNA translation. This inhibitor has not yet been found, but the sequencing of the 3′ UTR has shown that the *glp-1* mRNA has a 3′ UTR with the same sequences that recognize Pumilio and Nanos (Figure 12.18). Thus, the anterior cells of *C. elegans*, like those of *Drosophila*, are specified by translational gene regulation. ■

Mechanisms for the translational regulation of oocyte messages

In Chapter 4, we saw evidence that the oocyte contained messenger RNAs that were present but untranslated until the oocyte was fertilized or activated (as in the progesterone activation of frog oocytes immediately prior to fertilization). There are at present at least five mechanisms that regulate oocyte mRNA translation. Three of them involve the availability of mRNAs; the other two involve the efficiency of mRNA translation. Most species probably use more than one mechanism to regulate oocyte mRNA translation. The fundamental question is, How are messenger RNAs recruited into the polysomes? Although the oocyte and the early blastomeres contain the same population of messages, different subsets are being translated in the egg and embryo (Young and Raff, 1979; Mermod et al., 1980; Rosenthal et al., 1980; Taylor and Smith, 1985). The question then becomes, How do mRNAs that are lying dormant in the cytoplasm of the oocyte suddenly acquire their competence to be translated?

The Masked Maternal Message Hypothesis

This hypothesis contends that the oocyte messages are physically masked by proteins, preventing the mRNA from attaching to ribosomes. Upon oocyte maturation or fertilization, the masking protein would fall away, enabling the mRNA to be translated. Messenger RNA is never found devoid of proteins. However, the type of protein associated with the RNA can vary. In 1966, Spirin proposed that the mRNA of the oocyte is stored in *informosomes*, ribonucleoprotein complexes in which the mRNA is masked. The masked messages would be unable to bind to the ribosomes and thus would not be translated. At fertilization, the proteins masking the message would be released (possibly because of the ionic changes occurring during fertilization) and the message would be free to initiate translation. Support for this hypothesis followed shortly. In 1968, Infante and Nemer found that the unfertilized sea urchin egg contains RNP particles that sediment more slowly than ribosomes, and Gross and co-workers (1973) found that these particles contain various mRNAs.

Support for the masked maternal message hypothesis came from experiments showing that while the mRNA of unfertilized eggs stored in RNPs cannot be translated, the same RNAs can be translated if their RNPs are placed in solutions mimicking the changed ionic state of the egg after fertilization (Jenkins et al., 1978; Raff, 1980). It was proposed that the influx of sodium during fertilization might destabilize the RNP particle, thereby allowing its mRNA to be translated. Such unmasking might be occurring in the surf clam *Spisula*, where the mRNAs encoding the small subunit of ribonucleotide reductase and cyclin A are severely repressed in oocytes and are not translated until fertilization. Two procedures can "unmask" these messages. First, high salt concentrations (0.5 M KCl) allow these mRNAs to make proteins; so does the removal of a particular sequence of bases in the 3' untranslated regions of these messages (Figure 12.19; Standart et al., 1990). There are regions in the 3' UTRs of both these messages that are very similar and may constitute binding sites for an 82-kDa protein that binds to the 3' UTR of these mRNAs (Standart, 1992). Upon fertilization, this protein becomes phosphorylated, and it appears that the phosphorylated form of this protein can no longer block translation. The phosphorylation of this protein may be accomplished by the cdc2 kinase that is activated at fertilization (Walker et al., 1996).

Another support for the masked message hypothesis comes from the analysis of the translation of the message encoding *Xenopus* fibroblast growth factor receptor-1 (XFGFR1). This message is present, but not translated in growing oocytes. It begins to be translated when progesterone initiates the meiotic maturation. Robbie and colleagues (1995) have shown that this new translation is not dependent on the lengthening of the poly(A) tail or the translocation of the message. Rather, there appears to be a 43-kDa protein that is associated with the 3' UTR of the *Xfgfr1* mRNA and which proba-

(A)

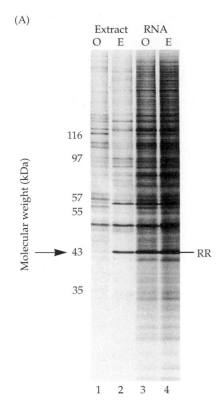

(B)

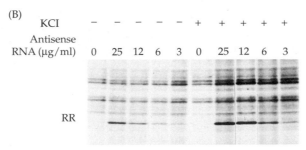

Figure 12.19
Unmasking the ribonucleotide reductase (RR) small subunit mRNA in clam oocytes. (A) Clam RR message is present but not translated in oocytes. Extracts from oocytes (lane 1) or activated eggs (lane 2) were mixed with ribosomes, translation factors, and radioactive amino acids and translated in vitro. The proteins were run on a gel, and autoradiographs were made. The RR protein is made in the egg extract but not in the oocyte extract. When the mRNA of the oocytes (lane 3) and the eggs (lane 4) were isolated and separated from all proteins (by phenol extraction), RR message was translated in both cases. (B) The degree of unmasking depends on the presence of a high salt concentration or the addition of antisense mRNA that blocks the 3′ UTR protein-binding site. (From Standart et al., 1990.)

bly is removed when progesterone stimulates oocyte maturation. This association stores the 5S RNA in an inactive form until it is later incorporated into new ribosomes. Amphibian oocytes contain specific proteins that bind to some mRNAs and not others (Richter and Smith, 1984; Audet et al., 1987; Swiderski and Richter, 1988), but it is not known if these proteins functionally mask endogenous RNAs. It is possible that these proteins facilitate the binding of a *general* RNA masking protein that would associate with the mRNA and render it untranslatable. The FRGY2 protein active in *Xenopus* oocytes may be such a general masking protein (Bouvet and Wolffe, 1994). This protein complexes with certain oocyte transcripts as they are being transcribed in the nucleus, and it is capable of rendering such messages translationally silent. The global "unpacking" of such messages at fertilization may involve ionic changes, the phosphorylation of certain proteins, or changes in RNP composition.

The Poly(A) Tail Hypothesis

Recent studies have demonstrated that altered polyadenylation is critical to the timing of oocyte mRNA translation and that this altered polyadenylation is regulated by the 3′ untranslated region.

The 3′ UTR can regulate the translational efficiency of oocyte messages by controlling the size of the poly(A) tail. In oocytes, the shortening of the poly(A) tail does not doom the message to extinction. It merely represses its ability to be translated (Hyman and Wormington, 1988). This repression is often temporary. In mouse oocytes, those mRNAs that are being used for oocyte growth and metabolism retain their long poly(A) tails and are immediately translated. However, those mRNAs that are to be stored in the oocyte for translation at meiotic maturation (just prior to ovulation) or at fertilization tend to lose most of their poly(A) tails upon entering the cytoplasm. These mRNAs retain only from 15 to 90 adenylate residues (Figure 12.20). At

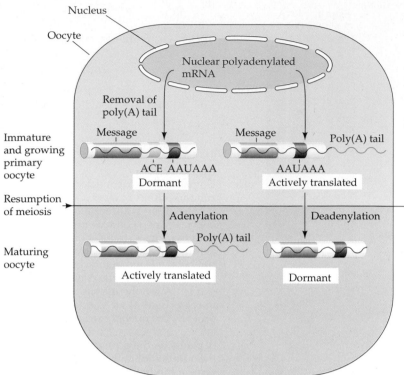

Figure 12.20
Model for the translational regulation of oocyte mRNAs in the mouse. Those mRNAs to be used in oocyte metabolism have polyadenylation sequences in their 3′ UTRs and retain their poly(A) tails. These mRNAs are translated until meiotic maturation (just prior to ovulation), when they lose their poly(A) tails. Those mRNAs that remain translationally dormant until meiotic maturation have cytoplasmic polyadenylation elements (CPEs) as well as the polyadenylation sequences, and they lose their poly(A) tails in the cytoplasm of the immature oocyte. When meiotic maturation begins, the tails are restored and translation of these messages is initiated.

meiotic maturation, an inversion occurs. Those mRNAs that had been actively translated lose their poly(A) tails and no longer function, while those poorly adenylated mRNAs that had been stored rapidly gain long (150 to 600 adenylates) poly(A) tails and are translated into proteins (Vassalli et al., 1989; Huarte et al., 1992).

In mammals, the messages that are translated in the immature oocyte have a standard AAUAAA polyadenylation sequence. These messages retain their poly(A) tails until the resumption of meiotic maturation. At that time, their tails are deadenylated, and they become translationally inactive. Those mRNAs that will be stored in the immature oocyte cytoplasm for translation after oocyte maturation get their poly(A) tails clipped off immediately after they leave the nucleus. These messages have *two* signals in their 3′ UTR: the polyadenylation sequence AAUAAA and a sequence known as **the cytoplasmic polyadenylation element (CPE)**, also called the adenylation control element, (ACE), whose consensus sequence in mice and frogs is UUUUUAU (Fox et al., 1989; Bachvarova, 1992; Huarte et al., 1992). Upon the resumption of oocyte maturation, these stored transcripts are polyadenylated again (probably by the same polyadenylation enzyme that was found in the nucleus) and become translationally active. The acquisition of a long tail is critical for the onset of translation from the stored oocyte mRNA, and the control of this lengthening depends on the presence or absence of a CPE.

In *Xenopus* oocytes, the story is similar, but with some variations.* Like the mammalian oocyte mRNAs, a long poly(A) tail is necessary for a message to be translated. A switch in RNA translation occurs during maturation. When the germinal vesicle (the haploid nucleus) breaks down to begin meiotic division, it releases deadenylation factors. Those mRNAs without CPEs are deadenylated, while those messages containing CPEs are able to become polyadenylated (Fox and Wickens, 1990; Varnum and Wormington, 1990; Varnum et al., 1992). The polyadenylation sequence and the CPE are both needed for translational activation of these messages, but in some instances, the presence *per se* of a poly(A) tail is not sufficient. In these cases, the *process* of polyadenylation is critical for the message to be translated. That is, an mRNA injected with a preexisting poly(A) tail will not be translated. It is possible that the process of polyadenylation also removes an inhibitor protein that otherwise masks the message (Fox et al., 1989; McGrew et al., 1989). There are some differences among the CPEs, and these differences may cause different patterns of polyadenylation in the messages containing them (Paris and Richter, 1990). For instance, a particular CPE with a stretch of 12 U bases inhibits the polyadenylation of those mRNAs containing it during the time of oocyte maturation. However, after fertilization, the mRNAs containing this CPE are polyadenylated and translated into proteins (Simon et al., 1992).

This suggests that there are specific factors that bind to these CPEs at various times. Joel Richter and colleagues (Paris et al., 1991; Hake and Richter, 1994) demonstrated that a 58-kDa oocyte protein, CPEB, binds to a specific CPE (UUUUUAAU). This protein was isolated by RNA affinity chromatography wherein *Xenopus* oocyte proteins were incubated with sepharose beads bound to RNAs containing a CPE with the sequence UUU-UUAAU. The binding of CPEB to this CPE prevents polyadenylation and may inhibit translation of these messages until oocyte maturation (immediately before fertilization). At that time, CPEB is phosphorylated by cdc2 kinase. (The cdc2 kinase is activated by progesterone, which stimulates the oocytes to resume meiosis prior to fertilization.) This phosphorylation appears to allow CPEB to recruit a cytoplasmic poly(A) polymerase to the message (Ballantyne et al., 1995; Gebauer and Richter, 1995). These messages become polyadenylated and subsequently translated.

The cdc2 kinase (as we remember from Chapter 5) is only active when it complexes with a cyclin. The cyclin proteins are also under translational regulation, and the 3′ UTRs of the cyclin mRNAs determine the times when they are translated. In *Xenopus* oocytes, the mRNAs for cyclins A1, B1, and B2 all have truncated poly(A) tails. Within 5 hours after the progesterone signal (at first meiotic metaphase), the poly(A) tails of these cyclin messages are elongated, and their translation begins (Sheets et al., 1994). This demonstrates that developmental cues can regulate which set of mRNAs become functional. It also shows that there are cascades of translational gene regulation during those hours prior to the nucleus becoming active.

The activation of messages by polyadenylation appears to be a critically important process in development. However, we still do not know why

---

*The functions of the poly(A) sequences and CPEs differ between mouse and frog oocytes. In frog oocytes, the deadenylation that occurs at maturation is the "default state," and deadenylation and translational inactivation occur unless a CPE is present. Polyadenylation will activate a masked message and maintain the translation of those mRNAs associated with polysomes. In mouse oocytes, the CPE controls both polyadenylation and deadenylation. In *immature* oocytes, messages lacking the CPE are translated immediately, while CPE-containing mRNAs are deadenylated and translationally inactivated. At maturation, the mouse system becomes similar to that of *Xenopus*, and the CPE-containing RNAs are now polyadenylated and active in translation (Huarte et al., 1992).

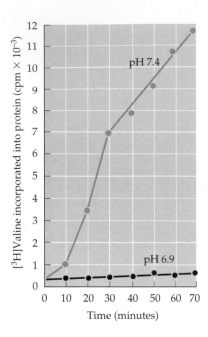

Figure 12.21
Evidence for the inefficiency of protein synthesis at prefertilization pH levels. The in vitro translation system made from unfertilized eggs is kept at pH 6.9 or dialyzed to pH 7.4. Endogenous messages are translated much more efficiently at the postfertilization pH. (After Winkler and Steinhardt, 1981.)

short poly(A) tails are unable to initiate translation, while larger ones can. One possibility (Kuge and Richter, 1995) is that 3′ poly(A) addition stimulates the methylation of the 5′ cap. They found that the progesterone-stimulation of meiotic maturation caused the methylation of mRNA caps, and that this methylation could be inhibited by preventing polyadenylation. Thus, the 3′ and 5′ ends of the message appear to interact. Another possibility (Hentze, 1997) is that the poly(A) tail binds to an initiation factor on the 40S ribosomal subunit to stimulate translation.

The Translational Efficiency Hypothesis

The preceding models of translational regulation assume that the translational apparatus is capable of efficiently translating any message, but that mRNA and the ribosomes are kept apart by physical or chemical means. This need not be the case. The initial low pH of the oocyte is itself able to impede protein synthesis. As discussed in Chapter 4, there is a dramatic release of hydrogen ions during sea urchin fertilization, resulting in an elevation of cytoplasmic pH. When Winkler and Steinhardt (1981) increased the pH of an oocyte lysate from oocyte levels (pH 6.9) to zygote levels (pH 7.4), they observed a burst of protein synthesis mimicking that seen during fertilization (Figure 12.21).

Hille and her colleagues (1985; Danilchik et al., 1986) have suggested that the change in pH activates the translational apparatus of the egg. Ribosomes and initiation factors derived from unfertilized eggs were less active in translation than ribosomes and factors derived from fertilized eggs. Moreover, the injection of exogenous globin message into unfertilized eggs did not increase the amount of protein being synthesized. The globin mRNA was translated at the expense of other messages, suggesting that there is a limiting amount of some portion of the translational apparatus. The limiting factor is probably a translational initiation factor. The addition of eIF2B (the GTP-binding recycling factor) or eIF4F (which contains cap-binding protein) to a lysate prepared from unfertilized eggs increased the translational efficiency of that lysate (Colin et al., 1987; Lopo et al., 1988). Alkalinization of egg cytoplasm may serve both to unmask the mRNA (either physically or through polyadenylation) and to activate initiation factors. Support for this notion also comes from Winkler and his colleagues (1985; Kelso-Winemiller and Winkler, 1991), who have seen a threefold increase in the binding of mRNA to ribosomes once the pH has been raised.

Other mRNA Activation Systems: Uncapped Messages and Sequestered Messages

UNCAPPED MRNA.   The 5′ and 3′ modified ends of messenger RNA are necessary for efficient translation. We have already seen how differences in the length of the 3′ poly(A) tail can effect differential RNA translation in *Xenopus* and *Spisula* oocytes. Certain moths use a mechanism for translational control involving changes at the 5′ cap (Kastern et al., 1982). To be efficiently translated, almost all eukaryotic messages need a 7-methylguanosine "cap" on their 5′ ends (Shatkin, 1976). The stored messages of the tobacco hornworm moth oocyte have a nonmethylated cap. The guanosine is present, but the methyl group has not been added to it. Such messages are not translated into proteins in a cell-free system. However, at fertilization, there is a burst of methylation in these oocytes, and the caps are completed. The mRNAs with the completed caps are then able to bind to the ribosomes and initiate translation. Data on artificial messages suggest that secondary structures (such as hairpin loops) in the 5′ untranslated region can also regulate the timing of RNA translation in the oocyte (Fu et al., 1991).

SEQUESTERED MRNA.   In some instances, the protein synthetic apparatus is compartmentalized, thereby preventing the mRNA (within the RNP) from getting close to the ribosomes (Moon et al., 1982). The histone mRNAs of sea urchin oocytes seem to be regulated by this type of restriction. The histone messages of the oocyte are not found in the cytoplasm. Rather, they are localized in the large pronucleus of the unfertilized egg. It is only when the pronucleus breaks down, at the end of fertilization, that the histone mRNA gets into the cytoplasm (Figure 12.22; DeLeon et al., 1983). This may not be the case for other messages. Less than 0.1 percent of the total mRNA of the unfertilized egg is found in pronuclei (Angerer and Angerer, 1981; Showman et al., 1982). The observation that some maternal messages as well as individual ribosomes are bound to the cytoskeleton (Moon et al., 1983) suggests that the cytoskeleton may also separate mRNAs from ribosomes. It is possible that all these mechanisms of translational control are utilized, even within the same oocyte. The egg has evolved numerous ways of regulating the translation of its stored mRNA, and species are able to use several of these mechanisms at once.

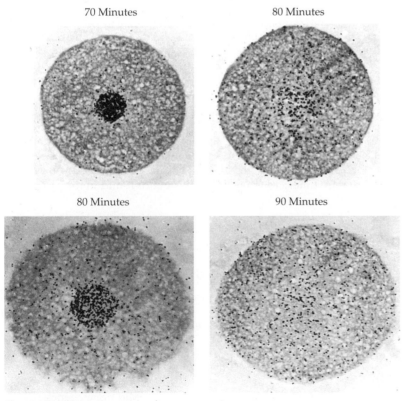

Figure 12.22
Sequestering of the sea urchin oocyte histone messages. A cDNA probe recognizing histone message is hybridized to sea urchin eggs fixed at various times after fertilization. Autoradiography shows the message to be sequestered in the maternal pronucleus until its breakdown 80–90 minutes after sperm entry. (From DeLeon et al., 1983, courtesy of L. and R. Angerer.)

The Activation of the Embryonic Genome

THE TIMING of developmental events differs enormously among animal species. At 24 hours after fertilization, *Drosophila* larvae have hatched and are busily eating; amphibian embryos are either at late-gastrula or early-neurula stages; and the sea urchin embryo is a late blastula or early gastrula with hundreds of cells. Mammalian embryos take their time. At 24 hours of development, a mouse zygote has divided only once, and a human egg still has about 6 hours until its first cleavage. Organisms also differ in the timing and abruptness of the transition from the cytoplasmic control of development to the regulation of development by nuclear transcription. In most species studied, the early embryo is an "RNA world" where the genome counts for naught (Wickens, 1992). In other embryos, nuclear transcription begins immediately upon fertilization, and new gene products are seen during the first cell cycle (Table 12.4).

Xenopus embryos appear to develop through the cleavage stage without the need for nuclear transcription. As mentioned in Chapter 5, the nucleus is essentially inactive until the "midblastula transition" at the end of the twelfth cell division (Figure 12.23; Newport and Kirschner, 1982). Until this midblastula

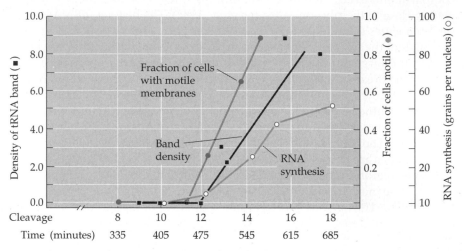

Figure 12.23 Activation of transcription and membrane motility in *Xenopus* after the twelfth cell division. Transcription was scored autoradiographically by the number of reduced silver grains over the nuclei of embryos immersed in radioactive uridine and by the activation of a cloned tRNA gene whose radioactive product could be assayed by measuring the density of the band of the autoradiographed gel. Motility was scored by determining the fraction of cells showing pseudopodia or blebbing in video recordings. (After Davidson, 1986.)

transition, development proceeds on the materials stored in the oocyte cytoplasm. Eventually, transcription is initiated in the nuclei of the embryo. After the midblastula transition, different genes get turned on at different times; but the genes that become activated earliest may be activated by the maternal factors in the oocyte. The OZ1 protein is a transcription factor that is made in the developing oocyte. It binds to a 14-base-pair DNA sequence found in the promoters

Table 12.4 Activation of embryonic genomes and duration of functional maternal mRNA

Organism	Time of first observable transcription from nucleus[a]	Time of major new nuclear transcription[a]	Longevity of functional maternal message
Mammal (*Mus musculus*)	Late 1-cell stage (11–17 hr)	8- to 16-cell early morula (day 3)	4-cell cleavage stage (day 2–4)
Amphibian (*Xenopus laevis*)	Early cleavage (≤32-cell) midblastula (3 hr)	12th cleavage (4000-cell) midblastula (7 hr)	Neurula stage (15–30 hr)
Echinoderm (*S. purpuratus* and other sea urchins)	Zygote (pronuclear stage) (≤0.5 hr)	Midblastula (~128 cells) (11 hr)	Late blastula (15 hr)
Insect (*Drosophila melanogaster*)	Syncytial blastoderm after 10th nuclear division (2.5 hr)	Cellular blastoderm after 14th nuclear division (3.5 hr)	Mid-organogenesis (~15 hr)

Source: Adapted from Wilt, 1964; Woodland and Ballantine, 1980; Clegg and Piko, 1983; Gilbert and Solter, 1985; Poccia et al., 1985; Weir and Kornberg, 1985; Davidson, 1986; Edgar and Schubiger, 1986; Shiokawa et al., 1989.
[a]Times indicate incubation at an appropriate temperature.

of several genes that are turned on at or shortly after the midblastula transition (Ovsenek et al., 1992). If other genes (such as the *Xenopus* β-globin gene) are linked to this sequence, they become expressed at the midblastula transition; but if the sequence is mutated, they are not expressed well (Figure 12.24). It is possible that this OZ1 protein is itself inactive until some other factor (perhaps associated with the lengthening of the cell cycle) activates it.

The concept that maternal proteins may activate the genome during the midblastula transition is supported by the investigations of the o mutant of the axolotl salamander. This is a maternal effect mutation wherein homozygous females produce eggs that are successfully fertilized and are totally normal until the late-cleavage and early-blastula stages (Briggs and Cassens, 1966). At midblastula, eggs shed by an *o/o* female have slower mitoses and go on to form a dorsal blastopore lip, but always arrest in gastrulation. Malacinski (1971) and Carroll (1974) have shown that in embryos from wild-type females, new RNA and protein synthesis starts at this midblastula stage. However, the midblastulae from eggs of *o/o* mothers do not undergo this burst of protein synthesis and have a pattern of proteins identical to that produced by enucleated zygotes (Figure 12.25). Briggs and Cassens (1966) demonstrated that the embryos from *o/o* mothers lacked a factor that activates the midblastula nuclear genome. In the absence of such a factor, the only development that occurs is that which can be supported by the stored oocyte mRNA. In amphibians, there is enough stored oocyte material to enable the embryo to enter gastrulation. Without new RNA synthesis, however, no further development can occur.

In *Drosophila*, there also appears to be a midblastula transition from the RNAs and proteins of the oocyte cytoplasm to nuclear transcription. This transition is first seen after the tenth nuclear division. This is the first cycle with a G2 phase, and the G2 phase increases in length from 10 minutes after the tenth cycle to 60 minutes following the fourteenth. In the G2 phase of the fourteenth cycle, the genome is transcribing at the highest level of activity seen during embryogenesis (Anderson and Lengyel, 1979; Weir and Kornberg, 1985). Edgar and Schu-

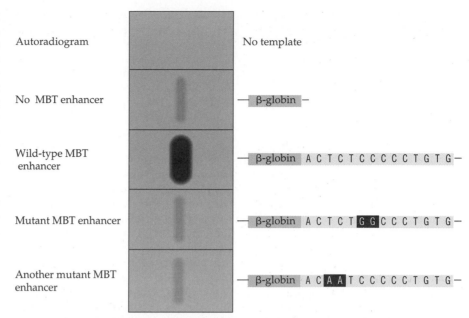

Figure 12.24 Effect of the OZ1-binding midblastula transition (MBT) enhancer. The DNA sequence that activates transcription at the midblastula transition for the *GS17* gene of *Xenopus laevis* was placed on a β-globin gene and injected into *Xenopus* oocytes. This globin construct became expressed at the midblastula stage. Globin genes without this enhancer or with a mutated MBT enhancer did not show significant expression at this stage. (After Ovsenek et al., 1992.)

biger (1986) have shown that *Drosophila* nuclei become competent to transcribe at cycle 10 but that most genes need a longer G2 phase to become activated. The high transcriptional activity of cycle 14 embryos can be induced prematurely by artificially extending the G2 period of younger embryos with cycloheximide. This activation can be accomplished with embryos as young as the tenth cycle, but not earlier. It appears, then, that most genes become capable of activation during cycle 10, but they do not initiate their transcription until cycle 14.

Sea urchins lack a stark midblastula transition. Although their enucleated eggs can develop through blastula stages, and although there is certainly a

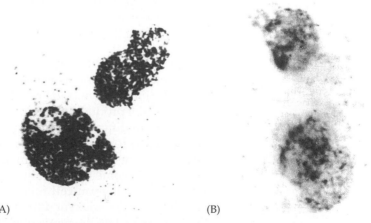

(A)   (B)

Figure 12.25 Incorporation of [³H]uridine into RNA of wild-type and *o/o*-mutant axolotl embryos. Blastula-stage embryos were incubated in the radioactive RNA precursor for 3 hours, washed, fixed, stained, and observed by autoradiography. (A) Normal embryo cells showing intense radioactivity, indicating RNA synthesis. (B) Embryo from an *o/o* female. Stain is present, but no significant labeling is seen, indicating that little or no transcription has occurred. (From Carroll, 1974.)

burst of nuclear transcription from midblastula nuclei, there seems to be no time in sea urchin development when the embryonic nucleus is not functioning. Low levels of transcription (including new histone messages) can be seen from pronuclei even before they fuse (Poccia et al., 1985). These newly transcribed messages join the larger pool of maternal mRNA. The chromatin of the first four cleavages is made primarily with histones stored in the oocyte cytoplasm and with histones synthesized from maternal messages. From the 16-cell stage onward, however, most histones are synthesized from messages transcribed from embryonic cell nuclei (Goustin and Wilt, 1981). This pattern is in marked contrast to that of *Xenopus* embryos, where a large pool of maternally stored histone protein and a large supply of stored oocyte histone message are utilized by thousands of cells.

Mammalian, ascidian, nematode, and molluscan embryos also appear to initiate transcription within the first cell cycle (Schauer and Wood, 1990). However, as in so many developmental events, mammals cannot be said to have evolved a uniform strategy. In the most-studied mammal group, mice, the embryonic genome is extremely active during the 2-cell stage. Between the 1-cell stage and the 2-cell embryo, over two-thirds of the proteins have a greater than fivefold alteration of their synthesis (Latham et al., 1991, 1992). If cultured with the transcriptional inhibitor a-amanitin (which blocks RNA polymerase II), mouse eggs are blocked at the 2-cell stage (Flach et al., 1982). In mice, the maternal mRNAs last about two days—roughly the same amount of time as those of the other phyla—and then, during the second day, the maternal messages are rapidly degraded (Clegg and Piko, 1983; Paynton et al., 1988). As the gene products encoded by the maternal messages decay, they are replaced by new proteins made from mRNA that is newly transcribed from the nucleus. In most instances, the sperm-derived chromosomes are probably activated simultaneously with the egg-derived chromosomes (Gilbert and Solter, 1985). Latham and colleagues (1992) have transplanted nuclei into different cytoplasms and demonstrated that the cytoplasm changes during the latter part of the 1-cell stage. The cytoplasm in the early 1-cell embryo will not support the transcription of genes from the nuclei of later embryos. However, the cytoplasm of late 1-cell embryos will. Since inhibitors of the cAMP-dependent protein kinase (PKA) inhibit the competence of the cytoplasm to support transcription, it is possible that the activation of PKA is essential for the cytoplasm's acquiring its transcriptionally permissive state. Other mammals do not necessarily follow the same schedule. Human mRNA synthesis is first seen at the 4-cell stage, and transcriptional inhibitors block development at the 4- to 8-cell stage. In cows and sheep, transcriptional activity is seen at the 8- to 16-cell stage (Braude et al., 1988; Telford et al., 1990).

In all animal species observed, there is a period of time when the phenomena of early development are controlled by messages and proteins stored in the oocyte cytoplasm. In most species (mammals being the exception), the nuclear genome is activated long before the maternal messages are degraded, so that both sets of mRNAs are being translated simultaneously. Eventually, as the maternal messages are degraded on day 1 or 2, the transcripts from the embryonic genome become more important. ∎

Translational gene regulation in larvae and adults

Translational control isn't only for eggs and their early embryos. Recent studies have shown widespread use of translational gene regulation for several critical processes later in development. As in studies on early embryogenesis, the 3' UTR has been seen to play a critical role. This region of the message, "long viewed as a wasteland of genetic information" (Wickens, 1992), is becoming one of the most interesting areas of developmental gene regulation.

Gamete Determination in C. elegans

A particularly dramatic role for the 3' UTR in masked mRNA is seen in *Caenorhabditis elegans*. This nematode worm has a female body but is hermaphroditic, producing both sperm and eggs at different times. The first germ cells to differentiate in the nematode become sperm, which are stored in the uterus for later use. After the fourth molt (from larva to adult), the germ cells cease making sperm and begin to make eggs. These eggs will eventually become fertilized by the stored sperm. The process determining which path the germ cell follows—to sperm or to egg—depends on the translational repression of different messages. The initiation of sperm formation is achieved by the repression of the *tra-2* message. The TRA-2 protein is essential for the development of eggs and female body cells, and repression of *tra-2* mRNA translation in germ cells causes them to become sperm. The 3' UTR of this message contains two regions of 28 nucleotides, each of which appears to bind a putative repressor protein that is synthesized during the larval stages associated with spermatogenesis. If these regions are mutated, the translation of the *tra-2* mRNA is not repressed, no sperm is made, and the nematode is functionally female instead of hermaphroditic (Evans et al., 1992). A protein that binds to these regions has been isolated and may mediate the translational repression (Figure 12.26; Goodwin et al., 1993).

The story does not end here. The switchover from spermatogenesis to oogenesis also requires suppressing the translation of the *fem-3* mRNA through its 3′ UTR. The FEM-3 protein is critical for specifying male body cells and sperm production. Transcription of the *fem-3* gene is inhibited by TRA-2 protein, but the repression of existing *fem-3* messages is also needed. This translational repression appears to be effected by the binding of a translational inhibitor by the 3′ UTR of the *fem-3* mRNA (see Figure 12.26; Ahringer and Kimble, 1991; Evans et al., 1992). Thus, the initiation of spermatogenesis in hermaphroditic nematodes and the transition from spermatogenesis to oogenesis appear to be regulated by translational repression through the 3′ UTR.

Natural Antisense RNA

It seems that whatever proteins can do, RNAs can do, too. If proteins can regulate translation by binding to specific sites on the 3′ UTR of messenger RNAs, so can small RNAs. "Translational control RNA" was originally proposed by Bester and co-workers in 1975. Since then, it has been discovered in *C. elegans* and in chicks.

Caenorhabditis elegans lives up to its name, having evolved a particularly elegant solution to the problem of controlling larval gene expression (Lee et al., 1993; Wightman et al., 1993). High levels of LIN-14 transcription factor specify the protein synthesis of early larval organs. Thereafter, the LIN-14 protein is not seen anymore, although *lin-14* messages are detected throughout development. *C. elegans* is able to inhibit the synthesis of LIN-14 from its mRNA by activating the *lin-4* gene. In loss-of-function *lin-4* mutations, the LIN-14 protein is synthesized continuously, and the nematode is arrested in early development. The *lin-4* gene does not encode any protein. Rather, it encodes two small RNAs (the most abundant being 25 nucleotides long, the other continuing for 40 more nucleotides) that are complementary to an imperfectly repeated site in the *lin-14* 3′ UTR. Figure 12.27 shows a hypothetical sketch of what might be happening. It appears that the binding of these

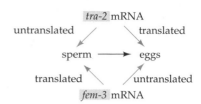

Figure 12.26
The transition from spermatogenesis to oogenesis during the fourth-instar larva of *C. elegans* is regulated by translation of the *tra-2* and *fem-3* messages. In both cases, the blocking of translation occurs through the binding of an inhibitory protein to the respective 3′ UTR.

Figure 12.27
Hypothetical model for the regulation of *lin-14* mRNA translation by the *lin-4* mRNAs. (This has not been experimentally confirmed.) The *lin-4* gene does not produce an mRNA. Rather, it produces small RNAs that do not make proteins. These RNAs are complimentary to a repeated sequence in the 3′ UTR of the *lin-14* mRNA. (After Wickens and Takayama, 1994.)

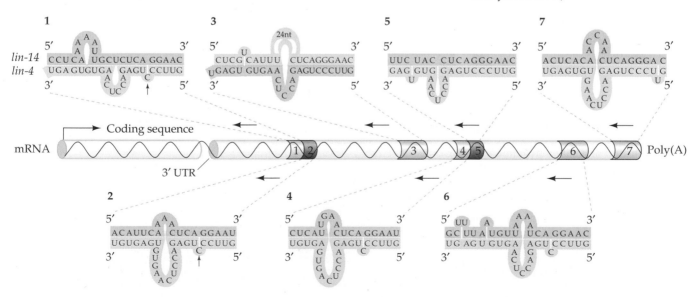

lin-4 transcripts to the *lin-14* mRNA 3′ UTR does not signal the destruction of the message, but rather prevents the message from being translated. [RNA5.html]

In the chick embryo, antisense mRNA is seen to regulate the synthesis of basic fibroblast growth factor (FGF2). Some tissues (such as the mesonephros) have *Fgf2* messages without the antisense transcript, while other tissues (such as undifferentiated limb mesoderm) contain both the *Fgf2* mRNA and its antisense complement. It is thought that the antisense RNA leads to the degradation of itself and the *Fgf2* transcript (Kimelman and Kirschner, 1989; Savage and Fallon, 1995).

Translational Control "Switches"

The coordinate and opposite regulation of the two major mammalian iron-binding proteins, ferritin and the transferrin receptor, has recently been elucidated (see Klausner and Harford, 1989; Klausner et al., 1993). Both the ferritin and the transferrin receptor mRNAs contain regions that bind an iron-responsive binding protein (IRE-BP). The ferritin message has this sequence on its leader sequence (5′ to the protein-encoding region), while the transferrin receptor message contains two of these sequences on its 3′ untranslated region. When cellular iron is in low supply, the iron-binding protein cannot bind iron and is in a conformation that binds to these mRNAs. When it binds to the leader sequence of the ferritin message, it blocks its translation, thus preventing the synthesis of this iron storage protein. Simultaneously, the protein binds to the 3′ end of the transferrin receptor message, thereby stabilizing it against degradation and enabling more transferrin receptors to be produced. The transferrin receptors bring more iron into the cell (Figure 12.28).

FERRITIN mRNA

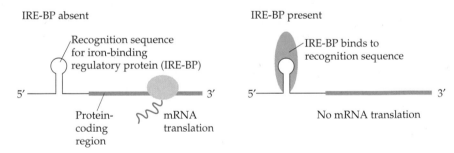

TRANSFERRIN RECEPTOR mRNA

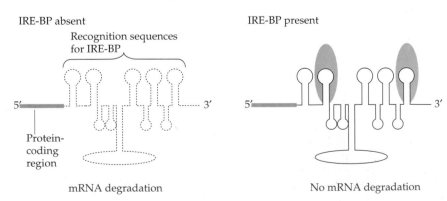

Figure 12.28
Coordinate and opposite translational regulation of ferritin and the transferrin receptor. Both messages contain regions that are recognized by an iron-binding regulatory protein (IRE-BP). In the absence of intracellular iron, this protein binds to these messages, inhibiting the translation of ferritin mRNA and stabilizing the mRNA for the transferrin receptor. (After Klausner and Harford, 1989.)

RNA Editing

One of the most unexpected mechanisms of translational control has recently been seen in the regulation of the apolipoprotein-B proteins. Apo-B proteins are components of serum lipid-carrier proteins and are thought to play a major role in the genesis of atherosclerosis. Apo-B48 (48-kDa) is synthesized in the intestines and becomes part of the chylomicron complexes necessary for the absorption and transport of dietary cholesterol and triglycerides. Apo-B100 (100-kDa) is made in the liver and is the major protein component of the very low-, low-, and intermediate-density lipid-carrier proteins. The Apo-B100 and Apo-B48 proteins are transcribed from the same gene, and no differential RNA processing is seen to occur to generate different mRNAs for these two proteins. Analysis of Apo-B cDNAs indicates that the *apo-B* message in the *liver* encodes the entire Apo-B100 peptide. The *intestinal* message, however, differs from the liver message by only one base. A C-to-U transition has occurred, changing a normal glutamine codon (CAA) into a terminator codon (UAA) at codon 2153. This difference results in the formation of the shorter Apo-B48 protein in the intestine (Chen et al., 1987; Powell et al., 1987). This **RNA editing** is an instance where a specific base change is made in an existing RNA, thereby changing the message. The primary transcript of the *apo-B* gene does not appear to be edited, and the C-to-U editing can be accomplished by a factor contained within the nucleus. Therefore, Lau and colleagues (1991) conclude that this RNA editing is done during the RNA-processing steps. The protein responsible for this editing is a cytidine deaminase (Navaratnam et al., 1995), and by altering the sequence structure of the RNA near the edited cytosine, Chen and co-workers (1990) discovered two regions that are critical for editing. One is a region of nucleotides conserved between various mammalian species, and the other is a species-specific sequence further downstream. They postulate an enzyme that recognizes these two regions and places its catalytic site upon the particular cytosine. The deamination of this cytosine converts it into a uridine residue (Figure 12.29). Up to this time, it had been an axiom of molecular biology that the nucleotide sequence of a message, once transcribed, could not be altered. Although RNA editing is an exceptionally rare event, it is also seen in certain organelle messages (see Scott, 1995; Simpson and Thiemann, 1995), in altering the calcium ion permeability of certain glutamate-gated ion channels during mammalian brain development (Sommer et al., 1991; Higuchi et al., 1993) and in altering the WT1 transcript factor (Sharma et al., 1994).

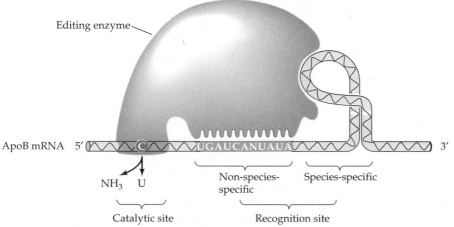

Figure 12.29
Model of an enzymatic mechanism that could enable deamination of a specific cytosine of *apo-B* mRNA. Two regions are needed for the RNA editing: a region that is conserved in several mammals and a species-specific element that has a hairpin loop structure that might be recognized by the enzyme. (After Chan, 1993.)

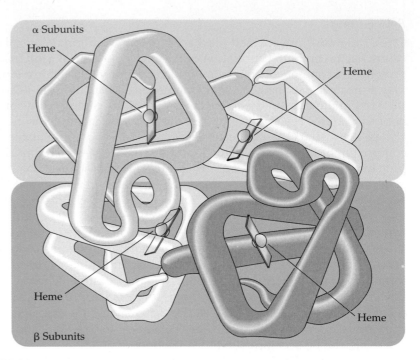

Figure 12.30
The structure of adult human hemoglobin, which has four polypeptide chains (two α, two β) and four heme molecules. (After Dickerson and Geis, 1983.)

Translational control and coordinated protein synthesis: Hemoglobin production

One of the major problems in genetic regulation is the coordinated production of several products from different regions of the genome. When a developing red blood cell synthesizes hemoglobin, it must ensure that the α-globin chains, β-globin chains, and heme molecules are in a 2:2:4 ratio (Figure 12.30). Any major deviation from this ratio results in severely debilitating diseases.

The heme molecule appears to regulate the proportional synthesis of the hemoglobin components. It accomplishes this feat in two ways. First, excess heme (i.e., heme that has not been bound to protein such as globin) shuts off its own synthesis (Karibian and London, 1965). It does this by inactivating δ-aminolevulinate synthase (DALA synthase), the first enzyme in the pathway for the production of heme (Figure 12.31). Thus, when there is more heme present than there are molecules to bind it, no more heme will be produced. Second, excess heme stimulates the production of globin proteins (Gribble and Schwartz, 1965; Zucker and Schulman, 1968). When heme (as its oxidized form, hemin) is added to a cell-free translation system that includes all the factors needed to translate mRNAs (Table 12.5), the synthesis of globin is greatly enhanced (Figure 12.32A). Therefore, if there is no globin to bind the heme, the excess heme shuts off its own synthesis and stimulates the production of more globin.

Several laboratories have investigated how a molecule as small as heme could regulate protein synthesis. In 1972, Adamson and colleagues demonstrated that the stimulatory effect of heme on globin synthesis could be mimicked by adding to the translation system those proteins that are loosely associated with the ribosomes. Because such solutions are rich in translation initiation factors, each factor was tested separately. It was found that eukaryotic initiation factor 2 (eIF2) restored protein synthesis to heme-deficient

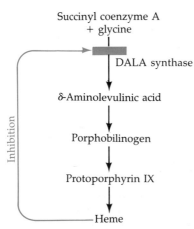

Figure 12.31
Feedback regulation of heme synthesis. (After Harris, 1975.)

Table 12.5 Components of the in vitro translation system containing rabbit reticulocyte lysate

Component	Concentration (in 100 µl)	Component	Concentration (in 100 µl)
Reticulocyte lysate (1:1)	50 µl	KCl	76 mM
tris-HCl (pH 7.6) buffer	10 mM	Proportional amino acid mixture	6–170 µM
ATP	1 mM	[14C]Leucine	0.8 µCi
GTP	0.2 mM	"Cold" leucine	26 µM
Creatine phosphate	5 mM	Hemin	10–30 µM
Creatine phosphokinase	10 µg	H$_2$O to bring the total reaction	
Magnesium acetate	2mM	to 100 µl volume	

Source: After London et al., 1976.

lysates in the translation system (Figure 12.32B). This initiation factor is responsible for combining with the initiator tRNA and complexing it to the 40S ribosomal subunit.

What, then, is the relationship between heme and eIF2? To answer this, London and his co-workers (Levin et al., 1976; Ranu et al., 1976; Ramaiah et al., 1992) added heme-deficient lysates to heme-supplemented translation systems. They found that a portion of the heme-deficient lysate could actually depress the synthesis of globin in the translation system to which it was added. This finding indicated that an inhibitor was present. This heme-responsive inhibitor protein, HRI, was isolated and was found to be a kinase capable of phosphorylating eIF2. Hemin binds to this kinase and inactivates it (see Chen and London, 1995).

Phosphorylated eIF2 eventually halts translation. Normally, once the ribosomal subunits come together, eIF2 is released as a complex with GDP (Raychaudhury et al., 1985). For eIF2 to be used again in initiation, it must complex with eIF2B (recycling factor). This eIF2B exchanges GTP for GDP

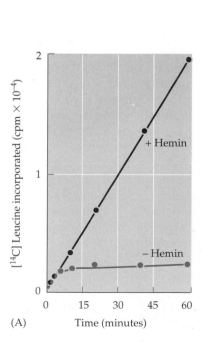

(A)

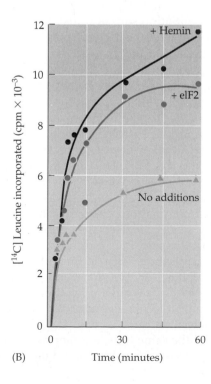

(B)

Figure 12.32
Translational regulation by hemin and eukaryotic initiation factor 2. (A) Translation of the globin mRNA in the rabbit reticulocyte in vitro protein synthesis system. Inclusion of hemin causes a dramatic elevation in protein synthesis. (B) Effect of additional eukaryotic initiation factor 2 on the rabbit reticulocyte in vitro translation system. The additional eIF2 elevated the level of protein synthesis close to that of the hemin-stimulated system. (A after London et al., 1976; B after Clemens et al., 1974.)

Figure 12.33
Scheme for the translational control of globin synthesis. As a result of inactivation by protein kinase, eIF2 is depleted unless heme inactivates the protein kinase.

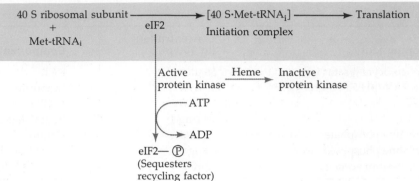

(see Figure 12.11), and the resulting eIF2-GTP complex is able to enter another round of initiation. However, if the α subunit of the eIF2 is phosphorylated, the eIF2B recycling factor binds but cannot let go (Gross et al., 1985; Thomas et al., 1985). Eventually, all the eIF2B (whose concentration is 10- to 20-fold lower than that of eIF2) gets bound on these complexes, and translation stops. Addition of eIF2B to heme-deficient lysates restores protein synthesis to the levels of heme-supplemented systems (Grace et al., 1984). Excess heme is able to bind to the protein kinase, inactivating it (Fagard and London, 1981). Inactivated kinase will not phosphorylate eIF2α, so translation proceeds. Thus, as long as heme is present, globin synthesis continues (Figure 12.33).

The story of the translational control of globin synthesis does not end here. As we discussed in Chapter 11, there are four active α-globin genes per diploid cell and only two active β-globin genes. If each gene were transcribed and translated at the same rate, one would expect twice as many α-globin molecules as β-globin molecules. This, of course, is not the case. One finds a 1.4:1 ratio of α:β mRNA, but a 1:1 ratio of the proteins (Lodish, 1971). The equalization of the proteins appears to involve translational regulation.

Kabat and Chappell (1977) have suggested that the equalization is done at the initiation step of translation. They showed that the α-globin mRNA competes with the β-globin message for initiation factors and that the β-globin message appears to be the better competitor. The β-globin message is recognized more efficiently by the initiation factors and is thus translated more frequently. When the two mRNAs were present in equal amounts with a severely limiting supply of initiation factors, only 3 percent of the resulting protein was α-globin. However, when the unfractionated mRNA (α- and β-globin messages from lysed cells) was added to an excess of such initiation factors, all mRNAs were translated with equal efficiency, and the resulting α to β ratio was 1.4:1. The cap-binding protein has been implicated as the initiation factor responsible for discriminating between the two types of globin message (Ray et al., 1983; Sarkar et al., 1984). While it is not yet known how this discrimination occurs, it is known that the secondary structure of the 5′ leader sequence affects the efficiency of translation (Pelletier and Sonenberg, 1985). As is seen in Figure 12.34, the 5′ ends of α- and β-globin messages differ significantly. Thus, at the initiation step of translation, the proper ratios of α-globin, β-globin, and heme are established. Although hemoglobin synthesis involves regulation at the transcriptional and RNA-processing levels, the final molecule is constructed through fine-tuned coordination at the level of translation.

At the same time, another remarkable example of translational regulation is occurring inside the red blood cell. The mRNA encoding the enzyme

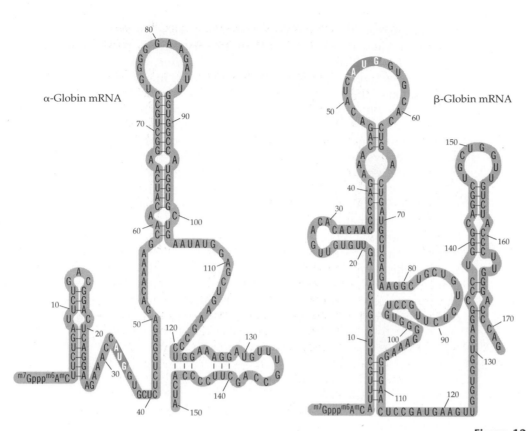

α-Globin mRNA

β-Globin mRNA

Figure 12.34
Probable secondary structures for the 5′ ends of mouse α- and β-globin chains. The translation-initiating AUG codons are in color. (After Pavlakis et al., 1980.)

15-lipoxygenase (15-LOX) is transcribed during early stages of red blood cell development in the bone marrow, but it is translated only when the red blood cell is about to enter the peripheral circulation. This enzyme is responsible for digesting the mitochondria during the last stages of red blood cell formation. The 3′ UTR of the *15-lox* mRNA has 10 tandem repeats of a pyrimidine-rich sequence that binds an erythrocyte-specific 48-kDa protein. This 48-kDa protein represses the translation of the *15-lox* message until the erythrocyte is ready to enter the circulation (Ostareck-Lederer et al., 1994). It is not yet known how this repressor protein is regulated during red blood cell development.

Epilogue: Posttranslational regulation

Translational control, then, is an important and widely used mechanism for regulating gene expression in development. It can be used to activate a certain set of existing mRNAs at a certain time or to regulate the ratio at which different, competing mRNAs can be translated. Animals have evolved several mechanisms whereby mRNAs can be stored in the oocyte for later use during early embryogenesis. The molecular bases for these translational regulatory mechanisms are now being studied.

However, when the peptide is synthesized, the story is still not over. Once a protein is made, it becomes part of a larger level of organization. It may become part of the structural framework of the cell, or it may become involved in one of the myriad enzymatic pathways for the synthesis or breakdown of cellular metabolites. In either case, the individual protein is now part of a complex "ecosystem" that integrates it into a relationship with numerous other proteins. Thus, several changes can still take place that determine whether or not the protein is active. First, some newly synthesized proteins are inactive without further modifications. These modifications can involve the cleaving away of certain inhibitory sections of the protein or the binding of a small compound to enhance its activity. Second, some proteins may be selectively inactivated. In some cases, inactivation involves the degradation of the protein itself; in other cases, inactivation may be brought about by the binding of an inhibitory ligand. Third, some proteins must be "addressed" to their specific intracellular destinations. The cell is not merely a sack of enzymes: proteins are often sequestered in certain regions, such as membranes, lysosomes, nuclei, or mitochondria. Fourth, some proteins need to assemble with other proteins to form a functional unit. The hemoglobin protein, the microtubule, and the ribosome are all examples of numerous proteins joining together to form a functional unit. Therefore, the expression of genetic information can still be influenced at the posttranslational level. Some of these cases (such as the phosphorylation of mitosis-promoting factor) have already been discussed, while others will be discussed as they appear. At this point, we will leave our discussion of the molecular aspects of gene expression and return to the dynamics of the developing embryo. We can now look at early developmental processes to study the molecular mechanisms for the determination of cell fate and tissue structure.

LITERATURE CITED

Adamson, S. D., Yau, P. M. P., Herbert, E. and Zucker, W. V. 1972. Involvement of hemin, a stimulatory fraction from ribosomes, and a protein synthesis inhibitor in the regulation of hemoglobin synthesis. *J. Mol. Biol.* 63: 247–264.

Ahringer, J. and Kimble, J. 1991. Control of the sperm-oocyte switch in *Caenorhabditis elegans* by the *fem-3* 3′ untranslated region. *Nature* 349: 346–348.

Anderson, K. W. and Lengyel, J. A. 1979. Rates of synthesis of major classes of RNA in *Drosophila* embryos. *Dev. Biol.* 70: 217–231.

Angerer, L. M. and Angerer, R. C. 1981. Detection of poly A+ RNA in sea urchin eggs and embryos by quantitative in situ hybridization. *Nucleic Acids Res.* 9: 2819–2840.

Audet, R. G., Goodchild, J. and Richter, J. D. 1987. Eukaryotic initiation factor 4A stimulates translation in microinjected *Xenopus* oocytes. *Dev. Biol.* 121: 58–68.

Axel, R., Feigleson, P. and Schutz, G. 1976. Analysis of the complexity and diversity of mRNA from chicken liver and oviduct. *Cell* 7: 247–254.

Bachvarova, R. F. 1992. A maternal tail of poly(A): The long and the short of it. *Cell* 69: 895–897.

Baker, B., Nagoshi, R. N. and Burtin, K. C. 1987. Molecular genetic aspects of sex determination in *Drosophila*. *BioEssays* 6: 66–70.

Ballantyne, S., Bilger, A. Åström, J., Virtanen, A. and Wickens, M. 1995. Poly(A) polymerases in the nucleus and cytoplasm of frog oocytes: dynamic changes during oocyte maturation and early development. *RNA* 1: 64–78.

Belote, J. M., McKeown, M., Boggs, R. T., Ohkawa, R. and Sosnowski, B. A. 1989. Molecular genetics of *transformer*, a genetic switch controlling sexual differentiation in *Drosophila*. *Dev. Genet.* 10: 143–154.

Berget, S. M. 1995. Exon recognition in vertebrate splicing. *J. Biol. Chem.* 270: 2411–2414.

Bester, A. J., Kennedy, D. S. and Heywood, S. M. 1975. Two classes of translational control RNA: Their role in the regulation of protein synthesis. *Proc. Natl. Acad. Sci. USA* 72: 1523–1527.

Boggs, R. T., Gregor, P., Idriss, S., Belote, J. M. and McKeown, M. 1987. Regulation of sexual differentiation in *D. melanogaster* via alternative splicing of RNA from the *transformer* gene. *Cell* 50: 739–747.

Bouvet, P. and Wolffe, A. P. 1994. A role for transcription and FRGY2 in masking maternal mRNA within *Xenopus* oocytes. *Cell* 77: 931–941.

Braude, P., Bolton, V. and Moore, S. 1988. Human gene expression first occurs between the four- and eight-cell stages of preimplantation development. *Nature* 332: 459–461.

Briggs, R. and Cassens, G. 1966. Accumulation in the oocyte nucleus of a gene product essential for embryonic development beyond gastrulation. *Proc. Natl. Acad. Sci. USA* 55: 1103–1109.

Carroll, C. R. 1974. Comparative study of the early embryonic cytology and nucleic acid synthesis of *Ambystoma mexicanum*

normal and *o* mutant embryos. *J. Exp. Zool.* 187: 409–422.

Chan, L. 1993. RNA editing: Exploring one mode with apolipoprotein B mRNA. *BioEssays* 15: 33–41.

Chen, C.-Y. A., You, Y. and Shyu, A.-B. 1992. Two cellular proteins bind specifically to a purine-rich sequence necessary for the destabilization function of a c-fos protein coding region determinant of mRNA instability. *Mol. Cell. Biol.* 12: 5748–5757.

Chen, C.-Y. A., Chen, T.-M. and Shyu, A.-B. 1994. Interplay of two functionally and structurally distinct domains of the c-fos AU-rich element specifies its mRNA-destabilizing function. *Mol. Cell. Biol.* 14: 416–426.

Chen, J.-J. and London, I. M. 1995. Regulation of protein synthesis by heme-regulated eIF-2α kinase. *Trends Bioch. Sci.* 20: 105–108.

Chen, S.-H. and 12 others. 1987. Apolipoprotein B48 is the product of a messenger RNA with an organ-specific inframe stop codon. *Science* 238: 363–366.

Chen, S.-H., Li, X., Liao, W. S. L., Wu, J. H. and Chan, L. 1990. RNA editing of apolipoprotein B mRNA: Sequence specificity determined by in vitro coupled transcription-editing. *J. Biol. Chem.* 265: 6811–6816.

Christofori, G. and Keller, W. 1988. 3′ cleavage and polyadenylation of mRNA precursors in vitro requires a poly(A) polymerase, a cleavage factor, and a snRNP. *Cell* 54: 875–889.

Clegg, K. B. and Piko, L. 1983. Poly(A) length, cytoplasmic adenylation, and synthesis of poly(A)+ RNA in early mouse embryos. *Dev. Biol.* 95: 331–341.

Clemens, M. J., Henshaw, E. C., Rahaminoff, H. and London, I. M. 1974. Met-tRNA$_{fmet}$ binding to 40S ribosomal units: A site for the regulation of initiation of protein synthesis by hemin. *Proc. Natl. Acad. Sci. USA* 71: 2946–2950.

Colin, A. M., Brown, B. D., Dholakia, J. N., Woodley, C. L., Wahba, A. J. and Hille, M. B. 1987. Evidence for simultaneous derepression of messenger RNA and the guanine nucleotide exchange factor in fertilized sea urchin eggs. *Dev. Biol.* 123: 354–363.

Cooper, G. M. 1996. *The Cell: A Molecular Approach.* Sinauer Associates, Sunderland, MA.

Coschigano, K. T. and Wensink, P. 1993. Sex-specific transcriptional regulation by the male and female doublesex proteins of *Drosophila. Genes Dev.* 7: 42–54.

Crossley, P. H. and Martin, G. R.1995. The mouse *Fgf8* gene encodes a family of polypeptides and is expressed in regions that direct outgrowth and patterning in the developing embryo. *Development* 121: 439–451.

Danilchik, M. V., Yablonka–Reuveniz, Z., Moon, R. T., Reed, S. K. and Hille, M. B. 1986. Separate ribosomal pools in sea urchin embryos: Ammonia activates a movement between pools. *Biochemistry* 25: 3696–3702.

Davidson, E. H. 1976. *Gene Activity in Early Development.* 1st Ed. Academic Press, New York.

Davidson, E. H. 1986. *Gene Activity in Early Development,* 3rd Ed. Academic Press, New York.

Davidson, E. H. and Britten, R. J. 1979. Regulation of gene expression: Possible role of repetitive sequences. *Science* 204: 1052–1059.

Decker, C. J. and Parker, R. 1994. Mechanisms of mRNA degradation in eukaryotes. *Trends Biochem.* 19: 336–340.

DeLeon, C. V., Cox, K. H., Angerer, L. M. and Angerer, R. C. 1983. Most early variant histone mRNA is contained in the pronucleus of sea urchin eggs. *Dev. Biol.* 100: 197–206.

Dickerson, R. E. and Geis, I. 1983. *Hemoglobin* Benjamin/Cummings, Menlo Park, CA.

Dubnau, J. and Struhl, G. 1996. RNA recognition and translational regulation by a homeodomain protein. *Nature* 379: 694–699.

Edgar, B. A. and Schubiger, G. 1986. Parameters controlling transcriptional activation during early *Drosophila* development. *Cell* 44: 871–877.

Evans, T. C., Goodwin, E. B. and Kimble, J. 1992. Translational regulation of development and maternal RNAs in *Caenorhabditis elegans. Semin. Dev. Biol.* 3: 381–389.

Evans, T. C., Crittenden, S. L., Kodoyianni, V. and Kimble, J. 1994. Translational control of maternal *glp-1* mRNA establishes an asymmetry in the *C. elegans* embryo. *Cell* 77: 183–194.

Fagard, R. and London, I. M. 1981. Relationship between phosphorylation and activity of heme-regulated eukaryotic initiation factor 2 kinase. *Proc. Natl. Acad. Sci. USA* 78: 866–870.

Ferrandon, D., Elphick, L., Nüsslein-Volhard C. and St. Johnon, D. 1994. Staufen protein associates with the 3′ UTR of bicoid to form particles that move in a microtubule-dependent manner. *Cell* 79: 1221–1232.

ffrench-Constant, C. and Hynes, R. O. 1989. Alternative splicing of fibronectin is temporally and spatially regulated in the chicken embryo. *Development* 106: 375–388.

Flach, G., Johnson, M. H., Braude, P. R., Taylor, R. A. S. and Bolton, V. N. 1982. The transition from maternal to embryonic control in the 2-cell mouse embryo. *EMBO J.* 1: 681–686.

Fox, C. A. and Wickens, M. 1990. Poly(A) removed during oocyte maturation: A default reaction selectively prevented by specific sequences in the 3′ UTR of certain maternal mRNAs. *Genes Dev.* 4: 2287–2298.

Fox, C. A., Sheets, M. D. and Wickens, M. 1989. Poly(A) addition during maturation of frog oocytes: distinct nuclear and cytoplasmic activities and regulation by the sequence UUUUUAU. *Genes Dev.* 3: 2151–2162.

Francke, C., Edstrom, J. E., McDowell, A. W. and Miller, O. L. 1982. Microscopic visualization of a discrete class of giant translation units in salivary gland cells of *Chironomus tentans. EMBO J.* 1: 59–62.

Fu, L., Ye, R., Browder, L. and Johnston, R. 1991. Translational potentiation of mRNA with secondary structure in *Xenopus. Science* 251: 807–810.

Gagnon, M. L., Angerer, L. M. and Angerer, R. C. 1992. Posttranscriptional regulation of ectoderm-specific gene expression in early sea urchin embryos. *Development* 114: 457–467.

Galau, G., Kelin, W. H., Davis, M. M., Wold, B., Britten, R. J. and Davidson, E. H. 1976. Structural gene sets active in embryos and adult tissues of the sea urchin. *Cell* 7: 487–505.

Gavis, E. R. and Lehmann, R. 1994. Translational regulation of *nanos* by RNA localization. *Nature* 369: 315–318.

Gebauer, F. and Richter, J. D. 1995. Cloning and characterization of a Xenopus poly(A) polymerase. *Mol. Cell. Biol.* 15:1422–1430.

Gilbert, S. F. and Solter, D. 1985. Onset of paternal and maternal *Gpi-2* expression in preimplantation mouse embryos. *Dev. Biol.* 109: 515–517.

Goodwin, E. B., Okkema, P. G., Evans, T. C. and Kimble, J. 1993. Translational regulation of *tra-2* by its 3′ untranslated region controls sexual identity in *C. elegans. Cell* 75: 329–339.

Goustin, A. S. and Wilt, F. H. 1981. Protein synthesis, polyribosomes, and peptide elongation in early development of *Strongylocentrotus purpuratus. Dev. Biol.* 82: 32–40.

Grace, M. and eight others. 1984. Protein synthesis in rabbit reticulocytes: Characteristics of the protein factor RF that reverses inhibition of protein synthesis in heme-deficient reticulocyte lysates. *Proc. Natl. Acad. Sci. USA* 79: 6517–6521.

Gribble, T. J. and Schwartz, H. C. 1965. Effect of protoporphyrin on hemoglobin synthesis. *Biochim. Biophys. Acta* 103: 333–338.

Gross, K. W., Jacobs-Lorena, M., Baglioni, G. and Gross, P. R. 1973. Cell-free translation of maternal messenger RNA from sea urchin eggs. *Proc. Natl. Acad. Sci. USA* 70: 2614–2618.

Gross, M., Redman, R. and Kaplansky, D. A. 1985. Evidence that the primary effects of phosphorylation of eukaryotic initiation factor 2a in rabbit reticulocyte lysate is inhibition of the release of eukaryotic initiation factor 2-GDP from 60S ribosomal subunits. *J. Biol. Chem.* 260: 9491–9500.

Guo, W., Mulligan, G. J., Wormsley, S. and Helfman, D. M. 1991. Alternative splicing of β-tropomyosin pre-mRNA: *Cis*-acting elements and cellular factors that block the use of a skeletal muscle exon in nonmuscle cells. *Genes Dev.* 5: 2096–2107.

Guyette, W. A., Matusik, R. J. and Rosen, J. M. 1979. Prolactin-mediated transcriptional and post-transcriptional control of casein gene expression. *Cell* 17: 1013–1023.

Hake, L.E. and Richter, J. D. 1994. CPEB is a specificity factor that mediates cytoplasmic polyadenylation during *Xenopus* oocyte maturation. *Cell* 79: 617–627.

Harris, H. 1975. *Principles of Human Biochemical Genetics.* Elsevier North-Holland, New York.

Hentze, M. W. 1997. eIF4G: A multipurpose ribosome adapter? *Science* 275: 500–501.

Hershey, J. W. B. 1989. Protein phosphorylation controls translation rates. *J. Biol. Chem.* 264: 20823–20826.

Higuchi, M., Single, F. N., Köhler, M., Sommer, B., Sprengel, R. and Seeburg, P. H. 1993. RNA editing of AMPA receptor subunit GluR-B: A base-pair intron-exon structure determines position and efficiency. *Cell* 75: 1361–1370.

Hille, M. B., Danilchik, M. V., Colin, A. M. and Moon, R. T. 1985. Translational control in echinoid eggs and early embryos. *In* R. H. Sawyer and R. M. Showman (eds.), *The Cellular and Molecular Biology of Invertebrate Development.* University of South Carolina Press, pp. 91–124.

Hodges, P. E. and Beggs. J. D. 1994. U2 fulfills a commitment. *Curr. Biol.* 4: 264–267.

Holt, J. T., Gopal, T. V., Moulton, A. D. and Nienhuis, A. W. 1986. Inducible production of c-*fos* antisense RNA inhibits 3T3 cell proliferation. *Proc. Natl. Acad. Sci. USA* 83: 4794–4798.

Horowitz, D. S. and Krainer, A. R. 1994. Mechanisms for selecting 5′ splice sites in mammalian pre-mRNA splicing. *Trend Genet.* 10: 100–106.

Hough-Evans, B. R., Wold, B. J., Ernst, S. G., Britten, R. J. and Davidson, E. H. 1977. Appearance and persistence of maternal RNA sequences in sea urchin development. *Dev. Biol.* 260: 258–277.

Huarte, J. and seven others. 1992. Transient translational silencing by reversible mRNA deadenylation. *Cell* 69: 1021–1030.

Hyman, L. E. and Wormington, W. M. 1988. Translational inactivation of ribosomal protein messenger RNAs during *Xenopus* oocyte maturation. *Genes Dev.* 2: 598–605.

Hynes, R. O. 1987. Fibronectins: A family of complex and versatile adhesive glycoproteins derived from a single gene. *Harvey Lect.* 81: 133–152.

Infante, A. and Nemer, M. 1968. Heterogeneous RNP particles in the cytoplasm of sea urchin embryos. *J. Mol. Biol.* 32: 543–565.

Jenkins, N. A., Kaumeyer, J. R., Young, E. M. and Raff, R. A. 1978. A test for masked message: The template activity of messenger ribonucleoprotein particles isolated from sea urchin eggs. *Dev. Biol.* 63: 279–298.

Jurnich, V. A. and Burtis, K. C. 1993. A positive role in differentiation of the male doublesex protein of *Drosophila*. *Dev. Biol.* 155: 235–249.

Kabat, D. and Chappell, M. R. 1977. Competition between globin messenger ribonucleic acids for a discriminating initiation factor. *J. Biol. Chem.* 252: 2684–2690.

Karibian, D. and London, I. M. 1965. Control of heme synthesis by feedback inhibition. *Biochem. Biophys. Res. Commun.* 18: 243–249.

Kastern, W. H., Swindlehurst, M., Aaron, C., Hooper, J. and Berry, S. J. 1982. Control of mRNA translation in oocytes and developing embryos of giant moths. I. Functions of the 5′ terminal "cap" in the tobacco hornworm *Manduca sexta*. *Dev. Biol.* 89: 437–449.

Kelso-Winemiller, L. and Winkler, M. M. 1991. "Unmasking" of stored maternal mRNAs and the activation of protein synthesis at fertilization in sea urchins. *Development* 111: 623–633.

Kiledjian, M., Wang, X. and Liebhaber, S. A. 1995. Identification of two KH domain proteins in the α-globin mRNP stability complex. *EMBO J.* 14: 4357–4364.

Kim-Ha, J., Kerr, K. and Macdonald, P. M. 1995. Translational regulation of oskar mRNA by bruno, a newly discovered RNA-binding protein, is essential. *Cell* 81: 403–412.

Kimelman, D. and Kirschner, M.W. 1989. An antisense messenger RNA directs the covalent modification of the transcript encoding fibroblast growth factor in *Xenopus* oocytes. *Cell* 59: 687–696.

Klausner, R. D. and Harford, J. B. 1989. *Cis-trans* models for post-transcriptional gene regulation. *Science* 246: 870–872.

Klausner, R. D., Rouault, T. A. and Harford, J. B. 1993. Regulating the fate of mRNA: The control of cellular iron metabolism. *Cell* 72: 19–28.

Kleene, K. C. and Humphreys, T. 1977. Similarity of hnRNA sequences in blastula and pluteus stage sea urchin embryos. *Cell* 12: 143–155.

Kleene, K. C. and Humphreys, T. 1985. Transcription of similar sets of rare maternal RNAs and rare nuclear RNAs in sea urchin blastulae and adult coelomocytes. *J. Embryol. Exp. Morphol.* 85: 131–149.

Kozak, M. 1986. Point mutations define a sequence flanking the AUG initiator codon that modulates translation by eukaryotic ribosomes. *Cell* 44: 283–292.

Kuge, H. and Richter, J. D. 1995. Cytoplasmic 3′ poly(A) addition induces 5′ cap ribose methylation: Implications for translational control of maternal mRNA. *EMBO J.* 14: 6301–6310.

Latham, K. E., Garrels, J. I., Chang, C. and Solter, D. 1991. Quantitative analysis of protein synthesis in mouse embryos. I. Extensive reprogramming at the one- and two-cell stages. *Development* 112: 821-932.

Latham, K. E., Solter, D. and Schultz, R. M. 1992. Acquisition of a transcriptionally permissive state during the 1-cell stage of mouse embryogenesis. *Dev. Biol.* 149: 457–462.

Lau, P. P., Xiong, W., Zhu, H.-J., Chen, S.-H. and Chan, L. 1991. Apolipoprotein B mRNA editing is an intranuclear event that occurs posttranscriptionally coincident with splicing and polyadenylation. *J. Biol. Chem.* 266: 20550–20554.

Lee, R. C., Feinbaum, R. L. and Ambros, V. 1993. The *C. elegans* heterochromatic gene *lin-4* encodes small RNAs with antisense complementarity to *lin-14*. *Cell* 75: 843–854.

Levin, D., Ranu, R., Ernst, V. and London, I. M. 1976. Regulation of protein synthesis in reticulocyte lysates: Phosphorylation of methionyl-tRNA$_f$ binding factor by protein kinase activity of translational inhibitor isolated from heme-deficient lysates. *Proc. Natl. Acad. Sci. USA* 73: 3112–3116.

Liang, L., Diehl-Jones, W. and Lasko, P. 1994. Localization of vasa protein to the *Drosophila* pole plasm is dependent on its RNA-binding and helicase activities. *Development* 120: 1201–1211.

Lieberfarb, M. E., Chu, T., Wreden, C., Theurkauf, W., Gergen, J. P. and Strickland, S. 1996. Mutations that perturb poly(A)-dependent maternal mRNA activation block the initiation of development. *Development* 122: 579–588.

Lodish, H. F. 1971. Alpha and beta globin messenger ribonucleic acid. Different amounts and rates of translation. *J. Biol. Chem.* 246: 7131–7138.

London, I. M., Clemens, M. J., Ranu, R. S., Levin, D. H., Cherbas, L. F. and Ernst, V. 1976. The role of hemin in the regulation of protein synthesis in erythroid cells. *Fed. Proc.* 35: 2218-2222.

Lopo, A. C., MacMillan, S. and Hershey, J. W. B. 1988. Translational control in early sea urchin embryogenesis: Initiation factor eIF4F stimulates protein synthesis in lysates from unfertilized eggs of *S. purpuratus*. *Biochemistry* 27: 351-357.

MacArthur, C. A., Lawshé, A., Xu, J., Santos-Ocampo, S., Heikinheimo, M., Chellaiah, C. and Ornitz, D. M. 1995. FGF-8 isoforms activate receptor splice forms that are expressed in mesenchymal regions of mouse development. *Development* 121: 3603–3613.

MacDougall, C., Harbison, D. and Bownes, M. 1995. The developmental consequences of alternative splicing in sex determination and differentiation in *Drosophila*. *Dev. Biol.* 172: 353–376.

Malacinski, G. M. 1971. Genetic control of qualitative changes in protein synthesis during early amphibian (Mexican axolotl) embryogenesis. *Dev. Biol.* 26: 442–451.

Mayeda, A. and Krainer, A. R. 1992. Regulation of alternative pre-mRNA splicing by hnRNP A1 and splicing factor SF2. *Cell* 68: 367–375.

McDevitt, M. A., Imperiale, M. J., Ali, H. and Nevins, J. R. 1984. Requirement of a downstream sequence for the generation of poly(A) addition site. *Cell* 37: 329–338.

McGrew, L., Dworkin-Rastl, E., Dworkin, M. B. and Richter, J. D. 1989. Poly(A) elongation during *Xenopus* oocyte maturation is required for translational recruitment and is mediated by a short sequence element. *Genes Dev.* 3: 803–815.

Meijlink, F., Curran, T., Miller, A. D. and Verma, I. M. 1985. Removal of a 67-base pair sequence in the non-coding region of protooncogene *fos* converts it to a transforming gene. *Proc. Natl. Acad. Sci. USA* 82: 4987–4991.

Melton, D. 1987. Translocation of a localized maternal mRNA to the vegetal pole of *Xenopus* oocytes. *Nature* 328: 80–82.

Mermod, J. J., Schatz, G. and Croppa, M. 1980. Specific control of messenger translation in *Drosophila* oocytes and embryos. *Dev. Biol.* 75: 177–186.

Moon, R. T., Danilchik, M. V. and Hille, M. 1982. An assessment of the masked messenger hypothesis: Sea urchin egg messenger ribonucleoprotein complexes are efficient templates for in vitro protein synthesis. *Dev. Biol.* 93: 389–403.

Moon, R. T., Nicosia, R. F., Olsen, C., Hille, M. B. and Jeffery, W. R. 1983. The cytoskeletal framework of sea urchin eggs and embryos: Developmental changes in the association of messenger RNA. *Dev. Biol.* 95: 447–458.

Morle, F., Lopez, B., Henni, T. and Godet, J. 1985. α-Thalassaemia associated with the deletion of two nucleotides at positions -2 and -3 preceding the AUG codon. *EMBO J.* 4: 1245–1250.

Mottes, J. R. and Iverson, L. E. 1995. Tissue-specific alternative splicing of hybrid *shaker/lacZ* genes correlates with kinetic differences in shaker K^+ currents in vivo. *Neuron* 14: 613–623.

Murata, Y. and Wharton, R. P. 1995. Binding of *pumillio* to maternal *hunchback* mRNA is required for posterior patterning in *Drosophila* embryos. *Cell* 80: 747–756.

Nagoshi, R. N., McKeown, M., Burtis, K. C., Belote, J. M. and Baker, B. S. 1988. The control of alternative splicing at genes regulating sexual differentiation in *D. melanogaster*. *Cell* 53: 229–236.

Navaratnam, N., Bhattacharya, S., Fujino, T., Patel, D., Jarmuz, A. L. and Scott, J. 1995. Evolutionary origins of apoB editing: catalysis by a cytidine deaminase that has acquired a novel RNA-binding motif at its active site. *Cell* 81: 187–195.

Newport, J. and Kirschner, M. 1982. A major developmental transition in early *Xenopus* embryos. II. Control of the onset of transcription. *Cell* 30: 687–696.

Orkin, S. H., Cheng, T.-C., Antonarakis, S. E. and Kazazian, H. H., Jr. 1985. Thalassemia due to a mutation in the cleavage-polyadenylation signal of the human β-globin gene. *EMBO J.* 4: 453–456.

d'Orval, B. C., Carafa, Y. d'A., Sirand-Pugnet, P., Gallego, M., Brody, E. and Marie, J. 1991. RNA secondary structure repression of a muscle-specific exon in HeLa cell nuclear extracts. *Science* 252: 1823–1828.

Ostareck-Lederer, A., Ostareck, D. H., Standart, N. and Thiele, B. 1994. Translation of 15-lipoxygenase mRNA is inhibited by a protein that binds to a repeated sequence in the 3' untranslated region. *EMBO J.* 13: 1476–1481.

Ovsenek, N., Zorn, A. M. and Krieg, P. A. 1992. A maternal factor, OZ-1, activates embryonic transcription of the *Xenopus laevis* GS17 gene. *Development* 115: 649–655.

Palatnik, C. M., Wilkins, C. and Jacobson, A. 1984. Translational control during early *Dictyostelium* development: Possible involvement of poly(A) sequences. *Cell* 36: 1017–1025.

Paris, J. and Richter, J. D. 1990. Maturation-specific polyadenylation and translational control: Diversity of polyadenylation elements, influence of poly(A) tail size and the formation of stable polyadenylation complexes. *Mol. Cell. Biol.* 10: 5634–5645.

Paris, J., Swensen, K., Piwnica-Worms, H. and Richter, J. D. 1991. Maturation-specific polyadenylation: In vitro activation by p34^cdc2 and phosphorylation of a 58-kD CPE-binding protein. *Genes Dev.* 5: 1697–1708.

Pavlakis, G. N., Lockard, R. E., Vamvakopolous, N., Rieser, L., Rajbhandary, U. L. and Vournakis, J. N. 1980. Secondary structure of mouse and rabbit α- and β-globin mRNAs: Differential accessibility of initiator AUG codons towards nucleases. *Cell* 19: 91–102.

Paynton, B. V., Rempel, R. and Bachvarova, R. 1988. Changes in states of adenylation and time course of degradation of maternal mRNAs during oocyte maturation and early embryonic development in the mouse. *Dev. Biol.* 129: 304–314.

Pelletier, J. and Sonenberg, N. 1985. Insertional mutagenesis to increase secondary structure within the 5' noncoding region of a eukaryotic mRNA reduces translational efficiency. *Cell* 40: 515–526.

Poccia, D., Wolff, R., Kragh, S. and Williamson, P. 1985. RNA synthesis in male pronuclei of the sea urchin. *Biochim. Biophys. Acta* 824: 349–356.

Powell, L. M., Wallis, S. C., Pease, R. J., Edwards, Y. H., Knott, T. J. and Scott, J. 1987. A novel form of tissue-specific RNA processing produces apolipoprotein-B48 in intestine. *Cell* 50: 831–840.

Proudfoot, N. J. and Brownlee, G. G. 1976. 3' Non-coding region sequences in eukaryotic messenger RNA. *Nature* 263: 211–214.

Raff, R. A. 1980. Masked messenger RNA and the regulation of protein synthesis in eggs and embryos. *In* D. M. Prescott and L. Goldstein (eds.), *Cell Biology: A Comprehensive Treatise*, Vol. 4. Academic Press, New York, pp. 107–136.

Ramaiah, K. V. A., Dhindsa, R. S., Chen, J. J., London, I. M. and Levin, D. 1992. Recycling and phosphorylation of eukaryotic initiation factor 2 on 60S subunits of 70S initiation complexes and polysomes. *Proc. Natl. Acad. Sci. USA* 89: 12063–12067.

Ranu, R. S., Levin, D. H., Delaunay, J., Ernst, U. and London, I. M. 1976. Regulation of protein synthesis in rabbit reticulocyte lysates: Characteristics of inhibition of protein synthesis by a translational inhibitor from heme-deficient lysates and its relationship to the initiation factor which binds Met-tRNAf. *Proc. Natl. Acad. Sci. USA* 73: 2720–2726.

Ray, B. K. and eight others. 1983. Role of mRNA competition in regulating translation: Further characterization of mRNA discriminatory initiation factors. *Proc. Natl. Acad. Sci. USA* 80: 663–667.

Raychaudhury, P., Chaudhuri, A. and Maitra, U. 1985. Formation and release of eukaryotic initiation factor 2 GDP complex during eukaryotic ribosomal polypeptide chain initiation complex formation. *J. Biol. Chem.* 260: 2140–2145.

Rebagliati, M. R., Weeks, D. L., Harvey, R. P. and Melton, D. A. 1985. Identification and cloning of localized maternal RNAs from *Xenopus* eggs. *Cell* 42: 769–777.

Restifo, L. L. and Guild, G. M. 1986. Poly(A) shortening of coregulated transcripts in *Drosophila*. *Dev. Biol.* 115: 507–510.

Richter, J. D. and Smith, L. D. 1984. Reversible inhibition of translation by *Xenopus* oocyte-specific proteins. *Nature* 309: 378–380.

Robbie, E. P., Peterson, M., Amaya, E. and Musci, T. J. 1995. Temporal regulation of the *Xenopus* FGF receptor in development: A translation inhibiting element in the 3' untranslated region. *Development* 121: 1775–1785.

Rodgers, W. H. and Gross, P. R. 1978. Inhomogeneous distribution of egg RNA sequences in the early embryo. *Cell* 14: 279–288.

Rosenthal, E., Hunt, T. and Ruderman, J. V. 1980. Selective translation of mRNA controls the pattern of protein synthesis during early development of the surf clam, *Spisula solidissima*. *Cell* 20: 487–494.

Ruskin, B., Zamore, P. D. and Green, M. R. 1988. A factor, U2AF, is required for U2 snRNP binding and splicing complex assembly. *Cell* 52: 207–219.

Safer, B. 1989. Nomenclature of initiation, elongation and termination factors for translation in eukaryotes. *Eur. J. Biochem.* 186: 1–3.

Sallés, F. J., Liebfarb, M. E., Wreden, C., Gergen, J. P. and Strickland, S. 1994. Coordinate initiation of *Drosophila* development by regulated polyadenylation of maternal messenger RNAs. *Science* 266: 1996–1999.

Sarkar, G., Edery, I., Gallo, R. and Sonenberg, N. 1984. Preferential stimulation of rabbit α-globin mRNA translation by a cap-binding protein complex. *Biochim. Biophys. Acta* 783: 122–129.

Savage, M. P. and Fallon, J. F. 1995. FGF-2 messenger RNA and its antisense message are expressed in a developmentally specific manner in the chick limb bud and mesonephros. *Dev. Dyn.* 202: 343–353.

Schauer, I. E. and Wood, W. B. 1990. Early *C. elegans* embryos are transcriptionally active. *Development* 110: 1303–1317.

Scott, J. 1995. A place in the world for RNA editing. *Cell* 81: 833–836.

Sharma, P. M., Bowman, M., Madden, S. L., Rauscher, F. J. III and Sukumar, S. 1994. RNA editing in the Wilms' tumor suppressor gene, WT-1. *Genes Dev.* 8: 720–731.

Shatkin, A. J. 1976. Capping of eukaryotic mRNAs. *Cell* 9: 645–653.

Shatkin, A. J. 1985. mRNA cap binding proteins: Essential factors for initiating translation. *Cell* 40: 223–224.

Shaw, G. and Kamen, R. 1986. A conserved AU sequence from the 3' untranslated region of GM-CSF mRNA mediates selective mRNA degradation. *Cell* 46: 659–667.

Sheets, M. D., Fox, C. A., Hunt, T., Vande Woude, G. and Wickens, M. 1994. The 3' untranslated region of c-*mos* and cyclin mRNAs stimulate translation by regulating cytoplasmic polyadenylation. *Genes Dev.* 8: 926–938.

Showman, R. M., Wells, D. E., Anstrom, J., Hursh, D. A. and Raff, R. A. 1982. Message-specific sequestration of maternal histone mRNA in the sea urchin egg. *Proc. Natl. Acad. Sci. USA* 79: 5944–5947.

Simon, R., Tassan, J.-P. and Richter, J. D. 1992. Translational control by poly(A) elongation during *Xenopus* development: Differential regressions and enhancement by a novel cytoplasmic polyadenylation element. *Genes Dev.* 6: 2580–2591.

Simpson, L. and Thiemann, O. H. 1995. Sense from nonsense: RNA editing in mitochondria of kinetoplastid protozoa and slime molds. *Cell* 81: 837–840.

Smibert, C. A., Wilson, J. E., Kerr, K. and Macdonald, P. M. 1996. Smaug protein represses translation of unlocalized *nanos* mRNA in the *Drosophila* embryo. *Genes Dev.* 10: 2600–2609.

Sommer, B., Köhler, M., Sprengel, R. and Seeburg, P. H. 1991. RNA editing in brain controls a determinant of ion flow in glutamate-gated channels. *Cell* 67: 11–19.

Sosnowski, B. A., Belote, J. M. and McKeown, M. 1989. Sex-specific alternative splicing of RNA from the *transformer* gene results from sequence-specific splice site blockage. *Cell* 58: 449–459.

Spirin, A. S. 1966. On "masked" forms of messenger RNA in early embryogenesis and in other differentiating systems. *Curr. Top. Dev. Biol.* 1: 1–38.

Standart, N. 1992. Masking and unmasking of maternal mRNAs. *Semin. Dev. Biol.* 3: 367-379.

Standart, N., Hunt, T. and Ruderman, J.V. 1986. Differential accumulation of ribonucleotide reductase subunits in clam oocytes: The large subunit is stored as a polypeptide, the small subunit as untranslated mRNA. *J. Cell Biol.* 103: 2129–2136.

Standart, N., Dale, M., Stewart, E. and Hunt, T. 1990. Maternal mRNA from clam oocytes can be specifically unmasked *in vitro* by antisense RNA complementary to the 3' untranslated region. *Genes Dev.* 4: 2157–2168.

Swiderski, R. E. and Richter, J. D. 1988. Photocrosslinking of proteins to maternal mRNA in *Xenopus* oocytes. *Dev. Biol.* 128: 349–358.

Tamkun, J. W., Schwartzbauer, J. E. and Hynes, R. O. 1984. A single rat fibronectin gene generates three different mRNAs by alternative splicing of a complex exon. *Proc. Natl. Acad. Sci. USA* 81: 5140–5144.

Taylor, M. A. and Smith, L. D. 1985. Quantitative changes in protein synthesis during oogenesis in *Xenopus laevis*. *Dev. Biol.* 110: 230–237.

Telford, N. A., Watson, A. J. and Schultz, G. A. 1990. Transition from maternal to embryonic control in early mammalian development: A comparison of several species. *Mol. Reprod. Dev.* 26: 90–100.

Thach, R. E. 1992. Cap recap: The involvement of eIF-4F in regulating gene expression. *Cell* 69: 177–180.

Thomas, N. S. B., Matts, R. L., Levin, D. H. and London, I. M. 1985. The 60S ribosomal subunit as a carrier of eukaryotic initiation factor 2 and the site of reversing factor activity during protein synthesis. *J. Biol. Chem.* 260: 9860–9866.

Tian, M. and Maniatis, T. 1993. Positive control of pre-mRNA splicing in vitro. *Science* 256: 237–240.

Tian, M. and Maniatis, T. 1993. A splicing enhancer complex controls alternative splicing of doublesex pre-mRNA. *Cell* 74: 105–114.

Varnum, S. M. and Wormington, W. M. 1990. Deadenylation of maternal mRNAs during *Xenopus* oocyte maturation does not require *cis* sequences: A default mechanism for translational control. *Genes Dev.* 4: 2278–2286.

Valcárcel, J., Singh, R., Zamore, P. D. and Greene, M. R. 1993. The protein Sex-lethal antagonizes the splicing factor U2AF to regulate alternative splicing of transformer pre-mRNA. *Nature* 362: 171–175.

Varnum, S., Hurney, C. A. and Wormington, W. M. 1992. Maturation-specific deadenylation in *Xenopus* oocytes requires nuclear and cytoplasmic factors. *Dev. Biol.* 153: 283–290.

Vassalli, J. D. and seven others. 1989. Regulated polyadenylation controls mRNA translation during meiotic maturation of mouse oocytes. *Genes Dev.* 3: 2163–2171.

Wagenaar, E. B. and Mazia, D. 1978. The effect of emetine on the first cleavage division of the sea urchin, *Strongylocentrotus purpuratus*. *In* E. R. Dirksen, D. M. Prescott and L. F. Fox (eds.), *Cell Reproduction: In Honor of Daniel Mazia*. Academic Press, New York, pp. 539–545.

Walker, J. Dale, M. and Standart, N. 1996. Unmasking messenger RNA in clam

oocytes: Role of phosphorylation of a 3'UTR masking element-binding protein at fertilization. *Dev. Biol.* 173: 292–305.

Wang, C., Dickinson, L. K. and Lehmann, R. 1994. Genetics of *nanos* localization in *Drosophila*. *Dev. Dyn.* 199: 103–115.

Weir, M. P. and Kornberg, T. 1985. Patterns of *engrailed* and *fushi tarazu* transcripts reveal novel intermediate stages of *Drosophila* segmentation. *Nature* 318: 433–439.

Wharton, R. P. and Struhl, G. 1991. RNA regulatory elements mediate control of *Drosophila* body pattern by the posterior morphogen, nanos. *Cell* 67: 955–967.

Wickens, M. and Stephenson, P. 1984. Role of the conserved AAUAAA sequence: Four AAUAAA point mutants prevent 3' end formation. *Science* 226: 1045–1051.

Wickens, M. 1992. Introduction: RNA and the early embryo. *Semin. Dev. Biol.* 3: 363–365.

Wickens, M. and Takayama, K. 1994. Deviants—or emissaries. *Nature* 367: 17–18.

Wightman, B., Ha, I. and Ruvkun, G. 1993. Posttranslational regulation of the heterochronic gene *lin-14* by *lin-4* mediates temporal pattern formation in *C. elegans*. *Cell* 75: 855–862.

Wilson, T. and Treisman, R. 1988. Removal of poly(A) and consequent degradation of c-*fos* mRNA facilitated by 3' AU-rich sequences. *Nature* 336: 396–399.

Winkler, M. M. and Steinhardt, R. A. 1981. Activation of protein synthesis in a sea urchin cell-free system. *Dev. Biol.* 84: 432–439.

Winkler, M. M., Nelson, E. M., Lashbrook, C. and Hershey, J. W. B. 1985. Multiple levels of regulation of protein synthesis at fertilization in sea urchin eggs. *Dev. Biol.* 107: 290–300.

Wold, B. J., Klein, W. H., Hough-Evans,B. R., Britten, R. J. and Davidson, E. H. 1978. Sea urchin embryo mRNA sequences expressed in nuclear RNA of adult tissues. *Cell* 14: 941–950.

Wu, J. and Maniatis, T. 1993. Specific interactions between proteins implicated in splice site selection and regulated alternative splicing. *Cell* 75: 1061–1070.

Young, E. M. and Raff, R. A. 1979. Messenger ribonucleoprotein particles in developing sea urchin embryos. *Dev. Biol.* 72: 24–40.

Zorn, A. M. and Krieg, P. A. 1992. Developmental regulation of alternative splicing in the mRNA encoding *Xenopus laevis* neural cell adhesion molecule (N-CAM). *Dev. Biol.* 149: 197–205.

Zucker, W. V. and Schulman, H. M. 1968. Stimulation of globin-chain initiation by hemin in the reticulocyte cell-free system. *Proc. Natl. Acad. Sci. USA* 59: 582–589.

Specification of Cell Fate and the Embryonic Axes

Autonomous cell specification by cytoplasmic determinants

13

I hold it probable that in the germ cells there exist fine internal differences which predetermine the subsequent transformation to a determinant substance; not differences which are mere potencies present in the germ cells, but actual material differences so fine that we have not as yet been able to demonstrate them.
R. VIRCHOW (1858)

Studying the period of cleavage we approach the source whence emerge the progressively branched streams of differentiation that end finally in almost quiet pools, the individual cells of the complex adult organism.
E. E. JUST (1939)

EACH METAZOAN ORGANISM is a complex assortment of specialized cell types. For example, red and white blood cells differ not only from each other but also from the heart cells that propel them through the body. They also differ from the outstretched neurons that conduct neural impulses from the brain to the heart, and from the glandular cells that secrete hormones into the blood. Table 13.1 presents a very incomplete list of specialized cell types, their characteristic products, and their functions.

Cell commitment and differentiation

The development of specialized cell types from the single fertilized egg is called **differentiation.** This overt change in cellular biochemistry and function is preceded by a process involving the covert commitment of cells to a particular fate or set of fates. Here, the cell does not *appear* phenotypically different from its uncommitted state, but somehow its developmental fate has become restricted. Although embryologists have routinely used the word *determination* to describe this hidden commitment, a particular tissue might be classified as determined or undetermined depending on which assay for determination was used (see Harrison, 1933). Slack (1991) has therefore divided this commitment into two stages, **specification** and **determination.** A cell or tissue is said to be *specified* when it is capable of differentiating autonomously when placed in a neutral environment such as a petri dish. (The environment is "neutral" with respect to the developmental pathway.) A cell or tissue is said to be *determined* when it is capable of differentiating autonomously even when placed in another region of the embryo. If it is able to differentiate according to its original fate even when placed in another region of the embryo, it is assumed that the commitment is irreversible.

We know of three major ways by which this commitment can take place (Table 13.2). The first mechanism of commitment involves the cytoplasmic segregation of determinative molecules during embryonic cleavage, whereby the cleavage planes separate qualitatively different regions of the zygote cytoplasm into different daughter cells. Each cell becomes specified by the type of cytoplasm it acquires during cleavage, and cell fate is thereby

505

Table 13.1 Some differentiated cell types and their major products

Cell type	Differentiated cell product	Specialized function
Keratinocyte (skin cell)	Keratin	Protection against abrasion, desiccation
Erythrocyte (red blood cell)	Hemoglobin	Transport of oxygen
Lens cell	Crystallins	Transmission of light
B lymphocyte	Immunoglobulins	Antibody synthesis
T lymphocyte	Cell-surface antigens (lymphokines)	Destruction of foreign cells; regulation of immune response
Melanocyte	Melanin	Pigment production
Pancreatic islet cells	Insulin	Regulation of carbohydrate metabolism
Leydig cell (♂)	Testosterone	Male sexual characteristics
Chondrocyte (cartilage cell)	Chondroitin sulfate; type II collagen	Tendons and ligaments
Osteoblast (bone-forming cell)	Bone matrix	Skeletal support
Myocyte (muscle cell)	Muscle actin and myosin	Contraction
Hepatocyte (liver cell)	Serum albumin; numerous enzymes	Production of serum proteins and numerous enzymatic functions
Neurons	Neurotransmitters (acetylcholine epinephrine, etc.)	Transmission of electrical impulses
Tubule cell (♀) of hen oviduct	Ovalbumin	Egg white proteins for nutrition and protection of embryo
Follicle cell (♀) of insect oviduct	Chorion proteins	Eggshell proteins for protection of embryo

determined without any reference to neighboring cells. This mechanism of committing cell fate is called **autonomous specification,** because the cells are specified by their own internal cytoplasmic components (Davidson, 1991). Thus, if a particular blastomere were removed early in development, that blastomere would produce the same cells that it would have if it were still part of the larger embryo, and the remaining embryo would lack those cells (and only those cells) that would have been formed by the missing cell (see Figure 1.29). Autonomous specification gives rise to a pattern of embryogenesis referred to as **mosaic development,** since the embryo appears to be constructed like a tile mosaic of self-differentiating parts.

A second way of committing cell fate involves interactions with neighboring cells. Here, the cells originally have the ability to follow more than one path of differentiation, and the interaction of these cells with other cells or tissues restricts the fates of one or both of the participants. This type of cell fate determination is sometimes called **conditional specification,** because the fate of a cell depends upon the conditions in which it finds itself. If a blastomere were removed from an early embryo of an organism having the conditional specification of its cells, the remaining embryonic cells could alter their normal fates so that the roles of the missing cell could be taken over. Thus, conditional specification gives rise to a pattern of embryogenesis called **regulative development.** As we will see, all organisms use both au-

Table 13.2 Modes of cell type specification and their characteristics

I. Autonomous specification

Characteristic of most invertebrates.

Specification by acquisition of certain cytoplasmic molecules present in the egg.

Invariant cleavages produce the same lineages in each embryo of the species. Blastomere fates are generally invariant.

Lineage "founder cells" are usually specified autonomously at poles of embryonic axes.

Cell type specification precedes any large-scale embryonic cell migration.

Produces "mosaic" ("determinative") development: cells cannot change fate if a blastomere is lost.

II. Conditional specification

Characteristic of all vertebrates and few invertebrates.

Specification by interactions between cells. Relative positions are important.

Variable cleavages produce no invariant fate assignments to cells.

Massive cell rearrangements and migrations precede or accompany specification.

Capacity for "regulative" development: allows cells to acquire different functions.

III. Syncytial specification

Characteristic of most insect classes.

Specification of body regions by interactions between cytoplasmic regions prior to cellularization of the blastoderm.

Variable cleavage produces no rigid cell fates for particular nuclei.

After cellularization, conditional specification is most often seen.

Source: After Davidson, 1991.

tonomous and conditional means to specify different cell types, and there is a spectrum ranging from mosaic to regulative development. However, most invertebrates have a predominantly autonomous mode of cell type specification, whereas vertebrates are characterized by their extensive use of conditional specification.

Many insects also utilize a third means to determine cell fate. Here, interactions between maternal components within the syncytial blastoderm occur before the cell membranes separating nuclei have formed. In **syncytial specification,** major cell fate decisions are made even before cells have formed. This chapter will focus on experiments that demonstrate autonomous specification, while the following chapters will cover conditional and syncytial modes of cell commitment during early embryogenesis.

Preformation and epigenesis

Any explanation of the differentiation of the various bodily cells from the fertilized egg has to explain (1) the constant morphology of each species (i.e., that chickens beget only chickens, not crocodiles) and (2) the diversity among the bodily parts of each organism. Indeed, one of the major characteristics of development is that each species reproduces a characteristic developmental pattern. Development involves the expression of the inherited properties of the species.

In the seventeenth century, a union of development and inheritance was achieved in the hypothesis of **preformationism.** According to this view, all the organs of the adult were prefigured in miniature within the sperm or (more usually) the ovum. Organisms were not seen to be "developed," but were "unrolled." This hypothesis had the backing of both science and philosophy (Gould, 1977; Roe, 1981). First, because all organs were prefigured, embryonic development merely required the growth of existing structures, not the formation of new ones. No extra mysterious force was needed for embryonic development. Second, just as the adult organism was prefigured in the germ cells, another generation already existed in a prefigured state within the germ cells of the first prefigured generation. This corollary, called *embôitment* (encapsulation), ensured that the species would always remain constant. Although certain microscopists claimed to see fully formed human miniatures within the sperm or egg, the major proponents of this hypothesis—Albrecht von Haller and Charles Bonnet—knew that organ systems develop at different rates and that embryonic structures need not be in the same place as those in the newborn.

The preformationists had no cell theory to provide a lower limit to the size of their preformed organisms, nor did they view humankind's tenure on Earth as potentially infinite. Rather, said Bonnet (1764), "Nature works as small as it wishes," and the human species existed in that finite time spanning the creation and the resurrection. This was in accord with the best science of its time, conforming to the French mathematician-philosopher René Descartes's principle of the infinite divisibility of a mechanical nature initiated, but not interfered with, by God.

Preformation was a conservative theory, emphasizing the lack of change between generations. Its principal failure was its inability to account for the variations known by the limited genetic evidence of the time. It was known, for instance, that matings between white and black parents produced children of intermediate skin color, an impossibility if inheritance and development were solely through either the sperm or the egg. In more controlled experiments, the German botanist Joseph Kölreuter (1766) had produced hybrid tobacco plants having the characteristics of both species. Moreover, by mating the hybrid to either the male or female parent, Kölreuter was able to "revert" the hybrid back to one or the other parental type after several generations. Thus, inheritance seemed to arise from a mixture of parental components. In addition, preformationism could not explain the generation of "monstrosities" and distinct deviations, such as hexadactylism (six fingers per hand), when both parents were normal.

There developed, then, an alternative hypothesis: **epigenesis.** According to this hypothesis, each adult organism develops anew from an undifferentiated condition. This view of development, having philosophical roots as far back as Aristotle, was revived by Kaspar Friedrich Wolff, a German embryologist working in St. Petersburg. By carefully observing the development of chick embryos, Wolff demonstrated that the embryonic parts develop from tissues that have no counterpart in the adult organism. The heart and blood vessels (which, according to preformationism, had to be present from the beginning to ensure embryonic growth) could be seen to develop anew in each embryo. Similarly, the intestinal tube was seen to arise by the folding of an originally flat tissue. This latter observation was explicitly detailed by Wolff, who proclaimed (1767), "When the formation of the intestine in this manner has been duly weighed, almost no doubt can remain, I believe, of the truth of epigenesis." However, to explain how an organism is created anew each generation, Wolff had to postulate an unknown force, the *vis essentialis* ("essential force"), which, acting like gravity or magnetism, would organize embryonic development.

Preformationism best explained the continuity between generations, whereas epigenesis best explained variation and the direct observations of organ formation. A reconciliation of sorts was attempted by the German philosopher Immanuel Kant (1724–1804) and his colleague, biologist Johann

Friedrich Blumenbach (1752–1840). Attempting to construct a scientific theory of racial descent, Blumenbach postulated a mechanical, goal-directed force called the *Bildungstrieb* ("development force"). Such a force, he said, was not theoretical but could be shown to exist by experimentation. A *Hydra*, when cut, regenerates its amputated parts from the rearrangement of existing elements. Some purposive organizing force could be observed in operation, and this force was a property of the organism itself. This *Bildungstrieb* was thought to be inherited through the germ cells. Thus, development could proceed epigenetically through a predetermined force inherent in the matter of the embryo (Cassirer, 1950; Lenoir, 1980). Moreover, such a force was believed to be susceptible to change, as demonstrated by the left-handed variant of snail coiling.

In this hypothesis, wherein epigenetic development is directed by preformed instructions, we are not far from the view held by some modern biologists that "the complete description of the organism is already written in the egg" (Brenner, 1979). However, until the rediscovery of Mendel's work at the beginning of the twentieth century, there was no consistent genetic theory in which to place such ideas of inherited variation, and each scientist was free to speculate on the mechanisms by which developmental patterns are inherited.

The French Teratologists

The attempts to find a hypothesis that would explain species constancy and epigenetic development led to the creation of modern embryology. The searches for such a hypothesis were undertaken in two different intellectual traditions. One, centered in France, sought to discover the mechanisms by which embryological errors cause some infants to be born with developmental abnormalities. This science became known as **teratology,** the study of congenital malformations. The second search was centered in Germany and focused on the physiology of developmental processes. Both research traditions began manipulating embryos to see how the developing organism would respond to these perturbations (Churchill, 1973; Fischer and Smith, 1984).

The French teratological experiments began in the 1820s with the studies of Etienne Geoffrey Saint-Hilaire and his son, Isadore. These investigators attempted to show that anomalous births were the products of disrupted fetal development rather than preformed aberrations. They sought to artificially produce developmental anomalies by altering the incubation conditions of chick egg development. Though largely unsuccessful in these attempts (their crude techniques either allowed normal development to continue or killed the embryos), they set the stage for Dareste's more refined analysis in 1877. Dareste performed thousands of experiments and traced developmental anomalies in chicks back to early stages of their development.

But the chick embryo was a poor choice of organism for studying the earliest stages of embryogenesis. If one wanted to examine whether perturbations at the earliest stages of development affected adult structures, one would have to use another organism. In 1886, a French medical student, Laurent Chabry, began studying teratogenesis in the more readily accessible tunicate embryo. This was a fortunate choice, because these embryos develop rapidly into larvae with relatively few cell types. Chabry set out to produce specific malformations by lancing specific blastomeres of the cleaving tunicate embryo. He discovered that each blastomere was responsible for producing a particular set of larval tissues. In the absence of those cells, the larva lacked just those structures normally formed by those cells. Moreover, he observed that when particular cells were isolated from the rest of the embryo, they formed their characteristic structure apart from the context of the other cells. Thus, each of the tunicate cells appeared to be developing autonomously.* As we discussed earlier, this ability of each cell to develop independently from the other embryonic cells is often referred to as *autonomous* or *mosaic development*, because the embryo seems to be a mosaic of self-differentiating parts.

Autonomous specification in tunicate embryos

More recent studies have shown that the tunicate embryo does indeed approximate a "mosaic of self-differentiated parts" constructed from information stored in the oocyte cytoplasm. As the embryo divides, different cells incorporate different regions of cytoplasm. These different cytoplasmic regions are thought to contain **morphogenetic determinants** that control the commitment of the cell to a particular cell type. Studies of tunicate cell determi-

Figure 13.1
Segregation of cytoplasmic determinants at the time of fertilization. (A) Fate map of the cytoplasmic regions of the tunicate *Halocynthia roretzi* shortly after the cytoplasmic movements of fertilization have ended. Anterior is to the left, posterior to the right. (B) The organs of the tunicate larva. (A after Nishida, 1987.)

*This was not the answer Chabry expected or the one he had hoped to find. In nineteenth-century France, conservatives favored preformationist views, which were interpreted to support hereditary inequalities between members of a community. What you were was determined by your lineage. Liberals, especially Socialists, favored epigenetic views, which were interpreted to indicate that everyone started off with an equal hereditary endowment, and no one had a "right" to a higher position than any other person. Chabry, a Socialist who hated the inherited rights of the aristocrats, took pains not to extrapolate his data to anything beyond tunicate embryos (see Fischer, 1991).

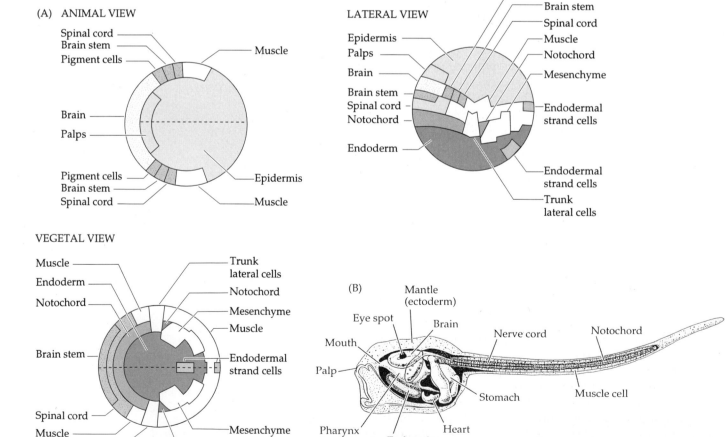

nation have been helped considerably by the eggs of certain species that segregate their cytoplasm into a series of colored regions immediately after fertilization (Plate 11).

The Muscle-Forming Determinant of the Yellow Crescent

In 1905, E. G. Conklin described how these colored plasms become apportioned into various blastomeres. The first cleavage separates the egg into right and left mirror images. From then on, each cell division on one side parallels a similar division on the other. By following the fate of *each* blastomere of the tunicate *Styela partita*, Conklin came to the astonishing conclusion that each of the colored regions of cytoplasm delineates a specific embryonic fate (Figure 13.1). The yellow cytoplasmic crescent gives rise to the muscle cells; the gray equatorial crescent produces the notochord and the neural tube; the clear animal cytoplasm becomes the larval epidermis; and the yolky gray vegetal region gives rise to the larval gut.

Reverberi and Minganti (1946) analyzed tunicate determination in a series of isolation experiments, and they, too, observed the self-differentiation of each isolated blastomere and the remaining embryo. The results of one of these experiments is shown in Figure 13.2. When the 8-cell embryo is separated into its four doublets (the right and left sides being equivalent), mosaic determination is the rule. The animal posterior pair of blastomeres gives rise to the ectoderm; the vegetal posterior pair produces endoderm, mesenchyme, and muscle tissue, just as expected from the fate map. Neural development, however, is an exception. The nerve-producing cells are generated from both the animal and the vegetal anterior quadrants, yet neither produces them alone. When these anterior pairs are reunited, though, the

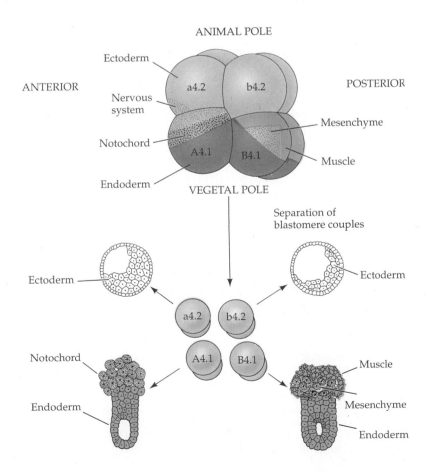

Figure 13.2
Mosaic determination in tunicates. When the four blastomere couples of the 8-cell embryo are dissociated, they develop as indicated, each forming separate structures. (After Reverberi and Minganti, 1946.)

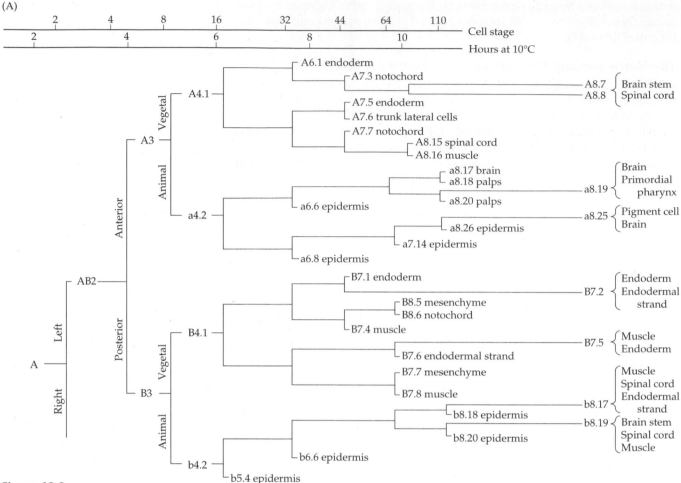

Figure 13.3
Determinative lineage of the tunicate blastomeres. (A) Lineage fate chart of embryonic development in the tunicate *H. roretzi*. Because both right and left halves develop identically, only half of the embryo is represented here. (B) Muscle cell lineages. (A after Nishida, 1987; B after Nishida, 1992a.)

brain and palp tissues arise. Even in an embryo as strictly determined as that of the tunicates, some inductive interactions take place between blastomeres. In fact, Ortolani (1959) has shown that this region of ectoderm is not determined for "neuralness" until the 64-cell stage, right before gastrulation. Thus, although most tissues are determined immediately by segregation of the egg cytoplasm, certain tissues in these embryos have a conditional determination by cell-cell interaction.

From the cell lineage studies of Conklin and others (Figures 13.2 and 13.3), it was known that only one pair of blastomeres (posterior vegetal; B4.1) in the 8-cell embryo is capable of producing tail muscle tissue. When cytoplasm is transferred from the B4.1 (muscle-forming) blastomere to the b4.2 (ectoderm-forming) blastomere of 8-cell tunicate embryos, the ecto-

Figure 13.4
Location of muscle-forming cytoplasm during early ascidian development. Regions of cytoplasm were transferred into the a4.2 (presumptive epidermis) blastomere and screened for muscle-specific proteins made from the a4.2-derived cells. The colored region represents the "yellow crescent" material thought to contain muscle-forming determinants. Percentages indicate the fraction of the specimen showing muscle gene expression. (A) 8-cell embryo. (B) Unfertilized egg. (C) Fertilized egg in the first phase of cytoplasmic movements. (D) Fertilized egg in the second phase of cytoplasmic movements. (After Nishida, 1992b.)

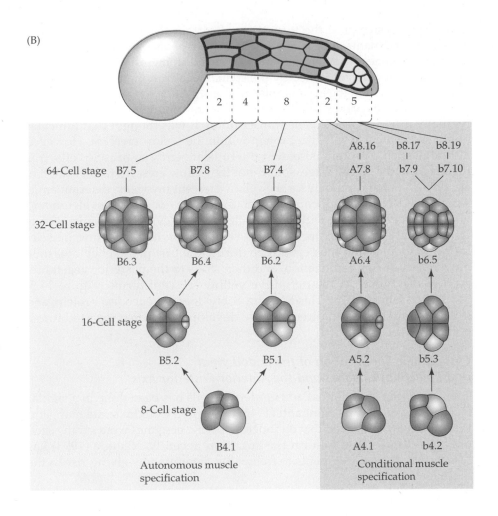

64-Cell stage   B7.5   B7.8   B7.4   A7.8   b7.9   b7.10
A8.16   b8.17   b8.19

32-Cell stage   B6.3   B6.4   B6.2   A6.4   b6.5

16-Cell stage   B5.2   B5.1   A5.2   b5.3

8-Cell stage   B4.1   A4.1   b4.2

Autonomous muscle
specification

Conditional muscle
specification

derm-forming blastomere generates muscle cells as well as its normal ecto-dermal progeny (Whittaker, 1982). Moreover, cytoplasm from the yellow plasm area of the fertilized egg can also cause the a4.2 blastomere to express muscle-specific proteins (Figure 13.4; Nishida, 1992a). Conversely, Tung and colleagues (1977) have shown that when larval *nuclei* are transplanted into enucleated tunicate egg fragments, the newly formed cells show the structures typical of those cells providing the cytoplasm, not of those cells providing the nuclei. We can conclude, then, that certain determinants that exist

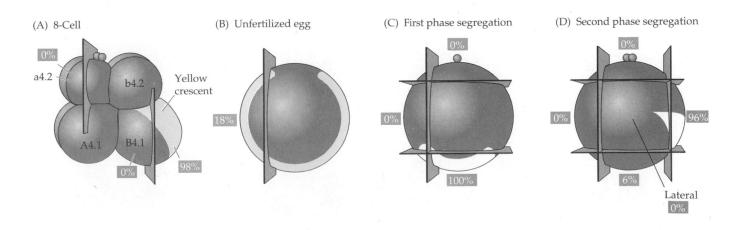

(A) 8-Cell

0%
a4.2   b4.2   Yellow
crescent
A4.1   B4.1
0%   98%

(B) Unfertilized egg

18%

(C) First phase segregation

0%
0%
100%

(D) Second phase segregation

0%
0%   96%
6%

Lateral
0%

in the cytoplasm cause the formation of certain tissues. These morphogenetic determinants appear to work by selectively activating (or inactivating) specific genes. The determination of the blastomeres and the activation of certain genes are controlled by the spatial localization of morphogenetic determinants within the egg cytoplasm. [cyto1.html]

It has been hypothesized that the yellow crescent myogenic determinant regulates muscle-specific gene transcription. It was thought that tunicates might segregate a MyoD-like protein within the yellow crescent. However, although such a protein is seen in the tunicate embryo muscle cells, it is turned on at the 32-cell stage, so it is not the yellow crescent factor (Satoh et al., 1995). A better candidate for the yellow crescent myogenic determinant is a maternal RNA that appears to be bound to the oocyte cytoskeleton and which segregates along with the muscle-forming cytoplasm. This RNA is found in the cortex of mature oocytes, segregates with the yellow, muscle-forming cytoplasm to the vegetal cap during the first phase of cytoplasmic movements during fertilization, and then shifts to the posterior vegetal region of the zygote as the definitive yellow crescent forms (Figure 13.5; Swalla and Jeffery, 1995). This RNA probably does not encode a protein, and it is not known if it can direct muscle development when inserted into a nonmuscle cell.

Cytoplasmic Specification of the Endodermal and Epidermal Lineages and the Anterior-Posterior Axis

Analysis of the endodermal and epidermal cells has been done in a similar fashion. Reverberi and Minganti (1946) confirmed Conklin's fate map, and Whittaker (1977) showed that endoderm-specific enzymes were synthesized only in cells destined to form the gut. More recently, Nishida (1993) has fused cells and cell fragments to follow the determinants giving rise to the

Figure 13.5
Spatial localization of an RNA (YC-RNA) that segregates with the yellow crescent muscle-forming cytoplasm. In situ hybridizations were performed at various stages of *Styela clava* development. (A) Unfertilized egg. (B) After first phase of oogenic cytoplasm movements of fertilization. (C) Frontal section through a 4-cell embryo showing expression in both vegetal blastomeres. (D) Frontal section of 32-cell embryo showing expression in six posterior muscle cells. (E) Tailbud embryo, showing YC-RNA expression in muscle progenitor cells on either side of the notochord. (From Swalla and Jefferey, 1995; photographs courtesy of the authors.)

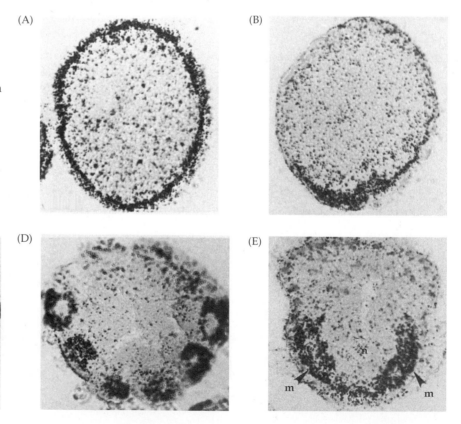

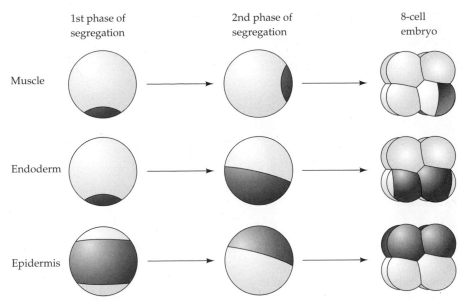

1st phase of segregation   **2nd phase of segregation**   **8-cell embryo**

Muscle

Endoderm

Epidermis

Figure 13.6
Comparison of the movements of the cytoplasmic determinants for three types of tunicate tissues. These figures represent the surface of the eggs only. (After Nishida, 1994a.)

epidermal and endodermal cell lineages. After fusing the cell or cell fragment with a cell from another lineage, Nishida used a biochemical or antigenic marker to determine if that cell took on a new fate. The epidermal determinants migrate to the apical region of the cell during fertilization and enter into the animal cap blastomeres (the a4.2 pair and b4.2 pair) of the 8-cell embryo. Conversely, the endodermal determinants were found to migrate to the vegetal hemisphere of the zygote and to become apportioned into the vegetal blastomeres (Figure 13.6; Nishida, 1994a).

The anterior-posterior axis is also determined during the migration of the oocyte cytoplasmic regions. When roughly 10 percent of the cytoplasm from the posterior vegetal region of the egg was removed after the second ooplasmic movement, most of the embryos failed to form their anterior-posterior axis. Rather, these embryos developed into radially symmetrical larvae with anterior fates. This posterior vegetal cytoplasm (PVC) was "dominant" to other cytoplasms in that when PVC was transplanted into the anterior vegetal region of zygotes that had their own PVC removed, the anterior of the cell became the new posterior, and the axis was reversed (Nishida, 1994b). These results suggest that posterior fate is determined through a particular determinant in the cytoplasm, while anterior fate is determined by the absence of the posterior vegetal cytoplasm. This correlates well with the observation that most of the posterior cell fates (such as muscle and endoderm) are cytoplasmically specified, but the anterior cell fates (such as brain and notochord) are generated by inductions (Figure 13.7).

In the tunicate embryo, the ooplasmic movements at fertilization create distinctively different cytoplasmic domains, which become apportioned into the blastomeres. The identity of these determinants and their mechanisms of action remain to be determined.

Cytoplasmic localization in mollusc embryos

The mosaic type of differentiation is widespread throughout the animal kingdom, especially in protostomal organisms such as ctenophores (comb jellies), annelids, nematodes, and molluscs, all of which initiate gastrulation at the future anterior end after only a few cell divisions. Molluscs provide some of the most impressive examples of "mosaic" development and of the

(A)

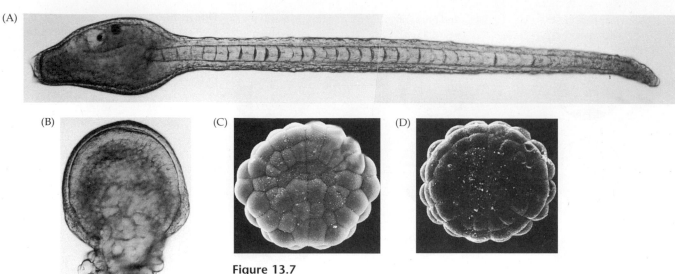

(B)

(C)

(D)

Figure 13.7
Comparison of normal tunicate embryos and embryos from which posterior vegetal cytoplasm had been removed. (A) Wild-type larva. (B) Radially symmetrical larva from egg in which PVC was removed. The larva has no anterior-posterior axis. These larvae consist of an outer epidermal layer, a central notochordal mass, and an intermediately placed endodermal layer. (C) Vegetal view of normal 76-cell embryo. (D) Vegetal view of radially symmetrical embryo whose PVC had been removed. (After Nishida, 1994b.)

phenomenon of **cytoplasmic localization,** wherein the morphogenetic determinants are found in a specific region of the oocyte. Moreover, these cytoplasmic factors become actively moved to one pole of the cell so that a blastomere having these factors can restrict their transmission to only one of its two daughter cells. The fate of the two daughter cells is thus changed by which one of them gets the morphogenetic determinant.

E. B. Wilson, the outstanding American embryologist at the beginning of this century, isolated early blastomeres from embryos of the mollusc *Patella coerulea* and compared their development with that of the same cells left within other *Patella* embryos. Figure 13.8 shows one set of results published by Wilson in 1904. Not only did the isolated blastomeres follow their normal developmental fates (in this case, to produce the ciliated trochoblast cells), but they also completed the normal number of cell divisions at precisely the

Figure 13.8
(A–C) Differentiation of trochoblast cells in the normal embryo of the mollusc *Patella*. (A) 16-cell stage seen from the side; the presumptive trochoblast cells are shaded. (B) 48-cell stage. (C) Ciliated larval stage, seen from the animal pole. Cilia are seen on trochoblast cells. (D–G) Differentiation of *Patella* trochoblast cells isolated and cultured in vitro. (D) Isolated trochoblast cell. (E,F) Results of first and second divisions in culture. (G) Ciliated product of (F). Even in isolated culture, cells become ciliated at the correct time. (After Wilson, 1904.)

Normal development of *Patella*

Presumptive trochoblast

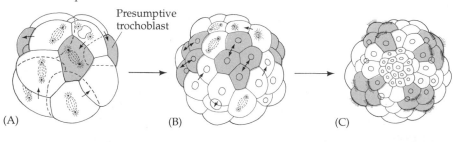

(A)

(B)

(C)

Isolated trochoblast development

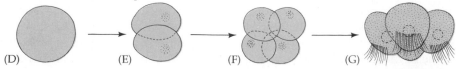

(D)

(E)

(F)

(G)

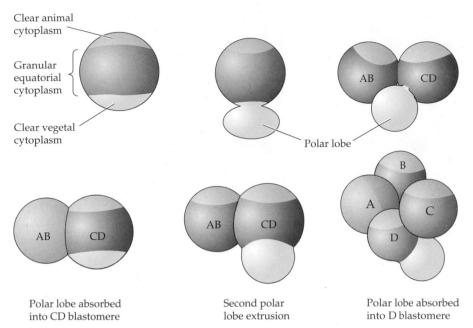

Clear animal cytoplasm

Granular equatorial cytoplasm

Clear vegetal cytoplasm

AB · CD

Polar lobe

AB · CD

AB · CD

B · A · C · D

Polar lobe absorbed into CD blastomere

Second polar lobe extrusion

Polar lobe absorbed into D blastomere

Figure 13.9
Cleavage in the mollusc *Dentalium.* Extrusion and reincorporation of the polar lobe occur twice. (After Wilson, 1904.)

same time as those cells remaining within the embryo. Their cleavages were in the correct orientation, and the derived cells became ciliated at the appropriate time. Wilson concluded from these experiments that these cells possess within themselves all the factors that determine the form and rhythm of cleavage and the characteristic and complex differentiation that they undergo, wholly independent of their relation to the remainder of the embryo. [cyto2.html]

The Polar Lobe

In his next experiment, Wilson was able to demonstrate that such development is predicated on the segregation of specific morphogenetic determinants into specific blastomeres. Certain spirally cleaving embryos (mostly in the mollusc and annelid phyla) extrude a bulb of cytoplasm immediately before first cleavage (see Figure 13.9). This protrusion is called the **polar lobe.** In certain species of snails, the region uniting the polar lobe to the rest of the egg becomes a fine tube. The first cleavage splits the zygote asymmetrically, so the polar lobe is connected only to the CD blastomere. In several species, nearly one-third of the total cytoplasmic volume is present in these anucleate lobes, giving them the appearance of another cell. This three-lobed structure is often referred to as the trefoil-stage embryo (Figure 13.10). The CD blastomere then absorbs the polar lobe material, but extrudes it again prior to second cleavage (Figure 13.9). After this division, the polar lobe is attached only to the D blastomere, which absorbs its material. Thereafter, no polar lobe is formed.

Wilson showed that if one removes the polar lobe at the trefoil stage, the remaining cells divide normally. However, instead of producing a normal trochophore (snail) larva, they produce an incomplete larva, wholly lacking its mesodermal organs—muscles, mouth, shell gland, and foot.* Moreover, Wilson demonstrated that the same type of abnormal embryo can be pro-

---

*The shell gland is an ectodermal organ formed through induction by mesodermal cells. Without the mesoderm, no cells are present to induce the competent ectoderm. Again, we see some limited induction within a mosaic embryo.

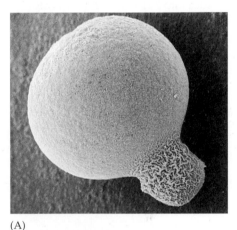

(A)

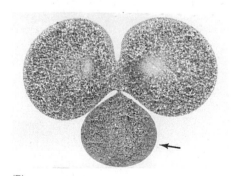

(B)

Figure 13.10
Polar lobes of molluscs. (A) Scanning electron micrograph of the extending polar lobe in the uncleaved egg of *Buccinum undatum.* The surface ridges are confined to the polar lobe region. (B) Section through the first cleavage, or trefoil-stage, embryo of *Dentalium.* The arrow points to the large polar lobe. (Courtesy of M. R. Dohmen.)

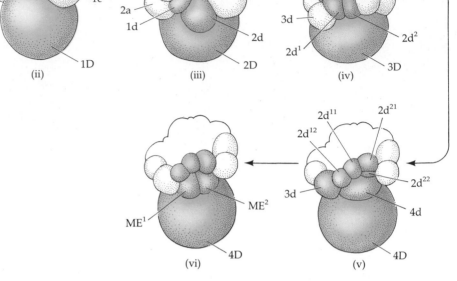

Figure 13.11
Development of the D blastomere. (A) Schematic diagrams of the lineage of the D blastomere in *Ilyanassa* embryos. (i) 4-cell embryo. (ii) 1D and 1d blastomeres at 8-cell stage. (iii) 16-cell stage containing 2D and 2d blastomeres (derived from 1D). The D-derived cells (color) often divide later than others. (iv) Division of the 2D macromere to generate 3D and 3d, cells while the 2d cell divides into 2d1 and 2d2. (v) 64-cell stage. The 3D blastomere produces the 4D and 4d cells. (vi) The 4d blastomere divides symmetrically to produce the two mesentoblasts ME^1 and ME^2. (B) 8-cell embryo. The small PB cell is the polar body and is not part of the embryo. (C) 12-cell embryo (1a–1d have not divided yet). (D) 32-cell embryo. (A after Clement, 1962; photographs from Craig and Morrill, 1986, courtesy of the authors.)

duced by removing the D blastomere from the 4-cell embryo. Wilson concluded that the polar lobe cytoplasm contains the mesodermal determinants and that these determinants give the D blastomere its mesoderm-forming capacity. Wilson also showed that the localization of the mesodermal determinants is established shortly after fertilization, thereby demonstrating that a specific cytoplasmic region of the egg, destined for inclusion into the D blastomere, contains whatever factors are necessary for the special cleavage rhythms of the D blastomere and for the differentiation of the mesoderm.

The morphogenetic determinants sequestered within the polar lobe are probably located in the cytoskeleton or cortex and not in the diffusible cytoplasm of the embryo. Evidence for this came from the studies of A. C. Clement (1968). When the animal hemisphere is separated from the vegetal hemisphere in the snail *Ilyanassa obsoleta*, the animal hemisphere forms ectodermal organs that resemble embryos formed from lobeless eggs. Clement took those embryos that had begun resorbing their second polar lobe and placed them into gelatin slabs. He then centrifuged the embedded embryos, forcing the fluid, yolky cytoplasm from the vegetal part of the cell into the animal hemisphere. By centrifuging these embryos in a second, viscous medium, he caused the separation of the animal and vegetal hemispheres. The animal halves from such centrifuged embryos did not develop any more mesodermal and endodermal structures than those of uncentrifuged eggs. Thus, the determinants of the polar lobe were not transferred to the animal hemisphere in the fluid contents of the vegetal hemisphere. Van den Biggelaar obtained similar results when he removed the cytoplasm from the polar lobe with a micropipette. Cytoplasm from other regions of the cell flowed into the polar lobe, replacing the portion that he had removed. The subsequent development of these embryos was normal. In addition, when he added the soluble polar lobe cytoplasm to the B blastomere, duplications of structures were not seen (Verdonk and Cather, 1983). Therefore, the diffusible part of the cytoplasm does not contain these morphogenetic determinants. They probably reside in the nonfluid cortical cytoplasm or on the cytoskeleton.

(B)

(C)

(D)

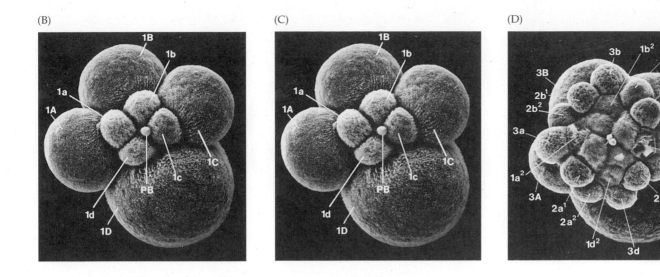

Clement also analyzed the further development of the D blastomere in order to observe the further appropriation of these determinants. The development of the D blastomere is illustrated in Figure 13.11. This macromere, having received the contents of the polar lobe, is larger than the other three. When one removes the D blastomere or its first or second macromere derivatives (1D or 2D), one obtains an incomplete larva, lacking heart, intestine, velum (the ciliated border of the larva), shell gland, eyes, and foot. When one removes the 3D blastomere (*after* the division of the 2D cell to form the 3d blastomere), one obtains an almost-normal embryo, having eyes, foot, velum, and some shell gland, but no heart or intestine (Figure 13.12). Therefore, some of the morphogenetic determinants originally present in the D blastomere were apportioned to the 3d cell. After the 4d cell is given off (by the division of the 3D blastomere), removal of the D derivative (the 4D cell) produces no qualitative difference in development. In fact, all the essential determinants for heart and intestine formation are now in the 4d blastomere, and removal of *that* cell results in a heartless and gutless larva (Clement, 1986). The 4d blastomere is responsible for forming (at its next division) the two **mesentoblasts,** the cells that give rise to both the mesodermal (heart) and endodermal (intestine) organs.

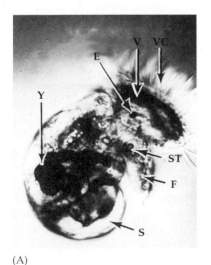

(A)

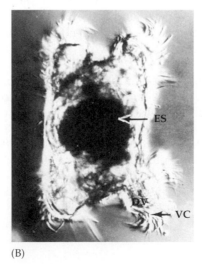

(B)

Figure 13.12
Importance of polar lobe in the development of *Ilyanassa*. (A) Normal veliger larva. (B) Aberrant larva, typical of those produced when the polar lobe of the D blastomere is removed. (E, eye; F, foot; S, shell; ST, statocyst, a balancing organ; V, velum; VC, velar cilia; Y, residual yolk; ES, everted stomodeum; DV, disorganized velum.) (From Newrock and Raff, 1975, courtesy of K. Newrock.)

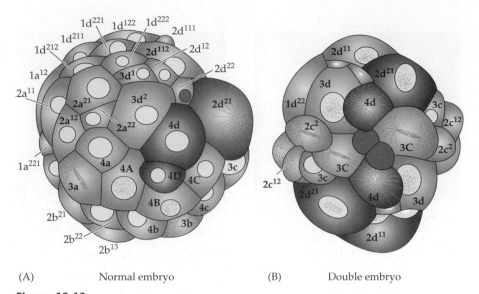

(A)   Normal embryo   (B)   Double embryo

Figure 13.13
Formation of twin embryos by the suppression of polar lobe formation in
Dentalium. (A) Normal embryo at sixth-cleavage stage. (B) Twin embryo formed
when low concentrations of cytochalasin inhibit polar lobe formation and the polar
lobe material is distributed to both the AB and CD blastomeres. (After Guerrier et
al., 1978.)

The material in the polar lobe is also responsible for organizing the dorsal-ventral (back-belly) polarity of the embryo. When polar lobe material is
allowed to pass into the AB blastomere as well as into the CD cell, twin larvae are formed that are joined at their ventral surfaces (Figure 13.13; Guerrier et al., 1978; Henry and Martindale, 1987).

Thus, experiments have demonstrated that the nondiffusible polar lobe
cytoplasm is extremely important for normal mollusc development because:

1. It contains the determinants for the proper rhythm and cleavage orientation of the D blastomere.
2. It contains certain determinants (those entering the 4d blastomere
   and hence leading to the mesentoblasts) for mesodermal and intestinal differentiation.
3. It is responsible for permitting the inductive interactions (through
   the material entering the 3d blastomere) leading to the formation of
   the shell gland and eye.
4. It contains determinants needed for specifying the dorsal-ventral
   axis of the embryo.

Although the polar lobe is clearly important for normal snail development,
we still do not know the mechanisms of its effects. There appear to be no
major differences in mRNA or protein synthesis between lobed and lobeless
embryos (Brandhorst and Newrock, 1981; Collier, 1983, 1984). One possible
clue has been provided by Atkinson (1987), who has observed differentiated
cells of the velum, digestive system, and shell gland within the lobeless embryo. Lobeless embryos can produce these cells, but they appear unable to
organize them into functional tissues and organs. Tissues of the digestive
tract can be found, but they are not connected; myocytes are scattered
around the lobeless larva but are not organized into a functional muscle tissue. Thus, the developmental functions of the polar lobe may be very complex. [cyto3.html], [evo2.html]

Cell specification in the nematode Caenorhabditis elegans

The ability to analyze development requires appropriate organisms. Sea urchins have long been a favorite organism of embryologists because their gametes are readily obtainable in large numbers, their eggs and embryos are transparent, and fertilization and development can occur under laboratory conditions. But sea urchins are difficult to rear in the laboratory for more than one generation, making their genetics difficult to study. Geneticists, on the other hand (at least those working with multicellular eukaryotes), favor *Drosophila*. The rapid life cycle, the readiness to breed, and the polytene chromosomes of the fly larva (which allow gene localization) make this animal superbly suited for hereditary analysis. But *Drosophila* development is very complex and difficult to study. A research program spearheaded by Sidney Brenner (1974) was set up to identify an organism wherein it might be possible to identify each gene involved in development as well as to trace the lineage of every single cell. Such an organism is *Caenorhabditis elegans*, a small (1 mm-long), free-living soil nematode (Figure 13.14A). It has a rapid period of embryogenesis (about 16 hours), which it can accomplish in petri dishes, and relatively few cell types. Moreover, its predominant form is hermaphroditic, each individual containing both eggs and sperm. These roundworms can reproduce either by self-fertilization or by cross-fertilization with the infrequently occurring males. The body of a hermaphroditic *C. elegans* contains exactly 959 somatic cells, whose entire lineage has been traced through its transparent cuticle (Figure 13.14B; Sulston and Horvitz, 1977; Kimble and Hirsch, 1979; Sulston et al., 1983). Furthermore, unlike vertebrate cell lineages, the cell lineage of *C. elegans* is almost entirely invariant

Figure 13.14
Caenorhabditis elegans. (A) Side view of adult hermaphrodite. Early in its development, sperm are formed. These sperm are stored during later stages, so that a mature egg passes through the sperm on its way to the vulva. In this manner, the hermaphrodite unites its own sperm with its own eggs. (B) Entire cell lineage chart for *C. elegans*. Each vertical line represents a cell; each horizontal line represents a cell division. (After Pines, 1992, based on Sulston and Horvitz, 1977, and Sulston et al., 1983.)

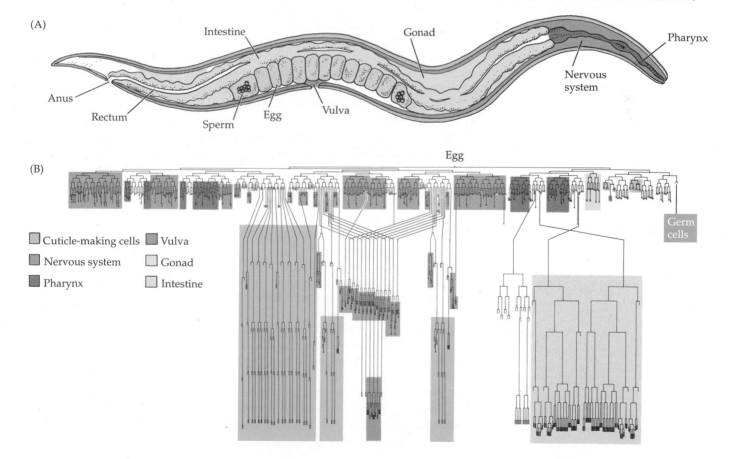

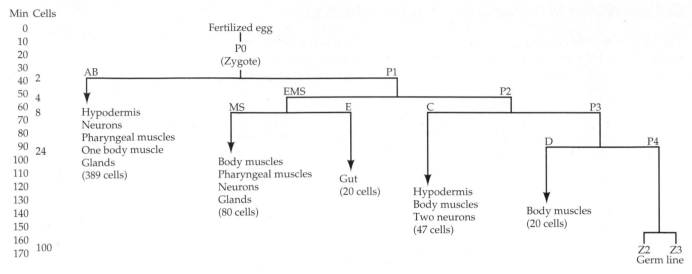

Min Cells

Figure 13.15
Abbreviated cell lineage chart for *C. elegans*, emphasizing the germ line precursors (P cells P0–P4) that receive the P granules. The number of cells (in parentheses) refers to those present in the newly hatched larva. Some of these continue to divide to produce the 959 somatic cells of the adult. (From Strome and Wood, 1983, courtesy of W. Wood.)

from one individual to the next. There is little room for randomness (Sulston et al., 1983). (This is a consequence of the spatial ordering of cytoplasm segregation.) *Caenorhabditis* also has a small number of genes for a multicellular organism—about 15,000 (Sulston et al., 1992).

The initial polarity appears to reside in the elongated egg, the anterior-posterior axis being the long axis of the egg. However, the decision as to which end will become the anterior and which the posterior seems to reside in the sperm. The position of the sperm nucleus entry defines the posterior pole (Goldstein and Hird, 1996).

The pattern of *C. elegans* division (Figure 13.15) resembles that of stem cell lines in that during early cleavage, asymmetrical divisions produce one "differentiating" daughter cell (collectively called the **founder cells** and denoted AB, MS, E, C, and D) and another stem cell (the P1–P4 lineage). The localization of cytoplasmic substances into specific blastomeres has been elegantly demonstrated in these asymmetrical divisions. Within the egg is the set of **germ line granules,** or **P granules,** that are redistributed in the zygote shortly after fertilization and become restricted to those cells capable of forming gametes. Using fluorescent antibodies to a component of the P granules, Strome and Wood (1983) discovered that during the pronuclear migration in the zygote, the randomly scattered P granules become localized in the posterior end of the zygote (toward the site of sperm entry), so that they only enter the blastomere (P1) formed from the posterior cytoplasm (Figure 13.16; Plate 10). Following cleavage, the P granules disperse throughout the P1 blastomere until the start of mitosis, when they once again migrate to the posterior end of the cell. Here they become apportioned to the P2 blastomere. Eventually, the P granules will reside in the P4 cell, whose progeny become the sperm and eggs of the adult. The localization of the P granules requires microfilaments but can occur in the absence of microtubules. Treating the zygotes with cytochalasin D (a microfilament inhibitor) prevents the segregation of these granules to the posterior of the cell, whereas demecolcine (a colchicine-like microtubule inhibitor) fails to stop this movement (Strome and Wood, 1983). Once within the posterior region of the zygote, the P granules remain there, even if the microfilaments are then disrupted (Hill and Strome, 1987, 1990). [other.html#cyto4]

The mechanisms for the movement and anchoring of these cytoplasmic granules remain unknown, but they are regulated by the *par* genes that control cytoplasm partitioning during the first cleavages of *C. elegans*. Mutations

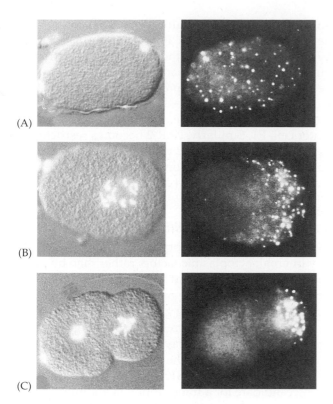

(A)

(B)

(C)

Figure 13.16
Asymmetrical localization of P granules during fertilization and first cleavage. Left-hand figures are stained to show DNA; right-hand figures are the same cells stained with fluorescent antibodies to P-granule protein. (A) A zygote prior to pronuclear migration shows random dispersal of P granules. (B) As pronuclei come together, granules become localized to the posterior periphery of the zygote. (C) A 2-cell embryo in which P1 is entering mitotic prophase; the P granules are now positioned at the posterior periphery to be transmitted to the P2 cell. (From Strome and Wood, 1983, courtesy of S. Strome.)

in six *par* (*par*tition-defective) genes are expressed as maternal effect mutants, where microfilament distributions are aberrant, and the P granules are distributed abnormally (Kemphues et al., 1988; Kirby et al., 1990). Early cleavages in these mutant embryos are symmetrical and synchronous, and the P granules are found in several blastomeres (Figure 13.17). The phenotypes of the *par-2* and *par-3* mutants resemble what occurs when wild-type

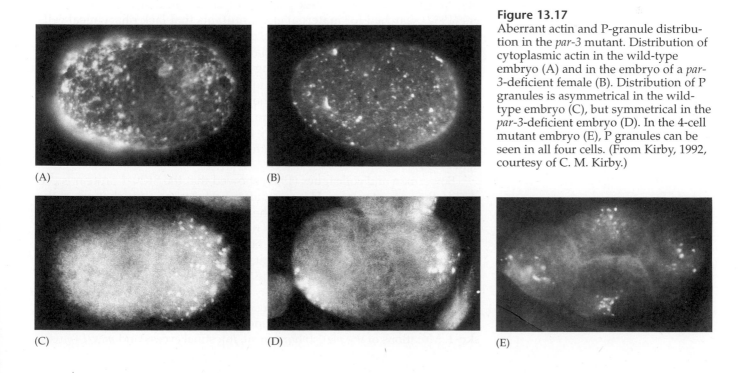

(A)

(B)

(C)

(D)

(E)

Figure 13.17
Aberrant actin and P-granule distribution in the *par-3* mutant. Distribution of cytoplasmic actin in the wild-type embryo (A) and in the embryo of a *par-3*-deficient female (B). Distribution of P granules is asymmetrical in the wild-type embryo (C), but symmetrical in the *par-3*-deficient embryo (D). In the 4-cell mutant embryo (E), P granules can be seen in all four cells. (From Kirby, 1992, courtesy of C. M. Kirby.)

embryos are exposed to a microfilament inhibitor for a 10-minute period during the first cell cycle (Hill and Strome, 1990). Furthermore, at least three of the proteins (PAR-1, PAR-2, and PAR-3) are themselves asymmetrically distributed in the zygote cortex (Etemad-Moghadam et al., 1995; Guo and Kemphues, 1995; Boyd et al., 1996). The PAR-2 protein is uniformly distributed throughout the oocyte cortex, but becomes localized to the posterior cortex in the one-cell embryo. At first cleavage, the PAR-2 protein enters only the posterior daughter cell, P1. Similarly, PAR-2 becomes restricted to the posterior pole of P1, P2, and P3. The PAR-2 protein may be critical for maintaining the PAR-1 protein in the posterior cortex, and PAR-1 appears to be involved in binding to the P granules (Boyd et al., 1996).

Maternal Control of Blastomere Identity: The Genetic Control of the Pharyngeal Progenitor Cells of C. elegans

The determination for much of the *C. elegans* embryo is autonomous, the cell fates being determined by internal, cytoplasmic factors rather than by interactions with neighboring cells. It is thought that the protein factors might determine cell fate by entering the nuclei of the particular blastomeres and activating or repressing specific fate-determining genes. Have any transcription factors been found in autonomously determined cell lineages? While the P granules of *C. elegans* are localized in a way consistent with a role as morphogenetic determinant, they do not enter the nucleus, and their role in development is still unknown. However, the **SKN-1 protein** of *C. elegans* embryos is a very promising candidate for a transcription factor morphogen.

The SKN-1 protein is a maternally specified polypeptide that may control the fate of the EMS blastomere, the cell that generates the posterior pharynx. After first cleavage, only the posterior blastomere, P1, has the ability to autonomously produce pharyngeal cells when isolated. After P1 divides, only the EMS is able to generate pharyngeal muscle cells, even when isolated from the other cells of the body (Priess and Thomson, 1987). Similarly, when the EMS cell divides, only one of its progeny, MS, has the intrinsic ability to generate pharyngeal tissue. This suggests that pharyngeal cell fate may be determined autonomously by maternal factors residing in the cytoplasm that get parceled to these particular cells. Bowerman and his co-workers (1992) searched for maternal effect mutants that lack pharyngeal cells, and they isolated a mutation in the *skn-1* gene. Embryos from homozygous *skn-1*-deficient mothers lack both pharyngeal and intestinal derivatives of EMS (Figure 13.18). Instead of making the normal intestine and pharyngeal structures, these embryos seem to make extra hypodermal (skin) tissue where their intestine and pharynx should be. Only those cells that are destined to form pharynx or intestine are affected by this mutation. Moreover, the protein that would be encoded by this message has a DNA-binding-site motif similar to that seen in the bZIP family transcription factors (Blackwell et al., 1994).

Bowerman and colleagues (1993) showed that the SKN-1 protein is present in the egg cytoplasm. However, after first cleavage, much more of this protein enters the P1 nucleus than the AB nucleus (Figure 13.19). After the second division, both P1 derivatives receive the SKN-1 protein in their nuclei. Thus, the SKN-1 protein might be a morphogen that activates certain genes in the P1 cell and its descendants. However, something else is needed to restrict the function of SKN-1 to the EMS cell and to prevent it from functioning in P2.

Restricting the EMS identity to a single blastomere of the 4-cell embryo requires the activity of two other genes, both of which appear to regulate skn-1. Mutations of the *pie-1* (*p*haryngeal, *i*ntestinal *excess*) and *mex-1* (*m*uscle

Wild-type   *skn*-1 mutant

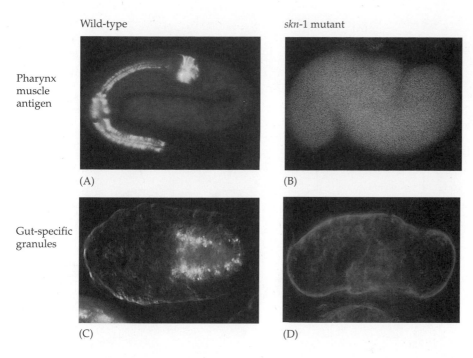

Pharynx muscle antigen

(A)   (B)

Gut-specific granules

(C)   (D)

Figure 13.18
Deficiencies of intestine and pharynx in *skn-1* mutants. Embryos derived from wild-type females (A,C) and females homozygous for mutant *skn-1* (B,D) were tested for the presence of pharyngeal muscles (A,B) and gut-specific granules (C,D). The pharyngeal muscle-specific antibody labels the pharynx musculature of those embryos derived from wild-type females but does not bind to any structure in the embryos from *skn-1* mutant females. Similarly, the birefringent gut granules characteristic of embryonic intestines are absent from the embryos derived from the *skn-1* mutant females. (From Bowerman et al., 1992, courtesy of B. Bowerman.)

*ex*cess) genes alter the determination of cells in the 8-cell *C. elegans* embryo such that several additional cells in the embryo become determined as MS cells (Mello et al., 1992). In embryos derived from *pie-1*-deficient females, the P3 and C sister blastomeres are converted into E and MS blastomeres, respectively, while in embryos from *mex-1*-deficient females, all the descendants of the AB cell are redefined as MS cells. Females simultaneously deficient in both *mex-1* and *pie-1* products generate embryos in which the anterior six cells are MS cells and the posterior two cells are E cells. Thus, PIE-1 and MEX-1 proteins act independently—MEX-1 during the first division, the PIE-1 protein during the second division (Figure 13.20; Bowerman et al., 1993). In all cases, the wild-type SKN-1 gene is needed for the extra MS cells to form, and embryos without *skn-1* lack any pharynx. This links the MEX-1 and PIE-1 proteins to the activation (rather than to the place-

Wild-type   *mex*-1

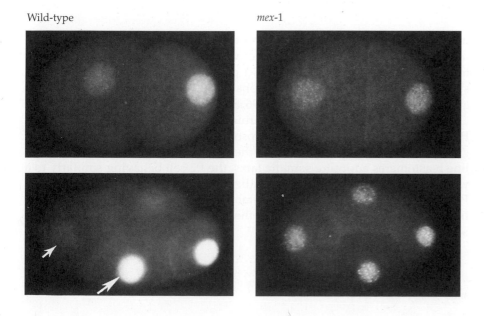

Figure 13.19
Cytoplasmic localization of the SKN-1 protein. Antibodies to the SKN-1 protein show that it is present predominantly in the P1 cell nucleus after first division. After second division, this protein accumulates in both cells derived from P1, but not in cells derived from AB (compare intensities of the nuclei marked by arrows). In *mex-1* mutants, the SKN-1 protein is distributed equally to all blastomeres. (From Bowerman et al., 1993, courtesy of B. Bowerman.)

Figure 13.20
Schematic model of MS cell determination in wild-type and mutant embryos. (A) The MS determinant (thought to be the product of the *skn-1* gene) is present in an inactive state within the egg. During first division in wild-type embryos, the MS determinant is localized into the posterior (P1) blastomere, and at second division, it becomes further localized into the EMS cell. At third division, the EMS cell divides into the MS cell (where the factor is activated) and the E cell. In embryos derived from *mex-1*-deficient females, the factor does not segregate at first division, but the segregation from P1 is normal at the next division. In embryos derived from *pie-1*-deficient females, the initial segregation of the MS determinant into P1 is normal, but the second asymmetrical distribution of the determinant (into the EMS cell) is defective. In the combined mutant, the patterns are superimposed, such that all four cells of the 4-cell embryo have the inactive MS determinant. (B) Summary of interactions involving *skn-1*, *pie-1*, and *mex-1*. (A after Mello et al., 1992; B after McGhee, 1995.)

(A)

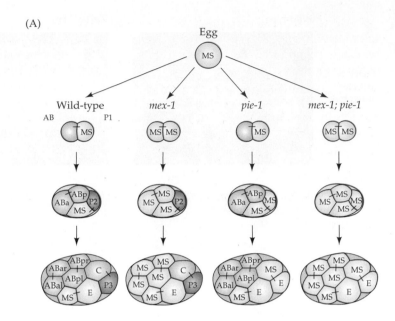

(B)

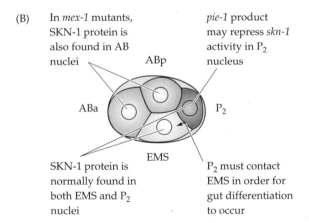

In *mex-1* mutants, SKN-1 protein is also found in AB nuclei

pie-1 product may repress *skn-1* activity in P2 nucleus

SKN-1 protein is normally found in both EMS and P2 nuclei

P2 must contact EMS in order for gut differentiation to occur

ment) of SKN-1*. Thus, the SKN-1 protein, placed into the egg cytoplasm, may be a transcription factor that activates specific genes in the MS blastomere to determine its fate. The PIE-1 protein, which prevents SKN-1 from specifying the pharynx in P2, is probably also a transcription factor that atagonizes its action (Mello et al., 1996; Seydoux et al., 1996).

PIE-1 probably plays a positive role as well. During each division, the PIE-1 protein is retained by the centrosome of the cell that becomes the next germ line blastomere. Mutations of the maternal *pie-1* gene result in germ line blastomeres adopting somatic fates, the P2 cell behaving similarly to a wild-type EMS blastomere. The localization and the genetic properties of PIE-1 suggest that it represses the establishment of somatic cell determination and preserves the totipotency of the germ cell lineage (Mello et al., 1996; Seydoux et al., 1996).

*The proper placement of SKN-1 appears to depend on the PAR proteins, especially PAR-3 and PAR-6 (Watts et al., 1996).

At the third division, the EMS blastomere gives rise to E (which forms the intestine) and MS (which predominantly forms the pharynx and muscle wall cells). The EMS cell contains SKN-1, and each of its descendants have equal amounts of SKN-1. So while SKN-1 is critical in determining which cell can give rise to the pharyngeal mesoderm (i.e., which blastomere becomes the EMS cell), something besides SKN-1 specifies MS to the exclusion of E. Studies by Lin and colleagues (1995) have shown that the POP-1 protein is critical for MS specification. In the absence of POP-1, the MS cell adopts the fate of another wild type E blastomere. In such maternal effect mutants, the MS cell produces neither pharyngeal nor muscle cells, producing intestinal cells instead. The POP-1 protein is probably a transcription factor, and it may interact with SKN-1 to specify MS development. [cyto5.html], [cyto6.html]

Regulation in C. elegans

The development of *C. elegans* is largely autonomous, but regulatory interactions between cells are also important for specifying cell fate. If the EMS blastomere is separated from all the other cells at the 4-cell stage shortly after its formation, it will not form gut-specific rhabditin granules. If it is recombined with the P2 blastomere, however, it will form such granules; but it will not do so when combined with ABa, ABp, or both AB derivatives (Figure 13.21; Goldstein, 1992). Cellular interactions are needed for this stage of intestinal determination.

Since the nematode has invariant cell lineages, it also has invariant cell-cell interactions. In the 4-cell embryo, sister blastomeres ABa (anterior) and ABp (posterior) have different developmental fates. ABa makes neurons, hypodermis, and the anterior pharynx cells, while ABp makes only neurons and hypodermal cells. However, if one experimentally reverses their positions, their fates are similarly reversed, and a normal embryo is formed. In other words, ABa and ABp are equivalent cells whose fate is determined by their positions within the embryo (Priess and Thomson, 1987). Under normal circumstances, however, the invariant embryonic cleavage pattern dictates that the descendants of ABa, not ABp, make 19 pharyngeal cells. ABa

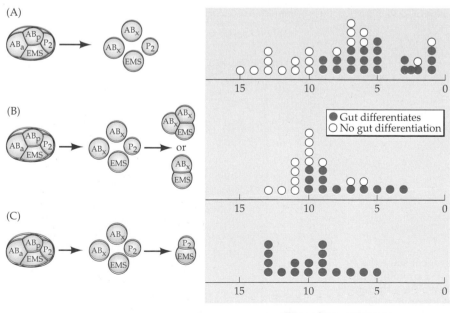

Time of separation
(min before EMS cleavage)

Figure 13.21
Results of isolation and recombination experiments, showing that cellular interactions are required for the EMS cell to form intestine lineage determinants. (A) When separated shortly after its formation, the EMS blastomere cannot produce gut-specific granules. If left there for longer periods, it can. (B) If the EMS cell is recombined with either or both derivatives of the AB blastomere, it will not form gut-specific granules. (C) If recombined with the P2 blastomere, the EMS cell will give rise to gut-specific structures. (After Goldstein, 1992.)

daughter cells differentiate into these pharyngeal muscle cells because of their interaction with the EMS blastomere or its descendants (which autonomously produce 18 pharyngeal muscle cells).

Genetic studies have shown that ABp becomes different from ABa through an interaction with the P2 cell. Moreover, these studies show that this interaction is mediated between the **GLP-1** protein on the ABp cell and the **APX-1** (**a**nterior **p**harynx e**x**cess) protein on the P2 blastomere. In an unperturbed embryo, both ABa and ABp contact the EMS blastomere, but only ABp contacts the P2 cell. If the P2 cell is killed at the early 4-cell stage, the ABp cell does not generate the intestinal valve cells as it normally would (Bowerman et al., 1992). Contact between ABp and P2 is essential for the specification of ABp cell fates, and the ABa cell can be made into an ABp-type cell if it is forced into contact with P2 (Hutter and Schnabel, 1994; Mello et al. 1994). The maternal product of the *glp-1* **gene** appears to be critical in distinguishing ABp from ABa. In embryos whose mothers had mutant *glp-1*, the ABp is transformed into an ABa cell (Hutter and Schnabel, 1994; Mello et al., 1994). Using temperature-sensitive alleles of *glp-1*, is was shown that the time for the GLP-1-dependent interaction is between the 4- and 12-cell stages, when P2 is required for the establishment of the ABp fates (Figure 13.22). The GLP-1 protein is a member of a widely conserved family called the Notch proteins, which serve as cell membrane receptors in many cell-cell interactions, and it is seen on the ABa and ABp cells (Evans et al., 1994).*

One of the most important ligands for Notch proteins such as GLP-1 is another cell-surface protein called Delta. In *C. elegans*, the Delta-like protein is APX-1, and it is found on the P2 cell (Mango et al., 1994; Mello et al., 1994). This APX-1 signal appears to break the symmetry between ABa and ABp, since it stimulates the GLP-1 protein solely on the AB descendant that it touches, namely, the ABp blastomere. In doing this, the P2 cell initiates the dorsal-ventral axis of *C. elegans*.

*As discussed in the previous chapter, the GLP1 protein is localized in the ABa and ABp blastomeres but the maternally encoded *glp-1* mRNA is found throughout the embryo. Evans and colleagues (1994) have postulated that there might be some translational determinant in the AB blastomere that enables the *glp-1* message to be translated in its descendants. The *glp-1* gene is also active in regulating postembryonic cell-cell interactions. It is used later by the distal tip cell of the gonad to control the number of germ cells entering meiosis; hence the name *germline* proliferation (see Chapters 17 and 22; Austin and Kimble, 1987).

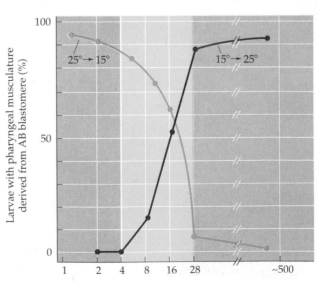

Figure 13.22
Temperature shift experiment to determine at what stage the maternal *glp-1* gene product is active. The GLP-1 protein in this mutant functions at 15° C, but not at 25° C. By shifting the temperature at different embryonic stages, it was found that the GLP-1 protein was needed between the 4- and 28-cell stage. (After Priess et al., 1987.)

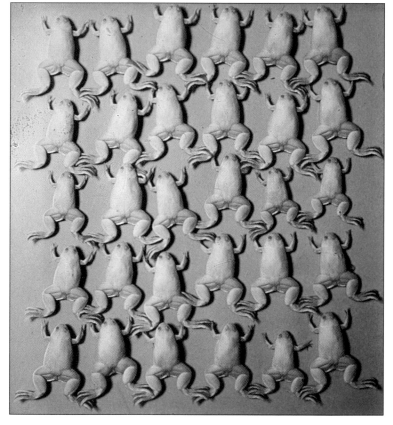

PLATE 1
A clone of *Xenopus* frogs
The nuclei for all the members of this clone came from a single individual—a female tailbud-stage tadpole whose parents (upper panel, right) were both marked by albino genes. The nuclei were transferred into activated enucleated unfertilized eggs from a wild-type female (upper panel). The resulting frogs were all female and albino (lower panel). Chapter 2. (Photograph courtesy of J. Gurdon.)

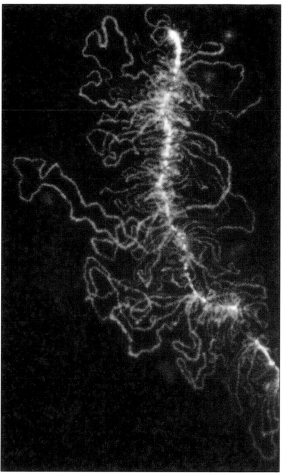

PLATE 2
Active transcription units on a newt chromosome
Oocytes from amphibians such as *Notophtalmus viridescens* have lampbrush chromosomes where the active RNA-synthesizing genes loop out. The DNA axis of these loops is stained with a white dye. The red stain is from an antibody that binds to RNA-binding proteins. Chapter 22. (Photograph courtesy of M. B. Roth and J. Gall.)

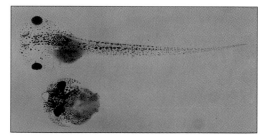

PLATE 3
Fibroblast growth factor is essential for ventral and lateral mesoderm production in *Xenopus*
When *Xenopus* eggs are injected with a dominant negative mutant receptor for fibroblast growth factor (FGF), the embryo is unable to respond to FGF. In the absence of the FGF signal, the ventral and lateral mesoderm fail to form, and the embryo lacks trunk and tail. Chapters 3 and 15. (Photograph courtesy of M. Kirschner.)

PLATE 4
Ability of dorsal blastopore lip to generate secondary neural axis in amphibians

The dorsal blastopore lip of amphibian embryos can organize a second embryonic axis when transplanted to the ventral side of another gastrula. This photograph is taken from an actual slide prepared by Hilde Mangold and shows that the secondary dorsal structures contain both host (unpigmented) and donor (pigmented) tissues. Chapter 15. (Photograph courtesy of P. Fässler and K. Sander.)

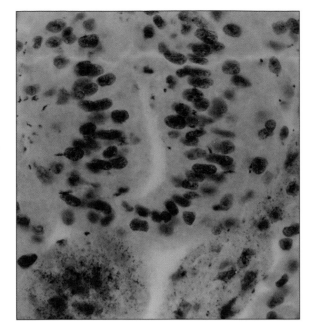

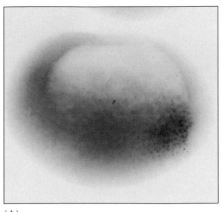

(A)

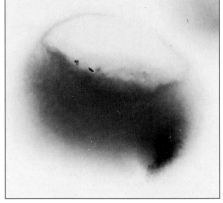

(B)

PLATE 5
Rescue of dorsal structures by the Noggin protein

The Noggin protein may be critical for inducing the dorsal mesoderm and the neural tube. When *Xenopus* eggs are exposed to UV irradiation prior to first cleavage, no dorsal structures form (top). If an early cell of such an embryo is injected with *noggin* RNA, the embryos form dorsal structures. If too much *noggin* message is injected, the embryos produce too much dorsal anterior tissue (bottom). Chapter 15. (Photographs courtesy of R. M. Harland.)

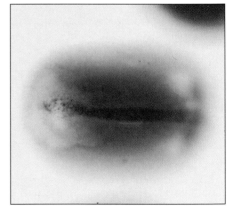

(C)

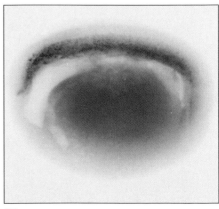

PLATE 6 (*right*)
The *noggin* gene is transcribed in the dorsal mesoderm and endoderm tissue

The *noggin* RNA accumulates in the region of the dorsal marginal zone (A) and is seen in the dorsal blastopore lip (B). When these cells involute, *noggin* expression is seen in the notochord and pharyngeal endoderm (C), which extend anteriorly in the center of the embryo (D). Chapter 15. (Photographs courtesy of R. M. Harland.)

(D)

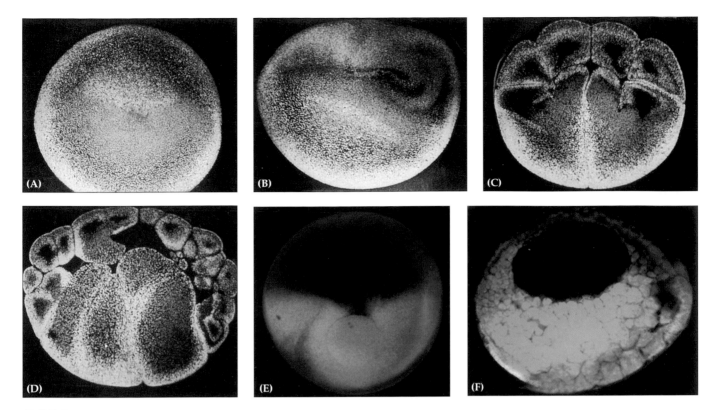

PLATE 7
Cytoplasm rearrangements in *Xenopus laevis*

(A) The unfertilized *Xenopus laevis* egg is radially symmetric. (B) Cytoplasmic movements are seen as the egg starts cleaving, 90 minutes after fertilization. The cytoplasm of the future dorsal side (right) differs from that of the future ventral side (left). These differences can be seen throughout embryonic cleavage (C,D) and result in the positioning of the dorsal morphogenetic determinants in the side of the embryo opposite the point of sperm entry. (E) The cytoplasmic movements correlate with the displacement of β-catenin. At the early 2-cell stage, β-catenin (stained orange) is localized predominantly in the future dorsal surface of the embryo. This pattern persists into the blastula stage (F). Chapters 4, 6, and 15. (A–D courtesy of M.V. Danilchik; E and F courtesy of R. T. Moon.)

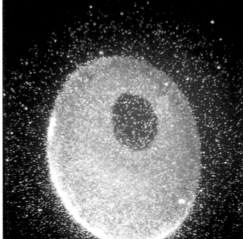

PLATE 8
Localization of a specific RNA in a region of the egg

The *vg1* RNA, which encodes a growth factor of the TGF-β family, is found by in situ hybridization to reside solely in the vegetal region of the *Xenopus* egg. The white crescent at the bottom of the egg is due to the radioactivity of the probe recognizing the RNA; the rest of the egg is green due to staining with Giemsa dye. Chapters 12, 15 and 22. (Photograph courtesy of D. A. Melton.)

PLATE 9
Effect of retinoic acid on limb regeneration

Retinoic acid (RA) causes regenerating cells to "forget" their original position. Regenerating salamander wrist tissue will usually form only a wrist. After treatment with RA, however, the regenerating tissue (here from a darkly pigmented salamander) regenerates an entire forearm (lower right limb) when grafted on to the severed hindlimb of a differently pigmented animal. Chapter 18. (Courtesy of K. Crawford.)

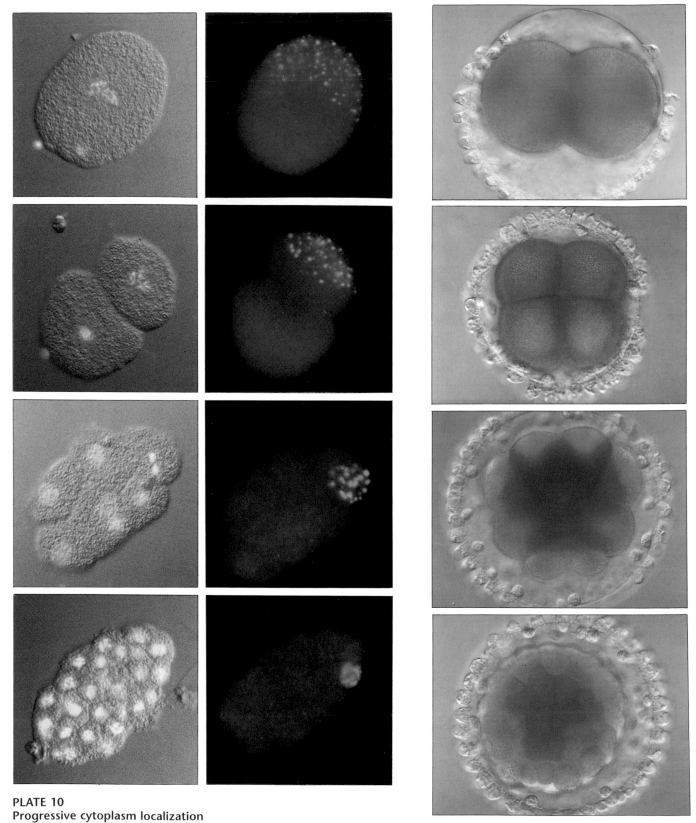

PLATE 10
Progressive cytoplasm localization

The segregation of certain cytoplasm granules ("P granules") is seen to progress into the posteriormost cells of the *Caenorhabditis elegans* embryo. These cells generate the sperm and egg of the nematode. When the pronuclei meet during fertilization, the P granules move to the posterior portion of the cell. This movement continues until they are found only in the P cell that gives rise to the gametes. The left-hand column is stained to show the position of the nuclei, while the right-hand column is stained to show the P granules. Chapter 13. (Photographs courtesy of S. Strome.)

PLATE 11
Cytoplasmic localization in tunicate embryos

Cleavage separates regions of cytoplasm into particular cells. The yellow crescent of the *Styela* embryo becomes localized into a small group of cells that will generate the larval musculature. This figure shows the 2-, 4-, 16- and 64-cell stages. Chapter 13. (Photographs courtesy of J. R. Whittaker.)

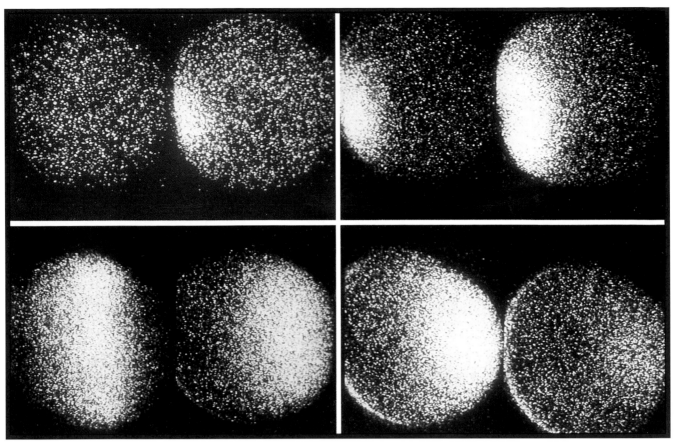

PLATE 12
Wave of calcium ions across sea urchin eggs during fertilization

When the sperm fuses with the egg, a wave of calcium begins at the site of sperm entry and propagates across the egg. This can be monitored by preloading the egg with a dye that fluoresces when it binds calcium. The wave takes 30 seconds to traverse the egg. Chapter 4. (Photographs courtesy of G. Schatten.)

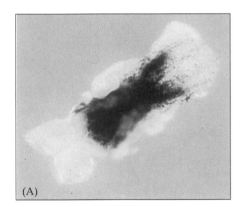

(A)

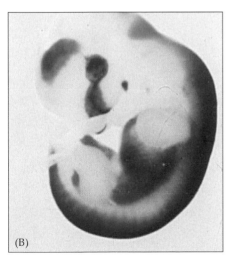

(B)

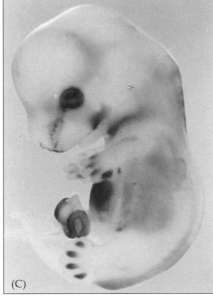

(C)

PLATE 13
Retinoic acid-responsive regions of the mouse embryo

A transgene consisting of a retinoic acid-responsive element fused to a β-galactosidase gene was inserted into a mouse embryo. Staining for β-galactosidase should reveal those cells that respond to endogenous concentrations of retinoic acid. (A) 3-somite stage showing retinoic acid responsiveness in the medial region of the embryo; (B) 11.5-day embryos showing staining in the frontonasal region and forebrain; (C) 14.5-day embryo showing staining in jaw, optic region, whisker pads, and interdigital regions of the limb. Chapters 11, 18 and 21. (Photographs courtesy of J. Rossant.)

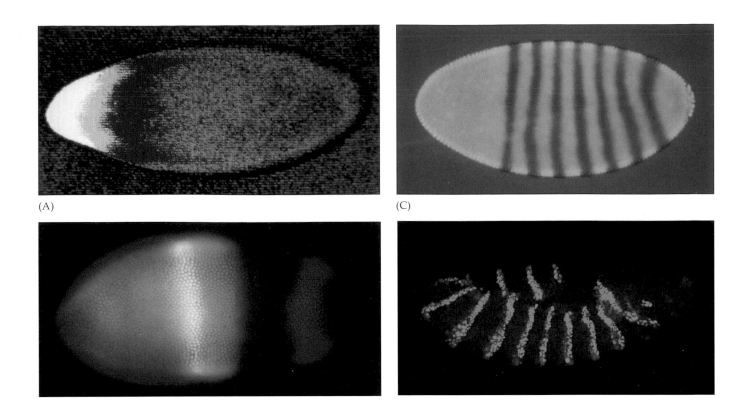

(A)

(C)

(B)

(D)

PLATE 14
Pattern formation in *Drosophila*

(A) The anterior-posterior axis is specified by cytoplasmic mRNAs and proteins. The gradient of Bicoid protein is especially important. High concentrations of this protein (yellow through red) cause head and thorax formation by activating the *hunchback* gene. (B) The gradients of proteins in the early embryo activate the gap genes. The protein products of the gap genes (such as *hunchback* and *Krüppel*) define large domains of the insect body. These proteins interact to form specific boundaries in the embryo. Here, Hunchback (orange) and Krüppel (green) proteins overlap to form a boundary (yellow). (C) The levels of the gap proteins cause the activation of specific pair rule genes (visible here as dark bands) that divide the embryo into segments along the anterior-posterior axis. (D) At the extended germ band stage, the 14 bands of segment polarity gene *engrailed* can be seen. Chapter 14. (Photgraphs courtesy of (A) W. Driever and C. Nüsslein-Volhard; (B) C. Rushlow and M. Levine; (C) T. Karr; and (D) S. Carroll and S. Paddock.)

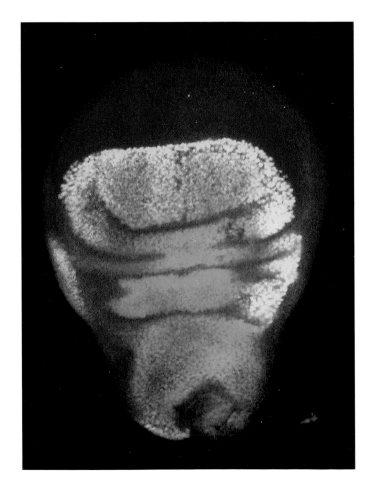

PLATE 15
Compartmentation of *Drosophila* wing imaginal disc

The red immunofluorescent stain labels the cells where the Vestigial protein is made (the future ventral wing); the green stain labels the cells expressing Apterous protein (necessary for dorsal wing formation). The area of overlap is yellow. Chapter 19. (Photograph courtesy of S. Carroll.)

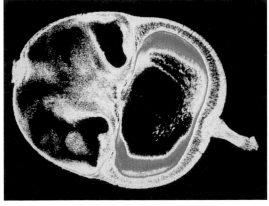

(A)

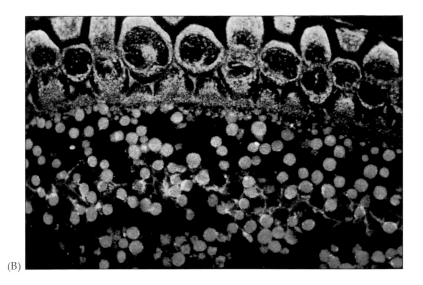

(B)

PLATE 16
RNA polymerase II localization in the oocytes of giant silk moths

(A) Fluorescence photomicrograph (using confocal lens) of *Hyalophora cecropia* egg chamber. Orange fluorescence indicates the presence of RNA polymerase II (stained with labeled amanitin). Green background indicates the location of actin. (B) Higher magnification of an *Antherea polyphemus* oocyte cortical region and follicle cells. Orange indicates RNA polymerase II. The other colors are background staining of yolk granules and follicle cells. Chapter 22. (Photographs courtesy of S. Berry.)

PLATE 17 (*above*)
Gynandromorph moth

A sexual mosaic (gynandromorph) of an *Io* moth, divided bilaterally in a rose-brown female half and a smaller-winged, yellow male half. Such sex mosaics are caused when an X chromosome is lost from a nucleus during an early mitotic division. Chapter 20. (Photograph by T. R. Manley; courtesy of *The Journal of Heredity*.)

PLATE 18 (*left*)
Control of development by the environment

Caterpillars of *Nemoria arizonaria* that hatch in the spring eat oak flowers and develop a cuticle that mimics the flowers. Caterpillars of the same species that hatch in the summer (after the flowers are gone) eat oak leaves; these caterpillars develop cuticle that resembles the oak twigs. Chemicals in the leaves appear to modify cuticle development. Chapter 21. (Photographs courtesy of E. Greene.)

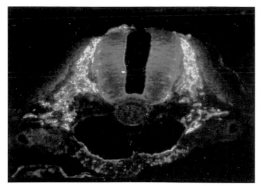

(A)

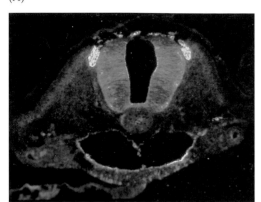

(B)

PLATE 19
Migration of chick neural crest cells
Chick neural crest cells can be followed in their migration by staining the cells with a fluorescently tagged monoclonal antibody. The neural crest cells (stained green) are found to migrate through the anterior (A) but not the posterior (B) regions of the somite tissue. This specific pattern of neural crest cell migration plays a role in determining the placement of peripheral neurons. Chapter 7. (Photographs courtesy of M. Bronner-Fraser.)

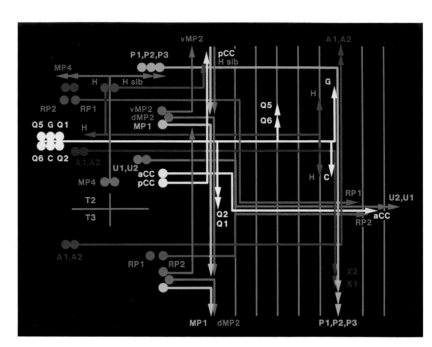

PLATE 20
Neural migration pathways in insects
Neural axons in insect embryos migrate in very specific patterns. Neurons derived from a common precursor (shown here in the same color) produce axons that selectively migrate along with other axons. The Q1 axon, for instance, travels until it meets the dMP2 axon and then travels with it, while the axon from the G neuron continues moving in a straight line until it meets the P1 axon. Chapter 8. (Photograph courtesy of C. Goodman.)

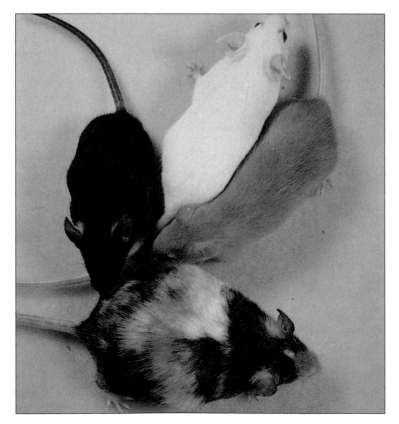

PLATE 21
A mouse with six parents
The multicolored mouse was formed by mixing together cells from three 4-cell stage embryos: an embryo from two black mice; an embryo from two white mice; and an embryo from two brown mice. Instead of forming a three headed monster, the embryo regulated to form one normal-size mouse with contributions from each of the three embryos. That each of the three embryos also provided germ line cells was shown by mating this mouse with a recessive (white) mouse; this mating produced offspring of all three colors. Chapter 5. (Photograph courtesy of C. Markert and *The Journal of Heredity*.)

The left-right axis is specified later, at the 12-cell stage, when the MS blastomere contacts half the progeny of the ABa and ABp cells, converting them into ABal (anterior left) and ABpl (posterior left), while the other two cells become their right-sided counterparts. The signal from the MS cell appears to activate GLP-1 on the AB progeny (Evans et al., 1994). However, the ligand giving this signal is different from APX-1 and has not yet been discovered (Hutter and Schnabel, 1995).

Sidelights & Speculations

"To Be or Not to Be: That Is the Phenotype"

GRANTED, we all are poised at life-or-death decisions, but this existential dichotomy is rarely as stark as that seen in the cell lineage of *C. elegans*. During the normal development of *C. elegans*, 131 cells kill themselves. This programmed cell death, or **apoptosis**, is an active event initiated by two genes, *ced-3* and *ced-4*. When these genes are expressed, the cells expressing them die. Loss-of-function mutations in either of these genes allows the survival of cells that normally would have undergone apoptosis. The products of *ced-3* and *ced-4* are thought either to be toxic to the cell or to cause the formation of toxic compounds from other metabolites (Ellis and Horvitz, 1986; Yuan and Horvitz, 1990).

What determines which cells shall live and which shall die? Studies in Robert Horvitz's laboratory (Hengartner et al., 1992) have demonstrated that the *ced-9* gene inhibits the activities of *ced-3* and *ced-4*. Mutations that inactivate the CED-9 protein cause numerous cells that normally survive to activate their *ced-3* and *ced-4* genes and die. This leads to the death of the entire embryo. Conversely, gain-of-function mutants of *ced-9* prevent those cells that normally undergo apoptosis from dying. (These are the same cells that survive in the *ced-3* and *ced-4* mutants.) Thus, the normal function of CED-9 is to prevent those cells that should live from initiating programmed cell death (Figure 13.23). The *ced-9* gene appears to function as a binary switch regulating the choice between life and death. This decision is made independently in every cell of the embryo; it is possible that every cell of the embryo is poised to die, and those cells that do survive do so because an active *ced-9* gene prevents programmed cell death from occurring. [cyto7.html]

It is not known how CED-9 prevents cell death, nor is it known how it becomes differentially regulated, although certain other genes produce products that turn it on. Similar genes are now being described in mammals as well. The gene *BCL-2* encodes an intracellular membrane protein that prevents or delays normal apoptosis of human neurons and lymphocytes (Hockenbery et al., 1990; Williams et al., 1990; Allsopp et al., 1993). Most lymphocytes and their precursors die during their maturation, and those that survive have a limited life span. They are prevented from dying by certain growth factors that appear to activate the *BCL-2* gene. Moreover, if *BCL-2* genes on a constitutively active promoter are transferred into cells that are soon going to die, BCL-2 protein is made and the life spans of the cells are significantly increased (Nuñez et al., 1990). This is done naturally by the Epstein-Barr virus (which causes mononucleosis). Lymphocytes infected with Epstein-Barr virus do not undergo their usual cell death because one of the viral proteins induces the activity of the *BCL-2* gene (Henderson et al., 1991).

The similarities between *ced-9* and *BCL-2* are so impressive that if an active human *BCL-2* gene is placed into *C. elegans* embryos, it prevents the normally occurring cell deaths (Vaux et al., 1992). This suggests that *BCL-2* functions in humans by acting on the human homologues of *ced-3* and *ced-4*. The human homologues of *ced-3* have been found, and they are a family of cysteine proteases that include **apopain** (CPP32). Apopain is able to inactivate the enzyme poly(ADP-ribose)polymerase, a protein thought to be necessary for genome structure and integrity, at the onset of apoptosis (Nicholson et al., 1995). Moreover, apopain is thought to be negatively regulated by BCL-2. Certain degenerative disorders (such as virus-induced lymphocyte apoptosis in AIDS or neurodegenerative disorders such as strokes) may arise from the inactivation or circumvention of genes such as *BCL-2*. If this is shown to be the case, those genes active in preventing cell death in development may be among the most medically important genes in our bodies. ∎

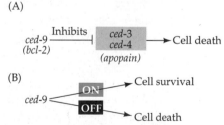

Figure 13.23 Model for the function of the *ced-9* gene in *C. elegans*. (A) *ced-9* acts as a negative regulator of *ced-3* and *ced-4*, the two genes whose activities cause cell death. In mammals, the homologue of ced-9 is Bcl-2, and the homologue of ced-3 is apopain. No homologue has yet been found for ced-4. (B) The binary switch effected by *ced-9*. When on, the activities of *ced-3* and *ced-4* are inhibited and the cell survives. If *ced-9* is not on, the *ced-3* and *ced-4* gene products kill the cell. (From Hengartner et al., 1992.)

Asymmetrical cell divisions in later development

Studies of *Drosophila* nervous system development have provided evidence that the segregation of morphogenetic determinants can occur in later development, after the main body plan has been established. In forming the central nervous system of the *Drosophila* larva, neural stem cells (neuroblasts) divide to form two distinct cell types. One daughter cell is another neuroblast (i.e., another stem cell to keep the population increasing), while the other daughter cell is a **ganglion mother cell** whose progeny are committed to become neurons (Chapter 8). This ganglion mother cell contains a distinct set of proteins, including the membrane protein **Numb** and the transcription factor **Prospero,** that specify its neuronal commitment. Interestingly, though, the Numb and Prospero proteins are *not* synthesized in the ganglion mother cell; they are synthesized in the neuroblast. This paradox was resolved when researchers found that while in the neuroblast, the Numb and Prospero proteins remain in the cytoplasm. However, as that cell begins to divide, these proteins become associated with the membrane that will form the ganglion mother cell. By the end of division, all the Numb and Prospero proteins have been partitioned into the ganglion mother cell where these proteins play their respective roles (Figure 13.24; Hirata et al., 1995; Knoblich et al., 1995; Spana and Doe, 1995). These two proteins share a common amino acid motif which is thought to be responsible for their asymmetric segregation. When this stretch of amino acids is deleted from these proteins, they become distributed randomly to both daughter cells. [cyto8.html]

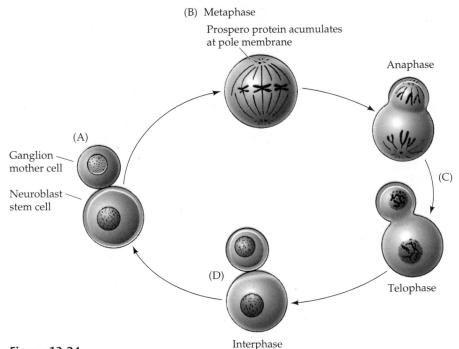

Figure 13.24
Asymmetrical distribution of the Prospero protein during development of the ganglion mother cell. (A) Neuroblast stem cell synthesizes the Prospero protein, which remains diffusely distributed in the cytoplasm. (B) At metaphase, all the Prospero protein has accumulated at one pole of the dividing neuroblast. (C) At anaphase and telophase, the Prospero protein enters into the ganglion mother cell and is excluded from the neuroblast. (D) The Prospero protein, being a transcription factor, enters the nucleus of the ganglion mother cell. The Numb protein colocalizes with Prospero in its leaving the neuroblast. However, it does not enter the nucleus of the neuroblast. (After Hirata et al., 1995.)

Cytoplasmic localization of germ cell determinants

Cytoplasmically localized determinants are found throughout the animal kingdom. The most frequently observed determinants are those responsible for the determination of germ cell precursors, that is, those cells that give rise to gametes. Even in many embryos in which other aspects of early development are regulative, those cells containing a certain region of egg cytoplasm are destined to become germ cell precursors.

Germ Cell Determination in Nematodes

Theodor Boveri (1862–1915) was the first person to look at an organism's chromosomes throughout its development. In so doing, he discovered a fascinating feature in the development of the roundworm *Parascaris aequorum* (formerly *Ascaris megalocephala*). This nematode has only two chromosomes per haploid cell, allowing detailed observations of the individual chromosomes. The cleavage plane of the first embryonic division is unusual in that it is equatorial, separating the animal half from the vegetal half of the zygote (Figure 13.25A). More bizarre, however, is the behavior of the chromosomes in the subsequent division of these first two blastomeres. The ends of the chromosomes in the animal-derived blastomere fragment into dozens of pieces just before cleavage of this cell. This phenomenon is called **chromosome diminution,** because only a portion of the original chromosome survives. Numerous genes are lost in these cells when the chromosomes fragment, and these genes are not included in the newly formed nuclei (Tobler et al., 1972). Meanwhile, in the vegetal blastomere, the chromosomes remain normal. During the second division, the animal cell splits meridionally while the vegetal cell again divides equatorially. Both vegetally derived cells have normal chromosomes. However, the chromosomes of the more animally located of these two vegetal blastomeres fragment before third division. Thus, at the 4-cell stage, only one cell—the most vegetal—contains a full set of genes. At successive cleavages, somatic nuclei are given off from

Figure 13.25
Distribution of germ plasm (color) during cleavage of (A) normal and (B) centrifuged zygotes of *Parascaris*. (A) The germ plasm is normally conserved in the most vegetal blastomere, as shown by the lack of chromosomal diminution in that particular cell. Thus, at the 4-cell stage, the embryo has one stem cell for its gametes. (B) When the first cleavage is displaced 90° by centrifugation, both resulting cells have vegetal germ plasm, and neither cell undergoes chromosome diminution. After the second cleavage, these two cells give rise to germinal stem cells. (After Waddington, 1966.)

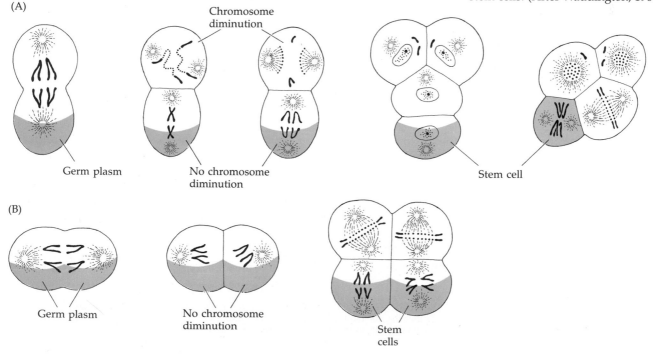

(A)

Chromosome diminution

Germ plasm   No chromosome diminution   Stem cell

(B)

Germ plasm   No chromosome diminution   Stem cells

this vegetalmost line, until the 16-cell stage when there are only two cells with undiminished chromosomes. One of these two blastomeres gives rise to the germ cells; the other eventually undergoes chromosome diminution and forms somatic cells. The chromosomes are kept intact only in those cells destined to form the germ line. If this were not the case, the genetic information would degenerate from one generation to the next. The cells that have undergone chromosome diminution generate the somatic cells.*

Boveri has been called the last of the great "observers" of embryology and the first of the great experimenters. Not content with observing the retention of the full chromosome complement by the germ cell precursors, he set out to test whether a specific region of cytoplasm protects the nuclei within it from diminution. If so, any nucleus happening to reside in this region should be protected. Boveri (1910) tested this by centrifuging *Parascaris* eggs shortly before their first cleavage. This treatment shifted the orientation of the mitotic spindle. When the spindle forms perpendicular to its normal orientation, both resulting blastomeres should contain some of the vegetal cytoplasm (see Figure 13.25). Indeed, Boveri found that after the first division, neither nucleus underwent chromosomal diminution. However, the next division was equatorial along the animal-vegetal axis. Here the resulting animal blastomeres both underwent diminution, whereas the two vegetal cells did not. Boveri concluded that the vegetal cytoplasm contains a factor (or factors) that protects nuclei from chromosomal diminution and determines them to be germ cells.

The pole plasm of nematodes, including that of *C. elegans*, remains largely uncharacterized. RNA helicase appears to localize in the germ plasm of both *C. elegans* and *Ascaris*, and antibody studies suggest that these enzymes may be part of the P granules (Roussell and Bennett, 1993; Kuznicki et al., 1996).

Germ Cell Determination in Insects

The germinal cytoplasm of insects is different from any other cytoplasm in the egg. Hegner (1911) found that when he removed or destroyed this region of beetle eggs before pole cell formation had occurred, the resulting embryos had no germ cells and were sterile. Geigy (1931) showed that irradiating *Drosophila* egg pole plasm with ultraviolet light produced sterile flies; Okada and co-workers (1974) extended this line of experimentation by showing that the addition of pole plasm from unirradiated donor embryos can cure the sterility of irradiated eggs (Figure 13.26). No other part of the cytoplasm could accomplish this reversal of sterility. This posterior pole plasm is conveniently marked with **polar granules** (Figure 13.27A). Although their role in germ cell determination is not known, their constant association with the pole plasm and the **pole cells** derived from it makes them a convenient marker for this region. This region of pole cells is easily identifiable under a scanning electron microscope (Figure 13.27B). [cyto9.html]

Recent work on the pole cell cytoplasm has focused primarily on *Drosophila* embryos. The nuclei of syncytial-stage *Drosophila* embryos are totipotent and can give rise to any cell type. Whichever nuclei land in the posterior pole are the first nuclei to form cells and incorporate the germ plasm. These cells become the precursors of the gametes (Schubiger and Wood, 1977). Nature has also provided confirmation of the importance of both pole plasm and its polar granules. *Drosophila* females homozygous for

---

*While these cases of chromosome diminution and elimination are exceptions to the general rule that the nuclei of differentiated cells retain unused genes, there is no evidence that different somatic cells in *Parascaris* retain different parts of the genome.

(A)

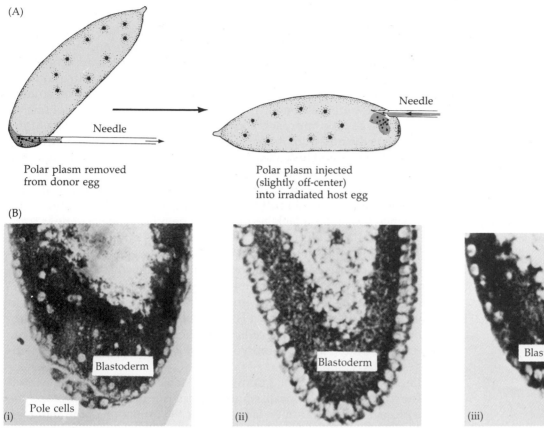

Polar plasm removed
from donor egg

Polar plasm injected
(slightly off-center)
into irradiated host egg

(B)

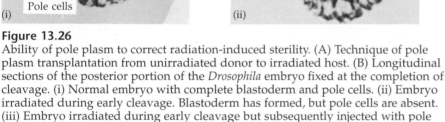

Blastoderm

Blastoderm

Blastoderm

(i) Pole cells

(ii)

(iii) Pole cells

Figure 13.26
Ability of pole plasm to correct radiation-induced sterility. (A) Technique of pole
plasm transplantation from unirradiated donor to irradiated host. (B) Longitudinal
sections of the posterior portion of the *Drosophila* embryo fixed at the completion of
cleavage. (i) Normal embryo with complete blastoderm and pole cells. (ii) Embryo
irradiated during early cleavage. Blastoderm has formed, but pole cells are absent.
(iii) Embryo irradiated during early cleavage but subsequently injected with pole
plasm from normal embryos. Pole cells and blastoderm are both seen. (From Okada
et al., 1974, courtesy of M. Okada.)

the *grandchildless* mutation produce normal but sterile offspring: $GG\male \times gg\female$
$\rightarrow Gg$ (sterile). Mahowald and colleagues (1979) have shown that when such
females are mated with normal males, the nuclei of the resulting embryos
never migrate into the pole plasm of the egg. No pole cells are formed, and
the resulting adults have no primordial germ cells with which to produce
gametes. Another maternal effect mutation—*agametic*—causes the absence of
germ cells in about half the gonads of offspring derived from homozygous
female flies. Here, the normal number of pole cells form, but the polar granules degenerate shortly after fertilization (Engstrom et al., 1982). Transplantation experiments demonstrate that the defect is in the polar cytoplasm and
not in the ovarian environment. Thus, we now have fairly strong evidence
that the pole plasm is directly concerned with germ cell determination.

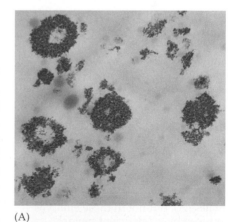

(A)

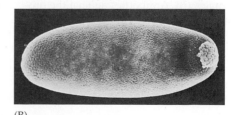

(B)

Figure 13.27
The pole plasm of *Drosophila*. (A) Electron micrograph of polar granules from particulate fraction from *Drosophila* pole cells. (B) Scanning electron micrograph of the
pole cells of a *Drosophila* embryo just prior to completion of cleavage. The pole cells
can be seen at the right of this picture. (Photographs courtesy of A. P. Mahowald.)

Components of the Drosophila *Pole Plasm*

What are the determinants of the *Drosophila* germ plasm, and how do they become localized in the posterior of the embryo? *Drosophila* polar granules have been isolated and appear to be composed of both protein and RNA (Mahowald, 1971a,b; Waring et al., 1978), but the identity of these macromolecules (and there are still some whose identity is unknown) was not uncovered until the genetic approach was used. One of the components of the germ plasm is the mRNA of the ***germ cell-less* (*gcl*)** gene. This gene was discovered by Jongens and his colleagues (1992) when they mutated *Drosophila* and screened for those females who did not have "grandchildren" ("grandoffspring"?). The rationale was that if a female did not place functional germ plasm in her eggs, she could still have offspring, but the offspring would be sterile (since they would lack germ cells). The wild-type *gcl* gene is transcribed in the nurse cells of the fly's ovary, and its mRNA is transported through the ring canals into the egg. Once inside the egg, it becomes transported to the posteriormost portion and resides within what will become the pole plasm (Figure 13.28A,B). This message gets translated into protein during the early stages of cleavage (Figure 13.28C,D). The *gcl*-encoded protein appears to enter the nucleus, and it is essential for pole cell production. Flies mutant for this gene lack germ cells, and when antisense RNA against the *gcl* message is placed into the embryo, the ability to make germ cells is similarly destroyed (Figure 13.29).

The second candidate for a germ plasm determinant is the Nanos protein. The *nanos* message is localized at the posterior pole of the egg, and the Nanos protein translated from it is needed for the formation of the abdomen in *Drosophila*. Recently, Kobayashi and colleagues (1996) also showed it to be essential for germ cell formation. Pole cells lacking Nanos do not migrate into the gonads and fail to become gametes.

A third candidate for a germ plasm component was a big surprise: **mitochondrial large ribosomal RNA.** Using the assay system of ultraviolet-irradiated eggs, Kobayashi and Okada (1989) showed that the injection of mitochondrial large ribosomal RNA (mtlrRNA) restores the ability of these irradiated embryos to form pole cells. Moreover, in normal fly eggs, mtlrRNA is located outside the mitochondria solely in the pole plasm of cleavage-stage embryos. Here it appears as a component of the polar granules (Kobayashi et al., 1993; Amikura et al., 1996). While mtlrRNA is involved in directing the formation of the pole cells, it does not enter them.

Figure 13.28
Localization of *germ cell-less* gene products in the posterior of the egg and embryo. The *gcl* mRNA can be seen in the posterior pole of early-cleavage embryos produced from wild-type females (A), but not in those embryos produced by *gcl*-deficient mutant females (B). Antibodies against the protein encoded by the *gcl* gene can be detected in the cellular blastoderm stage of those embryos produced by wild-type females (C), but not in the embryos from mutant females (D). (From Jongens et al., 1992, courtesy of T. A. Jongens.)

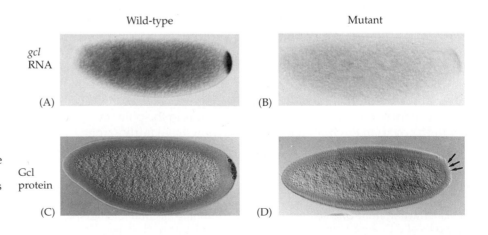

Wild-type   Mutant

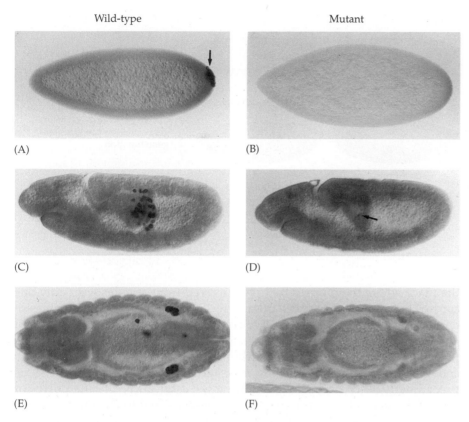

(A)   (B)

(C)   (D)

(E)   (F)

Figure 13.29
Migration of germ cells in embryos produced by wild-type females and in embryos produced by mutants that cannot synthesize the *germ cell-less* protein. Staining of the germ cells is accomplished by antibodies directed against Vasa, a polar granule component that is not mutated in either type of *Drosophila*. An early blastoderm-stage embryo from wild-type females (A) has pole cells at the posterior pole. Embryos from *gcl*-mutant females (B) do not. In embryos from wild-type females, these pole cells can be moved into the posterior midgut primordium (C) from whence they migrate into the gonads (E). Such cells are not seen in the embryos from mothers that lack *germ cell-less* activity (D,F). (From Jongens et al., 1992, courtesy of T. A. Jongens.)

A fourth component of *Drosophila* pole plasm (and one that becomes localized in the polar granules) is a non-translatable RNA called *Polar granule component (Pgc)*. While its exact function remains unknown, the pole cells of transgenic female flies making antisense RNA against *Pgc* fail to migrate to the gonads (Nakamura et al., 1996).

What directs *germ cell-less* mRNA, *nanos* message, and mtlrRNA (and presumably other pole plasm molecules) to the posterior of the egg? There are at least seven other mutants unable to form germ cells, and these mutants also have poorly formed abdomens. These mutants are in the *cappucino, spire, staufen, oskar, vasa, valois,* and *tudor* genes. Each of these genes is active in the ovary and places a gene product into the growing oocyte. By probing for the localization of the mRNA or protein for one gene in a mutant that lacks another gene, one can place the actions of these genes in a defined order (Figure 13.30). These studies (reviewed by Strome, 1992; Ephrussi and Lehmann, 1992) show that two proteins, those made by the *cappucino* and *spire* genes, are needed for the posterior localization of the Staufen protein. (That is, the Staufen protein will not be placed in the posterior of eggs from mothers whose ovaries cannot make Cappucino or Spire.) The Staufen protein is needed for the posterior localization of *oskar* mRNA. The protein made by the *oskar* message is a component of the polar granules and is critical for the posterior localization of the Vasa protein, another component of the polar granules. Mutants of *tudor* and *valoir* do not affect the positioning of Vasa, but appear to be critical for the maintenance of the pole plasm once it has formed (Hay et al., 1990; Lasko and Ashburner, 1990).

The assembly of pole plasm is organized by the *oskar* message. The amount and position of this mRNA determines the number of pole cells and

Figure 13.30
(A) Diagram of the rationale for determining the steps of the genetic pathway leading to the posterior localization of the germ cell determinants. (B) Summary of those steps.

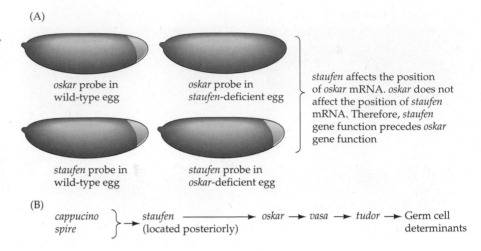

(A)

oskar probe in wild-type egg

oskar probe in *staufen*-deficient egg

staufen affects the position of *oskar* mRNA. *oskar* does not affect the position of *staufen* mRNA. Therefore, *staufen* gene function precedes *oskar* gene function

staufen probe in wild-type egg

staufen probe in *oskar*-deficient egg

(B)

cappucino *spire* → *staufen* (located posteriorly) ⟶ *oskar* → *vasa* → *tudor* → Germ cell determinants

the place where they form.* Embryos derived from females with only one copy of the *oskar* gene produce from 10 to 15 pole cells at the cellular blastoderm stage, whereas those with two copies produce around 35 pole cells. If one raises the number of *oskar* genes to four copies, about 50 pole cells are formed. Moreover, Ephrussi and Lehmann (1992) demonstrated that germ cells will form wherever the *oskar* message is localized and that the steps preceding it are crucial solely in getting the *oskar* mRNA to the posterior pole of the egg. If the *oskar* message becomes localized to the anterior of the embryo (which can be arranged experimentally), the germ plasm and germ cells will form in the anterior. The Oskar protein probably creates the first part of the scaffold of the polar granules. The Vasa and Tudor proteins bind to Oskar to create a more complex scaffold that can bind the germ cell determinants (Breitwieser et al., 1996). The localization of *gcl* mRNA and mtlr-RNA to the posterior pole of the egg is wiped out by any of the preceding mutations. In *valois* and *tudor* mutants, small amounts of *gcl* message can be seen localized to the posterior plasm in early-cleavage embryos, but this localization is lost by late cleavage (Jongens et al., 1992). Thus, the polar granules include both the germ cell determinants and the scaffolding that keeps them at the posterior of the egg and embryo. The scaffold will bind the *germ cell-less* mRNA (and presumably gene products for other germ cell determinants). These messages become translated into protein during early cleavage, enter the nucleus of the pole cells, and (in some as yet unknown way) determine the fate of these cells to be germ cells.

Germ Cell Determination in Amphibians

Cytoplasmic localization of germ cell determinants has also been observed in vertebrate embryos. Bounoure (1934) showed that the vegetal region of fertilized frog eggs contains a material with staining properties similar to those of *Drosophila* pole plasm (Figure 13.31). He was able to trace this corti-

---

*Named for neither the Grouch nor the Norse king, *oskar* received its appellation from the dwarf antihero of Günter Grass's novel *The Tin Drum*. The region-specific translation of *oskar* mRNA into specific isoforms is a complex matter. The *oskar* message is translocated through the egg to the posterior pole by a tropomyosin-containing network and is bound by the repressor protein Bruno to prevent its premature translation (Erdéyli et al., 1995; Kim-Ha et al., 1995). Upon the localization of the mRNA at the posterior pole, the Staufen protein allows it to become translated. Oskar protein is needed to retain the *oskar* mRNA (and the Oskar protein) at the posterior pole (Markussen et al., 1995; Rongo et al., 1995; Chapter 22).

Yolk platelets

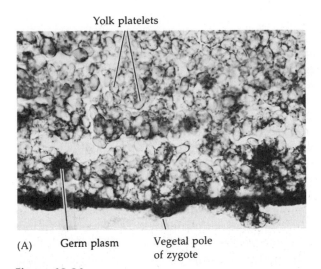

(A) Germ plasm   Vegetal pole of zygote

Germ plasm

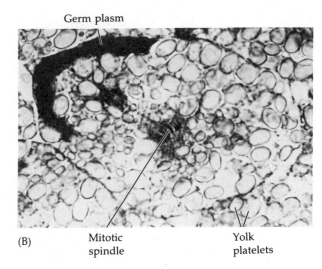

(B) Mitotic spindle   Yolk platelets

Figure 13.31
Germ plasm of frog embryos. (A) Germ plasm (dark regions) near the vegetal pole of a newly fertilized zygote. (B) Germ-plasm-containing cell in endodermal region of blastula in mitotic anaphase. Note the germ plasm entering into only one of the two yolk-laden daughter cells. (C) Primordial germ cell and somatic cells near the floor of the blastocoel in early gastrula. (Courtesy of A. Blackler.)

Somatic cell   Yolk platelets

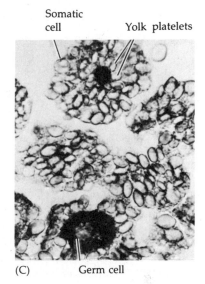

(C) Germ cell

cal cytoplasm into the few cells in the presumptive endoderm that would normally migrate into the genital ridge. By transplanting genetically marked cells from one embryo into another, differently marked strain, Blackler (1962) showed that these cells are the primordial germ cell precursors. The early movements of the germ plasm have been analyzed in detail by Savage and Danilchik (1993), who labeled the germ plasm with fluorescent dye. They found that the germ plasm of unfertilized eggs consists of tiny "islands" that appear to be tethered to the yolk mass near the vegetal cortex. These germ plasm islands move with this vegetal yolk mass during the cortical rotation of fertilization. After the rotation, the islands are released from the yolk mass and begin fusing together and migrating to the vegetal pole. This aggregation depends on microtubules, and the movement of these clusters to the vegetal pole is dependent on a kinesin-like protein that may act as the motor for germ plasm movement (Robb et al., 1996). Later, periodic surface contractions of the vegetal cell surface also appear to push this germ plasm along the furrows of the newly formed blastomeres, enabling it to enter the embryo.

When ultraviolet light is applied to the vegetal surface (but nowhere else) of the frog embryo, the resulting frogs are normal but lack germ cells in their gonads (Bounoure, 1939; Smith, 1966). Very few primordial germ cells reach the gonads, and those that do have about one-tenth the volume of normal primordial germ cells and have aberrantly shaped nuclei (Züst and Dixon, 1977). Savage and Danilchik (1993) found that UV light prevents the vegetal surface contractions and inhibits the migration of germ plasm to the vegetal pole. The *Xenopus* homologues of Nanos (a *Drosophila* protein essential for pole cell migration) and Vasa are specifically localized to this region (Forristal et al., 1995; Ikenishi et al., 1996; Zhou and King, 1996). So, like the *Drosophila* pole plasm, the cytoplasm from the vegetal region of frog zygotes contains the determinants for germ cell formation.

Coda

We have evidence in certain organisms that the determination of a cell's fate is due to the portion of egg cytoplasm that it acquires during cleavage. Such a cell differentiates independently of other cells, and organisms that utilize such a mechanism tend toward a mosaic, or determinate, type of development. Embryos exhibiting this form of development include those of molluscs, tunicates, and nematodes. The placement of morphogenetic determinant within the egg cytoplasm, their redistribution during egg development and fertilization, and the patterns of cell cleavage are important for determining which cell acquires which fate. Each of these phenomena is a function of the egg. Although most of the development of these organisms follows the mosaic pattern, some interactive determination is also seen. In tunicates, the nervous system and some of the muscles are formed by inductive interactions between blastomeres, and snails and nematodes also have certain organs formed in this interactive manner. In the next chapter, we will look at certain organisms in which interactions between molecules in the syncytial blastoderm of insect eggs constitute the primary mechanism of determining cell fate.

LITERATURE CITED

Allsopp, T. E., Wyatt, S., Paterson, H. F. and Davies, A. M. 1993. The proto-oncogene *bcl-2* can selectively rescue neurotrophic factor-dependent neurons from apoptosis. *Cell* 73: 295–307.

Amikura, R., Kobayashi, S., Saito, H. and Okada, M. 1996. Changes in subcellular localization of mtlrRna outside mitochondria in oogenesis and early embryogenesis of *Drosophila melanogaster*. *Dev. Growth Differ.* 38: 489–498.

Atkinson, J. W. 1987. An atlas of light micrographs of normal and lobeless larvae of the marine gastropod *Ilyanassa obsoleta*. *Int. J. Invert. Reprod. Dev.* 9: 169–178.

Austin, J. and Kimble, J. 1987. glp-1 is required in the germ line for regulation of the decision between mitosis and meiosis in *C. elegans*. *Cell* 51: 589–600.

Blackler, A. W. 1962. Transfer of primordial germ cells between two subspecies of *Xenopus laevis*. *J. Embryol. Exp. Morphol.* 10: 641–651.

Blackler, A. W. 1966. The role of a "germinal plasm" in the formation of primordial germ cells in *Rana pipiens*. *Dev. Biol.* 14: 330–347.

Blackwell. T. K., Bowerman, B., Priess, J. R. and Weintraub, H. 1994. Formation of a monomeric DNA binding domain by SKN bZIP and homeodomain elements. *Science* 266: 621–628.

Bonnet, C. 1764. *Contemplation de la Nature.* Marc-Michel Ray, Amsterdam.

Bounoure, L. 1934. Recherches sur la lignée germinale chez la grenouille rousse aux premiers stades du développement. *Ann. Sci. Natl. 10e Wer.* 17: 67–248.

Bounoure, L. 1939. *L'origine des Cellules Reproductries et le Probleme de la Lignée Germinale.* Gauthier-Villars, Paris.

Boveri, T. 1910. Über die Teilung centrifugierter Eier von *Ascaris megalocephala*. *Wilhelm Roux Arch. Entwicklungsmech. Org.* 30: 101–125.

Bowerman, B., Eaton, B. A. and Priess, J. R. 1992. *skn-1*, a maternally expressed gene required to specify the fate of ventral blastomeres in the early *C. elegans* embryo. *Cell* 68: 1061–1075.

Bowerman, B., Draper, B. W., Mello, C. C. and Priess, J. R. 1993. The maternal gene *skn-1* encodes a protein that is distributed unequally in early *C. elegans* embryos. *Cell* 74: 443–452.

Boyd, L., Guo, S., Levitan, D., Stinchcomb, D.T. and Kemphues, K.J. 1996. PAR-2 is asymmetrically distributed and promotes association of P granules and PAR-1 with the cortex in *C. elegans* embryos. *Development* 122: 3075–3084.

Brandhorst, B. P. and Newrock, K. M. 1981. Post-transcriptional regulation of protein synthesis in *Ilyanassa* embryos and isolated polar lobes. *Dev. Biol.* 83: 250–254.

Breitwieser, W., Marhussen, F.-H., Horstmann, H. and Eybrussi, A. 1996. Oskar protein interaction with Vasa represents an essential step in polar granule assembly. *Genes Dev.* 10: 2179–2188.

Brenner, S. 1974. The genetics of *Caenorhabditis elegans*. *Genetics* 77: 71–94.

Brenner, S. 1979. Cited in H. F. Judson, *The Eighth Day of Creation.* Simon and Schuster, New York, p. 219.

Cassirer, E. 1950. Developmental mechanics and the problem of cause in biology. In E. Cassirer (ed.), *The Problem of Knowledge.* Yale University Press, New Haven.

Chabry, L. M. 1887. Contribution a l'embryologie normale tératologique des ascidies simples. *J. Anat. Physiol. Norm. Pathol.* 23: 167–321.

Churchill, F. B. 1973. Chabry, Roux and the experimental method in nineteenth century embryology. In R. N. Giere and R. S. Westfall (eds.), *Foundations of Scientific Method: The Nineteenth Century.* Indiana University Press, Bloomington, pp. 161–205.

Clement, A. C. 1962. Deveopment of *Ilyanassa* following removal of the D macromere at successive cleavage stages *J. Exp. Zool.* 149: 193–215.

Clement, A. C. 1968. Development of the vegetal half of the *Ilyanassa* egg after removal of most of the yolk by centrifugal force, compared with the development of animal halves of similar visible composition. *Dev. Biol.* 17: 165–186.

Clement, A. C. 1986. The embryonic value of the micromeres in *Ilyanassa obsoleta*, as determined by deletion experiments. III. The third quartet cells and the mesentoblast cell, 4d. *Int. J. Invert. Reprod. Dev.* 9: 155–168.

Collier, J. R. 1983. The biochemistry of molluscan development. In N. H. Verdonk, J. A. M. van den Biggelaar and A. S. Tompa (eds.), *The Mollusca*, Vol. 3. Academic, New York, pp. 215–252.

Collier, J. R. 1984. Protein synthesis in normal and lobeless gastrulae of *Ilyanassa obsoleta. Biol. Bull.* 167: 371–377.

Conklin, E. G. 1905a. The organization and cell lineage of the ascidian egg. *J. Acad. Nat. Sci. Phila.* 13: 1–119.

Conklin, E. G. 1905b. Organ-forming substances in the eggs of ascidians. *Biol. Bull.* 8: 205–230.

Conklin, E. G. 1905c. Mosaic development in ascidian eggs. *J. Exp. Zool.* 2: 145–223.

Craig, M. M. and Morrill, J. B. 1986. Cellular arrangements and surface topology during early development in embryos of *Ilyanassa obsoleta. Int. J. Invert. Reprod. Dev.* 9: 209–228.

Dareste, C. 1877. *Recherches sur la Production Artificielle des Monstruosités ou Essais de Tératogenie Experimentale.* Reinwald, Paris.

Davidson, E. H. 1991. Spatial mechanisms of gene regulation in metazoan embryos. *Development* 113: 1–26.

Ellis, R. E. and Horvitz, H. R. 1986. Genetic control of programmed cell death in the nematode *C. elegans. Cell* 44: 817–829.

Engstrom, L., Caulton, J. H., Underwood, E. M. and Mahowald, A. P. 1982. Developmental lesions in the agametic mutant of *Drosophila melanogaster. Dev. Biol.* 91: 163–170.

Ephrussi, A. and Lehmann, R. 1992. Induction of germ cell formation by *oskar. Nature* 358: 387–392.

Erdélyi, M., Michon, A.-M., Guichet, A., Glotzer, J. B. and Ephrussi, A. 1995. Requirement for *Drosophila* cytoplasmic tropomyosin in *oskar* mRNA localization. *Nature* 377: 524–527.

Etemad-Moghadam, B., Guo, S. and Kemphues, K. J. 1995. Asymmetrically distributed PAR-3 protein contributes to cell polarity and spindle alignment in early *C. elegans* embryos. *Cell* 83: 743–752.

Evans, T. C., Crittenden, S. L., Kodoyianni, V. and Kimble, J. 1994. Translational control of maternal *glp-1* mRNA establishes an asymmetry in the *C. elegans* embryo. *Cell* 77: 183–194.

Fischer, J.-L. 1991. Laurent Chabry and the beginnings of experimental embryology in France. *In* S. Gilbert (ed.), *A Conceptual History of Modern Embryology.* Plenum, New York, pp. 31–41.

Fischer, J.-L. and Smith, J. 1984. French embryology and the "mechanics of development" from 1887 to 1910: L. Chabry, Y. Delage and E. Bataillon. *Hist. Phil. Life Sci.* 6: 25–39.

Forristall, C., Pondel, M., Chen, L. and King, M.L. 1995. Patterns of localization and cytoskeletal association of two vegetally localized RNAs, *Vg1* and *Xcat2. Development* 121: 201–208.

Geigy, R. 1931. Action de l'ultra-violet sur le pole germinal dans l'oeuf de *Drosophila melanogaster* (castration et mutabilité). *Rev. Suisse Zool.* 38: 187–288.

Goldstein, B. 1992. Induction of gut in *Caenorhabditis elegans* embryos. *Nature* 357: 255–257.

Goldstein, B. and Hird, S. N. 1996. Specification of the anterioposterior axis in *Caenorhabditis elegans. Development* 122: 1467–1474.

Gould, S. J. 1977. *Ontogeny and Phylogeny.* Belknap Press, Cambridge, MA.

Guerrier, P., van den Biggelaar, J. A. M., Dongen, C. A. M. and Verdonk, N. H. 1978. Significance of the polar lobe for the determination of dorsoventral polarity in *Dentalium vulgare* (da Costa). *Dev. Biol.* 63: 233–242.

Guo, S. and Kemphues, K. J. 1995. *par-1,* a gene required for establishing polarity in *C. elegans* embryos, encodes a putative ser/thr kinase that is asymmetrically distributed. *Cell* 81: 611–620.

Harrison, R. G. 1933. Some difficulties of the determination problem. *Am. Natur..* 67: 306–321.

Hay, B., Jan, L. Y. and Jan, Y. N. 1990. Localization of vasa, a component of *Drosophila* polar granules, in maternal effect mutants that alter embryonic anteroposterior polarity. *Development* 109: 425–433.

Hegner, R. W. 1911. Experiments with chrysomelid beetles. III. The effects of killing parts of the eggs of *Leptinotarsa decemlineata. Biol. Bull.* 20: 237–251.

Henderson, S. and seven others. 1991. Induction of *bcl-2* expression by Epstein-Barr virus latent membrane protein 1 protects infected B cells from programmed cell death. *Cell* 65: 1107–1115.

Hengartner, M. O., Ellis, R. E. and Horvitz, H. R. 1992. *Caenorhabditis elegans* gene *ced-9* protects cell from programmed cell death. *Nature* 356: 494–499.

Henry, J. J. and Martindale, M. Q. 1987. The organizing role of the D quadrant as revealed through the phenomenon of twinning in the polychaete *Chaetopterus variopedatus. Wilhelm Roux Arch. Dev. Biol.* 196: 449–510.

Hill, D. P. and Strome, S. 1987. An analysis of the role of microfilaments in the establishment and maintainance of asymmetry in *Caenorhabditis elegans* zygotes. *Dev. Biol.* 125: 75–84.

Hill, D. P. and Strome, S. 1990. Brief cytochalasin-induced disruption of microfilaments during a critical interval in 1-cell *C. elegans* embryos alters the positioning of developmental instructions to the 2-cell embryo. *Development* 108: 159–172.

Hirata, J., Nakagoshi, H., Nabeshima, Y-I. and Matsuzaki, F. 1995. Asymmetric segregation of the homeodomain protein Pros-

pero during *Drosophila* development. *Nature* 377: 627–630.

Hockenbery, D. M., Nuñez, G., Milliman, C., Schreiber, R. D. and Korsmeyer, S. J. 1990. Bcl-2 is an inner mitochondrial membrane protein that blocks programmed cell death. *Nature* 348: 334–336.

Hutter, H. and Schnabel, R. 1994. *glp-1* and inductions establishing embryonic axes in C. elegans. *Development* 120: 2051–2064.

Hutter, H. and Schnabel, R. 1995. Establishment of left-right asymmetry in the *Caenorhabditis elegans* embryo: A multistep process involving a series of inductive events. *Development* 121: 3417–3424.

Illmensee, K. 1968. Transplantation of embryonic nuclei into unfertilized eggs of *Drosophila melanogaster. Nature* 219: 1268–1269.

Ikenishi, K., Tanaka, T. S. and Komiya, T. 1996. Spatio-temporal distribution of the protein of the *Xenopus* vasa homologue *(Xenopus vasa-like gene-1, XVLG1)* in embryos. *Dev. Growth Differ.* 38: 527–535.

Jongens, T. A., Hay, B., Jan, L. Y. and Jan, Y. N. 1992. The *germ cell-less* gene product: A posteriorly localized component necessary for germ cell development in *Drosophila. Cell* 70: 569–584.

Kemphues, K. J., Priess, J. R., Morton, D. G. and Cheng, N. 1988. Identification of genes required for cytoplasmic localization in early *C. elegans* embryos. *Cell* 52: 311–320.

Kim-Ha, J., Kerr, K. and Macdonald, P. M. 1995. Translational regulation of *oskar* mRNA by Bruno, an ovarian RNA-binding protein, is essential. *Cell* 81: 403–412.

Kimble, J. and Hirsch, D. 1979. The postembryonic cell lineages of the hermaphrodite and male gonads in *Caenorhabditis elegans. Dev. Biol.* 70: 396–417.

Kirby, C. M. 1992. Cytoplasmic reorganization and the generation of asymmetry in *Caenorhabditis elegans,* with an emphasis on *par-3,* a maternal-effect gene essential for both processes. PhD dissertation, Cornell University, Ithaca, NY.

Kirby, C. M., Kusch, M. and Kemphues, K. 1990. Mutations in the *par* genes of *Caenorhabditis elegans* affect cytoplasmic reorganization during the first cell cycle. *Dev. Biol.* 142: 203–215.

Knoblich, J. A., Jan, L. Y. and Jan, Y. N. 1995. Asymmetric segregation of Numb and Prospero during cell division. *Nature* 377: 624–627.

Kobayashi, S. and Okada, M. 1989. Restoration of pole-cell forming ability to UV-irradiated *Drosophila* embryos by injection of mitochondrial lrRNA. *Development* 107: 733–742.

Kobayashi, S., Amikura, R. and Okada, M. 1993. Presence of mitochondrial large ribosomal RNA outside mitochondria in germ plasm of *Drosophila melanogaster. Science* 260: 1521–1524.

Kobayashi, S., Yamada, M., Asaoka, M. and Kitamura, T. 1996. Essential role of the posterior morphogen nanos for germline development in *Drosophila. Nature* 380: 708–711.

Kölreuter, J. G. 1766. *Vorläufige Narchricht von einigen das Geschlecht der Pflanzen beteffenden Versuchen und Beobachtungen, nebst Fortsetzugen 1, 2 und 3.* Leipzig.

Kuznicki, K., Gruidl, M., Smith, P., McCrone, S. and Bennett, K. 1996. *C. elegans* germline RNA helicases: Are they all components of the P granules? *Dev. Biol.* 175: 379

Lasko, P. F. and Ashburner, M. 1990. Posterior localization of vasa protein correlates with, but is not sufficient for, pole cell development. *Genes Dev.* 4: 905–921.

Lenoir, T. 1980. Kant, Blumenbach, and vital materialism in German biology. *Isis* 71: 77–108.

Lin, R., Thompson, S. and Priess, J. R. 1995. *pop-1* encodes an HMG box protein required for the specification of a mesoderm precursor in early *C. elegans* embryos. *Cell* 83: 599–609.

Mahowald, A. P. 1971a. Polar granules of *Drosophila.* III. The continuity of polar granules during the life cycle of *Drosophila. J. Exp. Zool.* 176: 329–343.

Mahowald, A. P. 1971b. Polar granules of *Drosophila.* IV. Cytochemical studies showing loss of RNA from polar granules during early stages of embryogenesis. *J. Exp. Zool.* 176: 329–343.

Mahowald, A. P., Caulton, J. H. and Gehring, W. J. 1979. Ultrastructural studies of oocytes and embryos derived from female flies carrying the *grandchildless* mutation in *Drosophila subobscura. Dev. Biol.* 69: 118–132.

Mango, S. E., Thorpe, C. J., Martin, P. R., Chamberlain, S. H. and Bowerman, B. 1994. Two maternal genes, *apx-1* and *pie-1*, are required to distinguish the fates of equivalent blastomeres in early *C. elegans* embryos. *Development* 120: 2305–2315.

Markussen, F.-H., Michon, A.-M., Breitwieser, W. and Eprussi, A. 1995. Translational control of *oskar* generates short OSK, the isoform that induces pole plasm assembly. *Development* 121: 3723–3732.

McGhee, J. D. 1995. Cell fate decisons in the early embryo of the nematode *Caenorhabditis elegans. Dev. Genet.* 17: 155–166.

Mello, G. C., Draper, B. W., Krause, M., Weintraub, H. and Priess, J. R. 1992. The pie-1 and *mex-1* genes and maternal control of blastomere identity in early *C. elegans* embryos. *Cell* 70: 163–176.

Mello, C. C., Draper, B. W. and Priess, J. R. 1994. The maternal genes *apx-1* and *glp-1* and establishment of dorsal-ventral polarity in the early *C. elegans* embryo. *Cell* 77: 95–106.

Mello, C. C., Schubert, C., Draper, B., Zhang, W., Lobel, R. and Priess, J. R. 1996. The PIE-1 protein and germline specification in *C. elegans* embryos. *Nature* 382: 710–712.

Nakamura, A., Amikura, R., Mukai, M., Kobayashi, S. and Lasko, P. F 1996. Requirement for a noncoding RNA in *Drosophila* polar granules for germ cell establishment. *Science* 274: 2075–2079.

Newrock, K. M. and Raff, R. A. 1975. Polar lobe specific regulation of translation in embryos of *Ilyanassa obsoleta. Dev. Biol.* 42: 242–261.

Nicholson and fifteen others. 1995. Identification and inhibition of the ICE/CED-3 protease necessary for mammalian apoptosis. *Nature* 376: 37–43.

Nishida, H. 1987. Cell lineage analysis in ascidian embryos by intracellular injection of a tracer enzyme. III. Up to the tissue restricted stage. *Dev. Biol.* 121: 526–541.

Nishida, H. 1990. Determinative mechanisms in secondary muscle lineages of ascidian embryos: Development of muscle-specific features in isolated muscle progenitor cells. *Development* 108: 559–568.

Nishida, H. 1992a. Determination of developmental fates of blastomeres in ascidian embryos. *Dev. Growth Differ.* 34: 253–262.

Nishida, H. 1992b. Regionality of egg cytoplasm that promotes muscle differentiation in embryo of the ascidian *Halocynthia roretzi. Development* 116: 521–529.

Nishida, H. 1993. Localized regions of egg cytoplasm that promote expression of endoderm-specific alkaline phosphatase in embryos of the ascidian *Halocynthia roretzi. Development* 118: 1–7.

Nishida, H. 1994a. Localization of egg cytoplasm that promotes differentiation to epidermis in embryos of the ascidian *Halocynthia roretzi. Development* 120: 235–243.

Nishida, H. 1994b. Localization of determinants for formation of the anterior-posterior axis in eggs of the ascidian *Halocynthia roretzi. Development* 120: 3093–3104.

Nuñez, G., London, L., Hockenbury, D., Alexander, M., McKearn, J. P. and Korsmeyer, S. J. 1990. Deregulation of *blc-2* gene expression selectively prolongs survival of growth factor-deprived hemopoietic cells. *J. Immunol.* 144: 3602–3610.

Okada, M., Kleinman, I. A. and Schneiderman, H. A. 1974. Restoration of fertility in sterilized *Drosophila* eggs by transplantation of polar cytoplasm. *Dev. Biol.* 37: 43–54.

Ortolani, G. 1959. Richerche sulla induzione del sistema nervoso nelle larve delle Ascidie. *Boll. Zool.* 26: 341–348.

Pines, M. (ed.). 1992. *From Egg to Adult.* Howard Hughes Med. Inst., Bethesda, pp. 30–38.

Priess, R. A. and Thomson, J. N. 1987. Cellular interactions in early *C. elegans* embryos. *Cell* 48: 241–250.

Priess, R. A., Schnabel, H. and Schnabel, R. 1987. The *glp-1* locus and cellular interactions in early *C. elegans* embryos. *Cell* 51: 601–611.

Reverberi, G. and Minganti, A. 1946. Fenomeni di evocazione nello sviluppo dell'uovo di Ascidie. Risultati dell'indagine spermentale sull'ouvo di *Ascidiella aspersa* e di *Ascidia malaca* allo stadio di 8 blastomeri. *Pubbl. Staz. Zool. Napoli* 20: 199–252. (Quoted in Reverberi, 1971, p. 537.)

Robb, D. L., Heasman, J., Raats, J. and Wylie, C. 1996. A kinesin-like protein is required for germ plasm aggregation in *Xenopus. Cell.* 87: 823–831.

Roe, S. 1981. *Matter, Life, and Generation: Eighteenth-Century Embryology and the Haller-Wolff Debate.* Cambridge University Press, Cambridge.

Rongo, C., Gavis, E. R. and Lehmann, R. 1995. Localization of *oskar* RNA regulates oskar translation and requires Oskar protein. *Development* 121: 2737–2746.

Roussell, D. L. and Bennett, K. L. 1993. glh, a germline putative RNA helicase from Caenorhabditis, has four zinc fingers. *Proc. Natl. Acad. Sci. USA* 90: 9300–9304.

Satoh, Y., Kusakabe, T., Araki, I. and Satoh, N. 1995. Timing of initiation of muscle-specific gene expression in the ascidian embryo precedes that of developmental fate restriction in lineage cells. *Dev. Growth Diff.* 37: 319–327.

Savage, R. M. and Danilchik, M. V. 1993. Dynamics of germ plasm localization and its inhibition by ultraviolet irradiation in early cleavage *Xenopus* eggs. *Dev. Biol.* 157: 371–382.

Schubiger, G. and Wood, W. J. 1977. Determination during early embryogenesis in *Drosophila melanogaster. Am. Zool.* 17: 565–576.

Seydoux, G., Mello, C. C., Pettitt, J., Wood, W. B., Priess, J. R. and Fire, A. 1996. Repression of gene expression in the embryonic germ lineage of *C. elegans. Nature* 382: 713–716.

Slack, J. M. W. 1991. *From Egg to Embryo: Regional Specification in Early Development.* Cambridge University Press, New York.

Smith, L. D. 1966. The role of a "germinal plasm" in the formation of primordial germ cells in *Rana pipiens. Dev. Biol.* 14: 330–347.

Strome, S. 1992. The germ of the issue. *Nature* 358: 368–369.

Strome, S. and Wood, W. B. 1983. Generation of asymmetry and segregation of germ-like granules in early *Caenorhabditis elegans* embryos. *Cell* 35: 15–25.

Spana, E. P. and Doe, C. Q. 1995. The prospero transcription factor is asymmetrically localized to the cell cortex during neuroblast mitosis in *Drosophila. Development* 121: 3187–3195.

Sulston, J. and Horvitz, H. R. 1977. Postembryonic cell lineages of the nematode *Caenorhabditis elegans. Dev. Biol.* 56: 110–156.

Sulston, J. E., Schierenberg, J., White, J. and Thomson, N. 1983. The embryonic cell lineage of the nematode *Caenorhabditis elegans. Dev. Biol.* 100: 64–119.

Sulston, J. and eighteen others. 1992. The *C. elegans* genome sequencing project: A beginning. *Nature* 356: 37–42.

Swalla, B. J. and Jeffery, W. R. 1995. A maternal RNA localized in the yellow crescent is segregated to the larval muscle cells during ascidian development. *Dev. Biol.* 170: 353–364.

Tobler, H., Smith, K. D. and Ursprung, H. 1972. Molecular aspects of chromatin elimination in *Ascaris lumbricoides. Dev. Biol.* 27: 190–203.

Tung, T. C., Wu, S. C., Yel, Y. F., Li, K. S. and Hsu, M. C. 1977. Cell differentiation in ascidians studied by nuclear transplantation. *Scientia Sinica* 20: 222–233.

Vaux, D. L., Weissman, I. L. and Kim, S. K. 1992. Prevention of programmed cell death in *Caenorhabditis elegans* by human *bcl-2. Science* 258: 1955–1957.

Verdonk, N. H. and Cather, J. N. 1983. Morphogenetic determination and differentiation. *In* N. H. Verdonk, J. A. M. van den Biggelaar and A. S. Tompa (eds.), *The Mollusca.* Academic Press, New York, pp. 215–252.

Waddington, C. H. 1966. *Principles of Development and Differentiation.* Macmillan, New York.

Waring, G. L., Allis, C. D. and Mahowald, A. P. 1978. Isolation of polar granules and the identification of polar granule-specific protein. *Dev. Biol.* 66: 197–206.

Watts, J. L. and seven others. 1996. *par-6,* a gene involved in the establishment of asymmetry in early *C. elegans* embryos, mediates the asymmetric localization of PAR-3. *Development* 122: 3133–3140.

Whittaker, J. R. 1977. Segregation during cleavage of a factor determining endodermal alkaline phosphatase development in ascidian embryos. *J. Exp. Zool.* 202: 139–153.

Whittaker, J. R. 1982. Muscle cell lineage can change the developmental expression in epidermal lineage cells of ascidian embryos. *Dev. Biol.* 93: 463–470.

Whittaker, J. R., Ortolani, G. and Farinella-Ferruzza, N. 1977. Autonomy of acetylcholinesterase differentiation in muscle lineage cells in ascidian embryos. *Dev. Biol.* 55: 196–200.

Williams, G. T., Smith, C. A., Spooncer, E., Dexter, T. M. and Taylor, D. R. 1990. Haemopoietic colony stimulating factors promote cell survival by suppressing apoptosis. *Nature* 343: 76–79.

Wilson, E. B. 1904. Experimental studies on germinal localization. I. The germ regions in the egg of *Dentalium.* II. Experiments on the cleavage-mosaic in *Patella* and *Dentalium. J. Exp. Zool.* 1: 1–72.

Wolff, K. F. 1767. De formatione intestinorum praecipue. *Novi Commentarii Academine Scientarum Imperialis Petropolitanae.*

Yuan, J. Y. and Horvitz, H. R. 1990. The *Caenorhabditis elegans* genes *ced-3* and *ced-4* act cell autonomously to cause programmed cell death. *Dev. Biol.* 138: 33–41.

Zhou, Y. and King, M. L. 1996. Localization of Xcat-2, a putative germ plasm component, to the mitochondrial cloud in *Xenopus* stage I oocytes. *Development.* 122: 2947–2953.

Züst, B. and Dixon, K. E. 1977. Events in the germ cell lineage after entry of the primordial germ cells into the genital ridges in normal and UV-irradiated *Xenopus laevis. J. Embryol. Exp. Morphol.* 41: 33–46.

The genetics of axis specification in Drosophila

14

When the spermatozoon enters the egg, it enters a cell system which has already achieved a certain degree of organization.
ERNST HADORN (1955)

Those of us who are at work on Drosophila *find a particular point to the question. For the genetic material available is all that could be desired, and even embryological experiments can be done....It is for us to make use of these opportunities. We have a complete story to unravel, because we can work things from both ends at once.*
JACK SCHULTZ (1935)

The chief advantage of Drosophila *initially was one that historians have overlooked: it was an excellent organism for student projects.*
ROBERT E. KOHLER (1994)

IN THE LAST CHAPTER, we discussed the specification of early embryonic cells by their acquiring different cytoplasmic determinants that had been stored in the oocyte. The cell membranes establish the region of cytoplasm incorporated into each cell, and it is thought that the morphogenetic determinants then direct differential gene expression in these blastomeres. During *Drosophila* development, cellular membranes do not form until after the thirteenth nuclear division. Prior to this time, all the nuclei share a common cytoplasm, and material can diffuse through the embryo. In these embryos, the specification of the cell types along anterior-posterior and dorsal-ventral axes is accomplished by the interactions of cytoplasmic materials within the single, multinucleated cell. Moreover, the initiation of the anterior-posterior and dorsal-ventral differences is controlled by the position of the egg within the mother's ovary. Whereas the sperm entry site may fix the axes in ascidians and nematodes, the fly's anterior-posterior and dorsal-ventral axes are specified by interactions between the egg and its surrounding follicle cells.

A summary of Drosophila development

As discussed in Chapter 3, *Drosophila* embryos develop very rapidly through a series of nuclear divisions that form a syncytial blastoderm. During the ninth division cycle, about five nuclei reach the surface of the posterior pole of the embryo. These nuclei become enclosed by cell membranes and generate the pole cells that give rise to the gametes of the adult. Most of the other nuclei arrive at the periphery of the embryo at cycle 10 and then undergo four more divisions at progressively slower rates. Following cycle 13, cell membranes grow between the nuclei to form a cellular blastoderm of about 6000 cells (Turner and Mahowald, 1977; Foe and Alberts, 1983). At cycle 14, the level of overall transcription, which had been very low, increases dramatically. At the same time, the 2- to 3-hour embryo begins gastrulation.

The first movements of *Drosophila* gastrulation segregate the presumptive mesoderm, endoderm, and ectoderm (Figure 14.1). The prospective mesoderm—about 1000 cells constituting the ventral midline—folds in to produce the **ventral furrow**. This furrow eventually pinches off from the sur-

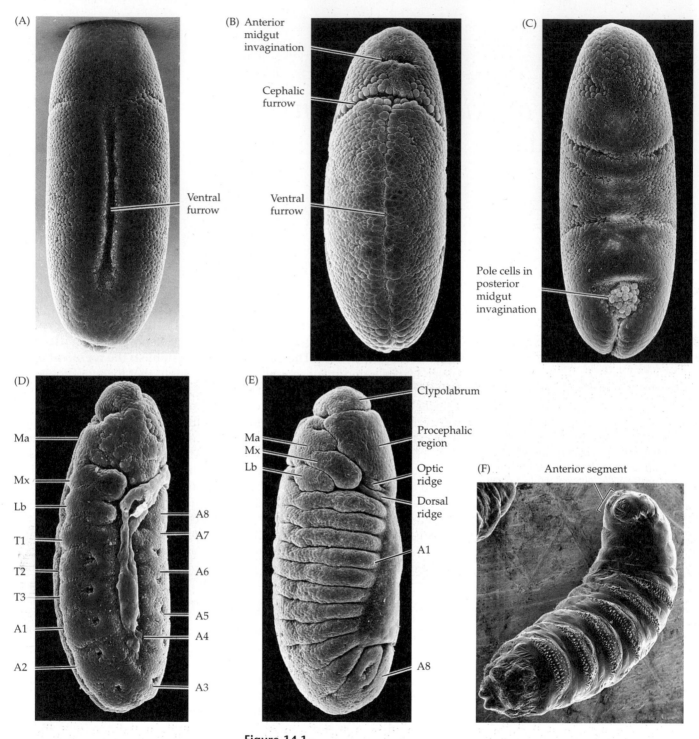

Figure 14.1
Gastrulation in *Drosophila*. (A) Ventral furrow beginning to form as cells flanking the ventral midline invaginate. (B) Furrow closes, with mesodermal cells placed internally and surface ectoderm flanking the ventral midline. (C) Dorsal view of a slightly older embryo, showing the pole cells and posterior endoderm sinking into the embryo. (D) Lateral view showing fullest migration of germ band. Subtle indentations mark the incipient segments along the germ band: Ma, Mx, and Lb correspond to the mandibular, maxillary, and labial head segments; T1–T3, thoracic segments; A1–A8, abdominal segments. (E) Germ band reversing direction. The true segments are now visible, as well as the other territories of the dorsal head, such as the clypolabrum, procephalic region, optic ridge, and dorsal ridge. (F) Newly hatched first-instar larva. (Courtesy of F. R. Turner.)

face to become a ventral tube within the embryo. It then flattens to form a layer of mesodermal tissue beneath the ventral ectoderm. The prospective endoderm invaginates as two pockets at the anterior and posterior ends of the ventral furrow. The pole cells are internalized along with the endoderm. At this time, the embryo bends to form the **cephalic furrow** and the **anterior** and **posterior transverse folds.** [other.html#droso1]

The cells remaining on the surface (the ectoderm) undergo convergence and extension, migrating toward the ventral midline to form the **germ band.** The germ band extends posteriorly, and perhaps because of the egg case, wraps around the top (dorsal) surface of the embryo. Thus, at the end of germ band formation, the cells destined to form the most posterior larval structures are located immediately behind the future head region. At this time, the body segments begin to appear, dividing the ectoderm and mesoderm. The germ band then retracts, placing the presumptive posterior segments into the posterior tip of the embryo.

While the germ band is in its extended position, several key morphogenetic processes occur: organogenesis, segmentation, and the segregation of the imaginal discs.* In addition, the nervous system forms from two regions of ventrally located of ectodermal cells. As described in Chapter 8, neuroblasts differentiate from this neurogenic ectoderm within each segment (and also from the nonsegmented region of the head ectoderm). Therefore, in insects like *Drosophila,* the nervous system is located ventrally, rather than being derived from a dorsal neural tube as in vertebrates.

THE ORIGINS OF ANTERIOR-POSTERIOR POLARITY

Overview

The general body plan of *Drosophila* is the same in the embryo, the larva, and the adult, each of which has a distinct head end and a distinct tail end, between which are repeating segmental units (Figure 14.2). Three of these segments form the thorax, while another eight segments form the abdomen.

*The details of imaginal disc differentiation will be discussed in Chapter 19. For more information on *Drosophila* developmental anatomy, see Bate and Martinez-Arias, 1993; Tyler and Schetzer, 1996; and Schwalm, 1997.

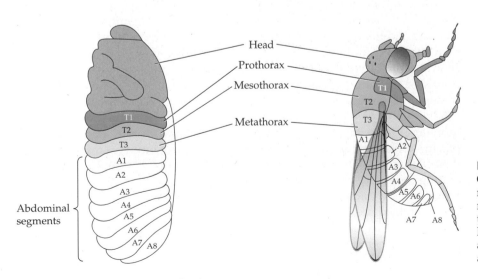

Figure 14.2
Comparison of larval and adult segmentation in *Drosophila.* The three thoracic segments can be distinguished by their appendages: T1 (prothoracic) has legs only; T2 (mesothoracic) has wings and legs; T3 (metathoracic) has halteres and legs.

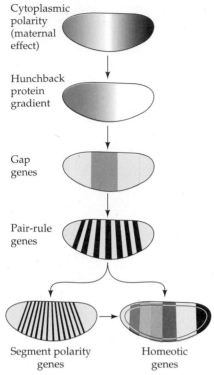

Cytoplasmic polarity (maternal effect)

Hunchback protein gradient

Gap genes

Pair-rule genes

Segment polarity genes   Homeotic genes

Figure 14.3
Generalized model of *Drosophila* pattern formation. The pattern is established by maternal effect genes that form gradients and regions of morphogenetic proteins. These morphogenetic determinants create a gradient of Hunchback protein that differentially activates the gap genes that define broad territories of the embryo. The gap genes enable the expression of pair-rule genes, each of which divides the embryo into regions about two segment primordia wide. The segment polarity genes then divide the embryo into segment-sized units along the anterior-posterior axis. The combination of these genes defines the spatial domains of the homeotic genes that define the identities of each of the segments. In this way, periodicity is generated from nonperiodicity, and each segment is given a unique identity.

Each segment of the adult fly has its own identity. The first thoracic segment, for example, has only legs; the second thoracic segment contains legs and wings. The third thoracic segment has legs and halteres (balancers). Thoracic and abdominal segments can also be distinguished by their cuticle. How does this pattern arise? During the past decade, the combined approaches of molecular biology, genetics, and embryology has lead to a detailed model describing how periodic pattern is generated along the anterior-posterior axis and how each segment is differentiated from the others.

The anterior-posterior polarity in the embryo, larva, and adult has its origin in the anterior-posterior polarity of the egg (Figure 14.3). The **maternal effect genes** in the fly ovaries produce messenger RNAs that are placed into different regions of the egg. These encode transcriptional and translational regulatory proteins that diffuse through the syncytial blastoderm and activate or repress the expression of certain zygotic genes. One pair of these proteins, Bicoid and Hunchback, regulates the production of anterior structures, while another pair of maternally specified proteins, Nanos and Caudal, regulates the formation of the posterior parts of the embryo. Next, the zygotic genes regulated by these maternal factors are expressed in certain broad (about three segments wide), partially overlapping domains. These genes are called **gap genes** (because mutations in them cause gaps in the segmentation pattern), and they are among the first genes transcribed in the embryo. The different concentrations of the gap gene proteins cause the transcription of **pair-rule genes** that divide the embryo into periodic units. The transcription pattern of each of these pair-rule genes gives a striped pattern of seven vertical bands perpendicular to the anterior-posterior axis. The stripes of the pair-rule gene proteins activate the transcription of the **segment polarity genes.** Their mRNA and protein products divide the embryo into 14 segment-wide units. This establishes the *periodicity* of the embryo. At the same time, proteins of the gap, pair-rule, and segment polarity genes interact to regulate another class of genes, the **homeotic genes,** whose transcription determines the developmental *fate* of each of these segments.

The maternal effect genes

Embryological Evidence of Polarity Regulation by Oocyte Cytoplasm

Classic embryological experiments demonstrated that there are at least two "organizing centers" in the insect egg. One is the anterior organizing center, the other the posterior organizing center. Klaus Sander (1975) postulated that these two organizing areas form two gradients, one initiated at the anterior end and the other at the posterior end. Each of these gradients forms its own structures at the poles and interacts with the other gradient to form the central portion of the embryo. Sander based this model on experiments that involved ligating the embryo at various times during development and transplanting regions of polar cytoplasm from one region of the egg to another (Figure 14.4). First, if he moved cytoplasm from the posterior pole more anteriorly, he obtained a small embryo anterior to the posterior pole plasm, while extra segments, not organized into an embryo, formed behind it (see Figure 14.4D). Second, if he ligated the egg early in development, separating the anterior from the posterior region, one half developed into an anterior embryo and one half developed into a posterior embryo, but neither half contained the middle segments of the embryo. The later in development the ligature was made, the fewer middle segments were missing. Thus, it appeared that there were indeed gradients emanating from the two poles dur-

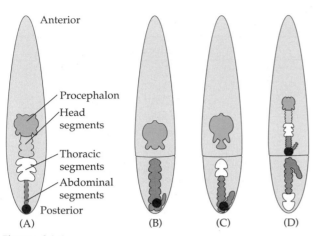

Anterior

Procephalon
Head
segments

Thoracic
segments

Abdominal
segments

Posterior

(A)   (B)   (C)   (D)

Figure 14.4
Sander's ligature experiments on the embryo of the leafhopper insect *Euscelis*. (A)
Normal embryo seen in ventral view. The black ball at the bottom represents a cluster of symbiotic bacteria that marks the posterior pole. (B) After ligating the early
embryo, partial embryos form, but the head and thoracic segments are missing
from both embryos. (C) When ligated later (at the blastoderm stage), more of the
missing segments are formed, but the embryos still lack the most central segments.
(D) When the posterior pole cytoplasm is transplanted into an embryo ligated at
the blastoderm stage, a small but complete embryo forms in the anterior half, while
the posterior half forms an inverted partial embryo. These results can be explained
in terms of gradients at the poles of the embryo that turn on one set of structures
and repress the formation of others. (After Sander, 1960, and French, 1988.)

ing cleavage and that these gradients interacted to produce the positional information determining the identity of each segment.

   The possibility that mRNA is responsible for generating the anterior gradient was suggested in a series of experiments by Kalthoff and Sander
(1968). They found that when the anterior portion of the *Smittia* (midge) egg
was exposed to ultraviolet light at wavelengths capable of inactivating RNA
(265 and 285 nm), the resulting embryo lacked its head and thorax. Instead,
the embryos developed two abdomens and telsons (tails) with mirror-image
symmetry: telson-abdomen-abdomen-telson (Figure 14.5). Further evidence
that RNA is important in specifying the anterior portion of the fly embryo
was obtained by Kandler-Singer and Kalthoff (1976), who submerged *Smittia* eggs in solutions containing various enzymes and then punctured the
eggs in specific regions. Double abdomens resulted when RNase was permitted to enter the anterior end. Other enzymes did not cause this abnormality, nor did RNase effect this change when it entered other regions of the
egg. Thus, Sander's laboratory postulated the existence of a gradient at either end of the egg, and it seemed likely that the egg sequestered an RNA
that generated a gradient of anterior-forming material.

The Molecular Model: Protein Gradients in the Early Embryo

In 1988, the gradient hypothesis was united with a genetic approach to the
study of *Drosophila* embryogenesis. If there were gradients, what were the
morphogens whose concentrations changed over space? What were the
genes that shaped these gradients? And did these substances act by activating or inhibiting certain genes in the areas where they were concentrated?
Christiane Nüsslein-Volhard led a research program that found that one set
of genes encoded gradient morphogens for the anterior part of the embryo,
another set of genes encoded the morphogens responsible for organizing the

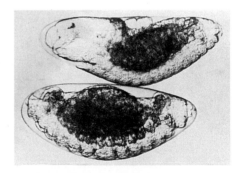

Figure 14.5
Normal and irradiated embryos of the
midge *Smittia*. The normal embryo (top)
shows a head on the left and abdominal
segments on the right. The UV-irradiated embryo has no head region but has
abdominal segments at both ends.
(From Kalthoff, 1969, courtesy of K.
Kalthoff.)

Table 14.1 Maternal effect genes that effect the anterior-posterior polarity of the *Drosophila* embryo

Gene	Phenotype	Proposed function and structure
ANTERIOR GROUP		
bicoid (bcd)	Head and thorax deleted, replaced by inverted telson	Graded anterior morphogen; contains homeodomain; represses *caudal*
exuperantia (exu)	Anterior head structures deleted	Anchors *bicoid* mRNA
swallow (swa)	Anterior head structures deleted	Anchors *bicoid* mRNA
POSTERIOR GROUP		
nanos (nos)	No abdomen	Posterior morphogen; represses *hunchback*
tudor (tud)	No abdomen, no pole cells	Localization of Nanos
oskar (osk)	No abdomen, no pole cells	Localization of Nanos
vasa (vas)	No abdomen, no pole cells; oogenesis defective	Localization of Nanos
valois (val)	No abdomen, no pole cells; cellularization defective	Stabilization of the Nanos localization complex
pumilio (pum)	No abdomen	Helps Nanos protein bind *hunchback* message
caudal (cad)	No abdomen	Activates posterior terminal genes
TERMINAL GROUP		
torso (tor)	No termini	Possible morphogen for termini
trunk (trk)	No termini	Transmits *torsolike* signal to *torso*
fs(1)Nasrat[fs(1)N]	No termini; collapsed eggs	Transmits *torsolike* signal to *torso*
fs(1)polehole[fs(1)ph]	No termini; collapsed eggs	Transmits *torsolike* signal to *torso*

Source: After Anderson, 1989.

abdominal region of the embryo, and a third set of genes encoded proteins that produced the terminal regions at both ends of the embryo (Figure 14.6; Table 14.1). This work resulted in a Nobel Prize for her and her colleague, Eric Wieschaus, in 1995. [droso1.html]

The anterior-posterior axis for the *Drosophila* embryo appears to be patterned before the nucleus even begins to function (Figure 14.7). The nurse cells of the ovary deposit mRNAs into the developing oocyte, and these mRNAs become apportioned to different regions of the cell. In particular, four messenger RNAs are critical to the formation of the anterior-posterior axis:

- *bicoid* and *hunchback* mRNAs, whose protein products are critical for head and thorax formation
- *nanos* and *caudal* mRNAs whose protein products are critical for the formation of the abdominal segments

The *bicoid* mRNAs are tethered to the anterior microtubules, while the *nanos* messages are bound to the posterior cortical cytoskeleton. The *hunchback* and *caudal* mRNAs are distributed throughout the oocyte. Upon fertilization, the mRNAs can be translated into proteins. At the anterior pole, the *bicoid* RNA is translated into Bicoid protein, which forms a gradient highest at the anterior. At the posterior pole, the *nanos* message is translated into Nanos pro-

Figure 14.6

Three independent genetic pathways interact to form the anterior-posterior axis of the *Drosophila* embryo. In each case, the initial asymmetry is established during oogenesis, and the pattern is organized by the maternal products soon after fertilization. The realization of the pattern comes about when the localized maternal products activate or repress specific zygotic genes in different regions of the embryo. (After St. Johnston and Nüsslein-Volhard, 1992.)

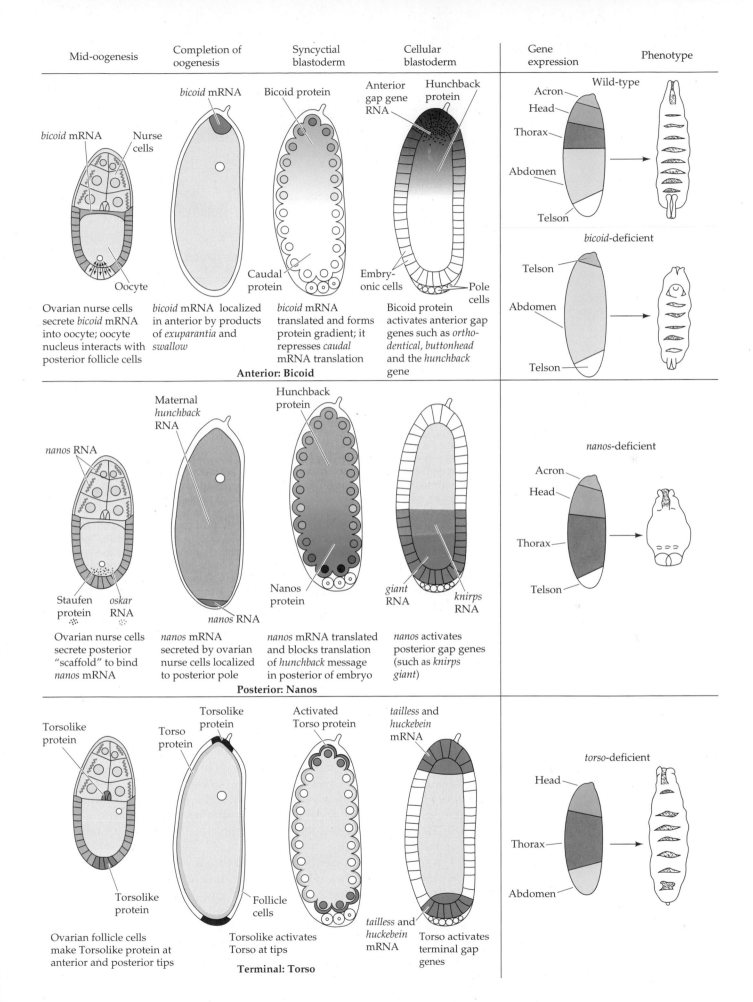

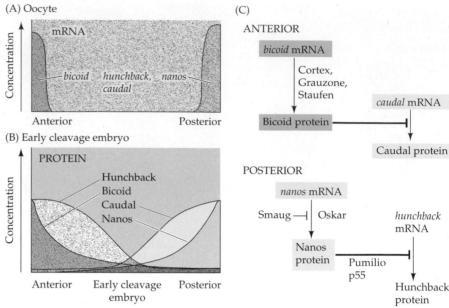

Figure 14.7
A model of anterior-posterior pattern generation by the maternal effect genes. (A) The *bicoid, nanos, hunchback,* and *caudal* messenger RNAs are placed into the oocyte by the ovarian nurse cells. The *bicoid* message is sequestered anteriorly. The *nanos* message is sent to the posterior pole. (B) Upon translation, the Bicoid protein gradient extends from anterior to posterior, and the Nanos protein gradient extends from posterior to anterior. Nanos inhibits the translation of the *hunchback* message (in the posterior), while Bicoid prevents the translation of *caudal* message (in the anterior). This results in opposing Caudal and Hunchback gradients. The Hunchback gradient is secondarily strengthened by the transcription of the *hunchback* gene from the anterior nuclei (since Bicoid acts as a transcription factor to activate *hunchback* transcription). (C) Parallel interactions whereby translational gene regulation establishes the anterior-posterior patterning of the *Drosophila* embryo. In the anterior of the embryo, *bicoid* mRNA is bound to the anterior cytoskeleton and is inhibited from translation by having a small polyadenylate tail. Upon fertilization, the tail is extended in a manner dependent on the Cortex, Grauzone, and Staufen proteins, and the *bicoid* mRNA is translated. The Bicoid protein suppresses the *caudal* mRNA from being translated. In the posterior region of the embryo, *nanos* mRNA is suppressed in the oocyte by the Smaug protein (which binds to its 3′ UTR). At fertilization, Oscar aids its translation, and the Nanos protein acts as a translational suppressor of *hunchback* mRNA. (C after Macdonald and Smibert, 1996.)

tein, which forms a gradient highest at the posterior. Bicoid protein inhibits the translation of the *caudal* RNA, thereby allowing Caudal protein to be synthesized only in the posterior of the cell. Conversely, the Nanos protein, in conjunction with the Pumilio protein, binds to *hunchback* RNA, preventing its translation in the posterior portion of the embryo. Bicoid also elevates the level of Hunchback protein in the anterior of the embryo by binding to the enhancers of the *hunchback* gene and stimulating its transcription (Figure 12.18). The result of these interactions is the creation of four protein gradients in the early embryo:

- An anterior-to-posterior gradient of Bicoid protein
- An anterior-to-posterior gradient of Hunchback protein
- A posterior-to-anterior gradient of Nanos protein
- A posterior-to-anterior gradient of Caudal protein

The stage is now set for the activation of the zygotic genes in those nuclei that had been busily dividing while this gradient was being established.

Gradient Models of Positional Information

HOW CAN CELLS be informed of their position in the embryo and then use that information to differentiate into the appropriate cell type? One explanation proposes gradients of morphogenetic substances (Boveri, 1901; Child, 1941; Wolpert, 1971). In these models, a soluble substance (morphogen) is posited to diffuse from a *source* (where it is produced) to a *sink* (where it is degraded), establishing a continuous range of concentrations within that region. Theoretical considerations (see Crick, 1970) suggest that such gradients could only function over relatively small distances, less than 100 cell diameters. In gradient models, the concentration of the morphogen changes over distance, the highest concentrations being near the source of the morphogen. The cells would have to have "sensors" that would respond differently to different concentrations of the gradient. If the morphogen were a transcription factor, then enhancer or promoter elements might bind the morphogen at different strengths (Figure 14.8). For example, if a morphogen were being made at the anterior of the body, the genes responsible for organizing head development might

have an enhancer that bound the morphogen poorly. Only when there was a large concentration of the morphogen present would that gene be active. The gene(s) responsible for thorax formation, on the other hand, might have an enhancer that bound the morphogen rather well, enabling it to respond to relatively low levels of that morphogen. The cells of the head would express *both* these genes, while the cells of the thorax would express only that gene whose enhancer could bind low amounts of the morphogen. The cells in the posterior portion of the body would not see any of this morphogen, and neither of these genes would be activated. In this way, cells could sense the presence of a morphogen and respond differentially, depending on the morphogen concentration. The sensor would not have to be an enhancer; it could just as well be a cell-surface receptor for a specific growth factor (see Chapter 17).

Most gradient models assume that all the cells that can respond to a gradient are equivalent. All these cells interpret the morphogen signal in the same way, and the concentration of morphogen that they receive determines their identity. However, the interpretation of gradients does not have to be linear. Take, for example, a series of exam grades that stretches uniformly from 100 to 60. In one scheme (a "linear" reading), a grade

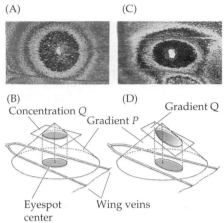

Figure 14.9 Gradient model of positional information proposed to explain butterfly wing spots. (A) Photograph of an eyespot on the wing of *Morpho peleides*. (B) Diagram of a two-gradient model that may explain the way the spot was generated. The origin of the morphogen is at the center of the spot and corresponds to the apex of a cone, the height of which reflects its concentration. Concentration *Q* represents the level of morphogen needed to reach the threshold sensitivity for the formation of color in those wing cells. (C) Photograph of the wing of *Smyrna blomfildia*, in which the eyespots are elliptical. (D) Different orientations of the sensitivity gradient *Q* could result in such elliptical eyespots. (After Nijhout, 1981, courtesy of H. F. Nijhout.)

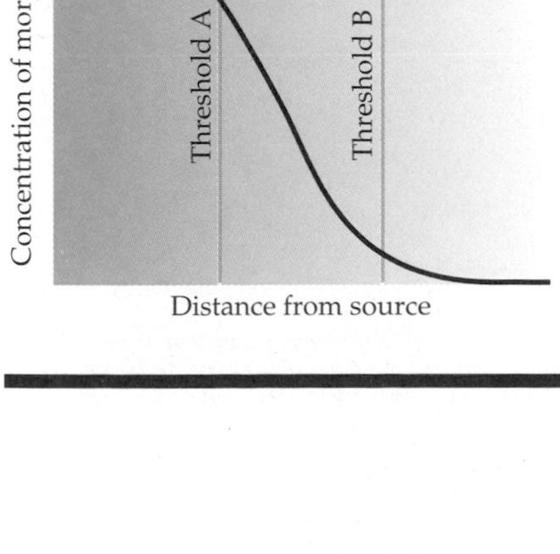

Figure 14.8 A hypothetical model for gradients establishing positional information. The concentration of the morphogen drops from the source. In this diagram, the receptors for the morphogen are enhancer elements of two genes that control cell fate, but the receptors could also be cytoplasmic receptors or membrane receptors. One of the receptors (in this case, the enhancer on gene A) needs a high concentration of morphogen in order to act. At high concentrations of morphogen, both genes A and B are active. In moderate concentrations, only gene B is active. Where the morphogen concentration falls below another threshold, neither gene is active. (After Wolpert, 1978.)

between 100 and 90 is A, 89–80 is B, 79–70 is C, and 69–60 is D. In another class (using a "curved" reading), 100–95 is A, 94–85 is B, 84–70 is C, and 69–60 is D. Nijhout (1981) has used a two-gradient model to explain the development of "eyespot" patterns on butterfly wings. One gradient consists of a linear diffusion of a morphogen. The second gradient involves the interpretation of this morphogen; in other words, the sensitivity threshold of the cells involved differs at different regions of the wing. The existence of the second gradient gives rise to an elliptical spot, not the circular spot that would result if the sensitivity gradient were absent (Figure 14.9). ■

Both genes active | Gene A inactive Gene B active | Both genes inactive

Gene A

Gene B

Concentration of morphogen

Threshold A

Threshold B

Distance from source

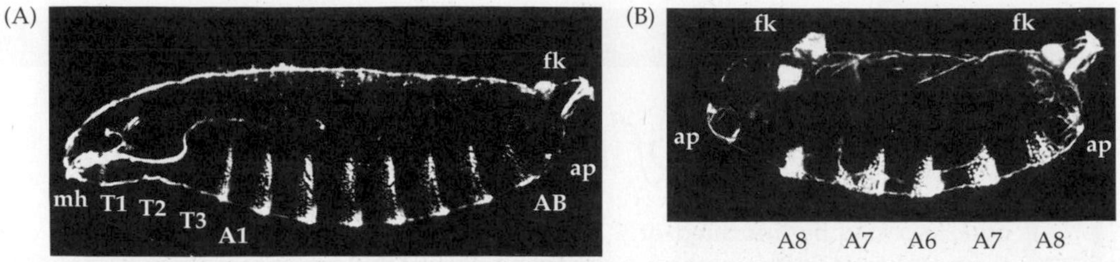

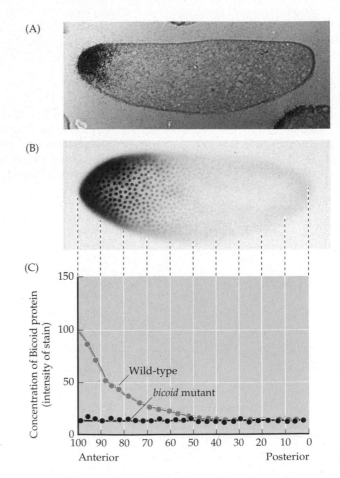

Figure 14.10
Phenotype of a strongly affected embryo from a female deficient in the *bicoid* gene. (A) Wild-type cuticle pattern. (B) *bicoid* mutant. The head and thorax have been replaced by a second set of posterior telson structures. Abbreviations: fk, filzkörper; ap, anal plates (both telson structures); T1–T3, thoracic segments; A1, A8, the two terminal abdominal segments; mh, cs, head structures. (From Driever et al., 1990, courtesy of W. Driever.)

Evidence that the Gradient of Bicoid Protein Constitutes the Anterior Organizing Center

In *Drosophila*, the phenotype of the *bicoid* mutant is very interesting in terms of gradients. Instead of having anterior structures (acron, head, and thorax) followed by abdominal structures and a telson, the structure of the *bicoid* mutant is telson-abdomen-abdomen-telson (Figure 14.10). It would appear that these embryos lack whatever morphogen is needed for anterior stuctures. Moreover, one could postulate that the substance that these mutants lack is the one postulated by Sander and Kalthoff to turn on genes for anterior structures and turn off genes for the telson structures. (Compare Figures 14.10 and 14.5.)

Further studies have strengthened the view that the product of the wild-type *bicoid* (*bcd*) gene is the morphogen that controls anterior development. First, *bicoid* is a maternal effect gene. The messenger RNA from the mother's

Figure 14.11
Gradient of Bicoid protein in the early *Drosophila* embryo. (A) Localization of *bicoid* mRNA to the anterior tip of the embryo. (B) Gradient of Bicoid protein shortly after fertilization. Note that the concentration is greatest anteriorly and trails off posteriorly. Notice also that Bicoid protein is concentrated in the nuclei of the embryo. (C) Densitometric scan of the Bicoid protein gradient. The upper curve represents the gradient of Bicoid protein in wild-type embryos. The lower curve represents Bicoid protein in embryos of *bicoid*-deficient mothers. (A from Kaufman et al., 1990; B and C from Driever and Nüsslein-Volhard, 1988a; photographs courtesy of the authors.)

bicoid genes is placed into the embryo by the mother's ovarian cells (Frigerio et al., 1986; Berleth et al., 1988; details in Chapter 22). The *bicoid* RNA is strictly localized in the anterior portion of the oocyte (Figure 14.11A). Driever and Nüsslein-Volhard (1988a) have shown that when Bicoid protein is translated from this RNA during early cleavage, it forms a gradient, with its highest concentration in the anterior of the egg and reaching background levels in the posterior third of the egg. Moreover, this protein soon becomes concentrated in the embryonic *nuclei* in the anterior portion of the embryo (Figure 14.11B,C; Plate 14A).

Further evidence that Bicoid protein is the anterior morphogen came from experiments that altered the steepness of the gradient. Two genes, *exuperantia* and *swallow*, are responsible for keeping the *bicoid* message at the anterior pole of the egg. In their absence, the *bicoid* message diffuses further into the posterior of the egg, and the gradient of Bicoid protein is shallower (Driever and Nüsslein-Volhard, 1988b). The phenotype produced by these two mutants is similar to that of *bicoid*-deficient embryos, but less severe. These embryos lack their most anterior structures and have an extended mouth and thoracic region. Thus, by altering the gradient of Bicoid protein, one correspondingly alters the fate of the embryonic regions.

Confirmation that the Bicoid protein is crucial for initiating head and thorax formation came from experiments in which purified *bicoid* RNA was injected into early-cleavage embryos (Figure 14.12; Driever et al., 1990). When injected into the anterior of *bicoid*-deficient embryos (whose mother lacked *bicoid* genes), the *bicoid* RNA rescued the embryos and caused them to have normal anterior-posterior polarity. Moreover, any location in the embryos where the *bicoid* messages were injected became the head. If *bicoid* RNA was injected into the center of the embryo, that middle region became the head and the regions on either side of it became thorax structures. If a large amount of *bicoid* RNA was placed in the *posterior* pole of a wild-type embryo (with its own endogenous *bicoid* message in its anterior pole), two heads emerged, one at either end. Therefore, the *bicoid* gene is now thought to encode the anterior morphogen of the *Drosophila* embryo.

The next question that emerged was, How is this localization of *bicoid* RNA accomplished? The current theories will be detailed in Chapter 22, but in brief, the anterior cytoskeleton anchors *bicoid* RNA through the message's 3' untranslated region. The posterior cytoskeleton has specific anchoring sites that will recognize the 3' UTR of the *nanos* message. Thus, the overall

Figure 14.12
Schematic representation of the experiments demonstrating that the *bicoid* gene encodes the morphogen responsible for head structures in *Drosophila*. The phenotypes of the *bicoid*-deficient and wild-type embryos are shown at the sides. When *bicoid*-deficient embryos are injected with *bicoid* mRNA, the point of injection forms the head structures. When the posterior pole of an early-cleavage wild-type embryo is injected with *bicoid* mRNA, head structures form at both poles. (After Driever et al., 1990.)

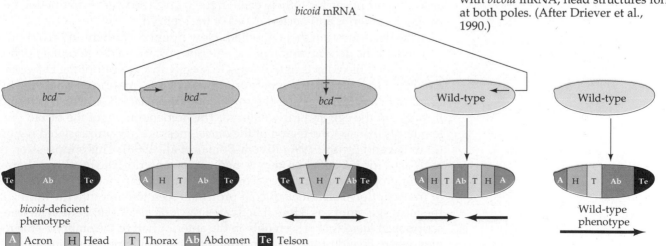

bicoid mRNA

bcd⁻ | bcd⁻ | bcd⁻ | Wild-type | Wild-type

bicoid-deficient phenotype

Wild-type phenotype

A Acron | H Head | T Thorax | Ab Abdomen | Te Telson

Figure 14.13
The importance of oocyte-follicle interactions in the formation of the anterior-posterior and dorsal-ventral axes of *Drosophila*. (A) The oocyte nucleus becomes localized to the posterior side of the egg. It localizes a factor (Gurken protein) that is received by the follicle cells at the posterior end of the egg chamber. (B,C) This causes the follicle cells to differentiate into posterior follicle cells and to secrete some factor that causes the oocyte to realign its microtubules. It is possible that this factor works by activating protein kinase A (PKA) in the oocyte cell membrane (see Chapter 22). (D) This reorganization allows the transport of Oskar protein and *nanos* mRNA to the posterior pole of the egg and retains *bicoid* message at the anterior pole of the egg. At the same time, the oocyte nucleus travels along the repolarized microtubules toward the anterior-dorsal region of the egg. Here, the same signal (Gurken protein) initiates the dorsal-ventral axis by signaling these cells to become dorsal follicle cells. (After Gonzáles-Reyes et al., 1995).

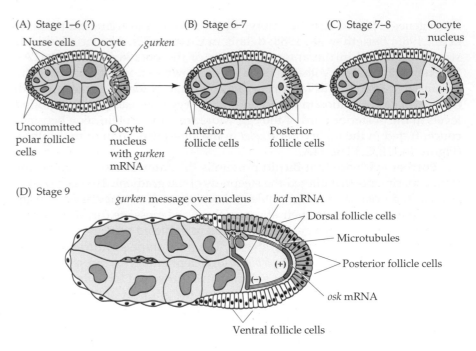

organization of the oocyte cytoskeleton is crucial for development. How does this cytoskeletal organization occur? During mid-oogenesis, the oocyte nucleus is positioned near the posterior pole of the oocyte (i.e. away from the nurse cells). The oocyte nucleus serves as the "collecting site" for the *gurken* RNA, a message that encodes a homologue of the epidermal growth factor and whose synthesis is not well understood. The *gurken* message collects directly above the nucleus, between the nucleus and the dorsal posterior follicle cells. Here, it becomes translated into Gurken protein and is secreted by the oocyte to those follicle cells nearest to the nucleus—the posterior follicle cells. This alters these follicle cells, causing them to secrete a factor that induces a reorganization of the microtubules in the oocyte. These microtubules initiate the reorganization of the oocyte cytoskeleton and enable the nucleus to move from its posterior position to an anterior-dorsal portion of the growing oocyte (Figure 14.13; Gonzáles-Reyes et al., 1995; Roth et al., 1995). Thus, the first signal for the anterior-posterior axis of the embryo comes from the maternal follicle cells. The distinction between anterior and posterior follicle cells in the ovary causes the distinction between the anterior and posterior axis of the embryo.

The next question then emerged: How might a gradient in Bicoid protein control the determination of the anterior-posterior axis? Recent evidence suggests that Bicoid acts in two ways to specify the anterior of the *Drosophila* embryo. First, it acts as a *repressor* of posterior formation. It does this by binding to and suppressing the translation of *caudal* RNA, which is found throughout the egg and early embryo. The homeodomain of the Bicoid protein binds to a specific region of the *caudal* message's 3' untranslated region (Dubnau and Struhl, 1996; Rivera-Pomar et al., 1996). This suppression is necessary, for if Caudal protein is made in the anterior, the head and thorax are not properly formed. The second mode of Bicoid function is at the level of transcriptional *activation*. Bicoid protein is seen to enter into the nuclei of the cleavage embryos. Here it activates the *hunchback* (*hb*) gene. The transcription of *hunchback* is seen only in the anterior half of the embryo—the region where Bicoid protein is seen. Mutants deficient in maternal and zygotic

Hunchback protein lack mouthparts and thorax structures. In the late 1980s, two laboratories independently demonstrated that Bicoid protein binds to and activates the *hunchback* gene (Driever and Nüsslein-Volhard, 1989; Struhl et al., 1989). The Hunchback protein derived from the synthesis of new *hunchback* mRNA joins the Hunchback protein synthesized by the translation of maternal messages in the anterior of the embryo. The Hunchback protein, also a transcription factor, is thought to repress abdominal-specific genes, thereby allowing the region of *hunchback* expression to form the head and thorax. Using DNase footprinting (in which proteins are bound to a segment of DNA, DNase is added, and the only DNA that remains is that protected by the DNA-binding protein), the researchers found that Bicoid protein binds to five sites in the upstream promoter region of the *hunchback* gene. These sites all have the consensus sequence 5'-TCTAATCCC-3'.

But binding does not necessarily mean activation. The activation of this gene by Bicoid protein was shown by fusing these *hunchback* promoter sites to chloramphenicol acetyltransferase (CAT) reporter genes and injecting these genes into early *Drosophila* embryos. In all cases, Bicoid protein was needed to activate the reporter genes. If injected into *bicoid*-deficient embryos, no CAT was produced (Figure 14.14). Also, while some activation was seen when only one of the five Bicoid-binding sequences was present, the full expression of the reporter gene (and presumably of *hunchback*) came when three of the five sites were present. Thus, the Bicoid protein gradient probably functions by activating *hunchback* gene transcription in the anterior portion of the embryo.

The Hunchback protein also works with Bicoid in generating the anterior pattern of the embryo. Driever and co-workers (1989) predicted that at least one other anterior gene besides *hunchback* must be activated by Bicoid. First, deletions of *hunchback* produce only some of the defects seen in the *bicoid* mutant phenotype. Second, as we saw in the *swallow* and *exuparentia* experiments, only moderate levels of Bicoid protein are needed to activate thorax formation (i.e., *hunchback* gene expression), but head formation needs higher concentrations. Driever and co-workers (1989) predicted that the promoters of such a head-specific gap gene would have low-affinity binding sites for Bicoid protein. This gene would be activated only at extremely high concentrations of Bicoid protein—that is, near the anterior tip of the embryo. Since then, three gap genes of the head have been discovered that are dependent on very high concentrations of Bicoid protein for their expression (Cohen and Jürgens, 1990; Finkelstein and Perrimon, 1990; Grossniklaus et al., 1994). The *buttonhead (btd), empty spiracles (ems),* and *orthodenticle (otd)* genes are needed to specify the progressively anterior regions of the head. In

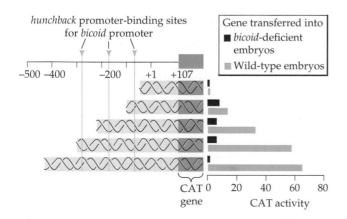

Figure 14.14
Influence of Bicoid protein in activating the *hunchback* gene. Different regions of the *hunchback* promoter were fused to the CAT reporter gene and injected into either wild-type embryos or embryos from *bicoid*-deficient mothers. The more Bicoid-binding sites there were in the promoter region, the more effective its expression in wild-type embryos. In embryos without Bicoid protein, no transcription resulted from any of the *hunchback* promoter-driven genes. (After Driever and Nüsslein-Volhard, 1989.)

addition to their needing high Bicoid levels for activation, these genes also require the presence of Hunchback protein to become transcribed (Simpson-Brose et al., 1994; Reinitz et al., 1995). The Bicoid and Hunchback proteins act synergistically as the enhancers of these "head genes" to promote their transcription.

The Posterior Organizing Center: Localizing and Activating the nanos Product

The posterior organizing center is defined by the activities of the *nanos* gene (Lehmann and Nüsslein-Volhard, 1991; Wang and Lehmann, 1991; Wharton and Struhl, 1991). The *nanos* RNA is produced in the ovary and is transported into the egg, where it becomes bound in the posterior region (farthest away from the ovarian nurse cells). The products of several other genes (*oskar*, *valois*, *vasa*, *staufen*, and *tudor*)—the same gene products that place the germ plasm determinant into the posterior pole plasm (see Chapter 13)—are needed to place the *nanos* RNA into the posterior part of the egg.* If *nanos* or any other of these maternal effect genes are absent in the mother, no embryonic abdomen forms (Lehmann and Nüsslein-Volhard, 1986; Schüpbach and Wieschaus, 1986).

The *nanos* message is translated into protein soon after fertilization, just as the *bicoid* message is. Tautz (1988) has shown that during normal abdomen formation, the protein product of the *nanos* gene represses the *translation* of *hunchback* RNA (see Figure 14.7). This *hunchback* RNA is initially present throughout the embryo, although more can be made from zygotic nuclei if they are activated by Bicoid protein. Thus, the combination of Bicoid and Nanos protein causes a gradient of Hunchback protein across the egg (Figure 14.15). The Bicoid protein activates *hunchback* gene transcription in the anterior part of the embryo, while the Nanos protein inhibits the translation of *hunchback* RNA in the posterior part of the embryo. If the *nanos* gene product were not present, Hunchback protein would be made throughout the embryo and would presumably inhibit the expression of abdomen-generating gap genes such as *knirps* (Hülskamp et al., 1989; Irish et al., 1989; Struhl, 1989). The *hunchback* gene, therefore, appears to be the focal point, under the regulation of both the anterior and the posterior organization centers long known to exist in insect development. These studies on *nanos* and *bicoid* function can now explain the embryological experiments. UV light or RNase treatment would destroy *bicoid* RNA, causing the loss of anterior structures and the duplication of the abdomen; ligation procedures might block the spread of Nanos, thereby allowing higher levels of Hunchback protein to accumulate.

Although Nanos is considered the major posterior morphogen, two other proteins, **Pumilio** and **Caudal,** are also important in constructing the posterior segments of *Drosophila*. Nanos protein does not bind directly to the *hunchback* message. Instead, Pumilio, a protein found throughout the embryo, binds to the 3' UTR of the *hunchback* message and forms a binding site on which Nanos is able to bind (Barker et al., 1992; Murata and Wharton, 1995). The binding of Nanos is critical in repressing *hunchback* message

---

*Like the placement of the *bicoid* message, the location of the *nanos* message is determined by its 3' untranslated region. If the *bicoid* 3' UTR is placed onto the protein-encoding region of *nanos* RNA, the *nanos* message gets placed in the anterior of the egg. When the RNA is translated, the Nanos protein inhibits the translation of *hunchback* and *bicoid* mRNAs, and the embryo forms two abdomens—one in the anterior of the embryo and one in the posterior (Gavis and Lehmann, 1992). The localization of *nanos* RNA is ultimately dependent on the interactions between the oocyte and the neighboring follicle cells that localize the *oskar* message to the posterior pole and localize *bicoid* RNA to the anterior pole (see Chapter 22).

(A)

MATERLAL

Transcription factors

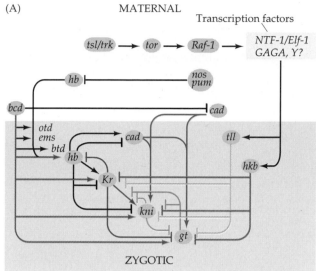

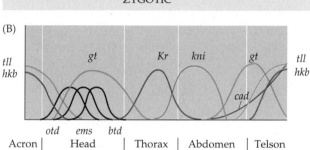

ZYGOTIC

(B)

tll
hkb

gt

Kr   kni

gt

tll
hkb

cad

otd   ems   btd

Acron | Head | Thorax | Abdomen | Telson

Figure 14.15

Conversion of maternal gradients into zygotic gap gene expression. (A) The gradients of maternal transcription factors Bicoid, Caudal, and Hunchback regulate the transcription of the gap genes. Hunchback and Caudal proteins come from both maternal messages and new zygotic transcription. (B) The concentration of Bicoid, Hunchback, and Caudal proteins is critical in specifying the positions where the gap genes are transcribed. These proteins diffuse, and the interactions between the proteins will be critical in activating the transcription of the pair-rule genes. At the two termini, the interaction between Torso and Torsolike activates the *tailless* and *huckebein* gap genes. (After Rivera-Pomar and Jäckle, 1996.)

translation. The Caudal protein is also important for forming posterior structures. Although embryos can form abdominal segments in the absence of Caudal, these segments are often fused together or are partially absent (Macdonald and Struhl, 1986; Mlodzik and Gehring, 1987).

The Terminal Gene Group

If both the anterior and the posterior organizing centers are nonfunctional, an embryo still can develop some anterior-posterior pattern (Nüsslein-Volhard et al., 1987). When females are made doubly mutant for both the anterior and the posterior morphogens, their embryos produce two telsons, one at each end of the embryo. Thus, there exists a third set of maternal effect genes that help to create the extremes of the anterior-posterior axis. Mutations in these *terminal genes* result in the loss of the unsegmented extremities of the organism: the anterior acron and the posterior telson. In the absence of these gene products, the segmented portion of the embryo expands to the extremities (Degelmann et al., 1986; Klingler et al., 1988). Therefore, the terminal gene set defines the boundaries of the segmented parts of the body.

The critical gene here appears to be **torso**, a gene encoding a receptor tyrosine kinase (see Figure 14.6). The *torso* RNA is synthesized by the ovarian cells, deposited into the oocyte, and translated after fertilization. The transmembrane Torso protein is not spatially restricted to the ends of the egg but is evenly distributed throughout the plasma membrane (Casanova and Struhl, 1989). A dominant mutation of *torso*, which imparts constitutive activity to the receptor, converts the entire anterior half of the embryo into an acron and the entire posterior half into a telson. Thus, Torso must normally be activated only at the ends of the egg. In fact, Stevens and her colleagues (1990) have shown that Torso protein is activated by follicle cells at either

Figure 14.16
Hypothetical model of Torso signaling. Torso-like protein, secreted from the anterior and posterior follicle cells is bound by the Torso receptor (which is found throughout the oocyte membrane). Binding of the ligand leads to the activation of torso and its autophosphorylation on specific tyrosine residues. The phosphotyrosine groups are recognized by the SH2 domain of Drk protein. The SH3 domain of Drk protein binds to SOS protein and thereby activates the GTPase activity of the Ras protein. This will activate the Raf protein which is the first member of a serine/threonine kinase cascade. This cascade usually functions by phosphorylating a transcription factor thereby allowing it to enter or function in the nucleus. That factor has not yet been identified. The end result is to stimulate the transcripton of the *huckebein* and *tailless* gap genes. (After Duffy and Perrimon, 1994).

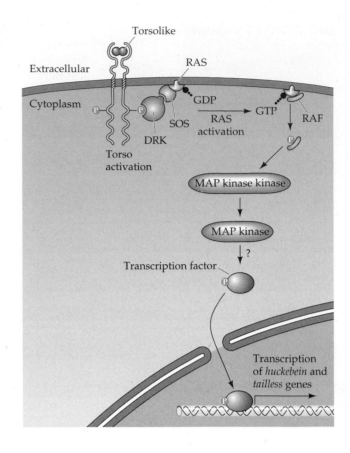

pole of the embryo. The activator of the Torso protein is probably **Torsolike,** since a loss-of-function mutation in the *torsolike* gene creates a phenotype almost identical to that produced by *torso*. The *torsolike* gene is expressed in the anterior and posterior follicle cells, and the secreted *torsolike* protein remains close to these cells (Martin et al., 1994). Stevens and colleagues showed that if the follicle cells at the poles of the egg chamber are deficient in the *torsolike* gene (even if the other follicle cells express the wild-type allele of this gene), the resulting embryo will have a phenotype similar to that of *torso*. It appears that Torsolike protein is secreted from the follicle cells and that it activates the Torso protein in the oocyte membrane.

Activation of Torso receptor tyrosine kinase involves the autophosphorylation of tyrosine residues and the subsequent activation of the Ras and Raf proteins (Figures 14.15 and 14.16; Duffy and Perrimon, 1994). These proteins activate the MAP kinase cascade (see Chapter 3), which (in some still unknown manner) stimulates the transcription of the *tailless* and *huckebein* gap genes. These genes then specify the termini of the embryo. The distinction between the anterior and posterior termini depends on the presence of Bicoid. If the terminal genes act alone, cells differentiate into telsons. However, if Bicoid is also present, the region forms an acron (Pignoni, et al., 1992).

The anterior-posterior axis of the embryo is therefore specified by three sets of genes: those that define the anterior organizing center, those that define the posterior organizing center, and those that define the terminal boundary region. The anterior organizing center is located at the anterior end of the embryo and acts through a gradient of Bicoid protein that activates anterior-specific gap genes and suppresses posterior-specific gap genes. The posterior organizing center is located at the posterior pole and acts through the formation of Nanos protein, which gets transported into the abdominal region. Here, Nanos inhibits the inhibitor of abdominal-specific gene expression and activates those genes that form the abdomen. The boundaries of the acron and telson are defined by the *torso* gene product, which is activated at the tips of the embryo.

The segmentation genes

An Overview

The commitment of cell fate in *Drosophila* appears to be a two-step process: specification and determination (Slack, 1983). Early in development, the fate of a cell depends on environmental cues such as those provided by the gradients mentioned earlier. This **specification** of cell fate is flexible and can still be altered in response to environmental signals. Eventually, the cells undergo a transition from this loose type of commitment to an irreversible **determination.** Here, the fate of a cell has become cell-intrinsic.* The transition from specification to determination in *Drosophila* is mediated by the **segmentation genes.** Segmentation genes divide the early embryo into a repeating series of segmental primordia along the anterior-posterior axis. Mutations in segmentation genes cause the embryo to lack certain segments or parts of segments, and these mutations show the existence of three classes of segmentation genes (Table 14.2). Often these mutations affect **parasegments,** regions of the embryo that are separated by mesodermal thickenings and ectodermal grooves and which divide the embryo into 14 regions (Martinez-Arias and Lawrence, 1985). The parasegments of the embryo do not become the segments of the larva or adult. Rather, they include the posterior part of an anterior segment and the anterior portion of the segment behind it (Figure 14.17). While the *segments* are the major anatomical divisions of the lar-

---

*Aficionados of information theory will recognize that the process by which the anterior-posterior information in morphogenetic gradients is transferred to discrete domains of homeotic selector genes represents a transition from analog to digital specification. Specification is analog, determination digital. This enables the transient information of the gradients in the syncytial blastoderm to be stabilized so that it can be utilized much later in development (Baumgartner and Noll, 1990).

TABLE 14.2 Major loci affecting segmentation pattern in *Drosophila*

Category	Loci	Category	Loci
Gap genes	*Krüppel (Kr)* *knirps (kni)* *hunchback (hb)* *giant (gt)* *tailess (tll)* *huckebein (hkb)* *buttonhead (btd)* *empty spiracles* *(ems)*	Pair-rule genes (secondary)	*fushi tarazu (ftz)* *odd-paired (opa)* *odd-skipped (odd)* *sloppy-paired (slp)* *paired (prd)*

		Segment
		   polarity genes
			engrailed (en)
			 wingless (wg)
			 cubitus interruptus^D
			   (ci^D)

Pair-rule genes (primary)	*orthodenticle (otd)* *hairy (h)* *even-skipped (eve)* *runt (run)*		*hedgehog (hh)* *fused (fu)* *armadillo (arm)* *patched (ptc)* *gooseberry (gsb)* *pangolin (pan)*

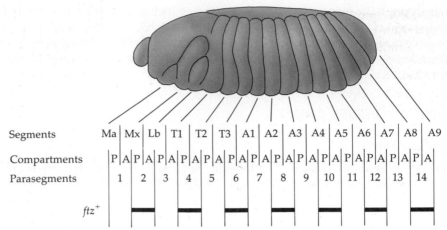

Figure 14.17
Segments and parasegments. A and P represent the anterior and posterior compartments of the segments. The parasegments are shifted one compartment forward. Ma, Mx, and Lb represent three of the head segments (mandibular, maxillary, and labial), T segments are thoracic, and A segments are abdominal. The parasegments are numbered 1 through 14. Underneath the map are the boundaries of gene expression observed by the in situ hybridization of radioactive cDNA from the pair-rule gene *fushi tarazu* (*ftz*). (After Martinez-Arias and Lawrence, 1985.)

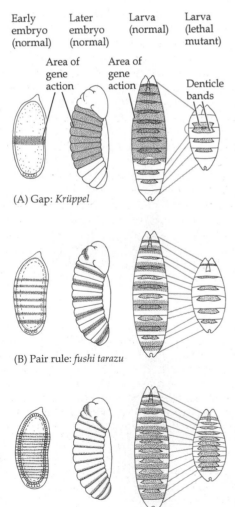

(A) Gap: *Krüppel*

(B) Pair rule: *fushi tarazu*

(C) Segment polarity: *engrailed*

val and adult body plan, these segments are built according to rules that use the *parasegment* as the basic unit of construction.

There are three classes of segmentation genes, each class expressed after the other (Figure 14.3). The transition from an embryo characterized by gradients of morphogens to an embryo having distinct units is accomplished by the products of the gap genes. The gap genes are activated or repressed by the maternal effect genes, and they divide the embryo into broad regions containing several parasegment primordia. The *Krüppel* gene, for example, is expressed primarily in parasegments 4–6 in the center of the *Drosophila* embryo (Figures 14.18A and 14.19; Plate 14A); the absence of *Krüppel* causes the embryo to lack these regions. The protein products of the gap genes interact with their neighboring gap-gene-encoded proteins and activate the transcription of the pair-rule genes. The transcription of these genes subdivides the broad gap gene domains into parasegments. Mutations of the pair-rule genes (as in *fushi tarazu*; Plate 14C) usually delete portions of every alternative segment. Figures 14.18 and 14.20 compare the morphology of the wild-type embryo with that of the *fushi tarazu* mutant. Finally, the segment polarity genes are responsible for maintaining certain repeated structures within each segment. Mutations in this group of genes cause a portion of each segment to be deleted and replaced by a mirror-image structure of another portion of the segment. For instance, in *engrailed* mutants, the portions of the posterior part of each segment are replaced by duplications of the anterior region of the subsequent segment (Figure 14.18C; Plate 14D). Thus, the segmentation genes are transcription factors that take the gradients of the early-cleavage embryo and transform the embryo into a periodic, parasegmental structure.

Figure 14.18
Three types of segmentation pattern mutants. The left panel shows the cleavage-stage embryo, with the region where the particular gene is normally transcribed in wild-type embryos shown in color. In the three panels on the right, the areas shown in color are deleted when these mutants develop. (After Mange and Mange, 1990.)

After the parasegmental boundaries are made, the pair-rule and gap genes interact to regulate the homeotic genes that determine the identity of each segment. By the end of the cellular blastoderm stage, each segment primordium has been given an individual identity by its unique constellation of gap, pair-rule, and homeotic gene products (Levine and Harding, 1989).

The Gap Genes

The gap genes were originally defined by a series of mutations whose embryos lacked groups of consecutive segments (Nüsslein-Volhard and Wieschaus, 1980). As shown in Figure 14.21, deletions caused by the *hunchback* (*hb*), *Krüppel* (*Kr*), and *knirps* (*kni*) genes span the entire segmented region of the *Drosophila* embryo. The *giant* (*gt*) gap gene overlaps with these three, and the phenotypes of the *tailless* and *huckebein* mutants delete portions of the unsegmented termini of the embryo.

The expression of these genes is dynamic. There is usually a low level of transcriptional activity across the entire embryo that becomes defined into discrete regions of high activity as cleavage continues (Jäckle et al., 1986). The critical element appears to be the expression of the Hunchback protein, which, by the end of nuclear division cycle 12, is at high levels across the anterior part of the embryo and then forms a steep gradient through about 15 nuclei. The last third of the embryo has undetectable Hunchback expression. The transcription patterns of the anterior gap genes are initiated by the different concentrations of the Hunchback and Bicoid proteins. High levels of

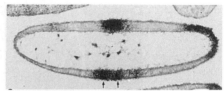

Figure 14.19
Expression of the *Krüppel* gene in the center and posterior of the *Drosophila* embryo (arrows). A 2.5-hour-old embryo was hybridized with cDNA that recognized *Krüppel* mRNA accumulations. (From Levine and Harding, 1989, courtesy of M. Levine.)

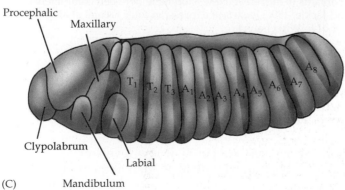

(A)

(B)

(C)

Procephalic
Maxillary
T_1 T_2 T_3 A_1 A_2 A_3 A_4 A_5 A_6 A_7 A_8
Clypolabrum
Labial
Mandibulum

Figure 14.20
Defects seen in the *ftz⁻* embryo. (A) Scanning electron micrograph of wild-type embryo seen in lateral view. (B) Same stage of an *ftz⁻* embryo. The white lines connect the homologous portions of the segmented germ band. (C) Diagram of wild-type embryonic segmentation. The shaded regions show the parasegments of the germ band that are missing in the *ftz⁻* embryo. (After Kaufman et al., 1990, photographs courtesy of T. Kaufman.)

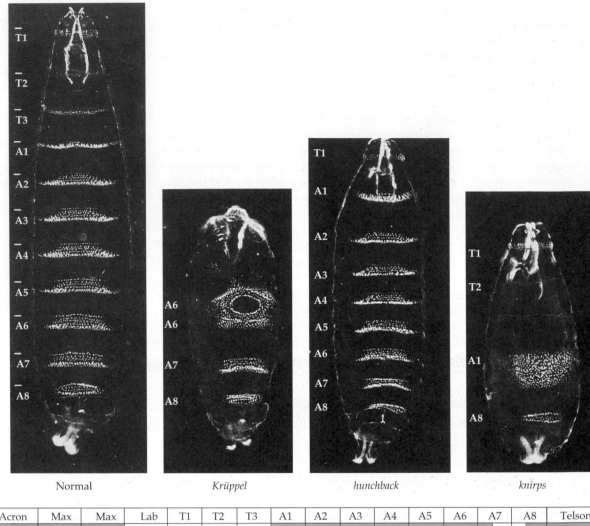

	Acron	Max	Max	Lab	T1	T2	T3	A1	A2	A3	A4	A5	A6	A7	A8	Telson
hunchback																
Krüppel																
knirps																
tailless																
giant																

Figure 14.21

Segment deletions in gap gene mutants. The table below the photographs indicates missing segmental regions with white bars. In *hunchback* mutants, the region is extended (lighter shading) if the mother as well as the zygote lacks *hunchback* gene activity. The actual domains of *huckebein* expression have not been completely defined. (After Gaul and Jäckle, 1990; *huckebein* expression after Weigel et al., 1990; photographs courtesy of E. Wieschaus.)

Hunchback protein induce the expression of *giant*, while the *Krüppel* transcript appears over the region where Hunchback begins to decline (see Figure 14.15). Also, high levels of Hunchback protein prevent the transcription of the posterior gap genes (such as *knirps*) in the anterior part of the embryo (Struhl et al., 1992).

In the posterior, Hunchback protein is at low levels or absent. It is thought that a gradient of the Caudal protein, highest at the posterior pole, is responsible for activating the abdominal gap genes *knirps* and *giant*. The *giant* gene has two methods for its activation, one for its anterior expression band and one for its posterior expression band (see Figure 14.15; Rivera-Pomar, 1995; Schulz and Tautz, 1995).

After the initial placement of these proteins by the maternal effect genes and Hunchback, they become stabilized and maintained by interactions between the different gap genes themselves.* For instance, *Krüppel* gene expression is negatively regulated on its anterior boundary by Hunchback protein and on its posterior boundary by Knirps and Tailless proteins (Jäckle et al., 1986; Harding and Levine, 1988; Hoch et al., 1992). If Hunchback activity is lacking, the domain of *Krüppel* expression extends anteriorly. If Knirps activity is lacking, *Krüppel* gene expression extends more posteriorly. The boundaries between the regions of gap gene transcription are probably created by mutual repression. Just as the Giant and Hunchback proteins can control the anterior boundary of *Krüppel* transcription, so can Krüppel determine the posterior boundaries of *giant* and *hunchback* transcription. If an embryo lacks the *Krüppel* gene, *hunchback* transcription continues into the area usually allotted to *Krüppel* (Jäckle et al., 1986; Kraut and Levine, 1991). These boundary-forming inhibitions are thought to be directly mediated by the gap gene products, because all four major gap genes (*hb, gt, Kr,* and *kni*) encode DNA-binding proteins that can activate or repress transcription (Knipple et al., 1985; Gaul and Jäckle, 1990; Capovilla et al., 1992).

Moreover, these interactions are highly specific, and the product of one gap gene can bind to the promoters of other gap genes. DNase I footprinting shows that the protein encoded by the wild-type *Krüppel* gene binds to the promoter region of the *hunchback* gene (which it inhibits) and to the promoter region of the *knirps* gene (which it stimulates). The *knirps* promoter region is also recognized by the protein product of the *tailless* gene, which inhibits *knirps* transcription. Hunchback protein (in addition to recognizing the *Krüppel* promoter) also recognizes its own promoter, suggesting that *hunchback* is involved in regulating its own expression (Pankratz et al., 1990; Štanojevíc et al., 1989; Treisman and Desplan, 1989).

The Pair-Rule Genes

The first indication of segmentation in the fly embryo comes when the pair-rule genes are expressed during the thirteenth division cycle. The transcription patterns of these genes are striking in that they each divide the embryo into the areas that are the precursors of the segmental body plan. As can be seen in Figure 14.22 and Plate 14C, one vertical stripe of nuclei (the cells are just beginning to form) expresses this gene, then another stripe of nuclei does not express it, and then another stripe of nuclei expresses this gene. The result is a "zebra stripe" pattern along the anterior-posterior axis, dividing the axis into 15 subunits (Hafen et al., 1984). Eight genes are currently known to be capable of dividing the early embryo in this fashion; they are listed in Table 14.2. It is important to note that not all nuclei express the same pair-rule genes. In fact, in each parasegment, each row of nuclei probably has its own constellation of pair-rule genes that distinguishes it from any other row.

How are some nuclei of the *Drosophila* embryo told to transcribe a particular gene while their neighbors are told not to transcribe it? The answer appears to come from the distribution of the protein products of the gap genes. Whereas the *RNA* of each of the gap genes has a very discrete distribution that defines abutting or slightly overlapping regions of expression, the

---

*The interactions between genes and gene products are facilitated by the fact that these reactions occur within a syncytium. The cell membranes have not yet formed.

(A)

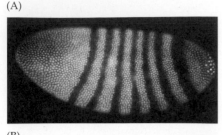

(B)

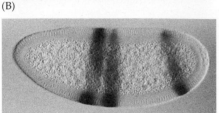

Figure 14.22
Specific promoter regions of the *even-skipped* gene control specific transcription bands in the embryo. (A) A reporter β-galactosidase gene was fused to regions of the *even-skipped* promoter and inserted into the fly genome. The full promoter region produces the seven normal transcription stripes. (B) If only the most proximal 480 base pairs of the promoter are present, only stripes 2, 3, and 7 form. (C) Partial map of the *eve* promoter, showing the regions responsible for the various stripes and for autoregulation. (Photographs courtesy of S. Carroll and M. Levine.)

(C)

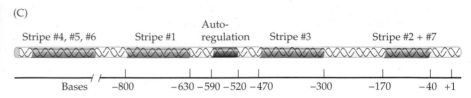

protein products of these genes extend more broadly. In fact, they overlap by at least 8–10 nuclei (which at this stage accounts for about two to three segment primordia). This was demonstrated in a striking manner by Štanojevíc and co-workers (1989). They fixed cellularizing blastoderms, stained the Hunchback protein with an antibody carrying a red dye, and simultaneously stained the Krüppel protein with an antibody carrying a green dye. Cellularizing regions that contained both proteins bound both antibodies and were stained bright yellow (Plate 14B). Similarly, Krüppel protein overlaps with Knirps protein in the posterior region (Pankratz et al., 1990).

Three genes are known to be the **primary pair-rule genes.** These genes—*hairy, even-skipped*, and *runt*—are essential for the formation of the periodic pattern, and they are the genes directly controlled by the gap proteins. The promoters of the primary pair-rule genes are recognized by gap gene proteins, and it is thought that the different concentrations of gap gene proteins determine whether the gene is transcribed or not. The promoters of these genes are often modular: the control over each stripe is located in a discrete region of the DNA. For example, one particular deletion in the promoter region of the *even-skipped* gene prevents the formation of the seventh *even-skipped* stripe, while a deletion slightly more downstream causes the loss of the second *even-skipped* stripe (Figure 14.23). DNase I footprinting of this latter region shows that it contains six binding sites for Krüppel protein, three binding sites for Hunchback protein, three binding sites for giant protein, and five binding sites for Bicoid protein. Genetic evidence shows that if some of these sites are deleted, the position of the second stripe moves. Štanojevíc and colleagues (1991) have shown that the second *even-skipped* stripe is repressed by both Giant and Krüppel proteins and is activated by Hunchback protein at low concentrations of Bicoid. This model is shown in Figure 14.23B,C. The region responsible for the third stripe of *even-skipped* transcription contains 20 Hunchback-binding sites and not one site for the Krüppel protein (Štanojevíc et al., 1989). This situation would enable the site to respond to very low levels of the *hunchback* gene product. Gap proteins activate transcription of some pair-rule genes while repressing the transcription of others. The result is the pattern of transcription stripes that emerges as the embryo develops.

Once initiated by the gap proteins, the transcription pattern of the primary pair-rule genes becomes stabilized by their interactions among themselves (Levine and Harding, 1989). The primary pair-rule genes also form

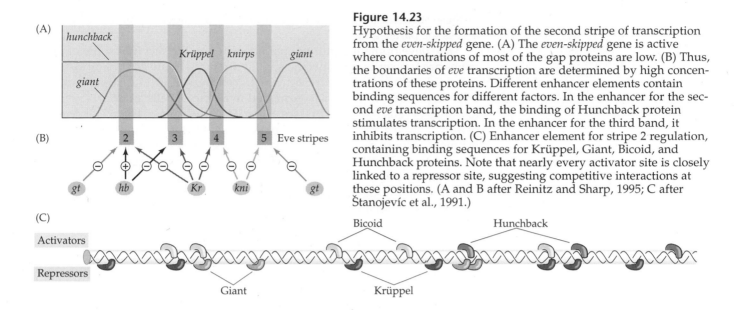

Figure 14.23
Hypothesis for the formation of the second stripe of transcription from the *even-skipped* gene. (A) The *even-skipped* gene is active where concentrations of most of the gap proteins are low. (B) Thus, the boundaries of *eve* transcription are determined by high concentrations of these proteins. Different enhancer elements contain binding sequences for different factors. In the enhancer for the second *eve* transcription band, the binding of Hunchback protein stimulates transcription. In the enhancer for the third band, it inhibits transcription. (C) Enhancer element for stripe 2 regulation, containing binding sequences for Krüppel, Giant, Bicoid, and Hunchback proteins. Note that nearly every activator site is closely linked to a repressor site, suggesting competitive interactions at these positions. (A and B after Reinitz and Sharp, 1995; C after Štanojevíc et al., 1991.)

the context that allows or inhibits the expression of the later-acting **secondary pair-rule genes.** One such secondary pair-rule gene is *fushi tarazu* (*ftz*; Japanese, "too few segments"). Early in cycle 14, *ftz* RNA and protein are seen throughout the segmented portion of the embryo. However, as the proteins from the primary pair-rule genes begin to interact with the *ftz* promoter, the *ftz* gene is repressed in certain bands of nuclei to create the interstripe regions. Meanwhile, the Ftz protein interacts with its own promoter to stimulate more transcription of the *ftz* gene (Figure 14.24; Edgar et al., 1986; Karr and Kornberg, 1989; Schier and Gehring, 1992).

The Segment Polarity Genes

So far, our discussion has identified interactions between molecules within the syncytial embryo. But once cells form, interactions take place between the *cells*. These intercellular interactions are mediated by the segment polarity genes, and they accomplish two important tasks. First, they reinforce the parasegmental periodicity established by the earlier transcriptional factors. Second, through this cell signaling, cell fates are established within each parasegment.

Many segment polarity genes encode proteins that are constituents of cell-signaling pathways. For instance, Wingless and Hedgehog are each secreted proteins that act as ligands, while Patched is a transmembrane protein that acts as a receptor (for Hedgehog). Other segment polarity genes, such as *disheveled*, *zeste white-3*, and *fused*, encode signal transducers (see Chapter 3), and some, such as *engrailed*, *armadillo*, and *cubitus interruptus*, are thought to be transcription factors activated by these pathways. Mutations

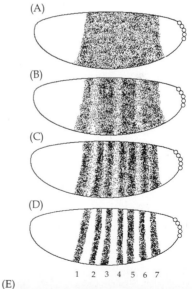

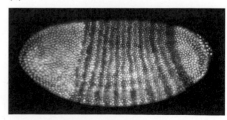

Figure 14.24
Transcription of the *ftz* gene. (A–D) At the beginning of cycle 14, there is low-level transcription in each of the nuclei in the segmented region of the *Drosophila* embryo. Within the next 30 minutes, the expression pattern alters as *ftz* transcription is enhanced in certain regions (which form the stripes) and repressed in the interstripe regions. (E) Double labeling of the *even-skipped* (darker bands) and *fushi tarazu* (lighter bands) transcripts, showing that *ftz* is expressed between the *eve* bands. (A–D after Karr and Kornberg, 1989; E courtesy of M. Levine.)

in these segment polarity genes lead to defects in segmentation and pattern across the parasegment.

The development of normal pattern relies on the fact that some of these genes are transcribed in specific spatial domains (Plate 14D). For instance, *wingless, engrailed,* and *hedgehog* are each expressed in 14 distinct bands of cells. In particular, *wingless* is expressed in a strip of cells anteriorly adjacent to a strip of cells that co-expresses *engrailed* and *hedgehog*. The misexpression of any of these genes destroys the patterning of the parasegment. The establishment of these restricted patterns of expression is determined by the pair-rule proteins. The transcription of the *wingless* gene is repressed by both Fushi tarazu and Even-skipped proteins, and it is stimulated by general activators found throughout the embryo. Meanwhile, the *engrailed* gene is active in those cells containing the Fushi tarazu protein (which stimulates *engrailed* transcription) and lacking Odd-skipped (which inhibits its transcription). This causes *wingless* to be transcribed solely in the cell directly anterior to the cells where *engrailed* is transcribed (Figure 14.25A).

Once *wingless* and *engrailed* expression is established in adjacent cells, this pattern must be maintained to retain the parasgmental periodicity of body plan established by the pair-rule genes. It should be remembered that the mRNAs and proteins involved in initiating these patterns are short-lived and that these patterns have to be maintained after the initiators of the pattern are no longer being synthesized. The *maintenance* of these patterns is regulated by interactions between cells expressing *wingless* and those expressing *engrailed*. The Wingless protein, secreted from the *wingless*-expressing cells, signals to adjacent cells by binding to the transmembrane protein D-Frizzled-2 (see Figure 3.38; Bhanot et al., 1996). This activates the signal transducer Disheveled, which will cause a reduction of Zeste white-3 kinase activity. The downregulation of this kinase is thought to allow the unphosphorylated Armadillo (β-catenin) protein to enter the nucleus, where it acts as a transcription factor positively regulating and thereby maintaining the expression of the *engrailed* gene (Siegfried et al., 1994).

This activation starts another portion of this reciprocal pathway. The Engrailed protein activates the transcription of the **hedgehog** (**hh**) gene. This gene encodes a secreted protein that activates a signal transduction pathway in anteriorly responding cells, leading to the maintenance of *wingless* gene transcription in its neighboring cell. The result is a reciprocal loop wherein the Engrailed-synthesizing cells secrete the Hedgehog protein, which maintains the expression of the *wingless* gene in the neighboring cell, while the Wingless-secreting cell causes the expression of the *engrailed* and *hedgehog* genes in the other cell (Heemskerk et al., 1991; Ingham et al., 1991; Mohler and Vani, 1992). In this way, the transcription pattern of these two cells is stabilized.

The second task accomplished by the segment polarity genes is to establish the cell fates across each parasegment. This is not understood completely, but the stabilized group of cells flanking the parasegment border expressing *wingless* and *hedgehog*, respectively, is essential. This can be seen in the dorsal epidermis, where the rows of cells produce different cuticular structures, depending on their position within the segment. The 1° row consists of large pigmented denticles. Posterior to these cells, the 2° row produces a smooth epidermal cuticle. The next two cell rows have a 3° fate, making small thick hairs, and these are followed by several rows of cells that adopt the 4° fate, producing fine hairs.

The *wingless*-expressing cells lie within the region differentiating the fine hairs, while the *hedgehog*-expressing cells are near the 1° row cells. The fates of the cells can be altered by experimentally increasing or decreasing the lev-

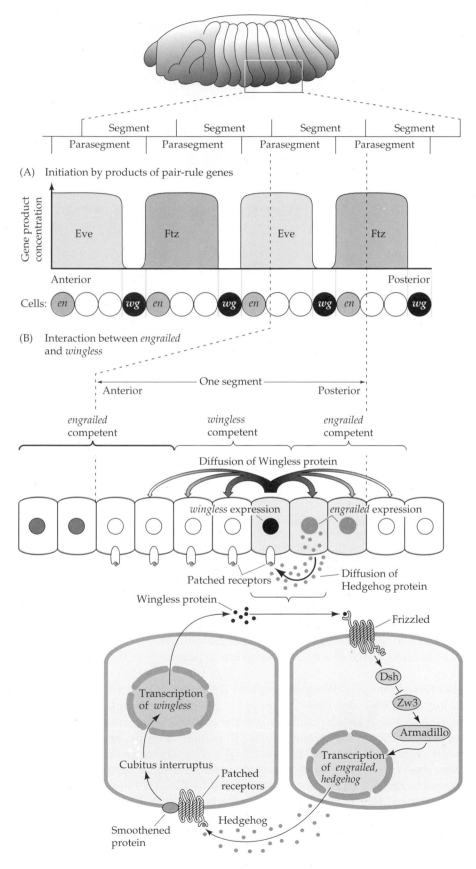

(A) Initiation by products of pair-rule genes

(B) Interaction between *engrailed* and *wingless*

Figure 14.25
Model for the transcription of segment polarity genes *engrailed* (*en*) and *wingless* (*wg*). (A) The initiation of *wg* and *en* expression is initiated by pair-rule genes. The *engrailed* gene is expressed when the cells contain high concentrations of either Even-skipped or Fushi tarazu proteins. The *wingless* gene is transcribed when neither *eve* or *ftz* genes are active, but a third gene (probably *odd-paired*) is expressed. (B) The continued expression of *wg* and *en* is maintained by interactions between the *engrailed*- and *wingless*-expressing cells. The Wingless protein is secreted and diffuses to the surrounding cells. In those cells competent to express *engrailed* (having Eve or Ftz proteins), Wingless protein is bound by the Frizzled receptor. This enables the activation of the *engrailed* gene. The Engrailed protein activates the transcription of the *hedgehog* gene and also activates its own (*engrailed*) gene transcription. Hedgehog protein diffuses from these cells and binds to the Patched protein. This binding prevents the Patched protein from inhibiting signaling from the Smoothened protein. The signal enables the transcription of the *wingless* gene and the subsequent secretion of the Wingless protein.

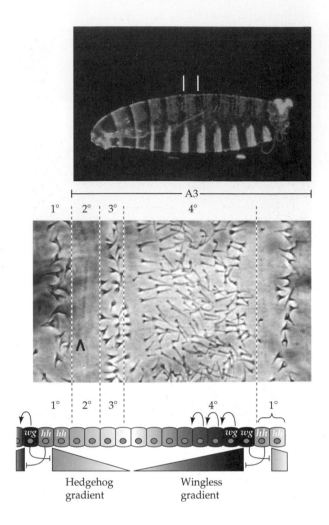

Figure 14.26
Cell specification by the Wingless/Hedgehog-signaling center. (A) Bright-field photograph of wild-type *Drosophila* embryo, showing the position of the third abdominal segment. (B) Close-up of the dorsal area of the A3 segment, showing the different cuticular structures made by the 1°, 2°, 3°, and 4° rows of cells. (C) Model for the role of Wingless and Hedgehog. Each signal is responsible for about half the pattern. Either each signal acts in a graded manner (shown here as gradients decreasing from their respective sources) to specify the fates of cells at a distance from these sources, or each signal might act locally on the neighboring cells, to initiate a cascade of inductions (shown here as sequential arrows). (After Heemskerk and DiNardo, 1994; photographs courtesy of the authors).

els of Hedgehog or Wingless protein (Heemskerek and DiNardo, 1994; Bokor and DiNardo, 1996; Porter et al, 1996). For example, if *hedgehog* is placed on a heat shock promoter and the embryos are grown at a temperature that activates the *hh* gene, more Hh protein is made, and the cells normally showing 3° fates will become 2° type cells. The rows of 4° cells farthest from the Wg-secreting cells can also become 3° or 2°. It seems that the cells closest to the Wg secreters cannot respond to Hh, and Hh cannot, by itself, specify the 1° fate. (This might require the expression of pair-rule gene products, especially Engrailed.) Thus, Hedgehog and Wingless appear necessary for elaborating the entire pattern of cell types across the parasegment. However, the mechanism by which they accomplish this specification is not clear. Either these signals act in a graded fashion, as morphogens, or they act locally to initiate a cascade of local signaling events, where each interaction uses a different ligand and receptor (Figure 14.26). The pattern of cell fates also changes the focus of patterning from parasegment to segment. One now has external markers, the *engrailed*-expressing cells becoming the most posterior cells of each segment.

The homeotic selector genes

Patterns of Homeotic Gene Expression

After the segmental boundaries have been established, the characteristic structures of each segment are specified. This specification is accomplished by the **homeotic selector genes** (Lewis, 1978). There are two regions of *Drosophila* chromosome 3 that contain most of these homeotic genes (Figure 14.27). One region, the **Antennapedia complex,** contains the homeotic genes *labial* (*lab*), *Antennapedia* (*Antp*), *Sex comb reduced* (*Scr*), *Deformed* (*Dfd*), and *proboscipedia* (*pb*). The *labial* and *Deformed* genes specify the head segments, while *Sex comb reduced* and *Antennapedia* contribute to giving the thoracic segments their identities. The *proboscipedia* gene appears to act only in adults, but in its absence, the labial palps of the mouth are transformed into legs (Wakimoto et al., 1984; Kaufman et al., 1990). The second region of homeotic genes is the **bithorax complex** (Lewis, 1978). There are three protein-coding genes found in this complex: *Ultrabithorax* (*Ubx*), which is required for the identity of the third thoracic segment; and *abdominal A* (*abdA*) and *Abdominal B* (*AbdB*), which are responsible for the segmental identities of the abdominal segments (Sánchez-Herrero et al., 1985). The lethal phenotype of the triple-point mutant *Ubx⁻, abdA⁻, AbdB⁻* is identical to that of a deletion of the entire bithorax complex (Casanova et al., 1987). The chromosome region containing both the Antennapedia complex and the bithorax complex is often referred to as the **homeotic complex (Hom-C).**

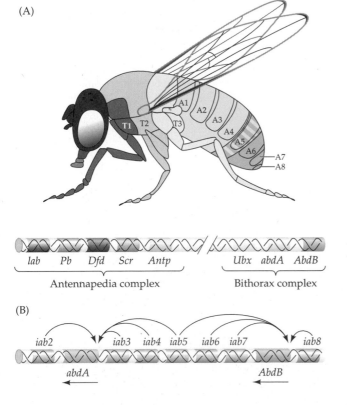

(A)

lab Pb Dfd Scr Antp   Ubx abdA AbdB

Antennapedia complex   Bithorax complex

(B)

iab2   iab3 iab4 iab5 iab6 iab7   iab8

abdA   AbdB

Figure 14.27
The functional domains of the bithorax complex and Antennapedia complex genes in *Drosophila*. (A) The bithorax complex has been divided into the three lethal complementation groups identified by E. B. Lewis. The Antennapedia complex genes are *labial* (*lab*), *Deformed* (*Dfd*), *Sex comb reduced* (*Scr*), and *Antennapedia* (*Antp*). (B) Summary of the control of the *abdA* and *AbdB* genes in *Drosophila*. The borders are controlled by the gap genes. The *infraabdominal* series of mutations controls the regulatory elements of these genes. (A after Dessain et al., 1992; B after Casares and Sánchez-Herrero, 1995.)

Figure 14.28
(A) Head of a wild-type fly. (B) Head of a fly containing the *Antennapedia* mutation that converts antennae into legs. (From Kaufman et al., 1990, courtesy of T. C. Kaufman.)

(A)

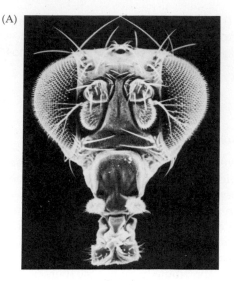

(B)

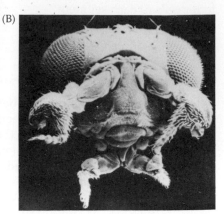

Because these genes are responsible for the specification of fly body parts, mutations in them lead to bizarre phenotypes. In 1894, William Bateson called these organisms "homeotic mutants," and they have fascinated developmental biologists for decades. The *Antennapedia* gene, for instance, is thought to specify the identity of the second thoracic segment. In the *dominant* mutation of *Antennapedia*, this gene is expressed in the head as well as in the thorax, and the imaginal discs of the head region are specified as thoracic. Thus, legs rather than antennae grow out of the head sockets (Figure 14.28). In the *recessive* mutant of *Antennapedia*, the gene fails to be expressed in the second thoracic segment, and antennae sprout out of the leg positions (Struhl, 1981; Frischer et al., 1986; Schneuwly et al., 1987). Likewise, when the *Ultrabithorax* gene is deleted, the third thoracic segment (which is characterized by halteres) becomes transformed into another *second* thoracic segment. The result (Figure 14.29) is a fly with four wings—an embarrassing situation for a classic dipteran.*

*Dipterans (two-winged insects such as flies) are thought to have evolved from normal four-winged insects; it is possible that this change arose via alterations in the bithorax complex. Chapter 23 includes more speculation on the relationship between bithorax genes and evolution.

Figure 14.29
This four-winged fruit fly was constructed by putting together three mutations in *cis* regulators of the *Ultrabithorax* gene. These mutations effectively transform the third thoracic segment into another second thoracic segment (i.e., halteres into wings). (Courtesy of E. B. Lewis.)

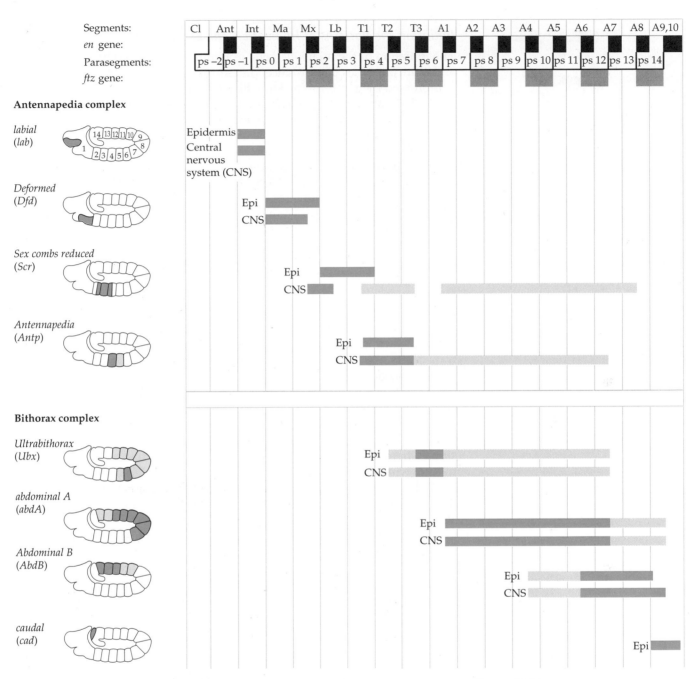

Figure 14.30
Regions of homeotic gene expression (both mRNA and protein) in the blastoderm of the *Drosophila* embryo and (a few hours later) in its central nervous system. The darker shaded areas are those segments or parasegments with the most product. The bars adjacent to the illustration represent gene expression within the parasegmental boundaries. (After Kaufman et al., 1990.)

These major homeotic selector genes have been cloned and their expression analyzed by in situ hybridization (Harding et al., 1985; Akam, 1987). The results of these experiments is summarized in Figure 14.30. Transcripts from each locus are detected in specific regions of the embryo and are especially prominent in the central nervous system. In homeotic mutants, this normal expression is altered. For instance, in the dominant *Antennapedia* alleles, the *Antennapedia* gene has been inverted on the chromosome, so that it has lost its own promoter and is under the control of a different promoter that is active in the head. This causes the ectopic expression of *Antp* in the head. Similarly, if the *Ultrabithorax* gene is placed on a new promoter and expressed in the head region, the antennae start making leg-specific structures and proteins (Mann and Hogness, 1990).

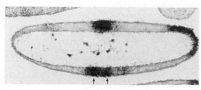

(A)

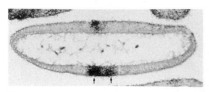

(B)

Figure 14.31
The initial expression of the homeotic gene *Antennapedia* (B) is predicated on the prior expression of *Krüppel* (A) in the same area. If the placement of *Krüppel* expression is altered, so is the expression of *Antennapedia*. (From Levine and Harding, 1989, courtesy of the authors.)

Initiating the Patterns of Homeotic Gene Expression

The initiation of the homeotic gene domains is influenced by the gap genes and pair-rule genes. For instance, the expression of the *abdA* and *AbdB* genes is repressed by gap proteins Hunchback and Krüppel. This inhibition prevents these abdomen-specifying genes from being active in the head and thorax (Casares and Sánchez-Herrero, 1995). Conversely, the *Ultrabithorax* gene is activated by certain levels of the Hunchback protein, so it is originally transcribed in a broad band in the middle of the embryo, and the Krüppel gap protein activates the transcription of *Antennapedia* (Figure 14.31; Harding and Levine, 1988; Struhl et al., 1992). The boundaries of homeotic gene expression are soon confined to parasegments as defined by the Fushi tarazu and Even-skipped proteins (Ingham and Martinez-Arias, 1986; Müller and Bienz, 1992).

Maintaining the Patterns of Homeotic Gene Expression

The expression of homeotic genes is a dynamic process. The *Antp* gene, for instance, although initially expressed in presumptive parasegment 4, soon appears in parasegment 5. As the germ band expands, *Antp* gene expression is seen in the presumptive neural tube as far posterior as parasegment 12. During further development, the pattern contracts again, and *Antp* transcripts are localized strongly to parasegments 4 and 5. Like other homeotic genes, *Antp* expression is negatively regulated by all the homeotic gene products posterior to it (Harding and Levine, 1989; González-Reyes and Morata, 1990). In other words, each of the bithorax complex genes represses the expression of *Antennapedia*. If *Ultrabithorax* is deleted, *Antp* activity extends through the region that would normally have expressed *Ubx* and stops where the *Abd* region begins. (This allows the third thoracic segment to form wings like the second thoracic segment, as in Figure 14.29.) If the entire bithorax complex is deleted, *Antp* expression extends throughout the abdomen. (The larva doesn't survive, but the cuticle pattern throughout the abdomen is that of the second thoracic segment.)

The gap proteins and the pair-rule proteins are transient, but the identities of the parasegments must be conserved so that specific differentiation can occur. Thus, once the transcription patterns of the homeotic genes have become stabilized, they are "locked" into place by alteration of the chromatin conformation in these genes. The repression of the homeotic genes appears to be maintained by the **Polycomb** family of proteins, while the active chromatin structure appears to be maintained by the **trithorax** proteins (Ingham and Whittle, 1980; McKeon and Brock, 1991; Simon et al., 1992).

REALISATOR GENES. The search is now on for the "realisator genes," genes that are the targets of the homeotic genes and that function to form the specified tissue or organ primordia. One method, pioneered in Walter Gehring's laboratory, has used "enhancer traps" to detect those genes regulated by *Antennapedia*. Here, a transposon containing a β-galactosidase reporter gene is linked to a weak promoter and is introduced randomly into the genome of different *Drosophila*. The expression of β-galactosidase (which can be readily detected by staining) becomes controlled by the enhancers in the vicinity of the promoter. If the enhancer is regulated by Antennapedia protein (which is present in the thoracic region of the embryo but not in the head), then β-galactosidase activity should be different when thoracic and head tissues are compared. Using this approach, Wagner-Bernholz and colleagues (1991) found what may be a critical gene that is regulated by *Antennapedia*. This gene, *salm*, is not active in leg imaginal discs from the thorax, but is ex-

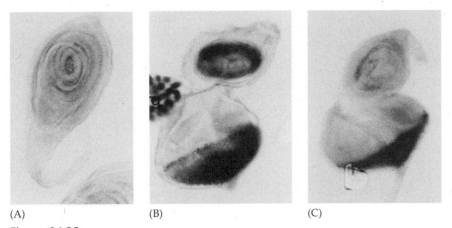

(A)   (B)   (C)

Figure 14.32
The "enhancer trap" transposon carries a β-galactosidase gene that becomes activated when placed near an enhancer. In one strain, the transposon became incorporated near a gene that was differentially regulated in the head and thorax. (A) Leg imaginal discs from wild-type larvae (at third-instar stage just prior to pupation) do not express a particular gene, *salm*. (B) The antennal discs from the same larva do express *salm*. (C) Antennal discs from an *Antennepedia* mutant show that this gene is repressed in this mutant. (From Wagner-Bernholz et al., 1991, courtesy of W. J. Gehring.)

pressed in the antennal imaginal disc (Figure 14.32). Thus, *salm* appears to be a gene that is repressed by the Antennapedia protein. The repression of the *salm* gene may be critical for the formation of leg tissue, rather than antennal tissue, from the thoracic imaginal discs.

Another approach to finding such genes has been sequencing. Gene sequencing has shown that some genes have enhancer elements that bind homeotic genes, thereby causing them to be regulated by homeotic gene expression patterns. One target gene, *decapentaplegic*, has a binding site for the Ultrabithorax protein in its enhancer. This enables the Decapentaplegic protein to be expressed in the visceral mesoderm of parasegment 7, where it is needed for the development of the midgut (Immergluck et al., 1990; Panganiban et al., 1990).

Another target of the homeotic proteins, the *Distal-less* gene (itself a homeobox-containing gene) is necessary for limb development and is active solely in the thorax. *Distal-less* expression is repressed in the abdomen, probably by a combination of Ubx and AbdA proteins that can bind to its enhancer and block transcription (Vachon et al., 1992; Castelli-Gair and Akam, 1995). This provides a paradox, since parasegment 5 (entirely thoracic and leg-producing) and parasegment 6 (which includes most of the legless first abdominal segement) both express *Ultrabithorax*. How can the two very different segments be specified by the same gene? Castelli-Gair and Akam (1995) have shown that the mere presence of Ubx protein in a group of cells is not sufficient for specification. Rather, the time and place of its expression within the parasegment can be critical. Before *Ubx* expression, parasegments 4–6 have similar potentials. At stage 10, *Ubx* expression in the anterior parts of parasegments 5 and 6 prevent those parasegments from forming structures (such as the anterior spiracle), characteristic of parasegment 4 . Moreover, in the posterior compartment of parasegment 6 (but not parasegment 5), Ultrabithorax protein blocks the formation of the limb primordium by repressing the *Distal-less* genes. At stage 11, by the time Ubx has pervaded all of parasegment 6, the *Distal-less* gene has become self-regulatory and cannot be repressed by Ultrabithorax (Figure 14.33).

Figure 14.33
Schematic representation of the differences between *Ubx* expression in parasegments 5 and 6. (A) Before *Ubx* expression, each parasegment is competent to make spiracles and legs. (B) At stage 10, early *Ubx* expression (shaded) blocks the formation of the anterior spiracle in PS5 and PS6, and it prevents limb formation in the posterior compartment of PS6. AbdA protein provides the same role in other abdominal segments. (C) At stage 11, the *Ubx* expression domain extends to the limb primordium of PS5 and PS6, but it is "too late" to repress *Distal-less* gene expression. (After Castelli-Gair and Akam, 1995.)

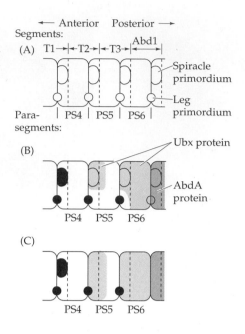

Cis-*Regulatory Elements and the Bithorax Complex*

The temporal and spatial differences in *Ubx* expression between parasegments 5 and 6 suggest that *Ubx* is regulated by different regulatory elements. Lewis and his colleagues (Lewis, 1978, 1985; Bender et al., 1983; Karch et al., 1985) have identified these *cis*-regulatory regions. At first (before the three genes of the bithorax complex were identified), these regions were thought to encode specific proteins. Now, it is known that they regulate the transcription of one of the three bithorax complex genes in specific parasegments. For instance, the *anterobithorax* (*abx*) and *bithorax* (*bx*) mutants cause the anterior compartment of the third thoracic segment (anterior balancers) to assume the identity of the anterior compartment of the second thoracic segment (anterior wings). Similarly, the *posterobithorax* (*pbx*) and *bithoraxoid* (*bxd*) mutants cause the posterior compartment of the third thoracic segment to resemble that of the second thoracic segment. The combination of *abx*, *pbx*, and *bxd* mutations in a single embryo causes the entire transformation of the third thoracic segment into another second thoracic segment. (The result is the fly shown in Figure 14.29.) Whereas these mutations were originally considered to be on separate genes, it now appears that they are mutations of enhancer elements that enable the position-specific expression of the *Ubx* gene (Lewis, 1985; Peifer et al., 1987).

The relationship between the *cis*-regulatory mutations and the three transcription units of the bithorax complex is shown in Figure 14.34. The protein-encoding regions of the bithorax complex take up less than one-tenth of the DNA in this complex. The regulatory mutations generally map to the flanking regions of these three genes or to introns within the genes. Further evidence that *abx*, *bx*, and *bxd* are *cis*-regulatory elements comes from analyzing specific mutations and deletions. The deletion of the *Ubx* gene results in the homeotic transformation of parasegment 5 (posterior T2 and anterior T3) and parasegment 6 (posterior T3 and anterior A1) into copies of parasegment 4 (posterior T1 and anterior T2). Such a transformation is lethal; the embryo dies before hatching. In *abx* and *bx* mutants, how-

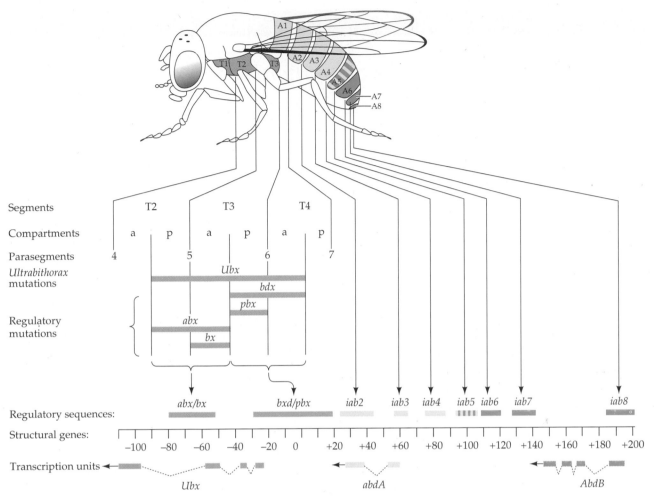

Figure 14.34
Regulatory mutations in the bithorax complex. The schematic adult fly is divided
into segments and anterior and posterior compartments. The regulatory regions of
the *Ultrabithorax* gene are shown below the fly. The shaded areas represent the
region specified by the particular regulatory domain. The continuous line below
that represents the 300,000-base-pair region of the complex. The three transcription
units that encode the three homeotic proteins of the bithorax complex are shown in
relation to the regulatory loci. Each of these genes is transcribed right to left. The
exons are shown as shaded boxes, the introns as dashed lines. Above the line are
the regulatory sequences defined by genetic mutations, and the color of the lines
corresponds to the gene that the sequence positively regulates. (After Peifer et al.,
1987; Beachy, 1990; Casares and Sánchez-Herrero, 1995.)

ever, only parasegment 5 is transformed into parasegment 4, as *Ubx* expression is reduced in parasegment 5 (Casanova et al., 1985; Peifer and Bender,
1986). Hence, the anterior wing emerges in what would otherwise have been
an anterior haltere. Similarly, the *bxd* mutations reduce *Ubx* expression in
parasegment 6 (Peifer et al., 1987). The *bithorax* regulatory element for Ubx
contains an enhancer that binds the proteins encoded by the *tailless*, *fushi
tarazu*, and *hunchback* segmentation genes (Qian et al., 1991). In the abdominal region, the *cis*-regulatory sequences *intra-abdominal* (*iab*) 2–8 direct the
expression of *abdA* or *AbdB* in the various segments (Boulet et al., 1991;
Sánchez-Herrero, 1991).

Molecular Regulation of Development: The Homeodomain Proteins

The Homeodomain

Homeodomain proteins are a family of transcription factors characterized by a 60-amino acid domain that binds to certain regions of DNA. The homeodomain was first seen in those proteins whose absence or misregulation causes homeotic transformations of *Drosophila* segments. It is thought that homeodomain proteins activate batteries of genes that specify the particular properties of that segment. Such homeodomain-containing proteins include the products of the eight homeotic genes of the homeotic complex, as well as other proteins, such as Fushi tarazu, Caudal, and Bicoid. Homeodomain transcription factors are important in determining the anterior-posterior axes of both invertebrates and vertebrates. In *Drosophila*, the presence of certain homeodomain-containing proteins is also necessary for the determination of specific neurons. Without these transcription factors, the fates of these neuronal cells are altered (Doe et al., 1988).

The homeodomain is encoded by the 180-base-pair **homeobox** (see Chapter 10). The homeodomains appear to specify the binding sites for these proteins and are critical in specifying cell fate. For instance, if a chimeric protein is con-structed mostly of Antennapedia but with the carboxy terminus (including the homeodomain) of Ultrabithorax, the protein can substitute for Ultrabithorax and specify the appropriate cells as parasegment 6 (Mann and Hogness, 1990). The isolated homeodomain of Antennapedia will bind to the same promoters as the entire Antennapedia protein, indicating that the binding of this protein is dependent on its home-odomain (Müller et al., 1988).

The homeodomain folds into three α helices, the latter two folding into a helix-turn-helix conformation that is characteristic of a family of transcription factors that bind DNA in the major groove of the double helix (Otting et al., 1990; Percival-Smith et al., 1990). The third helix is the recognition helix, and it is here that the amino acids make contact with the bases of the DNA. A four-base motif, TAAT, is conserved in nearly all sites recognized by homeodomains; it probably distinguishes those sites to which homeodomain proteins might bind. The 5' terminal T appears to be critical in this recognition, as mutating it destroys all homeodomain binding. The base pairs following the TAAT motif are important in distinguishing between similar recognition sites. For instance, the next base pair is recognized by the amino acid 9 within the recognition helix. Mutation studies have shown that Bicoid and Antennapedia homeodomain proteins use lysine or glutamine, respectively, at position 9 to distinguish related recognition sites. The lysine of the Bicoid homeodomain recognizes the G of CG pairs, while the glutamine of the Antennapedia homeodomain recognizes the A of an AT pair (Figure 14.35; Hanes and Brent, 1991). If that lysine is replaced by glutamine, a Bicoid protein will recognize Antennapedia-binding sites (Hanes and Brent, 1989, 1991). Other homeodomain proteins show a similar pattern, where one portion of the homeodomain recognizes the common sequence, while another portion recognizes a specific structure close to the TAAT.

Figure 14.35 Homeodomain-DNA interactions. (A) Homeodomain helix-turn-helix sequence within major groove of the DNA. (B) Proposed pairing between the Bicoid homeodomain lysine and the CG base pair of the recognition sequence, and between the glutamine of the Antennapedia homeodomain and the TA base pair of its recognition sequence. In both cases, the ninth amino acid of the helix bonds with the base pair immediately following the TAAT sequence. (A after Riddihough, 1992; B after Hanes and Brent, 1991.)

(A)

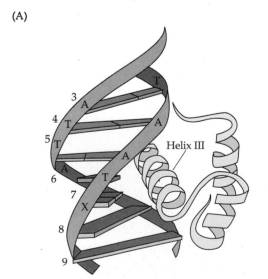

(B)

bicoid site, base pair 7

Cytosine   Guanine

Antp site, base pair 7

Thymine   Adenine

Cofactors for the Hom-C Genes

The homeotic genes of the *Drosophila* homeotic complex specify segmental fate, but they may need some help in doing it. The DNA-binding sites recognized by the homeodomains of the Hom-C proteins are very similar, and there is some overlap in their binding specificity. In 1990, Peifer and Wieschaus discovered that the product of the *Extradenticle* (*Exd*) gene interacts with several Hom-C proteins and may help specify the segmental identities. For instance, the Ubx protein is responsible for specifying the identity of the first abdominal segment (A1). Without Extradenticle protein, it will transform this segment into A3. Moreover, the Exd and Ubx proteins are both needed for the regulation of *decapentaplegic*, and the structure of the *decapentaplegic* promotor suggests that the Extradenticle protein may dimerize with the Ubx protein on the enhancer of this target gene (Raskolb and Wieschaus, 1994; van Dyke and Murre, 1994). The Extradenticle protein includes a homeodomain, and the human protein PBX1 resembles the Extradenticle protein and may play a similar role as a cofactor for human homeotic genes.

The product of the *teashirt* gene may also be an important cofactor. This zinc finger transcription factor is necessary for the functioning of the Scr product to distinguish between labial and first thoracic segments. It is critical for the specification of the anterior prothoracic (parasegment 3) identity, and it may be the gene that specifies the "groundstate condition" of the homeotic complex. If the bithorax complex and the *Antennapedia* gene are removed, all the segments become anterior prothorax. The function of the *teashirt* gene appears to be critical for working with Scr protein in distinguishing thorax from head and working throughout the trunk to prevent head structures from forming (Roder et al., 1992). [droso2.html] ∎

THE GENERATION OF DORSAL-VENTRAL POLARITY IN DROSOPHILA

In 1936, embryologist E. E. Just criticized those geneticists who sought to explain development by looking at specific mutations affecting eye color, bristle number, and wing shape. He said that he wasn't interested in the development of the bristles of a fly's back; rather, he wanted to know how the fly embryo makes the back itself. Fifty years later, embryologists and geneticists are finally answering that question.*

Dorsal protein: Morphogen for dorsal-ventral polarity

Dorsal-ventral polarity is established by the gradient of another transcription factor protein, **Dorsal**. Unlike Bicoid, whose gradient is established within a syncytium, the Dorsal gradient forms over a field of cells and is established as a consequence of cell-signaling events.

The specification of the dorsal-ventral axis can be divided into several steps. The critical step is the translocation of the Dorsal protein from the cytoplasm into the nuclei of the *ventral* cells during the fourteenth division cycle. Anderson and Nüsslein-Volhard (1984) have isolated 11 maternal effect genes, each of whose absence is associated with a lack of ventral structures (Figure 14.36). In addition, the absence of another maternal effect gene, *cactus*, causes the ventralization of all cells. The proteins encoded by these maternal genes are critical for making certain that the Dorsal protein only gets into the nuclei on the ventral surface of the embryo. The steps after the translocation of the Dorsal protein concern what the Dorsal protein does to specify the different regions of the embryo. Here, different concentrations of Dorsal protein in the nuclei appear to specify different fates in these cells.

Translocation of Dorsal Protein

The actual protein that distinguishes dorsum from ventrum is the protein product of the *dorsal* gene. The RNA from the mother's *dorsal* genes is put

---

*In a manner that Just could not have predicted, it turns out that some of the genes (such as *decapentaplegic*) that are involved in regulating bristle number or wing shape also have earlier functions regulating dorsoventral polarity.

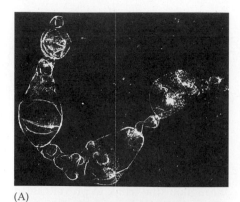

(A)

(B)

Figure 14.36
Rescue of larva by injection of wild-type mRNA into eggs destined to have the *snake* phenotype. (A) Deformed larva consisting entirely of dorsal cells. Larvae like these developed from eggs of a female homozygous for the *snake* allele. (B) Wild-type appearance of such larvae developing from *snake* eggs that received injections of mRNA from wild-type eggs. (From Anderson and Nüsslein-Volhard, 1984, courtesy of C. Nüsslein-Volhard.)

into the egg by the mother fly's ovarian cells. However, Dorsal protein is not synthesized from its maternal message until about 90 minutes after fertilization. When this protein is translated, it is found *throughout* the embryo, not just on the ventral or dorsal side. How, then, can this protein act as a morphogen if it is located everywhere in the embryo? In 1989, the surprising answer was found (Roth et al., 1989; Rushlow et al., 1989; Steward, 1989). While Dorsal protein can be found throughout the syncytial blastoderm of the early *Drosophila* embryo, it is transported into cell nuclei only in the *ventral* part of the embryo (Figure 14.37A,B). Here, Dorsal protein binds to certain nuclear genes to activate or suppress their transcription. If Dorsal protein does not enter the nucleus, the ventralizing genes (*snail* and *twisted*) are not transcribed, the dorsalizing genes (*decapentaplegic* and *zerknüllt*) are not repressed, and all the cells of the embryo are specified as dorsal cells. This hypothesis that the dorsal-ventral axis of *Drosophila* is specified by the selective transport of the Dorsal morphogen protein into the nucleus is strengthened by the analysis of mutations having an entirely dorsalized or an entirely ventralized phenotype (Figure 14.37C,D). In those mutants where all the cells are dorsalized (as is evident by their dorsal cuticle), Dorsal protein does not enter the nucleus in any cell. Conversely, in those mutants where all cells have a ventral phenotype, Dorsal protein is found in every cell nucleus.

Providing the asymmetrical signal for Dorsal protein translocation

Signal from the Oocyte Nucleus to the Follicle Cells

If Dorsal protein is found throughout the embryo but gets translocated into the nucleus of only ventral cells, something else must be providing the asymmetrical cues (Figure 14.38). It appears that the signal is mediated through a complex interaction between the oocyte and its surrounding follicle cells. The follicular epithelium surrounding the developing oocyte is ini-

Figure 14.37
Inclusion of Dorsal protein into the nuclei of ventral, but not lateral or dorsal, nuclei. (A) Fate map through the center of the *Drosophila* embryo. The most ventral part becomes the mesoderm; the next higher portion becomes the neurogenic (ventral) ectoderm. The lateral and epidermal ectoderm can be distinguished in the cuticle, and the dorsalmost region becomes the amnioserosa, the extraembryonic layer that surrounds the embryo. (B–D) Cross sections of embryos stained with antibody to show the presence of Dorsal protein. In all cases, the dark stain represents the Dorsal protein. (B) A wild-type embryo, showing Dorsal protein in the ventralmost nuclei. (C) A dorsalized mutant, showing no localization of Dorsal protein in any nucleus. (D) A ventralized mutant; Dorsal protein has entered the nucleus of every cell. (A from Rushlow et al., 1989; B–D from Roth et al., 1989, courtesy of the authors.)

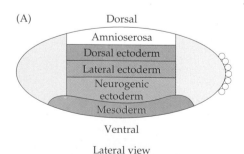

(A)

Dorsal

Amnioserosa
Dorsal ectoderm
Lateral ectoderm
Neurogenic ectoderm
Mesoderm

Ventral

Lateral view

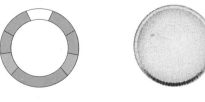

Transverse section

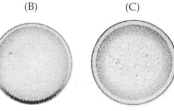

(B) (C)

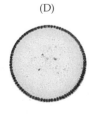

(D)

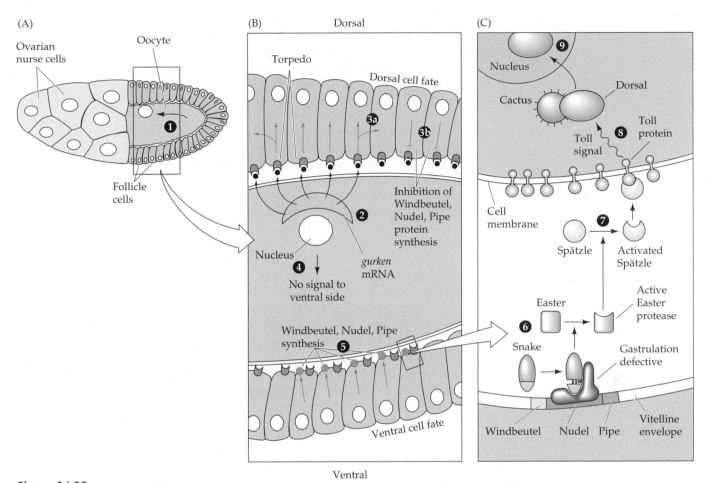

Figure 14.38
Schematic representation of a model for the generation of the dorsal-ventral polarity in *Drosophila*. (A) The oocyte develops an ovarian follicle consisting of 15 nurse cells (which supply maternal proteins and messages to the developing egg) and follicle cells. (B) The nucleus of the oocyte resides in what will become the dorsal side. The *cornichon* and *gurken* genes of the oocyte synthesize a signal that is received by the receptor made by the *torpedo* gene of the follicle cells. Given the short diffusibility of the signal, only the follicle cells closest to the oocyte nucleus (i.e., the dorsal follicle cells) receive this signal. The signal from the Torpedo receptor causes the follicle cells to differentiate a characteristic dorsal follicle morphology and (somehow) inhibits the synthesis of Windbeutel, Nudel, and Pipe proteins. Therefore, these proteins are made only by the ventral follicle cells. (C) The three ventral follicle proteins are thought to become incorporated into the vitelline membrane, but only on the ventral side. They split the products of the *snake* and *gastrulation defective* genes to create an active enzyme that will split the zymogen form of the Easter protein into an active Easter protease. The Easter protease splits the Spätzle protein into a form that can bind to the Toll receptor (which is found throughout the cell membrane). Thus, only the ventral side receives the Toll signal. This signal separates the Cactus protein from the Dorsal protein, allowing Dorsal protein to be translocated into the nucleus. The Dorsal protein enters the nucleus and ventralizes the cells. (After Schüpbach et al., 1991; Roth, 1994; Hong and Hashimoto, 1995.)

1. Oocyte nucleus travels to anterior dorsal side of oocyte. It collects *cornichon* and *gurken* mRNA

2. *cornichon* and *gurken* messages translated. The Gurken protein is received by Torpedo proteins during mid-oogenesis

3a. Torpedo signal causes follicle cells to differentiate to a dorsal morphology

3b. Synthesis of Windbeutel, Nudel, and Pipe proteins inhibited in dorsal follicle cells

4. Cornichon and Gurken proteins do not diffuse to ventral side

5. Ventral follicle cells synthesize Windbeutel, Nudel, and Pipe proteins

6. Ventral follicle proteins absorb Snake and Gastrulation-defective proteins to effect splitting of Easter zymogen, making active Easter protease only on ventral side

7. Easter splits Spätzle, which binds to Toll receptor protein

8. Toll signal causes phosphorylation and degradation of Cactus protein, releasing it from Dorsal.

9. Dorsal protein enters the nucleus and ventralizes the cell

tially symmetrical, but this symmetry is broken by a signal from the oocyte nucleus. As mentioned earlier in the chapter, the oocyte nucleus is originally located at the posterior end of the oocyte, away from the nurse cells. It then becomes translocated anteriorly, beneath one cortical surface of the oocyte, to an anterior dorsal position, along a track of microtubules. The oocyte nucleus then signals the overlying follicle cells and dorsalizes them (Montell et al., 1991; Schüpbach et al., 1991). The follicle cells above the nucleus assume a more columnar shape than the other follicle cells. These differences in shape and packing become accentuated as the egg matures, eventually distinguishing dorsal and ventral follicle cells. The dorsalizing signal from the oocyte nucleus appears to be made by the products of the *gurken* and *cornichon* genes (Schüpbach, 1987; Forlani et al., 1993). Mutations of these genes in the oocyte cause the ventralization of both the embryo and its surrounding follicle cells. (If the mutation is in the follicle cells and not in the egg, the embryo is normal.)

The dorsalizing signal appears to be received by the follicle cells through a receptor encoded by the *torpedo* gene. Molecular analysis has now established that *gurken* encodes a homologue of epidermal growth factor (EGF), while *torpedo* encodes a homologue of the vertebrate EGF receptor (Price et al., 1989; Neuman-Silberberg and Schüpbach, 1993). Maternal deficiency of *torpedo* causes the ventralization of the embryo. Moreover, the *torpedo* gene is active in the ovarian follicle cells, not in the embryo. This was discovered by making germ line/somatic chimeras. Schüpbach (1987) transplanted germ cell precursors from wild-type embryos to embryos whose mothers carried the *torpedo* mutation. Conversely, she transplanted those germ cells from the *torpedo* embryos to wild-type embryos (Figure 14.39). The results were surprising in that the wild-type eggs produced ventralized embryos when these eggs had developed within *torpedo* mutant follicles. The *torpedo* mutant eggs were able to produce normal embryos if the eggs developed within a wild-type ovary. Thus, unlike *gurken* and *cornichon* gene products, the wild-type torpedo gene is needed in the follicle cells, not in the egg itself.

Signal from the Follicle Cells to the Oocyte Cytoplasm

The genes *nudel (nd)*, *pipe (pip)*, and *windbeutel (wind)* are also needed in the follicle cell and not in the oocyte. If the mother lacks any of these three

Figure 14.39
Germ line chimeras made by interchanging pole cells (germ cell precursors) between wild-type embryos and embryos from mothers homozygous for the *torpedo* gene. These transplants produce wild-type females whose embryos come from eggs from the mutant mothers, and *torpedo*-deficient embryos with wild-type eggs. The eggs from the *torpedo*-deficient mothers produce normal embryos if the eggs develop in the wild-type ovary, while the wild-type eggs produce ventralized embryos if the eggs develop in the mutant ovary.

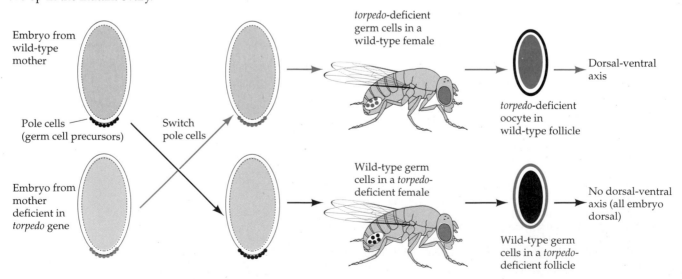

genes, the embryo forms a totally dorsalized phenotype. These genes are turned off by the activation of the Torpedo receptor (Stein et al., 1991). If they are allowed to be active (as is normally the case in the ventral follicle cells), their proteins are thought to be incorporated into the ventral portion of the vitelline envelope that is secreted around the egg by the follicle cells (Hecht and Anderson, 1992; Stein and Nüsslein-Volhard, 1992). In this way, an asymmetrical signal is now present on the envelope adjacent to the egg and separated from the egg by a perivitelline fluid. However, these proteins are not sufficient to create the signal for the translocation of Dorsal protein into the nucleus. Once again, we turn to the oocyte (now an embryo) to supply essential components that will generate the ventral signal from the follicle cells to the embryo.

The complex formed by the Nudel, Pipe, and Windbeutel proteins is thought to activate three serine proteases that are secreted by the embryo into the perivitelline fluid (see Figure 14.38; Hong and Hashimoto, 1995). These three serine proteases are the products of the *gastrulation defective* (*gd*), *snake* (*snk*), and *easter* (*ea*) genes. As with most extracellular proteases, they are secreted in an inactive form and become activated by peptide cleavage. It is thought that the Nudel-Pipe-Windbeutel complex first tethers and activates the Gastrulation-defective protein. This protein is a protease and cleaves the Snake protein. This cleavage activates the protease activity of the Snake protein, and the activated Snake protein cleaves the Easter protein. This cleavage activates the protease activity of the Easter protein, which cleaves the Spätzle protein (Chasan et al., 1992; Hong and Hashimoto, 1995).

Cleaved Spätzle protein is now able to bind to a receptor in the oocyte cell membrane, the product of the *Toll* gene. Toll protein is also a maternal product evenly distributed throughout the cell membrane of the egg (Hashimoto et al., 1988, 1991). The recessive mutation of *Toll* has a similar dorsalized phenotype, and injections of RNA from wild-type eggs will restore the dorsal-ventral polarity of eggs laid by *Toll⁻/Toll⁻* mothers. But unlike the case of *snake* or the other 10 maternal genes, the site of the injection is important. Whichever part of the egg is injected becomes the ventral region of the rescued embryo (Anderson et al., 1985). This suggests that the *Toll⁻/Toll⁻* eggs lack a dorsal-ventral axis (whereas in *snake*, the ventral region occurs in its normal place). In normal development, the Toll receptor is located throughout the oocyte cell membrane, but it becomes activated only by binding the Spätzle protein, which is produced on the ventral side of the egg. In this way, the Toll receptors on the ventral side of the egg are transducing a signal into the egg, while the Toll receptors on the dorsal side of the egg are not.

The Establishment of the Dorsal Protein Gradient

SEPARATION OF THE DORSAL AND CACTUS PROTEINS. The crucial outcome of signaling through the Toll receptor is the establishment of a gradient of Dorsal protein. How is this gradient established? It appears that the Cactus protein is sitting on the portion of the Dorsal protein that enables the Dorsal protein to get into nuclei. As long as this Cactus protein is bound to it, Dorsal protein remains in the cytoplasm. Thus, this entire complex signaling system is organized to split the Cactus protein from the Dorsal protein in the ventral part of the egg. When Spätzle binds to and activates the Toll protein, the Toll protein can activate the **Pelle** protein kinase. (The Tube protein is probably necessary for bringing Pelle up to the cell membrane, where it can be activated; Galindo et al., 1995.) The Pelle protein kinase can then phosphorylate the Cactus protein. Once phosphorylated, the Cactus protein is degraded, and the Dorsal protein can enter the nucleus (Kidd, 1992; Shelton and

Figure 14.40
Model of a conserved pathway for regulating nuclear transport of transcription factors in *Drosophila* and mammals. (A) In *Drosophila*, the Toll protein binds the signal from the Spätzle protein and activates the kinase region of the Pelle protein. The Pelle protein phosphorylates Cactus and Dorsal, causing the two proteins to separate from eachother. The Dorsal protein can then enter the nucleus and regulate the transcription of ventrally specific genes. (B) In mammalian lymphocytes, the IL-1 receptor can cause the phosphorylation of IκB (through a protein kinase that has not yet been identified). This enables the NF-κB protein to enter the nucleus and effect the transcription of several lymphocyte-specific genes. (After Shelton and Wasserman, 1993.)

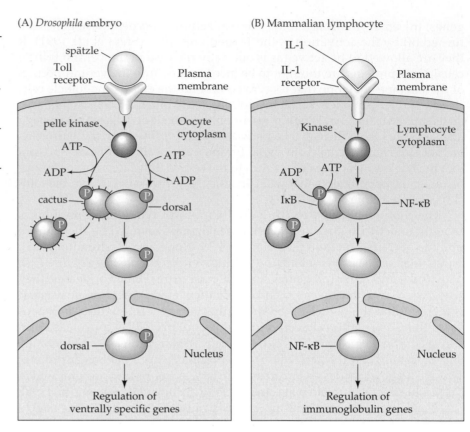

(A) *Drosophila* embryo

spätzle
Toll receptor
Plasma membrane
pelle kinase
Oocyte cytoplasm
ATP
ATP
ADP
ADP
cactus
dorsal
dorsal
Nucleus
Regulation of ventrally specific genes

(B) Mammalian lymphocyte

IL-1
IL-1 receptor
Plasma membrane
Kinase
Lymphocyte cytoplasm
ADP
ATP
IκB
NF-κB
NF-κB
Nucleus
Regulation of immunoglobulin genes

Wasserman, 1993; Whalen and Steward, 1993; Reach et al., 1996). The result is a gradient of Dorsal localization into the ventral cells of the embryo, with the highest concentrations of Dorsal protein in the most ventral cell nuclei.

The process described for the translocation of Dorsal protein into the nucleus is very similar to the process described in Chapter 10 for the translocation of the NF-κB transcription factor into the nucleus of mammalian lymphocytes. In fact, there is substantial homology between the NF-κB and Dorsal, between IκB and Cactus, between the Toll protein and the interleukin 1 receptor, between Pelle protein and an IL-1-associated protein kinase, and between the DNA sequences recognized by Dorsal and NF-κB (González-Crespo and Levine, 1994; Cao et al, 1996). Thus, the biochemical pathway used to specify the dorsal-ventral polarity in *Drosophila* appears to be the same as that used to differentiate lymphocytes in mammals (Figure 14.40).*

READOUT OF THE DORSAL PROTEIN GRADIENT. What does the Dorsal protein do once it is located in the nuclei of the ventral cells? Looking at the fate map of the cross section through the middle of the *Drosophila* embryo at the fourteenth division cycle (see Figure 14.37), it is obvious that the 16 cells with the highest concentration of Dorsal protein are those that generate the mesoderm. The next cell up from this region generates the specialized glial and neural cells of the midline. The next two cells are those that give rise to the ventral epidermis and ventral nerve cord, while the nine cells above them produce the dorsal epidermis. The most dorsal group of six cells do not di-

---

* Lemaitre and colleagues (1996) have shown that Toll and its ligand (Spätzle) are also involved in the *Drosophila* immune response against fungal infections.

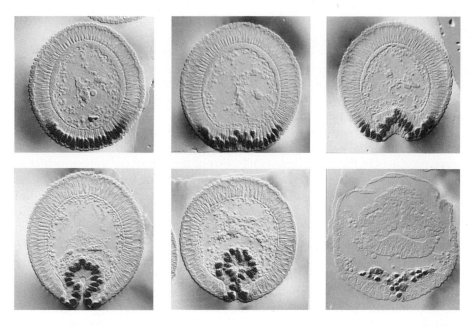

Figure 14.41
Gastrulation in *Drosophila*. In this cross-section, the mesodermal cells at the ventral portion of the embryo buckle inward, forming a tube that then flattens and generates the mesodermal organs. The nuclei are stained with antibody to the Twist protein. (From Leptin, 1991b, courtesy of M. Leptin.)

vide; they generate the amnioserosal covering of the embryo (Ferguson and Anderson, 1991).

This fate map is generated by the gradient of Dorsal protein in the nuclei. Large amounts specify the cells to be mesoderm, while lesser amounts specify the cells to be glial or ectodermal tissue (Jiang and Levine, 1993). The first morphogenetic event of *Drosophila* gastrulation is the invagination of the 16 ventralmost cells of the embryo (Figure 14.41). All the mesodermal derivatives of the body muscles, fat bodies, and gonads derive from these cells (Foe, 1989). The Dorsal protein specifies these cells to become mesoderm in two ways. First, Dorsal protein can activate specific genes that create the mesodermal phenotype. Three of the target genes for Dorsal protein are *twist, snail,* and *rhomboid* (Figure 14.42). These genes are transcribed only in the ventral cell nuclei that have received high concentrations of the Dorsal protein, since their enhancers do not bind Dorsal protein with a very high affinity (Thisse et al., 1988, 1991; Jiang, et al., 1991; Pan et al., 1991). The Twist protein *activates* mesodermal genes, while the Snail protein *represses*

Figure 14.42
Subdivision of the dorsal-ventral axis by the gradient of Dorsal protein in the nuclei. The Dorsal protein activates the zygotic genes *rhomboid, twist,* and *snail,* depending on its nuclear concentration. The Snail protein, formed most ventrally, inhibits the transcription of Rhomboid protein. The Dorsal protein inhibits the expression of *tolloid, decapentaplegic,* and *zerknüllt* in the ventral region. Differing concentrations of Zerknüllt protein determine the fates of the dorsal cells. (After Steward and Govind, 1993.)

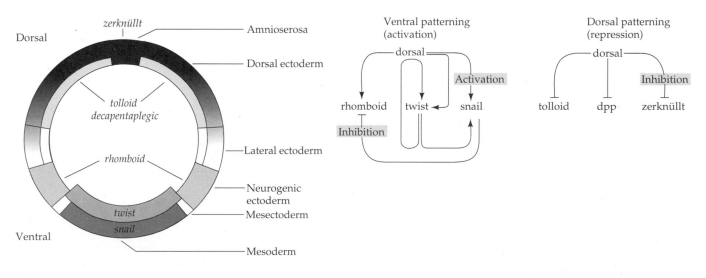

particular nonmesodermal genes that might otherwise be active. The *rhomboid* gene is interesting because it is activated by Dorsal but repressed by Snail. Thus, *rhomboid* expression is not found in the most ventral cells (i.e., the mesodermal precursors), but is expressed in the cells adjacent to the mesoderm that form the presumptive neuroectoderm (see Figure 14.42; Jiang and Levine, 1993). Both *snail* and *twist* are needed to get the entire mesodermal phenotype and proper gastrulation (Leptin et al., 1991a). The sharp border between the mesodermal cells and those cells adjacent to them that generate glial cells is produced by the presence of the *snail* and *twist* gene products in the ventralmost cells, but of only the *twist* gene product in the next dorsal cell (Kosman et al., 1991). In mutants of *snail*, the ventralmost cells still have the *twist* gene activated, and they resemble the more lateral cells (Nambu et al., 1990).

The Dorsal protein also determines the mesoderm directly. In addition to activating the mesoderm-stimulating genes (*twist* and *snail*), it directly inhibits the dorsalizing genes, *zerknüllt* (*zen*) and *decapentaplegic* (*dpp*). Thus, in the same cells, the Dorsal protein can act as an activator of some genes and a repressor of other genes. Whether the Dorsal protein functions to activate or repress depends in the structure of the genes' enhancers. The *zen* enhancer contains a binding site for a protein called DSP1 ("dorsal switch protein 1"). This protein is found throughout the embryo. When Dorsal protein is absent, it does not appear to have any effect on transcription. However, when

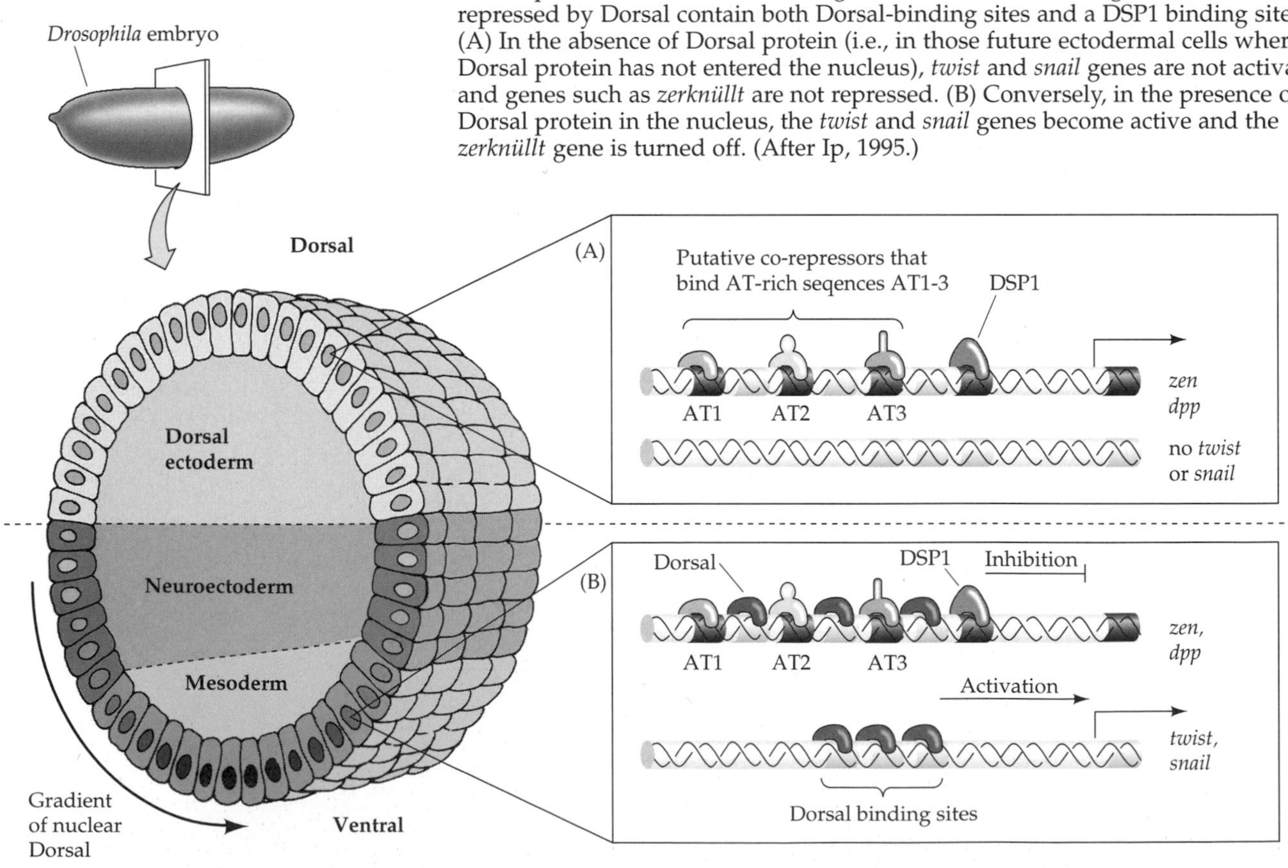

Figure 14.43
Activation and repression by the Dorsal protein. Enhancer in a gene activated by Dorsal protein (such as *twist* or *snail*) has multiple low-affinity binding sites for the Dorsal protein and no DSP1-binding site. Enhancers in those genes that are repressed by Dorsal contain both Dorsal-binding sites and a DSP1 binding site. (A) In the absence of Dorsal protein (i.e., in those future ectodermal cells where the Dorsal protein has not entered the nucleus), *twist* and *snail* genes are not activated, and genes such as *zerknüllt* are not repressed. (B) Conversely, in the presence of Dorsal protein in the nucleus, the *twist* and *snail* genes become active and the *zerknüllt* gene is turned off. (After Ip, 1995.)

Dorsal is also present at the enhancer site, it converts the activating function of Dorsal into a repressing function (Figure 14.43; Lehming et al., 1994; Ip, 1995). Mutants of *dorsal* express *dpp* and *zen* genes throughout the embryo (Rushlow et al., 1987), and embryos deficient in *dpp* and *zen* fail to form dorsal structures (Irish and Gelbart, 1987). Thus, in wild-type embryos, the mesodermal precursors express *twist* and *snail* (but not *zen* or *dpp*); precursors of the dorsal epidermis and amnioserosa express *zen* and *dpp* but not *twist* or *snail*; glial (mesectoderm) precursors express only *snail*; while the lateral neuroectodermal precursors do not express any of these four genes (Kosman et al., 1991; Ray et al., 1991). Thus, as a consequence of the responses to the Dorsal protein gradient, the axis becomes subdivided into mesoderm, mesectoderm, neurogenic ectoderm, epidermis, and amnioserosa. [droso3.html]

■ *AXES AND ORGAN PRIMORDIA*

The cartesian coordinate model and the specification of organ primordia

The anterior-posterior and dorsal-ventral axes of *Drosophila* embryos form a coordinate system that can be used to specify positions within the embryo. Theoretically, cells that are initially equivalent in developmental potential can respond to their coordinates by expressing different sets of genes. This has been seen in the formation of the salivary gland rudiments (Panzer et al., 1992). First, salivary glands only form in the strip of cells defined by the *Sex combs reduced* (*Scr*) gene along the anterior-posterior axis (parasegment 2). No salivary glands form in *Scr*-deficient mutants. Moreover, if *Scr* is caused to function throughout the embryo, salivary gland genes are expressed in a ventrolateral stripe along most of the length of the embryo. The position of the salivary gland along the dorsal-ventral axis is repressed by both Decapentaplegic and Dorsal. These proteins inhibit salivary gland formation both dorsally and ventrally. Thus, the salivary gland forms at the intersection of the vertical *Scr* expression band (second parasegment) and the horizontal region in the middle of the embryo's circumference that has neither *decapentaplegic* nor *dorsal* gene products (Figure 14.44). The cells that form the salivary gland are directed to do so by the intersecting gene activities of the anterior-posterior and dorsal-ventral axes.

A similar situation is seen with tissues that are found in every segment of the fly. Neuroblasts arise from 10 clusters of 4 to 6 cells each that form twice in every segment in the strip of neuroectoderm at the midline of the embryo (Skeath and Carroll, 1992). The potential to form neural cells is conferred on these cells by the expression of proneural genes from the achaete-scute gene complex: *achaete* (*ac*), *scute* (*sc*), and *lethal of scute* (*l'sc*). The cells in each cluster interact (in ways that are discussed in a Chapters 8 and 17) to generate a single neural cell from the cluster. Skeath and colleagues (1993) have shown that the pattern of *acheate* and *scute* transcription is imposed by a coordinate sytem. Their expression is repressed by Decapentaplegic and Snail proteins along the dorsal-ventral axis, while positive enhancement by pair-rule genes along the anterior-posterior axis causes their repetition in each half-segment. The enhancer recognized by these axis-specifying proteins lies between the *achaete* and *scute* genes and appears to regulate both of them. It is very likely, then, that the positions of organ primordia are specified throughout the fly through a two-dimensional coordinate system based on the intersection of the anterior-posterior and dorsal-ventral axes.

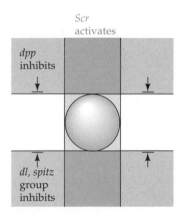

Figure 14.44
Cartesian coordinate system for the expression of genes giving rise to salivary glands. The genes are activated by the protein product of the *Sex combs reduced* homeotic gene along the anterior-posterior axis, and they are inhibited in the regions marked by *decapentaplegic* and *dorsal* gene products along the dorsal-ventral axis. This allows salivary glands to form in the midline of the embryo in the second parasegment. (After Panzer et al., 1992.)

Coda: Some principles of Drosophila development

We are beginning to learn how the genome influences the construction of the organism. The genes regulating pattern formation in *Drosophila* operate according to certain principles.

- There are morphogens—such as Bicoid and Dorsal—whose gradients determine the specification of different cell types. These morphogens can be transcription factors, or can act as signaling molecules.
- There is a temporal order wherein different classes of genes are transcribed, and the products of one gene often regulate the expression of another gene. In *Drosophila*, boundaries of gene expression can be created by the interaction between trascription factors and their gene targets. Here, the transcription factors transcribed earlier regulate the expression of the next set of genes.
- Translational control is extremely important in the early embryo, and localized mRNAs are critical in patterning the embryo.
- Individual cell fates are not defined immediately. Rather, there is a stepwise specification where a given field is divided and subdivided, eventually regulating individual cell fates.

Genetic studies on the *Drosophila* embryo have uncovered numerous genes that are responsible for the specification of the anterior-posterior and dorsal-ventral axes. We are far from a complete understanding of *Drosophila* pattern formation, but we are much more aware of its complexity than we were five years ago. The mutations of *Drosophila* have given us our first glimpses of the multiple levels of pattern regulation in a complex organism and have enabled the isolation of these genes and their gene products. Moreover, as we will see in the forthcoming chapters, these genes may provide clues to a general mechanism of pattern formation used throughout the animal kingdom.

LITERATURE CITED

Akam, M. E. 1987. The molecular basis for metameric pattern in the *Drosophila* embryo. *Development* 101: 1–22.

Anderson, K. V. and Nüsslein-Volhard, C. 1984. Information for the dorsal-ventral pattern of the *Drosophila* embryo is stored as maternal mRNA. *Nature* 311: 223–227.

Anderson, K., Bokla, L. and Nüsslein-Volhard, C. 1985. Establishment of dorsal-ventral polarity in the *Drosophila* embryo: The induction of polarity by the *Toll* gene product. *Cell* 42: 791–798.

Barker, D. D., Wang, C., Moore, J., Dickinson, L. K. and Lehmann, R. 1992. Pumillio is essential for function but not for distribution of the *Drosophila* abdominal determinant, Nanos. *Genes Dev.* 6: 2312–2326.

Bate, M. and Martinez-Arias, A. 1993. *The Development of Drosophila melanogaster*. Cold Spring Harbor Laboratory Press, Cold Spring Harbor, NY.

Bateson, W. 1894. *Materials for the Study of Variation*. Macmillan, London.

Baumgartner, S. and Noll, M. 1990. Networks of interaction among pair-rule genes regulating *paired* expression during primordial segmentation of *Drosophila*. *Mech. Dev.* 1: 1–18.

Beachy, P. A. 1990. A molecular view of the *Ultrabithorax* homeotic gene of *Drosophila*. *Trends Genet.* 6(2): 46–51.

Bender, W. and seven others. 1983. Molecular genetics of the bithorax complex in *Drosophila melanogaster*. *Science* 221: 23–29.

Berleth, T. and seven others. 1988. The role of localization of bicoid RNA in organizing the anterior pattern of the *Drosophila* embryo. *EMBO J.* 7: 1749–1756.

Bhanot, P. and eight others. 1996. A new member of the *frizzled* family from *Drosophila* functions a s a Wingless receptor. *Nature* 382: 225–230.

Bokor, P. and DiNardo, S. 1996. The roles of Hedgehog and Wingless in patterning the dorsal epidermis in *Drosophila*. *Development* 122: 1083–1092.

Boulet, A. M., Lloyd, A., Sakonju, S. 1991. Molecular definition of the morphogenetic and regulatory functions and the *cis*-regulatory elements of the *Drosophila Abd-B* gene. *Development* 111: 393–405.

Boveri, T. 1901. Über die Polarität des Seeigeleiers. *Eisverh. Phys. Med. Ges. Würzburg* 34: 145–175.

Cao, Z., Henzel, W. J. and Gao, X. 1996. IRAK: A kinase associated with the interleukin-1 receptor. *Science* 271: 1128–1131.

Capovilla, M., Eldon, E. D. and Pirrotta, V. 1992. The *giant* gene of *Drosophila* encodes a bZIP DNA-binding protein that regulates the expression of other segmentation gap genes. *Development* 114: 99–112.

Casanova, J. and Struhl, G. 1989. Localized surface activity of *torso*, a receptor tyrosine kinase, specifies body pattern in *Drosophila*. *Genes Dev.* 3: 2025–2038.

Casanova, J., Sanchez-Herrero, E. and Morata, G. 1985. Prothoracic transformation and functional structure of the *Ultrabithorax* gene of *Drosophila*. *Cell* 42: 663–669.

Casanova, J., Sanchez-Herrero, E., Busturia, A. and Morata, G. 1987. Double and triple mutant combination of the bithorax complex of *Drosophila*. *EMBO J.* 6: 3103–3109.

Casares, F. and Sánchez-Herrero, E. 1995. Regulation of the *infraabdominal* regions of the bithorax complex of Drosophila by gap genes. *Development* 121: 1855–1866.

Castelli-Gair, J. and Akam, M. 1995. How the Hox gene *Ultrabithorax* specifies two different segments: the significance of spatial and temporal regulation within metameres. *Development* 121: 2973–2982.

Chasan, R., Jin, Y. and Anderson, K. V. 1992. Activation of the *easter* zymogen is regulated by five other genes to define dorsal-ventral polarity in the *Drosophila* embryo. *Development* 115: 607–615.

Child, C. M. 1941. *Patterns and Problems of Development*. University of Chicago Press, Chicago.

Cohen, S. M. and Jürgens, G. 1990. Mutations of *Drosophila* head development by gap-like segmentation genes. *Nature* 346: 482–485.

Crick, F. H. C. 1970. Diffusion in embryogenesis. *Nature* 225: 420–422.

Degelmann, A., Hardy, P. A., Perrimon, N. and Mahowald, A. P. 1986. Developmental analysis of the *torso*-like phenotype in *Drosophila* produced by a maternal-effect locus. *Dev. Biol.* 115: 479–489.

Dessain, S., Gross, C. T., Kuziora, M. A. and McGinnis, W. 1992. *Antp*-type homeodomains have distinct DNA-binding specificities that correlate with their different regulatory functions in embryos. *EMBO Journal* 11: 991–1002.

Doe, C. Q., Hiromi, Y., Gehring, W. J. and Goodman, C. S. 1988. Expression and function of the segmentation gene *fushi tarazu* during *Drosophila* neurogenesis. *Science* 239: 170–175.

Driever, W. and Nüsslein-Volhard, C. 1988a. A gradient of bicoid protein in *Drosophila* embryos. *Cell* 54: 83–93.

Driever, W. and Nüsslein-Volhard, C. 1988b. The bicoid protein determines position in the *Drosophila* embryo in a concentration-dependent manner. *Cell* 54: 95–104.

Driever, W. and Nüsslein-Volhard, C. 1989. The bicoid protein is a positive regulator of *hunchback* transcription in the early *Drosophila* embryo. *Nature* 337: 138–143.

Driever, W., Thoma, G. and Nüsslein-Volhard, C. 1989. Determination of spatial domains of zygotic gene expression in the *Drosophila* embryo by the affinity of binding sites for the bicoid morphogen. *Nature* 340: 363–367.

Driever, W., Siegel, V. and Nüsslein-Volhard, C. 1990. Autonomous determination of anterior structures in the early *Drosophila* embryo by the bicoid morphogen. *Development.* 109: 811–820.

Dubnau, J. and Struhl, G. 1996. RNA recognition and translational regulation by a homeodomain protein. *Nature* 379: 694–699.

Duffy, J. B. and Perrimon, N. 1994. The torso pathway in *Drosophila*: Lessons on receptor tyrosine kinase signaling and pattern formation. *Dev. Biol.* 166: 380–395.

Edgar, B. A. and Schubiger, G. 1986. Parameters controlling transcriptional activation during early *Drosophila* development. *Cell* 44: 871–877.

Edgar, B. A., Weir, M. P., Schubiger, G. and Kornberg, T. 1986. Repression and turnover pattern of *fushi tarazu* RNA in the early *Drosophila* embryo. *Cell* 47: 747–754.

Ferguson, E. L and Anderson, K. V. 1991. Dorsal-ventral pattern formation in the *Drosophila* embryo The role of zygotically active genes. *Curr. Top. Dev. Biol.* 25: 17–43.

Finkelstein, R. and Perrimon, N. 1990. The *orthodenticle* gene is regulated by bicoid and torso and specifies *Drosophila* head development. *Nature* 346: 485–488.

Foe, V. E. 1989. Mitotic domains reveal early committment of cells in *Drosophila* embryos. *Development* 107: 1–22.

Foe, V. E. and Alberts, B. M. 1983. Studies of nuclear and cytoplasmic behavior during the five mitotic cycles that precede gastrulation in *Drosophila* embryogenesis. *J. Cell Sci.* 61: 31–70.

Forlani, S., Ferrandon, D., Saget, O. and Mohier, E. 1993. A regulatory function for K10 in the establishhment of dorsoventral polarity in the *Drosophila* egg and embryo. *Mech. Dev.* 41: 109–120.

French, V. 1988. Gradients and insect segmentation. *In* V. French, P. Ingham, J. Cooke and J. Smith (eds.), *Mechanisms of Segmentation*. Company of Biologists, Cambridge, pp. 3–16.

Frigerio, G., Burri, M., Bopp, D., Baumgartner, S. and Noll, M. 1986. Structure of the segmentation gene *paired* and the *Drosophila* PRD gene set as part of a gene network. *Cell* 47: 735–746.

Frischer, L. E., Hagen, F. S. and Garber, R. L. 1986. An inversion that disrupts the *Antennapedia* gene causes abnormal structure and localization of RNAs. *Cell* 47: 1017–1023.

Galindo, R. L., Edwards, D. N., Gillespie, S. K. H. and Wasserman, S. A. 1995. Interaction of the pelle kinase with the membrane-associated protein tube is required for transduction of the dorsoventral signal in *Drosophila* embryos. *Development* 121: 2209–2218.

Gaul, U. and Jäckle, H. 1990. Role of gap genes in early *Drosophila* development. *Annu. Rev. Genet.* 27: 239–275.

Gavis, E. R. and Lehmann, R. 1992. Localization of *nanos* RNA controls embryonic polarity. *Cell* 71: 301–313.

González-Crespo, S. and Levine, M. 1994. Related target enhancers for dorsal and NF-κB signalling pathways. *Science* 264: 255–258.

González-Reyes, A and Morata, G. 1990. The developmental effect of overexpressing a Ubx product in *Drosophila* embryos is dependent on its interactions with other homeotic products. *Cell* 61: 515–522.

Gonzáles-Reyes, A., Elliot, H. and St. Johnson, D. 1995. Polarization of both major body axes in *Drosophila* by *gurken*-torpedo signalling. *Nature* 375: 654–658.

Grossniklaus, U., Cadigan, K. M. and Gehring, W. J. 1994. Three maternal coordinate systems cooperate in the patterning of the *Drosophila* head. *Development* 120: 3155–3171.

Hafen, E., Levine, M. and Gehring, W. J. 1984. Regulation of *Antennapedia* transcript distribution by the bithorax complex in *Drosophila*. *Nature* 307: 287–289.

Hanes, S. D. and Brent, R. 1989. DNA specificity of the bicoid activator protein is determined by homeodomain recognition helix residue 9. *Cell* 57: 1275–1283.

Hanes, S. D. and Brent, R. 1991. A genetic model for interaction of the homeodomain recognition helix with DNA. *Science* 251: 426–430.

Harding, K. and Levine, M. 1988. Gap genes define the limits of *Antennapedia* and *Bithorax* gene expression during early development in *Drosophila*. *EMBO J.* 7: 205–214.

Harding, K. and Levine, M. 1989. *Drosophila*: The zygotic contribution. *In* D. M. Glover and B. D. Hames (eds.), *Genes and Embryos*. IRL, New York, pp. 38–90.

Harding, K., Wedeen, C., McGinnis, W. and Levine, M. 1985. Spatially regulated expression of homeotic genes in *Drosophila*. *Science* 229: 1236–1242.

Hashimoto, C., Hudson, K. L. and Anderson, K. V. 1988. The *Toll* gene of *Drosophila*, required for dorsal-ventral embryonic polarity, appears to encode a transmembrane protein. *Cell* 52: 269–279.

Hashimoto, C., Gerttula, S. and Anderson, K. V. 1991. Plasma membrane localization of the *Toll* protein in the syncytial *Drosophila* embryo: importance of transmembrane signalling for dorsal-ventral pattern formation. *Development* 11: 1021–1028.

Hecht, P. M. and Anderson, K. V. 1992. Extracellular proteases and embryonic pattern formation. *Trends Cell Biol.* 2: 197–202.

Heemskerk, J., DiNardo, S., Kostriken, R. and O'Farrell, P. H. 1991. Multiple modes of *engrailed* regulation in the progression towards cell fate determination. *Nature* 352: 404–410.

Heemskerk, J. and DiNardo, S. 1994. *Drosophila hedgehog* acts as a morphogen in cellular patterning. *Cell* 76: 449–460.

Hoch, M., Gerwin, N., Taubert, H. and Jäckle, H. 1992. Competition for overlapping sites in the regulatory region of the *Drosophila* gene *Krüppel*. *Science* 256: 94–97.

Hong, C. C. and Hashimoto, C. 1995. An unusual mosaic protein with a protease domain, encoded by the *nudel* gene, is involved in defining embryonic dorsoventral polarity in *Drosophila*. *Cell* 82: 785–794.

Hülskamp, M., Schröder, C., Pfeifle, C., Jäckle, H. and Tautz, D. 1989. Posterior segmentation of the *Drosophila* embryo in the absence of a maternal posterior organizer gene. *Nature* 338: 629–632.

Immergluck, K., Lawrence, P. A. and Bienz, M. 1990. Induction across germ layers in *Drosophila* mediated by a genetic cascade. *Cell* 62: 261–268.

Ingham, P. W. and Martinez-Arias, A. 1986. The correct activation of *Antennapedia* and bithorax complex genes requires the *fushi tarazu* gene. *Nature* 324: 592–597.

Ingham, P. W. and Whittle, R. 1980. *Trithorax*: A new homeotic mutation of *Drosophila* causing transformations of abdominal and thoracic imaginal segments. I. Putative role during embryogenesis. *Mol. Gen. Genet.* 179: 607–614.

Ingham, P. W., Taylor, A. M. and Nakano, Y. 1991. Role of *Drosophila patched* gene in positional signalling. *Nature* 353: 184–187.

Ip. Y. T. 1995. Converting an activator into a repressor. *Curr. Biol.* 5: 1–3.

Irish, V. F. and Gelbart, W. M. 1987. The *decapentaplegic* gene is required for dorsal-ventral patterning of the *Drosophila* embryo. *Genes Dev.* 1: 868–879.

Irish, V., Lehmann, R. and Akam, M. 1989. The *Drosophila* posterior-group gene *nanos* functions by repressing hunchback activity. *Nature* 338: 646–648.

Jäckle, H., Tautz, D., Schuh, R., Seifert, E. and Lehmann, R. 1986. Cross-regulatory interactions among the gap genes of *Drosophila*. *Nature* 324: 668–670.

Jiang, J. and Levine, M. 1993. Binding affinities and cooperative interactions with bHLH activators delimit threshold responses to the dorsal gradient morphogen. *Cell* 72: 741–752.

Jiang, J., Kosman, D., Ip, Y. T. and Levine, M. 1991. The *dorsal* morphogen gradient regulates the mesoderm determinant *twist*

in early *Drosophila* embryos. *Genes Dev.* 5: 1881–1891.

Kalthoff, K. 1969. Der Einfluss vershiedener Versuchparameter auf die Häufigkeit der Missbildung "Doppelabdomen" in UV-bestrahlten Eiern von *Smittia* sp. (Diptera,Chironomidae). *Zool. Anz. Suppl.* 33: 59–65.

Kalthoff, K. and Sander, K. 1968. Der Enwicklungsgang der Missbildung "Doppelabdomen" im partiell UV-bestrahlten Ei von *Smittia parthenogenetica* (Diptera, Chironomidae). *Wilhelm Roux Arch. Entwicklungsmech. Org.* 161: 129–146.

Kandler-Singer, I. and Kalthoff, K. 1976. RNase sensitivity of an anterior morphogenetic determinant in an insect egg (*Smittia* sp., Chironomidae, Diptera). *Proc. Natl. Acad. Sci. USA* 73: 3739–3743.

Karch, F. and seven others. 1985. The abdominal region of the bithorax complex. *Cell* 43: 81–96.

Karr, T. L. and Kornberg, T. B. 1989. *fushi tarazu* protein expression in the cellular blastoderm of *Drosophila* detected using a novel imaging technique. *Development* 105: 95–103.

Kaufman, T. C., Seeger, M. A. and Olsen, G. 1990. Molecular and genetic organization of the *Antennapedia* gene complex of *Drosophila melanogaster*. *Adv. Genet.* 27: 309–362.

Kidd, S. 1992. Characterization of the *Drosophila cactus* locus and analysis of interactions between cactus and dorsal proteins. *Cell* 71: 623–635.

Klingler, M., Erdélyi, M., Szabad, J. and Nüsslein-Volhard, C. 1988. Function of *torso* in determining the terminal anlagen of the *Drosophila* embryo. *Nature* 335: 275–277.

Knipple, D. C., Seifert, E., Rosenberg, U. B., Preiss, A. and Jäckle, H. 1985. Spatial and temporal patterns of *Krüppel* gene expression in early *Drosophila* embryos. *Nature* 317: 40–44.

Kosman, D., Ip, Y. T., Levine, M. and Arora, K. 1991. Establishment of the mesoderm-neuroectoderm boundary in the *Drosophila* embryo. *Science* 254: 118–122.

Kraut, R. and Levine, M. 1991. Mutually repressive interactions between the gap genes *giant* and *Krüppel* define middle body regions of the *Drosophila* embryo. *Development* 111: 611–621.

Lehmann, R. and Nusslein-Volhard, C. 1986. Abdominal segmentation, pole cell formation, and embryonic polarity require the localized activity of *oskar*, a maternal gene in *Drosophila*. *Cell* 47: 141–152

Lehmann, R. and Nüsslein-Volhard, C. 1991. The maternal gene *nanos* has a central role in posterior pattern formation of the *Drosophila* embryo. *Development* 112: 679–691.

Lehming, N., Thanos, D., Brickman, J. M., Ma, J., Maniatis, T. and Ptashne, M. 1994. An HMG-like protein that can switch a transcriptional activator to a repressor. *Nature* 371: 175–179.

Lemaitre, B., Nicolas, E., Michaut, L., Reichhart, J.-M. and Hoffmann, J. A. 1996. The dorsoventral regulatory gene casette spuatzle/Toll/cactus controls the potent antifungal response in *Drosophila* adults. *Cell* 86: 973–983.

Leptin, M. 1991a. *twist* and *snail* as positive and negative regulators during *Drosophila* mesoderm development. *Genes Dev.* 5: 1568–1576.

Leptin, M. 1991b. Mechanics and genetics of cell shape changes during *Drosophila* ventral furrow formation. *In* R. Keller et al. (eds.), *Gastrulation: Movements, Patterns, and Molecules*. Plenum, New York, pp. 199–212.

Levine, M. S. and Harding, K. W. 1989. *Drosophila*: The zygotic contribution. *In* D. M. Glover and B. D. Hames (eds.), *Genes and Embryos*. IRL, New York, pp. 39–94.

Lewis, E. B. 1978. A gene complex controlling segmentation in *Drosophila*. *Nature* 276: 565–570.

Lewis, E. B. 1985. Regulation of the genes of the bithorax complex in *Drosophila*. *Cold Spring Harbor Symp. Quant. Biol.* 50: 155–164.

Macdonald, P. M. and Smibert, C. A. 1996. Translational regulation of maternal mRNAs. *Curr. Opin. Genet. Dev.* 6: 403–407.

Macdonald, P. M. and Struhl, G. 1986. A molecular gradient in early *Drosophila* embryos and its role in specifying the body pattern. *Nature* 324: 537–545.

Mange, A. P. and Mange, E. J. 1990. *Genetics: Human Aspects*. Sinauer Associates, Sunderland, MA.

Mann, R. S. and Hogness, D. S. 1990. Functional dissection of *Ultrabithorax* proteins in *D. melanogaster*. *Cell* 60: 597–610.

Martin, J. R., Railbaud, A. and Ollo, R. 1994. Terminal elements in *Drosophila* embryo induced by *torso-like* protein. *Nature* 367: 741–745.

Martinez-Arias, A. and Lawrence, P. A. 1985. Parasegments and compartments in the *Drosophila* embryo. *Nature* 313: 639–642.

McKeon, J. and Brock, H. W. 1991. Interactions of the *Polycomb* group of genes with homeotic loci of Drosophila. *Roux Arch. Dev. Biol.* 199: 387–396.

Mlodzik, M. and Gehring, W. J. 1987. Expression of the *caudal* gene in the germ line of *Drosophila*: Formation of an RNA and protein gradient during early embryogenesis. *Cell* 48: 465–478.

Mohler, J. and Vani, K. 1992. Molecular organization and embryonic expression of the *hedgehog* gene involved in cell-cell communication in segmental patterning in *Drosophila*. *Development* 115: 957–971.

Montell, D. J., Keshishian, H. and Spradling, A. C. 1991. Laser ablation studies of the role of the *Drosophila* oocyte nucleus in pattern formation. *Science* 254: 290–293.

Müller, J. and Bienz, M. 1992. Sharp anterior boundary of homeotic gene expression conferred by the *fushi tarazu* protein. *EMBO J.* 11: 3653–3661.

Müller, M. Affolter, M., Leupin, W., Otting, G., Wüthrich, K. and Gehring, W. J. 1988. Isolation and sequence-specific DNA binding of the *Antennapedia* homeodomain. *EMBO J.* 7: 4299–4304.

Murata, Y. and Wharton, R. P. 1995. Binding of pumilio to maternal *hunchback* mRNA is required for posterior patterning in *Drosophila* embryos. *Cell* 80: 747–756.

Nambu, J. R., Franks, R. G., Hong, S. and Crews, S. 1990. The *single-minded* gene of *Drosophila* is required for the expression of genes important for the development of CNS midline cells. *Cell* 63: 63–75.

Neuman-Silberberg, F. S. and Schüpbach, T. 1993. The *Drosophila* dorsoventral patterning gene *gurken* produces a dorsally localized RNA and encodes a TGF-a-like protein. *Cell* 75: 165–174.

Nijhout, H. F. 1981. The color patterns of butterflies and moths. *Sci. Am.* 245 (5): 140-151.

Nüsslein-Volhard, C. and Wieschaus, E. 1980. Mutations affecting segment number and polarity in *Drosophila*. *Nature* 287: 795–801.

Nüsslein-Volhard, C., Fröhnhofer, H. G. and Lehmann, R. 1987. Determination of anterioposterior polarity in *Drosophila*. *Science* 238: 1675–1681.

Otting, G., Qian, Y. Q., Billeter, M., M ller, M., Affolter, M., Gehring, W. J. and Wüthrich, K. 1990. Protein-DNA contacts in the structure of a homeodomain-DNA complex determined by nuclear magnetic resonance spectroscopy in solution. *EMBO J.* 9: 3085–3092.

Pan, D., Huang, J.-D. and Courey, A. J. 1991. Functional analysis of the *Drosophila twist* promoter reveals a *dorsal*-binding ventral activator region. *Genes Dev.* 5: 1892–1901.

Panganiban, G. E. F., Reuter, R., Scott, M. P. and Hoffmann, F. M. 1990. A *Drosophila* growth factor homolog, *Decapentaplegic*, regulates homeotic gene expression within and across germ layers during midgut morphogenesis. *Development* 110: 1041–1050.

Pankratz, M. J., Seifert, E., Gerwin, N., Billi, B., Nauber, U. and Jäckle, H. 1990. Gradients of *Krüppel* and *knirps* gene products direct pair-rule gene stripe patterning in the posterior region of the *Drosophila* embryo. *Cell* 61: 309–317.

Panzer, S., Weigel, D. and Beckendorf, S. K. 1992. Organogenesis in *Drosophila melanogaster*: embryonic salivary gland determination is controlled by homeotic and dorsoventral patterning genes. *Development* 114: 49–57.

Peifer, M. and Bender, W. 1986. The *antero-bithorax* and *bithorax* mutations of the bithorax complex. *EMBO J.* 5: 2293–2303.

Peifer, M. and Wieschaus, E. 1990. Mutations in the *Drosophila* gene *extradenticle* affect the way specific homeodomain proteins regulate segment identity. *Genes Dev.* 4: 1209–1223.

Peifer, M., Karch, F. and Bender, W. 1987. The bithorax complex: Control of segmental identity. *Genes Dev.* 1: 891–898.

Percival-Smith, A., Müller, M., Affolter, M. and Gehring, W. J. 1990. The interaction with DNA of wild-type and mutant *fushi tarazu* homeodomains. *EMBO J.* 9: 3967–3974.

Pignoni, F., Steingrímsson, E. and Lengyel, J. A. 1992. *bicoid* and the terminal system activate *tailless* expression in the early *Drosophila* embryo. *Development* 115: 239–251.

Porter, J. A. and ten others. 1996. Hedgehog patterning activity. Role of a lipophilic modification by the carboxy-terminal autoprocessing domain. *Cell* 86: 21–34.

Price, J. V., Clifford, R. J. and Sch pbach, T. 1989. The maternal ventralizing gene *torpedo* is allelic to *faint little ball*, an embryonic lethal, and encodes the *Drosophila* EGF receptor homolog. *Cell* 56: 1085–1092.

Qian, S., Capovilla, M. and Pirrotta, V. 1991. The *bx* region enhancer, a distant *cis*-control element of the *Drosophila Ubx* gene and its regulation by *hunchback* and other segmentation genes. *EMBO J.* 10: 1415–1425.

Raskolb, C. and Wieschaus, E. 1994. Coordinate regulation of downstream genes by extradenticle and homeotic selector proteins. *EMBO J.* 15: 3561–3569.

Ray, R. Arora, K., Nüsslein-Volhard, C. and Gelbart, W. M. 1991. The control of cell fate along the dorsal-ventral axis of the *Drosophila* embryo. *Development* 113: 35–54.

Reach, M., Galindo, R. L., Towb, P., Allen, J. L., Karin, M. and Wasserman, S. A. 1996. A gradient of Cactus protein degradation establishes dorsoventral polarity in the *Drosophila* embryo. *Dev. Biol.* 180: 353–364.

Reinitz, J., Mjolsness, E. and Sharp, D. H. 1995. Model for cooperative control of positional information in *Drosophila* by bicoid and maternal hunchback. *J. Exp. Zool.* 271: 47–56.

Reinitz, J. and Sharp, D. H. 1995. Mechanism of *eve* stripe formation. *Mech. Dev.* 49: 133–158.

Riddihough, G. 1992. Homing in on the homeobox. *Nature* 357: 643–644.

Rivera-Pomar, R. and Jackle, H. 1996. From gradients to stripes in *Drosophila* embryogenesis: Filling in the gaps. *Trends Genet.* 12: 478–483.

Rivera-Pomar, R., Lu, X., Perrimon, N., Taubert, H. and Jäckle, H. 1995. Activation of posterior gap gene expression in the *Drosophila* blastoderm. *Nature* 376: 253–256.

Rivera-Pomar, R., Niessling, D., Schmidt-Ott, U., Gehring, W. J. and Häckle, H. 1996. RNA binding and translational suppression by bicoid. *Nature* 379: 746–749.

Roder, L., Vola, C. and Kerridge, S. 1992. The role of *teashirt* in trunk segmental identity in *Drosophila*. *Development* 115: 1017–1033.

Roth, S. 1994. Proteolytic generation of a morphogen. *Curr. Biol.* 4: 755–757.

Roth, S., Stein, D., Nüsslein-Volhard, C. 1989. A gradient of nuclear localization of the *dorsal* protein determines dorsoventral pattern in the *Drosophila* embryo. *Cell* 59: 1189–1202.

Roth, S., Neuman-Silberberg, F. S., Barcelo, G. and Schüpbach, T. 1995. cornichon and the EGF receptor signalling process are necessary for both anterior-posterior and dorsal-ventral pattern formation in *Drosophila*. *Cell* 81: 967–978.

Rushlow, C., Frasch, J., Doyle, H. and Levine, M. 1987. Maternal regulation of a homeobox gene controlling differentiation of dorsal tissues in *Drosophila*. *Nature* 330: 583–586.

Rushlow, C. A., Han, K., Manley, J. L. and Levine, M. 1989. The graded distribution of the *dorsal* morphogen is initiated by selective nuclear transport in *Drosophila*. *Cell* 59: 1165–1177.

Sánchez-Herrero, E. 1991. Control of the expression of the bithorax complex genes *abdominal-A* and *Abdominal-B* by *cis*-regulatory regions in *Drosophila* embryos. *Development* 111: 437–449.

Sánchez-Herrero, E., Verños, I., Marco, R. and Morata, G. 1985. Genetic organization of *Drosophila* bithorax complex. *Nature* 313: 108–113.

Sander, K. 1960. Analyse des ooplasmatischen Reaktionssystems von Euscelis plebajus Fall (Circadina) durch Isolieren und Kombinieren von Keimteilen. II. Die Differenzierungsleistungen nach Verlagern von Hinterpolmaterial. *Wilhelm Roux Arch. Entwicklungsmech. Org.* 151: 660–707.

Sander, K. 1975. Pattern specification in the insect embryo. In *Cell Patterning. CIBA Foundation Symp.* 29: 241–263.

Schier, A. F. and Gehring, A. J. 1992. Direct homeodomain-DNA interaction in the autoregulation of the *fushi tarazu* gene. *Nature* 356: 804–807.

Schneuwly, S., Kuroiwa, A. and Gehring, W. J. 1987. Molecular analysis of the dominant homeotic *Antennapedia* phenotype. *EMBO J.* 6: 201–206.

Schulz, C. and Tautz, D. 1995. Zygotic *caudal* regulation by hunchback and its role in abdominal segment formation of the *Drosophila* embryo. *Development* 121: 1023–1028.

Schüpbach, T. 1987. Germ line and soma cooperate during oogenesis to establish the dorsoventral pattern of egg shell and embryo in *Drosophila melanogaster*. *Cell* 49: 699–707.

Schüpbach, T. and Wieschaus, E. 1986. Maternal effect mutations altering the anterior-posterior pattern of the *Drosophila* embryo. *Roux Arch. Dev. Biol.* 195: 302–317.

Schüpbach, T., Clifford, R. J., Manseau, L. J. and Price, J. V. 1991. Dorsoventral signaling processes in *Drosophila* oogenesis. *In* J. Gerhart, (ed.) *Cell-Cell Interactions in Early Development*. Wiley–Liss, New York pp. 163–174.

Schwalm, F. 1997. Insects. In S.F. Gilbert and A. M. Raunio, (eds.), *Embryology: Constructing the Organism*. Sinauer Associates, Sunderland, MA.

Shelton, C. A. and Wasserman, S. A. 1993. *pelle* encodes a protein kinase required to establish dorsoventral polarity in the *Drosophila* embryo. *Cell* 72: 515–525.

Siegfried, E., Wilder, E. L. and Perrimon, N. 1994. Components of *wingless* signalling in *Drosophila*. *Nature* 367: 76–80.

Simon, J., Chiang, A. and Bender, W. 1992. Ten different *Polycomb* genes are required for spatial control of the abdA and AbdB homeotic products. *Development* 114: 493–505.

Simpson-Brose, M., Treisman, J. and Desplan, C. 1994. Synergy between the hunchback and bicoid morphogens is required for anterior patterning in *Drosophila*. *Cell* 78: 855–865.

Skeath, J. B. and Carroll, S. B. 1992. Regulation of proneural gene expression and cell fate during neuroblast segregation in the *Drosophila* embryo. *Development* 114: 939–946.

Skeath, J. B., Panganiban, G., Selegue, J. and Carroll, S. B. 1993. Gene regulation in two dimensions: The proneural achaete and scute genes are controlled by combinations of axis-patterning genes through a common intergenic control region. *Genes Dev.* 6: 2606–2619.

Slack, J. M. W. 1983. *From Egg to Embryo: Determinative Events in Early Development*. Cambridge University Press, Cambridge.

St. Johnston, D. and Nüsslein-Volhard, C. 1992. The origin of pattern and polarity in the *Drosophila* embryo. *Cell* 68: 201–219.

Štanojević, D., Hoey, T. and Levine, M. 1989. Sequence-specific DNA-binding activities of the gap proteins encoded by *hunchback* and *Krüppel* in *Drosophila*. *Nature* 341: 331–335.

Štanojević, D., Small, S. and Levine, M. 1991. Regulators of a segmentation stripe by overlapping activators and repressors in the *Drosophila* embryo. *Science* 254: 1385–1387.

Stein, D. and Nüsslein-Volhard, C. 1992. Multiple extracellular activities in *Drosophila* egg perivitelline fluid are required for establishment of embryonic dorsal-ventral polarity. *Cell* 68: 429–440.

Stein, D., Roth, S., Vogelsang, E., Nüsslein-Volhard, C. 1991. The polarity of the dorsoventral axis in the *Drosophila* embryo is defined by an extracellular signal. *Cell* 65: 725–735.

Stevens, L. M., Frohnhöfer, H. G., Klingler, M. and Nüsslein-Volhard, C. 1990. Localized requirement for *torso*-like expression in follicle cells for development of terminal anlagen of the *Drosophila* embryo. *Nature* 346: 660–662.

Steward, R. 1989. Relocalization of the dorsal protein from the cytoplasm to the nucleus correlates with its function. *Cell* 59: 1179–1188.

Steward, R. and Govind, S. 1993. Dorsal-ventral polarity in the *Drosophila* embryo. *Curr. Opin. Genet. Dev.* 3: 556–561.

Struhl, G. 1981. A homeotic mutation transforming leg to antenna in *Drosophila*. *Nature* 292: 635–638.

Struhl, G. 1989. Differing strategies for organizing anterior and posterior body pattern in *Drosophila* embryos. *Nature* 338: 741–744.

Struhl, G., Struhl, K. and Macdonald, P. M. 1989. The gradient morphogen *bicoid* is a concentration-dependent transcriptional activator. *Cell* 57: 1259–1273.

Struhl, G., Johnson, P. and Lawrence, P. 1992. Control of a *Drosophila* body pattern by the hunchback morphogen gradient. *Cell* 69: 237–249.

Tautz, D. 1988. Regulation of the *Drosophila* segmentation gene *hunchback* by two maternal morphogenetic centers. *Nature* 332: 281–284.

Thisse, B., Stoetzel, C., Gorostiza-Thisse, C. and Perrin-Schmidt, F. 1988. Sequence of the *twist* gene and nuclear localization of its protein in endomesodermal cells of early *Drosophila* embryos. *EMBO J.* 7: 2175–2183.

Thisse, C, Perrin-Schmidt, F., Stoetzel, C. and Thisse, B. 1991. Sequence-specific *trans* activation of the *Drosophila* twist gene by the *dorsal* gene product. *Cell* 65: 1191–1201.

Treisman, J. and Desplan, C. 1989. The products of the *Drosophila* gap genes *hunchback* and Krüppel bind to the *hunchback* promoters. *Nature* 341: 335–336.

Turner, F. R. and Mahowald, A. P. 1977. Scanning electron microscopy of *Drosophila melanogaster* embryogenesis. *Dev. Biol.* 57: 403–416.

Tyler, M. S. and Schetzer, J.W. 1996. *The Lives of a Fly*. Videocassette. ASAP Media Services, Orono; Sinauer Associates, Sunderland, MA.

Vachon, G., Cohen, B., Pfeifle, C., McGuffin, M. E., Botas, J. and Cohen, S. M. 1992. Homeotic genes of the bithorax complex repress limb development in the abdomen of the *Drosophila* embryo through the target gene *Distal-less*. *Cell* 71: 437–450.

Van Dyke, M. A. and Murre, C. 1994. Extradenticle raises the DNA binding specificity of homeotic selector gene products. *Cell* 78: 617–624.

Wagner-Bernholz, J. T., Wilson, C., Gibson, G., Schuh, R. and Gehring, W. J. 1991. Identification of target genes of the homeotic gene *Antennapedia* by enhancer detection. *Genes Dev.* 5: 2467–2480.

Wakimoto, B. T., Turner, F. R. and Kaufman, T. C. 1984. Defects in embryogenesis in mutants associated with the *Antennapedia* gene complex of *Drosophila melanogaster*. *Dev. Biol.* 102: 147–172.

Weigel, D., Jürgens, G., Klingler, M. and Jäckle, H. 1990. Two gap genes mediate maternal terminal information in *Drosophila*. *Science* 248: 495–498.

Whalen, A. M. and Steward, R. 1993. Dissociation of the dorsal-cactus complex and phosphorylation of the dorsal protein correlate with the nuclear localization of dorsal. *J. Cell. Biol.* 123: 523–534.

Wharton, R. P. and Struhl, G. 1991. RNA regulatory elements mediate control of *Drosophila* body pattern by the posterior morphogen *nanos*. *Cell* 67: 955–967.

Wolpert, L. 1971. Positional information and pattern formation. *Curr. Top. Dev. Biol.* 6: 183–224.

Wolpert, L. 1978. Pattern formation in biological development. *Sci. Am.* 239(4): 154–164.

Specification of cell fate by progressive cell-cell interactions

15

The study of the role of genes in ontogeny is a field for developmental physiology. Not that the geneticist will be excluded from the working out of this problem—he will become a genetic experimental embryologist. After a long journey which took him sometimes out of sight of his fellow biologists, he returns home again with some new concepts and tools.
CURT STERN (1936)

We are standing and walking with parts of our body which could have been used for thinking had they developed in another part of the embryo.
HANS SPEMANN (1943)

IN THE LAST CHAPTER, we saw that cell determination and axis specification can be caused by the interactions of specific cytoplasmic substances within a syncytial cell. Only later do cell-cell interactions occur to fix cell fate. But most types of organisms do not include a syncytial stage of early embryogenesis. In many species, including most vertebrates, cells become specified by their interactions with neighboring cells.

Regulative development

In deuterostomes such as sea urchins and vertebrates, the fate of a cell depends on its position in the embryo, rather than on what piece of cytoplasm it has acquired. Sidney Brenner (quoted in Wilkins, 1993) has remarked that animal development proceeds in either of two ways. Some organisms are specified predominantly in the "European style"; that is, each cell is determined by who its ancestors were. The lineage is the important thing. Conversely, the blastomeres of most vertebrates are specified predominantly in the "American style"; there is a great deal of mixing between cells, and each cell is determined by who its neighbors are. Each cell starts off with similar potentials and develops according to whom it meets. In these embryos, for at least part of cleavage, each cell is able to develop into the entire embryo if it is separated from the other cells, and the remaining cells are able to alter their fates to produce a whole embryo (as in twin formation). This type of commitment is called conditional (or dependent) specification, and it gives rise to **regulative development.**

During autonomous development, the axis of the embryo is determined by the placement of materials into each of the blastomeres. However, in regulative development, the axes form from the interactions of the constituent cells. In this chapter, we will trace the experiments that began over a century ago to understand how the nervous system is specified in amphibians.

Testing the germ plasm theory

August Weismann: The Germ Plasm Theory

The discovery of regulative determination has its roots in the failure of the mosaic theories of development formulated in Germany at the end of the nineteenth century. In 1883, August Weismann began proposing a theory that integrated such diverse biological phenomena as heredity, development, regeneration, sexual reproduction, and evolution by natural selection. This mechanical model for cellular differentiation was called the **germ plasm theory.** Based on the scant knowledge of fertilization available at that time, Weismann boldly proposed that the sperm and egg provided equal chromosomal contributions, both quantitatively and qualitatively, to the new organism. Moreover, chromosomes were postulated to carry the inherited potentials of this new organism and were said to be the basis for the continuity between generations.* However, not all the determinants on the chromosomes were thought to enter every cell of the embryo; instead of dividing equally, the chromosomes were hypothesized to divide in such a way that different nuclear determinants entered different cells. Whereas the fertilized egg would carry the full complement of determinants, certain cells would retain the "blood-forming" determinants while other cells would retain the "muscle-forming" determinants. Only in the nuclei of those cells destined to become gametes (the germ cells) were all types of determinants thought to be retained. The nuclei of all other cells would have only a fraction of the original determinant types.

Weismann's hypothesis proposed the continuity of the germ plasm and the diversity of the somatic lines. Differentiation was due to the "segregation of the nuclear determinants" into the various cell types. The chromosomes, while appearing equal in all cells, would be unequal in their qualities. Only the germ cell line would keep all the determinants, and this cell lineage would be totally *independent* of the somatic cells. Hence, there could be no inheritance of characteristics acquired by the somatic cells. Weismann obtained support for this model by cutting off the tails of newborn mice for 19 generations. The mice of each succeeding generation had tails of normal length, indicating that the germ line was insulated from insults to the somatic tissue.†

Weismann's germ plasm theory is depicted in Figure 15.1. It emphasizes the continuity and immortality of the germ line contrasted with the temporary nature of the adult organism, showing, as physiologist Michael Foster noted, that "the animal body is in reality a vehicle for ova." E. B. Wilson, who claimed that his remarkable textbook *The Cell in Development and Inheritance* (1896) grew out of Weismann's hypothesis, also saw the implications of Weismann's scheme:

---

*Embryologists were thinking in these terms some 15 years before the rediscovery of Mendel's work. Weismann (1892, 1893) also speculated that these nuclear determinants of inheritance functioned by elaborating substances that became active in the cytoplasm!

†At this time, the major alternative view was that of pangenesis. This theory, advocated as a "provisional hypothesis" by Charles Darwin, posited that each somatic cell contained particles (*pangenes*) that migrated back into the sex cells to provide for the transmission of that cell's characteristics. According to this theory, Weismann should have obtained mice with shorter tails. More recently, Thomas Jukes, commenting on Weismann's results, quoted Hamlet's intuition that "there's a divinity that shapes our ends, rough-hew them how we will." It should be noted that the independence of the germ line from the somatic line is not absolute in all organisms. In sponges, flatworms, hydroids, and colonial tunicates, germ cells can develop from somatic tissues, and genetic changes made to those somatic tissues can be inherited (Berrill and Liu, 1948; Buss, 1987).

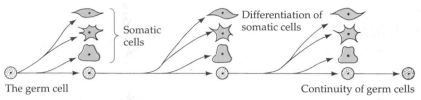

Figure 15.1
Weismann's theory of inheritance. The germ cell gives rise to the differentiating somatic cells of the body (indicated in color), as well as to new germ cells. (From Wilson, 1896.)

The death of the individual involves no breach of continuity in the series of cell divisions by which the life of the race flows onwards. The individual dies, it is true, but the germ-cells live on, carrying with them, as it were, the traditions of the race from which they have sprung, and handing them on to their descendants.

Wilhelm Roux: Mosaic Development

Weismann had intuited that chromosomes are the bearers of the inherited information for development. More importantly, though, he had proposed a hypothesis of development that could be tested immediately. Weismann had claimed that when the first cleavage division separated the future right half of the embryo from the future left half, there would be a separation of "right" determinants from "left" determinants in the resulting blastomeres. This assertion was tested by Wilhelm Roux, a young German embryologist. In 1888, Roux published the results of a series of experiments in which he took 2- and 4-cell frog embryos and destroyed some of the cells of each embryo with a hot needle. Weismann's hypothesis predicted the formation of right or left half embryos; Roux obtained half-morulae, just as Weismann predicted (Figure 15.2). These developed into half-neurulae having a complete right or left side, with one medullary fold, one ear pit, and so on. He therefore concluded that the frog embryo was a mosaic of self-differentiating parts and that it was likely that each cell received a specific set of determinants and differentiated accordingly. With this series of experiments, Roux inaugurated his program of developmental mechanics (*Entwicklungsmechanik*), an experimental and physiological approach to embryology (see Sander, 1991a,b). No longer, insisted Roux, would embryology merely be the servant of evolutionary studies. Rather, embryology would assume its role as an independent experimental science.

Figure 15.2
Roux's attempt to show mosaic development. Destroying one cell of a 2-cell frog embryo results in the development of only one-half of the embryo.

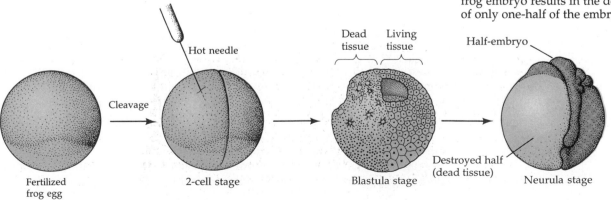

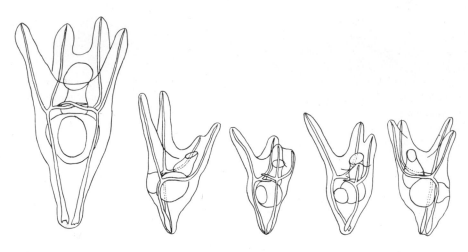

(A) Normal pluteus larva   (B) Plutei developed from single cells of 4-cell embryo

Figure 15.3
Driesch's demonstration of regulative development. (A) A normal pluteus larva. (B) Smaller, but normal, plutei that each developed from one blastomere of a dissected 4-cell embryo. (All larvae are drawn to the same scale.) (After Hörstadius and Wolsky, 1936.) Note that the four larvae derived in this way are not identical, despite their ability to generate all the necessary cell types. Such variations are also seen in those adult sea urchins formed in this way (Marcus, 1979).

Hans Driesch: Regulative Development

Nobody appreciated the experimental approach to embryology more than Hans Driesch. Driesch's goal was to explain development in terms of the laws of physics and mathematics. His initial investigations were similar to those of Roux. Roux's experiments were, technically, *defect* studies that answered the question of how the remaining blastomeres of an embryo would develop when a subset of them was destroyed. Driesch (1892) sought to extend this research by performing *isolation* experiments. Sea urchin blastomeres were separated from each other by vigorous shaking (or, later, by placing them in calcium-free seawater). To Driesch's surprise, each of the blastomeres from a 2-cell embryo developed into a complete larva. Similarly, when Driesch separated the blastomeres from 4- and 8-cell embryos, some of the cells produced entire pluteus larvae (Figure 15.3). Here was a result drastically different from that predicted by Weismann or Roux. Rather than self-differentiating into its future embryonic part, each blastomere could regulate its development so as to produce a complete organism. Here was the first experimentally observable instance of regulative development.

Regulative development was also demonstrated in another experiment by Driesch. In sea urchin eggs, the first two cleavage planes are meridional, passing through the animal and vegetal poles, whereas the third division is equatorial, dividing the embryo into four upper and four lower cells (see Figure 5.3). Driesch (1893) changed the direction of the third cleavage by gently compressing the early embryos between two glass plates, thus causing the third division to be meridional like the preceding two cleavages. After he released the pressure, the fourth division was equatorial. This procedure reshuffled the nuclei, causing a nucleus that normally would be in the region destined to form endoderm to now be in the presumptive ectoderm region. Some nuclei that would normally have produced dorsal struc-

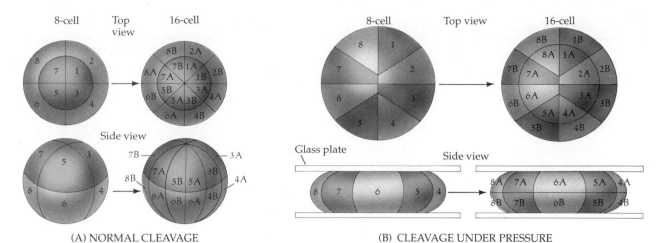

Figure 15.4
Driesch's pressure-plate experiment for altering the distribution of nuclei. (A) Normal cleavage from 8- to 16-cell sea urchin embryos, seen from the animal pole (upper sequence) and from the side (lower sequence). (B) Abnormal cleavage planes formed under pressure, as seen from the animal pole and from the side. (After Huxley and deBeer, 1934.)

tures were now found in the ventral cells (Figure 15.4). If the segregation of nuclear determinants had occurred (as had been proposed by Weismann and Roux), the resulting embryo should be strangely disordered. However, Driesch obtained normal larvae from these embryos. He concluded, "The relative position of a blastomere within the whole will probably in a general way determine what shall come from it."

The consequences of these experiments were momentous both for embryology and for Driesch personally. First, Driesch had demonstrated that the "prospective potency" of an isolated blastomere (those cell types it was possible for it to form) is greater than its "prospective fate" (the cell types it would normally give rise to over the unaltered course of its development). According to Weismann and Roux, the prospective potency and the prospective fate of a blastomere should be identical. Second, Driesch concluded that the sea urchin embryo is a "harmonious equipotential system," because all these potentially independent parts functioned together to form a single organism. Third, he concluded that the fate of a nucleus depended solely on its location in the embryo. Driesch (1894) hypothesized a series of events wherein development proceeded by the interactions of the nucleus and cytoplasm:

> *Insofar as it contains a nucleus, every cell, during ontogenesis, carries the totality of all primordia; insofar as it contains a specific cytoplasmic cell body, it is specifically enabled by this to respond to specific effects only...When nuclear material is activated, then, under its guidance, the cytoplasm of its cell that had first influenced the nucleus is in turn changed, and thus the basis is established for a new elementary process, which itself is not only the result but also a cause.*

This strikingly modern concept of nuclear-cytoplasmic interaction and nuclear equivalence eventually caused Driesch to abandon science. He could no longer envision the embryo as a physical machine, because it could be subdivided into parts that were each capable of re-forming the entire organism. In other words, Driesch had come to believe that development could

Table 15.1 Experimental procedures and results of Roux and Dreisch

Investigator	Organism	Type of experiment	Conclusion	Interpretation concerning potency and fate
Roux (1888)	Frog (*Rana fusca*)	Defect	Mosaic development (autonomous)	Prospective potency equals prospective fate
Driesch (1892)	Sea urchin (*Echinus microtuberculatus*)	Isolation	Regulative development (conditional)	Prospective potency is greater than prospective fate
Dreisch (1893)	Sea urchin (*Echinus* and *Paracentrotus*)	Recombination	Regulative development (conditional)	Prospective potency is greater than prospective fate

not be explained by physical forces. He was driven to invoke a vital force, *entelechy* ("internal goal-directed force"), to explain how development proceeds. Essentially, he believed that the embryo was imbued with an internal psyche and wisdom to accomplish its goals despite the obstacles embryologists placed in its path. Unable to explain his results in terms of the physics of his day, Driesch renounced the study of developmental physiology and became a philosophy professor, proclaiming vitalism until his death in 1941. Others, especially Oscar Hertwig (1894), were able to incorporate Driesch's experiments into a more sophisticated experimental embryology.*

The differences between Roux's experiments and those of Driesch are summarized in Table 15.1. The difference between isolation and defect experiments and the importance of the interactions provided by the destroyed blastomeres were highlighted in 1910, when J. F. McClendon showed that isolated frog blastomeres behave just like separated sea urchin cells. Therefore, the mosaic-like development of the first two frog blastomeres in Roux's study was an artifact of the defect experiment. Something in or on the dead blastomere still informed the live cells that it existed. We have already seen that early mammalian blastomeres have a regulative type of development. As we discussed in Chapter 5, each isolated blastomere of a mouse inner cell mass is capable of generating an entire fertile mouse. The ability of two or more early mouse embryos to fuse into one normal embryo (see Figure 5.28) and the phenomenon of identical twins (see Figure 5.27) also attest to the regulative ability of mammalian blastomeres. Therefore, even though Weismann and Roux pioneered the study of developmental physiology, their proposition that differentiation is caused by the segregation of nuclear determinants was soon shown to be incorrect.

---

*These experiments strengthened within embryology a type of mechanistic philosophy called *wholist organicism*. This philosophy refers to the views that (1) the properties of the whole cannot be predicted solely from the properties of the component parts, and (2) the properties of the parts are informed by their relationship to the whole. As an analogy, the meaning of a sentence obviously depends on the meanings of its component parts, words. However, the meaning of each word depends on the entire sentence. In the sentence, "The party leaders were split on the platform," the possible meanings of each noun and verb are limited by the meaning of the entire sentence and by the relationships to other words within the sentence. Similarly, a cell in the embryo develops its phenotype depending on its interactions within the entire embryo. The opposite materialist view is *reductionism*, which maintains that the properties of the whole can be known if all the properties of the parts are known. Embryology has traditionally espoused wholist organicism, while genetics has been characterized as being a reductionistic discipline (Haraway, 1976; Roll-Hansen, 1978; Allen, 1985; Tauber and Sarkar, 1992; Gilbert and Faber, 1996). Driesch became an outspoken opponent of the Nazis, and was one of the first non-Jewish professors to be forcibly retired when Hitler came to power (Harrington, 1996).

Sven Hörstadius: Potency and Oocyte Gradients

But Driesch was not 100 percent correct either. As we saw in the previous chapter, there are numerous animals that do develop largely as a mosaic of self-differentiating parts. More importantly, though, even the sea urchin embryo is not a collection of completely equipotential cells. In a series of experiments conducted from 1928 to 1935, Swedish biologist Sven Hörstadius separated various layers of early sea urchin embryos with fine glass needles and observed their subsequent development (Hörstadius, 1928, 1939). When the 8-cell embryo was divided meridionally through the animal and vegetal poles, both halves produced pluteus larvae, just as Driesch had foretold. But when embryos at the same stage were split equatorially (separating animal and vegetal poles), neither part developed into a complete larva (Figure 15.5). Rather, the animal half became a hollow ball of ciliated epidermal cells (called a **dauerblastula**), and the vegetal half developed into a slightly abnormal embryo with an expanded gut. Hörstadius was able to duplicate these results by cutting unfertilized sea urchin eggs in half and fertilizing the halves separately. In sea urchins, egg fragments (**merogones**) can divide and develop even if they have only a haploid nucleus. If a sperm enters the half that lacks the haploid egg nucleus, the merogone will still develop (Figure 15.6). When the egg has been split meridionally, normal embryos formed from both halves of the egg. However, when the oocyte has been equatorially cut, fertilization produced either the ciliated animal ball or an embryo with expanded vegetal gut. Therefore, even in sea urchin embryos, there appears to be some degree of mosaicism, at least along the animal-vegetal axis. This has been confirmed by Maruyama and co-workers (1985), who similarly split unfertilized sea urchin eggs either meridionally or equatorially. They found that when they separated the animal half from the vegetal half, only the fertilized vegetal half was able to form micromeres and to gastrulate. Therefore, the determinants that enable micromere formation and gastrulation appear to be localized in the vegetal portion of the egg. [regul1.html]

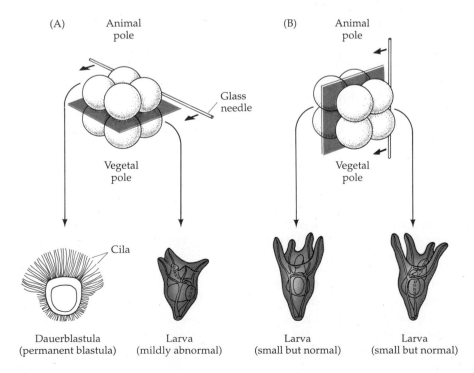

Figure 15.5
Early asymmetry in the sea urchin embryo. (A) When the four animal blastomeres are separated from the four vegetal blastomeres and each half is allowed to develop, the animal cells form a ciliated dauerblastula and the vegetal cells form a larva with an expanded gut. (B) When the 8-cell embryo is split so that each half contains animal and vegetal cells, small, normal-appearing larvae develop.

Figure 15.6
Asymmetry in the sea urchin egg. (A) When Hörstadius split the sea urchin egg meridionally so that both merogones contained animal and vegetal cytoplasm, small, normal-appearing plutei developed. (B) When the sea urchin egg was divided into animal and vegetal halves (merogones) and the halves were allowed to be fertilized by sperm, the animal half developed into a ciliated dauerblastula, and the vegetal half produced a pluteus with an expanded gut.

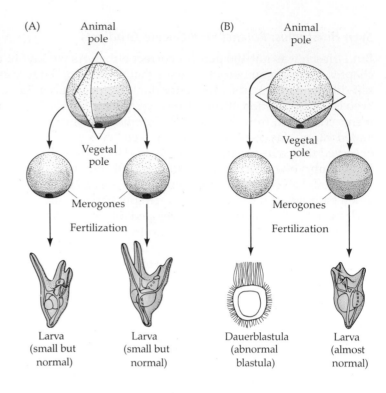

REGULATION OF CELL FATE IN LATE-CLEAVAGE EMBRYOS. These observations led Hörstadius to perform some of the most exciting experiments in the history of embryology. First, Hörstadius (1935) traced the normal development of each of the six tiers of cells of the 64-cell sea urchin embryo. As shown in Figure 15.7A, the animal cells and the first vegetal layer normally produce ectoderm; the second vegetal layer gives rise to endoderm and some larval mesoderm; and the micromeres generate the mesodermal skeleton.

Next, Hörstadius removed the fertilization membrane from the 64-cell embryos, separated the tiers with fine glass needles, and recombined them in various ways. The isolated animal hemisphere alone became a ball of ciliated ectoderm cells (Figure 15.7B). Such a ciliated dauerblastula was said to be "animalized." When Hörstadius recombined an isolated animal hemisphere with the veg_1 tier (Figure 15.7C), the resulting larva was less animalized. Ciliary development was suppressed, and a portion of the gut was formed. However, when the animal hemisphere was combined with the veg_2 tier (Figure 15.7D), a normal-looking pluteus larva developed. In this combination, the veg_2 cells, which normally form only the archenteron and its derivatives, now formed skeletal structures as well. Similarly, when the animal half was recombined with just the micromeres (Figure 15.7E), a small, normal-looking pluteus was formed, but in this case, the endoderm was completely derived from the animal cells. Here, the gut was formed by cells that would normally have created the ciliated ectoderm. These experiments showed that the animal cells have the genetic potential to be gut cells even at the 64-cell stage.

Forming an Integrated Organism: Restricting the Potency of Neighboring Cells

Driesch referred to the embryo as an "equipotential harmonious system" because each of the composite cells had surrendered most of its potential in order to form part of a single complete organism. Each cell could have be-

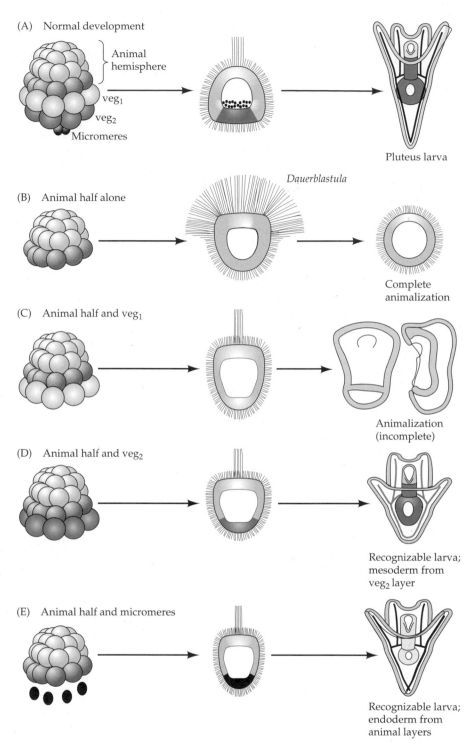

(A) Normal development

Animal hemisphere

veg₁
veg₂
Micromeres

Pluteus larva

(B) Animal half alone

Dauerblastula

Complete animalization

(C) Animal half and veg₁

Animalization (incomplete)

(D) Animal half and veg₂

Recognizable larva; mesoderm from veg₂ layer

(E) Animal half and micromeres

Recognizable larva; endoderm from animal layers

Figure 15.7
Hörstadius's demonstration of regulation in sea urchins. (A) Fate of each cell layer of the 64-cell sea urchin embryo through blastula to pluteus stage. The different cell layers are marked as in Figure 6.1. (B) Fate of the isolated animal half. (C) Recombination of animal half plus the veg₁ tier of cells. (D) Recombination of animal half plus the veg₂ tier of cells. (E) Recombination of animal half plus the micromeres. In each case, the original fates of the cells have been altered by the new neighbors. (After Hörstadius, 1939.)

come a complete animal on its own, yet didn't. What made the cells cooperate instead of becoming autonomous entities? In the case of snails and tunicates, the answer was simple. The maternal cytoplasm does not permit each cell to become autonomous; each cell can only develop into a portion of the embryo. In sea urchins and other embryos that show regulation, the answer is more complex.

Recent evidence suggests that the "harmonious equipotential system" is caused by negative induction events that mutually restrict the fate of neigh-

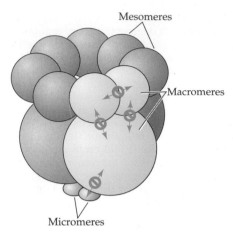

Figure 15.8
Summary of inhibitory inductions in the sea urchin blastula. Double-headed arrows illustrate the mutually restrictive interactions between adjacent cells. (After Henry et al., 1989.)

boring cells. Jon Henry and colleagues in Rudolf Raff's laboratory (1989) have shown that if one isolates pairs of cells from the animal cap of a 16-cell sea urchin embryo, these cells can give rise to both ectodermal and mesodermal components. However, their capacity to form mesoderm is severely restricted if they are aggregated with other animal cap pairs. Thus, the presence of neighbor cells, even of the same kind, restricts the potencies of both partners. Ettensohn and McClay (1988) have shown that potency is also restricted when a cell is combined with its neighbors along the animal-vegetal axis. First, they demonstrated that the number of primary mesenchyme cells appears to be fixed and can be regulated by changes in the macromeres. If all the 60 primary mesenchyme cells of *Lytechinus variegatus* are removed from the early gastrula, an equal number of secondary mesenchyme cells (from the archenteron that had been vegetal plate macromeres) convert into primary mesenchyme and start forming spicules. If one removes 20 primary mesenchyme cells, about 20 secondary mesenchyme cells become spicule-forming primary mesenchyme cells And so forth. Thus, the primary mesenchyme cells have a restrictive influence, preventing the formation of new primary mesenchyme cells from the archenteron, and there is negative induction occurring here. We do not yet know the mechanism by which the primary mesenchyme cells stop the archenteron from forming primary mesenchyme and set a limit for the number of such cells in the blastocoel.

By recombining cells from various layers, Khaner and Wilt (1990, 1991) found that in most cases, a cell from one layer *restricts* the ability of a cell from another layer to express all its potential fates (Figure 15.8). The major exception—as already mentioned above—is the recombining of animal pole mesomere cells with certain vegetal pole micromeres to form gut tissue from the mesomeres. However, in normal sea urchin development, these cells never associate with each other.

Regulation during amphibian development

Hans Spemann: Progressive Determination of Embryonic Cells

In the previous sections, we saw evidence for regulative development. We noted that the two major aspects of regulation—first, that an isolated blastomere has a potency greater than its normal embryonic fate, and second, that a cell's fate is determined by interactions between neighboring cells—hold true during the early stages of sea urchin cleavage. Eventually, however, the blastomeres become committed to certain fates. In 1918, Hans Spemann, of the University of Freiburg, discovered that a similar situation exists in the salamander egg. The experiments by which he and his colleagues analyzed this phenomenon over the next 20 years form the basis for much of our knowledge of embryonic physiology and won a Nobel Prize for Spemann in 1935.

Spemann, like Roux and Driesch, sought to test Weismann's hypothesis, and, by an ingenious method, he demonstrated that early newt blastomeres have identical nuclei, each capable of producing an entire larva. Shortly after fertilizing a newt egg, Spemann used a baby's hair to lasso the zygote in the plane of the first cleavage. He then partially constricted the egg, causing all the nuclear divisions to remain on one side of the constriction. Eventually, often as late as the 16-cell stage, a nucleus would escape across the constriction into the non-nucleated side. Cleavage then began on this side, too, whereupon the lasso was tightened until the two halves were completely separated. Twin larvae developed, one slightly older than the other (Figure 15.9). Spemann concluded from this that early amphibian nuclei are geneti-

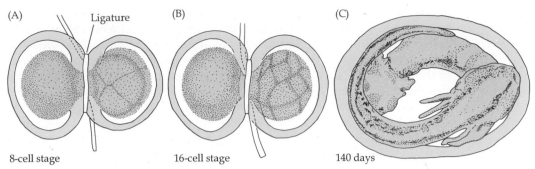

(A) Ligature

8-cell stage

(B)

16-cell stage

(C)

140 days

cally identical and that each cell is capable of giving rise to an entire organism. In this respect, amphibian blastomeres were similar to those of the sea urchin.

Moreover, when Spemann performed a similar experiment with the constriction still longitudinal but perpendicular to the plane of the first cleavage (separating the future dorsal and ventral regions rather than right and left sides), he obtained a different result altogether. The nuclei continued to divide on both sides of the constriction, but only one side—the future *dorsal* side of the embryo—would give rise to a normal larva. The other side would produce an unorganized tissue mass of ventral cells, which Spemann called the *Bauchstück*—the belly piece. This tissue mass was a ball of epidermal cells (ectoderm) containing blood and mesenchyme (mesoderm) and gut cells (endoderm), but no dorsal structures such as nervous system, notochord, or somites (Figure 15.10).

Why should these two experiments give different results? Could it be that when the egg is divided perpendicular to the first cleavage plane, some cytoplasmic substance is not equally distributed into the two halves? Fortu-

Figure 15.9
Spemann's demonstration of nuclear equivalence in newt cleavage. (A) When the fertilized egg of the newt *Triturus taeniatus* was constricted by a ligature, the nucleus was restricted to one-half of the embryo. The cleavage on that side of the embryo reached the 8-cell stage, while the other side remained undivided. (B) At the 16-cell stage, a single nucleus entered the as yet undivided half, and the ligature was constricted to complete the separation of the two halves. (C) After 140 days, each side had developed into a normal embryo. (After Spemann, 1938.)

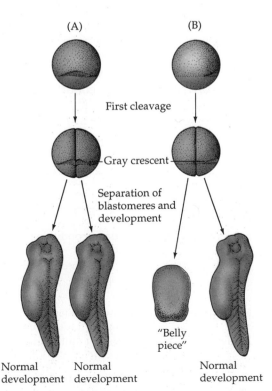

(A)

(B)

First cleavage

Gray crescent

Separation of blastomeres and development

Normal development

Normal development

"Belly piece"

Normal development

Figure 15.10
Asymmetry in an amphibian egg. (A) When the plane of first cleavage divides the egg into two blastomeres that each get one-half of the gray crescent, each experimentally separated cell develops into a normal embryo. (B) When only one of the two blastomeres receives the entire gray crescent, it alone forms a normal embryo. The other piece lacks dorsal structures and remains a mass of unorganized tissues. (After Spemann, 1938.)

nately, the salamander egg was a good place to look for answers. As we have seen in Chapters 4 and 6, there are dramatic movements in the cortical cytoplasm following the fertilization of amphibian eggs, and in some amphibians these movements expose a gray, crescent-shaped area of cytoplasm in the region directly opposite the point of sperm entry. Moreover, the first cleavage plane normally bisects this region equally into the two blastomeres. If these cells are then separated, two complete larvae develop. However, should this cleavage plane be aberrant (either in the rare natural event or in an experiment in which an investigator constricts a hair-loop lasso perpendicular to the normal cleavage plane), the gray crescent material passes into only one of the two blastomeres. Spemann found that when these two blastomeres are separated, only the blastomere containing the gray crescent develops normally.

It appears, then, that something in the gray crescent region is essential for proper embryonic development. But how does it function? What role does it play in normal development? The most important clue came from the fate map of this area of the egg, for it showed that the gray crescent region gives rise to the cells that initiate gastrulation. These cells form the dorsal lip of the blastopore. As was shown in Chapter 6, the cells of the dorsal blastopore lip are somehow committed to invaginate into the blastula, thus initiating gastrulation and the formation of the archenteron. Because all future amphibian development depends on the interaction of cells rearranged during gastrulation, Spemann speculated that the importance of the gray crescent material lies in its ability to initiate gastrulation and that crucial developmental changes occur during gastrulation.

In 1918, Spemann demonstrated that enormous changes in cell potency do indeed take place during gastrulation. He found that the cells of the *early* gastrula are uncommitted with respect to their eventual differentiation, but that the fates of the *late* gastrula cells are fixed. Spemann exchanged tissues between the early gastrulae of two differently pigmented species of newts (Figure 15.11). When a region of prospective epidermal cells was transplanted to an area where the neural plate formed, the transplanted cells gave rise to neural tissue. When prospective neural plate cells were transplanted to the region fated to become belly skin, these cells became epidermal (Table 15.2). Thus, these early newt gastrula cells were not yet committed to a specific type of differentiation. Their prospective potencies were still greater than their prospective fates. Such cells are said to exhibit conditional (i.e.

Table 15.2 Results of tissue transplantation during early- and late-gastrula stages in the newt

Donor region	Host region	Differentiation of donor tissue	Conclusion
EARLY GASTRULA			
Prospective neurons	Prospective epidermis	Epidermis	Dependent (conditional) development
Prospective epidermis	Prospective neurons	Neurons	Dependent (conditional) development
LATE GASTRULA			
Prospective neurons	Prospective epidermis	Neurons	Independent (autonomous) development (determined)
Prospective epidermis (determined)	Prospective neurons	Epidermis	Independent (autonomous) development

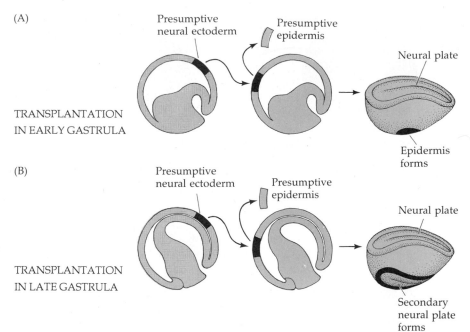

(A)

Presumptive neural ectoderm

Presumptive epidermis

Neural plate

TRANSPLANTATION IN EARLY GASTRULA

Epidermis forms

(B)

Presumptive neural ectoderm

Presumptive epidermis

Neural plate

TRANSPLANTATION IN LATE GASTRULA

Secondary neural plate forms

Figure 15.11
Determination of ectoderm during newt gastrulation. Presumptive neural ectoderm from one newt embryo is transplanted into a region in another embryo that normally becomes epidermis. (A) When the transfer is done in the early gastrula, the presumptive neural tissue develops into epidermis and only one neural plate is seen. (B) When the same experiment is performed on late-gastrula tissues, the presumptive neural cells form neural tissue, thereby causing two neural regions to form on the host. (After Saxén and Toivonen, 1962.)

regulative or dependent) development because their ultimate fates depend on their location in the embryo. However, when the same heteroplastic (interspecies) transplantation experiments were performed on *late* gastrulae, Spemann obtained completely different results. Rather than regulating their differentiation in accordance with their new location, the transplanted cells exhibited autonomous (or independent, or mosaic) development. Their prospective fate was fixed, and the cells developed independently of their new embryonic location. Specifically, prospective neural cells now developed into brain tissue even when placed in the region of prospective epidermis, and prospective epidermis formed skin even in the region of the prospective neural tube. Within the time separating early and late gastrulation, the potencies of these groups of cells had become restricted to their eventual paths of differentiation. Such cells are said to be determined: they can no longer regulate their differentiation into other cell types. It should be noted that the criteria for determination are completely operational. There are no obvious changes occurring in the cells, and no overt differentiation can be seen. The molecular basis of determination remains one of the major unsolved puzzles of development.

Hans Spemann and Hilde Mangold: Primary Embryonic Induction

The most spectacular transplantation experiments were published by Hans Spemann and Hilde Mangold in 1924.* They showed that when tissues are placed in new locations, the dorsal lip of the blastopore is the only self-differentiating region in the early gastrula. When dorsal blastopore lip tissue from an early gastrula was transplanted into the ventral ectoderm of another gastrula, it not only continued to be blastopore lip, but it also initiated gastrulation and embryogenesis in the surrounding tissue. In these experi-

*Hilde Proescholdt Mangold died in a tragic accident when her gasoline heater exploded. At the time she was 26 years old, and her paper was just being published. Hers is one of the very few doctoral theses in biology that have directly resulted in the awarding of a Nobel prize. For more information about Hilde Mangold and her times, see Hamburger (1984) and Fässler and Sander (1996).

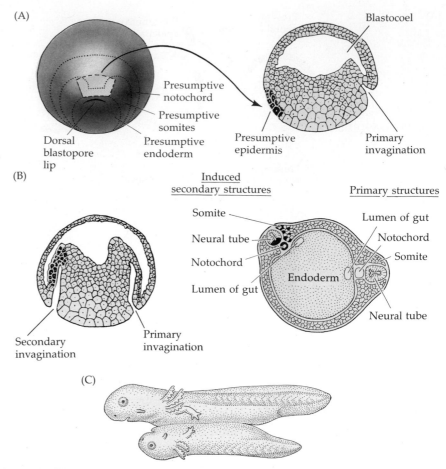

Figure 15.12
Self-differentiation of the dorsal blastopore lip tissue. (A) Dorsal blastopore lip from early gastrula is transplanted into another early gastrula in the region that normally becomes ventral epidermis. (B) Tissue invaginates and forms a second archenteron and then a second embryonic axis. Both donor and host tissues are seen in the new neural tube, notochord, and somites. (C) Eventually, a second embryo forms that is joined to the host.

ments, Spemann and Mangold used differently pigmented embryos from two species of newt: the darkly pigmented *Triturus taeniatus* and the nonpigmented *Triturus cristatus*. When Spemann and Mangold prepared these heteroplastic transplants, they were able to readily identify host and donor tissues on the basis of color. The dorsal blastopore lips (the dorsal marginal zone tissue) of early *T. cristatus* gastrulae were removed and implanted into the regions of early *T. taeniatus* gastrulae fated to become ventral epidermis (Figure 15.12). Unlike the other early-gastrula tissues, which developed according to their new location, the donor blastopore lip did not become belly skin. Rather, it invaginated just as it would normally have done (showing self-determination) and disappeared beneath the vegetal cells. The nonpigmented donor tissue then continued to self-differentiate into the chordamesoderm and other mesodermal structures that constituted the original fate of that blastopore tissue. As the axis was formed, host cells began to participate in the production of the new embryo, becoming organs that normally they never would have formed. Thus, a somite could be seen containing both colorless (donor) and pigmented (host) tissue. What was even more spectacular, the dorsal blastopore lip cells were able to interact with the host tissues

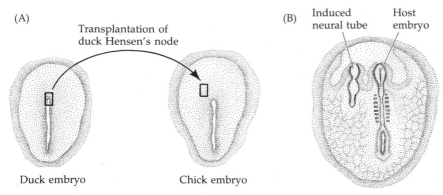

Figure 15.13
Induction of a new embryonic axis by Hensen's node. (A) Hensen's node tissue is removed from a duck embryo and implanted into a host chick embryo. (B) An accessory neural tube is induced at the graft site. (After Waddington, 1933.)

to form a complete neural plate from host ectoderm. Eventually, a secondary embryo formed, face to face with its host (see Figure 15.12; Plate 4). These technically difficult experiments have recently been repeated using nuclear markers, and the results of Spemann and Mangold have been confirmed (Gimlich and Cook, 1983; Smith and Slack, 1983; Jacobson, 1984; Recanzone and Harris, 1985).* [regul2.html]

Spemann (1938) referred to the dorsal blastopore lip cells as the **organizer** because (1) they induced the host's ventral tissues to change their fates to form a neural tube and dorsal mesodermal tissue, and (2) they organized these host and donor tissues into a secondary embryo with clear anterior-posterior and dorsal-ventral axes. He proposed that during normal development, these cells would organize the dorsal ectoderm into a neural tube and transform the flanking mesoderm into the body axis. It is now known (thanks largely to Spemann and his students) that the interaction of the chordamesoderm and ectoderm is not sufficient to "organize" the entire embryo. Rather, it initiates a series of sequential inductive events. The process by which one embryonic region interacts with a second region to influence that second region's differentiation or behavior is called **induction**. Because there are numerous inductions during embryonic development, this key induction wherein the blastopore lip cells induce the dorsal axis and the neural tube is traditionally called **primary embryonic induction.**[†]

We also know that the dorsal blastopore lip is active in organizing secondary embryos in *Amphioxus*, cyclostomes, and a variety of amphibians. In birds and mammals, the organizer originates at Koller's sickle (the posterior margin of the embryo), and Hensen's node acts as the dorsal blastopore lip. Cells migrating through Hensen's node become the head endoderm and chordamesoderm, while cells migrating through other portions of the primitive streak become lateral and ventral mesodermal cells. When Hensen's node from a young gastrula is transplanted into an epiblast of another young gastrula, it induces a complete secondary axis to form (Figure 15.13; Waddington, 1933; Storey et al., 1992; Khaner, 1995).

*Spemann's laboratory, and those of his students, usually used salamander embryos for their experiments. It turns out that *frog* ectoderm is much more difficult to induce than that of these urodeles.

[†]This classic term has been a source of confusion, because the induction of the neural tube by the notochord is no longer considered the first inductive process in the embryo. We will soon discuss inductive events that precede this "primary" induction.

The Nieuwkoop center

Despite the enormous amount of research that has been performed on amphibian embryos, we are just beginning to know the basic mechanisms of primary embryonic induction. In the past decade, numerous laboratories have focused their efforts on explaining embryonic induction in one amphibian—*Xenopus laevis*—and there is a consensus concerning the general outline of primary embryonic induction in this organism.

The data point to an orchestration of induction that has at least four stages. The first stage of induction takes place at fertilization. The unfertilized egg is radially symmetrical around the animal-vegetal axis. The entry of the sperm breaks this symmetry by causing the internal cytoplasm of the egg to rotate relative to the cortex (see Chapter 4). This asymmetry specifies the dorsal-ventral axis by mixing animal and vegetal cytoplasms in the vegetal cells that form opposite to the point of sperm entry. It appears that the cytoplasmic mixing activates dorsalizing determinants in these vegetal cells. These dorsalized vegetal cells are called the **Nieuwkoop center.** In the second stage, the descendants of these vegetal cells induce the cells above them to become the **Spemann-Mangold organizer.** The other vegetal cells induce the marginal cells above them to become the lateral and ventral mesoderm. Thus, there is an induction before "primary induction." In the third stage, the organizer converts neighboring mesoderm into dorsal mesoderm, and it instructs the dorsal ectoderm to become neural tissue. The fourth stage involves giving the induced neural tissue its regional characteristics (forebrain, hindbrain, spinal cord, etc.).

The Formation of the Nieuwkoop Center and Mesodermal Polarity

The endoderm is able to instruct those cells above it to become mesoderm. Moreover, the polarity of the endoderm is transferred to the mesodermal cells. Pieter Nieuwkoop (1969, 1973, 1977) demonstrated the importance of the vegetal (presumptive endoderm) cells in inducing the mesoderm. He removed the equatorial cells from the blastula and showed that neither the animal nor the vegetal caps produced mesodermal tissue. However, when the two caps were recombined, the animal cap cells were induced to form mesodermal structures such as notochord, muscles, kidney cells, and blood cells (Figure 15.14). The polarity of this induction (whether the region of animal cells formed notochord or muscles, etc.) depended on the dorsal-ventral polarity of the *endodermal* fragment. This set of factors capable of inducing the dorsal mesoderm has been called the Nieuwkoop center (Gerhart et al, 1989), and in *Xenopus laevis*, it resides in the dorsalmost vegetal cells of the blastula (Figure 15.15). [regu5.html]

The ventral and lateral vegetal cells also have roles in specifying the mesoderm. Whereas the ventral and lateral vegetal cells specify the intermediate (muscle, mesenchyme) and ventral (mesenchyme, blood, pronephric kidney) types of mesoderm, the dorsalmost vegetal cells specify the axial mesoderm components (notochord and somites; Figure 15.16). There appears, then, to be two signals: (1) a general "let's make mesoderm" signal from all the vegetal cells and (2) a specific "those cells above me are the dorsal (organizer) mesoderm" signal coming from the dorsalmost vegetal cells (D1). Dale and Slack (1987) provided evidence for a third inductive signal, one coming from the organizer cells (those marginal cells directly above the Nieuwkoop center) that "dorsalizes" the marginal mesodermal cells adjacent to them. When the ventral marginal cells are isolated, they give rise primarily to ventral mesodermal tissues. However, if they are cultured adjacent to dorsal marginal cells (i.e., the organizer), they generate intermediate

Dissected blastula fragments give rise to different tissue in culture:

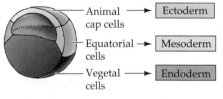

Animal and vegetal fragments give mesoderm

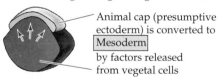

Figure 15.14
Summary of the experiments of Nieuwkoop and those of Nakamura and Takasaki, showing mesodermal induction by vegetal endoderm. Isolated animal cap cells become a mass of ciliated epidermis, isolated vegetal cells generate gutlike tissue, and isolated equatorial (marginal zone) cells become mesoderm. If the animal cap cells are combined with vegetal hemisphere cells, many of the cells of the animal cap generate mesodermal tissue.

mesodermal tissue. A fourth signal appears to be coming from the ventral region that opposes the signals from the organizer. Thus, there is evidence for a three-step specification of the mesoderm (Figure 15.17): (1) the induction of the organizer activity by the dorsalmost vegetal cells (the Nieuwkoop center), (2) the induction of the ventral mesoderm by the other vegetal cells, and (3) the dorsalization of the lateral marginal cells adjacent to the dorsal marginal cells to produce the intermediate mesoderm while the other marginal cells adopt ventral fates. The past decade has seen attempts to find the molecular interactions that generate this mesodermal patterning.

The Specification of Dorsoventral Polarity at Fertilization

As we saw in Chapters 4 and 6, dorsoventral specification is accomplished by the rotation of the inner cytoplasm of the egg relative to the cortex. If this rotation is inhibited by UV light, the embryo will not form dorsoanterior structures (Vincent and Gerhart, 1987). Render and Elinson (1986) and Wakahara (1989) cut eggs into fragments before and after this rotation. If the egg was cut before rotation, both sides developed dorsoanterior structures: head, notochord, neural tube. If the cut was made after the rotation had occurred, one fragment developed a head, heart, and some dorsal mesodermal structures, while the other fragment developed essentially into a *Bauchstück*, consisting almost solely of ventral cells, having little or no dorsal mesoderm and no nervous system. Sakai (1996) has shown that if the vegetal cytoplasm is deleted from the egg before rotation, the dorsal axis will not form, and that some dorsal determinants move from the vegetal cortex into the marginal zone on the future dorsal side. It appears, then, that this cytoplasmic rotation moves the dorsal-activating determinants to the future dorsal side of the egg.

By the 32-cell stage, the dorsoanterior determinants are contained within the dorsalmost (D1) blastomeres (see Figure 6.20; Gimlich and Gerhart, 1984;

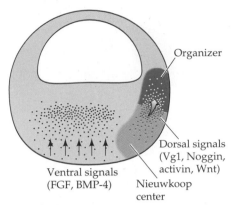

Figure 15.15
Model for mesoderm induction in *Xenopus*. A ventral signal (probably FGF2 or BMP4) is released throughout the vegetal region of the embryo. This induces the marginal cells to become mesoderm. BMP4 may specify the marginal cells to become posterior mesoderm. On the dorsal side (away from the point of sperm entry), a signal (probably initiated by Vg1 and propagated by activin, Noggin, and Wnt proteins) is released by the vegetal cells of the Nieuwkoop center. This dorsal signal induces the formation of the Spemann organizer in the overlying marginal zone cells. (After De Robertis et al., 1992.)

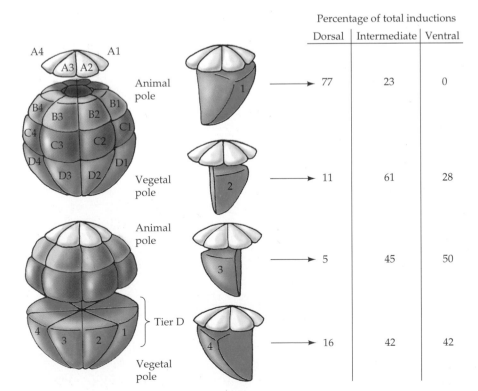

Figure 15.16
Regional specificity of mesoderm induction shown by recombining cells of 32-cell *Xenopus* embryos. The animal pole cells of 32-cell embryos were combined with individual vegetal blastomeres. The animal pole cells were labeled with fluorescent polymers to identify their descendants. The inductions resulting from these recombinations are summarized at right. (After Dale and Slack, 1987.)

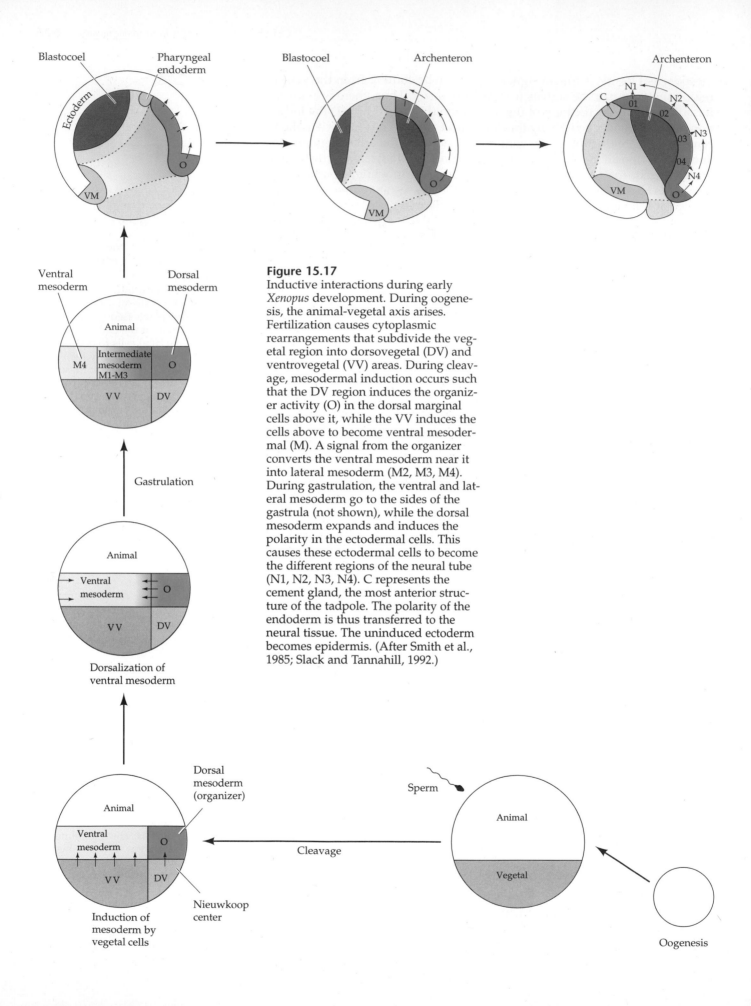

Figure 15.17
Inductive interactions during early *Xenopus* development. During oogenesis, the animal-vegetal axis arises. Fertilization causes cytoplasmic rearrangements that subdivide the vegetal region into dorsovegetal (DV) and ventrovegetal (VV) areas. During cleavage, mesodermal induction occurs such that the DV region induces the organizer activity (O) in the dorsal marginal cells above it, while the VV induces the cells above to become ventral mesodermal (M). A signal from the organizer converts the ventral mesoderm near it into lateral mesoderm (M2, M3, M4). During gastrulation, the ventral and lateral mesoderm go to the sides of the gastrula (not shown), while the dorsal mesoderm expands and induces the polarity in the ectodermal cells. This causes these ectodermal cells to become the different regions of the neural tube (N1, N2, N3, N4). C represents the cement gland, the most anterior structure of the tadpole. The polarity of the endoderm is thus transferred to the neural tissue. The uninduced ectoderm becomes epidermis. (After Smith et al., 1985; Slack and Tannahill, 1992.)

Gimlich, 1985, 1986). This localization was further demonstrated by recombination experiments (see Figure 15.16). Dale and Slack (1987) recombined single vegetal blastomeres from a 32-cell *Xenopus* embryo with the uppermost animal tier of a fluorescently labeled embryo of the same stage. The dorsalmost vegetal cell, as expected, induced the animal pole cells to become dorsal mesoderm. The remaining vegetal cells generally induced these animal cells to produce either intermediate or ventral mesodermal tissues. Thus, *dorsal vegetal cells* can induce animal cells to become dorsal mesodermal tissue.

It should be noted that in *Xenopus* (and other vertebrates), the formation of the anterior-posterior axis follows the formation of the dorsal-ventral axis. Once the dorsal portion of the embryo is established, the movement of the involuting mesoderm establishes the anterior-posterior axis. The mesoderm that migrates first through the dorsal blastopore lip gives rise to the anterior structures; the mesoderm in the ventral margin forms the posterior structures.

The molecular basis of mesoderm induction

Establishing Dorsal Regionalization: The Possible Role of β-catenin.

β-catenin is a multifunctional protein that can act as an anchor for cell membrane cadherins (Chapter 3) or as a nuclear transcription factor. In *Xenopus* embryos, the cortical rotation of fertilization is seen to move β-catenins to the future dorsal part of the egg. β-catenin continues to accumulate preferentially at the dorsal side throughout early cleavage, and this accumulation is seen in the nuclei of the dorsal cells (15.18A,B; Plate 7E,F; Schneider et al., 1996; Larabell et al., 1997). This region of β-catenin accumulation originally appears to cover both the Nieuwkoop center and organizer regions. During later cleavage, the cells with β-catenin may reside specifically in the Nieuwkoop center (Heasman et al., 1994; Guger and Gumbiner, 1995).

β-catenin is necessary for forming the dorsal axis, since depletion of β-catenin transcripts with antisense oligonucleotides results in the lack of dorsal structures (Heasman et al., 1994). Moreover, the injection of exogenous β-catenin into the ventral side of the embryo produces a secondary axis (Funayama et al., 1995; Guger and Gumbiner, 1995). β-catenin is part of the Wnt signaling transduction pathway and is negatively regulated by the glycogen synthase kinase 3 (GSK-3; Chapter 3). GSK-3 is also critical for axis formation, and activated GSK-3 blocks axis formation when added to the egg (Pierce and Kimelman, 1995; He et al., 1995; Yost et al., 1996). If endogenous GSK-3 is knocked out by a dominant negative mutation in the ventral cells of the early embryo, a second axis forms (Figure 15.18C). Labeling experiments (Yost, et al., 1996; Larabell et al., 1997) suggest that β-catenin is initially synthesized (from maternal messages) throughout the embryo, but that it gets degraded by GSK-3 phosphorylation specifically in the ventral cells. It is not known what causes these regional differences in GSK-3 activity. Experimentally, endogenous GSK-3 can be inhibited by adding Wnt proteins into the egg, and these Wnts have been found to induce secondary axes (McMahon and Moon, 1989; Sokol et al., 1991). But Wnts may not be the natural regulators of GSK-3 in the dorsal side of the embryo; dominant negative mutations of Wnt proteins and their receptors do not seem able to block the formation of the normal axis (Hoppler et al., 1996; Sokol, 1996). Current studies are underway to see if the cortical rotation in *Xenopus* eggs somehow regulates GSK-3 activity and if there is another agent (besides Wnt proteins) capable of inactivating GSK-3.

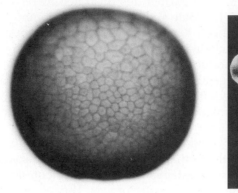

(A)

(B)

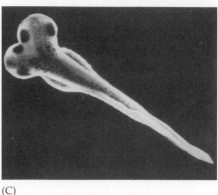

(C)

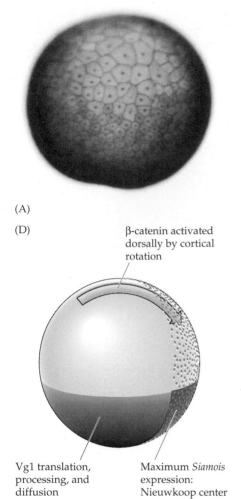

(D)

β-catenin activated dorsally by cortical rotation

Vg1 translation, processing, and diffusion

Maximum *Siamois* expression: Nieuwkoop center

Figure 15.18

Role of Wnt-pathway proteins in dorsal-ventral axis specification. (A,B) Differential translocation of β-catenin protein into *Xenopus* blastomere nuclei. (A) Presumptive dorsal side of a *Xenopus* blastula stained for β-catenin shows nuclear localization. (B) Such nuclear localization is not seen on the ventral side of the same embryo. (C) Dorsal axis formation caused by the injection of both blastomeres of a 2-cell *Xenopus* embryo with dominant inactive GSK-3β. Dorsal fate is actively suppressed by wild-type GSK-3β. (D) Irenic model whereby the Nieuwkoop center (characterized by *Siamois* gene expression and the ability to induce dorsal mesoderm) is created by the synergy of the activation of β-catenin dorsally and the activation of Vg1 vegetally. (A and B from Schneider et al., 1996, photographs courtesy of P. Hausen; C from Pierce and Kimelman, 1995, photograph courtesy of D. Kimelman.)

β-catenin is an HMG-box transcription factor that may be a DNA bending protein. It may predispose different cells to respond differently after gene expression is initiated at the midblastula transition. Once in the nucleus of the dorsal vegetal cells, it activates particular target genes, one of them being the homeobox-containing gene *Siamois*. This gene is expressed in the Nieuwkoop center immediately following the midblastula transition. If this gene is ectopically expressed in the ventral vegetal cells, a secondary axis emerges on the former ventral side of the embryo, and if cortical rotation is prevented, *Siamois* expression is eliminated (Lemaire et al., 1995; Brannon and Kimelman, 1996). Recent studies (Brannon and Kimelman, 1996) suggest that maximum *Siamois* expression occurs when there is synergism between the GSK-3/β-catenin pathway and a vegetally expressed TGF-β signal. The cortical rotation may activate the β-catenins and allow the expression of *Siamois* in the dorsal region of the embryo. At the same time, the translation of vegetally localized messages encoding a factor of the TGF-β family may generate a protein that permits the activation of β-catenin better in the vegetal cells than in the animal cells (Figure 15.18D). Cui and colleagues (1996) provide evidence that this TGF-β family member is the mature Vg1 protein, a protein expressed solely in the vegetal cells. The result is that *Siamois* would be expressed in the dorsalmost vegetal cells constituting the Nieuwkoop center.

The Functioning of the Nieuwkoop Center: Roles for Vg1 and Noggin

THE ACTIVATED VG1 PROTEIN. There may be several ways of inducing the dorsal mesoderm. First, the Vg1 protein is able to induce dorsal mesoderm formation in the cells above it. The mRNA for Vg1 is tethered to the vegetal yolk mass during oogenesis and remains in the vegetal hemisphere during cleavage (Chapters 4 and 12; Plate 8). After fertilization, Vg1 protein is made

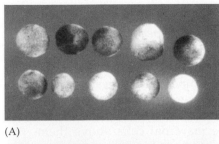

(A)

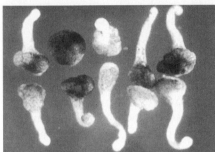

(B)

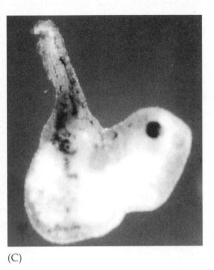

(C)

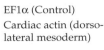

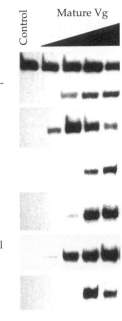

EF1α (Control)
Cardiac actin (dorso-lateral mesoderm)
Xbra (general mesoderm)
Gsc (dorsal anterior mesoderm)
Noggin (dorsal anterior mesoderm)
Xwnt8 (ventrolateral mesoderm)
NCAM (neural)

(D)

Figure 15.19
Mature Vg1 protein induces morphogenetic movements and dorsal mesodermal gene expression in ectodermal explants. Blastula-stage animal cap explants were cultured (A) in untreated medium or (B) in medium containing the mature (cleaved) Vg1 protein. The Vg1 protein induced convergent extension movements in the cap. When left in the treated medium for a longer time (C) the animal cap explants formed larva-like structures, including notochord, muscles, eyes, cement gland, and anterior-posterior axis. (D) As the concentration of Vg1 protein rises, it induces a more dorsal set of mesodermal markers. The lowest concentration is 0 (control), followed by 1, 3, 10, and 30 percent Vg1 supernatant. (After Kessler and Melton, 1995; photographs courtesy of D. A. Melton.)

throughout the vegetal hemisphere of the blastula, but it is in an inactive precursor form that must be cleaved to become active. Activated Vg1 protein is able to (1) induce dorsal mesoderm in animal cap cells; (2) induce an entire embryonic axis when microinjected into ventral vegetal cells; and (3) rescue the dorsal axis of UV-irradiated eggs when microinjected into the dorsal vegetal cells (Dale et al., 1993; Thomsen and Melton, 1993; Kessler and Melton, 1995).

Kessler and Melton (1995) have shown that activated Vg1 protein caused the active elongation of the notochord mesoderm as well as the dosage-dependent activation of mesodermal markers. When blastula-stage animal pole caps are placed in low concentrations of processed Vg1, the Vg1 protein induces the expression of genes such as *Brachyury*, that characterize general mesoderm. Slightly higher doses of Vg1 induce the expression of lateral mesodermal markers (*Xwnt8* and *actin*), and at higher concentrations, Vg1 induces these cells to express the dorsal mesoderm markers *goosecoid* and *noggin* (Figure 15.19). However, Cui and colleagues (1996) have found that Vg1, alone, is not able to cause the differentiation of the notochord in vivo. For this to occur, they find that the cells need both Vg1 and the Wnt-pathway products. (The Wnt-pathway was not sufficient alone to induce dorsal mesoderm.) It is possible that the combination of Vg1 and some product specified by the *Siamois* gene are able to induce the specification of the dorsal mesoderm and its differentiation into notochord.*

The mature (processed) Vg1 protein thus appears to be critical in the functioning (if not the establishment) of the Nieuwkoop center of amphibians. Vg1 is also seen in the homologous region of the chick embryo—the posterior marginal zone. Moreover, when processed Vg1 protein is experimentally introduced into lateral areas of the chick blastoderm, a new

---

*Alternatively, this may be another example of Spemann's (1938) notion of "double assurance." The embryo could specify the dorsal mesoderm by the synergism of Vg1 and β-catenin, (without a Nieuwkoop center). The same result could be achieved by a signal initiated by the *Siamois* gene of the Nieuwkoop center beneath it. Spemann likened double assurance to wearing both a belt and suspenders.

Nieuwkoop center is formed and a secondary axis is induced (Seleiro et al., 1996).

Induction of Ventral and Lateral Mesodermal Specificity

So far, we have been discussing the induction of the dorsal mesoderm by the dorsalmost vegetal cells. But there is more. The other vegetal cells are able to induce the cells above them to become ventral mesoderm. Experiments by Smith and his colleagues (1991) have shown that at midblastula, the ventrolateral and dorsal vegetal blastomeres of *Xenopus* induce the expression of the *Brachyury* gene in the marginal cells above them. The *Brachyury* mRNA encodes a transcription factor whose function is crucial to the formation of the mesoderm. It is expressed prior to α-actin and the other proteins that are the products of mesodermal cells, and if the *Brachyury* gene is expressed in cells where the transcription factor is normally inactive, those cells become mesodermal (Cunliffe and Smith, 1992). If the animal hemisphere containing the marginal zone cells is removed from the vegetal hemisphere at midblastula, no mesoderm is formed in the animal hemisphere. However, if vegetal cells are added back to such animal hemispheres, the *Brachyury* gene becomes expressed, and those cells expressing the gene become mesodermal. Thus, the vegetal cells induce the expression of mesodermal genes in the marginal zone cells. Without this interaction, the marginal zone cells remain ectodermal.

FIBROBLAST GROWTH FACTORS.   There is much debate as to the identities of the general mesodermal inducers found in the ventral and lateral vegetal cells. Fibroblast growth factors (and their messages) have been found in the *Xenopus* egg and embryo, and they are thought to allow the marginal cells to respond to the Vg1 (or another activin-like protein) (Cornell and Kimelman, 1994; LaBonne and Whitman, 1994). The functional significance of such secreted FGF molecules was demonstrated by destroying the embryo's receptors for FGF by dominant negative mutations (Chapter 3). When this experiment was done, the embryos that lacked functional FGF receptors had dramatically reduced amounts of posterior and lateral mesoderm (Plate 3). One possibility is that a graded amount of active Vg1 creates the pattern, with small amounts inducing ventral mesoderm, larger amounts inducing lateral mesoderm, and still larger concentrations inducing the dorsal mesoderm. Such gradients have been seen in culture.

BMP4.   Another molecule believed to be important for mesoderm specification is **bone morphogenetic protein 4 (BMP4)**. There appears to be an antagonistic relationship between BMP4 and the dorsal mesoderm. If the mRNA for BMP4 is injected into 1-cell *Xenopus* eggs, all the mesoderm in the embryo becomes ventrolateral mesoderm, and no involution occurs at the blastopore lip (Dale et al., 1992; Jones et al., 1992). Further evidence for the role of BMP4 in ventrolateral mesoderm induction came from implantation experiments. When animal caps from embryos injected with *bmp4* message were isolated and implanted into the blastocoels of young *Xenopus* blastulae, they caused the formation of an extra tail (Figure 15.20). Conversely, overexpression of a dominant-negative *bmp4* receptor resulted in the formation of two dorsal axes (Graff et al., 1994; Maeno et al., 1994). It is possible that BMP4 is inducing a set of transcription factors that specify the mesoderm to be lateral or posterior (Stennard et al., 1996; Zhang and King, 1996). Thus, the formation of the posterior (ventrolateral) mesoderm appears to be generated by the actions of FGF and BMP4.

Figure 15.20
The importance of BMP4 in producing posterior structures can be seen when *bmp4* mRNA was injected into embryos and the resulting animal cap cells were transplanted beneath the young gastrula ectoderm. The treated larvae usually developed an extra tail. (From Jones et al., 1992, courtesy of B. Hogan.)

The creation of organizer activity

Secreted Proteins from the Organizer

The organizer is induced by the Nieuwkoop center. While the Nieuwkoop center cells remain endodermal, the cells of the organizer become the dorsal mesoderm (head mesoderm, notochord, paraxial mesoderm) and become positioned underneath the dorsal ectoderm. There, they will induce the central nervous system to form. The properties of the organizer tissue can be divided into five major functions:

1. The ability to become dorsal mesoderm (notochord, etc.)
2. The ability to dorsalize the surrounding mesoderm into lateral mesoderm (when it would otherwise form ventral mesoderm)
3. The ability to dorsalize the ectoderm into neural ectoderm
4. The ability to initiate the movements of gastrulation
5. The ability to cause the neural plate to become the neural tube

The cells of the organizer ultimately contribute to four cell types—pharyngeal endoderm, head mesoderm, notochord, and the chordaneural hinge (Keller, 1976; Gont et al., 1993). The pharyngeal endoderm leads the migration of the organizer tissue and appears to induce the most anterior head structures. The head mesoderm induces the forebrain and midbrain, the notochord induces the hindbrain and trunk, and the chordaneural hinge induces the tip of the tail. Vodicka and Gerhart (1995) have recently correlated fluorescent-cell-marking techniques and in situ hybridization to obtain a map of those cells that give rise to the organizer. It was found that the most animal portion (10 percent) was derived from the A1 blastomeres of the 32-cell blastula; the central region (70 percent) was derived from the progeny of the B1 blastomeres; and around 20 percent (the vegetal and deep cells) was derived from the C1 blastomere directly above the D1 cells of the Nieuwkoop center. All six A1, B1, and C1 blastomeres produced deep and superficial cells. The progeny of the C1 blastomere form the most vegetal, leading part of the organizer, and these are the cells that form the head mesoderm.

When the organizer was first described, it started the first truly international scientific research program—the search for the organizer molecules. Researchers from Britain, Germany, France, the United States, Belgium, Finland, Japan, and the Soviet Union all tried to find these remarkable substances (see Gilbert and Saxén, 1993). R. G. Harrison (quoted by Twitty, 1966) referred to the amphibian gastrula as the "new Yukon to which eager miners were now rushing to dig for gold around the blastopore." Unfortunately, their picks and shovels proved too crude to uncover the molecules involved. The analysis of organizer molecules had to wait until recombinant DNA technologies enabled investigators to make cDNA clones from blastopore lip mRNA and to see which of these clones encoded factors that could dorsalize the embryo. The formation of the dorsal (organizer) mesoderm involves the activation of several genes. The secreted proteins of the Nieuwkoop center are thought to activate a set of transcription factors in the mesodermal cells above them. These transcription factors would then activate the genes encoding the secreted products of the organizer. Several organizer-specific proteins have been found and are listed in Table 15.3. Because the properties of the organizer depend on its secreted factors, we will start with these proteins. [regul3.html]

Table 15.3 Proteins expressed solely or almost exclusively in the organizer (partial list)

Nuclear proteins	Secreted proteins
Lim1	Chordin
XANF1	Noggin
Goosecoid	Follistatin
HNF3β-related proteins (e.g., Forkhead, Pintallavis)	Sonic hedgehog
	Cerberus
	Nodal-related proteins (several)

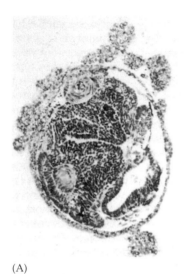

(A)

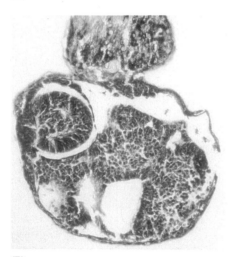

(B)

Figure 15.21
Soluble inducing factors and their identification. (A) Neural structures induced in presumptive ectoderm by newt dorsal lip, separated from the ectoderm by a Nucleopore filter with an average pore diameter of 0.05 μm. Anterior-type neural cells are evident, including some induced eyes. (B) Similar type of induction seen when *Xenopus* animal cap (presumptive ectoderm) is injected with *chordin* mRNA and treated with soluble FGF2. (A from Toivonen, 1979; B from Sasai et al., 1996; photographs courtesy of L. Saxén and E. De Robertis, respectively.)

Evidence for diffusible signals from the notochord came from several sources, the most critical being the transfilter studies of the Finnish investigators (Saxén, 1961; Toivonen et al., 1975; Toivonen and Wartiovaara, 1976). Newt dorsal lip was placed on one side of a filter thin enough so that no processes could fit through the pores, and competent gastrula ectoderm was placed on the other side. After several hours, neural structures were observed in the ectodermal tissue (Figure 15.21). The identities of these factors diffusing from the organizer took a quarter of a century to identify. There are several such molecules currently being studied: Chordin, Noggin, Follistatin, Sonic hedgehog, and Cerberus.

CHORDIN. One of the first roles of the organizer is to "protect itself" from being ventralized. BMP4 is produced throughout the *Xenopus* blastula and actively makes mesoderm ventral (Graff et al., 1994). In other words, the formation of ventral mesoderm is not merely due to the absence of dorsal cues; it is actively constructed. Moreover, as previously described, BMP4 can override the dorsal signals. The dorsalized mesoderm blocks the BMP4 signal by secreting **Chordin** and **Noggin** (Sasai et al., 1994; Holley et al., 1995). Chordin is a secreted protein that is activated by the homeodomain-containing transcription factors Goosecoid and Xnot2. It is originally detected in the dorsal marginal zone about an hour before gastrulation; and as gastrulation begins, *chordin* message is seen only in the dorsal blastopore lip (Figure 15.22). Thereafter, *chordin* is expressed in the prechordal plate (the head mesoderm that precedes the notchord anteriorly) and notochord. When the last inductions in the tail are occurring, Chordin is found in the chordoneural hinge, the last vestige of the organizer. Chordin can induce a secondary axis when microinjected into the ventral sides of *Xenopus* blastulae, and it is thought to do this by interfering with the action of BMP4.

BMP4 is initially expressed throughout the ectoderm and mesodermal regions of the late blastula. However, during gastrulation, *bmp4* transcripts are restricted to the ventrolateral marginal zone (Hemmati-Brivanlou and Thomsen, 1995; Northrop et al., 1995). The BMP4 protein induces the expression of several transcription factors (Xvent-1, Vox, Mix.1, Xom) that are key regulators of ventral mesodermal development. Thus, BMP4 activates ventral gene expression. The transcription factors induced by BMP4 repress *goosecoid* and other dorsal genes while at the same time activating ventrolateral mesodermal proteins (Gawantka et al., 1995; Hawley et al., 1995; Mead et al., 1996; Schmidt et al., 1996). In this way, BMP4 activates ventral mesodermal development and suppresses dorsal development. In *Xenopus*, *chordin* and *noggin* both directly bind and inactivate BMP4, thus preventing it from acting on cells near the organizer (Figure 15.23; De Robertis and Sasai, 1996; Piccolo et al., 1996; Sasai et al., 1996; Zimmerman et al., 1996).

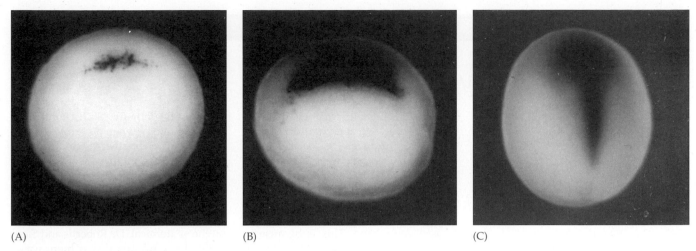

(A) (B) (C)

Figure 15.22
Chordin mRNA localization. (A) Whole-mount in situ hybridization shows that just prior to gastrulation, *chordin* message is expressed in the region that will become the dorsal blastopore lip. (B) As gastrulation begins, *chordin* is expressed at the dorsal blastopore lip, and (C) it is seen in the organizer tissues. (From Sasai et al., 1994; photographs courtesy of E. De Robertis.)

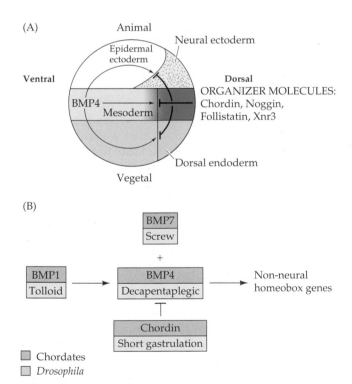

Figure 15.23
Model for the action of the organizer. (A) BMP4 (and certain other molecules) are powerful ventralizing factors. Organizer proteins such as Chordin and Noggin may block the action of BMP4. (Follistatin may inhibit the action of BMP7, which combines with BMP4 to activate it.) The antagonistic effects of these proteins can be seen in all three germ layers. (B) Homologous developmental pathways in the formation of the central nervous systems of a vertebrate (*Xenopus*) and an invertebrate (*Drosophila*). The vertebrate factor is in black, the homologous *Drosophila* protein is in color. (After De Robertis and Sasai, 1996; Sasai et al., 1996.)

BMP4 and Geoffroy's Lobster

RECENTLY, the laboratories of De Robertis and Kimelman have shown that the reactions that lead to the formation of the dorsal neural tube in *Xenopus* are the same reactions that lead to the formation of the ventral nerve cord in insects (see Figure 15.23B; Holley et al., 1995; Schmidt et al., 1995). The *Drosophila* homologue to the *bmp4* gene is *decapentaplegic* (*dpp*). As discussed in the previous chapter, the Dpp protein is responsible for the patterning of the dorsal-ventral axis of *Drosophila*, and it is present in the dorsal portion of the embryo and diffuses ventrally. Here, it is opposed by a protein called Short-gastrulation (Sog). Short-gastrulation is the *Drosophila* homologue of Chordin. Not only do the homologues appear to be similar; they can substitute for each other. When *short-gastrulation* mRNA is injected into ventral regions of *Xenopus* embryos, it induces the *Xenopus* notochord and neural tube. Injection of *chordin* mRNA into *Drosophila* gives ventral nervous tissue. Although the *Xenopus* Chordin usually functions to dorsalize the embryo, it ventralizes the *Drosophila* embryo. This is because in *Drosophila*, Dpp is made dorsally. In *Xenopus*, BMP4 is made ventrally. In both cases, Sog/Chordin makes neural tissue by blocking the effects of Dpp/BMP4. In *Drosophila*, Dpp interacts with the product of the *screw* gene to function. In *Xenopus*, the homologue of *screw*, *Bmp7*, appears to be essential for the ventralizing effect of BMP4 (Hawley et al., 1995).

In 1822, the French anatomist Etienne Geoffroy Saint-Hilaire provoked one of the most heated and critical confrontations in biology when he proposed that the lobster was but the vertebrate upside down. He claimed that the ventral side of the lobster (with its nerve cord) was homologous to the dorsal side of the vertebrate (Appel, 1987). It seems that he was correct on the molecular level, if not on the anatomical. De Robertis and Sasai (1996) have proposed that there was a common origin for all bilateral phyla—a hypothetical creature (dubbed *Urbilateria*) some 600 million years ago that was the ancestor for both the protostome and the deuterostome subkingdoms. The BMP4(Dpp)/Chordin(Sog) interaction is an example of "homologous processes," suggesting a unity of developmental principles among all animals (Gilbert et al., 1996). ■

NOGGIN.   One of the other agents of the organizer appears to be the product of the *noggin* gene. Smith and Harland (1991, 1992) isolated this gene by constructing a cDNA library from dorsalized (lithium-treated) gastrulae. RNAs synthesized from sets of these plasmids were injected into the ventralized embryos produced by UV irradiation. Those sets of plasmids whose RNAs rescued the dorsal axis were split into smaller sets, and so on, until single clones were isolated whose mRNAs were able to restore the dorsal axis in such embryos. One of these clones contained *noggin*. Smith and Harland (1992) have shown that newly transcribed (as opposed to maternal) *noggin* mRNA is first localized in the dorsal blastopore lip region and then becomes expressed in the notochord (Plate 6). Moreover, if the early embryo is treated with lithium chloride (LiCl) so that the entire mesodermal mantle becomes notochord-like organizer tissue, then *noggin* mRNA is found throughout the mesodermal mantle. Treatment of the early embryo with ultraviolet light (which prevents dorsal blastopore lip formation) inhibits the synthesis of *noggin* mRNA. Injection of *noggin* mRNA into 1-cell, UV-irradiated embryos completely rescues the dorsal axis and allows the formation of a complete embryo (Plate 5). If too much Noggin protein is synthesized at that time, the embryo becomes "hyperdorsal," forming *only* the head region (hence the name "noggin"). The mRNA for this Noggin protein is already present in the fertilized egg, and the sequence of the protein (as deduced from the gene) suggests strongly that Noggin is a secreted protein. It seems, then, that Noggin is an excellent candidate for mediating some of the functions of the organizer.

Recent evidence suggests that Noggin protein can accomplish two major functions of the Spemann-Mangold organizer: it *induces neural tissue* from the dorsal ectoderm, and it *dorsalizes the mesoderm cells* that would otherwise contribute to the ventral mesoderm. Smith and co-workers (1993) have shown that Noggin protein can dorsalize the ventral marginal zone cells at gastrulation and respecify their fate from ventral mesoderm (mesenchyme and blood cells) to more intermediate fates (muscle, heart, and pronephric kidney). When Smith and co-workers removed ventral marginal zones (the presumptive ventral mesoderm) from *Xenopus* gastrulae and placed them into medium containing soluble Noggin protein, these explants produced muscle-specific mRNA, something that is normally reserved for the dorsal marginal explants. These explants also became elongated (another characteristic of dorsal development). However, the elongated explants did not stain for notochordal tissue. These experiments show that soluble Noggin protein can induce gastrula ventral mesoderm cells to become muscle (but not notochord), and it therefore resembles the signal from the organizer that dorsalizes the lateral mesodermal tissue (see Figure 15.17).

The Noggin protein is also able to induce neural tissue in gastrula ectoderm without the presence of any dorsal mesoderm (Lamb et al., 1993). When Noggin is added to gastrula (or animal cap) ectoderm, the ectodermal cells are induced to express forebrain-specific neural markers. Moreover, the gene products for notochordal or muscle cells are not induced by the Noggin protein. Since Noggin is a secreted protein synthesized by the derivatives of the organizer (the head mesoderm and chordamesoderm) during gastrulation (when induction takes place), and since it inactivates BMP4 (which would ventralize the embryo), Noggin is thought to play a major role in both the dorsalization of the mesoderm and the neuralization of the dorsal ectoderm.*

FOLLISTATIN. Hemmati-Brivanlou and Melton (1994) demonstrated that the activin-binding protein Follistatin is present in the dorsal blastopore lip and later becomes restricted to the notochord. Although originally thought to bind only activin, there is now evidence (Yamashida et al., 1995) that Follistatin can inhibit the activities of BMP7. BMP7 is needed for the activation of BMP4, so by inhibiting BMP7, Follistatin can also prevent ventralization of the mesoderm. Follistatin also plays a role in dorsalizing the ectoderm. It appears that activin (or, more likely, an activin-like protein such as BMP7) is necessary for repressing neural induction. Follistatin, by binding this protein, releases the inhibition and allows the tissue to become neural (Hemmati-Brivanlou et al., 1994; Hawley et al., 1995).

It is interesting that Noggin, Chordin, and Follistatin are all inhibitors. Here we see a principle that underlies much of development: activation is often accomplished by inhibiting a repressor. This might be explained by the fact that in each nucleus, most of the genes are repressed. To activate a particular gene, an inhibitor of this repression is needed. Similarly, inhibition is often accomplished by the suppression of that inhibitor of the repressor. (Developmental biologists get used to talking in double and triple negatives.) In this instance, the "default" state of the ectoderm is to become neural, unless acted upon by BMP4. The proteins from the organizer mesoderm prevent to the action of BMP4 on the ectoderm.

*Noggin may also be functioning as part of the Nieuwkoop center. A material *noggin* mRNA is translated in the early blastula (Smith and Harland, 1992) and a recent investigation (Lustig et al., 1996) shows that Noggin works with a cofactor, *Xenopus nodal related-1,* to induce the early gastrula. Xnr1 may also be involved in forming the left-right axis in *Xenopus*. During neurulation, it is asymmetrically expressed in the lateral plate mesoderm, being present only in the left-hand side of the embryo. This expression pattern resembles that of *nodal* genes in chicks and mice, wherein the expression of *nodal* is critical for the establishment of the left-right axis (Chapter 16).

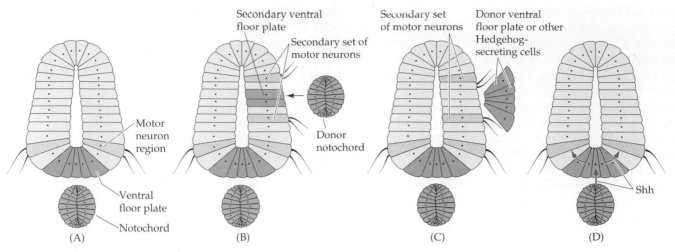

Figure 15.24
Cascade of inductions initiated by the notochord on the newly formed neural tube. (A) Two cell types in the newly formed neural tube. Those cells closest to the notochord become the ventral floor plate cells. Motor neurons emerge on the ventrolateral sides. (B) If a second notochord is transplanted adjacent to the neural tube, it induces a new set of floor plate cells and two new sets of motor neurons. (C) If the ventral floor plate cells are transplanted adjacent to the neural tube, new sets of motor neurons differentiate. (D) The inductive interactions between these cells. The red arrows represent secreted Sonic hedgehog protein. (After Placzek et al., 1990.)

SONIC HEDGEHOG. Sonic hedgehog is utilized after most of the inductive events of neurulation have taken place. It is used to pattern the newly formed neural tube. Sonic hedgehog is expressed in the notochord, and the amino-terminal portion of this protein is secreted (see Figure 7.11). If notochord fragments are taken from one embryo and transplanted to the lateral sides of a host neural tube, the host neural tube will form another set of floor plate cells at the sides of the neural tube. If a piece of notochord is removed from an embryo, the neural tube adjacent to the deleted region has no floor plate cells (Figure 15.24; Placzek et al., 1990; Yamada et al., 1991). These floor plate cells, once induced, induce the formation of motor neurons on either side of them. The same results can be obtained if the notochord fragments are replaced by pellets of cultured cells secreting Sonic hedgehog (Echelard et al., 1993; Roelink et al., 1994). Sonic hedgehog from the floor plate cells is able to further polarize the neural tube. It induces the motor neurons on the ventral lateral regions, and it prevents the dorsalization of the ventral neural tube by antagonizing the effects of BMP4 coming from the dorsal epidermis (see Chapter 7).*

CERBERUS. The induction of the most anterior head structures is accomplished by a secreted protein called Cerberus. Unlike the other secreted proteins, Cerberus promotes the formation of the cement gland, eyes, and olfactory placodes. Moreover, unlike Noggin or Chordin, Cerberus suppresses the formation of dorsal mesoderm, while inducing cardiac mesoderm and liver (a foregut endoderm derivative). When *cerberus* mRNA was injected into the vegetal ventral (D4) set of blastomeres at the 32-cell stage, ectopic head structures were formed (Figure 15.25; Bouwmeester et al., 1996). These head structures were made from the injected cell as well as from neighboring cells. The *cerberus* gene is expressed in those cells that lead the anterior movement of gastrulating cells into the embryo. These are the cells of the involuting endoderm (in the deep layer of the organizer) that give rise to the foregut and its derivatives and which underlie the head. The *cerberus* mes-

---

*BMP4 acts as a ventralizing agent in the formation of the neural tube (actively preventing its formation in the ventral part of the embryo), but once the neural tube is made, it can act as a dorsalizing agent, being secreted from the upper epidermis to dorsalize the neural tube (see Chapter 7). A versatile player, it will stimulate muscle development in the myotome, pattern the developing tooth, and even destroy the webbing between our fingers and toes. It is often paired with Sonic hedgehog in the formation of organ primordia.

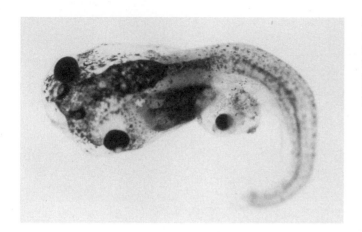

Figure 15.25
Cerberus mRNA injected into a single D4 (ventral vegetal) blastomere of a 32-cell *Xenopus* embryo induces head structures as well as a duplicated heart and liver. The secondary eye (a single cyclopic eye) and olfactory placode can be readily seen. (From Bouwmeester et al., 1996; photograph courtesy of E. M. De Robertis.)

sage is dependent on the rest of the organizer being active, and transcription of the *cerberus* gene is activated by Follistatin, Noggin, and Chordin. This may explain why *cerberus* transcription is limited to the region of the involuting endoderm closest to the organizer, a region that overlaps with *chordin* expression.

Transcription Factors Induced in the Organizer

The activities of the Nieuwkoop center are thought to activate a set of genes encoding transcription factors in the mesoderm above it. Several transcription factors have been found that are organizer-specific; that is, they are expressed solely in the dorsal blastopore lip and the resulting notochord. Two of these proteins are XANF-1 and Goosecoid.

XANF-1 is a homeodomain-containing transcription factor that may be one of the earliest expressed. By the beginning of gastrulation, XANF-1 is predominantly in the deep layers of the dorsal blastopore lip, the precursors of the head mesoderm, and injection of *XANF-1* mRNA into ventral blastomeres induces the formation of a secondary axis. Those injected cells become the anterior mesoderm of the secondary axis (Zaraisky et al., 1995). Thus, XANF-1 may control the migratory behavior of the deep cells of the dorsal blastopore lip and the differentiation of these cells into organiser tissue.

Goosecoid appears to function very much like XANF1. The message for Goosecoid was found by screening libraries of dorsal blastopore lip cDNA with probes to genes that are active in *Drosophila* axis formation (Blumberg et al., 1991; Cho et al., 1991a). The *goosecoid* transcripts are first detected at the late-blastula stage, indicating that this is a nuclear-controlled gene, and those transcripts accumulate in the area directly overlying the dorsal blastopore lip in the dorsal mesodermal precursor (C1 blastomere) cells. In animal cap cultures, either Vg1 protein or activin, but not FGF2 or Noggin, can induce the transcription of the *goosecoid* gene (Cho et al., 1991a; Thomsen and Melton, 1993).The expression of *goosecoid* mRNA also correlates with the organizer domain in experimentally treated animals. When LiCl is used to increase the dorsoanterior-inducing mesoderm in the marginal zone, the expression of *goosecoid* likewise is expanded. Conversely, when eggs are treated by UV light prior to first cleavage, both dorsoanterior induction and *goosecoid* expression are significantly inhibited. Injection of the full-length *goosecoid* message into the two ventral blastomeres of the 4-cell *Xenopus* embryo causes the progeny of these blastomeres to involute, undergo convergent extension, and form the dorsal mesoderm and head endoderm of the secondary axis (Figure 15.26; Niehrs et al., 1993). Moreover, labeling experi-

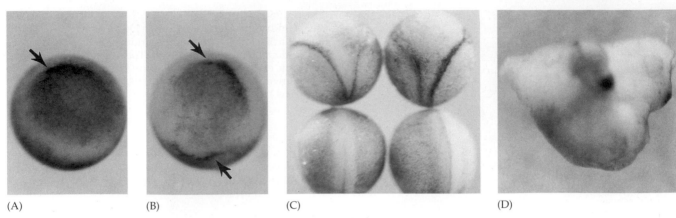

(A)   (B)   (C)   (D)

Figure 15.26
Ability of *goosecoid* mRNA to induce a new axis. (A) At gastrula, the control embryo (either uninjected or given an injection of *goosecoid*-like mRNA but lacking the homeobox) has one dorsal blastopore lip. (B) An embryo whose ventral vegetal blastomeres were injected at the 16-cell stage with *goosecoid* message. Note the secondary dorsal blastopore lip. (C) Top, two neurulae that had been injected with *goosecoid* mRNA, showing two axes; bottom, two control neurulae. (D) Twinned embryo produced by the *goosecoid* injection. Complete head structures have been induced. (After Cho et al., 1991a; Niehrs et al., 1993; courtesy of E. De Robertis.)

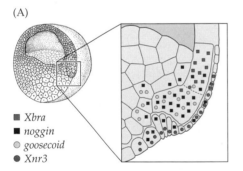

(A)

■ *Xbra*
■ *noggin*
○ *goosecoid*
● *Xnr3*

(B)

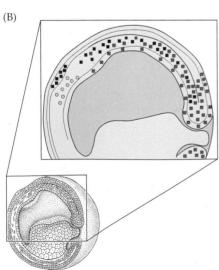

ments (Niehrs et al., 1993) show that the *goosecoid*-injected cells are able to recruit neighboring host cells into the dorsal axis as well. Thus, the Nieuwkoop center activates the *goosecoid* gene encoding a DNA-binding protein that (1) activates the migration properties (involution and convergent extension) of the dorsal blastopore lip cells, (2) autonomously determines the head endodermal and dorsal mesodermal fates of those cells expressing it, and (3) enables the *goosecoid*-expressing cells to recruit neighboring cells into the dorsal axis. Goosecoid has been found to activate the *Xotx2* gene in the anterior mesoderm and in the presumptive brain ectoderm (Blitz and Cho, 1995). *Xotx2* is the *Xenopus* homologue of the *orthodenticle* gene that is essential for brain development in flies and mice.

"Organizer-specific" gene expression can be used to subdivide the early organizer into regions having different combinations of these messages (Figure 15.27; Vodicka and Gerhart, 1995). At the beginning of gastrulation, as the organizer cells involute into the embryo, these configurations change. Within the deep cells, *goosecoid* is now seen in the most anterior portions (mostly constructed from C1 cells), especially the prechordal plate head mesoderm. The overlap of *noggin* and *Xbra* genes defines the notochord, and the region having *Xbra* without *noggin* defines the domain destined to become posterior endoderm. A second expression domain of *noggin* is seen in the anterior neural plate. [regu6.html]

Figure 15.27
Fine structure of the organizer. (A) At the beginning of gastrulation, *Xbra* is in the more animal cells of the organizer, while *noggin* is more vegetal. The vegetal cells involute first and become placed more anteriorly. (B) The same factors seen near the end of gastrulation. The expression zones are more discrete and less overlapping, and there is no correlation between the original cell placement and its later gene expression pattern. (After Vodicka and Gerhart, 1995.)

How Does the Organizer Neuralize the Ectoderm?

WHILE THE IDENTITY of the signaling molecules is being uncovered, the mechanism of their actions is still an enigma. It is probable that in addition to blocking the ventralizing signal (BMP4), the organizer must also activate the ectodermal cells to become the neural plate. While the molecule(s) doing this are not known, it appears that neuralization can be accomplished by the combination of two separate reactions: increasing intracellular cyclic AMP in the ectodermal cells and activating Protein Kinase C (PKC)

on their cell membranes. Several studies (Davids et al., 1987; Davids, 1988; Otte et al., 1988, 1989) have shown that if one and not the other of these events occurs, no neural tissue is formed. However, if one artificially activates Protein Kinase C *and* adenyl cyclase on the ectodermal cell membranes, neural tissue is generated. In this model, neural induction is accomplished by two reactions, and each reaction might be initiated by a different molecule. The participation of PKC in natural neural induction was given further support when Otte and co-workers

(1991; Otte and Moon, 1992) demonstrated that the PKC of the dorsal ectoderm differs from the PKC of the ventral ectoderm, both in its structure and in its ability to be activated by external compounds. Only the PKC found in the dorsal ectoderm correlates with the ability to respond to natural inducers. One possible reason nobody has been able to isolate *the* naturally occurring neural-inducing factor may be that there are several simultaneous factors at work. ■

The regional specificity of induction

The Determination of Regional Differences

One of the most fascinating phenomena in neural induction is the regional specificity of the neural structures that are produced. Forebrain (archencephalic), hindbrain (deuterencephalic), and spinocaudal regions of the neural tube must all be properly organized in an anterior-to-posterior direction. Thus, the organizer tissue not only induces the neural tube but also specifies the *regions* of the neural tube. This region-specific induction was shown by Otto Mangold (1933) in a series of experiments wherein various regions of the *Triturus* (newt) archenteron roof were transplanted into early-gastrula embryos (Figure 15.28). After the superadjacent neural plate was removed, four successive sections of the archenteron roof were excised from embryos that had just completed gastrulation and were placed into the blastocoels of early gastrulae. The most anterior portion of the archenteron roof induced balancers and portions of the oral apparatus (Figure 15.28A); the next most anterior section induced the formation of various head structures, including nose, eyes, balancers, and otic vesicles (Figure 15.28B); the third section induced the hindbrain structure (Figure 15.28C); and the most posterior segment induced the formation of dorsal trunk and tail mesoderm (Figure 15.28D). The induction of dorsal mesoderm—rather than the dorsal ectoderm of the nervous system—by the posterior end of the notochord was confirmed by Bijtel (1931) and Spofford (1945), who showed that the posterior fifth of the neural plate gives rise to tail somites and the posterior portions of the pronephric kidney duct.

Moreover, when dorsal blastopore lips from *early* salamander embryos (early gastrulae) were placed into other early salamander embryos, they formed secondary heads. When dorsal lips from *later-stage* embryos were transplanted into early salamander embryos, they induced the formation of secondary tails (Figure 15.29; Mangold, 1933). This means that the first cells of the organizer to enter the embryo induce the formation of brains and

Figure 15.28
Regional specificity of induction can be demonstrated by implanting different regions (color) of the archenteron roof into early *Triturus* gastrulae. The resulting animals have secondary parts. (A) Head with balancers. (B) Head with balancers, eyes, and forebrain. (C) Posterior part of head, deuterencephalon, and otic vesicles. (D) Trunk-tail segment. (After Mangold, 1933.)

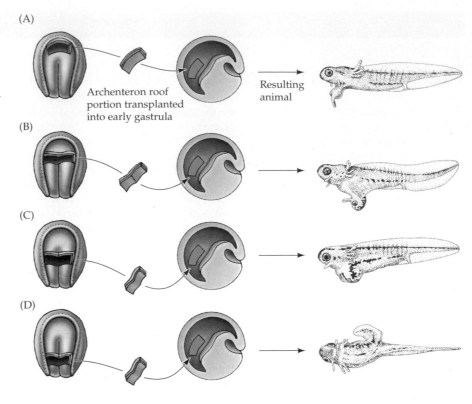

(A)

Archenteron roof portion transplanted into early gastrula

Resulting animal

(B)

(C)

(D)

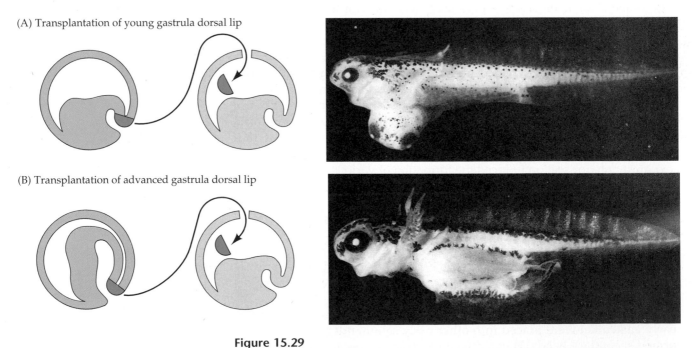

(A) Transplantation of young gastrula dorsal lip

(B) Transplantation of advanced gastrula dorsal lip

Figure 15.29
Regionally specific inducing action of the dorsal blastopore lip. (A) Young dorsal blastopore lips (which will form the anterior portion of the dorsal mesoderm) induce anterior structures when placed into young newt gastrulae. (B) Older dorsal blastopore lips placed into similar newt gastrulae produce more posterior structures. (From Saxén and Toivonen, 1962, photographs courtesy of L. Saxén.)

heads, while those cells that form the dorsal blastopore lip of later-stage embryos induce the cells above them to become spinal cords and tails. A similar phenomenon occurs in chick embryos (Storey at al., 1992).

The Double Gradient Model

In the 1950s, P. Nieuwkoop (1952) and Toivonen and Saxén (1955) proposed models of regional specificity involving two steps. In the first step, neural tissue was induced by the organizer. This neural tissue was "archencephalic," forebrain tissue. The second step consisted of a posteriorizing signal that was arranged as a gradient, with the highest concentration caudally. The posteriorizing signal would act on the anterior ectoderm, converting it into hindbrain and spinal cord tissue. Nieuwkoop's evidence came from the transplantation of folds of competent ectoderm at various positions along the anterior-posterior axis of the host gastrula. The proximal parts of these folds produced structures typical of the host's region of insertion, while the more distal part of the fold developed into neural structures of a more anterior nature than that of the insertion (Figure 15.30). Toivonen and Saxén's evidence came from studies involving artificial tissue-specific inducers. Guinea pig bone marrow, for instance, was found to induce mesodermal structures only. Pellets of guinea pig liver, however, could only cause the induction of forebrain structures. They implanted these inducers together within the blastocoel of the same early gastrula. Whereas the liver would only have induced forebrain and the bone marrow would only have induced mesoderm, the two together induced all the normal forebrain, hindbrain, spinal cord, and trunk mesoderm (Toivonen and Saxén, 1955). Thus, the regional specificity of neural induction may be due to opposing gradients of forebrain-inducing and spinal-cord-inducing substances (Figure 15.31). Similar results came from studies wherein anterior neural ectoderm was mixed with different amounts of posterior dorsal mesoderm (Figure 15.32; Toivonen and Saxén, 1968). Thus, the neural tissue was first determined to be forebrain,

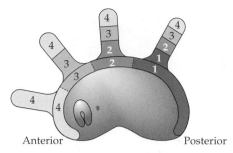

Figure 15.30
Evidence for a two-step model of neural induction: activation and transformation. A fold of gastrula ectoderm was implanted to a region of the neural plate. The more anterior structures are on the left-hand side, and 1–4 represent different neural structures. Folds of unspecified gastrula ectoderm tended to differentiate into anterior neural structures, but were posteriorized by material coming from the posterior of the embryo. (After Doniach, 1993.)

(A)

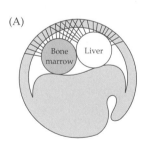

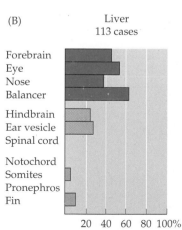

(B)

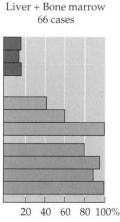

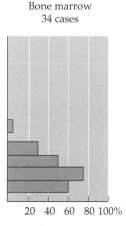

Figure 15.31
Evidence for the dual-gradient model of induction. (A) Simultaneous implantation of a neuralizing inducer (guinea pig liver) and a mesodermalizing inducer (guinea pig bone marrow) into the blastocoel of an early newt gastrula. (B) Results of such an implantation. Hindbrain and spinal cord structures, which were intermediate between forebrain and mesoderm on the fate map of the neural plate, were not induced well by either inducer. When the two inducers were implanted together, these structures were produced. (After Toivonen and Saxén, 1955.)

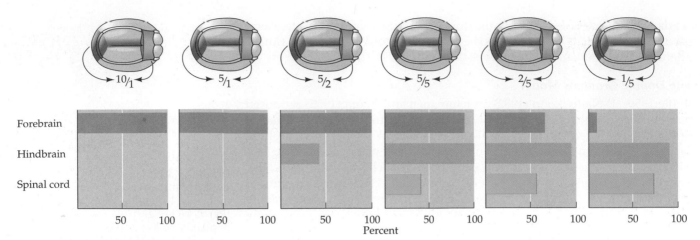

Figure 15.32
Evidence for the dual-gradient two-step induction within the amphibian embryo. The anterior region of the neural plate (i.e., cells that had already been naturally induced by a forebrain inducer, here seen in color) and cells from the posterior notochord were excised and mixed together in different ratios. The frequency of intermediate (hindbrain) structures increases as the ratio of anterior neural plate to mesodermal cells approaches 1:1. This suggests that the regional specification occurs after the neural plate cells have been determined to be neural. (From Gilbert and Saxén, 1993.)

but it then was posteriorized in a graded fashion by caudal substances. Most models of neural induction have converged on a scheme that includes (1) an initial "activation" step that determines that the cells are capable of becoming forebrain neural cells and (2) a "transformation" step in which a gradient of material from the posterior mesoderm causes the posteriorization of the neural specification (Figure 15.33). [regul4.html]

Molecular Correlates of Neural Caudalization

The neural inducers Chordin, Noggin, and Follistatin each induce exclusively anterior (forebrain-type) neural tissues. What, then, might be the factor(s) that posteriorize the neural tube? Several recent studies point to an FGF as the factor that specifies the neural ectoderm to become more caudal (Cox and Hemmati-Brivanlou, 1995; Lamb and Harland, 1995). When early-gastrula ectoderm (which had not yet been underlain by dorsal mesoderm) was isolated and neuralized by Noggin, Chordin, or Follistatin, anterior-type neural markers were found. When that tissue was incubated with a

Figure 15.33
The activation-transformation model of neural patterning. According to this model, the original neural induction ("activation") causes the neural ectoderm to be specified as the most anterior type of neural cells. The caudalization ("transformation") of these cells is accomplished by a gradient of some other substance, whose concentration is highest posteriorly. (After Doniach, 1995.)

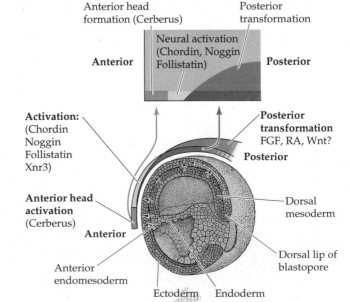

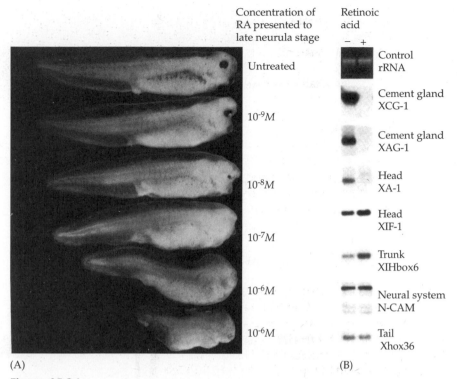

Concentration of
RA presented to
late neurula stage

Untreated

$10^{-9}M$

$10^{-8}M$

$10^{-7}M$

$10^{-6}M$

$10^{-6}M$

Retinoic
acid

− +

Control
rRNA

Cement gland
XCG-1

Cement gland
XAG-1

Head
XA-1

Head
XIF-1

Trunk
XIHbox6

Neural system
N-CAM

Tail
Xhox36

(A)   (B)

Figure 15.34
Retinoic acid causes posteriorization of neural structures. (A) Late-neurula embryos
exposed continuously to different concentrations of RA and allowed to grow until
controls reached tadpole stage. (B) Effect on neural marker mRNA expression when
blastulae are treated with 10^{-6} M RA for 2 hours (sufficient to give acephalic tadpoles). Inhibitory effect can be seen on the genes expressed most anteriorly. (A after
Ruiz i Altaba and Jessell, 1991; B after Sive et al., 1990.)

neural inducer plus FGF2, the ectoderm expressed more posterior neural
markers. Indeed, FGF2 will induce the forebrain to express hindbrain-specific genes. When FGF signaling is blocked in vivo by a dominant-negative
FGF receptor, the resulting tadpoles lack their posterior segments (Amaya et
al., 1991). FGF2 is probably not the natural posteriorizing FGF in *Xenopus*,
since it is not secreted and is not localized to any side of the embryo. However, an embryonic form of FGF (eFGF, a *Xenopus* FGF similar to mammalian
FGF4) is found in *Xenopus* posterior and tailbud mesoderm and has the
same effects as FGF2 (Isaacs et al., 1992). Overexpression of eFGF upregulates several posteriorly expressed genes, including the *Xenopus* homolo-
gogue of *caudal*. This, in turn, appears to activate the expression of more posterior *Hox* genes, leading to the more posterior specification of the nervous
system (Pownall et al., 1996).

In addition to the FGFs, other factors may be involved in patterning the
Xenopus nervous system. If early *Xenopus* gastrulae are treated with
nanomolar-to-micromolar concentrations of retinoic acid (RA), their forebrain and midbrain development is impaired in a concentration-dependent
fashion (Figure 15.34A; Papalopulu et al., 1991; Sharpe, 1991). When lower
concentrations are used, the actual induction of neural tissue does not appear to be inhibited, but fewer forebrain messages and structures are produced (Figure 15.34B; Durston et al., 1989, 1991; Sive et al, 1990). Retinoic
acid appears to affect both the mesoderm and the ectoderm. Ruiz i Altaba
and Jessell (1991) found that anterior dorsal mesoderm from RA-treated gas-

Figure 15.35
Xwnt3a can caudalize anterior neural tissue. Explants of competent ectoderm connected to dorsal blastopore lip were isolated as in Figure 15.30. The specific expressed mRNAs were identified by reverse transcriptase PCR. In this chart, the most anteriorly expressed neural markers are located highest. Overexpression of Wnt3a in the embryo with Xwnt3a wiped out the most anterior neural markers. Those ectodermal regions from uninjected embryos or those overexpressing a control protein (prolactin) were not affected. (From McGrew et al., 1995; photograph courtesy of R. T. Moon.)

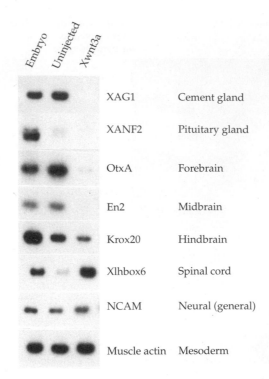

XAG1	Cement gland
XANF2	Pituitary gland
OtxA	Forebrain
En2	Midbrain
Krox20	Hindbrain
Xlhbox6	Spinal cord
NCAM	Neural (general)
Muscle actin	Mesoderm

trulae was unable to induce head structures in host embryos, and Sive and Cheng (1991) found that RA-treated ectoderm was unable to respond to the anterior-inducing mesoderm of untreated gastrulae. Another candidate for the caudalizing factor is *Xenopus* Wnt3a (McGrew et al., 1995). This protein is found in the neural ectoderm of the early neurula. When ectoderm is isolated from *Xenopus* gastrulae but remains connected to the dorsal blastopore lip, the ectoderm develops an anterior-posterior array of neural markers. If the embryo had been injected with *Xwnt3a* RNA (causing the overexpression of this protein), the anterior markers are lost (Figure 15.35).

Sidelights & Speculations

Vertical and Horizontal Signals from the Organizer

OUR DISCUSSION of the organizer so far has discussed those signals that go vertically from the organizer to the overlying ectoderm. We now know that there is a second set of signals that is produced by the dorsal blastopore lip and sent horizontally through the plane of the ectoderm (Figure 15.36). Recent evidence suggests that both vertical induction through the chordamesoderm and horizontal (planar) induction through the ectoderm are necessary for complete embryonic induction.

What roles might these signals play? First, there is some evidence that planar signals may be involved in the neuralizing activities of the organizer. Perhaps

Figure 15.36 Two modes of inducing the dorsal axis. In the *planar* mechanism, molecules are transferred from the dorsal blastopore lip tissue through the plane of the ectoderm. In the *vertical* mechanism, soluble molecules from the dorsal-blastopore-lip-derived chordamesoderm induce the cells above them to become neural tissue. (After Doniach, 1993.)

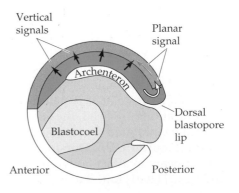

they provide the missing signals that activate neurulation (as opposed to the signals that block ventralization). If planar signals traveling from the dorsal blastopore lip through the ectoderm are responsible for neural induction, then the original source of such signals should be the *epithelium* of the dorsal marginal zone rather than the deep mesenchyme cells of that organizer zone. Shih and Keller (1992) have found this to be the case. They repeated the Spemann and Mangold experiment, but instead of using the entire dorsal marginal zone (DMZ), they transplanted either the epithelial cells or the deep cells of the DMZ (which they labeled with fluorescent dextran particles). The epithelial cells had all the inductive properties of the Spemann organizer and differentiated into mesodermal tissue. This epithelium was also able to rescue embryos that had been "ventralized" by UV irradiation. Neither the deep cells of the DMZ nor the ventral marginal cells could accomplish such organizer activities. A recently discovered inducer protein, *Xenopus nodal-related-3* (*Xnr3*) has been found in this superficial layer of the organizer, and it can convert animal pole caps into anterior neural ectoderm. Interestingly, unlike other inducers, it does not dorsalize the mesoderm. It is not yet known if this protein is part of the planar signaling system (Hansen et al., 1997).

However, the planar signals are not seen to be sufficient for neural induction. Nieuwkoop and Koster (1995) prevented vertical induction from occurring during *Xenopus* gastrulation and found that no neural differentiation took place. Furthermore, if the fibronectin binding fragment, RGD, is injected into the blastocoel of *Rana pipiens* gastrulae, the axial mesoderm fails to migrate toward the animal pole. Rather, it splits into two streams that involute horizontally along the equator of the embryo, forming two laterally located notochords. Each notochord induces a neural plate, but a neural plate does not form in the dorsal ectoderm, where the planar signals would have spread (Saint-Jeannet and Dawid, 1994). So in this model, the planar signals are redundant or may support the vertical signals from the notochord.

In the second model, the planar signals may be important in contributing to

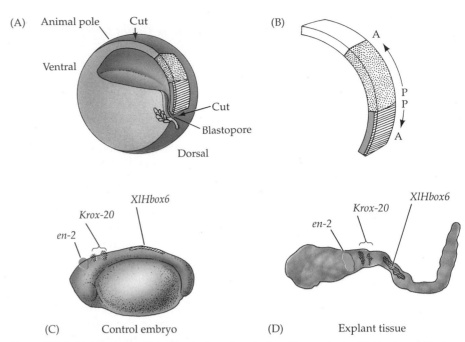

Figure 15.37 Expression pattern of neural markers induced by contact with the dorsal blastopore lip in plane of ectoderm. (A) Sagittal section of early *Xenopus* gastrula showing where cuts were made. (B) Explant depicting the anterior-posterior polarity expected from fate map: white region is epidermis; stippled region is presumptive mesenchyme; black region is dorsal mesoderm; striped region is archenteron roof. The explants were placed under coverslips to prevent migration of the mesoderm. (C) Expression of neural markers in the control embryo, stage 21. Homeobox genes *engrailed-2* and *XlHbox6* are expressed at the hindbrain-midbrain border and in the spinal cord, respectively; zinc finger protein gene *Krox-20* is expressed in rhombomeres 3 and 5 of the hindbrain. (D) The same order of expression is seen in the ectoderm of those explants retaining a connection to the dorsal blastopore lip. (After Doniach et al., 1992.)

the regional specificity of induction. Doniach and her colleagues (1992) showed that instructive, positionally specific information is provided by planar signals passing through the ectoderm. When explants are taken from early *Xenopus* gastrulae such that the ectoderm retains contact with the dorsal blastopore lip but never contacts the mesoderm, not only are the pan-neural markers NCAM and NF-3 induced in the ectoderm, but four position-specific neural markers—*engrailed-2, Krox-20, XlHbox1,* and *XlHbox6*—are expressed in the explant ectoderm in the appropriate anterior-posterior sequence (Figure 15.37). It appears, then, that the horizontally inductive signals from the dorsal blastopore lip are sufficient for inducing the anterior-posterior neural pattern. Ruiz i Altaba (1992) has also confirmed extensive neural patterning in these exogastrulae, showing that the pattern of neural markers in the exogastrulae reflects the nor-

mal pattern except in the forebrain and ventral regions. He also provides evidence that the transmission of these horizontal signals is through the notoplate (the ectoderm above the notchord).

In the third model, the planar signals complement the vertical signals in the creation of the neural tube. The planar signals appear to be involved in inducing the convergent extension of the hindbrain and spinal cord ectoderm adjacent to it (while the folding of the neural plate into a neural tube appears to be induced by the notochord). (Keller et al., 1992; Nieuwkoop and Koster, 1995). We are still trying to fit all the pieces of the induction puzzle together, while new pieces are being discovered. Spemann predicted that scientists would find that the embryo used more than one mechanism ("double assurance") to accomplish its ends. The embryo may well be using both planar and vertical signals to induce its nervous system. ∎

Homeobox Genes in Neural Specification

One of the most spectacular findings of this decade has been the discovery that flies and mice use the same homeobox genes for specifying regions along the anterior-posterior axis. However, the analysis of homeobox genes in *Xenopus* has not progressed as far owing to the inability to do gene knockouts on these frogs. As we will see in Chapter 16, retinoic acid is able to convert one part of the mouse body into a more posterior part by causing the expression of homeobox genes that are characteristic of the more posterior region. This is also seen in *Xenopus*. Both retinoic acid and eFGF have been shown to alter *Hox* gene expression. Pownall and colleagues (1996) have shown that eFGF promotes the expression of posterior *Hox* genes in *Xenopus* ectoderm, and both Cho and colleagues (1991b) and Sive and Cheng (1991) have shown that retinoic acid alters the expression of homeobox genes in a posterior direction in both ectoderm and mesoderm. Thus, in a variation of the two-step model proposed earlier, neural induction is proposed to lead to the creation of an anterior (forebrain-type) neural determination that is acted on by a posterior gradient of retinoic acid, eFGF or Wnt3a to create the regional specificities (Otte et al., 1991; Sharpe, 1991). Such an RA gradient (10-fold higher in the posterior than in the anterior) has been detected in the dorsal mesoderm of early *Xenopus* neurulae (Chen et al., 1994).

Competence and inductive cascades

The primary inductive interactions, although complex, cannot construct the entire embryo. However, the formation of the neural tube, dorsal mesoderm, pharyngeal endoderm, and other tissues creates the conditions for a cascade of further inductive events. Interactions by which one tissue interacts with another to specifically direct its fate are called **secondary inductions.***

Any system of embryonic induction has at least two components: a tissue capable of producing the inducing stimulus and a tissue capable of receiving and responding to it. So far, we have been looking at the specificity of production; now we must look at the specificity of responding cells. The ability to respond in a specific manner to a given stimulus is called **competence.** We have already seen that in the early gastrula, an implanted blastopore lip can induce a new neural plate and embryonic axis just about anywhere in the embryo it can meet ectoderm. However, with increasing embryonic age, the ectoderm loses this ability to respond, and the implantation of a dorsal blastopore lip beneath the prospective epidermis of a neurula-stage embryo will not cause it to form a new neural plate. It has lost its competence to respond to the new blastopore lip.

Although the late-neurula ectoderm is no longer competent to respond to the blastopore lip, it has become competent to respond to new inducers. This competence can be localized to particular areas. During gastrulation and early neurulation, the *head* ectoderm (but not the *trunk* ectoderm) becomes competent to form the lens, nose, and ear placodes. This competence is acquired by its being acted upon by the neural plate region (Henry and Grainger, 1990). Thus, the head region of the neurula is now competent to respond to contact from the optic vesicle (derived from the forebrain) to become lens.

---

*Although the inductions after "primary" embryonic induction have often been referred to as "secondary" inductions, there is no conceptual difference between them. We will return to secondary inductions in Chapter 17.

Moreover, once a tissue has been induced, it can induce other tissues. The D1 blastomeres of the Nieuwkoop center induce the cells above them to become the organizer. The organizer then induces the ectoderm above it to become the neural tube. The neural tube can then induce the head ectoderm to form lenses. And the inductions continue. Moreover, one tissue can induce several others. The organizer induces both mesoderm and ectoderm. Sonic hedgehog from the notochord not only induces the floor plate in the neural tube; coming from both the floor plate and the notochord, Sonic hedgehog induces the ventral medial somite to become the cartilage-forming sclerotome (see Figure 9.6; Fan and Tessier-Lavigne, 1994; Johnson et al., 1994). We will continue our discussion of secondary inductions in Chapter 17.

We are finally putting names to the "agents" and "soluble factors" of the experimental embryologists. We are finally delineating the intercellular pathways of paracrine factors and transcription factors that constitute the first steps in the processes of organogenesis. The international research program initiated by Spemann's laboratory in the 1920s is reaching toward a conclusion. But this research has found levels of complexity far deeper than Spemann would have conceived, and just as Spemann's experiments told us how much we didn't know, so today, we are faced with a whole new set of problems generated by our solutions to older ones: How is the Nieuwkoop center initiated? What is the activity of *Siamois*? How does the mesoderm become patterned? How are the signals from the notochord limited? How does the notochord differentiate? How does the ectoderm achieve its competence?

Surveying the field in 1927, Spemann remarked:

> We still stand in the presence of riddles, but not without hope of solving them. And riddles with the hope of solution—what more can a scientist desire?

The challenge still remains.

LITERATURE CITED

Allen, G. E. 1985. Thomas Hunt Morgan: Materialism and reductionism in the development of modern genetics. *Trends Genet.* 3: 151–154; 186–190.

Appel, T. A. 1987. *The Cuvier-Geoffroy Debate: French Biology in the Decades before Darwin*. Oxford University Press, NY.

Amaya, E., Musci. T. J. and Kirschner, M. W. 1991. Expression of a dominant negative mutant of the FGF receptor disrupts mesoderm formation in *Xenopus* embryos. *Cell* 66: 257–270.

Berrill, N. J. and Liu, C. K. 1948. Germplasm, Weismann, and hydrozoa. *Q. Rev. Biol.* 23: 124–132.

Bijtel, J. H. 1931. Über die Entwicklung des Schwanzes bei Amphibien. *Wilhelm Roux Arch. Entwicklungsmech. Org.* 125: 448–486.

Blitz, I. L. and Cho, K. W. Y. 1995. Anterior neurectoderm is progressively induced during gastrulation: the role of the Xenopus homeobox gene orthodenticle. *Development* 121: 993–1004.

Blumberg, B., Wright, C. V. E., De Robertis, E. M. and Cho, K. W. Y. 1991. Organizer-specific homeobox genes in *Xenopus laevis* embryos. *Nature* 253: 194–196.

Bouwmeester, T., Kim, S.-H., Sasai, Y., Lu, B. and De Robertis, E. M. 1996. Cerberus is a head-inducing secreted factor expressed in the anterior endoderm of Spemann's organizer. *Nature* 382: 595–601.

Brannon, M. and Kimelman, D. 1996. Activation of *Siamois* by the Wnt pathway. *Dev. Biol.* 180: 344–347.

Buss, L. 1987. *The Evolution of Individuality*. Princeton University Press, Princeton, NJ.

Chen, Y. P., Huang, L. and Solursh, M. 1994. A concentration gradient of retinoidsin the early *Xenopus laevis* embryo. *Dev. Biol.* 161: 70–76.

Cho, K. W. Y., Blumberg, B., Steinbeisser, H. and De Robertis, E. 1991a. Molecular nature of Spemann's organizer: The role of the *Xenopus* homeobox gene *goosecoid*. *Cell* 67: 1111–1120.

Cho, K. W. Y., Morita, A. A., Wright, C. V. E. and De Robertis, E. M. 1991b. Overexpression of a homeodomain protein confers axis-forming activity to uncommitted *Xenopus* embryonic cells. *Cell* 65: 55–64.

Cooke, J. and Webber, J. A. 1985. Dynamics of the control of body pattern in the development of *Xenopus laevis*. *J. Embryol. Exp. Morphol.* 88: 85–99.

Cornell, R. A. and Kimelman, D. 1994. Activin-mediated mesoderm induction requires FGF. *Development* 120: 453–462.

Cornell, R. A., Musci, T. J. and Kimelman, D. 1995. FGF is a prospective competence factor for early activin-type signals in *Xenopus* mesoderm induction. *Development* 121: 2429–2437.

Cox, W. G. and Hemmati-Brivanlou, A. 1995. Caudalization of neural fate by tissue recombination and bFGF. *Development* 121: 4349–4358.

Cui, Y., Tian, Q. and Christian, J. L. 1996. Synergistic effects of Vg1 and Wnt signals in the specification of dorsal mesoderm and endoderm. *Dev. Biol.* 180: 22–34.

Cunliffe, V. and Smith, J. C. 1992. Ectopic mesoderm formation in *Xenopus* embryos caused by widespread expression of a *Brachyury* homologue. *Nature* 358: 427–430.

Dale, L. and Slack, J. M. W. 1987. Regional specificity within the mesoderm of early embryos of *Xenopus laevis*. *Development* 100: 279–295.

Dale, L. Howes, G., Price, B. M. J. and Smith, J. C. 1992. Bone morphogenetic protein 4: a ventralizing factor in early *Xenopus* development. *Development* 115: 573–585.

Dale, L., Matthew, G. and Coleman, A. 1993. Secretion and mesoderm-inducing activity of the TGF-β related domain of *Xenopus* Vg1. *EMBO J.* 12: 4471–4480.

Davids, M. 1988. Protein kinases in amphibian ectoderm induced for neural differentiation. Wilhelm Roux Arch. *Dev. Biol.* 197: 339–344.

Davids, M., Loppnow, B., Tiedemann, H. and Tiedemann, H. 1987. Neural differentiation of amphibian gastrula ectoderm exposed to phorbol ester. *Wilhelm Roux Arch. Dev. Biol.* 106: 137–140.

De Robertis, E. M. and Sasai, Y. 1996. A common plan for dorsoventral patterning in Bilateria. *Nature* 380: 37–40.

De Robertis, E. M., Blum, M., Niehrs, C. and Steinbeisser, H. 1992. Goosecoid and the organizer. *Development 1992 Suppl.*: 167–171.

Doniach, T. 1993. Planar and vertical induction of anteroposterior pattern during the development of the amphibian central nervous system. *J. Neurobiol.* 24: 1256–1275.

Doniach, T. 1995. Basic FGF as an induced of anteroposterior neural pattern. *Cell* 83: 1067–1070.

Doniach, T., Phillips, C. R. and Gerhart, J. C. 1992. Planar induction of anteroposterior pattern in the developing central nervous system of *Xenopus laevis*. *Science* 257: 542–545.

Driesch, H. 1892. The potency of the first two cleavage cells in echinoderm development. Experimental production of partial and double formations. *In* B. H. Willier and J. M. Oppenheimer (eds.), *Foundations of Experimental Embryology*. Hafner, New York.

Driesch, H. 1893. Zur Verlagerung der Blastomeren des Echinideneies. *Anat. Anz.* 8: 348–357.

Driesch, H. 1894. *Analytische Theorie de organischen Entwicklung*. W. Engelmann, Leipzig.

Durston, A. and Otte, A. P. 1991. A hierarchy of signals mediates neural induction in *Xenopus laevis*. *In* J. Gerhart (ed.), *Cell-Cell Interactions in Early Development*. Wiley-Liss, New York, pp. 109–127.

Durston, A., Timmermans, A., Hage, W. J., Hendriks, H. F. J., de Vries, N. J., Heideveld, M. and Nieuwkoop, P. D. 1989. Retinoic acid causes an anteroposterior transformation in the developing central nervous system. *Nature* 330: 140–144.

Echelard, Y., Epstein, D. J., St-Jacques, B., Shen, L., Mohler, J., McMahon, J. A. and McMahon, A. 1993. Sonic hedgehog, a member of a family of putative signaling molecules, is implicated in the regulation of CNS polarity. *Cell* 75: 1417–1430.

Ettensohn, C. A. and McClay, D. R. 1988. Cell lineage conversion in the sea urchin embryo. *Dev. Biol.* 125: 396–409.

Fan, C.-M. and Tessier-Lavigne, M. 1994. Patterning of mammalian somites by surface ectoderm and notochord: Evidence for sclerotome induction by sonic hedgehog. *Cell* 79: 1175–1186.

Fässler, P. E. and Sander, K. 1996. Hilde Mangold (1898–1924) and Spemann's organizer: Achievement and tragedy. *Roux Arch. Dev. Biol.* 205: 323–332.

Funayama, N., Fagotto, F., McCrea, P. and Grumbiner, B. M. 1995. Embryonic axis induction by the armadillo repeat domain of β-catenin: Evidence for intracellular signalling. *J. Cell Biol.* 128: 959–968.

Gawantka, V., Delius, H., Hirschfeld, K., Blumenstock, C. and Niehrs, C. 1995. Antagonizing the Spemann organizer: Role of the homeobox gene *Xvent-1*. *EMBO J.* 14: 6268–6279.

Geoffroy Saint-Hilaire, E. 1822. Considérations gén´rales sur la vertèbre. *Mém. Mus. Hist. Natur.* 9: 89–119.

Gerhart, J. C., Danilchik, M., Doniach, T., Roberts, S., Browning, B. and Stewart, R. 1989. Cortical rotation of the *Xenopus* egg: Consequences for the anteroposterior pattern of embryonic dorsal development. *Development* [Suppl.] 107: 37–51.

Gilbert. S. F. and Faber, M. 1996. Looking at embryos: The visual and conceptual aesthetics of embryology. *In* A. I. Tauber (ed.) *The Elusive Synthesis.: Aesthetics and Science*. Kluwer Press, Dordecht. pp. 125–151.

Gilbert, S. F., Opitz, J. and Raff, R. A. 1996. Resynthesizing evolutionary and developmental biology. *Dev. Biol.* 173: 357–372.

Gilbert, S. F. and Saxén, L. 1993. Spemann's Organizer: Models and molecules. *Mech. Dev.* 41: 73–89.

Gimlich, R. L. 1985. Cytoplasmic localization and chordamesoderm induction in the frog embryo. *J. Embryol. Exp. Morphol.* 89: 89–111.

Gimlich, R. L. 1986. Acquisition of developmental autonomy in the equatorial region of the *Xenopus* embryo. *Dev. Biol.* 116: 340–352.

Gimlich, R. L. and Cook, J. 1983. Cell lineage and the induction of nervous systems in amphibian development. *Nature* 306: 471–473.

Gimlich, R. L. and Gerhart, J. C. 1984. Early cellular interactions promote embryonic axis formation in *Xenopus laevis*. *Dev. Biol.* 104: 117–130.

Gont, L. K., Steinbeisser, H., Blumberg, B. and De Robertis, E. M. 1993. Tail formation as a continuation of gastrulation: the multiple tail populations of the *Xenopus* tailbud derive from the late blastopore lip. *Development* 119: 991–1004.

Graff, J. M., Thies, R. S., Song, J. J., Celeste, A. J. and Melton, D. A. 1994. Studies with a *Xenopus* BMP receptor suggest that ventral mesoderm-inducing signals override dorsal signals *in vivo*. *Cell* 79: 169–179.

Guger, K. A. and Gumbiner, B. M. 1995. β-catenin has wnt-like activity and mimics the Nieuwkoop signaling center in Xenopus dorsal-ventral patterning. *Dev. Biol.* 172: 115–125.

Hamburger, V. 1984. Hilde Mangold, co-discoverer of the organizer. *J. Hist. Biol.* 17: 1–11.

Hamburger, V. 1988. *The Heritage of Experimental Embryology: Hans Spemann and the Organizer*. Oxford University Press, Oxford.

Hansen, C. S., Marion, C. D., Steele, K., George, S. and Smith, W. C. 1997. Direct neural induction and selective inhibition of mesoderm and epidermis inducers by Xnr3. *Development* 124: 483–492.

Haraway, D. J. 1976. *Crystals, Fabrics and Fields: Metaphors of Organicism in Twentieth-Century Biology*. Yale University Press, New Haven.

Harrington, A. 1996. *Reenchanted Science: Holism in German Culture from Wilheim II to Hitler*. Princeton University Press, Princeton, NJ.

Hawley, S. H. B. Wünnenberg,-Stapleton, K., Hashimoto, C., Laurent, M. N., Watabe, T., Blumberg, B. W. and Cho, K. W. Y. 1995. Disruption of BMP signals in embryonic *Xenopus* ectoderm leads to direct neural induction. *Genes Dev.* 9: 2923–2935.

He, X., Saint-Jeannet, J.-P., Woodgett, J. R., Varmus, H. E. and Dawid, I. B. 1995. Glycogen synthase kinase-3 and dorsoventral patterning in *Xenopus* embryos. *Nature* 374: 617–622.

Heasman, J. M. and eight others. 1994. Overexpression of cadherins and underexpression of β-catenin inhibit dorsal mesoderm induction in early Xenopus embryos. *Cell* 79: 791–803.

Hemmati-Brivanlou, A. and Melton, D. A. 1992. A truncated activin receptor inhibits mesoderm induction and formation of axial structures in *Xenopus* embryos. *Nature* 359: 609–614.

Hemmati-Brivanlou, A. and Melton, D. 1994. Inhibition of activin signalling promotes neuralization in *Xenopus*. *Cell* 77: 273–281.

Hemmati-Brivanlou, A. and Thomsen, G. H. 1995. Ventral mesodermal patterning in *Xenopus* embryos: Expression patterns and activities of BMP-2 and BMP-4. *Dev. Genet.* 17: 78–89.

Henry, J. J. and Grainger, R. M. 1990. Early tissue interactions leading to embryonic lens formation in *Xenopus laevis*. *Dev. Biol.* 141: 149–163.

Henry, J. J., Amemiya, S., Wray, G. A. and Raff, R. A. 1989. Early inductive interactions are involved in restricting cell fates of mesomeres in sea urchin embryos. *Dev. Biol.* 136: 140–153.

Hertwig, O. 1894. *Zeit- und Streitfragen der Biologie I. Präformation oder Epigenese? Grundzüge einer Entwicklungstheorie der Organismen.* Gustav Fischer, Jena.

Holley S. A., Jackson, P. D., Sasai, Y., Lu, B., De Robertis, E. M., Hoffmann, F. M. and Ferguson, E. L. 1995. A conserved system for dorsal-ventral patterning in insects and vertebrates involving *sog* and *chordin*. *Nature* 376: 249–253.

Hoppler, S, Brown, J. D. and Moon, R. T. 1996. Expression of a dominant negative Wnt blocks induction of *MyoD* in Xenopus embryos. *Genes Dev.* 10: 2805–2817.

Hörstadius, S. 1928. Über die Determination des Keimes bei Echinodermen. *Acta Zool.* 9: 1–191.

Hörstadius, S. 1935. Über die Determination im Verlaufe der Eiachse bei Seeigeln. *Publ. Staz. Zool. Napoli* 14: 251–479.

Hörstadius, S. 1939. The mechanics of sea urchin development studied by operative methods. *Biol. Rev.* 14: 132–179.

Hörstadius, S. and Wolsky, A. 1936. Studien über die Determination der Bilateralsymmetrie des jungen Seeigelkeimes. *Wilhelm Roux Arch. Entwicklungsmech. Org.* 135: 69–113.

Huxley, J. S. and deBeer, G. R. 1934. *Elements of Experimental Embryology*. Cambridge University Press, Cambridge.

Isaacs, H. V., Tannahill, D. and Slack, J. M. W. 1992. Expression of a novel FGF in the *Xenopus* embryo. A new candidate inducing factor for mesoderm formation and anteroposterior specificiation. *Development* 114: 711–720.

Jacobson, M. 1984. Cell lineage analysis of neural induction: Origins of cells forming the induced nervous system. *Dev. Biol.* 102: 122–129.

Johnson, R. L., Laufer, E., Riddle, R. D. and Tabin, C. 1994. Ectopic expression of sonic hedgehog alters dorsal-ventral patterning of somites. *Cell* 79: 1165–1173.

Jones, C. M., Lyons, K. M., Lapan, P. M., Wright, C. V. E. and Hogan, B. L. M. 1992. DVR-4 (bone morphogenetic protein-4) as a posterior-ventralizing factor in *Xenopus* mesoderm induction. *Development* 115: 639–647.

Kageura, H. and Yamana, K. 1983. Pattern regulation in isolated halves and blastomeres of early *Xenopus laevis*. *J. Embryol. Exp. Morphol.* 74: 221–234.

Kageura, H. and Yamana, J. 1986. Pattern formation in 8-cell composite embryos of *Xenopus laevis*. *J. Embryol. Exp. Morphol.* 91: 79–100.

Keller, R. E. 1976. Vital dye mapping of the gastrula and neurula of *Xenopus laevis* II. Prospective areas and morphogenetic movements of the deep layer. *Dev. Biol.* 51: 118–137.

Keller, R., Shih, J., Sater, A. K. and Moreno, C. 1992. Planar induction of convergence and extension of the neural plate by the organizer of *Xenopus*. *Dev. Dyn.* 193: 218–234.

Kessler, D. S. and Melton, D. A. 1995. Induction of dorsal mesoderm by soluble, mature Vg1 protein. *Development* 121: 2155–2164.

Khaner, O. 1995. The rotated hypoblast of the chicken embryo does not initiate an ectopic axis in the epiblast. *Proc. Natl. Acad. Sci. USA* 92: 10733–10737.

Khaner, O. and Wilt, F. 1990. The influence of cell interactions and tissue mass on differentiation of sea urchin mesomeres. *Development* 109: 625–634.

Khaner, O. and Wilt, F. 1991. Interactions of different vegetal cells with mesomeres during early stages of sea urchin development. *Development* 112: 881–890.

Ku, M. and Melton, D. A. 1993. *Xwnt-11*, a maternally expressed *Xenopus wnt* gene. *Development* 119: 1161–1173.

LaBonne, C. and Whitman, M. 1994. Mesoderm induction by activin requires FGF-mediated intracellular signals. *Development* 120: 463–472.

Lamb, T. M. and Harland, R. M. 1995. Fibroblast growth factor is a direct neural inducer, which combined with noggin generates anterior-posterior pattern. *Development* 121: 3627–3636.

Lamb, T. M. and seven others. 1993. Neural induction by the secreted polypeptide noggin. *Science* 262: 713–718.

Larabell, C. A. and seven others. 1997. Establishment of the dorsal-ventral axis in *Xenopus* embryos is presaged by early asymmetries in β-catenin which are modulated by the Wnt signaling pathway. *J. Cell Biol.* 136: 1123–1136.

Lemaire, P., Garrett, N. and Gurdon, J. B. 1995. Expression cloning of *Siamois*, a Xenopus homeobox gene expressed in dorsal-vegetal cells of blastulae and able to induce a complete secondary axis. *Cell* 81: 85–94.

Lustig, K. D., Kroll, K., Sun, E., Ramos, R., Elmendorf, H. and Kirschner, M. W. 1996. A *Xenopus* nodal-related gene that acts in synergy with noggin to induce complete secondary axis and notochord formation. *Development* 122: 3275–3282.

Maeno, M., Ong, R. C., Suzuki, A., Ueno, N. and Kung, H. F. 1994. A truncated bone morphogenesis protein-4 receptor alters the fate of ventral mesoderm to dorsal mesoderm—role of animal pole tissue in the development of ventral mesoderm. *Proc. Natl. Acad. Sci. USA* 91: 10260–10264.

Mangold, O. 1933. Über die Induktionsfahigkeit der verschiedenen Bezirke der Neurula von Urodelen. *Naturwissenschaften* 21: 761–766.

Marcus, N. H. 1979. Developmental aberrations associated with twinning in laboratory-reared sea urchins. *Dev. Biol.* 70: 274–277.

Maruyama, Y. K., Nakaseko, Y. and Yagi, S. 1985. Localization of the cytoplasmic determinants responsible for primary mesenchyme formation and gastrulation in the unfertilized eggs of the sea urchin *Hemicentrotus pulcherrimus*. *J. Exp. Zool.* 236: 155–163.

McClendon, J. F. 1910. The development of isolated blastomeres of the frog's egg. *Am. J. Anat.* 10: 425–430.

McGrew, L. L., Lai, C.-J. and Moon, R. T. 1995. Specification of the anteroposterior neural axis through synergistic interaction of the wnt signaling cascade with noggin and follistatin. *Dev. Biol.* 172: 337–342.

McMahon, A. P. and Moon, R. T. 1989. Ectopic expression of the proto-oncogene *int-1* in *Xenopus* leads to duplication of the embryonic axis. *Cell* 58: 1075–1084.

Mead, P. E., Brivanlou, I. H., Kelly, C. M. and Zon, L. I. 1996. BMP-4-responsive regulation of dorsal-ventral patterning by the homeobox protein Mix.1. *Nature* 382: 357–360.

Nakamura, O. and Takasaki, H. 1970. Further studies on the differentiation capacity of the dorsal marginal zone in the morula of *Triturus pyrrhogaster*. *Proc. Japan Acad.* 46: 700–705.

Niehrs, C., Keller, R., Cho, K. W. Y. and De Robertis, E. M. 1993. The homeobox gene *goosecoid* controls cell migration in *Xenopus* embryos. *Cell* 72: 491–503.

Nieuwkoop, P. D. 1952. Activation and organization of the central nervous system in amphibians. III. Synthesis of a new working hypothesis. *J. Exp. Zool.* 120: 83–108.

Nieuwkoop, P. D. 1969. The formation of the mesoderm in urodele amphibians. I. Induction by the endoderm. *Wilhelm Roux Arch. Entwicklungsmech. Org.* 162: 341–373.

Nieuwkoop, P. D. 1973. The "organisation center" of the amphibian embryo: Its origin, spatial organisation and morphogenetic action. *Adv. Morphogen.* 10: 1–39.

Nieuwkoop, P. D. 1977. Origin and establishment of embryonic polar axes in amphibian development. *Curr. Top. Dev. Biol.* 11: 115–132.

Nieuwkoop, P. D. and Koster, K. 1995. Vertical versus planar induction in amphibian early development. *Develop. Growth Differ.* 37: 653–688.

Northrop, J., Woods, A., Seger, R., Suzuki, A., Ueno, N., Krebs, E. and Kimelman, D. 1995. BMP-4 regulates the dorsal-ventral differences in FGF/MAPKK-mediated mesoderm induction in *Xenopus. Dev. Biol.* 172: 242–252.

Otte, A. P. and Moon, R. T. 1992. Protein kinase C isozymes have distinct roles in neural induction and competence in *Xenopus. Cell* 68: 1021–1029.

Otte, A. P., Kramer, I. J. M. and Durston, A. J. 1991. Protein kinase C and regulation of the local competence of *Xenopus* ectoderm. *Science* 251: 570–573.

Otte, A. P., Koster, C. H., Snoek, G. T. and Durston, A. J. 1988. Protein kinase C mediates neural induction in *Xenopus laevis. Nature* 334: 818–620.

Otte, A. P., van Run, P., Heideveld, M., van Driel, R. and Durston, A. J. 1989. Neural induction is mediated by cross-talk between the protein kinase C and cyclic AMP pathways. *Cell* 58: 641–648.

Papalopulu, N., Clarke, J. D. W., Bradley, L., Wilkinson, D., Krumlauf, R. and Holder, N. 1991. Retinoic acid causes abnormal development and segmental patterning of the anterior hindbrain in *Xenopus* embryos. *Development* 113: 1145–1158.

Piccolo, S., Sasai, Y. Lu, B. and De Robertis, E. M. 1996. Dorsoventral patterning in Xenopus: Inhibition of ventral signals by direct binding of chordin to BMP-4. *Cell* 86: 589–598.

Pierce, S. B. and Kimelman, D. 1995. Regulation of Spemann organizer formation by the intracellular kinase Xgsk-3. *Development* 121: 755–765.

Placzek, M., Tessier-Lavigne, M., Yamada, T., Jessell, T. and Dodd, J. 1990. Mesodermal control of neural cell identity: Floor plate induction by the notochord. *Science* 250: 985–988.

Pownall, M. E., Tucker, A. S., Slack, J. M. W. and Isaacs, H. V. 1996. eFGF, Xcad3 and *Hox* genes form a molecular pathway that establishes the anteropostior axis in *Xenopus. Development* 122: 3881–3892.

Recanzone, G. and Harris, W. A. 1985. Demonstration of neural induction using nuclear markers in *Xenopus. Wilhelm Roux Arch. Dev. Biol.* 194: 344–354.

Render, J. and Elinson, R. P. 1986. Axis determination in polyspermic *Xenopus* eggs. *Dev. Biol.* 115: 425–433.

Roelink, H. and ten others. 1994. Floor plate and motoneuron induction by *vhh-1*, a vertebrate homolog of *hedgehog* expressed by the notochord. *Cell* 76: 761–775.

Roll-Hansen, N. 1978. *Drosophila* genetics: A reductionist research program. *J. Hist. Biol.* 11: 159–210.

Roux, W. 1888. Contributions to the developmental mechanics of the embryo. On the artificial production of half-embryos by destruction of one of the first two blastomeres and the later development (postgeneration) of the missing half of the body. *In* B. H. Willier and J. M. Oppenheimer (eds.), 1974, *Foundations of Experimental Embryology.* Hafner, New York, pp. 2–37.

Ruiz i Altaba, A. 1992. Planar and vertical signals in the induction and patterning of the *Xenopus* nervous system. *Development* 115: 67–80.

Ruiz i Altaba, A. and Jessell, T. 1991. Retinoic acid modifies mesodermal patterning in early Xenopus embryos. *Genes Dev.* 5: 175–187.

Saint-Jeannet, J.-P. and Dawid, I. B. 1994. Vertical versus planar neural induction in *Rana pipiens* embryos. *Proc. Natl. Acad. Sci. USA* 91: 3049–3053.

Sakai, M. 1996. The vegetal determinants required for the Spemann organizer move equatorially during the first cell cycle. *Development* 122: 2207–2214.

Sander, K. 1991a. Wilhelm Roux and his programme for developmental biology. *Wilhelm Roux Arch. Dev. Biol.* 200: 1–3.

Sander, K. 1991b. "Mosaic work" *and* "assimilating effects" in embryogenesis: Wilhelm Roux's conclusions after disabling frog blastomeres. *Wilhelm Roux Arch. Dev. Biol.* 200: 237–239.

Sasai, Y., Lu, B., Steinbeisser, H., Geissert, D., Gont, L. K. and De Robertis, E. M. 1994. *Xenopus chordin*: A novel dorsalizing factor activated by organizer-specific homeobox genes. *Cell* 79: 779–790.

Sasai, Y., Lu, B., Piccolo, S. and De Robertis, E. M. 1996. Endoderm induction by the organizer-secreted factors chordin and noggin in *Xenopus* animal caps. *EMBO J.* 15: 4547–4555.

Saxén, L. 1961. Transfilter neural induction of amphibian ectoderm. *Dev. Biol.* 3: 140–152.

Saxén, L. and Toivonen, S. 1962. *Embryonic Induction.* Prentice-Hall, Englewood Cliffs, NJ.

Schmidt, J., Francoise, V., Bier, E. and Kimelman, D. 1995. *Drosophila* short gastrulation induces an ectopic axis in *Xenopus*: Evidence for conserved mechanisms of dorso-ventral patterning. *Development* 121: 4319–4328.

Schmidt, J. E., Dassow, G. van and Kimelman, D. 1996. Regulation of dorsal-ventral patterning: the ventralizing effects of the novel *Xenopus* homobox gene *Vox. Development* 122: 1711–1721.

Schneider, S., Steinbeisser, H., Warga, R. M. and Hausen, P. 1996. β-catenin translocation into nuclei demarcates the dorsalizing centers in frog and fish embryos. *Mech. Dev.* 57: 191–198.

Seleiro, E. A. P., Connolly, D. J. and Cooke, J. 1996. Early developmental expression and experimental axis determination by the chick *Vg1* gene. *Curr. Biol.* 6: 1476–1486.

Sharpe, C. R. 1991. Retinoic acid can mimic endogenous signals involved in transformation of the *Xenopus* nervous system. *Neuron* 7: 239–247.

Shih, J. and Keller, R. 1992. The epithelium of the dorsal marginal zone of *Xenopus* has organizer properties. *Development* 116: 887–899.

Sive, H. L. and Cheng, P. F. 1991. Retinoic acid perturbs the expression of *Xhox.lab* genes and alters mesodermal determination in *Xenopus laevis. Genes Dev.* 5: 1321–1332.

Sive, H. L., Draper, B. W., Harland, R. M. and Weintraub, H. 1990. Identification of a retinoic acid sensitive period during primary axis formation in *Xenopus laevis. Genes and Devel.* 4: 932–942.

Slack, J. M. W. and Tannahill, D. 1992. Mechanism of anteroposterior axis specification in vertebrates: Lessons from the amphibians. *Development* 114: 285–302.

Smith, J. C. and Slack, J. M. W. 1983. Dorsalization and neural induction: Properties of the organizer in *Xenopus laevis. J. Embryol. Exp. Morphol.* 78: 299–317.

Smith, J. C., Dale, L. and Slack, J. M. W. 1985. Cell lineage labels and region-specific markers in the analysis of inductive interactions. *J. Embryol. Exp. Morphol.* [Suppl.] 89: 317–331.

Smith, J. C., Price, B. M. J., Green, J. B. A., Weigel, D. and Herrmann, B. G. 1991. Expression of a *Xenopus* homolog of *Brachyury* (*T*) is an immediate-early response to mesoderm induction. *Cell* 67: 79–87.

Smith, W. C. and Harland, R. M. 1991. Injected *wnt-8* RNA acts early in *Xenopus* embryos to promote formation of a vegetal dorsalizing center. *Cell* 67: 753–765.

Smith, W. C. and Harland, R. M. 1992. Expression cloning of *noggin*, a new dorsalizing factor localized to the Spemann organizer in *Xenopus* embryos. *Cell* 70: 829–840.

Smith, W. C., Knecht, A. K., Wu, M. and Harland, R. M. 1993. Secreted noggin mimics the Spemann organizer in dorsalizing *Xenopus* mesoderm. *Nature* 361: 547–549.

Sokol, S, Y. 1996. Analysis of Disheveled signalling pathways during *Xenopus* development. *Curr. Biol.* 6: 1456–1467.

Sokol, S., Wong, G. and Melton, D. A. 1990. A mouse macrophage factor induces head structures and organizes a body axis in *Xenopus. Science* 249: 561–564.

Sokol, S.. Christian, J. L., Moon, R. T. and Melton, D. A. 1991. Injected wnt RNA induces a complete body axis in *Xenopus* embryos. *Cell* 67: 741–752.

Spemann, H. 1918. Über die Determination der ersten Organanlagen des Amphibienembryo. *Wilhelm Roux Arch. Entwicklungsmech. Org.* 43: 448–555.

Spemann, H. 1927. Neue Arbieten über Organisatoren in der tierischen Entwicklung. *Naturwissenschaften* 15: 946–951.

Spemann, H. 1938. *Embryonic Development and Induction*. Yale University Press, New Haven.

Spemann, H. and Mangold, H. 1924. Induction of embryonic primordia by implantation of organizers from a different species. *In* B. H. Willier and J. M. Oppenheimer (eds.), *Foundations of Experimental Embryology*. Hafner, New York, pp. 144–184.

Spofford, W. R. 1945. Observations on the posterior part of the neural plate in *Amblystoma. J. Exp. Zool.* 99: 35–52.

Stennard, F., Carnac G. and Gurdon, J. B. 1996. The *Xenopus* T-box gene, Antipodean, encodes a vegetally localised maternal mRNA and can trigger mesoderm formation. *Development* 122: 4179–4188.

Storey, K., Crossley, J. M., De Robertis, E., Norris, W. E. and Stern, C. D. 1992. Neural induction and regionalisation in the chick embryo. *Development* 114: 729–741.

Tauber, A. I. and Sarkar, S. 1992. The human genome project: Has blind reductionism gone too far? *Perspec. Biol. Med.* 35: 220–235.

Thomsen, G. H. and Melton, D. A, 1993. Processed Vg1 protein is an axial mesoderm inducer in *Xenopus. Cell* 74: 433–441.

Toivonen, S. 1979. Transmission problem in primary induction. *Differentiation* 15: 177–181.

Toivonen, S. and Saxén, L. 1955. The simultaneous inducing action of liver and bone marrow of the guinea pig in implantation and explantation experiments with embryos of *Triturus. Exp. Cell Res.* [Suppl.] 3: 346–357.

Toivonen, S. and Saxén, L. 1968. Morphogenetic interaction of presumptive neural and mesodermal cells mixed in different ratios. *Science* 159: 539–540.

Toivonen, S. and Wartiovaara, J. 1976. Mechanism of cell interaction during primary induction studied in transfilter experiments. *Differentiation* 5: 61–66.

Toivonen, S., Tarin, D., Saxén, L., Tarin, P. J. and Wartiovaara, J. 1975. Transfilter studies on neural induction in the newt. *Differentiation* 4: 1–7.

Twitty, V. C. 1966. *Of Scientists and Salamanders*. Freeman, San Francisco. [p. 39]

Vincent, J.-P. and Gerhart, J. C. 1987. Subcortical rotation in *Xenopus* eggs: An early step in embryonic axis specification. *Dev. Biol.* 123: 526–539.

Vodicka, M. A. and Gerhart, J. C. 1995. Blastomere derivation and domains of gene expression in the Spemann Organizer of *Xenopus laevis. Development* 121: 3505–3518.

Waddington, C. H. 1933. Induction by the primitive streak and its derivatives in the chick. *J. Exp. Biol.* 10: 38–46.

Wakahara, M. 1989. Specification and establishment of dorsal-ventral polarity in eggs and embryos of *Xenopus laevis. Dev. Growth Diff.* 31: 197–207.

Weismann, A. 1892. *Essays on Heredity and Kindred Biological Problems*. Translated by E. B. Poulton, S. Schoenland and A. E. Shipley. Clarendon, Oxford.

Weismann, A. 1893. *The Germ-Plasm: A Theory of Heredity*. Translated by W. Newton Parker and H. Ronnfeld. Walter Scott Ltd., London.

Wilkins, A. S. 1993. *Genetic Analysis of Animal Development*, 2nd ed. Wiley-Liss, New York.

Wilson, E. B. 1896. *The Cell in Development and Inheritance*. Macmillan, New York.

Yamada, T., Placzek, M., Tanaka, H., Dodd, J. and Jessell, T. M. 1991. Control of cell pattern in the developing nervous system: Polarizing activity of floor plate and notochord. *Cell* 64: 635–647.

Yamana, K. and Kageura, H. 1987. Reexamination of the "regulative development" of amphibian embryos. *Cell Differ.* 20: 3–10.

Yamashida, H. and seven others. 1995. Osteogenic protein-1 binds to activin type II receptors and induces certain activin-like effects. *J. Cell Biol.* 130: 217–226.

Yost, C., Torres, M., Miller, J. R., Brown, CJ. D., Lai, C.-J. and Moon, R. T. 1996. The axis-inducing ability, stability, and subcellular localization of β-catenin is regulated in *Xenopus* embryos by glycogen synthase kinase 3. *Genes Dev.* 10: 1443–1454.

Zaraisky, A. G., Ecochard, V., Kazanskaya, O. V., Lukyanov, S. A., Fesenko, I. V. and Duprat-A.-M. 1995. The homeobox-containming gene *XANF-1* may control development of the Spemann organizer. *Development* 121: 3839–3847.

Zhang, J. and King, M. L. 1996. *Xenopus VegT* RNA is localized to the vegetal cortex during oogenesis and encodes a novel T-box transcription factor involved in mesodermal patterning. *Development* 122: 4119–4129.

Zimmerman, L. B., De Jesús-Escobar, J. and Harland, R. M. 1996. The Spemann organizer signal noggin binds and inactivates bone morphogenetic protein 4. *Cell* 86: 599–606.

Establishment of body axes in mammals and birds

16

Between the fifth and tenth days the lump of stem cells differentiates into the overall building plan of the [mouse] embryo and its organs. It is a bit like a lump of iron turning into the space shuttle. In fact it is the profoundest wonder we can still imagine and accept, and at the same time so usual that we have to force ourselves to wonder about the wondrousness of this wonder.
MIROSLAV HOLUB (1990)

It is known that nature works constantly with the same materials. She is ingenious to vary only the forms. As if, in fact, she were restricted to the same primitive ideas, one sees her tend always to cause the same elements to reappear, in the same number, in the same circumstances, and with the same connections.
E. GEOFFROY SAINT-HILAIRE (1807)

THIS CHAPTER DETAILS some of the research that has provided insights into how the body axes of mammals and birds become established. Much of this work utilizes data provided by the analysis of embryos whose development has been malformed through mutations or disrupted by particular chemicals.

Despite the attempts of the quoted *savant* Geoffroy Saint-Hilaire (who believed that *all* the animals of the world shared a common body plan), most developmental biologists would not have predicted that the body plans of flies and mammals would be specified through the same set of genes. Having diverged over 500 million years ago, the fly's body and the vertebrate body seem exceptionally different. Most insects specify their axes in the common cytoplasm of the syncytial blastoderm, while vertebrate axes are specified by inductive interactions between groups of cells. In the *Drosophila* embryo, the general body plan is specified while the cells are a cylindrical monolayer surrounding the yolk; in mammals, cells have undergone extensive movements by the time their body parts are specified. The formation of appendages in insects results from the extension of ectodermal imaginal discs, while the mammalian limb is generated by complex inductive interactions between the ectodermal and mesodermal cells that have migrated to these areas. However, recent studies have shown that the anterior-posterior axis of developing mammals is specified by the same homeotic genes that specify the *Drosophila* body axis. Indeed, the homeobox has been called the Rosetta stone of developmental biology (Riddihough, 1992; Slack and Tannahill, 1992), as it enables us to transfer our genetic knowledge of *Drosophila* embryos into the less known realm of mammalian development.

Initiating the anterior-posterior axis

Establishing a Nieuwkoop Center

The establishment of the organizer appears to be similar throughout vertebrates. In teleost fishes, the cells of the blastoderm remain relatively coherent until gastrulation, and the mesodermal precursors form a belt around the margin, adjacent to the yolk cell (Wilson et al., 1995). There is evidence

Figure 16.1
A common strategy for establishing the Nieuwkoop center in vertebrates. During oogenesis, inactive maternal determinants (open circles) are transported (perhaps with the yolk) to a new cytoplasmic environment. This results in their conversion to an active form (filled circles). In the amphibian egg, this is accomplished by cytoplasmic rotation. In the teleost and bird embryos, the mechanism is not known, but the activated determinants become concentrated close to a group of cells at the mesodermal margin. (After Grunwald and Wilson, 1996.)

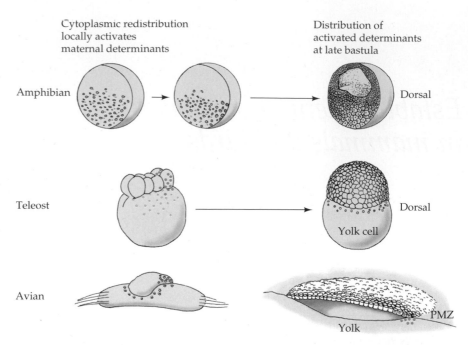

(Oppenheimer, 1936; Tung et al., 1945; Grunwald and Wilson, 1996) that the future dorsal side of the yolk cell acts like a Nieuwkoop center, transferring maternal factors to the blastoderm (Figure 16.1). In avian embryos (and presumably in mammals, as well) the posterior marginal zone (PMZ) may be the equivalent of the Nieuwkoop center (Eyal-Giladi and Khaner, 1989; Khaner and Eyal-Giladi, 1989). Transplantation experiments demonstrated that this is the site where cells collect to form the primitive streak. It had been thought that the hypoblast had axis-inducing ability, but recent studies (Khaner, 1995) suggest that this capacity resides solely in the PMZ. The hypoblast appears merely to direct the streak's subsequent movements. The identification of the chick posterior marginal zone with the Nieuwkoop center is strengthened by the finding that the chick homologue of *Vg1* is transcribed in this region. Moreover, when cultured cells secreting the mature (processed) chick Vg1 protein are placed along the lateral edges of the blastoderm, they induce new primitive streaks to form (Seleiro et al., 1996). Like the amphibian Nieuwkoop center, the future position of the PMZ is fixed shortly after fertilization and depends on gravity and rotation.

Gene Expression in the Organizer Tissues

As mentioned in Chapter 6, the mammalian homologue to the amphibian's dorsal blastopore lip is its node at the anterior end of the primitive streak. In birds, this is called Hensen's node, and in mammals (despite the fact that Hensen first described it in rabbits), this structure is often just called the node. The primitive streak serves as the lateral blastopore lips. The node contains many of the same proteins found in the frog organizer, including Goosecoid, Nodal, Lim-1, and HNF3β. The *nodal* gene is essential for the initiation of the primitive streak and for its continued maintenance. Its expression is first seen in the ventral margin (where gastrulation begins). Afterward, Nodal protein is seen in the anteriormost region of the streak (Figure 16.2; Conlon et al., 1994). When this gene is deleted, the developing embryo has a defective streak and cannot gastrulate. Later in gastrulation (as we will see), *nodal* gene expression is important in forming the left-right axis of the embryo.

Similarly, *goosecoid* mRNA is first seen in the cells in Koller's sickle, when the cells that form the primitive streak collect there. It is then detected in Hensen's node, as the streak moves forward. However, when the node regresses, *goosecoid*-expressing cells remain in the head mesoderm and the pharyngeal endoderm (prechordal plate) just as in amphibians; as the head region forms, *goosecoid* expression occurs in the most anterior cells (Izpisúa-Belmonte et al., 1993). Its expression in these cells appears to be critical for inducing genes involved in head formation. If the *goosecoid* genes are deleted from mouse embryos, the axis forms normally, but the head does not form properly (Rivera-Pérez et al., 1995). Another gene, *Lim-1*, is also expressed in these cells, and mice having their *Lim-1* genes knocked out do not develop heads (see Figure 7.17).

The *HNF-3β* gene resembles similar genes in the *Xenopus* organizer (*XFH1*, *XFD1/1*, and *pintallavis*). HNF-3β is found in the prechordal mesoderm that is thought to induce regional specificity in the forebrain and midbrain, and when this gene is deleted, no node is formed. The embryos have severe deficiencies in their head mesoderm and notochord, fail to gastrulate properly, and lack forebrain and midbrain structures (Ang and Rossant, 1994; Weinstein et al., 1994).

While Goosecoid, Lim-1, and HNF-3β appear to be required for specifying the cells of the anterior dorsal mesoderm, the middle and posterior dorsal axis appears to be specified by the Brachyury (T) protein (MacMurray and Shin, 1988; Yanagisawa, 1990; Stott et al., 1993). Notochord formation and differentiation require expression of the *T* gene (Gluecksohn-Schoenheimer, 1938; Herrmann, 1991; Rashbass et al., 1991), and mutations of the *Brachyury* gene cause posterior axial malformations (Figure 2.25C).

As the node begins to form, it begins to secrete **scatter factor.** This protein appears to promote the ability of the epiblast cells to respond to inducing signals from the node and/or notochord (Streit et al., 1995). As discussed in Chapter 6, the primitive streak elongates, and the neural tube is formed along the midline of the embryo. The prechordal mesoderm is thought to induce the head structures, while the notochord can induce the hindbrain and spinal cord. As the neural tube is laid down, it becomes specified as to the type of neural tube it will be—forebrain, midbrain, hindbrain, or spinal cord. The mesoderm and endoderm are similarly patterned. Recent studies now suggest that this specification is accomplished by the same homeobox-containing genes that specify the anterior-posterior axis in *Drosophila*.

Figure 16.2
Organizer gene expression in developing mouse embryos. Expression of the *nodal* gene during the extension of the primitive streak. (Photograph courtesy of M. R. Kuehn.)

Specifying the mammalian anterior-posterior axis: The Hox code hypothesis

Homology of the Homeotic Gene Complexes Between Drosophila and Mammals

The *Drosophila* homeotic gene complex (HOM-C) on chromosome 3 contains the Antennapedia and Bithorax classes of homeotic genes and can be seen as a single functional unit. (Indeed, in other insects, such as the flour beetle *Tribolium*, it is a single unit.) The HOM-C genes are arranged in the same general order as their expression pattern along the anterior-posterior axis, the most 3' gene (*labial*) being required for producing the most anterior structures, the most 5' gene (*AbdB*) specifying the development of the posterior abdomen. Mouse and human genomes contain four copies of HOM-C per haploid set (*Hox A* through *D* in the mouse, *HOX A* through *D* in humans; Boncinelli et al., 1988; McGinnis and Krumlauf, 1992; Scott, 1992). Not only are the same general *types* of homeotic genes found in both flies and mammals, but the *order* of these genes on the respective chromosomes is remarkably similar. And if this similarity isn't enough to argue for a common scheme of axis formation, it was soon discovered that the *expression pattern* of these genes follows the same pattern: those mammalian genes homologous to the *Drosophila labial, proboscipedia,* and *Deformed* genes are expressed anteriorly, while those genes homologous to the *Drosophila Abdominal-B* gene are expressed posteriorly. The mammalian *Hox/HOX* genes are numbered from 1 to 13, starting from that end of the complex being expressed most anteriorly. Figure 16.3 shows the relationships between the *Drosophila* and mouse homeotic gene sets. The equivalent genes in each mouse complex (such as *Hoxa-1, Hoxb-1,* and *Hoxd-1*)are called a **paralogous group.** It is thought that the four mammalian *Hox* complexes were formed from chromosome duplications. Because there is not a one-to-one correspondence between the *Drosophila* HOM-C genes and the mammalian *Hox* genes, it is likely that independent gene duplications have occurred since these two animal branches diverged (Hunt and Krumlauf, 1992).

Expression of Hox Genes in the Vertebrate Nervous System and its Derivatives

Hox gene expression can be seen along the dorsal axis (neural tube, neural crest, paraxial mesoderm, and surface ectoderm) from the anterior boundary of the hindbrain through the tail. It is also seen in the derivatives of these tissues, especially the derivatives of the neural crest cells. For instance, the hindbrain region of the head gives rise not only to the hindbrain and its cranial ganglia, but also to the cartilage of the ears, jaws, and neck, the aortic arches, and to organs such as the thymus, thyroid, and parathyroid glands. As discussed in Chapter 7, the hindbrain neural tube becomes divided into segmental units called **rhombomeres.** The migration of the cranial neural crest cells also appears to be organized on the rhombomeric pattern such that a specific cranial ganglion and the branchial arch it innervates originate from the crest arising from the same rhombomere (Lumsden et al., 1991). These neural crest cells appear to retain positional information from their original place along the anterior-posterior axis. When avian premigratory neural crest cells that would normally migrate to the *first* branchial arch (to form jaw cartilage) are placed into the region of the crest whose cells normally migrate to the *second* branchial arch (to form the hyoid cartilage), the grafted neural crest cells migrate into the *second* branchial arch, but they form the structures (jaw cartilage) characteristic of the *first* arch. Moreover, they will interact with the surface ectoderm and paraxial mesoderm to form the *first* arch musculature (the beak and jaw muscles). This strongly suggests that before they migrate, the neural crest cells are already committed to form at least some of the structures appropriate to their level on the anterior-posterior axis (Noden, 1988).

This positional commitment may be the result of these cells' expressing particular combinations of *Hox* genes. For instance, the *Hox-B* genes are expressed in the presumptive mouse neural tube before neural crest formation, and when the neural crest cells migrate, they retain the *Hox-B* gene expression pattern characteristic of their place of origin (Hunt et al., 1991a). With only one known exception (*Hoxb-1*), the anterior boundary of each *Hox* gene stops at the rhombomere border two rhombomeres away from the anterior border of the next *Hox* gene (Wilkinson et al., 1989; Keynes and Lumsden, 1990). As represented in Figure 16.4, the homeobox genes *Hoxb-2, -3,* and *-4* are found throughout the spinal cord, but *Hoxb-2* stops at the border of rhombomeres 2 and 3; *Hoxb-3* stops at the 4/5 border, and *Hoxb-4* stops at

(A)

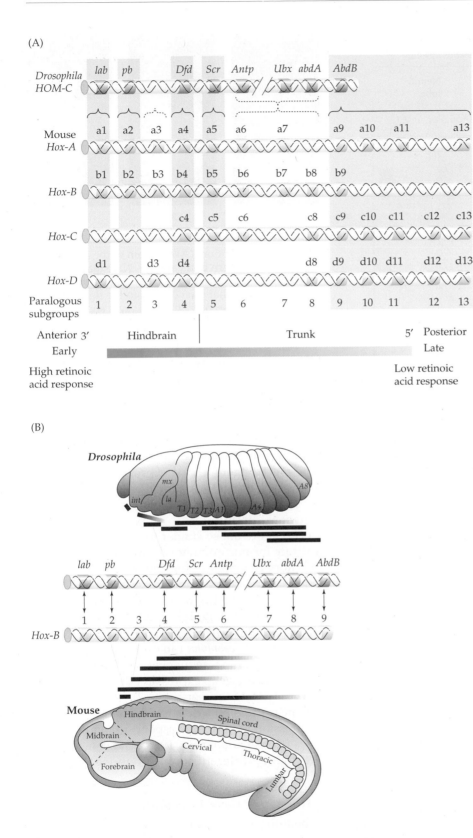

Figure 16.3
Evolutionary conservation of homeotic gene organization and transcriptional expression in flies and mice. (A) Conservation between the homeobox cluster on *Drosophila* chromosome 3 and the four *Hox* gene clusters in the mouse genome. The shaded regions show particularly strong structural similarities between species, and one can see that the order on the chromosomes has been conserved. Those genes at the 5′ end (since all mouse homeobox genes are transcribed in the same direction) are those that are expressed more posteriorly, are expressed later, and can be induced only by high doses of retinoic acid. Genes having similar structures, the same relative positions on each of the four chromosomes, and similar expression patterns belong to the same paralogous group. (B) Comparison of the transcription patterns of the HOM-C and *Hox-B* genes of *Drosophila* (10 hours) and mice (12 days), respectively. Another set of genes that controls the formation of the fly head (*orthodonticle* and *empty spiracles*) have homologues in the mouse that show expression in midbrain and forebrain. The homologous human genes are called (capitalized) HOX genes. (A after Krumlauf, 1993; B after McGinnis and Krumlauf, 1992.)

(B)

(A)

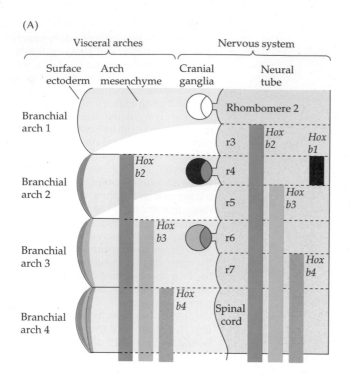

(B)

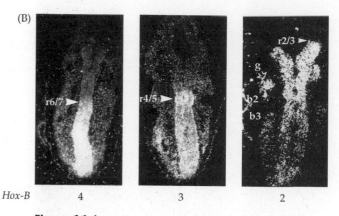

Hox-B

Figure 16.4
Hox gene transcription. (A) Diagram of the *Hox* gene transcription pattern in the mouse. Note that the pattern is staggered between the neural tube and the mesoderm (such that crest cells from the third rhombomere enter the second branchial arch) and that the *Hox* gene expression borders coincide with rhombomere boundaries. (B) Transcription patterns of *Hox-B* homeotic genes in the 9.5-day mouse hindbrain. (A from McGinnis and Krumlauf, 1992; B from Hunt et al., 1991a.)

the border between the sixth and seventh rhombomere. The more 5′ *Hox* genes are found solely in the posterior regions of the neural tube, where they also form a "nested" set. The more 5′ genes have expression boundaries more posterior than the less 5′ genes. When the neural crest cells contact the surface ectoderm, they cause the ectodermal cells to express the same set of *Hox* genes (Figure 16.4A; Hunt et al., 1991b).

Some of the mammalian *Hox* genes are so similar to their *Drosophila* homologues that they can substitute for one another. The mouse *Hoxb-6* gene can perform some of the regulatory functions of the *Drosophila Antennapedia* gene when the murine gene is transfected into *Drosophila*. The human *HOXD-4* gene can also perform some of the regulatory functions of its *Drosophila* homologue, *Deformed* (Malicki et al., 1990; McGinnis et al., 1990). Moreover, the enhancer region of the *Drosophila Deformed* gene (a gene specifying head-specific gene expression in *Drosophila*) can cause gene expression in the mouse hindbrain; and the regulatory sequences of a human homologue of *Deformed* provide head-specific gene expression in *Drosophila* embryos (Awgulewitsch and Jacobs, 1992; Malicki et al., 1992).

A similar pattern of *Hox* gene expression seems to exist within the trunk as well. Here the *Hox* gene expression patterns correspond to somite (rather than rhombomere) boundaries (Kessel and Gruss, 1991), and some paralogous genes are expressed at slightly different somite boundaries (Figure 16.5).

Experimental Analysis of a Hox Code: Gene Targeting

The expression patterns of the murine *Hox* genes suggest a code whereby certain combinations of *Hox* genes specify a particular region of the anterior-posterior axis (Hunt and Krumlauf, 1991). Particular sets of paralogous genes would provide segmental identity along the anterior-posterior axis of the body. Evidence for such a code comes from three sources:

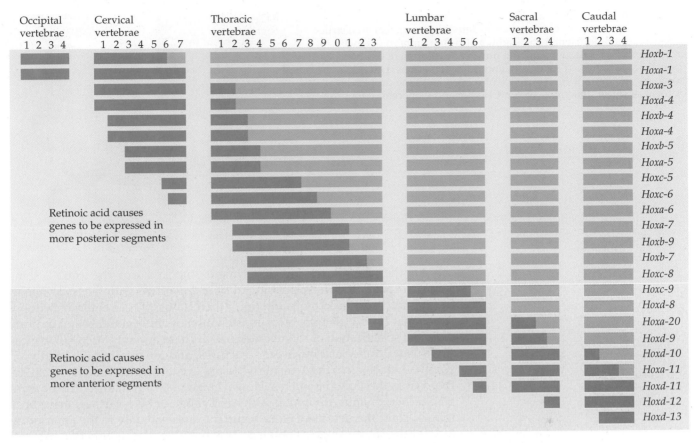

Occipital vertebrae 1 2 3 4	Cervical vertebrae 1 2 3 4 5 6 7	Thoracic vertebrae 1 2 3 4 5 6 7 8 9 0 1 2 3	Lumbar vertebrae 1 2 3 4 5 6	Sacral vertebrae 1 2 3 4	Caudal vertebrae 1 2 3 4	
						Hoxb-1
						Hoxa-1
						Hoxa-3
						Hoxd-4
						Hoxb-4
						Hoxa-4
						Hoxb-5
						Hoxa-5
						Hoxc-5
						Hoxc-6
						Hoxa-6
						Hoxa-7
						Hoxb-9
						Hoxb-7
						Hoxc-8
						Hoxc-9
						Hoxd-8
						Hoxa-20
						Hoxd-9
						Hoxd-10
						Hoxa-11
						Hoxd-11
						Hoxd-12
						Hoxd-13

Retinoic acid causes genes to be expressed in more posterior segments

Retinoic acid causes genes to be expressed in more anterior segments

Figure 16.5
The somite *Hox* code in the trunk and neck of the mouse embryo. Areas of major expression are indicated in darker color, while the posterior regions of expression are not as defined as might be suggested by the lighter color. The effect of retinoic acid is to push anterior gene expression more posteriorly and posterior gene expression more anteriorly. (After Kessel, 1992.)

- *Gene targeting* or "knockout" experiments (see Chapter 2), in which mice are constructed that lack both copies of one or more particular *Hox* genes.
- *Retinoic acid-induced homeosis,* in which mouse embryos given retinoic acid have a different pattern of *Hox* gene expression along the anterior-posterior axis and have abnormal differentiation of their axial structures.
- *Comparative anatomy,* in which the types of vertebrae in different species are correlated with the constellation of *Hox* genes in these vertebrae.

When Chisaka and Capecchi (1991) knocked out the *Hoxa-3* gene from inbred mice, the homozygous mutant *Hoxa-3* mice died soon after they were born. Autopsies of these mice revealed that their neck cartilage was abnormally short and thick and that they had severely deficient or absent thymuses, thyroids, and parathyroid glands. Their heart and major blood vessels were also malformed (Figure 16.6). This set of malformations is extremely similar to the human congenital disorder DiGeorge syndrome, in which these same deficiencies of neural-crest-derived structures are found. Further analysis showed that the number and migration of the neural crest cells that form these structures are normal. Rather, it appears that the *Hoxa-3* genes are responsible for specifying cranial neural crest cell fate and for enabling these cells to differentiate and proliferate into neck cartilage and the fourth and sixth pharyngeal arch derivatives (Manley and Capecchi, 1995).

Figure 16.6

Deficient development of neural-crest-derived pharyngeal arch structures in *Hoxa-3*-deficient mice. Right, a 10.5-day embryo of a heterozygous *Hoxa-3* mouse showing normal development of thymus (pouch 3), parathyroid (pouch 4), and other structures. Left, a homozygous mutant *Hoxa-3*-deficient mouse lacks the proper development of those structures. (From Chisaka and Capecchi, 1991.)

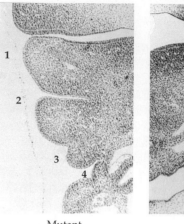
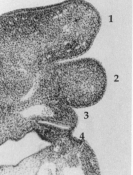

Mutant   Wild-type

Another gene-targeting experiment knocked out the *Hoxa-1* gene (Lufkin et al., 1991). The expression of *Hoxa-1* overlaps with the *Hoxa-3* gene, but it is also expressed more anteriorly than *Hoxa-3*. Those embryos without functional *Hoxa-1* genes show a constellation of abnormalities that indicate deficient specification of rhombomeres 4–7. These mutants often fail to close their neural tube, lack inner ear structures, and are missing the hindbrain ganglia (which form the acoustic, glossopharyngeal, and vagus nerves) derived from these rhombomeres. However, no malformations of the pharyngeal arches, thymus, thyroid, parathyroid gland, or neck cartilage have been found. Thus, the defects of *Hoxa-1* mutants are seen only in the anterior region of the *Hoxa-1* expression area. (It is possible that its functions are not needed or are redundant in the posterior portion of its range.) Unlike the defects of the *Hoxa-3*-deficient mice (which are confined to the neural crest), the *Hoxa-1* defects are seen in the central nervous system and placode-derived tissue, as well as in the paraxial mesoderm. Knockout of the *Hoxa-2* gene also produces mice whose neural crest cells have been respecified. Cranial elements normally formed by the neural crest cells of the second branchial arch (stapes, styloid bones) are missing and are replaced by duplication of the structures of the first branchial arch (incus, malleus, etc.) (Gendron-Maguire et al., 1993; Rijli et al., 1993). Thus, without certain *Hox* genes, some regionally specific organs along the anterior-posterior axis fail to form or become respecified as other regions. The initial evidence supports the notion that different sets of *Hox* genes are necessary for the complete specification of any region of the axis and that a set of paralogous genes may be responsible for different subsets of organs within these regions.

Partial Transformations of Segments by Knockout of **Hox** *Genes Expressed in the Trunk*

If the *Hox* genes do indeed form a code that specifies the anterior-posterior axis, one would expect that changing the constellation of *Hox* genes expressed in any particular region of the embryo might change one structure into another structure along the anterior-posterior axis. This has been shown to be the case when the *Hoxc-8* gene is deleted from the embryo by gene targeting (Le Mouellic et al., 1992). In such mice, several axial skeletal segments resemble more anterior segments, much like what is seen in *Drosophila* loss-of-function homeotic mutations. As can be seen in Figure 16.7, the first lumbar vertebra of this mouse has formed a rib—something characteristic of the thoracic vertebrae anterior to it. The knockout of the *Hoxb-4* gene partially converts the second cervical vertebra (the axis vertebra) into a copy of the

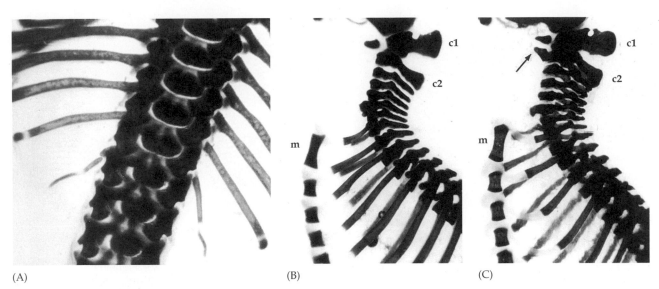

(A) (B) (C)

Figure 16.7
Homeotic transformations in the mouse induced by gene knockouts of homeobox genes expressed in the trunk. (A) Partial transformation of the first lumbar vertebra into a thoracic vertebra by the knockout of the *Hoxc-8* gene. Thoracic vertebrae, but not lumbar vertebrae, have ribs associated with them. (B,C) Partial transformation of the second cervical vertebra into a second copy of the first cervical vertebra by the knockout of the *Hoxb-4* gene. (B) The wild-type mouse has a first cervical vertebra characterized by a ventral tubercle. (C) In the mutant mouse, the second cervical vertebra also has this tubercle (arrow). (A from Le Mouellic et al., 1992; B and C from Ramirez-Solis et al., 1993; photographs courtesy of the authors.)

first cervical vertebra (the atlas), and deletion of the *Hoxa-5* gene causes the posterior transformation of the seventh cervical (neck) vertebra into a rib-forming thoracic vertebra (Jeannotte et al., 1993; Ramirez-Solis et al., 1993).

One can get severe axial transformations by knocking out two or more genes of a paralogous set. Mice homozygous for the *Hoxd-3* deletion have mild abnormalities of the craniocervical joint (the atlas is reduced in size), while mice homozygous for the *Hoxa-3* deletion have no abnormality of this joint (see the previous discussion of this mutant). When the two mutants are bred together, both sets of problems become more severe. The mice with neither *Hoxa-3* nor *Hoxd-3* gene sets have no atlas bone at all, and the hyoid and thyroid cartilage is so reduced in size that there are holes in the skeleton (Condie and Capecchi, 1994). It appears that there are synergistic interactions occurring between the products of the *Hox* genes and that in some functions, one of the paralogues can replace the other.

The regulation of the vertebrate *Hox* genes appears to be controlled by factors similar to those that regulate the Hom-C genes in flies. In *Drosophila*, there is a homeobox gene, *caudal*, that resides outside the Hom-C complex. This maternal effect gene functions in *Drosophila* to direct the expression of the most posterior (*AbdB*) HOM-C genes. A mammalian homologue to that gene, *Cdx1*, plays a similar role in the paraxial mesoderm. It becomes expressed in the primitive streak during gastrulation, when the specification of the anterior-posterior axis is being made, and it is turned off shortly thereafter. If this gene is deleted from the mouse embryo, the expression patterns of the *Hox* genes shift one somite posteriorly, and anterior skeletal structures are found more posteriorly (Subramanian et al., 1995). Similarly, the repression of *Drosophila* Hom-C genes is mediated by a set of genes that includes *extra sex combs* (*esc*). If the mouse homologue of this gene (*embryonic ectoderm development*; *eed*) plays the same role, then one would expect mutations in *eed* to result in the derepression of *Hox* genes and the homeotic transformation of anterior structures into posterior structures. This is indeed what happens. Mutant *eed* genes cause the transformation of anterior skeletal structures into posterior ones (Schumacher et al., 1996).

Experimental Analysis of the Hox Code: Retinoic Acid Teratogenesis

Such homeotic changes are also seen when mouse embryos are given teratogenic doses of retinoic acid (RA). Exogenous retinoic acid given to mouse embryos in utero can cause certain *Hox* genes to become expressed in groups

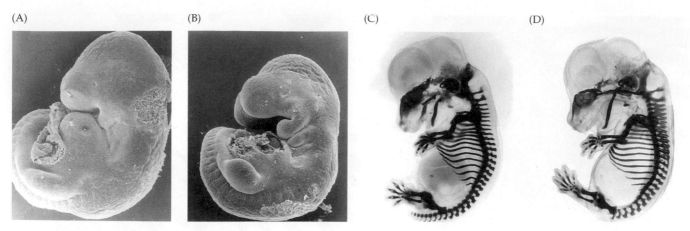

(A) (B) (C) (D)

Figure 16.8

Mouse embryo cultured at day 8 in control medium (A,C) or in medium containing teratogenic retinoids (B,D). At day 2 (A,B), the first pharyngeal arch of the treated embryos has a shortened and flattened appearance and has apparently fused with the second pharyngeal arch. At day 17 (C,D), craniofacial malformations can be seen in the neural-crest-derived cartilage of the treated embryos. Meckel's cartilage is completely displaced from the mandibular (lower jaw) to the maxillary (upper mouth) region. The malleus and incus cartilages are also not formed. (A and B from Goulding and Pratt, 1986; C and D from Morriss-Kay, 1993; photographs courtesy of the authors.)

of cells that usually do not express these genes (Conlon and Rossant, 1992; Kessel, 1992). Moreover, the craniofacial abnormalities of mouse embryos from mothers given teratogenic doses of RA (Figure 16.8) can be mimicked by causing the expression of *Hoxa-7* throughout the embryo (Balling et al., 1989). If high doses of RA can activate *Hox* genes in inappropriate cells along the anterior-posterior axis, and if the constellation of active *Hox* genes specifies the region of the anterior-posterior axis, then mice given RA in utero should show homeotic transformations manifested as malformations occurring along that axis. Kessel and Gruss (1991) found this to be the case. Wild-type mice have 7 cervical (neck) vertebrae, 13 thoracic vertebrae, and 6 lumbar vertebrae (in addition to the sacral and tail vertebrae). When exposed to RA on day 8 of gestation, the first one or two lumbar vertebrae were transformed into thoracic vertebrae, while the first sacral vertebra often became a lumbar vertebra (Figure 16.9). In some cases, the entire posterior region of the mouse embryo failed to form. These changes in structure were correlated with changes in the constellation of *Hox* genes expressed in these tissues. For example, when RA was given to embryos on day 8 (during gastrulation), *Hoxa-10* expression was shifted posteriorly, and an additional set of ribs formed on what had been the first lumbar vertebra. When posterior *Hox*

(A) (B) (C)

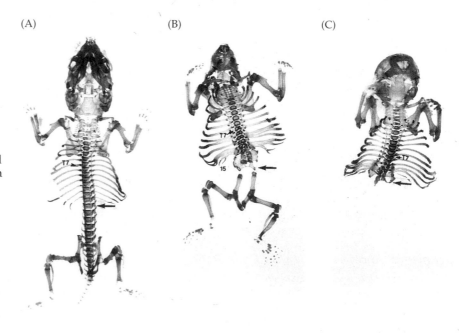

Figure 16.9

Retinoic acid administered to pregnant mice alters *Hox* gene expression and phenotype in fetuses. The figure shows changes to axial skeleton (vertebrae and ribs) caused by RA exposure in utero on day 8. (A) The wild-type has 7 cervical vertebrae, 13 thoracic vertebrae, 6 lumbar vertebrae, 4 fused sacral vertebrae, and tail vertebrae. This arrangement is altered by RA given to the mothers. In some cases (B,C), RA caused the loss of the lumbar, sacral, and caudal vertebrae. (A and B after Kessel and Gruss, 1991; C from Kessel, 1992; photographs courtesy of the authors.)

Figure 16.10
Retinoic acid mediates homeotic transformation of hindbrain regions. In untreated mouse embryos on day 8.5, *Hoxb-1* expression is limited to the r4 rhombomere. When exposed to retinoic acid at this time, *Hoxb-1* expression expands anteriorly toward the midbrain. After 2 days, *Hoxb-1* in normal embryos is expressed in the descendant cells of the r4 rhombomere and in the midline cells of r5, which generate the facial motor nerve (mnVII). In the RA-treated embryos, the normal pattern of r4/5 has been duplicated in r2/3. The neural crest expression of *Hoxb-2* is also duplicated, and a second facial motor nerve is formed. This suggests that retinoic acid mediates the homeotic transformation of r2/3 to r4/5. (After Krumlauf, 1993.)

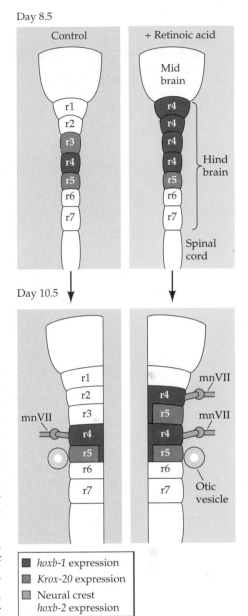

genes did not become expressed at all, the caudal part of the embryo failed to form.*

In the central nervous system, retinoic acid induces the anterior expression of *Hox* genes that are usually expressed only more posteriorly, and they cause rhombomeres 2 and 3 to assume the identity of rhombomeres 4 and 5 (Figure 16.10; Marshall et al., 1992; Kessel, 1993). In this instance, the trigeminal nerve (which arises from rhombomere 2) is transformed into another facial nerve (characteristic of rhombomere 4), and abnormalities of the first branchial arch indicate that the neural crest cells of the second and third rhombomeres have been transformed into more posterior phenotypes.

Retinoic acid probably plays a role in axis specification during normal development, and the source of retinoic acid is probably Hensen's node (Hogan, 1992; Maden et al., 1996). Since the early node appears to contain the precursors of both the anterior and the posterior structures, it is possible that the specification of these cells depends on the amount of time spent within the high retinoic acid concentrations of the node. The more time spent in the node, the more posterior the specification. This is seen to occur in culture, as embryonal carcinoma cells express more "posterior" *Hox* genes the longer they are exposed to retinoic acid (Simeone et al., 1990). Moreover, *Hoxa-1*, *Hoxb-1*, and *Hoxd-4* each have retinoic acid response element in their upstream regulatory regions (see Chapter 21). Giving RA exogenously would mimic the situation normally encountered only by the posterior cells. Avantaggiato and colleagues (1996) have shown that when retinoic acid is given to embryos during the mid-streak stages, the anteriormost regions of the neural tube do not form and are replaced by tissue resembling the hindbrain. This correlates with a loss of forebrain and midbrain gene expression (*Emx1*, *Emx2*) in this region and their replacement by hindbrain-specific *Hox* genes such as *Hoxb-1*. The evidence points to a *Hox* code wherein different constellations of *Hox* genes specify the regional characteristics along the anterior-posterior axis. Moreover, as these expression patterns are similar for mammals and insects, it appears that there is a common developmental plan on which the anterior-posterior axis of most animals is constructed.

Evidence for a Hox Code from Comparative Anatomy

A new type of comparative embryology is now emerging. Gaunt (1994) and Burke and her collaborators (1995) have compared the vertebrae of the mouse and chick. Although the mouse and chick have a similar number of vertebrae, they apportion them differently. Mice (like all mammals, be they giraffes or whales) have only 7 cervical (neck) vertebrae. These are followed

---

*Hoxa-10 is also important in specifying the axial pattern of the genital ducts. Knockouts of *Hoxa-10* create mice wherein the upper region of the uterus is transformed into tissue resembling the oviduct. This region coincides with the anterior limit of *Hoxa-10* expression in the wild-type Müllerian duct (Benson et al., 1996).

Figure 16.11
Schematic representation of the mouse and chick vertebral pattern along the anterior-posterior axis. The boundaries of certain *Hox* genes have been placed onto these domains. (After Burke et al., 1995.)

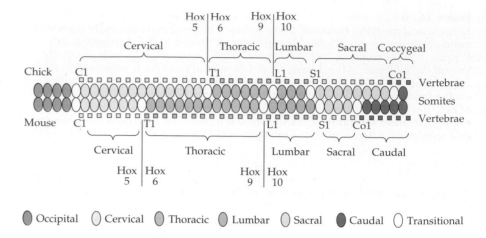

by 13 thoracic (ribbed) vertebrae, 6 lumbar vertebrae, 4 sacral vertebrae, and a variable (20+) number of caudal (tail) vertebrae (Figure 16.11). The chick, on the other hand, has 14 cervical vertebrae, seven thoracic vertebrae, 12 or 13 (depending on the strain) lumbosacral vertebrae, and 5 coccygeal vertebrae. The question is, Does the constellation of *Hox* genes correlate with the type of vertebra (e.g., cervical or thoracic) or with the relative position of the vertebrae (e.g., number 8 or 9)? The answer is that the constellation of *Hox* genes predicts the *type* of vertebra. In the mouse, the transition between cervical and thoracic vertebrae is between vertebrae 7 and 8; in the chick it is between vertebrae 13 and 14. In both cases, the *Hox-5* paralogues are seen in the last cervical vertebrae, while the anterior boundary of the *Hox-6* paralogues extends to the first thoracic vertebra. Similarly, in both cases, the thoracic/lumbar transition is seen at the boundary between the *Hox-9* and *Hox-10* paralogue groups. It appears that there is a code of *Hox* gene expression that determines the type of vertebra along the anterior-posterior axis.

Sidelights & Speculations

Animals as Variations on the Same Developmental Theme

ONE OF THE MOST CELEBRATED (and acrimonious) debates in biology was fought in Paris at the height of the French Revolution. There, in the Académie des Sciences, E. Geoffroy Saint-Hilaire contested with Georges Cuvier over the nature of the animal kingdom. Cuvier, the eminent comparative anatomist who had "made zoology a French science," emphasized the *differences* that separate the phyla from one another. There could be no "Chain of Being" linking all organisms, nor could there be any way that the parts of an insect could be seen as homologous to those of a mollusc or vertebrate. The only thing linking an insect leg, a mollusc foot, and a vertebrate leg was their locomotive function. Anatomically and embryologically, they were distinct and noncomparable entities.

Geoffroy Saint-Hilaire emphasized the *similarities* among all phyla. He claimed that all animals were organized according to the same basic principles, and an insect was but a vertebrate turned upside down. A head was formed at one end, a tail at the other, and all the animals had neural tubes, whether dorsal or ventral. Instead of nature composed of intrinsically different species, all animals were united in a kind of brotherhood, reminiscent of the Revolution's *egalité* and *fraternité* (Appel, 1987).

Since that time, different biological traditions have emphasized either the differences or the similarities among organisms. Comparative anatomy (following Cuvier) emphasizes the differences, while morphology (following Geoffrey Saint-Hilaire) celebrates the "underlying unities." Genetics and cell biology look at all animals (and plants) as composed in basically the same way and following the same laws, while embryology has traditionally seen each species as developing in a different manner.

Recently, though, embryology is providing evidence for the underlying unity of animal nature. Jonathan Slack and his colleagues (1993) have defined an animal as an organism that displays a particular spatial pattern of *Hox* gene expression. They propose that the body plan of each phylum is typified at a particular "phylotypic" stage during its development. For vertebrates, this would be the tailbud stage (where, despite their differences in cleavage and gastrulation, vertebrate embryos converge and have tailbuds and pharyngeal pouches); for insects, the fully segmented germ band is where the embryos converge. At this stage, the homeotic gene expression pattern of the *Hox*/HOM-C genes is seen most clearly and is remarkably similar in all animals. Those genes resembling *Deformed* and *labial* are expressed in the anterior of the embryo; those resembling *Abdominal B* are expressed in the posterior. Even nematodes and hydras have homeotic gene clusters that appear to be expressed in the same anterior-posterior fashion (Schummer et al., 1992; Wang et al. 1993). Although fungi and plants have homeobox genes, they are not homologous with the ones in animals, nor are they arranged in the same chromosomal order or expressed in the same anterior-posterior pattern. In this way, the spatial pattern of *Hox* gene expression is being used as the primary underlying characteristic defining animal existence. This observation has not yet been tested in several phyla, and it will be very interesting to see whether this general pattern is seen throughout the animal kingdom.

Getting a Head: More Vertebrate and Invertebrate Homologies

In *Drosophila*, the brain is composed of three neuromeres. The neuromeres are specified by two homeobox-containing genes that are not linked to the HOM-C region. These genes are *orthodenticle* (*otd*), which is expressed predominantly in the anteriormost neuromere, and *empty spiracles* (*ems*), which is expressed in the posterior two brain neuromeres. Loss-of-function mutations of *otd* eliminate the anteriormost neuromere of the developing *Drosophila* embryo, and loss-of-function mutations of *ems* eliminate the second and third neuromeres (Hirth et al., 1995). In frogs and mice, the homologues of these genes (*Otx-1, Otx-2, Emx-1, Emx-2*) are also expressed in the brain (Simeone et al., 1992), although the exact patterns of transcription are not identical (Figure 16.12). The *Otx-2* gene has been knocked out by gene targeting (Acampora et al., 1995; Matsuo et al., 1995; Ang et al., 1996), and the resulting mice have neural and mesodermal head deficiencies anterior to the r3 rhombomere. In humans, mutations of *EMX-2* lead to a rare condition known as schizencephaly, in which there are clefts ripping through the entire cerebral cortex (Brunelli et al., 1996). Even though the *Drosophila otd* and *ems* genes are specified by the Bicoid and Hunchback gradients, and the mammalian *Otx* and *Emx* transcripts are induced by the anterior dorsal mesoderm, it appears that these same genes are used for specifying the brain regions.* ■

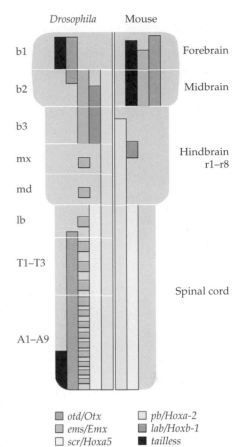

Figure 16.12 Expression of the regulatory genes in *Drosophila* and mouse, emphasizing those genes expressed in the head. A1–9 are abdominal segments; b1-3 are neuromere ("brain") segments; 1b, md, and mx are the labia, mandibular, and maxillary segments, respectively; r, rhombomere; T1-3, thoracic segments. (After Thor, 1995.)

*In addition to expressing the homologues of homeobox-containing genes *ems* and *otd*, the mammalian brain also expresses the homologue of the *tailless* gene. This gene is expressed in the most anterior and posterior portions of the *Drosophila* embryo and is a member of the steroid receptor family (Monaghan et al., 1995).

Dorsal-ventral and left-right axes in mammals and birds

Very little is known about how mammals form their dorsal-ventral axis. In chicks, the dorsal-ventral axis is determined by gravity placing the hypoblast on the ventral side (see Chapter 5). In mice and humans, the hypoblast forms on the side of the inner cell mass that is exposed to the blastocyst fluid. As development proceeds, the notochord maintains dorsal-ventral polarity by inducing specific dorsal-ventral patterns of gene expression in the neural tube (Goulding et al., 1993). [mamaxis1.html]

Similarly, we know very little about the formation of the left-right axis. The mammalian body is not symmetrical. The heart is assigned to the left

(A)

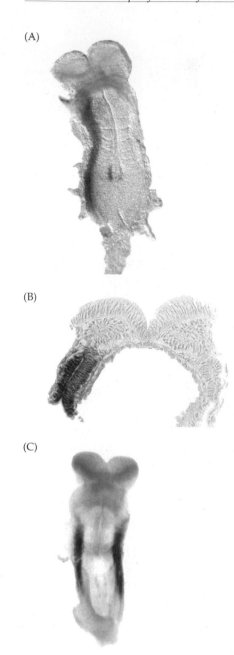

(B)

(C)

Figure 16.13
Asymmetry of gene expression in the mouse embryo. (A) In situ hybridization for *nodal* mRNA in the 5-somite mouse embryo. The *nodal* gene expression is confined to the lateral plate mesoderm on the left side of the embryo. (B) Cross section through the embryo at the same stage as (A). (C) In mice with the *inverted (iv)* mutation, *nodal* expression is seen in the lateral plate mesoderm on both sides of the embryo. The heart has an equal chance of looping to either side. (After Lowe et al., 1996; photographs courtesy of M. R. Kuehn.)

side of the chest cavity, even though it forms in the center. The spleen is found solely on the left side of the abdomen, while the major lobe of the liver is on the right side of the abdomen. It is not known what regulates these asymmetries. However, recent findings suggest two levels in regulating the left-right axis: a global level and an organ-specific level.

In the mouse, two genes are known whose mutations destroy normal right-left asymmetry. The first gene, *situs inversus viscerum (iv)*, randomizes the left-right axis for each asymmetrical organ (Hummel and Chapman, 1959; Layton, 1976). This means that the heart may loop to the left in one homozygous animal, but loop to the right in another. Moreover, the direction of heart looping is not coordinated with the placement of the spleen or the stomach. This can cause serious problems, even death. The second gene, *inversion of embryonic turning (inv)*, causes a more global phenotype. Mice homozygous for an insertion mutation at this locus were found to have *all* their asymmetrical organs on the wrong side of the body (Yokoyama et al., 1993).* Since all the organs are reversed, this asymmetry does not have dire consequences for the mice.

While we do not know the proteins encoded by *iv* and *inv*, some of the components of this pathway have been recently discovered. In the mouse, the *lefty* and *nodal* genes are expressed only in the left lateral plate mesoderm, and their expression precedes the characteristic right-handed looping of the heart and the right-handed rotation of the embryo (Figure 16.13; Collignon et al., 1996; Lowe et al., 1996; Meno et al., 1996.). In mice homozygous for the *inversion of embryonic turning* mutation, these genes become expressed solely on the *right* side of the lateral plate mesoderm, whereas in mice with the randomizing *situs inversus viscerum* mutation, the expression of *nodal* and *lefty* is either normal, switched, or absent. These genes encode paracrine factors of the TGF-β family, and it is not known what tissues they influence. One possible site of influence is the symmetrical heart tube that forms in the embryonic midline. Between the inner endocardium and the outer myocardium of this double-walled tube is an extracellular matrix (cardiac jelly) that contains the protein Flectin. In the chick embryo, this protein is asymmetrically expressed at the time of cardiac looping, accumulating predominantly in the left side of the matrix (Plate 33; Tsuda et al., 1996).

The mechanisms that would yield asymmetrical transcription of *nodal* and *lefty* are still unclear, but hints are coming from studies of the chick embryo (Levin et al., 1995). The critical observation is that during midgastrulation, the **sonic hedgehog (shh)** message is transcribed symmetrically throughout Hensen's node. A few hours later, however, transcription on the right side stops, and *shh* transcription is only seen on the left side of the node. At the same time as this asymmetry develops, the **activin receptor IIa (cActRIIa)** gene is expressed only on the right side of the node (Figure 16.14). This receptor can be induced by activin. The *sonic hedgehog* expression on the left side does not last long, disappearing at about 24 hours incubation, and the expression of the chick **nodal** gene, becomes expressed solely on the left side (Plate 25). It is tempting to put these genes into a common pathway where activin (or an activin-like molecule) would be made only on the right side of Hensen's node in the chick embryo. It would induce the

*This gene was discovered accidentally when Yokoyama and colleagues (1993) made transgenic mice wherein the transgene (for the tyrosinase enzyme) inserted randomly into the genome. In one instance, this gene inserted itself into a region of chromosome 4, knocking out the existing gene.

Figure 16.14
Pathway for left-right asymmetry in the chick embryo. (A) Top: Expression pattern of the *sonic hedgehog, activin receptor IIa,* and *cNR-1* genes with respect to Hensen's node. The activin receptor is on earliest, followed by sonic hedgehog and last by *cNR-1*. Bottom: A day later, the asymmetry is seen in the "right-handed" heart loop. The hypothetical pathway between these genes is shown beneath them. (B,C) Dorsal and close-up views of the in situ hybridization of the *sonic hedgehog* mRNA. (D,E) Dorsal and close-up views of the *activin receptor IIa* message. (A after Roush, 1995, and Wolpert and Brown, 1995; B–E from Levin, et al., 1995, courtesy of C. Tabin and C. Stern.)

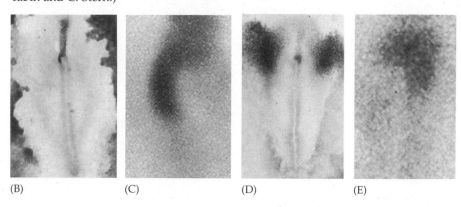

(B)   (C)   (D)   (E)

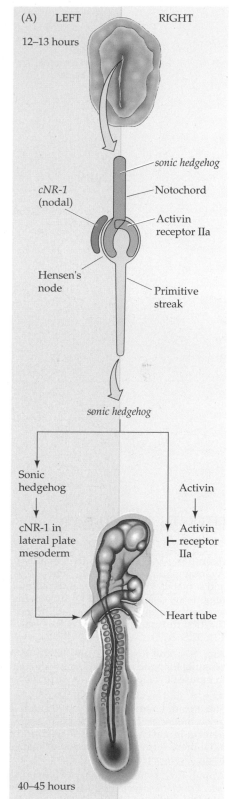

synthesis of the activin receptor IIa and then function through that receptor to block *sonic hedgehog* expression wherever that receptor was located. Therefore, it would block *shh* transcription on the right side of the node. *Sonic hedgehog*, then, would only become expressed on the left side of the node. Sonic hedgehog protein would then be secreted on the left side of the embryo and activate the *nodal* gene in the lateral plate mesoderm that contains the precursors of the heart. There, they might cause accumulation of flectin protein in the left side of the extracellular matrix. Experiments suggest that this pathway is a good approximation. Activin is indeed synthesized at the appropriate time and only in the right-hand side of Hensen's node. If blocked by the experimental addition of Follistatin, the asymmetry of *sonic hedgehog* expression vanishes, and the heart has an equal chance of looping either way (Levin et al., 1997). When activin-soaked beads were placed into the left side of Hensen's node, they induced the synthesis of cActRIIa on that side, and the *shh* gene (usually expressed only on the left side) was repressed. This, in turn, suppressed the transcription of *nodal*. In this situation, the heart tube formed randomly, having an equal probability of being left or right. A similar condition was produced when cells secreting Sonic hedgehog were implanted into the right side of the node. In these cases, Nodal was induced symmetrically in the lateral plate mesoderm, and the heart had a 50 percent chance of having a left-handed tube (Figure 16.15). The formation of the right-left axis in the mouse also appears to use activin receptors and nodal protein, but it does not appear to link the two through Sonic hedgehog (Collignon et al., 1996). The chick and mouse appear to have subtle variations on how they construct their axes. [mamaxis2.html]

Several different paths—teratogenesis, gene knockouts, organizer-specific gene studies, clinical genetics, even fruit fly genetics—are leading us to an understanding of a fundamental mystery: how the vertebrate embryo begins to know up from down, mouth from anus, and right from left. We have learned more about this in the past five years than in all the years preceding them.

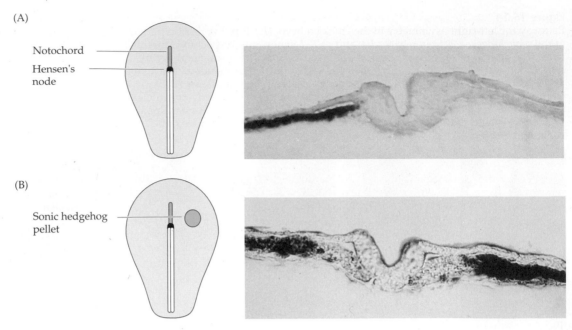

Figure 16.15
Ectopic expression of *sonic hedgehog* leads to symmetrical *cNR-1 (nodal)* expression and randomization of heart looping. (A) Wild-type expression of *cNR-1* gene, showing expression on the left side. Nearly all hearts develop right-sided loops. This pattern is also seen when pellets containing control substances are implanted into the right side of the node or when a Sonic-hedgehog-containing pellet is implanted on the left side (where *shh* is usually expressed). (B) When Sonic hedgehog pellets are implanted on the right side of the node, the expression of *cNR-1* becomes bilaterally symmetrical. (From Levin et al., 1995; photographs courtesy of the authors.)

LITERATURE CITED

Acampora, D., Mazan, S., Lallemand, Y., Avantaggiato, V., Maury, M., Simeone, A. and Brulet, P. 1995. Forebrain and midbrain regions are deleted in *Otx2⁻/⁻* mutants due to a defective anterior neuroectoderm specification during gastrulation. *Development* 121: 3279–3290.

Ang, S. L. and Rossant, J. 1994. HNF-3β is essential for node and notochord formation in mouse development. *Cell* 78: 561–574.

Ang, S. L., Jin, O., Rhinn, M., Daigle, N., Stevenson, L. and Rossant, J. 1995. A targeted mouse *otx2* mutation leads to severe defects in gastrulation and formation of axial mesoderm and to deletion of rostral brain. *Development* 122: 243–252.

Appel, T. A. 1987. *The Cuvier-Geoffroy Debate: French Biology in the Decades before Darwin*. Oxford University Press, New York.

Avantaggiato, V. Acampora, D., Tuorto, F. and Simeone, A. 1996. Retinoic acid induces stage-specific repatterning of the rostral central nervous system. *Dev. Biol.* 175: 347–357.

Awgulewitsch, A. and Jacobs, D. 1992. *Deformed* autoregulatory element from *Drosophila* functions in a conserved manner in transgenic mice. *Nature* 358: 341–344.

Balling, R., Mutter, G., Gruss, P. and Kessel, M. 1989. Craniofacial abnormalities induced by ectopic expression of the homeobox gene *Hox-1.1* in transgenic mice. *Cell* 58: 337–347.

Benson, G. V., Lim, H., Paria, B. C., Satokata, I., Dey, S. K. and Maas, R. L. 1996. Mechanisms of reduced fertility in Hoxa-10 mutant mice: Uterine homeosis and loss of maternal Hoxa-10 expression. *Development* 122: 2687–2696.

Boncinelli, E., Somma, R., Acampora, D., Pannese, M., D'Esposito, M., Faiella, A. and Simeone, A. 1988. Organization of human homeobox genes. *Hum. Reprod.* 3: 880–886.

Brunelli, S., Faiella, A., Capra, V., Nigro, V., Smeone, A., Cama, A. and Boncinelli, E. 1996. Germline mutations in the homeobox gene *Emx2* in patients with severe schizencephaly. *Nat. Genet.* 12: 94–96.

Burke, A. C., Nelson, A. C., Morgan, B. A. and Tabin, C. 1995. Hox genes and the evolution of vertebrate axial morphology. *Development* 121: 333–346.

Chisaka, O. and Capecchi, M. 1991. Regionally restricted developmental defects resulting from targeted disruption of the mouse homeobox gene *Hox-1.5. Nature* 350: 473–479.

Collignon, J., Varlet, I. and Robertson, E. J. 1996. Relationship between asymmetric *nodal* expression and the direction of embryonic turning. *Nature* 381: 155–158.

Condie, B. G. and Capecchi, M. R. 1994. Mice with targeted disruptions in the paralogous genes *hoxa-3* and *hoxd-3* reveal synergistic interactions. *Nature* 370: 304–307.

Conlon, F. L., Lyons, K. M., Takaesu, N., Barth, K. S., Herrmann, B. and Robertson, E. J. 1994. A primary requirement for *nodal* in the formation of the primitive streak in the mouse. *Development* 120: 1919–1928.

Conlon, R. A. and Rossant, J. 1992. Exogenous retinoic acid rapidly induces anterior ectopic expression of murine *Hox-2* genes in vivo. *Development* 116: 357–368.

Eyal-Giladi, H. and Khaner, O. 1989. The chick's marginal zone and primitive streak formation. II. Quantification of marginal zone's potencies—temporal and spatial aspects. *Dev. Biol.* 134: 215–221.

Gaunt, S. J. 1994. Conservation in the Hox code during morphological evolution. *Int. J. Dev. Biol.* 38: 549–552.

Gendron-Maguire, M., Mallo, M., Zhang, M. and Gridley, T. 1993. *Hoxa-2* mutant mice exhibit homeotic transformation of skeletal elements derived from cranial neural crest. *Cell* 75: 1317–1331.

Glueksohn-Schoenheimer, S. 1938. The development of two tailless mutants in the house mouse. *Genetics* 23: 573–584.

Goulding, E. H. and Pratt, R. M. 1986. Isotretinoin teratogenicity in mouse whole embryo culture. *J. Craniofac. Genet. Dev. Biol.* 6: 99–112.

Goulding, M. D., Lumsden, A. and Gruss, P. 1993. Signals from the notochord and floor plate regulate the region-specific expression of two *Pax* genes in the developing spinal cord. *Development* 117: 1001–1016.

Grunwald, D. J. and Wilson, E. T. 1996. A unifying model for the origin of the vertebrate organizer. *Netherl. J. Zool.* 46: 22–46.

Herrmann, B. G. 1991. Expression pattern of the *Brachyury* gene in whole-mount *Twis/Twis* mutant embryos. *Development* 113: 913–917.

Hirth, F., Therianos, S., Loop, T., Gehring, W. J., Reichart, H. and Furukubo-Tokunaga, K. 1995. Developmental defects in brain segmentation caused by mutations of the homeobox genes *orthodenticle* and *empty spiracles* in *Drosophila*. *Neuron* 15: 769 –778.

Hogan, B. L. M., Thaller, C. and Eichele, G. 1992. Evidence that Hensen's node is a site of retinoic acid synthesis. *Nature* 359: 237–241.

Hummel, K. P. and Chapman, D. B. 1959. Visceral inversion and associated anomalies in the mouse. *J. Hered.* 50: 9–13.

Hunt, P. and Krumlauf, R. 1991. Deciphering the *Hox* code: Clues to patterning branchial regions of the head. *Cell* 66: 1075–1078.

Hunt, P. and Krumlauf, R. 1992. *Hox* codes and positional specification in vertebrate embryonic axes. *Annu. Rev. Cell Biol.* 8: 227–256.

Hunt, P. and seven others. 1991a. A distinct *Hox* code for the branchial region of the head. *Nature* 353: 861–864.

Hunt, P., Wilkinson, D. and Krumlauf, R. 1991b. Patterning the vertebrate head: Murine *Hox-2* genes mark distinct subpopulations of premigratory and migratory neural crest. *Development* 112: 43–50.

Izpisúa-Belmonte, J. C., De Robertis, E. M., Storey, K. G. and Stern, C. D. 1993. The homeobox gene *goosecoid* and the origin of organizer cells in the early chick blastoderm. *Cell* 74: 645–659.

Jeannotte, L., Lemieux, M., Cherron, J., Poirier, F. and Robertson, E. J. 1993. Specification of axial identity in the mouse: Role of the *Hoxa-5 (Hox 1.3)* gene. *Genes Dev.* 7: 2085–2096.

Kessel, M. 1992. Respecification of vertebral identities by retinoic acid. *Development* 115: 487–501.

Kessel, M. 1993. Reversal of axonal pathways from rhombomere 3 correlates with extra *Hox* expression domains. *Neuron* 10: 379–393.

Kessel, M. and Gruss, P. 1991. Homeotic transformations of murine vertebrae and concomitant alteration of *Hox* codes induced by retinoic acid. *Cell* 67: 89–104.

Keynes, R. and Lumsden, A. 1990. Segmentation and the origin of regional diversity in the vertebrate central nervous system. *Neuron* 2: 1–9.

Khaner, O. 1995. The rotated hypoblast of the chicken embryo does not initiate an ectopic axis in the epiblast. *Proc. Natl. Acad. Sci. USA* 92: 10733–10737.

Khaner, O. and Eyal-Giladi, H. 1989. The chick's marginal zone and primitive streak formation. I. Coordinative effects of induction and inhibition. *Dev. Biol.* 134: 206–214.

Krumlauf, R. 1993. *Hox* genes and pattern formation in the branchial region of the vertebrate head. *Trends. Genet.* 9: 106–112.

Layton, W. M. Jr. 1976. Random determination of a developmental process. *J. Hered.* 67: 336–338.

Le Mouellic, H., Lallemand, Y. and Brûlet, P. 1992. Homeosis in the mouse induced by a null mutation in the *Hox-3.1* gene. *Cell* 69: 251–264.

Levin, M., Johnson, R. L., Stern, C., Kuehn, M. and Tabin, C. 1995. A molecular pathway determining left-right asymmetry in chick embryogenesis. *Cell* 82: 803–814.

Levin, M., Pagan, S., Roberts, D. J., Cooke, J., Kuehn, M. R. and Tabin, C. J. 1997. Different aspects of laterality are independently controlled by an apparently streak-autonomous signaling pathway initiating by activin. In press.

Lowe, L. A. and eight others. 1996. Conserved left-right asymmetry of nodal expression and alterations in murine situs inversus. *Nature* 381: 158–161.

Lufkin, T., Dierich, A., LeMeur, M., Mark, M. and Chambon, P. 1991. Disruption of the *Hox-1.6* homeobox gene results in defects in a region corresponding to its rostral domain of expression. *Cell* 66: 1105–1119.

Lumsden, A., Sprawson, N. and Graham, A. 1991. Segmental origin and migration of neural crest cells in the hindbrain region of the chick embryo. *Development* 113: 1281–1291.

MacMurray, A. and Shin, H.-S. 1988. The antimorphic nature of the *Tc* allele at the mouse *T* locus. *Genetics* 120: 545–550.

Maden, M., Gale, E., Kostetski, I. and Zile, M., 1996. Vitamin A-deficient quail embryos have half a hindbrain and other neural defects. *Curr. Biol.* 6: 417–426.

Malicki, J., Schugart, K. and McGinnis, W. 1990. Mouse *Hox-2.2* specifies thoracic segmental identity in *Drosophila* embryos and larvae. *Cell* 63: 961–967.

Malicki, J. Cianetti, L. C., Peschle, C. and McGinnis, W. 1992. Human HOX4B regulatory element provides head-specific expression in *Drosophila* embryos. *Nature* 358: 345–347.

Manley, N. R. and Capecchi, M. R. 1995. The role of *Hoxa-3* in mouse thymus and thyroid development. *Development* 121: 1989–2003.

Marshall, H., Nonchev, S., Sham, M. H., Muchamore, I., Lumsden, A. and Krumlauf, R. 1992. Retinoic acid alters hindbrain *Hox* code and induces transformation of rhombomeres 2/3 into a 4/5 identity. *Nature* 360: 737–741.

Matsuo, I., Kuratani, S., Kimura, C., Takeda, N. and Aizawa, S. 1995. Mouse *Otx2* functions in the formation and patterning of the rostral head. *Genes Dev.* 9: 2646–2658.

McGinnis, W. and Krumlauf, R. 1992. Homeobox genes and axial patterning. *Cell* 68: 283–302.

McGinnis, N., Kuziora, M. A. and McGinnis, W. 1990. Human *Hox 4.2* and *Drosophila Deformed* encode similar regulatory specificities in *Drosophila* embryos and larvae. *Cell* 63: 969–976.

Meno, C. and seven others. 1996. Left-right asymmetric expression of the TGFβ-family member lefty in mouse embryos. *Nature* 381: 151–155.

Monaghan, P. , Grau, E., Bock, D. and Schulz, G. 1995. The mouse homolog of the orphan nuclear receptor tailless is expressed in the developing brain. *Development* 121: 839–851.

Morriss-Kay, G. 1993. Retinoic acid and craniofacial development: Molecules and morphogenesis. *BioEssays* 15: 9–15.

Nature Genetics (editorial). 1995. Risk assessment and religion. *Nat. Genet.* 11: 105–106.

Noden, D. 1988. Interactions and fates of avian craniofacial mesenchyme. *Development* [Suppl.] 103: 121–140.

Oppenheimer, J. M. 1936. The development of isolated blastoderms of *Fundulus heteroclitus. J. Exp. Zool.* 72: 247–269.

Ramirez-Solis, R., Zheng, H., Whiting, J., Krumlauf, R. and Bradley, A. 1993. *Hoxb-4* (*Hox 2.6*) mutant mice show homeotic transformation of a cervical vertebra and defects in the closure of the sternal rudiments. *Cell* 73: 279–294.

Rashbass, P. R., Cook, L. A., Herrmann, B. G. and Beddington, R. S. P. 1991. A cell autonomous function of *Brachyury* in *T/T* embryonic stem cell chimeras. *Nature* 353: 348–351.

Riddihough, G. 1992. Homing in on the homeobox. *Nature* 357: 643–644.

Rijli, F. M., Mark, M., Lakkaraju, S., Dierich, A., Dollé, P. and Chambon, P. 1993. A homeotic transformation is generated in the rostral branchial region of the head by disruption on *Hoxa-2*, which acts as a selector gene. *Cell* 75: 1333–1349.

Rivera-Pérez, J. A., Mallo, M., Gendron-Maguire, M., Gridley, T. and Behringer, R. R. 1995. Goosecoid is not an essential component of the mouse gastrula organizer but is required for craniofacial and rib development. *Development* 121: 3005–3012.

Roush, W. 1995. Embryos travel forking path as they tell left from right. *Science* 269: 1514–1515.

Schumacher, A., Faust, C. and Magnuson, T. 1996. Positional cloning of a global regulator of anterior-posterior patterning in mice. *Nature* 383: 250–253.

Schummer, M., Scheurlen, I., Schaller, C. and Galliot, B. 1992. HOM/HOX homeobox genes are present in hydra (*Chlorohydra viridissima*) and are differentially expressed during regeneration. *EMBO J.* 11: 1815–1825.

Scott, M. 1992. Vertebrate homeobox gene nomenclature. *Cell* 71: 551–553.

Seleiro, E. A. P., Connolly, D. J. and Cooke, J. 1996. Early developmental expression and experimental axis determination by the chicken *Vg1* gene. *Curr. Biol.* 6: 1476–1486.

Simeone, A., Acampora, D., Arcioni, L., Boncinelli, E. and Mavilio, F. 1990. Sequential activation of *Hox* 2 genes by retinoic acid in human embryonal carcinoma cells. *Nature* 34: 763–766.

Simeone, A., Gulisano, M., Acampora, Stornaiuolo, A., Rambaldi, M. and Boncinelli, E. 1992. Two vertebrate homeobox genes related to *Drosophila* empty spiracles are expressed in the embryonic cerebral cortex. *EMBO J.* 11: 2541–2550.

Slack, J. M. W. and Tannahill, D. 1992. Mechanism of anteroposterior axis specification in vertebrates: Lessons from the amphibians. *Development* 114: 285–302.

Slack, J. M. W., Holland, P. W. H. and Graham, C. F. 1993. The zootype and the phylotypic stage. *Nature* 361: 490–492.

Stott, D., Kisbert, A. and Herrmann, B. G. 1993. Rescue of the tail defect of *Brachyury* mice. *Genes Dev.* 7: 197–203.

Streit, A., Stern, C. D., Théry, C., Ireland, G. W., Aparacio, S., Sharpe, M. J. and Gherardi, E. 1995. A role for HGF/SF in neural induction and its expression in Hensen's node during gastrulation. *Development* 121: 813–824.

Subramanian, V., Meyer, B. I. and Gruss, P. 1995. Disruption of the murine homeobox gene *Cdx1* affects axial skeletal identities by altering the mesodermal expression domains of *Hox* genes. *Cell* 83: 641–653.

Thor, S. 1995. The genetics of brain development: Conserved programs in flies and mice. *Neuron* 15: 975–977.

Tsuda, T., Philp N., Zile, M. H. and Linask, K. K. 1996. Left-right asymmetric localization of flectin in the extracellular matrix during heart looping. *Dev. Biol.* 173: 39–50.

Tung, T.-C., Vhang, C.-Y. and Tung, Y.-F.-Y. 1945. Experiments on the developmental potencies of blastoderms and fragments of telestean eggs separated latitudinally. *Proc. Zool. Soc. Lond.* 115: 175–188.

Wang, B. B., Müller-Immergluck, M. M., Aystin, J., Robinson, N. T., Chisholm, A. and Kenyon, C. 1993. A homeotic gene cluster patterns the anteroposterior body axis of *C. elegans. Cell* 74: 29–42.

Weinstein, D. C., Ruiz i Altaba, A., Chen, W. S., Hoodless, P., Prezioso, V. R., Jessell, T. M. and Darnell, J. E. Jr. 1994. The winged-helix transcription factor HNF-3β is required for notochord development in the mouse embryo. *Cell* 78: 575–588.

Wilkinson, D. G., Bhatt, S., Cook, M., Boncinelli, E. and Kruflauf, R. 1989. Segmental expression of *Hox-2* homeobox-containing genes in the developing mouse hindbrain. *Nature* 341: 405–409.

Wilson, E. T., Cretekos, C. J. and Helde, K. A. 1995. Cell mixing during epiboly in the zebrafish embryo. *Dev. Genet.* 17: 6–15.

Wolpert, L. and Brown, N. A. 1995. hedgehog keeps to the left. *Nature* 377: 103–104.

Yanagisawa, K. O. 1990. Does the *T* gene determine the anterior-posterior axis of the mouse embryo? *Japan. J. Genet.* 65: 287–297.

Yokoyama, T., Copeland, N .G., Jenkins, N. A., Montgomery, C. A., Elder, F. F. B. and Overbeek, P. A. 1993. Reversal of left-right symmetry: A *situs inversus* mutation. *Science* 260: 679–682.

Cellular Interactions During Organ Formation

V

Proximate tissue interactions: Secondary induction

17

In dealing with such a complex system as the developing embryo, it is futile to inquire whether a certain organ rudiment is "determined" and whether some feature of its surroundings, to the exclusion of others, "determines" it. A score of different factors may be involved and their effects most intricately interwoven. In order to resolve this tangle we have to inquire into the manner in which the system under consideration reacts with other parts of the embryo at successive stages of development and under as great a variety of experimental conditions as it is possible to impose.
R. G. HARRISON (1933)

The aspiration to truth is more precious than its assured possession.
G. E. LESSING (1778)

ORGANS ARE COMPLEX STRUCTURES composed of numerous types of tissues. In the vertebrate eye, for example, light is transmitted through the transparent corneal tissue and focused by the lens tissue (the diameter of which is controlled by muscle tissue), eventually impinging on the tissue of the neural retina. The precise arrangement of tissues in this organ cannot be disturbed without damaging its function. Such coordination in the construction of organs is accomplished by one group of cells changing the behavior of an adjacent set of cells, thereby causing them to change their shape, mitotic rate, or differentiation. This action at close range, sometimes called **proximate interaction** or **secondary induction,** enables one group of cells to respond to a second group of cells and, in changing, often to become able to alter a third set of cells.

Instructive and permissive interactions

Howard Holtzer (1968) distinguished two major modes of proximate tissue interaction. In **instructive interaction** a signal from the inducing cell is necessary for initiating new gene expression in the responding cell. Without the inducing cell, the responding cell would not be capable of differentiating in that particular way. For example, in Chapter 15, we discussed the ability of the notochord to induce the formation of floor plate cells in the neural tube. All neural tube cells are capable of responding to the notochord signal, but only those closest to the notochord are induced. The other cells become non–floor plate cells. Moreover, if one removes the notochord from the embryo, cells that would normally become floor plate cells will not differentiate into that type of cell, and if one adds an additional notochord laterally to the neural plate, this new notochord will induce a secondary set of floor plate cells. The responding neural tube cells would somehow be told to express a set of genes different from the set they would express had they not been in contact with the notochord. The notochord is said to be an inducing tissue acting *instructively.* Wessells (1977) has proposed four general principles characteristic of most instructive interactions:

1. In the presence of tissue A, responding tissue B develops in a certain way.

2. In the absence of tissue A, responding tissue B does not develop in that way.
3. In the absence of tissue A, but in the presence of tissue C, tissue B does not develop in that way.
4. In the presence of tissue A, a tissue D, which would normally develop differently, is changed to develop like B.

The second type of proximate tissue interaction is **permissive interaction.** Here, the responding tissue contains all the potentials needed to be expressed, and it only needs an environment that allows the expression of these traits. For instance, many developing tissues need a solid substrate containing fibronectin or laminin in order to develop. The fibronectin or laminin does not alter the type of cell that is to be produced, but only enables its expression.*

Competence and receptors

It should be noted that in the above principles, the responding tissue must be competent to respond. **Competence** is the ability to respond to an inductive signal (Waddington, 1940). It is not a passive state, but an actively acquired condition. When we detailed the induction of the neural tube, we observed that the gastrula ectoderm is capable of being induced by the dorsal blastopore lip or its mesodermal derivatives. Thus, the gastrula ectoderm is said to be *competent* to respond to the inductive stimuli. This competence for neural induction is acquired during late cleavage and lost during the late-gastrula stages. As this competence to respond to dorsal lip induction diminishes, some regions of ectoderm gain the competence to respond to lens inducers. Still later, the competence to lens inducers is lost, but the ectoderm can respond to ear placode inducers (Servetnick and Grainger, 1991). Therefore, competence is itself a differentiated phenotype that distinguishes cells spatially and temporally.

It is generally thought that competence can be attained in several ways. First, a cell can become competent by synthesizing a receptor for the inducer molecule. As we will see later in this chapter, a B cell is not competent to respond to induction by T cells until it has bound antigens. When the antigens are bound, they create a set of receptors that enable them to respond to the inducing molecules secreted by the T cells. This mechanism of competence is also seen in the induction of sympathetic neuron differentiation (Birren and Anderson, 1990; Cattanco and McKay, 1990). Since the early 1960s, it has been known that the differentiation of sympathetic neurons depends on nerve growth factor (NGF); but when the progenitor cells for these neurons were actually isolated, they did not respond to NGF. Moreover, they lacked the receptors capable of binding NGF. Rather, to differentiate, these cells first had to be exposed to fibroblast growth factor (FGF). The exposure to FGF resulted in these cells expressing the NGF receptor on their cell membranes. These FGF-treated cells could respond to NGF (Figure 17.1). The original progenitor cell was not competent to be induced by NGF because it lacked the NGF receptor. When it was induced by FGF, it became competent to respond to NGF.

---

*It is easy to distinguish permissive and instructive relationships by an analogy with a more familiar situation. This textbook is made possible by permissive and instructive interactions. The reviewers can convince me to change the material in the chapter. This is an *instructive interaction*, as the information is changed from what it would have been. However, the information in the book could not be expressed without *permissive interactions* by the publisher and printer.

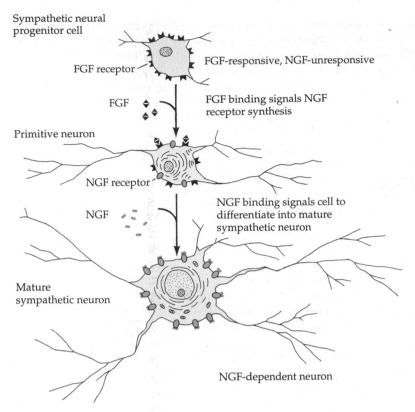

Sympathetic neural
progenitor cell

FGF receptor

FGF-responsive, NGF-unresponsive

FGF

FGF binding signals NGF
receptor synthesis

Primitive neuron

NGF receptor

NGF

NGF binding signals cell to
differentiate into mature
sympathetic neuron

Mature
sympathetic neuron

NGF-dependent neuron

Figure 17.1
Induction and competence of a sympathetic neuron precursor lineage. The original
stem cell is a mitotically active cell that has no NGF receptors but can respond to
FGF. This gives rise to a primitive neural cell that has processes but still divides.
This primitive neuron has receptors for NGF. The NGF-responsive cell can differentiate into the mature nondividing sympathetic neuron (characterized by its large
soma, prominent nucleolus, extensive processes, and dependence on NGF for survival). (After Birren and Anderson, 1990.)

Second, a cell might achieve competence by synthesizing a molecule
that allows the receptor to function. Receptors alone may bind the inducer,
but that does not mean that the receptors are functional. Often, a receptor
acts by sending a signal to the nucleus. As we have seen in Chapter 3, once a
receptor has bound a ligand, it activates enzymes that manufacture the signal for division or differentiation. If one of these enzymes is not present, the
signal is not transmitted. So a cell might achieve competence by synthesizing a missing link in the signaling pathway.

Third, competence may be acquired by the repression of an inhibitor. If
the inhibitor is present, a cell might bind the inducer, send the signal to the
nucleus, and still not be able to be induced. For example, inducers often
cause cell shape changes (as in the induction of the neural tube). If the cell
were inhibited from changing its shape, it would not be able to respond.

Paracrine factors

Proximate interactions are usually mediated by proteins that can diffuse
over small distances to induce changes in their neighboring cells. These proteins are often called **paracrine factors** or **growth and differentiation factors**

(**GDFs**).* Whereas endocrine factors (hormones) go through the blood to exert their effects, paracrine factors (such as the FGF and NGF mentioned previously) are secreted into the immediate spaces around the cell producing them. During the past decade, developmental biologists have discovered that the formation of numerous organs is actually effected by a relatively small population of proteins. The embryo inherits a rather compact "tool kit" and uses many of the same proteins to construct the heart, the kidneys, the teeth, the eye, and other organs. Moreover, the same proteins are utilized throughout the animal kingdom, and the factors active in creating the *Drosophila* eye or heart are very similar to those used in generating mammalian organs. These proteins can be grouped into four major families on the basis of their structures. These families are the fibroblast growth factor (FGF) family, the Hedgehog family, the Wingless (Wnt) family, and the TGF-β superfamily.

The Fibroblast Growth Factors

The fibroblast growth factor (FGF) family currently has nine structurally related members. FGF1 is also known as acidic FGF, FGF2 is sometimes called basic FGF, and FGF7 sometimes goes by the name of keratinocyte growth factor. Although there are nine distinct known FGF genes, they can generate a variety of protein isoforms by varying their RNA splicing or initiation codon in different tissues (Lappi, 1995). FGFs activate a set of receptor tyrosine kinases called the fibroblast growth factor receptors. The reactions initiated by the activated fibroblast growth factors have been discussed in Chapter 3. FGFs are associated with several developmental functions, including angiogenesis, mesoderm formation, and axon extension. While FGFs can often substitute for each other, their expression patterns give them separate functions. FGF2 is especially important in angiogenesis, and FGF8 is important for the development of the midbrain (Crossley et al., 1996). In the mouse, targeted disruptions of certain FGF genes produce specific abnormalities. Absence of *Fgf3* leads to disorganized somite formation, abnormal tail vertebrae, and inner ear defects, while the absence of *Fgf4* results in early embryonic death due to the failure of the inner cell mass to grow. The only problem that *Fgf5*-deficient mice seem to have is abnormally long hair (Figure 17.2; Hebert et al., 1994; Wilkie et al., 1995).

FGFs also are active in the growth plates of long bones and in the sutures of the skull bones (Muenke and Schell, 1995). Mutations causing the premature activation of FGF receptors are the leading cause of dwarfism (early maturation of the long bone growth plates) and craniosynostoses (premature fusion of the skull bones) (Figure 9.19). [cell7.html]

---

*Physiologists have described three major modes by which soluble molecules effect changes in cells. *Paracrine factors* are soluble molecules that effect changes in cells adjacent to or nearby the secreting cell. In embryology, such factors have also been called morphogens. *Endocrine factors* (hormones) are soluble molecules that travel via the blood to effect changes in cells at a distance from the secreting cell. *Autocrine factors* are molecules that effect changes in the cell that secreted them. For autocrine effects to happen, the cell synthesizes a molecule for which it has its own receptor. Although autocrine stimulation is not common, it is seen in placental cytotrophoblast cells, which synthesize and secrete the platelet-derived growth factor whose receptor is on the cytotrophoblast cell membrane (Goustin et al., 1985). The result is the explosive proliferation of that tissue. There is considerable debate as to how far paracrine factors can operate. Activin, for instance, can diffuse over many cell diameters and can induce different sets of genes at different concentrations (Gurdon et al., 1994, 1995). The Vg1, BMP4 and Nodal proteins, however, probably work only on their adjacent neighbors (Jones et al., 1996; Reilly and Melton, 1996). These factors may induce the expression of other short-range factors from these neighbors, and a cascade of paracrine inductions can be initiated.

(A)

(B)

Figure 17.2
Roles of fibroblast growth factors and their receptors. Hair growth in a mouse deficient for FGF5 (the *angora* mutation) (A) is much longer than in control littermates (B). (Photographs courtesy of C. Peterson.)

The Hedgehog Family

In *Drosophila*, Hedgehog protein plays several critical roles in patterning the developing fly. In the early embryo, it acts in a concentration-dependent manner to specify each embryonic parasegment, and as we will see in later chapters, hedgehog also works later in development to specify the axes of the leg and wing imaginal discs (Basler and Struhl, 1994; Heemskerk and DiNardo, 1994). Vertebrates have at least three homologues of the *Drosphila hedgehog* gene: *sonic hedgehog* (*shh*), *desert hedgehog* (*dhh*), and *indian hedgehog* (*ihh*). *Desert hedgehog* is expressed in Schwann and Sertoli cells, and mice homozygous for a null allele of *dhh* have defective spermatogenesis. *Indian hedgehog* is expressed in the gut and in cartilage (Bitgood and McMahon, 1995; Bitgood et al., 1996).

Sonic hedgehog is the most widely used of the three vertebrate homologues. Made by the notochord, it is the protein that is responsible for inducing the floor plate cells and the motor neurons in the neural tube (Placzek et al., 1990; Yamada et al., 1993; see Chapter 8). The Hedgehog protein secreted by the notochord (actually, the N-terminal two-thirds of this protein) is also responsible for inducing the sclerotome in the somites (Fan and Tessier-Lavigne, 1994; Johnson et al., 1994). Sonic hedgehog has been shown to mediate the formation of the left-right axis in chicks, to initiate the anterior-posterior axis in limbs, and to induce the polarized axis of the gut (Riddle et al., 1993; Levin et al., 1995; Roberts et al., 1995). Often, Sonic hedgehog works with other paracrine factors, such as Wnt and FGF. As we will see in the next chapter, *shh* in the limb bud induces the expression of FGF4 in the posterior mesoderm, and the combination of FGF4 and Wnt7a is needed to maintain *shh* expression. In the developing tooth, Sonic hedgehog, FGF4, and other paracrine factors are concentrated in the region where cell interactions are taking place (Figure 17.3; Vaahtokari et al., 1996a).

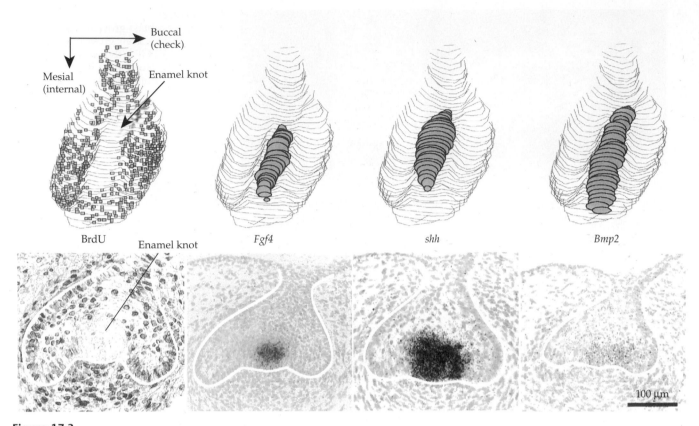

Figure 17.3
Concentration of paracrine growth and differentiation factors in the region where morphogenesis and differentiation are occurring in the 14-day embryonic mouse lower molar. (The boundary of the tooth epithelium is shown in white.) The paracrine factors are being secreted by the nondividing epithelial cells, the enamel knot. (The panel on the left shows that the cells of the enamel knot are not replicating DNA.) Above each of the in situ hybridizations is a serial reconstruction of the area of expression. See page 682 for details. (From Jernvall, 1995; photographs courtesy of A. Vaahtokari, J. Jernvall, and I. Thesleff.)

The Wnt Family

The Wnts comprise a family of cysteine-rich glycoproteins, and there are at least 15 members of this family in vertebrates. Their name comes from fusing the name of the *Drosophila* segment polarity gene *wingless* with the name of one of its vertebrate homologues, *integrated*. As we have seen in Chapter 7, Wnt1 appears to be active in inducing the myotome in the somites and in establishing the boundaries of the midbrain (McMahon and Bradley, 1990; Ku and Melton, 1993; Stern et al., 1995). As we will see in subsequent chapters, the *Wnt* genes also are important in establishing the polarity of vertebrate limbs, just as *wingless* establishes polarity during insect limb development. Interestingly, in both cases, there are interactions with hedgehog family members. During mouse gastrulation, *Wnt3a*, *Wnt5a*, and *Wnt5b* are all expressed in distinct yet overlapping regions within the primitive streak. Wnt3a is the only Wnt protein seen in the regions of the streak that will generate the dorsal (somite) mesoderm, and mice homozygous for a null allele of the *Wnt3a* gene have no somites caudal to the forelimbs (Figure 17.4; Takada et al., 1994).

The Wnt signaling pathway is intimately connected to the hedgehog pathway. As shown in Figure 3.38, hedgehog stimulates the expression of *wg* and Wingless protein stimulates the expression of *hedgehog*. In *Drosophila*, one of the things that Hedgehog does to activate *wingless* gene expression is to counter the repression of the Patched protein. Once the *patched* gene's repression is eliminated, *wingless* can be expressed. Ectopic expression of the *patched* gene inhibits cell growth. A similar pathway is thought to exist in humans, and each of the molecules in the *Drosophila* pathway has a mammalian homologue. In humans, sporadic loss-of-function mutations of the *patched* gene in somatic tissues cause **basal cell carcinomas,** the most com-

(A)

(B)

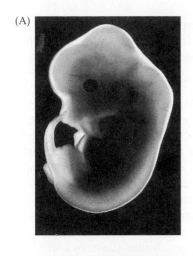

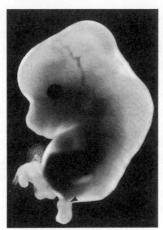

(C)

(D)

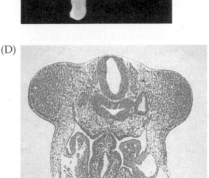

Figure 17.4
Absence of caudal somites in mouse embryos homozygous for a null allele of *Wnt3a*. (A) Wild-type 12.5-day mouse embryo. (B) Same stage embryo of a *Wnt3a* mutant mouse, showing no tail-bud and a truncated axis. (C) Cross section through the forelimb area of a 9.5-day wild-type mouse embryo. (D) Cross section through the same stage of a *Wnt3a* mutant embryo. No somites are seen (The clump of cells near the spinal cord are most likely from the neural crest.) (From Takada et al., 1994; photographs courtesy of A. P. McMahon.)

mon type of human cancer. Hereditable mutations of the *patched* gene give rise to **basal cell nevus syndrome,** an autosomal dominant condition characterized by developmental anomalies (rib and craniofacial alterations, fused fingers) and malignant tumors (medulloblastomas and basal cell carcinomas) (Hahn et al., 1996; Johnson et al., 1996).*

The TGF-β Superfamily

There are over 30 structurally related members of the TGF-β superfamily, and they regulate some of the most important interactions in development (Figure 17.5). The peptides encoded by TGF-β superfamily genes become processed such that the carboxyl terminal region contains the mature peptide. These peptides are dimerized into homodimers (with themselves) or heterodimers (with other TGF-β peptides) and secreted from the cell. The TGF-β superfamily includes the TGF-β family, the activin family, the bone morphogenesis proteins (BMPs), the Vg1 family, and other proteins, including Dorsalin (which is active in patterning the neural tube; see Chapter 7), glial-derived neurotrophic factor (necessary for kidney and enteric neuron differentiation), and Müllerian inhibitory factor (which is involved in mammalian sex determination; see Chapter 20). In *Drosophila*, the decapentaplegic protein is homologous to the vertebrate BMP4.

The receptors that bind the TGF-β superfamily members transmit the signal to the nucleus by activating specific **smad** proteins. The smad proteins reside in the cytoplasm, but when the receptors bind the TGF-β superfamily

---

* Basal cell carcinomas, tumors of the basal cell layer of the epidermis, afflict about 750,000 people every year in the United States, with most of these cancers arising from exposure to sunlight in people of northern European ancestry. On the other hand, basal cell nevus syndrome (sometimes called Gorlin's syndrome) is extremely rare.

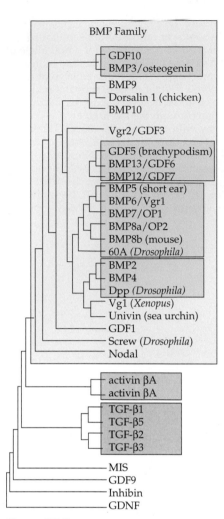

Figure 17.5
Relationships between members of the TGF-β superfamily. (After Hogan, 1995.)*

*Unfortunately, different laboratories capitalize and/or hyphenate the names of these and other factors differently. Our particular orthography is not designed to give priority to any laboratories or conventions. However, one is reminded of Cohen's (1982) dictum that "Academicians are more likely to share each other's toothbrush than each other's nomenclature."

members, they activate (probably by phosphorylation) one of these 50-kDa polypeptides. This converts the smad protein into a transcription factor that can enter the nucleus and activate specific genes (Graff et al., 1996; Hoodless et al., 1996; Liu et al., 1996).

THE TGF-β FAMILY. This family includes TGF-β1, 2, 3, and 5. TGF-β1 appears be important in the formation of branched organs. Exogenous TGF-β1 has been found to inhibit ductal growth in adult mouse mammary glands (Daniel, 1989; Silberstein et al., 1992), to cause malformations of embryonic murine salivary glands (Hardman et al., 1994), and to prevent the branching of embryonic kidneys (Ritvos et al., 1995). Thus, TGF-β1 may be critical in the normal branching process, perhaps mediating these and other processes by enhancing the production of extracellular matrix components such as fibronectin, collagens I and IV (Ignotz and Massagué, 1986; Penttinen et al., 1988), osteonectin (Wrana et al., 1991), and proteoglycans (Bassols and Massagué, 1988; Morales and Roberts, 1988), while inhibiting cell matrix proteolysis (Edwards et al., 1987; Saksela et al., 1987). This could have a net effect of stabilizing tissue structure. The exact effects of TGF-β proteins are often dependent on the cell type encountering them, and the same TGF-β can have opposite effects (such as stopping or accelerating cell division) in different cell types.

The effects of TGF-β are difficult to sort out, because members of the TGF-β family appear to function similarly and can compensate for losses of the others when expressed together. Moreover, targeted deletions of the *Tgfβ1* gene are difficult to interpret, since the mother can supply this factor through the placenta and milk (Letterio et al., 1994).

THE BMP FAMILY. Although originally discovered for their ability to induce bone growth, the BMPs regulate developmental processes as diverse as cell proliferation, apoptosis, cell migration, cell differentiation, and morphogenesis (Hogan, 1996). (As it turned out, BMP1 is not a member of this family.) BMPs are distinguished from the other TGF-β superfamily groups by their having seven, rather than nine, conserved cysteines in the mature peptide. We have already seen BMP proteins such as Nodal (active in axis formation in both *Xenopus* and mice), BMP4 (important in mesoderm specification, neural tube polarity, and somite patterning), and Decapentaplegic (which determines dorsoventral polarity in *Drosophila*). BMP4 is also implicated in inducing apoptosis in the neural crest cells migrating from odd-numbered rhombomeres (Graham et al., 1993) and in the webbing between the toes of chick embryos (see Chapter 23). BMP4 and Decapentaplegic are extremely similar, and human *BMP4* genes can rescue fly embryos lacking *dpp* (Padgett et al., 1993). The absence of some BMPs causes specific skeletal abnormalities (Kingsley et al., 1994; Storm et al., 1994). Mutations of the *BMP5* gene result in a small skeleton and small ears owing to reduced precartilage condensations, while mutations of *gdf5* cause short limbs and a reduced number of toes. Like the other TGF-β members, the BMPs function by dimerizing receptors on the target cells and activating their serine/threonine kinases (Liu et al., 1995).

Juxtacrine Signaling

While most known regulators of induction are diffusible proteins, some proteins might remain bound to the cell surface. Certain Wnt proteins, for instance, lack a secretion signal and may interact with their neighbor's receptors while bound to their cell membrane. Similarly, the Hedgehog proteins can exist in a membrane-bound form prior to their proteolytic processing. In

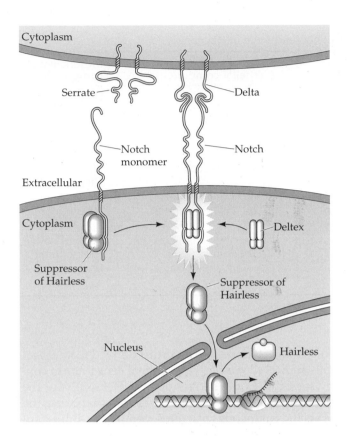

Figure 17.6
Cell-cell signaling between two juxtaposed cells. This speculative model for Delta-Notch signaling is based on genetic evidence from *Drosophila* crosses. The Notch receptor protein can bind to the Serrate or the Delta proteins of the adjacent cell through its extracellular domains. The Delta protein acts as a ligand and dimerizes the Notch protein in the latter's membrane. This dimerization is stabilized by interactions between the proteins. This dimerization may allow the exchange of the Suppressor of Hairless protein with Deltex. The Suppressor of Hairless protein had been bound to the cytoplasmic side of the Notch molecule, but once released, it becomes a transcription factor. This factor can control the fate of the cell, directing it to become skin rather than neural tissue. (After Artavanis-Tsakonas et al., 1995.)

such signaling, the cells would have to be in direct contact for the signal to be effective. Such a signaling pathway has been seen for the Delta signal received by the Notch protein, a signal whose developmental functions in *Drosophila* will be discussed later. Notch extends through the cell membrane, and its external surface contacts Delta or Serrate proteins that extend out from the adjacent cell. When the delta proteins connect to Notch, they stabilize its dimerization and allow conformational changes to occur on the cytoplasmic side of the Notch protein. These changes allow the deltex protein to exchange with a protein called Suppressor of Hairless. When separated from the Notch protein, Suppressor of Hairless protein enters the nucleus to become a transcription factor (Figure 17.6; Artavanis-Tsakonas et al., 1995). Thus, the Notch protein is able to receive the signal from Delta only when the cells are juxtaposed. Hence, this type of signaling is sometimes called **juxtacrine signaling.** [prox1.html]

Epithelial-mesenchymal interactions

Some of the best-studied cases of secondary induction are those involving the interactions of epithelial sheets with adjacent mesenchymal cells. These are called **epithelial-mesenchymal interactions.** The epithelium can come from any germ layer, whereas the mesenchyme is usually derived from loose mesodermal or neural crest tissue. Examples of epithelial-mesenchymal interactions are listed in Table 17.1.

Regional Specificity of Induction

Using the induction of cutaneous structures as our examples, we will look at the properties of epithelial-mesenchymal interactions. The first phenomenon is the regional specificity of induction. Skin is composed of two main tissues:

Table 17.1 Some epithelio-mesenchymal interactions

Organ	Epithelial component	Mesenchymal component
Cutaneous structures (hair, feathers, sweat glands, mammary glands)	Epidermis (ectoderm)	Dermis (mesoderm)
Limb	Epidermis (ectoderm)	Mesenchyme (mesoderm)
Gut organs (liver, pancreas, salivary glands)	Epithelium (endoderm)	Mesenchyme (mesoderm)
Pharyngeal and respiratory associated organs (lungs, thymus, thyroid)	Epithelium (endoderm)	Mesenchyme (mesoderm)
Kidney	Ureteric bud epithelim (mesoderm)	Mesenchyme (mesoderm)
Tooth	Jaw epithelium (ectoderm)	Mesenchyme (neural crest)

an outer epidermis derived from ectoderm and a dermis derived from mesoderm. The chick epidermis signals the underlying dermal cells to form condensations (probably by secreting Sonic hedgehog and TGF-β2), and the condensed dermal mesoderm responds by secreting factors that cause the formation of regionally specific cutaneous structures, comprised almost entirely of ectodermal cells (Nohno et al., 1995, Ting-Berreth and Chuong, 1996; Plate 23). These are the broad wing feathers, the narrow thigh feathers, and the scales and claws of the feet. After separating the embryonic epithelium and mesenchyme from each other, one can recombine them in different ways (Saunders et al., 1957). Some of the recombinations are illustrated in Figure 17.7. As you can see, the mesenchyme is responsible for the specificity of induction in the competent ectoderm. This same type of ectoderm develops according to the region from which the mesoderm was taken. Here, the mesenchyme has an instructive role, calling into play different sets of genes in the responding cells.

This regional specificity of induction is critical during the development of the digestive and respiratory systems. In the morphogenesis of the endo-

Figure 17.7
Regional specificity of induction. When cells of the dermis (mesoderm) are recombined with the epidermis (ectoderm) in the chick, the type of cutaneous structure made by the ectoderm is determined by the original location of the mesoderm. (Adapted from Saunders, 1980.)

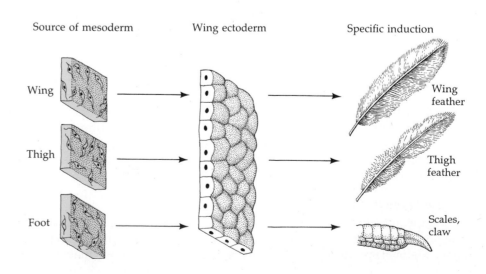

Source of mesoderm   Wing ectoderm   Specific induction

Wing   Wing feather

Thigh   Thigh feather

Foot   Scales, claw

dermal tubes, the endodermal epithelium is able to respond differently to different regionally specific mesenchymes. This enables the digestive tube and respiratory tube to develop different structures in different regions of the tube. Thus, as the digestive tube meets new mesenchymes, it differentiates into esophagus, stomach, small intestine, and colon (Gumpel-Pinot et al., 1978; Fukumachi and Takayama, 1980). This regional specificity of mesenchymal induction is dramatically apparent in the formation of the respiratory system. In the developing mammal, the respiratory epithelial tube responds in two distinct fashions. When in the region of the neck, it grows straight, forming the trachea. After entering the thorax, it branches, forming two bronchi and then the lungs. The respiratory epithelium can be isolated soon after it has split into two bronchi, and the two sides can be treated differently. Figure 17.8 shows the result of such an experiment. The right bronchial epithelium retained its lung mesenchyme, whereas the left bronchus was surrounded by tracheal mesenchyme (Wessells, 1970). The right bronchus proliferated and branched under the influence of the lung mesenchyme, whereas the left side continued to grow in an unbranched manner. Thus, respiratory epithelium is extremely malleable and can differentiate according to its mesenchymal instructions.

The specificity of the mesoderm is thought to be controlled by its interactions with the endodermal tube during earlier stages of development. Roberts and colleagues (1995) have implicated Sonic hedgehog in this specification. Early in development, the expression of *shh* is limited to the posterior endoderm of the hindgut. This appears to be necessary for the induction in the mesoderm of a nested set of *Hox* genes that resemble the posterior *HOM-C* gene set of *Drosophila*. Like the situation in the vertebrae, the anterior borders of the expression pattern delineate the morphological boundaries of the regions that will form the cloaca, large intestine, ceca, mid-ceca (at the midgut/hindgut border), and the posterior portion of the midgut (Plate 22; Figure 17.9). Thus, the endoderm expression of Sonic appears to induce a nested expression of *Hox* genes in the mesoderm. These *Hox* genes

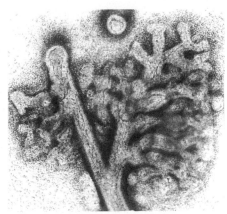

Figure 17.8
Ability of presumptive lung epithelium to differentiate with respect to the source of the inducing mesenchyme. After embryonic mouse lung epithelium has branched into two bronchi, the entire rudiment is excised and cultured. The right bronchus is left untouched, while the tip of the left bronchus is covered with tracheal mesenchyme. The tip of the right bronchus forms the branches characteristic of the lung, whereas no branching occurs in the tip of the left bronchus. (From Wessells, 1970, courtesy of N. Wessells.)

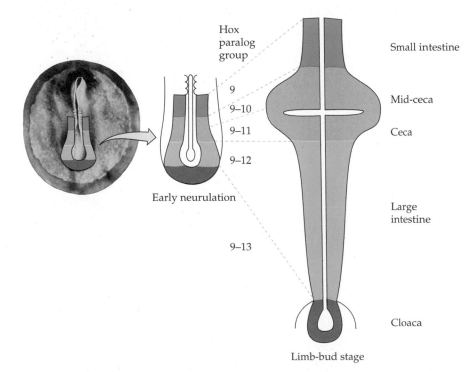

Figure 17.9
Regional specification of the visceral mesoderm through interactions with the posterior gut endoderm. The expression and secretion of Sonic hedgehog in the endoderm generates a nested set of *Hox* gene expression in the adjacent mesoderm. After the mesoderm is specified, it can act on the endodermal tube to induce specific morphological regions. (After Roberts et al., 1995.)

probably specify the mesoderm so that they can interact with the endodermal tube and specify its regions.

Genetic Specificity of Induction

Whereas the mesenchyme may instruct the epithelium as to what sets of genes to activate, the responding epithelium can comply with this information only so far as its genome permits. In a classic experiment, Hans Spemann and Oscar Schotté (1932) transplanted flank ectoderm from an early *frog* gastrula to the region of a *newt* gastrula destined to become parts of the mouth. Similarly, the presumptive flank ectodermal tissue of *newt* gastrula was placed into the presumptive oral regions of *frog* embryos. The structures of the mouth region differ greatly between salamander and frog larvae. The *Triturus* salamander larva has club-shaped balancers beneath its mouth, whereas the frog tadpoles produce mucus-secreting glands, and suckers (Figure 17.10). The frog tadpoles also have a horny jaw without teeth, whereas the salamander has a set of calcareous teeth in its jaw. The larvae resulting from the transplants were chimeras. The salamander larvae had froglike mouths, and the frog tadpoles had salamander teeth and balancers. In other words, the mesodermal cells instructed the ectoderm to make a mouth, but the ectoderm responded by making the only mouth it "knew" how to make, no matter how inappropriate.*

The same genetic specificity is seen in combinations of chicken skin and mouse skin (Coulombre and Coulombre, 1971). When ectoderm normally destined to become cornea is isolated from chicken embryos and combined

---

*Spemann is reported to have put it this way: "The ectoderm says to the inducer, 'you tell me to make a mouth; all right, I'll do so, but I can't make your kind of mouth; I can make my own and I'll do that.'" (Quoted in Harrison, 1933.)

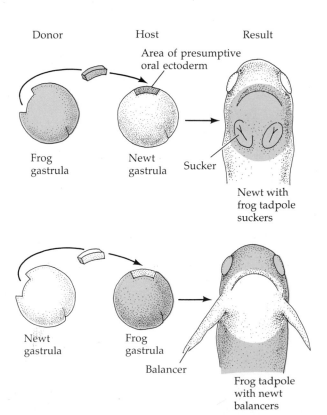

Figure 17.10
Genetic specificity of induction. Reciprocal transplantation between the presumptive oral ectoderm regions of newt and frog gastrulae leads to newt larvae with tadpole suckers and frog tadpoles with newt balancers. (After Hamburgh, 1970.)

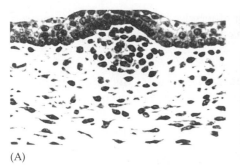

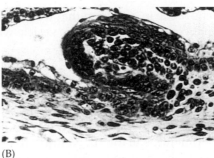

(A)　　　　　　　　　　　　(B)

Figure 17.11
Genetic specificity of cutaneous induction. (A) Section of the corneal region of a 17-day chick embryo. At 5 days of incubation, the lens of this eye had been replaced by the flank dermis of an early mouse embryo. A condensation of the mouse embryo cells is located directly beneath the chick epithelium. (B) Feather forming from the corneal epithelium from such a specimen. Mouse cells are present in the feather rudiment. (From Coulombre and Coulombre, 1971, courtesy of A. J. Coulombre.)

with chick skin mesoderm, the ectoderm produces feather buds typical of chick skin. Moreover, when the same tissue—presumptive cornea ectoderm—is combined with *mouse* skin mesoderm, feather buds also appear (Figure 17.11). The mouse mesoderm has instructed the chick cornea to make a cutaneous structure. In the mouse, this would normally be hair. The competent chick ectoderm, however, does the best it can, developing its cutaneous structures—namely, feathers.

Thus, the instructions sent by the mesenchymal tissue can cross species barriers. Salamanders respond to frog signals, and chick tissue responds to mammalian inducers. The response of the epithelium, however, is species-specific. So, whereas organ-type specificity (feather or claw) is usually controlled by the mesenchyme within a species, species specificity is usually controlled by the responding epithelium. [prox2.html]

Cascades of embryonic induction: Lens induction

The Phenomena of Lens Induction

Proximate cell interactions provide a mechanism whereby coordinated organ development can occur, for a responding tissue can also become an inducing tissue. Recent studies have shown that secondary induction is a very complex process. Indeed, what we have traditionally been calling "secondary" inductions are usually only the last induction in a cascade that began much earlier in embryogenesis, and many tissues acquire their competence through a previous induction. Although these tissues may look unchanged through a microscope, they have been induced such that they can respond to a new inducer. This is probably true of the epidermal inductions mentioned earlier, and it is certainly true for that most intensively studied secondary induction, the formation of the lens.

THE OPTIC CUP MODEL OF LENS INDUCTION. As discussed in Chapter 7, the cells that form the lens are derived from a region of head ectoderm that is contacted by the optic vesicle of the forebrain. This work was pioneered by Hans Spemann, and his review of these studies in 1938 has made lens induction the paradigm of secondary inductive events. The basic experiments were as follows. First, when Spemann (1901) destroyed the optic vesicle primordium of the frog *Rana temporaria*, lenses failed to develop. Thus, Spemann concluded that contact of the optic vesicle with the overlying ectoderm was essential for inducing the formation of the lens. Second, Warren Lewis (1904, 1907) confirmed and extended this conclusion. He removed optic vesicles from late-stage neurulae and transplanted them to the head ectoderm of regions that would not usually form lenses. He found that the head ectoderm of this region would then form lens-like structures, and he concluded that the optic vesicle was sufficient to induce the formation of lens tissue in ectoderm that would not otherwise have formed it. It appeared that contact with the optic vesicle was all that was needed to induce lenses in the overlying ectoderm.

DISAGREEMENTS WITH THE OPTIC CUP MODEL. There were dissenters from this view, however. Mencl (1908) noticed that certain fish have congential defects wherein no eye forms. Nevertheless, these fish have lenses in their head ectoderm. More importantly, when King (1905) tried to repeat Spemann's experiments, she found, contrary to her expectations, that lenses still formed even when the optic vesicle rudiments had been obliterated. These and other investigators began to find that lenses could form without contact with the optic vesicle.*

As more data began accumulating, it seemed that there was a great deal of species diversity. Some species appeared to form lenses without the need for optic vesicles, while the lenses appeared to be completely dependent on the optic vesicle contact in other species. Spemann (1938) reconciled these results by arguing that an organism could evolve a margin of safety by developing two ways of forming a particular tissue. Thus, the lens would normally arise by contact with the optic vesicle, but, failing this, could arise separately if it had to. This concept was called the "double assurance" hypothesis. In 1966, Jacobson integrated more data into this model. He noted that the lens-forming ectoderm sequentially comes into contact with presumptive foregut endoderm, presumptive heart mesoderm, and the optic vesicle. Each of these tissues, he predicted, would act in an additive way to induce the formation of lenses in this tissue. In some species, the threshold for lens induction would be low, and contact with the endoderm would be sufficient. In other species, the threshold would be high, and all three inducers would have to be active. Here, lens formation would seem to depend on the optic vesicle, but in actuality, the optic vesicle would be only the last of three inducers.

The Cellular Basis of Lens Induction

While not ruling out the role of the mesoderm and endoderm, recent studies in *Xenopus* stress the importance of the anterior neural plate as an early inducer of lens ectoderm. These experiments indicate that the presumptive lens ectoderm receives its ability to become lens very early in development (during late-gastrula to midneurula stages) and that the optic vesicle merely localizes the differentiation of this already autonomous tissue. In other words, the head ectoderm will form lenses without the contact of the optic cup, but the optic cup is needed for the full differentiation of the lens and its proper positioning with respect to the rest of the eye. This model (Figure 17.12; Saha et al., 1989; Grainger, 1992) divides the determination of lens ectoderm into four stages: competence, bias, determination, and final differentiation. Competence to respond to the initial inducing signal is seen as an autonomous process within the ectoderm, and the bias to produce lenses is provided by the anterior neural plate. The specification of the lens occurs around the time of neural tube closure, as the optic vesicle approaches the head ectoderm, and final determination is induced by the optic vesicle.

---

*The interpretation of these experiments has been extremely difficult because of species differences in the induction mechanisms, the temperatures at which maximal induction takes place, and the difficulty in getting uncontaminated pieces of tissue for transplantation. See Jacobson and Sater (1988) and Saha et al. (1989, 1991) for reviews of these dissenters and their experiments.

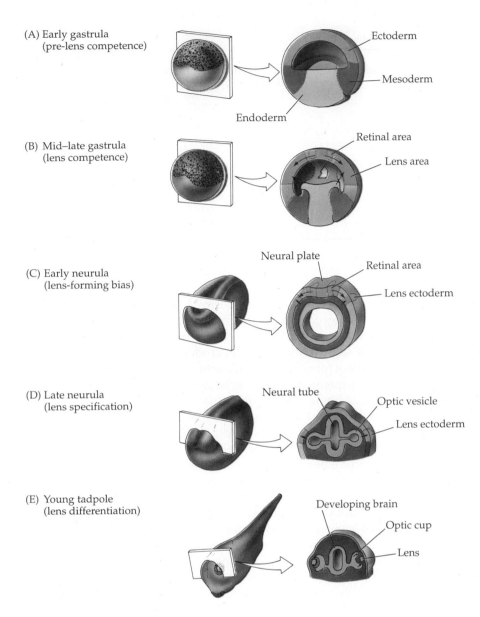

(A) Early gastrula
(pre-lens competence)

Ectoderm

Mesoderm

Endoderm

(B) Mid–late gastrula
(lens competence)

Retinal area

Lens area

(C) Early neurula
(lens-forming bias)

Neural plate

Retinal area

Lens ectoderm

(D) Late neurula
(lens specification)

Neural tube

Optic vesicle

Lens ectoderm

(E) Young tadpole
(lens differentiation)

Developing brain

Optic cup

Lens

Figure 17.12
A current model for lens induction. The inductive signals are indicated by arrows. (A) At early gastrula, the ectoderm has not achieved the competence to become lens (although it does have competence to become neural tissue). (B) During midgastrula, the lens-forming ectoderm becomes competent to respond to the lens-inducing signal from the presumptive neural plate (possibly the presumptive retinal cells). During late gastrula, this signal from neuralized ectoderm (most likely the presumptive eye region) induces the presumptive lens-forming ectoderm. An additional inductive signal may be coming from the prospective mesoderm or foregut endoderm. (C) At early neurula, the signals from the anterior neural region have caused a lens-forming bias in the head ectoderm. This signal may be reinforced by induction from the anterior lateral mesoderm. (D) At the late-neurula stage, the optic vesicle contacts the lens-forming ectoderm, signaling the final determination of this tissue into lens. (E) At the tadpole stage, the presumptive lens ectoderm differentiates into lens tissue. (After Saha et al., 1989; Grainger, 1992.)

ECTODERMAL COMPETENCE AND LENS BIAS. In 1987, Henry and Grainger demonstrated that the determination of lens-forming ability occurs very early in *Xenopus* development. They transplanted ectoderm from *Xenopus* embryos into the lens-forming region of the neurula. Would that ectoderm be able to form a lens when contacted hours later by the optic vesicle? Ectoderm from very young gastrulae was not competent to form lenses. However, when late-gastrula ectoderm was transplanted into neurulae, it was capable of responding to the optic vesicle by forming a lens (Table 17.2). No other tissue responded in this way. Other ectodermal regions from gastrulae also had a limited ability to form lenses, but this ability was lost as development proceeded.

It appeared, then, that the lens-forming ectoderm achieved the competence to form lenses long before the optic vesicle contacted it. When and how was this lens-forming state achieved? Experiments by Nieuwkoop (1952) had suggested that a signal from the neural plate could travel through the ectoderm. Could the neural plate induce the presumptive epidermis lat-

Table 17.2 Increasing responsiveness of prospective lens ectoderm with age

Donor stage	Operation Donor	Operation Host neurula	Number examined	Induced lens (%)	Lens-like body (%)	Ectodermal thickening (%)	Nonlens body (%)	No response (%)	Total positive
Midgastrula			24	0	4	38	8	50	1 (4%)
Late gastrula			21	10	14	42	10	24	5 (24%)
Early neurula			24	75	8	0	4	13	20 (83%)
Late neurula			20	95	5	0	0	0	20 (100%)

Source: After Henry and Grainger, 1987.

eral to it to become lens-forming ectoderm? Henry and Grainger (1990) tested this hypothesis by combining the prospective anterior neural plate region of late-gastrula embryos with ectoderm from the region that would eventually become lens. While isolated ectoderm from the potential lens-forming region did not form lens proteins when cultured alone, the same region did form lens proteins when cultured next to the prospective anterior neural plate tissue. Although lens differentiation was often rudimentary, it was very specific. Lens proteins were not made when the gastrula ectoderm was combined with other tissues, including foregut endoderm or cardiac mesoderm (Figure 17.13A). These experiments show that the anterior portion of the prospective neural plate (which contains the future retinal regions) provides a signal that biases this tissue to become lens.

But are all tissues able to respond to the signal coming from the anterior neural plate? Servetnick and Grainger (1991) showed that only mid- to late-gastrula ectoderm is competent to respond to these signals. They removed animal cap ectoderm from various gastrula stages and then transplanted them into the presumptive lens region of neural-plate-stage embryos (Figure 17.13B). Ectoderm from early gastrulae showed little or no competence to form lenses (as assayed by the production of crystallin proteins), but ectoderm from slightly later stages was able to form lenses. By the end of gastrulation, this ability to respond to the neural plate signal had been lost. This competence was seen to be inherent within the ectoderm itself and was not induced by other surrounding tissues. The animal cap ectoderm from various embryonic stages could be removed, cultured in glass for a certain period of time, and then placed back into neural-plate-stage embryos. Such ectoderm showed the same pattern of competence, even though it had spent part of its development inside a petri dish. It appears, then, that the ectoderm acquires the competence to respond to inducing signals from the anterior neural plate at the early midgastrula stages, and during late gastrula, the anterior neural plate induces a lens-forming bias in this tissue. This lens-forming bias could be demonstrated by transplanting the tissue into other regions of the head and having it become lens (while ectoderm from the earlier stages could not).

DETERMINATION AND DIFFERENTIATION OF THE LENS. Lens determination can be shown by isolating ectoderm and culturing it apart from the embryo. At the time of neural tube closure, the ectoderm from the lateral regions of the forebrain will give rise to small lenses even under these conditions. The optic

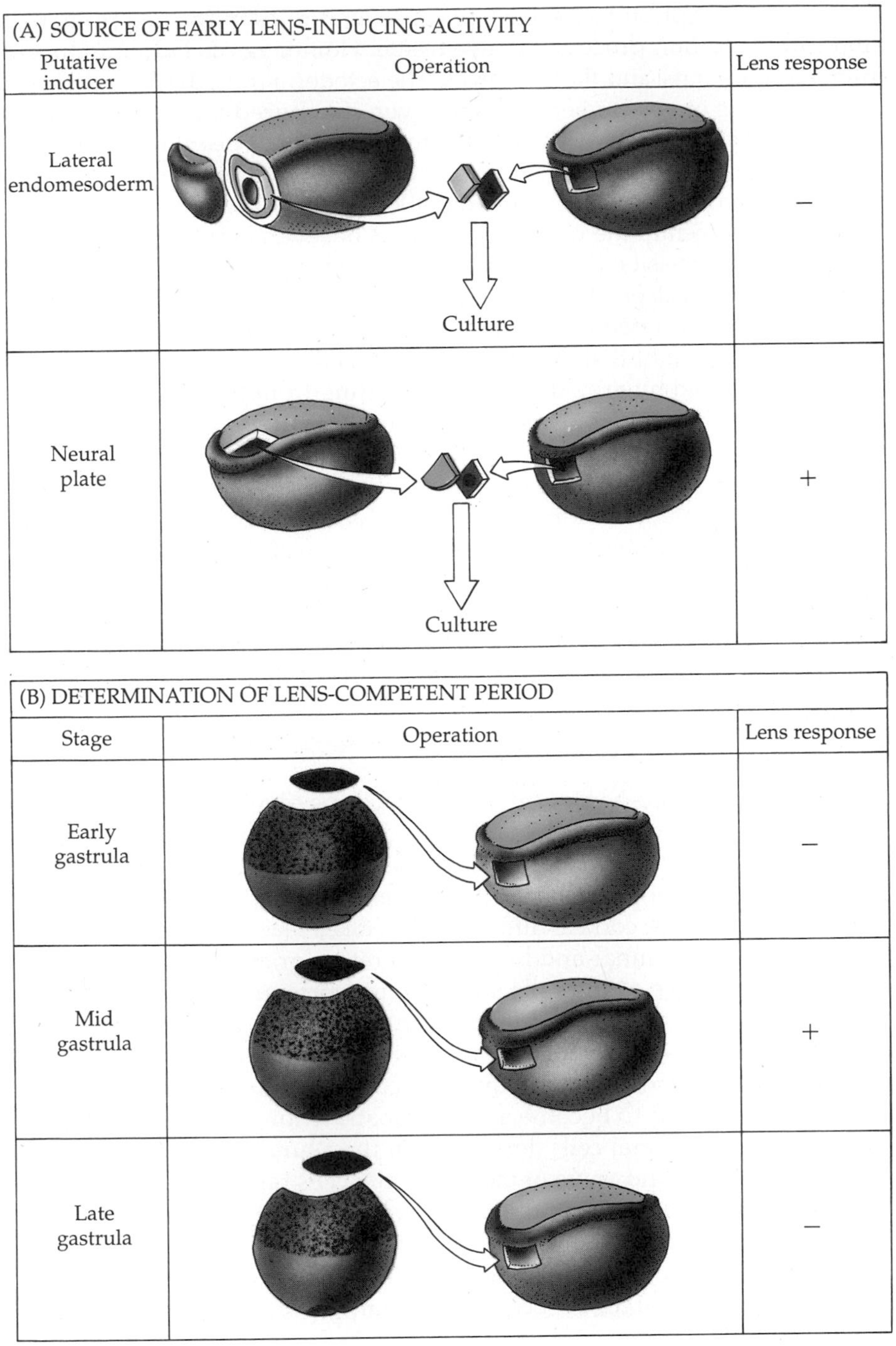

(A) SOURCE OF EARLY LENS-INDUCING ACTIVITY

Putative inducer	Operation	Lens response
Lateral endomesoderm	Culture	–
Neural plate	Culture	+

(B) DETERMINATION OF LENS-COMPETENT PERIOD

Stage	Operation	Lens response
Early gastrula		–
Mid gastrula		+
Late gastrula		–

Figure 17.13
Early determination of lens-forming ability in *Xenopus* ectoderm. (A) The source of lens-determining signal was found to be the anterior neural plate. Presumptive lens ectoderm was cultured with either lateral endoderm/ mesoderm or with the anterior neural plate (the two major tissues adjacent to it). The ectoderm formed lens proteins only when cultured with the neural plate. (B) The time at which the anterior neural plate cells could induce competence in the ectoderm was determined by transplanting presumptive ectoderm from different-stage donor gastrulae into the lens-forming region of the neurula. Only the ectoderm from mid-gastrula embryos was competent to respond to the signals. (After Grainger, 1992.)

cup has not yet contacted this tissue, showing that it is not critical in initiating lens induction in *Xenopus*. However, it does play a role in enabling the complete lens phenotype to become expressed. The lenses that form in the absence of the optic vesicle are usually very rudimentary. It is not known whether the influence of the optic vesicle is a directly positive one, promoting the differentiation of the lens placode into a fully differentiated lens, or if the influence is to remove an inhibitor of lens differentiation. It has been proposed (von Woellwarth, 1961; Henry and Grainger, 1987) that the cranial neural crest cells prevent lens differentiation and that contact with the optic vesicle shields the lens placode from these inhibiting signals.

The transcription factor **Pax6** plays several roles in the determination and differentiation processes of eye tissue. Homozygous *Pax6* mutants of humans, mice, rats, and flies lack eyes. The ectoderm from the *Pax6*-deficient rat embryo is unable to become lens, even if cultured with optic vesicles from wild-type embryos. The head ectoderm has not been determined by the earlier signals from the neural plate or mesoderm (Fujiwara et al., 1994). Pax6 is also critical for the expression of lens crystallins. Not only are Pax6-binding sites seen in the regulatory regions of several crystallin genes, but lens-specific expression of these proteins depends on Pax6 expression (Cvekl et al., 1995; Richardson et al., 1995).

The lens is situated between the anterior and vitrous chambers of the eye, and it is thought that the differentiation of the lens (discussed in Chapter 7) is mediated by growth factors emanating from these two chambers. The anterior chamber appears to concentrate a mitogenic protein (whose identity remains unknown) that is specific for causing mitosis and inhibiting differentiation in the lens-forming epithelium. This protein is thought to come into the anterior chamber from the blood capillaries. In the vitreous chamber, FGF1 and 2 both stimulate the elongation and differentiation of the lens cells and block the mitogenic activity of the anterior chamber growth factor (Hyatt and Beebe, 1993; Schulz et al., 1993). The result is the elongation of those lens cells on the dorsal surface of the lens placode and the continued proliferation of cells on the ventral side of the lens placode (Figure 17.14).

Cornea Formation

Once the lens placode has invaginated, it becomes covered by two cell layers from the adjacent ectoderm. Now the developing lens can act as an inducer. The ectoderm that is to become the cornea has probably also been determined during an earlier stage of development (Meier, 1977). Now, under the influence of the lens, corneal differentiation takes place. The overlying ectoderm becomes columnar and fills with secretory granules. These granules migrate to the base of the cells and secrete a primary stroma containing about 20 layers of types I and II collagen (see Figure 17.14). Neighboring capillary endothelial cells migrate into this region (on the primary stroma) and secrete hyaluronic acid into this matrix. The hyaluronic acid causes the matrix to swell and to become a good substrate for the migration of two waves of mesenchymal cells derived from the neural crest. Upon entering the matrix, the second wave of mesenchymal cells remains there, secreting type I collagen and hyaluronidase. The hyaluronidase causes the stroma to shrink. Under the influence of thyroxine from the developing thyroid gland, this secondary stroma is dehydrated, and the collagen-rich matrix of epithelial and mesenchymal tissues becomes the transparent cornea (see Hay, 1980; Bard, 1990).

We can see, then, that "simple" inductive interactions are actually well-coordinated dramas in which the actors must come on stage and speak their lines at the correct times and positions. In acquiring new information, they can also impart information for others to use. With this in mind, we can now study some principles about secondary induction obtained from other developing organs.

Formation of parenchymal organs

Epithelial-mesenchymal interactions are also seen in the formation of duct-forming organs, such as the kidney, liver, lung, mammary gland, and pancreas. In the formation of these organs, we see the reciprocal induction of the

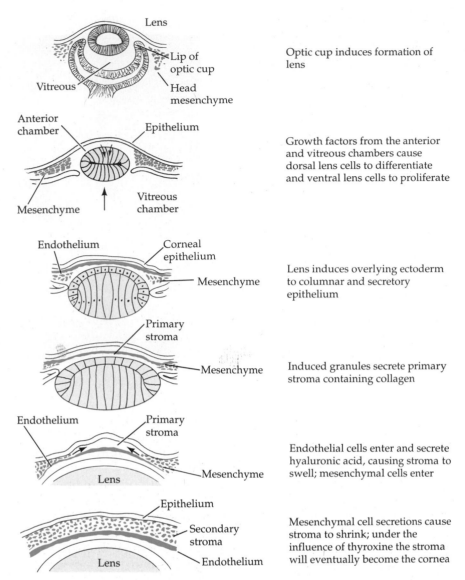

Optic cup induces formation of lens

Growth factors from the anterior and vitreous chambers cause dorsal lens cells to differentiate and ventral lens cells to proliferate

Lens induces overlying ectoderm to columnar and secretory epithelium

Induced granules secrete primary stroma containing collagen

Endothelial cells enter and secrete hyaluronic acid, causing stroma to swell; mesenchymal cells enter

Mesenchymal cell secretions cause stroma to shrink; under the influence of thyroxine the stroma will eventually become the cornea

Figure 17.14
Lens and corneal development. The optic cup induces the final determination of the lens. Mitogenic proteins (black arrow) in the anterior chamber maintain a line of proliferating cells in the ventral surface of the lens, while fibroblast growth factors (colored arrows) stimulate the differentiation of the dorsal lens epithelium. Under the inductive influence of the lens, the corneal epithelium differentiates and secretes a primary stroma consisting of collagen layers. Endothelial cells then secrete hyaluronic acid into this region, enabling mesenchyme cells from the neural crest to enter. Afterward, hyaluronidase (secreted by either the mesenchyme or endothelium) digests the hyaluronic acid, causing the primary stroma to shrink. (After Hay and Revel, 1969; Hyatt and Beebe, 1993.)

mesenchyme upon the epithelium and the epithelium upon the mesenchyme.

Morphogenesis of the Mammalian Kidney

THE PROGRESSION OF RENAL TUBULES.   Like the eye, the mammalian kidney is an exceedingly intricate structure. Its functional unit, the nephron, contains over 10,000 cells and at least 12 different cell types, each cell type located in a particular place in relation to the other cell types along the length of the nephron. The development of the mammalian kidney progresses through three major stages. Early in development (day 22 in humans; day 8 in mice), the **pronephric duct** arises in the intermediate mesoderm just ventral to the anterior somites. The cells in this duct migrate caudally, and the anterior region of the duct induces the adjacent mesenchyme to form the **pronephric kidney tubules** (Figure 17.15A). While the pronephric tubules form functioning kidneys in fish and in amphibian larvae, they are not thought to be active in mammalian amniotes. In mammals, the pronephric tubules and the anterior portion of the pronephric duct degenerate, but the more caudal por-

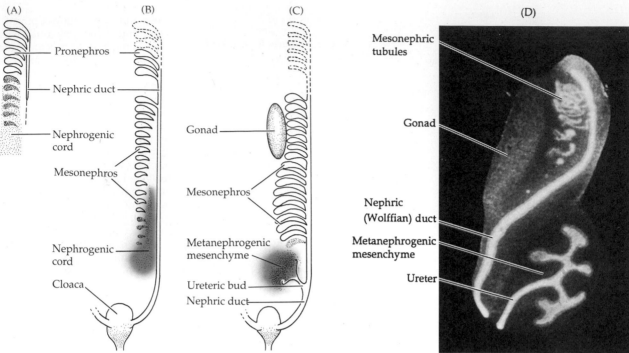

(A)

Pronephros

Nephric duct

Nephrogenic
cord

Mesonephros

Nephrogenic
cord

Cloaca

(B)

(C)

Gonad

Mesonephros

Metanephrogenic
mesenchyme

Ureteric bud
Nephric duct

(D)

Mesonephric
tubules

Gonad

Nephric
(Wolffian) duct

Metanephrogenic
mesenchyme

Ureter

Figure 17.15
General scheme of development in the
vertebrate kidney. (A) The original
tubules, constituting the pronephric kidney, are induced from the nephrogenic
mesenchyme by the pronephric duct as
it migrates caudally. (B) As the pronephros degenerates, the mesonephric
tubules form. (C) The final mammalian
kidney, the metanephros, is induced by
the ureteric bud. (D) Section of a mouse
kidney showing the initiation of the
metanephric kidney (bottom) while the
mesonephros is still apparent. The duct
tissue is stained with a fluorescent antibody to a cytokeratin found in the
pronephric duct and its derivatives.
(A–C after Saxén, 1987; D courtesy of S.
Vainio.)

tions of the pronephric duct persist and become the central component of
the excretory system throughout its development (Toivonen, 1945; Saxén,
1987). This remaining duct is often referred to as the **nephric** or **Wolffian
duct.**

As the pronephric tubules degenerate, the midportion of the nephric
duct initiates a new set of kidney tubules in the adjacent mesenchyme. This
set of tubules constitutes the **mesonephros,** or **mesonephric kidney.** In humans, about 30 mesonephric tubules form, beginning around day 25. However, as more tubules are induced caudally, the anterior mesonephric tubules
begin to regress (although in mice, the anterior tubules remain while the
posterior ones regress; Figure 17.15B). In female mammals, this regression is
complete. However, as we will discuss in Chapter 20, some of these
mesonephric tubules persist in male mammals to become the sperm-carry-
ing tubes (the vas deferens and efferent ducts) of the testes.

The permanent kidney of amniotes, the **metanephros,** is generated by
some of the same components as the earlier, transient kidney types, and it is
thought to originate by a complex interaction between epithelial and mesenchymal components of the intermediate mesoderm. In the first two steps,
the **metanephrogenic mesenchyme** forms in posteriorly located regions of
the intermediate mesoderm, and it induces the formation of a branch from
each of the paired nephric ducts. These epithelial tubes are called the
ureteric buds. These buds eventually separate from the nephric duct to become the ureters that take the urine to the bladder. When the ureteric buds
emerge from the nephric duct, they enter the metanephrogenic mesenchyme. In the third and fourth steps, the ureteric buds induce this mesenchymal tissue to condense around the buds and to differentiate into the
nephrons of the mammalian kidney. The fifth step of kidney initiation occurs
when this nephron-forming tissue induces the further branching of the
ureteric bud (Figure 17.15C,D).

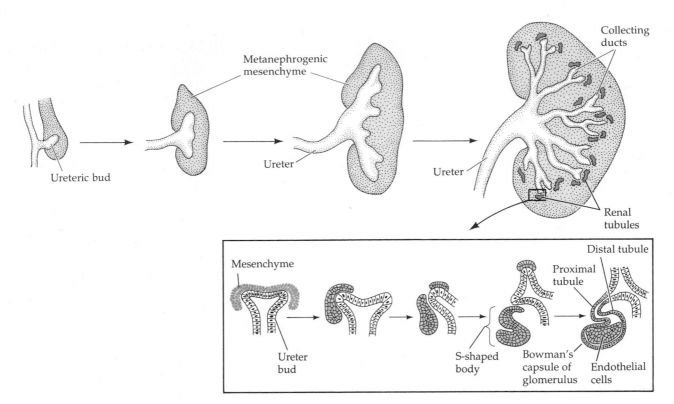

Figure 17.16
Reciprocal induction in the development of the mammalian kidney. As the ureteric bud enters the metanephrogenic mesenchyme, the mesenchyme induces the bud to branch. At the tips of the branches, the epithelium induces the mesenchyme to aggregate and cavitate to form the renal tubules. The formation of the nephron from the mesenchyme cells is shown in the insert. After aggregating at the branches, the mesenchyme cells form an epithelial node that extends into an S-shaped tube, which fuses with the ureteric bud epithelium. (Insert after Romanoff, 1960.)

RECIPROCAL INDUCTION DURING KIDNEY DEVELOPMENT. These two mesodermal tissues, the ureteric bud and the metanephrogenic mesenchyme, interact and reciprocally induce each other (Figure 17.16). The metanephrogenic mesenchyme causes the ureteric bud to elongate and branch. At the tip of these branches, the ureteric bud induces the loose mesenchyme cells to form an epithelial aggregate. Each aggregate of about 20 cells will proliferate and differentiate into the intricate structure of the renal nephron. First, each node elongates into a "comma" shape and then forms a characteristic S-shaped tube. Soon after the S-shaped tube is formed, the cells of this epithelium begin to differentiate into regionally specific cell types, such as the capsule cells, the podocytes, and the distal and proximal tubule cells. At this time, a connection develops between the ureteric bud and the newly formed tube, thereby enabling material to pass from one to the other. The newly formed tubes derived from the mesenchyme form the secretory nephrons of the functioning kidney, and the branched ureteric bud gives rise to the renal collecting ducts and to the ureter, which drains the urine from the kidney.

Clifford Grobstein (1955, 1956) documented this **reciprocal induction** in vitro. He separated the ureteric bud from the mesenchyme and cultured them either individually or together. In the absence of mesenchyme, the ureteric bud does not branch. In the absence of the ureteric bud, the mesenchyme soon dies. When they are placed together, however, the ureteric bud grows and branches, and tubules form throughout the mesenchyme (Figure 17.17). Although certain other tissues (notably neural tube) will enable the metanephrogenic mesenchyme to form kidney tubules, the ureteric bud branches only under instructions from the metanephrogenic mesenchyme. Mesenchymes that induce branching in other epithelia (such as salivary gland mesenchyme) will not cause the ureteric bud to branch (Bishop-Calame, 1966).

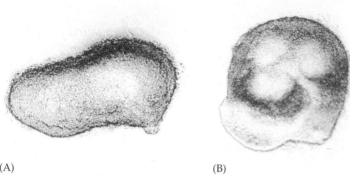

(A)   (B)

Renal tubules   Collecting ducts

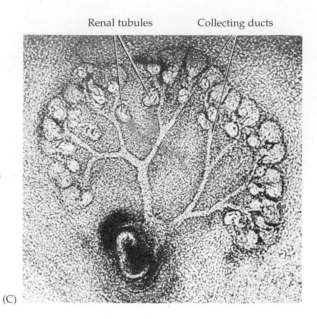

(C)

Figure 17.17
Kidney induction observed in vitro. (A) An 11-day mouse metanephric rudiment includes both ureteric bud and metanephrogenic mesenchyme. (B) After the first day of culture, tubules can be seen at the tips of the branching ureters. (C) The branching collecting ducts formed by the ureteric bud and the kidney tubules formed by the mesenchymal condensations at the tips of these buds can be clearly seen after 8 days of culture. (A and B from Saxén and Sariola, 1987; C from Grobstein, 1955; all photographs courtesy of the authors.)

The first metanephric nephrons, then, are immediately joined to the collecting ducts. These nephrons are carried outward into the metanephrogenic mesenchyme as the ureter continues to grow (Figure 17.18A). The ends of the ureteric buds, however, retain their ability to induce tubule formation in this mesenchyme, and the result is the formation of tubular arcades (Figure 17.18B). As the ureteric branch migrates further through the mesenchyme, new nephrons are formed and become joined to the same collecting duct (Figure 17.18C; Osathanondh and Potter, 1963).

The Mechanisms of Kidney Organogenesis

There appear to be at least six sets of signals operating in the reciprocal induction of the metanephros.

SIGNAL 1: FORMATION OF THE METANEPHROGENIC MESENCHYME.   It is one thing to say that the ureteric bud induces the metanephrogenic mesenchyme to become the epithelium of the nephrons. It is another thing to understand how this process occurs. Like the development of the lens by the optic vesicle, it is thought that induction of the metanephrogenic mesenchyme by the ureteric bud is only the final step that triggers a cascade of events in the competent mesenchyme. Only the metanephrogenic mesenchyme has the ability to respond to the ureteric bud to form kidney tubules, and if induced by other tissues (such as embryonic salivary gland or neural tube), the metanephrogenic mesenchyme will respond by forming kidney tubules and no other structures (Saxén, 1970; Sariola et al., 1982). Thus, the metanephrogenic mesenchyme cannot become any tissue other than kidney tubules. The competence to respond to ureteric bud inducers is thought to be regulated by **WT1,** a transcription factor found in the metanephrogenic mesenchyme, and if this mesenchyme lacks this factor, the uninduced cells die (Kriedberg et al., 1993). In situ hybridization shows that WT1 is normally first expressed in the intermediate mesoderm prior to kidney formation and is then expressed in the developing kidney, gonad, and mesothelium (Pritchard-Jones et al., 1990; van Heyningen et al., 1990; Armstrong et al., 1992). Although this mesenchyme appears homogeneous, the metanephrogenic mesenchyme

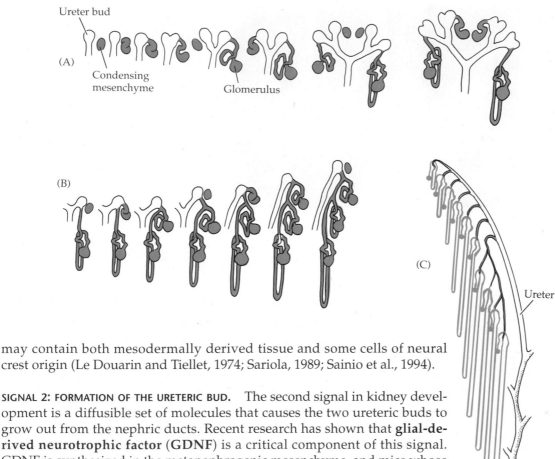

Figure 17.18
Schematic representation of human nephron development. (A) Formation of early nephrons directly attached to the ureteric bud epithelium. (B) Formation of nephron arcades where several nephrons are joined to the same collecting duct. (C) General arrangement of human nephrons at birth. The deeper nephrons constitute an arcade, while the nephrons closer to the surface are directly connected to the collecting duct of the ureter. (After Osathanondh and Potter, 1963.)

may contain both mesodermally derived tissue and some cells of neural crest origin (Le Douarin and Tiellet, 1974; Sariola, 1989; Sainio et al., 1994).

SIGNAL 2: FORMATION OF THE URETERIC BUD. The second signal in kidney development is a diffusible set of molecules that causes the two ureteric buds to grow out from the nephric ducts. Recent research has shown that **glial-derived neurotrophic factor** (**GDNF**) is a critical component of this signal. GDNF is synthesized in the metanephrogenic mesenchyme, and mice whose *gdnf* genes had been knocked out died soon after birth from their lack of kidneys (Moore et al., 1996; Pichel et al., 1996; Sánchez et al., 1996). The **GDNF receptor** (the c-Ret protein) is synthesized in the Wolffian ducts and later becomes concentrated in the growing ureteric buds (Figure 17.19; Schuchardt et al., 1996; Trupp et al., 1996). Mice lacking the GDNF receptor also die of renal agenesis. Another protein synthesized by the metanephrogenic mesenchyme is **hepatocyte growth factor** (**HGF;** scatter factor), and the receptor for HGF is made by the ureteric buds. Antibodies to HGF will block ureteric bud outgrowth in cultured kidney rudiments (Santos et al., 1994; Woolf et al., 1995). The synthesis of GDNF and HGF by the mesenchyme is thought to be regulated by the *WT1* gene.

In another mouse mutation, the *Danforth short-tail* mutant, the ureteric bud is initated but doesn't enter into the metanephrogenic mesenchyme (Gluecksohn-Schoenheimer, 1943). Here, too, the kidney does not form. The failure of the ureteric bud to grow has been correlated with the absence of *Wnt11* expression in the tips of the ureteric bud. *Wnt11* expression is maintained by proteoglycans made by the mesenchyme. It appears that once the ureteric bud has entered into the region of the mesenchyme, the mesenchymal proteoglycans stimulate its continued growth by maintaining the expression and secretion of Wnt11 (Davies et al., 1995; Kispert et al., 1996).

SIGNAL 3: PREVENTION OF MESENCHYMAL APOPTOSIS. The third signal is sent from the ureteric bud to the mesenchyme, and it alters the fate of the mesenchyme cells. If left uninduced by the ureteric bud, the mesenchyme cells undergo apoptosis (Koseki et al., 1992). However, if induced by the ureteric

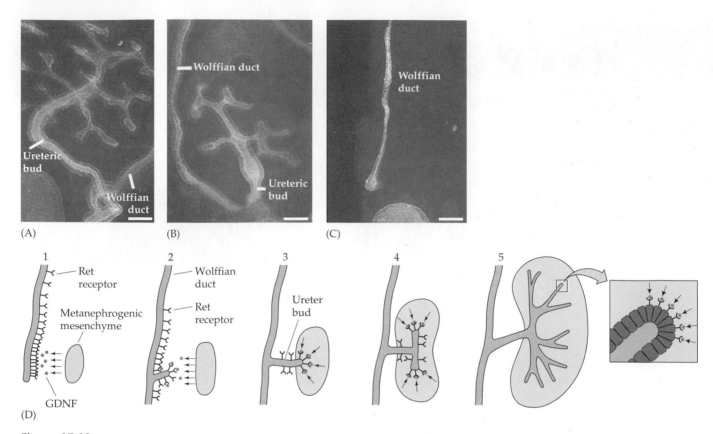

Figure 17.19

Ureteric bud growth is dependent on GDNF and its receptor. (A) The ureteric bud from a 11.5-day wild-type mouse embryonic kidney cultured for 72 hours has a characteristic branching pattern. (B) In embryonic mice heterozygous for the genes encoding GDNF, the size of the ureteric bud and the number and length of its branches are reduced. (C) In mouse embryos missing both copies of the *gdnf* genes, the ureteric bud does not form from the Wolffian duct. (D) The receptors for GDNF are concentrated in the posterior portion of the Wolffian duct. GDNF secreted by the metanephrogenic mesenchyme stimulates the growth of the ureteric bud from this duct. At later stages, the GDNF receptor is found exclusively at the tips of the ureteric buds. Scale bars equal 100 μm. (A–C from Pichel et al., 1996; photographs courtesy of J. G. Pichel and H. Sariola; D after Schuchardt et al, 1995.)

bud, the mesenchyme cells are rescued from the precipice of death and are converted into proliferating stem cells (Bard and Ross, 1991; Bard et al., 1996). The factors secreted from the ureteric bud include **fibroblast growth factor 2** (**FGF2**) and **bone morphogenetic protein 7** (**BMP7**). FGF2 has three modes of action in that it inhibits apoptosis, promotes the condensation of mesenchyme cells, and maintains the synthesis of WT1 (Perantoni et al., 1995). BMP7 has similar effects, and in the absence of BMP7, the mesenchyme of the kidney undergoes apoptosis (Figure 17.20; Dudley et al.,

Figure 17.20

Kidney malformation in *BMP7*-deficient mouse embryo. At embryonic day 19, the mutant kidneys are significantly smaller than those of wild-type embryos. (From Dudley et al., 1995; photograph courtesy of E. J. Robinson.)

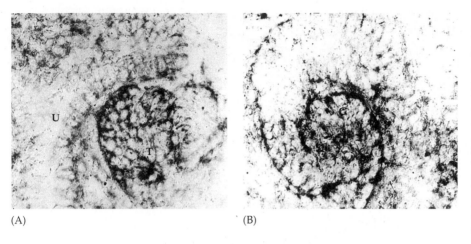

(A)   (B)

Figure 17.21
The extracellular matrix proteoglycan syndecan is not synthesized or secreted by mesenchyme cells until after induction. This molecule is probably involved in structuring the new tubule epithelium, and it distinguishes the cells of the tubule from the remaining mesenchyme. (A) Immunological staining of syndecan shows its presence in the newly induced mesenchyme cells (T) that are becoming epithelial. Some staining (U) is also seen on the ureteric bud epithelium. (B) Intense syndecan staining is seen in the region of the developing tubule that is to become the renal glomerulus (G). (From Vainio et al., 1989, courtesy of L. Saxén.)

1995; Luo et al., 1995). The induced mesenchyme cells also synthesize receptors for epidermal growth factor and neural growth factor and can respond to these proteins by proliferating.

SIGNAL 4: CONVERSION OF MESENCHYME CELLS INTO AN EPITHELIUM. The ureteric bud causes dramatic changes in the extracellular matrix of the metanephrogenic mesenchyme cells. The uninduced mesenchyme secretes an extracellular matrix consisting largely of fibronectin and collagen types I and III. Upon induction, these proteins disappear and are replaced by an epithelial basal lamina made of laminin and type IV collagen. The changes in the extracellular matrix appear to be critical in tubule formation, since the induced mesenchyme secretes a receptor for laminin that enables its participation in epithelium formation (Ekblom et al., 1994). The cytoskeleton also changes from one characteristic of mesenchyme cells to one typical of epithelia (Ekblom et al., 1983; Lehtonen et al., 1985). In this manner, the loose mesenchyme cells are linked together as a polarized epithelium on a basal lamina.

Prior to these changes, the newly induced metanephrogenic mesenchyme synthesizes two adhesive proteins, E-cadherin and Syndecan. **Syndecan** is an adhesive proteoglycan first seen around the mesenchyme cells surrounding the ureteric bud as the bud first enters the region of mesenchyme. As the ureteric bud initiates its first branch, the entire mesenchymal region around the branch stains positively for syndecan (Figure 17.21). Syndecan *mRNA* is present in the uninduced kidney mesenchyme, but it is not translated into protein unless the mesenchyme is induced (Figure 17.22; Vainio et al., 1989a, 1992). Not only might syndecan regulate the *condensation* of the mesenchyme into an epithelium, but it might also promote the *proliferation* of these cells. By labeling proliferating cells with bromodeoxyuridine (which is incorporated into DNA only if the cells are dividing) and labeling the syndecan-expressing cells with fluorescent antibodies to syndecan, Vainio and colleagues (1992) demonstrated a close correlation between the dividing cells and those expressing syndecan.

In addition, the Pax2 transcription factor is synthesized in the induced mesenchyme.* When antisense RNA to Pax2 prevents the translation of the *Pax2* mRNA that is transcribed as a response to induction, the mesenchyme

---

*Pax2 plays several roles during kidney development. Its most critical function occurs even earlier than mesenchyme conversion, as it appears that Pax2 may be important in specifying the intermediate mesoderm. In mouse mutants of Pax2, no urogenital system forms (Torres et al., 1995).

Figure 17.22
Syndecan expression in induced and uninduced kidney mesenchymes. (A) In situ hybridization localizing *syndecan* mRNA in the mesenchymal aggregates of a 15-day embryonic mouse kidney. Visualization of the autoradiograph is by dark-field illumination. (B) Isolated kidney mesenchyme (M) induced by spinal cord (SPC) shows intense syndecan expression when stained with fluorescent antibodies to syndecan. Uninduced mesenchyme does not. (C) The amount of syndecan (labeled with radioactive sulfur) isolated from an induced kidney mesenchyme is 10-fold greater than that isolated from a similar amount of uninduced mesenchyme. (After Vainio et al., 1992, courtesy of S. Vainio.)

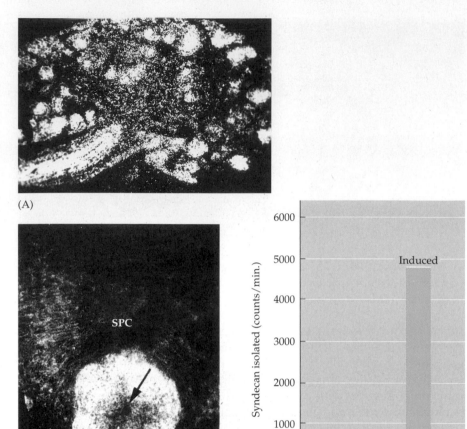

(A)

(B)

(C)

cells of cultured kidney rudiments fail to condense (Rothenpieler and Dressler, 1993). [prox3.html]

Once induced, and after it has started to condense, the mesenchyme begins to secrete Wnt4, which acts in an autocrine fashion to complete the transition from mesenchymal mass to epithelium (Stark et al., 1994). *Wnt4* expression is detected in the condensing mesenchyme cells and in the comma-shaped aggregates. In the S-shaped aggregate, it is found at the region where the newly epithelialized cells fuse with the ureteric bud tips. In mice lacking the *Wnt4* genes, the mesenchyme remains morphologically undifferentiated, and no pretubular aggregates are formed.

SIGNAL 5: CONVERSION OF THE AGGREGATED CELLS INTO A NEPHRON. In the fifth stage of kidney formation, the condensed epithelium is specified into the different cell types of the nephron. In this stage, the genes responsible for cell specification are activated. In recent years, three genes have been found whose products may be important for this specification. The first is the gene for gap junction protein, Connexin 43. This protein is seen in the condensed mesenchyme and connects the cells of the S-shaped body (Sainio et al., 1992). The second gene is *Pax2*. *Pax2* is active in the condensed mesenchyme and is turned off as the cells differentiate. If it remains active, the podocytes, glomerulus, and proximal tubule cells form abnormally (Dressler et al., 1993). The third gene encodes the low-affinity receptor for nerve growth factors, NGFR. NGFR is absent on uncondensed mesenchyme, but becomes abundantly present on the condensed cells that later form the nephrons.

When antisense oligonucleotides to NGFR were added to rat kidney rudiments in culture, the condensed cells failed to form kidney tubules (Figure 17.23; Sariola et al., 1991). It is not known what signal causes the conversion of the aggregates into nephrons.

SIGNAL 6: THE CONTINUED GROWTH OF THE URETERIC BUD AND THE DIFFERENTIATION OF THE NEPHRON. After the initial interactions create the first aggregates, the metanephrogenic mesenchyme cells near the kidney border begin to proliferate to form stem cells. These stem cells can interact with the ureteric bud branches to form new nephrons or they can produce stromal cells. The stromal cells migrate to the central portion of the kidney and produce factors (as yet unknown) that (1) enable the continued growth of the ureteric bud and (2) stimulate the differentiation of the nephron into the convoluted renal tubules, Henle's loop, the glomerulus, and the juxtaglomerular apparatus. The transcription factor BF2 is synthesized in these stromal cells, and when it is knocked out in mouse embryos, the resulting kidney lacks the branched ureteric tree (it branches only three or four times instead of seven or eight, resulting in an 8- to 16-fold reduction in the number of branches), and the aggregates do not differentiate into nephrons (Hatini et al., 1996). So it appears that the factors necessary for these two functions are synthesized by the stromal cells and are regulated by transcription factor BF2.

There is also evidence that reciprocal interactions between the ureteric bud and the metanephrogenic mesenchyme may be critical in maintaining these stromal cells. The combination of FGF2 and conditioned medium from rat ureteric bud cell lines is capable of inducing the complete differentiation of nephrons in isolated metanephrogenic mesenchyme. FGF2 is needed to induce the aggregation of the mesenchyme cells, but the substances secreted into the culture medium by the ureteric bud cells is capable of turning these aggregates into nephrons (Karavonova et al., 1996). It is probable that the ureteric bud factors (which remain to be identified) stimulate the stromal cells to produce their factors (which also remain to be identified) so that the aggregate can differentiate into the nephron and so that the branches can continue to grow. The identification of these factors has become one of the important new foci of kidney developmental biology (Bard, 1996).

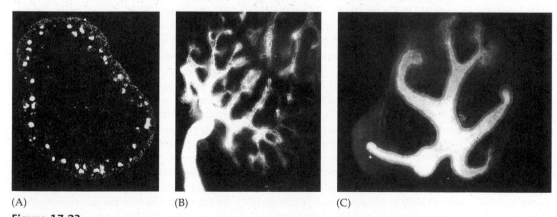

(A) (B) (C)

Figure 17.23
The role of the low-affinity NGF receptor in kidney morphogenesis. (A) In situ hybridization shows the localization of *NGFR* mRNA in the condensed mesenchymes of an 18-day rat embryonic kidney. (B) Higher magnification of the branching pattern of the ureteric bud (stained with antibodies to an epithelial specific cytokeratin) in a 13-day kidney cultured 5 days in vitro. (C) Ureteric bud from a kidney similar to that in (B) but cultured in antisense oligonucleotides to *NGFR* mRNA. (From Sariola et al., 1991, courtesy of H. Sariola.)

Coordinated Differentiation and Morphogenesis in the Tooth

DURING THE MORPHOGENESIS of any organ, numerous dialogues are occurring between the interacting tissues. In epithelial-mesenchymal interactions, the mesenchyme influences the epithelium; the epithelial tissue, once changed by the mesenchyme, can secrete factors that change the mesenchyme. Such interactions continue until an organ is formed with organ-specific mesenchyme cells and organ-specific epithelia. The identification of the chemicals involved in these intertissue conversations is under way in several laboratories. Some of the most extensively studied interactions are those that form the mammalian tooth. Here, the jaw epithelium differentiates into the enamel-secreting **ameloblasts,** while the neural-crest-derived mesenchyme cells become the dentin-secreting **odontoblasts.**

First, the epithelium causes the mesenchyme to aggregate at specific sites. At this time, the *epithelium* possesses the potential to generate tooth structures out of several types of mesenchyme cells (Mina and Kollar, 1987; Lumsden, 1988). However, this tooth-forming potential soon becomes transferred to the mesenchyme that aggregated beneath it. These *mesenchymal* cells form the dental papilla and are now able to induce tooth morphogenesis in other epithelia (Kollar and Baird, 1970). At this stage, the jaw epithelium has lost its ability to instruct tooth formation in other mesenchymes. Thus, the "odontogenic potential" has shifted from the epithelium to the mesenchyme. At the basement membrane that separates the epithelium from the mesenchyme, the epithelium induces the mesenchyme to become the odontoblasts, while the mesenchyme induces the epithelium to become the ameloblast cells (Figure 17.24; Thesleff et al., 1989).

This shift in the odontogenic potential coincides with a shift in the synthesis of bone morphogenetic protein 4 (BMP4). During the earliest phases of tooth development, BMP4 is synthesized in the epithelium. This epithelial BMP4 induces the differentiation of the mesenchyme and stimulates the mesenchyme to express three transcription factors, including homeodomain-containing proteins Msx1 and Msx2 The induction of mesenchymal differentiation can be mimicked by placing BMP4 on agarose beads and applying them to the mesenchymel mass (Vainio et al., 1993). Thus, BMP4 appears to be a critical morphogenetic signal from the epithelium to the mesenchyme.

A critical event in the analysis of tooth development was the discovery that the signaling center for tooth development is an obscure group of epithelial cells referred to as the **enamel knot** (Jernvall et al., 1994). This group of cells, first seen at the outset of the cap stage, appears as a nondividing population of cells in the center of the growing cusps (see Figure 17.3). Moreover, in situ hybridization demonstrated that this enamel knot is the source of sonic hedgehog, FGF4, BMP7, BMP4, and BMP2 secretion (Koyoma et al., 1996; Vaahtokari et al., 1996a). As a nondividing population secreting growth factors capable of being received by both the epithelium and the mesenchyme, the enamel knot is thought to direct the cusp morphogenesis of the tooth and to be critical in directing the evolutionary changes of tooth structure in mammals (Jernvall, 1995).

A summary of recent research correlating mesenchyme induction and differentiation is shown in Figure 17.24. As can be seen, the mesenchyme at one stage is different from the mesenchyme at other stages. The mesenchyme cells are first induced (by epithelial expression of BMP4, BMP2, BMP7, and probably FGF8) to express a set of transcription factors that includes Msx1 and Lef1. If the genes for either of these two proteins are knocked out, the developing mouse lacks teeth. In the human condition caused by a mutation of *MSX1*, the patients are missing teeth (Satokata and Maas, 1994; Kratochwil et al., 1996; Vas-

tardis et al., 1996). As the dental mesenchyme cells condense, they are induced to synthesize the membrane protein syndecan and the extracellular matrix protein tenascin. These proteins (which can bind each other) appear at the time the epithelium induces mesenchymal aggregation, and Thesleff and colleagues (1990) have proposed that these two molecules may interact to bring about this condensation. As in the kidney, syndecan expression also correlates with the proliferation of the clustered mesenchyme cells, suggesting that it is regulating cell division as well as aggregation (Vainio et al., 1991).

After the mesenchyme cells have aggregated, they begin to secrete FGF3, BMP3, BMP4, HGF, and activin (Wilkinson et al., 1989; Thesleff and Sahlberg, 1996). These signals presumably induce the formation of the enamel knot in the epithelium. The enamel knot then secretes its potent cocktail of growth and differentiation factors which cause the growth and differentiation of both the mesoderm and epithelium. The mesenchyme cells begin to differentiate into odontoblasts, and tenascin is induced to be expressed at much higher levels and at the same sites as alkaline phosphatase. Both these proteins have been associated with bone and cartilage differentiation, and they may promote the mineralization of the extracellular matrix (Mackie et al., 1987).

Finally, as the odontoblast phenotype emerges, osteonectin and type I collagen are secreted as components of the extracellular matrix. The enamel knot disappears through apoptosis (Vaahtokari et al., 1996b). By this steplike process, the cranial neural crest cells of the jaw can be transformed into the dentin-secreting odontoblasts. These interactions occur at specific times of development and are correlated to the maturation of the epithelium. Under normal circumstances, two independent phenomena—morphogenesis and cell differentiation—are coordinated in organ formation. ■

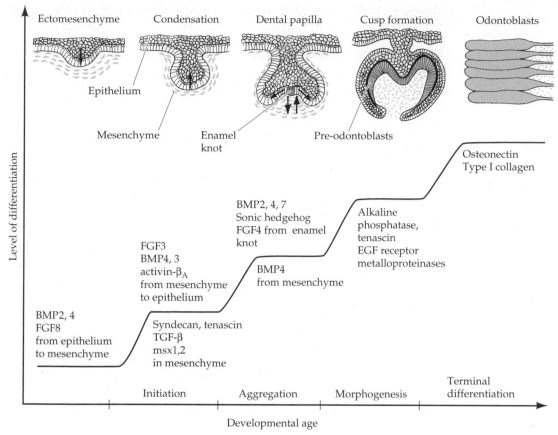

Figure 17.24 Coordinated differentiation and morphogenesis in the mammalian tooth. As development progresses, the neural crest-derived mesenchyme of the jaw undergoes stepwise differentiation as it interacts with jaw epithelium. (After Thesleff et al., 1990; Thesleff and Sahlberg, 1996.)

Mechanisms of branching in the formation of parenchymal organs

The generation of organ-specific epithelial branching patterns has remained a largely unexplored area. Earlier studies (for reviews, see Bard, 1990; Mizuno and Yasugi, 1990) revealed three major patterns whereby the mesenchyme regulates branching specificity. In the kidney, only one type of mesenchyme can cause branching to occur (Saxén, 1987). In the salivary and mammary glands, the mesenchyme specifies the branching pattern, but the differentiation of the epithelium is determined autonomously by the epithelium (Lawson, 1974; Sakakura et al., 1976). In epithelial tubes that form tracts containing several differently branching regions (such as the respiratory, digestive, and reproductive tracts), the regional mesenchymes specify both the branching patterns *and* the types of proteins in each region (Wessells, 1970; Cunha et al., 1976a,b; Hilfer et al., 1985; Haffen et al., 1987). For instance, in the region of the endodermal tube that is going to become liver, the mRNA for albumin (a liver-specific protein) is synthesized in the hepatic region of the epithelium even before the cells aggregate to form the liver rudiment. All that is needed for albumin mRNA synthesis is that the epithelium be in close contact with the mesenchyme cells of that area. In both liver and pancreas, these early interactions with the region-specific mesenchyme produce a low level of specific gene expression in that region of the proliferating endodermal tube (Rutter et al., 1964; Cascio and Zaret, 1991). This initial pattern will be amplified when the organs form their morphological structures.

The Extracellular Matrix as a Critical Element in Branch Formation

The mechanisms for such branching may have both general and specific components (Grobstein, 1967) and may depend on the interaction between those forces promoting cell growth and those forces promoting intercellular cohesion. The general components are thought to involve the selective degradation of the epithelial basement membrane at the branching sites (Bernfield et al., 1984; Mizuno and Yasugi, 1990).

As seen in the kidney and many other organs, the mesenchyme can interact with an epithelial tube by causing it to branch. Branching occurs when the epithelial outgrowths are divided by clefts, yielding lobules on either side of the cleft. These lobules grow to create branches. The branching of epithelial buds depends upon the presence of the mesenchyme. In some cases, such as the interaction of respiratory epithelium with several different mesenchymes, the interaction is instructive. In most cases, however, the interactions are merely permissive. The buds are prepared to branch and form acini, but they need support from the mesenchyme. It is now thought that the mesenchyme causes cleft formation and branching by splitting the lobule and selectively digesting part of the epithelial tissue's basal lamina.

The control of cleft formation appears, in part, to be a function of collagen molecules. Collagen III fibrils are produced by the mesenchyme cells but accumulate only within the clefts of lobules (Figure 17.25; Grobstein and Cohen, 1965; Nakanishi et al., 1988). Moreover, the extent of branching can be artificially regulated by preserving or removing collagen molecules (Nakanishi et al., 1986a). Figure 17.26 shows the branching of a 12-day rudiment of a submandibular salivary gland under conditions that prevent the degradation of collagen fibrils (collagenase inhibitor was added to the medium) or accelerate its degradation (exogenous collagenase was added to the medium). Without the collagen, no clefts are seen, but when the endogenous collagenase is unable to remove excess collagen, supernumerary clefts appear.

Mesenchymal cells

Epithelial cell

Collagen in cleft between epithelial cells

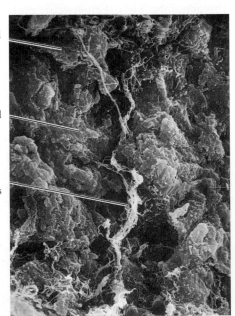

Figure 17.25
Scanning electron micrograph of collagen fibril accumulation within an early cleft of a 12-day mouse salivary gland. (From Nakanishi et al., 1986b, courtesy of Y. Nakanishi.)

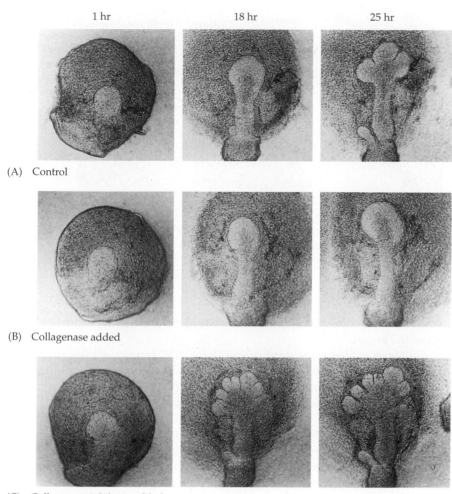

1 hr   18 hr   25 hr

(A) Control

(B) Collagenase added

(C) Collagenase inhibitor added

Figure 17.26
Control of epithelial cleft formation by mesenchymal collagen. The 12-day mouse salivary gland rudiments were cultured and observed at 1, 18, and 25 hours. (Row A) Normal development, showing three major lobes. (Row B) Growth of lobule but no branching when exogenous collagenase (5 μg/ml) was added to the medium. (Row C) Supernumerary branches when collagenase inhibitor (5 μg/ml) was added to the medium to suppress endogenous collagenase activity. (From Nakanishi et al., 1986a, courtesy of Y. Nakanishi.)

The mechanism by which collagen initiates this branching is still unclear. Nakanishi and co-workers (1986b) have proposed that the mesenchyme cells align collagen fibrils by their traction to form ridges that cut into the lobular epithelium to form clefts. These clefts then become clearly defined as more migrating mesenchyme cells further deform the lobule by their traction. This initial cleft formation does not depend on epithelial cell proliferation (Nakanishi et al., 1987). The collagen fibrils may also be responsible for the development of the cleft into distinct branches. Bernfield and Banerjee (1982) have proposed that collagen can protect the basal lamina of the epithelial cells against hyaluronidase secreted by the mesenchyme cells. They showed that the mesenchyme cells do indeed digest glycosaminoglycan (GAG) from the lobule (Banerjee and Bernfield, 1979) and that the GAGs at the tips are more susceptible than those in the clefts. When heparan sulfate GAGs are removed from cultured salivary gland rudiments, branching ceases (Nakanishi et al., 1993). The breakdown of the basal lamina would enable the expansion of the branch by the increased mitoses stimulated in that area. In this model, shown in Figure 17.27, the mesenchyme promotes epithelial growth, degrades the GAG, and deposits collagen fibers in the cleft. The epithelium synthesizes the basal lamina materials and stimulates mesenchymal collagen synthesis. The result is a differential breakdown of the basal lamina at the tips of the lobes, thus enabling the dividing cells of the lobe to form branches. Here, the interaction of mesenchyme cells with the

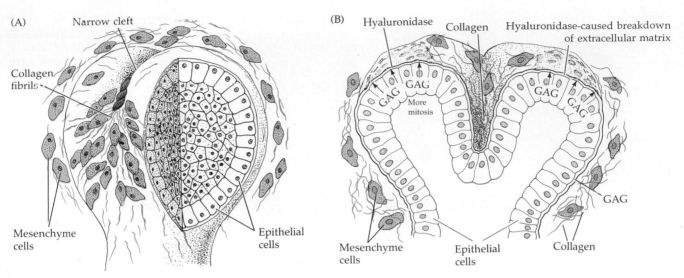

Figure 17.27
A model for cleft formation and branching in the mouse salivary gland rudiment. (A) A furrow is made in the lobule by contracting a bundle of collagen fibrils (shown here as a twisted, rope-like structure) by the traction of the mesenchyme cells. As shown in Figure 17.25, the fibrils extend between two groups of mesenchyme cells. (B) The elongation of the two separated lobules into branches may ensue, as the GAGs at the tips of the lobules are more sensitive to hyaluronidase, since they do not have the protection of the collagen fibrils. The stalk of the lobule is stabile, while the increased division (stimulated by the mesenchyme) at the tips pushes the lobule forward. (A from Nakanishi et al., 1986b; B after Wessells, 1977.)

extracellular matrix of the epithelium would determine the branching pattern of the organ.

Collagen is also important in stabilizing the branches once they have formed. If one adds collagenase to salivary gland rudiments after branching has occurred, the collagen is removed, and the branches coalesce into a globe (Grobstein and Cohen, 1965; Wessels and Cohen, 1968).

Paracrine Factors Effecting Branching Patterns

We still do not know for certain the identities of those molecules secreted by the mesenchyme that are responsible for inducing these epithelial branching patterns. Recent evidence has implicated several paracrine factors in these events. The first candidate is **transforming growth factor β1** (TGF-β1). This molecule is abundant in embryonic organs. When exogenous TGF-β1 is added to cultured mammary glands or to embryonic salivary gland, lung, or kidney rudiments, it prevents the epithelium from branching (Figure 17.28; Silberstein et al., 1990; Hardman et al., 1994; Serra et al., 1994; Ritvos et al., 1995). TGF-β1 is known to promote the synthesis of extracellular matrix proteins and to inhibit the metalloproteinases that can digest these matrices (Penttinen et al., 1988; Nakamura et al., 1990). It is possible that TGF-β1 plays a role in stabilizing the branches after they have forked.

A second molecule that may be important in epithelial branching is **activin**. Activin is known to be important in specifying the right/left axis in chicks, and it has been detected in the salivary glands, pancreas, and kidneys of embryonic mice. When activin is added exogenously to embryonic mouse kidney, salivary, or pancreatic rudiments, it severely distorts the normal branching pattern (Figure 17.29; Ritvos et al., 1995). The epithelial cells are not dead and are still capable of inducing the mesenchyme cells to form nephrons, but the branches are grossly disordered. The similarities between the salivary gland rudiments treated with collagenase and those treated with activin suggest that activin may trigger digestion of the extracellular matrix at the site of a new branch and that adding it exogenously causes the breakdown of the extracellular matrix throughout the epithelium.

Several additional paracrine factors appear to be responsible for inducing the branching of lung epithelia. One form of platelet-derived growth factor can induce lung branching, and antisense RNA against its message inhibits it (Souza et al., 1995). Cultured lung epithelium can also be stimulated

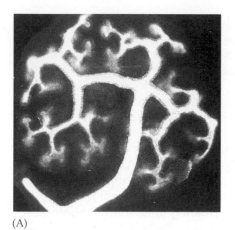

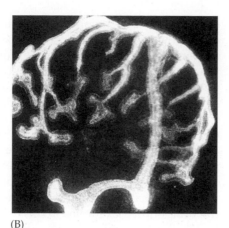

(A)   (B)

Figure 17.28
The effect of TGF-β1 on the morphogenesis of kidney epithelium. (A) An 11-day mouse kidney cultured 4 days in control medium has normal branching pattern. (B) An 11-day mouse kidney cultured in TGF-β1 shows no branching until reaching the periphery of the mesenchyme, and the branches formed are elongated. (After Ritvos et al., 1995.)

to branch by exposing it to amphiregulin, a paracrine factor similar to epidermal growth factor. Antibodies to amphiregulin will inhibit branching in these cultures (Schugar et al., 1996). The mesenchyme of the embryonic mouse lung secretes FGF7, while the lung epithelium synthesizes the FGF7 receptor. Antisense oligonucleotides to either FGF7 or its receptor block epithelial branching in cultured lung rudiments, as will loss-of-function mutations of this receptor* (Peters et al., 1994; Post et al., 1996). In addition to the secretion of amphiregulin and FGF7 by the mesenchyme, Sonic hedgehog appears to be secreted by the distal tips of the lung buds (Bellusci et al., 1996).

Induction at the single-cell level

Embryonic induction occurs when interactions between inducing and responding cells bring about changes in the developmental pathway of the responding cell (Jacobson and Sater, 1988). Without the induction, the responding cell would become one cell type; with the induction, it becomes

*In a remarkable coincidence, the formation of the *Drosophila* tracheal system also depends on FGF (Glazer and Shilo, 1991; Samakoulis et al., 1996). The lungs and tracheae in vertebrates are evolutionary novelties having no anatomical or embryonic similarities to insect tracheae.

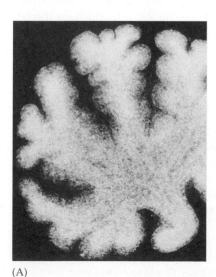

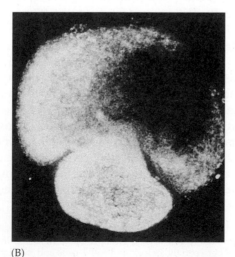

(A)   (B)

Figure 17.29
The effects of activin on the morphogenesis of salivary gland epithelium. Embryonic salivary gland rudiments were cultured 4 days in control medium (A) and in medium containing 7.5 nM activin (B). After these 4 days, the organs were fixed and stained for epithelial cytokeratin. (From Ritvos et al., 1995.)

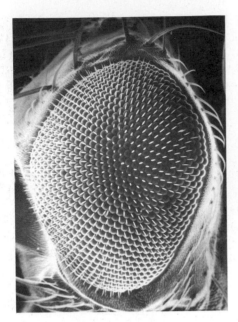

Figure 17.30
Scanning electron micrograph of a compound eye in *Drosophila*. Each facet is a single ommatidium. A sensory bristle projects from each ommatidium. (Courtesy of T. Venkatesh.)

another. Our discussions of induction have usually concerned tissues, not cells. However, induction can also occur at the single-cell level. The first examples of this phenomenon came from studies of the immune system. Here, the reception of antigens (foreign substances) by the B cell gave it the competence to respond to paracrine and juxtacrine factors synthesized by the helper T cell. There is a reciprocal dialogue between the B cell and the T cell such that both of them differentiate and proliferate in the presence of the foreign antigen (Clark and Ledbetter, 1994; Essen et al., 1995). Indeed, AIDS is a disease of induction, wherein the helper T cell has been destroyed and cannot induce the differentiation of B cells and macrophages.* [prox5.html]

Recent research into the development of *Drosophila* and *Caenorhabditis* has shown that induction does indeed occur on the cell-to-cell level. Some of the best-studied examples involve the formation of the retinal photoreceptors in the *Drosophila* eye. The retina consists of about 800 units called **ommatidia** (Figure 17.30). Each ommatidium is composed of 20 cells arranged in a precise pattern. The eye develops in the flat epithelial layer of the eye imaginal disc of the larva. There are no cells directly above or below this layer, so the interactions are confined to neighboring cells in two dimensions. The differentiation of the randomly arranged epithelial cells into retinal photoreceptors and their surrounding lens tissue occurs during the last (third) larval stage. An indentation forms at the posterior margin of the imaginal disc, and this **morphogenetic furrow** begins to travel forward toward the anterior of the epithelium (Figure 17.31). The movement of the furrow depends on the tag-team proteins, Hedgehog and Decapentaplegic. Hedgehog is expressed by the cells immediately posterior to the furrow (i.e., those that have just differentiated), and it induces the expression of the decapentaplegic protein within the furrow (Heberlein et al., 1993; Ma et al., 1993). As the retinal cells begin to differentiate behind the furrow, they secrete the hedgehog protein, which drives the furrow anteriorly (Brown et al., 1995). As the furrow passes through a region of cells, those cells begin to differentiate in a specific order. The first cell to develop is the central (R8) photoreceptor. (It is not yet known how the furrow instructs certain cells to become R8 photoreceptors, but it is possible that the Dpp and Hedgehog proteins in the furrow region induce R8 determination.) The R8 cell is thought to induce the cell before it and the cell after it (with respect to the furrow) to become the R2 and R5 photoreceptors, respectively. The R2 and R5 photoreceptors are functionally equivalent, so the signal from R8 is probably the same to both cells (Tomlinson and Ready, 1987). Signals from these cells induce four more adjacent cells to become the R3, R4, and then the R1 and R6 photoreceptors. Last, the R7 photoreceptor appears. The other cells around these photoreceptors become the lens cells. Lens determination is the "default" condition if the cells are not induced. [prox5.html]

A series of mutations has been found that blocks some of the steps of this induction cascade. The *rough* (*ro*) mutation, for instance, blocks the induction of the R3 and R4 photoreceptors. The *sevenless* (*sev*) mutation and the *bride of sevenless* (*boss*) mutations can each prevent the R7 cell from differentiating. (This cell becomes a lens cell instead.) Analysis of these mutations has shown that they are involved in the inductive process. The *sevenless* gene is required in the R7 cell itself. If mosaic embryos are made such that some of the cells of the eye disc are heterozygous (normal) and some are homozygous for the *sevenless* mutation, the R7 photoreceptor is seen to develop only

*In humans, these T cells are called *helper/inducer T cells*, a name that recognizes their developmental role. The CD4 glycoprotein is normally involved in mediating nonspecific cell adhesion between the helper/inducer T cells and B lymphocytes (Doyle and Strominger, 1987).

Figure 17.31
Differentiation of photoreceptors in the late larval eye imaginal disc. The morphogenetic furrow (arrow) crosses the disc from posterior (left) to anterior (right). Behind the furrow, the photoreceptor cells differentiate in a defined sequence (shown below). The first photoreceptor cell to differentiate is R8. R8 appears to induce the differentiation of R2 and R5, and the cascade of induction continues until the R7 photoreceptor is differentiated. (After Tomlinson, 1988, photograph courtesy of T. Venkatesh.)

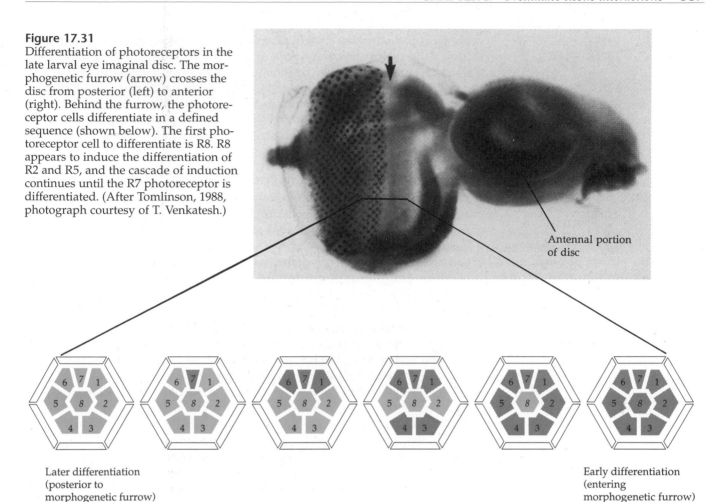

Antennal portion of disc

Later differentiation (posterior to morphogenetic furrow)

Early differentiation (entering morphogenetic furrow)

if the R7 precursor cell has the wild-type *sevenless* allele (Basler and Hafen, 1989; Bowtell et al., 1989). Antibodies to this protein find it in the cell membrane, and the sequence of the *sevenless* gene suggests that it is a transmembrane protein with a tyrosine kinase site in its cytoplasmic domain (Banerjee et al., 1987; Hafen et al., 1987). This is consistent with the protein's being a receptor for some signal.

This signal for the R7 precursor to differentiate into the R7 photoreceptor probably comes directly from a protein encoded by the wild-type allele of *bride of sevenless* (*boss*). Flies homozygous for the *boss* mutation also do not have the R7 photoreceptors. Gene mosaic studies wherein some of the cells of the imaginal disc are normal and some of the cells are homozygous for the *boss* mutation show that the wild-type *boss* gene is not needed in the R7 precursor cell. Rather, the R7 photoreceptor only differentiates if the wild-type *boss* gene is expressed in the *R8* cell. Thus, the *bride of sevenless* gene is encoding some protein whose existence in the R8 cell is necessary for the differentiation of the R7 cell.* The signal produced by the Boss protein proba-

*All the photoreceptor precursors synthesize the Sev protein, and the Boss signal given by the R8 photoreceptor is probably given to and received by all the surrounding cells. What, then, prevents R1–R6 cells from also becoming R7 cells? The restrictive agent is probably the product of the *seven-up* (*sup*) gene. In *sup*-deficient mutants, the R1, R3, R4, and R6 precursors all develop the R7 phenotype. The *sup* gene encodes a transcription factor of the steroid-receptor family (Mlodzik et al., 1990). This is not the entire story, either. There is most likely a parallel pathway, wherein the Sevenless receptor also activates the Corkscrew protein. Corkscrew activates the Daughter-of-sevenless (dos) protein. The Dos protein facilitates activation of Ras (Herbst et al., 1996).

Figure 17.32
Summary of genes known to be involved in the induction of *Drosophila* photoreceptors. For development to continue beyond the differentiation of the R8, R2, and R5 photoreceptors, the *rough* gene (*ro*) must be present in both the R2 and R5 cells. For the differentiation of the R7 photoreceptor, the *sevenless* gene (*sev*) has to be active in the R7 precursor cell, while the *bride of sevenless* gene (*boss*) must be active in the R8 photoreceptor. (After Rubin, 1989.)

bly works by cell contact. Wild-type *boss* genes in an R8 cell in one ommatidium will not correct the deficiency of a mutant *boss* allele in the adjacent ommatidia, and the extracellular domain of the Boss protein is sufficient to activate the *sevenless* tyrosine kinase in a neighboring cell (Reinke and Zipursky, 1988; Hart et al., 1993). A summary of the known cell-to-cell inductions in the *Drosophila* retina (Figure 17.32) shows that individual cells are able to induce other individual cells to create the precise arrangement of cells in particular tissues.

Vulval Induction in the Nematode Caenorhabditis elegans

The vulva of *Caenorhabditis elegans* is a case where one inductive signal may generate a variety of cell types. This organ forms during the larval stage from six blast cells called the **vulval precursor cells** (**VPCs**). The cell connecting the overlying gonad to the vulval precursor cells is called the **anchor cell**. It secretes the LIN-3 protein, a relative of epidermal growth factor (Hill and Sternberg, 1992). If the anchor cell is destroyed (or if the *lin-3* gene is mutated), the VPCs will not form a vulva; they will become part of the hypodermis (skin) (Kimble, 1981). The six vulval precursor cells influenced by the anchor cell form an **equivalence group.** Each member of this group is competent to become induced by the anchor cell and can assume any of three fates, depending on its proximity to the anchor cell (Figure 17.33). The cell directly beneath the anchor cell divides to form the *central* vulval cells. The two cells flanking that central cell divide to become the *lateral* vulval cells, while the three cells farther away from the anchor cell generate *hypoblast* cells. If the anchor cell is destroyed, all six cells of the equivalence group divide once and contribute to the hypodermal tissue. If the three central cells are destroyed, the three outer cells, which normally form hypodermal cells, generate vulval cells instead. The LIN-3 protein is received by the LET-23 receptor tyrosine kinase on the VPCs, and the signal is transferred to the nucleus through the Ras-MAP kinase pathway (see Chapter 3).

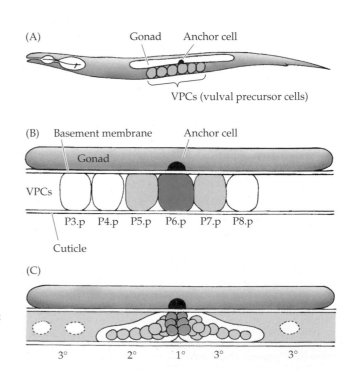

Figure 17.33
The VPCs and their descendants. (A) Location of the gonad, anchor cell, and VPCs in the second instar larva of a *C. elegans* hermaphrodite. (B,C) Relationship of the anchor cell to the six VPCs and their subsequent lineages. 1° lineages result in the central vulval cells; 2° lineages constitute the lateral vulval cells; 3° lineages generate hypodermal cells. The outline of the vulva is shown in the fourth instar larva, the circles representing the positions of the nuclei. (After Katz and Sternberg, 1996.)

There are three mechanisms by which such inductions can take place (Katz and Sternberg , 1996):

1. *The graded signal hypothesis.* Here, the VPC closest to the anchor cell receives the highest concentration of LIN-3 protein and generates the central vulval cells. The two VPCs adjacent to it receive a low amount of LIN-3 and become the lateral vulval cells. The VPCs farther away from the anchor cell do not receive enough LIN-3 to matter, so they become hypodermis (Katz et al., 1995).

2. *The sequential induction model.* Here, LIN-3 protein works only on the cell directly beneath it. This cell will generate the central vulval lineage. It will also signal laterally to the two adjacent cells and instruct them to generate the lateral vulval lineages. These cells will not instruct the peripheral cells of VPCs to do anything, so they become hypodermis (Koga and Oshima, 1995; Simske and Kim, 1995).

3. *The non-equivalence model.* Here, the VPCs can substitute for one another, but they are not identical. They have their biases and can respond to even small concentrations of LIN-3 protein. However, the biases cause the cell beneath the anchor cell to generate the central vulval lineage (Sternberg, 1989; Sternberg and Horvitz, 1989).

Interestingly, there is evidence that all three models function during normal development (Kenyon, 1995; Katz and Sternberg, 1996). There is probably a graded LIN-3 signal from the anchor cell, which reinforces the already existing biases of the VPCs. Moreover, once the VPC beneath the anchor cell becomes determined to form the central vulval lineage, it signals the cells adjacent to it to prohibit them from also forming the central vulval cells. This **lateral inhibition** of the "secondary" vulval precursor cells by the "primary" VPC is accomplished through the LIN-12 proteins (Figure 17.34; Sternberg, 1988). If all these systems are operating during normal development, notes Kenyon (1995), "then together they could produce the ever-perfect tiny vulvae that *C. elegans* is so famous for."

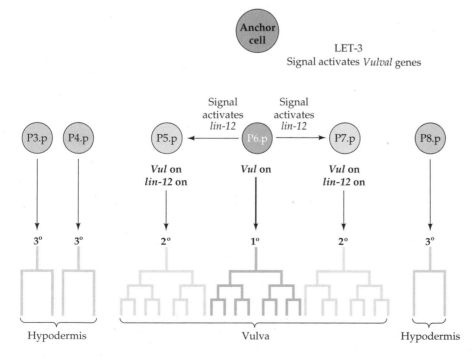

Figure 17.34
Model for the determination of vulval cell lineages in *C. elegans*. The LIN-3 signal from the anchor cell causes the determination of the P6.p cell to generate the central vulval lineage. Lower doses of LIN-3 cause the P5.p and P7.p cells to form the lateral vulval lineages. The P6.p (central lineage) cell also secretes a short-range signal that induces the neighboring cells to activate the LIN-12 protein. This also prevents the P5.p and P7.p cells from generating the primary, central vulval cell lineage.

Cell-Cell Interactions and Chance in the Determination of Cell Types

THE DEVELOPMENT of the vulva in *C. elegans* shows several instances of induction on the cellular level. The first instance concerns induction of the gonadal anchor cell. The formation of the anchor cell is mediated by the *lin-12* gene, which encodes a cell-surface receptor protein. In wild-type hermaphrodites, two adjacent cells, Z1.ppp and Z4.aaa, have the potential to become the gonadal anchor cell. They interact in a manner that causes one of them to be the anchor cell while the other one becomes the precursor of the uterine tissue. In recessive *lin-12* mutants, *both* cells become anchor cells, while in dominant mutations, both cells become uterine precursors (Greenwald et al., 1983). Studies using genetic mosaics and cell ablations have shown that this decision is made in the second larval stage and that the *lin-12* gene only needs to function in that cell destined to become the uterine precursor cell. The presumptive anchor cell does not need it. Seydoux and Greenwald (1989) speculate that these two cells originally synthesize both the signal for uterine differentiation (the LAG-2 protein) and the receptor for this molecule (the LIN-12 protein) (Figure 17.35; Wilkinson et al, 1994). During a particular time in larval development, the cell that, by chance, is secreting more of this differentiation signal causes its neighbor to cease its production of the signal molecule and to increase its production of LIN-12 protein. The cell secreting the signal becomes the gonadal anchor cell, while the cell receiving the signal through its LIN-12 protein becomes the ventral uterine precursor cell. Thus, the two cells are thought to determine each other prior to their respective differentiation events.

The anchor cell/ventral uterine precursor decision illustrates two important aspects of determination in two originally equivalent cells. First, the initial difference between them is created by chance. Second, these initial differences are reinforced by feedback. Such deter-

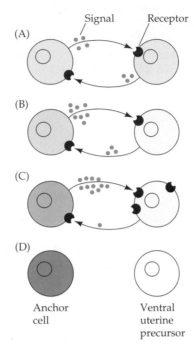

Figure 17.35   Model for the generation of two cell types (anchor cell and ventral uterine precursor) from two equivalent cells (Z1.ppp and Z4.aaa). (A) The cells start off as equivalent, with fluctuating amounts of signal (arrow) and receptor (inverted arrow). The *lag-2* gene is thought to encode the signal; the *lin-12* gene is thought to encode the receptor. Reception of the signal turns down LAG-2 production and upregulates LIN-12. (B) A stochastic (chance) event causes one cell to produce more signal substance than the other cell at some particular critical time. This stimulates more LIN-12 activity in the neighboring cell. (C) This difference is amplified, since the cell with more LIN-12 doesn't produce as much signal. (D) Eventually, one cell delivers the signal, and the other cell receives it. The signaling cell becomes the anchor cell; the receiving cell becomes the ventral uterine precursor. (After Greenwald and Rubin, 1992.)

mination is also seen in the determination of which of the originally equivalent epidermal cells of the insect embryo generate the neurons of the peripheral nervous system. Here, the choice is between becoming a skin (hypodermal) cell or a neuroblast. The *Notch* gene of *Drosophila* also channels a bipotential cell into one

of two alternative paths. Soon after gastrulation, a region of about 1800 ectodermal cells lies along the ventral midline of the *Drosophila* embryo. These cells have the potential to form the ventral nerve cord of the insect, and about one-quarter of these cells become neuroblasts, while the rest become the precursors of the hypodermis. The cells that give rise to neuroblasts are intermingled with those cells destined to give rise to hypodermal precursors. Thus, each ectodermal cell in the nerve-forming regions of the fly embryo can give rise to either hypodermal or neural precursor cells (Hartenstein and Campos-Ortega, 1984). In the absence of *Notch* gene transcription in the embryo, the cells develop into neural precursors rather than into a mixture of hypodermal and neural precursor cells (Figure 17.36; Artavanis-Tsakonis et al., 1983; Lehmann et al., 1983). These embryos die, having a gross excess of neural cells at the expense of the ventral and head hypodermis (Poulson, 1937; Hoppe and Greenspan, 1986). The *Notch* gene has been cloned (Kidd et al., 1983; Yedvobnick et al., 1985) and found to be transcribed during the early half of embryogenesis (and later in the early pupal stage). Both Notch and LIN-12 proteins share remarkable sequence homologies to one another. They are both transmembrane proteins that may act as receptors for signals from adjacent cells (Yochem et al., 1988).

Heitzler and Simpson (1991) have proposed that the Notch protein, like LIN-12, serves as a receptor for intercellular signals involved in distinguishing equivalent cells. Moreover, they provide evidence that another transmembrane protein, the product of the *delta* gene (whose absence creates a phenotype very similar to that of *Notch* deficiencies), is the ligand for Notch. Genetic mosaics show that whereas *Notch* is needed in the cells that are to become epidermis, the *delta* gene is needed in the cells that induce the epidermal phenotype.

Greenwald and Rubin (1992) have

Figure 17.36 Representation of the effect of the *Notch* mutation. In wild-type embryos, the neurogenic ectoderm cells generate both neuroblasts and skin (hypodermal) cells. In *Notch*-deficient embryos, however, all the neurogenic ectoderm generates neuroblasts. The proportion of neuroblasts to hypodermal cells differs between regions of the embryo.

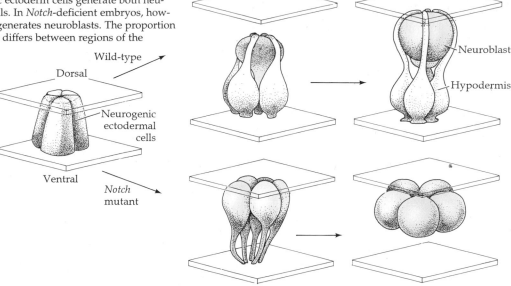

proposed a model based on the LIN-12 hypothesis to explain the spacing of neuroblasts in the proneural clusters of epidermal and neural precursors (Figure 17.37). Initially, all the cells have equal potentials and signaling. However, when one of the cells, by chance, produces more signal (say, *delta* product), it activates the receptors on adjacent cells and reduces their signaling level. Since the signaling levels on adjacent cells are low, the neighbors of those low-signaling cells will tend to be high-level signalers. In this way, a spacing of neuroblasts is produced.

The role of chance in cell determination is not as uncommon as may be supposed. As we will discuss in Chapter 22, the maturation of only one ovum per month in humans is determined largely on the chance number of hormone receptors on the follicle cells. Similarly, the decision as to whether or not a cell becomes part of the embryo or part of the trophoblast—certainly a major decision in mammalian development—is also determined by the chance position of the cell during compaction. Such chance factors can cause interactions that are amplified, eventually distinguishing two cell types from what had been a homogeneous cell population. ■

Figure 17.37 Model to explain the spacing patterns of neuroblasts among the initially equivalent neurogenic ectoderm cells. Based on the model for two cells shown in Figure 17.36, each cell both gives and receives the same signal. (A) A field of equivalent cells, all of which signal and receive equally. (B) A chance event causes one of the cells (darker shading) to produce more signal. Its surrounding cells receive this higher amount of signal and reduce their own signal level (lighter shading). (C) The rest of the pattern is now constrained. Those cells that have downregulated their own signaling (in response to the events in B) are less likely to express more signal than their neighboring cells. The cells surrounded by more downregulated signalers are more likely to become signalers. (D,E) The fates of the cells throughout the field become specified as the amplification of the signals creates populations of signalers surrounded by populations of receivers. In the case of the neurogenic genes, the signal is thought to be from the Delta protein, the receiver being the Notch protein. (After Greenwald and Rubin, 1992.)

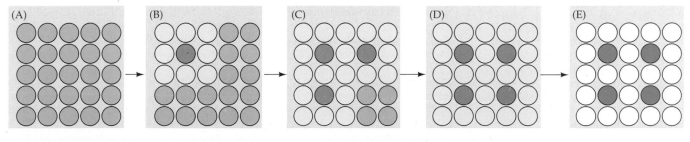

Induction is the process initiated when one cell or group of cells signals a neighboring cell or group of cells to change its developmental fate. In organisms as complex as mammals, reciprocal inductive interactions are essential to coordinate the parts into a coherent whole.

LITERATURE CITED

Armstrong, J. F., Pritchard-Jones, K., Bickmore, W. A., Hastie, N. D. and Bard, J. B. L. 1992. The expression of the Wilms' tumor gene, *WT-1*, in the developing mammalian embryo. *Mech. Dev.* 40: 85–97.

Artavanis-Tsakonis, S., Muskavitch, M. A. T. and Yedvobnick, Y. 1983. Molecular cloning of *Notch*, a locus affecting neurogenesis in *Drosophila melanogaster*. *Proc. Natl. Acad. Sci. USA* 80: 1977–1981.

Artavanis-Tsakonas, S., Matsuno, K. and Fortini, M. E. 1995. Notch signaling. *Science* 268: 225–232.

Banerjee, S. and Bernfield, M. 1979. Developmentally regulated neutral hyaluronidase activity during epithelial mesenchymal interaction. *J. Cell Biol.* 83: 469a.

Banerjee, U., Renfranz, P. J., Pollock, J. A. and Benzer, S. 1987. Molecular characterization and expression of *sevenless*, a gene involved in neuronal pattern formation in the *Drosophila* eye. *Cell* 49: 281–291.

Bard, J. B. L. 1990. *Morphogenesis: The Cellular and Molecular Processes of Developmental Anatomy*. Cambridge University Press, Cambridge.

Bard, J. B. L. 1996. A new role for the stromal cells in kidney development. *BioEssays* 18: 705–707.

Bard, J. B. L. and Ross, A. S. A. 1991. LIF, the ES cell inhibition factor, reversibly blocks nephrogenesis in cultured mouse kidney rudiments. *Development* 113: 193–198.

Bard, J. B. L. , Davies, J. A., Karavanova, I., Lehtonen, E., Sariola, H. and Vainio, S. 1996. Kidney development: the inductive interactions. *Semin. Cell Dev. Biol.* 7: 195–202.

Basler, K. and Hafen, E. 1989. Ubiquitous expression of *sevenless*: Position-dependent specification of cell fate. *Science* 243: 931–934.

Basler, K. and Struhl, G. 1994. Compartment boundaries and the control of *Drosophila* limb pattern by hedgehog protein. *Nature* 368: 208–214.

Bassols, A. and Massagué, J. 1988. Transforming growth factor β regulates the expression and structure of extracellular matrix chondroitin/dermatan sulfate-proteoglycans. *J. Biol. Chem.* 263, 3039–3045.

Bellusci, S., Henderson, R., Winnier, G., Oikawa, T. and Hogan, B. L. M. 1996. Evidence from normal expression and targeted misexpression that bone morphogenesis protein-4 (BMP-4) plays a role in mouse embryonic lung morphogenesis. *Development* 122: 1693–1702.

Bernfield, M. and Banerjee, S. D. 1982. The turnover of basal lamina glycosaminoglycan correlates with epithelia morphogenesis. *Dev. Biol.* 90: 291–305.

Bernfield, M., Banarjee, S. D., Koda, J. E. and Rapraeger, A. C. 1984. Remodeling of basement membrane as a mechanism of morphogenetic tissue interaction. *In* R. L. Trelstad (ed.), *The Role of Extracellular Matrix in Development*. Alan R. Liss, New York, pp. 545–547.

Birren, S. J. and Anderson, D. J. 1990. A v-*myc*-immortalized sympathoadrenal progenitor cell line in which neuronal differentiation is initiated by FGF but not NGF. *Neuron* 4: 189–201.

Bishop-Calame, S. 1966. Étude experimentale de l'organogenese du systéme urogénital de l'embryon de poulet. *Arch. Anat. Microsc. Morphol. Exp.* 55: 215–309.

Bitgood, M. J. and McMahon, A. P. 1995. Hedgehog and BMP genes are coexpressed at many diverse sites of cell-cell interaction in the mouse embryo. *Dev. Biol.* 172: 126–158.

Bitgood, M. J., Shen, L. and McMahon, A. P. 1996. Sertoli cell signalling by desert hedgehog regulates the male germline. *Curr. Biol.* 6: 298–304.

Bowtell, D. D. L., Simon, M. A. and Rubin, G. M. 1989. Ommatidia in the developing *Drosophila* eye require and can respond to sevenless for only a restricted period. *Cell* 56: 931–936.

Brown, N. L., Sattler, C. A., Paddock, S. W. and Carroll, S. B. 1995. Hairy and Emc negatively regulate morphogenetic furrow progression in the *Drosophila* eye. *Cell* 80: 879–887.

Cascio, S. and Zaret, K. S. 1991. Hepatocyte differentiation initiates during endodermal-mesenchymal interactions prior to liver formation. *Development* 113: 217–225.

Cattanco, E. and McKay, R. D. G. 1990. Proliferation and differentiation of neuronal stem cells regulated by nerve growth factor. *Nature* 347: 762–765.

Clark, E. A. and Ledbetter, J. A. 1994. How B and T cells talk to each other. *Nature* 367: 425–428.

Cohen, M. M. Jr. 1982. *The Child with Multiple Birth Defects*. Raven Press, NY.

Coulombre, J. L. and Coulombre, A. J. 1971. Metaplastic induction of scales and feathers in the corneal anterior epithelium of the chick embryo. *Dev. Biol.* 25: 464–478.

Crossley, P. H., Martinez, S. and Martin, G. R. 1996. Midbrain development induced by FGF8 in the chick embryo. *Nature* 380: 66–68.

Cunha, G. R. 1976a. Epithelial-stromal interactions in the development of the urogenital tract. *Int. Rev. Cytol.* 47, 137–194.

Cunha, G. R. 1976b. Stromal induction and specification of morphogenesis and cyto-differentiation of the epithelia of the Müllerian ducts and urogenital sinus during

development of the uterus and vagina in mice. *J. Exp. Zool.* 196, 361–370.

Cvekl, A., Sax, C. M., Li, X., McDermott, J. B. and Piatigorsky, J. 1995. Pax-6 and lens-specific transcription of the chicken δ1-crystallin gene. *Proc. Natl. Acad. Sci. USA* 92: 4681–4685.

Daniel, C. W. (1989). TGF-β1 induced inhibition of mouse mammary ductal growth: developmental specificity and characterization. *Dev. Biol.* 134: 20–30.

Davies, J. A., Lyon, M., Gallagher, J. and Garrod, D. R. 1995. Sulphated proteoglycan is required for collecting duct growth and branching but not nephron formation during kidney development. *Development.* 121: 1507–1517.

Deuchar, E. M. 1975. *Cellular Interactions in Animal Development*. Chapman and Hall, London.

Doyle, C. and Strominger, J. L. 1987. Interaction between CD4 and class II MHC molecules mediates cell adhesion. *Nature* 330: 256–259.

Dressler, G. R., Wilkinson, J. E., Rothenpieler, V. W., Patterson, L. T., Williams-Simons, L. and Westphal, H. 1993. Deregulation of *Pax-2* expression in transgenic mice generates severe kidney abnormalities. *Nature* 362: 65–67.

Dudley, A. T., Lyons, K. M. and Robertson, E. J. 1995. A requirement for bone morphogenesis protein-7 during development of the mammalian kidney and eye. *Genes Dev.* 9: 2795–2807.

Edwards, D. R. and seven others. 1987. Transforming growth factor beta modulates the expression of collagenase and metalloproteinase inhibitor. *EMBO J.* 6, 1899–1904.

Ekblom, P., Thesleff, I., Saxén, L., Miettinen, A. and Timpl, R. 1983. Transferrin as a fetal growth factor: Acquisition of responsiveness related to embryonic induction. *Proc. Natl. Acad. Sci. USA* 80: 2651–2655.

Ekblom, P. and eight others. 1994. Role of mesenchymal nidogen for epithelial morphogenesis in vitro. *Development* 120: 2003–2014.

Essen, D. van, Kikutani, H. and Gray, D. 1995. CD40 ligand-transduced co-stimulation of T cells in the development of helper function. *Nature* 378: 620–623.

Fan, C. M. and Tessier-Lavigne, M. 1994. Patterning of mammalian somites by surface ectoderm and notochord: Evidence for sclerotome induction by a hedgehog homolog. *Cell* 79: 1175–1186.

Fujiwara, M., Uchida, T., Osumi-Yamashita, N. and Eto, K. 1994. Uchida rat (*rSey*): a new mutant rat with craniofacial abnormalities resembling those of the mouse *Sey* mutant. *Differentiation* 57: 31–38.

Fukumachi, H. and Takayama, S. 1980. Epithelial-mesenchymal interaction in differentiation of duodenal epithelium of fetal rats in organ culture. *Experientia* 36: 335–336.

Glazer, L. and Shilo, B. Z. 1991. The *Drosophila* FGF-R homolog is expressed in the embryonic tracheal system and appears to be required for directed tracheal cell extension. *Genes Dev.* 5: 697–705.

Glueckschn-Schoenheimer, S. 1943. The morphological manifestations of a dominant mutation in mice affecting tail and urogenital system. *Genetics* 28: 341–348.

Goustin, A. S. and nine others. 1985. Coexpression of the *sis* and *myc* proto-oncogenes in developing human placenta suggests autocrine control of trophoblast growth. *Cell* 41: 301–312.

Graff, J. M., Bansal, A. and Melton, D. A. 1996. Xenopus Mad proteins transduce distinct subsets of signals for the TGF-β superfamily. *Cell* 85: 479–487.

Graham, A., Heyman, I. and Lumsden, A. 1993. Even-numbered rhombomeres control the apoptotic elimination of neural crest cells from odd-numbered rhombomeres in the chick hindbrain. *Development* 119: 233–245.

Grainger, R. M. 1992. Embryonic lens induction: Shedding light on vertebrate tissue determination. *Trends Genet.* 8: 349–355.

Greenwald, I. and Rubin, G. M. 1992. Making a difference: The role of cell-cell interactions in establishing separate identities for equivalent cells. *Cell* 68: 271–281.

Greenwald, I., Sternberg, P. W. and Horvitz, H. R. 1983. The *lin-12* locus specifies cell fates in *Caenorhabditis elegans*. *Cell* 34: 435–444.

Grobstein, C. 1955. Induction interaction in the development of the mouse metanephros. *J. Exp. Zool.* 130: 319–340.

Grobstein, C. 1956. Trans-filter induction of tubules in mouse metanephrogenic mesenchyme. *Exp. Cell Res.* 10: 424–440.

Grobstein, C. 1967. Mechanisms of organogenetic tissue interaction. *Natl. Cancer Inst. Monogr.* 26: 279–299.

Grobstein, C. and Cohen, J. 1965. Collagenase: Effect on the morphogenesis of embryonic salivary epithelium in vitro. *Science* 150: 626–628.

Gumpel-Pinot, M., Yasugi, S. and Mizuno, T. 1978. Différenciation d'épithéliums endodermiques associés au mésoderme splanchnique. *Comp. Rend. Acad. Sci.* (Paris) 286: 117–120.

Gurdon, J. B., Harger, P., Mitchell, A. and Lemaire, P. 1994. Activin signalling and response to a morphogen gradient. *Nature* 371: 487–492.

Gurdon, J. B., Mitchell, A. and Mahony, D. 1995. Direct and continuous assessment by cells of their position in a morphogen gradient. *Nature* 376: 520–521.

Hafen, E., Basler, K., Edstrom, J. E. and Rubin, G. M. 1987. *sevenless*, a cell-specific homeotic gene of *Drosophila*, encodes a putative transmembrane receptor with a tyrosine kinase domain. *Science* 236: 55–63.

Haffen, K., Kedinger, M. and Simonassmann, P. 1987. Mesenchyme-dependent differentiation of epithelial progenitor cells in the gut. *J. Pediat. Gastroent.* 6: 15–23.

Hahn, H. and twenty others. 1996. Mutations of the human homolog of *Drosophila* patched in the nevoid basal cell carcinoma syndrome. *Cell* 85: 841–851.

Hamburgh, M. 1970. *Theories of Differentiation*. Elsevier, New York.

Hardman, P., Landels, E., Woolf, A. S. and Spooner, B. S. 1994. TGF–β1 inhibits growth and branching morphogenesis in embryonic mouse submandibular and sublingual glands *in vitro*. *Devel. Growth Differ.* 36: 567–577.

Harrison, R.G. 1933. Some difficulties of the determination problem. *Am. Nat.* 67: 306–321.

Hart, A. C., Krämer, H. and Zipursky, S. L. 1993. Extracellular domain of the boss transmembrane ligand acts as an antagonist of the the sev receptor. *Nature* 361: 732–736.

Hartenstein, V. and Campos-Ortega, J. A. 1984. Early neurogenesis in wild-type *Drosophila melanogaster*. *Wilhelm Roux Arch. Dev. Biol.* 193: 308–325.

Hatini, V., Huh, S. O., Herzlinger, D., Soares, V. C. and Lai, E. 1996. Essential role of stromal morphogenesis in kidney morphogenesis revealed by targeted disruption of winged helix transcription factor, BF-2. *Genes Dev.* 10: 1467–1478.

Hay, E. D. 1980. Development of the vertebrate cornea. *Int. Rev. Cytol.* 63: 263–322.

Hay, E. D. and Revel, J. -P. 1969. Fine structure of the developing avian cornea. *In* A. Wolsky and P. S. Chen (eds.), *Monographs in Developmental Biology*. Karger, Basel.

Heberlein, U., Wolff, T. and Rubin, G. M. 1993. The TGF-β homolog *dpp* and the segment polarity gene *hedgehog* are required for propagation of a morphogenetic wave in the *Drosophila* retina. *Cell* 75: 913–926.

Hebert, J., Rosequist, T., Gotz, J. and Martin, G. 1994. FGF-5 as regulator of hair growth cycle: Evidence from targeted and spontaneous mutations. *Cell* 78: 1–20.

Heemskerk, J. and DiNardo, S. 1994. *Drosophila hedgehog* acts as a morphogen in cellular patterning. *Cell* 76: 449–460.

Heitzler, P. and Simpson, P. 1991. The choice of cell fate in the epidermis of *Drosophila*. *Cell* 64: 1083–1092.

Henry, J. J. and Grainger, R. M. 1987. Inductive interactions in the spatial and temporal restriction of lens-forming potential in embryonic ectoderm. *Dev. Biol.* 124: 200–214.

Henry, J. J. and Grainger, R. M. 1990. Early tissue interactions leading to embryonic lens formation in *Xenopus laevis*. *Dev. Biol.* 141: 149–163.

Herbst, R., Carroll, P. M., Allard, J. D., Schilling, J., Raabe, T. and Simon, M. A. Daughter of sevenless is a substrate of the phosphotyrosine phosphatase corkscrew and functions during sevenless signaling. *Cell* 85: 899–909.

Hilfer, S. R., Rayner, R. M. and Brown, J. W. 1985. Mesenchymal control of branching pattern in the fetal mouse lung. *Tissue Cell* 17: 523–538.

Hill, R. J. and Sternberg, P. W. 1992. The gene *lin-3* encodes an inductive signal for vulval development in *C. elegans*. *Nature* 358: 470–476.

Hogan, B. L. M. 1996. Bone morphogenesis proteins: multifunctional regulators of vertebnrate development. *Genes Dev.* 10: 1580–1594.

Holtzer, H. 1968. Induction of chondrogenesis: A concept in terms of mechanisms. *In* R. Gleischmajer and R. E. Billingham (eds.), *Epithelial-Mesenchymal Interactions*. Williams & Wilkins, Baltimore, pp. 152–164.

Hoodless, P. A., Haerry, T., Abdollah, S., Stapleton, M., O'Connor, M. B., Attisano, L. and Wrana, J. L. 1996. MADR1, a MAD-related protein that functions in BMP2 signaling pathways. *Cell* 85: 489–500.

Hoppe, P. E. and Greenspan, R. J. 1986. Local function of the *Notch* gene for embryonic ectodermal pathway choice in *Drosophila*. *Cell* 46: 773–783.

Hyatt, G. A. and Beebe, D. C. 1993. Regulation of lens cell growth and polarity by embryonic-specific growth factor and by inhibitors of lens cell proliferation and differentiation. *Development* 117: 701–709.

Ignotz, R. A. and Massague, J. (1986). Transforming growth factor beta stimulates the expression of fibronectin and collagen and their incorporation into the extracellular matrix. *J. Biol. Chem.* 261, 4337–4345.

Jacobson, A. G. 1966. Inductive processes in embryonic development. *Science* 152: 25–34.

Jacobson, A. G. and Sater, A. K. 1988. Features of embryonic induction. *Development* 104: 341–359.

Jernvall, J. 1995. Mammalian molar cusp patterns: Developmental mechanisms of diversity. *Acta Zool. Fennica* 198: 1–61.

Jernvall, J., Kettunen, P., Karavanova, I., Martin, L. B. and Theseleff,I. 1994. Evidence for the role of the enamel knot as a control center in mammalian tooth cusp formation: non-dividing cells express growth stimulating *Fgf-4* gene. *Int. J. Dev. Biol.* 38: 463–469.

Johnson, R. L., Laufer, E., Riddle, R. D. and Tabin, C. 1994. Ectopic expression of *Sonic hedgehog* alters dorsal-ventral patterning of somites. *Cell* 79: 1165–1173.

Johnson, R. L. and ten others. 1996. Human homolog of *patched*, a candidate gene for the basal cell nevus syndrome. *Science* 272: 1668–1671.

Jones, C. M., Armes, N. and Smith, J. C. 1996. Signalling by TGF-β family members: Short-range effects of Xnr-2 and BMP4 contrast with the long range effects of activin. *Curr. Biol.* 6: 1468–1475.

Karavanova, I. D., Dove, L. F., Resau, J. H. and Perantoni, A. O. 1996. Conditioned medium from a rat ureteric bud cell line in combination with βFGF induces complete differentiation of isolated metanephric mesenchyme. *Development.* 122: 4159-4167.

Katz, W. S. and Sternberg, P. W. 1996. Intercellular signalling in *Caenorhabditis elegans* vulval pattern formation. *Semin. Cell Dev. Biol.* 7: 175–183.

Katz, W., Hill, R. J., Clandenin, T. R. and Sternberg, P. W. 1995. Different levels of the *C. elegans* growth factor LIN-3 promote distinct vulval precursor fates. *Cell* 82: 297–307.

Kenyon, C. 1995. A perfect vulva every time: gradients and signaling cascades in C. elegans. *Cell* 82: 171–174.

Kidd, S., Lockett, T. J. and Young, M. W. 1983. The *Notch* locus of *Drosophila melanogaster. Cell* 34: 421–433.

Kimble, J. 1981. Alterations in cell lineage following laser ablation of cells in the somatic gonad of *Caenorhabditis elegans. Dev. Biol.* 87: 286–300.

King, H. D. 1905. Experimental studies on the eye of the frog embryo. *Wilhelm Roux Arch. Entwicklungsmech. Org.* 19: 85–107.

Kingsley, D. M., Bland, A. E., Grubber, J. M., Marker, P. C., Russell, L. B., Copeland, N. G. and Jenkins, N. A. 1994. The mouse *short ear* skeletal morphogenesis locus is associated with defects in a bone morphogenesis protein of the TGF-β superfamily. *Cell* 71: 399–410.

Kispert, A., Vainio, S., Shen, L., Rowitch, D. R. and McMahon, A. P. 1996. Proteoglycans are required for maintenance of Wnt-11 expression in the ureter tips. *Development* 122: 3627-3637.

Koga, M. and Ohshima, Y. 1995. Mosaic analysis of the *let-23* gene function in vulval induction of *Caenorhabditis elegans. Development* 121: 2655–2666.

Kollar, E. J. and Baird, G. 1970. Tissue interaction in developing mouse tooth germs. II. The inductive role of the dental papilla. *J. Embryol. Exp. Morphol.* 24: 173–186.

Koseki, C., Herzlinger, D. and Al-Auqati, Q. 1992. Apoptosis in metanephric development. *J. Cell Biol.* 119: 1327–1333.

Koyama, E. and ten others. 1996. Polarizing activity, *sonic hedgehog*, and tooth development in embryonic and postnatal mouse. *Dev. Dyn.* 206: 59–72.

Kratochwil, K., Dull, M., Farinas, I., Galceran, J. and Grosschedl, R. 1996. *Lef1* expression is activated by BMP-4 and regulates inductive interactions in tooth and hair development. *Genes Dev.* 10: 1382–1394.

Kreidberg, J. A., Sariola, H., Loring, J. M., Maeda, M., Pelletier, J., Housman, D. and Jaenisch, R. 1993. *WT-1* is required for early kidney development. *Cell* 74: 679–691.

Ku, M. and Melton, D. A. 1993. *Xwnt-11,* a maternally expressed *Xenopus wnt* gene. *Development* 119: 1161–1173.

Lappi, D. A. 1995. Tumor targeting through fibroblast growth factor receptors. *Semin. Cancer Biol.* 6: 279–288.

Lawson, K. A. 1974. Mesenchyme specificity in rodent salivary gland development: the response of salivary epithelium to lung mesenchyme in vitro. *J. Embryol. Exp. Morphol.* 32: 469–493.

Le Douarin, N. and Tiellet, M.-A. 1974. Experimental analysis of the migration and differentiation of neuroblasts of the autonomic nervous system and of neuroectodermal derivatives, using a biological cell marking technique. *Dev. Biol.* 41: 162–184.

Lehmann, R., Jimenez, F., Dietrich, U. and Campos-Ortega, J. A. 1983. On the phenotype and development of mutants of early neurogenesis in *Drosophila melanogaster. Wilhelm Roux Arch. Dev. Biol.* 192: 62–74.

Lehtonen, E., Virtanen, I. and Saxén, L. 1985. Reorganization of the intermediate cytoskeleton in induced mesenchyme cells is independent of tubule morphogenesis. *Dev. Biol.* 108: 481–490.

Letterio, J. J., Geiser, A. G., Kulkarni, A. B., Roche, A. B., Sporn, N. S. and Roberts, A. B. 1994. Maternal rescue of TGF-β1-null mice. *Science* 264: 1936–1938.

Levin, M., Johnson, R. L., Stern, C., Kuehn, M. and Tabin, C. 1995. A molecular pathway determining left-right asymmetry in chick embryogenesis. *Cell* 82: 803–814.

Lewis, W. 1904. Experimental studies on the development of the eye in amphibia. I. On the origin of the lens, *Rana palustris. Am. J. Anat.* 3: 505–536.

Lewis, W. 1907. Experimental studies on the development of the eye in amphibia. III. On the origin and differentiation of the lens. *Am. J. Anat.* 6: 473–509.

Liu, A., Ventura, F., Doody, J. and Maasague, J. 1995. Human Type II receptor for bone morphogenesis proteins (BMPs): Extension of the two-kinase receptor model to the BMPs. *Mol. Cell Biol.* 15: 3479–3486.

Liu, F., Hata, A., Baker, J. C., Doody, J., Cárcamo, J., Harland, R. M. and Massagué, J. 1996. A human Mad protein acting as a BMP-regulated transcription factor. *Nature* 381: 620–623.

Lumsden, A. G. S. 1988. Spatial organization of the epithelium and the role of neural crest cells in the initiation of the mammalian tooth germ. *Development* 103 [Suppl.]: 155–169.

Luo, G., Hofmann, C., Bronckers, A. L. J. J., Sohocki, M., Bradley and Karsenty, G. 1995. BMP-7 is an inducer of nephrogenesis and is also required for eye development and skeletal patterning. *Genes Dev.* 9: 2808–2820.

Ma, C., Zhou, Y., Beachy, P. A. and Moses, K. 1993. The segment polarity gene hedgehog is required for progression of the morphogenetic furrow in the developing *Drosophila* retina. *Cell* 75: 927–938.

Mackie, E. J., Thesleff, I. and Chiquet-Ehrismann, R. 1987. Tenascin is associated with chondrogenic and osteogenic differentiation *in vivo* and promotes chondrogenesis in vivo. *J. Cell Biol.* 105: 2569–2579.

McCormick, F. 1989. *ras* GTPase activating protein: Signal transmitter and signal terminator. *Cell* 56: 5–8.

McMahon, A. P. and Bradley, A. 1990. The *Wnt-1 (int-1)* proto-oncogene is required for the development of a large region of the mouse brain. *Cell* 62: 1073–1085.

Meier, S. 1977. Initiation of corneal differentiation prior to cornea-lens association. *Cell Tissue Res.* 184: 255–267.

Mencl, E. 1908. Neue Tatsachen zur Selbstdifferenzierung der Augenlinse. *Wilhelm Roux Arch. Entwicklungsmech. Org.* 25: 431–450.

Mina, M. and Kollar, E. J. 1987. The induction of odontogenesis in non-dental mesenchyme combined with early murine mandibular arch epithelium. *Arch. Oral Biol.* 32: 123–127.

Mizuno, T. and Yasugi, S. 1990. Susceptibility of epithelia to directive influences of mesenchymes during organogenesis: Uncoupling of morphogenesis and cytodifferentiation. *Cell Differ. Devel.* 31, 151–159.

Mlodzik, M., Hiromi, Y., Weber, U., Goodman, C. S. and Rubin, G. M. 1990. The *Drosophila seven-up* gene, a member of the steroid receptor gene superfamily, controls photoreceptor cell fates. *Cell* 60: 211–224.

Morales, T. I. and Roberts, A. B. 1988. Transforming growth factor β regulates the metabolism of proteoglycans in bovine cartilage organ cultures. *J. Biol. Chem.* 263, 12828–12831.

Moore, M. W. and nine others. 1996. Renal and neuronal abnormalities in mice lacking GDNF. *Nature* 382: 76–79.

Muenke, M and Schell, U. 1995. Fibroblast growth factor receptor mutations in human skeletal disorders. *Trends Genet.* 11: 308–313.

Nakamura, T. Okuda, S., Miller, D., Ruoslahti, E. and Border, W. 1990. Transforming growth factor-β (TGF-β) regulates production of extracellular matrix (ECM) components by glomerular epithelial cells. *Kidney Int.* 37, 221.

Nakanishi, Y., Sugiura, F., Kishi, J.-I. and Hayakawa, T. 1986a. Collagenase inhibitor stimulates cleft formation during early morphogenesis of mouse salivary gland. *Dev. Biol.* 113: 201–206.

Nakanishi, Y., Sugiura, F., Kishi, J.-I. and Hayakawa, T. 1986b. Scanning electron microscopic observations of mouse embryonic submandibular glands during initial branching: Preferential localization of fibrillar structures at the mesenchymal ridges participating in cleft formation. *J. Embryol. Exp. Morphol.* 96: 65–77.

Nakanishi, Y., Morita, T. and Nogawa, H. 1987. Cell proliferation is not required for the initiation of early cleft formation in mouse embryonic submandibular epithelium in vitro. *Development* 99: 429–437.

Nakanishi, Y., Nogawa, H., Hashimoto, Y., Kishi, J.-I. and Hayakawa, T. 1988. Accumulation of collagen III at the cleft points of developing mouse submandibular gland. *Development* 104: 51–59.

Nakanishi, Y., Uematsu, J., Takamatsu, H., Fukuda, Y. and Yoshida, K. 1993. Removal of heperan sulfate chains halted epithelial branching morphogenesis of developing mouse submandibular gland in vitro. *Dev. Growth Differ.* 35: 371–384.

Nieuwkoop, P. 1952. Activation and organization of the central nervous system in amphibians. *J. Exp. Zool.* 120: 1–108.

Nohno, T. W. Kawakami, Y., Ohuchi, H., Fujiwara, A., Yoshioka, H. and Noji, S. 1995. Involvement of the *sonic hedgehog* gene in chick feather formation. *Biochem. Biophys. Res. Comm.* 206: 33–39.

Osathanondh, V. and Potter, E. 1963. Development of human kidney as shown by microdissection. III. Formation and interrelationships of collecting tubules and nephrons. *Arch. Pathol.* 76: 290–302.

Padgett, R. W., Wozney, J. M. and Gelbart, W. M. 1993. Human BMP sequences can confer normal dorsal-ventral patterning in the *Drosophila* embryo. *Proc. Natl. Acad. Sci. USA* 90: 2905–2909.

Penttinen, R. P., Kobayashi, S. and Bornstein, P. 1988. Transforming growth factor-β increases mRNA for matrix proteins in the presence and in the absence of the changes in mRNA stability. *Proc. Natl. Acad. Sci. USA* 85: 1105–1108.

Perantoni, A. O., Dove, L. F. and Karavanova, I. 1995. Basic fibroblast growth factor can mediate the early inductive events in renal development. *Proc. Natl. Acad. Sci USA* 92: 4696–4700.

Peters, K., Werner, S., Liao, S., Wert, S., Whitsett, J. A. and Williams, L. T. 1994. Targeted expression of a dominant negative FGF receptor blocks branching morphogenesis and epithelial differentiation of the mouse lung. *EMBO J.* 13: 3296–3301.

Pichel, J. G.and eleven others. 1996. Defects in enteric innervation and kidney development in mice lacking GDNF. *Nature* 382: 73–76.

Placzek, M., Tessier-Lavigne, M., Yamada, T., Jessell, T. and Dodd, J. 1990. Mesodermal control of neural cell identity: Floor plate induction by the notochord. *Science* 250: 985–988.

Post, M., Souza, P., Liu, J., Tseu, I., Wang, J., Kuliszewski, M. and Tanswell, A. K. 1996. Keratinocyte growth factor and its receptor are involved in regulating early lung branching. *Development* 122: 3107–3115.

Poulson, D. F. 1937. Chromosomal deficiencies and the embryonic development of *Drosophila melanogaster*. *Proc. Natl. Acad. Sci. USA* 23: 133–137.

Pritchard-Jones, K. and eleven others. 1990. The candidate Wilms' tumour gene is involved in genitourinary development. *Nature* 346: 194–197.

Reilly, K. M. and Melton, D. A. 1996. Short-range signaling by candidate morphogens of the TGF-β family and evidence for a relay mechanism of induction. *Cell* 86: 743–754.

Reinke, R. and Zipursky, A. L. 1988. Cell-cell interaction in the *Drosophila* retina: The *bride of sevenless* gene is required in photoreceptor cell R8 for R7 cell development. *Cell* 55: 321–330.

Richardson, J., Cvekl, A and Wistow, G. 1995. Pax-6 is essential for lens-specific expression of ζ-crystallin. *Proc. Natl. Acad. Sci. USA* 92: 4674–4680.

Riddle, R. D., Johnson, R. L., Laufer, E. and Tabin, C. 1993. *Sonic hedgehog* mediates the polarizing activity of the ZPA. *Cell* 75: 1401–1416.

Ritvos, O., Tuuri, T., Erämaa, M., Sainio, K., Hilden, K., Saxén, L. and Gilbert, S. F. 1995. Activin disrupts epithelial branching morphogenesis in developing murine kidney, pancreas, and salivary gland. *Mech. Dev.* 50: 229–245.

Roberts, D. J., Johnson, R.L. Burke, A. C., Nelson, C. E., Morgan, B. A. and Tabin, C. 1995. Sonic hedgehog is an endodermal signal inducing *Bmp-4* and *Hox* genes during induction and regionalization of the chick hindgut. *Development* 121: 3163–3174.

Romanoff, A. L. 1960. *The Avian Embryo*. Macmillan, New York.

Rothenpieler, U. W. and Dressler, G. R. 1993. Pax-2 is required for mesenchyme-to-epithelium conversion during kidney development. *Development* 119: 711–720.

Rubin, G. M. 1989. Development of the *Drosophila* retina: Inductive events studied at single cell resolution. *Cell* 57: 519–520.

Rutter, W. J., Wessells, N. K. and Grobstein, C. 1964. Controls of specific synthesis in the developing pancreas. *Natl. Cancer Inst. Monogr.* 13: 51–65.

Saha, M. S. 1991. Spemann seen through a lens. *In* S. F. Gilbert (ed.), *A Conceptual History of Modern Embryology*. Plenum, New York, pp. 91–108.

Saha, M. S., Spann, C. L. and Grainger, R. M. 1989. Embryonic lens induction: More than meets the optic vesicle. *Cell Differ. Dev.* 28: 153–172.

Sainio, K., Nonclercq, D., Saarma, M., Palgi, J., Saxén, L. and Sariola, H. 1994. Neuronal characteristics of embryonic renal stroma. *Int. J. Dev. Biol.* 38: 77–84.

Sainio, K. and seven others. 1992. Differential expression of gap junction mRNAs and proteins in the developing murine kidney and in experimentally induced nephric mesenchymes. *Development* 115: 827–837.

Sakakura, T., Nishizuka, Y. and Dawe, C. J. 1976. Mesenchyme-dependent morphogenesis and epithelium-specific cytodifferentiation in mouse mammary gland. *Science* 194, 1439–1441.

Saksela, O., Moscatelli, D. and Rifkin, D. B. 1987. The opposing effects of basic fibroblast growth factor and transforming growth factor β on the regulation of plasminogen activator activity in capillary endothelial cells. *J. Cell Biol.* 105, 957–963.

Samakoulis, C., Hacohen, N., Manning, G., Sutherland, D. C., Guillemin, K. and Krasnow, M. A. 1996. Development of the *Drosophila* tracheal system occurs by a series of morphologically distinct but genetically coupled branching events. *Development* 122: 1395–1407.

Sánchez, M. P., Silos-Santiago, I., Frisén, J., He, B., Lira, S. A. and Barbacid, M. 1996. Renal agenesis and the absence of enteric neurons in mice lacking GDNF. *Nature* 382: 70–73.

Santos, O. F. P., Baras, E. J. G., Yang, X.-M., Matsumoto, K., Nakamura, T., Park, M. and Nigam, S. K. 1994. Involvement of hepatocyte growth factor in kidney development. *Dev. Biol.* 163: 525–529.

Sariola, H., Ekblom, P. and Saxén, L. 1982. Restricted developmental options of the metanephric mesenchyme. *In* M. Burger and R. Weber (eds.), *Embryonic Development*, Part B: *Cellular Aspects*. Alan R. Liss, New York, pp. 425–431.

Sariola, H., Holm-Sainio, K. and Henke-Fahle, S. 1989. The effect of neuronal cells on kidney differentiation. *Int. J. Dev. Biol.* 33: 149–155.

Sariola, H. and seven others. 1991. Dependence of kidney morphogenesis on the expression of nerve growth factor receptor. *Science* 254: 571–573.

Satokata, I. and Maas, R. 1994. *Msx1* deficient mice exhibit cleft palate and abnormalities of craniofacial and tooth development. *Nat. Genet.* 6: 348–355.

Saunders, J. W., Jr. 1980. *Developmental Biology.* Macmillan, New York.

Saunders, J. W., Jr., Cairns, J. M. and Gasseling, M. T. 1957. The role of the apical ectodermal ridge of ectoderm in the differentiation of the morphological structure of and inductive specificity of limb parts of the chick. *J. Morphol.* 101: 57–88.

Saxén, L. 1970. Failure to demonstrate tubule induction in heterologous mesenchyme. *Dev. Biol.* 23: 511–523.

Saxén, L. 1987. *Organogenesis of the Kidney.* Cambridge University Press, Cambridge.

Saxén, L. and Sariola, H. 1987. Early organogenesis of the kidney. *Pediat. Nephrol.* 1: 385–392.

Schuchardt, A., D-Agati, V., Pachnis, V. and Constantini, F. 1996. Renal agenesis and hypodysplasia in *ret-k⁻* mutant mice result from defects in ureteric bud development. *Development* 122: 1919–1929.

Schugar, L., Johnson, G. R., Gilbride, K., Plowman, D. D. and Mandel, R. 1996. Amphiregulin in lung branching morphogenesis: interaction with heparan sulfate proteoglycan modulates cell proliferation. *Development.* 122: 1759–1767.

Schulz, M. W., Chamberlain, C. G., de Longh, R. U. and McAvoy, J. W. 1993. Acidic and basic FGF in ocular media and lens: Implications for lens polarity and growth patterns. *Development* 118: 117–126.

Serra, R., Pelton, R. W. and Moses, H. 1994 TGFβ1 inhibits branching morphogenesis and N-myc expression in lung bud organ cultures. *Development* 120: 2153–2161.

Servetnick, M. and Grainger, R. M. 1991. Changes in neural and lens competence in *Xenopus* ectoderm: Evidence for an autonomous developmental timer. *Development* 112: 177–188.

Seydoux, G. and Greenwald, I. 1989. Cell autonomy of *lin-12* function in a cell fate decision in *C. elegans. Cell* 57: 1237–1245.

Silberstein, G. B., Flanders, K. C., Roberts, A. B. and Daniel, C.W. 1992. Regulation of mammary morphogenesis: Evidence for extracellular matrix inhibition of ducted budding by transforming growth factor β-1. *Dev. Biol.* 152: 354–362.

Silberstein, G. B., Strickland, P., Coleman, S. and Daniel, C. W. 1990. Epithelium-dependent extracellular matrix synthesis in transforming growth factor-β1-growth-inhibited mouse mammary gland. *J. Cell Biol.* 110: 2209–2219.

Simske, J. S. and Kim, S. K. 1995. Sequential signaling during Caenorhabditis elegans vulval induction. *Nature* 375: 142–146.

Souza, P., Kuliszewski, M., Wang, J., Tseu, I., Tanswell, A. K. and Post, M. 1995.

PDGF-AA and its receptor influence early lung branching via an epithelial-mesenchymal interaction. *Development* 121: 2559–2567.

Spemann, H. 1901. Über Correlationen in der Entwicklung des Auges. *Verh. Anat. Ges. 15 Vers. Bonn.* 61–79.

Spemann, H. 1938. *Embryonic Development and Induction.* Yale University Press, New Haven.

Spemann, H. and Schotté, O. 1932. Über xenoplatische Transplantation als Mittel zur Analyse der embryonalen Induktion. *Naturwissenschaften* 20: 463–467.

Stark, K., Vainio, S., Vassileva, G. and McMahon, A. P. 1994. Epithelial transformation of metanehric mesenchyme in the developing kidney regulated by Wnt-4. *Nature* 372: 679–683.

Stern, H. M., Brown, A. M. C. and Hauschka, S. D. 1995. Myogenesis in paraxial mesoderm: preferential induction by dorsal neural tube and by cells expressing Wnt-1. *Development* 121: 3675–3686.

Stern, M. J. and eight others. 1993. The human *GRB2* and *Drosophila drk* genes can functionally replace the *Caenorhabditis elegans* cell signalling gene *sem-5. Mol. Biol. Cell.* 4: 1175–1188.

Sternberg, P. W. 1988. Lateral inhibition during vulval induction in *Caenorhabditis elegans. Nature* 335: 551–554.

Sternberg, P. W. and Horvitz, H. R. 1989. The combined action of two intercellular signalling pathways specifies three cell fates during vulval induction in *C. elegans. Cell* 58: 679–693.

Storm, E. E., Huynh, T. V., Copeland, N. G., Jenkins, N. A., Kingsely, D. M. and Lee, S. J. 1994. Limb alterations of *brachyopodism* mice due to mutations in a new member of the TGF-β superfamily. *Nature* 368: 639–643.

Takada, S., Stark, K. L., Shea, M. J., Vassileva, G., McMahon, J. A. and McMahon, A. P. 1994. *Wnt-3a* regulates somite and tailbud formation in the mouse embryo. *Genes Dev.* 8: 174–189.

Thesleff, I. and Sahlberg, C. 1996. Growth factors as inductive signals regulating tooth morphogenesis. *Semin. Cell Dev. Biol.* 7: 185–193.

Thesleff, I., Vainio, S. and Jalkanen, M. 1989. Cell-matrix interaction in tooth development. *Int J. Dev. Biol.* 33: 91–97.

Thesleff, I., Vaahtokari, A. and Vainio, S. 1990. Molecular changes during determination and differentiation of the dental mesenchyme cell lineage. *J. Biol. Bucalle* 18: 179–188.

Ting-Berreth, S. A. and Chuong, C.-M. 1996. Local delivery of TGFβ2 can substitute for placode epithelium to induce mesenchymal condensation during skin morphogenesis. *Dev. Biol.* 179: 347–359.

Toivonen, S. 1995. Über die Entwicklung der Vor und Uriniere biem Kaninchen. *Ann. Acac. Sci. Fenn. ser. A* 8: 1–27.

Tomlinson, A. 1988. Cellular interactions in the developing *Drosophila* eye. *Development* 104: 183–193.

Tomlinson, A. and Ready, D. F. 1987. Cell fate in the *Drosophila* ommatidium. *Dev. Biol.* 123: 264–275.

Torres, M., Gomez-Pardo, E., Dressler, G. R. and Gruss, P. 1995. Pax2 controls multiple steps of urogenital development. *Development* 121: 4057–4065.

Trupp, M. and twelve others. 1996. Functional receptor for GDNF encoded by the c-*ret* proto-oncogene. *Nature* 381: 785–789.

Vaahtokari, A., Aberg, T., Jernvall, J., Keränen, S. and Thesleff, I. 1996a. The enamel knot as a signalling center in the developing mouse tooth. *Mech. Dev.* 54: 39–43.

Vaahtokari, A., Aberg, T. and Thesleff, I. 1996b. Apoptosis in the developing tooth: association with an embryonic signaling center and suppression by EGF and FGF-4. *Development* 122: 121–129.

Vainio, S., Lehtonen, E., Jalkanen, M., Bernfield, M. and Saxén, L. 1989. Epithelial-mesenchymal interactions regulate the stage-specific expression of a cell surface proteoglycan, syndecan, in the developing kidney. *Dev. Biol.* 134: 382–391.

Vainio, S., Jalkanen, M., Vaahtokari, A., Sahlberg, C., Mali, M., Bernfield, M. and Thesleff, I. 1991. Expression of syndecan gene is induced early, is transient, and correlates with changes in mesenchymal proliferation during tooth organogenesis. *Dev. Biol.* 147: 322–333.

Vainio, S., Jalkanen, M., Bernfield, M. and Saxén, L. 1992. Transient expression of syndecan in mesenchymal cell aggregates of the embryonic kidney. *Dev. Biol.* 152: 221–232.

Vainio, S. Karavanova, I., Jowett, A. and Thesleff, I. 1993. Identification of BMP-4 as a signal mediating secondary induction between epithelial and mesenchymal tissues during early tooth development. *Cell* 75: 45–58.

van Heyningen, V. and eleven others. 1990. Role for Wilms tumor gene in genital development? *Proc. Natl. Acad. Sci. USA* 87: 5383–5386.

Vastardis, H., Karimbux, N., Guthua, S. W., Seidman, J. G. and Seidman, C. E. 1996. A human *MSX1* homeodomain missense mutation causes selective tooth agenesis. *Nat. Genet.* 13: 417–421.von Woellwarth, V. 1961. Die Rolle des neuralleistenmaterials und der Temperatur bei der Determination der Augenlinse. *Embriologia* 6: 219–242.

Waddington, C. H. 1940. *Organisers and Genes.* Cambridge University Press, Cambridge.

Wessells, N. K. 1970. Mammalian lung development: Interactions in formulation and morphogenesis of tracheal buds. *J. Exp. Zool.* 175: 455–466.

Wessells, N. K. 1977. *Tissue Interaction and Development*. Benjamin, Menlo Park, CA.

Wessells, N. K. and Cohen, J. H. 1968. Effects of collagenase on developing epithelia in vitro: lung, ureteric bud and pancreas. *Dev. Biol.* 18: 294–309.

Wilkie, A. O. M., Morriss-Kay, G. M., Jones, E. Y. and Heath, J. K. 1995. Functions of fibroblast growth factors and their receptors. *Curr. Biol.* 5: 500–507.

Wilkinson, D. G., Bhatt, S. and McMahon, A. P. 1989. Expression of the FGF-related proto-oncogene *int-2* suggests multiple roles in fetal development. *Development* 105: 131–136.

Wilkinson, H. A., Fitzgerald, K. and Greenwald, I. 1994. Reciprocal changes in expression of the receptor lin-12 and its ligand lag-2 prior to commitment in a *C. elegans* cell fate decision. *Cell* 79: 1187–1198.

Woolf, A. S. and eight others. 1995. Roles of hepatocyte growth factor/scatter factor and the Met receptor in the early development of the metanephros. *J. Cell Biol.* 128: 171–184.

Wrana, J. L., Overall, C. M. and Sodek, J. 1991. Regulation of a secreted acidic protein rich in cysteine (SPARC) in human fibroblasts by transforming growth factor-β. *Eur. J. Biochem.* 197: 519–528.

Yamada, T., Placzek, M., Tanaka, H., Dodd, J. and Jessell, T. M. 1991. Control of cell pattern in the developing nervous system: Polarizing activity of floor plate and notochord. *Cell* 64: 635–647.

Yamada, T., Pfaff, S. L., Edlund, T. and Jessell, T. M. 1993. Control of cell pattern in the neural tube: Motor neuron induction by diffusible factors from notochord and floor plate. *Cell* 73: 673–686.

Yedvobnick, B., Muskavitch, M. A. T., Wharton, K. A., Halpern, M. E., Paul, E., Grimwade, B. G. and Artavanis-Tsakonas, S. 1985. Molecular genetics of *Drosophila* neurogenesis. *Cold Spring Harbor Symp. Quant. Biol.* 50: 841–854.

Yochem, J., Weston, K. and Greenwald, I. 1988. The *Caenorhabditis elegans lin-12* gene encodes a transmembrane protein with overall similarity to *Drosophila Notch*. *Nature* 335: 547–550.

Development of the tetrapod limb

18

My arms are longer than my legs...I am my own sculptor: I am shaping myself from within with living, wet, malleable materials: what other artist has ever had available to him as perfect a design as the one possessed by my hammers and chisels: the cells move to the exact spot for building an arm: it's the first time they've ever done it, never before and never again, do your mercies benz understand what I'm saying? I will never be repeated.
CARLOS FUENTES (1989)

What can be more curious than that the hand of a man, formed for grasping, that of a mole for digging, the leg of a horse, the paddle of the porpoise, and the wing of the bat should all be constructed on the same pattern and should include similar bones, and in the same relative positions?
CHARLES DARWIN (1859)

Pattern formation in the limb

PATTERN FORMATION is the process by which embryonic cells form ordered spatial arrangements of differentiated tissues. The ability to carry out this process is one of the most dramatic properties of developing organisms and one that has provoked a sense of awe in scientists and laypeople alike. How is it that the embryo is able not only to generate the different cell types of the body, but also to produce them in a way that forms functional tissues and organs? It is one thing to differentiate the chondrocytes and osteocytes that synthesize the cartilage and bone matrices, respectively; it is another thing to produce these cells in a temporal-spatial orientation that generates a functional bone. It is still another thing to make that bone a humerus and not a pelvis or a femur. The ability of limb cells to sense their relative positions and to differentiate with regard to their position has been the subject of intense debate and experimentation. How are the cells that differentiate into the cartilage of the embryonic bone specified so as to form digits at one end and a shoulder at the other? (It would be quite a useless appendage if the order were reversed.) Here the cell types are the same, but the patterns they form are different.

The vertebrate limb is an extremely complex organ with an asymmetrical pattern of parts. The bones of the forelimb, be it wing, hand, flipper, or fin, consist of a proximal humerus (adjacent to the body wall), a radius and an ulna in the middle region, and the distal bones of the wrist and the digits (Figure 18.1). Originally, these structures are cartilaginous, but eventually, most of the cartilage is replaced by bone. The position of each of the bones and muscles in the limb is precisely organized. Polarity exists in other dimensions as well. In humans, it is obvious that each hand develops as a mirror image of the other. It is possible for other arrangements to exist—such as the thumb developing on the left side of both hands—but this is not generally seen. Similarly, the palm (ventral) is readily distinguishable from the wrist (dorsal). In some manner, the three-dimensional pattern of forelimb is routinely produced. The fundamental problem of morphogenesis—how specific structures arise in particular places—is exemplified in limb development. How is it that one part of the lateral plate mesoderm develops limb-forming capacities? How is it that the fingers form at one end of the limb

Figure 18.1
Skeletal pattern of the chick wing. According to convention, digits are numbered II, III, and IV. Digits I and V are not found in chick wings. (After Saunders, 1982.)

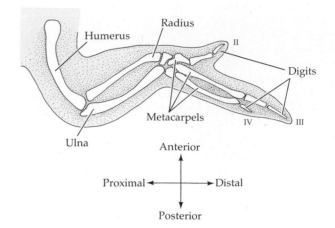

and nowhere else? How is it that the little finger develops at one edge of the limb and the big finger at the other?

The basic "morphogenetic rules" for forming a limb appear to be the same in all tetrapods (see Hinchliffe, 1991). Fallon and Crosby (1977) showed that grafted pieces of reptile or mammalian limb bud can direct the formation of chick limbs, and Sessions and co-workers (1989) have found that regions of frog and salamander limb buds can direct the patterning of each other's limbs. Moreover, the *regeneration* of salamander limbs appears to follow the same rules as those of developing limbs (Muneoka and Bryant, 1982). But what are these morphogenetic rules?

The positional information needed to construct a limb has to function in a three-dimensional coordinate system.* During the past five years, particular proteins have been identified that play a role in the formation of each of the limb axes. The proximal-distal (shoulder-finger; hip-toe) growth appears to be regulated by the **fibroblast growth factor** (**FGF**) family of proteins. The anterior-posterior (thumb-pinky) axis seems to be regulated by the **Sonic hedgehog** protein, and the dorsal-ventral (knuckle-palm) axis is regulated, at least in part, by **Wnt7a**. The interactions of these proteins determine the differentiation of the cell types and also mutually support each other.

Formation of the limb bud

The Limb Field

A **morphogenetic field** can be described as a group of cells whose position and fate are specified with respect to the same set of boundaries (Weiss, 1939; Wolpert, 1977). A particular field of cells will give rise to its particular organ (forelimb, eye, tail, etc.) when transplanted to a different part of the embryo, and the cells of the field can regulate their fates to make up for missing cells in the field (Huxley and De Beer, 1934; Opitz, 1985; De Robertis et al., 1991). One of the first fields identified was the limb field.

The mesodermal cells that give rise to the vertebrate limb can be identified by (1) removing certain groups of cells and observing whether a limb develops in their absence (Detwiler, 1918; Harrison, 1918), (2) transplanting certain groups of cells to new locations and observing whether they form a limb (Hertwig, 1925), and (3) marking groups of cells with dyes or radioactive

---

*Actually, it is a four-dimensional system in which time is the fourth axis. Developmental biologists get used to seeing nature in four dimensions.

precursors and observing which descendants of marked cells partake in limb development (Rosenquist, 1971). By these procedures, the prospective limb area has been precisely localized in many vertebrate embryos. Figure 18.2 shows the prospective forelimb area in the tailbud stage of the salamander *Ambystoma maculatum*. The center of this disc is normally destined to give rise to the limb itself. Adjacent to it are the cells that will form the peribrachial flank tissue and the shoulder girdle. These two regions encompass the classic "limb disc" used in the experiments cited in this chapter. However, if all these cells are extirpated from the embryo, a limb will still form, albeit somewhat later, from an additional ring of cells that surrounds this area. If this last ring of cells is included in the extirpated tissue, no limb will develop. This larger region, representing all the cells in the area capable of forming a limb, is called the **limb field.**

The limb field originally has the ability to regulate for lost or added parts. In the tailbud-stage *Ambystoma*, any half of the limb disc is able to regenerate the entire limb when grafted to a new site (Harrison, 1918). This potential can also be shown by splitting the limb disc vertically into two or more segments and placing thin barriers between these segments to prevent their reunion. When this is done, each part develops into a full limb. The regulative ability of the limb bud has recently been highlighted by a remarkable experiment of nature. In a pond in Santa Cruz, California, numerous multilegged frogs and salamanders have been found (Figure 18.3). The presence of these extra appendages has been linked to the infestation of the larval abdomen with parasitic trematode worms. The eggs of these worms apparently split the limb bud in several places while the tadpole was first forming these structures (Sessions and Ruth, 1990). Thus, like an early sea urchin embryo, the limb field represents a "harmonious equipotential system" wherein a cell can be instructed to form any part of the limb.

Specification of the Limb Fields: **Hox** *Genes and Retinoic Acid*

Limbs will not form just anywhere along the body axis. Rather, there are very discrete positions where the limb fields are generated. Interestingly, in all vertebrates there are only four limb buds per embryo, and they are always opposite each other with respect to the midline. Although limbs of different vertebrates may differ with respect to which somite level they arise from, their position is constant with respect to the level of *Hox* gene expression along the anterior-posterior axis. For instance, in fish (where the pectoral and pelvic fins correspond to the anterior and posterior limbs, respectively), amphibians, birds, and mammals, the forelimb buds are found at the most anterior expressing region of *Hoxc-6*, the position of the first thoracic vertebra (Oliver et al., 1988; Molven et al., 1990; Burke et al., 1995). The lateral plate mesoderm in the region of the limbs is also special in that it will induce myoblasts to come out from the somites and enter into the limb bud. No other region of the lateral plate mesoderm will do that (Hayashi and Ozawa, 1995). [limb1.html], [mesend1.html]

Retinoic acid appears to be critical for the initiation of limb bud outgrowth, since blocking the synthesis of retinoic acid with certain drugs prevents limb bud initiation (Stratford et al., 1996). Bryant and Gardiner (1992) suggest that a gradient of retinoic acid along the anterior-posterior axis might activate certain homeotic genes in particular cells and thereby specify them to become included in the limb field. The source of this retinoic acid would be Hensen's node (Hogan et al., 1992). The specification of a limb field by retinoic-acid-activated *Hox* genes might explain a bizarre observation made by Mohanty-Hejmadi and colleagues (1992) and repeated by Maden (1993). When *tails* of tadpoles were amputated and the stump ex-

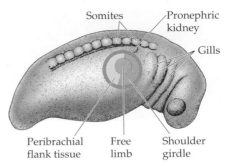

Figure 18.2
Prospective forelimb field of the salamander *Ambystoma maculatum*. The central area contains those cells destined to form the limb per se; the cells surrounding the free limb are those that give rise to the peribrachial flank tissue and the shoulder girdle. The cells outside these regions usually are not included in limbs but can form a limb if the more central tissues are extirpated. (After Stocum and Fallon, 1982.)

Figure 18.3
Regulative ability of the limb field, seen when the early hindlimb fields of a *Hyla regila* tadpole were split by numerous trematode eggs. (Courtesy of S. Sessions.)

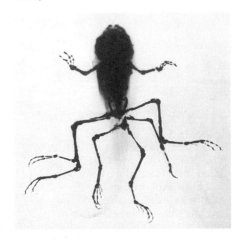

Figure 18.4
Legs regenerating from the tail blastema of a marbled balloon frog tadpole. The tail blastema had been treated with retinoic acid after amputation. (From Mohanty-Hejmadi et al., 1992, courtesy of P. Mohanty-Hejmadi.)

posed to retinoic acid during the first days of regeneration, these tadpoles regenerated several *legs* from their tail stump (Figure 18.4). It appears that the retinoic acid may have caused a homeotic transformation in the regenerating tail by respecifying the tail tissue into limb fields (Müller et al., 1996).

Growth of the Early Limb Bud: Fibroblast Growth Factors as Inducers of the Limb Bud

Limb development begins when mesenchyme cells proliferate from the somatic layer of the limb field lateral plate mesoderm (limb *skeletal* precursors) and from the somites (limb *muscle* precursors) (Figure 18.5). The cells accumulate under the epidermal tissue of the neurula. The circular bulge on the surface of the embryo is called the **limb bud.** The mesenchyme cells in the limb bud multiply to create a bulge that will proliferate to form a limb. The initial stages of this proliferation may be under the regulation of nearby intermediate mesoderm such as the mesonephros (the primitive kidney). If the mesonephros on one side of the embryo is removed at this stage or if a thin impermeable membrane is imposed between the mesonephros and a limb bud, the limb mesenchyme cells of that particular limb bud stop dividing (Stephens et al., 1991; Geduspan and Solursh, 1992).

Recent experiments (Crossley et al., 1996) suggest that the molecule coming from the intermediate mesoderm is **fibroblast growth factor 8** (**FGF8**). In the chick mesonephric mesenchyme, at stages 14 and 15 (when the prospective limb buds are first apparent), the regions of FGF8 expression in the mesonephros coincide with the regions where limb buds will form (Figure 18.6). Moreover, FGFs can induce limb formation. A bead soaked in FGF8 (or related FGFs) can be inserted into the "interlimb" region (opposite somites 21–25; see Figures 18.6A and 18.7) at stage 15. After a week's incubation, an ectopic limb forms there. [limb2.html]

Induction of the Apical Ectodermal Ridge

The ability of FGF8 (or other FGFs) to induce the mesodermal outgrowths of the early limb bud may only be a permissive, rather than an instructive, activity. The formation of the limb bud needs not only an active mesodermal inducer but also a competent ectoderm. The ectoderm competent to form a limb bud appears to reside solely at the border between the dorsal and ventral surfaces of the embryo.

Figure 18.5
Limb bud formation. Proliferation of mesodermal cells from the somatic region of the lateral plate mesoderm causes the limb bud in the amphibian embryo to bulge outward. These cells generate the skeletal elements of the limb. (Migration of somitic cells into the limb bud generates limb musculature.)

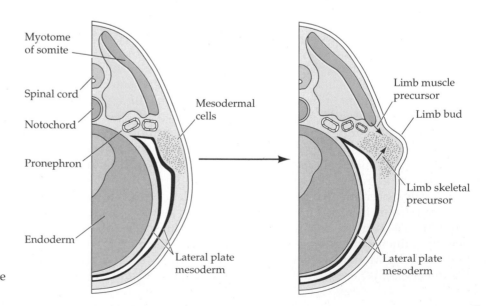

As the limb bud forms, the mesodermal cells induce the overlying ectoderm to form a structure called the **apical ectodermal ridge** (**AER**; Figure 18.8; Kieny, 1960; Saunders and Reuss, 1974). This ridge runs along the distal margin of the limb bud and will become a major signaling center for the developing limb. Its roles include (1) maintaining the mesoderm beneath it in a plastic, proliferating phase that enables the linear (proximal-distal) growth of the limb; (2) maintaining the expression of those molecules that generate the anterior-posterior (thumb-pinky) axis; and (3) interacting with the proteins specifying the anterior-posterior and dorsal-ventral axes so that each cell is given instructions on how to differentiate.

The AER is positioned at the boundary between the dorsal and ventral ectoderm. Only the ectoderm at this junction in the early limb bud has the ability to form an AER (Goetinck, 1964; Fraser and Abbott, 1971). In mutants where the limb bud ectoderm is dorsalized (as in the chick mutant *limbless*), the AER fails to form, and limb development ceases (Carrington and Fallon, 1988). Moreover, FGF-soaked beads will not induce an AER when placed beneath the purely dorsal or ventral ectoderm of the back or belly. The dorsal-ventral boundary appears to be critical. Recent experiments (Laufer et al.,1997; Rodriguez and Izpisúa-Belmonte, 1997; Tanaka et al., 1997) demonstrate that the apposition of dorsal and ventral ectoderm from the chick limb bud is necessary to cause the formation of an AER. When dorsal limb bud ectoderm was grafted into the ventral ectoderm of another limb bud, a new AER was formed in addition to the original one (Figure 18.9). It appears that at stage 15 (just prior to limb bud formation), the dorsal ectoderm is synthesising a secreted protein called **Radical fringe.*** As the limb buds emerge (at

---

*So named after the *fringe* gene of *Drosophila*. The search for the vertebrate homologues of the *fringe* gene was motivated by studies (to be discussed in the next chapter) showing that the formation of the wing margin in *Drosophila* is dependent on the expression boundary of this gene. Since *hedgehog* and *wingless* genes appear to be playing roles in both vertebrate and insect limb formation, several laboratories sought the vertebrate *fringe* genes to see if they might create the equivalent of the wing margin, namely, the AER. It had been predicted that expression boundaries between dorsal and ventral regions would be critical in vertebrate and invertebrate limb formation (Bryant et al., 1981; Meinhardt, 1984; Javois and Iten, 1986), but the molecules involved are just now being identified.

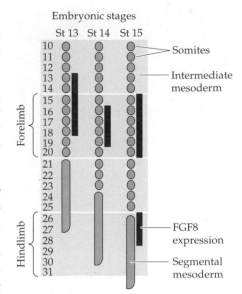

Figure 18.6
FGF8 expression in the intermediate mesoderm of the chick embryo at stages 13–15. Schematic diagram representing a lateral half of the embryo during the time of limb bud induction. The numbers on the left indicate somite levels. (The somites are represented as circles breaking off from the segmental mesoderm, which is portrayed as a white bar.) The shaded stripe indicates the position of the intermediate mesoderm, and FGF8 expression in this intermediate mesoderm is shown by the darker regions of this stripe. The positions of the prospective forelimb and hindlimb buds have been shaded in gray. (After Crossley et al., 1996.)

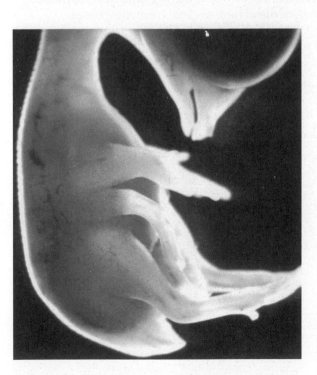

Figure 18.7
Ectopic limb formed by the implantation of an FGF-soaked bead into the interlimb mesoderm at stage 15. Late embryo showing forelimb, hindlimb, and the intermediate limb induced by the FGF-soaked bead (Photograph courtesy of G. R. Martin.)

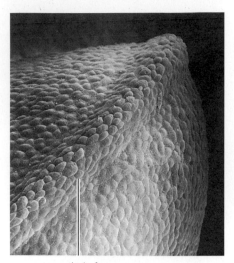

Apical
ectodermal ridge

Figure 18.8
Scanning electron micrograph of an
early chick forelimb bud, with its apical
ectodermal ridge in the foreground.
(Courtesy of K. W. Tosney.)

Figure 18.9
Formation of an ectopic AER when ventral tissue is transplanted into dorsal
limb bud tissue. (A) Operation whereby
ventral ecoderm from a chick hindlimb
bud is transplanted into the dorsal surface of a host hindlimb bud at the same
stage. (B) After 26 hours incubation, an
ectopic AER has formed (the original
AER is denoted by an arrow, the ectopic
AER by an arrowhead). (C) As the AER
forms, *radical fringe* expression (arrowhead) in the limb bud becomes confined
to the dorsal cells at the D/V junction
which will form the AER. (From Laufer
et al., 1997; photographs courtesy of E.
Laufer.)

stage 17), there is a sharp demarcation between the dorsal cells that express
the *radical fringe* gene and the ventral ectoderm cells that do not. As the bud
continues to grow, the expression of *radical fringe* becomes restricted almost
exclusively to those dorsal ectoderm cells at the dorsal/ventral limb bud border. These cells begin to express *Fgf8*, and they become the AER. (As we will
see the FGF8 secreted from the AER is thought to be critical in its ability to
maintain the proliferation of the mesoderm beneath it and to sustain the expression of the *sonic hedgehog* gene to organize the anterior-posterior axis; see
Figure 18.10.)

The importance of the *radical fringe*-expressing and non-expressing border is confirmed by studies wherein this gene is expressed ectopically on
retroviruses. If ventral limb bud cells are infected with a retrovirus expressing *radical fringe*, a new boundary between cells expressing *radical fringe* and
those not expressing it is created, and a new AER is generated there. Conversely, if the ectopic expression of *radical fringe* destroys the existing boundary between expressing and non-expressing cells, that region of the original
AER does not form.

The formation of the AER may involve the interaction between the secretion of FGFs (such as FGF8) from the mesoderm and the boundary of *radical fringe* expression along the dorsal-ventral ectoderm border. The limited
secretion of FGFs may be critical in localizing which cells along the dorsal-
ventral flank of the embryo produce the limb buds. It is still not known how
the *radical fringe* expression/non-expression border and FGFs induce the formation of the AER.

Generation of the proximal distal axis of the limb

The Apical Ectodermal Ridge: The Ectodermal Component

The proximal-distal growth and differentiation of the limb bud is made possible by a series of interactions between the limb bud mesenchyme and the
AER (Figure 18.11; Harrison, 1918; Saunders, 1948):

1. When the AER is removed at any time during limb development,
    further development of distal limb skeletal elements ceases.

Stage 18/19 host   Stage 18/19 donor
leg bud   ectodermal jacket

(A)

27

32

V   D

Ectoderm graft   AER

Somites

(B)

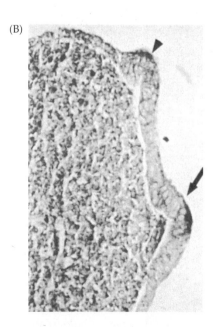

(C)

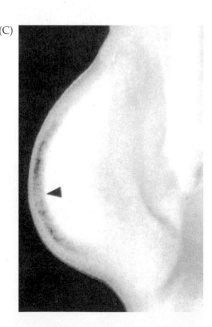

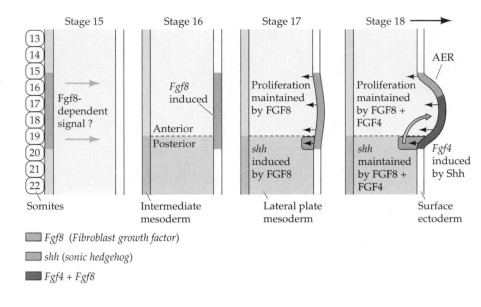

Somites   Intermediate mesoderm   Lateral plate mesoderm   Surface ectoderm

■ *Fgf8* (Fibroblast growth factor)
■ *shh* (sonic hedgehog)
■ *Fgf4* + *Fgf8*

Figure 18.10
A molecular model for the initiation of the limb bud. FGF8 secreted by the intermediate mesoderm and/or *radical fringe* expression in the ectodermal margin induces FGF8 expression in the overlying surface ectoderm. The anterior-posterior boundary is present at stage 16 (and perhaps earlier). The FGF8 secretion by the ectoderm induces the proliferation in the mesenchyme cells and induces Sonic hedgehog expression in the posterior region of the limb bud. Sonic hedgehog induces FGF4 expression in the posterior portion of the limb bud ectoderm. FGF2 is also made by the ectoderm, although it is not yet clear if it is induced by the FGF8 from the mesonephric mesoderm. (After Crossley et al., 1995.)

2. When an extra AER is grafted onto an existing limb bud, supernumerary structures are formed, usually toward the distal end of the limb.
3. When leg mesenchyme is placed directly beneath the wing AER, distal hindlimb structures (toes) develop at the end of the limb. (However, if this mesenchyme is placed farther from the AER, the hindlimb mesenchyme becomes integrated into wing structures.)
4. When nonlimb mesoderm is grafted beneath the AER, the AER regresses and limb development ceases.

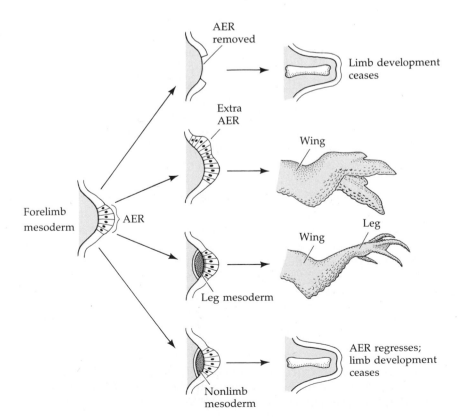

Figure 18.11
Summary of the effect of the apical ectodermal ridge (AER) on the underlying mesenchyme. (Modified from Wessells, 1977.)

Figure 18.12
Cross section through the distal region of a chick limb 3 days after a wedge of the AER was removed from the area that would form interdigital tissue. Instead of degenerating, the remaining interdigital tissue formed an extra digit. (From Hurle et al., 1989, courtesy of the authors.)

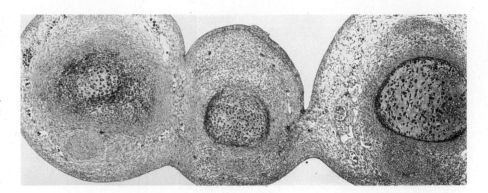

Thus, although the mesenchyme cells induce and sustain the AER and determine the type of limb to be formed, the AER is responsible for the sustained outgrowth and development of the limb (Zwilling, 1955; Saunders et al., 1957; Saunders, 1972; Krabbenhoft and Fallon, 1989). The AER keeps the mesenchyme directly beneath it in a state of mitotic proliferation and prevents the mesenchyme cells from forming cartilage. Hurle and co-workers (1989) found that if they cut away a small portion of the AER in a region that would normally fall between the digits of the chick leg, an extra digit emerged at that place (Figure 18.12). It thus appears that one function of the AER is to keep the mesenchyme cells proliferating and thereby inhibit their forming cartilage.

The Progress Zone: The Mesodermal Component

The proximal-distal axis is defined only after the induction of the apical ectodermal ridge by the underlying mesoderm. The limb bud elongates by the proliferation of the mesenchyme cells underneath the AER. This region of cell division is called the **progress zone,** and it extends about 200 μm in from the AER. Molecules from the AER are thought to keep these progress zone mesenchyme cells dividing, and it is now thought that FGFs are the molecules responsible (Savage and Fallon, 1995; Crossley et al., 1996). When the mesenchyme cells leave the progress zone, they differentiate in a regionally specific manner. The first cells leaving the progress zone form proximal structures; those cells that have undergone numerous divisions in the progress zone become the more distal structures (Saunders, 1948; Summerbell, 1974). Therefore, when the AER is removed from an early-stage wing bud, the cells of the progress zone stop differentiating, and only a humerus forms. When the AER is removed slightly later, humerus, radius, and ulna form (Figure 18.13; Rowe et al., 1982).

Proximal-distal polarity resides in the mesodermal compartment of the limb. If the AER provides the positional information—somehow instructing the undifferentiated mesoderm beneath it as to what structures to make—then older AERs combined with younger mesoderm should produce limbs with deletions in their middle, while younger AERs combined with older mesoderm should produce duplications of structures. This was not found to be the case, however (Rubin and Saunders, 1972). Rather, normal limbs formed in both experiments. But when the entire progress zone, including the mesoderm and AER, from an early embryo was placed on the limb bud of a later-stage embryo, new proximal structures were produced beyond those already present. Conversely, when old progress zones were added to young limb buds, distal structures immediately developed, so that digits were seen to emerge from the humerus without the intervening ulna and radius (Figure 18.14; Summerbell and Lewis, 1975).

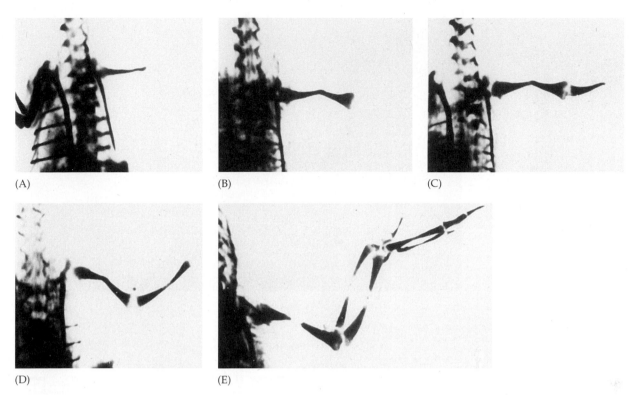

(A)   (B)   (C)

(D)   (E)

Figure 18.13
Dorsal view of chick skeletal pattern after removal of the entire AER from the right wing bud of embryos at various stages. The last photo (E) is of a normal wing skeleton. (From Iten, 1982, courtesy of L. Iten.)

Hox *Genes and the Specification of the Proximal-Distal Limb Axis*

The analysis of naturally occurring and experimentally induced mutations has given rise to the hypothesis that the 5′ (*Abdominal-B*-like) *Hox* genes specify individual portions of the proximal-distal axis of the limb. The 5′ ends of the *Hoxa* and *Hoxd* series of paralogous genes (paralogues 9–13) appear to be active in the forelimb bud of mice. Davis and colleagues (1995) have knocked out all four loci for the paralogous genes *Hoxa-11* and *Hoxd-11*. (There are no *Hoxb-11* genes in mice, and *Hoxc-11* is not well expressed in the forelimb, although it is expressed in the hindlimb.) The resulting mice lacked the ulna and radius of their forelimbs (Figure 18.15A,B). Based on the expression patterns of the *Hoxa* and *Hoxd* series of genes, where the most 5′ genes of the *Hox* clusters are expressed most distally, these researchers pro-

Figure 18.14
Control of proximal-distal specification by the cells of the progress zone (PZ). (A) Extra set of ulna and radius formed when early-bud PZ is transplanted to late wing bud that has already formed ulna and radius. (B) Lack of intermediate structures seen when late-bud PZ is transplanted to early limb bud. The hinges indicate the location of the grafts. (From Summerbell and Lewis, 1975, courtesy of D. Summerbell.)

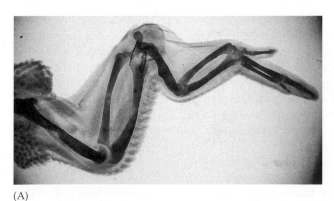

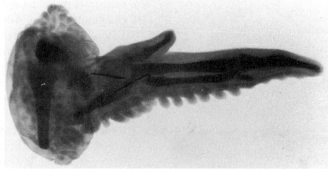

(A)   (B)

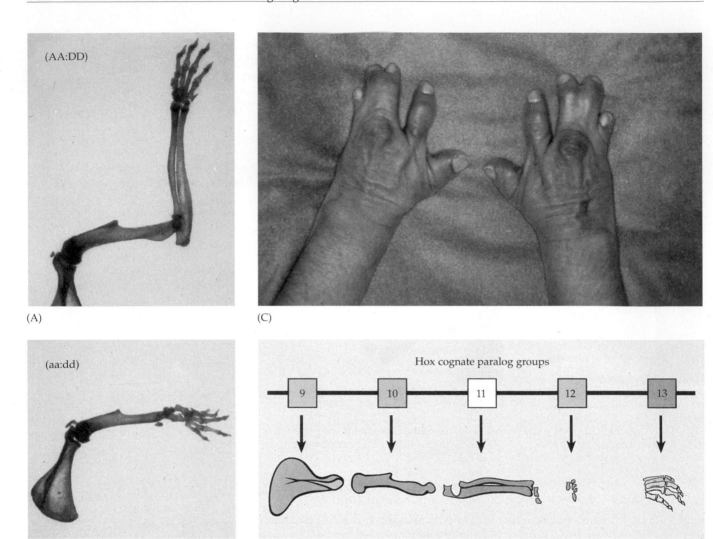

(A)

(AA:DD)

(B)

(aa:dd)

(C)

(D)

Hox cognate paralog groups

9   10   11   12   13

Figure 18.15
Deletion of limb bone elements by the deletion of paralogous *Hox* genes. (A) Wild-type mouse forelimb. (B) Forelimb of mouse made doubly mutant such that it lacked functional *Hoxa-11* and *Hoxd-11* genes. The ulna and radius are absent. (C) Synpolydactyly resulting from homozygosity at the HOXD-13 loci. (D) Hypothesis that the 5′ paralogues of *Hox* genes could specify particular regions of the forelimb. (A,B, and D after Davis et al., 1995; photographs courtesy of M. Capecchi. C from Muragaki et al., 1996, courtesy of B. Olsen.)

posed a model wherein the paralogous genes specify the identity of a limb region (Figure 18.15D). This model makes obvious predictions concerning the phenotype of other doubly and triply knockout mice (when paralogues 13 or 12 are deleted), and research is under way to test this model.

This model is supported by the analysis of two naturally occurring mutations. Mice homozygous for a loss-of-function allele of *Hoxa-13* have a severe malformation of all four paws, which develop only one digit, a malformed version of digit 4 (Mortlock et al., 1996). Humans homozygous for a loss-of-function mutation of *Hoxd-13* show abnormalities of the hands and feet wherein metacarpal and metatarsal bones are transformed into short carpal and tarsal bones. This results in the fusion of digits (Figure 18.15C; Muragaki et al., 1996). In both cases, the autopod (the most distal portion of the limb) is affected by the loss-of-function of the most 5′ *Hox* gene. The mechanism by which *Hox* genes could specify the proximal-distal axis is not yet understood, but one clue comes from the analysis of chicken *Hoxa-13*. Ectopic expression of this gene (which is usually expressed in the distal ends of the developing chick limbs) appears to make the cells expressing it stickier. This, in turn, would cause the cartilaginous nodules to condense in specific ways (Yokouchi et al., 1995; Newman, 1996).

Interactions Between the AER and the Progress Zone

The molecular signals by which the AER and the progress zone mesenchyme interact are starting to be identified. The division of the mesenchyme cells in the progress zone appears to be regulated by secretion of members of the FGF family, such as FGF2 (Fallon et al., 1994), FGF4 (Niswander et al., 1993), and FGF8 (Mahmood et al., 1995; Crossley et al., 1996; Vogel et al., 1996). These fibroblast growth factors are seen to be secreted from the AER into the underlying mesenchyme (see Cover; Figure 18.16). Moreover, if the AER is removed, the implantation of beads filled with FGF2, FGF8 or FGF4 will replace it (Figure 18.17). Thus, it appears that the AER acts to promote growth by secreting fibroblast growth factors (Crossley et al., 1996; Vogel et al., 1996). FGF8 is one of the first molecules seen in the region of the ectoderm that becomes the AER, and its expression is critical in the growth of the limb bud (Figure 18.16).

Mutations in the Interactions between the Progress Zone and the AER

The relationships between the AER and the limb bud mesenchyme can best be seen by mutations of chick limb development. The *polydactylous* mutation, as the name implies, confers extra digits on each limb. By recombining mu-

(A)   (B)   (C)   (D)

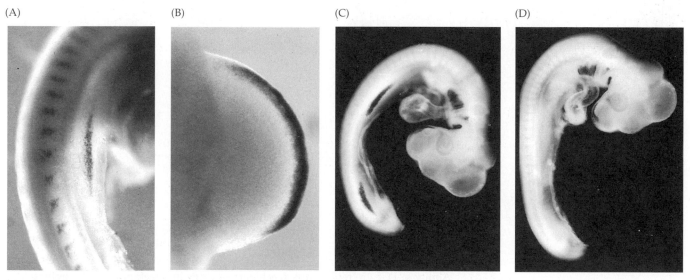

Figure 18.16
FGF8 and limb morphogenesis. (A) In situ hybridization showing expression of *Fgf8* message in the ectoderm as the limb bud begins to form. (B) Expression of *Fgf8* RNA in the apical ectodermal ridge, the source of mitotic signals to the underlying mesoderm. (C) In normal chick embryos (stage 17; roughly 24 hours), FGF8 is expressed in the apical ectodermal ridge of both the forelimb and hindlimb buds. It is also expressed in several other places in the embryo. (D) In the *limbless* mutant of chickens, FGF8 is not expressed in the limb buds, although it is not lost in other embryonic regions. Here, the limb buds form but fail to develop into limbs. (A and B courtesy of J. C. Izpisúa Belmonte; C and D courtesy of A. López-Martínez and J. F. Fallon.)

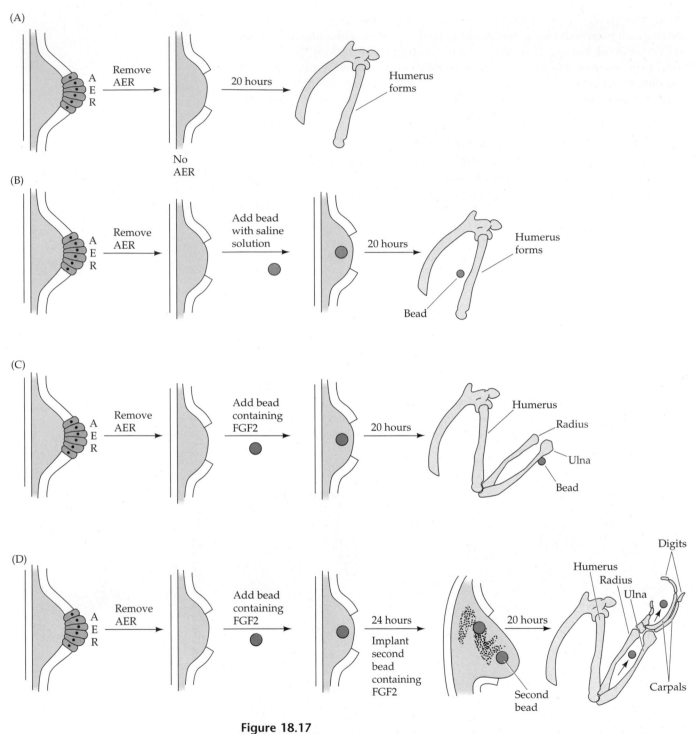

Figure 18.17
Ability of FGF2 to replace apical ectodermal ridge in developing chick forelimb bud. (A) When the AER is removed from stage 20 chick wing buds, only the humerus is able to form. (B) When a slow-release gel bead soaked in saline solution is placed into the progress zone mesenchyme, the limb is still truncated and forms only a humerus. (C) When a bud soaked in FGF2 is placed in the progress zone, growth of the limb bud continues, and the ulna and radius are formed. (D) If a second FGF2-containing bead is placed into the progress zone after most of the FGF2 of the first bead has dissipated, the limb bud continues to grow and to make the metacarpals and digits. (After Fallon et al., 1994.)

Table 18.1 Mutations affecting the reciprocal interactions between AER and its underlying mesenchyme[a]

Mesoderm	Epidermis	Result	Conclusion
POLYDACTYLOUS			
Polydactylous	Wild-type	Polydactylous	Mesoderm is affected
Wild-type	Polydactylous	Wild-type	by the mutation
EUDIPLOPODIA			
Eudiplopodia	Wild-type	Wild-type	Ectoderm is affected
Wild-type	Eudiplopodia	Eudiplopodia	by the mutation
LIMBLESS			
Limbless	Wild-type	Wild-type	Ectoderm is affected
Wild-type	Limbless	Limbless	by the mutation

[a]By reciprocal transplantation between wild-type and mutant AER and mesenchyme, the aberrant compartment of the induction can be identified.

tant and wild-type tissues (Table 18.1), the defect can be traced to mesodermal cells that induce too broad an AER. In the mutant *eudiplopodia* (Greek, "two good feet"), there are not only extra digits, but two complete rows of toes on each hindlimb (Figure 18.18). Similar reconstitution experiments show that here the defect is in the ectodermal tissue. Chick embryos homozygous for the *limbless* mutation initiate the limb bud formation, but the AER fails to form. Recombination experiments show that the *limbless* ectoderm is unable to form an AER, even when placed on wild-type limb mesoderm; a normal ridge can form when normal ectoderm is grafted to the limb field in place of the mutant ectoderm (Figure 18.19; Carrington and Fallon, 1988).

In addition, there are some naturally limbless vertebrates whose lack of limb formation can be traced to deficiencies in the AER-mesenchyme interactions. The curse against snakes in the Book of Genesis seems to have been directed at the distal end of the limb bud, since the AER of these reptiles degenerates prematurely and at the same time as cell death in the underlying mesenchyme (Lande, 1978). It is not known whether the initial defect is in the mesenchyme or the AER. [limb3.html]

(A)

(B)

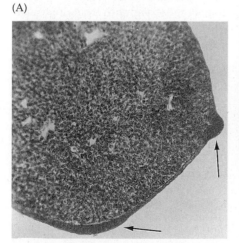 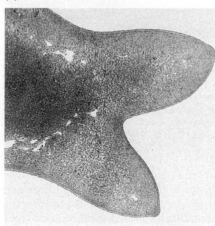

Figure 18.18
Cross sections of hindlimb buds from *eudiplopodia* chick embryos. (A) Two AERs on hindlimb bud; extra outgrowth on the dorsal side will form an extra set of toes. (B) Both outgrowth regions are covered by an AER. It has recently been shown (Laufer et al., 1997) that two patches of *radical fringe* emerge on the limb buds of this mutant, and each becomes associated with the new AER. (From Goetinck, 1964, courtesy of P. Goetinck.)

Figure 18.19
The limbless embryo fails to form AER, and the defect appears to reside in the ectoderm. If wild-type quail ectoderm replaces the mutant chick ectoderm in the region that forms a forelimb, the wing will develop on that side of the embryo. No other limb forms. (After Carrington and Fallon, 1988; photograph courtesy of J. Fallon.)

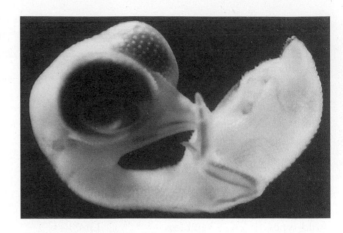

Sidelights & Speculations

The Regeneration of Salamander Limbs and the Retention of the Proximal-Distal Axis

IT IS OFTEN USEFUL to find adult models of embryonic development. For two centuries, the regeneration of the amphibian limb has been not only one of the most awesome instances of regulation, but also a model for vertebrate development of the tetrapod limb. When a salamander limb is amputated, the remaining cells are able to reconstruct a complete limb with all its differentiated cells arranged in the proper order. It is remarkable that not only has the limb regenerated, but the remaining cells retain the information specifying their position and the position of the cells that have been removed. In other words, the new cells construct only the missing structures and no more; for example, when a wrist is amputated, the salamander forms a new wrist and not a new elbow (Figure 18.20). In some way, the salamander limb "knows" where the proximal-distal axis has been severed and is able to regenerate from that point on. Oscar Schotté declared that he would give his right arm to know the secret of limb regeneration (in Goss, 1991).

Upon limb amputation, a plasma clot forms; and within 6–12 hours, epidermal cells from the remaining stump migrate to cover the wound surface, forming the **wound epidermis.** This single-layered

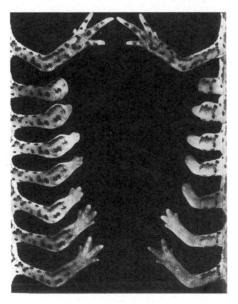

Figure 18.20 Regeneration of a salamander forelimb. On the left, the amputation was made below the elbow; the amputation shown on the right cut through the humerus. In both cases, the correct positional information is respecified. (From Goss, 1969, courtesy of R. J. Goss.)

structure is required for the regeneration of the limb. It proliferates to form the **apical ectodermal cap.** The nerves innervating the limb degenerate for a short distance proximal to the plane of amputation (see Chernoff and Stocum, 1995). During the next 4 days, the cells beneath the developing cap undergo a dramatic **dedifferentiation:** bone cells, cartilage cells, fibroblasts, myocytes, and neural cells lose their differentiated characteristics and become detached from each other. Genes that are expressed in differentiated tissues (such as the *MRF4* and *Myf5* genes expressed in the muscle cells) are downregulated, while there is a dramatic increase in the expression of genes such as *msx1*, which are associated with the proliferating progress zone mesenchyme (Simon et al., 1995). The well-structured limb region at the cut edge of the stump thus forms a proliferating mass of indistinguishable, dedifferentiated cells just beneath the apical ectodermal cap. This dedifferentiated cell mass is called the **regeneration blastema**, and these cells will continue to proliferate and differentiate to form the new structures of the limb. If the blastema cells are destroyed, no regeneration takes place (Butler, 1935). Moreover, once the cells have dedifferentiated to form a blastema, they have regained their embryonic plasticity.

In most instances, however, neural tissue is essential for the formation of the new limb by other cells. Singer (1954) demonstrated that a minimum number

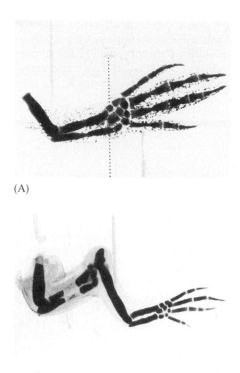

(A)

(B)

Figure 18.21 Effects of vitamin A (a retinoid) on regenerating salamander limbs. (A) Normal regenerated axolotl limb (9×) with humerus, paired radius and ulna, carpals, and digits. Dotted line shows plane of amputation. (B) Regeneration after amputation through the carpal area, but after the limb blastema had been placed in retinol palmitate for 15 days. A new humerus, ulna, radius, carpal set, and digit set emerge (5×). (From Maden et al., 1982; photographs courtesy of M. Maden.)

of nerve fibers must be present for regeneration to take place. It is thought that the neurons release a mitosis-stimulating factor that increases the proliferation of the blastema cells (Singer and Caston, 1972; Mescher and Tassava, 1975). After an initial neuron-dependent phase, regeneration can proceed without neural stimulation. One candidate for this crucial neural substance is **glial growth factor (GGF)**. This peptide is known to be produced by newt neural cells, is present in the blastema, and is lost upon denervation. When GGF is added to a denervated blastema, the mitotically arrested cells are able to divide again (Brockes and Kinter, 1986). Another candidate is **transferrin,** an iron-transport protein that is necessary for mitosis in all dividing cells (since ribonucleotide reductase, the rate-limiting enzyme of DNA synthesis, re-

quires a ferric ion in its active site). When hindlimbs are severed, the sciatic nerve transports transferrin along the axon and releases large quantities of this protein into the blastema (Munaim and Mescher, 1986; Mescher, 1992). Neural extracts and transferrin are both able to stimulate cell division in denervated limbs, and chelation of ferric ions from the neural extracts abolishes their mitotic activity (Munaim and Mescher, 1986; Albert and Boilly, 1988). A third candidate is **FGF2**. Mullen and colleagues (1996) have shown that the apical ectodermal cap transcribes large amounts of *Dlx3*, an amphibian homologue of *Drosophila Distal-less*. During the neuron-dependent stages of regeneration, ectodermal *Dlx3* expression is dependent on innervation. The presence of *Dlx3* correlates with an outgrowth-permissive epidermis. In the late stages of regeneration, *Dlx3* expression does not depend on neurons. If amputated limbs are denervated at a neuron-dependent stage, *Dlx3* expression and regeneration can be maintained by beads containing FGF2 (Mullen et al., 1996).

Retinoic acid appears to play an important role both in the dedifferentiation of the cells to form the regeneration blastema and in the respecification processes as the cells redifferentiate. If the blastemas of regenerating salamander limbs are soaked in sufficient concentrations of retinoic acid (or other retinoids), the regenerating limbs have duplications along the proximal-distal axis (Figure 18.21; Niazi and Saxena, 1978; Maden, 1982). A complete limb (starting at the most proximal bone) grows out of the limb stump regardless of the original level of amputation. It appears that the retinoic acid causes the cells to be respecified to the most proxi-

mal position (Figure 18.22; Plate 9; Crawford and Stocum, 1988b).

Retinoic acid is synthesized in the regenerating limb wound epidermis and is seen to form a gradient along the proximal-distal axis of the blastema (Brockes, 1992; Scadding and Maden, 1994). This gradient of retinoic acid can activate genes differentially in different regions of the blastema. One of these retinoic-acid-responsive genes is the *msx1* gene that is associated with mesenchyme proliferation (Shen et al., 1994; Viviano et al., 1995). Another set of genes that may be respecified by retinoic acid are the *HoxA* genes. Gardiner and colleagues (1995) have shown that the expression pattern of certain *HoxA* genes in the *distal* cells of the regeneration blastema is changed by exogenous retinoic acid into an expression pattern characteristic of more proximal cells. It is probable that during normal regeneration, the wound epidermis/apical ectodermal cap secretes retinoic acid, which activates those genes needed for cell proliferation, downregulates the genes that are specific for differentiated cells, and activates a set of *Hox* genes that tells the cells where they are in the limb and how much they need to grow. The mechanisms by which these *Hox* genes do this is not known, but changes in cell-cell adhesion and other surface qualities of the cells have been seen (Nardi and Stocum, 1983; Stocum and Crawford, 1987; Bryant and Gardiner, 1992).

So we are faced with a situation wherein the adult cells of an organism can return to an "embryonic" condition and begin the formation of the limb anew. Just as in embryonic development, the blastema forms successively more distal structures (Rose, 1962). Thus, the blastema must contain some

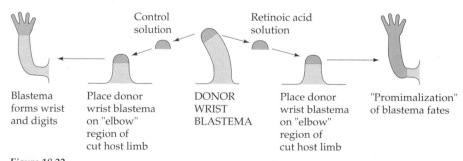

Control solution

Retinoic acid solution

Blastema forms wrist and digits

Place donor wrist blastema on "elbow" region of cut host limb

DONOR WRIST BLASTEMA

Place donor wrist blastema on "elbow" region of cut host limb

"Promimalization" of blastema fates

Figure 18.22
Wrist blastemas from recently cut axolotl limbs will generate the wrist when placed onto host limbs cut at the elbow area (see Chapter 3). However, if placed into retinoic acid, these blastema will begin regenerating at the place they have joined the host tissue, and they generate structures proximal to those of the wrist. (Data from Crawford and Stocum, 1998a,b.)

positional information that directs a blastema on a stump containing a humerus neither to make another humerus nor to start immediately producing digits. Not only does the blastema regenerate those structures beginning at the ap-

propriate proximal-distal level in the limb, but the polarities of the anterior-posterior ("thumb-pinky") and dorsal-ventral ("wrist-palm") axes also correspond to those of the stump. ■

Specification of the limb anterior-posterior axis

The Zone of Polarizing Activity

The self-differentiation of the anterior-posterior axis is the first change from the pluripotent condition. In chicks, this axis is specified long before a limb bud is recognizable. Hamburger (1938) showed that as early as the 16-somite stage, prospective wing mesoderm transplanted to the flank area develops into a limb with the anterior-posterior and dorsal-ventral polarities of the donor graft and not those of the host tissue (Figure 18.23).

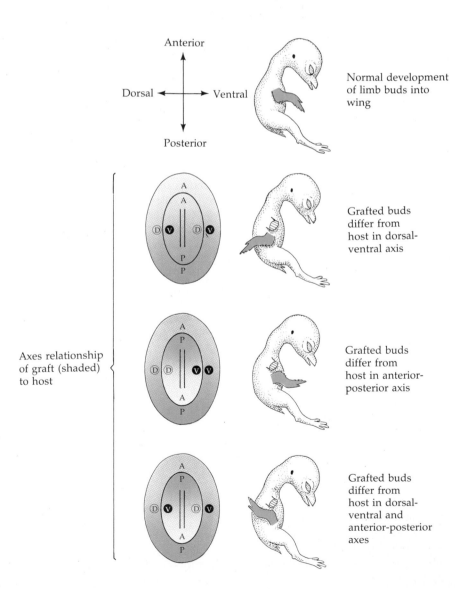

Figure 18.23
Specification of the anterior-posterior and dorsal-ventral axes in the chick wing. The grafted limb bud develops in accordance with its own polarity and does not adopt the polarity of its host. Wings that develop from grafted limb buds are shown in color. For the sake of clarity, the host's normally developed wing is not shown. (After Hamburger, 1938.)

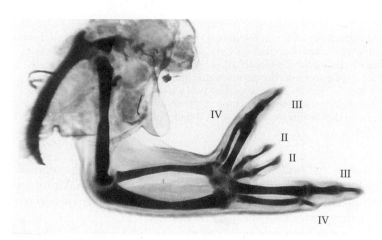

Figure 18.24
Duplicated digits emerge as mirror image of normal digits when ZPA is grafted to anterior limb bud mesoderm. (From Honig and Summerbell, 1985, photograph courtesy of D. Summerbell.)

Although the differentiation of the proximal-distal structures is thought to depend on how many divisions a cell undergoes while in the progress zone, positional information instructing a cell as to its position on the anterior-posterior and dorsal-ventral axes must come from other sources. Several experiments (Saunders and Gasseling, 1968; Tickle et al., 1975; Summerbell, 1979) suggest that the anterior-posterior axis is specified by a small block of mesodermal tissue near the posterior junction of the young limb bud and the body wall. When this tissue from a young limb bud is transplanted into a position on the anterior side of another limb bud (Figure 18.24), the number of digits of the resulting wing is doubled. Moreover, the structures of the extra set of digits are mirror images of the normally produced structures. The polarity has been maintained, but the information is now coming from both an anterior and a posterior direction. This region of the mesoderm has been called the **zone of polarizing activity** (**ZPA**).

The distribution and strength of the ZPA's positional signaling activity in the chick wing and leg buds have been mapped (Hinchliffe and Sansom, 1985; Honig and Summerbell, 1985). As shown in the drawings in Figure 18.25, the polarizing activity (measured after grafting the posterior marginal cells into the anterior margin of the limb bud) is highest in a particular region of the posterior margin and tapers off from there. It weakens as development progresses.

Sonic Hedgehog as Defining the ZPA

The search for the molecule(s) conferring this polarizing activity of the chick limb bud became one of the most intensive quests in developmental biology. The present candidates for the ZPA factors came from studies that assumed an evolutionary homology in the developmental regulatory systems between *Drosophila* and vertebrates. As you will recall from Chapter 16, the homeotic genes of *Drosophila* have vertebrate counterparts that play critical

Figure 18.25
Map of positional signaling activity as the limb develops. The colors represent the intensity of *sonic hedgehog* expression. The numbers represent the percentage of grafts showing complete duplications where these regions were transplanted into the anterior margin of the early limb bud. (Drawings after Honig and Summerbell, 1985, expression data from Riddle et al., 1993.)

Stage 17

Stage 19

Stage 21

Stage 23

Stage 25

Stage 27

Stage 29

Transfect *shh*-expressing virus
and allow viral spread

Infectable strain
of chick embryo
fibroblast cells

Centrifuge pellet cells

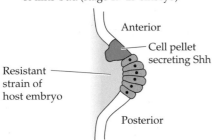

Implant in anterior portion
of limb bud (stage 19–23 embryo)

Anterior

Cell pellet
secreting Shh

Resistant
strain of
host embryo

Posterior

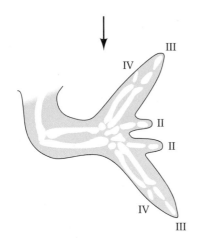

III

IV

II

II

IV

III

Figure 18.26

Assay for polarizing activity of Sonic hedgehog. The *sonic hedgehog* gene was insert-ed within an active promoter of a chicken virus, and the recombinant virus was placed into cultured chick embryo fibroblast cells (CEF). The virally infected cells were pelleted and implanted into the anterior margin of a limb bud of a chick embryo resistant to infection by this virus. The virus was therefore unable to infect the host, but it was able to express and secrete high levels of Sonic hedgehog. The resulting limbs showed that the secreted material had polarizing activity. (After Riddle et al., 1993.)

developmental functions. And as was mentioned in Chapter 14, the segment polarity gene *hedgehog* is thought to encode a diffusible protein that interacts with neighboring cells. Would it be too much to ask that there be a verte-brate homologue that performs a similar function?

Using the known sequence of the *Drosophila hedgehog* gene, Riddle and his co-workers (1993) used the polymerase chain reaction to identify a *hedge-hog*-like message in chick limb buds. They dubbed the gene *sonic hedgehog*.* In situ hybridization showed that *sonic hedgehog* expression was not found throughout the entire limb bud, but was localized exactly to the region pre-viously shown by Honig and Summerbell to contain the highest ZPA activ-ity (Figure 18.25).

Riddle and co-workers showed that in all probability, the secretion of Sonic hedgehog protein would be sufficient for ZPA activity. They trans-fected embryonic chick fibroblasts (which normally would never synthesize this protein) with a viral vector containing the *sonic hedgehog* gene (Figure 18. 26). The gene became expressed and translated in these fibroblasts, which were then inserted into the anterior ridge of an early chick limb bud. The ZPA-like digit polarity reversals were seen. More recently, beads containing Sonic hedgehog protein were shown to cause the same duplications (López-Martinez et al., 1995). Thus, Sonic hedgehog looks like the active agent of the ZPA.

Interactions between the AER and the ZPA to Integrate Growth and Pattern

How does Sonic hedgehog bring about the anterior-posterior pattern of the limb? Recent investigations suggest that it does not act alone and that coop-eration with signals from the AER are crital for its function. These interac-tions may establish patterns of *Hox* gene expression that could specify the anterior-posterior axis.

SONIC HEDGEHOG AS INITIATOR OF MORPHOGEN SECRETION.   It is still not known how the ZPA specifies the anterior-posterior axis. One model suggests that short-range signals specify the cells that will produce the digits and that these cells migrate across the limb bud (Figure 18.27A). However, no such migration has been observed. This leaves two other models. In the inductive cascade model (Figure 18.27B), a progression of short-range signals is suces-sively propagated from the ZPA to the responding tissues. Thus, Sonic hedgehog does not diffuse across the limb bud in a steadily decreasing gra-dient. Rather, Sonic hedgehog diffuses a small distance and induces those re-

*Yes, after the Sega cartoon character. The *hedgehog* gene in *Drosophila*, as in most genes, is named for its mutant phenotype. (This causes a lot of confusion. The genes "for" eyelessness or limblessness are actually those genes whose products prevent these deficiencies.) In *Drosophila*, the deficiency of *hedgehog* expression results in the cuticle having more pointy denticles, hence appearing like a hedgehog.

(A)   Short-range signaling and displacement

(B)   Sequential short-range signaling

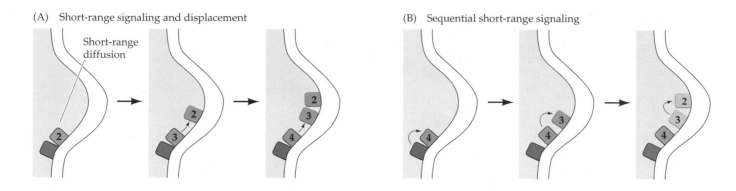

(C)   Progressive spread of graded signal and one-way promotion

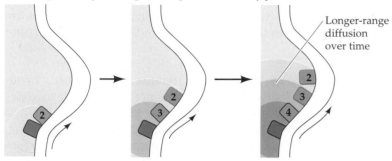

Figure 18.27
Models of ZPA activity. (A) Model of ZPA function by short-range signaling and subsequent displacement of specified tissue. (B) Model of ZPA function by sequential short-range signals. (C) Model of ZPA function by the progressive spread of a graded signal, wherein the responding tissue responds to concentration gradient. (After Tickle, 1995.)

sponding cells to secrete another protein. This second protein diffuses a small distance to activate those cells near them, and these cells secrete a third protein, and so forth. In this manner, a series of signals is sent from the source of Sonic hedgehog toward the anterior of the limb bud (see López-Martínez et al., 1995). The third model is called the soluble morphogen model (Figure 18.27C; Wolpert, 1969, 1977; Tickle, 1981) wherein the tissue responds differently to different concentrations of soluble molecules secreted by the ZPA. At first, the tissue nearest the ZPA gets low concentrations of the morphogen and is specified to be the most distal digit. However, as secretion continues, that tissue is exposed to a higher concentration and is respecified as a more proximal (posterior) digit. The next region of cells slightly farther away from the ZPA gets a low concentration of morphogen and becomes specified for the more distal (anterior) structures. This continues until all the digits are specified across the limb bud.

Neither of the last two models has been excluded, but there is evidence that while Sonic hedgehog appears to define and "empower" the ZPA, it is not the soluble morphogen responsible for specifying the digits. There may be a cascade of inductive signals. The active part of the Sonic hedgehog protein has been shown to be its N-terminal region, which is split from the rest of the protein in chick limbs. This active end can diffuse from the cell, but it is not seen to diffuse far from its source in the posterior limb bud (López-Martínez et al., 1995). When it does diffuse, it appears to activate bone mor-

Figure 18.28
Some of the molecular interactions by which the limb bud formation is initiated and maintained. Dynamic pattern of *HoxD* gene expression during a portion of chick wing morphogenesis. Some of the major loops include (1) the maintainance of Sonic hedgehog (Shh) by the combination of Wnt7a and FGF4; (2) the maintenance of Shh by the combination of retinoic acid and FGF4; (3) the reciprocal induction of FGF4 and Shh to maintain each other; (4) the interaction between FGF4 and Shh to activate the expression of *HoxD* genes and to maintain cell division in the progress zone mesenchyme. (After Nelson et al., 1996; Niswander et al., 1994.)

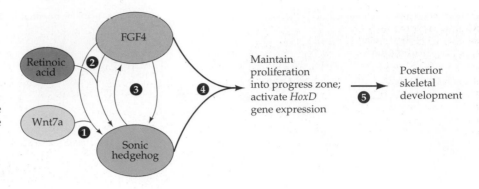

phogenetic proteins, especially BMP2 (Francis et al., 1994; Laufer et al., 1994). These proteins do not diffuse very far either, and investigators are searching for other molecules that may be activated by the bone morphogenetic proteins.

SONIC HEDGEHOG AS CO-ACTIVATOR OF *HOX* GENES AND CELL PROLIFERATION. In addition to activating the genes for bone morphogenetic proteins (especially BMP2), there are two other important targets for sonic hedgehog. The first set of targets may be the 5′ *Hoxd* genes (Figure 18.28; *Hoxd-9* through *Hoxd-13*). During normal chick or mouse limb development, there evolves a characteristic "concentric nested" pattern of *Hoxd* gene expression centered at the posterior margin that had been defined as the ZPA (Dollé et al., 1989; Nelson et al., 1996). The region closest to the center has all of these 5′ *Hoxd* genes expressed, while expression of these genes falls sequentially as the cells get progressively further removed from the ZPA. Moreover, the transplantation of either the ZPA or the Sonic-hedgehog-secreting cells to the anterior margin leads to the formation of mirror-image patterns of *Hoxd* gene expression and mirror-image digit patterns (Izpisúa-Belmonte et al., 1991; Nohno et al., 1991; Riddle et al., 1993).

While it originally appeared that there might be a code wherein the expression of different *HoxD* genes specified the anterior-posterior digit pattern, recent studies show the problem to be more complex. Sonic hedgehog may be specifying pattern by acting in conjunction with signals from the AER. First, *Hoxd* gene expression is controlled by cooperation between the AER and ZPA. In the absence of an AER, Sonic hedgehog is unable to induce the expression of the *Hoxd* genes (Laufer et al., 1994). However, the addition of retinoic acid can replace the missing AER (Helms et al., 1996; Ogura et al., 1996).* Retinoic acid has long been known to induce limb polarization. Beads soaked in retinoic acid can mimic ZPA tissue and induce a mirror-image reversal of anterior-posterior polarity (Tickle et al., 1982, 1985), and a single retinoic acid-soaked bead could replace a ZPA when the normal ZPA tissue was removed (Eichele, 1989). However, the retinoic acid content of the ZPA does not seem high enough to activate retinoic-acid-responsive genes (Plate 13; Noji et al., 1991; Rossant et al., 1991), and theoretical considerations (see Wanek et al., 1991) make it unlikely that retinoic acid is the active agent of the ZPA. Recent studies, however, suggest that retinoic acid induces a co-

---

*The morphogenetically active retinoic acid in the limb bud may differ between species. In the chick limb, the active RA appears to be didehydroretinoic acid. However, this form is not seen in the mouse limb bud (Stratford et al., 1996).

factor of Sonic hedgehog. While Sonic hedgehog alone could induce *Hoxd-9* through *Hoxd-11*, the induction of the most 5′ *HoxD* genes, *Hoxd-12* and *Hoxd-13*, could be accomplished only if retinoic acid were present (Ogura et al., 1996). Grafting experiments show that such retinoic-acid-induced factors are produced in the AER (Helms et al., 1996).

One candidate for the retinoic-acid-induced factor is FGF4. Retinoic acid induces the expression of *Fgf4* in the AER, and it does so independently of detectable Sonic hedgehog (Niswander et al., 1994). Moreover, when a bead of FGF4 replaces the AER, *HoxD* gene expression is promoted (Laufer et al., 1994). Duprez and her colleagues (1996) have found that retinoic acid will induce BMP2 and that BMP2 will induce both FGF4 in the AER and the expression of Hoxd11 and 13 in the mesoderm. Therefore, the normal induction of the most 5′ *Hox* genes (which cannot be accomplished by Shh alone) is done by the combination of BMP2 and FGF4.

This creates an interesting situation because Sonic hedgehog and FGF4 mutually activate one another. Sonic hedgehog activates the expression of the *Fgf4* gene in the posterior region of the AER (see Figure 18.9), while *Fgf4* expression is necessary for the normal expression of the *sonic hedgehog* gene. (Such a relationship had been suggested by the studies of Todt and Fallon, 1987, who showed that the AER was necessary for ZPA function.) There is, then, a positive feedback loop wherein Sonic hedgehog from the posterior mesoderm activates *Fgf4* in the AER, and FGF4 (probably in conjunction with FGF8) from the AER maintains *sonic hedgehog* expression (see Figure 18.28B; Laufer et al., 1994; Niswander et al., 1994).

Specifying the ZPA

We do not yet know what causes the activation of the *sonic hedgehog* genes specifically in cells of the posterior limb bud and not in the cells located more anteriorly. It is possible that the *sonic hedgehog* gene is being activated by an FGF protein coming from the newly formed apical ectodermal ridge, and FGF8 is present in the AER and capable of activating *sonic hedgehog*. But why doesn't it activate all the mesenchyme cells beneath the ridge? The answer may reside in the differential competence of certain mesenchyme cells to *respond* to the FGF signal. Charité and colleagues (1994) have suggested that the Hoxb-8 protein may be critical in providing this restricted competence. They observed that the *Hoxb-8* gene was usually expressed in the *posterior* half of the mouse forelimb bud. They then constructed transgenic mice in which the *Hoxb-8* gene was placed under the control of a new promoter that would cause its expression throughout the forelimb buds. This resulted in the expression of *sonic hedgehog* in the anterior portion of the limb buds, the creation of a new ZPA, a new region of *HoxD* gene expression, and mirror-image forelimb duplications. This evidence suggests that the Hoxb-8 protein is involved in specifying the expression of *sonic hedgehog* and thus establishing the ZPA.

The generation of the dorsal-ventral axis

The third axis of the limb defines the dorsal limb (knuckles, nails) from the ventral limb (pads, soles). In 1974, MacCabe and co-workers demonstrated that the dorsal-ventral polarity of the limb bud is determined by the ectoderm encasing it. If the ectoderm is rotated 180° with respect to the limb bud mesenchyme, the dorsal-ventral axis is partially reversed; the distal elements (digits) are "upside down." This suggested that the late specification of the dorsal-ventral axis of the limb is regulated by its ectodermal component. The **Wnt7a** gene is expressed in the dorsal (but not ventral) ectoderm of the

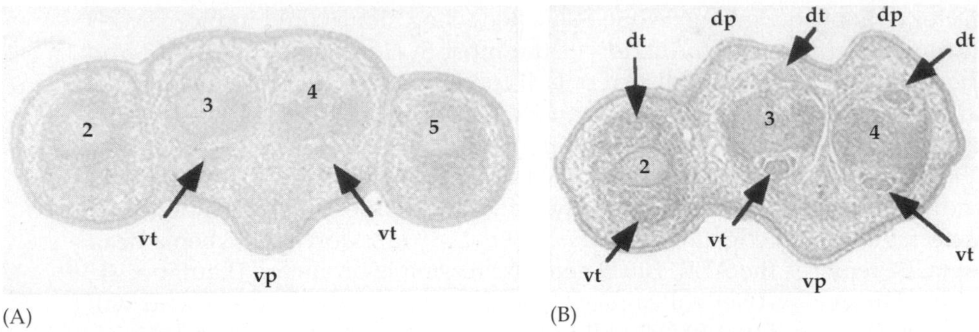

Figure 18.29
Dorsal-to-ventral transformations of limb regions in mice deficient for both *Wnt7a* genes. (A) Histological section (stained with hemotoxylin and eosin) of wild-type 15.5-day embryonic mouse forelimb paw. The ventral tendons and ventral footpads are readily seen. (B) Same section through a mutant embryo deficient in *Wnt7a*. Tendons and footpads are now duplicated on what would be the dorsal surface of the paw. dt, dorsal tendons; dp, dorsal footpad; vp, ventral footpad; vt, ventral tendon. Numbers indicate digit identity. (From Parr and McMahon, 1995; photographs courtesy of the authors.)

chick and mouse limb buds (Deally, 1993; Parr et al., 1993). In 1995, Parr and MacMahon genetically deleted *Wnt7a* from the mouse embryo. The resulting embryos had sole pads on both surfaces of their paws, showing that Wnt7a was needed for the dorsal patterning of the limb (Figure 18.29). Wnt7a induces the *Lmx1* gene in the dorsal mesenchyme, and this gene encodes a transcription factor that appears to be essential for specifying the dorsal cell fates in the limb (Riddle et al., 1995; Vogel et al., 1995). If this factor is expressed in the ventral mesenchyme cells, they develop a dorsal phenotype.

Such *Wnt7a*-deficient mice also lacked posterior digits, suggesting that *Wnt7a* was also needed for the anterior-posterior axis. Yang and Niswander (1995) made a similar set of observations in chick embryo. These investigators removed the dorsal ectoderm from the developing limb and found that such an operation resulted in the loss of posterior skeletal elements from the limbs. The reason that these limbs lacked posterior digits was that *sonic hedgehog* and *Fgf4* expression was missing. Viral-induced expression of *Wnt7a* could replace the dorsal ectoderm and restore *sonic hedgehog* expression and posterior pattern. The synthesis of Sonic hedgehog is stimulated by the combination of FGF4 and Wnt7a proteins. The three axes of the chick embryo are all interelated and coordinated.

Distinguishing the forelimb from the hindlimb

So far, we have been treating the forelimb and the hindlimb as if they were the same. Indeed, they do follow the same rules of pattern formation. Yet their pattern is different. A foot is not a hand, and a chick leg certainly is not a wing. So how do they become different? It appears that the precartilage cells of the chick wing and leg respond very differently to growth factors, and this causes them to associate with one another and differentiate in different ways (Downie and Newman, 1994). In cultures of embryonic cells, retinoic acid will enhance chondrogenesis in wing mesenchyme and inhibit chondrogenesis in leg mesenchyme. TGF-β1 will convert leg cartilage-forming nodules into sheets, whereas it has no effect on wing cartilagenous nodules except to promote chondrogenesis. The wing precartilage cells also produce a different pattern of fibronectin than their corresponding leg cartilage cells (Figure 18.30).

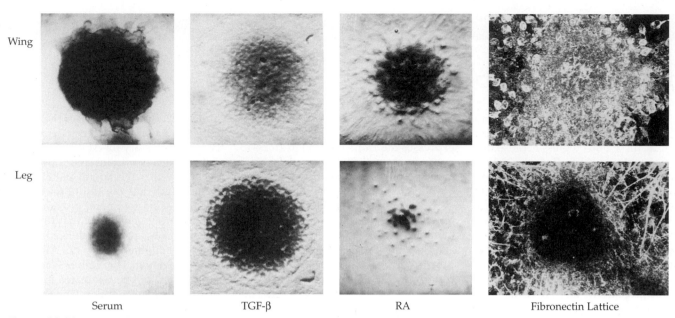

Figure 18.30
Differential response of wing and leg precartilage cells (stage 24) to specific morphogenetic factors. Pictures of the cells in serum, TGF-β, and retinoic acid are macroscopic photographs of the cell colonies. The photographs of the fibronectin lattices deposited by the cells are fluorescence photomicrographs at 40×. (From Downie and Newman, 1994.)

This difference in fibronectin deposition may be very important in regard to wing/leg differences. The placement, timing, and architecture of cartilage deposition in the cultures of limb bud tissue closely parallel the deposition of fibronectin. In wing cultures, the cartilage condensations were broad and flat; in leg cultures they were compact and spheroidal. In both cases, the deposition of cartilage paralleled the placement of fibronectin (Downie and Newman, 1995). Thus, there are inherent differences in the forelimb and hindlimb precartilaginous mesenchyme cells that cause them to respond differently to growth factors and to deposit fibronectin differently. The placement of fibronectin is critical in directing the placement and extent of chondrogenesis. The mechanism by which this cartilage formation occurs and bifurcates to form the limb skeleton is a subject of great interest. [limb4.html]

Recently, pairs of related transcription factors have been shown to be differentially expressed in the hindlimb and forelimb buds. The chick *Hoxc-4* and *Hoxc-5* are expressed in the wing buds, while *Hoxc-9*, *Hoxc-10*, and *Hoxc-11* are expressed exclusively in the leg buds (Nelson et al., 1996). The *Brachyury*-like gene *tbx5* is transcribed in the mouse forelimbs, while its closely related *tbx4* is expressed in the hindlimbs (Gibson-Brown et al., 1996). It has yet to be seen if any of these genes is causally involved in directing forelimb versus hindlimb specificity.* However, the loss of human TBX5 results in the Holt-Oram Syndrome, characterized by abnormalities of the heart and upper limbs (Basson et al., 1996; Li et al., 1996). The legs are not affected.

---

*When referring to the hand, one has an orderly set of names to specify each digit (*digitus pollicis, d. indicis, d. medius, d. annularis, and d. minimus,* respectively, from thumb to little finger). No such nomenclature exists for the pedal digits, but the plan proposed by Phillips (1991) has much merit. The pedal digits, from hallux to small toe, would be named *porcellus fori, p. domi, p. carnivorus, p. non voratus,* and *p. plorans domi,* respectively.

Lessons from limbless

A S NOTED EARLIER, the *limbless* mutant can form limb buds, but these buds regress because they do not form an AER. Recent work on gene expression in this mutant shows that limb budding is accompanied by "normal" gene expression patterns that are not set up by the three signaling centers. These data suggest that the gene expression patterns normally associated with development of the tetrapod limb would have occurred anyway and represent a "prepattern" that directs development of the tetrapod limb. (Ros et al., 1996).The signaling centers only reinforce and sustain this pattern.

First, the *limbless* limb bud expresses neither FGF4 nor FGF8, implying that these proteins are not required for normal budding. Second, the limbless mesoderm expresses the *Hoxd-11* through *Hoxd-13* genes in a posteriorly nested fashion, along with the asymmetrical expression of *BMP4* and *Wnt5a*. It does this in the absence of detectable *sonic hedgehog* expression or an AER (Grieshammer et al., 1996; Noramly et al., 1996; Ros et al., 1996).

Experimental analysis of the *limbless* limb bud reveals that *limbless* buds will form AERs and limbs if given beads secreting FGFs. Moreover, such beads induce Sonic hedgehog in the posterior region, showing that there is a polarity to the limb bud, even in the absence of the AER. Interestingly, the limb formed is "bi-dorsal," expressing *Wnt7a* through-

out the ectoderm. This raises the possibility that a dorsoventral ectodermal interface is necessary for AER induction.

The *limbless* mutant also raises the possibility that the lateral plate mesoderm already has the ability to express the anterior-posterior and proximal-distal patterning genes, and that this prepattern is subsequently stabilized, maintained, or augmented by the AER and Sonic hedgehog. The limb is a complicated organ, and new research makes it seem even more complex. The analysis of development of the tetrapod limb has provided biologists with some of its greatest successes in understanding development, but it also keeps posing some of our greatest challenges. ∎

Cell death and the formation of the digits

Cell death also plays a role in sculpting the limb. Indeed, it is essential if joints are to form and if our fingers are to become separate (Zaleske, 1985). The death (or lack of death) in specific cells in the vertebrate limb is genetically programmed and has been selected for during evolution. One such case involves the webbing or nonwebbing of feet. The difference between a chicken's foot and that of a duck is the presence or absence of cell death between the digits (Figure 18.31A,B). Saunders and co-workers (1962; Saunders and Fallon, 1966) have shown that after a certain stage, chick cells between the digit cartilage are destined to die and will do so even if transplanted to another region of the embryo or placed into culture. Before that time, however, transplantation to a duck limb will save them. Between the time when the cell's death is determined and when death actually takes place, levels of DNA, RNA, and protein synthesis decrease dramatically (Pollak and Fallon, 1976).

In addition to the **interdigital necrotic zone,** there are three other regions that are "sculpted" by cell death. The ulna and radius are separated from each other by an **interior necrotic zone,** and two other regions, the **anterior** and **posterior necrotic zones,** further shape the end of the limb (Figure 18.31B; Saunders and Fallon, 1966). Although these zones are said to be "necrotic," this is a holdover from the days when no distinction was made between "necrotic" cell death and "apoptotic" cell death. These cells die by apoptosis, and the death of the interdigital tissue is associated with the fragmentation of DNA (Mori et al., 1995). In humans, there are several syndromes characterized by webbed digits ("syndactyly" syndromes), but the mutation responsible is known in only one instance (Vortkamp et al., 1991; Hui and Joyner, 1993), in which the gene encodes a transcription factor that

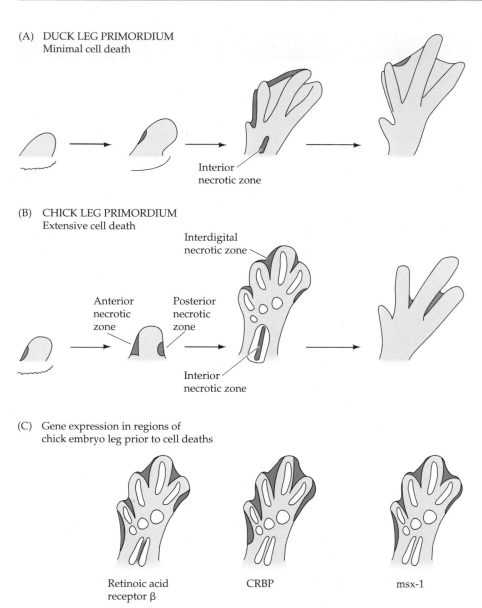

(A) DUCK LEG PRIMORDIUM
Minimal cell death

Interior
necrotic zone

(B) CHICK LEG PRIMORDIUM
Extensive cell death

Interdigital
necrotic zone

Anterior
necrotic
zone

Posterior
necrotic
zone

Interior
necrotic zone

(C) Gene expression in regions of
chick embryo leg prior to cell deaths

Retinoic acid
receptor β

CRBP

msx-1

Figure 18.31
Patterns of cell death in leg primordia of (A) duck and (B,C) chick embryos.
Shading indicates areas of cell death. In the duck, cell death is very slight, whereas
there are regions of extensive cell death in the interdigital tissue of the chicken leg.
(After Saunders and Fallon, 1966.)

is expressed in the interdigital mesenchyme. The signal for apoptosis in
chick limbs may be the BMP4 protein. The expression of BMP4 is seen in the
interdigital spaces in chick limbs, and inhibiting BMP4 signaling prevents
the interdigital cells from undergoing apoptosis. Interestingly, this BMP expression is not seen in embryonic duck interdigital mesenchyme at this time
(Gañan et al., 1996; Zou and Niswander, 1996; see Chapter 23).

The limb has been a cornerstone in the study of pattern generation in
vertebrates. It arises from numerous interrelated processes that involve the
placement and growth of the limb bud, the induction of the AER, the maintenance of the progress zone mesenchyme, the formation and the mutual
maintenance of the ZPA and AER, the formation of the dorsal-ventral axis,

Evolution of the Tetrapod Limb

Of Fins and Limbs

Macroevolution, the generation of morphological novelties in the evolution of new species and higher taxa, results from alterations of development. Some of the most important macroevolutionary changes resulted from the transition of aquatic animals onto land. One of the most obvious changes is that of the fish fin to the amphibian leg. As Richard Owen (1849) pointed out, there is considerable homology between the bones of the fin and the tetrapod limb, the pectoral and pelvic fins of the fish being homologous to the forelimb and hindlimb, respectively. While specific homologies were able to be made between the proximal elements (zeugopod; tibia and fibula) of the fin and limb, the homologies proposed between the autopod of the limb (the hand or foot at the distal end) and the rays of the fins "did not hold water." This was true even when one compared the tetrapod limb to the fins of crossopterygian (lobe-finned) fishes thought to have been closely related to the ancestors of the amphibians (see Coates, 1994; Hinchliffe, 1994). The problem becomes even more vexing when one looks at the limbs of the first tetrapods known. Rather than having the canonical five digits, these primitive amphibians had six (*Tulerpedon*), seven (*Ichthyostega*), or even eight (*Acanthostega*) digits on their limbs. There are no fin-like rays associated with these limbs, and they are thought to have functioned as paddles for a shallow-water niche rather than for weight-bearing on land. Again, while there seems to be homology for the proximal elements of the limb, the autopod seems to be something new—what evolutionary biologists call a "neomorphic" structure.

Recent studies have strongly suggested that the placement of the 5′ end ("*Abd*-like") *Hox* genes of the *HoxD* group may be crucial in the change from fin to limb. In the early limb bud of chicks and mice, the 5′ genes (*Hoxd-11, 12, 13*) are restricted to the posterior end

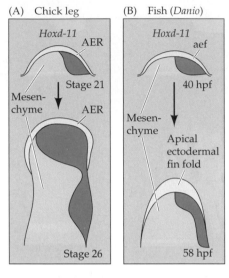

Figure 18.32 Differences in *Hoxd-11* expression in fish and chick embronic apendages. (A) Regions of *Hoxd-11* expression in the mouse hindlimb from an early bud stage to a later one. During the later stages, the *Hoxd-11* expression pattern extends across the anterior-posterior border of the progress zone. (B) In the zebrafish pectoral fin, *Hoxd-11* expression continues to be expressed posteriorly, but it does not extend anteriorly. hpf; hours past fertilization. (After Sordino et al., 1995.)

of the limb bud. This is similar to the situation in the zebrafish fin bud (Figure 18.32). However, in tetrapods, there is a second phase, where the expression of the *Abd*-like *HoxD* genes changes. Instead of being restricted to the posterior of the limb bud, the expression of the 5′ *HoxD* genes sweeps across the distal mesenchyme, just beneath the AER. This band of expression is coincident with the "digital arch" from which the digits form (Morgan and Tabin, 1994; Sordino et al., 1995; Nelson et al, 1996). These studies show that while the *HoxD* gene expression pattern is homologous in the proximal regions, the expression in the late bud distal mesenchyme is new. It also confirms the paleontological-developmental studies of Shubin and Alberch (1986), who proposed that the path of

digit formation was not (as previously believed) through the fourth digit (making the fin rays homologous to the other digits), but through an arch of distal wrist condensations (metapterygia) that begins posteriorly and turns anteriorly across the distal mesenchyme (Figure 18.33). Thus, the border of 5′ *HoxD* gene expression follows the metapterygial axis that Shubin and Alberch hypothesized as being the origin of digits. Sordino and colleagues propose that the proximal localization of the *HoxD* gene transcripts represents the original pattern and is common to all vertebrates. The reoriented, distal, phase of *HoxD* gene expression represents a new and "derived" condition. This, in turn, might have evolved owing to changes in the regulation of the 5′ *HoxD* genes. The early *HoxD* pattern is formed independently of Sonic hedgehog. The distalized expression pattern

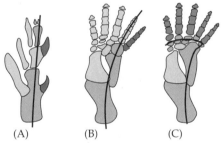

Figure 18.33 Origin of digits (autopod) as an evolutionary novelty dependent on 5′ (*AbdB*-like) *HoxD* gene expression. (A) Representation of a primitive fish fin, showing a central axis (black) with rays radiating anteriorly (light gray) and posteriorly (dark gray). (B) Representation of the older view of tetrapod Development of the tetrapod limb. The central axis curved through the fourth digit. The fifth digit was seen to be homologous to a posterior ray; the first, second, and third digits were seen as being homologous to anterior rays. (C) Current view of autopod formation. The axis originally extends posteriorly, but then curves anteriorly across the metapterygial cartilage. The tibia is considered to branch anteriorly, but the digits are not homologous to any rays. (After Nelson and Tabin, 1995.)

appears to be governed by Sonic hedgehog expression. The foot and hand, then, appear to be new structures in evolution, and they appear to have been formed by the repositioning of *HoxD* gene expression during fin development. Needless to say, this is not the only change that has occurred to create digits. Other *Hox* genes and probably *sonic hedgehog* have changed their expression patterns as well. We are finally coming to a point in biology where changes in gene expression can be linked to large evolutionary changes.

Of Fly Legs and Chick Legs

Evolution involves modification with descent. This was often documented with homologies. The flipper of a seal, the wing of a bat, the arm of a squirrel, and the arm of a human are all based on the same homologous "plan," but with modifications. Each is a different modification of the reptilian limb plan, which is itself homologous to that of mammals. One of the most important findings in modern developmental biology concerns the homologies of processes as well as those of structures. As we will see in Chapter 23, certain developmental pathways and interactions have been conserved over evolutionary time and have been modified by different animal groups. This can be seen with the development of the limb. Fly limbs and vertebrate limbs are familiar examples of analogy (as opposed to homology). Where **homologous** structures are seen to be modifications of an original structure and may now serve different functions, **analogous** structures have the same function but are not seen as being derived from a common structure.

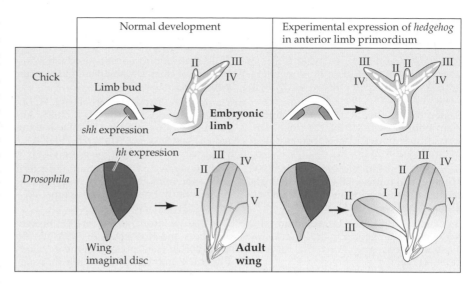

Figure 18.34 Homology of process in the formation of the anterior-posterior axes in *Drosophila* and chick appendages. (A) A chick limb bud expresses *sonic hedgehog* in its posterior region. If *sonic hedgehog* is also expressed in an anterior region, the developing limb develops a mirror-image duplication of the anterior-posterior axis. (B) A *Drosophila* wing disc expresses *hedgehog* in its posterior compartment. If *hedgehog* is expressed in the anterior compartment as well, the wing develops a mirror-image duplication of the anterior-posterior axis. (After Ingham, 1994.)

Fly limbs and vertebrate limbs have little in common except their function. However, both fly and vertebrate limbs appear to be formed through the same developmental pathway. (There appears to be a homology of process underlying the analogy of structures.) As we have seen, *sonic hedgehog* is usually expressed in the posterior part of the limb bud. If it is expressed in the anterior part of the bud, mirror-image duplications arise (Riddle et al., 1993). In the *Drosophila* wing disc (the field of cells that gives rise to the wing during metamorphosis), the Hedgehog protein is usually expressed in the posterior portion of the disc. If it is expressed anteriorly, mirror-image duplications of the wing will form (Figure 18.34; Basler and Struhl, 1993; Ingham, 1994). Furthermore, certain genes regulated by Hedgehog have been conserved as well (Marigo et al., 1996), and the ventral limb compartments of both insects and vertebrates appear to be regulated by the expression of the *engrailed* gene (Davis et al., 1991; Loomis et al., 1996). Thus, it seems like nature figured out how to make a limb only once, and both arthropods (*Drosophila*) and vertebrates (chicks and mice) use that process to this day. ∎

the generation of spaced precartilaginous condensations of mesenchyme that will form the cartilage and bone tissue, and the imposition of asymmetry on the limb bud by the ZPA. There are still lively debates concerning the mechanisms by which these processes occur, and there is considerable controversy over the molecules that might regulate such phenomena.

LITERATURE CITED

Albert, P. and Boilly, B. 1988. Effect of transferrin on amphibian limb regeneration: A blastema cell culture study. *Roux Arch. Dev. Biol.* 197: 193–196.

Basler, K. and Struhl. G. 1993. Compartment boundaries and the control of *Drosophila* limb pattern by *hedgehog* protein. *Nature* 368: 208–214.

Basson, C. T., and thirteen others. 1996. Mutations in human TBX5 cause limb and cardiac malformation in Holt-Oram syndrome. *Nat. Genet.* 15: 30–35.

Brockes, J. P. 1992. Introduction of a retinoid reporter gene into the urodele limb blastema. *Proc. Natl. Acad. Sci. USA* 89: 11386–11390.

Brockes, J. P. and Kinter, C. R. 1986. Glial growth factor and nerve-dependent proliferation in the regeneration blastema of urodele amphibians. *Cell* 45: 301–306.

Bryant, S. V. and Gardiner, D. M. 1992. Retinoic acid, local cell-cell interactions, and pattern formation in vertebrate limbs. *Dev. Biol.* 152: 1–25.

Bryant, S. V., French V. and Bryant, P. J. 1981. Distal regeneration and symmetry. *Science* 21: 993–1002.

Burke, A. C., Nelson, C. E., Morgan, B. A. and Tabin, C. 1995. Hox genes and the evolution of vertebrate axial morphology. *Development* 121: 333–346.

Butler, E. G. 1935. Studies on limb regeneration in X-rayed *Ambystoma* larvae. *Anat. Rec.* 62: 295–307.

Carrington, J. L. and Fallon, J. F. 1988. Initial limb budding is independent of apical ectodermal ridge activity: Evidence from a *limbless* mutant. *Development* 104: 361–367.

Charité, J., Graaff, W. de, Shen, S. and Duchamps, J. 1994. Ectopic expression of *Hoxb-8* causes duplications of the ZPA in the forelimb and homeotic transformation of axial structures. *Cell* 78: 559–601.

Chernoff, E. A. G. and Stocum, D. 1995. Developmental aspects of spinal cord and limb regeneration. *Dev. Growth Diff.* 37: 133–147.

Coates, M. I. 1994. The origin of vertebrate limbs. *Development 1994 Suppl.* 169–180.

Crawford, K. and Stocum, D. L. 1988a. Retinoic acid coordinately proximalizes regenerate pattern and blastema differential affinity in axolotl limbs. *Development* 102: 687–698.

Crawford, K. and Stocum, D. L. 1988b. Retinoic acid proximalizes level-specific properties responsible for intercalary regeneration in axolotl limbs. *Development* 104: 703–712.

Crossley, P. H., Monowada, G., MacArthur, C. A. and Martin, G. R. 1996. Roles for FGF8 in the induction, initiation, and maintenance of chick Development of the tetrapod limb. *Cell* 84: 127–136.

Davis, A. P., Witte, D. P., Hsieh-Li, H. M., Potter, S. and Capecchi, M. R. 1995. Absence of radius and ulna in mice lacking *hoxa-11* and *hoxd-11. Nature* 375–791–795.

Davis, C. A., Holmyard, D. P., Millen, K. J. and Joyner, A. L. 1991. Examining pattern formation in mouse, chicken and frog embryos with an En-specific antiserum. *Development* 111: 287–298.

Dealy, C. N., Roth, A., Ferrari, D., Brown, A. M. C. and Kosher, R. A. 1993. *Wnt-5a* and *Wnt-7a* are expressed in the developing chick limb bud in a manner that suggests roles in pattern formation along the proximodistal and dorsoventral axes. *Mech. Dev.* 43: 175–186.

De Robertis, E. M., Morita, E. A. and Cho, K. W. Y. 1991. Gradient fields and homeobox genes. *Development* 112: 669–678.

Detwiler, S. R. 1918. Experiments on the development of the shoulder girdle and the anterior limb of *Amblystoma punctatum. J. Exp. Zool.* 25: 499–538.

Dollé, P., Izpisúa-Belmonte, J.-C., Falkenstein, H., Renucci, A. and Duboule, D. 1989. Coordinate expression of the murine *Hox-5* complex homeobox-containing genes during limb pattern formation. *Nature* 342: 767–772.

Downie, S. A. and Newman, S. A. 1994. Morphogenetic differences between fore and hind limb precartilage mesenchyme: Relation to mechanisms of skeletal pattern formation. *Dev. Biol.* 162: 195–208.

Downie, S. A. and Newman, S.A. 1995. Different roles for fibronectin in the generation of fore and hind limb precartilage condensations. *Dev. Biol.* 172: 519–530.

Duprez, D. M., Kostakopoulou, K., Francis-West, P. H., Tickle, C. and Brickell, P. M. 1996. Activation of Fgf-4 and HoxD gene expression by BMP-2 expressing cells in the developing chick limb. *Development* 122: 1821–1828.

Eichele, G. 1989. Retinoic acid induces a pattern of digits in anterior half wing buds that lack the zone of polarizing activity. *Development* 107: 863–867.

Fallon, J. F. and Crosby, G. M. 1977. Polarizing zone activity in limb buds of amniotes. *In* D. A. Ede, J. R. Hinchliffe and M. Balls (eds.), *Vertebrate Limb and Somite Morphogenesis.* Cambridge University Press, Cambridge.

Fallon, J. F., Lopez, A., Ros, M. A., Savage, M. P., Olwin, B. B. and Simandl, B. K. 1994. FGF-2: Apical ectodermal ridge growth signal for chick Development of the tetrapod limb. *Science* 264: 104–107.

Francis, P. H., Richardson, M. K., Brickell, P. M. and Tickle, C. 1994. Bone morphogenesis proteins and a signalling pathway that controls patterning in developing chick limb. *Development* 120: 209–218.

Fraser, R. A. and Abbott, U. K. 1971. Studies on limb morphogenesis VI. Experiments with early stages of the polydactylous mutant *Eudiplopodia. J. Exp. Zool.* 176: 237–248.

Gañan, Y., Macias, D., Duterque-Coquillaud, M., Ros, M. A. and Hurle, J. M. 1996. Role of TGF-βs and BMPs as signals controlling the position of the digits and the areas of interdigital cell death in the developing chick autopod. *Development* 122: 2349–2357.

Gardiner, D. M., Blumberg, B., Konine, Y. and Bryant, S. V. 1995. Regulation of *HoxA* expression in developing and regenerating axolotl limbs. *Development* 121: 1731–1741.

Geduspan, J. S. and Solursh, M. 1992. A growth-promoting influence from the mesonephros during limb outgrowth. *Dev. Biol.* 151: 242–250.

Gibson-Brown, J. J., Agulnik, S. I., Chapman, D. L., Alexiou, M., Garvey, N., Silver, L. M. and Papaioannou, V. E. 1996. Evidence of a role for T-box genes in the evolution of limb morphogenesis and the specification of forelimb/hindlimb identity. *Mech. Dev.* 56: 93–101.

Goetinck, P. 1964. Studies on limb morphogenesis. III. Experiments with the polydactylous mutant *Eudiplopodia. Dev. Biol.* 10: 71–91.

Goss, R. J. 1969. *Principles of Regeneration.* Academic Press, New York.

Goss, R. J. 1991. The natural history (and mystery) of regeneration. *In* C. E. Dinsmore (ed.), *A History of Regeneration Research.* Cambridge University Press, New York, pp. 7–23.

Grieshammer, U., Minowada, G., Pisenti, J. M., Abbott, U. and Martin, G. R. 1996. The *limbless* mutation causes abnormalities in limb dorsal ventral patterning implication for the mechanism of apical ridge formation. *Development* 122: 3851–3861.

Hamburger, V. 1938. Morphogenetic and axial self-differentiation of transplanted limb primordia of 2-day chick embryos. *J. Exp. Zool.* 77: 379–400.

Hayashi, K. and Ozawa, E. 1995. Myogenic cell migration from somites is induced by tissue contact with medial region of the presumptive limb mesoderm in chick embryos. *Development* 121: 661–669.

Harrison, R. G. 1918. Experiments on the development of the forelimb of *Amblystoma*, a self-differentiating equipotential system. *J. Exp. Zool.* 25: 413–461.

Helms, J. A., Kim, C. H., Eichele, G. and Thaller, C. 1996. Retinoic acid signaling is required during early chick Development of the tetrapod limb. *Development* 122: 1385–1394.

Hertwig, O. 1925. Haploidkernige Transplante als Organisatoran diploidkeniger Extremitaten be *Triton. Anat. Anz.* [Suppl.] 60: 112–118.

Hinchliffe, J. R. 1991. Developmental approaches to the problem of transformation of limb structure in evolution. *In* J. R. Hinchliffe (ed.), *Developmental Patterning of the Vertebrate Limb*. Plenum, New York, pp. 313–323.

Hinchliffe, J. R. 1994. Evolutionary biology of the tetrapod limb. *Development 1994 Suppl.* 163–168.

Hinchliffe, J. R. and Sansom, A. 1985. The distribution of the polarizing zone (ZPA) in the legbud of the chick embryo. *J. Embryol. Exp. Morphol.* 86: 169–175.

Hogan, B. L. M., Thaller, Thaller, C. and Eichele, G. 1992. Evidence that Hensen's node is a source of retinoic acid synthesis. *Nature* 359: 237–241.

Honig, L. S. and Summerbell, D. 1985. Maps of strength of positional signaling activity in the developing chick wing bud. *J. Embryol. Exp. Morphol.* 87: 163–174.

Hui, C.-C. and Joyner, A. L. 1993. A mouse model of Grieg cephalopolysyndactyly syndrome: the *extra-toes* mutation contains an intragenic deletion of the *Gli3* gene. *Nat. Genet.* 3: 241–246.

Hurle, J. M., Gañan, Y. and Macias, D. 1989. Experimental analysis of the in vivo chondrogenic potential of the interdigital mesenchyme of the chick limb bud subjected to local ectodermal removal. *Dev. Biol.* 132: 368–374.

Huxley, J. S. and De Beer, G. R. 1934. *The Elements of Experimental Embryology*. Cambridge University Press, Cambridge.

Ingham, P. W. 1994. *Hedgehog* points the way. *Curr. Biol.* 4: 345–350.

Iten, L. E. 1982. Pattern specification and pattern regulation in the embryonic chick limb bud. *Am. Zool.* 22: 117–129.

Izpisúa-Belmonte, J.-C., Tickle, C., Dollé, P., Wolpert, L. and Duboule, D. 1991. Expression of the homeobox *Hox-4* genes and the specification of position in chick wing development. *Nature* 350: 585–589.

Javois, L. C. and Iten, L. E. 1986. The handedness and origin of supernumerary limb structures following 180° rotation of the chick limb bud on its stump. *J. Embryol. Exper. Morphol.* 91: 135–152.

Kieny, M. 1960. Rôle inducteur du mésoderme dans la différenciation précoce du bourgeon de membre chez l'embryon de poulet. *J. Embryol. Exp. Morphol.* 8: 457–467.

Krabbenhoft, K. M. and Fallon, J. F. 1989. The formation of leg or wing specific structures by leg bud cells grafted to the wing bud is influenced by proximity to the apical ridge. *Dev. Biol.* 131: 373–382.

Lande, R. 1978. Evolutionary mechanisms of limb loss in tetrapods. *Evolution* 32: 73–92.

Laufer, E., Nelson, C. E., Johnson, R. L., Morgan, B. A. and Tabin, C. 1994. *Sonic hedgehog* and *Fgf-4* act through a signalling cascade and feedback loop to integrate growth and patterning of the developing limb bud. *Cell* 79: 993–1003.

Laufer, E. and seven others. 1997. The *Radical fringe* expression boundary in the limb bud ectoderm regulates AER formation. *Nature.* 386: 366–367.

Li, Q. Y. and sixteen others. 1996. Holt-Oram syndrome is casued by mutations in TBX5, a member of the *Brachyury* (T) gene family. *Nat. Genet.* 15: 21–29.

Loomis, C. A., Harris, E., Michaud, J. Wurst, J., Hanks, W. and Joyner, A. L. 1996. The mouse *Engrailed-1* gene and ventral limb patterning. *Nature* 382: 360–363.

López-Martínez, A. and seven others. 1995. Limb-patterning activity and restricted posterior localization of the amino-terminal product of sonic hedgehog cleavage. *Curr. Biol.* 5: 791–796.

MacCabe, J. A., Errick, J. and Saunders, J. W. Jr. 1974. Ectodermal control of dorsoventral axis in leg bud of chick embryo. *Dev. Biol.* 39: 69–82.

Maden, M. 1982. Vitamin A and pattern formation in the regenerating limb. *Nature* 295: 672–675.

Maden, M. 1993. The homeotic transformation of tails into limbs in *Rana temporaria* by retinoids. *Dev. Biol.* 159: 379–391.

Mahmood, R. and nine others. 1995. A role for FGF-8 in the initiation and maintenance of vertebrate limb outgrowth. *Curr. Biol.* 5: 797–806.

Marigo, V., Johnson, R. L., Vortkamp, A., and Tabin, C. J. 1996. Sonic hedgehog differentially regulates expression of GLI and GLI3 during Development of the tetrapod limb. *Development* 180: 273–283.

Meinhardt, H. 1984. Models for positioning signaling, the threefold subdivision of segments and the pigmentation pattern of molluscs. *J Embryol. Exp. Morphol.* 83 Suppl.: 289–311.

Mescher, A. L. 1992. Trophic activity of regenerating peripheral nerves. *Comments Dev. Neurobiol.* 1: 373–390.

Mescher, A. L. and Tassava, R. A. 1975. Denervation effects on DNA replication and mitosis during the initiation of limb regeneration in adult newts. *Dev. Biol.* 44: 187–197.

Mohanty-Hejmadi, P., Dutta, S. K. and Mahapatra, P. 1992. Limbs generated at the site of tail amputation in marbled balloon frog after vitamin A treatment. *Nature* 355: 352–353.

Molven, A., Wright, C. V. E., Bremiller, R., De Robertis, E. M. and Kimmel, C. B. 1990. Expression of a homeobox gene product in normal and mutant zebrafish embryos: Evolution of the tetrapod body plan. *Development* 109: 279–288.

Morgan, B. A. and Tabin, C. 1994. *Development 1994 Suppl.*, p. 181–186.

Mori, C., Nakamura, N., Kimura, S., Irie, H., Takigawa, T. and Shiota, K. 1995. Programmed cell death in the interdigital tissue of the fetal mouse limb is apoptosis with DNA fragmentation. *Anat. Rec.* 242: 103–110.

Mortlock, D. P., Post, L. C. and Innis, J. W. 1996. The molecular basis of *hypodactyly* (*Hd*): a deletion in *Hoxa13* leads to arrest of digital arch formation. *Nat. Genet.* 13: 284–289.

Mullen, L. M., Bryant, S.V., Torok, M. A., Blumberg, B. and Gardiner, D. M. 1996. Nerve dependency of regeneration: the role of *Distal-less* and FGF signaling in amphibian limb regeneration. *Development.* 122: 3487–3497.

Müller, G., Streicher, J. and Müller, R. 1996. Homeotic duplicate of the pelvic body segment in regenerating tadpole tails induced by retinoic acid. *Dev. Genes Evol.* 206: 344–348.

Munaim, S. I. and Mescher, A. L. 1986. Transferrin and the trophic effect of neural tissue on amphibian limb reneration blastemas. *Dev. Biol.* 116: 138–142.

Muneoka, K. and Bryant, S. V. 1982. Evidence that patterning mechanisms in developing and regenerating limbs are the same. *Nature* 298: 369–371.

Muragaki, Y., Mundlos, S., Upton, J. and Olsen, B. 1996. Altered growth and branching patterns in synpolydactyly caused by mutations in HOXD13. *Science* 272: 548–551.

Nardi, J. B. and Stocum, D. L. 1983. Surface properties of regenerating limb cell: Evidence for gradation along the proximodistal axis. *Differentiation* 25: 27–31.

Nelson, C. E. and Tabin, C. 1995. Footnote on limb evolution. *Nature* 375: 630–631.

Nelson, C. E. and nine others. 1996. Analysis of *Hox* gene expression in the chick limb bud. *Development* 122: 1449–1466.

Newman, S. A. 1996. Sticky fingers: *Hox* genes and cell adhesion in vertebrate Development of the tetrapod limb. *BioEssays* 18: 171–174.

Niazi, I. A. and Saxena, S. 1978. Abnormal hindlimb regeneration in tadpoles of the toad *Bufo andersonii* exposed to excess vitamin A. *Folia Biol.* (Krakow) 26: 3–8.

Niswander, L. and Martin, G. M. 1993. FGF-4 and BMP-2 have opposite effects on limb growth. *Nature* 361: 68–71.

Niswander, L., Tickle, C., Vogel, A., Booth, I. and Martin, G. R. 1993. FGF-4 replaces the apical ectodermal ridge and directs outgrowth and patterning of the limb. *Cell* 75: 579–587.

Niswander, L., Jeffrey, S., Martin, G. R. and Tickle, C. 1994. A positive feedback loop coordinates growth and patterning in the vertebrate limb. *Nature* 371: 609–612.

Nohno, T. and seven others. 1991. Involvement of the *Chox-4* chicken homeobox genes in determination of anteroposterior axial polarity during Development of the tetrapod limb. *Cell* 64: 1197–1205.

Noji, S. and ten others. 1991. Retinoic acid induces polarizing activity but is unlikely to be a morphogen in the chick limb bud. *Nature* 350: 83–86.

Noramly, S., Pisenti, J., Abbott, U. and Morgan, B. 1996. Gene expression in the *limbless* mutant: Polarized gene expression in the absence of Shh and AER. *Dev. Biol.* 179: 339–346.

Ogura, T., Alvarez, I. S., Vogel, A., Rodrigez, C., Evans, R. M. and Belmonte, J. C. I. 1996. Evidence that *Shh* cooperates with a retinoic acid inducible co-factor to establish ZPA-like activity. *Development* 122: 537–542.

Oliver, G., Wright, C. V. E., Hardwicke, J. and De Robertis, E. M. 1988. A gradient of homeodomain protein in developing forelimbs of *Xenopus* and mouse embryos. *Cell* 55: 1017–1024.

Opitz, J. M. 1985. The developmental field concept. *Am. J. Med. Genet.* 21: 1–11.

Owen, R. 1849. *On the Nature of Limbs.* J. Van Voor, London.

Parr, B. A. and McMahon, A. P. 1995. Dorsalizing signal wnt-7a required for normal polarity of D-V and A-P axes of the mouse limb. *Nature* 374: 350–353.

Parr, B. A., Shea, M. J., Vassileva, G. and McMahon, A. P. 1993. Mouse *Wnt* genes exhibit discrete domains of expression in early embryonic CNS and limb buds. *Development* 119: 247–261.

Phillips, J. 1991. Higgledy, piggledy. *N. Engl. J. Med.* 324: 497.

Pollak, R. D. and Fallon, J. F. 1976. Autoradiographic analysis of macromolecular synthesis in prospectively necrotic cells of the chick limb bud. II. Nucleic acids. *Exp. Cell Res.* 100: 15–22.

Riddle, R. D., Johnson, R. L., Laufer, E. and Tabin, C. 1993. *Sonic hedgehog* mediates the polarizing activity of the ZPA. *Cell* 75: 1401–1416.

Riddle, R. D., Ensini, M., Nelson, C., Tsuchida, T., Jessell, T. M. and Tabin, C. 1995. Induction of the LIM homeobox gene *Lmx1* by WNT7a establishes dorsoventral

patter in the vertebrate limb. *Cell* 83: 631–640.

Rodriguez-Esteban, C., Schwabe, J. W. R., De La Peña, J., Foys, B., Eshelman, B. and Izpisúa-Belmonte, J. C. 1997. Radical fringe positions the apical ectodermal ridge at the dorsoventral boundary of the vertebrate limb. *Nature* 386: 360–366.

Ros, M. A., Lopez-Martinez, A., Simandl, B. K., Rodriguez, C., Belmonte, J. C. I., Dahn, R. and Fallon, J. F. 1996. The limb field mesoderm determines initial limb bud anteroposterior asymmetry and budding independent of sonic hedgehog or apical ectodermal gene expressions. *Development* 122: 2319–2330.

Rose, S. M. 1962. Tissue-arc control of regeneration in the amphibian limb. *In* D. Rudnick (ed.), *Regeneration.* Ronald, New York, pp. 153–176.

Rosenquist, G. C. 1971. The origin and movement of the limb-bud epithelium and mesenchyme in the chick embryo as determined by radioautographic mapping. *J. Embryol. Exp. Morphol.* 25: 85–96.

Rossant, J., Zirngibl, R., Cado, D., Shago, M. and Giguère, V. 1991. Expression of a retinoic acid response element-*hsplacZ* transgene defines specific domains of transcriptional activity during mouse embryogenesis. *Genes Dev.* 5: 1333–1344.

Rowe, D. A., Cairnes, J. M. and Fallon, J. F. 1982. Spatial and temporal patterns of cell death in limb bud mesoderm after apical ectodermal ridge removal. *Dev. Biol.* 93: 83–91.

Rubin, L. and Saunders, J. W., Jr. 1972. Ectodermal–mesodermal interactions in the growth of limbs in the chick embryo: Constancy and temporal limits of the ectodermal induction. *Dev. Biol.* 28: 94–112.

Saunders, J. W., Jr. 1948. The proximal-distal sequence of origin of the parts of the chick wing and the role of the ectoderm. *J. Exp. Zool.* 108: 363–404.

Saunders, J. W., Jr. 1972. Developmental control of three-dimensional polarity in the avian limb. *Ann. N.Y. Acad. Sci.* 193: 29–42.

Saunders, J. W., Jr. 1982. *Developmental Biology.* Macmillan, New York.

Saunders, J. W., Jr. and Fallon, J. F. 1966. Cell death in morphogenesis. *In* M. Locke (ed.), *Major Problems of Developmental Biology.* Academic Press, New York, pp. 289–314.

Saunders, J. W., Jr. and Gasseling, M. T. 1968. Ectodermal-mesodermal interactions in the origin of limb symmetry. *In* R. Fleischmajer and R. E. Billingham (eds.), *Epithelial-Mesenchymal Interactions.* Williams & Wilkins, Baltimore, pp. 78–97.

Saunders, J., Jr. and Reuss, C. 1974. Inductive and axial properties of prospective wing-bud mesoderm in the chick embryo. *Dev. Biol.* 38: 4150.

Saunders, J. W., Jr., Cairns, J. M. and Gasseling, M. T. 1957. The role of the apical ridge of ectoderm in the differentiation of the morphological structure and inductive specificity of limb parts of the chick. *J. Morphol.* 101: 57–88.

Saunders, J. W., Jr., Gasseling, M. T. and Saunders, L.C. 1962. Cellular death in morphogenesis of the avian wing. *Dev. Biol.* 5: 147–178.

Savage, M. P. and Fallon, J. F. 1995. FGF-2 mRNA and its antisense message are expressed in a developmentally specific manner in the chick limb bud and mesonephros. *Dev. Dyn.* 202: 343–353.

Scadding, S. R. and Maden, M. 1994. Retinoic acid gradients during limb regeneration. *Dev. Biol.* 162: 608-617.

Sessions, S. and Ruth, S. B. 1990. Explanation for naturally occurring supernumary limbs in amphibians. *J. Exp. Zool.* 254: 38–47.

Sessions, S. K., Gardiner, D. M. and Bryant, S. V. 1989. Compatible limb patterning mechanisms in urodeles and anurans. *Dev. Biol.* 131: 294–301.

Shen, R. Q., Chen, Y. P., Huang, L., Vitale, E. and Solursh, M. 1994. Characterization of the human *msx-1* promoter and an enhancer responsible for retinoic acid induction. *Cell. Mol. Biol. Res.* 40: 297–312.

Shubin, N. H. and Alberch, P. 1986. A morphogenetic approach to the origin and basic organization of the tetrapod limb. *Evol. Biol.* 20: 319–387.

Simon, H. G., Nelson, C., Goff, D., Laufer, E., Morgan, B. A. and Tabin, C. 1995. The differential expression of myogenic regulatory genes and *msx-1* during dedifferentiation and redifferentiation of regenerating amphibian limbs. *Dev. Dyn.* 202: 1–12.

Singer, M. 1954. Induction of regeneration of the forelimb of the postmetamorphic frog by augmentation of the nerve supply. *J. Exp. Zool.* 126: 419–472.

Singer, M. and Caston, J. D. 1972. Neurotrophic dependence of macromolecular synthesis in the early limb regenerate of the newt, *Triturus. J. Embryol. Exp. Morphol.* 28: 1–11.

Sordino, P., Hoeven, F. van der and Duboule, D. 1995. *Hox* gene expression in teleost fins and the origin of the vertebrate digits. *Nature* 375: 678–681.

Stephens, T. D. and seven others. 1991. Axial and paraxial influences on limb morphogenesis. *J. Morphol.* 208: 367–379.

Stocum, D. L. and Crawford, K. 1987. Use of retinoids to analyse the cellular basis of memory in regenerating amphibian limbs. *Biochem. Cell Biol.* 65: 750–761.

Stocum, D. L. and Fallon, J. F. 1982. Control of pattern formation in urodele limb ontogeny: A review and a hypothesis. *J. Embryol. Exp. Morphol.* 69: 7–36.

Stratford, T., Horton, C. and Maden, M. 1996. Retinoic acid is required for the initiation of outgrowth in the chick limb bud. *Curr Biol.* 6: 1124–1133.

Summerbell, D. 1974. A quantitative analysis of the effect of excision of the AER from the chick limb bud. *J. Embryol. Exp. Morphol.* 32: 651–660.

Summerbell, D. 1979. The zone of polarizing activity: Evidence for a role in abnormal chick limb morphogenesis. *J. Embryol. Exp. Morphol.* 50: 217–233.

Summerbell, D. and Lewis, J. H. 1975. Time, place and positional value in the chick limb bud. *J. Embryol. Exp. Morphol.* 33: 621–643.

Tabin, C. J. 1991. Retinoids, homeoboxes, and growth factors: Toward molecular models for Development of the tetrapod limb. *Cell* 66: 199–217.

Tabin, C. J. 1992. Why we have (only) five fingers per hand: *Hox* genes and the evolution of paired limbs. *Development* 116: 289–296.

Tanaka, M., Tamura, K., Noji, S., Nohno, T. and Ide, H. 1997. Induction of additional limb at the dorsal-ventral boundary of a chick embryo. *Dev. Biol.* 182: 191–203.

Tickle, C. 1981. The number of polarizing region cells required to specify additional digits in the developing chick wing. *Nature* 289: 295–298.

Tickle, C. 1995. Vertebrate Development of the tetrapod limb. *Curr. Biol.* 5: 478–484.

Tickle, C., Summerbell, D. and Wolpert, L. 1975. Positional signaling and specification of digits in chick limb morphogenesis. *Nature* 254: 199–202.

Tickle, C., Alberts, B., Wolpert, L. and Lee, J. 1982. Local application of retinoic acid to the limb bud mimics the action of the polarizing region. *Nature* 296: 564–566.

Tickle, C., Lee, J. and Eichele, G. 1985. A quantitative analysis of the effect of all-*trans*-retinoic acid on the pattern of chick wing development. *Dev. Biol.* 109: 82–95.

Todt, W. L. and Fallon, J. F. 1987. Posterior apical ectodermal ridge removal in chick wing bud triggers a series of events resulting in defective anterior pattern. *Development* 101: 505–515.

Tosney, K. W. and Landmesser, L. T. 1985. Development of the major pathways for neurite outgrowth in the chick hindlimb. *Dev. Biol.* 109: 193–214.

Viviano, C. M., Horton, C. E., Maden, M. and Brockes, J. P. 1995. Synthesis and release of 9-*cis* retinoic acid by the urodele wound epidermis. *Development* 121: 3753–3762.

Vogel, A., Rodriguez, C., Warnken, W. and Izpisúa-Belmonte, J.-C. 1995. Dorsal cell fate specified by chick *Lmx1* during vertebrate Development of the tetrapod limb. *Nature* 378: 716–720.

Vogel, A., Rodriguez, C. and Izpisúa-Belmonte, J.-C. 1996. Involvement of FGF-8 in initiation, outgrowth, and patterning of the vertebrate limb. *Development* 122: 1737–1750.

Vortkamp, A., Gessler, M. and Grzeschik, K.-H. GLI3 zinc-finger gene interrupted by translocations in Grieg syndrome family. *Nature* 352: 539–540.

Wanek, N., Gardiner, D. M., Mueoka, K. and Bryant, S. V. 1991. Conversion by retinoic acid of anterior cells into ZPA cells in the chick wing bud. *Nature* 350: 81–83.

Weiss, P. 1939. *Principles of Development.* Holt, Rinehart & Winston, New York.

Wessells, N. K. 1977. *Tissue Interaction and Development.* Benjamin, Menlo Park, CA.

Wolpert, L. 1969. Positional information and the spatial pattern of cellular formation. *J. Theoret. Biol.* 25: 1–47.

Wolpert, L. 1977. *The Development of Pattern and Form in Animals.* Carolina Biological, Burlington, NC.

Yang, Y. Z. and Niswander, L. 1995. Interaction between signaling molecules Wnt7a and Shh during vertebrate limb development: Dorsal signals regulate anteroposterior patterning. *Cell* 80: 939–947.

Yokouchi, Y., Nakazato, S., Yamamoto, M., Goto, Y., Kameda, T., Iba, H. and Kuroiwa, A. 1995. Misexpression of *Hoxa-13* induces cartilage homeotic transformation and changes in adhesiveness in chick limb buds. *Genes Dev.* 9: 2509–2522.

Zaleske, D. J. 1985. Development of the upper limb. *Hand Clin.* 1985(3): 383–390.

Zou, H. and Niswander, L. 1996. Requirement for BMP signaling in interdigital apoptosis and scale formation. *Science* 272: 738–741.

Zwilling, E. 1955. Ectoderm-mesoderm relationship in the development of the chick embryo limb bud. *J. Exp. Zool.* 128: 423–441.

Cell interactions at a distance: Hormones as mediators of development

19

The old order changeth, yielding place to the new.
ALFRED LORD TENNYSON (1886)

The earth-bound early stages built enormous digestive tracts and hauled them around on caterpillar treads. Later in the life-history these assets could be liquidated and reinvested in the construction of an entirely new organism—a flying-machine devoted to sex.
CARROLL M. WILLIAMS (1958)

ORGAN FORMATION IN ANIMALS is accomplished by the interactions of numerous cell types. In the preceding two chapters, we have seen how developmental interactions can be mediated by adjacent cell populations. In this chapter, we will discuss the regulation of development by diffusible molecules that travel long distances from one cell type to another. Diffusible regulators of development that travel through the blood to cause changes in the differentiation or morphogenesis of other tissues are called **hormones.**

Metamorphosis: The hormonal directing of development

Because minute quantities of hormones are enough to accomplish their actions, it is exceedingly difficult to isolate them from embryos. Some of the most thorough analysis of the hormonal control of development has therefore centered upon the dramatic "reprogramming" of development known as **metamorphosis.**

In many species of animals, embryonic development leads to a larval stage with characteristics very different from those of the adult organism. Very often, the larval forms are specialized for some function, such as growth or dispersal. The pluteus larva of the sea urchin, for instance, can travel on ocean currents, whereas the adult urchin leads a sedentary existence. The caterpillar larvae of butterflies and moths are specialized for feeding, whereas their adult forms are specialized for flight and reproduction and often lack the mouth parts necessary for eating. The division of functions between larva and adult is often remarkably distinct (Wald, 1981). Mayflies hatch from eggs and develop for several months. All this development enables them to spend one day as fully developed winged insects, mating quickly before they die. As might be expected from this discussion, the larval form and the adult form often live in different environments. Moreover, as first noted by Weismann (1875), the larvae must have their own adaptation to help them survive. The adult viceroy butterfly mimics the more unpalatable monarch butterfly, but the viceroy caterpillar does not resemble the beautiful larva of the monarch. Rather, the viceroy larva escapes detection by resembling bird droppings (Begon et al., 1986).

During metamorphosis, developmental processes are reactivated by specific hormones, and the entire organism changes to prepare itself for its new mode of existence. These changes are not solely ones of form. In amphibian tadpoles, metamorphosis causes the developmental maturation of liver enzymes, hemoglobin, and eye pigments, as well as the remodeling of the nervous, digestive, and reproductive systems. Thus, metamorphosis is a time of dramatic developmental change affecting the entire organism.

This chapter focuses on three cases in which hormones reactivate the developmental processes after birth: amphibian metamorphosis, insect metamorphosis, and mouse breast development.

Amphibian metamorphosis

In amphibians, metamorphosis is generally associated with the changes that prepare an aquatic organism for a terrestrial existence. In urodeles (salamanders), the changes include the resorption of the tail fin, the destruction of the external gills, and the change of skin structure. In anurans (frogs and toads), the metamorphic changes are most striking, and almost every organ is subject to modification (Table 19.1). The changes in form are very obvious (Figure 19.1). Regressive changes include the loss of the tadpole's horny teeth and internal gills, as well as the destruction of the tadpole's tail. At the same time, constructive processes such as limb development and dermoid gland

Table 19.1 Summary of some metamorphic changes in anurans

System	Larva	Adult
Locomotory	Aquatic; tail fins	Terrestrial; tailless tetrapod
Respiratory	Gills, skin, lungs; larval hemoglobins	Skin, lungs; adult hemoglobins
Circulatory	Aortic arches; aorta; anterior, posterior, and common cardinal veins	Carotid arch; systemic arch; jugular veins
Nutritional	*Herbivorous*: long spiral gut—intestinal symbionts; small mouth—horny jaws, labial teeth	*Carnivorous*: Short gut—proteases; large mouth—long tongue
Nervous	Lack of nictitating membrane; porphyropsin, lateral line system—Mauthner's neurons	Development of ocular muscles, nictitating membrane, rhodopsin, loss of lateral line system—degeneration of Mauthner's neurons; tympanic membrane
Excretory	Largely ammonia, some urea (ammonotelic)	Largely urea, high activity of enzymes of ornithine-urea cycle (ureotelic)
Integumental	Thin, bilayered epidermis with thin dermis; no mucous glands or granular glands	Stratified squamous epidermis with adult keratins; well-developed dermis contains mucous glands and granular glands secreting antimicrobial peptides

Source: Data from Turner and Bagnara, 1976; Reilly et al., 1994.

construction are also evident. For locomotion, the paddle tail recedes while the hindlimbs and forelimbs differentiate. The cartilaginous skull of the tadpole is replaced by the predominantly bony skull of the froglet. The horny teeth constructed for tearing pond plants disappear as the mouth and jaw take a new shape, and the tongue muscle develops. Meanwhile the large intestine characteristic of herbivores shortens to suit the more carnivorous diet of the adult frog. The gills regress, and the gill arches degenerate. The lungs enlarge, and muscles and cartilage develop for pumping air in and out of the lungs. The sensory apparatus changes, too, as the lateral line system of the tadpole degenerates and the eye and ear undergo further differentiation. In the ear, the middle ear develops, as does the tympanic membrane so characteristic of frogs and toads. In the eye, both nictitating membranes and eyelids emerge. Moreover, the eye pigment changes. In tadpoles, as in freshwater fishes, the major retinal photopigment is porphyropsin, a complex between the protein opsin and the aldehyde of vitamin A_2. In adult frogs, the pigment changes to rhodopsin, the characteristic photopigment of terrestrial and marine vertebrates. Rhodopsin consists of opsin conjugated to the aldehyde of vitamin A_1 (Wald, 1945, 1981; Smith-Gill and Carver, 1981; Hanken and Hall, 1988).

Other biochemical events are also associated with metamorphosis. Tadpole hemoglobin binds oxygen faster and releases it more slowly than does adult hemoglobin (McCutcheon, 1936). Moreover, Riggs (1951) showed that the binding of oxygen by tadpole hemoglobin is independent of pH, whereas frog hemoglobin (like most other vertebrate hemoglobins) shows increased oxygen binding as the pH rises (the Bohr effect). Another biochemical change in the metamorphosis of some frogs is the induction of those enzymes necessary for the production of urea. Tadpoles, like most freshwater fishes, are **ammonotelic;** that is, they excrete ammonia. Many adult frogs (such as the genus *Rana*, but not *Xenopus*) are **ureotelic,** excreting urea, like most terrestrial vertebrates. During metamorphosis, the liver develops enzymes necessary to create urea from carbon dioxide and ammonia. These enzymes constitute the urea cycle, and each of them is seen to arise during metamorphosis (Figure 19.2).

Hormonal Control of Amphibian Metamorphosis

All these diverse changes are brought about by the secretion of the hormones **thyroxine (T_4)** and **triiodothyronine (T_3)** from the thyroid during metamorphosis (Figure 19.3). It is now believed that T_3 is the active hormone, as it will cause the metamorphic changes in thyroidectomized tadpoles in much lower concentrations than will T_4 (Kistler et al., 1977; Robinson et al., 1977). The control of metamorphosis by thyroid hormones was shown by Gudernatsch (1912), who found that tadpoles metamorphosed prematurely when fed powdered sheep thyroid gland. Allen (1916) and Hoskins and Hoskins (1917) found that when they removed the thyroid rudiment from early tadpoles, the larvae never metamorphosed, becoming giant tadpoles instead.

REGIONALLY SPECIFIC CHANGES.   The various organs of the body respond differently to hormonal stimulation. The same stimulus causes certain tissues to degenerate while causing others to develop and differentiate. For instance, tail degeneration is clearly associated with the increasing levels of thyroid hormones. The degeneration of tail structures is relatively rapid, as the bony skeleton does not extend to the tail, which is supported only by the notochord (Wassersug, 1989). This degeneration can be shown in vitro (Weber, 1967) when isolated tail pieces are placed in agar dishes and subjected to

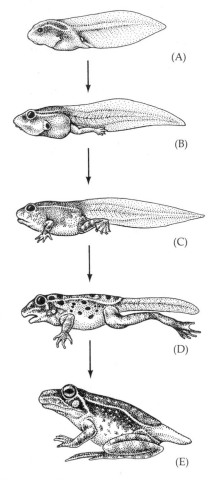

Figure 19.1
Sequence of metamorphosis in the frog *Rana pipiens*. (A) Premetamorphic tadpole. (B) Prometamorphic tadpole showing hindlimb growth. (C) Onset of metamorphic climax as forelimbs emerge. (D,E) Climax stages.

Figure 19.2
Development of the urea cycle during anuran metamorphosis. (A) The major features of the urea cycle, by which nitrogenous wastes can be detoxified and excreted. (B) Emergence of urea cycle enzyme activities correlated with metamorphic changes in the frog *Rana catesbeiana*. (After Cohen, 1970.)

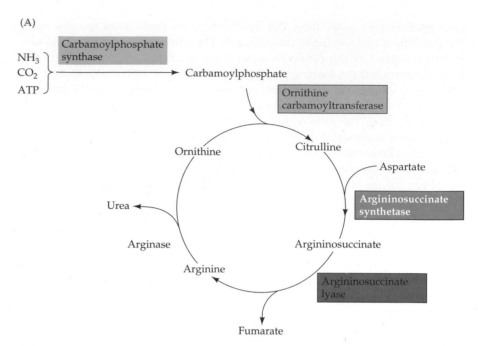

(A)

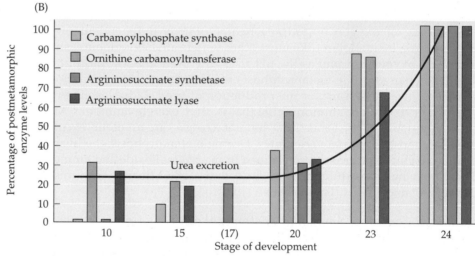

(B)

chemical treatments. Those tails grown in untreated medium remain healthy, whereas those placed into medium containing thyroid hormones undergo characteristic regression. Moreover, prolactin inhibits the tail degeneration induced by thyroid hormones (Brown and Frye, 1969). The regression of the tail is thought to occur in four stages. First, protein synthesis decreases in the striated muscle cells of the tail (Little et al., 1973). Next, there is an increase in the lysosomal enzymes. Concentrations of proteases, RNase, DNase, collagenase, phosphatase, and glycosidases all rise in the epidermis, notochord, and nerve cord cells (Fox, 1973). Cell death is probably caused by the release of these enzymes into the cytoplasm. The epidermis helps to digest the muscle tissue, probably by releasing these digestive enzymes. If the epidermis is surgically removed from the tail tips, the tips will not regress when cultured in thyroxine (Eisen and Gross, 1965; Niki et al., 1982). After this cell death, macrophages collect in the tail region, digesting the debris with their own proteolytic enzymes (Kaltenbach et al., 1979). The result is that the tail becomes a large sac of proteolytic enzymes (Figure

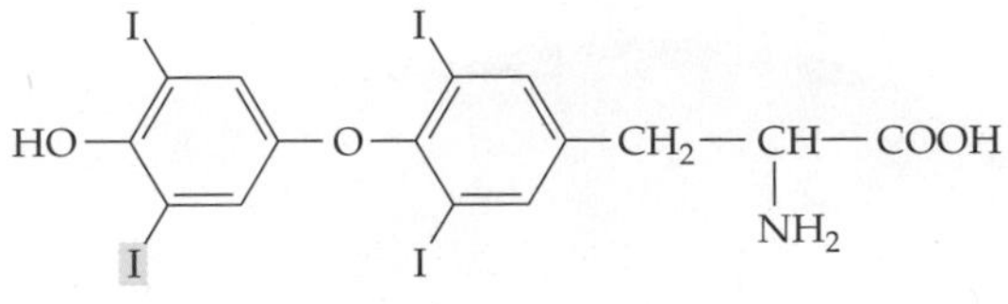

HO—⬡—I ... —O—⬡— CH₂—CH—COOH

$$HO-\bigcirc-O-\bigcirc-CH_2-CH(NH_2)-COOH$$

Thyroxine (T₄)

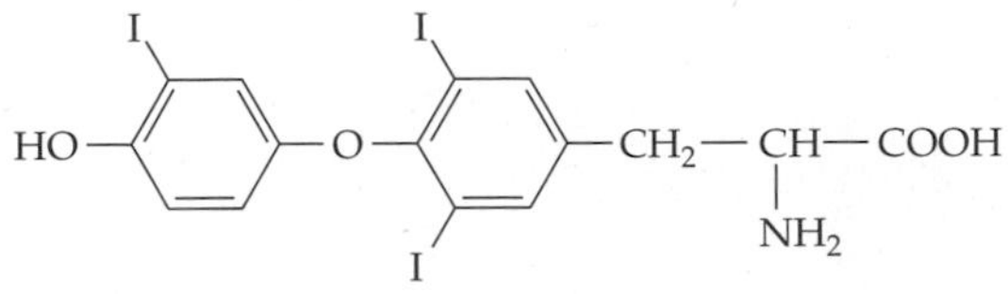

$$HO-\bigcirc-O-\bigcirc-CH_2-CH(NH_2)-COOH$$

Triiodothyronine (T₃)

Figure 19.3
Formulas of thyroxine (T₄) and triiodothyronine (T₃).

19.4). The major proteolytic enzymes appears to be collagenases and other metalloproteinases whose synthesis depends on thyroid hormones. If a metalloproteinase inhibitor (TIMP) is added to the tails, it prevents the thyroid hormone-induced tail regression (Oofusa and Yoshizato, 1991; Patterson et al., 1995).

The response to thyroid hormones is intrinsic to the organ itself and is not dependent on surrounding tissues. In the epidermis, the response to thyroid hormones depends on which part of the body the epidermis covers. Tadpole head and body epidermal cells usually undergo a small turnover (as expected in skin), and T_3 does not change this rate. In the *tail*, however, T_3 causes a rapid increase in the keratinization and death of these cells. It also causes a tail-specific suppression of stem cell divisions that could give rise to more epidermal cells. The result is the death of the tail epidermal cells

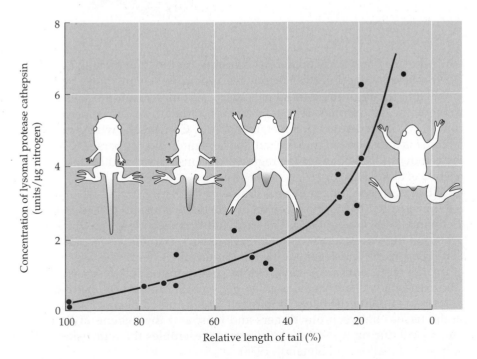

Figure 19.4
Increase in lysosomal protease activity during tail regression in *Xenopus laevis*. The lysosomal enzymes are thought to be responsible for digesting the tail cells. (After Karp and Berrill, 1981.)

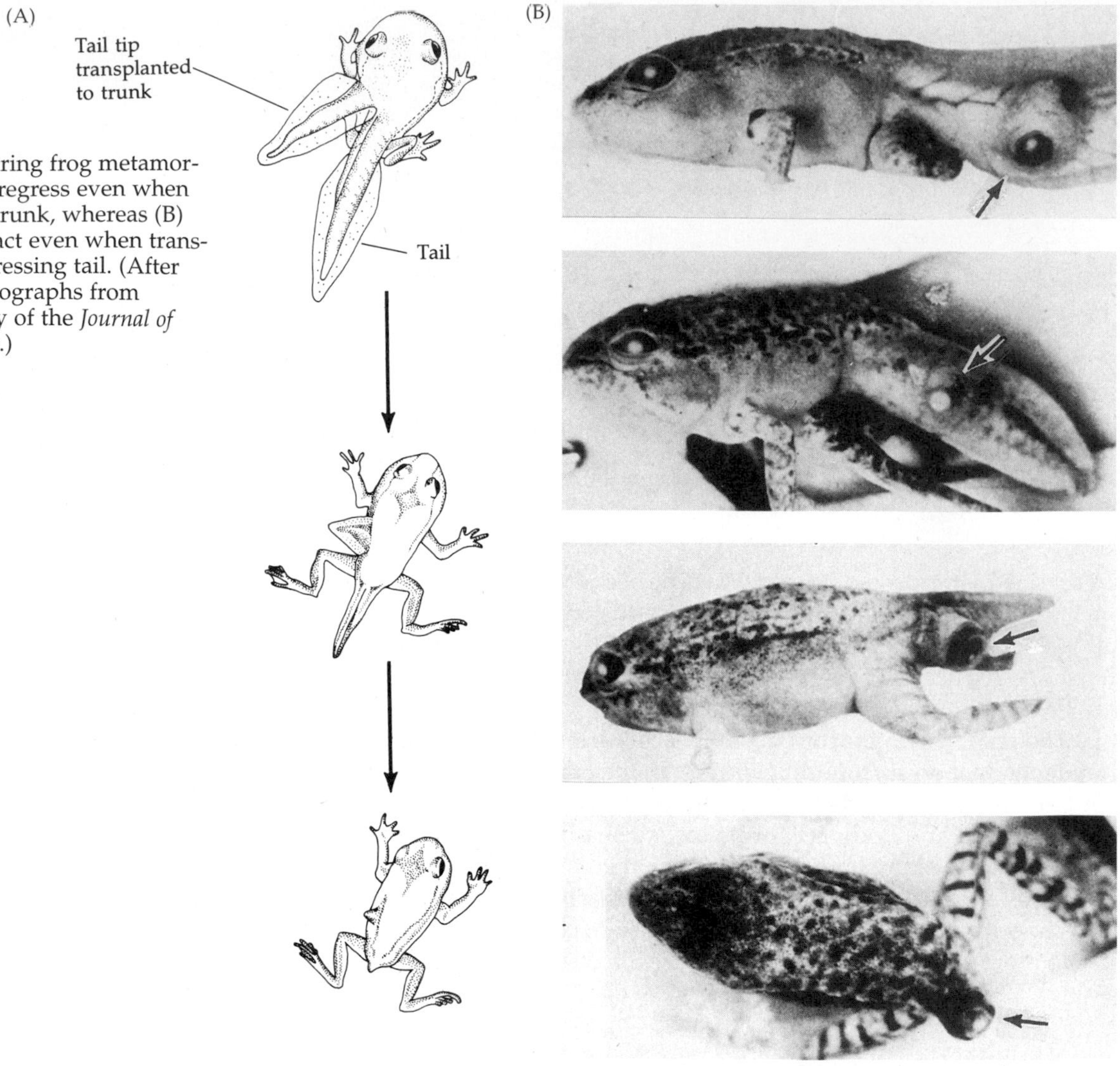

(A)

Tail tip transplanted to trunk

Tail

Figure 19.5
Organ specificity during frog metamorphosis. (A) Tail tips regress even when transplanted to the trunk, whereas (B) eye cups remain intact even when transplanted into the regressing tail. (After Schwind, 1933; photographs from Geigy, 1941, courtesy of the *Journal of Experimental Zoology*.)

(B)

while the head and body epidermis continues to function (Nishikawa et al., 1989). These local epidermal responses appear to be controlled by the regional specificity of the dermal mesoderm. If tail dermatome cells (which generate the tail dermis) are transplanted into the trunk, the epidermis they contact will degenerate upon metamorphosis. Conversely, when trunk dermatome is transplanted into the tail, those regions of skin persist. Changing the ectoderm does not alter the regional response to thyroid hormones (Kinoshita et al., 1989).

This organ-specific response is dramatically demonstrated when tail tips are transplanted to the trunk region or when eye cups are placed in the tail (Schwind, 1933; Geigy, 1941). The extra tail tip placed into the trunk is not protected from degeneration, but the eye retains its integrity despite the fact that it lies within the degenerating tail (Figure 19.5). Thus, the degeneration of the tail represents a cell-autonomous programmed cell death. Only specific tissues die when a signal is given. Such programmed cell deaths are important in molding the body. In humans, programmed degeneration occurs in the tissues between our fingers and toes, and the degeneration of the human tail during week 4 of development resembles the regression of the tadpole tail (Fallon and Simandl, 1978).

COORDINATION OF DEVELOPMENTAL CHANGES. One of the major problems of metamorphosis is the coordination of developmental events. The tail should not degenerate until some other means of locomotion—the limbs—has developed, and the gills should not regress until the animal can utilize its newly developed lung muscles. The means of coordinating the metamorphic events appears to be different amounts of hormone that produce different specific effects (Kollros, 1961). This model is called the **threshold concept.** As the concentration of thyroid hormones gradually builds up, different events occur at different concentrations of the hormone. If tadpoles are deprived of their thyroids and are placed in a dilute solution of thyroid hormones, the only morphological effects are the shortening of the intestines and accelerated hindlimb growth. However, at higher concentrations of thyroid hormones, tail regression is seen before the hindlimbs are formed. These experiments suggest that as thyroid hormone levels rise, the hindlimbs develop first and then the tail regresses. Similarly, when T_3 is given to tadpoles, it induces the earliest-forming bones at the lowest dosages and the last bones at the higher doses, mimicking the natural situation (Hanken and Hall, 1988). Thus, the timing of metamorphosis is regulated by the competency of different tissues to respond to thyroid hormones.

NEURONAL CHANGES. But what happens to the nervous system when the animal is constructing a new organism from the old? The adaptational anatomy of a frog certainly differs from that of its tadpole. One readily observed consequence of anuran metamorphosis is the movement of the eyes forward from their originally lateral position (Figure 19.6).* The lateral eyes of the tadpole are typical of preyed-upon herbivores, whereas the frontally located

*One of the most spectacular movements of the eyes during metamorphosis occurs in flatfish such as flounder. Originally, the eyes are on opposite sides of the face. However, during metamorphosis, one of the eyes migrates dorsally to meet the other at the top of the head, allowing the fish to dwell on the bottom, looking upward (Martin and Drewry, 1978).

Figure 19.6
Eye migration and associated neuronal changes during metamorphosis of the *Xenopus laevis* tadpole. The eyes of the tadpole are laterally placed, so there is relatively little binocular field. Eyes migrate dorsally and rostrally during metamorphosis, creating a large binocular field for the adult frog. Below the metamorphosing tadpole is a representation of the optic region of its brain. When horseradish peroxidase is injected into the retina, the optic neurons translocate it to the contralateral (opposite) side of the brain (small arrow), but not to the ipsilateral side. As metamorphosis continues, the ipsilateral projections (involved in binocular vision) begin to be seen (large arrow). (From Hoskins and Grobstein, 1984, courtesy of P. Grobstein.)

eyes of the frog befit its more predatory lifestyle. To catch its prey, the frog needs to see in three dimensions. That is, it has to acquire a binocular field of vision wherein input from both eyes converges in the brain. In the tadpole, the right eye innervates the left side of the brain and vice versa. There are no ipsilateral (same-side) projections of the retinal neurons. During metamorphosis, however, these additional ipsilateral pathways emerge, enabling input from both eyes to reach the same area of the brain (Currie and Cowan, 1974; Hoskins and Grobstein, 1985a). In *Xenopus*, these new neuronal pathways result not from the remodeling of existing neurons, but from the formation of new neurons that differentiate in response to thyroid hormones (Hoskins and Grobstein, 1985a,b). Both the movement of the eyes to their new positions and the differentiation of new neurons that extend processes ipsilaterally to the brain are thyroid-hormone-dependent changes.

Other neurons also undergo profound changes. Some nerve cells, such as those that innervate the tadpole tail muscles, die (Forehand and Farel, 1982). This neuronal death does not seem to be caused by the death of the target tissue but appears to be a separate response to thyroid hormones. Other neurons, such as certain motor neurons in the tadpole jaw, switch their allegiances from larval muscle to the newly formed adult muscle (Alley and Barnes, 1983). Still other neurons, such as those innervating the tongue (a newly formed muscle not present in the larva), have lain dormant during the tadpole stage and first form connections during metamorphosis (Grobstein, 1987). The brain also undergoes changes in its structure during metamorphosis. Thus, the anuran nervous system undergoes enormous restructuring during metamorphosis. Some neurons die, others are born, and others change their specificity.

BEHAVIORAL CHANGES. Metamorphosis also brings behavioral changes; obviously, the behavior of a frog differs from that of its tadpole. The study of tropical frogs has recently demonstrated surprising behaviors involving tadpole-frog interrelationships. The poison arrow frog, *Dendrobates,* is found in the rain forests of Central America. Most of the time, these highly toxic frogs live in the leaf litter of the forest floor. After laying eggs in a damp leaf, a parent (sometimes male, sometimes female) stands guard over the eggs. When the eggs mature into tadpoles, the guarding frog allows them to wriggle onto its back (see Plate 34). The frog then climbs into the canopy until it finds a bromeliad with a small pool of water in its leaf base. Here it deposits one of its tadpoles, then goes back for another, and so on, until the brood has been placed in numerous small pools. Then each day the female returns to these pools and deposits a small number of unfertilized eggs into them, replenishing the dwindling food supply for the tadpoles until they finish metamorphosis (Mitchell, 1988; vanWijngaarden and Bolanos, 1992; Brust, 1993). It is not known how the female frog remembers—or is informed about—where the tadpoles have been deposited.

Molecular Responses to Thyroid Hormones During Metamorphosis

Evidence from inhibitor experiments suggested that thyroid hormones controlled metamorphosis at the level of transcription. Weber (1967) demonstrated that the injection of actinomycin D into normal prometamorphic tadpoles inhibited tail regression and head remodeling. In the liver (which is remodeled during metamorphosis rather than being destroyed or replaced), the metamorphic changes are accompanied by dramatic increases in ribosomal and messenger RNA synthesis, the rate of protein synthesis increasing nearly 100-fold within 4 hours of thyroid hormone stimulation (Cohen et al., 1978). Many of these new mRNAs are those coding for the new enzymes of the adult liver. Mori and co-workers (1979) have shown that much of the increase in carbamoylphosphate synthase can be attributed to the increased transcription from that gene.

Dot blots wherein radioactive mRNA from premetamorphic and metamorphosing bullfrog tadpoles is hybridized to cloned genes have demonstrated three types of responses to thyroid hormones. The transcription of one set of genes increases in response to either natural or experimentally induced metamorphosis; the transcription of another set of genes is dramatically decreased; and a third set of genes remains unaffected by thyroid hormones (Lyman and White, 1987; Mathison and Miller, 1987). The transcription of mRNAs for albumin, adult globin, adult skin keratin, and the *Xenopus* homologue of Sonic hedgehog is controlled by T_3. The transcription of the *sonic hedgehog* gene is interesting, since it suggests that the regional patterning of the organs forming during metamorphosis might be generated by the reappearance of some of the same molecules that had structured the embryo (Stolow and Shi, 1995).

But these are relatively late gene responses to T_3. The earliest reponse to T_3 is the transcriptional activation of the **thyroid hormone receptor (TR)** genes (Yaoita and Brown, 1990; Kawahara et al., 1991). Thyroid hormone receptors are members of the steroid hormone receptor superfamily of transcription factors. There are two major types of TR, TRα and TRβ, and the mRNAs of both are present at relatively low levels before metamorphosis begins (Table 19.2; Kawahara et al., 1991; Baker and Tata, 1992). However, the synthesis of these mRNAs accelerates dramatically as metamorphosis begins. The injection of exogenous T_3 causes a 2- to 5-fold increase in TRα message and a 20- to 50-fold increase in the mRNA for TRβ. This "autoinduction" of T_3 receptor message by T_3 may play a significant role in the acceleration of metamorphosis (Figure 19.7). The more T_3 receptors a tissue has, the more competent it should be to respond to small amounts of T_3. Thus, **metamorphic climax,** that time when the visible changes of metamorphosis occur rapidly, may be brought about by the enhanced production and induction of more T_3 receptors. The mechanism for this induction is not known, but Kanamori and Brown (1992) have shown that the upregulation of TRβ mRNA is significantly blocked by inhibitors of protein synthesis. Thus, other proteins are probably involved in the T_3-responsiveness of the TR genes. The TR does not work alone, however, but forms a dimer with the retinoid receptor, RX. This dimer binds thyroid hormones and can enter the nucleus to effect transcription (Wong and Shi, 1995).

Table 19.2 Relative accumulation of TRα and β mRNA in *Xenopus* tadpoles following treatment with T_3 and prolactin

	Relative units	
Treatment	**TRα**	**TRβ**
None	505	24
T_3	1290	368
Prolactin + T_3	799	<10
Prolactin	405	43

Source: After Baker and Tata, 1992.

(A) PREMETAMORPHOSIS

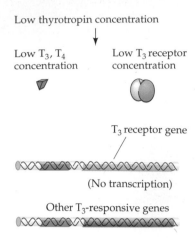

Low thyrotropin concentration

Low T_3, T_4
concentration

Low T_3 receptor
concentration

T_3 receptor gene

(No transcription)

Other T_3-responsive genes

(B) EARLY METAMORPHOSIS
(PROMETAMORPHOSIS)

Thyrotropin concentration increases

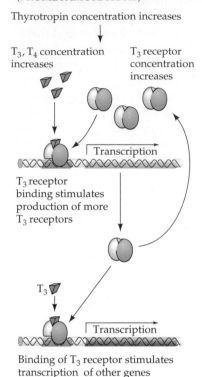

T_3, T_4 concentration
increases

T_3 receptor
concentration
increases

Transcription

T_3 receptor
binding stimulates
production of more
T_3 receptors

T_3

Transcription

Binding of T_3 receptor stimulates
transcription of other genes

(C) METAMORPHIC CLIMAX

High thyrotropin concentration

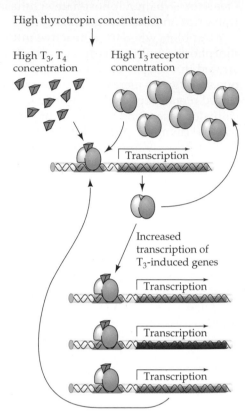

High T_3, T_4
concentration

High T_3 receptor
concentration

Transcription

Increased
transcription of
T_3-induced genes

Transcription

Transcription

Transcription

Some T_3- induced protein
stimulates more T_3 message

Figure 19.7
Hypothetical model for the acceleration of metamorphosis in *Xenopus* by the auto-induction of T_3 receptors by T_3. (A) In the tadpole, premetamorphosis is character-ized by low levels of thyrotropin (thyroid hormone-releasing factor), thyroid hor-mones, and T_3 receptors. (B) At the onset of metamorphosis, the levels of thy-rotropin are increased (probably as a part of the developmental maturation of the pituitary gland). This increases the amount of T_3. The T_3 binds to the small amount of T_3 receptor to stimulate the transcription of more T_3 receptor mRNA. Some other T_3-induced proteins are also needed for the transcription of more T_3 message. (C) At metamorphic climax, the large concentrations of T_3 induce the synthesis of still more T_3 receptors, which cause a more rapid response to T_3.

The hormone prolactin has also been found to inhibit the upregulation of TRα and TRβ mRNAs. Moreover, if the upregulation of the thyroid recep-tors is blocked by prolactin, the tail is not resorbed, and the adult-specific keratin gene is not activated (Tata et al., 1991; Baker and Tata, 1992). Injec-tions of prolactin stimulate larval growth and inhibit metamorphosis (Bern et al., 1967; Etkin and Gona, 1967), but there is dispute as to whether this re-flects the natural role of prolactin (Takahashi et al., 1990; Buckbinder and Brown, 1993). We still do not know the mechanisms by which levels of thy-roid hormone are regulated in the tadpole, nor do we know how the recep-tion of thyroid hormone elicits different responses (proliferation, differentia-tion, cell death) in different tissues.

Heterochrony

OST SPECIES of animals develop through a larval phase. However, some species have modified their life cycles by either greatly extending or shortening their larval period. The phenomenon wherein animals change the relative time of appearance and rate of development of characters already present in their ancestors is called **heterochrony.** Here we will discuss three extreme types of heterochrony. **Neoteny** refers to the retention of the juvenile form owing to the retardation of body development relative to the germ cells and gonads, which achieve maturity at the normal time. **Progenesis** also involves the retention of the juvenile form, but in this case, the gonads and germ line develop at a faster rate than normal, and they become sexually mature while the rest of the body is still in a juvenile phase. In **direct development,** the embryos abandon the stages of larval development entirely and proceed to construct a small adult.

Neoteny

In certain salamanders, sexual maturity occurs in what is usually considered a larval state. The reproductive system and germ cells mature, while the rest of the body retains its juvenile form throughout its life. In most instances, metamorphosis fails to occur and sexual maturity takes place in a "larval" body.

The Mexican axolotl, *Ambystoma mexicanum*, does not undergo metamorphosis in nature because its pituitary gland does not release an active thyroid-stimulating hormone (TSH) to activate T_3 synthesis in its thyroid glands (Prahlad and De-Lanney, 1965; Norris et al., 1973; Taurog et al., 1974). Thus, when investigators gave *A. mexicanum* either thyroid hormones or TSH, they found that the salamander metamorphosed into an adult not seen in nature (Huxley, 1920). Other species, such as *A. tigrinum*, metamorphose only if given cues from the environment. Otherwise, they become neotenic, successfully mating as larvae.

(A) (B)

Figure 19.8 Metamorphosis induced in the axolotl. (A) Normal condition of the axolotl. (B) Specimen treated with thyroxine to induce metamorphosis. (Courtesy of G. Malacinski.)

In part of its range, *A. tigrinum* is a neotenic salamander, paddling its way through the cold ponds of the Rocky Mountains. However, in the warmer region of its range, the larval form of *A. tigrinum* is transitory, leading to the land-dwelling tiger salamander. Neotenic populations from the Rockies can be induced to undergo metamorphosis simply by placing them in water at higher temperatures. It appears that the hypothalamus of this species cannot produce TSH-releasing factor at the low temperatures.

Some salamanders, however, are permanently neotenic, even in the laboratory. Whereas thyroxine is able to produce the long-lost adult form of *A. mexicanum* (Figure 19.8), the neotenic species of *Necturus* and *Siren* remain unresponsive to thyroid hormones (Frieden, 1981); their neoteny is permanent. Yaoita and Brown (1990) noted that the mRNA for thyroid hormone receptor β is absent in *Necturus* and thus cannot be induced by T_3. The genetic lesions thought to be responsible for neoteny in several species are shown in Figure 19.9.

De Beer (1940) and Gould (1977) have speculated that neoteny is a major factor in the evolution of more complex taxa. By retarding the development of somatic tissues, natural selection is given a flexible substrate. According to Gould, neoteny would "provide an escape from specialization. Animals can relinquish their highly specialized adult forms, return to the lability of youth, and prepare themselves for new evolutionary directions."

Progenesis

In progenesis, gonadal maturation is accelerated while the rest of the body develops normally to a certain stage. Progenesis has enabled some salamander species to find new ecological niches. *Bolitoglossa occidentalis* is a tropical salamander that, unlike other members of its genus, lives in trees. This salamander has webbed feet and a small body size that suit it for arboreal existence, the webbed feet producing suction for climbing and the small body size making such traction efficient. Alberch and

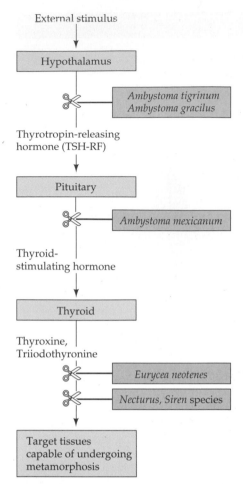

External stimulus

Hypothalamus

Ambystoma tigrinum
Ambystoma gracilus

Thyrotropin-releasing
hormone (TSH-RF)

Pituitary

Ambystoma mexicanum

Thyroid-
stimulating hormone

Thyroid

Thyroxine,
Triiodothyronine

Eurycea neotenes

Necturus, Siren species

Target tissues
capable of undergoing
metamorphosis

Figure 19.9 Stages along the hypothalamus-pituitary-thyroid axis of salamanders where various species are thought to have blocked metamorphosis. *Eurycea, Necturus,* and *Siren* appear to have a receptor defect in the responsive tissues. *Eurycea* will metamorphose when exposed to extremely high concentrations of thyroxine, while *Necturus* and *Siren* do not respond to any dose. (After Frieden, 1981.)

Alberch (1981) have shown that *B. occidentalis* resembles juveniles of the related species *B. subpalmata* and *B. rostrata* (whose young are small, with digits that have not yet grown past their webbing). It is thought that *B. occidentalis* became a sexually mature adult at a much smaller size than its predecessors. This gave it a phenotype that made tree-dwelling a possibility.

Direct Development

While some animals have extended their larval period of life, others have "accelerated" their development by abandoning their "normal" larval forms. This latter phenomenon, called direct development, is typified by frog species that lack tadpoles and by sea urchins that lack pluteus larvae. Elinson and his colleagues (del Pino and Elinson, 1983; Elinson, 1987) have studied *Eleutherodactylus coqui*, a small frog that is one of the most populous animals on the island of Puerto Rico. Unlike eggs of *Rana* and *Xenopus*, the eggs of *E. coqui* are fertilized while they are still within the female's body. Each egg is about 3.5 mm in diameter (roughly 20 times the volume of *Xenopus* eggs). After the eggs are laid, the male gently sits on the developing embryos, protecting them from predators and desiccation (Taigen et al., 1984). Early development is like that of most frogs. Cleavage is holoblastic, gastrulation is initiated at a subequatorial position (Figure 19.10A), and the neural folds become elevated from the surface (Figure 19.10B). However, shortly after the neural tube closes, limb buds appear on the surface (Figure 19.10C). This early emergence of limb buds is the first indication that development is direct and will not pass through a limbless tadpole stage. Moreover, the emergence of the limbs does not depend on thyroid hormones (Lynn and Peadon, 1955). What emerges from the egg jelly three weeks after fertilization is not a tadpole but a little frog (Figure 19.10D). The froglet has a tail for the first part of its life, but it is used for respiration rather than locomotion. Such direct-developing frogs do not need ponds for their larval stages and can therefore colonize new regions inaccessible to other frogs.

Raff (1987) has studied direct development in sea urchins. In the "typical" sea urchin, the primary mesenchyme cells invaginate and secrete the calcium carbonate skeleton of pluteus larvae. These larvae feed and grow until coelomic vesicles (also derived from the micromeres) form at the sides of the gut (Pehrson and Cohen, 1986). The left coelom grows further to produce a hydrocoel, which induces the overlying ectoderm to invaginate to form a vestibule. The hydrocoel and vestibule form a rudiment that grows within the larva until it is released at metamorphosis to become a juvenile sea urchin (Figure 19.11).

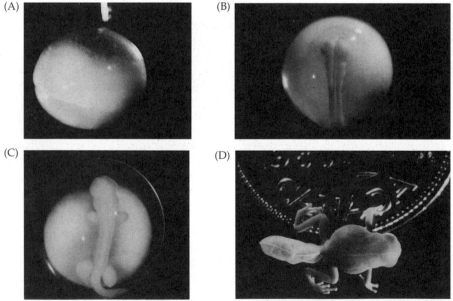

(A)   (B)   (C)   (D)

Figure 19.10 Direct development of the frog *Eleutherodactylus coqui*. (A) Early gastrula showing blastopore lip. (B) Dorsal view of neurula showing raised neural folds. (C) A day after the closure of the neural folds, limb buds are seen. (D) Three weeks after fertilization, a tiny froglet hatches, seen here beside a Canadian penny (the inflation of the tail is an artifact caused by the chemical fixatives used to prepare the specimen). (From Elinson, 1987, courtesy of R. P. Elinson.)

(A) (B)

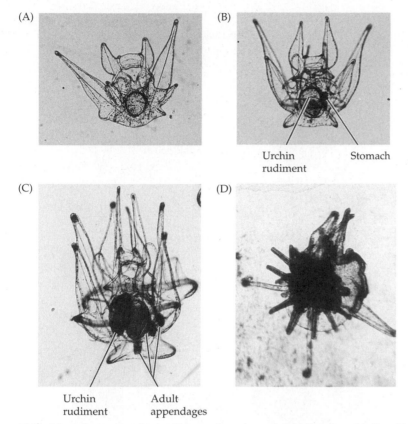

Urchin Stomach
rudiment

(C) (D)

Urchin   Adult
rudiment   appendages

Figure 19.11 Normal metamorphosis of pluteus larva into adult in the sea urchin *Lytechinus pictus.* (A) Pluteus larva 8 days after fertilization. (B) An 11-day pluteus larva with urchin rudiment on left coelomic sac. (C) A 19-day pluteus with developing sea urchin rudiment. (D) About 11 minutes after attaching to substrate, the larval arms are being resorbed. (From Hinegardner, 1969, courtesy of R. T. Hinegardner.)

Several sea urchin species have suppressed stages of the pluteus larvae while accelerating the development of the adult rudiment. As in direct development in frogs, direct development in sea urchins depends on a large, yolky egg. In fact, Raff has found a correlation between egg volume and the extent of direct development (Table 19.3). North American and European sea urchins have eggs whose diameters range from 60 to 200 μm. These species undergo indirect development through the pluteus larvae. Eggs in the 300–350 μm range produce partial pluteus larvae that have larval skeletons but no gut (and therefore cannot feed). These species show an accelerated growth of the adult rudiment, so a feeding juvenile urchin is quickly formed. There are some yolky eggs that reach 2 mm in diameter

(about the same volume as *Xenopus* eggs). These embryos develop directly without any pluteus stage. The feeding stage is not needed because nutrition can be provided by the yolk.

Nature has provided an excellent comparison in two Australian species of the sea urchin genus *Heliocidaris. Heliocidaris erythrogramma* and *H. tuberculata* are common species that, according to morphological and DNA-sequencing data, are very closely related. They live side by side and spawn during the same time in the summer. However, *H. erythrogramma* has an egg with a diameter of 425 μm and is a direct developer; *H. tuberculata* produces an egg with a diameter of 95 μm and develops through a typical pluteus larva. The comparison between these species reveals that the direct developer has eliminated the larval stages and proceeds directly to coelom formation and the construction of the juvenile sea urchin (Figure 19.12). The pluteus larva is designed for swimming and feeding (Strathmann, 1971, 1975), using its arms as supports for bands of cilia that sweep food particles into the mouth. The cells of the direct developer have changed their fates such that no larval skeleton or mouth forms. In the gastrulation of direct-developing sea urchins, one does not see the descendants of the micromere cells invaginating to form the larval skeleton. Rather, these cells are immediately involved in forming the calcareous spine of the young adult. Also, the tip of the archenteron of the direct-developing urchin forms an extensive hydrocoel that interacts with the ectodermal vestibule to form the sea urchin rudiment at gastrulation. In indirect developers, only two cells initiate the formation of the vestibule, and these interact with the hydrocoel after the pluteus structure has been established (Wray and Raff, 1990, 1991). Thus, we have an interesting paradox. In one sense, the development of larval stages seems highly constrained. Larvae from different classes of echinoderms are very similar, and the tadpoles of different groups of frogs are also very much the same. However, these constraints can be eliminated by abandoning the need for a feeding larval stage. Increasing the amount of yolk available to the embryo seems to make this possible. ■

Table 19.3 Relationship of developmental mode to egg size in sea urchins

Number of species	Egg size range (μm)	Developmental mode
83	60–345	Feeding pluteus larva
1	280	Facultative feeding pluteus
2	300–350	Abbreviated pluteus, nonfeeding
19	400–2000	Pluteus lost; direct development

Source: After Raff, 1987.

Figure 19.12 Changes in cell fate and gastrulation in an indirect-developing sea urchin and a direct-developing sea urchin. The fate maps at the 32-cell stage show the differences in cell fate. The vegetal fates (indicated by shading) include coelom (C), gut (G), pigment cells (P), and skeletogenic mesenchyme (S). Those cells giving rise to neural tissues are denoted as N. Note that the direct developer has not produced separate micromeres and macromeres. The indirect developer forms a pluteus, and within this larval structure, interactions form the rudiment of the juvenile sea urchin (color). In the direct developer, such interactions between coelom and vestibule cells occur immediately at gastrulation, and the juvenile rudiment (color) is formed without any feeding larval stage. Both types of development generate the same adult structures. (After Raff, 1994.)

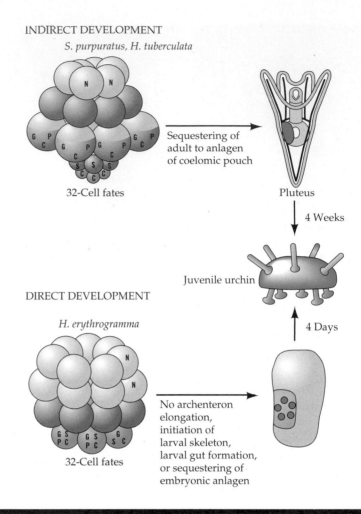

INDIRECT DEVELOPMENT

S. purpuratus, H. tuberculata

32-Cell fates

Sequestering of adult to anlagen of coelomic pouch

Pluteus

4 Weeks

Juvenile urchin

4 Days

DIRECT DEVELOPMENT

H. erythrogramma

32-Cell fates

No archenteron elongation, initiation of larval skeleton, larval gut formation, or sequestering of embryonic anlagen

Metamorphosis in insects

Eversion and Differentiation of the Imaginal Discs

Whereas amphibian metamorphosis is characterized by the remodeling of existing tissues, insect metamorphosis often involves the destruction of larval tissues and their replacement by an entirely different population of cells.

There are three major patterns of insect development. A few insects, such as springtails, have no larval stage and undergo direct development. Other insects, notably grasshoppers and bugs, undergo a gradual, **hemimetabolous** metamorphosis (Figure 19.13A). Adult organs are formed without any profound discontinuity. The rudiments of the wing, genital organs, and other adult structures are present at hatching, and they become more mature with each molt. At the last molt, the emerging insect is a winged and sexually mature adult. The larval form of a hemimetabolous insect is called a **nymph.**

In the **holometabolous** insects (flies, beetles, moths, and butterflies), there is a dramatic and sudden transformation between the larval and adult stages (Figure 19.13B). The juvenile larva (caterpillar, grub, maggot) undergoes a series of molts as it becomes larger. The newly hatched insect larva is covered by a hard **cuticle.** To grow, the insect must produce a new, larger cuticle and shed the old one. Thus, the postembryonic development of these

(A) HEMIMETABOLOUS
DEVELOPMENT

(B) HOLOMETABOLOUS
DEVELOPMENT

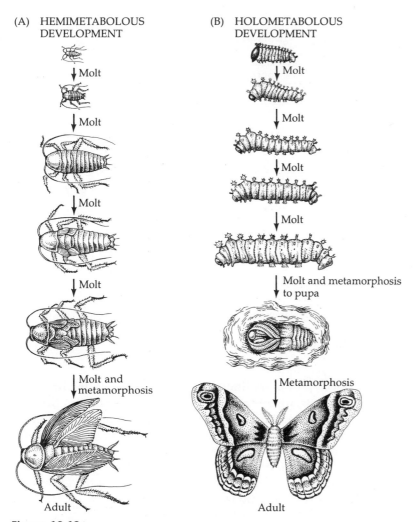

Figure 19.13
(A) Hemimetabolous (incomplete) metamorphosis. (B) Holometabolous (complete)
metamorphosis.

insects consists of a succession of molts. The number of molts before becoming an adult is characteristic for the species, although environmental factors
can increase or decrease the number. The stages between these molts are
called **instars.** The instar stages grow in a stepwise fashion, each instar stage
being qualitatively larger than the previous one. After the last instar stage,
the larva undergoes a metamorphic molt to become a **pupa.** The pupa does
not feed, and its energy must come from those foods ingested while a larva.
During pupation, the adult structures are formed and replace the larval
structures. Eventually, an **imaginal molt** enables the adult to shed the pupal
case and emerge.

Drosophila undergoes four molts in its life cycle. The embryo develops
into the first-instar larva and then molts to become the second-instar larva.
Subsequent molts separate the second instar from the third instar, the third
instar from the pupa, and the pupa from the adult. At each molt, the epidermal cells separate from the cuticle and secrete a "molting fluid" into the intervening space. When the epidermal cells have secreted a new cuticle, they
degrade the old one by activating enzymes in the molting fluid (Hepburn,
1985). It is within the pupal cuticle that the transformation of juvenile into
adult occurs. Most of the old body of the larva is systematically destroyed as

Figure 19.14
The locations and developmental fates of the imaginal discs in *Drosophila melanogaster*. (After Fristrom et al., 1969.)

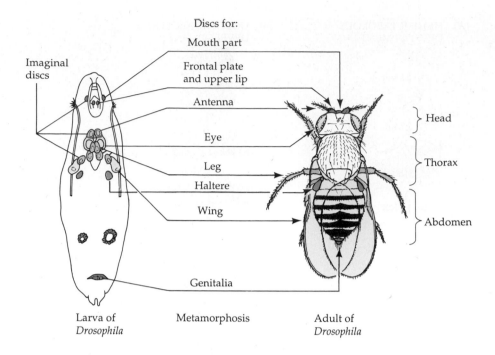

new adult organs develop from undifferentiated nests of cells, the **imaginal discs** (and, in some insects, **histoblasts**). When the adult organism (**imago**) is developed, the imaginal molt results in the shedding of the pupal cuticle, and the mature insect emerges. In holometabolous larvae, then, there are two cell populations: the larval cells, which are used for the functions of the juvenile, and the thousands of imaginal cells, which lie in clusters awaiting the signal to differentiate. Figure 19.14 shows the location of the *Drosophila* imaginal discs and the structures into which they develop.

In *Drosophila*, there are 10 major pairs of imaginal discs, which reconstruct the entire adult (except for the abdomen), and a genital disc, which forms the reproductive structures. The abdominal epidermis forms from a small group of imaginal cells called histoblasts, which lie in the region of the larval gut, and other nests of histoblasts located throughout the larva form the internal organs of the adult. The imaginal discs can be seen in the newly hatched larva as local thickenings of the epidermis. In *Drosophila*, these newly hatched eye-antenna, wing, haltere, leg, and genital discs contain 70, 38, 20, 36–45, and 64 cells, respectively (Madhavan and Schneiderman, 1977). Whereas most of the larval cells have a very limited mitotic capacity, the imaginal discs divide rapidly at specific characteristic times. As the cells proliferate, they form a tubular epithelium that folds in upon itself in a compact spiral (Figure 19.15A). The largest disc, that of the wing, contains some 60,000 cells, whereas the leg and haltere discs contain around 10,000 (Fristrom, 1972). At metamorphosis, these cells differentiate and elongate (Figure 19.15B).

The fate map and elongation sequence of the leg disc are shown in Figure 19.16. At the end of larval development, the leg disc is an epithelial sac connected by a thin stalk to the larval epidermis. On one side of the sac, the epithelium is folded into a series of concentric folds "reminiscent of a Danish pastry" (Kalm et al., 1995). At the end of the larval period, the cells at the center of the disc telescope out to become the most distal portions of the leg—the claws and the tarsus. The outside cells become the proximal structures—the coxa and the adjoining epidermis. After differentiating, the cells

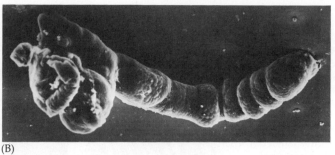

(A) (B)

Figure 19.15
Imaginal disc elongation. Scanning electron micrograph of *Drosophila* third-instar leg disc before (A) and after (B) elongation. (From Fristrom et al., 1977, courtesy of D. Fristrom.)

of the appendages and epidermis secrete a cuticle appropriate for the specific region. Although the discs are composed primarily of epidermal cells, a small number of **adepithelial** cells migrate into the disc early in development. During the pupal period, these cells give rise to the muscles and nerves that serve that structure.

The process of elongation can be initiated in culture by placing imaginal discs in a solution containing the molting hormone, **20-hydroxyecdysone.** Moreover, such eversion can be inhibited by adding any of three sets of drugs. (1) Inhibitors of RNA synthesis and protein synthesis inhibit eversion when added to cultured imaginal discs at the same time as 20-hydroxyecdysone. It is known that RNA and protein synthesis occurs prior to elongation and that some of these proteins are needed for elongation to occur. (2) Cytochalasin B, an inhibitor of microfilament function, also inhibits elongation, thereby indicating a need for actin microfilaments. (3) Protease inhibitors also inhibit elongation (Pino-Heiss and Schubiger, 1989), as cell-surface proteases are required for the releasing of constraints upon cell shape. Taken together, these data suggest that the eversion of the imaginal discs requires new protein synthesis, a well-developed system of actin microfilaments, and some cellular communication by the cell surface (Fristrom et al., 1977; Kalm, 1995).

Studies by Condic and her colleagues (1990) have demonstrated that the elongation of the imaginal disc is due primarily to cell shape change within

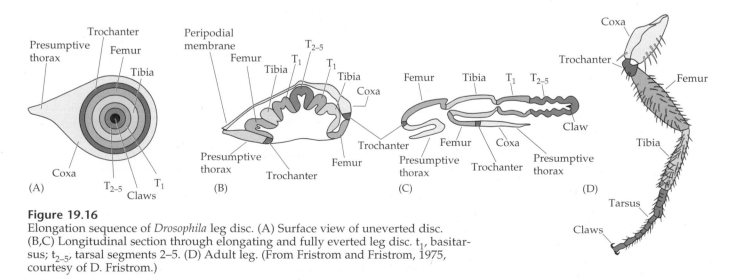

Figure 19.16
Elongation sequence of *Drosophila* leg disc. (A) Surface view of uneverted disc. (B,C) Longitudinal section through elongating and fully everted leg disc. t_1, basitarsus; t_{2-5}, tarsal segments 2–5. (D) Adult leg. (From Fristrom and Fristrom, 1975, courtesy of D. Fristrom.)

Figure 19.17
Changes in cell shape during the elongation of a *Drosophila* leg imaginal disc. Top: Optical sections through the elongating second leg disc. The arrows mark the basitarsal segments, and the calibration bar represents 100 μm. Bottom: Higher magnification (calibration bar represents 10 μm) of cell apices through the basitarsal area. The cell boundaries are stained by fluorescently labeled phalloidin. (A) Beginning prepupal stage. (B) A 6-hour prepupa. (C) Leg disc from a beginning prepupa treated with trypsin. The basitarsal cells are initially compressed along the proximal-distal axis. Upon hydroxyecdysone treatment or trypsinization, the compression is released, and the cells expand to elongate the tissue. (From Condic et al., 1990, courtesy of the authors.)

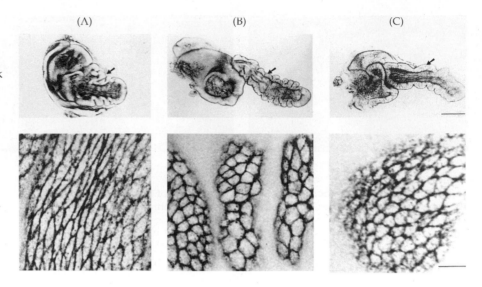

the disc epithelium. Using fluorescently labeled phalloidin to stain the peripheral microfilaments of the leg disc cells, they showed that the cells of the early third-instar discs are tightly compressed along the proximal-distal axis. This compression is maintained through several rounds of cell division. Then, when the tissue begins elongating, the compression is removed and the cells "spring" into their rounder state (Figure 19.17). This conversion of an epithelium of compressed cells into a longer epithelium of noncompressed cells represents a novel mechanism for the extension of an organ during development.

Sidelights & Speculations

The Determination of the Leg and Wing Imaginal Discs

Determination of Discs from Ectoderm

The molecular biology of insect metamorphosis begins with the specification of certain epidermal cells to become imaginal disc precursors. As we discussed in Chapter 14, the organ rudiments in *Drosophila* are specified on an orthagonal grid by intersecting anterior-posterior and dorsal-ventral signals. In most segments, homeobox gene products prevent *Distal-less* gene expression and the establishment of the limb primordia; but in those segments that are specified to be thoracic, limb formation is permitted (see Figure 14.33). Cohen and colleagues (1993) have demonstrated that the leg and wing originate

from the same set of imaginal precursors, specified at the intersection between the anteroposterior stripes of Wingless (Wg) protein expression and the horizontal band of cells expressing the Decapentaplegic (Dpp) protein. Both proteins are soluble and have a limited range. In the early *Drosophila* embryo (at germ band extension about 4.5 hours after fertilization), a single group of cells at the intersection of these domains forms the imaginal disc precursors in the abdomen. These cells (and only these cells) express the Distal-less protein. As the cells expressing dpp are moved dorsally, these Distal-less-expressing cells move to establish a secondary cluster of imaginal cells (derived from the original ventral cluster). The initial clusters form

the leg imaginal discs, while the secondary cluster forms the wing and haltere discs. Thus, the leg and wing discs have a common origin (Figure 19.18).

Determination of Disc Identity

Despite their common origin, it is obvious that the leg and wing discs are determined to become different structures. As we detailed earlier, the specification of these discs into their particular fates is probably accomplished by the interactions of homeotic genes. Even so, we still do not know the molecules that specify the leg discs to be different from the wing discs, or the eye discs to be different from the antenna discs. We do know that if certain homeotic genes are expressed in the wrong places (such as the

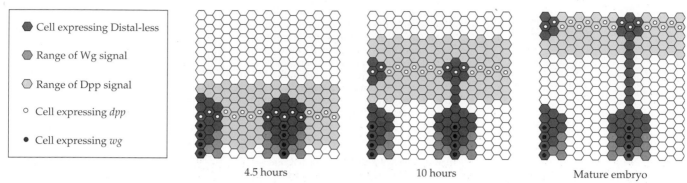

- ⬡ Cell expressing Distal-less
- ⬡ Range of Wg signal
- ⬡ Range of Dpp signal
- ○ Cell expressing *dpp*
- ● Cell expressing *wg*

4.5 hours   10 hours   Mature embryo

Figure 19.18 Schematic model for the allocation and separation of the leg-wing disc in the *Drosophila* thorax. The embryo is divided into an orthagonal grid with vertical stripes of Wingless (Wg) and a horizontal band of Decapentaplegic (Dpp) synthesis and secretion. The initial disc forms at the intersection of these secretory domains. The Dpp-secreting cells migrate dorsally, bringing with them some of the imaginal disc cells. These dorsal disc cells generate the wing disc, while the cells remaining form the leg disc. (After Cohen et al., 1993.)

expression of *Antennapedia* in the eye-antenna disc), the discs become respecified (so that legs grow from the antenna disc). The determination of the wing disc appears to be regulated by the *vestigial* gene. The *vestigial* gene regulates wing disc identity. Using a targeted gene expression system, Kim and colleagues (1996) have caused the *vestigial* gene to be expressed in eye, antenna, and leg discs (Figure 19.19). When this happens, regions of the normal structure are converted into wing.

Determination of the Disc Polarity

Recent evidence suggests that the leg and wing axes become specified by interactions at their compartmental boundaries (Meinhardt, 1980; Causo, 1993; Tabata, 1995). After these initial in-

teractions, a polar coordinate system (similar to those discussed in the previous chapter for vertebrate limb development) may subdivide the regions more finely (Held, 1995).

Anterior-posterior axis. During the *first* larval instar, the leg and wing imaginal discs acquire their anterior-posterior (A/P) axis. The discs become split into two **compartments** representing the future anterior and posterior regions of the appendage (i.e., from the front of the wing to the rear of the wing). The posterior compartment is defined by the expression of the *engrailed* gene in the posterior cells of the disc (Figure 19.20; Garcia-Bellido et al., 1973; Lawrence and Morata, 1976). If *engrailed* function is absent, all the disc cells become anterior-

ized. The boundary between the posterior and anterior compartments is strictly observed. Cells from one side cannot produce descendants that cross over the boundary to the other.

In the wing disc, posterior cells express Hedgehog protein which acts as a short-range signal to induce the expression of Dpp in adjacent anterior cells, while the expression of *engrailed* in the posterior cells render them non-responsive to the Hedgehog they secrete. The Dpp protein acts as a long-range signal to establish the anterior-posterior axis of the wing (Guillen et al., 1995; Tabata et al., 1995; Nellen et al., 1996).

In the leg disc, the posterior compartment also secretes Hedgehog protein. However, here, Hedgehog induces the *dorsal* cells of the anterior compartment to secrete Dpp while it induces the *ventral* cells of the anterior compartment to secrete Wingless. (Jiang and Struhl, 1996).

Dorsal-ventral axis. In the *second*-instar larva, a second axis, the dorsal-ventral axis, is determined in the wing disc. The D/V boundary lies at the future margin of the wing blade, thus separating the upper surface of the wing from the lower (Bryant, 1970; Garcia-Bellido et al., 1973). The gene involved in this compartmentation event is *apterous*. Cells expressing *apterous* become the dorsal cells (Plate 15; Frontispiece; Blair, 1993; Diaz-Benjumea et al., 1993). When *apterous* is deleted, all cells in the disc acquire ven-

(A)

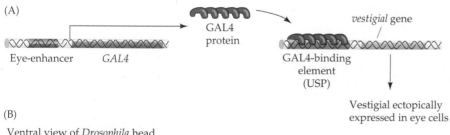

Eye-enhancer   *GAL4*   GAL4 protein   GAL4-binding element (USP)   *vestigial* gene

Vestigial ectopically expressed in eye cells

(B)

Ventral view of *Drosophila* head

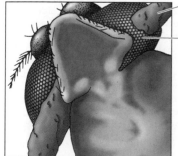

Wing-like outgrowth from ventral eye

Eye

Figure 19.19 The *vestigial* gene determines wing disc identity. (A) Kim and colleagues constructed lines of *Drosophila* that carry the yeast transcriptional activator protein GAL4 coupled to an enhancer, such as the eye enhancer shown in this figure. Although GAL4 protein is expressed in the eyes of these flies, it does not bind to any *Drosophila* DNA. However, if the fly is mated to another strain that carries the *vestigial* gene downstream from the GAL4-binding element (the UAS—upstream activator sequence), the GAL4 protein activates this gene. Thus, in these flies, the GAL4 protein is made in the eye disc and activates the expression of the *vestigial* gene. (B) The resulting eye contains regions of wing tissue. (After Kim et al., 1996.)

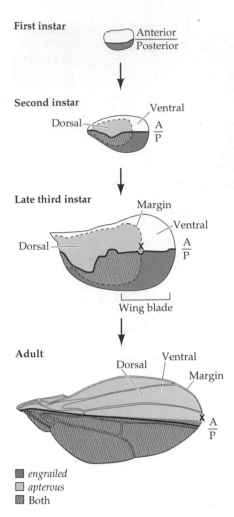

First instar

Anterior / Posterior

Second instar

Dorsal — Ventral — A/P

Late third instar

Margin — Ventral — Dorsal — X — A/P

Wing blade

Adult

Dorsal — Ventral — Margin — X — A/P

■ *engrailed*
□ *apterous*
▨ Both

Figure 19.20 Compartmentation and gene expression in the wing disc. (A) In the first-instar larva, the anterior-posterior axis has been formed and is manifest by the expression of the *engrailed* gene in the posterior compartment. In the second instar, the dorsal-ventral axis forms and is seen by the expression of the *apterous* gene in the future dorsal surface. In the third-instar larva, the borders of *engrailed* expression extend slightly beyond the A/P boundary. Where secreted and membrane proteins interact at the junction of the D/V and A/P axes, cells are determined to become the distal tip of the wing (X). (After Blair, 1995.)

tral fates. Both *engrailed* and *apterous* are considered **selector genes,** since they regulate the fate of a compartment. Like the homeotic selector genes discussed in Chapter 14, these genes contain homeoboxes and are thought to encode transcription factors. In the wing, there is no growth along the D/V axis, since the ectoderm remains one cell layer thick on either side of the wing margin. It is not known what causes the initial D/V polarity in the leg disc.

Proximal-distal axis. The interaction between the D/V and A/P axes at their boundaries is critical for the outgrowth along the proximal-distal axis. During metamorphosis, the "distalization" of the proximal-distal axis from the base of the thorax outward to the tip of the wing or leg is accomplished by cell interactions at the boundaries between the D/V and A/P axes.

In the leg, Hedgehog protein from the

posterior compartment induces the closest cells of the dorsal anterior compartment to secrete Decapentaplegic protein and induces Wingless protein from the closest cells of the ventral anterior compartment. Both Decapentaplegic and Wingless proteins activate the *optomotor-blind* gene, whose protein product promotes the outgrowth of the limb appendages (Wilder and Perrimon, 1995; Grimm and Pflugfelder, 1996). Moreover, where these three diffusible proteins meet defines the distalmost tip of the limb (i.e., the claw). This region begins to express the *Distal-less* and *arista-less* genes that characterize the distal tip re-

gion and stimulate growth and differentiation of the cells (Figure 19.21A; Campbell et al., 1993; Basler and Struhl, 1994; Diaz-Benjumea et al., 1994). If Dpp protein is made by a cluster of cells in the ventral anterior compartment or if Wingless protein is expressed by a small group of cells in the dorsal anterior compartment (by switching on genes) an entire new proximal-distal axis will be formed at that site of expression (Figure 19.21B; Plate 27).

The situation in the wing is a bit more difficult to understand. Hedgehog protein from the posterior compartment induces the adjacent cells of both the anterior dorsal and anterior ventral compartments to secrete Dpp. This establishes the conditions of cell growth and patterning along the A/P axis. At the cells that give rise to the margin, the *apterous*-expressing cells of the dorsal surface meet the ventral cells that do not express *apterous*. The Apterous transcrip-

(A)

dpp

Distalless

hh

Anterior — Posterior

wg — hh

(B)

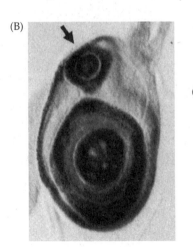

(C)

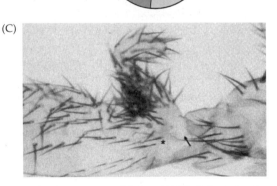

Figure 19.21 Model of axis formation in the developing *Drosophila* leg. (A) Hedgehog protein is only synthesized and secreted by the Engrailed-synthesizing cells of the posterior side of the disc. Hedgehog protein diffuses over a few cell diameters and induces the strip of adjacent posterior cells in the dorsal region of the disc to express *decapentaplegic* genes. The Dpp protein then diffuses and patterns the anterior dorsal side of the disc. The Hedgehog secretion by the posterior cells instructs the adjacent ventral anterior cells adjacent to the posterior cells to synthesize and secrete the Wingless protein. This will help pattern the ventral anterior wing. (B) When a clone of Wingless-expressing cells is made to form in the dorsal region of the disc (by manipulating a *wingless* transgene), it organizes a new limb axis to form. Here, this axis can be seen when stained for the presence of *Distalless* gene expression. (C) New limb axis formed when a clone of Dpp-expressing cells is ectopically expressed. (B and C from Diaz-Benjumea et al., 1994; photographs courtesy of the authors.)

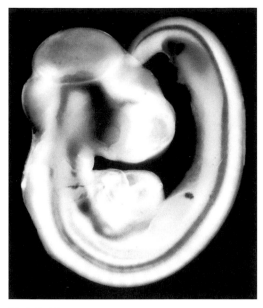

PLATE 22
sonic hedgehog **expression in the 3-day chick embryo.**
Sonic hedgehog is involved in numerous inductive interactions wherein one tissue influences the differentiation of another tissue. Wholemount in situ hybridization finds *sonic hedgehog* mRNA in the notochord, neural floor plate cells, foregut and midgut, and in the posterior limb bud mesoderm. Chapters 7, 8, 15, and 18. (Photograph courtesy of C. Tabin.)

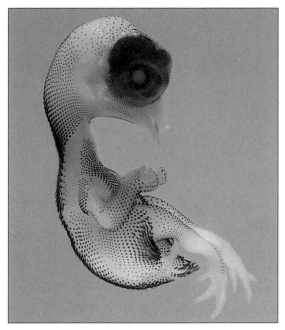

PLATE 23
sonic hedgehog **expression in the 10-day chick embryo.**
After mediating several important interactions during organ formation, *sonic hedgehog* becomes expressed in the ectoderm of the developing feather germs and foot scales. This wholemount in situ hybridization shows the hexagonal array of feather pattern. Chapter 17. (Photograph courtesy of Won-Sun Kim and John F. Fallon.)

PLATE 24
The Myf-5 protein is expressed in muscle cell precursors.
The genetic elements regulating the temporal and spatial expression the *Myf-5* gene can be discerned by fusing a β-galactosidase gene to the sequences surrounding the *Myf-5* locus. Here a particular sequence upstream from the *Myf-5* gene causes the expression of the gene (black stain) in the neck muscles, pharyngeal arches, ocular muscles, forelimb muscles, and segmented myotomes of a 13.5-day mouse embryo. Chapters 2 and 9. (Photograph courtesy of A. Patapoutian, G. Lyons, J. Miner, and B. Wold.)

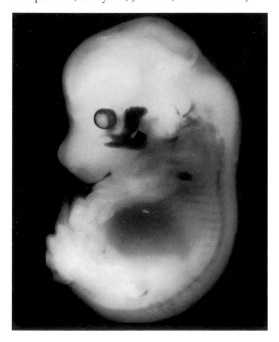

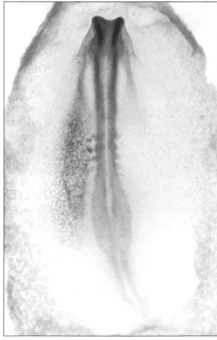

PLATE 25
Asymmetrical expression of the *nodal* gene in the 24-hour chick embryo.
Wholemount in situ hybridization using probes for the chick *nodal* gene finds it to be expressed in the lateral plate mesoderm only on the left-hand side. It can be seen here as the region stained purple. This gene is important in establishing the left-right axis of the chick. Chapter 16. (Courtesy of C. Stern.)

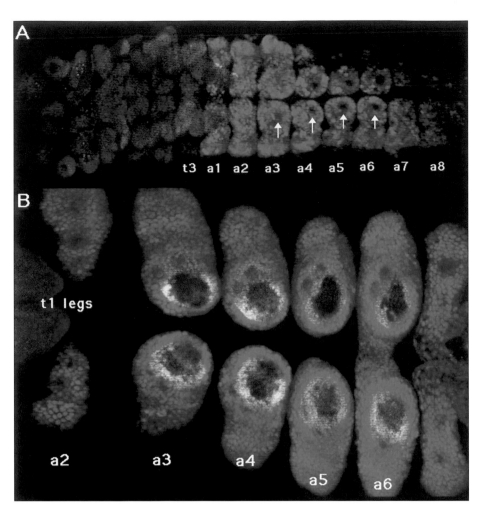

PLATE 26
Regulation of homeotic gene expression in the formation of insect legs.
The larvae of flies do not have prolegs, but those of butterflies do. Here the products of the homeotic genes *Ultrabithorax* and *abdominal-A* are stained green and the Distal-less protein (needed for limb development) is stained orange. In the early caterpillar of the buckeye butterfly *Precis*, the thoracic limbs (from T1-3) are readily seen. Some abdominal segments (A3-6) begin to make "holes" in their expression domain of the homeotic proteins. Below, when the caterpillar has grown, Distal-less expression can be seen in these regions. (The yellow indicates overlap of expression domains.) Chapters 14, 19 and 23. (Photographs courtesy of B. Warren, S. Paddock, and S. Carroll.)

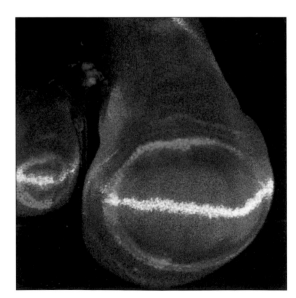

PLATE 27
The Wingless protein plays critical roles in organizing the wing imaginal disc of *Drosophila*.
Cells at the junction between the dorsal and ventral compartments of the wing disc induce the expression of Wingless protein in a thin band of cells straddling that boundary. Wingless protein then induces the expression of other proteins such as Vestigial (here stained red) several diameters away. Chapter 19. (Photograph courtesy of K. Basler.)

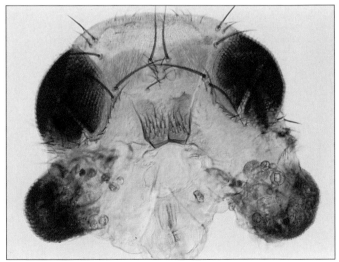

PLATE 28
Ectopic expression of the *Drosophila eyeless* gene causes the formation of new eyes in other regions of the adult.
Here, the *eyeless* gene was experimentally activated in the regions of the fly larvae that form head cuticle. Upon metamorphosis, pigmented compound eyes emerged from this tissue. Chapter 23. (Photograph courtesy of W. Gehring and *Science*.)

PLATE 29
Seasonal polyphensim of *Araschina levana*, the European map butterfly.
Several species of butterflies develop differently at different seasons. In *A. levana,* the summer form is represented at the top, the spring form is represented below. In this case, the developmental differences are elicited by the environment, specifically the differences in daylength. Chapter 21. (Photograph courtesy of H. F. Nijhout.)

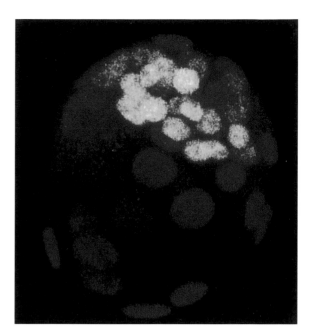

PLATE 30
Expression of the transcription factor Oct4 in the mouse blastocyst.
The Oct4 transcription factor is found in those cells that will form the embryo, while it is absent from those cells that will form the placenta. Chromatin is stained with propidium iodide (red) while the Oct4 protein is stained green. The overlap is indicated by the yellow color which shows the presence of Oct4 only in the cells of the inner cell mass. Chapters 5 and 22. (Photograph courtesy of H. R. Schöler.)

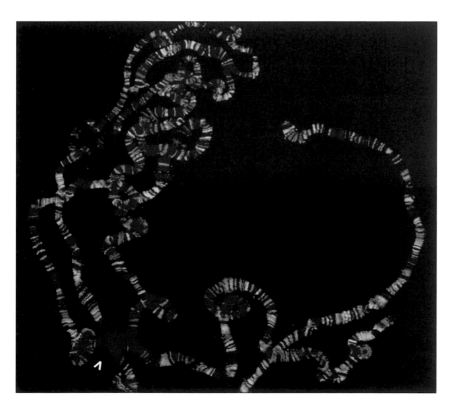

PLATE 31
Insulators of gene expression.
The BEAF-32 protein binds to hundreds of sites in the *Drosophila* polytene chromosomes, dividing the chromosomes into functional domains. It is suspected that regulatory signals in one domain do not cross the boundary into the next. The DNA has been stained red with propidium iodide. The antibody to the BEAF-32 protein is stained green and the overlap appears yellow. Chapter 11. (Photograph courtesy of U. K. Laemmli.)

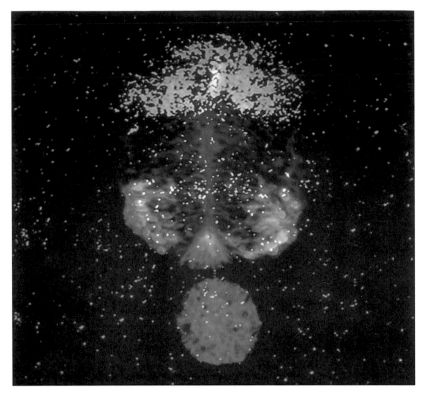

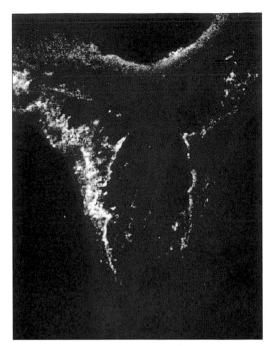

PLATE 32
Dorsal-ventral polarity of the chick neural tube.
Diffusible signals from the notochord (green tube at bottom) induce the formation of the floor plate at the ventral side of the neural tube (green). The floor plate cells induce the formation of the two regions of motor neurons (gold) on the ventrolateral sides. The notochord also restricts the expression of Dorsalin protein (needed for neural crest cell development) to the most dorsal region of the neural tube (blue). Chapter 7 and 17. (Photograph courtesy of T. M. Jessell.)

PLATE 33
Asymmetrical expression of the Flectin protein in the developing chick heart.
This extracellular matrix protein (stained yellow) accumulates predominantly on the left-hand side of the stage 10 chick embryo. Chapters 9 and 16. (Scanning laser confocal photograph courtesy of K. Linask.)

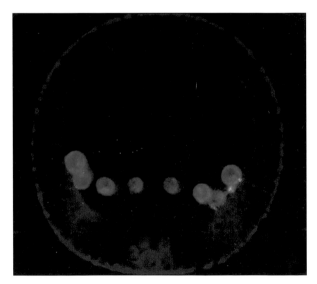

PLATE 34
Parental care of frog tadpoles.
Tadpoles of the reticulated poison-dart frog are carried on their parents' backs to small pools of water at the base of bromeliad leaves in the rain forest canopy. The female of the Peruvian Amazon species will then supply unfertilized eggs as food for the developing tadpoles. Chapter 19. (Photograph by M. Fogden/DRK Photo.)

PLATE 35
Localization of primary mesenchyme cells in the sea urchin embryo.
Only part of the mesenchyme blastula is shown in this confocal immunofluorescence micrograph. The primary mesenchyme cells are stained green and β-catenin is stained red. β-catenin is seen in the adherens junctions of the embryonic cell membranes, and it also is found in the cytoplasm and nuclei of those cells serving as targets for the migration of the primary mesenchyme cells. Chapter 6. (Photograph courtesy of J. R. Miller and D. McClay.)

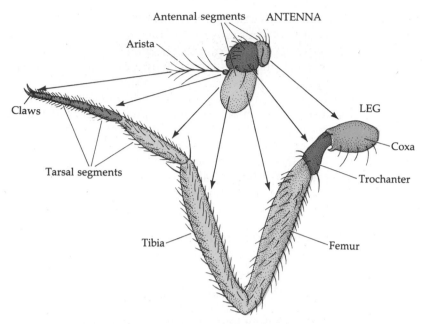

Figure 19.22 Correspondence between portions of the antenna and portions of the leg. In the mutant *Antennapedia*, regions of the antenna are transformed into leg structures. The arrows show the portions of the antenna that form specific corresponding portions of the leg. Such correspondence has also been seen in the transcription patterns of genes such as *salm*. (Ater Postlethwait and Schneiderman, 1971).

tion factor activates the expression of the *fringe* and *serrate* genes in the dorsal cells (Irvine and Wieschaus, 1994; Williams et al., 1994; Kim et al., 1995). The Fringe and Serrate proteins act to promote the transcription of the *vestigial* and *wingless* genes in cells lining the D/V boundary (Frontispiece). The Vestigial transcription factor activates the ventral-specific wing genes, while the Wingless protein diffuses from the cell to signal the adjacent dorsal cell to express its dorsal-specific wing genes. In this way, the growth and differentiation of the dorsal and ventral wing surfaces are coordinated. The dorsal and ventral wing surfaces become glued together by integrins on both epithelia (Brower and Jaffe, 1989; Kim et al., 1996). [meta1.html]

The boundary hypothesis discussed here does not explain certain observations involving the D/V polarity of the leg or the distalization of the appendages. Held (1995) suggests that there is a gradient of Dpp protein that stimulates the synthesis of the (still unidentified) molecules needed for extending the appendage and establishing the polarity in all three dimensions.

Homologous Specification. The molecules used by imaginal discs to specify positional information may be the same throughout the fly. That is, the discs may specify the respective fates of their cell by the same mechanisms. This is called **homologous specification.** Thus cells in the eye disc may respond to the same positional cues as cells in the leg disc. Homologous specification can be seen with certain homeotic mutants such as *Antennapedia*, in which antennal structures are transformed into legs (Postlethwait and Schneiderman, 1971). Occasionally, the entire antenna becomes an entire leg, but is is more common that only a portion of the antenna becomes leg-like. In the latter cases, the replacement is absolutely position-specific. The cells of the antenna disc that normally would have formed the distal tip of the antenna (arista) are transformed into the most distal portion of the leg (claw); cells specified to give rise to the second portion of the antenna are transformed into the second portion (trochanter) of the leg. The corresponding parts of the two structures are shown in Figure 19.22. It is apparent, then, that the two differently determined discs must use a common mechanism for the specification of cell fates within the respective discs.* ∎

*Yes, it's complex and likely to get more so. This is not without its humor, however. Sidney Brenner (1996) recalls his fellow Nobel Laureate Francis Crick being frustrated by this complexity and saying, "God knows how these imaginal discs work." Brenner fantasized a meeting wherein Crick asks the Deity how He constructed these entities, only to have God bewildered by the complexity as well. Eventually, all God could do was to reassure Crick that "we've been building flies here for 200 million years and we have had no complaints."

Remodeling of the Nervous System

As in anuran metamorphosis, insect metamorphosis causes a major restructuring of the organism's nervous system. Some nerves die, while other nerves take on new functions. In Chapter 17, we saw the development of photoreceptors from the epithelial cells of the eye disc. Here, a new set of neurons is generated to take on a new function. The neurons that have been connected to dying tissues either die with the tissue or are respecified for new functions. The nerve that innervates the proleg muscle of the caterpillar of the *Manduca* moth is independently sensitive to ecdysone and perishes simultaneously with its larval target tissue. However, the motor neuron innervating the second oblique muscle of the larva survives the death of its target to innervate a newly formed adult muscle (the fourth dorsal external muscle) that differentiates during metamorphosis (Truman et al., 1985).

In some instances, larval functions are taken over by different regions in the adult. The larval firefly has its paired lanterns in the eighth (last) abdominal segment; the neurons from this segment control the larva's luminescence. During pupation, the sixth and seventh segments also develop the light-producing photocytes and the nerves to control the timing of the flash. By the end of pupation, only the sixth and seventh segments have functional lanterns. Moreover, if the larval lanterns are removed, the adult lanterns will still form (Strause et al., 1979). Thus, what had been a neural function of the eighth-segment ganglia has become a function of the ganglia of the sixth and seventh segments.

Hormonal Control of Insect Metamorphosis

The hormonal control of insect metamorphosis was shown by the dramatic experiments of Wigglesworth (1934), who studied *Rhodnius prolixus*, a blood-sucking bug that has five instars before undergoing a striking metamorphosis. When a first-instar larva of *Rhodnius* was decapitated and fused to a molting fifth-instar larva, the minute first instar developed the cuticle, body structure, and genitalia of the adult. This showed that blood-borne hormones are responsible for the induction of metamorphosis. Wigglesworth also showed that the **corpora allata,** near the insect brain, produce a hormone that counteracts this tendency to undergo metamorphosis. If the corpora allata were removed from a third-instar larva, the next molt turned the larva into a precocious adult. Conversely, if the corpora allata from fourth-instar larvae were implanted into fifth-instar larvae, these larvae would molt into extremely large "sixth-instar" larvae rather than into adults. We now know that the corpora allata secrete **juvenile hormone,** a natural inhibitor of metamorphosis (which will be discussed shortly).

Transplantation of insect tissues carried out in several laboratories eventually generated an integrated account of how metamorphosis takes place. Although the detailed mechanisms of metamorphosis differ between species, the general pattern of hormone action is usually very similar (Figure 19.23). Like amphibian metamorphosis, the metamorphosis of insects appears to be regulated by effector hormones controlled by neurosecretory peptide hormones in the brain (for reviews, see Gilbert and Goodman, 1981; Granger and Bollenbacher, 1981). The molting process is initiated in the brain, where neurosecretory cells release **prothoracicotropic hormone (PTTH)** in response to neural, hormonal, or environmental factors. PTTH is a family of peptide hormones with a molecular weight of approximately 40,000, and it stimulates the production of ecdysone by the prothoracic gland (Figure 19.24). Ecdysone, however, is not an active hormone, but a prohormone that must be converted into an active form. This conversion is accomplished by a heme-containing oxidase in the mitochondria and microsomes of peripheral tissues such as the fat body. Here the ecdysone is changed to the active hormone **20-hydroxyecdysone** (Figure 19.25).*

Each molt is occasioned by one or more pulses of 20-hydroxyecdysone. For a molt from a larva, the first pulse produces a small rise in the hydroxyecdysone concentration in the larval hemolymph (blood) and elicits a change in cellular commitment. The second, large pulse of hydroxyecdysone

---

*Since its discovery in 1954, when Butenandt and Karlson isolated 25 mg of ecdysone from 500 kg of silkworm moth pupae, 20-hydroxyecdysone has gone under several names, including β-ecdysone, ecdysterone, and crustecdysone.

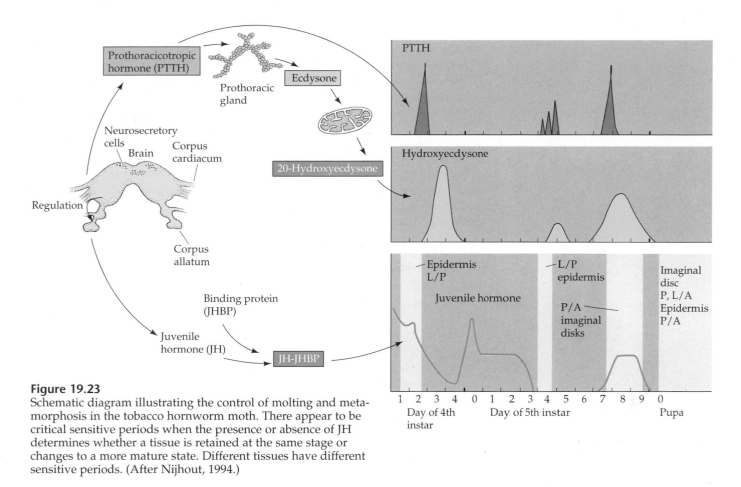

Figure 19.23
Schematic diagram illustrating the control of molting and metamorphosis in the tobacco hornworm moth. There appear to be critical sensitive periods when the presence or absence of JH determines whether a tissue is retained at the same stage or changes to a more mature state. Different tissues have different sensitive periods. (After Nijhout, 1994.)

initiates the differentiation events associated with molting. The hydroxyecdysone produced by these pulses commits and stimulates the epidermal cells to synthesize enzymes that digest and recycle the components of the cuticle. In some cases, environmental conditions can control molting, as in the case of the silkworm moth *Hyalophora cecropia*. Here, PTTH secretion ceases after the pupa has formed. The pupa remains in this suspended state, called **diapause,** throughout the winter. If not exposed to cold weather, diapause lasts indefinitely. Once exposed to two weeks of cold, however, the pupa can molt when returned to a warmer temperature (Williams, 1952, 1956; see Chapter 21).

The second major effector hormone in insect development is **juvenile hormone (JH)**. The structure of a common juvenile hormone active in butterfly and moth caterpillars is shown in Figure 19.25A. JH is secreted by the corpora allata. The secretory cells of the corpora allata are active during larval molts but inactive during the metamorphic molt. This hormone is responsible for preventing metamorphosis. As long as JH is present, the hydroxyecdysone-stimulated molts result in a new larval instar. In the last larval instar, the medial nerve from the brain to the corpora allata inhibits the gland from producing juvenile hormone, and there is a simultaneous increase in the body's ability to degrade existing JH (Safranek and Williams, 1989). Both these mechanisms cause JH levels to drop below a critical threshold value. This triggers the release of PTTH from the brain (Nijhout and Williams, 1974; Rountree and Bollenbacher, 1986). PTTH, in turn, stimulates the prothoracic glands to secrete a small amount of ecdysone. The resulting hydrox-

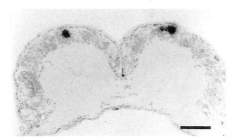

Figure 19.24
Cellular localization of PTTH mRNA in the *Bombyx mori* (silkworm moth) larva. In situ hybridization of a radioactive cloned gene for the 224-amino-acid peptide localizes the PTTH mRNA to two neurosecretory cells in the left hemisphere of the brain and two neurosecretory cells in the right hemisphere. In this section, one PTTH-secreting cell can be seen on each side. The bar represents 100 μm. (From Kawakami et al., 1990, courtesy of H. Ishizaki and A. Kawakami.)

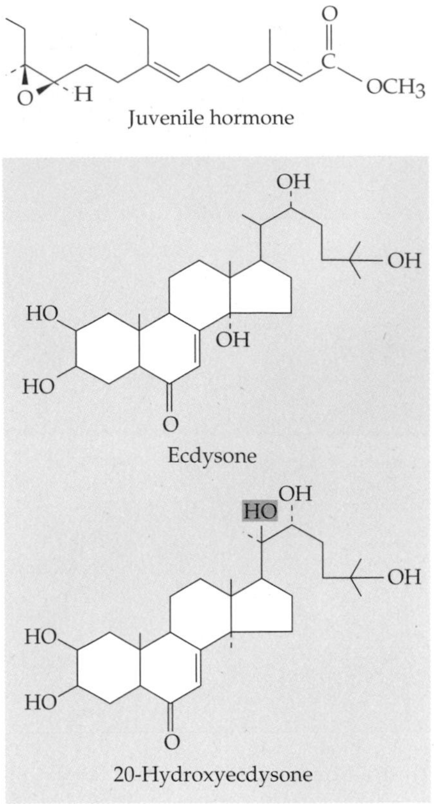

Figure 19.25
Structures of a commonly occurring juvenile hormone, ecdysone, and the active molting hormone 20-hydroxyecdysone.

yecdysone, in the absence of JH, commits the cells to pupal development. Larva-specific mRNAs are not replaced, and new mRNAs are synthesized whose protein products inhibit the transcription of the larval messages. After the second ecdysone pulse, new pupa-specific gene products are synthesized (Riddiford, 1982), and the subsequent molt shifts the organism from larva to pupa. It appears, then, that the first ecdysone pulse during the last larval instar triggers the processes that inactivate the larva-specific genes and prepare the pupa-specific genes to be transcribed. The second ecdysone pulse transcribes the pupa-specific genes and initiates the molt (Nijhout, 1994).

From the 1950s until recently, it had been thought that the type of molt was determined by the juvenile hormone concentration at the time of the ecdysone pulses. High levels of JH induced larvae, intermediate levels of JH produced pupae, and low levels of JH produced adults (see Piepho, 1951; also see Chapter 20 of the Fourth Edition of this book). However, when the titer of JH could actually be determined, it was found that it fluctuated during the final instar period, having specific peaks and troughs. Metamorphosis is not correlated with or caused by a progressive decline in JH activity. The control of metamorphosis appears to be more complex.

In the tobacco hornworm moth *Manduca sexta*, there are specific times when different cells are sensitive to juvenile hormone (see Figure 19.23). As a general rule, if JH is present during a JH-sensitive period, the current developmental state is maintained, whereas if JH is absent during that period, this tissue will progress to a more mature developmental state. The onset and duration of the JH-sensitive period appears to be an autonomous state of the cell and is not controlled by hormones (Nijhout, 1994). (It has been hypothesized that this may be a time when JH receptors are available in these tissues.) In each larval instar, there is a period where the presence of JH prevents the larval epidermis from transforming into pupal epidermis. If JH is present, the epidermis continues to be larval; if JH is absent, it becomes pupal. During the penultimate instar larva, JH titers are able to retain the epidermis in its larval condition. During the last instar, there are two windows of JH sensitivity. The first is for the epidermis. At this time, though, JH levels have dropped significantly. Thus, the epidermis will be transformed from larval epidermis to pupal epidermis. The second JH-sensitive period concerns the imaginal disc tissue. At this time, however, the JH titer has risen again, so that the imaginal discs are not instructed to evert and differentiate. The molt transforms the larva into a pupa (Nijhout and Wheeler, 1982). The next time the ecdysone pulses occur, no JH is seen during the critical periods. The epidermis transforms from pupal to adult, and the imaginal discs are allowed to evert and differentiate. Injection of JH into the pupa at this time can cause it to molt again into a second pupa (Williams, 1959).

Like frog metamorphosis, the timing of ecdysis has to be meticulously coordinated. Many of the behaviors seen during metamorphosis are characteristic of that stage, and failure to perform then leaves the insect fatally trapped in its old cuticle. The coordination of movements and cuticle changes is probably regulated by a cascade of hormones, wherein **eclosion hormone** from the brain activates the secretion of ecdysis-triggering hormone from cells at the base of each spiracle. The **ecdysis-triggering hormone** would then signal the abdominal ganglia of each segment to initiate the movements that allow the larva to shed its old shell (Žitňan et al., 1996).

In *Drosophila*, there is a variation on this general theme (Riddiford, 1993). The ecdysone is released by the ring gland (a structure having regions similar to both the corpus allatum and the prothoracic gland). A high titer pulse of ecdysone at the end of the third instar period signals the onset of metamorphosis. The larva ceases movement, everts its spiracles, and allows the larval cuticle to harden into a **puparium** (pupal case) that surrounds the organism during its metamorphosis. At this time, the imaginal discs evert to form the basic outline of the adult body, but with the head still tucked within the body cavity. After 12 hours (at 25°C), a brief pulse of ecdysone triggers the eversion of the head from the thorax and the transition from "prepupa" to pupa. The head is pushed out by the contraction of abdominal muscles, which push an air bubble up to the anterior to make room for the head to evert (Fristrom and Fristrom, 1993). A subsequent burst of ecdysone brings about the final differentiation of adult form in the *Drosophila* pupa immediately prior to **eclosion,** the "hatching" of the adult from the pupal case. Like other insects, *Drosophila* has an eclosion hormone that initiates the movements and behaviors that enable the adult to wiggle from its pupal case out into the larger world.

The Molecular Biology of Hydroxyecdysone Activity

BINDING OF HYDROXYECDYSONE TO DNA.   During molting and metamorphosis, certain regions of the polytene chromosomes of *Drosophila* become puffed out in certain cells (see Figure 2.13; Clever, 1966; Ashburner, 1972; Ashburner and Berondes, 1978). These **chromosome puffs** represent areas where the DNA is being actively transcribed. Moreover, the organ-specific pattern of puffing can be reproduced by culturing the larval tissue and adding hormones to the medium or by adding hydroxyecdysone to an earlier-stage larva. When hydroxyecdysone is added to larval salivary glands, certain puffs are produced and others regress (Figure 19.26). This puffing is medi-

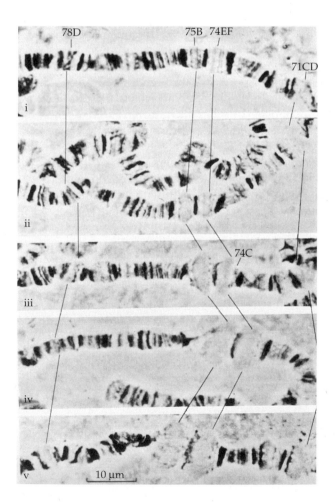

Figure 19.26
Ecdysone-induced puffs in cultured salivary gland cells of *D. melanogaster*. The chromosome region here is the same as in Figure 2.13. Puffing is induced by hydroxyecdysone. (i) Uninduced control. (ii–v) Hydroxyecdysone-stimulated chromosomes at 25 minutes, 1 hour, 2 hours, and 4 hours. (Courtesy of M. Ashburner.)

ated by the binding of hydroxyecdysone to specific places on the chromosomes; fluorescent antibodies against hydroxyecdysone find this hormone localized to the regions of the genome that are sensitive to it (Gronemeyer and Pongs, 1980).

DIFFERENT HYDROXYECDYSONE RECEPTORS IN DIFFERENT TISSUES.   The tissues of late-instar larvae can be grossly divided into three types based on their responses to hydroxyecdysone: (1) the strictly larval tissues (such as larval salivary glands, muscle, and gut) that undergo cell death in response to hydroxyecdysone; (2) the imaginal tissues that divide and differentiate to produce adult structures when exposed to hydroxyecdysone; and (3) tissues that undergo extensive modification or remodeling, such as the fat body or central nervous system. It is not known how one group of cells proliferates while another group of cells degenerates when given the same signal, but recent studies (Talbot et al., 1993; Truman et al., 1994) suggest that not all ecdysone receptors are the same in every tissue. The gene for the **ecdysone receptor** (**EcR**) can be alternatively spliced into three mRNAs that will yield three different, but related, proteins: EcR-A, EcR-B1, and EcR-B2 (Figure 19.27). All cells appear to have some of each, but the strictly larval tissues and regressing neurons are characterized by their great abundance of EcR-B1 compared with EcR-A. Imaginal discs and differentiating neurons, on the other hand, show a preponderance of the EcR-A isoform over EcR-B1. It is therefore possible that the different receptors activate different sets of genes when they bind hydroxyecdysone.

THE BROAD-COMPLEX.   Another reason for the tissue-specific response to ecdysone may be the presence of other transcription factors in these tissues. One of the early genes stimulated by ecdysone is the *Broad-Complex* (*BR-C*) gene. This is a complex gene composed of overlapping transcription units that create several transcription factor proteins through differentially spliced messages. In some *BR-C* mutants, the salivary gland fails to undergo its normal death during metamorphosis. In other mutations, the head fails to evaginate or the CNS fails to be remodeled. Staining with isoform-specific antibodies shows a fascinating correlation between the type of BR-C protein in the nucleus and the type of response to ecdysone. Organs such as the salivary gland that are destined for histolysis during metamorphosis express the Z1 isoform; imaginal discs destined for cell differentiation express the Z2 isoform; and the central nervous system (which undergoes marked remodeling during metamorphosis) expresses all isoforms, with Z3 predominating (Figure 19.28; Emery et al., 1994). Transgenic flies have demonstrated that these differences are functionally important. Ecdysone-dependent gene tran-

Figure 19.27
Formation of the ecdysone receptors. Alternative mRNA splicing of the ecdysone receptor (EcR) transcript creates three types of EcR mRNAs. These generate proteins having the same DNA-binding site and hydroxyecdysone-binding site, but with very different amino termini. (After Talbot et al., 1993.)

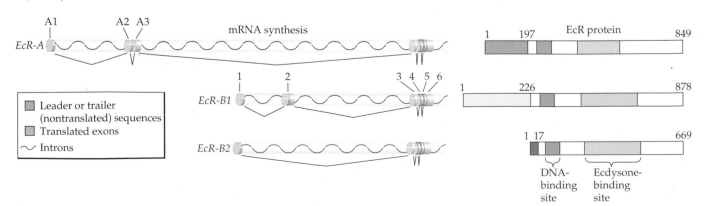

(A)

Anti-Z3

Fat body   Salivary gland

(B)

DNA stain

Fat body   Salivary gland

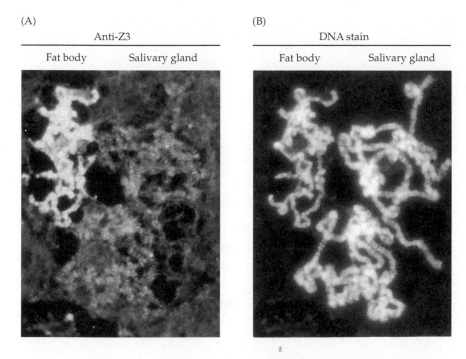

Figure 19.28
Specifity of Broad-Complex isoforms. Fluorescent antibodies locate Z3 isoform in fat body nuclei but not in nuclei of salivary glands. (A) Chromosome preparations of fat body (left) and salivary gland (right) stained with antibodies specific to the Z3 isoform of Broad-Complex. (B) The same preparations stained for DNA. (Photographs courtesy of I. Emery.)

scription in the salivary gland (eventually leading to its destruction) involves the early ecdysone-dependent expression of the Z1 isoform from the *Broad-Complex.* The Z2, Z3, or Z4 proteins will not suffice (Crossgrove et al., 1996). These isoforms also correlate with the types of mutation generated by mutant alleles at this locus. Thus, the specificity of response might be controlled by a specific isoform of the *Broad-Complex* that is stimulated by ecdysone. However, some other factor in the larval tissue must interact with the splicing machinery of the cell to generate the particular exon structure of the *BR-C* message.

DIFFERENT ECDYSONE RECEPTORS WITHIN A SINGLE CELL. The responses to hydroxyecdysone must be coordinated temporally as well as spatially. So in addition to the heterogeneity of responses to hydroxyecdysone between tissues, there is also a heterogeneity of responses to hydroxyecdysone within an individual cell. Hydroxyecdysone-sensitive puffs occurring during the late stages of the third-instar larva (as it prepares to form the pupa) can be grossly divided into three categories: puffs that hydroxyecdysone causes to regress; puffs that hydroxyecdysone induces rapidly; and puffs that are first seen several hours after stimulation. For example, in the larval salivary gland, about six puffs emerge within a few minutes of hydroxyecdysone treatment. These genes do not need protein synthesis to be active. A much larger set of genes was found to be induced later in development, and these genes do need protein synthesis to become transcribed. Ashburner (1974, 1990) predicted that the "early" genes make a protein product that is essential for the activation of the "late" genes. Moreover, this protein itself would turn off the transcription of the early gene (Figure 19.29).

Recent investigations support this view and suggest that the early genes represent transcription factors that can mediate the ecdysone effect. The ecdysone receptors (EcRs) constitute a family of transcription factors derived from a single gene, and they bind this steroid hormone and bring it to the specific region of DNA. Like the vertebrate steroid-binding receptors, EcRs form heterodimers. The ecdysone receptors do not bind either ecdysone or

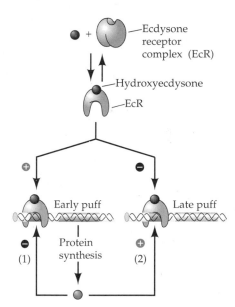

Figure 19.29
The Ashburner model of hydroxyecdysone regulation of transcription. Hydroxyecdysone binds to its receptor, and this compound binds to an early puff gene and a late puff gene. The early puff gene is activated, and its protein product (1) represses the transcription of its own gene and (2) activates the late puff gene, perhaps by displacing the ecdysone receptor. (After Richards, 1992.)

Figure 19.30
Patterns of ecdysone-regulated gene expression in *Drosophila* metamorphosis. (A) Temporal pattern of gene expression. The ecdysone pulses are the bars on top, the height corresponding to the intensity of the pulse. Development proceeds from left to right, starting with the third instar, the molts being represented by dotted lines. (B) Interactions underlying the temporal transcription patterns. Arrows represent activation, while blunted lines indicate repressive effects. (After Thummel, 1996.)

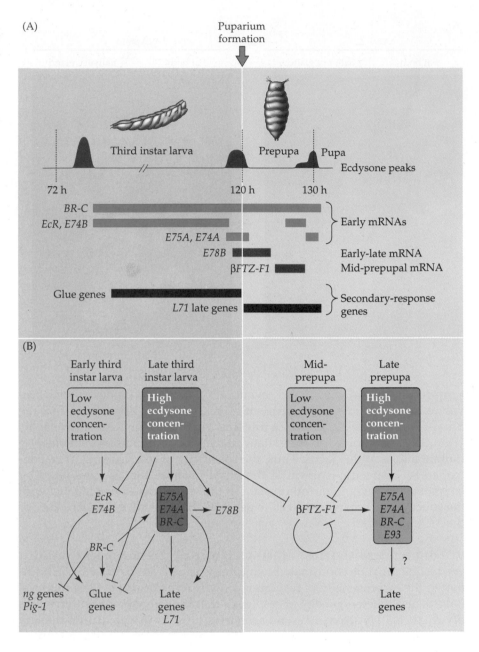

their respective DNA sequences without first forming a heterodimer with the product of the *ultraspiracle* (*USP*) gene (the *Drosophila* analogue of the retinoid receptor; Yao et al., 1992; Thomas et al., 1993). When the EcR/USP heterodimer has been formed, it binds hydroxyecdysone and activates the earliest ecdysone-responsive genes.

Some of these interactions are being elucidated. As shown in Figure 19.30, the *EcR*, *BR-C*, and *E74B* genes are expressed under low concentrations of ecdysone, such as those found toward the end of the third instar period. The BR-C proteins are required to maintain the transcription of the glue protein genes (whose protein glue allows the *Drosophila* pupa to stick to its substrate) and to repress earlier larval genes. E74B is needed both to maintain the activation of the glue genes and to repress genes such as *L71* whose proteins help form the puparium. At the end of the third instar period, there is a characteristically high ecdysone pulse. These higher concentrations of ecdysone repress the glue genes and switch the transcription of the *E74* gene from synthesizing E74B to transcribing the related protein E74A.* Whereas E74B inhibited *L71* gene expression, E74A stimulates it (Urness and Thummel, 1995). In this and other cases, a transition from larva to pupa is occurring.

In addition, the cascade of transcriptional activation and repression may generate new ecdysone receptors. When the *EcR* gene is downregulated at puparium formation, the products of the *E75* or *E78* genes may take over its functions (Koelle et al., 1991; Stone and Thummel, 1993). In this manner, ecdysone induces a cascade of transcription factors that can activate or repress different sets of genes.

So it appears that ecdysone can initiate "waves" of transcriptional activation, and different levels of ecdysone can activate different sets of genes. In this way, *Drosophila* development apears to be like amphibian development, the coordination of the changes being orchestrated by different concentrations of the hormone. The "targets" of these transcription factors are now starting to be identified. Some of these targets appear to be "competence factors" that give other genes the ability to be induced later in development. For instance, during the mid-prepupal stage, the ecdysone titer is lowered. This enables another transcription factor, βFTZ-F1, to be transcribed. The gene encoding βFTZ-F1 needs to have the earlier ecdysone pulse to become potentially active, but it begins transcription only when the ecdysone titer is diminished. Other targets may include the *reaper* and *hid* genes, which become activated in those tissues (such as salivary gland) that undergo ecdysone-dependent cell death.

Molecular biology is beginning to unravel one of the most fascinating networks of interaction known in developmental biology and certainly one of the first examples of animal development that we become aware of—the metamorphosis from larva to adult insect.

---

*E74A and E74B proteins arise from the same gene by the activation of different promoters. They both share the same carboxy-terminal end with its DNA-binding region. However, the E74A protein has a longer amino terminus. The *E74B* mRNAs are transcribed at ecdysone concentrations tenfold lower than those needed to activate the transcription of the *E74A* messages (Karim and Thummel, 1991).

Sidelights & Speculations

Environmental Control Over Larval Form and Function

MOST DISCUSSIONS of development are limited to within the developing organism's body. However, an organism's development can sometimes be regulated by environmental factors outside the body. There are several types of developmental phenomena where chemicals produced by one organism (often of another species) induce changes in the development of another organism.

When Karel Sláma came from Czechoslovakia to work in Carroll Williams's laboratory at Harvard, he brought with him his chief experimental animal, the European plant bug *Pyrrhocoris apterus*. To the consternation of the entire laboratory, these bugs failed to undergo metamorphosis at the end of the fifth instar. Rather, they became large sixth-instar larvae—something never before observed in nature or in the labora-

tory—and ultimately died before becoming adults. After many variables were tested, the paper towels lining the dishes were tested for their effect on the larvae. The results were as conclusive as they were surprising: larvae reared on European paper (including pages of the journal *Nature*) underwent metamorphosis as usual, while larvae reared on American paper (such as shredded copies of the journal *Science*) did not undergo

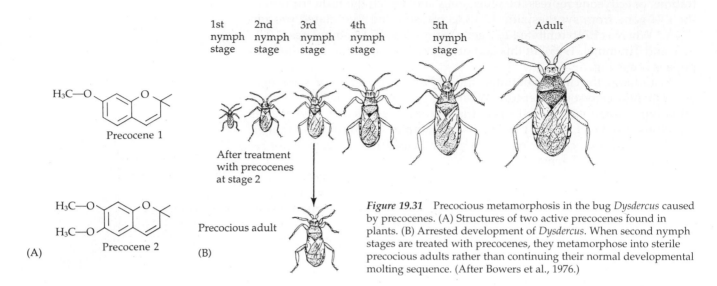

Figure 19.31   Precocious metamorphosis in the bug *Dysdercus* caused by precocenes. (A) Structures of two active precocenes found in plants. (B) Arrested development of *Dysdercus*. When second nymph stages are treated with precocenes, they metamorphose into sterile precocious adults rather than continuing their normal developmental molting sequence. (After Bowers et al., 1976.)

metamorphosis. It was eventually determined that the source of the American paper was the balsam fir, a tree indigenous to the northern United States and Canada. This tree synthesizes a compound that closely resembles juvenile hormone (Bowers et al., 1966; Sláma and Williams, 1966; Williams, 1970), and it probably employs this juvenile hormone analogue to get rid of certain insect predators.

Some other plants have compounds that produce the same effect—the death of insect predators—but do so by eliciting metamorphosis too early. Two compounds that have been isolated from composite herbs have been found to cause the premature metamorphosis of certain insect larvae into sterile adults (Bowers et al., 1976). These compounds are called **precocenes;** their chemical structures are shown in Figure 19.31A. When the larvae or nymphs of these insects are dusted with either of these compounds, they undergo one more molt and then metamorphose into the adult form (Figure 19.31B). Precocenes accomplish this by causing the selective death of the corpus allatum cells in the immature insect (Schooneveld, 1979; Pratt et al., 1980). These cells are responsible for synthesizing juvenile hormone. Without juvenile hormone, the larva commences its metamorphic and imaginal molts. Moreover, juvenile hormone is also responsible for the maturation of the insect egg (Chapter 21). Without this hormone, females are sterile. So the precocenes are able to protect the plant by causing the premature metamorphosis of certain insect larvae into sterile adults.* ∎

*Many more of these environmentally induced changes in larval development will be discussed in Chapter 21.

Multiple hormonal interactions in mammary gland development

The development of the breast is initiated during embryonic development, but it is completed only in the lactating mammal at the end of pregnancy. During breast development, different hormones provide different information to the rudimentary tissue. Mammary development can be divided into four stages: the embryonic stage, the adolescent stage, pregnancy, and lactation. The differentiated products of the mammary glands, casein and other milk proteins, are made only during the final stage (Topper and Freeman, 1980).

Embryonic Stage

In the normal development of the female mouse, two bands of raised epidermal tissue appear on both sides of the ventral midline on day 11 of gestation. This tissue is called the **mammary ridge.** Within each ridge, cells collect at centers of concentration and remain there, forming the **mammary buds** (Figure 19.32). In the mouse, there are five of these buds on each side; in humans, only one per side. In the days immediately prior to birth, the epithelial cells at these places proliferate rapidly, giving rise to the **mammary cord.**

(A)

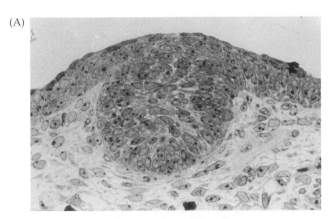

Figure 19.32
Sequence of early mammary gland development in the female mouse. (A) Mammary bud of 12-day fetus. Epithelial ectoderm cells protrude into mesenchyme. (B) Mammary cord of 15-day fetus. A small cleft at the bottom signals the initiation of branching. (C) Cord cavity extending to form a hollow lumen in the 20-day fetus. (From Hogg et al., 1983, courtesy of C. Tickle.)

(B)

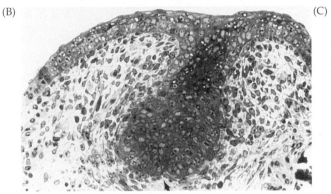

(C)

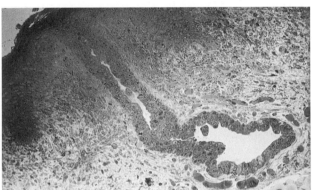

This cord opens at the skin at one end, forming the nipple, while its other end begins branching into ducts. Here development ceases until puberty.

The development of mammary tissue in male mice is identical to that of females until days 13–15 of gestation. At this time, the mesenchyme condenses around the center of the mammary bud, and the cells of the cord die. Thus, a small cord of epithelial cells is detached from the skin (Figure 19.33), and the mammary gland does not extend to the surface. No further development occurs.

This cell death in the mammary cord of males has been studied by culturing the mouse mammary buds in vitro. Such buds from female mice normally develop lobes connected to the surface (Figure 19.34). However, when testosterone is added to the culture medium, the buds degenerate. Mammary buds from male mice also produce lobes if they are cultured in the absence of testosterone; thus, the hormone testosterone prevents mammary development in the male. Testosterone causes this specific cell death by instructing the mesenchyme cells to destroy the epithelial cord. This was shown by a series of recombination experiments. There exists in mice (and in humans as well) a mutation called **androgen insensitivity syndrome,** in which chromosomally male (XY) individuals do not make a functional testosterone receptor. Thus, even though these individuals have testes that are actively secreting testosterone, they are unable to respond to it. One of the results is that these individuals have female-type breast development (see Figure 19.9). Kratochwil and Schwartz (1976) isolated mesenchyme and epithelial cells from normal and mutant mammary buds and cultured them in various combinations. Some cultures were given testosterone and some were not. The results are shown in Figure 19.35. When both mesenchyme and epithelium were wild-type, the rudiment developed into breast tissue. When testosterone was added, the mesenchyme condensed around the bud,

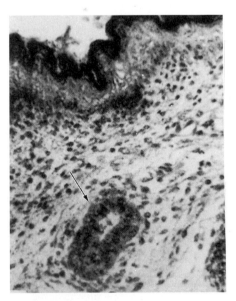

Figure 19.33
Mammary rudiment in a male mouse fetus. The rudiment (arrow) has separated from the epidermis. (From Raynaud, 1961.)

Figure 19.34
Role of testosterone in mediating the detachment of the mammary cord. (A) Female mouse mammary tissue, either in vivo or in culture, will grow downward from the epidermis and branch. (B) When female mouse mammary tissue is cultured in the presence of testosterone, the bud elongates, but mesenchyme cells aggregate around the stalk, and the lower portion is cut off, just as in normal male development. (C) When male mouse mammary tissue is cultured in the absence of testosterone, it develops as it would in the female mouse. (After Kratochwil, 1971.)

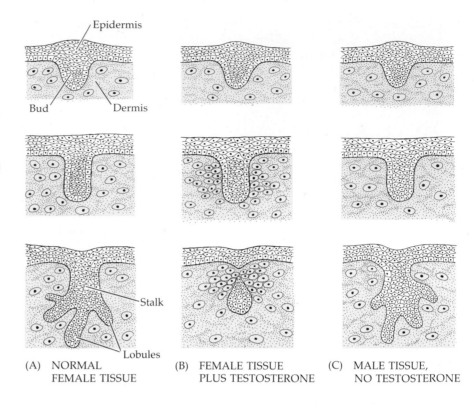

(A) NORMAL FEMALE TISSUE   (B) FEMALE TISSUE PLUS TESTOSTERONE   (C) MALE TISSUE, NO TESTOSTERONE

and the cord was severed. When normal epithelium was cultured with mutant mesenchyme (which could not respond to testosterone), normal breast development occurred in the presence of testosterone. However, when the mesenchyme was normal and the epithelium was mutant, testosterone was able to cause the degeneration of the mammary cord. Thus, the target of

Figure 19.35
Evidence that the mesenchyme cell is the target of testosterone in the arrest of mammary development. (A) Cultured mammary rudiment from 14-day female embryo. (B) Mammary rudiment from 14-day male embryo beginning its response to testosterone. (C) Recombined mammary bud containing wild-type epithelial cells and androgen-insensitive mesenchyme, cultured with testosterone. No androgen response is seen. (D) Recombined mammary bud containing androgen-insensitive epithelial cells and wild-type mesenchyme, cultured with testosterone. Mesenchyme cells are condensing at the neck of the bud. (From Kratochwil and Schwartz, 1976, courtesy of K. Kratochwil.)

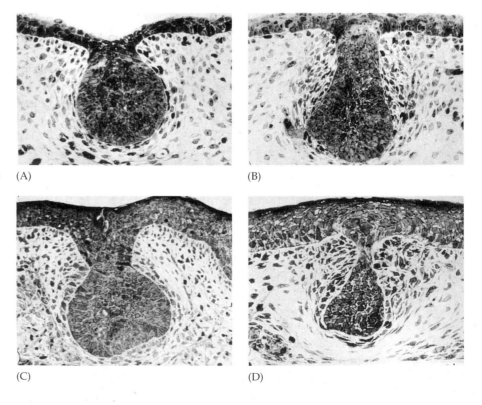

testosterone is the mesenchyme, not the epithelium. The mesenchyme must be responsive to testosterone for its action to occur. In males, the testosterone induces the mammary mesenchyme to destroy its adjacent epithelium. The effect is specific for the organ in that no other mesenchyme will kill the mammary epithelium, and no other epithelium can be destroyed by mammary mesenchyme (Dürnberger and Kratochwil, 1980).

Adolescence

During adolescence (which in the mouse occurs from week 4 to week 6), the duct system of the mammary gland proliferates extensively. The milk-secreting alveolar cells at the tips of the ducts have not differentiated yet, and no milk is produced. The extensive cell division is under the control of estrogen and growth hormones and appears to be concentrated at the ductal tips. Studies by Coleman and her colleagues (1988) implicate epidermal growth factor (EGF) as the responsible factor controlling ductal growth during this period. They implanted EGF in slow-release plastic pellets into the mammary glands of 5-week mice. These mice had had their ovaries removed, and their mammary development was thus halted. The ducts adjacent to the EGF implant reinitiated their growth and morphological development, while the ducts farther away from the implant did not (Figure 19.36). Moreover, when sections of the mammary gland were incubated with radioactive EGF, the EGF was seen to bind to the tip of the ducts and to be associated with the cells that were undergoing mitosis. EGF probably acts directly to cause the growth of the mammary glands during adolescence.*

Pregnancy and Lactation

Between adolescence and pregnancy, mouse breast cells are mitotically dormant and undifferentiated. This state is changed during the second half of pregnancy. Under the influence of the hormones estrogen and progesterone (the latter from the placenta), new ducts are formed, and the distal cells of the ducts begin to develop the characteristics of secretory tissue.

---

*The receptor of the epidermal growth factor (EGFR) may be a key factor in the etiology of breast cancers (which affect one out of every eight women in the United States). It is thought some breast cancers may develop if estrogen induces TGF-α, an alternative ligand for the EGFR. The activation of the EGFR would cause the continued proliferation of the mammary tissue (Sainsbury et al. 1985; Klijn et al., 1992; McIntyre et al., 1995).

Figure 19.36
EGF-dependent mammary gland growth in the absence of estrogen. (A) No duct growth or differentiation is seen in estrogen-deficient mice when a pellet of bovine serum albumin (∗) is implanted into the mammary gland. (B) When a pellet containing EGF is implanted into the estrogen-deficient mammary gland, nearby ducts increase their size and develop lobular tissue at their tips (arrows). (C) Normal mammary duct development in a control 5-week virgin mouse. (From Coleman et al., 1988, courtesy of S. Coleman.)

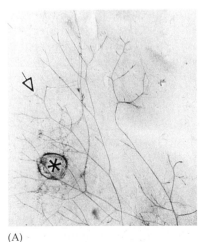

(A)

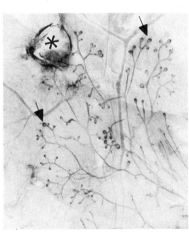

(B)

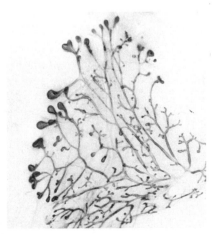

(C)

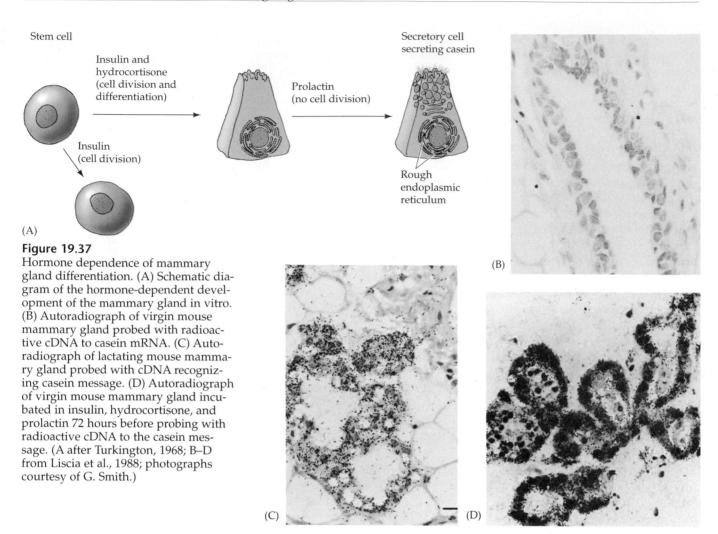

Figure 19.37
Hormone dependence of mammary gland differentiation. (A) Schematic diagram of the hormone-dependent development of the mammary gland in vitro. (B) Autoradiograph of virgin mouse mammary gland probed with radioactive cDNA to casein mRNA. (C) Autoradiograph of lactating mouse mammary gland probed with cDNA recognizing casein message. (D) Autoradiograph of virgin mouse mammary gland incubated in insulin, hydrocortisone, and prolactin 72 hours before probing with radioactive cDNA to the casein message. (A after Turkington, 1968; B–D from Liscia et al., 1988; photographs courtesy of G. Smith.)

When midpregnancy mammary glands are cultured in vitro, most of the cells have little rough endoplasmic reticulum and Golgi apparatus and no casein granules. When insulin or another promoter of DNA synthesis is added to these cultures, the cells become responsive to other hormones (Turkington et al., 1965). (It is likely that the insulin is merely mimicking the effects of placental lactogens, hormones that have a similar structure that are made during pregnancy). Glucocorticoids then induce the formation of the rough endoplasmic reticulum, where casein and the other proteins are synthesized. When the mouse gives birth, prolactin is secreted. Prolactin causes the casein gene to be transcribed and stabilizes the casein message once it is made (Figure 19.37). During the period of lactation (when the young are suckling), a female mouse can produce about 10 percent of her body weight in milk per day. About 80 percent of the proteins in that milk are caseins, and of these, β-casein is the most abundant.

The promoter of the mouse β-casein gene is located immediately upstream from the β-casein gene, and it is bound by a transcription factor, Mammary gland factor (MGF). High levels of this transcription factor accumulate during late pregnancy and lactation, but the factor is inactive unless it is phosphorylated. The phosphorylation of MGF occurs when prolactin binds its two prolactin receptors together at the cell surface. This activates their tyrosine kinase domains, and they phosphorylate a particular tyrosine on the MGF molecule. This phosphorylated MGF can enter the nucleus and

Figure 19.38
The *β-casein* mRNA levels in cultured mouse mammary gland cells placed in different culture conditions. (A) Endogenous *β-casein* mRNA when cells were cultured for 6 days on either extracellular matrix or plastic in medium containing such hormones as insulin, hydrocortisone, or prolactin. The extracellular matrix and prolactin were essential. (B) Reporter gene *CAT* expression when fused to a construct containing the *β-casein* enhancer and promoter. The fused gene was transfected into cultured mouse mammary cells and maintained 6 days in culture under various substrate and hormonal conditions. The fused gene was expressed only in the presence of prolactin and the extracellular matrix. Without the enhancer (having only the promoter), no transcription was seen under any conditions. (After Schmidhauser et al., 1992.)

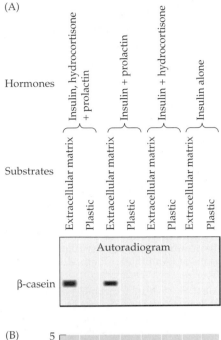

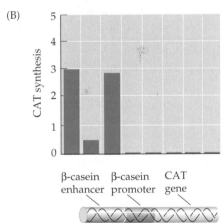

bind to the promoter region of the milk protein genes (Groner and Gouilleux, 1995). Withdrawal of pups from their mothers during lactation results in a rapid decrease in MGF activity. Restoration of the suckling pups restores activity to its maximum level within 4 hours. The effect may be mediated by pituitary or hypothalamic hormones that are responsive to suckling (Schmitt-Ney et al., 1992).

Casein is synthesized in competent mammary cells in response to prolactin only when the cells are anchored to an extracellular matrix (Figure 19.38). The enhancer of β-casein is responsive to both prolactin and the extracellular matrix. Using a reporter gene (*CAT*) linked to different regions of the 5′ flanking sequence, Schmidhauser and colleagues (1992) found a 160-base-pair sequence 1517 base pairs from the transcription initiation site (Figure 19.39). This enhancer site only functions in mammary cells, and it is responsive to both prolactin and the presence of the extracellular matrix (see Figure 19.38B).

The development of the mammary gland, then, involves a complex interplay of several hormones, paracrine proteins, and environmental factors at four different stages of life: embryonic, adolescent, pregnant, and lactating. The mammary gland never develops in normal males and does not become a fully differentiated organ in females until the middle of pregnancy in the adult organism. Studies of this organ have given us insights into the complexity of local and hormonal control in mammalian development.

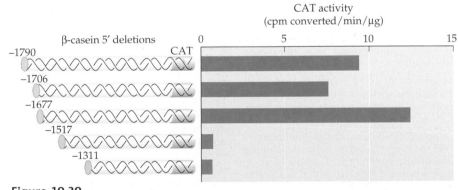

Figure 19.39
Constructs important for identifying the enhancer of the mouse *β-casein* gene. The *CAT* gene was used as a reporter and was fused to the 5′ end of the mouse *β-casein* gene. Exonuclease chopped off successively larger pieces from the 5′ flanking region of the gene. While the fused gene containing 1677 base pairs of the 5′ flanking sequence gave full activity, the sequence containing only 1517 base pairs gave very little activity. Thus, the enhancer was postulated to exist within these 160 base pairs. (After Schmidhauser et al., 1992.)

We have seen that diffusible regulation of cell-cell interactions is also important in regulating development. In studying the reactivation of development that occurs during metamorphosis and breast development, we can identify the roles of hormones in eliciting new patterns of differentiation and morphogenesis. We can also see the interactions between the development of the organism and the ecosystem of which it is part. In the next chapter, we will study the roles of cell-autonomous and diffusible factors in the processes responsible for gonad development and sex determination.

LITERATURE CITED

Alberch, P. and Alberch, J. 1981. Heterochronic mechanisms of morphological diversification and evolutionary change in the neotropical salamander *Bolitoglossa occidentalis* (Amphibia: Plethodontidae). *J. Morphol.* 167: 249–264.

Allen, B. M. 1916. Extirpation experiments in *Rana pipiens* larva. *Science* 44: 755–757.

Alley, K. E. and Barnes, M. D. 1983. Birthdates of trigeminal motor neurons and metamorphic reorganization of the jaw myoneural system in frogs. *J. Comp. Neurol.* 218: 395–405.

Ashburner, M. 1972. Patterns of puffing activity in the salivary glands of *Drosophila*. VI. Induction by ecdysone in salivary glands of *D. melanogaster* cultured in vitro. *Chromosoma* 38: 255–281.

Ashburner, M. 1974. Sequential gene activation by ecdysone in polytene chromosomes of *Drosophila melanogaster*. II. Effects of inhibitors of protein synthesis. *Dev. Biol.* 39: 141–157.

Ashburner, M. 1990. Puffs, genes, and hormones revisited. *Cell* 61: 1–3.

Ashburner, M. and Berondes, H. D. 1978. Puffing of polytene chromosomes. In *The Genetics and Biology of Drosophila*, Vol. 2B. Academic Press, New York, pp. 316–395.

Baker, B. S. and Tata, J. R. 1992. Prolactin prevents the autoinduction of thyroid hormone receptor mRNAs during amphibian metamorphosis. *Dev. Biol.* 149: 463–467.

Basler, K. and Struhl, G. 1994. Compartment boundaries and the control of *Drosophila* limb pattern by hedgehog protein. *Nature* 368: 208–214.

Begon, M., Harper, J. L. and Townsend, C. R. 1986. *Ecology: Individuals, Populations, and Communities*. Blackwell Scientific, Oxford.

Bern, H. A., Nicoll, C. S. and Strohman, R. C. 1967. Prolactin and tadpole growth. *Proc. Soc. Exp. Biol. Med.* 126: 518–521.

Blair, S. S. 1993. Mechanisms of compartment formation: Evidence that non-proliferating cells do not play a role in defining the D/V lineage restriction in the developing wing of *Drosophila*. *Development* 119: 339–351.

Blair, S. S. 1995. Compartments and appendage development in *Drosophila*. *Bio. Essays* 17: 299–309.

Bowers, W. S., Fales, H. M., Thompson, M. J. and Uebel, E. C. 1966. Identification of an active compound from balsam fir. *Science* 154: 1020–1021.

Bowers, W. S., Ohta, T., Cleere, J. S. and Marsella, P. A. 1976. Discovery of insect anti-juvenile hormones in plants. *Science* 193: 542–547.

Brenner, S. 1996. Francisco Crick in Paradiso. *Curr. Biol.* 6: 1202.

Brower, D. L. and Jaffe, S. M. 1989. Requirements for integrins during *Drosophila* wing development. *Nature* 342: 285–287.

Brown, P. S. and Frye, B. E. 1969. Effect of prolactin and growth hormone on growth and metamorphosis of tadpoles of the frog *Rana pipiens*. *Gen. Comp. Endocrinol.* 13: 139–145.

Brust, D. G. 1993. Maternal brood care by *Dendrobates pumilio*: A frog that feeds its young. *J. Herpatol.* 26: 102–105.

Bryant, P. J. 1970. Cell lineage relationships in the imaginal wing disc of *Drosophila melanogaster*. *Dev. Biol.* 22: 389–411.

Buckbinder, L. and Brown, D. D. 1993. Expression of the *Xenopus* prolactin and thyrotropin genes during metamorphosis. *Proc. Natl. Acad. Sci. USA* 90: 3820–3824.

Butenandt, A. and Karlson, P. 1954. Über die Isolierung eines Metamorphosen-Hormons der Insekten in kristallisierter Form. *Z. Naturforsch., Teil B* 9: 389–391.

Campbell, G., Weaver, T. and Tomlinson, A. 1993. Axis specification in the developing Drosophila appendage: The role of *wingless*, *decapentaplegic*, and the homeobox gene *aristaless*. *Cell* 74: 1113–1123.

Causo, J. P., Bate, M. and Martánez-Arias, A. 1993. A *wingless*-dependent polar coordinate system in *Drosophila* imaginal discs. *Science* 259: 484–489.

Clever, U. 1966. Induction and repression of a puff in *Chironomus tentans*. *Dev. Biol.* 14: 421–438.

Cohen, B., Simcox, A. A. and Cohen, S. M. 1993. Allocation of the thoracic imaginal primordia in the *Drosophila* embryo. *Development* 117: 597–608.

Cohen, P. P. 1970. Biochemical differentiation during amphibian metamorphosis. *Science* 168: 533–543.

Cohen, P. P., Brucker, R. F. and Morris, S. M. 1978. Cellular and molecular aspects of thyroid-hormone action during amphibian metamorphosis. *Horm. Prot. Peptides* 6: 273–381.

Coleman, S., Silberstein, G. B. and Daniel, C. W. 1988. Ductal morphogenesis in the mouse mammary gland: Evidence supporting a role for epidermal growth factor. *Dev. Biol.* 127: 304–315.

Condic, M. L., Fristrom, D. and Fristrom, J. W. 1990. Apical cell shape changes during *Drosophila* imaginal leg disc elongation: A novel morphogenetic mechanism. *Development* 111: 23–33.

Crossgrove, K., Bayer, C. A., Fristrom, J. W. and Guild, G. M. 1996. The *Drosophila Broad Complex* early gene directly regulates late gene transcription during the ecdysone-induced puffing cascade. *Dev. Biol.* 180: 745–758.

Currie, J. and Cowan, W. M. 1974. Evidence for the late development of the uncrossed retinothalamic projections in the frog *Rana pipiens*. *Brain Res.* 71: 133–139.

De Beer, G. 1940. *Embryos and Ancestors*. Clarendon Press, Oxford.

del Pino, E. M. and Elinson, R. P. 1983. A novel development pattern for frogs: Gastrulation produces an embryonic disk. *Nature* 306: 589–591.

Diaz-Benjumea, F. J. and Cohen, S. M. 1993. Interaction between dorsal and ventral cells in the imaginal disc directs wing development in *Drosophila*. *Cell* 75: 741–752.

Diaz-Benjumea, F. J., Cohen, B and Cohen, S, M. 1994. Cell interaction between compartments establishes the proximal-distal axis of *Drosophila* wings. *Nature* 372: 175–179.

Dürnberger, H. and Kratochwil, K. 1980. Specificity of tissue interaction and origin of mesenchymal cells in the androgen response of the embryonic mammary gland. *Cell* 19: 465–471.

Eisen, A. Z. and Gross, J. 1965. The role of epithelium and mesenchyme in the production of a collagenolytic enzyme and a hyaluronidase in the anuran tadpole. *Dev. Biol.* 12: 408–418.

Elinson, R. P. 1987. Change in developmental patterns: Embryos of amphibians with large eggs. *In* R. A. Raff and E. C. Raff (eds.), *Development as an Evolutionary Process.* Alan R. Liss, New York, pp. 1–21.

Emery, I. F., Bedian, V. and Guild, G. M. 1994. Differential expression of *Broad-Complex* transcription factors may forecast tissue-specific developmental fates during *Drosophila* metamorphosis. *Development* 120: 3275–3287.

Etkin, W. and Gona, A. G. 1967. Antagonism between prolactin and thyroid hormone in amphibian development. *J. Exp. Zool.* 165: 249–258.

Fallon, J. F. and Simandl, B. K. 1978. Evidence of a role for cell death in the disappearance of the embryonic human tail. *Am. J. Anat.* 152: 111–130.

Forehand, C. J. and Farel, P. B. 1982. Spinal cord development in anuran larvae. I. Primary and secondary neurons. *J. Comp. Neurol.* 209: 386–394.

Fox, H. 1973. Ultrastructure of tail degeneration in *Rana temporaria* larva. *Folia Morphol.* 21: 103–112.

Frieden, E. 1981. The dual role of thyroid hormones in vertebrate development and calorigenesis. *In* L. I. Gilbert and E. Frieden (eds.), *Metamorphosis: A Problem in Developmental Biology.* Plenum, New York, pp. 545–564.

Fristrom, D. and Fristrom, J. W. 1975. The mechanisms of evagination of imaginal disks of *Drosophila melanogaster.* I. General considerations. *Dev. Biol.* 43: 1–23.

Fristrom, D. and Fristrom, J. W. 1993. The metamorphic development of the adult epidermis. *In* M. Bate and A. Martinez-Arias, (eds.) *The Development of Drosophila melanogaster.* Cold Spring Harbor Laboratory Press, Cold Spring Harbor, pp. 843–897.

Fristrom, J. W. 1972. The biochemistry of imaginal disc development. *In* H. Ursprung and R. Nothiger (eds.), *The Biology of Imaginal Discs.* Springer-Verlag, Berlin, pp. 109–154.

Fristrom, J. W., Raikow, R., Petri, W. and Stewart, D. 1969. In vitro evagination and RNA synthesis in imaginal discs of *Drosophila melanogaster. In* E. W. Hanley, *Problems in Biology: RNA in Development.* University of Utah Press, Salt Lake City.

Fristrom, J. W., Fristrom, D., Fekete, E. and Kuniyuki, A. H. 1977. The mechanism of evagination of imaginal discs of *Drosophila melanogaster. Am. Zool.* 17: 671–684.

Garcia-Bellido, A., Ripoll, P. and Morata, G. 1973. Developmental compartmentalization of the wing disc of *Drosophila. Nat. New Biol.* 245: 251–253.

Geigy, R. 1941. Die metamorphose als Folge gewebsspezifischer determination. *Rev. Suisse Zool.* 48: 483–494.

Gilbert, L. I. and Goodman, W. 1981. Chemistry, metabolism, and transport of hormones controlling insect metamorphosis. *In* L. I. Gilbert and E. Frieden (eds.), *Metamorphosis: A Problem in Developmental Biology.* Plenum, New York, pp. 139–176.

Gould, S. J. 1977. *Ontogeny and Phylogeny.* Harvard University Press, Cambridge, MA, p. 283.

Granger, N. A. and Bollenbacher, W. E. 1981. Hormonal control of insect metamorphosis. *In* L. I. Gilbert and E. Frieden (eds.), *Metamorphosis: A Problem in Developmental Biology.* Plenum, New York, pp. 105–138.

Grimm, S. and Pflugfelder, G. O. 1996. Control of the gene *optomotor-blind* in *Drosophila* wing development by *decapentaplegic* and *wingless. Science* 271: 1601–1604.

Grobstein, P. 1987. On beyond neuronal specificity: Problems in going from cells to networks and from networks to behavior. *In* P. Shinkman (ed.), *Advances in Neural and Behavioral Development*, Vol. 3, Ablex, Norwood, NJ, pp. 1–58.

Gronemeyer, H. and Pongs, O. 1980. Localization of ecdysterone on polytene chromosomes of *Drosophila melanogaster. Proc. Natl. Acad. Sci. USA* 77: 2108–2112.

Groner, B. and Gouilleux F. 1995. Prolactin-mediated gene activation in mammary epithelial cells. *Curr. Opin. Genes Dev.* 5: 587–594.

Gudernatsch, J. F. 1912. Feeding experiments on tadpoles. I. The influence of specific organs given as food on growth and differentiation. A contribution to the knowledge of organs with internal secretion. *Wilhelm Roux Arch. Entwicklungsmech. Org.* 35: 457–483.

Guillen, I., Mullor, J. L., Capdevilla, J., Sanchez-Herrero, E., Morata, G. and Guerrero, I. 1995. The function of *engrailed* and the specification of *Drosophila* wing pattern. *Development* 121: 3447–3456.

Hanken, J. and Hall, B. K. 1988. Skull development during anuran metamorphosis II. Role of thyroid hormones in osteogenesis. *Anat. Embryol.* 178: 219–227.

Held, L. I. Jr. 1995. Axes, boundaries and coordinates: the ABCs of fly leg development. *BioEssays* 18: 721–732.

Hepburn, H. R. 1985. Structure of the integument. *In* G. A. Kerkut and L. I. Gilbert (eds.), *Comprehensive Insect Physiology, Biochemistry, and Pharmacology,* Vol 3. Pergamon Press, Oxford, pp. 1–58.

Hinegardner, R. T. 1969. Growth and development of the laboratory cultured sea urchin. *Biol. Bull.* 137: 465–475.

Hogg, N. A. S., Harrison, D. J. and Tickle, C. 1983. Lumen formation in the mammary gland. *J. Embryol. Exp. Morphol.* 73: 39–57.

Hoskins, E. R. and Hoskins, M. M. 1917. On thyroidectomy in amphibia. *Proc. Soc. Exp. Biol. Med.* 14: 74–75.

Hoskins, S. G. and Grobstein, P. 1984. Thyroxine induces the ipsilateral retinothalamic projection in *Xenopus laevis. Nature* 307: 730–733.

Hoskins, S. G. and Grobstein, P. 1985a. Development of the ipsilateral retinothalamic projection in the frog *Xenopus laevis.* II. Ingrowth of optic nerve fibers and production of ipsilaterally projecting cells. *J. Neurosci.* 5: 920–929.

Hoskins, S. G. and Grobstein, P. 1985b. Development of the ipsilateral retinothalamic projection in the frog *Xenopus laevis.* III. The role of thyroxine. *J. Neurosci.* 5: 930–940.

Huxley, J. 1920. Metamorphosis of axolotl caused by thyroid feeding. *Nature* 104: 436.

Irvine, K. D. and Wieschaus, E. 1994. *fringe,* a boundary-specific signaling molecule, mediates interactions between dorsal and ventral cells during Drosophila wing development. *Cell* 79: 595–606.

Jiang, J. and Struhl, G. 1996. Complementary and mutually exclusive activities of *decapentaplegic* and *wingless* organize axial patterning during Drosophila leg development. *Cell* 86: 401–409.

Kalm, L., von, Fristrom, D. and Fristrom, J. 1995. The making of a fly leg: a model for epithelial morphogenesis. *BioEssays* 17: 693–702.

Kaltenbach, J. C., Fry, A. E. and Leius, V. K. 1979. Histochemical patterns in the tadpole tail during normal and thyroxine-induced metamorphosis. II. Succinic dehydrogenase, Mg- and Ca-adenosine triphosphatases, thiamine pyrophosphatase, and 5' nucleotidase. *Gen. Comp. Endocrinol.* 38: 111–126.

Kanamori, A. and Brown, D. D. 1992. The regulation of thyroid hormone receptor β genes by thyroid hormone in *Xenopus laevis. J. Biol. Chem.* 267: 739–745.

Karim, F. D. and Thummel, C. S. 1991. Ecdysone coordinates the timing and amounts of *E74A* and *E74B* transcription in *Drosophila. Genes Dev.* 5: 1067–1079.

Karp, G. and Berrill, N. J. 1981. *Development.* McGraw-Hill, New York.

Kawahara, A., Baker, B. S. and Tata, J. R. 1991. Developmental and regional expression of thyroid hormone receptor genes during *Xenopus* metamorphosis. *Development* 112: 933–943.

Kawakami, A. and 9 others 1990. Molecular cloning of the *Bombyx mori* prothoracicotropic hormone. *Science* 247: 1333–1335.

Kim, J., Irvine, K. D. and Carroll, S. B. 1995. Cell recognition, signal induction, and symmetrical gene activation at the dorsalventral boundary of the developing *Drosophila* wing. *Cell* 82: 795–802.

Kim, J., Sebring, A., Esch, J. J., Kraus, M. E., Vorwrk, K., Magee, J. and Carroll, S. B. 1996. Integration of positional information and identity by *Drosophila vestigial* gene. *Nature* 382: 133–138.

Kinoshita, T., Takahama, H., Sasaki, F. and Watanabe, K. 1989. Determination of cell death in the developmental process of anuran larval skin. *J. Exp. Zool.* 251: 37–46.

Klijn, I. G. M., Berns, P. M. J. J., Schmitz, P. I. M. and Foekens, J. A. 1992. The clinical significance of EGF-R in human breast cancer: A review of 5232 patients. *Endocr. Rev.* 13: 3–17.

Kistler, A., Yoshizato, K. and Frieden, E. 1977. Preferential binding of tri-substituted thyronine analogs by bullfrog tadpole tail fin cytosol. *Endocrinology* 100: 134–137.

Koelle, M. R., Talbot, W. S., Segraves, W. A., Bender, M. T. , Cherbas, P. and Hogness, D. S. 1991. The *Drosophila* EcR gene encodes an ecdysone receptor, a new member of the steroid receptor superfamily. *Cell* 67: 59–77.

Kollros, J. J. 1961. Mechanisms of amphibian metamorphosis: Hormones. *Am. Zool.* 1: 107–114.

Kratochwil, K. 1971. In vitro analysis of the hormonal basis for sexual dimorphism in the embryonic development of the mouse mammary gland. *J. Embryol. Exp. Morphol.* 25: 141–153.

Kratochwil, K. and Schwartz, P. 1976. Tissue interaction in androgen response of embryonic mammary rudiment of mouse: Identification of target tissue for testosterone. *Proc. Natl. Acad. Sci. USA* 73: 4041–4044.

Lawrence, P. A. and Morata, G. 1976. Compartments of the wing of *Drosophila*: a study of the *engrailed* gene. *Dev. Biol.* 50: 321–337.

Liscia, D. S., Doherty, P. J. and Smith, G. H. 1988. Localization of α-casein gene transcription in sections of epoxy resin-embedded mouse mammary tissues by *in situ* hybridization. *J. Histochem. Cytochem.* 36: 1503–1510.

Little, G., Atkinson, B. G. and Frieden, E. 1973. Changes in the rates of protein synthesis and degradation in the tail of *Rana catesbeiana* tadpoles during normal metamorphosis. *Dev. Biol.* 30: 366–373.

Lyman, D. F. and White, B. A. 1987. Molecular cloning of hepatic mRNAs in *Rana catesbeiana* response to thyroid hormone during induced and spontaneous metamorphosis. *J. Biol. Chem.* 262: 5233–5237.

Lynn, W. G. and Peadon, A. M. 1955. The role of the thyroid gland in direct development of the anuran *Eleutherodactylus martinicenis. Growth* 19: 263–286.

Madhavan, M. M. and Schneiderman, H. A. 1977. Histological analysis of the dynamics of growth of imaginal disc and histioblast nests during the larval development of *Drosophila melanogaster. Wilhelm Roux Arch. Dev. Biol.* 183: 269–305.

Martin, F. D. and Drewry, G. E. 1978. *Development of Fishes of the Mid-Atlantic Bight,* Vol. 6. U. S. Department of the Interior, Washington, D.C.

Mathison, P. M. and Miller, L. 1987. Thyroid hormone induction of keratin genes: A two-step activation of gene expression during development. *Genes Dev.* 1: 1107–1117.

McCutcheon, F. H. 1936. Hemoglobin function during the life history of the bullfrog. *J. Cell. Comp. Physiol.* 8: 63–81.

McIntyre, B. S., Birkenfeld, H. P. and Sylvester, P. W. 1995. Relationship between EGFR levels, autophosphorylation, and mitogenic resposiveness in normal mouse mammary epithelial cells in vitro. *Cell Prolif.* 28: 45–56.

Meinhardt, H. 1980. Cooperation of compartments for the generation of positional information. *Z. Naturforsch.* 35c: 1086–1091.

Mitchell, A. W. 1988. *The Enchanted Canopy.* Macmillan, New York.

Mori, M., Morris, S. M., Jr. and Cohen, P. P. 1979. Cell-free translation and thyroxine induction of carbamylphosphate synthetase I messenger RNA in tadpole liver. *Proc. Natl. Acad. Sci. USA* 76: 3179–3183.

Nellen, D., Burke, R., Struhl, G. and Basler, K. 1996. Direct and long-range action of a Dpp morphogen gradient. *Cell* 85: 357–368.

Nijhout, H. F. 1994. *Insect Hormones.* Princeton University Press, Princeton, NJ.

Nijhout, H. F. and Wheeler, D. E. 1982. Juvenile hormone and the physiological basis of insect polymorphisms. *Quart. Rev. Biol.* 57: 109–133.

Nijhout, H. F. and Williams, C. M. 1974. Control of moulting and metamorphosis in the tobacco hornworm, *Manduca sexta*: Cessation of juvenile hormone secretion as a trigger for pupation. *J. Exp. Biol.* 61: 493–501.

Niki, K., Namiki, H., Kikuyama, S. and Yoshizato, K. 1982. Epidermal tissue requirement for tadpole tail regression induced by thyroid hormone. *Dev. Biol.* 94: 116–120.

Nishikawa, A., Kaiho, M. and Yoshizato, K. 1989. Cell death in the anuran tadpole tail:

Thyroid hormone induces keratinization and tail-specific growth inhibition of epidermal cells. *Dev. Biol.* 131: 337–344.

Norris, D. O., Jones, R. E. and Criley, B. B. 1973. Pituitary prolactin levels in larval, neotenic, and metamorphosed salamanders (*Ambystoma tigrinum*). *Gen. Comp. Endocrinol.* 20: 437–442.

Oofusa, K. and Yoshizato, K. 1991. Biochemical and immunological characterization of collagenase in tissues of metamorphosing bullfrog tadpoles. *Dev. Growth Differ.* 33: 329–339.

Patterson, D., Hayes, W. P. and Shi, Y. B. 1995. Transcriptional activation of the metalloproteinase gene *stromelysin*-3 coincides with thyroid hormone-induced cell death during frog metamorphosis. *Dev. Biol.* 167: 252–262.

Pehrson, J. R. and Cohen, L. H. 1986. The fate of the small micromeres in sea urchin development. *Dev. Biol.* 113: 522–526.

Piepho, H. 1951. Über die Lenkung der Insektenmetamorphose durch Hormone. *Verh. dtsch. Zool. Gessel.* 62–76.

Pino-Heiss, S. and Schubiger, G. 1989. Extracellular protease production by *Drosophila* imaginal discs. *Dev. Biol.* 132: 282–291.

Postlethwait, J. H. and Schneiderman, H. A. 1971. Pattern formation and determination in the antenna of the homeotic mutant *Antennapedia* of *Drosophila melanogaster. Dev. Biol.* 25: 606–640.

Prahlad, K. V. and DeLanney, L. E. 1965. A study of induced metamorphosis in the axolotl. *J. Exp. Zool.* 160: 137–146.

Pratt, G. E., Jennings, R. C., Hammett, A. F. and Brooks, G. T. 1980. Lethal metabolism of precocene-I to a reactive epoxide by locust corpora allata. *Nature* 284: 320–323.

Raff, R. A. 1987. Constraint, flexibility, and phylogenetic history in the evolution of direct development in sea urchins. *Dev. Biol.* 119: 6–19.

Raff, R. A. 1994. Developmental mechanisms in the evolution of animal form: Origins and evolvability of body plans. *In* S. Bengtson, (ed.), *Early Life on Earth,* Columbia University Press, New York, pp. 489–500.

Raynaud, A. 1961. Morphogenesis of the mammary gland. *In* S. K. Kon and A. T. Cowrie (eds.), *Milk: The Mammary Gland and Its Secretion,* Vol. 1. Academic Press, New York, pp. 3–46.

Richards, G. 1992. Switching partners? *Curr. Biol.* 2: 657–658.

Riddiford, L. M. 1982. Changes in translatable mRNAs during the larval-pupal transformation of the epidermis of the tobacco hornworm. *Dev. Biol.* 92: 330–342.

Riddiford, L. M. 1993. Hormones and *Drosophila* development. *In* M. Bate and A.

Martinez-Arias (eds.), *The Development of Drosophila melanogaster*, Cold Spring Harbor Laboratory Press, Cold Spring Harbor, pp. 899–939

Riggs, A. F. 1951. The metamorphosis of hemoglobin in the bullfrog. *J. Gen. Physiol.* 35: 23–40.

Robinson, H., Chaffee, S. and Galton, V. A. 1977. Sensitivity of *Xenopus laevis* tadpole tail tissue to the action of thyroid hormones. *Gen. Comp. Endocrinol.* 32: 179–186.

Rountree, D. B. and Bollenbacher, W. E. 1986. The release of the prothoracicotropic hormone in the tobacco hornworm, *Manduca sexta*, is controlled intrinsically by juvenile hormone. *J. Exp. Biol.* 120: 41–58.

Safranek, L. and Williams, C. M. 1989. Inactivation of the corpora allata in the final instar of the tobacco hornworm, *Manduca sexta*, requires integrity of certain neural pathways from the brain. *Biol. Bull.* 177: 396–400.

Sainsbury, J. R. C., Malcolm, A. J., Appleton, D. R., Farndon, J. R. and Harris, A. L. 1985. Presence of epidermal growth factor receptor as an indicator of poor prognosis in patients with breast cancer. *J. Clin. Pathol.* 38: 1225–1228.

Schmidhauser, C., Casperson, G. F., Myers, C. A., Sanzo, K. T., Bolten, S. and Bissell, M. J. 1992. A novel transcriptional enhancer is involved in the prolactin- and extracellular matrix-dependent regulation of β-casein gene expression. *Mol. Biol. Cell* 3: 699–709.

Schmitt-Ney, M., Happ, B., Ball, R. K. and Groner, B. 1992. Developmental and environmental regulation of a mammary gland-specific nuclear factor essential for the transcription of the gene encoding β-casein. *Proc. Natl. Acad. Sci. USA* 89: 3130–3134.

Schooneveld, H. 1979. Precocene-induced collapse and resorption of corpora allata in nymphs of *Locusta migratoria*. *Experientia* 35: 363–364.

Schwind, J. L. 1933. Tissue specificity at the time of metamorphosis in frog larvae. *J. Exp. Zool.* 66: 1–14.

Sláma, K. and Williams, C. M. 1966. The juvenile hormone. V. The sensitivity of the bug, *Pyrrhocoris apterus*, to a hormonally active factor in American paper-pulp. *Biol. Bull.* 130: 235–246.

Smith-Gill, S. J. and Carver, V. 1981. Biochemical characterization of organ differentiation and maturation. *In* L. I. Gilbert and E. Frieden (eds.), *Metamorphosis: A Problem in Developmental Biology*. Plenum, New York, pp. 491–544.

Stolow, M. A. and Shi, Y. B. 1995. *Xenopus* sonic hedgehog as a potential morphogen during embryogenesis and thyroid hormone dependent metamorphosis. *Nucleic Acids Res.* 23: 2555–2562.

Stone, B. L. and Thummel, C. S. 1993. *Drosophila* 78C early-late puff contains E78, an ecdysone-inducible gene that encodes a novel member of the nuclear hormone superfamily. *Cell* 75: 1–20.

Strathmann, R. R. 1971. The feeding behavior of planktotrophic echinoderm larvae: Mechanisms, regulation and rates of suspension feeding. *Exp. Mar. Biol. Ecol.* 6: 109–160.

Strathmann, R. R. 1975. Larval feeding in echinoderms. *Am. Zool.* 15: 717–730.

Strause, L. G., DeLuca, M. and Case, J. F. 1979. Biochemical and morphological change accompanying light organ development in the firefly, *Photuris pennsylvanica*. *J. Insect Physiol.* 125: 339–347.

Tabata, T., Schwartz, E., Gustavson, E., Ali, Z. and Kornberg, T. B. 1995. Creating a Drosophila wing de novo, the role of *engrailed*, and the compartment border hypothesis. *Development* 121: 3359–3369.

Taigen, T. L., Plough, F. H. and Stewart, M. M. 1984. Water balance of terrestrial anuran (*Eleutherodactylus coqui*) eggs: Importance of paternal care. *Ecology* 65: 248–255.

Takahashi. N., Yoshihama, K., Kikuyama, S., Yamamoto, K., Wakabayashi, K. and Kato, Y. 1990. Molecular cloning and nucleotide sequence of complementary DNA for bullfrog prolactin. *J. Mol. Endocrinol.* 5: 281–287.

Talbot, W. S., Swyryd, E. A. and Hogness, D. S. 1993. *Drosophila* tissues with different metamorphic responses to ecdysone express different ecdysone receptor isoforms. *Cell* 73: 1323–1337.

Tata, J. R., Kawahara, A. and Baker, B. S. 1991. Prolactin inhibits both thyroid hormone-induced morphogenesis and cell death in cultured amphibian larval tissues. *Dev. Biol.* 146: 72–80.

Taurog, A., Oliver, C., Porter, R. L., McKenzie, J. C. and McKenzie, J. M. 1974. The role of TRH in the neoteny of the Mexican axolotl (*Ambystoma mexicanum*). *Gen. Comp. Endocrinol.* 24: 267–279.

Thomas, H. E., Stunnenberg, H. G. and Stewart, A. F. 1993. Heterodimerization of the *Drosophila* ecdysone receptor with retinoid X receptor and *ultraspiracle*. *Nature* 362: 471–475.

Thummel, C. S. 1996. Flies on steroids: *Drosophila* metamorphosis and the mechanisms of steroid hormone action. *Trends Genet.* 12: 306–310.

Topper, Y. J. and Freeman, C. S. 1980. Multiple hormone interactions in the developmental biology of the mammary gland. *Physiol. Rev.* 60: 1049–1106.

Truman, J. W., Weeks, J. and Levine, R. B. 1985. Developmental plasticity during the metamorphosis of an insect nervous sys-

tem. *In* M. J. Cohen and F. Strumwasser (eds.), *Comparative Neurobiology*. Wiley, New York, pp. 25–44.

Truman, J. W., Talbot, W. S., Fahrbach, S. E. and Hogness, D. S. 1994. Ecdysone receptor expression in the CNS correlates with stage-specific responses to ecdysteroids during *Drosophila* and *Manduca* development. *Development* 120: 219–234.

Turkington, R. W. 1968. Hormone-dependent differentiation of mammary gland in vitro. *Curr. Top. Dev. Biol.* 3: 199–218.

Turkington, R. W., Juergens, W. G. and Topper, Y. J. 1965. Hormone-dependent synthesis of casein *in vitro*. *Biochem. Biophys. Acta* 111: 573–576.

Turner, C. D. and Bagnara, J. T. 1976. *General Endocrinology*, 6th ed. Saunders, Philadelphia.

Urness, L.D. and Thummel, C. D. 1995. Molecular analysis of a steroid-induced regulatory hierarchy: The *Drosophila* E74A protein directly regulates L71–6 transcription. *EMBO J.* 14: 6239–6246.

van Wijngaarden, R. and Bolanos, F. 1992. Parental care in *Dendrobates granuliferus* (Anura, Dendrobatidae) with a description of the tadpole. *J. Herpetol.* 26: 102–105.

Wald, G. 1945. The chemical evolution of vision. *Harvey Lect.* 41: 117–160.

Wald, G. 1981. Metamorphosis: An overview. *In* L. I. Gilbert and E. Frieden (eds.), *Metamorphosis: A Problem in Developmental Biology*. Plenum, New York, pp. 1–39.

Wassersug, R. J. 1989. Locomotion in amphibian larvae (or why aren't tadpoles built like fish). *Am. Zool.* 29: 65–84.

Weber, R. 1967. Biochemistry of amphibian metamorphosis. *In* R. Weber (ed.), *The Biochemistry of Animal Development*, Vol. 3. Academic Press, New York, pp. 227–301.

Weismann, A. 1875. Über den Saison-Dimorphismus der Schmetterlinge. In *Studien Zur Descendenz-Theorie*. Engelmann, Leipzig.

Wigglesworth, V. B. 1934. The physiology of ecdysis in *Rhodnius prolixus* (Hemiptera). II. Factors controlling moulting and metamorphosis. *Q. J. Microsc. Sci.* 77: 121–222.

Wilder, E. L. and Perrimon, N. 1995. Dual function of wingless in the *Drosophila* leg imaginal disc. *Development* 121: 477–488.

Williams, C. M. 1952. Physiology of insect diapause. IV. The brain and prothoracic glands as an endocrine system in the *Cecropia* silkworm. *Biol. Bull.* 103: 120–138.

Williams, C. M. 1956. The juvenile hormone of insects. *Nature* 178: 212–213.

Williams, C. M. 1959. The juvenile hormone. I. Endocrine activity of the corpora allata of the adult *Cecropia* silkworm. *Biol. Bull.* 116: 323–338.

Williams, C. M. 1970. Hormonal interactions between plants and insects. *In* E. Sondheimer and J. B. Simeone (eds.), *Chemical Ecology*. Academic Press, New York, pp. 103–132.

Williams, J. A., Paddock, S. W., Vorwek, K. and Carroll, S. B. 1994. Organization of wing formation and induction of a wing-patterning gene at the dorsal/ventral compartment boundary. *Nature* 368: 299–305.

Wong, J. M. and Shi, Y. B. 1995. Coordinated regulation and transcriptional activation of *Xenopus* thyroid hormone and retinoid-X receptors. *J. Biol. Chem.* 270: 18479–18483.

Wray, G. A. and Raff, R. A. 1990. Novel origins of lineage founder cells in the direct-developing sea urchin, *Heliocidaris erythrogramma. Dev. Biol.* 141: 41–54.

Wray, G. A. and Raff, R. A. 1991. Rapid evolution of gastrulation mechanisms in a sea urchin with lecithotrophic larvae. *Evolution* 45: 1741–1750.

Yao, T.-P., Segraves, W. A., Oro, A. E., McKeown, M. and Evans, R. M. 1992. *Drosophila ultraspiracles* modulates ecdysone receptor function via heterodimer formation. *Cell* 71: 63–72.

Yaoita, Y. and Brown, D. D. 1990. A correlation of thyroid hormone receptor gene expression with amphibian metamorphosis. *Genes Dev.* 4: 1917–1924.

Žitňan, D., Kingan, T. G., Hermesman, J. L. and Adams, M. E. 1996. Identification of ecdysis-triggering hormone from an epitracheal endocrine system. *Science* 271: 88–91.

Sex determination

20

Sexual reproduction is…the masterpiece of nature.
ERASMUS DARWIN (1791)

It is quaint to notice that the number of speculations connected with the nature of sex have well-nigh doubled since Drelincourt, in the eighteenth century, brought together two hundred and sixty-two "groundless hypotheses," and since Blumenbach caustically remarked that nothing was more certain than that Drelincourt's own theory formed the two hundred and sixty-third.
J. A. THOMSON (1926)

T HE MECHANISMS by which an individual's sex is determined has been one of the great questions of embryology since antiquity. Aristotle, who collected and dissected embryos, claimed that sex was determined by the heat of the male partner during intercourse. The more heated the passion, the greater the probability of male offspring. (Aristotle counseled elderly men to conceive in the summer if they wished to have male heirs.) Aristotle (ca. 335 B.C.E.) promulgated a very straightforward hypothesis of sex determination: women were men whose development arrested too early. The female was "a mutilated male" whose development stopped because the coldness of the mother's womb overcame the heat of the man's semen. Women were colder and more passive than men, and female sexual organs had not matured to the point where they could provide active seeds. This view was accepted by the Christian Church and by Galen (whose anatomy texts were to be the standard for over 1000 years). Around the year 200 C.E., Galen wrote:

> *Just as mankind is the most perfect of all animals, so within mankind, the man is more perfect than the woman, and the reason for this perfection is his excess heat, for heat is Nature's primary instrument…the woman is less perfect than the man in respect to the generative parts. For the parts were formed within her when she was still a fetus, but could not because of the defect in heat emerge and project on the outside.*

The view that women were but poorly developed men and that their genitalia were like men's, only turned inside out, was a very popular view for over a thousand years. Even in 1543, Andreas Vesalius, the Paduan anatomist who overturned much of Galen's anatomy (and who risked censure by the church for arguing that men and women have the same number of ribs), held this view. The illustrations from his two major works, *De Humani Corporis Fabrica* and *Tabulae Sex*, show that he saw the female genitalia as internal representations of the male genitalia (Figure 20.1). Nevertheless, Vesalius's books sparked a revolution in anatomy, and by the end of the 1500s, anatomists dismissed Galen's representation of female anatomy. During the 1600s and 1700s, females were seen as producing eggs that could transmit parental traits, and the physiology of the sex organs began to be

(A)

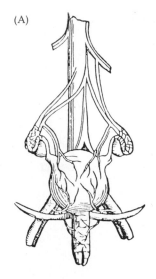

(B)

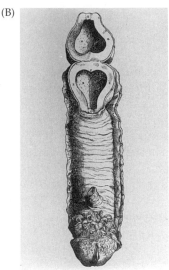

Figure 20.1
Vesalius's representations (1538, 1543) of the female reproductive organs. (A) Vesalius's rendering of Galen's conception of the female tract from the vagina to the uterus. (B) Vesalius's rendering of the female reproductive system. (Reprinted in Schiebinger, 1989.)

studied. Still, there was no consensus about how the sexes became determined (see Horowitz, 1976; Tuana, 1988; Schiebinger, 1989). [sex8.html]

During that time, the environment—heat and nutrition, in particular—was believed to be important in determining sex. In 1890, Geddes and Thomson summarized all available data on sex determination and came to the conclusion that the "constitution, age, nutrition, and environment of the parents must be especially considered" in any such analysis. They argued that factors favoring the storage of energy and nutrients predisposed one to have female offspring, whereas factors favoring the utilization of energy and nutrients influenced one to have male offspring.

This environmental view of sex determination remained the only major scientific theory until the rediscovery of Mendel's work in 1900 and the rediscovery of the sex chromosome by McClung in 1902. Based on his knowledge of Mendelism, Correns speculated that the 1:1 sex ratio of most species could be achieved if the male was heterozygous and the female homozygous for some sex-determining factor. It was not until 1905, however, that the correlation (in insects) of the female sex with XX sex chromosomes and the male sex with XY or XO chromosomes was established (Stevens, 1905; Wilson, 1905). This suggested strongly that a specific nuclear component was responsible for directing the development of the sexual phenotype. Thus, evidence accumulated that sex determination occurs by nuclear inheritance rather than by environmental happenstance.

Today we find that both environmental and internal mechanisms of sex determination can operate in different species. We will first discuss the chromosomal mechanisms of sex determination and then consider the ways by which the environment regulates the sexual phenotype.

Chromosomal sex determination in mammals

Primary Sex Determination

Primary sex determination concerns the determination of the gonads. In mammals, the determination of sex is strictly chromosomal and is not usually influenced by the environment. In most cases, the female is XX and the male is XY. Every individual must have at least one X chromosome. Since the female is XX, each of her eggs has a single X chromosome. The male, being XY, can generate two types of sperm: half bear the X chromosome, half the Y. If the egg receives another X chromosome from the sperm, the resulting individual is XX, forms ovaries, and is female; if the egg receives a Y chromosome from the sperm, the individual is XY, forms testes, and is male. The Y chromosome carries a gene that encodes a testis-determining factor. This factor organizes the gonad into a testis rather than an ovary. Unlike the case in *Drosophila* (discussed later), the mammalian Y chromosome is a crucial factor for determining sex in mammals. A person with five X chromosomes and one Y chromosome (XXXXXY) would be male. Furthermore, an individual with only a single X chromosome and no second X or Y (i.e., XO) develops as a female and begins making ovaries, but is unable to maintain the ovarian follicles.

Secondary Sex Determination

Secondary sex determination concerns the bodily phenotype outside the gonads. A male mammal has a penis, seminal vesicles, a prostate gland, and often sex-specific size, vocal cartilage, and musculature. A female mammal has a vagina, cervix, uterus, oviducts, mammary glands, and often sex-specific size, vocal cartilage, and musculature. The secondary sexual character-

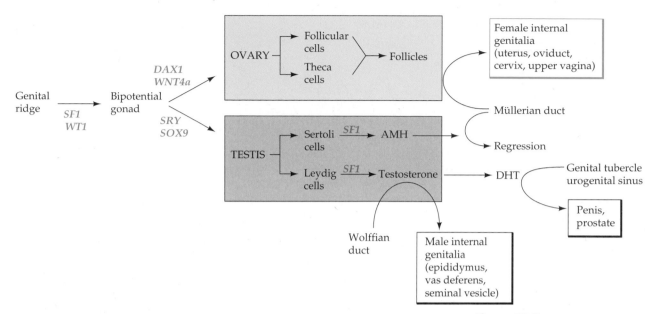

Figure 20.2
Postulated cascades leading to the formation of the sexual phenotypes in mammals. The conversion of the genital ridge into the bipotential gonad needs the *SF1* and *WT1* genes, since mice lacking either of these genes lack gonads. The bipotential gonad appears to be moved into the female pathway by the *Wnt4* or *DAX1* genes and into the male pathway by the *SRY* gene (on the Y chromosome) in conjunction with autosomal genes such as *SOX9*. The ovary makes thecal cells and granulosa cells, which together are capable of synthesizing estrogen. Under estrogen (first from the mother, then from the gonads), the Müllerian duct differentiates into the female genitalia and the offspring develop the secondary sex characteristics of females. The testis makes two major hormones. The first, anti-Müllerian duct factor (AMH), causes the Müllerian duct to regress. The second, testosterone, causes the differentiation of the Wolffian duct into the male internal genitalia. In the urogenital region, testosterone is converted into dihydrotestosterone (DHT), and this hormone causes the morphogenesis of the penis and prostate. (After Marx, 1995.)

istics are usually determined by *hormones* secreted from the gonads. However, in the *absence* of gonads, the female phenotype is generated. When Jost (1953) removed fetal rabbit gonads before they had differentiated, the resulting rabbits were female, regardless of whether they were XX or XY. They each had oviducts, a uterus, and a vagina, and each lacked a penis and male accessory structures.

The scheme of mammalian sex determination is shown in Figure 20.2. If the Y chromosome is absent, the gonadal primordia develop into ovaries. The estrogenic hormones produced by the ovary enable the development of the **Müllerian duct** into the uterus, oviducts, and upper end of the vagina. If the Y chromosome is present and the testes form, they secrete two major hormones. The first hormone—**anti-Müllerian duct hormone (AMH;** also referred to as Müllerian-inhibiting substance, MIS)—destroys the Müllerian duct. The second hormone—**testosterone**—masculinizes the fetus, stimulating the formation of the penis, scrotum, and other portions of the male anatomy, as well as inhibiting the development of the breast primordia. Thus, the body has the female phenotype unless it is changed by the two hormones elaborated from the fetal testes. We will now take a more detailed look at these events.

The Developing Gonads

The development of gonads is a unique embryological situation. All other organ rudiments can normally differentiate into only one type of organ. A lung rudiment can become only a lung, and a liver rudiment can develop only into a liver. The gonadal rudiment, however, has two normal options. When it differentiates, it can develop into either an ovary or a testis. The type of differentiation taken by this rudiment determines the future sexual development of the organism. But, before this decision is made, the mammalian gonad first develops through an **indifferent (bipotential) stage,** during which time it has neither female nor male characteristics. In humans, the gonadal rudiment appears in the intermediate mesoderm during week 4 and remains sexually indifferent until week 7. During this indifferent stage, the epithelium of the genital ridge proliferates into the loose connective mesenchymal tissue above it (Figure 20.3A,B). These epithelial layers form the sex cords, which will surround the germ cells that migrate into the human

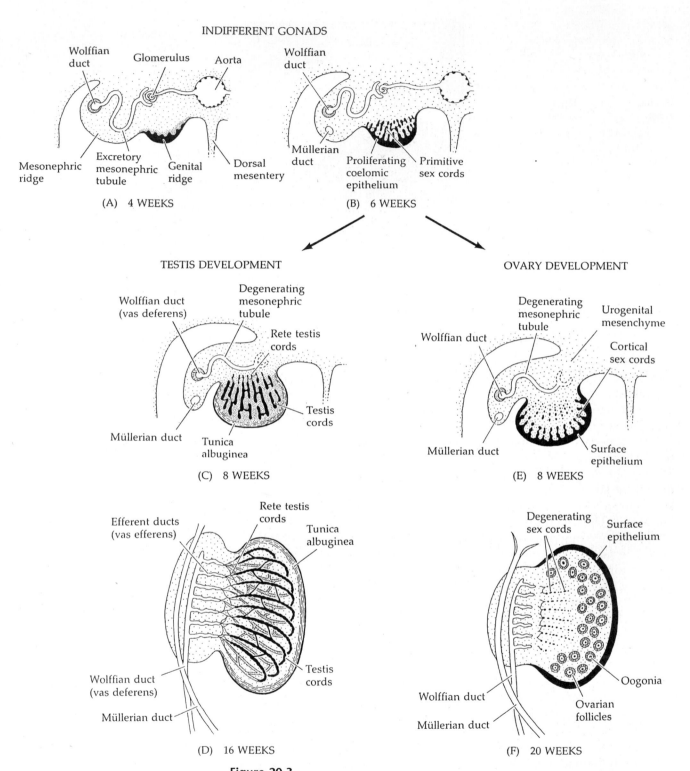

Figure 20.3
Differentiation of human gonads shown in transverse section. (A) Genital ridge of a 4-week embryo. (B) Genital ridge of a 6-week indifferent gonad showing primitive sex cords. (C) Testis development in the eighth week. The sex cords lose contact with the cortical epithelium and develop the rete testis. (D) By the sixteenth week of development, the testis cords are continuous with the rete testis and connect with the Wolffian duct. (E) Ovary development in an 8-week human embryo, as primitive sex cords degenerate. (F) The 20-week human ovary does not connect to the Wolffian duct, and new cortical sex cords surround the germ cells that have migrated into the genital ridge. (After Langman, 1981.)

gonad during week 6. In both XY and XX gonads, the sex cords remain connected to the surface epithelium.

If the fetus is XY, then the sex cords continue to proliferate through the eighth week, extending deeply into the connective tissue.* These cords fuse with each other, forming a network of internal (medullary) sex cords and, at its most distal end, the thinner **rete testis** (Figure 20.3C,D). Eventually, the testis cords lose contact with the surface epithelium and become separated from it by a thick extracellular matrix, the **tunica albuginea.** Thus, the germ cells are found in the cords *within* the testes. During fetal life and childhood, these cords remain solid. At puberty, however, the cords hollow out to form the seminiferous tubules, and the germ cells begin sperm production. The sperm are transported from the inside of the testis through the rete testis, which joins the **efferent ducts.** These efferent tubules are the remnants of the mesonephric kidney, and they link the testis to the Wolffian duct. This duct used to be the collecting tube of the mesonephric kidney. In males, the Wolffian duct differentiates to become the **vas deferens,** the tube through which the sperm pass into the urethra and out of the body. Meanwhile, during fetal development the interstitial mesenchyme cells of the testes have differentiated into **Leydig cells,** which make testosterone. The cells of the testis cords differentiate into **Sertoli cells,** which nurture the sperm and secrete anti-Müllerian duct hormone.

In females, the germ cells will reside near the outer surface of the gonad. Unlike the sex cords in males, which continue their proliferation, the initial sex cords of XX gonads degenerate. However, the epithelium soon produces a new set of sex cords, which do not penetrate deeply into the mesenchyme but stay near the outer surface (cortex) of the organ. Thus, they are called cortical sex cords. These cords are split into clusters, each cluster surrounding a germ cell (Figure 20.3E,F). The germ cell will become the ova, and the surrounding epithelial sex cords will differentiate into the **granulosa cells.** The mesenchyme cells of the ovary differentiate into the **thecal cells.** Together, the thecal and granulosa cells form the **follicles** that envelop the germ cells and secrete steroid hormones. Each follicle will contain a single germ cell. In females, the Müllerian duct remains intact, and it differentiates into the oviducts, uterus, cervix, and upper vagina; the Wolffian duct, deprived of testosterone, degenerates. A summary of the development of mammalian reproductive systems is shown in Figure 20.4. [sex1.html]

Mammalian primary sex determination: Y-chromosomal genes for testis determination

Several genes have been found whose function is necessary for normal sexual differentiation. Unlike other developing organs, the genes involved in sex determination differ extensively between phyla, so one cannot look at *Drosophila* sex-determining genes and expect to see their homologues directing mammalian sex determination. However, since the phenotype of mutations in sex-determining genes is often sterility, clinical studies have been used to identify those genes active in determining whether humans become male or female. The experimental manipulations to confirm the functions of these genes can be done in mice.

---

*In mice and rabbits, some cells of the mesonephros (the primitive kidney) migrate into the genital ridge and become part of the interstitial cell population. These appear to be necessary for establishing normal cord structure (Buehr et al., 1993).

Figure 20.4
Summary of the development of gonads and their ducts in mammals. Note that both the Wolffian and Müllerian ducts are present at the indifferent-gonad stage. The regional development of the Wolffian ducts depends on the mesenchyme that they encounter. Lower portions of the Wolffian duct that would normally form the epididymis will form seminal vesicle tissue if cultured with mesenchyme associated with the upper (seminal vesicle) portions of the duct. (Higgins et al., 1989.)

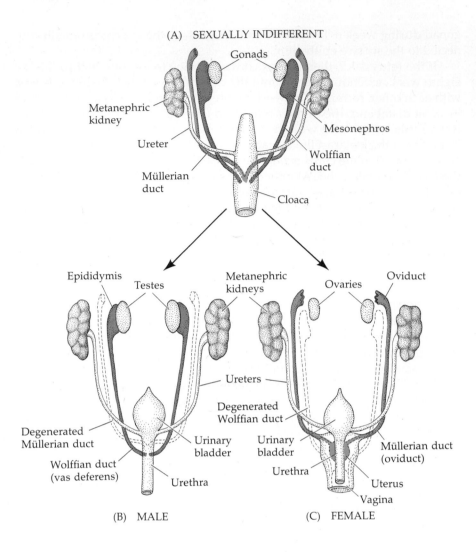

	GONADS	
Gonadal type	Testis	Ovary
Sex cords	Medullary (internal)	Cortical (external)
	DUCTS	
Remaining duct for germ cells	Wolffian	Müllerian
Duct differentiation	Vas deferens, epididymis, seminal vesicle	Oviduct, uterus, cervix, upper portion of vagina

SRY: *The Y-Chromosome Sex Determinant*

In humans, the major gene for the testis-determining factor resides on the short arm of the Y chromosome. Individuals who are born with the short arm but not the long arm of the Y chromosome are male, while individuals born with the long arm of the Y chromosome but not the short arm are female. By analyzing the DNA of XX *men* and XY *women,* the position of the testis-determining gene has been narrowed down to a 35,000-base-pair region of the Y chromosome located near the tip of the small arm. [sex2.html] In this region, Sinclair and colleagues (1990) found a male-specific DNA sequence that could encode a peptide of 223 amino acids. Such a peptide would probably be a transcription factor, since it would contain a DNA-binding domain called the **HMG box.** This HMG (high-mobility group) box is found in several transcription factors and nonhistone chromatin proteins, and it induces bending in the region of DNA to which it binds (Figure 20.5; Giese et al., 1992). This gene was called **SRY** (*s*ex-determining *r*egion of the *Y*), and there is evidence that it is indeed the gene that encodes the human testis-determining factor. *SRY* is found in XY males and in the rare XX males, and it is absent from normal XX females and from many of the XY females. Another group of XY females was found to have point or frameshift mutations in the *SRY* gene, and these mutations prevent the SRY protein from binding to or bending the DNA (Pontiggia et al., 1994; Werner et al., 1995). At least two genes involved in secondary sex determination (the genes for AMH and the P450 aromatase involved in steroid synthesis) contain SRY-binding sites in their promoters (Haqq et al., 1993), and the sequence-specific binding of SRY to another testis-specific gene leads to that gene's activation (Cohen et al., 1994).

If *SRY* actually does encode the major testis-determining factor, one would expect that it would act in the genital ridge immediately before or during testis differentiation. This prediction has been met in studies of the homologous gene found in mice. The mouse gene (*Sry*) also correlates with the presence of testes; it is present in XX males and absent in XY females (Gubbay et al., 1990; Koopman et al., 1990). The *Sry* gene is expressed in the somatic cells of the indifferent mouse gonad immediately before or during its differentiating into a testis; its expression then disappears (Hacker et al., 1995).

The most impressive evidence for *Sry* being the gene for testis-determining factor comes from transgenic mice. If *Sry* induces testes, then inserting *Sry* DNA into the genome of a normal XX mouse zygote should cause that XX mouse to form testes. Koopman and colleagues (1991) took the 14-kilobase region of DNA that includes the *Sry* gene (and presumably its regulatory elements) and microinjected this sequence into the pronuclei of newly fertilized mouse zygotes. In several instances, the XX embryos injected with this *Sry*-containing sequence developed testes, male accessory organs, and penises (Figure 20.6). (Functional sperm were not formed; but they were not

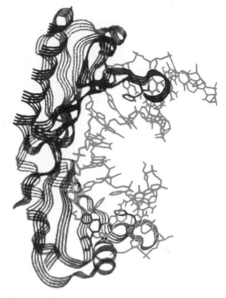

Figure 20.5
Association of DNA with the SRY protein can cause the DNA to bend 70°–80°. The black structure represents the HMG box of the SRY protein. The red coil is the double helix of DNA specifically bound by SRY. In this case, it is a region from the promoter of the anti-Müllerian hormone gene. (After Haqq et al., 1994; Werner et al., 1995.)

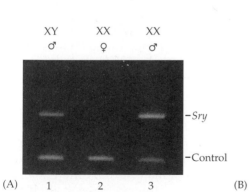

Figure 20.6
XX mouse transgenic for *Sry* is male. (A) Polymerase chain reaction followed by electrophoresis shows the presence of the *Sry* gene in normal XY males and in a transgenic XX *Sry* mouse. The gene is absent in the female XX littermate. (B) The external genitalia of the transgenic mouse are male (right) and essentially the same as in an XY male (left). (From Koopman et al., 1991, photograph courtesy of the authors.)

expected because the presence of two X chromosomes prevents sperm formation in XXY mice and men.) Therefore, there are good reasons to think that *Sry/SRY* is the major gene on the Y chromosome for testis determination in mammals.

The testis-determining gene of the Y chromosome is necessary but not sufficient for the development of the mammalian testis. Studies on mice (Eicher and Washburn, 1983; Washburn and Eicher, 1989) had shown that the SRY of some strains of mice failed to produce testes when placed in a different autosomal background. When the SRY protein binds to its sites on DNA, it probably creates large conformational changes. It unwinds the double helix in its vicinity and bends the DNA as much as 80° (Pontiggia et al., 1994; Werner et al., 1995). This bending may bring distantly bound proteins of the transcription apparatus into close contact, enabling them to interact and influence transcription. The identities of these proteins are not yet known. [sex3.html]

Mammalian primary sex determination: Autosomal genes in testis determination

SOX9: *Autosomal Sex Reversal in Campomelic Dysplasia*

If SRY is a transcription factor, it would be expected to activate or repress a battery of genes in the genital ridge. One candidate for such a gene is the **SOX9** gene in humans. The *SOX9* gene encodes a putative transcription factor that also contains an HMG box. Individuals missing a functional copy of this gene have a syndrome called campomelic dysplasia, a disease involving numerous skeletal and organ systems; they die soon after birth from respiratory distress arising from defective bronchia and tracheas (Foster et al., 1994; Wagner et al., 1994; Mansour et al., 1995). However, about 75 percent of the XY patients with this syndrome develop as phenotypic females or hermaphrodites. It appears that *SOX9* is essential for testis formation. Moreover, the mouse homologue of this gene, *Sox9*, is expressed only in male (XY) but not in female (XX) genital ridges, and in the same genital ridge cells as *Sry*. *Sox9* is expressed just slightly after *Sry* expression (Wright et al., 1995; Kent et al., 1996).

SF1: *The Link Between SRY and the Male Developmental Pathways*

Another protein that might be activated by SRY and be a cofactor with SRY is the transcription factor **SF1**. SF1 (steroidogenic *f*actor 1) is a protein that activates several genes involved in steroid synthesis. Indeed, it works in the Leydig cells of the testes to activate those genes encoding enzymes of the testosterone pathway. However, SF1 has recently been shown to have two other critical functions (Figure 20.7). First, deleting the *Sf1* genes from mice causes them to develop without adrenal glands or gonads (Luo et al., 1994). (The gonads develop, but then degenerate. The mice die from lack of corticosterone.) Second, SF1 appears to be involved in testis development. While SF1 levels decline in the genital ridge of XX mouse embryos, SF1 stays on in the developing testis. It is thought that SRY activates the *Sf1* gene, and the SF1 protein then activates both components of male sexual differentiation (the Sertoli AMH and the Leydig testosterone pathway) (Shen et al., 1994). Both SRY and SF1 may be needed to activate the *AMH* gene, suggesting that interactions between these proteins may be important (Haqq et al., 1994; Shen et al., 1994).

In mice, research into sex reversal has shown that the Y chromosome from one type of mouse does not necessarily make testes in another strain of mouse. It seems that the SRY proteins have diverged so greatly that they can

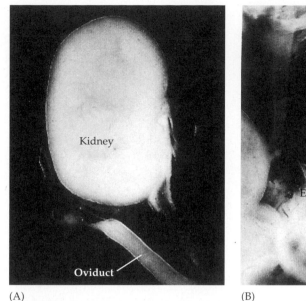

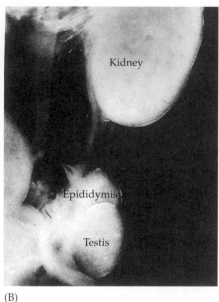

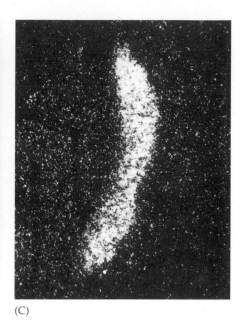

(A)   (B)   (C)

Figure 20.7
Functions of SF1 during gonadogenesis. (A) Knockout of the *SF1* gene from the mouse embryo causes the loss of both adrenals and testes. (The Müllerian duct persists and becomes the oviduct.) (B) A control showing the epididymis and testis. (C) In situ hybridization showing the activation of the *Sf1* gene throughout the developing testis of a 12.5-day mouse embryo. (A and B from Luo et al., 1994; C from Shen et al., 1994.)

no longer interact with the other proteins of the transcription apparatus (Coward et al., 1994; Eicher, 1994).

Mammalian primary sex determination: Ovary development

DAX1: *A Potential Ovary-Determining Gene on the X Chromosome*

In 1980, Bernstein and his colleagues (1980) reported two sisters who were genetically XY. Their Y chromosomes were normal, but they had duplicated a small portion of the small arm of the X chromosome (Xp21). Subsequent cases were found, and it was concluded that if there were two copies of this region on the active X chromosome, the SRY signal would be reversed (Figure 20.8). A double dosage of this region disrupted testis formation, but the absence of this region was compatable with testis formation. Bardoni and her colleagues (1994) proposed that this region contained a gene that competed with the SRY factor and that was important in directing the development of the ovary. In testicular development, this gene would be suppressed, but having two active copies of the gene would override this repression. This gene, *DAX1,* has been cloned and shown to encode a member of the nuclear hormone receptor family (Muscatelli et al., 1994; Zanaria, 1994). Preliminary data (reported in Zanaria, 1994) suggests that *DAX1* is expressed in the genital ridges of the mouse embryo.

Wnt4a: *A Potential Ovary-Determining Gene on an Autosome*

The *Wnt4a* gene is another gene that may be critical in ovary determination. This gene is expressed in the mouse genital ridge while it is still in its indifferent stage. It then becomes undetectable in XY gonads (which become

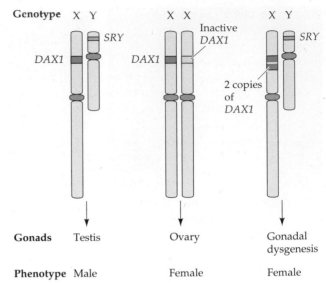

Figure 20.8
Phenotypic sex reversal in humans having two copies of the *DAX1* locus. *DAX1* (on the X chromosome) plus *SRY* (on the Y) produces testes. *DAX1* without the *SRY* (since the other *DAX1* locus is on the inactive X chromosome) produces ovaries. Two active copies of *DAX1* (on the active X chromosome) plus an *SRY* (from the Y chromosome) lead to a poorly formed gonad. Since the gonad makes neither AMH nor testosterone, the phenotype is female. (After Genetics Review Group, 1995.)

testes), whereas Wnt4a expression is maintained in XX gonads as they begin to form ovaries. If mice are bred that lack the *Wnt4a* genes, the ovary fails to properly form, and its cells express testis-specific markers, including AMH and testosterone producing enzymes (Vainio and McMahon, 1996).

It is possible that SRY may form testes by repressing *Wnt4a* expression in the genital ridge as well as by promoting *SF1*. It should be realized that both testis and ovary development are active processes. In mammalian primary sex determination, neither is a "default state" (Eicher and Washburn, 1986). Although remarkable progress has been made in recent years, we still do not know what the testis- or ovary-determining genes are doing, and the problem of primary sex determination remains (as it has since prehistory) one of the great unsolved problems of biology. [sex4.html]

Secondary sex determination in mammals

Hormonal Regulation of the Sexual Phenotype

Primary sex determination involves the formation of either an ovary or a testis from the indifferent gonad. This, however, does not give the complete sexual phenotype. Secondary sex determination concerns the development of the female and male phenotypes from the hormones secreted by the ovaries and testes. Both female and male secondary sex determination have two major temporal components. The first occurs within the embryo during organogenesis; the second occurs during adolescence.

As mentioned earlier, if the indifferent gonads are removed from an embryonic animal, the female phenotype is realized. The Müllerian ducts develop while the Wolffian duct degenerates. This also is seen in certain humans who are born without functional gonads. Individuals whose cells have only one X chromosome (and no Y chromosome) originally develop ovaries, but these ovaries atrophy before birth, and the germ cells die before puberty. However, under the influence of estrogen derived first from the ovary but then from the mother and placenta, these infants are born with a female genital tract (Langman and Wilson, 1982).*

The formation of the male phenotype involves the secretion of testicular hormones that promote Wolffian duct development and cause the Müllerian duct to atrophy. The first of these hormones is the anti-Müllerian duct hormone, the Sertoli cell hormone that causes the degeneration of the Müllerian duct. The second of these hormones is the steroid testosterone, which is secreted from the fetal testicular Leydig cells. This hormone causes the Wolffian duct to differentiate into the epididymis, vas deferens, and seminal vesicles, and it causes the urogenital swellings and sinus to develop into the scrotum and penis. The existence of these two independent systems of masculinization is demonstrated by people having **androgen insensitivity syndrome.** These XY individuals have the testis-determining factor gene and so have testes that make testosterone and AMH. However, these people lack the testosterone receptor protein and therefore cannot respond to the testosterone made in their testes (Meyer et al., 1975). Because they are able to respond to estrogen made in their adrenal glands, they are distinctly female in appearance (Figure 20.9). However, despite their female appearance, these individuals do have testes, and even though they cannot respond to testosterone, they do respond to AMH. Thus, their Müllerian ducts degenerate. These people develop as normal but sterile women, lacking a uterus and oviducts and having testes in their abdomen.†

Testosterone and Dihydrotestosterone

So there are two distinct masculinizing hormones, testosterone and AMH. There is evidence, though, that testosterone might not be the active hormone in certain tissues. Testosterone appears to be responsible for promoting the formation of male reproductive structures (the epididymis, seminal vesicles, and vas deferens) from the Wolffian duct primordium. However, it does not directly masculinize the male urethra, prostate, penis, or scrotum. These latter functions are controlled by **5α-dihydrotestosterone.** Siiteri and Wilson (1974) showed that testosterone is converted to 5α-dihydrotestosterone in the urogenital sinus and swellings, but not in the Wolffian duct.

Imperato-McGinley and her colleagues (1974) found a small community in the Dominican Republic in which several inhabitants had a genetic deficiency of the enzyme 5α-ketosteroid reductase 2, the enzyme that converts

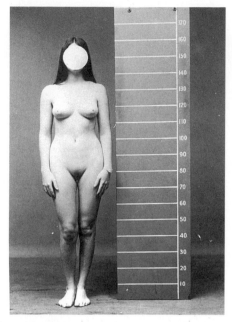

Figure 20.9
An XY individual with androgen insensitivity syndrome. Despite the XY karyotype and the presence of testes, the individual develops female secondary sex characteristics. Internally, however, the woman lacks the Müllerian duct derivatives and has undescended testes. (Courtesy of C. B. Hammond.)

---

*The mechanisms by which estrogen might promote the differentiation of the Müllerian ducts are not well understood. During embryonic development, the duct is extremely sensitive to estrogenic compounds, as is known from the teratogenic effects of diethylstilbesterol (DES). This compound is a synthetic estrogen that was given to women from the 1940s through the 1960s to help maintain pregnancies. Daughters born to women who used this drug have a high incidence of Müllerian duct anomalies, including malformation of the vaginal and cervical epithelia, structural abnormalities of the oviducts and uterus, and a higher than normal incidence of vaginal cancer (Robboy et al., 1982; Bell, 1986).

†Androgen insensitivity syndrome is one of several conditions called *pseudohermaphroditism*. True hermaphrodites (rare in humans and most mammals, but the norm in some invertebrates) contain both male and female gonadal tissues. Mammalian true hermaphrodites have abnormalities of *primary* sex determination and can occur when the Y chromosome is translocated to the X chromosome. If the translocated X is inactivated, the Y will be shut off. Some of the gonadal cells will be XX and others will be XY (Berkovitz et al., 1992). In a pseudohermaphrodite condition, there is only one type of gonad, but the *secondary* sexual characteristics differ from what is expected from the gonadal sex. In humans, male peseudohermaphrodites can result from androgen insensitivity syndrome or from the failure to produce testosterone because of a gene defect in one of the enzymes leading to its synthesis (Geissler et al., 1994). Female pseudohermaphroditism can occur when the body has an overproduction of testosterone.

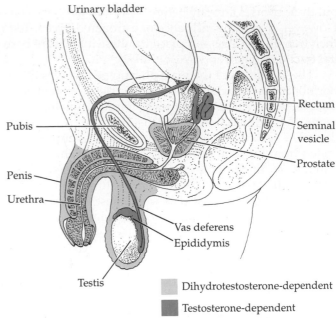

Figure 20.10
Testosterone- and dihydrotestosterone-dependent regions of the fetal human male genital system. (After Imperato-McGinley et al., 1974.)

testosterone to dihydrotestosterone. Affected individuals lack a functional gene for this enzyme (Andersson et al., 1991; Thigpen et al., 1992). Although these XY individuals have functioning testes, they have a blind vaginal pouch and an enlarged clitoris. They appear to be girls and are raised as such. Their internal anatomy, however, is male: testes, Wolffian duct development, and Müllerian duct degeneration. Thus, it appears that the formation of external genitalia is under the control of dihydrotestosterone, whereas the Wolffian duct differentiation is controlled by testosterone itself (Figure 20.10). Interestingly enough, the external genitalia become responsive to testosterone at puberty, causing obvious masculinization in a person originally thought to be a girl.

Anti-Müllerian Hormone

Anti-Müllerian duct hormone (AMH) is a 560-amino-acid glycoprotein (Cate et al., 1986) made in the Sertoli cells (Tran et al., 1977). When fragments of fetal testes or isolated Sertoli cells are placed adjacent to cultured segments containing portions of the Wolffian and Müllerian ducts, the Müllerian duct atrophies even though no change occurs in the Wolffian duct (Figure 20.11). This atrophy is caused both by cell death and by the epithelial cells of the duct becoming mesenchymal and migrating away (Trelstad et al., 1982). The mouse *AMH* gene has a promoter sequence that is bound by both the SF1 protein and SRY (Haqq et al., 1994; Shen et al., 1994). [sex5.html]

We see, then, that once testes are formed, they secrete two hormones that cause the masculinization of the fetus. One of these hormones—testosterone—may be converted into a more active form by the tissues that create the external genitalia. In females, estrogen secreted from the fetal ovaries appears sufficient to induce the differentiation of the Müllerian duct into the uterus, oviducts, and cervix. In such a manner, the sex chromosomes control the sexual phenotype of an individual.

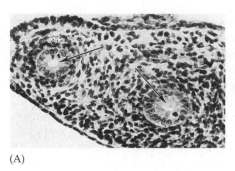
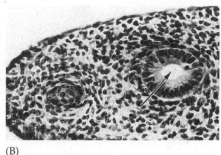

(A)   (B)

Figure 20.11
Assay for anti-Müllerian duct hormone activity in the anterior segment of a 14.5-day fetal rat reproductive tract after 3 days in culture. (A) Both the Müllerian duct (arrow at left) and Wolffian duct (arrow at right) are open. (B) The Wolffian duct (arrow) is open, but the Müllerian duct has degenerated and closed. (Courtesy of N. Josso.)

The Central Nervous System

One of the most controversial areas of secondary sex determination involves the development of sex-specific behaviors. In songbirds, testosterone is seen to regulate the growth of male-specific neuronal clusters in the brain. Male canaries and zebra finches sing eloquently, whereas females sing little, if ever. These songs are used to mark territories and to attract mates. The ability to sing is controlled by six different clusters of neurons (nuclei) in the avian brain (Figure 20.12). Neurons connect each of these regions to one another. In male canaries, these nuclei are several times larger than the corresponding cluster of neurons in female canaries; and in zebra finches, the females may lack one of these regions entirely (Arnold, 1980; Konishi and Akutagawa, 1985).

Testosterone plays a major role in song production. In adult male zebra finches, Pröve (1978) demonstrated a linear correlation between the amount of song and the concentration of serum testosterone. It has been shown that the seasonal changes in testosterone levels are correlated with the seasonal singing patterns of these birds. When testosterone levels are low, there is not only a decrease in bird song but also a decrease in the size of the male-specific brain nuclei (Nottebohm, 1981). In adult chaffinches, castration eliminates song, but injection of testosterone induces such birds to sing even in November, when they are normally silent (Thorpe, 1958). In several species of birds, the females can be induced to sing by injecting them with testosterone (Nottebohm, 1980). Four of the song-controlling regions of the brain grow 50–69 percent in such birds, whereas other brain regions show no such growth. Autoradiographic studies (Arnold et al., 1976) have shown that the neurons of the song-controlling nuclei incorporate radioactive testosterone, whereas other regions of the brain do not. It is apparent, then, that gonadal hormones can play a major role in the development of the regions of the nervous system that generate sex-specific behaviors.

Figure 20.12
Sexual dimorphism in the avian brain. Schematic diagram indicates the major neural areas thought to be involved in the production of bird song in the zebra finch. Circles represent specific brain areas; the size of each circle is proportional to the volume occupied by that region. Circles with dashed lines are estimated volumes. The numbers within each circle represent the percentage of cells therein that incorporate radiolabeled testosterone. The volume differences in three of these regions (HVc, RA, and nXIIts) are significant between the sexes, and area X has not been observed in the brains of female finches. The differences in testosterone binding in regions HVc and MAN are significant, and no sex differences in steroid hormone binding have been observed in other regions of the brain. The arrows indicate the axonal paths connecting the regions in the male finch. (After Arnold, 1980.)

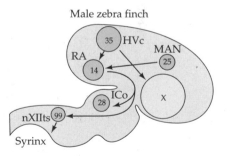

Male zebra finch

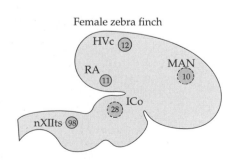

Female zebra finch

The situation is not as clear in mammals, for there are fewer behaviors that exclusively characterize one sex. Penile thrusting in rats is one such behavior, and it is controlled by motor neurons to the levator ani and bulbocavernosus muscles. Both these neurons originate from a spinal nucleus that can specifically concentrate testosterone. In female rats, these muscles are vestigial, and the volume of the controlling neurons is greatly reduced (Breedlove and Arnold, 1980; Tobin and Joubert, 1992). Testosterone appears to cause two types of changes in these responsive neurons. In fetal and newborn rats, testosterone prevents a "normally" occurring death of neurons in this region. Female rats lose up to 70 percent of the neurons of this spinal nucleus, while male rats lose only 25 percent. In adult rats, testosterone acts on this nucleus to maintain the size of the nerve cells and their dendrites. The soma area and dendrite length of this spinal nucleus are reduced by half when an adult rat is castrated. This reduction is reversed by the injection of testosterone (Nordeen et al., 1985; Kurz et al., 1986). In humans, the different growth rates between the sexes produce slightly different anatomical features. Although human female brains are about 10 percent smaller than male brains, the granular layer of some cortical regions contains more densely packed neurons than similar regions in male brains (Witelson et al., 1995).

Testosterone is not the only steroid capable of mediating behavior. In the mammalian brain, estrogen-sensitive neurons also are seen. These neurons are located at positions in neural circuits that are known to mediate reproductive behavior: the hypothalamus, the pituitary, and the amygdala (Figure 20.13; McEwen, 1981). Pfaff and McEwen (1983) demonstrated that estrogen alters the electrical and chemical features of those hypothalamic neurons capable of binding estrogen in their chromatin. Terasawa and Sawyer (1969) previously had found that the electrical activity of these neurons varies during the seasonal estrogen cycle of the rat, becoming elevated at the time of ovulation. Moreover, estrogen appears to stimulate those neurons in the regions that induce female reproductive behavior. Ovariectomized rats given estrogen injections directly to the hypothalamus displayed **lordosis,** a position that stimulates mounting behavior in male mice, whereas control ovariectomized rats did not show that behavior (Barfield and Chen, 1977; Rubin and Barfield, 1980). The mechanism by which estrogen causes the specific neuronal activity at these times is thought to involve increasing permeability of these neurons to potassium (Nabekura et al., 1986).

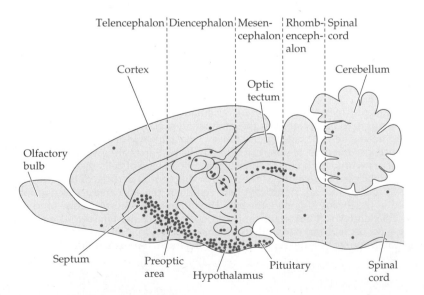

Figure 20.13
Representation of the estrogen-binding regions in a female rat brain. (After Kandel and Schwartz, 1985.)

The Development of Sexual Behaviors

The Organization/Activation Hypothesis

Does prenatal (or neonatal) exposure to particular hormones impose permanent sex-specific changes in the central nervous system? Such sex-specific neural changes have been shown in regions of the brain that regulate "involuntary" sexual physiology. The cyclical secretion of luteinizing hormone by the adult female rat pituitary is dependent on the lack of testosterone during the first week of its life. The luteinizing hormone secretion of female rats can be made noncyclical by giving them testosterone at 4 days after birth; conversely, the luteinizing hormone secretions of males can be made cyclical by removing their testes within a day of their birth (Barraclough and Gorski, 1962). It is thought that sex hormones can act during the fetal or neonatal stage of a mammal's life to organize the nervous system and that during adult life, the same hormones may have transitory, activational effects. This is called the **organization/activation hypothesis.**

Interestingly, the hormone chiefly responsible for determining the male brain pattern is estradiol.* Testosterone in fetal or neonatal blood can be converted into estradiol by P450 aromatase, and this conversion occurs in the hypothalamus and limbic system—two areas of the brain known to regulate reproductive and hormonal behaviors (Reddy et al. 1974; McEwen et al., 1977). Thus, testosterone is able to cause its effects by being converted into estradiol. But the fetal environment is rich in estrogens from the gonads and placenta. What stops the estrogens from masculinizing the nervous system of a female fetus? Fetal estrogen (both male and female) is bound by α-fetoprotein. This protein is made in the fetal liver and becomes a major compo-

nent of the fetal blood and cerebrospinal fluid. It will bind estrogen but not testosterone.

Attempts to extend the organization/activation hypothesis to "voluntary" sexual behaviors are more controversial because there is no truly sex-specific behavior that distinguishes the two sexes of many mammals and because there are multiple effects of hormone treatment on the developing mammal. For instance, injecting testosterone into a week-old female rat will increase her pelvic thrusts and diminish her amount of lordosis (Phoenix et al., 1959; Kandel et al., 1995). These changes can be ascribed to testosterone-mediated changes in the central nervous system, but they also can be due to hormonal effects on other tissues. Testosterone enables the growth of the muscles that allow pelvic thrusting. And since testosterone causes females to grow larger and to close their vaginal orifices, one cannot conclude that the lack of lordosis is due solely to testosterone-mediated changes in the neural circuitry. (Harris and Levine, 1965; De Jonge et al., 1988; Moore, 1990; Moore et al., 1992; Fausto-Sterling, 1995).

Extrapolating from rats to humans is a very risky business, as no sex-specific behavior has yet been identified in humans, and what is "masculine" in one culture may be considered "feminine" in another (see Jacklin, 1981; Bleier, 1984; Fausto-Sterling, 1992). As one review (Kandel et al., 1995) concludes:

> There is ample evidence that the neural organization of reproductive behaviors, while importantly influenced by hormonal events during a critical prenatal period, does not exert an immutable influence over adult sexual behavior or even over an individual's sexual orientation. Within the life of an individual, religious, social, or psychological motives can prompt biologically similar persons to diverge widely in their sexual activities.

Male Homosexuality

Certain behaviors are often said to be part of the "complete" male or female

phenotype. The brain of a mature man is said to be formed such that it causes the man to desire mating with a mature woman, and the brain of a mature woman causes her to desire to mate with a mature man. However, as important as desires are in our lives, they cannot be detected by in situ hybridization or isolated by monoclonal antibodies. We do not yet know if sexual desires are instilled in us by our social education or are "hardwired" in our brains by genes or hormones during our intrauterine development or by other means.

In 1991, Simon LeVay proposed that part of the anterior hypothalamus of homosexual men has the anatomical form typical of women rather than of heterosexual men. The hypothalamus is thought to be the source of our sexual urges, and rats have a sexually dimorphic area in their anterior hypothalamus that appears to regulate sexual behavior. Thus, this study generated a great deal of publicity and discussion. The major results are shown in Figure 20.14. The interstitial nuclei of the anterior hypothalamus (INAH) were divided into four regions. Three of them showed no signs of sexual dimorphism. One of them, INAH3, showed a significant statistical difference between males and females; it was claimed that the male INAH3 is, on average, more than twice as large as the female INAH3. Moreover, LeVay's data suggested that the INAH3 of homosexual males was similar to that of the females and less than half the size of the heterosexual men's INAH3. This finding, LeVay claimed, "suggests that sexual orientation has a biological substrate."

There have been several criticisms of LeVay's interpretation of the data. First, the data are from populations, not individuals. One can also say that there is a statistical range and that men and women have the same general range. Indeed, one of the INAH3s from a "homosexual male" was greater than all but one from the 16 "heterosexual males." Second, the "heterosexual men" were not necessarily heterosexual, nor were

---

*The terms *estrogen* and *estradiol* are often used interchangeably. However, estrogen refers to a class of steroid hormones responsible for establishing and maintaining specific female characteristics. Estradiol is one of these hormones, and in most mammals (including humans), it is the most potent of the estrogens.

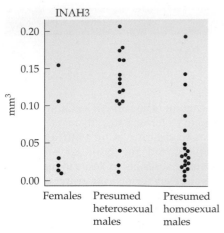

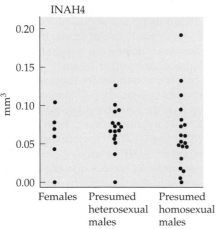

Figure 20.14 A portion of the data that may suggest a biological basis for homosexuality. INAH4 and INAH3 are two groups of hypothalamic neurons. INAH4 shows no sexual dimorphism in volume, while INAH3 shows a statistically significant clustering, although the range is similar. The INAH3 from autopsies of "homosexual" male brains clusters toward the female distribution. However, no cause-and-effect relationship can be posited. (After LeVay, 1991.)

the "homosexual men" necessarily homosexual; the brains came from corpses of people whose sexual preferences were not known. This brings up another issue: homosexuality has many forms and is thus not a phenotype in the usual sense. Third, the brains of the "homosexual men" were taken from patients who had died of AIDS. AIDS affects the brain, and its effect on the hypothalamic neurons is not known. Fourth, as the study was done on the brains of dead subjects, one cannot infer cause and effect. As mentioned in Chapter 1, such data show only correlations, not causation. It is as likely that behaviors can affect regional neuronal density size as it is that the regional neuronal density can affect behaviors. If one interprets the data as indicating that the INAH3 of male homosexuals is smaller than that of male heterosexuals, one still does not know whether that is a *result* of homosexuality or a *cause*. Fifth, even if a difference does exist, there is no evidence that the difference has anything to do with sexuality. Sixth, these studies do not indicate when such differences (if they exist) emerge. The question of whether differences in male, female, and homosexual male INAH3 occur during embryonic development, shortly after birth, during the first few years of life, during adolescence, or at some other time was not addressed.

In 1993, a correlation was made between a particular DNA sequence on the X chromosome and a particular subgroup of male homosexuals (homosexual men who had a homosexual brother). Out of 40 pairs of homosexual brothers wherein one had inherited a particular region of the X chromosome from his mother, 33 of them had brothers who also inherited this region (Hamer et al., 1993). One would have expected only 20 of them to have done so, on average. Again, this is only a statistical concordance, and one that could be coincidental. Moreover, the control (seeing if the same marker existed in the "nonhomosexual" males of these families) was not reported, and the statistical bias of the observations has been called into question, especially since other laboratories have not been able to repeat the result (Risch et al., 1993; Marshall, 1995). In a more recent study from the same laboratory, Hu and colleagues (1995) found little or no increase in this region when homosexual men were compared to their nonhomosexual brothers. They conclude that this region is "neither necessary nor sufficient for a homosexual orientation." Thus, despite the reports of these studies in the public media, no "gay gene" has been found.

It is worth remembering that genes encode RNAs and proteins, not behaviors. While genes may bias behavioral outcomes, we have no evidence for their "controlling" them. The existence of people with "multiple personality syndrome" indicates that one genotype can support a wide range of personalities. This is certainly a problem with any definition of a "homosexual phenotype," since many people alternate between homosexual and heterosexual behavior. Thus, the question as to whether homosexual desires are formed by genes within the nucleus, by sex hormones during fetal development, or by experiences after birth is still an open question. ■

Chromosomal sex determination in Drosophila

The Sexual Development Pathway

The sex-determining mechanisms in mammals and insects such as *Drosophila* are very different. In mammals, the Y chromosome plays a pivotal role in determining the male sex. Thus, XO mammals are *females*, with ovaries, a uterus, and oviducts (but usually very few, if any, ova). In *Drosophila*, sex determination is achieved by a balance of female determinants on the X chromosome and male determinants on the autosomes (non-sex chromosomes). If there is but one X chromosome in a diploid cell (1X:2A), the fly is male. If there are two X chromosomes in a diploid cell

Table 20.1 Ratios of X chromosomes to autosomes in different sexual phenotypes in *Drosophila melanogaster*

X chromosomes	Autosome sets (A)	X:A ratio	Sex
3	2	1.50	Metafemale
4	3	1.33	Metafemale
4	4	1.00	Normal female
3	3	1.00	Normal female
2	2	1.00	Normal female
2	3	0.66	Intersex
1	2	0.50	Normal male
1	3	0.33	Metamale

Source: After Strickberger, 1968.

(2X:2A), the fly is female (Bridges, 1921, 1925). Thus, XO *Drosophila* are sterile *males*. Table 20.1 shows the different X-to-autosome ratios and the resulting sex.

In *Drosophila*, and in insects in general, one can observe **gynandromorphs**—animals in which certain regions are male and other regions are female (Figure 20.15; Plate 17). This can happen when an X chromosome is lost from one embryonic nucleus. The cells descended from that cell, instead of being XX (female), are XO (male). Because there are no sex hormones in insects to modulate such events, each cell makes its own sexual "decision." The XO cells display male characteristics, whereas the XX cells display female traits. This situation provides a beautiful example of the association between X chromosomes and sex. As can be seen from this example, the Y chromosome plays no role whatsoever in *Drosophila* sex determination. It is necessary only to ensure fertility in males. The Y chromosome is active only late in development, during sperm formation.

Any theory of *Drosophila* sex determination must explain how the X-to-autosome ratio is read and how this information is transmitted to the genes controlling the male or female phenotypes. Although we do not yet know the intimate mechanisms by which the X:A ratio is made known to the cells, research in the past decade has revolutionized our view of *Drosophila* sex determination. Much of this research has concerned the identification and analysis of genes that are necessary for sexual differentiation and the placing of those genes in a developmental sequence. Loss-of-function mutations in most of these genes—*Sex-lethal (Sxl)*, *transformer (tra)*, and *transformer-2 (tra2)*—

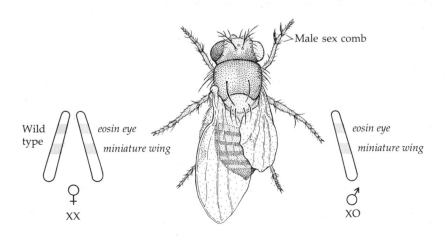

Figure 20.15
Gynandromorph of *D. melanogaster* in which the left side is female (XX) and the right side is male (XO). The male side has lost an X chromosome bearing the wild-type alleles of eye color and wing shape, thereby allowing the expression of the recessive alleles *eosin eye* and *miniature wing* on the remaining X chromosome. (After Morgan, 1919.)

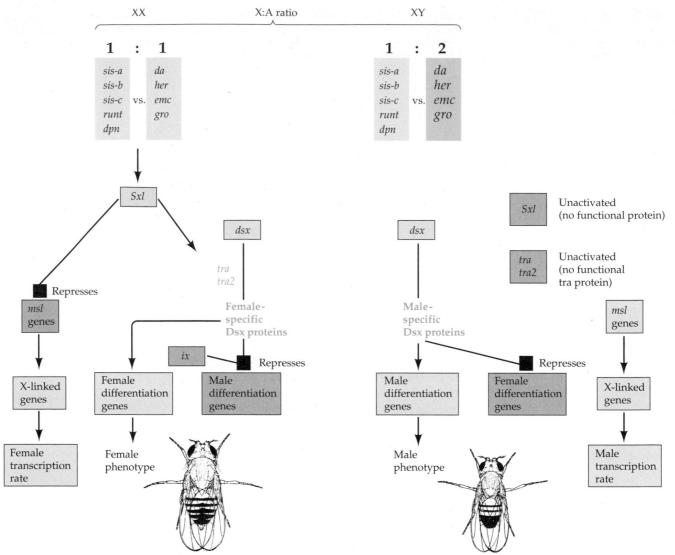

Figure 20.16
Proposed regulation cascade for *Drosophila* somatic sex determination. Arrows represent activation, while a block at the end of a line indicates suppression. The *msl* loci, under the control of the *Sxl* gene, regulate the dosage compensatory transcription of the male X chromosome. (After Baker et al., 1987.)

transform XX individuals into males. Such mutations have no effect on sex determination in XY males. Homozygosity of the *intersex* (*ix*) gene causes XX flies to develop an intersex phenotype having portions of male and female tissue in the same organ. The *doublesex* (*dsx*) gene is important for the sexual differentiation of both sexes. If *dsx* is absent, both XX and XY flies turn into intersexes (Baker and Ridge, 1980; Belote et al., 1985a).

The position of these genes in a developmental pathway is based on (1) the interpretation of genetic crosses resulting in flies bearing two or more of these mutations and (2) the determination of what happens when there is a complete absence of the products of one of these genes. Such studies have generated the model of regulatory cascade seen in Figure 20.16.

The Sex-lethal *Gene as Pivot for Sex Determination*

INTERPRETING THE X:A RATIO.   The first phase of *Drosophila* sex determination involves reading the X:A ratio. What elements on the X chromosome are "counted," and how is this information used? It appears that high values of the X:A ratio are responsible for activating the feminizing switch gene *Sex-lethal* (*Sxl*). At low values (males), *Sxl* remains inactive during the early stages of development (Cline, 1983; Salz et al., 1987). In XX *Drosophila*, *Sxl* is activated during the first 2 hours after fertilization, and this gene transcribes a particular embryonic type of *Sxl* mRNA that is only found for about 2 hours more (Salz et al., 1989). Once activated, the *Sxl* gene remains active despite any further changes in the X:A ratio (Sánchez and Nöthiger, 1983). Early *Sxl* function is necessary for XX embryos to initiate the female developmental pathway and to maintain an appropriate level of transcription from the two X chromosomes.

This female-specific activation of *Sxl* is thought to be stimulated by "numerator elements" on the X chromosome that constitute the X part of the X:A ratio. Cline (1988) has demonstrated that two of these numerator elements are the X chromosome genes *sisterless-a* and *sisterless-b*. The *Sxl* gene does not seem to sense these numerator elements without the presence of products of the *runt* and *daughterless (da)* genes. The lack of Daughterless protein prevents the activation of *Sxl*. This does not affect XY embryos (since they do not activate *Sxl* anyway), but it is lethal in female embryos, since the mechanism for dosage compensation causes the two X chromosomes to be transcribed at the faster (male) rate (Cline, 1986; Cronmiller and Cline, 1987; Duffy and Gergen, 1991)—hence the name of the mutation. Thus, shortly after fertilization, the *sis-a, sis-b, runt,* and *da* genes enable *Sxl* to be transcriptionally active only in female embryos.

The "denominator elements" are those genes that get counted from the autosomes. One of the major denominator elements appears to be the *deadpan* gene (Younger-Shepherd et al., 1992). Males with too high a ratio of *sis-b* to *deadpan* activate *Sxl* and die, while females with too low a ratio of *sis-b* to *deadpan* fail to activate *Sxl* and die. Another denominator gene encodes Extramachrochaetae, a protein that competes with the binding of Daughterless to the *Sxl* promoter (Van Doren et al., 1991). The *daughterless, sis-a, sis-b,* and *deadpan* genes are all helix-loop-helix (HLH) transcription factors, and it is possible that the denominator and numerator proteins can form heterodimers with each another. Presumably, the denominator proteins are able to form heterodimers that block those of the activator (Sis and Daughterless) proteins. (Figure 20.17A). It appears, then, that the X: Autosome ratio is measured by the competition of X-encoded activation and autosomally encoded repressors on the promoter of the *Sxl* gene. [sex6.html]

MAINTENANCE OF SXL FUNCTION.   Shortly after *Sxl* transcription has taken place, a second promoter on the *Sex-lethal* gene is activated, and this gene is transcribed in both males and females. However, analysis of the cDNA from *Sxl* mRNA shows that the *Sxl* mRNA of males differs from the *Sxl* mRNA of females (Bell et al., 1988). This is the result of differential RNA processing. Moreover, the Sxl protein appears to bind to its own mRNA precursor to splice it in the female manner. Since males do not have any available Sxl protein, their new *Sxl* transcripts are processed in the male manner (Keyes et al., 1992). The male *Sxl* mRNA is nonfunctional. While the female-specific *Sxl* message encodes a protein of 354 amino acids, the male-specific *Sxl* transcript contains a translation termination codon (UGA) after amino acid 48. The differential RNA processing that puts this termination codon into the male-specific mRNA is shown in Figures 20.17B and 20.18. In males, the nuclear transcript is spliced in a manner that yields three exons, and the termination codon is within the central exon. In females, RNA processing yields only two exons, and the male-specific central exon is now spliced out as a large intron. Thus, the female-specific mRNA lacks the termination codon.

The protein made by the female-specific *Sxl* transcript can be predicted from its nucleotide sequence. This protein would contain two regions that

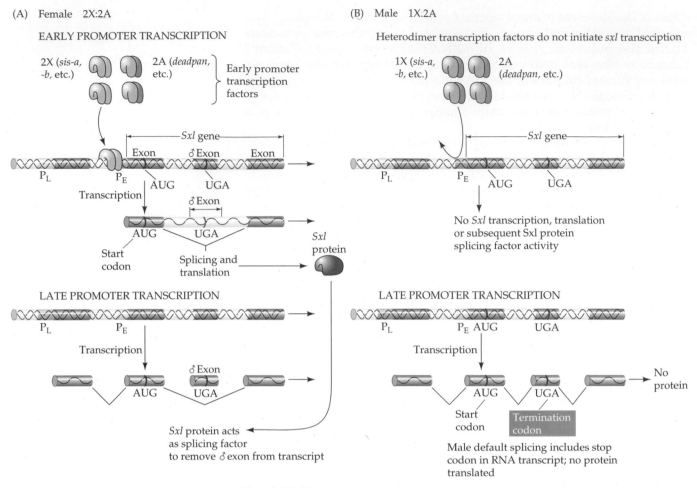

Figure 20.17
The differential activation of the *Sxl* gene in females and males. (A) In wild-type *Drosophila* with two X chromosomes and two sets of autosomes (XX; AA), the numerator transcription factor subunits (*sis-a, sis-b*, etc.) are not fully complexed by inhibitory subunits derived from genes (such as *deadpan*) on the autosomes. These numerator factors activate the early promoter of the *Sxl* gene, which produces a transcript that is automatically spliced to a female-specific mRNA that encodes functional Sxl protein. Eventually, constitutive transcription of *Sxl* starts from the late promoter. If Sxl is already available (i.e., from early transcription), the *Sxl* pre-mRNA is spliced to form the functional female-specific message. (B) In wild-type *Drosophila* with one X chromosome and two sets of autosomes (XO; AA), the numerator transcription factors are bound by the denominator subunits and cannot activate the early promoter. When the *Sxl* gene is transcribed from the late promoter, RNA splicing does not exclude the male-specific exon in the mRNA. The resulting message encodes a truncated and nonfunctional peptide, since the male-specific exon contains a translation termination codon. (After Keyes et al., 1992.)

are important for binding to RNA. These regions are shared with nuclear RNA-binding proteins such as those in snRNPs. Bell and colleagues (1988) have proposed that there are two targets for the RNA-binding protein encoded by *Sxl*. One of these targets is the pre-mRNA of *Sxl* itself. This would be the mechanism that would maintain the female state of the pathway after the initial activating event had passed. The second target of the female-specific Sxl protein would be the pre-mRNA of the next gene on the pathway, *transformer*.

The transformer *Genes*

The *Sxl* gene regulates somatic sex determination by controlling the processing of the *transformer* gene transcript. As we saw in Chapter 12, the *transformer* gene (*tra*) is alternatively spliced in males and females. There is a female-specific mRNA and also a nonspecific mRNA that is found in both
females and males. Like the male *Sxl* message, the nonspecific *tra* mRNA
contains a termination codon early in the message, making the protein nonfunctional (Boggs et al., 1987). In *tra*, the second exon of the nonspecific
mRNA has the termination codon. This exon is not utilized in the female-
specific message (see Figure 20.18). How is it that the females make a different transcript than the males? It is thought that the female-specific protein
from the *Sxl* gene activates a female-specific 3′ splice site in the *transformer*
pre-mRNA, causing it to be processed in a way that splices out the second
exon. To do this, the Sxl protein blocks the binding of splicing factor U2AF
to the nonspecific splice site by specifically binding to the polypyrimidine
tract adjacent to it. This causes U2AF to bind to the lower-affinity (female-
specific) 3′ splice site and generate a female-specific mRNA (Valcárcel et al.,
1993). The protein encoded by this message is critical in female sex determination. If the female-specific transcript is artificially produced in XY flies,
those flies become female. The nonspecific transcript has no effect on either
males or females (McKeown et al., 1988).

The female-specific *tra* product acts in concert with the *transformer-2*
(*tra2*) gene to help generate the female phenotype. (The *tra2* gene is not
needed for male sex determination, although it is needed for spermatogenesis later in development.) The *tra2* gene is constitutively active and makes
the same protein product in both males and females. This Tra2 protein, like
that of the female-specific Sxl protein, contains an RNA-binding domain
(Amrein et al., 1988; Goralski et al., 1988). It is proposed that the *tra2* gene
can bind to the transcript of the *doublesex* gene, but only in the presence of
the female-specific Tra protein (Baker, 1989).

doublesex: *The Switch Gene of Sex Determination*

The *doublesex* gene is active in both males and females, but its primary transcript is processed in a sex-specific manner (see Figure 12.9; Baker et al.,
1987). Male and female transcripts are identical through the first three exons.
The 3′ exons differ markedly. What is an exon for the female-specific transcripts is part of the untranslated 3′ end of the male-specific message. Moreover, molecular analyses of the dominant *dsx* mutations reveal them to con-

Figure 20.18
The pattern of sex-specific RNA splicing
in three major *Drosophila* sex-determining genes. The pre-mRNAs are located
in the center of the diagram and are
identical in both male and female
nuclei. In each case, the female-specific
transcript is shown at the left, while the
default transcript (whether male or nonspecific) is shown to the right. Exons are
numbered, and the positions of the termination codons and poly(A) sites are
marked. (After Baker, 1989.)

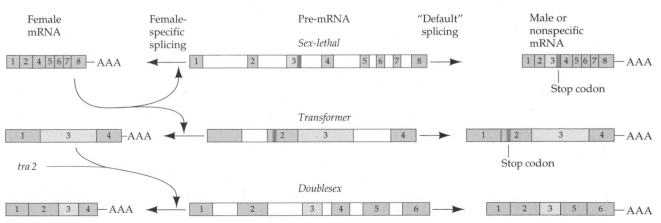

tain insertions in the female-specific exon. If a dominant *dsx* allele exists in an XX individual, the fly becomes male.

The alternative RNA processing appears to be the result of the *transformer* genes (see Figure 20.18). The Tra2 and female-specific Tra1 proteins bind specifically to a DNA sequence adjacent to the female-specific 3' splice site of the *dsx* pre-mRNA, and they recruit nonspecific splicing factors to this site (Tian and Maniatis, 1993). If *tra* is not produced, the *doublesex* transcript is spliced in the male-specific manner. The downstream 3' splice site is used, and a male-specific transcript is made. This encodes an active protein that inhibits female traits and promotes male traits. On the other hand, if the *transformer* gene is making its active, female-specific protein, a different type of processing is done (Ryner and Bruce, 1991). The Transformer proteins bind to sequences within the female-specific exon and activate the female-specific 3' splice site. (The alternative would have been their blocking the male-specific 3' splice site.) This activation of an otherwise unused female-specific 3' splice site produces an mRNA encoding a female-specific protein that activates female-specific genes (such as those of the yolk proteins) and inhibits male development.

The functions of the Doublesex proteins can be seen in the formation of the *Drosophila* genitalia. Male and female genitalia in *Drosophila* are derived from separate cell populations. In male (XY) flies, the female primordium is repressed and the male primordium differentiates into the adult genital structures. In female (XX) flies, the male primordium is repressed, and the female primordium differentiates. If the *doublesex* gene is absent (and thus neither transcript is made), *both* the male and the female primordia develop and intersexual genitalia are produced. Thus, one of the roles of the sex-specific *doublesex* transcripts is to actively inhibit the development of the inappropriate genitalia. Male *dsx* transcripts inhibit female development; female-specific *dsx* transcripts inhibit male development (Nöthiger et al., 1977; Schüpbach et al., 1978). According to this model (Baker, 1989), the sexual determination cascade comes down to what type of mRNA is going to be processed from the *doublesex* transcript. If the X:A ratio is 1, then *Sxl* makes a female-specific splicing factor that causes the *tra* gene transcript to be spliced in a female-specific manner. This female-specific protein interacts with the Tra2 splicing factor to cause the *doublesex* pre-mRNA to be spliced in a female-specific manner. If the *doublesex* transcript is not acted on in this way, it will be processed in a "default" manner to make the male-specific message.

Target Genes for the Sex Determination Cascade

Numerous proteins in *Drosophila* are present in one sex and not the other. In females, these include yolk proteins and eggshell (chorion) proteins. In males, the sex combs of the legs are sex-specific structures. Coschigano and Wensink (1993) have shown that both the male and female *doublesex* transcripts bind to three sites within the 127-base-pair enhancer of the *yolk protein* genes. Their binding and mutagenesis studies demonstrate that the male-specific Doublesex product inhibits transcription by its binding to these sites, whereas the female-specific Doublesex protein activates gene transcription from the same sites. In addition to inhibiting female-specific gene products and organs, the male Doublesex protein may also have a positive role in promoting the differentiation of the male sex combs (Jursnich and Burtis, 1993).

Temperature-sensitive mutations of sex-determining genes can enable researchers to determine the critical times at which certain target genes are sensitive to a sex-determining switch. When *temperature-sensitive* (*ts*) alleles of the *tra2* gene were used, the sexual development pathways in *Drosophila* were shown to be active from the late larval stages through the adult period. The *tra2^{ts}* gene is a temperature-sensitive allele in which the female phenotype is expressed at permissive (colder) temperatures and the male phenotype is expressed at nonpermissive (warmer) temperatures. During late larval and pupal stages, raising the temperature from permissive to nonpermissive causes an XX larva or pupa to develop into a male. Moreover, when adult mutants are kept at low temperatures, the adult fat body makes yolk proteins that will enter the oocyte. When shifted to a higher, nonpermissive temperature, transcription of the *yolk protein* genes ceases (Belote et al., 1985b). One remarkable finding has been that if adult XX *tra-2^{ts}* flies are kept at the nonpermissive temperature for several days, they begin to exhibit male courtship behaviors (Belote and Baker, 1987).

Hermaphroditism

Hermaphroditism in the Nematode C. elegans

The nematode *Caenorhabditis elegans* usually has two sexual types: hermaphrodite and male. Most individuals of this species are **hermaphroditic,*** having both testes and ovaries. As larvae, these hermaphrodites make sperm, which is stored in the nematode's genital tract (Figure 20.19). The adult ovary produces eggs, and these eggs become fertilized as they migrate into the uterus. (The sperm is already present in the hermaphroditic adult.) Self-fertilization almost always produces more hermaphrodites. Only 0.2 percent of the progeny are males. These males, however, can mate with hermaphrodites; and because their sperm has a competitive advantage over endogenous hermaphroditic sperm, the sex ratio resulting from such matings is about 50 percent hermaphrodites to 50 percent males (Hodgkin, 1985).

In *C. elegans*, the hermaphrodite is XX and the male is XO. As in *Drosophila*, sex is determined by the ratio of X chromosomes to autosomes. In closely related species of nematodes, XX females are found, suggesting that the hermaphrodites evolved from females. Somatically, the females and hermaphrodites are identical, the only difference being that the hermaphrodites make sperm during their early development before switching over to egg production. In *C. elegans*, there even exists a dominant mutation (*tra-1^D*)

*Hermaphrodites are named after the son of Hermes (Mercury) and Aphrodite (Venus). Having inherited the beauty of both parents, he excited the love of the nymph of the Salmacis fountain. As he bathed in this fountain, she embraced him, praying to the gods that they might forever be united. She got her wish in the most literal of fashions.

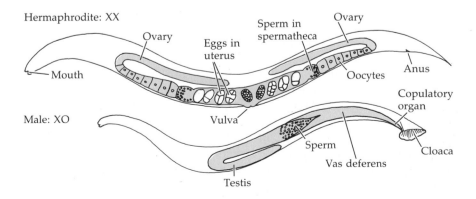

Figure 20.19
Schematic diagrams of the hermaphrodite and male *Caenorhabditis elegans*, emphasizing their reproductive systems. (From Hodgkin, 1985.)

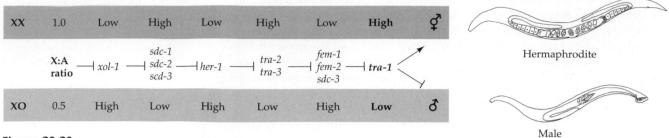

XX	1.0	Low	High	Low	High	Low	**High**	♀

$$\text{X:A ratio} \dashv xol\text{-}1 \dashv \begin{array}{c} sdc\text{-}1 \\ sdc\text{-}2 \\ scd\text{-}3 \end{array} \longrightarrow her\text{-}1 \longrightarrow \begin{array}{c} tra\text{-}2 \\ tra\text{-}3 \end{array} \dashv \begin{array}{c} fem\text{-}1 \\ fem\text{-}2 \\ sdc\text{-}3 \end{array} \dashv tra\text{-}1$$

XO	0.5	High	Low	High	Low	High	**Low**	♂

Hermaphrodite

Male

Figure 20.20
Schematic model of somatic sex determination in *C. elegans*. The *sdc-1* gene is postulated to be involved in transmitting the X/A ratio. It controls X chromosome dosage compensation as well as suppressing the *her-1* gene if the ratio is 1. The high/low designation reflects functional gene activity. The activity of the *sdc* genes eventually leads to the activity of the *tra-1* gene, whose activity promotes the hermaphroditic phenotype. The *scd* genes can be inhibited by the *xol* gene, which is only active in XO (males). (After Hodgkin, 1985; Miller et al., 1988.)

that transforms XX or XO individuals into fertile females. In colonies with such an allele, three sexes are possible and functioning (Hodgkin, 1980).

As in *Drosophila*, sex determination in *C. elegans* involves several autosomal genes that read and respond to the X:A ratio. The gene that integrates the numerators and denominators of *C. elegans* development is *xol-1* (*XO-lethal*). High levels of XOL-1 during gastrulation turn off the pathway for hermaphroditic development, thereby turning the animal into a male (Rhind et al., 1995). XOL-1 appears to accomplish this by repressing the *sdc* (sex determination control) genes, whose activities make the animal a hermaphrodite (Miller et al., 1988).

The pathway for *C. elegans* sex determination was deciphered by finding mutations in genes necessary for hermaphrodite development (the *tra* genes), as well as others necessary for the expression of the male phenotype (the *her* and *fem* genes). By creating genotypes carrying different combinations of these mutations, Hodgkin (1980) and others were able to construct a model for this developmental pathway (Figure 20.20). For instance *tra-2* mutations all suppressed the *her-1* mutation, indicating that *her-1* is later in the pathway.

The crucial gene in the pathway for sex determination appears to be *tra-1*. If the wild-type *tra-1* gene is active, the individual is a hermaphrodite. If *tra-1* is not functional, the individual is a male. The other genes appear to regulate this single switch gene.

But what does this linear genetic pathway have to do with the actual cellular events leading to sex determination? Recent studies indicate that some of these genes encode proteins of a signaling pathway between cells. Analysis of genetic mosaics suggests that *sdc-1* and *her-1* are not necessarily acting in the cells that make them. Rather, these genes appear to make secreted products. In contrast, *tra-1* acts in a cell-autonomous fashion and is therefore likely to be part of a signal-receiving apparatus. The sequence of the *tra-1* gene suggests that it encodes a zinc finger transcription factor (Hunter and Wood, 1990; Zarkower and Hodgkin, 1992; Perry et al., 1993). Kuwabara and Kimble (1992) have recently proposed a model that integrates this genetic pathway with the cellular biology of sex determination. The HER-1 protein is thought to promote male development in XO nematodes by inhibiting TRA-2. The protein encoded by *tra-2*, however, is not a transcription factor or a splicing factor, but an integral membrane protein with multiple transmembrane domains. Moreover, its mRNA is found (in different amounts) in both males and females. According to this speculative model (Figure 20.21), the FEM proteins combine to create one large FEM protein complex, and this complex is bound by the TRA-2 membrane protein. In XX individuals, the FEM protein complex is bound to the membrane, and the TRA-1 protein can enter the nucleus. In XO nematodes, however, the HER-1 protein binds to

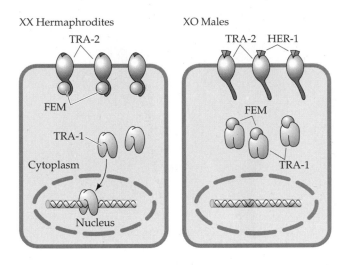

XX Hermaphrodites

XO Males

Figure 20.21
Hypothetical scheme for the actions of sex-determining genes in *C. elegans*. In XX individuals, the FEM proteins are sequestered near the cell membrane by the products of the *tra-2* genes. In the absence of the FEM proteins, TRA-1 protein enters the nucleus to transcribe genes needed for hermaphroditic development. In XO individuals, the HER-1 protein binds to the TRA-2 product, causing the TRA-2 product to release the FEM proteins. Once free in the cytoplasm, the FEM proteins can bind the TRA-1 product, preventing it from entering the nucleus. (After Kuwabara and Kimble, 1992.)

the extracellular region of the TRA-2 protein, causing the TRA-2 protein to release the FEM complex. This FEM complex, once free in the cytoplasm, can bind the TRA-1 protein and prevent its entering the nucleus. Since the TRA-1 protein (a putative transcription factor) cannot enter the nucleus, it cannot activate the hermaphrodite-specific genes. More studies need to be done to confirm or disprove this model, but it is useful both for suggesting new research and for visualizing how the genes might generate the pathway for sex determination in *C. elegans*.

One of the most interesting problems of *C. elegans* is its hermaphroditism. How did such a condition arise in an organism that probably had a male/female sex system? What gene changes arose, and were there other solutions that could have prevailed? The sex-determining genes of the closely related species *C. ramanei* (with male and female individuals) are now being identified so that such questions may be answered. [sex7.html]

Hermaphroditism in Fishes

While hermaphroditism is not uncommon in worms and insects, it is rarely seen in vertebrates. In birds and mammals, hermaphroditism is usually a pathological condition causing infertility. The most common vertebrate hermaphrodites are fishes, which display several types of hermaphroditism (Yamamoto, 1969). Some fishes, however, are **gonochoristic;** that is, they have a chromosomally determined sex that is either male or female. Hermaphroditic fish species can be divided into three groups. The first are the *synchronous* hermaphrodites, in which ovaries and testicular tissues exist at the same time and in which both sperm and eggs are produced. One such species is *Servanus scriba*. In nature and in aquaria, these fish form spawning pairs. As soon as one of the fish spawns its eggs, the other fish fertilizes them. Then the fish reverse their roles, and the fish that was formerly male spawns its eggs so that they can be fertilized by the sperm of its partner (Clark, 1959).

In other hermaphroditic species, an individual undergoes a genetically programmed sex change during its development. In these cases, the gonads are dimorphic, having both male and female areas. One or the other is predominant during a certain phase of life. In *protogynous* ("female-first") hermaphrodites, an animal begins its life as a female, but later becomes male. The reverse is the case in *protandrous* ("male-first") species. Figure 20.22 shows the gonadal changes of the protandrous hermaphroditic fish *Sparus*

Figure 20.22
Gonadal changes in the hermaphroditic fish *Sparus auratus*, shown in section through the gonad of (A) the male phase, (B) the transitory phase, and (C) the final, female phase. (Courtesy of the family of T. Yamamoto.)

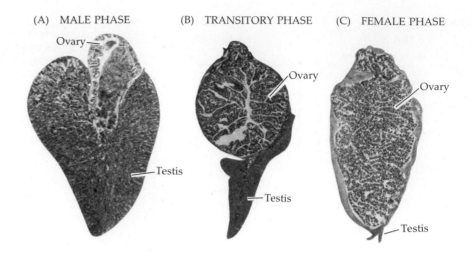

(A) MALE PHASE (B) TRANSITORY PHASE (C) FEMALE PHASE

auratus. At first, testicular tissue predominates, but after a transition period during which both testicular and ovarian tissues are seen, the ovarian cells take over.

Environmental sex determination

Temperature-Dependent Sex Determination in Reptiles

While the sex of most snakes and most lizards is determined by sex chromosomes at the time of fertilization, the sex of most turtles and all species of crocodilians is determined by the environment after fertilization. In these reptiles, the temperature of the eggs during a certain period of development is the deciding factor in determining sex (Bull, 1980), and small changes in temperature can cause dramatic changes in the sex ratio. Generally, eggs incubated at low temperatures (22°–27°C) produce one sex, whereas eggs incubated at higher temperatures (30°C and above) produce the other. There is only a small range of temperatures that permits both males and females to hatch from the same brood of eggs. Figure 20.23 shows the abrupt temperature-induced change in sex ratios for certain species of turtles. If eggs are incubated below 28°C, all the turtles hatching from them will be male. Above 32°C, every egg gives rise to a female. At temperatures in between, the broods will give rise to individuals of either sex. Variations on this theme also exist. The eggs of snapping turtles, for instance, become female at either cold (20°C or lower) or hot (30°C or above) temperatures. Between these extremes, males predominate.

One of the best-studied reptiles is the European pond turtle, *Emys obicularis*. In laboratory studies, incubating *Emys* eggs at temperatures above 30°C produces females, while temperatures below 25°C produce all male broods. The threshold temperature (where the sex ratio is even) is 28.5°C (Pieau et al., 1994). The developmental period during which sex determination occurs can be studied by incubating eggs at the male-producing temperature for a certain amount of time and then shifting the eggs to an incubator of the female-producing temperature (and vice-versa). In *Emys*, the last third of development appears to be the most critical for sex determination. It is not thought that turtles can reverse their sex after this period.

The pathways toward maleness and femaleness are just being delineated. Estrogen induces ovarian differentiation at masculinizing temperatures, and the sensitive time for the effects of estrogens coincides with the

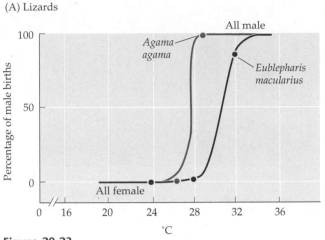

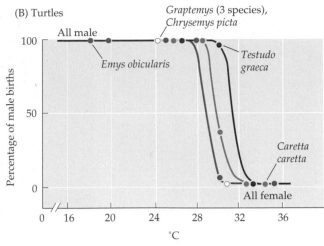

Figure 20.23
Relationship between sex ratio and incubation temperature in reptiles. (A) Two species of lizards in which higher temperatures result in the generation of male off-spring. (B) Seven species of turtles in which higher temperatures result in female offspring. (After Bull, 1980.)

time when sex determination usually occurs (Bull et al., 1988; Gutzke and Chymiy, 1988). It appears that the enzyme aromatase (which can convert testosterone into estrogen) is important. The aromatase activity of *Emys* is very low at the "male" temperature of 25°C. At the "female" temperature of 30°C, the aromatase activity increases dramatically during the critical period for sex determination (Desvages et al., 1993; Pieau et al., 1994). Temperature-dependent aromatase activity is also seen in diamondback terrapins, and its inhibition masculinizes their gonads (Jeyasuria et al., 1994). It is possible that the regulator of aromatase activity is the anti-Müllerian hormone. AMH is known to decrease aromatase activity in the *Emys* gonads (Desvages and Pieau, 1992).

Ferguson and Joanen (1982) have studied sex determination in the Mississippi alligator, both in the laboratory and in the field; they have concluded that sex is determined between 7 and 21 days of incubation. Eggs raised at 30°C or below produce female alligators, whereas those incubated at 34°C or above produce all males. Moreover, nests constructed on levees (close to 34°C) give rise to males, whereas those built in wet marshes (close to 30°C) produce females. The evolutionary advantages and disadvantages to temperature-dependent sex determination are discussed in Chapter 21.

Location-Dependent Sex Determination in Bonellia viridis and Crepidula fornicata

The sex of the echiuroid worm *Bonellia* depends on where a larva settles. The female *Bonellia* is a marine, rock-dwelling animal, with a body about 10 cm long (Figure 20.24). It has a proboscis, however, that can extend to over a meter in length. This proboscis serves two functions. First, it sweeps food from the rocks into the digestive tract of the female *Bonellia*. Second, should a larva land on the proboscis, it enters the female's mouth, migrates to the uterus, and differentiates into a 1- to 3-mm-long symbiotic male. Thus, when a larva settles on a rocky surface, it becomes a female, but if that same larva settles on the proboscis of a female, it becomes a male. The male *Bonellia* spends its life within the body of the female, fertilizing her eggs.

Baltzer (1914) demonstrated that when larvae were cultured in the absence of adult females, about 90 percent of them became females. However,

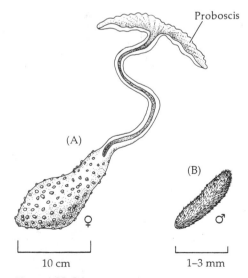

Figure 20.24
Extreme sexual dimorphism in *Bonellia viridis*. (A) Female, around 10 cm, with a proboscis capable of extending over a meter. (B) Symbiotic male (highly magnified compared with the female), 1–3 mm long. (After Barnes, 1968.)

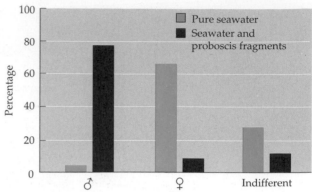

Figure 20.25
In vitro analysis of *Bonellia* differentiation. Larval *Bonellia* were placed either in normal seawater or in seawater containing fragments of the female proboscis. A majority of the animals cultured in the presence of the proboscis fragments became males, whereas normally they would have become females. (After Leutert, 1974.)

when these larvae were cultured in the presence of an adult female or its isolated proboscis, 70 percent of them adhered to the proboscis and developed the male structures. These results have been more recently confirmed by Leutert (1974; Figure 20.25).

The molecule(s) responsible for masculinizing the larvae can be extracted from the proboscis of adult females. When larvae are cultured in normal seawater in the absence of adult females, most become females. When cultured in seawater containing aqueous extracts of proboscis tissue, most become either male or an intermediate form, neither completely male nor completely female (Nowinski, 1934; Agius, 1979). The compound or compounds that attract the larvae to the proboscis and cause its masculinization are currently being purified.

Another example in which sex determination is affected by the position of the organism is the case of the slipper snail *Crepidula fornicata*. Here, individuals pile up on top of each other to form a mound (Figure 20.26). Young individuals are always male. This phase is followed by the degeneration of the male reproductive system and a period of lability. The next phase can be either male or female, depending on the animal's position in the mound. If the snail is attached to a female, it will become male. If such a snail is removed from its attachment, it will become female. Similarly, the presence of large numbers of males will cause some of the males to become females. However, once the individual becomes female, it will not revert to being male (Coe, 1936).

Coda

Nature has provided many variations on her masterpiece. In some species, sex is determined solely by chromosomes, whereas in other species, sex is a matter of environmental conditions. Within these two large categories, numerous variations also exist. A complete catalogue of known sex-determining mechanisms would require a separate (and very interesting) volume.

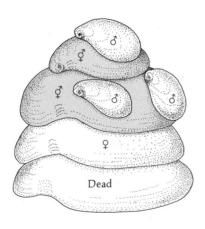

Figure 20.26
Cluster of *Crepidula* snails. Two individuals are changing from male to female. After these molluscs become female, they will be fertilized by the male above them. (After Coe, 1936.)

LITERATURE CITED

Agius, L. 1979. Larval settlement in the echiuran worm *Bonellia vividis*: Settlement on both the adult proboscis and body trunk. *Mar. Biol.* 53: 125–129.

Amrein, H., Gorman, M. and Nöthiger, R. 1988. The sex-determining gene *tra-2* of *Drosophila* encodes a putative RNA binding protein. *Cell* 55: 1025–1035.

Andersson, S., Berman, D. M., Jenkins, E. P. and Russell, D. W. 1991. Deletion of steroid 5a-reductase 2 gene in male pseudohermaphroditism. *Nature* 354: 159–161.

Aristotle. The generation of animals. Translated by A. Platt. *In* J. Barnes (ed.), 1984, *The Complete Works of Aristotle*, Vol. 8. Princeton University Press, Princeton, NJ.

Arnold, A. P. 1980. Sexual differences in the brain. *Am. Sci.* 68: 165–173.

Arnold, A. P., Nottebohm, F. and Pfaff, D. W. 1976. Hormone concentrating cells in vocal control and other brain regions of the zebra finch (*Poephila guttata*). *J. Comp. Neurol.* 165: 487–512.

Baker, B. S. 1989. Sex in flies: The splice of life. *Nature* 340: 521–524.

Baker, B. S. and Ridge, K. A. 1980. Sex and the single cell. I. On the action of major loci affecting sex determination in *Drosophila melanogaster*. *Genetics* 94: 383–423.

Baker, B. S., Nagoshi, R. N. and Burtis, K. C. 1987. Molecular genetic aspects of sex determination in *Drosophila*. *BioEssays* 6: 66–70.

Baltzer, F. 1914. Die Bestimmung und der Dimorphismus des Geschlechtes bei *Bonellia*. *Sber. Phys.-Med. Ges. Würzb.* 43: 1–4.

Bardoni, B. and eleven others. 1994. A dosage sensitive locus at chromosome Xp21 is involved in male to female sex reversal. *Nat. Genet.* 7: 497–501.

Barfield, R. J. and Chen, J. J. 1977. Activation of estrous behavior in ovariectomized rats by intracerebral implants of estradiol benzoate. *Endocrinology* 101: 1716–1725.

Barnes, R. D. 1968. *Invertebrate Zoology*. Saunders, Philadelphia.

Barraclough, C. A. and Gorski, R. A. 1962. Studies on mating behavior in the androgen-sterilized female rat in relation to the hypothalamic regulation of sexual behavior. *J. Endocrinol.* 25: 175–182.

Bell, S. E. 1986. A new model of medical technology development: A case study of DES. *Sociol. Health Care* 4: 1–32.

Bell, L. R., Maine, E. M., Schedl, P. and Cline, T. W. 1988. *Sex-lethal*, a *Drosophila* sex determination switch gene, exhibits sex-specific RNA splicing and sequence similarity to RNA binding proteins. *Cell* 55: 1037–1046.

Belote, J. M. and Baker, B. S. 1987. Sexual behavior: Its genetic control during development and adulthood in *Drosophila melanogaster*. *Proc. Natl. Acad. Sci. USA* 84: 8026–8030.

Belote, J. M., McKeown, M. B., Andrew, D. J., Scott, T. N., Wolfner, M. F. and Baker, B. S. 1985a. Control of sexual differentiation in *Drosophila melanogaster*. *Cold Spring Harbor Symp. Quant. Biol.* 50: 605–614.

Belote, J. M., Handler, A. M., Wolfner, M. F., Livak, K. L. and Baker, B. S. 1985b. Sex-specific regulation of yolk protein gene expression in *Drosophila*. *Cell* 40: 339–348.

Berkovitz, G. D. and seven others. 1992. The role of the sex-determining region of the Y chromosome (SRY) in the etiology of 46, XX true hermaphrodites. *Hum. Genet.* 88: 411–416.

Bernstein, R., Jenkins, T., Dawson, B., Wagner, J., Devald, G., Koo, G. C. and Wachtel, S. S. 1980. Female phenotype and multiple abnormalities in sibs with a Y chromosome and partial X-chromosome duplication: H-Y antigen and Xg blood group findiungs. *J. Med. Genet.* 17: 291–300.

Bleier, R. 1984. *Science and Gender*. Pergamon, New York, pp. 80–114.

Boggs, R. T., Gregor, P., Idriss, S., Belote, J. M. and McKeown, M. 1987. Regulation of sexual differentiation in *D. melanogaster* via alternative splicing of RNA from the *transformer* gene. *Cell* 50: 739–747.

Breedlove, S. M. and Arnold, A. P. 1980. Hormone accumulation in a sexually dimorphic motor nucleus of the rat spinal cord. *Science* 210: 565–566.

Bridges, C. B. 1921. Triploid intersexes in *Drosophila melanogaster*. *Science* 54: 252–254.

Bridges, C. B. 1925. Sex in relation to chromosomes and genes. *Am. Nat.* 59: 127–137.

Buehr, M., Gu, S. and McLaren, A. 1993. Mesonephric contribution to testis differentiation in the fetal mouse. *Development* 117: 273–281.

Bull, J. J. 1980. Sex determination in reptiles. *Q. Rev. Biol.* 55: 3–21.

Bull, J. J., Gutzke, W, H. N. and Crews, D. 1988. Sex reversal by estradiol in three reptilian orders. *Gen. Comp. Endocrinol.* 70: 425–428.

Cate, R. L. and eighteen others. 1986. Isolation of the bovine and human genes for Müllerian inhibiting substance and expression of the gene in animal cells. *Cell* 45: 685–698.

Clark, E. 1959. Functional hermaphroditism and self-fertilization in a serranid fish. *Science* 129: 215–216.

Cline, T. W. 1983. The interaction between *daughterless* and *Sex-lethal* in triploids: A novel sex-transforming maternal effect linking sex determination and dosage compensation in *Drosophila melanogaster*. *Dev. Biol.* 95: 260–274.

Cline, T. W. 1986. A female-specific lethal lesion in an X-linked positive regulator of the *Drosophila* sex determination gene, *Sex-lethal*. *Genetics* 113: 641–663.

Cline, T. W. 1988. Evidence that *sisterless-a* and *sisterless-b* are two of several discrete "numerator elements" of the X/A sex determination signal in *Drosophila* that switch *Sxl* between two alternative stable expression states. *Genetics* 119: 829–862.

Coe, W. R. 1936. Sexual phases in *Crepidula*. *J. Exp. Zool.* 72: 455–477.

Cohen, D. R., Sinclair, A. H. and McGovern, J. D. 1994. SRY protein enhances transcription of Fos-related antigen 1 promoter constructs. *Proc. Natl. Acad. Sci. USA* 91: 4372–4376.

Coschigano, K. T. and Wensink, P. C. 1993. Sex-specific transcriptional regulation of the male and female doublesex proteins of *Drosophila*. *Genes Dev.* 7: 42–54.

Coward, P., Nagai, K., Chen, D., Thomas, H. D., Nagamine, C. M. and Lau, Y-F. C. 1994. Polymorphism of a CAG trinucleotide repeat within *Sry* correlates with B6Y^{Dom} sex reversal. *Nat. Genet.* 6: 245–250.

Cronmiller, C. and Cline, T. W. 1987. The *Drosophila* sex determination gene *daughterless* has different functions in the germ line versus the soma. *Cell* 48: 479–487.

De Jonge, F. H., Muntjewerff, J.-W., Louwerse, A. L. and Van de Poll, N. E. 1988. Sexual behavior and sexual orientation of the female rat after hormonal treatment during various stages of development. *Horm. Behav.* 22: 100–115.

Desvages, G. and Pieau, C. 1992. Time required for temperature-induced changes in gonadal aromatase activity and related gonadal structure in turtle embryos. *Differentiation* 52: 13–18.

Desvages, G., Girondot, M. and Pieau, C. 1993. Sensitive stages for the effects of temperature on gonadal aromatase activity in embryos ofthe marine turtle *Dermochelys coriacea*. *Gen. Comp. Endocrinol.* 92: 54–61.

Duffy, J. B. and Gergen, J. P. 1991. The *Drosophila* segmentation gene *runt* acts as a position-specific numerator element necessary for the uniform expression of the sex determining gene *sex-lethal*. *Genes Dev.* 5: 2176–2187.

Eicher, E. M. 1994. Sex and trinucleotide repeats. *Nat. Genet.* 6: 221–223.

Eicher, E. M. and Washburn, L. L. 1983. Inherited sex reversal in mice: Identification of a new sex-determining gene. *J. Exp. Zool.* 228: 297–304.

Fausto-Sterling, A. 1992. *Myths of Gender*. Basic Books, New York.

Fausto-Sterling, A. 1995. Animal models for the development of human sexuality: A critical evaluation. *J. Homosexuality* 28: 217–236.

Ferguson, M. W. J. and Joanen, T. 1982. Temperature of egg incubation determines sex in *Alligator mississippiensis*. *Nature* 296: 850–853.

Foster, J. W. and eleven others. 1994. Campomelic dysplasia and autosomal sex reversal caused by mutations in an *SRY*-related gene. *Nature* 372: 525–530.

Galen, C. *On the Usefulness of the Parts of the Body*. Translated by M. May, 1968. Cornell University Press, Ithaca, NY.

Geddes, P. and Thomson, J. A. 1890. *The Evolution of Sex*. Walter Scott, London.

Genetics Review Group. 1995. One for a boy, two for a girl? *Curr. Biol.* 5: 37–39.

Giese, K., Cox, J. and Grosschedl, R. 1992. The HMG domain of lymphoid enhancer factor 1 bends DNA and facilitates the assembly of functional nucleoprotein structures. *Cell* 69: 185–195.

Geissler, W. M. and nine others. 1994. Male pseudohermaphroditism caused by mutations of testicular 17b-hydroxysteroid dehydrogenase 3. *Nat. Genet.* 7: 34–39.

Goralski, T. J., Edström, J.-E. and Baker, B. S. 1988. The sex determination locus *transformer-2* of *Drosophila* encodes a polypeptide with similarity to RNA binding proteins. *Cell* 56: 1011–1018.

Gubbay, J. and eight others. 1990. A gene mapping to the sex-determining region of the mouse Y chromosome is a member of a novel family of embryonically expressed genes. *Nature* 346: 245–250.

Gutzke, W. H. N. and Chymiy, D. B. 1988. Sensitive periods during embryology for hormonally induced sex determination in turtles. *Gen. Comp. Endocrinol.* 71: 265–267.

Hacker, A., Capel, B., Goodfellow, P. and Lovell-Badge, R. 1995. Expression of *Sry*, the mouse sex determining gene. *Development* 121: 1603–1614.

Hamer, D. H., Hu, S., Magnuson, V. L., Hu, N. and Pattatucci, A. M. L. 1993. A linkage between DNA markers on the X chromosome and male sexual orientation. *Science* 261: 321–327.

Haqq, C. M., King, C. Y., Donahoe, P. K. and Weiss, M. A. 1993. Sry recognizes conserved DNA sites in sex-specific promoters. *Proc. Natl. Acad. Sci. USA* 90: 1097–1101.

Haqq, C., King, C.-Y., Ukiyama, E., Falsafi, S., Haqq, N., Donahoe, P. K. and Weiss, M. A. 1994. Molecular basis of mammalian sexual determination: Activation of Müllerian inhibiting substance gene expression by SRY. *Science* 266: 1494–1500.

Harris, G. W. and Levine, S. 1965. Sexual physiology of the brain and its experimental control. *J. Physiol.* 181: 379–400.

Higgins, S. J., Young, P. and Cunha, G. R. 1989. Induction of functional cytodifferentiation in the epithelium of tissue recombinants II. Instructive induction of Wolffian duct epithelia by neonatal seminal vesicle mesenchyme. *Development* 106: 235–250.

Hodgkin, J. 1980. More sex-determination mutants of *Caenorhabditis elegans*. *Genetics* 96: 649–664.

Hodgkin, J. 1985. Males, hermaphrodites, and females: Sex determination in *Caenorhabditis elegans*. *Trends Genet.* 1: 85–88.

Horowitz, M. C. 1976. Aristotle and women. *J. Hist. Biol.* 9: 183–213.

Hu, S., Pattatucci, A. M. L., Patterson, C., Li, L., Fulker, D. W., Cherny, S. S., Kruglyak, L. and Hamer, D. H. 1995. Linkage between sexual orientation and chromosome Xq28 in males but not in females. *Nat. Genet.* 11: 248–256.

Hunter, C. P. and Wood, W. B. 1990. The *tra-1* gene determines sexual phenotype cell-autonomously in *C. elegans*. *Cell* 63: 1193–1204.

Imperato-McGinley, J., Guerrero, L., Gautier, T. and Peterson, R. E. 1974. Steroid 5a-reductase deficiency in man: An inherited form of male pseudohermaphroditism. *Science* 186: 1213– 1215.

Jacklin, D. 1981. Methodological issues in the study of sex-related differences. *Dev. Rev.* 1: 266–273.

Jeyasuria, P., Roosenburg, W. M. and Place, A. R. 1994. Role of P-450 aromatase in sex determination of the diamondback terrapin, *Malaclemys terrapin*. *J. Exper. Zool.* 270: 95–111.

Josso, N., Picard, J.-Y. and Tran, D. 1977. The anti-Müllerian hormone. *Recent Prog. Horm. Res.* 33: 117–167.

Jost, A. 1953. Problems of fetal endocrinology: The gonadal and hypophyseal hormones. *Recent Prog. Horm. Res.* 8: 379–418.

Jursnich, V. A. and Burtis, K. C. 1993. A positive role in differentiation for the male doublesex protein of *Drosophila*. *Dev. Biol.* 155: 235–249.

Kandel, E. R. and Schwartz, J. H. 1985. *Principles of Neural Science*. Elsevier, NY.

Kandel, E. R., Schwartz, J. H. and Jessell, T. M. 1995. *Essentials of Neural Science and Behavior*. Appleton and Lange, Norwalk, CT.

Kent, J.,Wheatley, S. C., Andrews, J. E., Sinclair, A. H. and Koopman, P. 1996. A male-specific role for SOX9 in vertebrate sex determination. *Development* 122: 2813–2822.

Keyes, L. N., Cline, T. W. and Schedl, P. 1992. The primary sex determination signal of *Drosophila* acts at the level of transcription. *Cell* 68: 933–943.

Konishi, M. and Akutagawa, E. 1985. Neuronal growth, atrophy, and death in a sexu-

ally dimorphic song nucleus in the zebra finch brain. *Nature* 315: 145–147.

Koopman, P., Münsterberg, A., Capel, B., Vivian, N. and Lovell-Badge, A. 1990. Expression of a candidate sex-determining gene during mouse testis differentiation. *Nature* 348: 450–452.

Koopman, P., Gubbay, J., Vivian, N., Goodfellow, P. and Lovell-Badge, R. 1991. Male development of chromosomally female mice transgenic for *Sry*. *Nature* 351: 117–121.

Kurz, E. M., Sengelaub, D. R. and Arnold, A. P. 1986. Androgens regulate the dendritic length of mammalian motoneurons in adulthood. *Science* 232: 395–397.

Kuwabara, P. E. and Kimble, J. 1992. Molecular genetics of sex determination in *C. elegans*. *Trends Genet.* 8: 164–168.

Langman, J. 1981. *Medical Embryology*, 4th Ed. Williams & Wilkins, Baltimore.

Langman, J. and Wilson, D. B. 1982. Embryology and congenital malformations of the female genital tract. *In* A. Blaustein (ed.), *Pathology of the Female Genital Tract*, 2nd Ed. Springer-Verlag, New York, pp. 1–20.

Leutert, T. R. 1974. Zur Geschlechtsbestimmung und Gametogenese von *Bonellia vividis* Rolando. *J. Embryol. Exp. Morphol.* 32: 169–193.

LeVay, S. 1991. A difference in hypothalamic structure between heterosexual and homosexual men. *Science* 253: 1034–1037.

Luo, X., Ikeda, Y. and Parker, K. L. 1994. A cell-specific nuclear receptor is essential for adrenal and gonadal development and sexual differentiation. *Cell* 77: 481–490.

Mansour, S., Hall, C. M., Pembrey, M. E. and Young, I. D. 1995. A clinical and genetic study of campomelic dysplasia. *J. Med. Genet.* 32: 415–420.

Marshall, E. 1995. NIH's "gay gene" study questioned. *Science* 268: 1841.

Marx, J. 1995. Mammalian sex determination: Snaring the genes that divide sexes for mammals. *Science* 269: 1824–1825.

McClung, C. E. 1902. The accessory chromosomesex determinant? *Biol. Bull.* 3: 72–77.

McEwen, B. S. 1981. Neural gonadal steroid actions. *Science* 211: 1303–1311.

McEwen, B. S., Leiberburg, I., Chaptal, C. and Krey, L. C. 1977. Aromatization: Important for sexual differentiation of the neonatal rat brain. *Horm. Behav.* 9: 249–263.

McKeown, M., Belote, J. M. and Boggs, R. T. 1988. Ectopic expression of the female *transformer* gene product leads to female differentiation of chromosomally male *Drosophila*. *Cell* 53: 887–895.

Meyer, W. J., Migeon, B. R. and Migeon, C. J. 1975. Locus on human X chromosome for dihydrotestosterone receptor and androgen insensitivity. *Proc. Natl. Acad. Sci. USA* 72: 1469–1472.

Miller, L. M., Plenefisch, J. D., Casson, L. P. and Meyer, B. 1988. *xol-1*: A gene that controls the male mode of both sex determination and X chromosome dosage compensation in *C. elegans*. *Cell* 55: 167–183.

Moore, C. L. 1990. Comparative development of vertebrate sexual behavior: Levels, cascades, and webs. *In* D. A. Dewsbury (ed.), *Issues in Comparative Psychology*. Sinauer Associates, Sunderland, MA, pp. 278–299.

Moore, C. L., Dou, H. and Juraska, J. M. 1992. Maternal stimulation affects the number of motor neurons in a sexually dimorphic nucleus of the lumbar spinal cord. *Brain Res.* 572: 52–56.

Morgan, T. H. 1919. *The Physical Basis of Heredity*. Lippincott, Philadelphia.

Muscatelli, F. and fourteen others. 1994. Mutations in the DAX-1 gene give rise to both X-linked adrenal hypoplasia congenita and hypogonadotropic hypogonadism. *Nature* 372: 672–634.

Nabekura, J., Oomura, Y., Minami, T., Mizuno, Y. and Fukuda, A. 1986. Mechanism of the rapid effect of 17β-estradiol on medial amygdala neurons. *Science* 233: 226–228.

Nordeen, E. J., Nordeen, K. W., Sengelaub, D. R. and Arnold, A. P. 1985. Androgens prevent normally occurring cell death in a sexually dimorphic spinal nucleus. *Science* 229: 671–673.

Nöthiger, R., Dübendorfer, A. and Epper, F. 1977. Gynandromorphs reveal two separate primordia for male and female genitalia in *Drosophila melanogaster*. *Wilhelm Roux Arch.* 181: 367–373.

Nottebohm, F. 1980. Testosterone triggers growth of brain vocal control nuclei in adult female canaries. *Brain Res.* 189: 429–436.

Nottebohm, F. 1981. A brain for all seasons: Cyclical anatomical changes in song control nuclei of the canary brain. *Science* 214: 1368–1370.

Nowinski, W. 1934. Die vermännlichende Wirkung fraktionierter Darmextrakte des Weibchens auf die Larven der *Bonellia viridis*. *Pubbl. Staz. Zool. Napoli* 14: 110–145.

Perry, M. D., Li, W., Trent, C., Robertson, B., Fire, A., Hageman, J. M. and Wood, W. B. 1993. Molecular characterization of the *her-1* gene suggests a direct role in cell signaling during *Caenorhabditis elegans* sex determination. *Genes Dev.* 7: 216–228.

Pfaff, D. W. and McEwen, B. S. 1983. The actions of estrogens and progestins on nerve cells. *Science* 219: 808–814.

Phoenix, C. H., Goy, R. W., Gerall, A. A. and Young, W. C. 1959. Organizing action of prenatally administered testosterone proprionate on the tissues mediating mating behavior in the female guinea pig. *Endocrinology* 65: 369–382.

Pieau, C., Girondot, N., Richard-Mercier, G, Desvages, M., Dorizzi, P and Zaborski, P. 1994. Temperature sensitivity of sexual differentiation of gonads in the European pond turtle. *J. Exper. Zool.* 270: 86–93.

Pontiggia, A., Rimini, R., Goodfellow, P. N., Lovell-Badge, R. and Bianchi, M. E. 1994. Sex-reversing mutations affect the architecture of Sry/DNA complexes. *EMBO J.* 13: 6115–6124.

Pröve, E. 1978. Courtship and testosterone in male zebra finches. *Z. Tierpsychol.* 48: 47–67.

Reddy, V. R., Naftolin, F. and Ryan, K. J. 1974. Conversion of androstenedione to estrone by neural tissues from fetal and neonatal rats. *Endocrinology* 94: 117–121. ·

Rhind, N. B., Miller, L. M., Kopczynski, J. B. and Meyer, B. J. 1995. *xol-1* acts as an early switch in the *C. elegans* male/hermaphrodite decision. *Cell* 80: 71–82.

Risch, N., Squires-Wheeler, E. and Keats, B. J. B. 1993. Male sexual orientation and genetic evidence. *Science* 262: 2063–2065.

Robboy, S. J., Young, R. H. and Herbst, A. L. 1982. Female genital tract changes related to prenatal diethylstilbesterol exposure. *In* A. Blaustein (ed.), *Pathology of the Female Genital Tract*, 2nd Ed. Springer-Verlag, New York, pp. 99–118.

Rubin, B. S. and Barfield, R. J. 1980. Priming of estrus responsiveness by implants of 17β-estradiol in the ventromedial hypothalamic nuclei of female rats. *Endocrinology* 106: 504–509.

Ryner, L. C. and Bruce, B. S. 1991. Regulation of *doublesex* pre-mRNA processing occurs by 3′-splice site activation. *Genes Dev.* 5: 2071–2085.

Salz, H. K., Cline, T. W. and Schedl, P. 1987. Functional changes associated with structural alterations induced by mobilization of a P element inserted into the *Sex-lethal* gene of *Drosophila*. *Genetics* 117: 221–231.

Salz, H. K., Maine, E. M., Keyes, L. N., Samuels, M. E., Cline, T. W. and Schedl, P. 1989. The *Drosophila* female-specific sex-determination gene, *Sex-lethal*, has stage-, tissue-, and sex-specific RNAs suggesting multiple modes of regulation. *Genes Dev.* 3: 708–719.

Sánchez, L. and Nöthiger, R. 1983. Sex determination and dosage compensation in *Drosophila melanogaster*: Production of male clones in XX females. *EMBO J.* 1: 485–491.

Schiebinger, L. 1989. *The Mind Has No Sex?* Harvard University Press, Cambridge, MA.

Schüpbach, T., Wieschaus, E. and Nöthiger, R. 1978. The embryonic organization of the genital disc studied in genetic mosaics of *Drosophila melanogaster*. *Wilhelm Roux Arch.* 185: 249–270.

Shen, W-H., Moore, C. C. D., Ikeda, Y., Parker, K.L. and Ingraham, H. A. 1994. Nuclear receptor steroidogenic factor 1 regulates the Müllerian inhibiting substance gene: a link to the sex determination cascade. *Cell* 77: 651–661.

Siiteri, P. K. and Wilson, J. D. 1974. Testosterone formation and metabolism during male sexual differentiation in the human embryo. *J. Clin. Endocrinol. Metab.* 38: 113–125.

Sinclair, A. H. and nine others. 1990. A gene from the human sex-determining region encodes a protein with homology to a conserved DNA-binding motif. *Nature* 346: 240–244.

Stevens, N. M. 1905. Studies in spermatogenesis with especial reference to the "accessory chromosome." *Carnegie Inst. Washington Rep.* 36.

Terasawa, E. and Sawyer, C. H. 1969. Changes in electrical activity in rat hypothalamus related to electrochemical stimulation of adenohypophyseal function. *Endocrinology* 85: 143–149.

Thigpen, A. E., Davis, D. L., Imperato-McGinley, J. and Russell, D. W 1992. The molecular basis of steroid 5α-reductase deficiency in a large Dominican kindred. *N. Engl. J. Med.* 327: 1216–1219.

Thorpe, W. H. 1958. The learning of song patterns by birds with especial reference to the song of the chaffinch, *Fringilla coelebs*. *Ibis* 100: 535–570.

Tian, M. and Maniatis, T. 1993. A splicing enhancer complex controls alternative splicing of *doublesex* pre-mRNA. *Cell* 74: 105–114.

Tobin, C. and Joubert, Y. 1992. Testosterone-induced development of the rat levator ani muscle. *Dev. Biol.* 146: 131–138.

Tran, D., Meusy-Dessolle, N. and Josso, N. 1977. Anti-Müllerian hormone is a functional marker of foetal Sertoli cells. *Nature* 269: 411–412.

Trelstad, R. L., Hayashi, A., Hayashi, K. and Donahoe, P. K. 1982. The epithelial-mesenchymal interface of the male rate Müllerian duct: Loss of basement membrane integrity and ductal regression. *Dev. Biol.* 92: 27–40.

Tuana, N. 1988. The weaker seed. *Hypatia* 3: 35–59.

Vainio, S. and McMahon, A. 1996. Wnt-4a as a signal regulating sex organogenesis. WNT Meeting Abstracts, Stanford, CA.

Valcárcel, J., Singh, R., Zamore, P. and Greene, M. R. 1993. The protein Sex-lethal antagonizes the splicing factor U2AF to regulate alternative splicing of *transformer* pre-mRNA. *Nature* 362: 171–175.

Van Doren, Ellis, H. M. and Posakony, J. W. 1991. The *Drosophila* extramacrochaetae protein antagonizes sequence-specific DNA binding by daughterless/achaete-scute protein complexes. *Development* 113: 245–255.

Wagner, T. and thirteen others. 1994. Autosomal sex reversal and campomelic dysplasia are caused by mutations in and around the *SRY*-related gene *SOX9*. *Cell* 79: 1111–1120.

Washburn. L. L. and Eicher, E. M. 1989. Normal testis determination in the mouse depends on genetic interaction of a locus on chromosome 17 and the Y chromosome. *Genetics* 123: 173–179.

Werner, M. H., Huth, J. R., Groneborn, A. M. and Clore, G. M. 1995. Molecular basis of human 46X,Y sex reversal revealed from the three-dimensional solution structure of the human SRY–DNA complex. *Cell* 81: 705–714.

Wilson, E. B. 1905. The chromosomes in relation to the determination of sex in insects. *Science* 22: 500–502.

Witelson, S. F., Glezer, I. I. and Kigar, D. L. 1995. Women have greater density of neurons in posterior temporal cortex. *J. Neurosci.* 15: 3418–3428.

Wright, E. and eight others. 1995. The *Sry*-related gene *Sox9* is expressed during chondrogenesis in mouse embryos. *Nat. Genet.* 9: 15–20.

Yamamoto, T.-O. 1969. Sex differentiation. *In* W. S. Hoar and D. J. Randall (eds.), *Fish Physiology*, Vol. 3. Academic Press, New York, pp. 117–175.

Younger-Shepherd, S., Vaessin, H., Bier, E., Jan, L. Y. and Jan, Y. N. 1992. *deadpan*, an essential pan-neural gene encoding an HLH protein, acts as a denominator in *Drosophila* sex determination. *Cell* 70: 911–922.

Zanaria, E. and thirteen others. 1994. An unusual member of the nuclear hormone receptor superfamily responsible for X-linked adrenal hypoplasia congenita. *Nature* 372: 635–641.

Zarkower, D. and Hodgkin, J. 1992. Molecular analysis of the *C. elegans* sex-determining gene *tra-1*: A gene encoding two zinc finger proteins. *Cell* 70: 237–249.

Environmental regulation of animal development

21

We may now turn to consider adaptations towards the external environment; and firstly the direct adaptations...in which an animal, during its development, becomes modified by external factors in such a way as to increase its efficiency in dealing with them.

C. H. WADDINGTON (1957)

I N THE FIRST HALF of the nineteenth century, "biology" was the study of "the organism in relation to its conditions of existence," and the investigation of the living organism was usually performed in its original habitat. Only in the 1850s did "physiology" emerge as an attempt to quantitate biological phenomena in the laboratory. Embryology remained within the realm of biology, while physiology investigated the structures and functions of adult organisms irrespective of their original environments (Nyhart, 1995).

Within this biological context, embryology was seen as the motor of evolutionary change, and development was seen as being conditioned by the environment. August Weismann (1875), for instance, noted that butterflies of the same species hatching at different seasons could be colored differently, and he could turn the summer morph into the spring form by cooling the pupae. Carl Siebold showed that some parthenogenetic aphids could give rise to males and sexual females late in the breeding season to produce an overwintering egg (which would invariably hatch as a parthenogenetic female), and several investigators studied environmental sex determination in *Bonellia* and in insect hives (see Hertwig, 1894). The first generation of experimental embryologists pursued several investigations into the environmental effects on development, including studies on the effects of ion or nutrient deprivation on sex determination and morphogenesis (Selenka, 1876; Born, 1881; Herbst, 1893). (Born's studies showing that the sex of frog embryos could be altered by environmental factors were featured prominently in the film *Jurassic Park*.)

But the tide was changing. In the 1870s and 1880s, young zoologists were moving away from these "biological questions" toward questions of internal physiology and anatomy. Older embryologists, such as Carl Siebold and Ernst Haeckel, who had put their work in an evolutionary or environmental context, despaired that "the next generation of 'scientific zoologists' will only know *cross-sections* and *stained* tissues, but neither the *entire* animal nor its mode of life" (Haeckel, 1881). They were astonished that these younger investigators cared not at all for studying the living embryo in its natual habitat.* Siebold excused these excesses by noting that publication demands on these young scientists were forcing them to do research that could be performed in a few months rather than the years it took for his type of investigation (Nyhart, 1995). When Wilhelm Roux attempted to ally experimental embryology with physiology, he postulated that development was caused by internal factors, especially those inside the nucleus. Experimental embryology moved away from environmental explanations and concentrated on those forces within the fertilized egg that enable the embryo to develop. This has been the general direction of developmental biology.

Now, with new interest in the relationship between development and evolution and with dismay over the loss of organismal diversity and the effects of environmental pollutants, there is renewed concern about the regulation of development by the environment (see Weele, 1995). Some people are convinced that "DNA provides the programme which controls the development of the embryo" (Wolpert, 1991) or that everything that is needed to form the embryo is within the fertilized egg. However, there are numerous examples (and *Homo sapiens* provides some of the best) wherein the *environment* plays a critical role in determining the organism's phenotype. We have already discussed such environmental regulation of development when we encountered sex determination in *Bonellia, Crepidula,* and many reptiles (see Chapter 20). The genetic ability to respond to such environmental factors has to be inherited, of course, but it is the environment that can give different phenotypes from the same nuclear genotype in these cases.

■ ENVIRONMENTAL REGULATION OF NORMAL DEVELOPMENT

Environmental cues used by organisms to complete their development

Certain environmental cues, such as a 1G gravitational field and a 0.85 percent saline ocean, can be utilized during development. Thus, it is not surprising that many eggs use gravity as a force to ensure the polarity of their oocytes, and the perturbation of gravity can disrupt development in frogs and birds (Pflüger, 1883; Born, 1884). Similarly, we saw in Chapter 4 that sea urchin oocytes (and no doubt the oocytes of many other species) use sodium ions from the seawater to substitute for hydrogen ions and help activate the egg (Jaffe, 1980). Mammalian embryos are intimately connected to their food, oxygen, and ion sources during their entire prenatal development. In these instances, the cues for normal development are not in the oocyte, but are assumed to be present in the environment in which the egg develops.

Larval Settlement

The inclusion of environmental cues into normal development occurs during the settling of marine larvae. Here, the cues may not be universal, but they need to be part of the environment if further development is to occur. A planktonic larva often needs to settle near a source of food or a firm substrate on which to metamorphose. If prey or anchors give off soluble molecules, these molecules can be used by the larvae as cues to begin their settle-

---

*These fears and the rhetoric expressing them are remarkably similar to those of today's older developmental biologists despairing that the younger investigators are only gene cloners having no knowledge of the structure of whole embryos (see Nyhart, 1995).

Table 21.1 Specific settlement substrates of molluscan larvae

Molluscan species	Substrate
GASTROPODA (SNAILS, NUDIBRANCHS)	
Nassarius obsoletus	Mud from adult habitat
Philippia radiata	*Porites lobata* (a cnidarian)
Adalaria proxima	*Electra pilosa* (a bryozoan)
Doridella obscura	*Electra crustulenta* (a bryozoan)
Phestilla sibogae	*Porites compressa* (a cnidarian)
Rostanga pulchra	*Ophlitaspongia pennata* (a sponge)
Trinchesia aurantia	*Tubularia indivisa* (a cnidarian)
Elysia chlorotica	Primary film of microorganisms from adult habitat
Haminoea solitaria	Primary film of microorganisms from adult habitat
Aplysia californica	*Laurencia pacifica* (a red alga)
Aplysia juliana	*Ulva* spp. (green algae)
Aplysia parvula	*Chondrococcus hornemanni* (a red alga)
Stylocheilus longicauda	*Lyngbya majuscula* (a cyanobacterium)
Onchidoris bilamellata	Living barnacles
AMPHINEURA (CHITONS)	
Tonicella lineata	*Lithophyllum* sp. and *Lithothamnion* sp. (red algae)
LAMELLIBRANCHIA (BIVALVES)	
Teredo sp.	Wood
Bankia gouldi	Wood
Mercenaria mercenaria	Clam liquor; sand
Placopecten magellanicus	Adult shell; sand; etc.
Mytilus edulis	Filamentous algae; other nonbiological silk material
Crassostrea virginica	Shell liquor; body extract; "shellfish glycogen"

ment. In molluscs, there are often very specific cues for settlement (Table 21.1). Most nudibranch (sea slug) larvae undergo metamorphosis only if triggered by living adult prey (which differs from species to species). In some cases, the soluble product of the prey that triggers metamorphosis has been identified (Hadfield, 1977). The larva of the shipworm *Teredo navalis* is induced to settle by compounds released by wood, and soluble material eluted from oyster shells induces the settlement of oyster larvae.*

The red abalone *Haliotis rufescens* has larvae that only settle when they physically contact coralline red algae. A brief contact is all that is required for the competent larvae to stop swimming and begin metamorphosis. The chemical agent responsible for this change has not yet been isolated, but recognition of an algal peptide induces metamorphosis in competent larvae. Larvae that are not competent to induce metamorphosis do not appear to have this receptor. The receptor is thought to be linked to a G protein similar

---

*In 1880, William Keith Brooks, an embryologist at Johns Hopkins University (and thesis adviser to T. H. Morgan, E. B. Wilson, R. G. Harrison, and E. G. Conklin), was asked to help the ailing oyster industry of Chesapeake Bay. For decades, oysters had been dredged from the bay, and there had always been a new crop to take their place. But recently, each year brought fewer oysters. What was responsible for the decline? Experimenting with larval oysters, Brooks discovered that the American oyster (unlike its better-studied European cousin) needed a hard substrate on which to metamorphose. For years, oystermen had thrown the shells back into the sea, but with the advent of suburban sidewalks, the oystermen were selling the shells to the cement factories. Brooks's solution: throw the shells back into the bay. The oyster population responded, and the Baltimore wharves still sell their descendants.

to those found in vertebrates, and the activation of this G protein may be necessary for inducing larval settlement and metamorphosis (Morse et al., 1984; Baxter and Morse, 1992; Degnan and Morse, 1995).

Food is not the only cue used in larval settling. The larvae of the black-fly, for instance, cling to hard surfaces in streams and passively feed from the suspended particles in the flow. These larvae actively seek areas of high stream velocity. Once in a high-velocity zone, these larvae are relatively immune from predation by flatworms. In laboratory experiments (Hansen et al., 1991), flatworms could not capture blackfly larvae when flow rates were greater than 35 cm/sec. The reason is probably that for flatworms to eat their prey, they must raise their head off the surface. This exposes their frontal surface to the flow and reduces the amount of surface area adhering to the substrate. Thus, at high stream velocities, flatworms are at risk of being washed downstream if they try to eat. So the blackflies are able to survive, metamorphose, and annoy another season's campers.

Blood Meals

In many mosquitoes, egg production is triggered by a blood meal. (In *Drosophila*, the environmental cue for egg production appears to be the photoperiod.) Only female mosquitoes bite, and they make no vitellogenin before this meal. In *Aedes aegypti*, the digested products of the blood meal stimulate the brain to secrete **egg development neurosecretory hormone** (**EDNH,** also known as ovarian ecdysteroidogenic hormone, OEH). This stimulates the ovary to make ecdysteroids, which instruct the fat body cells to make vitellogenin for the oocytes (Fallon et al., 1974; Hagedorn, 1983; Borovsk et al., 1990). The vitellogenin is critical for egg production. Thus, without the blood meal, there is no vitellogenin and no eggs (Figure 21.1).

In the blood-sucking bug *Rhodinus prolixus*, adult females produce a new batch of eggs each time they drink blood. This blood meal serves two functions. The blood proteins supply the amino acids needed for vitellogenin synthesis, and the physical stretching of the abdomen by the blood initiates the endocrine stimuli that activates juvenile hormone secretion by the corpora allata. The juvenile hormone stimulates vitellogenin synthesis in the ovary and fat body (see Nijhout, 1994). Moreover, the stretching caused by the single blood meal induces the larval molt. If this bug takes many small meals, it will survive, but it will not molt or grow. In these instances, mammals have been used for part of the insect's development.

Developmental Symbiosis

In some of the above examples, the development of one individual is made possible by the presence of another individual of a different species. In some organisms, this relationship has become symbiotic (Sapp, 1994). Here, these symbionts have been so tightly integrated into the host organism that the host cannot develop without them. The adult squid *Euprymna scolopes* is equipped with a light organ composed of sacs containing the luminous bacteria *Vibrio fischeri*. The juvenile squid, however, does not contain these light-emitting symbionts; nor does it have a structure to house them. Rather, the squid acquires the bacteria from the seawater pumped through its mantle cavity. The bacteria bind to a ciliated epithelium that extends into this cavity. These bacteria induce the death of these cells, their replacement by a nonciliated epithelium, and the differentiation of the surrounding epithelial cells to become storage receptacles for the bacteria (Figure 21.2; McFall-Ngai and Ruby, 1991; Montgomery and McFall-Ngai, 1995).

Symbiosis between the egg masses and photosynthetic algae are critical for the development of several species. The supply of oxygen limits the rate

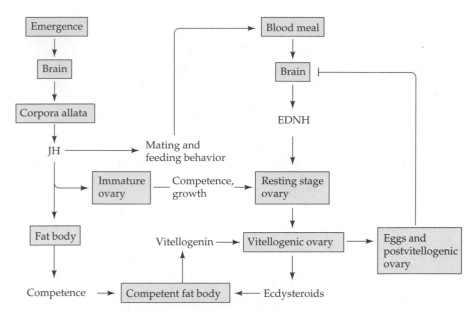

Figure 21.1
Flow chart of interactions enabling the production of eggs in the mosquito *Aedes aegypti*. (After Hagedorn, 1983; Nijhout, 1994.)

of development when eggs are packed together in tight masses, and the development of those embryos on the inside of the cluster is retarded compared with those near the surface (Strathmann and Strathmann, 1995). While there is a steep gradient of oxygen from the outside of the cluster to deep within it, the embryos seem to have gotten around this problem by coating themselves with a thin film of photosynthetic algae. In clutches of amphibian and snail eggs, photosynthesis from algal "fouling" enables net oxygen production in the light, while respiration exceeds photosynthesis in the dark (Bachmann et al., 1986; Pinder and Friet, 1994; Cohen and Strathmann, 1996). Thus, the algae rescue the eggs by photosynthesis.

An even tighter link between morphogenesis and symbiosis is afforded by the leafhopper *Euscelis incisus*. Here, the symbiosis occurs within the egg. There are symbiotic bacteria in this species that are within the egg cytoplasm and which are transferred through the generations, just like mitochondria. These bacteria have become so specialized that they can only multiply inside this organism's cytoplasm, and the host embryo has become so dependent

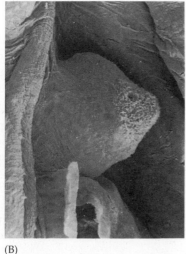

(A)   (B)

Figure 21.2
Scanning electron micrograph of a light organ primordium of a 3-day-old juvenile squid *E. scolopes*. (A) Light organ in an uninfected juvenile. (B) Light organ of a juvenile infected with the symbiotic *V. fischeri* bacteria. Regression of the epithelium is obvious in (B). (After Montgomery and McFall-Ngai, 1995; photographs courtesy of M. McFall-Ngai.)

Figure 21.3
Microbial symbionts are necessary for gut formation in the leafhopper *Euscelis incisus*. (A) Control embryo with symbionts has normal gut formation. (B) Abnormal gut-deficient embryo forming when antibiotics have eliminated most of the symbiotic bacteria from the egg. (After Schwemmler, 1974; photographs courtesy of W. Schwemmler.)

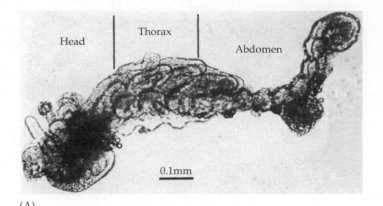

(A)

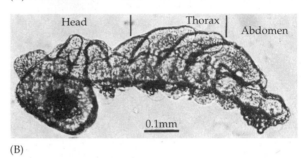

(B)

on the bacterium that it cannot complete embryogenesis without it. In fact, it is thought that the bacterial symbionts are essential for the formation of the embryonic gut. If the bacteria are surgically or metabolically removed (by feeding antibiotics to the larvae or adults), the bacteria can be eliminated from the developing eggs. These symbiont-free oocytes develop into embryos that lack an abdomen (Figure 21.3; Sander, 1968; Schwemmler, 1974, 1989). The endosymbiont may be secreting a factor that enters into the egg cytoplasm.

There is even a developmental symbiosis in the mammalian gut. Bacteria colonize the gut from the moment of birth, and the ecological succession in the human intestine progresses through a series of colonizations involving over 400 bacterial species. The intestinal epithelial cells of mice leading a germ-free existence fail to synthesize certain mRNAs that encode particular glycosylation enzymes (Bry et al., 1996). However, if a particular strain of bacteria is allowed to colonize their guts, these microbes induce the mRNA to become expressed. [env1.html]

Predictable environmental differences as cues for development

Seasonality and Sex: Aphids and Volvox

As already mentioned, several species of parthenogenetic aphids have a fascinating lifestyle wherein the hatched egg gives rise to several generations of asexually reproducing females. During the autumn, however, a particular type of female is produced whose eggs can give rise to both males and sexual females. These sexual forms mate, and the egg that forms is able to survive the winter. When it hatches, a new generation of asexual females is produced. One of the great mysteries of this type of parthenogenesis was solved in 1909 by Thomas Hunt Morgan (before he started working on fruit flies). Morgan analyzed the chromosomes of the hickory aphid through several

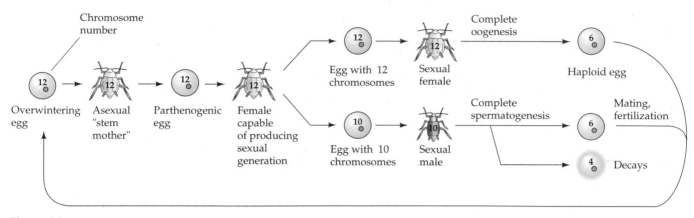

Figure 21.4
Chromosomal changes during the life cycle of the phylloxeran aphid. Fall weather
induces the production of males and females, which mate to produce the overwintering egg.

generations (Figure 21.4). He found that the diploid number of the female
aphids is 12. During oogenesis, only one polar body is extruded from the developing ovum, so the diploid number of 12 is retained in the egg. This egg
develops parthenogenetically without being fertilized. In the females that
can give rise to eggs that become male or female, a modification of this oogenesis occurs. In the female-producing eggs, six chromosome pairs enter
the sole polar body. The diploid number of 12 is thereby retained. In male-
producing eggs, however, an extra chromosome pair enters the polar body.
The male diploid number is 10. These males and females are sexual and
have complete meiotic divisions. The female produces oocytes with a haploid set of 6 chromosomes. The males, however, divide their 10 chromosomes to produce some sperm with a haploid number of 4 chromosomes
and other sperm with a haploid number of 6 chromosomes. The sperm with
4 chromosomes degenerate. The sperm with 6 chromosomes fertilize the egg
with its 6 chromosomes to restore the diploid chromosome number 12.
When the egg hatches after the winter, it is a female.

That solved one riddle. The riddle of how the autumn weather regulates
whether the female is sexual or parthenogenetic or whether the organism is
winged or apterous is still unsolved. Similarly, we do not know what regulates whether the diploid oocyte gives rise to male- or female-producing
eggs. Moreover, the environmental factors are used differently by different
species. Figure 21.5 shows one type of life cycle found in aphids. In the hickory aphids and *Megoura viciae*, there is an alternation of sexual and asexual
generations. In *Megoura*, temperature determines the sex early in development (the extreme temperatures favoring the production of females). In female development, photoperiod and temperature determine whether the female will reproduce sexually or parthenogenetically, and a combination of
temperature and population density will determine whether the female is
winged or wingless (Beck, 1980). It appears that juvenile hormone controls
the parthenogenetic/sexual switch (addition of juvenile hormone to adults
producing sexual offspring causes them to have parthenogenetic offspring)
and inhibits the formation of wings (Hardie, 1981; Hardie and Lees, 1985).
But it is not known how the environmental changes become transformed
into titers of juvenile hormone or how the fall weather or sunlight causes the
differential movement of chromosomes into the polar body.

Figure 21.5
Environmental effects on the life cycle of the aphid *Megoura viciae*. (A) Alternation of sexual and asexual generations, wherein the sexual generation is produced in the autumn. (B) Developmental alternatives provided by environmental factors in the life cycle of *Megoura*. (A after Nijhout, 1994; B after Beck, 1980.)

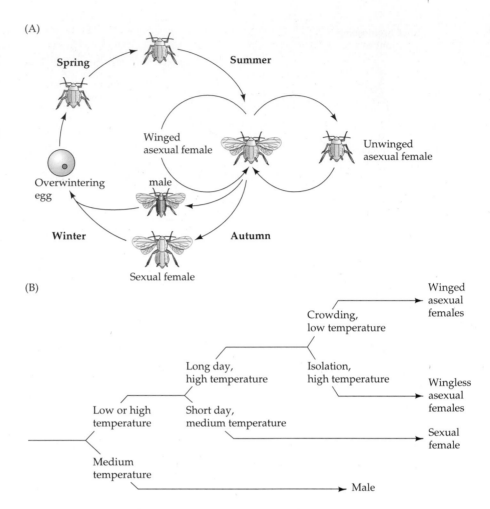

In Chapter 1, we discussed the volvox life cycle and its dependency on temperature. Here, too, temperature is responsible for switching asexual forms of the organism into sexual forms. The asexually reproducing female gives rise to offspring that make either sperm or eggs. The result of this fertilization is the zygote whose outer coat can protect it from desiccation and cold as the pond dries up and winter comes.

Diapause

Many species of insects have evolved a strategy called **diapause.** Diapause is a suspension of development that can occur at the embryonic, larval, pupal, or adult stage, depending on the species. In some species, diapause is facultative and occurs only when induced by environmental conditions; in other species, diapause has become an obligatory part of the life cycle. The latter is often seen in temperate-zone insects, where diapause is induced by changes in the photoperiod (the relative lengths of day and night). The day length when 50 percent of the population has entered diapause is called the **critical day length,** and it is usually quite sudden (Figure 21.6). Insects entering diapause when the day length falls below this threshold are called long-day insects. Those insects that develop normally when there are only a few hours of sunlight and which enter diapause when exposed to longer days are called short-day insects. The critical day length is a genetically determined property (Danilevskii, 1965; Tauber et al., 1986).

Diapause is not a physiological response brought about by harsh conditions. Rather, it is brought about by token stimuli that presage a change in the environment, beginning before the severe conditions actually arise. Diapause is especially important for temperate-zone insects, enabling them to survive the winter. Embryos of the silkworm moth *Bombyx mori* overwinter as embryos, entering diapause just before segmentation. The gypsy moth *Lymantia dispar* initiates its diapause as a fully formed larva, ready to hatch as soon as diapause ends. Other insects experience diapause as eggs, pupae, or even adults.

In the silkworm *Bombyx*, embryonic diapause appears to be regulated by **diapause hormone,** a 24-amino-acid peptide that is produced in the subesophageal ganglion (Fukuda, 1952; Hasegawa, 1952). This hormone acts on the maturing oocytes in the pupal stage and causes their development to stop later, once the embryo has reached about 12,000 cells (Kitazawa et al., 1963). Larval diapause, however, appears to be controlled by inhibition of PTTH production (see Chapter 19). This prevents the larvae from molting and entering pupation. In many butterflies, inhibition of PTTH is due to a continued elevated titer of juvenile hormone. Similarly, the lack of PTTH and ecdysone secretion once pupation has occurred will cause diapause during this part of development. Diapausing pupae can be reactivated by adding back 20-hydroxyecdysone. However, under normal conditions, the brain of a diapausing pupa (such as that of the moth *Hyalophora*) is activated by the exposure to cold weather for a particular duration. Moth pupae kept in warm conditions will remain in diapause until they die (see Nijhout, 1994). The mechanism by which these temperature and day length changes regulate hormone production remains to be elucidated. [env2.html]

Phenotypic plasticity: Polyphenism and reaction norms

The ability of an individual to express one phenotype under one set of circumstances and another phenotype under another set of environmental conditions is called **phenotypic plasticity.** There are two main types of phenotypic plasticity: polyphenism and reaction norms. **Polyphenism** refers to discontinuous ("either/or") phenotypes elicited by the environment. Migratory locusts, for instance, exist in two mutually exclusive forms: the short-winged, uniformly colored, solitary phase and the long-winged, brightly colored, gregarious phase. The environment (mainly population density) determines which morphology the young locust will take (see Pener, 1991). Similarly, the nymphs of planthoppers can develop in two ways, depending on their environment. High population density and certain plant communities lead to the production of migratory insects, wherein the third thoracic segment produces a large hindwing. Low population densities and other food plants lead to the development of flightless plantsuckers, wherein the third throracic segment develops into a haltere-like vestigial wing (Figure 21.7; Raatikainen, 1967; Denno et al., 1985). The seasonal coat color changes in arctic animals are other examples of polyphenism.*

In certain cases, the genome encodes a potential *range* of phenotypes, and the environment selects the phenotype that is usually the most adaptive. For instance, constant and intense labor can make our muscles grow larger; but there is a genetically defined limit to how much hypertrophy is

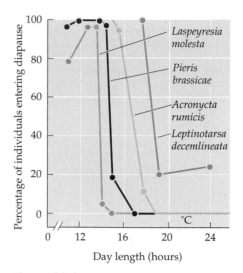

Figure 21.6
The photoperiodic response of long-day insects, which are induced to enter diapause when the daylight hours falls below a certain level. The four species shown here, (*Laspeyresia molesta, Pieris brassicae, Acronycta rumicis,* and *Leptinotarsa decemlineata*) each leaves diapause when daylight is 14–17 hours. (After Danilevskii, 1965.)

---

*Although seasonal polyphenism is usually considered adaptive, there are times when it does not increase the fitness of the organism. For instance, the photoperiod can cause a hare's color to change from brown to white, but if it doesn't snow, the hare will be conspicuous against the dark background.

Figure 21.7
Composite diagram showing the short-winged (left) and long-winged (right) forms of the planthopper *Prokelisia marginata*. The long-winged form is an excellent flier; the short-winged form is flightless. (After Denno et al., 1985.)

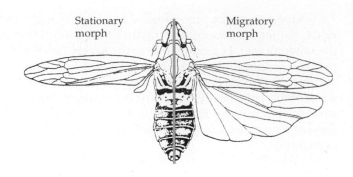

Stationary morph   Migratory morph

possible. Similarly, the microhabitat of a young salamander can cause its color to change (again, within genetically defined limits). This continuous range of phenotypes expressed by a single genotype across a range of environmental conditions is called the **reaction norm** (Woltereck, 1909; Schmalhausen, 1949; Stearns et al., 1991). The reaction norm is thus a property of the genome and can also be selected. Different genotypes will be expected to differ in the direction and amount of plasticity that they are able to express (Gotthard and Nylin, 1995; Via et al., 1995). The extent to which reaction norms can be inherited forms the basis for the evolution of phenotypic plasticity.

Seasonal Polyphenism in Butterflies

A dramatic example of polyphenism occurs in the moth *Nemoria arizonaria*. This moth has a fairly typical life cycle. Eggs hatch in the spring, and the caterpillars feed on young oak flowers (catkins). These larvae metamorphose in the late spring, mate in the summer, and produce another brood of caterpillars on the oak trees. These caterpillars eat the oak leaves, metamorphose, and mate. Their eggs overwinter to start the cycle over again next spring. What is remarkable is that the caterpillars that hatch in the spring look nothing like their progeny that hatch in the summer (Plate 18). The caterpillars that hatch in the spring and eat oak catkins are yellow-brown, rugose, and beaded, resembling nothing else but an oak catkin. They are magnificently camouflaged against predation. But what of the caterpillars that hatch in the summer, after all the catkins are gone? They, too, are well camouflaged, appearing like year-old oak twigs. What controls this? By doing reciprocal feeding experiments, Greene (1989) was able to convert the spring morphs into the summer forms by feeding them oak leaves. The reciprocal experiment did not turn the later morphs into catkin-like caterpillars. Thus, it appears that the catkin form is the "default state" and that something induces the twiglike morphology. This substance is probably a tannin that is concentrated in the oak leaves as they mature.

Another example of seasonal polyphenism is the European map butterfly, *Araschnia levana*, which has two seasonal phenotypes so different that Linnaeus classified them as two different species (Weele, 1995). The spring morph is bright orange with black spots, while the summer form is mostly black with a white band (Figure 21.8; Plate 29). The change from spring to summer morph is controlled by changes in both day length and temperature during the larval period. These are seen to control the timing of ecdysone release, which initiates the last metamorphic molts (Shapiro, 1976; Koch and Buchmann, 1987). When diapausing pupae are injected with 20-hydroxyecdysone so that they restart development within 3 days after pupation, the summer forms emerge. If the injection is done more than 10 days after pupation, the spring forms are produced.

Figure 21.8
Seasonal polyphenism in the butterfly *Araschnia laevana*. (A) The summer morph that emerges from the nondiapausing pupa. (B) The orange and brown spring morph that emerges from the diapausing pupa. (See Plate 29 for color photographs.) (Photographs courtesy of H. F. Nijhout.)

Throughout much of the Northern Hemisphere, one can see a polyphenism in the *Colias* and *Pieris* butterflies (the cabbage whites and sulphurs) between those that eclose during the long days of summer and those that eclose at the end of the season, in the short days of autumn. The hindwing pigments of the short-day forms are darker than those of the long-day butterflies. This has a functional advantage during the colder months of autumn; the darker short-day butterflies use their pigments to heat themselves up between flights. The darker pigments absorb sunlight more efficiently, raising the body heat faster than lighter pigments (Shapiro, 1968, 1978; Watt, 1968, 1969; Hoffmann, 1973; see Nijhout, 1991). [env3.html]

In tropical parts of the world, there is often a wet and a dry season. In Africa, the Malawian butterfly *Bicyclus anynana* has a polyphenism that is adaptive to seasonal changes. The cold-and-dry-season morph is cryptic, looking like the dead brown leaves of its habitat. The hot-and-wet-season morph is more active, and it has ventral hindwing eyespots that deflect attacks from predatory birds or lizards (Figure 21.9). The determining factor appears to be the temperature during pupation. Low temperatures produce the dry-season morph; high temperatures produce the wet-season morph (Brakefield and Reitsma, 1991). The development of butterfly eyespots begins in late larval stages, when the transcription of the *Distal-less* gene is restricted to a small focus that will become the center of each eyespot. During the early pupal stage, *Distal-less* expression is seen in a wider area, and this is thought to constitute the activating signal that determines the size of the spot. Last, the cells receiving the signal determine the color they will take. The seasonal *Bicyclus* morphs appear to diverge at the later stages of signal activation and color differentiation (Figure 21.10; Brakefield et al., 1996).

Figure 21.9
The two seasonal morphs of the *Bicyclus anynana* butterfly from Malawi. (A) Dry-season form that blends into dead, brown, leaf litter. (B) Wet-season form with conspicuous ventral hindwing eyespots. The wet-season form can be mimicked by raising larvae at high (23°C) temperatures; whereas larvae grown in lower temperatures (17°C, approximating the temperatures in the transition to the dry season) develop into the dry-season morph. (After Brakefield et al., 1996; photographs courtesy of S. Carroll and P. Brakefield.)

Figure 21.10
Developmental stages leading to the formation of eyespots.
(A) *Distal-less* gene expression occurs in regions of the wing imaginal disc where eyespots have the potential to form.
(B) Foci of *Distal-less* expression are stabilized in particular regions of the wing. (C) In the pupa, *Distal-less* foci expand.
(D) The surrounding cells respond to the signal by producing particular pigments, depending on their distance from the focus and their position in the wing. In *Bicyclus*, the two morphs are indistinguishable until the signaling stage (C).
(After Brakefield et al., 1996.)

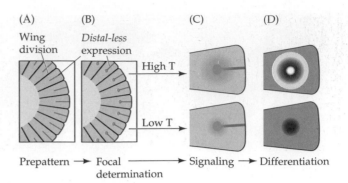

Nutritional Polyphenism

Not all polyphenism is controlled by the seasons. In bees, the size of the female larva at its pupal molt determines whether the individual is to be a worker or a queen. A larva fed nutrient-rich "royal jelly" retains the activity of its corpora allata during its last instar stage. The juvenile hormone secreted by these organs delays pupation, allowing the resulting bee to emerge larger and (in some species) more specialized in her anatomy (Figure 21.11A; Brian, 1974, 1980; Plowright and Pendrel, 1977). The juvenile hormone levels of larvae destined to become queens is 25 times the titer of larvae destined to become workers, and application of juvenile hormone onto worker larvae can transform them into queens as well (Wirtz, 1973; Rachinsky and Hartfelder, 1990).

Similarly, ant colonies are predominantly female, and the females can be extremely polymorphic (Figure 21.11A). The two major types of females are the worker and the gyne. The gyne is a potential queen. In more specialized species, a larger worker, the soldier, is also seen. In *Pheidole bicarinata*, these castes are determined by the levels of juvenile hormone in the developing larvae. Larvae given protein-rich food have an elevated juvenile hormone titer which causes an abrupt developmental switch that "reprograms" the

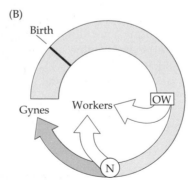

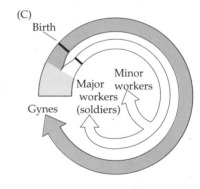

Figure 21.11
(A) Photograph of remarkable dimorphism of worker (left) and queen ant (right) in the species *Pheidologeton diversus*. The two are sisters, but one has been fed such that its larvae keeps growing and metamorphoses eventually into a fertile "queen."
(B,C) Gyne (queen) and worker formation in ants. Lightly colored areas represent bipotentiality to become either workers or gynes. The circled N represents a nutritional switch controlled by the larva's environment. (B) *Myrmica rubra*, wherein only the larvae that overwinter (OW) remain bipotential. In the last instar, the nutritional switch determines caste. (C) *Pheidole pallidula*, wherein the queen controls gyne determination through hormones that act during embryogenesis.
(Copyrighted photograph courtesy of Mark W. Moffett at the National Geographic Society; B and C after Wheeler, 1986.)

size at which the larvae will begin metamorphosis. This causes a large and discontinuous size difference between the soldier and worker castes, the head and jaws growing at a faster rate than the rest of the body. This reprogramming also involves changes in gene activity, since the cuticular proteins of the workers and soldiers are different (Passera, 1985; Wheeler, 1991).

In different species, caste determination can be environmental, hormonal, or a combination of both. These developmental patterns of caste determination have been analyzed by Diana Wheeler (1986, 1991) and are summarized in Figure 21.11B,C. In most species, ant larvae are bipotential until near pupation. In *Myrmica rubra*, only larvae that overwinter remain bipotential. After winter, the queen stimulates workers to *under*feed the last-instar larvae. This means that as long as there is a queen, no new queens can result. If the larvae are fed, they can becomes gynes. Thus, larvae remain bipotential until late in their last instar. In other species such as *Pheidole pallidula*, the queen controls gyne formation through chemicals that act during embryogenesis, so that no new queens are formed. However, the workers remain bipotential and can become minor or major workers, depending on nutrition.

Environmental-Dependent Sex Determination

There are many species in which the environment determines whether the individual is to be male or female. The temperature dependence of sex determination in fish and reptiles has provided the best-studied cases. Figure 21.12 displays the major patterns of temperature-dependent sex determination in reptiles. This type of environmental sex determination has advantages and disadvantages. One advantage is that it probably gives the species the benefits of sexual reproduction without tying the species to a 1:1 sex ratio. In crocodiles, where temperature extremes produce females while moderate temperatures produce males, the sex ratio may be as great as 10 females to each male (Woodward and Murray, 1993). The major disadvantage of temperature-dependent sex determination may be in its narrowing the temperature limits within which a species can exist. This would mean that thermal pollution (either locally or by "global warming") could conceivably eliminate a species in a given area (Janzen and Paukstis, 1991). Ferguson and Joanen (1982) speculate that dinosaurs may have had temperature-dependent sex determination and that their sudden demise may have been caused by a slight change in temperature that created conditions wherein only males or females hatched from their eggs.

Charnov and Bull (1977) have argued that environmental sex determination would be adaptive in certain habitats characterized by patchiness, having certain regions where it is more advantageous to be male and other regions where it is more advantageous to be female. Conover and Heins (1987)

Figure 21.12
Patterns of temperature-dependent sex determination. In the first three panels, different temperatures give a predominance of males or females. In the last panel, temperature does not play a role. (After Bull, 1980.)

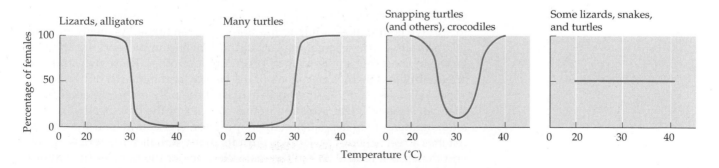

Figure 21.13
Relationship between temperature and sex ratio F:(F+M) during the period of sex determination in *Menidia menidia*. In those fish collected from the northernmost portion of its range (Nova Scotia), temperature had little effect on sex determination. When embryos were collected from fish at more southerly locations (especially from Virginia through South Carolina), the environment had a large effect. (After Conover and Heins, 1987.)

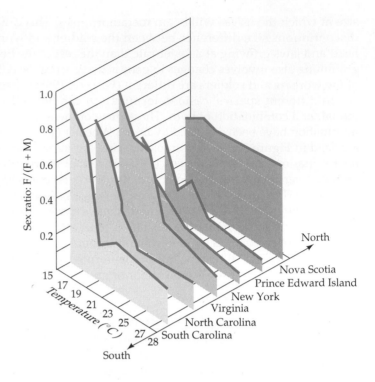

provide evidence that in certain fish, females benefit from being larger, since size translates into higher fecundity. It is an advantage to be born early in the breeding season if you are a female *Menidia*, since you would have a longer feeding season and thus grow larger. In the males, size is of no importance. Conover and Heins showed that in the southern range of *Menidia*, females are indeed born early in the breeding season. Temperature appears to play a major role. However, in the northern reaches of its range, the same species shows no environmental sex determination. Rather, a 1:1 ratio is generated at all temperatures (Figure 21.13). The authors speculate that the more northern populations have a very short feeding season, so there is no advantage for a female to be born earlier. Thus, this species of fish has environmental sex determination in those regions where it is adaptive and genotypic sex determination in those regions where it is not. Here again, one sees that the environment can induce sexual phenotype, or sexual phenotype can be a property of the genome, as it is with most mammals.

Unpredictable environmental factors controlling animal development

Most studies of adaptations concern the roles that adult structures play in enabling the individual to survive in otherwise precarious or hostile environments. However, the embryo, too, has to survive in its habitat, and it has to survive before those adult adaptations are even made. As mentioned earlier, protective coloration of the larva is one such example, and the ability of larvae to injest foods toxic to their predators is another. These two strategies are exemplified by the caterpillars of the viceroy and monarch butterflies, respectively (see page 733). Temperature isn't the only environmental factor that can effect sex determination in fish. The sex of the blue-headed wrasse, a Panamanian fish, depends on the population it encounters. If the embryo reaches a reef where a male lives with its many females, the wrasse grows and mates as a female. When the male dies, one of the females (usually the largest) becomes a male. Within a day, its ovaries shrink and its testes grow. If the same embryo had landed on a reef that had no males or that had territory undefended by a male, the embryo would develop into a male wrasse (Warner, 1993).

Inducible Defenses Against Predation

Some embryos are protected from environmental conditions by materials secreted into or around the egg. In other cases, the environment induces a particular developmental pathway rather than allowing the normal pathway to be followed. In the *Nemoria* caterpillar, the diet alters the phenotype and protects the individual from predation. Some animals have taken this a step further: The development of a juvenile is changed by chemicals released by the predator itself, enabling the juveniles to better escape those same predators. This is sometimes called **predator-induced defense** (or **predator-induced polyphenism**).

To demonstrate predator-induced defense, one has to show that the phenotypic change is caused by the predator (usually from soluble chemicals released by the predator) and that the phenotypic modification increases the fitness of its bearers when the predator is present (Adler and Harvell, 1990).* For instance, several *Daphnia* and rotifer species will alter their morphology when they develop in pond water in which their predators were cultured (Figure 21.14; Dodson, 1989; Adler and Harvell, 1990). The predatory rotifer *Asplanchna* releases into its water a soluble compound that induces the eggs of a prey species, *Keratella slacki*, to develop into individuals with slightly larger bodies, but with anterior spines 130 percent longer than they would otherwise develop. These changes make them more difficult to eat. The snail *Thais lamellosa* develops a thickened shell and a "tooth" in its aperture when exposed to the effluent of the crab species that preys on it. In a mixed population, crabs won't attack the thicker snails until over 50 percent of the normal snails are devoured (Palmer, 1985).

Predator-involved polyphenism is not constrained to invertebrates. Mc-Collum and Van Buskirk (1996) have shown that in the presence of their predators, the tail fin of the gray treefrog *Hyla chrysoscelis* grows larger and

*The phenomenon of **cyclomorphosis,** in which there is a cyclical variation of morphology in certain species of *Daphnia* (Woltereck, 1909), has not been correlated with any particular predator. It may be due to other factors (Dodson, 1989).

Figure 21.14
Predator-induced polyphenism. Typical (upper row) and predator-induced (lower row) morphs of various organisms. The numbers beneath each column represents the percentage of organisms surviving predation when both induced and uninduced individuals were presented with predators (in various assays). (Data from Adler and Harvell, 1980 and references cited therein.)

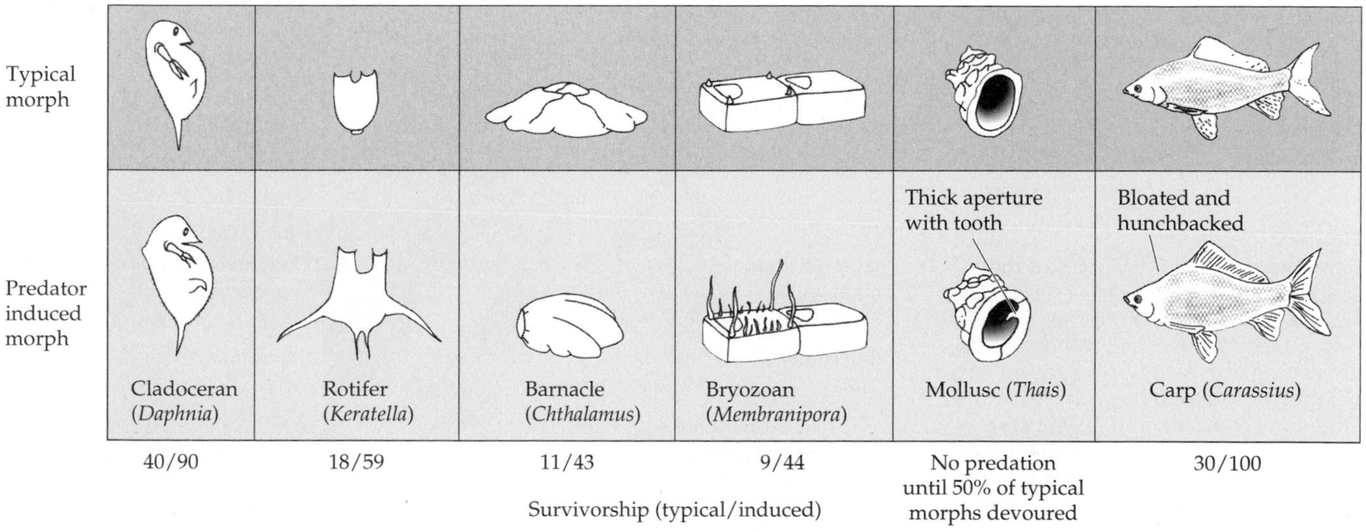

Typical morph					
				Thick aperture with tooth	Bloated and hunchbacked
Predator induced morph					
Cladoceran (*Daphnia*)	Rotifer (*Keratella*)	Barnacle (*Chthalamus*)	Bryozoan (*Membranipora*)	Mollusc (*Thais*)	Carp (*Carassius*)
40/90	18/59	11/43	9/44	No predation until 50% of typical morphs devoured	30/100

Survivorship (typical/induced)

becomes bright red. This allows the tadpole to swim away faster and to deflect strikes toward the tail region. The carp *Carassius carassius* is able to respond to the presence of a predatory pike only if the pike had already eaten fish. The carp grows into a pot-bellied, hunched-back morph that will not fit into the pike's jaws. As in most predator-induced defenses, there is a trade-off (otherwise one would expect the induced morph to become the normal phenotype). In this case, the induced morphology puts a drag on swimming conditions, and the fatter fish cannot swim as efficiently (Brönmark and Pettersson, 1994). Figure 21.14 shows the typical and predator-induced morphs for several species. In each case, soluble filtrate from water surrounding the predator is able to induce these changes. As Figure 21.14 shows, the induced morph is more successful at surviving the predator. [env4.html]

Phenotypic Plasticity and Changes in the Environment

The spadefoot toad, *Scaphiopus couchii*, has a remarkable life cycle. The toads are called out from hibernation by the thunder that accompanies the first spring storm in the Sonoran desert. (Unfortunately, motorcycles will produce the same sounds, causing these toads to come out from hibernation and die in the scorching Arizona sunlight.) The toads breed in the temporary ponds caused by the rain, and the embryos develop quickly into larvae. After the larvae metamorphose, the new toads then return to the desert, burrowing into the sand until the next year's storms bring them out.

The desert ponds are ephemeral pools that can dry up quickly or persist depending on the initial depth and the frequency of the rainfall. One might think that there are only two alternative scenarios confronting such a toad embryo: either (1) the pond persists until you have time to metamorphose and you live, or (2) the pond dries up before metamorphosis, and you die. These toads (and numerous other amphibians), however, have evolved a third alternative. The time of metamorphosis is controlled by the pond. If the pond does not dry out, development continues at its normal rate, and the algae-eating tadpoles eventually develop into juvenile spadefoot toads. However, if the pond is drying, overcrowding occurs, and some of the tadpoles embark on an alternative developmental pathway. They develop a wider mouth and need more powerful jaw muscles which enable them to eat, among other things, other *Scaphiopus* tadpoles. These carnivorous tadpoles metamorphose quickly, albeit into a smaller version of the spadefoot juvenile. But they survive while the other *Scaphiopus* tadpoles perish from desiccation or ingestion by their pond-mates (Figure 21.15; Newman, 1989; 1992).

Figure 21.15
Polyphenism in the tadpoles of the spadefoot toad, *Scaphiopus couchii*. The typical morph is an omnivore, usually eating insects and algae. When ponds are drying, the carnivorous (cannibalistic) morph forms. It has a wider mouth, larger jaw muscles, and an intestine modified for a carnivorous diet. (Photograph and drawing courtesy of R. Ruibel.)

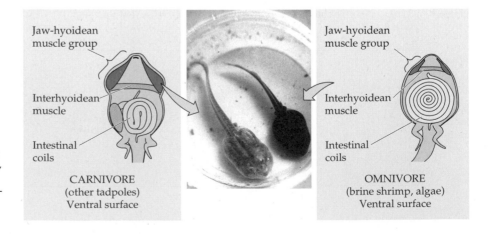

Jaw-hyoidean muscle group

Interhyoidean muscle

Intestinal coils

CARNIVORE
(other tadpoles)
Ventral surface

Jaw-hyoidean muscle group

Interhyoidean muscle

Intestinal coils

OMNIVORE
(brine shrimp, algae)
Ventral surface

Such phenotypic plasticity is also seen in echinoderm larvae. When food is scarce, the ciliated arms of the pluteus larva grow longer and increase the ability of the larva to obtain food. But this is done at a cost to the adult rudiment growing within the larva, and it takes longer for those long-armed plutei (even if they can acquire food) to metamorphose (Hart and Strathmann, 1994).

Phenotypic plasticity gives an individual the ability to respond to different environmental conditions. Different phenotypes are more fit in different environments. In spadefoot toad, the faster-developing form is more fit in the quickly drying ponds, but the slower-developing tadpoles (which develop into larger, more robust toads) are more fit in nondry conditions. There is a "trade-off" in evolving this phenotypic plasticity, but it helps ensure that there will always be some animals surviving in each condition.

Sidelights & Speculations

Genetic Assimilation

IN DISCUSSING the trade-offs between uninduced and induced morphs, we mentioned that if the induced morph did not have a significant trade-off, one would expect it to become the predominant form of the species. This was predicted independently by C. H. Waddington and I. I. Schmalhausen to explain how some species could rapidly evolve in particular directions (see Gilbert, 1994). Both were impressed by the calluses found on ostrich feet. Most mammalian skin has the ability to form calluses on those areas that abrade the ground or some other surface.* Here, the skin cells respond to friction by proliferating. While such examples of environmentally induced callus formation are widespread, the ostrich is *born* with calluses where it will touch the ground (Figure 21.16). Waddington and Schmalhausen hypothesized that since the skin cells are already *competent* to be induced by friction, they could be induced by other things as well. As ostriches evolved, a mutation appeared that enabled the skin cells to respond to a substance within the embryo. Waddington (1942) wrote:

Presumably its skin, like that of other animals, would react directly to external pressure and rubbing by becoming thicker…This capacity to react must itself be dependent upon genes…It may then not be too difficult for a gene mutation to occur which will modify some other area in the embryo in such a way that it takes over the function of external pressure, interacting with the skin so as to "pull the trigger" and set off the development of callosities.

By this transfer of induction from an external inducer to an internal inducer, a trait that had been induced by the environment became part of the genome of the organism and could be selected. Waddington called this phenomenon "genetic assimilation," while Schmalhausen (1949) called it "stabilizing selection." Both scientists had used orthodox embryology and orthodox genetics to explain examples that had been considered cases of a Lamarckian "inheritance of acquired characteristics."

The transfer of environmental stimulus to genetic stimulus might be seen in sex determination in *Menidia* and caste determination in ants. Similarly, preexisting developmental plasticity in the feeding echinoderm larvae may have bridged the transition from pluteus (feeding) larvae to larvae that lack the ciliated arms. The change in the allocation of resources between larval and juvenile structures parallels that seen where the food re-

Figure 21.16 Ventral side of an ostrich; arrows mark the calluses. (From Waddington, 1942).

serves are stored in the egg. Thus, the changes already present as adaptations to external food resources may have become genetically fixed in those species whose larvae do not need to hunt their food (Strathmann et al., 1992).

---

*And until this century, writers were recognized by the calluses on their fingers. (Thus, from observing his fingers, Sherlock Holmes correctly surmised that the red-headed man had been hired as a scrivener.)

If genetic assimilation indicates the genetic fixation of one of the phenotypes that had been adaptively expressed, then butterflies would be a good place to look for further examples. Brakefield and colleagues (1996) have shown that they could genetically fix the different morphs of the adaptive polyphenism of *Bicyclus*, and Shapiro (1976) has shown that the short-day (cold-weather) adaptive phenotype of several butterflies is the same as the single genetically produced phenotype of related species or subspecies living at higher altitudes or latitudes. One can also produce the cold-weather phenotype by incubating the larvae or pupae of warm-season butterflies in the refrigerator. [env5.html]

Genetic assimilation may play an important role in providing a bias for evolutionary change. If an organism inherits a reaction norm, the developmental pathways leading to a particular phenotype are already in place, and all that evolution need do is supply a constant initiator of those pathways. ∎

The continuing plasticity of development

The ability of an organism to monitor and respond to environmental change is critical to survival in complex habitats. Our two major sensory systems, the nervous system and the immune system, enable us to regulate our bodies developmentally in response to environmental stimuli.

The Immune System: Development in the Adult

If predator-induced polyphenism is an adaptive response to potential threats, the mammalian immune system is its highest achievement. The mammalian immune system is an incredibly elaborate mechanism for sensing and destroying materials that are foreign to the body. When we are exposed to a foreign molecule (called an **antigen**), we manufacture **antibodies** and secrete them into our blood serum (see Chapters 10 and 17 for details). These antibodies combine with the antigen to inactivate or eliminate the antigen. The basis for the immune response is summarized in the clonal selection hypothesis (Burnett, 1959). It contains five major postulates:

1. Each B lymphocyte (B cell) can make one and only one type of antibody. That is, one B cell may be making an antibody that binds to poliovirus, while a neighboring B cell might be making an antibody that binds to diphtheria toxin.
2. Each B cell will take the antibodies it makes and place them into its cell membrane with the specificity-bearing side outward.
3. Antigens are presented to the B cells (usually on the surfaces of macrophages).
4. Only those B cells that bind to the antigen can complete their development into antibody-secreting **plasma cells.** The B cells divide repeatedly, produce an extensive rough endoplasmic reticulum, and synthesize enormous amounts of antibody molecules. These antibodies are secreted into the blood.
5. The specificity of the antibody is exactly the same as that which was on the cell surface of the B cells.

The type of antibody molecule on the cell surface of the B cell is determined by chance. Out of the 10 million types of antibody proteins the cell can possibly synthesize, each B cell makes only one type. These B cells are continually being created and destroyed. However, when an antigen binds to a set of B cells, these cells are stimulated to divide and differentiate into plasma cells (that secrete the antibody) and memory cells (that populate lymph nodes and respond rapidly when exposed to the same antigen later in life) (Figure 21.17). Thus, each person's constellation of plasma cells and memory cells differs depending on which antigens he or she has encountered. Identical twins have different populations of B cell descendants in their spleens and lymph nodes.

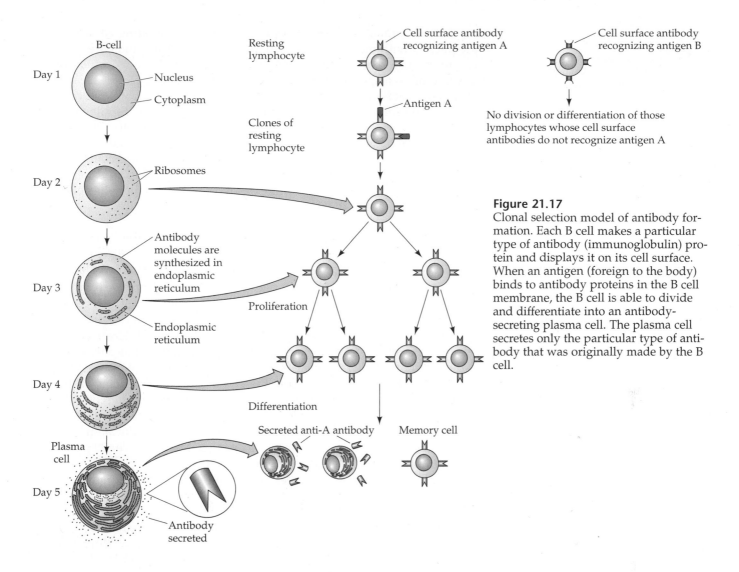

Figure 21.17
Clonal selection model of antibody formation. Each B cell makes a particular type of antibody (immunoglobulin) protein and displays it on its cell surface. When an antigen (foreign to the body) binds to antibody proteins in the B cell membrane, the B cell is able to divide and differentiate into an antibody-secreting plasma cell. The plasma cell secretes only the particular type of antibody that was originally made by the B cell.

Learning: An Environmentally Adaptive Nervous System

In Chapter 8, we discussed how activity can be a critical factor in deciding which neuronal synapses are retained by the adult organism. Here we will extend that discussion to highlight those remarkable instances in which new experiences modify the original set of neuronal connections, causing the creation of new neurons or the formation of new synapses between existing neurons. Since neurons, once formed, do not divide, the "birthday" of a neuron can be identified by treating the organism with radioactive thymidine. Normally, very little radioactive thymidine is taken up into the DNA of a neuron that has already been formed. However, if a new neuron differentiates by cell division during the treatment, it will incorporate radioactive thymidine into its DNA. Such new neurons are seen to be generated when male songbirds learn their songs. Juvenile zebra finches memorize a model song and then learn the pattern of muscle contractions necessary to sing a particular phrase. In this learning and repetition process, new neurons are generated in the hyperstriatum of the finch's brain. Many of these new neurons send axons to the archistriatum, which is responsible for controlling the vocal musculature (Nordeen and Nordeen, 1988). These changes are not seen in males who are too old to learn the song, nor are they seen in juvenile females (who do not sing these phrases); this is discussed more fully in Chapter 20.

The cerebral cortices of young rats reared in stimulating environments are packed with more neurons, synapses, and dendrites than are found in rats reared in isolation (Turner and Greenough, 1983). Even the adult brain is developing in response to new experiences. When adult canaries learn new songs, they generate new neurons whose axons project from one vocal region of the brain to another (Alvarez-Buylla et al., 1990). Similarly, when adult rats learn to keep their balance on dowels, their cerebellar Purkinje cell neurons develop new synapses (Black et al., 1990). Thus, the nervous system continues to develop in adult life, and the pattern of neuronal connections is a product of inherited patterning and patterning produced by experiences. This interplay between innate and experiential development has been detailed most dramatically in studies on mammalian vision.

EXPERIENTIAL CHANGES IN INHERENT MAMMALIAN VISUAL PATHWAYS. Some of the most interesting research on mammalian neuronal patterning concerns the effects of sensory deprivation on the developing visual system in kittens and monkeys. The paths by which electrical impulses pass from the retina to the brain in mammals are shown in Figure 21.18. Axons from the retinal ganglion cells form the two optic nerves, which meet at the optic chiasm. As in *Xenopus* tadpoles, some fibers go to the opposite (contralateral) side of the brain, but unlike most other vertebrates, mammalian retinal cells also send inputs into the same (ipsilateral) side of the brain. These nerves end at the two **lateral geniculate nuclei.** Here the input from each eye is kept separate, the uppermost and anterior layers receiving the axons from the contralateral eye, and the middle of the bodies receiving input from the ipsilateral eye. The situation becomes more complicated as neurons from the lateral geniculate nuclei connect with the neurons of the **visual cortex.** Over 80 percent of the neural cells in the cortex receive input from both eyes. The result is binocular vision and depth perception. Another remarkable finding is that the retinocortical projection is the same for both eyes. If a cortical neuron is stimulated by light flashing across a region of the left eye 5° above and 1° to the left of the fovea,* it will also be stimulated by a light flashing across a region of the right eye 5° above and 1° to the left of the fovea. Moreover, the response evoked in the cortical cell when both eyes are stimulated is greater than the response when either retina is stimulated alone.

Hubel, Wiesel, and their co-workers (see Hubel, 1967) demonstrated that the development of the nervous system depends to some degree on the experience of the individual during a critical period of development. In other words, all of neuronal development is not encoded in the genome: some is learned. Experience appears to strengthen or stabilize some neuronal connections that are already present at birth and to weaken or eliminate other connections. These conclusions come from studies of partial sensory deprivation. Hubel and Wiesel (1962, 1963) sewed shut the right eyelids of newborn kittens and left them closed for 3 months. After this time, they unsewed the lids of the right eye. The cortical cells of such kittens could not be stimulated by shining light in the right eye. Almost all the inputs into the visual cortex came from the left eye only. The behavior of the kittens revealed the inadequacy of their right eyes: when only the left eyes of these animals were covered, the kittens became functionally blind. Because the lateral geniculate neurons appeared to be stimulated from both right and left eyes in these kittens, the physiological defect appeared to be between the lateral geniculate nuclei and the visual cortex. In rhesus monkeys, where similar

---

*The fovea is a depression in the center of the retina where only cones are present and the rods and blood vessels are absent. Here it serves as a convenient landmark.

(A)

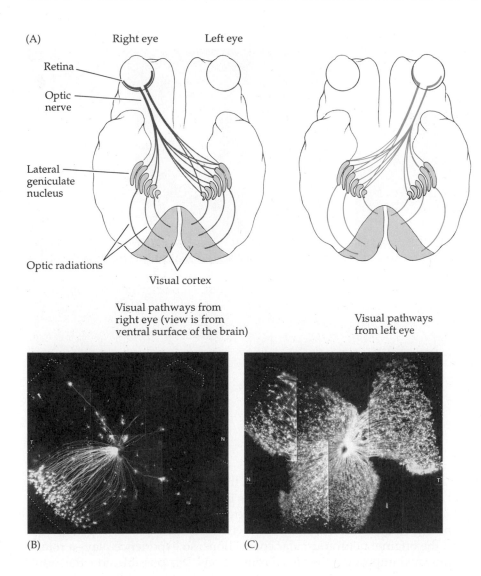

Right eye   Left eye

Retina

Optic nerve

Lateral geniculate nucleus

Optic radiations

Visual cortex

Optic chiasm

Visual pathways from right eye (view is from ventral surface of the brain)

Visual pathways from left eye

Combined left and right visual pathways

(B)

(C)

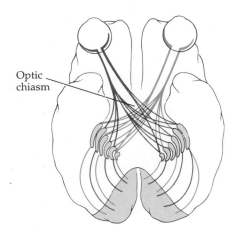

Figure 21.18
Major pathways of the mammalian visual system. (A) In mammals, the optic nerve from each eye branches, sending nerve fibers to a lateral geniculate nucleus on each side of the brain. On the ipsilateral side, a particular part of the retina goes to a particular part of the lateral geniculate nucleus. On the contralateral side, the lateral geniculate nucleus receives input from all parts of the retina. Neurons from each lateral geniculate nucleus innervate the visual cortex on the same side. (B,C) Isolated (and filleted) retinas showing ipsilateral (B) and contralateral (C) projections from the mouse 16-day embryonic retinal ganglial cells. The fluorescent carbocyanine dye DiI was inserted behind the optic chiasm, and the dye was allowed to enter the retinal axons. The dye diffuses along the axons, thereby labeling them to their origin. Ipsilateral projections mostly come from a single part of the retina (in this case, the ventrotemporal region). Contralateral projections to the same site come from all over the retina. (B and C from Colello and Guillery, 1990, courtesy of the authors.)

phenomena are observed, the defect has been correlated with a lack of protein synthesis in the lateral geniculate neurons innervated by the covered eye (Kennedy et al., 1981).

Although it would be tempting to conclude that the resulting blindness was due to a failure to form the proper visual connections, this is not the case. Rather, when a kitten is born, axons from lateral geniculate neurons receiving input from each eye overlap extensively in the visual cortex (Hubel and Wiesel, 1963). However, when one eye is covered early in the kitten's life, its connections into the visual cortex are taken over by those of the other eye (Figure 21.19). Competition occurs, and experience plays a role in strengthening and stabilizing the connections from each lateral geniculate nucleus to the visual cortex. Thus, when both eyes of a kitten are sewn shut for 3 months, most cortical cells can still be stimulated by appropriate illumination of one eye or the other. The critical time in kitten development for this validation of neuronal connections begins between the fourth and the sixth week of the kitten's life. Monocular deprivation up to the fourth week produces little or no physiological deficit, but after 6 weeks, it produces all the characteristic neuronal changes. If a kitten has had normal visual experience for the first 3 months, any subsequent monocular deprivation (even for a year or more) is without effect. The synapses have been stabilized.

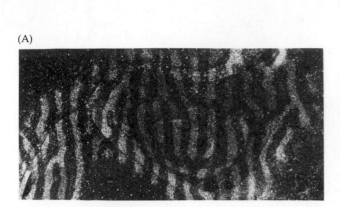

(A)

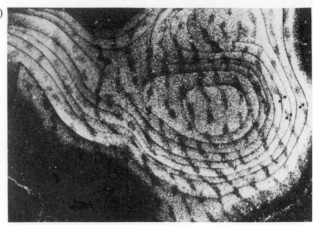

(B)

(C) (D) Cortical layer 3

Cortical layer 4

Figure 21.19
Dark-field autoradiographs of monkey striate cortex 2 weeks after one eye was injected with [³H]proline in the vitreous humor. Each retinal neuron takes up the radioactive label and transfers it to the cells with which it forms synapses. (A) Normal labeling pattern. The white stripes indicate that roughly half the columns took up the label, while the other half did not, a pattern reflecting that half the cells were innervated by the labeled eye and half were innervated by the unlabeled eye. (B) Labeling pattern when the unlabeled eye was sutured shut for 18 months. The axonal projections from the normal (labeled) eye take over the regions that would normally have been innervated by the sutured eye. (C,D) Drawings of axons from kitten geniculate nuclei in which one eye was occluded for 33 days. The terminal branching of the axons in the occluded eye (C) were far less extensive than those of the nonoccluded eye (D). (A and B from Wiesel, 1982, courtesy of T. Wiesel; C and D after Antonini and Stryker, 1993.)

Two principles, then, can be seen in the patterning of the mammalian visual system. First, neuronal connections involved in vision are present even before the animal sees; and second, experience plays an important role in determining whether or not certain connections remain.* Just as experience refines the original neuromuscular connections, so experience plays a role in refining and improving the visual connections. It is possible, too, that adult functions such as learning and memory arise from the establishment and/or strengthening of different synapses by experience. As Purves and Lichtman (1985) remark:

The interaction of individual animals and their world continues to shape the nervous system throughout life in ways that could never have been programmed. Modification of the nervous system by experience is thus the last and most subtle developmental strategy.

*Recent studies (Colman et al., 1997) have shown that divergence in neurotransmitter release results in changes in synaptic adhesivity and causes the withdrawal of the axon providing the weaker stimulation. Those of you who have studied neurobiology will recall (if properly potentiated) that the concept of the Hebbian synapse is predicated on the notion that experience influences neural pathways. If an axon from neuron A activates neuron B, so that the firing of neuron B is always associated with the firing of neuron A, then the synapse between neuron A and neuron B is strengthened. There are many means by which this strengthening could occur, but most hypotheses focus on changes that would enable calcium ions to more readily enter neuron B. This type of synapse could explain the phenomenon of long-term potentiation, which is thought to be the basis of correlative memory (where one sensation recalls others). Such Hebbian mechanisms may mediate the competition between lateral geniculate nuclei axons for cells in the visual cortex (Stent, 1973; Reiter and Stryker, 1988).

■ ENVIRONMENTAL DISRUPTION OF NORMAL DEVELOPMENT

Malformations and disruptions

From the first part of this chapter, it is clear that the instructions for development do not reside wholly in the genes or even in the zygote. The organism is sensitive to cues from the environment. However, this makes the organism vulnerable to environmental changes that can disrupt development.

If it seems amazing that any one of us survives to be born, you are correct; it is estimated that from one-half to two-thirds of all human conceptions do not develop successfully to term (Figure 21.20). Many of these embryos express their abnormality so early that they fail to implant in the uterus. Others implant but fail to establish a successful pregnancy. Thus, most abnormal embryos are spontaneously aborted before the woman even knows she is pregnant (Boué et al., 1985). Edmonds and co-workers (1982), using a sensitive immunological test that can detect the presence of human chorionic gonadotropin (hCG) 8 or 9 days after fertilization, monitored 112 pregnancies in normal women. Of these hCG-determined pregnancies, 67 failed to be maintained.

It appears, then, that many human embryos are impaired early in development and do not survive long in utero. Defects in the lungs, limbs, face, or mouth, however, would not be deleterious to the fetus (which does not depend on those organs while inside the mother), but can seriously threaten life once the baby is born. About 5 percent of all human births have a recognizable malformation, some of them mild, some very severe (McKeown, 1976).

Congenital ("at birth") abnormalities and the demise of embryos and fetuses prior to birth are caused both intrinsically and extrinsically. Those abnormalities caused by genetic events (mutations, aneuploidies, translocations) are called **malformations.** For instance, aniridia (absence of the iris), caused by a mutation of the *PAX6* gene, is a malformation. Down's syndrome, caused by trisomy of chromosome 21, is likewise a malformation. Most early embryonic and fetal demise is probably due to chromosomal abnormalities that interfere with normal developmental processes.

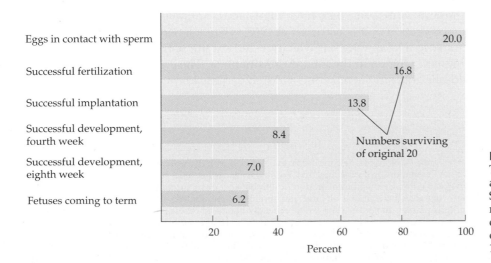

Figure 21.20
The hypothetical fates of 20 eggs that are naturally fertilized in the United States and western Europe. Under normal conditions, only 6.2 eggs of the original 20 would be expected to develop successfully to term. (After Volpe, 1987.)

Table 21.2 Some agents thought to cause disruptions in human fetal development[a]

DRUGS AND CHEMICALS

Alcohol

Aminoglycosides (Gentamycin)

Aminopterin

Antithyroid agents (PTU)

Bromine

Cigarette smoke

Cocaine

Cortisone

Diethylstilbesterol (DES)

Diphenylhydantoin

Heroin

Lead

Methylmercury

Penicillamine

Retinoic acid (Isotretinoin, Accutane)

Streptomycin

Tetracycline

Thalidomide

Trimethadione

Valproic acid

Warfarin

IONIZING RADIATION (X-RAYS)

HYPERTHERMIA

INFECTIOUS MICROORGANISMS

Coxsackie virus

Cytomegalovirus

Herpes simplex

Parvovirus

Rubella (German measles)

Toxoplasma gondii (toxoplasmosis)

Treponema pallidum (syphilis)

METABOLIC CONDITIONS IN THE MOTHER

Autoimmune disease (including Rh incompatibility)

Diabetes

Dietary deficiencies, malnutrition

Phenylketonuria

Source: Adapted from Opitz, 1991.

[a]This list includes known and possible teratogenic agents and is not exhaustive.

Abnormalities due to exogenous agents (certain chemicals or viruses, radiation, or hyperthermia) are called **disruptions.** The agents responsible for these disruptions are called **teratogens** (Greek "monster-formers"), and the study of how environmental agents disrupt normal development is called **teratology.*** Teratogens work during certain critical periods of development. The most critical time for any organ is when it is growing and forming its structures. Different organs have different critical periods, although the time from day 15 to day 60 is critical for many organs. The heart forms primarily during weeks 3 and 4, while the external genitalia are most sensitive during weeks 8 and 9. The brain and skeleton are always sensitive, from the beginning of week 3 to the end of pregnancy and beyond.

Teratogenic agents

Different agents are teratogenic in different organisms. A partial list of agents that are teratogenic in humans is given in Table 21.2.

The major class of teratogens includes drugs and environmental chemicals. Some chemicals that are naturally found in the environment can cause birth defects. Even in the pristine alpine meadows of the Rocky Mountains, teratogens are found. Here grows the skunk cabbage *Veratrum californicum*, upon which sheep sometimes feed. If pregnant ewes eat this plant, their fetuses tend to develop severe neurological damage, including cyclopia, the fusion of two eyes in the center of the face (Figure 21.21). This condition also occurs in humans, pigs, and many other mammals; the affected organism dies shortly after birth (as a result of severe brain defects, including the lack of a pituitary gland).

Quinine and alcohol, two substances derived from plants, can also cause congenital malformations. Quinine can cause deafness, and alcohol (when more than 2–3 ounces per day are imbibed by the mother) can cause physical and mental retardation in the infant. Nicotine and caffeine have not been proved to cause congenital anomalies, but women who are heavy smokers (20 cigarettes a day or more) are likely to have infants that are smaller than those born to women who do not smoke. Smoking also significantly lowers the number and motility of sperm in the semen of males who smoke at least four cigarettes a day (Kulikauskas et al., 1985).

In addition, our industrial society produces hundreds of new artificial compounds that come into general use each year. Pesticides and organic mercury compounds have caused neurological and behavioral abnormalities in infants whose mothers have ingested them during pregnancy. A tragic demonstration of this occurred in 1965, when a Japanese firm dumped mercury into a lake, where it was ingested by the fish, which were eaten by pregnant women in the village of Minamata. The congenital brain damage and blindness in the resulting children became known as Minamata disease.

---

*In some cases, the same condition can be caused by a disruption (from an exogenous agent) or a malformation (from the nucleus). For instance, certain axial malformations in mice can be produced either by the administration of retinoic acid or by mutations in certain *Hox* genes. In some instances, the mutation and the teratogen are known to affect the same enzyme. Chondroplasia punctata is a congenital defect of bone and cartilage, characterized by abnormal bone mineralization, underdevelopment of nasal cartilage, and shortened fingers. It is caused by a defective gene on the X chromosome. An identical phenotype is produced by the ingestion of the rat-killing compound, warfarin. It appears that the defective gene is normally responsible for producing an arylsulfatase protein necessary for cartilage growth. The warfarin compound inhibits this same enzyme (Franco et al., 1995).

Retinoic Acid as a Teratogen

In some instances, a compound used for development in the body can have deleterious effects if given in large amounts at particular times. Retinoic acid is important in forming the anterior-posterior axis of the mammalian embryo and also in forming the limbs. In these cases, the retinoic acid is made from discrete cells and works in a small area. However, if retinoic acid is provided from the mother in large amounts, cells can respond to it that normally would not receive such high concentrations of this molecule. In Chapter 16, we discussed the effect of retinoic acid on mouse development. Inside the body, vitamin A and 13-*cis*-retinoic acid become isomerized to the developmentally active forms of retinoic acid, all-*trans*-retinoic acid and 9-*cis*-retinoic acid (Creech Kraft, 1992). Retinoic acid cannot bind directly to genes. To effect gene regulation, RA must bind to a group of transcription factors called the retinoic acid receptors (RARs). These proteins have the same general structure as the steroid and thyroid hormone receptors, and they are active only when they have bound retinoic acid (Linney, 1992). The retinoic acid receptors bind to specific enhancer elements in the DNA called retinoic acid response elements. Retinoic acid response elements contain at least two copies of the sequence GGTCA (Ruberte et al., 1990, 1991a). Some of the *Hox* genes have retinoic acid response elements in their promoters (Yu et al., 1991; Pöpperl and Featherstone, 1993; Studer et al., 1994). There are three major types of retinoic acid receptors: RAR-α, RAR-β, and RAR-γ. They each bind both forms of retinoic acid, and they each bind to the same retinoic acid response element.

Retinoic acid has been useful in treating severe cystic acne and has been available (under the name Accutane) since 1982. Because the deleterious effects resulting from the administration of large amounts of vitamin A or its analogues to various species of pregnant animals have been known since the 1950s (Cohlan, 1953; Giroud and Martinet, 1959; Kochhar et al., 1984), the drug contains a label warning that it should not be used by pregnant women. However, about 160,000 women of childbearing age (15 to 45 years) have taken this drug since it was introduced, and some of them have used it during pregnancy. Lammer and his co-workers (1985) studied a group of women who inadvertently exposed themselves to retinoic acid and who elected to remain pregnant. Of the 59 fetuses, 26 were born without any noticeable anomalies, 12 aborted spontaneously, and 21 were born with obvious anomalies. The malformed infants had a characteristic pattern of anomalies, including absent or defective ears, absent or small jaws, cleft palate, aortic arch abnormalities, thymic deficiencies, and abnormalities of the central nervous system.*

This pattern of multiple congenital anomalies is similar to that seen in rat and mouse embryos whose pregnant mothers have been given these drugs. Goulding and Pratt (1986) have placed 8-day mouse embryos in a solution containing 13-*cis*-retinoic acid at very low concentrations ($2 \times 10^{-6} M$). Even at this concentration, approximately one-third of the embryos developed a very specific pattern of anomalies, including dramatic reduction in

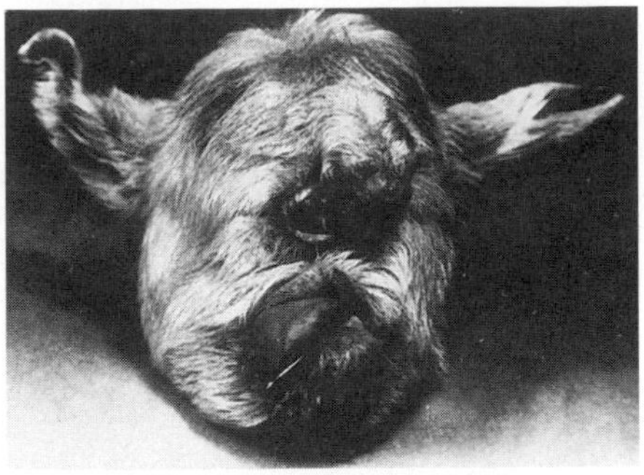

Figure 21.21
Head of a cyclopic lamb born of a ewe who had eaten *Veratrum californicum* early in pregnancy. The cerebral hemispheres fused, forming only one eye and no pituitary gland. (From Binns et al., 1964, courtesy of J. F. James and the USDA-ARS Poisonous Plant Laboratories.)

---

*Public health is a critical factor, since there is significant overlap between the population using acne medicine and the population of women of childbearing age. It is claimed that half of the pregnancies in America are unplanned (Nulman et al., 1997). Vitamin A is itself teratogenic when injected in megadose amounts. Rothman and colleagues (1995) found that pregnant women who took more than 10,000 international units of preformed vitamin A per day (in the form of vitamin supplements) had about a 2 percent chance of having a baby born with disruptions similar to those produced by retinoic acid.

Figure 21.22
Normal 17-day mouse embryo (A) and 17-day mouse embryo whose mother had been given retinoic acid on day 8 of pregnancy (B). Craniofacial malformations can be seen in the neural-crest-derived cartilage of the treated embryos. Meckel's cartilage is completely displaced from the mandibular (lower jaw) to the maxillary (upper mouth) region. The malleus and incus cartilages are also not formed. (From Morriss-Kay, 1993; photograph courtesy of G. Morriss-Kay.)

(A)   (B)

the size of the first and second pharyngeal arches (Figure 21.22). In normal mice, the first arch eventually forms the maxilla and mandible of the jaw and two ossicles of the middle ear, while the second arch forms the third ossicle of the middle ear as well as other facial bones.

The basis for this developmental disruption appears to reside in the drug's ability to alter the expression of the *Hox* genes and thereby respecify portions of the anterior-posterior axis and inhibit neural crest cell migration from the cranial region of the neural tube (Moroni et al., 1994; Studer et al., 1994). Radioactively labeled retinoic acid binds to the cranial neural crest cells and arrests both their proliferation and their migration (Johnston et al., 1985; Goulding and Pratt, 1986). The binding seems to be specific to the *cranial* neural-crest-derived cells, and the teratogenic effect of the drug is confined to a specific developmental period (days 8–10 in mice; days 20–35 in humans). Animal models of retinoic acid teratogenesis have been extremely successful in elucidating the mechanisms of teratogenesis at the cellular level. [env6.html]

Thalidomide as a Teratogen

Before 1961, there was very little evidence for drug-induced malformations in humans. But in that year, Lenz and McBride independently accumulated evidence that a mild sedative, **thalidomide,** caused an enormous increase in a previously rare syndrome of congenital anomalies. The most noticeable of these anomalies was **phocomelia,** a condition in which the long bones of the limbs are absent (amelia) or severely deficient (peromelia), thus causing the resulting appendage to resemble a seal flipper (Figure 21.23). Over 7000 affected infants were born to women who had taken this drug, and a woman need only have taken one tablet to produce children with all four limbs deformed (Lenz, 1962, 1966; Toms, 1962). Other abnormalities induced by the ingestion of thalidomide included heart defects, absence of the external ears, and malformed intestines. The drug was withdrawn from the market in November, 1961.

Nowack (1965) documented the **period of susceptibility** during which thalidomide caused these abnormalities. The drug was found to be teratogenic only during days 34–50 after the last menstruation (about 20 to 36 days postconception). The specificity of thalidomide action is shown in Figure 21.23C. From day 34 to day 38, no limb abnormalities are seen. During this period, thalidomide can cause the absence or deficiency of ear components. Malformations of upper limbs are seen before those of the lower limbs, since the arms form slightly before the legs during development.

(A)

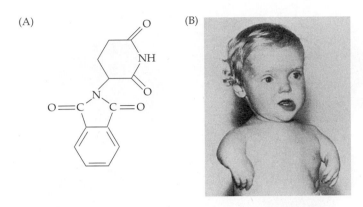

(B)

Figure 21.23
Thalidomide structure and effect. (A) Chemical structure of thalidomide. (B) Phocomelia in an infant whose mother took thalidomide during the first 2 months of pregnancy. (C) Timing of susceptibility to the teratogenic effects of thalidomide. (After Nowack, 1965.)

(C)

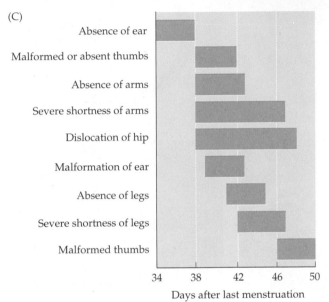

Days after last menstruation

The thalidomide tragedy showed the limits of animal models as tests of the potential teratogenic effects of drugs. Different species (and strains within species) metabolize thalidomide differently. Pregnant mice and rats—the animals usually used to test such compounds—do not generate malformed pups when given thalidomide. Rabbits produce some malformed offspring, but the defects are different from those seen in affected human infants. Primates such as the marmoset appear to have a susceptibility similar to that of humans, and affected marmoset fetuses have been studied in an attempt to discover how thalidomide causes these disruptions. McCredie (1976a,b) proposed that thalidomide might affect the differentiation of neural-crest-derived cells, and McBride and Vardy (1983) reported that the most noticeable difference seen before limb malformations concerns the size of the dorsal root ganglia and their neurons. The number of neurons in these ganglia is markedly reduced (Figure 21.24). These authors speculate that the neurons from these ganglia are necessary for maintaining limb development and that thalidomide works by interfering with or destroying these neurons.

Another hypothesis (Neubert et al., 1995; Geitz, et al., 1996) proposes that the initial target of thalidomide is the adhesion molecules of the limb bud and its capillaries. The addition of low doses of thalidomide to marmosets or to cultured endothelial cells results in the downregulation of several cell-cell and cell substrate adhesion molecules.

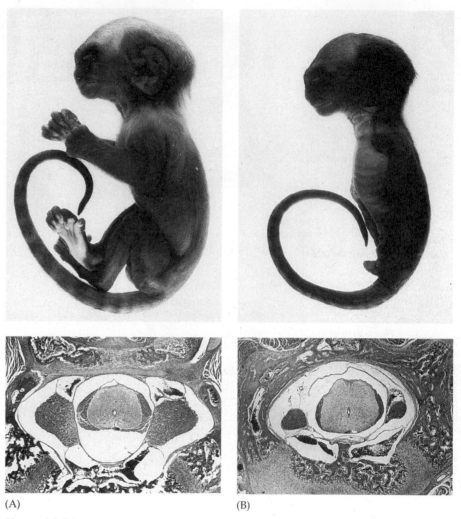

(A)   (B)

Figure 21.24
Thalidomide effects on fetal marmosets. Top figures show phenotypes of marmoset fetuses late in gestation. Bottom figures show spinal cord cross sections at the level of the forelimbs. (A) Fetus of a control marmoset. (B) Fetus of a marmoset treated with 25 mg thalidomide per kilogram of body weight between days 38 and 46 of pregnancy. (From McBride and Vardy, 1983, courtesy of W. G. McBride.)

A third mechanism to explain thalidomide teratogenicity has been advanced by Lash and Saxén (1972). Lash (1963) had noticed that the primitive kidney, the mesonephros, induced cartilage growth in cultured limb tissue. Lash and Saxén observed that thalidomide inhibited this mesonephros-induced cartilage growth in cultures of human organs obtained from electively aborted embryos. Moreover, radioactive thalidomide appeared to bind specifically to the human mesonephros. The molecular mechanism of this selective thalidomide teratogenicity is still not known, and this will remain a difficult problem to study so long as our only animal models are other primates.

The thalidomide tragedy also underscores another important principle: the metabolism of embryos is different from that of adults, and the *construction* of an organ can be affected by chemicals that have no deleterious effect on the normal *functioning* of that organ. Several medicines for adults are teratogenic to embryos. These include methotrexate (a drug used to stop tumor cell growth), anticonvulsants such as trimethadione and phenytoin, and anti-coagulants such as warfarin. Smoking cigarettes during pregnancy has been associated with fetal growth retardation, but neither the drinking of coffee nor the intake of triocyclic antidepressants has been seen to produce significant developmental abnormalities (see Friedman, 1992; Nulman et al., 1997).

Alcohol as a Teratogen

In terms of frequency and cost to society, the most devastating teratogen is undoubtedly ethanol. In 1968, Lemoine and colleagues noticed a syndrome of birth defects in the children of alcoholic mothers. Such a **fetal alcohol syndrome (FAS)** was also noted by Jones and Smith (1973). Babies with FAS were characterized as having small head size, an indistinct philtrum (the pair of ridges that run between the nose and mouth above the center of the upper lip), a narrow upper lip, and a low nose bridge. The brain of such a child may be dramatically smaller than normal and often shows defects in neuronal and glial migration (Figure 21.25; Clarren, 1986). There is also prominent extra cell death in the frontonasal process and in cranial nerve ganglia (Sulik et al., 1988). Fetal alcohol syndrome is the third most prevalent type of mental retardation (behind fragile X syndrome and Down syndrome) and affects one out of every 500 to 750 children born in the United States (Abel and Sokol, 1987).

Children with fetal alcohol syndrome are developmentally and mentally retarded, with a mean IQ of around 68 (Streissguth and LaDue, 1987). Patients with a mean chronological age of 16.5 years were found to have the functional vocabulary of 6.5-year-olds and to have the mathematical abilities of fourth-graders. Most of the adults and adolescents with FAS cannot handle money or their own lives, and they have difficulty learning from past experiences. Moreover, in many instances of FAS, the behavioral abnormalities exist without any gross physical changes in the head size or IQ (J. Opitz, personal communication, 1996). There is great variation in the ability of mothers and fetuses to metabolize ethanol, and it is thought that 30–40 percent of the children born to alcoholic mothers who drink during pregnancy will have FAS. It is also thought that lower amounts of ethanol ingestion by the mother can lead to *fetal alcohol effect*, a less severe form of FAS, but a condition that lowers the functional and intellectual abilities of the sufferer.*

---

*For a remarkable account of raising a child with fetal alcohol syndrome, as well as an analysis of FAS in Native American culture in the United States, see Dorris (1989). The personal and sociological effects of FAS are well integrated with the scientific and economic data.

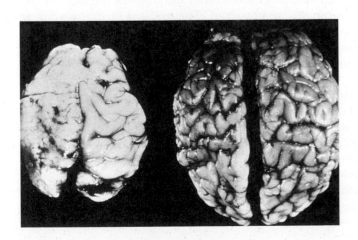

Figure 21.25
Comparison of a brain from an infant with fetal alcohol syndrome (left) with that of a normal infant of the same age (right). The brain from the infant with FAS is significantly smaller, and the pattern of convolutions is obscured by glial cells that have migrated over the top of the brain. (Photograph courtesy of S. Clarren.)

Figure 21.26
Possible mechanisms producing fetal alcohol syndrome. (A–C) Cell death by ethanol-induced superoxide radicals. Staining with Nile blue sulfate reveals areas of cell death. (A) Control 9-day mouse embryo head region. (B) Head region of ethanol-treated embryo, showing areas of cell death. (C) Head region of 9-day embryo treated with both ethanol and superoxide dismutase, an inhibitor of superoxide radicals. The superoxide inhibitor prevents the alcohol-induced cell death. (D) Graph of the inhibition of L1-mediated cell adhesion by ethanol. (A–C from Kotch et al., 1995; photographs courtesy of K. Sulik; D after Ramanathan et al., 1996.)

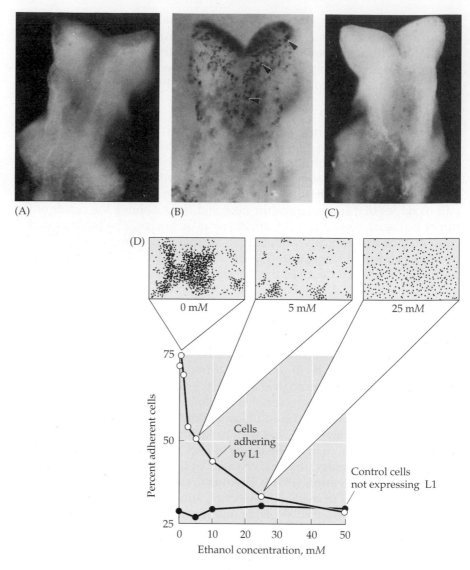

A mouse model system has been used to explain the effects of alcohol on the face and nervous system. When ethanol is given to mice at the time of their gastrulation, it induces the same range of developmental defects as in humans. As early as 12 hours after the mother has injested the alcohol, abnormalities of development are observed. The midline structures fail to form, allowing the abnormally close proximity of the medial processes of the face. Forebrain anomalies are also seen, and the more severely affected fetuses lack a forebrain entirely (Sulik et al., 1988). In the ethanol-treated embryos, cell death can be seen prominently in the frontonasal (facial) process, as well as in the cranial nerve ganglia and visceral arch mesoderm. Recent studies suggest that ethanol may induce its teratogenic effects by more than one mechanism. First, anatomical evidence suggests that neural crest migration is severely impaired. Second, cell death may be caused by the production of superoxide radicals that can oxidize cell membranes and lead to cytolysis (Figure 21.26A–C; Davis et al., 1990; Kotch et al., 1995). Third, alcohol may directly prevent the ability of cell adhesion molecule L1 to function in holding cells together. Ramanathan and colleagues (1996) have shown that ethanol can block the adhesive functions of L1 proteins in vitro at levels as low as 7 mM, a concentration of ethanol produced in the blood or brain with a single drink (Figure 21.26D). Moreover, mutations in human *L1* genes cause a syndrome of mental retardation and malformations similar to that seen in severe cases of fetal alcohol syndrome. [env7.html]

Other Teratogenic Agents

Drugs and chemicals are not the only extrinsic agents capable of causing disruptions in development. Another class of teratogens includes viruses. Gregg (1941) first documented the fact that women who had rubella (German measles) during the first third of their pregnancy had a 1 in 6 chance of giving birth to an infant with eye cataracts, heart malformations, or deafness. This was the first evidence that the mother could not fully protect the fetus from the outside environment. The earlier the rubella infection occurred during the pregnancy, the greater the risk that the embryo would be malformed. The first 5 weeks appear to be the most critical, because this is when the heart, eyes, and ears are being formed. The rubella epidemic of 1963–1965 probably resulted in about 20,000 fetal deaths and 30,000 infants with birth defects in the United States. Two other viruses, *Cytomegalovirus* and *Herpes simplex*, are also teratogenic. *Cytomegalovirus* infection of early embryos is nearly always fatal, but infection of later embryos can lead to blindness, deafness, cerebral palsy, and mental retardation.

Bacteria and protists are rarely teratogenic, but two of them can damage human embryos. *Toxoplasma gondii*, a protozoan carried by rabbits and cats (and their feces), can cross the placenta and cause brain and eye defects in the fetus. *Treponema pallidum*, the cause of syphilis, can kill early fetuses and produce congenital deafness in older ones.

Ionizing radiation can break chromosomes and alter DNA structure. For this reason, pregnant women are told to avoid unnecessary X-rays, even though there is no evidence for congenital anomalies resulting from diagnostic radiation (Holmes, 1979). Heat from high fevers is also a possible teratogen. [env8.html], [env11.html]

While we know the causes of certain malformations, most congenital abnormalities are not yet able to be explained. For instance, congenital cardiac anomalies occur in about 1 in every 200 live births. Genetic causes are responsible for about 8 percent of these heart abnormalities, and about 2 percent can be explained by known teratogens. That leaves 90 percent of the them unexplained (O'Rahilly and Müller, 1992). We still have a great deal of research to do. In fact, we have not screened most chemicals for their teratogenic effects. There are over 50,000 artificial chemicals currently used in our society and about 200 to 500 new materials being made each year (Johnson, 1980). The problem of screening these chemicals is of major importance, and standard protocols are expensive, long, and subject to interspecies differences in metabolism. There is still no consensus on how to test a substance's teratogenicity for human embryos.

In the former Soviet Union, the unregulated "industrial production at-all-costs" approach leaves a legacy of soaring birth defects. In some regions of Kazakhstan, teratogens such as lead, mercury, and zinc are found in high concentrations in drinking water, vegetables, and the air. In these places, nearly half the people tested have extensive chromosome breakage. In some areas, the incidence of birth defects has doubled since 1980 (Edwards, 1994). Although teratogenic compounds have always been with us, the fetus is placed at ever greater risk as more untested compounds enter our environment each year.

Environmental Estrogens

THERE IS PROBABLY no bigger controversy in the field of environmental toxicology than whether pesticides are responsible for breast cancer in women, the decline of sperm counts in men and the congenital malfunctions in wild animals. Americans use some 2 billion pounds of pesticides each year. Moreover, some pesticide residues stay in the food chain for decades. Although banned in the United States in 1972, DDT has an environmental half-life of about 100 years (Nature, 1995). Recent evidence has shown that DDT [dichloro-diphenyl-trichloroethane] and its chief metabolic byproduct, DDE (which lacks one of the chlorine atoms), can act as estrogenic compounds, either by mimicking the female sex hormone estrogen or by inhibiting androgen effectiveness (Davis et al., 1993; Kelce et al., 1995). These compounds have been linked to such environmental problems as the decrease in the alligator populations in Florida, the feminization of fish in Lake Superior, the rise in breast cancers, and the worldwide decline of human sperm counts (Carlsen et al., 1992; Keiding and Skakkebaek, 1993; Stone, 1994). Guillette and co-workers (1994) have linked a DDT spill in Florida's Lake Apopka to a 90 percent decline in the birthrate of alligators and to the reduced penis size in the young males. Dioxin, another pesticide ingredient, has been linked to cancer; and male offspring of pregnant rats exposed to dioxin have reduced sperm counts, smaller testes, and less male-specific sexual behaviors.

Some scientists, however, say that these claims are exaggerated. Although the pesticides can mimic estrogen, they do so only in large amounts, and they bind to the estrogen receptor a 1000-fold more weakly than do normal estrogens. Another argument is that some of the pesticide compounds have weak anti-estrogenic effects, that would cancel out the weak estrogenic effects. Critics of pesticides point out, however, that even

though these compounds bind weakly to the estrogen receptor, they are present in the blood serum over 100 times the concentration of normal estrogens. They also argue that some of the small amounts of other active compounds can act synergistically to stimulate estrogenic activity of these otherwise weakly estrogenic molecules (see Hansen and Jansen, 1994; Stone, 1994). Arnold and colleagues (1996), for example, have shown that a combination of two weak environmental estrogens gives an effect 1000-times stronger than either chemical alone. The recent paper by Kelce and colleagues (1995) suggests that the critical mode of operation might not be the weakly estrogenic effects of DDT, but the potent antitestosterone effect of its metabolite, DDE. DDE is able to inhibit androgen-responsive transcription at doses comparable to those found in contaminated soil in the United States and other countries.

Some estrogenic compounds may be in the food we eat and in the wrapping that surrounds them, for some of the chemicals used to set plastics have been found to be estrogenic. The discovery of the estrogenic effect of plastic stabilizers was made in a frightening way. Investigators at Tufts University Medical School had been studying estrogen-responsive tumor cells. These cells require estrogen in order to proliferate. Their experiments were going well until 1987, when the experiments went awry. The control cells were growing as well as the estrogen-treated cells. It was as if someone had contaminated the medium by adding estrogen to it. What was the source of contamination? After four months testing all the components of their experimental system, they discovered that the source of estrogen was the plastic tubes that held their water and serum. The company that made the tubes refused to tell the investigators about their new process for stabilizing the polystyrene plastic, so the scientists

had to discover it themselves. It turned out to be *p*-nonylphenol, a chemical used to harden the PVC plastic of the plumbing tubes that bring water and to stabilize the polystyrene plastics that hold water, milk, orange juice, and other liquids (Soto et al., 1991; Colburn et al., 1996). This compound is also the degradation product of detergents and household cleaners. A related compound, 4-*tert*-pentylphenol, has a potent estrogenic effect on human cultured cells and can cause male carp (*Cyprinus carpis*) to develop oviducts, ovarian tissue, and oocytes (Gimeno et al., 1996).

Some other environmental estrogens are polychlorinated biphenyls (PCBs). These compounds were widely used as refrigerants before they were banned in the 1970s as being responsible for causing cancer in rats. They remain in the food chain, though, and have been blamed for the widespread decline in the reproductive capacities of otters, seals, mink, and fish. PCBs resemble diethylstilbesterol in shape, and they may affect the estrogen receptor as DES does, perhaps by binding to another site on the estrogen receptor. The structure of these compounds does resemble that of thyroid hormones (Figure 21.27). Thyroid hormones are critical for the growth of the cochlea of the inner ear, and rats whose mothers were exposed to PCBs had poorly developed cochleas and hearing defects (Goldey and Crofton in Stone, 1995). [env9.html]

Throughout the northern United States and southern Canada there appears to be a dramatic increase in the number of developmentally deformed frogs in what seem to be pristine woodland ponds. The main abnormalities are extra and malformed limbs. It is not known what is causing these disruptions, but speculation (see Hilleman, 1996) is that pesticides (sprayed for mosquito and tick control) might be activating the retinoic acid receptors and respecifying issues as limbs.

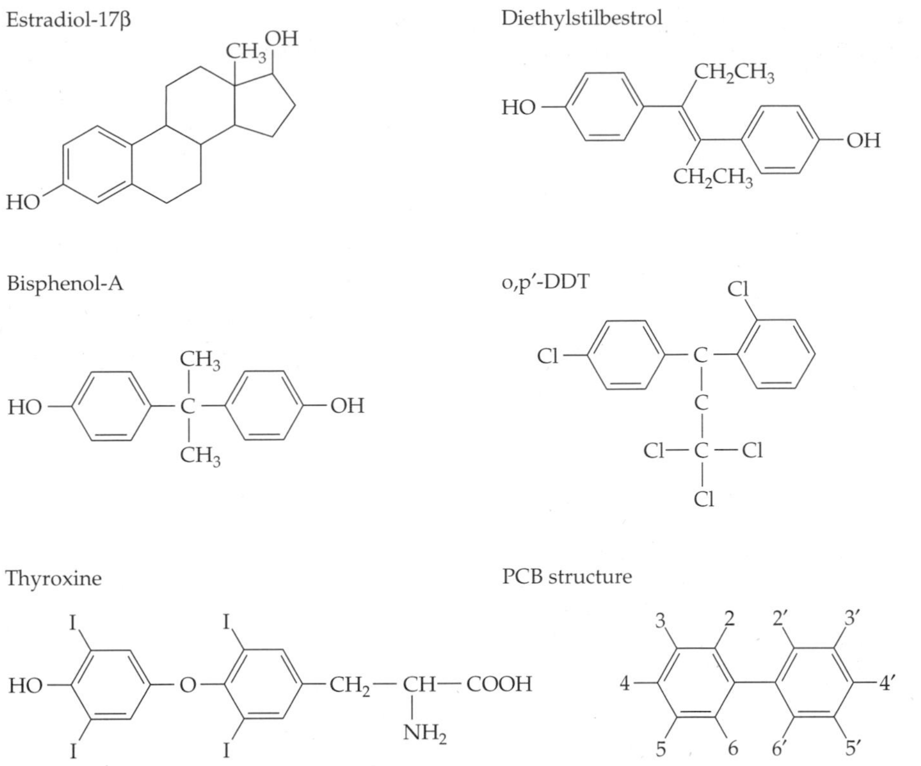

Estradiol-17β

Diethylstilbestrol

Bisphenol-A

o,p'-DDT

Thyroxine

PCB structure

Figure 21.27 Structures of hormones and hormonally disruptive compounds.

It is difficult to document the effects of environmental compounds on humans, and it is even more difficult to determine the effects of "cocktails" consisting of different compounds ingested at different times. There is a great deal more research that needs to be done on the biochemistry of these compounds, their effects on development, and the epidemiology of developmental abnormalities. At the moment, there seems to be evidence coming from animal studies that humans and natural animal populations may be at risk from these hormonal modulators, but all the needed data are not in. [env10.html] ∎

Genetic-environmental interactions

The observation that a substance may be teratogenic in one species but not in another strongly suggests that there is a genetic component to whether or not a substance can produce changes in normal development. Recent evidence suggests that different alleles in the human population can influence whether a substance is benign or dangerous to the fetus. For example, there is only a slight risk in the general population that heavy smoking by the mother will cause facial malformations in her fetus. However, if the fetus has a particular allele (A2) of the gene for growth factor TGF-α, the smoke absorbed through the placenta can raise the risk of cleft lip and palates 10-fold (Shaw et al., 1996). Similarly, different alleles encoding the enzyme alcohol dehydrogenase-2 have differing abilities to degrade ethanol. Whether heavy maternal alcohol consumption leads to fetal alcohol syndrome or fetal alcohol effect may be due to the types of alcohol dehydrogenase isozymes in the mother and fetus (McCarver-May, 1996). Thus, whether or not a compound is "teratogenic" depends on many things, including the genes of the individuals exposed to it.

Coda

Development usually occurs in a rich environmental milieu, and most animals are sensitive to environmental cues. The environment can determine sexual phenotype, can induce remarkable structural and chemical adaptations according to the season, can induce specific morphological changes that allow an individual to escape predation, and can induce caste determination in insects. The environment can also alter the structure of our neurons and the specificity of our immunocompetent cells. Unfortunately, the environment can also be the source of chemicals that disrupt normal developmental processes.

While development usually occurs in a complex natural environment, development can most easily be studied in the laboratory. Indeed, our "model systems" are animals that are readily domesticated and whose development is least affected by environmental factors (Bolker, 1995). However, as we become aware of the complexity of development, we realize that development is critically keyed to the environment. It can take a community to develop an embryo. The exploration of environmental regulation of development is just beginning.

LITERATURE CITED

Abel, E. L. and Sokol, R. J. 1987. Incidence of fetal alcohol syndrome and economic impact of FAS-related anomalies. *Drug Alcohol Depend.* 19: 51–70.

Adler, F. R. and Harvell, C. D. 1990. Inducible defenses, phenotypic variability, and biotic environments. *Trends Ecol. Evol.* 5: 407–410.

Alvarez-Buylla, A., Kirn, J. R. and Nottebohm, F. 1990. Birth of projection neurons in adult avian brain may be related to perceptual or motor learning. *Science* 249: 1444–1446.

Antonini, A. and Stryker, M. P. 1993. Rapid remodeling of axonal arbors in the visual cortex. *Science* 260: 1818–1821.

Arnold, S. F., Klotz, D. M. Collins, B. M., Vonier, P. M., Guilllete, L. J. Jr. and McLachlan, J. A. 1996. Synergistic activation of estrogen receptor with combinations of environmental chemicals. *Science* 272: 1489–1492.

Bachmann, M. D., Carlton, R. G., Burkholder, J. M. and Wetzel, R. G. 1986. Symbiosis between salamander eggs and green algae: Microelectrode measurements inside eggs demonstrate effects of photosynthesis on oxygen concentrations. *Can. Zool.* 64: 1586–1588.

Baxter, G. T. and Morse, D. E. 1992. Cilia from abalone larvae contain a receptor-dependent G-protein transduction system similar to that in mammals. *Biol. Bull.* 183: 147–154.

Beck, S. D. 1980. *Insect Photoperiodism.* 2nd ed. Academic Press, NY.

Begon, M., Harper, J. L. and Townsend, C. R. 1986. *Ecology: Individuals, Populations, and Communities.* Blackwell Scientific, Oxford.

Binns, W., James, L. F. and Shupe, J. L. 1964. Toxicosis of *Veratrum californicum* in ewes and its relationship to a congenital deformity in lambs. *Ann. N.Y. Acad. Sci.* 111: 571–576.

Black, J. E., Issacs, K. R. anderson, B. J. Alcantara, A. A. and Greenough, W. T. 1990. Learning causes synaptogenesis, whereas motor activity causes angiogenesis, in cerebellar cortex of adult rats. *Proc. Natl. Acad. Sci. USA* 87: 5568–5572.

Bolker, J. A. 1995. Model systems in developmental biology. *BioEssays* 17: 451–455.

Born, G. 1881. Experimentelle Untersuchungen über die Entstehung der Geschlechtsunterschiede. *Jahres-Bericht d. Schleischen Gesell f. väterländ. Culture 21 Jan.* pp. 2–23.

Born, G. 1884. Über die Einflussder Schwere uaf die Froschie. *Verh. Med. Sect. Schles. Gessell. f. väterländ. Culture 4 April 1884.*

Borovsk, D., Carlson, D. A., Griffin, P. R., Shabanowitz, J. and Hunt, D. F. 1990. Mosquito oostatic factor: A novel decapeptide modulating trypsin-like enzyme biosynthesis in the midgut. *FASEB J.* 4: 3015–3020.

Boué, A., Boué, J. and Gropp, A. 1985. Cytogenetics of pregnancy wastage. *Adv. Hum. Genet.* 14: 1–57.

Brakefield, P. M. and Reitsma, N. 1991. Phenotypic plasticity, seasonal climate, and the population biology of *Bicyclus* butterflies (Satyridae) in Malawi. *Ecol. Entomol.* 16: 291–303.

Brakefield, P. M. and seven others. 1996. Development, plasticity, and evolution of butterfly eyespot patterns. *Nature* 384: 236–242.

Brian, M. V. 1974. Caste differentiation in *Myrmica rubra*: The role of hormones. *J. Insect Physiol.* 20: 1351–1365.

Brian, M. V. 1980. Social control over sex and caste in bees, wasps and ants. *Biol. Rev.* 55: 379–415.

Brönmark, C. and Pettersson, L. 1994. Chemical cues from piscivores induce a change in morphology in crucian carp. *Oikos* 70: 396–402.

Bry, L., Falk, P. G., Midtvedt, T. and Gordon, J. I, 1996. A model of host-microbial interactions in an open mammalian ecosystem. *Science* 273: 1380–1383.

Bull, J. J. 1980. Sex determination in reptiles. *Q. Rev. Biol.* 55: 3–21.

Burnett, F. M. 1959. *The Clonal Selection Theory of Immunity.* Vanderbilt University Press, Nashville.

Carlsen, E., Giwercman, A., Keiding, N. and Skakkebaek, N. E. 1992. Evidence for decreasing quality of semen during past 50 years. *Brit. Med. J.* 305: 609–613.

Charnov, E. L. and Bull, J. J. 1977. When is sex environmentally determined? *Nature* 266: 828–830.

Clarren, S. K. 1986. Neuropathology in the fetal alcohol syndrome. *In* J. R. West (ed.), *Alcohol and Brain Development.* Oxford University Press, New York.

Cohen, C. S. and Strathmann, R. R. 1996. Embryos at the edge of tolerance: Effects of environment and structure of egg masses on supply of oxygen to embryos. *Biol. Bull.* 190: 8–15.

Cohlan, S. Q. 1953. Excessive intake of vitamin A as a cause of congenital anomalies in the rat. *Science* 117: 535–537.

Colburn, T., Dumanoski, D., and Myers, J. P. 1996. *Our Stolen Future.* Dutton, New York.

Colello, R. J. and Guillery, R. W. 1990. The early development of retinal ganglion cells with uncrossed axons in the mouse: retinal position and axon course. *Development* 108: 515–523.

Colman, H., Nabekura, J. and Lichtman, J. W. 1997. Alterations in synaptic strength preceding axon withdrawal. *Science* 275: 356–361

Conover, D. O. and Heins, S. W. 1987. Adaptive variation in environmental and genetic sex determination in a fish. *Nature* 326: 496–498.

Creech Kraft, J. 1992. Pharmacokinetics, placental transfer, and teratogencity of 13-*cis* retinoic acid, its isomer and metabolites. *In* G. M. Morriss-Kay (ed.), *Retinoids in Normal Development and Teratogenesis.* Oxford University Press, Oxford, pp. 267–280.

Danilevskii, A. S. 1965. *Photoperiodism and Seasonal Development of Insects.* Oliver and Boyd, Edinburgh.

Davis, D. L., Bradlow, H. L., Wolff, M., Woodruff, T., Hoel, D. G. and Anton-Culver, H. 1993. Xenoestrogens as preventable causes of breast cancer. *Environ. Health Perspect.* 101: 372–377.

Davis, W. L., Crawford, L. A., Cooper, O. J., Farmer, G. R., Thomas, D. and Freeman, B. L. 1990. Ethanol induces the generation of reactive free radicals by neural crest cells in culture. *J. Craniofac. Genet. Dev. Biol.* 10: 277–293.

Degnan, B. M. and Morse, D. E. 1995. Developmental and morphogenetic gene regulation in *Haliotis rufescens* larvae at metamorphosis. *Amer. Zool.* 35: 391–398.

Denno, R. F., Douglass, L. W. and Jacobs, D. 1985. Crowding and host plant nutrition: environmental determinants of wing form in *Prokelisia marginata. Ecology* 66: 1588–1596.

Dodson, S. 1989. Predator-induced reaction norms. *BioScience* 39: 447–452.

Dorris, M. 1989. *The Broken Cord.* Harper and Row, New York.

Edmonds, D. K., Lindsay, K. S., Miller, J. F., Williamson, E. and Wood, P. J. 1982. Early embryonic mortality in women. *Fertil. Steril.* 38: 447–453.

Edwards, M. 1994. Pollution in the former Soviet Union: Lethal legacy. *Natl. Geog.* 186 (2): 70–115.

Fallon, A. M., Hagedorn, H. H., Wyatt, G. R. and Laufer, H. 1974. Activation of vitellogenin synthesis in the mosquito *Aedes aegypti* by ecdysone. *J. Insect Physiol.* 26: 829–1823.

Ferguson, M. W. J. and Joanen, T. 1982. Temperature of egg incubation determines sex in *Alligator mississippiensis. Nature* 296: 850–853.

Franco, B. and twelve others. 1995. A cluster of sulfatase genes on Xp22.3 mutations in chondrodysplasia punctata (CDPX) and implications for warfarin embryopathy. *Cell* 81: 15–21.

Friedman, J. M. 1992. Effects of drugs and other chemicals on fetal growth. *Growth Genet. Horm.* 8(4): 1–5.

Fukuda, S. 1952. Function of the pupal brain and subesophageal ganglion in the production of non-diapause and diapause eggs in the silkworm. *Annot. Zool. Japan* 25: 149–155.

Geitz, H., Handt, S. and Zwingenberger, K. 1996. Thalidomide selectively modulates the density of cell surface molecules involved in the adhesion cascade. *Immunopharmacology* 31: 213–221.

Gilbert, S. F. 1994. Dobzhansky, Waddington and Schmalhausen: Embryology and the Modern Synthesis. *In* M. B. Adams (ed.), *The Evolution of Theodosius Dobzhansky: Essays on His Life and Thought in Russia and America.* Princeton University Press, Princeton, pp. 143–154.

Gimeno, S., Gerritsen, A., Bowmer, T. and Komen, H. 1996. Feminization of male carp. *Nature* 384: 221–222.

Giroud, A. and Martinet, M. 1959. Teratogenese pur hypervitaminose A chez le rat, la souris, le cobaye, et le lapin. *Arch. Fr. Pediatr.* 16: 971–980.

Gotthard, K. and Nylin, S. 1995. Adaptive plasticity and plasticity as an adaptation: A selective review of plasticity in animal morphology and life history. *Oikos* 74: 3–17.

Goulding, E. H. and Pratt, R. M. 1986. Isotretinoin teratogenicity in mouse whole embryo culture. *J. Craniofac. Genet. Dev. Biol.* 6: 99–112.

Greene, E. 1989. A diet-induced developmental polymorphism in a caterpillar. *Science* 243: 643–646.

Gregg, N. M. 1941. Congenital cataract following German measles in the mother. *Trans. Opthalmol. Soc. Aust.* 3: 35.

Guillette, L. J., Gross, T. S., Masson, G. R., Matter, J. M., Percival, H. F. and Woodward, A. R. 1994. Developmental abnormalities of the gonad and abnormal sex hormone concentrations in juvenile alligators from contaminated and control lakes in Florida. *Environ. Health Perspect.* 102: 680–688.

Hadfield, M. G. 1977. Metamorphosis in marine molluscan larvae: An analysis of stimulus and response. *In* R.-S. Chia and M. E. Rice (eds.), *Settlement and Metamorphosis of Marine Invertebrate Larvae.* Elsevier, New York, pp. 165–175.

Haeckel, E. 1891. Quoted in Nyhart, L., 1995.

Hagedorn, H. H. 1983. The role of ecdysteroids in the adult insect. *In* G. Downer and H. Laufer (eds.), *Endocrinology of Insects.* Alan R. Liss, New York, pp. 241–304.

Hansen, L. G. and Jansen, H. T. 1994. Environmental estrogens (Letter to editor in response to Stone, 1994). *Science* 266: 526.

Hansen, R. A., Hart, D. D. and Merz, R. A. 1991. Flow mediates predator-prey interaction between triclad flatworms and larval blackflies. *Oikos* 60: 187–196.

Hardie, J. 1981. Juvenile hormone and photoperiodically controlled polymorphism in *Aphis fabae*: Postnatal effects on presumptive gynoparae. *J. Insect Physiol.* 27: 347–355.

Hardie, J. and Lees, A. D. 1985. Endocrine control of polymorphism and polyphenism. *In* G. A. Kerkut and L. I. Gilbert (eds.), *Comprehensive Insect Physiology, Biochemistry, and Pharmacology.* Vol. 8, pp. 441–490.

Hart, M. W. and Strathmann, R. R. 1994. Functional consequences of phenotypic plasticity in echinoid larvae. *Biol. Bull.* 186: 291–299.

Hasegawa, K. 1952. Studies on voltinism of the silkworm, Bombyx mori L., with special reference to the organs controlling determination of voltinism. *J. Fac. Agric. Tottori Univ.* 1: 83–124.

Herbst, C. 1893. Experimentelle Untersuchungen über den Einfluss der veränderten chemischen Zusammensetzung des umgebenden Mediums auf die Entwicklung der Thiere. II. Wierteres über die morphologische Wirkung der Lithiumsalze und ihre theoretische Bedeutung. *Mitt. d. zool. Station Neapel.* 11: 136–220.

Hertwig, O. 1894. *The Biological Problem of To-day: Preformed or Epigenesis* (P. C. Mitchell, translator). Macmillan, New York.

Hilleman, B. 1996. Frog deformities pose a mystery. *Chem. Engin. News* 74: 24.

Hoffmann, R. J. 1973. Environmental control of seasonal variation in the butterfly *Colias eurytheme* I. Adaptive aspects of a photoperiodic response. *Evolution* 27: 387–397.

Holmes, L. B. 1979. Radiation. *In* V. C. Vaughan, R. J. McKay and R. D. Behrman (eds.), *Nelson Textbook of Pediatrics,* 11th Ed. Saunders, Philadelphia.

Hubel, D. H. 1967. Effects of distortion of sensory input on the visual system of kittens. *Physiologist* 10: 17–45.

Hubel, D. H. and Wiesel, T. N. 1962. Receptive fields, binocular interaction and functional architecture in the cat's visual cortex. *J. Physiol.* 160: 106–154.

Hubel, D. H. and Wiesel, T. N. 1963. Receptive fields of cells in striate cortex of very young, visually inexperienced kittens. *J. Neurophysiol.* 26: 944–1002.

Jaffe, L. A. 1980. Electrical polyspermy block in sea urchins: Nicotine and low sodium experiments. *Dev. Growth Differ.* 22: 503–507.

Janzen, F. J. and Paukstis, G. L. 1991. Environmental sex determination in reptiles: Ecology, evolution, and experimental design. *Q. Rev. Biol.* 66: 149–179.

Johnson, E. M. 1980. Screening for teratogenic potential: Are we asking the proper questions? *Teratology* 21: 259.

Johnston, M. C., Sulik, K. K., Webster, W. S. and Jarvis, B. L. 1985. Isotretinoin embryopathy in a mouse model: Cranial neural crest involvement. *Teratology* 31: 26A.

Jones, K. L. and Smith, D. W. 1973. Recognition of the fetal alcohol syndrome. *Lancet* 2: 999–1001.

Keiding, N and Skakkebaek, N. E. 1993. Are estrogens involved in falling sperm counts and disorders of the male reproductive tract? *Lancet* 341: 1392–1395.

Kelce, W. R., Stone, C. R., Laws, S. C., Gray, L. E., Kemppainen, J. A. and Wilson, E. M. 1995. Persistent DDT metabolite *p,p'*-DDE is a potent androgen receptor antagonist. *Nature* 375: 581–585.

Kennedy, C., Suda, S., Smith, C. B., Miyaoka, M., Ito, M. and Sokoloff, L. 1981. Changes in protein synthesis underlying functional plasticity in immature monkey visual system. *Proc. Natl. Acad. Sci USA* 78: 3950–3953.

Kitazawa, T., Kanda, T. and Takami, T. 1963. Changes of mitotic activity in the silkworm egg in relation to diapause. *Bull. Seric. Exp. Sta.* 18: 283–295.

Koch, P. B. and Buchmann, D. 1987. Hormonal control of seasonal morphs by the timing of ecdysteroid release in *Araschnia levana* (Nymphalidae: Lepidoptera). *J. Insect Physiol.* 36: 159–164.

Kochhar, D. M., Penner, J. D. and Tellone, C. I. 1984. Comparative teratogenic activities of two retinoids: Effects on palate and limb development. *Teratogen. Carcinogen. Mutagen.* 4: 377–387.

Kotch, L. E., Chen, S.-Y. and Sulik, K. K. 1995. Ethanol-induced teratogenesis: free radical damage as a possible mechanism. *Teratology* 52: 128–136.

Kulikauskas, V., Blaustein, A. B. and Ablin, R. J. 1985. Cigarette smoking and its possible effects on sperm. *Fertil. Steril.* 44: 526–528.

Lammer, E. J. and eleven others. 1985. Retinoic acid embryopathy. *N. Engl. J. Med.* 313: 837–841.

Lash, J. W. 1963. Studies on the ability of embryonic mesonephros explants to form cartilage. *Dev. Biol.* 6: 219–232.

Lash, J. W. and Saxén, L. 1972. Human teratogenesis: In vitro studies of thalidomide-inhibited chondrogenesis. *Dev. Biol.* 28: 61–70.

Lemoine, E. M., Harousseau, J. P., Borteyru, J. P. and Menuet, J. C. 1968. Les enfants de parents alcooliques: Anomalies observées *Oest. Med.* 21: 476–482.

Lenz, W. 1962. Thalidomide and congenital abnormalities. *Lancet* 1: 45. (First reported at a 1961 symposium.)

Lenz, W. 1966. Malformations caused by drugs in pregnancy. *Am. J. Dis. Child.* 112: 99–106.

Linney, E. 1992. Retinoic acid receptors: transcription factors modulating gene expression, development, and differentiation. *Curr. Top. Dev. Biol.* 27: 309–350.

McBride, W. G. 1961. Thalidomide and congenital abnormalities. *Lancet* 2: 1358.

McBride, W. G. and Vardy, P. H. 1983. Pathogenesis of thalidomide teratogenesis in the marmot (*Callithrix jacchus*): Evidence suggesting a possible trophic influence of cholinergic nerves in limb morphogenesis. *Dev. Growth Differ.* 25: 361–373.

McCarver-May, D. G. 1996. Genetic differences in alcohol dehydrogenase and fetal alcohol effects. Abstracts of the Ninth International Congress of Human Genetics *Brazil J. Genet.* 19: 73.

McCollum, S. A. and Van Buskirk, J. 1996. Costs and benefits of a predator induced polyphenism on the gray treefrog *Hyla chrysoscelis*. *Evolution* 50: 583–593.

McCredie, J. 1976a. Neural crest defects: A neuroanatomic basis for classification of multiple malformations related to phocomelia. *J. Neurol. Sci.* 28: 373–387.

McCredie, J. 1976b. The pathogenesis of congenital malformations. *Australas. Radiol.* 19: 348–355.

McKeown, T. 1976. Human malformations: An introduction. *Br. Med. Bull.* 32: 1–3.

McFall-Ngai, M. J. and Ruby, E. G. 1991. Symbiont recognition and subsequent morphogenesis as early events in an animal-bacterial mutualism. *Science* 254: 1491–1494.

Montgomery, M. K. and McFall-Ngai, M. J. 1995. The inductive role of bacterial symbionts in the morphogenesis of a squid light organ. *Amer. Zool.* 35: 372–380.

Morgan, T. H. 1909. Sex determination and parthenogenesis in phylloxerans and aphids. *Science* 29: 234–237.

Moroni, M. C., Vigano, M. A. and Mavilio, F. 1994. Regulation of human *Hoxd-4* gene by retinoids. *Mech. Dev.* 44: 139–154.

Morris-Kay, G. 1993. Retinoic acid and craniofacial development: Molecules and morphogenesis. *BioEssays* 15: 9–15.

Morse, A. N. C., Froyd, C. A. and Morse, D. E. 1984. Molecules from cyanobacteria and red algae that induce larval settlement and metamorphosis in the mollusc *Haliotis rufescens*. *Marine Biol.* 81: 293–298.

Nature Genetics (editorial). 1995. Risk assessment and religion. *Nat. Genet.* 11: 105–106.

Neubert, R., Hinz, N., Thiel, R. and Neubert, D. 1995. Down-regulation of adhesion receptors on cells of primate embryos as a probable mechanism of the teratogenic action of thalidomide. *Life Sci.* 58: 295–316.

Newman, R. A. 1989. Developmental plasticity of *Scaphiopus couchii* tadpoles in an unpredictable environment. *Ecology* 70: 1775–1787.

Newman, R. A. 1992. Adaptive plasticity in amphibian metamorphosis. *BioScience* 42: 671–678.

Nijhout, H. F. 1991. *The Development and Evolution of Butterfly Wing Patterns*. Smithsonian Institution Press, Washington, D. C.

Nijhout, H. F. 1994. *Insect Hormones*. Princeton University Press, Princeton.

Nordeen, K. W. and Nordeen, E. J. 1988. Projection neurons within a vocal pathway are born during song learning in zebra finches. *Nature* 334: 149–151.

Nowack, E. 1965. Die sensible Phase bei der Thalidomide-Embryopathie. *Humangenetik* 1: 516–536.

Nulman, I. and eight others. 1997. Neurodevelopment of children exposed in utero to antidepressant drugs. *N. Engl. J. Med.* 336: 258–262.

Nyhart, L. K. 1995. *Biology Takes Form: Animal Morphology and the German Universities, 1800–1900*. University of Chicago Press, Chicago.

O'Rahilly, R. and Müller, F. 1992. *Human Embryology and Teratology*. Wiley-Liss, New York.

Opitz, J. M. and Paul, N. W. (eds.) 1993. *Blastogenesis: Normal and Abnormal, March of Dimes Birth Defects Foundation Original Article Series*. Wiley-Liss, New York.

Palmer, A. R. 1985. Adaptive value of shell variation in *Thais lamellosa*: Effect of thick shells on vulnerability to and preference by crabs. *Veliger* 27: 349–356.

Passera, L. 1985. Soldier determination in ants of the genus *Pheidole*. *In* J. A. L. Watson, B. M Okot-Kotber and C. Noirot, (eds.) *Caste Determination in Social Insects*. Pergmon, Oxford, pp. 331–346.

Pener, M. P. 1991. Locust phase polymorphism and its endocrine relations. *Adv. Insect Physiol.* 3: 1–79.

Pflüger, E. 1883. Über den Einfluss der Schwerkraft auf die Theilung der Zellen. I, II, III. *Pflüger's Arch.* 32.

Pinder, A. W. and Friet, S. C. 1994. Oxygen transport in egg masses of the amphibians *Rana sylvatica* and *Ambystoma maculatum*: Convection, diffusion, and oxygen production by algae. *J. Exp. Biol.* 197: 17–30.

Plowright, R. C. and Pendrel, B. A. 1977. Larval growth in bumble-bees. *Can. Entomol.* 109: 967–973.

Purves, D. and Lichtman, J. W. 1985. *Principles of Neural Development*. Sinauer Associates, Sunderland, MA.

Pöpperl, H. and Featherstone, M. S. 1993. Identification of retinoic acid response element upstream from the mouse *Hox-4.2* gene. *Mol. Cell. Biol.* 13: 257–265.

Raatikainen, M. 1967. Bionomics, enemies, and population dynamics of *Javesella pellucida* (F.) (Homoptera, Delphaidae). *Annales Agric. Fenniae* 6: 1–49.

Rachinsky, A. and Hartfelder, K. 1990. Corpora allata activity, a prime regulating element for caste-specific juvenile hormone titre in honey bee larvae (*Apis mellifera carnica*). *J. Insect Physiol.* 36: 329–349.

Ramanathan, R., Wilkemeyer, M. F., Mittel, B., Perides, G. and Charness, M. E. 1996. Alcohol inhibits cell-cell adhesion mediated by human L1. *J. Cell Biol.* 133: 381–390.

Reiter, H. O. and Stryker, M. P. 1988. Neural plasticity without postsynaptic action potentials: Less-active inputs become dominant when kitten visual cortical cells are phamacologically inhibited. *Proc. Natl. Acad. Sci. USA* 85: 3623–3627.

Rothman, K. J., Moore, L. L., Singer, M. R., Nguyen, U. -S. D. T., Mannino, S. and Milunsky, A. 1995. Teratogenicity of high vitamin A intake. *N. Engl. J. Med.* 333: 1369–1373.

Ruberte, E., Dollé, P., Krust, A., Zalent, A., Morriss-Kay, G. and Chambon, P. 1990. Specific spatial and temporal distribution of retinoic acid receptor γ transcripts during mouse embryogenesis. *Development* 108: 213–222.

Ruberte, E. and seven others. 1991. Retinoic acid receptors in the embryo. *Semin. Dev. Biol.* 2: 153–159.

Sander, K. 1968. Entwicklungsphysiologische Untersuchungen am embryonalen Mycetom von Euscelis plebejus F. (Homoptera, Ciciadina). I. *Dev. Biol.* 17: 16–38.

Sapp, J. 1994. *Evolution by Association: A History of Symbiosis.* Oxford University Press, NY.

Schmalhausen, I. I. 1949. *Factors of Evolution: The Theory of Stabilizing Selection.* University of Chicago Press, Chicago.

Schwemmler, W. 1974. Endosymbionts: factors of egg patterning. *J. Insect Physiol.* 20: 1467–1474.

Schwemmler, W. 1989. Insect symbiosis as a model system for egg cell differentiation. *In* W. Schwemmler and G. Gassner, eds. *Insect Endosymbiosis.* CRC Press, Boca Raton, pp. 37–53.

Selenka, E. 1876. Zur Entwicklung der Holothurien: Ein Beitrage zur Keimblättertheorie. *Zeitschr. wissensch. Zool.* 27: 155–178.

Shapiro, A. M. 1968. Photoperiodic induction of vernal phenotype in *Pieris protodice* Boisduval and Le Conta (Lepidoptera: Pieridae). *Wasmann J. Biol.* 26: 137–149.

Shapiro, A. M. 1976. Seasonal polyphenism. *Evol. Biol.* 9: 259–333.

Shapiro, A. M. 1978. The evolutionary significance of redundancy and variability in phenotypic induction mechanisms of pierid butterflies (Lepidoptera). *Psyche* 85: 275–283.

Shaw, G. M., Wasserman, C. R., Lammer, E. J., O'Malley, C. D., Murray, J. C., Basart, A. M. and Tolarova, M. M. 1996. Orofacial clefts, parental cigarette smoking, and transforming growth factor-alpha gene variants. *Am. J. Hum. Genet.* 58: 551–561.

Soto, A., Justicia, H., Wray, J. and Sonnenschein, C. 1991. p-nonylphenol: an estrogenic xenobiotic released from "modified" polystyrene. *Environ. Health Perspect.* 92: 167–173.

Stearns, S. C., de Jong, G. and Newman, R. A. 1991. The effects of phenotypic plasticity on genetic correlations. *Trends Ecol. Evol.* 6: 122–126.

Stent, G. S. 1973. A physiological mechanism for Hebb's postulate of learning. *Proc. Natl. Acad. Sci. USA* 70: 997–1001.

Stone, R. 1994. Environmental estrogens stir debate. *Science* 265: 308–310.

Stone, R. 1995. Environmental toxicants under scrutiny at Baltimore meeting. *Science* 267: 1770–1771.

Strathmann, R. R. and Strathmann, M. F. 1995. Oxygen supply and limits on aggregation of embryos. *J. Mar. Biol. Assoc. U. K.* 75: 413–428.

Strathmann, R. R., Fenaux, L. and Strathmann, M. F. 1992. Heterochronic developmental plasticity in larval sea urchins and its implication for evolution on nonfeeding larvae. *Evolution* 46: 972–986.

Streissguth, A. P. and LaDue, R. A. 1987. Fetal alcohol: Teratogenic causes of developmental disabilities. *In* S. R. Schroeder (ed.), *Toxic Substances and Mental Retardation.* American Association of Mental Deficiency, Washington, DC, pp. 1–32.

Studer, M., Popperl, H., Marshall, H., Kuroiwa, A. and Krumlauf, R. 1994. Role of a conserved retinoic acid response element in rhombomere restriction of *Hoxb-1. Science* 265: 1728–1732.

Sulik, K. K., Cook, C. S. and Webster, W. S. 1988. Teratogens and craniofacial malformations: relationships to cell death. *Development* 103 (Suppl.): 213–231.

Tauber, M. J., Tauber, C. A. and Masaki, S. 1986. *Seasonal Adaptations of Insects.* Oxford University Press, Oxford.

Toms, D. A. 1962. Thalidomide and congenital abnormalities. *Lancet* 2: 400.

Turner, A. M. and Greenough, W. T. 1983. Synapses per neuron and synaptic dimensions in occipital cortex of rats reared in complex, social, or isolation housing. *Acta Stereologica* 2 (Suppl. 1): 239–244.

Via, S., Gomulkiewicz, R., De Jong, G., Scheiner, S. M., Schlichting, C. D. and Van Tienderen, P. H. 1995. Adaptive phenotypic plasticity: Consensus and controversy. *Trends Ecol. Evol.* 10: 212–217.

Voet, D. and Voet, J. G. 1995. *Biochemistry I,* 2nd ed. John Wiley, NY.

Volpe, E. P. 1987. Developmental biology and human concerns. *Am. Zool.* 27: 697–714.

Waddington, C. H. 1942. Canalization of development and the inheritance of acquired characteristics. *Nature* 150: 563–565.

Warner, R. R. 1993. Mating behavior and hermaphroditism in coral reef fishes. *In* P. W. Sherman and J. Alcock, (eds), *Exploring Animal Behavior.* Sinauer Associates, Sunderland, MA, pp. 188–196.

Watt, W. B. 1968. Adaptive significance of pigment polymorphism in *Colias* butterflies, I. Variation of melanin in relation to thermoregulation. *Evolution* 22: 437–458.

Watt, W. B. 1969. Adaptive significance of pigment polymorphism in *Colias* butterflies, II. Thermoregulation and periodically controlled melanin production in *Colias eurytheme. Proc. Natl. Acad. Sci. USA* 63: 767–774.

Weele, C. van der, 1995. *Images of Development: Environmental Causes in Ontogeny.* Elinkwijk, Utrecht.

Weismann, A. 1875. "Über den Saison-Dimorphismus der Schmetterlinge. In *Studien zur Descendenz-Theorie.* Engelmann, Leipzig.

Wheeler, D. 1986. Developmental and physiological determinants of caste in social hymenoptera: Evolutionary implications. *Am. Nat.* 128: 13–34.

Wheeler, D. 1991. The developmental basis of worker caste polymorphism in ants. *Amer. Nat.* 138: 1218–1238.

Wiesel, T. N. 1992. Postnatal development of the visual cortex and the influence of environment. *Nature* 299: 583–591.

Wirtz, P. 1973. Differentiation in the honeybee larva. *Meded. Landb. Hogesch. Wagningen.* 73–75, 1–66.

Wolpert, L. 1991. *The Triumph of the Embryo.* Oxford University Press. Oxford.

Woltereck, R. 1909. Weitere experimentelle Untersuchungen über Artveränderung, speziell über das Wesen quantitativer Artunderscheide bei Daphniden. *Versuch. Deutsch. Zool. Ges.* 1909: 110–172.

Woodward, D. E. and Murray, J. D. 1993. On the effect of temperature-dependent sex determination on sex ratio and survivorship in crocodilians. *Proc. R. Soc. Lond.* [B] 252: 149–155.

Yu, V. C. and nine others. 1991. RXRb: A coregulator that enhances binding of retinoic acid, thyroid hormone, and vitamin D receptors to their cognate response elements. *Cell* 67: 1251–1266.

The saga of the germ line

22

*And the end of all our exploring
Will be to arrive where we started
And know the place for the first time.*
T. S. ELIOT (1942)

WE BEGAN OUR ANALYSIS of animal development by discussing fertilization, and we will finish our studies of individual development by investigating **gametogenesis,** the processes by which the sperm and the egg are formed. Germ cells provide the continuity of life between generations, and the mitotic ancestors of our own germ cells once resided in the gonads of reptiles, amphibians, fishes, and invertebrates. In many animals, such as insects, roundworms, and vertebrates, there is a clear and early separation of germ cells from somatic cell types. In several animal phyla (and in the entire plant kingdom), this division is not as well established. In these species (which include cnidarians, flatworms, and tunicates), somatic cells can readily become germ cells even in adult organisms. The zooids, buds, and polyps of many invertebrate phyla testify to the ability of somatic cells to give rise to new individuals.

In those organisms where there is an established germ line that separates early in development, the germ cells do not arise within the gonad itself. Rather, their precursors—the **primordial germ cells** (**PGCs**)—migrate into the developing gonads. The first step in gametogenesis, then, involves forming the PGCs and getting them into the genital ridge as the gonad is forming. The initiation of the germ cell lineage (the germ line) in amphibians, insects, and roundworms was discussed in Chapter 13. We resume our story of the germ line with the migration of the PGCs from their place of origin into the gonads.

Germ cell migration

Germ Cell Migration in Amphibians

As discussed in Chapter 13, the germ plasm of anuran amphibians—frogs and toads—collects around the vegetal pole in the 1-cell embryo. During cleavage, this material is brought upward through the yolky cytoplasm, and the RNA-rich granules become associated with the endodermal cells lining the floor of the blastocoel (Figure 22.1; Bounoure, 1934; Ressom and Dixon, 1988; Kloc et al., 1993). The PGCs become concentrated in the posterior region of the larval gut, and as the abdominal cavity forms, the anuran PGCs

Figure 22.1
Changes in the position of germ plasm (color) in an early frog embryo. Originally located near the vegetal pole of the uncleaved egg (A), the germ plasm advances along the cleavage furrows (B) until it becomes localized at the floor of the blastocoel (C). (After Bounoure, 1934.)

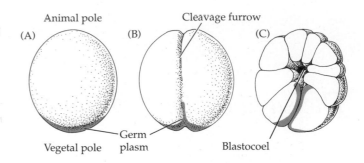

emigrate along the dorsal side of the gut, first along the dorsal mesentery (which connects the gut to the region where the mesodermal organs are forming) and then along the abdominal wall and into the genital ridges. They migrate up this tissue until they reach the developing gonads (Figure 22.2). *Xenopus* PGCs move by extruding a single filopodium and then streaming their yolky cytoplasm into the filopodium while retracting their tail. Contact guidance in this migration seems likely as both the cells and the extracellular matrix over which they migrate are oriented in the direction of that migration (Wylie et al., 1979). Furthermore, PGC adhesion and migration can be inhibited if the mesentery is treated with antibodies against *Xenopus* fibronectin (Heasman et al., 1981). Thus, the pathway for germ cell migration in these frogs appears to be composed of an oriented fibronectin-containing extracellular matrix. The fibrils over which the PGCs travel lose this polarity soon after migration has ended.* As they migrate, *Xenopus* PGCs divide about three times, and approximately 30 PGCs colonize the gonads (Whitington and Dixon, 1975; Wylie and Heasman, 1993). These will divide to form the germ cells.

The primordial germ cells of urodele amphibians (salamanders) have an apparently different origin, which has been traced by reciprocal transplantation experiments to the regions of the mesoderm that involute through the ventrolateral lips of the blastopore. Moreover, there does not seem to be any particular localized "germ plasm" in salamander eggs. Rather, the interaction of the dorsal endoderm cells and animal hemisphere cells creates the conditions needed to form germ cells in the particular areas that involute through the ventrolateral lips (Sutasurya and Nieuwkoop, 1974). So in salamanders, the PGCs are formed by induction within the mesodermal region and presumably follow a different path into the gonad.

Germ Cell Migration in Mammals

There is no obvious germ plasm in mammals, and mammalian germ cells are not morphologically distinct during early development. However, by using monoclonal antibodies that recognize cell-surface differences between the PGCs and their surrounding cells, Hahnel and Eddy (1986) showed that mouse PGCs originally reside in the epiblast of the gastrulating embryo. Ginsburg and her colleagues (1990) localized this region to the extraembryonic mesoderm just posterior to the primitive streak of the 7-day mouse embryo. Here about eight large, alkaline-phosphatase-staining cells are seen. If this area is removed, the remaining embryo becomes devoid of germ cells, while the isolated segment develops a large number of primordial germ cells. In normal mouse embryos, the germ cell precursors in the extraembry-

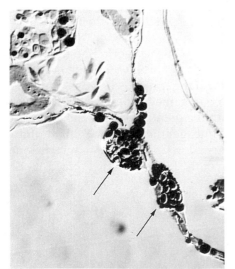

Figure 22.2
Migration of primordial germ cells in a frog. This phase-contrast photomicrograph of a section through the body wall and dorsal mesentery of a *Xenopus* embryo shows the migration of two large primordial germ cells (arrows) along the dorsal mesentery. (From Heasman et al., 1977, courtesy of the authors.)

*This does not necessarily hold true for all anurans. In the frog *Rana pipiens*, the germ cells follow a similar route but may be passive travelers along the mesentery rather than actively motile cells (Subtelny and Penkala, 1984).

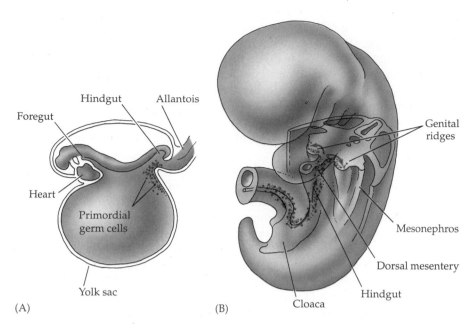

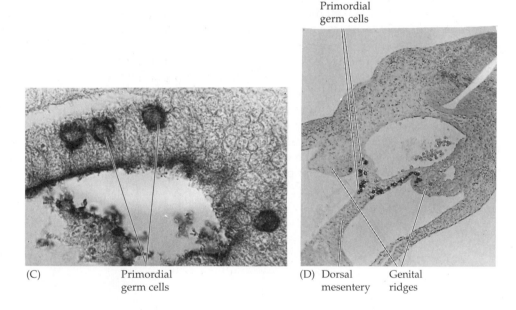

Figure 22.3
Pathway for the migration of mammalian primordial germ cells. (A) Primordial germ cells seen in the yolk sac near the junction of the hindgut and allantois. (B) Migration through gut and, dorsally, up the dorsal mesentery into the genital ridge. (C) Four large PGCs in the hindgut of mouse embryo (near the allantois and yolk sac) stain positively for high levels of alkaline phosphatase. (D) Such alkaline-phosphatase-staining cells can be seen migrating up the dorsal mesentery and entering the genital ridges. (A and B from Langman, 1981; C from Heath, 1978; D from Mintz, 1957; photographs courtesy of the authors.)

onic mesoderm then migrate back into the embryo, first into the mesoderm of the primitive streak and then to the endoderm by way of the allantois. The route of the mammalian PGC migration from the allantois (Figure 22.3) resembles that of the anuran PGC migration. After collecting at the allantois by day 7.5 (Chiquoine, 1954; Mintz, 1957), the mammalian PGCs migrate to the adjacent yolk sac (Figure 22.3A,C). By this time, they have already split into two populations that will migrate to either the right or the left genital ridge. The PGCs then move caudally from the yolk sac through the newly formed hindgut and up the dorsal mesentery into the genital ridge (Figure 22.3B,D). Most of the PGCs have reached the developing gonad by the eleventh day after fertilization. During this trek, they have proliferated from an initial population of 10 to 100 cells to the 2500 to 5000 PGCs present in the gonads by day 12. Like the PGCs of *Xenopus*, mammalian PGCs appear to be closely associated with the cells over which they migrate, and they move by extending filopodia over the underlying cell surfaces. These cells are also capable of penetrating cell monolayers and migrating through the cell sheets (Stott and Wylie, 1986). The mechanism by which the primordial germ cells know the route of this journey is still unknown. Fibronectin is likely to be an important substrate on which the PGCs migrate (ffrench-Constant et al., 1991), and in vitro evidence suggests that the genital ridges of 10.5-day mouse embryos secrete a diffusible TGF-β1-like protein that is capable of attracting mouse primordial germ cells (Godin et al, 1990; Godin and Wylie, 1991). Whether the genital ridge is able to provide such cues in vivo still must be tested.

Although no germ plasm has been found, the retention of totipotency has been correlated with the expression of a nuclear transcription factor, Oct4. This factor is expressed in the early-cleavage blastomere nuclei and is then expressed in the inner cell mass. During gastrulation, it becomes expressed solely in those posterior epiblast cells thought to give rise to the primordial germ cells. After that, this protein is seen only in the primordial germ cells and oocytes (Figure 22.4; Yeom et al., 1996; Plate 30).

The proliferation of the PGCs appears to be promoted by the **stem cell factor,** the same growth factor needed for the proliferation of neural-crest-derived melanoblasts and hematopoietic stem cells (see Chapter 7). The stem cell factor is produced by the cells along the migration pathway and remains bound to their cell membranes. It appears that the presentation of this protein on membranes is important for its activity. Mice homozygous for the *White* mutation (*W*) are deficient in germ cells (and melanocytes and blood cells), as their stem cells lack the receptor for stem cell growth factor. Mice homozygous for the *Steel* mutation have a similar phenotype, as they lack the ability to make this growth factor. Mice homozygous for the *Steel-Dickie* (Sl^d) allele have reduced numbers of germ cells, as these mice can make the stem cell growth factor, but it does not remain bound to their membranes (Dolci et al., 1991; Matsui et al., 1991). The addition of stem cell factor to PGCs taken from 11-day mice will stimulate their proliferation for about 24 hours and appears to prevent programmed cell death that would otherwise occur (Pesce et al., 1993).

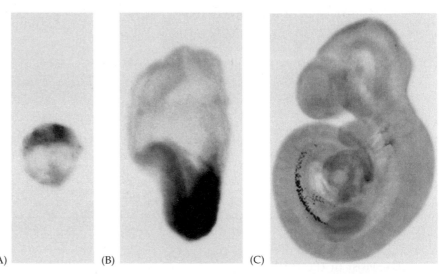

(A)　　　　　　(B)　　　　　　(C)

Figure 22.4
Expression of *Oct4* mRNA correlates with totipotency and ability to form germ cells. An *Oct4/lacZ* transgene driven by the *Oct4* promotor region shows its expression in (A) inner cell mass, (B) posterior epiblast in 8.5-day embryo, and (C) migrating PGCs in 10.5-day embryo. (After Yeom et al., 1996; permission courtesy of H. R. Schöler.)

Teratocarcinomas and Embryonic Stem Cells

STEM CELL FACTOR increases the proliferation of migrating mouse primordial germ cells in culture, and this proliferation can be further increased by adding another growth factor, leukemia inhibition factor (LIF). However, the life span of these cells is short, and the cells soon die. But if an additional mitotic regulator—basic fibroblast growth factor—is added, a remarkable change takes place. The cells continue to proliferate, producing a pluripotent embryonal stem cell with characteristics resembling the cells of the inner cell mass (Matsui et al., 1992). We have discussed such embryonic stem cells earlier, as these are the cells that can be transfected with recombinant genes and inserted into the blastocyst to create transgenic mice.

Such a mammalian germ cell or stem cell contains within it all the information needed for subsequent development. What would happen if such a cell became malignant? In one type of tumor, the germ cells become embryonic stem

cells, as in the experiment already referred to. This type of tumor is called a **teratocarcinoma.** Whether spontaneous or experimentally produced, a teratocarcinoma contains an undifferentiated stem cell population that has biochemical and developmental properties remarkably similar to those of cells of the inner cell mass (Graham, 1977). Moreover, these stem cells not only divide but

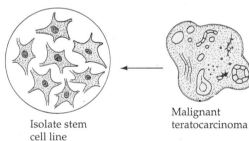

Figure 22.5 Photomicrograph of a section through a teratocarcinoma, showing numerous differentiated cell types. (From Gardner, 1982; photograph from C. Graham, courtesy of R. L. Gardner.)

can also differentiate into a wide variety of tissues, including gut and respiratory epithelia, muscle, nerve, cartilage, and bone (Figure 22.5). Once differentiated,

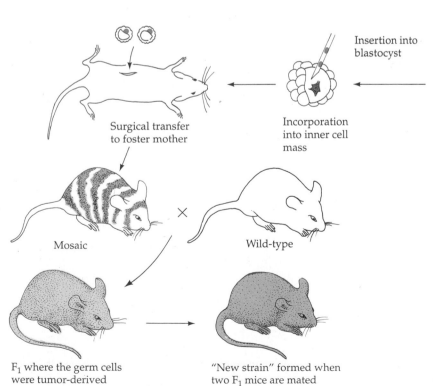

Figure 22.6 Protocol for breeding mice whose genes are derived largely from tumor cells. Stem cells are isolated from a mouse teratocarcinoma and inserted into blastocysts from a different strain of mouse. The chimeric blastocysts are placed into a foster mother. If the tumor cells are integrated into the blastocyst, the mouse that develops will have many of its cells derived from the tumor. If the tumor has given rise to germ cells, the mosaic mice can be mated to normal mice to produce an F_1 generation. The F_1 mice should be heterozygous for all the chromosomes of the tumor cell. Matings between F_1 mice produce F_2 mice having some homozygous genes derived from tumor cells. (After Stewart and Mintz, 1981.)

these cells no longer divide and are therefore not malignant. Such tumors can give rise to most of the tissue types in the body. Thus, the teratocarcinoma stem cells mimic early mammalian development, but the tumor they form is characterized by random, haphazard development.

In 1981, Stewart and Mintz formed a mouse from cells derived in part from a teratocarcinoma stem cell. Stem cells that had arisen in a teratocarcinoma of an agouti (yellow-tipped) strain of mice were cultured for several cell generations and were seen to maintain the characteristic chromosome complement of the parental mouse. Individual stem cells of this type were injected into the blastocysts of black mice. The blastocysts were then transferred to the uterus of a foster mother, and live mice were born. Some of these mice had coats of two colors, indicating that the tumor cell had integrated itself into the embryo. Moreover, when mated to a mouse carrying an appropriate marker, the chimeric mouse was able to generate mice having some of the phenotypes of the tumor "parent." The malignant embryonal carcinoma cell had produced many, if not all, types of normal somatic cells and even had produced normal, functional germ cells! When mice having a tumor cell for one parent were mated together, the resultant litter contained mice that were homozygous for a large number of genes from the tumor cell (Figure 22.6). ∎

Germ Cell Migration in Birds and Reptiles

In birds and reptiles, primordial germ cells are derived from epiblast cells that migrate from the central region of the area pellucida to a crescent-shaped zone in the hypoblast at the anterior border of the area pellucida (Figure 22.7; Eyal-Giladi et al., 1981; Ginsburg and Eyal-Giladi, 1987). This extraembryonic region is called the **germinal crescent,** and the primordial germ cells multiply in this region. Unlike the PGCs in amphibians and mammals, germ cells in birds and reptiles migrate primarily by means of the bloodstream (Figure 22.8). When the blood vessels form in the germinal crescent, the PGCs enter the vessels and are carried by the circulation to the region where the hindgut is forming. Here, they exit from the circulation, become associated with the mesentery, and migrate into the genital ridges (Swift, 1914; Kuwana, 1993). The PGCs of the germinal crescent appear to enter the blood vessels by **diapedesis,** a type of movement common to lymphocytes and macrophages that enables cells to squeeze between the endothelial cells of small blood vessels.

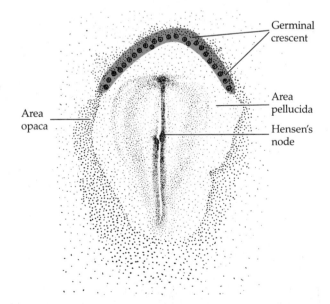

Figure 22.7
Dorsal view of a primitive-streak-stage embryo, showing the region, called the germinal crescent, in which the germ cells arise. (After Swift, 1914.)

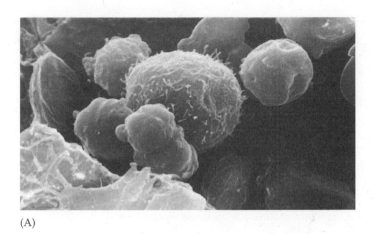

(A)

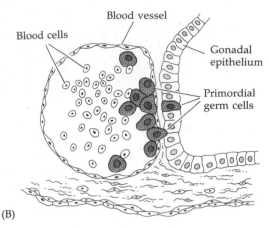

(B)

Figure 22.8
Primordial germ cells in the chick embryo. (A) Scanning electron micrograph of chick PGC in a capillary of a gastrulating embryo. The PGC can be identified by its large size and the microvilli on its surface. (B) Transverse section near the prospective gonadal region of a chick embryo. Several PGCs within the blood vessel cluster next to the epithelium. One PGC is crossing through the blood vessel endothelium, and another PGC is already located within the epithelium. (A from Kuwana, 1993, courtesy of T. Kuwana; B after Romanoff, 1960.)

The PGCs thus enter the embryo by being transported in the blood (Pasteels, 1953; Dubois, 1969). The PGCs must also "know" to get out of the blood when they reach the developing gonad (see Figure 22.8B). When the germinal crescent of a chick embryo is removed, and the circulation of that embryo is joined with that of a normal chick embryo, the primordial germ cells from the normal embryo will migrate into both sets of gonads (Simon, 1960). It is not known what causes this attraction for the genital ridges. One possibility is that the developing gonad produces a chemotactic substance that attracts PGCs and retains them in the capillaries bordering the gonad (Rogulska, 1969). (Such substances are known to be secreted by lymphocytes at the sites of infection to attract macrophages to that area and permit them to pass through the capillary wall by diapedesis.) Evidence for such chemotaxis came from studies (Kuwana et al., 1986) in which circulating chick PGCs were isolated from the blood and cultured between gonadal rudiments and other embryonic tissues. The PGCs migrated into the gonadal rudiments during a 3-hour incubation.

Another possibility is that the endothelial cells of the gonadal capillaries have a cell-surface compound that causes the PGCs to adhere there specifically. Using monoclonal antibodies that recognize different cell-surface molecules, Auerbach and Joseph (1984) have shown that the endothelial cells of several capillary networks have different cell membrane components and that the endothelial cells from ovarian capillaries differ from all others tested.* Both chemotaxis and differential cell adhesion mechanisms may be working. Whatever these factors may be, they are not species-specific. The chicken gonad attracts circulating PGCs from the turkey and even the mouse (Reynaud, 1969; Rogulska et al., 1971).

Primordial Germ Cell Migration in Drosophila

During *Drosophila* embryogenesis, germ cells move from the posterior pole to the gonads. The first step is a passive phase, wherein the germ cells are displaced by the movements of the embryonic cells during gastrulation. The differentiation of the endoderm triggers active amoeboid movement in the

---

*A similar situation is thought to occur when lymphocytes migrate through the bloodstream and leave the circulation when they enter the capillary bed of a particular lymphoid organ. The mechanism for this "homing" and organ specificity involves the lymphocytes' ability to specifically adhere to the blood vessel endothelial cells in these organs. Peripheral lymph node endothelial cells contain a glycoprotein, a **selectin,** in their cell membranes that is essential for the binding and exit of those lymphocytes that can recognize it. For each selectin on these endothelial cells, there is a complementary molecule on the lymphocytes that can recognize it (Gallatin et al., 1983, 1986).

(A)

(B)

(C)

(D)

(E)

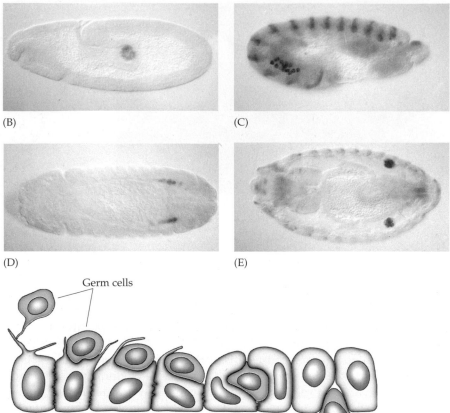

Germ cells

(F)

Figure 22.9
Primordial germ cell migration in *Drosophila*. (A) Germ cells stained with antibodies to the Vasa protein show germ cells originating at the posterior pole. (B) During germ band extension, the germ cells are moved into the posterior midgut. (C) Germ cells migrate through the gut wall (the embryo is counterstained for Engrailed protein) and (D) migrate in two single files through the mesoderm, where they (E) aggregate in the developing gonads. (F) Process of migrating through the gut wall, which is triggered by endodermal differentiation. (A–E from Warrior, 1994, permission courtesy of R. Warrior; F after Jaglarz and Howard, 1995.)

primordial germ cells, and they travel through the gut endothelium, migrating toward the mesoderm. The primordial germ cells then split into two groups, each of which becomes associated with a developing gonad primordium (Figure 22.9; Warrior, 1994).

The product of the *wuwen* gene appears to be responsible for directing the migration of the PGCs from the endoderm into the mesoderm. This protein is expressed in the endoderm immediataly before PGC migration, and it repels the PGCs. In loss-of-function mutants of this gene, the PGCs wander randomly (Zhang et al., 1997).

Meiosis

Once in the gonad, primordial germ cells continue to divide mitotically, producing millions of potential gametes. The PGCs of both male and female gonads are then faced with the necessity of reducing their chromosome number from the diploid to the haploid condition. In the haploid condition, each chromosome is represented only once, whereas diploid cells have two copies of each chromosome. To accomplish this reduction, the male and female germ cells undergo meiosis.

After the last mitotic division, a period of DNA synthesis occurs, so that the cells initiating meiosis have twice the normal amount of DNA in their nuclei. In this state, each chromosome consists of two sister chromatids attached at a common centromere. (In other words, the diploid nucleus contains four copies of each chromosome, but the chromosomes are seen as two chromatids bound together.) Meiosis (shown in Figure 1.13) entails two cell divisions. In the first division, homologous chromosomes (e. g., the chromosome 3 pair in the diploid cell) come together and are then separated into different cells. Hence, the first meiotic division separates homologous chromosomes into two daughter cells such that each cell has only one copy of each chromosome. But each of the chromosomes has already replicated. The second meiotic division then separates the two sister chromatids from each other. Consequently, each of the four cells produced by meiosis has a single (haploid) copy of each chromosome.

The first meiotic division begins with a long prophase, which is subdivided into five parts. During the **leptotene** (Greek, "thin thread") stage, the chromatin of the chromatids is stretched out very thinly, and it is not possible to identify individual chromosomes. DNA replication has already occurred, however, and each chromosome consists of two parallel chromatids. At the **zygotene** (Greek, "yoked threads") stage, homologous chromosomes pair side by side. This pairing is called **synapsis,** and it is characteristic of meiosis. Such pairing does not occur during mitotic divisions. Although the mechanism whereby each chromosome recognizes its homologue is not known, pairing seems to require the presence of the nuclear membrane and the formation of a proteinaceous ribbon called the **synaptonemal complex.** This complex is a ladderlike structure with a central element and two lateral bars (von Wettstein, 1984; Schmekel and Daneholt, 1995). The chromatin is associated with the two lateral bars and the chromatids are thus joined together (Figure 22.10). Examinations of the meiotic cell nuclei with the electron microscope (Moses, 1968; Moens, 1969) suggest that paired chromosomes are bound to the nuclear membrane, and Comings (1968) has suggested that the nuclear envelope helps bring together the homologous chromosomes. The configuration formed by the four chromatids and the synaptonemal complex is referred to as a **tetrad** or a **bivalent.**

During the next stage of meiotic prophase, the chromatids thicken and shorten. This stage has therefore been called the **pachytene** (Greek, "thick thread") stage. Individual chromatids can now be distinguished under the light microscope, and crossing-over may occur. Crossing-over represents exchanges of genetic material whereby genes from one chromatid are ex-

(A)

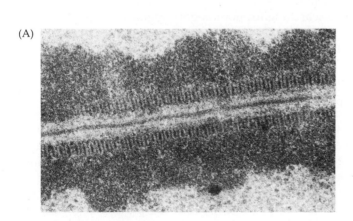

(B)

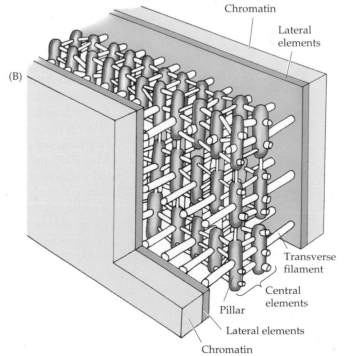

Figure 22.10
The synaptonemal complex. (A) Homologous chromosomes held together at the first meiotic prophase in the *Neottiella* oocyte. (B) Interpretive diagram of the synaptonemal complex structure. (A from von Wettstein, 1971, courtesy of D. von Wettstein; B after Schmekel and Daneholt, 1995.)

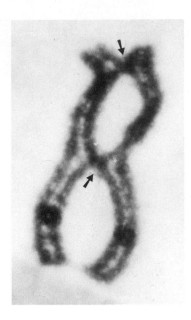

Figure 22.11
Chiasmata in diplotene bivalent chromosomes of salamander oocytes. Centromeres are visible as darkly stained circles; arrows point to the two chiasmata. (Courtesy of J. Kezer.)

changed with homologous genes from another chromatid. This crossing-over continues into the next stage, the **diplotene** (Greek, "double threads"). Here, the synaptonemal complex breaks down, and the two homologous chromosomes start to separate. Usually, however, they remain attached at various places called **chiasmata,** which are thought to represent regions where crossing-over is occurring (Figure 22.11). The diplotene stage is characterized by a high level of gene transcription. In some species, the chromosomes of both male and female germ cells take on the "lampbrush" appearance characteristic of chromosomes that are actively making RNA. During the next stage, **diakinesis** (Greek, "moving apart"), the centromeres move away from each other, and the chromosomes remain joined only at the tips of the chromatids. This last stage of meiotic prophase ends with the breakdown of the nuclear membrane and the migration of the chromosomes to the metaphase plate.

During anaphase I, homologous chromosomes are separated from each other in an independent fashion. This stage leads to telophase I, during which two daughter cells are formed, each cell containing one partner of the homologous chromosome pair. After a brief **interkinesis,** the second division of meiosis takes place. During this division, the centromere of each chromosome divides during anaphase so that each of the new cells gets one of the two chromatids, the final result being the creation of four haploid cells. Note that meiosis has also reassorted the chromosomes into new groupings. First, each of the four haploid cells has a different assortment of chromosomes. In humans, where there are 23 different chromosome pairs, there can be 2^{23} (nearly 10 million) different types of haploid cells formed from the genome of a single person. In addition, the crossing-over that occurs during the pachytene and diplotene stages of prophase I further increases genetic diversity and makes the number of different gametes incalculable.

The mechanism for homologue pairing is unknown. It is thought that the first events involve searching for homologous regions of chromatin and that this process might utilize DNA repair enzymes (Baker et al., 1996). Mouse mutants lacking such repair enzymes have abnormal synapses. After the homologous regions are aligned, synapsis is initiated at local regions. In *Drosophila*, recent evidence suggests that synapsis is initiated at regions of heterochromatin (Dernburg et al., 1996). Moreover, interactions between the **Mei-S332** protein and the heterochromatin around the kinetochore are critical in holding the sister chromatids together (Karpen et al., 1995, 1996; Kerrebrock et al., 1995). The gene for the Mei-S332 protein was discovered by screening mutations for their inability to complete meiosis. In these mutants, sister chromatids separated precociously about 90 percent of the time. In vertebrates, the **Rad51** protein corresponds to the lateral elements of the synaptonemal complex and appears to mediate the pairing of the homologues (Ashley et al., 1995). At the end of meiosis, this protein is localized only at the sites where the homologues are still attached. It would be important to know how these and other proteins interact during human gametogenesis, since most nondisjunction events (such as those leading to trisomies) are thought to be defects of meiotic pairing (see Yoon et al., 1996).

Sidelights & Speculations

Big Decisions: Mitosis or Meiosis? Sperm or Egg?

IN MANY SPECIES, the germ cells migrating into the gonad are bipotential and can differentiate into sperm or ova, depending on their gonadal environment. When the ovaries of salamanders are experimentally transformed into testes, the resident germ cells cease their oogenic differentiation and begin developing as sperm (Burns, 1930; Humphrey, 1931). Similarly, in the housefly and mouse, the gonad is able to direct the differentiation of the germ cell (McLaren, 1983; Inoue and Hiroyoshi, 1986). Thus, in most organisms, the sex of the gonads and its germ cells is the same.

But what about hermaphroditic animals, in which the change from sperm production to egg production is a naturally occurring physiological event? How is the same animal capable of producing sperm during one part of its life and oocytes during another part? Using *Caenorhabditis elegans*, Kimble and her colleagues have identified two "decisions" that presumptive germ cells have to make. The first involves whether to enter meiosis or to remain a mitotically dividing stem cell. The second decision concerns whether the meiotic cell is to become an egg or a sperm. Recent evidence shows that these decisions are intimately linked. The mitotic/meiotic decision is controlled by a single nondividing cell at the end of each gonad, the **distal tip cell**. The germ cell precursors near this cell divide mitotically, forming the pool of germ cells; but as these cells get farther away from the distal tip cell, they enter meiosis. If the distal tip cells are destroyed by a focused laser beam, *all* the germ cells enter meiosis, and if the distal tip cell is placed into a different location in the gonad, germ line stem cells are generated near this new position (Figure 22.12; Kimble, 1981; Kimble and White, 1981). It appears that the distal tip cells secrete some substance that maintains these cells in mitosis and inhibits their meiotic differentiation. Austin and Kimble (1987) have isolated

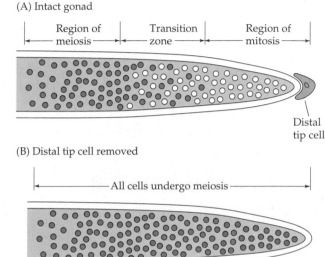

(A) Intact gonad

Region of meiosis | Transition zone | Region of mitosis

Distal tip cell

(B) Distal tip cell removed

← All cells undergo meiosis →

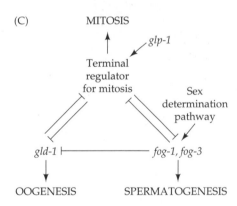

(C)

MITOSIS

glp-1

Terminal regulator for mitosis

Sex determination pathway

gld-1 ⊢   *fog-1, fog-3*

OOGENESIS   SPERMATOGENESIS

a mutation that mimics the phenotype obtained when the distal tip cells are removed. All the germ cell precursors of nematodes homozygous for the recessive mutation *glp-1* initiate meiosis, leaving no mitotic population. Instead of the 1500 germ cells usually found in the fourth larval stage of hermaphroditic development, these mutants produce only 5 to 8 sperm cells. When genetic chimeras are made in which wild-type germ cell precursors are found within a mutant larva, the wild-type cells are able to respond to the distal tip cells and undergo mitosis. However, when mutant germ cell precursors are found within

Figure 22.12 Regulation of the mitosis-or-meiosis decision by the distal tip cell of the *C. elegans* ovotestis. (A) Intact gonad early in development with regions of mitosis (shaded cells) and meiosis. (B) Gonad after laser ablation of the distal tip cell. All germ cells enter meiosis. (C) Model for the interactions by which germ cells adopt a single fate. The *gld-1* gene is active (and oogeneisis occurs) unless inhibited by either the mitotic signal (if the cell is activated by GLP-1) or the spermatogenic signal (if the sex determination genes such as *tra-1* and *fem-3* are active). The mitotic signal can inhibit both the oogenic signal (GLD-1) and the spermatogenic signal (FOG-1, FOG-3). Both the spermatogenic and oogenic signals inhibit the mitotic signal, and the spermatogenic signal can inhibit the oogenic signal. Mitosis is promoted by the activation of germ cell GLP-1, while the sex determination pathway can activate the *fog-1* and *fog-3* genes. (C after Ellis and Kimble, 1995.)

wild-type larvae, they each enter meiosis. Thus, the *glp-1* gene appears to be responsible for enabling the germ cells to respond to the distal tip cell's signal.

*The *glp-1* gene appears to be involved in a number of inductive interactions in *C. elegans*. You will no doubt recall that *glp-1* is also needed by the AB blastomere to receive inductive signals from the EMS blastomere to form pharyngeal muscles (see Chapter 13).

(A) Wild-type: Sperm and oocytes

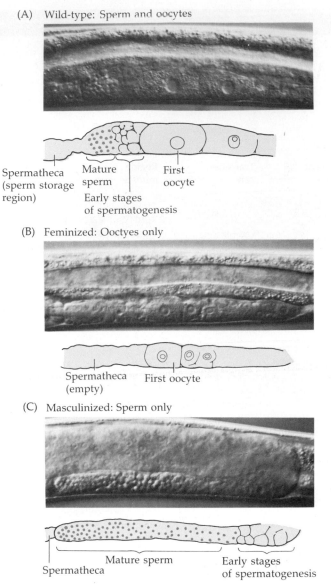

Spermatheca (sperm storage region)

Mature sperm

Early stages of spermatogenesis

First oocyte

(B) Feminized: Oocytes only

Spermatheca (empty)

First oocyte

(C) Masculinized: Sperm only

Spermatheca

Mature sperm

Early stages of spermatogenesis

Figure 22.13 Gonads of wild-type and mutant *C. elegans*. (A) Wild-type hermaphrodite producing first sperm and then eggs. (B) Female animal produced by the *fem-1* mutation produces only oocytes. (C) Masculinized hermaphrodite produced by loss-of-function mutations of *mog* genes (or mutations of the 3′ UTR of *fem-3*) produces only sperm. (Photographs courtesy of J. Kimble.)

fem genes are responsible for the sperm/oocyte decision (Figure 22.14).

The laboratories of Hodgkin (1985) and Kimble (1986) have isolated several genes needed for germ cell pathway selection. Figure 22.14 presents a scheme for how these genes might function in the switch from sperm formation to oocyte formation. During early development, the *fem* genes, especially *fem-3*, are critical for the specification of sperm cells. Loss-of-function mutations of these genes convert XX nematodes into females (i.e., spermless hermaphrodites). As long as the FEM proteins are made in the germ cells, sperm are produced. The active *fem* genes are thought to activate the *fog* genes (whose loss-of-function mutations cause the feminization of the germ line and eliminate spermatogenesis). The *fog* gene products activate the genes involved in transforming the germ cell into sperm and also inhibit those genes that would otherwise direct the germ cells to initiate oogenesis. Oogenesis can begin only when *fem* activity is suppressed. This suppression appears to act at the level of RNA translation. The 3′ untranslated region (3′ UTR) of the *fem-3* mRNA contains a sequence that binds a repressor during normal development. If this region is mutated such that the repressor protein cannot bind, the *fem-3* mRNA remains translatable, and oogenesis never occurs. The result is a hermaphrodite body that produces only sperm (Ahringer and Kimble, 1991; Ahringer et al., 1992). The *trans*-acting repressor factor has not yet been identified, but it is likely to be the product of one of the *mog* genes (Graham and Kimble, 1993). It is thought that proteins or messages stored in the oocyte may control the timing of this process, such that spermatogenesis occurs as long as there are repressors of *mog* expression. When these maternal inhibitors of *mog* expression decay, the MOG proteins are able to inhibit the synthesis of FEM proteins, thereby switching the gametogenesis from sperm to eggs. Another gene, *gld-1* (defective in *germ line development*), is essential for oogenesis to take place. Entry of presumptive oocytes into the meiotic pathway correlates with a dramatic increase in GLD-1. In loss-of-function mutants for *gld-1*, oogenesis is absent and the germ line cells continue to

After the cells begin their meiotic divisions, they still must become either sperm or ova. Generally, in each ovotestis, the most proximal germ cells produce sperm, while the most distal (near the tip) become eggs (Hirsh et al., 1976). The genetics of this switch are currently being analyzed. As discussed in Chapter 20, the genes for sex determination generate either a female body that is functionally hermaphroditic or a male body. In the germ line, the sex determination pathway activates or represses cer-

tain genes that are critical to the cell's becoming egg or sperm. For example, *mog* (*masculinization of germ line*) mutant homozygotes develop as sperm-producing males, and *fem-1* homozygous mutants develop as egg-producing females (Figure 22.13). The double mutants homozygous for both *tra-1* and *fem-1* have a unique phenotype. They are somatically male but are female in the germ line (Doniach and Hodgkin, 1984). This suggests that *tra-1* is the pivotal sex-determining gene of the *somatic* tissues, but that the

(A) Somatic sex determination

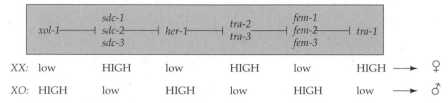

	sdc-1			tra-2		fem-1		
xol-1	sdc-2	her-1		tra-3		fem-2	tra-1	
	sdc-3					fem-3		

XX:	low	HIGH	low	HIGH	low	HIGH	→	♀
XO:	HIGH	low	HIGH	low	HIGH	low	→	♂

(B) Germline sex determination

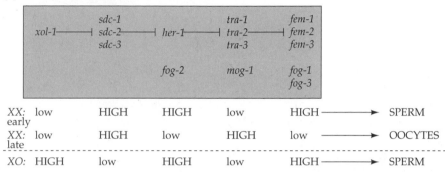

	sdc-1			tra-1		fem-1	
xol-1	sdc-2	her-1		tra-2		fem-2	
	sdc-3			tra-3		fem-3	
		fog-2		mog-1		fog-1	
						fog-3	

XX: early	low	HIGH	HIGH	low	HIGH	→	SPERM
XX: late	low	HIGH	low	HIGH	low	→	OOCYTES
XO:	HIGH	low	HIGH	low	HIGH	→	SPERM

Figure 22.14   Model of sex determination in the germ line of *C. elegans* hermaphrodites, based on analysis of mutations. (A) Sex determination in somatic tissues, showing a hierarchy of negative regulation. (B) Control of sex determination in the germ line. The genes *fog-2* and *mog-1* regulate sex determination in the germ line. The *fog-1* and *fog-3* genes act downstream to initiate spermatogenesis. (After Ellis and Kimble, 1995.)

proliferate, forming germ line tumors (Francis et al., 1995; Jones et al., 1996).

In *Drosophila*, the germ cells are instructed to differentiate into either sperm or eggs by the gonadal cells. Female gonadal cells make a product that is received by the germ cell and which activates a series of proteins whose activity is critical for the early transcription of the germ cell *Sxl* gene. The proper X:autosome ratio is also needed. By this mechanism, the XX flies get to make eggs while the XY flies make sperm (Burtis, 1993; Oliver et al., 1993). ■

Spermatogenesis

Spermatogenesis is the production of sperm from the primordial germ cells. Once the vertebrate primordial germ cells arrive at the genital ridge of male embryos, they become incorporated into the sex cords. They remain there until maturity, at which time the sex cords hollow out to form the seminiferous tubules, and the epithelium of the tubules differentiates into the Sertoli cells. During his lifetime, a human male can produce 10^{12} to 10^{13} gametes (Reijo et al., 1995). The spermatogenic cells are bound to the Sertoli cells by N-cadherin molecules on their respective cell surfaces and by galactosyltransferase molecules on the spermatogenic cells that bind a receptor on the Sertoli cells (Newton et al., 1993; Pratt et al., 1993). The Sertoli cells nourish and protect the developing sperm cells, and **spermatogenesis**—the developmental pathway from spermatogonial stem cell to mature sperm—occurs in the recesses of the Sertoli cells (Figure 22.15). The processes by which the PGCs generate sperm have been studied in detail in several organisms, but we will focus here on spermatogenesis in mammals. After reaching the gonad, the PGCs divide to form **type A$_1$ spermatogonia.** These cells are smaller than the PGCs and are characterized by an ovoid nucleus that contains chromatin associated with the nuclear membrane. The A$_1$ spermatogonia are found adjacent to the outer basement membrane of the sex cords. At maturity, these spermatogonia are thought to divide so as to make another type A$_1$ spermatogonium as well as a second, paler type of cell, the **type A$_2$ spermatogonium.** Thus, each type A$_1$ spermatogonium is a stem cell capable of regenerating itself as well as producing a new cell type. The A$_2$ spermatogonia divide to produce the A$_3$ spermatogonia, which then beget the type A$_4$ spermatogonia. It is possible that each of the type A spermatogonia are stem cells, capable of self-renewal. The A$_4$ spermatogonium has three options. It can form another A$_4$ spermatogonium (self-renewal); it can undergo cell

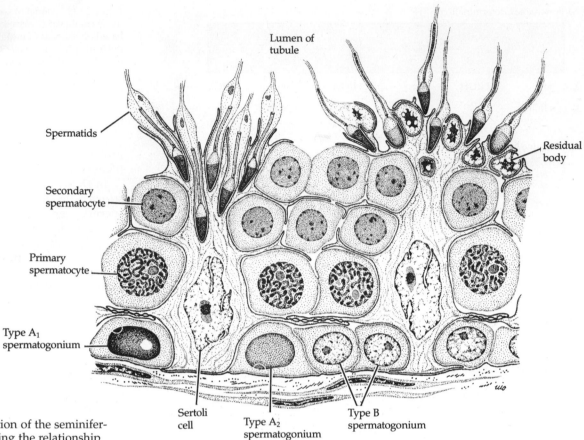

Lumen of
tubule

Spermatids

Secondary
spermatocyte

Primary
spermatocyte

Type A₁
spermatogonium

Residual
body

Sertoli
cell

Type A₂
spermatogonium

Type B
spermatogonium

Figure 22.15
Drawing of a section of the seminiferous tubule, showing the relationship between Sertoli cells and the developing sperm. As cells mature, they progress toward the lumen of the seminiferous tubule. (After Dym, 1977.)

death (apoptosis); or it can differentiate into the first committed stem cell, the **intermediate spermatogonium.** Intermediate spermatogonia are committed to become spermatozoa, and they mitotically divide once to form the **type B spermatogonia.** These cells are the precursors to the spermatocytes and are the last cells that undergo mitosis. These cells divide once to generate the **primary spermatocytes**—the cells that enter meiosis. It is not known what causes the spermatogonia to take the path toward differentiation rather than self-renewal; nor is it known what stimulates the cells to enter meiotic rather than mitotic division (Dym, 1994).

Looking at Figure 22.16, we find that during the spermatogonial divisions, cytokinesis is not complete. Rather, the cells form a syncytium whereby each cell communicates with the other via cytoplasmic bridges about 1 μm in diameter (Dym and Fawcett, 1971). The successive divisions produce clones of interconnected cells, and because ions and molecules readily pass through these intercellular bridges, each cohort matures synchronously.

Each primary spermatocyte undergoes the first meiotic division to yield a pair of **secondary spermatocytes,** which complete the second division of meiosis. The haploid cells formed are called **spermatids,** and they are still connected to each other through their cytoplasmic bridges. The spermatids that are connected in this manner have haploid nuclei but are functionally diploid, since the gene product made in one cell can readily diffuse into the cytoplasm of its neighbors (Braun et al., 1989). During the divisions from type A₁ spermatogonium to spermatid, the cells move farther and farther away from the basement membrane of the seminiferous tubule and closer to its lumen (see Figure 22.15). Thus, each type of cell can be found in a particular layer of the tubule. The spermatids are located at the border of the

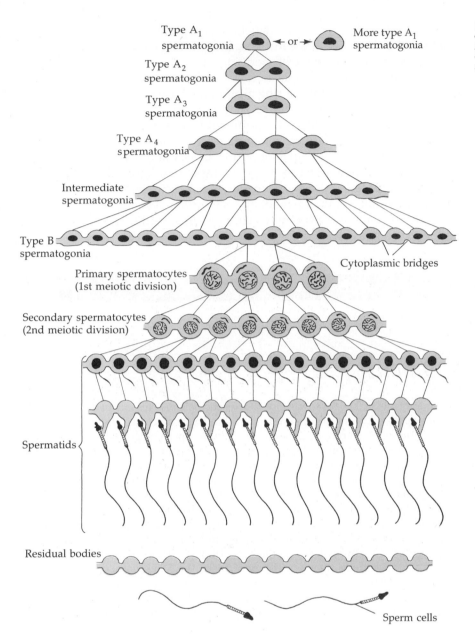

Figure 22.16
The formation of syncytial clones of human male germ cells. (After Bloom and Fawcett, 1975.)

lumen, and here they lose their cytoplasmic connections and differentiate into sperm cells. In humans, the progression from spermatogonial stem cell to mature sperm takes 65 days (Dym, 1994).

Spermiogenesis

The haploid spermatid is a round, unflagellated cell that looks nothing like the mature vertebrate sperm. The next step in sperm maturation, then, is **spermiogenesis** (or **spermateliosis**), the differentiation of the sperm cell. For fertilization to occur, the sperm has to meet and bind with the egg, and spermiogenesis differentiates the sperm for these functions of motility and interaction. The processes of mammalian sperm differentiation can be seen in Figure 4.2. The first steps involve the construction of the acrosomal vesicle from the Golgi apparatus. The acrosome forms a cap that covers the sperm nucleus. As the cap is formed, the nucleus rotates so that the acrosomal cap will then be facing the basal membrane of the seminiferous tubule. This rotation is necessary because the flagellum is beginning to form from the centriole on the other side of the nucleus, and this flagellum will extend into the lumen. During the last stage of spermiogenesis, the nucleus flattens and condenses, the remaining cytoplasm (the "cytoplasmic droplet") is jettisoned, and the mitochondria form a ring around the base of the flagellum. The resulting sperm then enter the lumen of the tubule.

In the mouse, the entire development from stem cell to spermatozoan takes 34.5 days. The spermatogonial stages last 8 days, meiosis lasts 13 days, and spermiogenesis takes up another 13.5 days. In humans, spermatic development takes nearly twice as long to complete. Because the type A_1 spermatogonia are stem cells, spermatogenesis can occur continuously. Each day, some 100 million sperm are made in each human testicle, and each ejaculation releases 200 million sperm. Unused sperm are either resorbed or passed out of the body in urine.

Sidelights & Speculations

Gene Expression During Sperm Development

Gene Expression Before Male Meiosis

Gene expression in the sperm is stage-specific, and even the haploid cells are able to synthesize certain products. The initiation of spermatogenesis during puberty is probably regulated by the synthesis of **BMP8B** by the spermatogonia. When BMP8B reaches a critical concentration, the spermatogonia can differentiate into round spermatids. These cells produce high levels of BMP8B, which can then further stimulate the spermatogonia to differentiate. Mice lacking BMP8B do not initiate spermatogenesis at puberty (Zhao et al., 1996). In humans, the *DAZ* gene located on the long arm of the Y chromosome is deleted in many infertile men, many of whom make no sperm at all. The *DAZ* gene is expressed exclusively in male germ cells, especially in the spermatogonin, and it appears to encode an RNA-binding protein (Reijo et al., 1995; Menke et al., 1997). *DAZ* is homologous to two *Drosophila* genes, *Rb97D* and *boule*, which also encode RNA-binding proteins and which are both essential for spermatogenesis. Spermatogonia degenerate in male flies deficient in *Rb97D*, while the germ cells of male flies lacking the *boule* gene do not enter meiosis (Karsch-Mizrachi and Haynes, 1993; Eberhart et al., 1996). RNA-binding proteins are critical in

spermatogenesis, because many of the genes expressed in the sperm are regulated at the level of translation (Schäfer et al., 1995). Indeed, in some animals, much of spermatogenesis occurs without any new gene transcription. The synthesis of protamine, the basic protein that replaces histones in the haploid sperm nucleus, is regulated by the phosphorylation of an 18-kDa binding protein that recognizes the 3′ untranslated region of the mouse protamine message (Kwon and Hecht, 1993).

In *Drosophila*, the *roughex* gene transcribed by premeiotic *Drosophila* spermatogonia controls the numbers of meiotic divisions. Males lacking functional copies of the *roughex* gene undergo an extra meiotic metaphase in addition to the two normal ones. Increasing the concentrations of Roughex results in the failure to execute meiosis II (Gönczy et al., 1994).

Gene Expression During Male Meiosis

Much of gene transcription during spermatogenesis takes place during the diplotene stage of meiotic prophase. The genes that are transcribed specifically during spermatogenesis are often those whose products are necessary for sperm motility or binding to the egg. In *Drosophila melanogaster*, one of the sperm-

specific genes transcribed is for β2-tubulin. This isoform of β-tubulin is seen only during spermatogenesis, and it is responsible for forming the meiotic spindles, the axoneme, and the microtubules associated with the lengthening mitochondria.* Hoyle and Raff (1990) have shown that another β-tubulin isoform, β3-tubulin (which is normally expressed in mesodermal cells and epidermis), cannot substitute for the β2-tubulin. When they fused the 5′ regulatory region from the β2-tubulin gene to the coding sequences of the β3-tubulin gene, the β3-tubulin gene was able to be expressed in the developing sperm. When this gene was expressed in the absence of the β2-tubulin gene, the resulting germ cells failed to undergo meiosis, axoneme assembly, or nuclear shaping. Only the mitochondrial elongation occurred. This indicates that the formation of the meiotic spindles and axoneme of sperm cells cannot be accomplished by just any β-tubulin and that the

---

*Making the sperm axoneme in *Drosophila* is a large undertaking. The sperm tail is 2 mm long—as long as the entire male fly. The sperm of the related species, *D. bifurca*, is 58.3 mm long, approximately 20 times longer than the flies producing them. Just as remarkable, the *D. melanogaster* egg incorporates the entire sperm (Karr, 1991). Only about 3 mm of the *D. bifurca* sperm is taken into the egg (Pitnick et al., 1995).

transcription of sperm-specific isoforms is important.

Those genes whose products are necessary for the binding of the sperm and the extracellular matrices of the egg are also transcribed during spermatogenesis. The gene for sea urchin bindin is transcribed relatively late in spermatogenesis, and its mRNA is translated into bindin shortly after being made (Nishioka et al., 1990). The bindin accumulates in vesicles that fuse together to form the single acrosomal vesicle of the mature sea urchin sperm. Figure 22.17 shows the localization of the bindin protein in the acrosomal vesicle of the sperm while it is still in the testis.

Haploid Gene Expression in Spermatocytes

In addition to gene transcription in diploid cells during meiotic prophase, certain genes are transcribed in the spermatids (reviewed in Palmiter et al., 1984). This evidence for **haploid gene expression** comes from studies involving heterozygous mice in which two different populations of sperm are seen to exist—one population expressing the mutant phenotype and one population expressing the wild-type trait. If the synthesis of the RNA or protein were to occur while the cells were still diploid, all the sperm would show the same phenotype. Transcription of the gene for protamine is seen in the early haploid cells (round spermatids) although their translation is delayed several days (Peschon et al., 1987). The gene for the β1,4-galactosyltransferase that binds the sperm to the zona pellucida is transcribed only during the haploid phase of mouse sperm maturation (Hardvin-Lepers et al., 1993). These genes expressed in the haploid stages may be regulated by follicle-stimulating hormone from the pituitary gland (Foulkes et al., 1993; Blendy et al., 1996; Nantel et al., 1996).*

Paternal Effect Genes

In some species, sperm provide important developmental information that can-

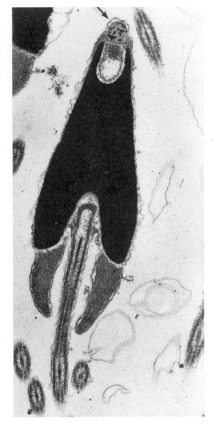

Figure 22.17   Localization of bindin in sperm acrosome by gold-labeled antibindin antibodies. The gold atoms enable the antibodies to show up as black dots in the electron micrograph. These sperm are still inside the sea urchin testis. (Courtesy of D. Nishioka.)

not be compensated for by the egg. We have discussed the imprinting of mammalian chromosomes wherein the sperm and egg DNA differ in their methylation patterns (see Chapters 4 and 11). There are also cases of **paternal effect genes.** Here, homozygous recessive alleles in the male cause abnormal development in the embryo, even if the female is homozygous for the wild-type allele, while the reciprocal cross, where the father is wild-type and the mother is homozygous for the mutant allele, leads to normal embryos. One such paternal effect gene is *spe-11* in *C. elegans.* The sperm containing mutant alleles at this locus are unable to

direct chromosomal movements that orient the mitotic spindle of the embryo, suggesting that the mutation affects the microtubule-organizing regions, such as the centrioles (Figure 22.18; Hill et al., 1989). Paternal effect mutations have been identified in *Drosophila* and these may also involve the structure of the zygote mitotic spindle (Karr, 1996). [fert10.html]

Terminating Gene Expression

Eventually, the haploid genome is condensed as the histones are replaced by protamines or by specifically modified histones. Many of the sperm histones become modified in the late spermatid stage during spermiogenesis. These modifications (such as dephosphorylating the *N*-terminal regions of certain histones) cause the chromatin to condense. Condensation results in severely reduced transcription. Thus, transcription from the male genome is not detected again until it is reactivated sometime during development (Poccia, 1986; Green and Poccia, 1988). ■

(A)

(B)

Figure 22.18   Immunofluorescence photomicrographs of mitotic spindles in a first-cleavage *C. elegans* embryo when sperm is (A) from a wild-type male and (B) from a male homozygous for the paternal effect gene *spe-11*. In (B), three microtubule-organizing centrioles can be seen instead of the usual two poles of mitosis. (From Hill et al., 1989, courtesy of S. Strome.)

---

*This mechanism seems unduly complex. The postmeiotic genes appear to be regulated by the CREM transcription factor. This gene for transcription factor, the cyclic-AMP-responsive element modulator, is transcribed during early spermatogenesis, but the message decays rapidly. The protein it makes inhibits the transcription of the postmeiotic genes. However, reception of FSH by the meiotic cells causes the alternative splicing of the *CREM* mRNA precursor, causing it to become a stable message for an activating isoform of the molecule. Gene targeting of the mouse *CREM* gene results in the lack of postmeiotic gene expression and the death of the spermatocytes.

Oogenesis

Oogenic Meiosis

Oogenesis—the differentiation of the ovum—differs from spermatogenesis in several ways. Whereas the gamete formed by spermatogenesis is essentially a motile nucleus, the gamete formed by oogenesis contains all the factors needed to initiate and maintain metabolism and development. Therefore, in addition to forming a haploid nucleus, oogenesis also builds up a store of cytoplasmic enzymes, mRNAs, organelles, and metabolic substrates. While the sperm becomes differentiated for motility, the oocyte develops a remarkably complex cytoplasm.

The mechanisms of oogenesis vary more than those of spermatogenesis. This difference should not be surprising, since the patterns of reproduction vary so greatly among species. In some species, such as sea urchins and frogs, the female routinely produces hundreds or thousands of eggs at a time, whereas in other species, such as humans and most mammals, only a few eggs are produced during the lifetime of an individual. In those species that produce thousands of ova, the oogonia are self-renewing stem cells that endure for the lifetime of the organism. In those species that produce fewer eggs, the oogonia divide to form a limited number of egg precursor cells. In humans, the thousand or so oogonia divide rapidly from the second to the seventh month of gestation to form roughly 7 million germ cells (Figure 22.19). After the seventh month of embryonic development, however, the number of germ cells drops precipitously. Most oogonia die during this period, while the remaining oogonia enter prophase of the first meiotic division (Pinkerton et al., 1961). These latter cells, called the **primary oocytes,** progress through the first meiotic prophase until the diplotene stage, at which point they are maintained until puberty. With the onset of adolescence, groups of oocytes periodically resume meiosis. Thus, in the human female, the first part of meiosis is begun in the embryo, and the signal to resume meiosis is not given until roughly 12 years later. In fact, some oocytes are maintained in meiotic prophase for nearly 50 years. As Figure 22.19 indicates, primary oocytes continue to die even after birth. Of the millions of primary oocytes present at birth, only about 400 mature during a woman's lifetime.

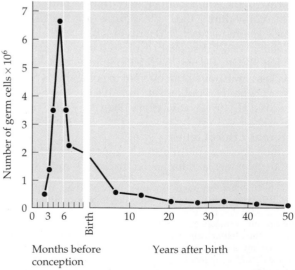

Figure 22.19
Changes in the number of germ cells in the human ovary. (After Baker, 1970.)

Figure 22.20
Polar body formation in the oocyte of the whitefish *Coregonus.* (A) Anaphase of first meiotic division, showing the first polar body pinching off with its chromosomes. (B) Metaphase (within the oocyte, arrow) of the second meiotic division, with the first polar body still in place. The first polar body may or may not divide again. (From Swanson et al., 1981, courtesy of C. P. Swanson.)

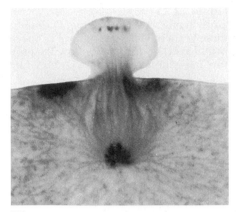

(A)

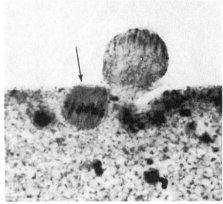

(B)

Oogenic meiosis also differs from spermatogenic meiosis in its placement of the metaphase plate. When the primary oocyte divides, the nucleus of the oocyte, called the **germinal vesicle,** breaks down and the metaphase spindle migrates to the periphery of the cell. At telophase, one of the two daughter cells contains hardly any cytoplasm, whereas the other cell has nearly the entire volume of cellular constituents (Figure 22.20). The smaller cell is called the **first polar body,** and the larger cell is referred to as the **secondary oocyte.** During the second division of meiosis, a similar unequal cytokinesis takes place. Most of the cytoplasm is retained by the mature egg (ovum), and a second polar body receives little more than a haploid nucleus. Thus, oogenic meiosis serves to conserve the volume of oocyte cytoplasm in a single cell rather than splitting it equally among four progeny.

In a few species of animals, meiosis is severely modified such that the resulting gamete is diploid and need not be fertilized to develop. Such animals are said to be **parthenogenetic.** In the fly *Drosophila mangabeirai,* one of the polar bodies acts as a sperm and "fertilizes" the oocyte after the second meiotic division. In other insects (such as *Moraba virgo*) and the lizard *Cnemidophorus uniparens,* the oogonia double their chromosome number before meiosis, so that the halving of chromosomes restores the diploid number. The germ cells of the grasshopper *Pycnoscelus surinamensis* dispense with meiosis altogether, forming diploid ova by two mitotic divisions (Swanson et al., 1981). In the preceding examples, the species consist entirely of females. In other species, haploid parthenogenesis is widely used not only as a means of reproduction, but also as a mechanism of sex determination. In the Hymenoptera (bees, wasps, and ants), unfertilized eggs develop into males, whereas fertilized eggs, being diploid, develop into females. The haploid males are able to produce sperm by abandoning the first meiotic division, thereby forming two sperm cells through second meiosis.

Maturation of the Oocyte in Amphibians

The egg is responsible for initiating and directing development, and in some species (as seen earlier), fertilization is not even necessary. The accumulated material in the oocyte cytoplasm includes energy sources and organelles (the yolk and mitochondria); the enzymes and precursors for DNA, RNA, and protein syntheses; stored messenger RNAs; structural proteins; and morphogenetic regulatory factors that control early embryogenesis. A partial catalogue of the materials stored in the oocyte cytoplasm is shown in Table 22.1. Most of this accumulation takes place during meiotic prophase I, and this stage is often subdivided into **previtellogenic** (Greek, "before yolk formation") and **vitellogenic** (yolk-forming) phases.

Eggs of fishes and amphibians are derived from a stem cell oogonial population that can generate a new cohort of oocytes each year. In the frog *Rana pipiens,* oogenesis lasts three years. During the first two years, the oocyte increases its size very gradually. During the third year, however, the rapid accumulation of yolk in the oocyte causes the egg to swell to its characteristically large size (Figure 22.21). Eggs mature in yearly batches, the first cohort maturing shortly after metamorphosis; the next group matures a year later.

Table 22.1 Cellular components stored in the mature oocyte of *Xenopus laevis*

Component	Approximate excess over amount in larval cells
Mitochondria	100,000
RNA polymerases	60,000–100,000
DNA polymerases	100,000
Ribosomes	200,000
tRNA	10,000
Histones	15,000
Deoxyribonucleoside triphosphates	2,500

Source: After Laskey, 1979.

Figure 22.21
Growth of oocytes in the frog. During the first three years of life, three cohorts of oocytes are produced. The drawings follow the growth of the first-generation oocytes. (After Grant, 1953.)

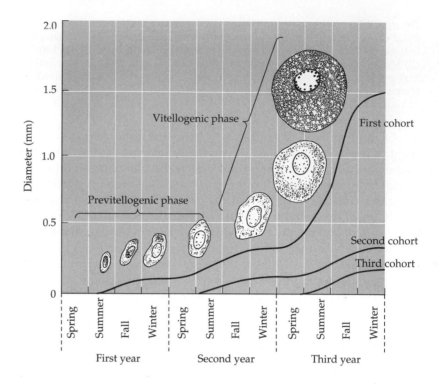

Vitellogenesis occurs when the oocyte reaches the diplotene stage of meiotic prophase. Yolk is not a single substance, but a mixture of materials used for embryonic nutrition. The major yolk component is a 470-kDa protein called **vitellogenin.** It is not made in the frog oocyte (as are the major yolk proteins of organisms such as annelids and crayfish), but is synthesized in the liver and carried by the bloodstream to the ovary (Flickinger and Rounds, 1956). This large protein passes between the follicle cells of the ovary and is incorporated into the oocyte by **micropinocytosis,** the pinching off of membrane-bounded vesicles at the base of microvilli (Dumont, 1978). In the mature oocyte, vitellogenin is split into two smaller proteins: the heavily phosphorylated **phosvitin** and the lipoprotein **lipovitellin.** These two proteins are packaged together into membrane-bounded **yolk platelets** (Figure 22.22A). Glycogen granules and lipochondrial inclusions store the carbohydrate and lipid components of the yolk, respectively.

Most eggs are highly asymmetrical, and it is during oogenesis that the animal-vegetal axis of the egg is specified. Danilchik and Gerhart (1987) have shown that although the concentration of yolk in *Xenopus* oocytes increases nearly 10-fold as one goes from the animal to the vegetal poles of the mature egg, vitellogenin uptake is uniform around the surface of the oocyte. What differs is its movement *within* the oocyte, and this depends on where the yolk proteins enter. When yolk platelets are formed in the future animal hemisphere, they move inward toward the center of the cell. The vegetal yolk platelets, however, do not actively move, but remain at the periphery for long periods of time, enlarging as they stay there. They are slowly displaced from the cortex as new yolk platelets come in from the surface. As a result of this differential intracellular transport, the amount of yolk steadily increases in the vegetal hemisphere, until the vegetal half of a mature *Xenopus* oocyte contains nearly 75 percent of the yolk (Figure 22.22B–E). The mechanism of this translocation remains unknown.

As the yolk is being deposited, the organelles also become arranged asymmetrically. The cortical granules begin to form from the Golgi appara-

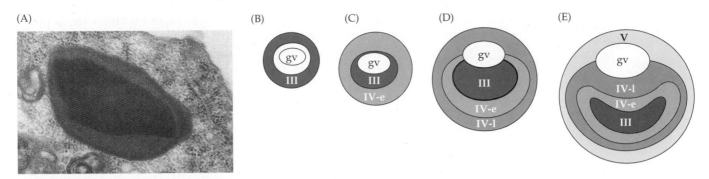

Figure 22.22
Yolk distribution in *Xenopus*. (A) An amphibian yolk platelet. (B-E) Establishment of animal-vegetal polarity of yolk platelets in the *Xenopus* oocyte. (B) In the late stage III (600 µm) oocyte, yolk platelets enter the cell equally at all points of the surface. (C,D) As the oocyte grows, the platelets at the future animal pole are displaced toward the vegetal pole, while those at the vegetal pole remain there. More yolk still enters the egg on all sides. (E) By the end of vitellogenesis, the earliest platelets (III) are all in the vegetal hemisphere, which has now concentrated roughly 75 percent of the oocyte yolk. Time of entry of yolk into oocyte platelets is indicated by shading and Roman numerals: III, stage III platelets; IV-e, early stage IV platelets; IV-l, late stage IV platelets; V, stage V platelets; gv, germinal vesicle. (After Danilchik and Gerhart, 1987; photograph courtesy of L. K. Opresko.)

tus and are originally scattered randomly through the oocyte cytoplasm. They later migrate to the periphery of the cell. The mitochondria replicate at this time, dividing to form millions of organelles that will be apportioned to the different cells during cleavage. (In *Xenopus*, new mitochondria are not formed until after gastrulation is initiated.) As vitellogenesis nears an end, the oocyte cytoplasm becomes stratified. The cortical granules, mitochondria, and pigment granules are found at the periphery of the cell, within the actin-rich oocyte cortex. Within the inner cytoplasm, distinct gradients emerge. While the yolk platelets become more heavily concentrated at the vegetal pole of the oocyte, the glycogen granules, ribosomes, lipid vesicles, and endoplasmic reticulum are found more toward the animal pole. Even specific mRNAs stored in the cytoplasm become localized to certain regions of the oocyte. [germ1.html]

While the precise mechanisms for establishing these gradients remain unknown, studies using inhibitors have shown that the cytoskeleton is critically important in localizing specific RNAs and morphogenetic factors. There seem to be two routes to get mRNAs localized into the vegetal cortex (Forristall et al., 1995; Kloc and Etkin, 1995). Messages such as those encoding the Vg1 protein are initially present throughout the oocyte and are translocated into the vegetal cortex in two-step process (Yisraeli et al., 1990). In the first phase, microtubules are needed to bring the *Vg1* mRNA into the vegetal hemisphere. In the second phase, microfilaments are responsible for anchoring the *Vg1* message to the cortex. The portion of the *Vg1* mRNA that binds to these cytoskeletal elements resides in the 3' untranslated region. When a specific 340-base sequence is placed onto a *β-globin* message, that *β-globin* mRNA is similarly localized to the vegetal cortex (see Chapter 12; Mowry and Melton, 1992). Other mRNAs, such as *Xlsirt* (a family of RNAs that do not encode proteins but which may be necessary for maintaining Vg1 at the cortex), *Xwnt11* and *Xcat2* (which encodes an RNA-binding protein related to Nanos), leave the germinal vesicle to become localized in the mitochondrial "cloud" located at the vegetal pole of the nucleus. These mes-

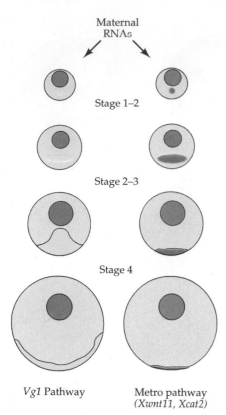

Figure 22.23
Schematic representations of the two pathways for localizing mRNAs to the vegetal region of the *Xenopus* oocyte. The METRO (message transport organizer) pathway accumulates messages in the mitochondrial cloud, and islands of them become transported into the vegetal pole cortex. In the *Vg1* pathway, messages are seen throughout the egg, but they are translocated through a microtubule-driven system to the microfilaments of the vegetal cortex. (After Kloc and Etkin, 1995.)

sages get compartmentalized into clusters associated with the germ plasm and transported to the vegetal cortex in a manner that appears to be independent of the cytoskeleton (Figure 22.23; Kloc et al., 1996).

Completion of Meiosis: Progesterone and Fertilization

Amphibian oocytes can remain for years in the diplotene stage of meiotic prophase. Resumption of meiosis in the amphibian primary oocyte requires progesterone. This hormone is secreted by follicle cells in response to the gonadotropic hormones secreted by the pituitary gland. Within 6 hours of progesterone stimulation, **germinal vesicle breakdown** (**GVBD**) occurs, the microvilli retract, the nucleoli disintegrate, and the lampbrush chromosomes contract and migrate to the animal pole to begin division. Soon afterward, the first meiotic division occurs, and the mature ovum is released from the ovary by a process called **ovulation.** The ovulated egg is in second meiotic metaphase when it is released.

How does progesterone enable the egg to break its dormancy and resume meiosis? To understand the mechanisms by which this activation is accomplished, it is necessary to briefly review the model for early blastomere division that was presented in Chapter 5. Maturation-promoting factor (MPF) is responsible for the resumption of meiosis. Its activity is cyclical, being high during cell division and undetectable during interphase. MPF is a protein kinase that contains an enzymatic subunit (cyclin). Since all the components of MPF are present in the amphibian oocyte, it is generally thought that progesterone somehow converts a pre-MPF complex into active MPF, perhaps by the activation of the cdc25 phosphatase (see Chapter 5; Minishull, 1993).

The mediator of the progesterone signal is probably the **c-mos** protein. Progesterone reinitiates meiosis by causing the egg to polyadenylate the maternal c-*mos* mRNA that has been stored in its cytoplasm (Sagata et al., 1988, 1989; Sheets et al, 1995). This message translates into a 39-kDa phosphoprotein, pp39^{mos}. This protein is detectable only during oocyte maturation and is destroyed quickly upon fertilization. Yet during its brief lifetime, it plays a major role in releasing the egg from its dormancy. If the translation of pp39^{mos} is inhibited (by injecting *mos*-antisense mRNA into the oocyte), pp39^{mos} does not appear, and germinal vesicle breakdown and the renewal of oocyte maturation do not occur. Having stimulated the reinitiation of meiosis, pp39^{mos} enables the oocyte to undergo a meiotic division, but freezes the second meiotic cycle in metaphase. This metaphase block is caused by the combined actions of pp39^{mos} and another protein, cyclin-dependent kinase 2 (cdk2; Gabrielli et al., 1993). These two proteins are now thought to constitute the **cytostatic factor** (**CSF**) that is found in mature frog eggs and which can block cell cycles in metaphase (Masui, 1974). It is thought that CSF prevents the degradation of cyclin.

The next question concerns the mechanisms by which fertilization enables the oocyte that is at second metaphase to complete that division to form a haploid gamete. Recent evidence suggests that the calcium ion flux attending fertilization enables the calcium-binding protein **calmodulin** to become active. Calmodulin, in turn, can activate the **calmodulin-dependent protein kinase II.** Calmodulin-dependent protein kinase II is necessary and sufficient to inactivate cdc2 kinase and stimulate c-mos degradation (Lorca et al., 1993). Calpain II, a calcium-dependent protease, degrades the pp39^{mos} (Watanabe et al., 1989). Thus, the two components of CSF are inactivated or destroyed. Without CSF, cyclin can be degraded, and the meiotic division can be completed (Figure 22.24).

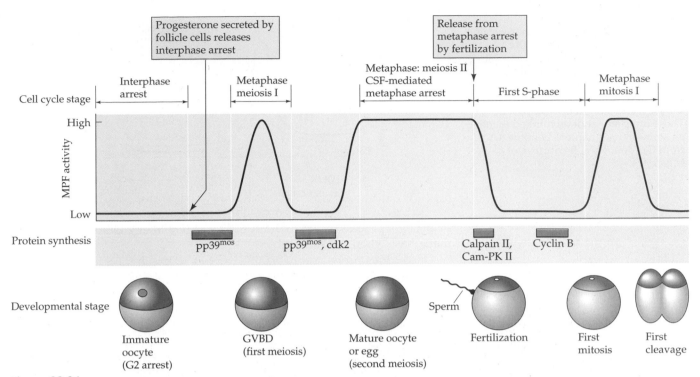

Figure 22.24
Schematic representation of *Xenopus* oocyte maturation, showing the regulation of meiotic cell division by pp39mos, cdk2, and calpain II. The solid line on the graph represents the relative levels of active MPF. The bars below the graph show the periods when the synthesis of particular proteins are needed for entry into the next M phase. GVBD is the point of germinal vesicle breakdown. Oocyte morphology is diagrammed below. (After Minishull, 1993.)

Gene Transcription in Oocytes

In most animals (insects being a major exception), the growing oocyte is active in transcribing genes where products are either (1) necessary for cell metabolism, (2) necessary for oocyte-specific processes, or (3) needed for early development before the nuclei begin to function. In mice, for instance, the growing diplotene oocyte is actively transcribing the genes for zona pellucida proteins ZP1, ZP2, and ZP3. Moreover, these genes are *only* transcribed in the oocyte and not in any other cell (Epifano et al., 1995; see Chapter 2).

The amphibian oocyte has certain periods of very active RNA synthesis. During the diplotene stage, certain chromosomes stretch out large loops of DNA, causing the chromosome to resemble a lampbrush (a handy instrument for cleaning test tubes in the days before microfuges). These **lampbrush chromosomes** (Plate 2) can be seen as the sites of RNA synthesis by in situ hybridization. Oocyte chromosomes can be prepared, denatured, and incubated with radioactive RNA that encodes a specific protein. After the unbound RNA is washed away, autoradiography visualizes the precise location of the gene. Figure 22.25 shows diplotene chromosome I of the newt *Triturus cristatus* after incubation with radioactive histone mRNA. It is obvious that a histone gene (or set of histone genes) is located on one of these loops of the lampbrush chromosome (Old et al., 1977). Electron micrographs of gene transcripts from lampbrush chromosomes also enable one to see chains of mRNA coming off each gene as it is transcribed (Hill and MacGregor, 1980).

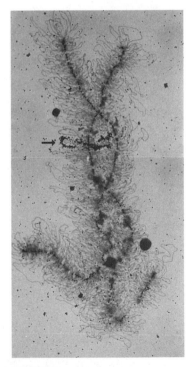

Figure 22.25
Localization (arrowhead) of histone genes on a lampbrush chromosome in an amphibian oocyte. Genes have been visualized by in situ hybridization and autoradiography. (From Old et al., 1977, courtesy of H. G. Callan.)

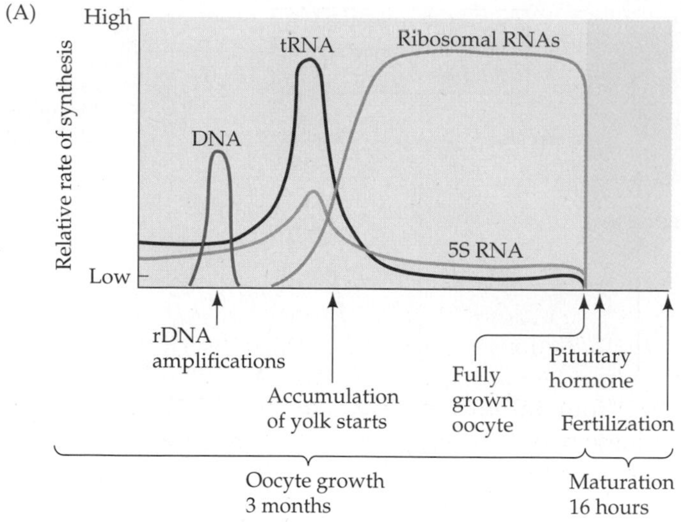

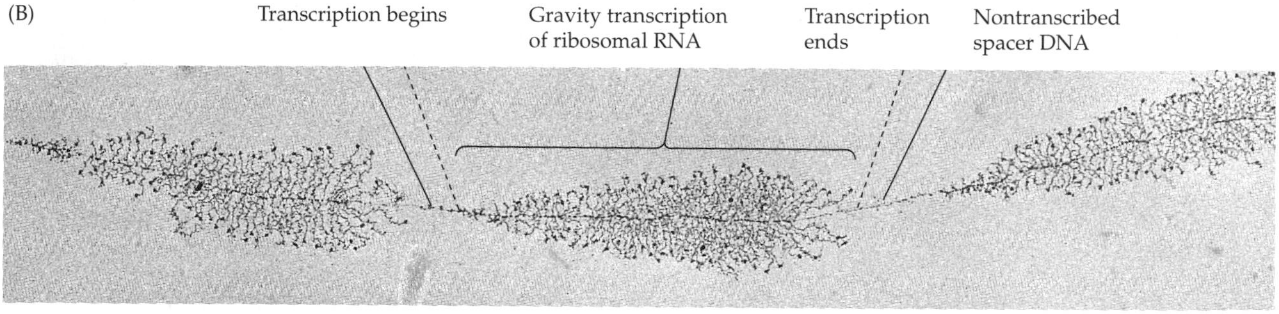

Figure 22.26
Ribosomal RNA production in *Xenopus* oocytes. (A) Relative rates of DNA, tRNA, and rRNA synthesis in amphibian oogenesis during the last three months before ovulation. (B) The transcription of the RNA precursor of the 28S, 18S, and 5.8S ribosomal RNAs. These units are tandemly linked together, some 450 per haploid genome. (A from Gurdon, 1976; B courtesy of O. L. Miller, Jr.)

(A)

(B)

In addition to mRNA synthesis, the patterns of rRNA and tRNA transcription are also regulated during oogenesis. Figure 22.26A shows ribosomal and transfer RNA synthesis during the *Xenopus* oogenesis. Transcription appears to begin in early (stage I, 25-40 μm) oocytes, during the diplotene stage of meiosis. At this time, all the ribosomal and transfer RNAs needed for protein synthesis until the midblastula stage are made, and all the maternal mRNAs for early development are transcribed. This stage lasts months in *Xenopus*. The rate of ribosomal RNA production is prodigious. The *Xenopus* oocyte genome has over 1800 genes encoding 18S and 28S rRNA, and these genes are selectively amplified such that there are over 500,000 genes making these ribosomal RNAs (Figure 22.26B; Brown and Dawid, 1968). After reaching a certain size, the chromosomes of the mature (stage VI) oocyte condense, and the genes are not actively transcribing. This "mature oocyte" condition can also be months long. Upon hormonal stimulation, the oocyte completes its first meiotic division and is ovulated. The mRNAs stored by the oocyte now join with the ribosomes to initiate protein synthesis. Within hours, the second meiotic division has begun, and the secondary oocyte has been fertilized. The embryo's genes do not begin active transcription until the midblastula transition (Davidson, 1986). [germ2.html]

As we have seen in Chapter 12, the oocytes of several species make two classes of mRNAs—those for immediate use in the oocyte and those that are stored for use during early development. In sea urchins, the translation of stored maternal messages is initiated by fertilization, while in frogs the signal for such translation is initiated by progesterone as the egg is about to be ovulated. One of the actions of the MPF kinase activity induced by progesterone may be the phosphorylation of CPE-binding proteins on the stored oocyte mRNAs. The phosphorylation of these factors is associated with the lengthening of the poly(A) tails in the stored messages and with translation of the stored mRNAs (Paris et al., 1991).

Meroistic Oogenesis in Insects

There are several types of oogenesis in insects, but most studies have focused on those insects, such as *Drosophila* and moths, that undergo **meroistic** oogenesis. In meroistic oogenesis, cytoplasmic connections remain between the cells produced by the oogonium. In *Drosophila*, each oogonium divides four times to produce a clone of 16 cells connected to each other through **ring canals.** The production of these interconnected cells (called **cystocytes**) involves a highly ordered array of cell divisions (Figure 22.27). Only those two cells having four interconnections are capable of developing into oocytes, and of these two, only one becomes the egg. The other begins meiosis but does not complete it. Thus, only one of the 16 cystocytes can become an ovum. All the other cells become **nurse cells.** As it turns out, the cell destined to become the oocyte is that cell residing at the most posterior tip of the egg chamber that encloses the 16-cell clone. However, since the nurse cells are connected to the oocyte by their cytoplasmic bridges, the entire complex can be seen as one egg-producing unit.

The meroistic ovary confronts us with some interesting problems. If all the cells are connected so that proteins and RNAs shuttle freely between them, why should they have different developmental fates? Why should one cell become the oocyte while the others become "RNA-synthesizing factories," sending mRNAs, ribosomes, and even centrioles into the oocyte? Why is the flow of protein and RNA in one direction only? As the cystocytes divide, a large spectrin-rich structure called the **fusosome** forms and spans the cytoplasmic bridges between the cells (Figure 22.27B). It is constructed asymmetrically, as it always grows from the spindle pole that remained in one of the cells (Lin and Spradling, 1995). The cell that retained the fusosome during the first division becomes the oocyte. It is not yet known if the fusosome contains oogenic determinants or if it directs the traffic of materials into this particular cell.

Once the patterns of transport are established, the cytoskeleton becomes actively involved in transporting mRNAs from the nurse cells into the oocyte cytoplasm (Cooley and Therkauf, 1994). The microtubular array is critical for oocyte determination. If this lattice is disrupted (either chemically

Figure 22.27
The formation of 16 interconnected cystocytes in *Drosophila*. (A) Diagram of adult ovariole showing sequence of oogenesis as younger germinal cysts mature within the ovariole. (B) Division of the cystocyte-forming cells (cystoblasts). The cells are represented schematically as dividing in a single plane. The stem cell divides to produce another stem cell plus a cell that is committed to form the cystocytes. Only one of the 16 cystocytes becomes an oocyte; the others become nurse cells, connected to the oocyte by ring canals (cytoplasmic bridges). The centriole of cystocyte 1 retains the fusosome (in red), which grows through the ring canal toward its mitotic sister. The arrow shows the polarity, pointing to the cell from which the fusosome grew. After three more mitotic divisions, the 16-cell cyst is formed. If intracellular transport is coordinated by the fusosome, the transport of mRNAs and proteins would be toward cystocyte 1, which would thus become the oocyte. (A after Ruohola et al., 1991; B after Lin and Spradling, 1995.)

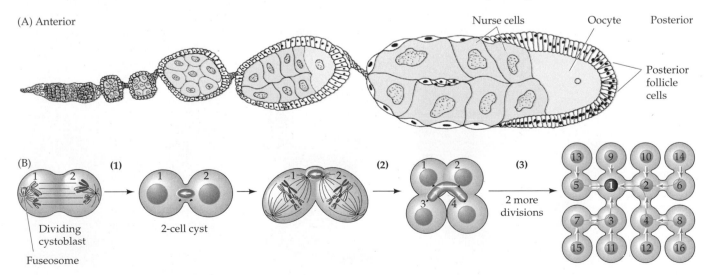

(A) Anterior

Nurse cells   Oocyte   Posterior

Posterior follicle cells

(B)   (1)   2-cell cyst   (2)   (3)

Dividing cystoblast

Fuseosome

2 more divisions

or by mutations such as *bicaudal-D* or *egalitarian*), gene products are transmitted in all directions and all 16 cells differentiate into nurse cells (Gutzeit, 1986; Theurkauf et al., 1992, 1993; Spradling, 1993). It is possible that some compounds transported from the nurse cells into the oocyte become associated with transport proteins such as kinesin that would enable them to travel along the tracks of microtubules extending through the ring canal (Theurkauf et al., 1992; Sun and Wyman, 1993). Actin may become important for maintaining this distinction during later stages of oogenesis. Mutations that prevent actin microfilaments from lining the ring canals prevent the transport of mRNAs from nurse cell to oocyte, and the disruption of the actin microfilaments randomizes the distribution of mRNA (Cooley et al., 1992; Watson et al., 1993). Thus, the microtubule and microfilamentous cytoskeleton appears to control the movement of organelles and RNAs between nurse cells and oocyte such that developmental cues are exchanged only in the appropriate direction.

TRANSPORT OF RNA FROM NURSE CELLS TO OOCYTE. The oocytes of meroistic insects do not pass through a transcriptionally active stage, nor do they have lampbrush chromosomes. Rather, autoradiographic evidence shows that RNA synthesis is largely confined to the nurse cells and that the RNA made by these cells is actively transported into the oocyte cytoplasm. This can be seen in Figure 22.28. When the egg chambers of a housefly are incubated in radioactive cytidine, the nuclei of the nurse cells show intense labeling. When the labeling is stopped and the cells are incubated for 5 more hours in nonradioactive media, the labeled RNA is seen to enter the oocyte from the nurse cells (Bier, 1963). Oogenesis takes place in only 12 days, so the nurse cells are very metabolically active during this time. They are aided in their transcriptional efficiency by becoming polytene. Instead of having two copies of each chromosome, they replicate their chromosomes until they have produced 512 copies. The 15 nurse cells are known to pass ribosomal and messenger RNAs as well as proteins into the oocyte cytoplasm, and entire ribosomes may be transported as well (Plate 16). The mRNAs do not associate with polysomes, which suggests that they are not immediately active in protein synthesis (Paglia et al., 1976; Telfer et al., 1981).

Figure 22.28
Transport of mRNA from nurse cells into fly oocytes. (A,B) Autoradiographs of the follicle cell of the housefly, *Musca domestica*, after incubation with [³H]cytidine. (A) Egg chamber fixed immediately after label was introduced. The nuclei of the nurse cells are heavily labeled, indicating that they are synthesizing new RNA. The oocyte remains unlabeled except where some RNA is escaping into the oocyte through the cytoplasmic connection between it and a nurse cell (arrow). (B) A similar egg chamber fixed 5 hours later. Label is gone from the nurse cell nuclei but has moved into the cytoplasm. Moreover, radioactive RNA can be seen passing into the oocyte cytoplasm through the two channels between the nurse cells and the oocyte. (C) Autoradiograph of *Drosophila* egg chamber stained by a radioactive probe to the *bicoid* mRNA. This message is transported from the nurse cells and remains in the anteriormost portion of the oocyte. (A and B from Bier, 1963, courtesy of D. Ribbert; C from Stephanson et al., 1988, courtesy of E. C. Stephanson.)

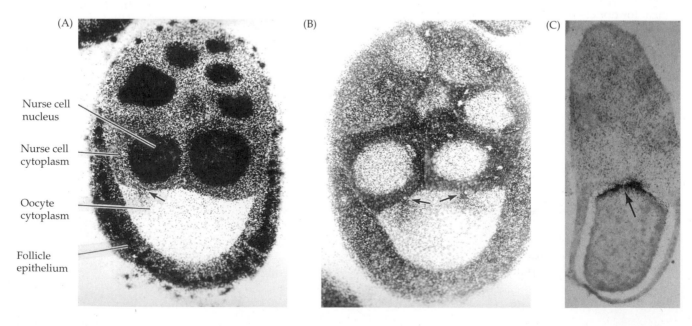

Nurse cell nucleus

Nurse cell cytoplasm

Oocyte cytoplasm

Follicle epithelium

The Origin of Drosophila *Embryonic Axes During Oogenesis*

THE ANTERIOR-POSTERIOR and dorsal-ventral axes are both established during mid-oogenesis (González-Rayes et al., 1995; Roth et al., 1995). The mRNA for the anterior determinant, *bicoid*, is placed in the anterior region of the egg; the mRNAs for the posterior determinants, *oskar* and *nanos*, are sent to the posterior pole; and the *gurken* message becomes concentrated at one region of the egg, initiating the reactions that establish that side as the dorsal surface of the embryo. The mechanisms for constructing these axes involve complex interactions between the nurse cells, oocyte, and follicle cells (see Figure 14.13). First, the *gurken* message is produced by the nurse cells, and it gathers around the oocyte nucleus, positioning itself between the nucleus and the plasma membrane. The nucleus is then at the posterior region of the egg, and the newly translated Gurken protein activates its receptor in the posterior pole follicle cells. (The Gurken protein resembles epidermal growth factor.) The posterior pole follicle cells respond by sending out a signal (perhaps cyclic AMP) that activates **protein kinase A** (**PKA**) in the oocyte cell membrane. As a result of the PKA activation, the microtubules of the oocyte are reoriented (Lane and Kalderon, 1994).* Instead of having their plus ends pointed toward the nurse cells (i.e., anteriorly), they reverse so that their plus ends are posterior (where the nucleus had been).

---

*PKA also is known to organize microtubules in axonal outgrowth (Shea et al., 1992), and as we saw in Chapter 1, it can mediate stalk cell differentiation in *Dictyostelium* (Williams et al., 1993).

The reorientation of the microtubules is a critical event. The *oskar* and *nanos* mRNAs are synthesized by nurse cells and are initially seen at the future anterior of the egg. These mRNAs can be transported to the posterior pole along the microtubules toward the plus end (but not toward the minus end). Thus, these messages can now become transported to the posterior pole. The *oskar* message is critical for organizing the pole plasm, and if it is translated before reaching the posterior pole, it can establish abdomens and germ cells elsewhere. During its journey to the posterior, the *oskar* message is repressed by the **Bruno** protein (which binds to the 3' UTR of the *oskar* message). Once at the posterior, this repression is abrogated, and Oskar protein can be made (Kim-Ha et al., 1995; Rongo et al., 1995). Conversely, the *bicoid* message is kept at the anterior by the minus ends of the microtubules. If the microtubules are disaggregated, *bicoid* message diffuses into the cytoplasm, and if the polarity of the microtubules is retained in its original conformation (as in flies that lack PKA), the *bicoid* message is transported to the posterior pole (Figure 22.29; Macdonald et al., 1991; Marcey et al., 1991; Pokrywka and Stephenson, 1991). This positioning of the *bicoid* mRNA in the future anterior and the *oskar* and *nanos* messages in the future posterior sets up the conditions for organizing the anterior-posterior axis (see Chapter 15).

The realignment of microtubules facilitates the movement of the nucleus, with its Gurken signal, along the egg plasma membrane to the anterior dorsal corner (Roth et al., 1995; Gozález-Reyes et al., 1995). Here, the Gurken protein

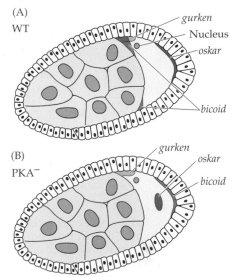

Figure 22.29   RNA localization in the oocytes of wild-type and PKA-deficient mutants of *Drosophila*. (A) In the wild-type oocyte (stage 9), *oskar* mRNA is at the posterior pole, *bicoid* mRNA is at the anterior margins, and *gurken* mRNA is localized at the dorsal anterior corner. (B) In the PKA-deficient oocytes, the distribution of *gurken* message is not affected, but *oskar* mRNA fails to be localized to the posterior pole and accumulates centrally, while *bicoid* mRNA is transported to the posterior pole. (After Lasko, 1995.)

causes the adjacent follicle cells to become the dorsal cells. (The pole and lateral follicle cells make different proteins and respond differently to the Gurken signal. The pole follicle cells activate oocyte PKA; the lateral follicle cells become dorsalized and repress the synthesis of ventralizing proteins). Thus, the anerior-posterior and dorsal-ventral axes in *Drosophila* are initiated before fertilization even takes place. ∎

TRANSPORT OF YOLK PROTEINS INTO THE EGG.   The three major yolk proteins in *Drosophila* are made in the fat body and ovary, but not in the oocyte itself (Bownes, 1982; Brennen et al., 1982). Yolk synthesis is controlled by several interacting agents, including sex, juvenile hormone levels, ecdysone, and

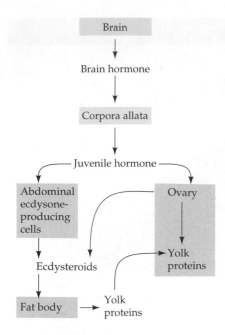

Figure 22.30
Model for the hormonal regulation of yolk peptide synthesis in *D. melanogaster*. In response to a brain hormone, the corpora allata produce juvenile hormone, which causes the ovary to make yolk proteins and ecdysteroids. JH also induces ecdysteroid synthesis in abdominal cells. These ecdysteroids cause the fat body to produce yolk proteins that are transported to the ovary. (After Bownes, 1982.)

nutrition. These physiological agents are "integrated" by the enhancer region between the two yolk protein genes of *Drosophila* (see Figure 10.13; Bownes et al., 1988). These genes are only active in female flies, and this is regulated by the binding of the female-specific Doublesex protein to this enhancer region. It is thought that a brain hormone, responding to environmental cues,* stimulates the corpora allata to secrete juvenile hormone (Figure 22.30). Juvenile hormone (1) regulates the uptake of the yolk peptides at the surface of the oocyte, (2) stimulates the synthesis of ovarian yolk proteins (which are identical to those made by the fat body), and (3) causes the ovarian follicles and other abdominal cells to secrete ecdysone. Ecdysone is metabolized into its active form—20-hydroxyecdysone—and stimulates the fat body to produce yolk proteins, just as estradiol stimulates the amphibian liver to do so. Similarly, the administration of ecdysone to adult males causes their fat bodies to secrete yolk proteins as well (Postlethwait et al., 1980), and yolk protein gets taken into insect oocytes through receptor-mediated endocytosis (Raikhel and Dhadialla, 1992). The receptors for vitellogenin are located in the regions of oocyte membrane at the base of and between the microvilli. The receptor-vitellogenin complexes are internalized and the vitellogenin freed from the receptor inside the endocytic vacuole. This vacuole fuses with other such endosomes to form the yolk storage granule full of vitellogenin.

Oogenesis in Mammals

Ovulation of the mammalian egg follows one of two basic patterns, depending on the species. One type of ovulation is stimulated by the physical act of intercourse itself. Physical stimulation of the cervix triggers the release of gonadotropins from the pituitary. These gonadotropins signal the egg to resume meiosis and initiate the events expelling the ovum from the ovary. This method ensures that most copulations lead to fertilized ova, and animals that utilize this method of ovulation—rabbits and minks—have a reputation for procreative success.

Most mammals, however, have a periodic type of ovulation. The female ovulates only at specific times of the year, called **estrus** (or its English equivalent, "heat"). In these cases, environmental cues, most notably the amount and type of light during the day, stimulate the hypothalamus to release gonadotropin-releasing factor. This factor stimulates the pituitary to release its gonadotropins—follicle-stimulating hormone (FSH) and luteinizing hormone (LH)—which cause the follicle cells to proliferate and secrete estrogen. The estrogen subsequently enters certain neurons and evokes the pattern of mating behavior characteristic of the species. Gonadotropins also stimulate follicular growth and the initiation of ovulation. Thus, estrus and ovulation occur close together.

Humans have a variation on the theme of periodic ovulation. Although human females have a cyclical ovulation (averaging about 29.5 days) and no definitive yearly estrus, most of human reproductive physiology is shared with other primates. The characteristic primate periodicity in maturing and releasing ova is called the **menstrual cycle** because it entails the periodic

---

*In *Drosophila*, the environmental cue appears to be the photoperiod. In the common mosquito, the cue is the blood meal. Only female mosquitoes bite, and they make no vitellogenin before this meal. Some factor in the blood stimulates the mosquito's brain to release juvenile hormone and the corpus-cardiacum-stimulating factor. The latter factor causes the release of the egg development neurosecretory hormone (EDNH). EDNH stimulates the ovary to secrete ecdysone, which, in concert with JH, stimulates the fat body to synthesize vitellogenin (Hagedorn, 1983; Borovsk et al., 1990). (See Chapter 21.)

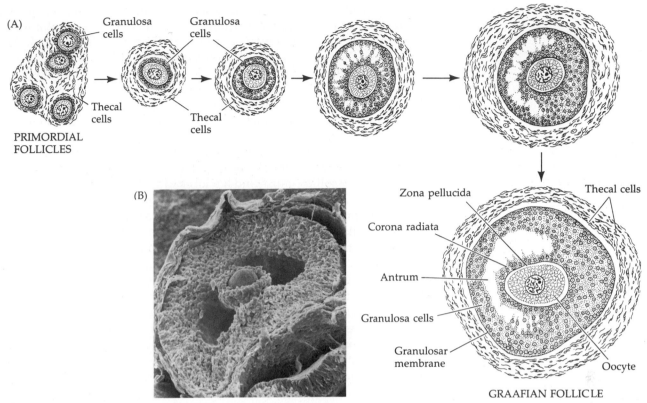

Figure 22.31
The ovarian follicle of mammals. (A) Maturation of the ovarian follicle. When
mature, it is often called a Graafian follicle. (B) Scanning electron micrograph of a
mature follicle in the rat. The oocyte (center) is surrounded by the smaller granulosa cells that will make the corona. (A after Carlson, 1981; B courtesy of P.
Bagavandoss.)

shedding of blood and cellular debris from the uterus at monthly intervals.*
The menstrual cycle represents the integration of three very different activities: (1) the ovarian cycle, the function of which is to mature and release an
oocyte, (2) the uterine cycle, the function of which is to provide the appropriate environment for the developing blastocyst to implant, and (3) the cervical cycle, the function of which is to allow sperm to enter the female reproductive tract only at the appropriate time. These three functions are
integrated through the hormones of the pituitary, hypothalamus, and ovary.

The majority of the oocytes within the adult human ovary are maintained in the prolonged diplotene stage of the first meiotic prophase (often
referred to as the **dictyate** state). Each oocyte is enveloped by a primordial
follicle consisting of a single layer of epithelial granulosa cells and a less-or-
ganized layer of mesenchymal thecal cells (Figure 22.31). Periodically, a
group of primordial follicles enters a stage of follicular growth. During this
time, the oocyte undergoes a 500-fold increase in volume (corresponding to
an increase in oocyte diameter from 10 μm in a primordial follicle to 80 μm
in a fully developed follicle). Concomitant with oocyte growth is an increase

*The periodic shedding of the uterine lining is an active process seen throughout mammals.
Moreover, the uterus has intricate circulatory adaptations (such as spiral arteries) that allow
blood to flow freely for a time without clotting and then to stop flowing (to prevent hemorrhage). Profet (1993) has proposed that menstruation serves a crucial immunological function,
protecting the uterus against infections from semen or other environmental agents.

in the number of follicular granulosa cells, which form concentric layers about the oocyte. This proliferation of granulosa cells is mediated by a paracrine factor, GDF-9, a member of the TGF-β family (Dong et al., 1996). Throughout this growth period, the oocyte remains in the dictyate stage. The fully grown follicle thus contains a large oocyte surrounded by several layers of granulosa cells. Many of these cells will stay with the ovulated egg, forming the **cumulus,** which surrounds the egg in the oviduct. In addition, during the growth of the follicle, an **antrum** (cavity) forms, which becomes filled with a complex mixture of proteins, hormones, cAMP, and other molecules. At any given time, a small group of follicles is maturing. However, after progressing to a more mature stage, most oocytes and their follicles die. To survive, the follicle must find a source of gonadotropic hormones and, "catching the wave" at the right time, must ride it until it peaks. Thus, for oocyte maturation to occur, the follicle needs to be at a certain stage of development when the waves of gonadotropin arise.

Day 1 of the menstrual cycle is considered to be the first day of "bleeding" (Figure 22.32). This bleeding from the vagina represents the sloughing off of extrauterine tissue and blood vessels that would have aided the implantation of the blastocyst. In the first part of the cycle (called the **proliferative** or **follicular phase**), the pituitary gland starts secreting increasingly large amounts of FSH. The group of maturing follicles, which have already undergone some development, respond to this hormone by further growth and cellular proliferation. FSH also induces the formation of LH receptors on the granulosa cells. Shortly after this period of initial follicle growth, the pituitary begins secreting LH. In response to LH, the meiotic block is broken. The nuclear membranes of competent oocytes break down, and the chromosomes assemble to undergo the first meiotic division. One set of chromosomes is kept inside the oocyte, and the other is given to the small polar body. Both are encased by the zona pellucida, which has been synthesized by the growing oocyte. It is in this stage that the egg will be ovulated.

The two gonadotropins, acting together, cause the follicle cells to produce increasing amounts of estrogen, which has at least five major activities in regulating the further progression of the menstrual cycle:

1. It causes the uterine mucosa to begin its proliferation and to become enriched with blood vessels.
2. It causes the cervical mucus to thin, thereby permitting sperm to enter the inner portions of the reproductive tract.
3. It causes an increase in the number of FSH receptors on the follicular granulosa cells (Kammerman and Ross, 1975) while causing the pituitary to lower its FSH production. It also stimulates the granulosa cells to secrete the peptide hormone inhibin, which also suppresses pituitary FSH secretion (Rivier et al., 1986; Woodruff et al., 1988).
4. At low concentrations, it inhibits LH production, but at high concentrations, it stimulates it.
5. At very high concentrations and over long durations, estrogen interacts with the hypothalamus, causing it to secrete gonadotropin-releasing factor.

As estrogen levels increase as a result of follicular production, FSH levels decline. LH levels, however, continue to rise as more estrogen is secreted. As estrogens continue to be made (days 7–10), the granulosa cells continue to grow. Starting at day 10, estrogen secretion rises sharply. This rise is followed at midcycle by an enormous surge of LH and a smaller burst of FSH.

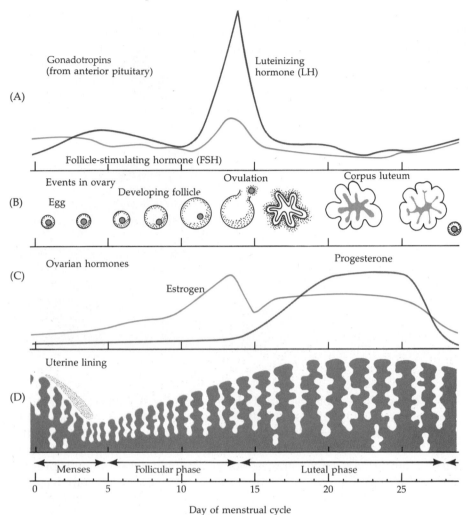

(A)

Gonadotropins
(from anterior pituitary)

Luteinizing
hormone (LH)

Follicle-stimulating hormone (FSH)

(B)

Events in ovary

Egg

Developing follicle

Ovulation

Corpus luteum

(C)

Ovarian hormones

Estrogen

Progesterone

(D)

Uterine lining

Menses

Follicular phase

Luteal phase

0   5   10   15   20   25

Day of menstrual cycle

Figure 22.32
The human menstrual cycle. The coordination of (B) ovarian and (D) uterine cycles is controlled by (A) the pituitary and (C) the ovarian hormones. During the follicular phase, the egg matures within the follicle, and the uterine lining is prepared to receive the embryo. The mature egg is released around day 14. If an embryo does not implant into the uterus, the uterine wall begins to break down, leading to menses.

Experiments with female monkeys have shown that exposure of the hypothalamus to greater than 200 pg of estrogen per milliliter of blood for more than 50 hours results in the hypothalamic secretion of gonadotropin-releasing factor. This factor causes the subsequent release of FSH and LH from the pituitary. Within 10 to 12 hours after the gonadotropin peak, the egg is ovulated (Figure 22.33; Garcia et al., 1981). Although the detailed mechanism of ovulation is not yet known, the physical expulsion of the mature oocyte from the follicle appears to be due to an LH-induced increase in collagenase, plasminogen activator, and prostaglandin within the follicle (Lemaire et al., 1973). The mRNA for plasminogen activator has been dormant in the oocyte cytoplasm. LH causes this message to be polyadenylated and translated into this powerful protease (Huarte et al., 1987). Prostaglandins may cause localized contractions in the smooth muscles in the ovary and may also increase the flow of water from the ovarian capillaries (Diaz-Infante et al., 1974; Koos and Clark, 1982). If ovarian prostaglandin synthesis is inhibited, ovulation does not take place. In addition to the prostaglandin-induced pressure, collagenases and the plasminogen activator protease loosen and digest the extracellular matrix of the follicle (Beers et al., 1975; Downs and Longo, 1983). The result of LH, then, would be increased follicular pressure coupled with the degradation of the follicle wall. A hole would be digested through which the ovum could burst.

Figure 22.33
Ovulation in the rabbit. The ovary of a living, anesthetized rabbit was exposed and observed. When the follicle started to ovulate, the ovary was removed, fixed, and stained. (Courtesy of R. J. Blandau.)

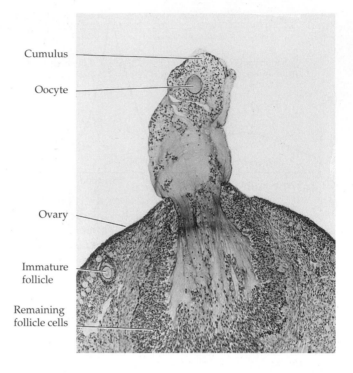

Cumulus

Oocyte

Ovary

Immature follicle

Remaining follicle cells

Following ovulation, the **luteal phase** of the menstrual cycle begins. The remaining cells of the ruptured follicle under the continued influence of LH become the **corpus luteum.** (They are able to respond to this LH because the surge in FSH stimulates them to develop even more LH receptors.) The corpus luteum secretes some estrogen, but its predominant secretion is **progesterone.** This steroid hormone circulates to the uterus, where it completes the job of preparing the uterine tissue for blastocyst implantation, stimulating the growth of the uterine wall and its blood vessels. Blocking the progesterone receptor with the synthetic steroid mifepristone (RU486) stops the uterine wall from thickening and prevents the implantation of the blastocyst into the uterus (Couzinet et al., 1986).* Progesterone also inhibits the production of FSH, thereby preventing the maturation of any more follicles and ova. (For this reason, such a combination of estrogen and progesterone has been used in birth control pills. The growth and maturation of new ova are prevented so long as FSH is inhibited.)

If the ovum is not fertilized, the corpus luteum degenerates, progesterone secretion ceases, and the uterine wall is sloughed off. With the decline in serum progesterone levels, the pituitary secretes FSH again, and the cycle is renewed. However, if fertilization occurs, the trophoblast secretes a new hormone, **luteotropin,** which causes the corpus luteum to remain active and serum progesterone levels to remain high. Thus, the menstrual cycle enables the periodic maturation and ovulation of human ova and allows the uterus to periodically develop into an organ capable of nurturing a developing organism for nine months.

*RU486 is thought to compete for the progesterone receptor inside the nucleus. RU486 can bind to the progesterone site in the receptor, and the receptor-RU486 complex appears to form heterodimers with the normal progesterone-carrying progesterone receptor. When this RU486-progesterone complex binds to the progesterone-responsive enhancer elements on the DNA, transcription from this site is inhibited (Vegeto et al., 1992; Spitz and Bardin, 1993). In Europe, RU486 has become a widely used alternative to surgical abortion (Palka, 1989; Maurice, 1991).

The Reinitiation of Meiosis
in Mammalian Oocytes

IF NUMEROUS FOLLICLES are capable of maturing when follicle-stimulating hormone is secreted, how is it that usually only one follicle and its oocyte prevail? It appears that the follicle capable of producing the most estrogen in response to FSH is the one that matures, while all the others die. Those sets of follicles initially receiving FSH not only begin to proliferate, but also produce new luteinizing hormone receptors on their thecal cells (Figure 22.34). The reception of LH causes these thecal cells to initiate estrogen production. As we have seen, estrogen has two disparate effects involving the future reception of FSH. At one level, it turns down the pituitary secretion of FSH, while at another level, it increases the FSH receptors on the follicle cells. Thus, the more estrogen a follicle produces, the more FSH receptors it has while less FSH remains in circulation. As FSH concentrations get progressively lower, only one follicle can bind

the available FSH. Only this follicle can still grow, and the other follicles die.

What does LH do that causes the reinitiation of meiosis? To address this question, the nature of the meiotic block has been intensely studied. As with amphibian oocytes, the dictyate stage is extremely important, because that is the time during which oocytes grow, differentiate the structures specific to oocytes, and acquire the ability to resume meiosis (Sorensen and Wassarman, 1976). Early experiments demonstrated that follicle-enclosed oocytes, in vivo or in vitro, do not undergo maturation unless exposed to gonadotropins, whereas oocytes removed from the follicle spontaneously resume meiosis even without the hormonal stimulus (Pincus and Enzmann, 1935).

It appears, then, that meiosis is normally inhibited by the follicle cells and can be reinitiated by gonadotropins. This hypothesis—that the follicle cells are important regulators of meiosis—is strengthened by observations that granulosa cells communicate with the oocyte through processes extending to the oocyte through the zona. These processes have gap junctions that enable small molecules to pass between the oocyte and the granulosa cells of the follicle (Figure 22.35; Anderson and Albertini, 1976; Gilula et al., 1978).

Because the elevation of cAMP levels inhibits oocyte maturation (Cho et al., 1974), it has been proposed that the meiotic arrest is maintained by the transfer of cAMP through the gap junctions from the follicular granulosa cell to the oocyte (Dekel and Beers, 1978, 1980). The luteinizing hormone surge could trigger maturation by terminating the gap junction communication, thereby inhibiting the transfer of cAMP into the oocyte. Several lines of evidence now support this hypothesis. First, the decline of cAMP appears critical for the resumption of meiosis. Germinal vesicle breakdown can be prevented by inhibiting the degradation of cAMP in follicle-free eggs

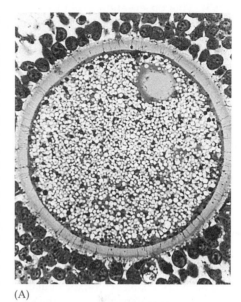

(A)

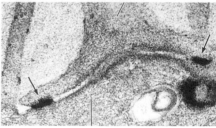

(B)　　　　　Oocyte

Figure 22.35 Communication between oocyte and granulosa cells. (A) Sheep oocyte surrounded by the zona pellucida and follicle cells. The granulosa cells of the follicle are extending processes through the zona pellucida and touching the oocyte. (B) Electron micrograph of follicle cell processes establishing gap junction connections with a rhesus monkey oocyte. Gap junctions (arrows) are stained with ionic lanthanum. (A from Moor and Cran, 1980, courtesy of the authors; B from Anderson and Albertini, 1976, courtesy of D. Albertini.)

or by providing such eggs directly with cAMP (Bornslaeger et al., 1986). The decline in oocyte cAMP concentration occurs immediately prior to the resumption of meiosis (Figure 22.36; Schultz et al., 1983).

Figure 22.34 Positive feedback cycle in mammalian follicle cells. Reception of follicle-stimulating hormone (FSH) leads to the production of more luteinizing hormone (LH) receptors. The follicle cells secrete estrogen when stimulated by LH; the estrogen causes both an increase in the number of FSH receptors and a decrease in the amount of pituitary FSH production. Eventually, very few follicles are able to receive the small amounts of FSH produced, thereby amplifying their ability to receive LH. These few follicles are able to mature.

Second, the gonadotropins can cause the loss of communication between the follicle cells and the oocyte. The follicle cells appear to be important sources of oocyte cAMP, and changes of cAMP concentration in the follicle cells are reflected in oocyte cAMP levels (Bornslaeger and Schultz, 1985; Racowsky, 1985). This observation explains why oocytes remain in meiotic arrest when they are surrounded by the follicle cells but resume meiosis when they are removed from them.

The gonadotropin surge, however, can elevate the *follicle cell* cAMP concentrations to new levels. In response to this elevation, the mature follicle cells synthesize hyaluronic acid, which causes a physical disruption of the contact between the follicle cell processes and the oocyte (Eppig, 1979; Larsen et al., 1986). The bridges through which cAMP flows from follicular granulosa cell to oocyte have thereby been removed, allowing the mammalian oocyte to resume meiosis (Dekel and Sherizly, 1985; Racowsky and Satterlie, 1985).

Like amphibian oocytes, the ovulated mouse oocyte is suspended in the second meiotic metaphase and is fertilized in that state. Paules and his co-workers (1989) have shown that maturing mouse

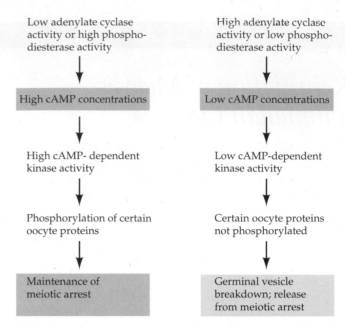

Figure 22.36 Summary of proposed mechanism whereby the oocyte cAMP level regulates resumption of meiosis by the oocyte. The cAMP levels within the oocyte are provided, at least in part, by cAMP from the follicle cells. Cyclic AMP cannot cross cell membranes, but it can enter the oocyte through gap junctions connecting the oocyte with its follicle cells. When the connections are released, oocyte cAMP levels decline, leading to the release of meiotic arrest.

oocytes also contain the pp39^{mos} cytostatic factor responsible for the arrest of meiosis in metaphase II. Female mice deficient for the *mos* gene do not stop their division in metaphase II, and the eggs often try to develop parthenogenetically (Colledge et al., 1994; Hashimoto et al., 1994). Evidently, similar events must take place for the maturation of the amphibian and mammalian oocytes. ∎

The egg and the sperm will both die if they do not meet. We are now back where we began. The stage is set for fertilization to take place. As F. R. Lillie recognized in 1919, "The elements that unite are single cells, each on the point of death; but by their union a rejuvenated individual is formed, which constitutes a link in the eternal process of Life."

LITERATURE CITED

Ahringer, J. and Kimble, J. 1991. Control of the sperm-oocyte switch in *Caenorhabditis elegans* hermaphrodites by the fem-3 3'untranslated region. *Nature* 349: 346–348.

Ahringer, J., Rosequist, T. A., Lawson, D. N. and Kimble, J. 1992. The *C. elegans* sex determining gene, *fem-3*, is regulated posttranscriptionally. *EMBO J.* 11: 2303–2310.

Anderson, E. and Albertini, D. F. 1976. Gap junctions between the oocyte and companion follicle cells in the mammalian ovary. *J. Cell Biol.* 71: 680–686.

Ashley, T., Plug, A. W., Xu, J. H., Solari, AJ., Reddy, G., Golub, E. I. and Ward, D. C.

1995. Dynamic changes in Rad51 distribution on chromatin during meiosis in male and female vertebrates. *Chromosoma* 104: 19–28.

Auerbach, R. and Joseph, J. 1984. Cell surface markers on endothelial cells: A developmental perspective. *In* E. A. Jaffe (ed.), *The Biology of Endothelial Cells.* Nijhoff, The Hague, pp. 393–400.

Austin, J. and Kimble, J. 1987. *glp-1* is required in the germ line for regulation of the decision between mitosis and meiosis in C. *elegans. Cell* 51: 589–599.

Baker, S. M. and eleven others. 1996. Involvement of the mouse Mlh1 in DNA mismatch repair and meiotic crossing over. *Nat. Genet.* 13: 336–342.

Baker, T. G. 1970. Primordial germ cells. *In* C. R. Austin and R. V. Short (eds.), *Reproduction in Mammals,* Vol. 1: *Germ Cells and Fertilization.* Cambridge University Press, Cambridge, pp. 1–13.

Beers, W. H., Strickland, S. and Reich, E. 1975. Ovarian plasminogen activator: Relationship to ovulation and hormonal regulation. *Cell* 6: 387–394.

Bier, K. 1963. Autoradiographische Untersuchungen über die Leistungen des Follikelepithels und der Nahrzellen bei der Dottbildung und Eiweissynthese im Fliegenova. *Wilhelm Roux Arch. Entwicklungsmech. Org.* 154: 552–575.

Blendy, J. A., Kaestner, K. H., Weinbauer, G. F., Nieschlag, E. and Schütz, G. 1996. Severe impairment of spermatogenesis in mice lacking the CREM gene. *Nature* 380: 162–165.

Bloom, W. and Fawcett, D. W. 1975. *Textbook of Histology*, 10th Ed. Saunders, Philadelphia.

Bornslaeger, E. A. and Schultz, R. M. 1985. Regulation of mouse oocyte maturation: Effect of elevating cumulus cell cAMP on oocyte cAMP levels. *Biol. Reprod.* 33: 698–704.

Bornslaeger, E. A., Mattei, P. and Schultz, R. M. 1986. Involvement of cAMP-dependent protein kinase and protein phosphorylation in regulation of mouse oocyte maturation. *Dev. Biol.* 114: 453–462.

Borovsk, D., Carlson, D. A., Griffin, P. R., Shabanowitz, J. and Hunt, D. F. 1990. Mosquito oostatic factor: A novel decapeptide modulating trypsin-like enzyme biosythesis in the midgut. *FASEB J.* 4: 3015–3020.

Bounoure, L. 1934. Recherches sur lignée germinale chez la grenouille rousse aux premiers stades au développement. *Ann. Sci. Zool. Ser.* 17, 10: 67–248.

Bownes, M. 1982. Hormonal and genetic regulation of vitellogenesis in *Drosophila*. *Q. Rev. Biol.* 57: 247–274.

Bownes, M., Scott, A. and Shirras, A. 1988. Dietary components modulate yolk protein transcription in *Drosophila melanogaster*. *Development* 103: 119–128.

Braun, R. E., Behringer, R. R., Peschon, J. J., Brinster, R. L. and Palmiter, R. D. 1989. Genetically haploid spermatids are phenotypically diploid. *Nature* 337: 373–376.

Brennen, M. D., Weiner, A. J., Goralski, T. J. and Mahowald, A. P. 1982. The follicle cells are a major site of vitellogenin synthesis in *Drosophila melanogaster*. *Dev. Biol.* 89: 225–236.

Brown, D. D. and Dawid, I. B. 1968. Specific gene amplification in oocytes. *Science* 160: 272–280.

Burns, R. K., Jr. 1930. The process of sex-transformation parabiotic *Amblystoma*. I. Transformation from female to male. *J. Exp. Zool.* 55: 123–169.

Burtis, K. C. 1993. The regulation of sex determination and sexually dimorphic differentiation in *Drosophila*. *Curr. Opin. Cell Biol.* 5: 1006–1014.

Carlson, B. M. 1981. *Patten's Foundations of Embryology*. McGraw-Hill, New York.

Chiquoine, A. D. 1954. The identification, origin, and migration of the primordial germ cells in the mouse embryo. *Anat. Rec.* 118: 135–146.

Cho, W. K., Stern, S. and Biggers, J. D. 1974. Inhibitory effect of dibutyryl cAMP on mouse oocyte maturation in vitro. *J. Exp. Zool.* 187: 383–386.

Colledge, W. H., Carleton, M. B. L., Udy, G. B. and Evans, M. J. 1994. Disruption of c-mos causes parthenogenetic development of unfertilized mouse eggs. *Nature* 370: 65–68.

Comings, D. E. 1968. The rationale for an ordered arrangement of chromatin in the interphase nucleus. *Am. J. Hum. Genet.* 20: 440–460.

Cooley, L. and Theurkauf, W. E. 1994. Cytoskeletal functions during *Drosophila* oogenesis. *Science* 266: 590–596.

Cooley, L., Verheyen, E. and Ayers, K. 1992. *chickadee* encodes a profilin required for intercellular cytoplasm transport during *Drosophila* oognesis. *Cell* 69: 173–184.

Couzinet, B., Le Strat, N., Ulmann, A., Baulieu, E. E. and Schaison, G. 1986. Termination of early pregnancy by the progesterone antagonist RU486 (Mifepristone). *N. Engl. J. Med.* 315: 1565–1570.

Danilchik, M. V. and Gerhart, J. C. 1987. Differentiation of the animal-vegetal axis in *Xenopus laevis* oocytes: Polarized intracellular translocation of platelets establishes the yolk gradient. *Dev. Biol.* 122: 101–112.

Davidson, E. 1986. *Gene Activity in Early Development*, 3rd Ed. Academic Press, Orlando, FL.

Dekel, N. and Beers, W. H. 1978. Rat oocyte maturation in vitro: Relief of cyclic cAMP inhibition by gonadotropins. *Proc. Natl. Acad. Sci. USA* 75: 4369–4373.

Dekel, N. and Beers, W. H. 1980. Development of the rat oocyte in vitro: Inhibition and induction of maturation in the presence or absence of the cumulus oophorus. *Dev. Biol.* 75: 247–254.

Dekel, N. and Sherizly, I. 1985. Epidermal growth factor induces maturation of rat follicle-enclosed oocytes. *Endocrinology* 116: 406–409.

Dernburg, A. F., Sedat, J. W. and Hawley, R. S. 1996. Direct evidence of a role for heterochromatin in meiotic chromosome segregation. *Cell* 86: 135–146.

Diaz-Infante, A., Wright, K. H. and Wallach, E. E. 1974. Effects of indomethacin and PGF2a on ovulation and ovarian contraction in the rabbit. *Prostaglandins* 5: 567–581.

Dolci, S. and eight others. 1991. Requirement for mast cell growth factor for primordial germ cell survival in culture. *Nature* 352: 809–811.

Dong, J., Albertini, D. F., Nishimori, K., Kumar, T. R., Lu, N., amd Matzuk, M. 1996. Growth differentiation factor-9 is required during early ovarian folliculogenesis. *Nature* 383: 531–534.

Doniach, T. and Hodgkin, J. 1984. A sex-determining gene, *fem-1*, required for both male and hermaphroditic development in *C. elegans*. *Dev. Biol.* 106: 223–235.

Downs, S. and Longo, F. J. 1983. Prostaglandins and preovulatory follicular maturation in mice. *J. Exp. Zool.* 228: 99–108.

Dubois, R. 1969. Le mécanisme d'entrée des cellules germinales primordiales dans le réseau vasculaire, chez l' embryon de poulet. *J. Embryol. Exp. Morphol.* 21: 255–270.

Dumont, J. N. 1978. Oogenesis in *Xenopus laevis*. VI. Route of injected tracer transport in follicle and developing oocyte. *J. Exp. Zool.* 204: 193–200.

Dym, M. 1977. The male reproductive system. *In* L. Weiss and R. O. Greep (eds.), *Histology*, 4th Ed. McGraw-Hill, New York, pp. 979–1038.

Dym, M. 1994. Spermatogonial stem cells of the testis. *Proc. Natl. Acad. Sci. USA* 91: 11287–11289.

Dym, M. and Fawcett, D. W. 1971. Further observations on the number of spermatogonia, spermatocytes, and spermatids connected by intercellular bridges in the mammalian testis. *Biol. Reprod.* 4: 195–215.

Eberhart, C. G., Maines, J. Z. and Wasserman, S. A. 1996. Meiotic cell cycle requirement for a fly homologue of human *Deleted in Azoospermia*. *Nature* 381: 783–785.

Ellis, R. E. and Kimble, J. 1995. The *fog-3* gene and regulation of cell fate in the germ line of *Caenorhabditis elegans*. *Genetics* 139: 561–677.

Epifano, O., Liang, L.-f., Familari, M., Moos, M. C. Jr. and Dean, J. 1995. Coordinate expression of the three zona pellucida genes during mouse oogenesis. *Development* 121: 1947–1956.

Eppig, J. J. 1979. FSH stimulates hyaluronic acid synthesis by oocyte-cumulus cell complexes from mouse preovulatory follicles. *Nature* 281: 483–484.

Eyal-Giladi, H., Ginsburg, M. and Farbarou, A. 1981. Avian primordial germ cells are of epiblastic origin. *J. Embryol. Exp. Morphol.* 65: 139–147.

ffrench-Constant, C., Hollingsworth, A., Heasman, J. and Wylie, C. 1991. Response to fibronectin of mouse primordial germ cells before, during, and after migration. *Development* 113: 1365–1373.

Flickinger, R. A. and Rounds, D. E. 1956. The maternal synthesis of egg yolk proteins as demonstrated by isotopic and serological means. *Biochem. Biophys. Acta* 22: 38–72.

Forristall, C., Pondel, M., Chen, L. and King, M. L. 1995. Patterns of localization and cytoskeletal association of two vegetally localized RNAs, Vg1 and Xcat-2. *Development* 121: 201–208.

Foulkes, N., Schlotter, F., Pévet, P. and Sassone-Corsi, P. 1993. Pituitary FSH directs the CREM functional switch during spermatogenesis. *Nature* 362: 264–267.

Francis, R., Barton, M. K., Kimble, J. and Schedl, T. 1995. gld-1, a tumor suppressor gene required for oocyte development in *Caenorhabditis elegans*. *Genetics* 139: 579–606.

Gabrielli, B., Roy, L. M. and Maller, J. L. 1993. Requirement for cdk2 in cytostatic factor-mediated metaphase II arrest. *Science* 259: 1766–1769.

Gallatin, W. M., Weissman, I. L. and Butcher, E. C. 1983. A cell surface molecule involved in organ-specific homing of lymphocytes. *Nature* 304: 30–35.

Gallatin, W. M., St. John, T. P., Siegelman, M., Reichert, R., Butcher, E. C. and Weissman, I. L. 1986. Lymphocyte homing receptors. *Cell* 44: 673–680.

Garcia, J. E., Jones, G. S. and Wright, G. L. 1981. Prediction of the time of ovulation. *Fert. Steril.* 36: 308–315.

Gardner, R. L. 1982. Manipulation of development. *In* C. R. Austin and R. V. Short (eds.), *Embryonic and Fetal Development*, Cambridge University Press, Cambridge, pp. 159–180.

Gilula, N. B., Epstein, M. L. and Beers, W. H. 1978. Cell-to-cell communication and ovulation. A study of the cumulus-oocyte complex. *J. Cell Biol.* 78: 58–75.

Ginsburg, M. and Eyal-Giladi, H. 1987. Primordial germ cells of the young chick blastoderm originate from the central zone of the area pellucida irrespective of the embryo-forming process. *Development* 101: 209–219.

Ginsburg, M., Snow, M. H. L. and McLaren, A. 1990. Primordial germ cells in the mouse embryo during gastrulation. *Development* 110: 521–528.

Godin, I. and Wylie, C. C. 1991. TGFβ1 inhibits proliferation and has a chemotactic effect on mouse primordial germ cells in culture. *Development* 113: 1451–1457.

Godin, I., Wylie, C. and Heasman, J. 1990. Genital ridges exert long-range effects on primordial germ cell numbers and direction of migration in culture. *Development* 108: 357–363.

Gönczy, P., Thomas, B. J. and DiNardo, S. 1994. roughex is a dose-dependent regulator of the second meiotic division during *Drosophila* spermatogenesis. *Cell* 77: 1015–1025.

González-Reyes, A., Elliot, H. and St. Johnson, D. 1995. Polarization of both major body axes in Drosophila by gurken-torpedo signalling. *Nature* 375: 654–658.

Graham, C. E. 1977. Teratocarcinoma cells and normal mouse embryogenesis. *In* M. I. Sherman (ed.), *Concepts of Mammalian Embryogenesis*. M.I.T. Press, Cambridge, MA, pp. 315–394.

Graham, P. L. and Kimble, J. 1993. The mog-1 gene is required for the switch from spermatogenesis to oogenesis in *C. elegans*. *Genetics* 133: 919–931.

Grant, P. 1953. Phosphate metabolism during oogenesis in *Rana temporaria*. *J. Exp. Zool.* 124: 513–543.

Green, G. R. and Poccia, D. L. 1988. Interaction of sperm histone variants and linker DNA during spermiogenesis in the sea urchin. *Biochemistry* 27: 619–625.

Gurdon, J. B. 1976. *The Control of Gene Expression in Animal Development*. Harvard University Press, Cambridge, MA.

Gutzeit, H. O. 1986. The role of microfilaments in cytoplasmic streaming in *Drosophila* follicles. *J. Cell Sci.* 80: 159–169.

Hagedorn, H. H. 1983. The role of ecdysteroids in the adult insect. *In* G. Downer and H. Laufer (eds.), *Endocrinology of Insects*. Alan R. Liss, New York, pp. 241–304.

Hahnel, A. C. and Eddy, E. M. 1986. Cell surface markers of mouse primordial germ cells defined by two monoclonal antibodies. *Gamete Res.* 15: 25–34.

Hardvin-Lepers, A., Shaper, J. and Shaper, N. L. 1993. Characterization of two cis-regulatory regions in the murine β1,4-galactosyltransferase gene. *J. Biol. Chem.* 268: 14348–14359.

Hashimoto, N. and ten others. 1994. Parthenogenetic activation of oocytes in c-mos-deficient mice. *Nature* 370: 68–71.

Heasman, J., Mohun, T. and Wylie, C. C. 1977. Studies on the locomotion of primordial germ cells from *Xenopus laevis* in vitro. *J. Embryol. Exp. Morphol.* 42: 149–162.

Heasman, J., Hynes, R. D., Swan, A. P., Thomas, V. and Wyle, C. C. 1981. Primordial germ cells of *Xenopus* embryos: The role of fibronectin in their adhesion during migration. *Cell* 27: 437–447.

Heath, J. K. 1978. Mammalian primordial germ cells. *Dev. Mammals* 3: 272–298.

Hill, D. P., Shakes, D. C., Wards, S. and Strome, S. 1989. A sperm-supplied product essential for initiation of normal embryogenesis in *Caenorhabditis elegans* is encoded by the paternal effect embryonic-lethal gene, spe-11. *Dev. Biol.* 136: 154–166.

Hill, R. S. and MacGregor, H. C. 1980. The development of lampbrush chromosome-type transcription in the early diplotene oocytes of *Xenopus laevis*: An electron microscope analysis. *J. Cell Sci.* 44: 87–101.

Hirsh, D., Oppenheim, D. and Klass, M. 1976. Development of the reproductive system of *Caenorhabditis elegans*. *Dev. Biol.* 49: 200–219.

Hodgkin, J., Doniach, T. and Shen, M. 1985. The sex determination pathway in the nematode *Caenorhabditis elegans*: Variations on a theme. *Cold Spring Harbor Symp. Quant. Biol.* 50: 585–593.

Hoyle, H. D. and Raff, E. C. 1990. Two *Drosophila* β-tubulin isoforms are not functionally equivalent. *J. Cell Biol.* 111: 1009–1026.

Huarte, J., Belin, D., Vassalli, A., Strickland, S. and Vassalli, J. -D. 1987. Meiotic maturation of mouse oocytes triggers the translatio and polyadenylation of dormant tissue-type plasminogen activator mRNA. *Genes Dev.* 1: 1201–1211.

Humphrey, R. R. 1931. Studies of sex reversal in *Amblystoma*. III. Transformation of the ovary of *A. tigrinum* into a functional testis through the influence of a testis resident in the same animal. *J. Exp. Zool.* 58: 333–365.

Inoue, H. and Hiroyoshi, T. 1986. A maternal-effect sex-transformation mutant of the housefly, *Musca domestica* L. *Genetics* 112: 469–481.

Jaglarz, M. K. and Howard, K. R. 1995. The active migration of *Drosophila* primordial germ cells. *Development* 121: 3495–3503.

Jones, A. R., Francis, R. and Schedl, T. 1996. GLD-1, a cytoplasmic protein essential for oocyte differentiation, shows stage- and sex-specific expression during *Caenorhabditis elegans* germline development. *Dev. Biol.* 180: 165–183.

Kammerman, S. and Ross, J. 1975. Increase in numbers of gonadotropin receptors on granulosa cells during follicle maturation. *J. Clin. Endocrinol.* 41: 546–550.

Karpen, G. H., Le, M.-H. and Le, H. 1996. Centric heterochromatin and the efficiency of achiasmatic disjunction in *Drosophila* female meiosis. *Science* 273: 118–121.

Karr, T. L. 1991. Intracellular sperm-egg interaction in *Drosophila*: A three-dimensional structural analysis of a paternal product in the developing egg. *Mech. Dev.* 34: 101–111.

Karr, T. L. 1996. Paternal investment and intracellular sperm-egg interaction during and following fertilization in *Drosophila*. *Curr. Top. Dev. Biol.* 34: 89–115.

Karsch-Mizrachi, I. and Haynes, S. R. 1993. The Rb97D gene encodes a potential RNA-binding protein required for spermatogenesis in *Drosophila*. *Nucl. Acids Res.* 21: 2229–2235.

Kerrebrock, A. W., Moore, D. P., Wu, J. S. and Orr-Weaver, T. L. 1995. Mei-S332, a *Drosophila* protein required for sister-chromatid cohesion, can localize to meiotic centromere regions. *Cell* 83: 247–256.

Kim-Ha, J., Kerr, K. and Macdonald, P. M. 1995. Translational regulation of oskar mRNA by bruno, an ovarian RNA-binding protein, is essential. *Cell* 81: 403–412.

Kimble, J. E. 1981. Strategies for control of pattern formation in *Caenorhabditis elegans*. *Philos. Trans. R. Soc. Lond.* [B] 295: 539–551.

Kimble, J. E. and White, J. G. 1981. Control of germ cell development in *Caenorhabditis elegans*. *Dev. Biol.* 81: 208–219.

Kimble, J., Barton, M. K., Schedl, T. B., Rosenquist, T. A. and Austin, J. 1986. Controls of postembryonic germ line development in *Caenorhabditis elegans*. *In* J. Gall (ed.), *Gametogenesis and the Early Embryo*. Alan R. Liss, New York, pp. 97–110.

Kloc, M. and Etkin, L. 1995. Two distinct pathways for the localization of RNAs at the vegetal cortex in *Xenopus* oocytes. *Development* 121: 287–297.

Kloc, M., Larabell, C. and Etkin, L. 1996. Elaboration of the messenger transport organizer pathway for localization of RNA to the vegetal cortex of *Xenopus* oocytes. *Dev. Biol.* 180: 119–130.

Kloc, M., Spohr, G. and Etkin, L. 1993. Translocation of repetitive RNA sequences with the germ plasm in *Xenopus* oocytes. *Science* 262: 1712–1714.

Koos, R. D. and Clark, M. R. 1982. Production of 6-keto-prostaglandin $F_{1\alpha}$ by rat granulosa cells in vitro. *Endocrinology* 111: 1513–1518.

Kuwana, T. 1993. Migration of avian primordial germ cells toward the gonadal anlage. *Dev. Growth Differ.* 35: 237–243.

Kuwana, T., Maeda-Suga, H. and Fujimoto, T. 1986. Attraction of chick primordial germ cells by gonadal anlage in vitro. *Anat. Rec.* 215: 403–406.

Kwon, Y. K. and Hecht, N. B. 1993. Binding of a phosphoprotein to the 3′ untranslated region of the mouse protamine 2 mRNA temporally represses its translation. *Mol. Cell Biol.* 13: 6547–6557.

Lane, M. E. and Kalderon, D. 1994. RNA localization along the anteroposterior axis of the *Drosophila* oocyte requires PKA-mediated signal transduction to direct normal microtubule organization. *Genes Dev.* 8: 2986–2995.

Langman, J. 1981. *Medical Embryology*, 4th Ed. Williams & Wilkins, Baltimore.

Larsen, W. J., Wert, S. E. and Brunner, G. D. 1986. A dramatic loss of cumulus cell gap junctions is correlated with germinal vesicle breakdown in rat oocytes. *Dev. Biol.* 113: 517–521.

Laskey, R. A. 1979. Biochemical processes in early development. *In* A. T. Bull, J. R. Lagnado, J. O. Thomsen and K. F. Tipton, (eds.), *Companion to Biochemistry*, Vol. 2. Longman, London, pp. 137–160.

Lasko, P. 1995. Cell-cell signalling, microtubule organization and RNA localization: Is PKA a link? *BioEssays* 17: 105–107.

Lin, H. and Spradling, A. C. 1995. Fusosome asymmetry and oocyte determination in *Drosophila*. *Dev. Genet.* 16: 6–12.

Lemaire, W. J., Yang, N. S. T., Behram, H. H. and Marsh, J. M. 1973. Preovulatory changes in concentration of prostaglandin in rabbit graafian follicles. *Prostaglandins* 3: 367–376.

Lillie, F. R. 1919. *Problems of Fertilization*. University of Chicago Press, Chicago.

Lorca, T., Cruzalegui, F. H., Fesquet, D., Cavadore, J.-C., Méry, J., Means, A. and Dorée, M. 1993. Calmodulin-dependent protein kinase II mediates inactivation of MPF and CSF upon fertilization of *Xenopus* eggs. *Nature* 366: 270–273.

Manseau, L. J. and Schüpbach, T. 1989. *cappuccino* and *spire*: two unique maternal-effect loci required for both the anteroposterior and dorsoventral patterns of the *Drosophila* embryo. *Genes Dev.* 3: 1437–1452.

Marcey, D., Watkins, W. S. and Hazelrigg, T. 1991. The temporal and spatial distribution pattern of maternal *exuparentia* protein: Evidence for a role in establishment but not maintainance of *bicoid* mRNSA localization. *EMBO J.* 10: 4259–4266.

Masui, Y. 1974. A cytostatic factor in amphibian: Its extraction and partial characterization. *J. Exp. Zool.* 187: 141–147.

Matsui, Y., Toksoz, D., Nishikawa, S., Nishikawa, S.-I., Williams, D., Zsebo, K. and Hogan, B. L. M. 1991. Effect of Steel factor and leukemia inhibitory factor on murine primordial germ cells in culture. *Nature* 353: 750–752.

Matsui, Y., Zsebo, K. and Hogan, B. L. M. 1992. Derivation of pluripotential embryonic stem cells from murine primordial germ cells in culture. *Cell* 70: 841–847.

Maurice, J. 1991. Improvements seen for RU-486 abortions. *Science* 254: 198–200.

McLaren, A. 1983. Does the chromosomal sex of a mouse cell affect its development? *Symp. Br. Soc. Dev. Biol.* 7: 225–227.

Menke, D. B., Mutter, G. I. and Page, D. C. 1997. Expression of DAZ, an azoospermia factor candidate, in human spermatogonia. *Am. J. Hum. Genet.* 60: 237–241.

Minishull, J. 1993. Cyclin synthesis: Who needs it? *BioEssays* 15: 149–155.

Mintz, B. 1957. Embryological development of primordial germ cells in the mouse: Influence of a new mutation. *J. Embryol. Exp. Morphol.* 5: 396–403.

Moens, P. B. 1969. The fine structure of meiotic chromosome polarization and pairing in *Locusta migratoria*. *Chromosoma* 28: 1–25.

Moor, R. M. and Cran, D. G. 1980. Intercellular coupling of mammalian oocytes. *Dev. Mammals* 4: 3–38.

Moses, M. J. 1968. Synaptonemal complex. *Annu. Rev. Genet.* 2: 363–412.

Mowry, K. L. and Melton, D. A. 1992. Vegetal messenger RNA localization directed by a 340-nt RNA sequence element in *Xenopus* oocytes. *Science* 255: 991–994.

Nantel, F. and eight others. 1996. Spermiogenesis deficiency and germ-cell apoptosis in CREM-mutant mice. *Nature* 380: 159–162

Newton, S. C., Blaschuk, O. W. and Millette, C. F. 1993. N-cadherin mediates Sertoli cell-spermatogenic cell adhesion. *Dev. Dyn.* 197: 1–13.

Nishioka, D., Ward, D., Poccia, D., Costacos, C. and Minor, J. E. 1990. Localization of bindin expression during sea urchin spermatogenesis. *Mol. Reprod. Dev.* 27: 181–190.

Old, R. W., Callan, H. G. and Gross, K. W. 1977. Localization of histone gene transcripts in newt lampbrush chromosomes by in situ hybridization. *J. Cell Sci.* 27: 57–80.

Oliver, B., Kim, Y.-J. and Baker, B. S. 1993. *Sex-lethal*, master and slave: A hierarchy of germ-line sex determination in *Drosophila*. *Development* 119: 897–908.

Paglia, L. M., Berry, J. and Kastern, W. H. 1976. Messenger RNA synthesis, transport, and storage in silkmoth ovarian follicles. *Dev. Biol.* 51: 173–181.

Palka, J. 1989. The pill of choice? *Science* 245: 1319–1323.

Palmiter, R. D., Wilkie, T. M., Chen, H. Y. and Brinster, R. L. 1984. Transmission distortion and mosaicism in an unusual transgenic mouse pedigree. *Cell* 36: 869–877.

Paris, J., Swenson, K., Piwnice-Worms, H. and Richter, J. D. 1991. Maturation-specific polyadenylation: In vitro activation by p34cdc2 and phosphorylation of a 58-kD CPE-binding protein. *Genes Dev.* 5: 1697–1708.

Pasteels, J. 1953. Contributions à l'étude du developpement des reptiles. I. Origine et migration des gonocytes chez deux Lacertiens. *Arch. Biol.* 64: 227–245.

Paules, R. S., Buccione, R., Moscel, R. C., Vande Woude, G. F. and Eppig, J. J. 1989. Mouse *mos* protoncogene product is present and functions during oogenesis. *Proc. Natl. Acad. Sci. USA* 86: 5395–5399.

Pesce, M., Farrace, M. G., Piacentini, M., Dolci, S. and De Felici, M. 1993. Stem cell factor and leukemia inhibitory factor promote primordial germ cell survival by suppressing programmed cell death (apoptosis). *Development* 118: 1089–1094.

Peschon, J. J., Behringer, R. R., Brinster, R. L. and Palmiter, R. D. 1987. Spermatid-specific expression of protamine-1 in transgenic mice. *Proc. Natl. Acad. Sci. USA* 84: 5316–5319.

Pincus, G. and Enzmann, E.V. 1935. The comparative behavior of mammalian eggs in vivo and in vitro. I. The activation of ovarian eggs. *J. Exp. Med.* 62: 665–675.

Pinkerton, J. H. M., McKay, D. G., Adams, E. C. and Hertig, A. T. 1961. Development of the human ovary: A study using histochemical techniques. *Obstet. Gynecol.* 18: 152–181.

Pitnick, S., Spicer, G. S. and Markow, T. A. 1995. How long is a giant sperm? *Nature* 375: 109.

Poccia, D. 1986. Remodeling of nucleoproteins during gametogenesis, fertilization, and early development. *Int. Rev. Cytol.* 105: 1–65.

Pokrywka, N. J and Stephenson, E. C. 1991. Microtubules mediate the localization of *bicoid* RNA during *Drosophila* oogenesis. *Development* 113: 55–66.

Postlethwait, J. H., Brownes, M. and Jowett, T. 1980. Sexual phenotype and vitellogenin synthesis in *Drosophila melanogaster. Dev. Biol.* 79: 379–387.

Pratt, S. A., Scully, N. F. and Shur, B. D. 1993. Cell surface β-1,4-galactosyltransferase on primary spermatocytes facilitates their initial adhesion to Sertoli cells in vitro. *Biol. Reprod.* 49: 470–482.

Profet, M. 1993. Menstruation as a defense against pathogens transported by sperm. *Q. Rev. Biol.* 68: 335–385.

Racowsky, C. 1985. Effect of forskolin on the spontaneous maturation and cyclic AMP content of hamster and oocyte-cumulus complexes. *J. Exp. Zool.* 234: 87–96.

Racowsky, C. and Satterlie, R. A. 1985. Metabolic, fluorescent dye and electrical coupling between hamster oocytes and cumulus cells during meiotic maturation in vivo and in vitro. *Dev. Biol.* 108: 191–202.

Raikhel, A. S. and Dhadialla, T. S. 1992. Accumulation of yolk proteins in insect oocytes. *Annu. Rev. Entomol.* 37: 217–251.

Reijo, R. and twelve others. 1995. Diverse spermatogenic defects in humans caused by Y chromosome deletions encompassing a novel RNA-binding protein gene. *Nat. Genet.* 10: 383–393.

Reynaud, G. 1969. Transfert de cellules germinales primordiales de dindon à l'embryon de poulet par injection intravasculaire. *J. Embryol. Exp. Morphol.* 21: 485–507.

Ressom, R. E. and Dixon, K. E. 1988. Relocation and reorganization of germ plasm in *Xenopus* embryos after fertilization. *Development* 103: 507–518.

Rivier, C., Rivier, J. and Vale, W. 1986. Inhibin-mediated feedback control of follicle-stimulating hormone secetion in the female rat. *Science* 234: 205–208.

Rogulska, T. 1969. Migration of chick primordial germ cells from the intracoelomically transplanted germinal crescent into the genital ridge. *Experientia* 25: 631–632.

Rogulska, T. Ozdzenski, W. and Komer, A. 1971. Behavior of mouse primordial germ cells in chick embryo. *J. Embryol. Exp. Morphol.* 25: 155–164.

Romanoff, A. L. 1960. *The Avian Embryo.* Macmillan, New York.

Rongo, C., Gavis, E. R. and Lehmann, R. 1995. Localization of *oskar* RNA regulates *oskar* translation and requires Oskar protein. *Development* 121: 2737–2746.

Roth, S., Neuman-Silbergger, F. S., Barcelo, G. and Schüpbach, T. 1995. *cornichon* and the EGF-receptor signalling process are necessary for both anterior-posterior and dorsal-ventral pattern formation in *Drosophila. Cell* 81: 967–978.

Ruohola, H., Bremer, K. A., Baker, D., Swedlow, J. R., Jan, L. Y. and Jan, Y. N. 1991. Role of neurogenic genes in establishment of follicle cell fate and oocyte polarity during oogenesis in *Drosophila. Cell* 66: 433–449.

Sagata, N., Watanabe, N., Vande Woude, G. F. and Ikawa, Y. 1989. The c-*mos* proto-oncogene product is a cytostatic factor responsible for meiotic arrest in vertebrate eggs. *Nature* 342: 512–518.

Sagata, N., Oskarsson, M., Copeland, T., Brumbaugh, J. and Vande Woude, G. F. 1988. Function of c-*mos* proto-oncogene product in meiotic maturation in *Xenopus* oocytes. *Nature* 335: 519–525.

Schäfer, M., Nayernia, K., Engel, W. and Schäfer, U. 1995. Translational control of spermatogenesis. *Dev. Biol.* 172: 344–352.

Schmekel, K. and Daneholt, B. 1995. The central region of the synaptonemal complex revealed in three dimensions. *Trends Cell Biol.* 5: 239–242.

Schultz, R. M., Montgomery, R. R. and Belanoff, J. R. 1983. Regulation of mouse oocyte maturation: Implications of a decrease in oocyte cAMP and protein dephosphorylation in commitment to resume meiosis. *Dev. Biol.* 97: 264–273.

Sheets, M. D., Wu, M. and Wickens, M. 1995. Polyadenylation of c-*mos* mRNA as a control point in *Xenopus* meiotic maturation. *Nature* 374: 511–516.

Shea, T. B., Beermann, M. L., Leli, U. and Nixon, R. A. 1992. Opposing influences of protein kinase activities on neurite outgrowth in human neuroblastoma cells: Initiation by kinase A and restriction by kinase C. *J. Neurosci. Res.* 33: 398–407.

Simon, D. 1960. Contribution à l'étude de la circulation et du transport des gonocytes primaires dans les blastodermes d'oiseau cultivé in vitro. *Arch. Anat. Microsc. Morphol. Exp.* 49: 93–176.

Sorensen, R. and Wassarman, P. M. 1976. Relationship between growth and meiotic maturation of the mouse oocyte. *Dev. Biol.* 50: 531–536.

Spitz, I. M. and Bardin, C. W. 1993. Mifeprisone (RU486): A modulator of progestin and glucocorticoid action. *N. Engl. J. Med.* 329: 404–412.

Spradling, A. C. 1993. Germine cysts: Communes that work. *Cell* 72: 649–651.

Stephanson, E. C., Chao, Y.-C. and Fackenthal, J. D. 1988. Molecular analysis of the *swallow* gene of *Drosophila melanogaster. Genes Dev.* 2: 1655–1665.

Stewart, T. A. and Mintz, B. 1981. Successful generations of mice produced from an established culture line of euploid teratocarcinoma cells. *Proc. Natl. Acad. Sci. USA* 78: 6314–6318.

Stott, D. and Wylie, C. C. 1986. Invasive behaviour of mouse primordial germ cells in vitro. *J. Cell Sci.* 86: 133–144.

Subtelny, S. and Penkala, J. E. 1984. Experimental evidence for a morphogenetic role in the emergence of primordial germ cells from the endoderm of *Rana pipiens. Differentiation* 26: 211–219.

Sun, Y.-A. and Wyman, R. J. 1993. Reevaluation of electrophoresis in the *Drosophila* egg chamber. *Dev. Biol.* 155: 206–215.

Sutasurya, L. A. and Nieuwkoop, P. D. 1974. The induction of primordial germ cells in the urodeles. *Wilhelm Roux Arch. Entwicklungsmech. Org.* 175: 199–220.

Swanson, C. P., Merz, T. and Young, W. J. 1981. *Cytogenetics: The Chromosome in Division, Inheritance and Evolution.* Prentice-Hall, Englewood Cliffs, NJ.

Swift, C. H. 1914. Origin and early history of the primordial germ-cells in the chick. *Am. J. Anat.* 15: 483–516.

Telfer, W. H., Woodruff, R. I. and Huebner, E. 1981. Electrical polarity and cellular differentiation in meroistic ovaries. *Am. Zool.* 21: 675–686.

Theurkauf, W. E., Smiley, S., Wong, M. L. and Alberts, B. M. 1992. Reorganization of the cytoskeleton during *Drosophila* oogenesis: Implications for axis specification and intercellular transport. *Development* 115: 923–936.

Theurkauf, W. E., Alberts, B. M., Jan, Y. N. and Jongens, T. A. 1993. A control code for microtubules in the differentiation of *Drosophila* oocytes. *Development* 118: 1169–1180.

Vegeto, E., Allan, G. F., Schrader, W. T., Tsai, M.-J., McDonnell, D. P. and O'Malley, B. W. 1992. The mechanism of RU486 antagonism is dependent on the conformation of the carboxy-terminal tail of the human progesterone receptor. *Cell* 69: 703–713.

von Wettstein, D. 1971. The synaptonemal complex and four-strand crossing over. *Proc. Natl. Acad. Sci. USA* 68: 851–855.

von Wettstein, D. 1984. The synaptonemal complex and genetic segregation. *In* C. W. Evans and H. G. Dickinson (eds.), *Controlling Events in Meiosis.* Cambridge University Press, Cambridge, pp. 195–231.

Warrior, R. 1994. Primordial germ cell migration and the assembly of the *Drosophila* embryonic gonad. *Dev. Biol.* 166: 180–194.

Watanabe, N., Vande Woude, G. F., Ikawa, Y. and Sagata, N. 1989. Specific proteolysis of the c-*mos* proto-oncogene product by calpain on fertilization of *Xenopus* eggs. *Nature* 342: 505–517.

Watson, C. A., Sauman, I. and Berry, S. J. 1993. Actin is a major structural and functional element of the egg cortex of giant silkmoths during oogenesis. *Dev. Biol.* 155: 315–323.

Whitington, P. M. and Dixon, K. E. 1975. Quantitative stuidies of germ plasm and germ cells during early embryogenesis of *Xenopus laevis. J. Embryol. Exp. Morphol.* 33: 57–74.

Williams, J. and seven others. 1993. Interacting signalling pathways regulating prestalk cell differentiation and movement during the morphogenesis of *Dictyostelium. Development* Suppl.: 1–7.

Woodruff, T. K., D'Agostino, J., Schwartz, N. B. and Mayo, K. E. 1988. Dynamic changes in inhibin messenger RNAs in rat ovarian follicles during the reproductive cycle. *Science* 239: 1296–1299.

Wylie, C. C. and Heasman, J. 1993. Migration, proliferation, and potency of primordial germ cells. *Semin. Dev. Biol.* 4: 161–170.

Wylie, C. C., Heasman, J., Swan, A. P. and Anderton, B. H. 1979. Evidence for substrate guidance of primordial germ cells. *Exp. Cell Res.* 121: 315–324.

Yeom, Y. I. and seven others. 1996. Germline regulatory element of Oct-4 specific for the totipotent cycle of embryonal cells. *Development* 122: 881–894.

Yisraeli, J. K., Sokol, S. and Melton, D. A. 1990. A two-step model for the localization of a maternal mRNA in *Xenopus* oocytes: Involvement of microtubules and microfilaments in translocation and anchoring of Vg1 mRNA. *Development* 108: 289–298.

Yoon, P. W. and eight others. 1996. Advanced maternal age and risk of Down syndrome characterized by the meiotic stage of the chromosomal error: A population-based study. *Amer. J. Hum. Genet.* 58: 628–633.

Zhang, N., Zhang, J., Purcell, K. J., Cheng, Y. and Howard, K. 1997. The *Drosophila* protein Wuwen repels migratory germ cells. *Nature* 385: 64–67.

Zhao, G.-Q., Deng, K., Labosky, P. A., Liaw, L. and Hogan, B. L. M. 1996. The gene encoding bone morphogenetic protein 8B is required for the initiation and maintenance of spermatogenesis in the mouse. *Genes Dev.* 10: 1657–1669.

Developmental mechanisms of evolutionary change

23

How does newness come into the world? How is it born? Of what fusions, translations, conjoinings is it made? How does it survive, extreme and dangerous as it is? What compromises, what deals, what betrayals of its secret nature must it make to stave off the wrecking crew, the exterminating angel, the guillotine?
SALMAN RUSHDIE (1988)

The first Bird was hatched from a Reptile's egg.
WALTER GARSTANG (1922)

CHARLES DARWIN was heir to centuries of speculation concerning the origins of the diversity of animal life. Darwin's own education was steeped in the British tradition of natural theology that held that God's wisdom, omnipotence, and benevolence could be seen in the works of His creation. The dominant part of this tradition was an account of Creation proclaiming that the species were intricately designed works of the Creator. The fingers of the human hand were seen as exquisitely (some said perfectly) designed contrivances that allowed humans mastery of their environment. The shovel-like claw of the mole was, again, perfectly adapted for its "office of existence," as were the wings of a bird and the fins of a fish. A more sophisticated form of natural theology, championed in Britain by anatomist and embryologist Richard Owen, held that adaptations were merely of secondary importance. Rather, homologies were critical. Homologous structures were those organs that had the same underlying parts arranged in similar ways, the differences being their secondary modifications. What was really important was that the human hand, the mole's claw, the bird's wing, and the fish's fin were each based on the same plan. In abstracting the plan of the limb, we could determine the grand design upon which God had constructed all vertebrate appendages. To Owen (1848), the homologies underlying animal diversity were what counted, not the secondary adaptations of these basic unities.

"Unity of Type" and "Conditions of Existence"

Charles Darwin's Synthesis

Darwin acknowledged his debt to these earlier debates when he wrote (1859), "It is generally acknowledged that all organic beings have been formed on two great laws—Unity of Type, and Conditions of Existence." Darwin went on to explain that his theory would explain unity of type by descent. The changes creating these types and causing the marvelous adaptations to the conditions of existence, moreover, were explained by natural selection. Darwin called this "descent with modification." After reading Johannes Müller's summary of von Baer's laws in 1842, Darwin saw that em-

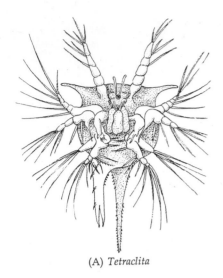

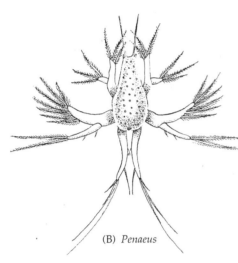

Figure 23.1
Nauplius larvae of (A) a barnacle (*Tetraclita*, seen in ventral view) and (B) a shrimp (*Penaeus*, seen in dorsal view). The shrimp and barnacle share a similar larval stage despite their radical divergence in later development. (After F. Müller, 1864.)

bryonic resemblances would be a very strong argument in favor of the genetic connectedness of different animal groups. "Community of embryonic structure reveals community of descent," he would conclude in *Origin of Species*.

Larval forms had been used for taxonomic classification even before Darwin. J. V. Thompson, for instance, had demonstrated that larval barnacles were almost identical to larval crabs, and he therefore counted barnacles as arthropods, not molluscs (Figure 23.1; Winsor, 1969). Darwin, an expert on barnacle taxonomy, celebrated this finding: "Even the illustrious Cuvier did not perceive that a barnacle is a crustacean, but a glance at the larva shows this in an unmistakable manner." Darwin's evolutionary interpretation of von Baer's laws set a paradigm that was to be followed for many decades, namely, that relationships between groups can be discovered by finding common larval forms. Kowalevsky (1871) would soon make a similar discovery (publicized in Darwin's *Descent of Man*) that tunicate larvae have notochords and form their neural tube and other organs in a manner very similar to that of the primitive chordate amphioxus. The tunicates, another enigma of classification schemes (usually placed, along with barnacles, as a mollusc), thereby found a home with the chordates. Darwin also noted that embryonic organisms sometimes make structures that are inappropriate for their adult form but that show their relatedness to other animals. He pointed out the existence of eyes in embryonic moles, pelvic rudiments in embryonic snakes, and teeth in embryonic baleen whales. In this book, we noted that mammalian embryos form a rudimentary yolk sac, send blood vessels to this yolk sac, and undergo gastrulation in a manner resembling that of birds and reptiles, whose development is constrained by the yolk.

Darwin also argued that adaptations that depart from the "type" and allow an organism to survive in its particular environment develop late in the embryo. He noted that the differences between species and genera are, as predicted by von Baer's laws, produced only late in development; he even chloroformed pigeons (with great reluctance) to prove to himself that this was indeed the case. Thus, Darwin recognized two ways of looking at "descent with modification." One could emphasize the common *descent* by pointing out embryonic homologies between two or more groups of animals, or one could emphasize the *modifications* by showing how development was altered to produce structures that enabled animals to adapt to particular conditions.

Darwin did not attempt to construct complete phylogenies from embryological data, but his work influenced many of his contemporaries to do so. One of the first scientists to realize the evolutionary importance of von Baer's studies was Elie Metchnikoff. Metchnikoff appreciated that evolution consists of modifying embryonic organisms, not adult ones. He wrote (1891):

> *Man appeared as a result of a one-sided, but not total, improvement of organism, by joining not so much adult apes, but rather their unevenly developed fetuses. From the purely natural historical point of view, it would be possible to recognize man as an ape's "monster," with an enormously developed brain, face and hands.*

Thus, organisms were seen to evolve through changes in their embryonic development. (In the early 1900s, this fusion of evolution and embryology was wrongly interpreted to support a linear (as opposed to a branched) model of evolution. The interpretation of Ernst Haeckel was that every organism evolved by the terminal addition of a new stage to the end of the last. Thus, he saw the entire animal kingdom as representing truncated steps of human development (see Gasman, 1971; Gould, 1977). [evo1.html]

E. B. Wilson and F. R. Lillie

If changes in embryonic development effected evolutionary changes, how did these developmental changes take place? During the late 1800s, many investigators attempted to link development to phylogeny through the analysis of cell lineages. They meticulously observed each cell in developing embryos and compared the ways that different organisms formed their tissues. In 1898, two eminent embryologists gave cell lineage lectures at the Marine Biology Laboratories at Woods Hole, Massachusetts, and their lectures served to emphasize the two ways that embryology was being used to support evolutionary biology. The first lecture, presented by E. B. Wilson, was a landmark in the use of embryonic homologies to establish phylogenetic relationships. Wilson had observed the spiral cleavage patterns of flatworms, molluscs, and annelids, and he had discovered that in each case, the same organs came from the same groups of cells. For him this meant that these phyla all had a common ancestor. The various groups of cleavage-stage cells in flatworms, molluscs, and annelids

> *Show so close a correspondence both in origin and in fate that it seems impossible to explain the likeness save as a result of community of descent. The very differences, as we shall see, give some of the most interesting and convincing evidence of genetic affinity; for processes which in the lower forms play a leading role in the development are in the higher forms so reduced as to be no more than vestiges or reminiscences of what they were, and in some cases seem to have disappeared as completely as the teeth of birds or the limbs of snakes.*

The next lecturer was F. R. Lillie, who had also done his research on the development of mollusc embryos and on modifications of cell lineage. He stressed the modifications, not the similarities, of cleavage. His research on *Unio*, a mussel whose cleavage is altered to produce the "bear-trap" larva that enables it to survive in flowing streams, was highlighted in Chapter 5. Lillie argued that "modern" evolutionary studies would do better to concentrate on changes in embryonic development that allowed for survival in particular environments rather than to focus on ancestral homologies that united animals into lines of descent.

In 1898, then, the two main avenues of approach to evolution and development were clearly defined: finding underlying unities that unite disparate groups of animals, and detecting the differences in development that enable species to adapt to particular environments. (Indeed, these same lines of argument characterized the two types of natural theologies before Darwin.) Darwin had thought these to be temporally distinguished—that is, one would find underlying unities in the earliest stages while the later stages would diverge to allow specific adaptations (see Ospovat, 1981). However, Wilson and Lillie were both discussing the cleavage stage of embryogenesis. These two ways of characterizing evolution and development are still the major approaches today.

The evolution of early development: E. Pluribis Unum

The Emergence of Embryos

In the evolution and development of living organisms, one sees the emergence of multicellularity from single-celled organisms. A new whole is formed from component cellular parts. This is a fundamental step in the emergence of a new level of complexity. The volvocaceans and the dictyostclids mentioned in Chapter 1 represent but 2 of the 17 types of protists in which multicellularity was attained (Buss, 1987). However, only three groups (those that generated fungi, plants, and animals) evolved the ability to form multicellular aggregates that could differentiate into particular cell types, i.e., an embryo.

The first embryos had to solve a fundamental problem. Since each of the component cells had the genetic apparatus and the cytoplasmic architecture needed to divide, why shouldn't each cell continue its own proliferation? What would cause these cells to sacrifice their proliferative capacity to form a collective individual? There may have been more than one solution. Buss suggests that in these early embryos, there was a sharp dichotomy between proliferation and differentiation and that our protist ancestors never learned the trick of dividing once they had differentiated cilia. While some other protist groups (especially ciliates) could make more microtubule-organizing centers, our ancestors could not. To this day, no ciliated metazoan cell divides (although ciliated metazoan cells can lose their cilia and then divide). Buss speculates that the ancestors of today's metazoans stopped their cellular proliferation by differentiating into a blastula of ciliated cells. (The early embryos of the first metazoan phyla—sponges and cnidarians—are characterized as balls of ciliated cells, just like the sea urchin embryos discussed in Chapter 5.) These ciliated blastulae could move, but it would appear that all their development had ceased, for ciliated cells do not divide, nor do they become any other differentiated cell type. To develop into an organism, this dilemma had to be solved.

This problem was solved by retaining or producing a population of nonciliated cells. These nonciliated cells could proliferate new cells, while the ciliated cells allowed the embryo to move. But these dividing cells could not go just anywhere. They could not grow on top of the ciliated cells or movement would cease. They could not grow into the water or they would be a drag on the embryo's movement. Rather, they would have to migrate *inside* the blastocoel (Figure 23.2). This movement and proliferation of cells is thought to be the origin of gastrulation. Thus, the blastula arose as a means of joining autonomous cells into a federation. The gastrula arose as a compromise within this federation that allowed the embryo to develop while moving (Buss, 1987).*

The earliest embryos probably developed in this mosaic fashion. However, induction provided a second mechanism to ensure that totipotent blastomeres remained together to form a single individual. Here, each cell sacrificed its autonomy to create a coherent community. Henry and co-workers (1989) have found that whereas individual sea urchin blastomeres can be totipotent, aggregates made of these same cells are not. Rather, each cell restricts the potency of its neighbor (see Chapter 15). This restrictive regulation is also seen in chimeric mice (see Chapter 5), wherein mammalian blastomeres combine to form a single chimeric mouse rather than two individual mice. There appear to be very important restrictions on cell potency once

---

*This is a modification of the theory originally proposed by Metchnikoff (1886) to account for the origin of multicellular organisms. Using hydroid and sponge embryos, Metchnikoff pointed out that certain cells from the wall of the blastula "draw in their flagellum, become amoeboid and mobile, multiply by division, fill the cavity of the blastula, and become capable of digesting." This embryonic state, he felt, "is therefore entitled to be considered the prototype of multicellular beings." Metchnikoff attempted to make a phylogeny of all organisms on the basis of their germ layers, and he believed that all mesodermal cells could be characterized by their ability to phagocytize foreign substances. His discoveries in comparative embryology eventually allowed him to formulate the conceptual foundations of a new science, immunology. (For details of Metchnikoff's theory of multicellular origins, see Chernyak and Tauber, 1988, 1991.)

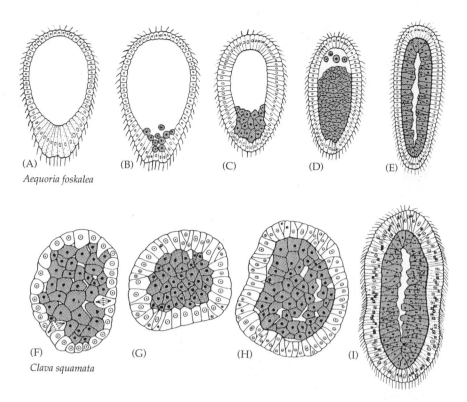

(A)

Aequoria foskalea

(F) (G) (H) (I)

Clava squamata

Figure 23.2
Gastrulation in two hydroid cnidarians.
(A–E) Gastrulation in *Aequoria foskalea*,
wherein a ciliated blastula is formed.
Cells at the vegetal pole lose their cilia
and migrate into the blastocoel to form
a mitotically dividing population. (F–I)
Gastrulation in *Clava squamata*, wherein
a cell-filled stereoblastula is formed and
the outer layer then becomes ciliated.
Both plans converge on the ciliated
planula larva characteristic of cnidarians. (Epiboly of a nonciliated ectoderm
is not seen in embryos that are free-
swimming.) (After Buss, 1987.)

cells are brought together. Moreover, once the inside population can interact with the outside population of cells and with other parts of the inside popu-lation, inductive events can occur to give rise to new organs.

Whichever way this community of cells has been formed, their integra-tion into a unified embryo is accomplished by maternal input into the egg cytoplasm. It is this set of instructions that causes cells to cleave in certain arrangements, to stick to one another, and to differentiate at particular times. As we saw in Chapter 12, the sea urchin embryo becomes a ciliated blastula even in the absence of nuclear transcription. Only at gastrulation does the nucleus begin to regulate development. Thus, selection at the level of cell propagation (which had been the rule of survival among the protists) has been superseded by selection at the level of the individual multicellular or-ganism.

Formation of New Phyla: Modifying Developmental Pathways

Only about three dozen animal body plans are currently being used on this planet (Margulis and Schwartz, 1988; Brusca and Brusca, 1990). These consti-tute the animal phyla. This is not to say that these body plans are the only possible ones. The Burgess Shale, a repository of early Cambrian soft-body fossils, is interpreted as containing representatives of 20 more phyla that never evolved descendants in the upper strata (Figure 23.3). In addition, this small band of sediment, about the size of a city block, contains about a dozen previously unknown classes of arthropods. These animals are not "primitive" members of existing phyla or classes, but are specialized exam-ples of their own groups (Whittington, 1985; Gould, 1989). There are also two specimens in the Burgess Shale that may be related to ancestral forms of existing phyla. One is a peripatus-like animal that may be close to the ances-tral form of insects; the other appears to be a well-preserved chordate called *Pikaia gracilens* that may be related to the ancestral chordates (see Figure 23.3B). This latter fossil has several features that recommend its being classi-

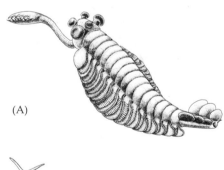

(A)

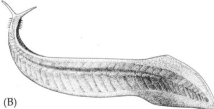

(B)

Figure 23.3
Two fossil organisms from the mid-Cambrian Burgess Shale. (A) *Opabina*, an organism having five eyes on its head, a frontal appendage with a termi-nal claw, body segments with dorsal gills, and a three-segment tail piece. (B) *Pikaia gracilens*, a possible chordate. (From Gould, 1989.)

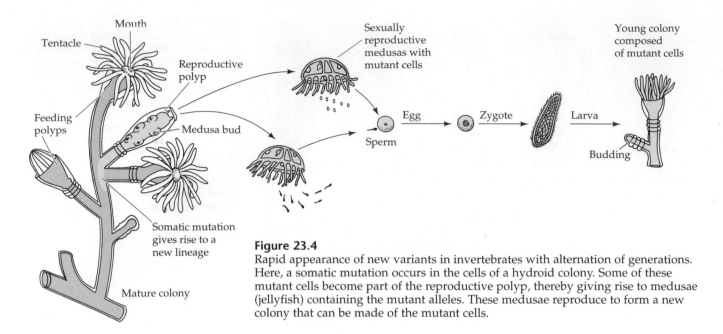

Figure 23.4
Rapid appearance of new variants in invertebrates with alternation of generations. Here, a somatic mutation occurs in the cells of a hydroid colony. Some of these mutant cells become part of the reproductive polyp, thereby giving rise to medusae (jellyfish) containing the mutant alleles. These medusae reproduce to form a new colony that can be made of the mutant cells.

fied in our phylum: it appears to have a notochord, and the zigzag bands along its sides look very much like the somite-derived musculature found in *Amphioxus* (Conway Morris and Whittington, 1979).* Thus, all the known metazoan phyla (and many heretofore unknown ones) appear to have been formed by the Cambrian radiation about 540 million years ago (Bowring et al., 1993; Wray et al., 1996).

How is it that no new phylum has emerged in the past half-billion years? Kauffman (1993) proposes a mathematical model that predicts that any evolving system (whether it be a phylum, species, automobile, or religion) displays this pattern of divergence followed by the locking in on a particular subset of the original diversity. Kauffman uses the metaphor of a rugged fitness landscape wherein there are peaks and valleys of fitness, and all organisms start out with the same average fitness value (in the middle of a peak). If they take large jumps, they have a 50 percent chance of becoming fitter organisms. Eventually, the chance of finding a fitter body plan decreases if an organism takes a jump far from where it is situated. Long jumps become risky, and the chance that these higher peaks are already occupied increases. Instead, small jumps (on the same peak) may make the organism somewhat fitter than the surrounding population. Then what we see is a diversification around a few successful models. In general, the duration between successful long jumps doubles with each attempt. During the early Cambrian, it is possible that the genome had not become stabilized into the sets of interactions that we see today. Moreover, in many invertebrate groups, there is an alternation of generations, wherein a sexual form generates an asexual form (zooid, polyp, bud) that then gives rise to the sexual form again. In such cases, somatic mutations in the asexual form can enter into the body of the sexual form and be propagated very rapidly (Figure 23.4; Buss, 1987).

*An even earlier fossil, *Yunnanazoon lividum*, from the early Cambrian, some 525 million years ago, was first reported as being a chordate (Chen et al., 1995). However, the interpretation of the fossil notochord has been challenged by Shu and colleagues (1996), who view *Yunnanazoon* as the earliest known hemichordate.

How, then, can one modify one *Bauplan* to create another *Bauplan*? The first way would be to modify the earliest stages of development. According to von Baer (see Chapter 7), animals of different species but of the same genus diverge very late in development. The more divergent the species are from one another, the earlier one can distinguish their embryos. Thus, embryos of the snow goose are indistinguishable from those of the blue goose until the very last stages. However, snow goose development diverges from chick development a bit earlier, and goose embryos can be distinguished from lizard embryos at even earlier stages. It appears, then, that mutations that create new *Baupläne* could do so by altering the earliest stages of development.

These early developmental changes can be effected by changing the localization of cytoplasmic determinants, changing the rate of cell division of one cell or group of cells relative to the others, or changing the positions of the cells as they divide. In Chapter 5, we saw that a modification of molluscan cleavage can give the bulk of cytoplasm to the ectodermal cells that form the larval shell. This is due to changing the manner in which the blastomeres divide and apportion cytoplasm. In annelid worms, the differences between polychaetes and oligochaetes stem from differences in the cytoplasmic localization of morphogens within the egg (Figure 23.5). Although they both un-

Figure 23.5
Comparison of the development of two classes of annelid worms, (A) the polychaete *Podarke* and (B) the oligochaete *Tubifex*. Their cleaving embryos, blastula fate maps, and products of gastrulation are seen. In *Podarke*, gastrulation leads to the formation of a trochophore larva. In *Tubifex*, there is no larval stage, and the embryo develops directly into a segmented body. (After Anderson, 1973.)

(A) *Podarke*

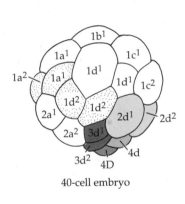

40-cell embryo

Fate map

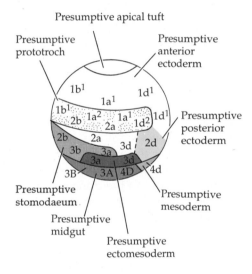

Trochophore larva

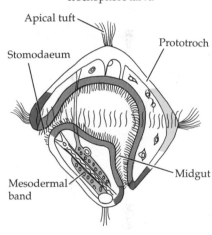

(B) *Tubifex*

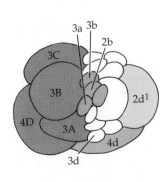

Cleaving embryo

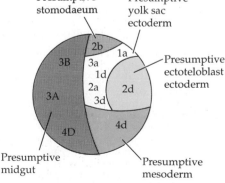

Fate map

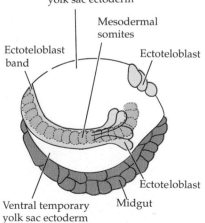

Gastrulation

Polygordius

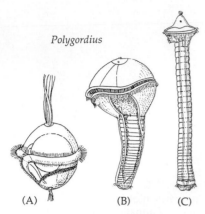

(A)   (B)   (C)

Patella

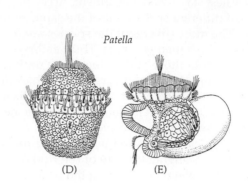

(D)   (E)

Vestimentiferan

(F)

Figure 23.6
Divergence of development after the trochophore larval stage. (A–C) Metamorphosis of the polychaete annelid *Polygordius* from its free-swimming trochophore larva shows the formation of a segmented trunk. Eventually, the larval structures shrink at the anterior end as the head forms. (D–E) Metamorphosis of the prosobranch (clam) mollusc *Patella*. After the trochophore stage, it develops a molluscan foot, shell gland, and visceral hump. (F) Scanning electron micrograph of the trochophore larva of a vestimentiferan. (A–E after Grant, 1978; F from Jones and Gardiner, 1989, courtesy of the authors.)

dergo spiral cleavage, they apportion their morphogens into different cells. Polychaetes undergo a relatively standard spiral cleavage to give rise to the trochophore larva. Oligochaetes, however, put most of their cytoplasm into those cells destined to form adult, rather than larval, structures. This group then skips the larval stage. If a mutation were to place a certain cytoplasmic morphogen into one region of the egg instead of another, or if a mutation caused a change in the axis of cell division so that different sets of cells acquired these determinants, then a radically different phenotype could be produced. As E. G. Conklin wrote in 1915, "We are vertebrates because our mothers were vertebrates and produced eggs of the vertebrate pattern."

Another way of evolving new phyla may involve modifying the larva. Darwin and others thought that similarities of larval form signified common descent. However, this can be reinterpreted to mean that changes that give rise to different phyla may occur in larvae. Snails, echiuroids, and polychaetes have very similar patterns of division and form trochophore larvae (Figure 23.6). In fact, the placement of the newly discovered phylum Vestimentifera (the bright red, gutless invertebrates found in the deep ocean trenches) near the annelids was made in part on the basis of vestimentiferans' having trochophore larvae (Jones and Gardiner, 1989; Young et al., 1996). Thus, one of the principal mechanisms for establishing new phyla and classes may be the rearrangement of development during the larval stage so that metamorphosis brings about new types of organization. Garstang (1928) showed how the veliger larva of certain snails could have arisen by mutation and then been selected because the new arrangement of head and shell allowed the head to retract beneath the shell for safety. He also framed the hypothesis that chordates arose from ancestral tunicate larvae that had become neotenic. Unfortunately, soft-bodied larvae rarely fossilize, so we know very little about the mechanisms by which chordates and other phyla may have arisen from early Cambrian larvae.*

*Larval forms often bridge the gap between the different adult forms. The larval form is seen either as being ancestral to two groups or as "breaking away" by neoteny and forming a different type of organism. This has often been hypothesized as the mechanism by which chordates emerged from invertebrates and vertebrates arose from chordates. The tornaria larva of hemichordates is formed in a deuterostome manner similar to that of echinoderm larvae and looks enough like echinoderm larvae to have been originally mistaken for them. This would link echinoderms and chordates. Garstang (1928) and Berrill (1955) hypothesized that the larvae of certain tunicates could have evolved into chordates such as amphioxus by neotenic development. In this way, the tunicates would keep the notochord, larval musculature, and feeding apparatus of the larval tunicate while becoming sexually mature. There are, in fact, neotenic free-swimming tunicates (such as *Larvacea*). Modifications of this view (using a different protochordate stock) have been suggested by Jefferies (1986). The origin of chordates remains a difficult problem.

Modularity: The prerequisite for changing evolution through development

There are only 35 or so *Baupläne*, but there are several million different species, each with its own pattern of development. Therefore, most evolution has occurred within the framework of an existing *Bauplan*. How is this done? How can the development of an embryo change when development is so finely tuned and complex? It was once thought that the only way to promote evolution was to add a step to the end of embryonic development, but we now know that even early stages can be altered to produce evolutionary novelties. The reason why changes can be made during development is that the embryo, like the adult organism, is composed as a series of interacting modules (Riedl, 1978; Bonner, 1988).

Modularity

Development occurs through discrete and interacting modules (Riedl, 1978; Gilbert et al., 1996; Raff, 1996; Wagner, 1996). Organisms are constructed of units that are coherent within themselves and yet part of a larger unit. Thus, cells are parts of tissues, which are parts of organs, which are parts of systems, and so on. Such a hierarchically nested system has been called a level-interactive modular array (Dyke, 1988). In development, such modules include morphogenetic fields (for example, those described for the limb or eye), imaginal discs, cell lineages (such as the inner cell mass or trophoblast), insect parasegments, and vertebrate organ rudiments. Modular units allow different parts of the body to change without interfering with other functions.

The fundamental principle of modularity allows three processes to alter development: dissociation, duplication and divergence, and co-option (Raff, 1996). Since the modules are on all levels from molecular to organismal, it is not surprising that one sees these principles operating at all levels of development.

Dissociation: Heterochrony and Allometry

Not all parts of the embryo are connected to one another. One can dissect out the limb field of a salamander neurula, and the eyes are not affected. By mutation or environmental perturbation, one part of the embryo can change without the other part. This modularity of development can allow changes that are either spatial or temporal. **Heterochrony** is a shift in the relative timing of two developmental processes during embryogenesis from one generation to the next. In other words, one module can change its temporal expression relative to the other modules of the embryo. We have come across this concept in our discussions of neoteny and progenesis in salamanders (see Chapter 19). Heterochrony can be caused in different ways. In salamander heterochronies, where the larval stage is retained, heterochrony is caused by gene mutations in the induction competence system. Other heterochronic phenotypes, however, are caused by the heterochronic expression of certain genes. The direct development of the adult sea urchin rudiment (see Chapter 19) involves the early activation of adult genes and the suppression of larval gene expression (Raff and Wray, 1989). Heterochrony can "return" an organism to a larval state, free from the specialized adaptations of the adult. Heterochrony can also give larval characteristics to an adult organism, as in the small size and webbed feet of arboreal salamanders or the fetal growth rate of human newborn brain tissue. [evo2.html]

Another consequence of modularity is allometry. **Allometry** occurs when different parts of the organism grow at different rates. Allometry can

Figure 23.7
Allometric growth in the whale head. The jaw has pushed forward, causing the nose to move to the top of the skull. (The premaxilla is present in the early human fetus, but it fuses with the maxilla by the end of the third month of gestation. The human premaxilla was discovered by Wolfgang Goethe, among others, in 1786.) (After Slijper, 1962.)

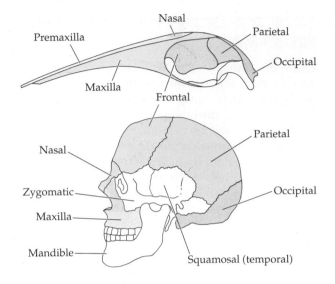

Figure 23.8
Transverse section through the anterior region of the pocket gopher (*Thomomys*) embryo, showing the anterior opening of the pouch (AP) and the continuity between the pouch at this stage and the buccal cavity (BC) across the developing lip area. (EP, epithelial cells; MC, Meckel's cartilage; T, tongue.) (From Brylski and Hall, 1988, courtesy of the authors.)

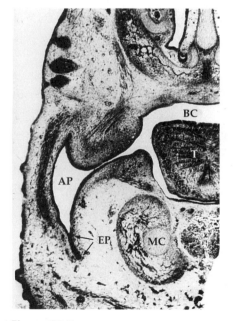

be very important in forming variant body plans within a *Bauplan*. Such differential growth changes can involve altering a target cell's sensitivity to growth factors or altering the amounts of growth factors produced. Again, the vertebrate limb can provide a useful illustration. Local differences in chondrocytes cause the central toe of the horse to grow at a rate 1.4 times that of the lateral toes (Wolpert, 1983). This means that as the horse grew larger during evolution, this regional difference caused the five-toed horse to become a one-toed horse. A particularly dramatic example of allometry in evolution comes from skull development. In the very young (4- to 5-mm) whale embryo, the nose is in the usual mammalian position. However, the enormous growth of the maxilla and premaxilla (upper jaw) pushes over the frontal bone and forces the nose to the top of the skull (Figure 23.7). This new position of the nose (blowhole) allows the whale to have a large and highly specialized jaw apparatus and to breathe while parallel to the water's surface (Slijper, 1962).

Allometry can also generate evolutionary novelty by small, incremental changes that eventually cross some developmental threshold (sometimes called a bifurcation point). Eventually, a change in quantity becomes a change in quality when such a threshold is crossed. It has been postulated that this type of mechanism produced the external fur-lined "neck" pouches of pocket gophers and kangaroo rats that live in deserts. External pouches differ from internal ones in that (1) they are fur-lined and (2) they have no internal connection to the mouth. They are very useful in that they allow these animals to store seeds without running the risk of desiccation. Brylski and Hall (1988) have dissected the heads of pocket gopher and kangaroo rat embryos and have looked at the way the external cheek pouch is constructed. When data from these animals were compared with data from animals that form internal cheek pouches (such as hamsters), the investigators found that the pouches are formed in very similar manners. In both cases, the pouches are formed within the embryonic cheeks by outpocketings of the cheek (buccal) epithelium into the facial mesenchyme (Figure 23.8). In animals with internal cheek pouches, these evaginations stay within the cheek. However, in animals that form external pouches, the elongation of the snout draws up the outpocketings into the region of the lip. As the lip epithelium rolls out of the oral cavity, so do the outpockets. What had been internal becomes external. The fur lining is probably derived from the external pouches' coming in contact with dermal mesenchyme, which can induce hair to form in epithelia (see Chapter 17). Such a pouch has no internal opening to the mouth. Indeed, the transition from internal to external pouch is one of threshold. The placement of the evaginations anteriorly or posteriorly determines whether the pouch is internal or not. There is no "transition stage" having two openings, one internal and one external. One could envision this externalization occurring by a chance mutation that shifted the outpocketing to a slightly more anterior location. Such a trait would be selected for in the desert. As Van Valen reflected in 1976, evolution can be defined as "the control of development by ecology."

Duplication and Divergence

Modularity also allows duplication and divergence to occur. The duplication part of this process allows the formation of redundant structures, and the divergence part of the process allows these structures to assume new roles. One of the copies can maintain the original role while the others are free to mutate and diverge functionally. This can happen at numerous levels. The TGF-β family, the MyoD family, and the globins each probably started as a single gene that duplicated several times. After the duplication, mutations caused the divergences that gave the members of each family new functions. On the tissue level, one sees duplication and divergence in the somites that give rise to the cervical, thoracic, and lumbar skeletons.

There are also duplications and divergences of particular developmental patterns. Epithelial-mesenchymal interactions appear to be variations on a single theme (Figure 23.9; Maderson, 1975; Burke, 1989a). The secretory glands of the epidermis are modifications of the same type of induction—

Figure 23.9
The interrelationships of epidermal-mesenchymal inductions. During morphogenesis, the mesenchyme can cause the adjacent epidermis to invaginate (A–C) or evaginate (D–H). In some cases, as in the formation of limbs or turtle carapace, the mesenchyme causes the formation of an apical ectodermal ridge (G,H). (After Burke, 1989a.)

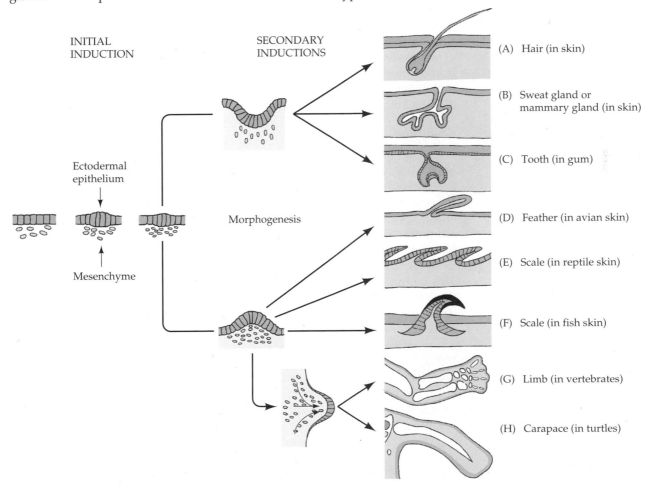

INITIAL INDUCTION

SECONDARY INDUCTIONS

Ectodermal epithelium

Mesenchyme

Morphogenesis

(A) Hair (in skin)

(B) Sweat gland or mammary gland (in skin)

(C) Tooth (in gum)

(D) Feather (in avian skin)

(E) Scale (in reptile skin)

(F) Scale (in fish skin)

(G) Limb (in vertebrates)

(H) Carapace (in turtles)

Figure 23.10
Midtrunk cross section through the embryo of the turtle *Chelydra serpentina*. (A) The carapacial ridge (arrow) is formed at the boundary of the somitic and lateral plate mesoderm and now represents the dorsal-ventral boundary. The thickened mesodermal bands extending from the center into the carapace area are the rib condensations. (B) Higher magnification of the carapacial ridge. (From Burke, 1989b, courtesy of the author.)

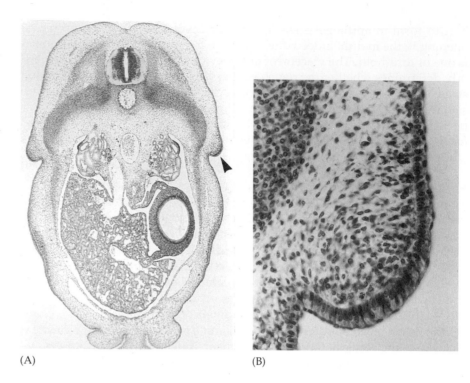

(A)   (B)

mammary glands are embryologically modified sweat glands. Likewise, the fearsome rows of sharks' teeth are modifications of its body scales. Changes in induction can change scales into feathers (as in the case of bantam chicks) and are responsible for such remarkable adaptations as the avian lung, the ruminant stomach, the tusks of elephants (modified incisors), and the tusks of walruses (modified upper canine teeth). The carapace (shell) of the turtle is an evolutionary novelty that appears to form in a manner reminiscent of limbs. There is even a carapacial ridge that organizes the mesenchyme much like the apical ectodermal ridge of the limb bud (Figure 23.10; Burke, 1989b).

Co-option

No one structure is destined for any particular purpose. A pencil can be used for writing, but it can also be used as a toothpick, a dagger, a hole-puncher, or a drumstick. On the molecular level, one finds that the gene *engrailed* is used for segmentation in *Drosophila* embryos, is used later to specify its neurons, and is used in the larval stages to provide an anterior-posterior axis to imaginal discs. Similarly, a protein that functions as an enzyme in the liver can function as a structural crystalin protein in the lens (Piatigorsky and Wistow, 1991). In other words, preexisting units can be recruited for new functions. This co-option is also seen on the morphological level. Wings have evolved three times during vertebrate evolution, and in each case, different forearm structures were modified for an entirely new function.

One of the most celebrated cases of co-option is the use of embryonic jaw parts for the creation of the mammalian middle ear (reviewed in Gould, 1990). Neural crest cells distinguish vertebrates from the protochordates and invertebrates. The protochordates have a dorsal neural tube and notochord, but no real "head." The cranial neural crest cells are largely responsible for the creation of the face, skull, and branchial arches. It is thought that the development of the head originally allowed for more efficient predation, placing the sensory structures adjacent to the prey-capturing jaws (Gans and Northcutt, 1983; Langille and Hall, 1989; Hall, 1992). Two remarkable transitions have occurred in the evolution of the vertebrate jaw. The first is the creation of jaws from the gill arches of unjawed fish. The second is the use of the bones that had articulated the upper and lower jaws in reptiles to become the malleus and incus (hammer and stirrup) bones of the middle ear. In the first vertebrates, a series of gills opened behind the jawless mouth. When gill slits became supported by cartilaginous elements, the first set of these gill supports surrounded the mouth to form the jaw. There is ample evidence that jaws are modified gill supports. First, both these sets of bones are made from neural crest cells. (Most other bones come from mesodermal tissue.) Second, both structures form from upper and lower bars that bend forward and are hinged in the middle. Third, the jaw musculature seems to be homologous to the original gill support musculature. Thus, the first transformation of the first branchial arch cartilage was from gill apparatus to jaw apparatus. But the story does not end here.

The upper portion of the second branchial arch supporting the gill becomes the hyomandibular bone of jawed fishes. This element supports the skull and links the jaw to the cranium (Figure 23.11A). As we saw in Chapter 7, this hyomandibular bone functions in mammals as the stapes, one of the middle ear bones. But fish do not use this bone for hearing, so how did a bone used for gill support and then for cranial support become part of the mammalian auditory apparatus? As fish came up onto land, they had a new problem: how to hear in a medium as thin as air. The hyomandibular bone happens to be near the otic capsule, and bony material is excellent for transmitting sound. Thus, while still functioning as a cranial brace, the hyomandibular bone of the first amphibians also began functioning as a sound transducer (Clack, 1989). As the terrestrial vertebrates altered their locomotion, jaw structure, and posture, the cranium became firmly attached to the rest of the skull and did not need the hyomandibular brace. It then seems to have become specialized into the stapes bone of the middle ear. What had been this bone's secondary function became its primary function.

The original jaw bones changed also. The first branchial arch generates the jaw apparatus. In amphibians, reptiles, and birds, the posterior portion of this cartilage forms the quadrate bone of the upper jaw and the articular bone of the lower jaw. These bones connect to one another and are responsible for articulating the upper and lower jaws. However, in mammals, this articulation occurs at another region (the dentary and squamosal bones), thereby "freeing" these bony elements to acquire new functions. The quadrate bone of the reptilian upper jaw evolved into the mammalian incus bone, and the articular bone of the reptile's lower jaw has become our malleus. This latter process was first described by Reichert in 1837, when he observed in the pig embryo that the mandible (jawbone) ossifies on the side of Meckel's cartilage, while the posterior region of Meckel's cartilage ossifies, detaches from the rest of the cartilage, and enters into the region of the middle ear to become the malleus (Figure 23.11B,C).*

*The lack of transition forms is often cited by Creationists as a criticism of evolution. For instance, in the transition from reptiles to mammals, three of the bones of the reptilian jaw became the incus and malleus, leaving only one bone (the dentary) in the lower jaw. Gish (1973), a Creationist, says that this is an impossible situation, since no fossil has been discovered showing two or three jaw bones and two or three ear ossicles. Such an animal, he claims, would have dragged its jaw on the ground. However, such a specific transition form (and there are over a dozen documented transition forms between reptile and mammalian skulls) need never have existed. Hopson (1966) has shown on embryological grounds how the bones of the jaw could have divided and been used for different functions, and Romer (1970) has found reptilian fossils wherein the new jaw articulation was already functional while the older bones were becoming useless. There are several species of therapsid reptiles that had two jaw articulations, with the stapes brought into close proximity with the upper portion of the quadrate bone (which would become the incus). [evo3.html]

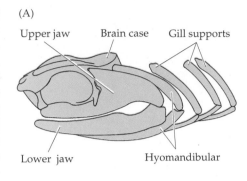

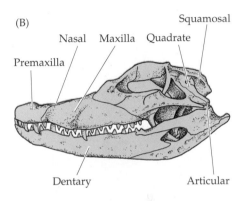

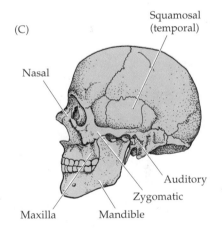

Figure 23.11
Jaw evolution in fish (A), reptile (B), and mammal (C). (A) Homologies of the jaws and the gill arches as seen in the skull of the paleozoic shark *Cobeledus aculentes*. (B) Lateral view of an alligator skull. The articular portion of the lower jaw articulates with the quadrate bone of the skull. (C) In mammals, the quadrate becomes internalized to form the incus of the middle ear. The articular bone retains its contact with the quadrate, becoming the malleus of the middle ear. Lateral view of the human skull, showing the junction of the lower jaw with the squamosal (temporal) region of the skull. (A after Zangerl and Williams, 1975.)

The existence of discrete developmental modules allows the principles of dissociation, duplication and divergence, and co-option to form new types of organisms.

Correlated Progression

One evolutionary consequence of the modular nature of development is **correlated progression.** Here, changes in one part of the embryo induce changes in another. Skeletal cartilage informs the placement of muscles, and muscles induce the placement of nerve axons. In such cases, if one structure changes, it will induce other structures to change with it (Thomson, 1988). The dramatic changes in bone arrangement from agnathans to jawed fish, from jawed fish to amphibians, and from reptiles to mammals were coordinated with changes in jaw structure, jaw musculature, tooth deposition and shape, and modifications of the cranial vault and ear (Kemp, 1982; Thomson, 1988). In 1995, Rowe made the case that the migration of the reptilian jaw cartilage to form the middle ear cartilage is itself a case of correlated progression, namely, a consequence of the enlarging braincase, which freed the cartilage precursors to migrate caudally.

One can also see correlated progression over a shorter time in domesticated animals. Humans have a great talent for selecting hereditary variants in domestic animals that involve those neural crest cells forming the frontonasal and mandibular processes. In some cases, such as that of bulldogs, the breed is selected for a wide face with very little angle between head and jaw. Other breeds, such as the collie, are selected for narrow snouts with a long jaw protruding away from the head. All breeds can move their jaws, shake their heads, and bark, despite the differences in the way their bones are shaped or positioned. Each variation is genetically determined, and it is important to note that each represents a harmonious rearrangement of the different bones with each other and with their muscular attachments. As the skeletal elements were selected, so were the muscles that moved them, the nerves that controlled these movements, and the blood vessels that fed them.*

The mechanism through which the jaw apparatus has maintained its integrity from lampreys to amniotes is a remarkable example of embryonic modules. The neural crest-derived structures of the vertebrate head include the pharyngeal arches (the precursors of the jaw, middle ear, tongue skeleton, etc.) as well as the dermal bones of the face, and the facial musculature (see Chapter 7). The braincase is produced from mesodermal tissues. By replacing individual chick rhombomeres with those of quails, Köntges and Lumsden (1996) were able to map the fates of the neural crest cells associated with the quail rhombomeres (Figure 23.12). The antibodies staining the quail neural crest cells showed that each rhombomere gives rise to particular skeletal elements *and to the muscles attached to them.* Moreover, the muscle-and-skeleton modules from each rhombomere were found to be innervated by a particular cranial nerve. For instance, the neural crest cells from rhombomere 4 generated four skeletal tissues—the retroarticular process of the lower jaw (found in birds but not mammals), a portion of the tongue skeleton, the stapes bone of the middle ear, and, surprisingly, the small portion of the braincase where the jaw-opening muscle attaches to the otherwise mesodermally derived skull. The muscles connecting these four skeletal elements also came from the rhombomere 4 neural crest cells. These muscles are all in-

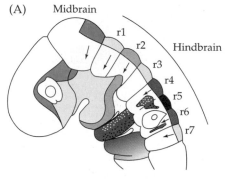

(A) Midbrain

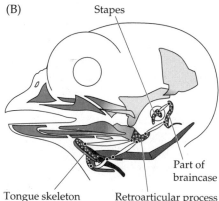

(B)

Figure 23.12
Chick embryo rhombomere neural crest cells and their musculoskeletal "packets." (A) Two-day chick embryo showing the contribution of the rhombomeric crest cells to the pharyngeal arches. (Most of the neural crest cells from r3 and r5 undergo apoptosis, while the rest of these cells contribute to the larger population of r4 neural crest cells.) (B) 10-day embryo showing the bones of the upper and lower jaws, tongue skeleton, and middle ear derived from the rhombomeric crest cells. The muscles derived from r4 are attached to bones from the same rhombomere, and the part of the braincase attached to the r4-derived jaw-opening muscle is also derived from rhombomere 4. (Other muscles have been omitted for clarity.) (After Ahlberg, 1997.)

---

*This coordination is not quite universal, however. In dogs with greatly shortened faces (such as bulldogs), the skin has not coordinated its development with the bones and therefore hangs in folds from the head (Stockard, 1941).

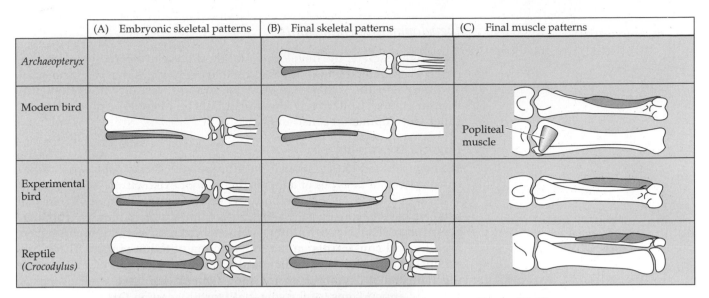

	(A) Embryonic skeletal patterns	(B) Final skeletal patterns	(C) Final muscle patterns
Archaeopteryx			
Modern bird			Popliteal muscle
Experimental bird			
Reptile (*Crocodylus*)			

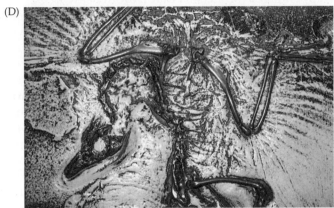

(D)

Figure 23.13
Experimental "atavisms" produced by altering embryonic fields in the limb. (A–C) Results of Müller's experiments wherein gold foil split the chick hind-limb field. (A,B) The embryonic and final bone pattern, indicating that the fibulare structure was retained by the experimental chick limb, as it is in extant reptiles and as it is thought to have been in *Archaeopteryx*. (C) Some of the correlated muscle changes in the experimental chick embryos. The popliteal muscle is present in the chick, but is absent from reptile limbs and from the experimental limb. The fibularis brevis muscle, which normally originates from both the tibia and fibula in chicks, takes on the reptilian pattern of originating solely from the fibula in the operated limbs. (D) Fossil *Archaeopteryx* in limestone. Imprints of feathers can be clearly seen. Were it not for the feathers, this toothed organism would probably have been classified as a reptile. (A–C after Müller, 1989; photograph courtesy of B. A. Miller/Biological Photo Service.)

nervated by cranial nerve VII. The rhombomeres form a modular unit, then, comprised of pharyngeal arch skeletal elements, the muscles that move them, the attachment site of the muscles to the braincase, and the nerves that innervate the muscles. Because these muscles and bones are formed from the same cells, their relationships can be maintained despite the dramatic changes in position and function that these elements might have over time.

Correlated progression has also been shown experimentally. Repeating earlier experiments of Hampé (1959), Gerd Müller (1989) inserted barriers of gold foil into the prechondrogenic hindlimb buds of a 3.5-day chick embryo. This barrier separated the regions of tibia formation and fibula formation. The results of these experiments are twofold. First, the tibia is shortened and the fibula bows and retains its connection to the fibulare. Such relationships between the tibia and fibula are not usually seen in birds, but they are characteristic of reptiles (Figure 23.13). Second, the musculature of the hindlimb undergoes parallel changes with the bones. Three of the muscles that attach to these bones now show characteristic reptilian patterns of insertion. It seems, therefore, that experimental manipulations that alter the development of one part of the mesodermal limb-forming field also alter the development of other mesodermal components. As with the correlated progression seen in face development, these changes all appear to be due to interactions within a field, in this case, the chick hindlimb field. These are not global effects and can occur independently of the other portions of the body.

Developmental constraints

Although discrete, the developmental modules can interact with one another. These interactions limit the possible phenotypes that can be created, and they also allow change to occur in certain directions more efficiently than in others.* Collectively, these restraints on phenotype production are called **developmental constraints.**

Physical Constraints

We have alluded to the fact that there are relatively few *Baupläne*, and one can easily imagine types of animals that do not exist within existing phyla. Why aren't there more major body types among the animals? To answer this, we have to consider the constraints imposed on evolution. There are three major classes of constraints on morphogenetic evolution. First, there are **physical constraints** on the construction of the organism. These constraints of diffusion, hydraulics, and physical support allow only certain mechanisms of development to occur. One cannot have a vertebrate on wheeled appendages (of the sort that Dorothy saw in Oz) because blood cannot circulate to a rotating organ; this entire possibility of evolution has been closed off. Similarly, structural parameters and fluid dynamics forbid the existence of 5-foot-tall mosquitoes. [evo4.html]

Morphogenetic Constraints

There are also constraints involving **morphogenetic construction rules** (Oster et al., 1988). Bateson (1894) noted that when organisms depart from their normal development, they do so in only a limited number of ways. Research in this area attempts to find the architectural parameters on which organisms are constructed and seeks to show how these parameters can be modified during evolution. Some of the best examples of these types of constraints come from the analysis of limb formation in vertebrates. Holder (1983) pointed out that although there have been many modifications of the vertebrate limb over 300 million years, some modifications (such as a middle digit shorter than its surrounding digits) are not found. Moreover, analyses of natural populations suggest that there is a relatively small number of ways that limb changes can occur (Wake and Larson, 1987). If a longer limb is favorable in a given environment, the humerus may become elongated. One never sees two smaller humeri joined together in tandem, although one could imagine the selective advantages that such an arrangement might have. This indicates a construction scheme that has certain rules.

The principal rules of vertebrate limb formation have been summarized by Oster and his co-workers (1988). They find that a reaction-diffusion mechanism can explain the known morphologies of the limb and can explain why other morphologies are forbidden. This model posits that the aggregations of cartilage actively recruit more cells from the surrounding area and laterally inhibit the formation of other foci of condensation. The number of foci depends on the geometry of the tissue and the strength of the lateral inhibition. If the inhibition remains the same, the size of the tissue volume must increase to get two foci forming where one had been allowed before. At a certain threshold (called a bifurcation threshold), this size is reached, and the limb can branch into two foci.

---

*Leibniz, probably the philosopher who most influenced Darwin, noted that existence must be limited not only to the possible but to the *compossible*. That is, whereas numerous things *can* come into existence, only those that are mutually compatible will actually exist (see Lovejoy, 1964). So although many developmental changes are possible, only those that can integrate into the rest of the organism (or which can cause a compensatory change in the rest of the organism) will be seen.

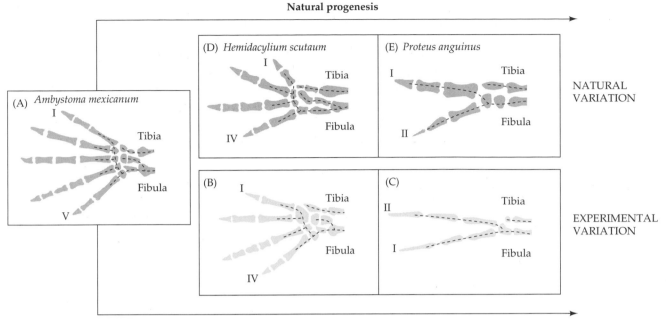

Natural progenesis

NATURAL VARIATION

EXPERIMENTAL VARIATION

Experimental decrease in cell number

Figure 23.14
Relationship between cell number and number of digits in salamanders. (A) The hindlimb of an axolotl (*Ambystoma mexicanum*) with its five symmetrical digits. (B,C) Digits on axolotl hindlimb after hindlimb bud is incubated in colchicine to reduce cell number. (D,E) Two wild salamanders formed by progenesis, each having a smaller limb bud. (D) *Hemidactylium scutatum.* (E) *Proteus anguinus.* The parallels between the experimental variation and the natural variation can be seen, and the common denominator is reduced cell numbers in the limb buds. (After Oster et al., 1988.)

Evidence for this mathematical model comes from experimental manipulations and comparative anatomy. When an axolotl limb bud is treated with the antimitotic drug colchicine, the dimensions of the limb are reduced. In such limbs, there is not only a reduction of digits, but a reduction of certain digits in a certain order, as expected by the mathematical model and from the "forbidden" morphologies. Moreover, these reductions of specific digits are very similar to those limbs of progenetic salamanders, those species that reach maturity at a smaller stage than did their ancestors and whose limbs develop from smaller limb buds (Figure 23.14; Alberch and Gale, 1983, 1985). Thus, the use of reaction-diffusion mechanisms to construct limbs may constrain the possibilities that can be generated during development, because only certain types of limbs are possible using these rules.

Phyletic Constraints

Phyletic constraints comprise the third set of constraints on the evolution of new types of structures (Gould and Lewontin, 1979). These are the historical restrictions based on the genetics of an organism's development. For instance, once a structure is generated by inductive interactions, it is difficult to start over again. The notochord, which is still functional in adult protochordates (Berrill, 1987), is considered vestigial in adult birds and mammals. Yet it may be transiently necessary in the embryo to specify the neural tube. Similarly, Waddington (1938) noted that although the pronephric kidney of the chick embryo is considered vestigial (since it has no ability to concentrate urine), it is the source of the ureteric bud that induces the formation of a functional kidney during chick development.

This type of phyletic constraint has recently been reviewed by Raff and colleagues (1991). Until recently, it was thought that the earliest stages of development would be the hardest to change, because altering them would either destroy the embryo or generate a radically new phenotype. But recent work (and the reappraisal of older work) has shown that alterations can be made to early cleavage without upsetting the final form. Modifications of morphogens in mollusc embryos can give rise to new types of larvae that still metamorphose into molluscs, and changes in sea urchin cytoplasmic morphogens can generate sea urchins that develop without larvae but still become sea urchins. In fact, looking at vertebrates, one can see that there is an entire history leading up to the famous diagram of von Baer's law shown in Chapter 7. All the vertebrates arrive at this particular stage of development (called the *pharyngula*), but they can do so by different means (Figure 23.15). Birds, reptiles, and fish arrive there after meroblastic cleavages of different sorts; amphibians get to the same stage by way of radial holoblastic cleavage; and mammals reach the same place after constructing a blastocyst, chorion, and amnion. The earliest stages of development, then, appear to be extremely

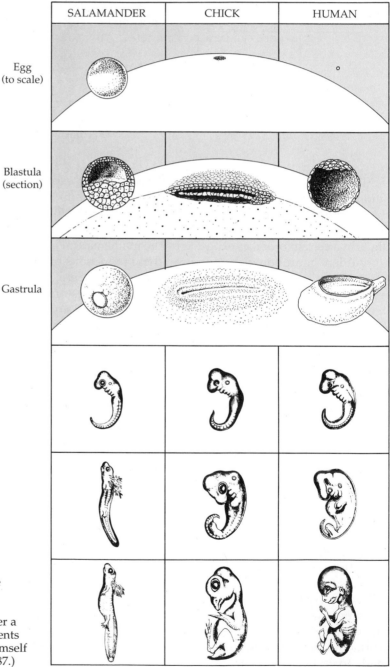

Figure 23.15
The bottleneck at the "pharyngula" stage of vertebrate development. The bottom of this chart is the standard depiction of von Baer's law (as shown in Chapter 7), demonstrating the divergence of vertebrate classes after a common embryonic stage. The top of this chart represents the divergent beginnings of development. Von Baer himself (1886) was aware of this bottleneck. (After Elinson, 1987.)

plastic. Similarly, the later stages are very different, as the different pheno-types of mice, sunfish, snakes, and newts amply demonstrate. There is some-thing in the *middle* of development that appears to be invariant.

Raff argues that the formation of new *Baupläne* is inhibited by the need for global sequences of induction during the neurula stage (Figure 23.16). Before that stage, there are few inductive events. After that period, there are a great many inductive events, but almost all of them done within discrete modules. During early organogenesis, however, there are several inductive events occurring simultaneously that are global in nature. At this stage, the modules overlap and interact with one another. In vertebrates, to use von Baer's example, the earliest stages involve specifying axes and undergoing gastrulation. Induction has not happened on a large scale. Moreover, as Raff and colleagues have shown (Henry et al., 1989), there is a great deal of regu-lative ability here, so small changes in morphogen distribution or the posi-tion of cleavage planes can be accommodated. After the major body plan is fixed, inductions occur all over the body but are compartmentalized into discrete organ-forming systems. The lens induces the cornea, but if it fails to do so, only the eye is affected. Similarly, there are inductions in the skin that form feathers, scales, or fur. If they do not occur, the skin or patch of skin may lack these structures. But during early organogenesis, the interactions are more global (Slack, 1983). Failure to have the heart at a certain place can affect the induction of eyes (see Chapter 17). Failure to induce the mesoderm in a certain region leads to malformations of the kidneys, limbs, and tail. It is this stage that constrains evolution and that typifies the vertebrate phylum. Thus, once a vertebrate, it is difficult to evolve into anything else.

Coevolution of Ligand and Receptor: Reproductive Isolation

Another developmental constraint concerns the ability of one tissue to inter-act with another. In development, things have to fit together if the organism is to survive. Ligands have to fit with receptors, and they have to be ex-pressed at the right place and at the right time. Changes in the ligand must be accommodated by complementary changes in the receptor if the receptor is to function. However, if the change in the ligand (or receptor) structure produces too big a change, it will not bind to its receptor (or ligand), and de-velopment will stop. These complementary changes can lead to a separation of functions, as can be seen in the evolution of hormone families and their receptors (Moyle et al., 1994).

Such separation of functions can cause reproductive isolation and the separation of species when the receptor and ligand are proteins on the sperm and egg. While most proteins of related marine species are very simi-lar, the proteins responsible for fertilization are often extremely different (Metz et al., 1994). In sea urchins, the bindin of the sperm and the comple-mentary receptors of the egg have coevolved so that the bindin of one species often does not recognize the bindin receptors on the oocyte. Hof-mann and Glabe (1994) have proposed a model whereby there would be several distinct recognition sites between bindin and its receptor. Mutations would cause some of these sites to be altered, and this would select comple-mentary alterations on the opposite gamete. There would be a stage wherein some sperm could bind, albeit weakly to the eggs, but eventually, this process of alteration and accommodation would produce two reproduc-tively isolated groups within the species (Figure 23.17). In abalones, muta-tions of a small region of the lysin protein and its corresponding receptor ap-pear to be responsible for the species specificity of fertilization. Moreover, the evolution of these changes in lysin and bindin proteins appears to be

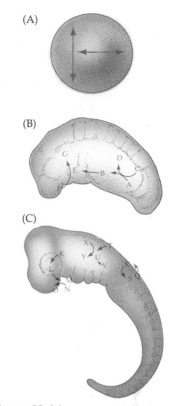

Figure 23.16
Mechanism for the bottleneck at the pharyngula stage of vertebrate develop-ment. (A) In the cleaving embryo, glob-al interactions exist, but there are very few of them (mainly to specify the axes of the organism). (B) At the neurula to pharyngula stages, there are many glob-al interactions. (C) After the pharyngula stage, there are even more inductive interactions, but these are primarily local in effect, confined to their own fields. (After Raff, 1994.)

Figure 23.17
Hypothetical model for the recognition pattern of sperm and egg in two related sea urchin species. The sperm of *Strongylocentrotus purpuratus* can bind the receptor in the egg. Similarly, the sperm of *S. franciscanus* can bind with its egg receptor. *S. purpuratus* sperm will not bind to *S. franciscanus* eggs, while the *S. franciscanus* sperm will weakly bind to the eggs of *S. purpuratus*. It is postulated that repeated elements in the bindin protein each interact with complementary sites in their respective bindin receptors. A coevolution between bindin and its receptor may have separated the two species. (After Shaw et al., 1994.)

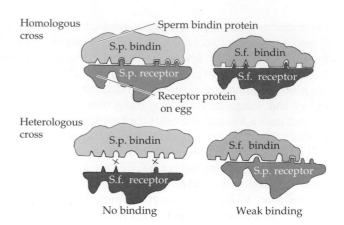

rapid and correlates with speciation (Shaw et al., 1994; Lee et al., 1995; Metz and Palumbi, 1996).*

The developmental genetic mechanisms of evolutionary change: Homologous regulatory genes

Descent with modification can now be demonstrated on the molecular level. Moreover, the modifications can be shown to involve regulatory genes. Roth (1984) defined homology as "the sharing of pathways of development, which are controlled by genealogically-related genes." But when Roth made that influential definition, the developmental pathways had not been elucidated. We can now say a great deal more about evolution and development, and we can give substance to the concept that we inherit pathways of development and that evolution can occur by changing elements of these pathways.

Pax6 *and Eye Development*

A major assumption of the population genetics approach to evolution was that "the search for homologous genes is quite futile except in very close relatives" (Dobzhansky, 1955; Mayr, 1966). However, molecular biology and developmental genetics has shown this assumption to be invalid. The remarkable conclusion is that the genes responsible for particular developmental functions have been conserved over hundreds of millions of years. Moreover, modifications of these genes and their targets can cause much of the diversity of living organisms. One of the most striking of these findings has been that the *Pax6* gene orchestrates eye development in species as distant as flies and humans.

The development of the mammalian eye, the insect eye, and the molluscan eye is each strikingly different from the other. The fly eye contains numerous ommatidia and develops from a morphogenetic furrow sweeping across an imaginal disc. The cephalopod eye develops through the separation of lens-forming and retina-forming regions of a common placode. The mammalian eye develops through a series of inductive interactions involv-

---

*Another example of a developmental mutation causing reproductive isolation involves a more mechanical function. The snail shell coiling mutations discussed in Chapter 5 are mutations that act during early development to change the position of the mesodermal organs. Mating between left-coiling and right-coiling snails is mechanically very difficult, if not impossible, in some species (Clark and Murray, 1969). As this mutation is inherited as a maternal effect gene, a group of related snails would emerge that could mate with one another but not with other members of the original population. These reproductively isolated snails could expand their range and, by the accumulation of new mutations, form a new species (Alexandrov and Sergievsky, 1984).

ing a bulge from the diencephalon contacting the surface ectoderm (see Chapter 17). It had been thought that the three types of eyes showed convergent evolution and that the eye had evolved independently in each of these three groups. However, recent research shows that the eyes of insects and vertebrates did not come from separate origins, but arose in the distant past from a common ancestor.

Mutant alleles of the human *PAX6* gene are responsible for malformations of the eye (Hanson et al., 1994). Heterozygotes are conspicuous since they lack the iris. One human fetus thought to have homozygous PAX6 mutations has been reported, and it lacks eyes completely and has several craniofacial abnormalities (Hodgson and Saunders, 1980). In the mouse and the rat, this gene is called *Small eyes*, owing to the phenotype of the heterozygote. Homozygous mouse and rat fetuses die soon after birth, and they have neither nose nor eyes (Hogan et al., 1986; Grindley et al., 1995). The close similarity between the structure and function of *Pax6* genes in mice and humans was expected. However, in 1994, Quiring and her colleagues in Walter Gehring's laboratory showed that the *Drosophila* genome contained a homologue of *Pax6* that encoded a protein whose sequence showed 94 percent identity to the human Pax6 protein. Loss-of-function mutations of *Drosophila Pax6* map to the *eyeless* gene, a gene characterized by small eyes in heterozygotes and eyelessness in homozygotes. It appears, then, that there is a common gene—*Pax6*—that is necessary for the development of eyes in both insects and vertebrates.*

Since positive evidence is stronger than negative evidence, Gehring's laboratory expressed the *Drosophila Pax6* (i.e., the wild-type *eyeless* gene) in imaginal discs that usually do not express it. Halder and his colleagues (1995) placed the gene encoding the yeast GAL4 transcriptional activator protein downstream from an enhancer that would function in a specific nonneural portion of the fly, such as the leg or wing imaginal disc. They then constructed a transposon, placing the cDNA for the *eyeless* gene downstream from a sequence composed of five GAL4-binding sites. The GAL4 protein would only be made in a particular imaginal disc, and when that protein was made, it would cause the transcription of the *eyeless* cDNA in those particular cells (Figure 23.18A). In flies where the *eyeless* cDNA was being expressed in the antennal discs, the antennae became the pigmented and bristled ommatidia characteristic of the *Drosophila* eye (Plate 28). When the *eyeless* cDNA was expressed in the wing disc, part of the wing cuticle gave rise to eyes (Figure 23.18B). Even more remarkable, when the *Drosophila eyeless* cDNA was replaced by mouse *Pax6* cDNA and placed under the control of the *GAL4* expression system, the murine *Pax6* protein caused the formation of ectopic *Drosophila* eyes (Figure 23.18C)! The *Pax6* gene appears to be a regulator of the eye-forming pathway in both vertebrates and insects. But the eyes are not the same. It remains to be seen how these pathways diverged during evolution to produce the different types of eyes now seen.

It appears that the *Pax* gene is conserved throughout the animal kingdom and that it encodes a transcription factor that binds to eye-forming genes throughout the animal kingdom. The *Pax6* gene is not the only developmental regulator that seems to be homologous in insects and mammals. Another such gene is the homeobox-containing gene *tinman*. The *tinman*

---

*The model before this research was that eyes had independently evolved as many as 40 times. Gehring's laboratory has mentioned the cloning of *Pax6* homologues from flatworms and cephalopods as well. A second *Drosophila* gene, *dachshund (dac)* can also give rise to ectopic eyes when expressed in the wrong imaginal disc. Since it appears that *eyeless* can activate *dachshund* expression and vice-versa, the two genes may have evolved a self-reinforcing positive feedback loop (Shen and Mardon, 1997).

Figure 23.18
Pax-6 as a homologous gene for eye development in insects and vertebrates. (A) The targeted expression of the *Pax6* cDNA in a *Drosophila* non-eye imaginal disc. A strain of *Drosophila* is constructed wherein the gene for the yeast GAL4 protein is placed downstream from an enhancer sequence that stimulates expression in the wing, leg, or antennal imaginal disc. Usually, the yeast protein can find no sequence to activate. However, if a transposon is added to the embryo that carries the cDNA for *Pax6* downstream from the GAL4-binding sites, that cDNA will be expressed in whichever imaginal disc the GAL4 protein is made. (B) *Drosophila* ommatidia emerging in the wing of a fruit fly when *eyeless* cDNA was expressed in the *Drosophila* wing disc. (C) *Drosophila* ommatidia emerging in the leg of a fruit fly when mouse *Pax6* cDNA was expressed in the *Drosophila* leg disc. (From Halder et al., 1995; photographs courtesy of W. J. Gehring.)

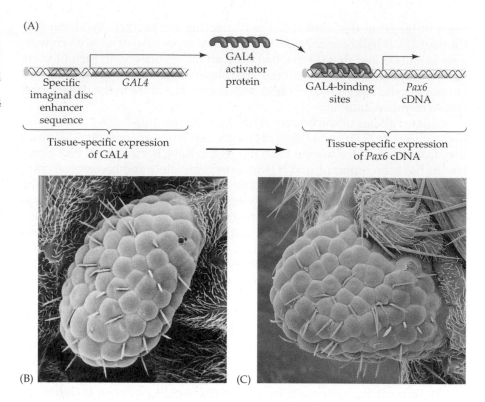

gene is expressed in the *Drosophila* splanchnic mesoderm, eventually residing in the region of the cardiac mesoderm. Loss-of-function mutants of *tinman* lack a heart (hence its name after the *Wizard of Oz* character) (Bodmer, 1993). In mice, the homologous gene is called *Cardiac-specific homeobox* (*Csx*), and it, too, is originally expressed in the splanchnic mesoderm and then continues to be expressed in those cells that form the heart tubes (Manak and Scott, 1994). Thus, although the heart of vertebrates and the heart of insects have hardly anything in common except their ability to pump fluids, they both appear to be predicated on the expression of the same gene, *Csx/tinman*. The difference between the hearts must be in the genes regulated by the CSX/Tinman protein.

BMP4 and Limb Morphogenesis

In some instances, a homologous gene can take on a new function if expressed in a new place. The expression of *Bmp4* in the chick limb is a good example of how a small developmental change can create an evolutionarily important morphological alteration. Most people would agree that the duck and the chick are not alike, although their embryogenesis is extremely similar until the last days. At that time, the bill of the duck becomes distinguishable from the beak of the chick, and the webbed feet of the duck are retained, while webbing is lost in the chick hindlimbs.

BMP4 is known to induce cells to undergo apoptosis in cranial neural crest cells, lung mesenchyme, and tooth buds. It also causes apoptosis in the interdigital soft tissue in the chick limb. Not only is *Bmp4* expressed in the interdigital tissue, but if chick limbs are infected with a virus expressing a dominant negative form of the BMP receptor, the interdigital tissue will not undergo apoptosis when given the BMP4 signal (Figure 23.19; Yokouchi et al., 1996; Zou and Niswander, 1996). The chick and the duck show very similar patterns of BMP expression. However, duck embryos do not express any

(A)

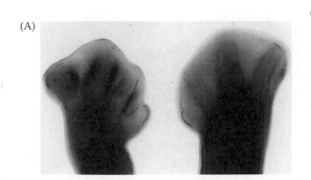

(B)

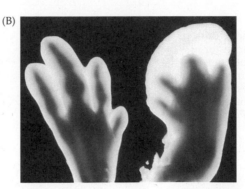

Figure 23.19
Expression of BMP needed to induce apoptosis in the interdigital webbing of chick embryos. (A) BMP4 is seen in the interdigital webbing of the chick hindlimb (left) but not in that of the duck (right) at the same developmental stage. (B) When the BMP signal is blocked by a dominant negative receptor infected into the hindlimb, the interdigital apoptosis does not occur and the digits are shorter. (From Zou and Niswander, 1996; photographs courtesy of L. Niswander.)

Bmp4 (or related *Bmp2* or *7*) in their interdigital tissues. Therefore, by slightly changing the regulation of *Bmp4*, a new morphology is produced that can be selected or rejected by natural selection. Changes in development may produce the *arrival* of the fittest. Their *survival* depends on their environment.

Hox *Genes and the Evolution of Vertebrates*

One of the most remarkable pieces of evidence for deep homologies between all the animals of the world is provided by the *Hox* genes. As mentioned in Chapter 16, the *Hom-C* genes of the fruit fly are homologous to those of the mammal. Not only are the genes, themselves, homologous, but they are in the same order on their respective chromsomes. The expression patterns are also remarkably similar; the expression of the genes at the 3′ end are expressed anteriorly, while those at the 5′ end are expressed more posteriorly. If this evidence of homology were not enough, Malicki and colleagues (1992) demonstrated that the human *HOX4B* gene could mimic the function of its *Drosophila* homologue, *Deformed*, when introduced into *Dfd*-deficient *Drosophila* embryos. Slack and his colleagues (1993) postulated that the *Hox* gene expression pattern defines the development of all animals and that the pattern of *Hox* gene expression is constant for all phyla, the *labial*-type *Hox* genes being expressed anteriorly, the *Ubx*-type *Hox* genes being expressed in the center, and the *AbdB*-type genes expressed posteriorly. The global regulation of these *Hox* genes is also similar from species to species. The Caudal protein is used to induce the posterior domains of *Drosophila*, and it seems to be doing the same thing in mice and nematodes (Subramanian et al., 1995). If the underlying *Hox* gene expression is uniform, it is thought that the differences in the phyla emerge from differences in how these genes are regulated and what genes the *Hox*-derived proteins regulate.*

In vertebrates, there are four *Hox* complexes. In amphioxus, a nonvertebrate chordate that lacks a true head, brain, neural crest tissues, and spinal chord, there is only one *Hox* complex, looking very much like that of the insect (Figure 23.20; Holland and Garcia-Fernández, 1996). By the time fish evolved, there were four *Hox* complexes. The *Hox* genes appear to interpret the positional information along the anterior-posterior body axis, and the importance of these genes in relating evolution and development was suggested by certain "atavistic" structures that resulted from the loss of particu-

---

*The reason for this remarkable conservation of structure in the *Hox* gene complex is thought to be the sharing of *cis*-regulatory regimes by neighboring genes. If a *Hox* gene is moved to a different region within the complex, its regulation is altered. The critical regulatory regimes might be the binding sites for the Polycomb proteins. These proteins are also conserved throughout evolution, and they silence the *Hox* genes at specific times and places. Here, then, we see a "phyletic constraint" at the molecular level (Chiang et al., 1995; Müller et al., 1995; van der Hoeven et al., 1996).

Figure 23.20
Postulated ancestry of the homeotic genes from a hypothetical ancestor of both deuterostomes and prototomes. Amphioxus has only one cluster, similar to that of insects. Vertebrates have four clusters, none of which is complete. (After Holland and Garcia-Fernández, 1996.)

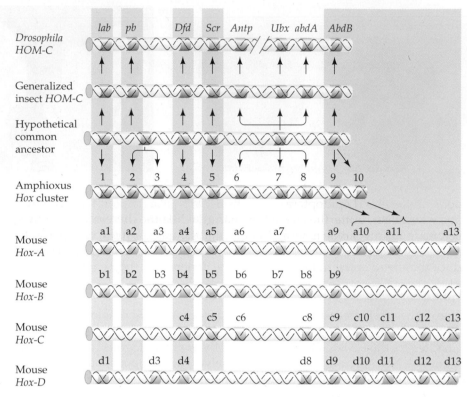

lar *Hox* genes. Disruption of the *Hoxa-2* genes results in a partial transformation of the second pharyngeal arch into a copy of the first pharyngeal arch. The mutant fetuses lack the stapes and styloid bones formed from the second arch, but have extra malleus, incus, tympanic, and squamosal bones. They also possess a rodlike cartilage that is fused to the alisphenoid element and whose caudal end is in contact with the supernumerary incus. This cartilage has no counterpart in normal mice, but its anatomical relationships suggest that it is homologous to the pterygoquadrate cartilage seen in reptiles. The complex formed by the pterygoquadrate cartilage and the incus is thought to have been present in therapsids, the reptile stock that gave rise to the mammals (Rijli et al., 1993; Mark et al., 1995). When the *Hoxa-2* gene is misregulated by knocking out the retinoic acid receptors, a distinct pterygoquadrate cartilage develops that links the incus and alisphenoid bones (Figure 23.21; Lohnes et al., 1994).

The question remained, however, whether the *Hox* genes specify the axis according to a counting system or by a code wherein different *Hox* genes specify different vertebrae. This is an important question because it gives insight into how the same *Hox* genes can specify different bodies. By comparing the *Hox* gene expression patterns with the type of vertebrae, it was shown that the type of vertebrae was specified by the constellation of *Hox* genes expressed in the somites (Gaunt, 1994; Burke et al., 1995). For example, the mouse has 5 occipital, 7 cervical, 13 thoracic, 6 lumbar, and 4 sacral vertebrae. The chick, on the other hand, has 5 occipital, 14 cervical, 7 thoracic, 9 lumbar, and 4 sacral vertebrae. Although the total number of presacral vertebrae differs only by 1 (34 versus 35), there are obvious transpositions between the species (Goodrich, 1930). In both animals, *Hoxc-5* is expressed at the end of the cervical vertebrae, while *Hoxc-6* comes on at the beginning of the thoracic series. In the mouse this occurs at the boundary between the twelfth and thirteenth vertebrae, and in chicks between the nine-

Figure 23.21
Representation of skeletal elements derived from the first pharyngeal arch (in gray) and the second pharyngeal arch (in black). (AS, alisphenoid; I, incus; I2, duplicated incus; P and P2, normal and duplicated pteroid cartilage; PQ, pterygoquadrate cartilage; SQ, squamosal; SQ2, duplicated squamosal.) (After Mark et al., 1995.)

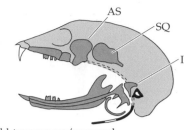

Wild-type mouse/mammal

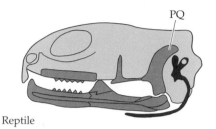

Reptile

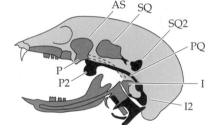

Hoxa-2 null mutant

teenth and twentieth. So in vertebrates, changes in morphology can be brought about by changing the domains of *Hox* gene expression.

Hox *Genes and the Evolution of Arthropods*

The same question yielded a different answer when it was asked of arthropods. Butterflies (Lepidoptera) differ from *Drosophila* (Diptera) in two obvious ways. First, butterflies have four wings, while the dipterans have two. Second, butterfly larvae have abdominal limbs, called **prolegs,** which are absent in fly maggots. The most likely way of creating these differences would be to change the pattern of homeotic gene expression (Lewis, 1978). In *Drosophila*, *Ultrabithorax* (*Ubx*) is expressed in the halteres, but not in the wings. Loss-of-function mutations of *Ubx* convert the halteres into mesothoracic wings, while ectopic expression of Ubx in the wing discs cause them to form into halteres (see Chapter 14). One would expect, then, that *Ubx* would be inactive in the butterfly hindwing discs. This is not the case. Warren and colleagues (1994) showed high levels of *Ubx* expression in the hindwing discs of the buckeye butterfly, *Precis coenia*. In fact, the pattern of Hom-C gene expression in *Precis* was essentially the same as the pattern in *Drosophila*. In the butterfly, *Ubx* modifies wing morphology to make a hindwing (rather than a forewing). In the fly, it modifies the wing into a haltere. The present hypothesis is that the target genes of Ubx may have been changed, not the pattern of *Ubx* expression.

THE EVOLUTION OF WING NUMBER.   Insect wings are thought to have evolved from the multibranched gill appendages of ancestral crustaceans. Specifically, the pattern of apterous expression in the dorsal osmoregulatory flaps (epipods) of crustacea resembles its expression in developing insect wings (Kukalova-Peck, 1978; Averof and Cohen, 1997). Carroll and his colleagues (1995) suggest that the orginal insect had wings coming from all segments (like the crustacean's gills). In modern insects, different homeotic genes suppress this potential in most segments. In other words, wing formation arose independently of homeotic genes in an organism that had been using the homeotic genes for segment identity or neural patterning. Only later did the wing formation program come under the control of the homeotic genes. Several observations point to this conclusion. First, although *Antennapedia* is expressed in the two segments (second and third thorax) capable of making wings, it is not needed for wing formation. The winged mesothoracic (T2) segment may therefore represent the "ground state" that used to be in all segments before homeotic genes began regulating wing formation. Second, instead of *Antennapedia* positively regulating wing development in T2 and T3, it appears that other Hom-C genes repress wing development in other primordia. Loss-of-function mutations of the Hom-C genes cause the formation of ectopic wing primordia in the segments in which they are expressed (see Figure 14.29 for a fly where *Ubx* is removed). Therefore, with the possible exception of the lower abdominal segments controlled by *Abd-b*, the potential for wing development exists in all segments and is repressed by the homeotic genes. Third, if *Scr* expression is induced in wing discs (by using the GAL4 system mentioned earlier), wing development is aborted in the early stages.

(A) Apterygoyte

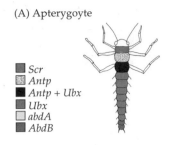

■ *Scr*
▧ *Antp*
■ *Antp + Ubx*
■ *Ubx*
□ *abdA*
■ *AbdB*

(B) Palaeodictyopteran nymph

(C) Palaeozoic mayfly nymph

(D) Primitive neoptera adult

(E) Modern endopterygote (Lepidoptera)

(F) Modern endopterygote (Diptera)

Figure 23.22
Evolutionary schema of wing development. Apterogotes (A) already had the standard insect Hom-C pattern. When wings emerged (B,C), all segments had them, irrespective of the Hom-C genes expressed in them. (D,E) In most insects, *AbdB*, *Scr* and *abdA* prevented wing formation. (F) In dipterans such as *Drosophila*, *Ubx* also acquired the ability to repress wing development. (After Carroll et al., 1995.)

By combining developmental genetics and the fossil record (Kukalova-Peck, 1978), Carroll and colleagues have proposed the following scenario (Figure 23.22): when wings first arose, they were found on all segments, and there was no homeotic regulation of their number or character. Assuming that the pattern of Hom-C gene expression has remained the same, the HOM-C proteins acquired the ability to regulate wing formation through the evolution of Scr-, AbdA-, and Ubx-sensitive sites in the regulatory regions of the wing-forming genes. The evolution of Scr-responsive elements would lead to the modification or reduction of the prothoracic (T1) wings, while AbdA-responsive elements would lead to the reduction of abdominal wings. In four-winged insects, *Ubx* represses wing formation in the first abdominal segment (A1), and in two-winged insects, it controls wing development in both A1 and T3. Thus, unlike the situation in mammals, the evolution of insect segment identity does not appear to correspond with changes in the Hom-C genes. Rather, the proteins encoded by these homeotic genes acquired new regulatory "targets."

EVOLUTION OF INSECT LEG NUMBER. Another important evolutionary lesson is that the Hom-C genes are not all-powerful regulators. Rather, they can be regulated locally by the products of other genes. In arthropods, many groups are distinguished by the number of limbs. Insects have six legs as adults, three pairs arising from each of the three thoracic segments. In *Drosophila*, the *Distal-less* (*Dll*) gene is critical for providing proximal-distal axis of the appendages (see Figures 14.33 and 19.21). *Distal-less* expression occurs in the cephalic and thoracic limb-forming discs (both for legs, mandibles, and wings), but it is excluded in the abdomen by the AbdA and Ubx proteins. Thus, the appendages grow into legs and wings in the thorax and into mandibles in the head. The *Drosophila* larva never develops limbs in its abdomen.

However, butterfly and moth caterpillars are characterized by rudimentary abdominal legs called prolegs. Panganiban and her colleagues (1994) cloned the *Distal-less* homologue from the buckeye butterfly and mapped its expression during butterfly development. During the early portion of *Precis* embryogenesis, *Dll* expression is the same as it is in *Drosophila*. It is seen first in the head regions during gastrulation (antennal, maxilla, and labial segments), and in the thoracic regions that will give rise to the leg imaginal discs (Figure 23.23A). However, as development procedes, the *Dll* gene of *Precis* becomes expressed in the third through sixth abdominal segments (Figure 23.23B). Whereas *Dll* expression is seen in both the proximal ring and "socks" of the true thoracic legs, the expression of *Distal-less* in the abdomen is restricted to the proximal ring.

Thus, the lepidopteran prolegs appear to be homologous to the proximal portion of the thoracic legs. The expression in the maxilla and labial segments in both *Drosophila* and *Precis* is interesting because it is consistent with recent paleontological evidence (Kukalova-Peck et al., 1992) that, although these jaw structures originated from limb primordia, distal limb elements are lost from all arthropod jaws.

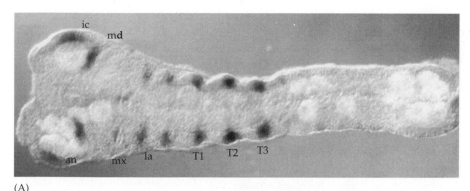

(A)

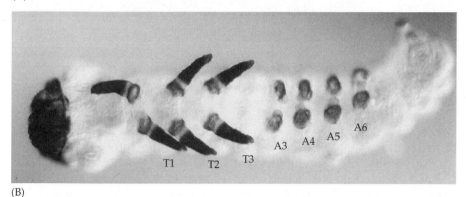

(B)

Figure 23.23
Distal-less gene expression in *Precis*. (A) At 12 percent embryogenesis, *Dll* transcripts appear in the three thoracic segments (T1, T2, T3) as well as in the antennal (an), maxilla (mx), and labial (la) segments. (B) By 40 percent of embryonic development, *Dll* expression in *Precis* has diverged significantly from that of *Drosophila* in that *Dll* expression is also seen in abdominal segments 3–6. (A and B from Panganiban et al., 1994, courtesy of the authors.)

The presence of larval prolegs and *Distal-less* expression in the *Precis* abdominal segments suggests that *Distal-less* is regulated differently in dipterans and lepidopterans. Two possibilities come to the fore. (1) The *Distal-less* genes of *Precis* are not repressed by the AbdA and Ubx homeodomain proteins, or (2) the expression of the repressing homeodomain genes is somehow abrogated in the abdominal regions of *Precis*. Warren and co-workers (1994) showed that the *Drosophila* and *Precis* embryos had the same initial pattern of Hom-C gene expression. However, at about 20 percent of the way through *Precis* embryogenesis, Hom-C gene expression is lost in small patches of segments A3–A6. Neither *Ubx* nor *abdA* is expressed in the region of the abdominal segments that give rise to the prolegs (Plate 26). Shortly thereafter, the *Distal-less* and *Antennapedia* genes are expressed in those "holes." It is not known what molecules are used to downregulate the *abdA* and *Ubx* gene expression in the regions of *Distal-less* expression. The *Polycomb* group genes are the best suspects, since they can repress both genes in *Drosophila*.

Homologous pathways of development

One of the most exciting findings of the past decade has been not only homologous regulatory genes but also homologous developmental pathways (Zuckerkandl, 1994; Gilbert, 1996; Gilbert et al., 1996). Two of these pathways have already been discussed in earlier chapters. First, as seen in Chapter 15, the chordin/BMP4 pathway demonstrates that in both vertebrates and invertebrates, chordin/short-gastrulation inhibits the lateralizing effects of BMP4/decapentaplegic, thereby allowing the ectoderm protected by chordin/short-gastrulation to become the neurogenic ectoderm. The reactions are so similar that *Drosophila* decapentaplegic protein can induce ventral fates in *Xenopus* and can substitute for the short-gastrulation protein (Holley et al., 1995). Second, in Chapter 18, we saw that the interactions be-

Figure 23.24
Similar regulation of the alcohol dehydrogenase gene in *Drosophila* and humans. CREB/ATF and C/EBP are positive regulators of the alcohol dehydrogenase gene. AEF is a negative regulator. (After Abel et al., 1992; Zuckerkandl, 1994.)

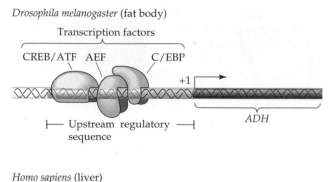

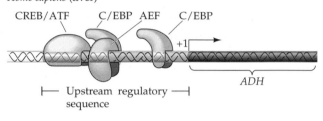

Figure 23.25
The widely used RTK-RAS pathway. The outline of the pathway is shown on the left side, along with the names of the elements in different species. The ligand, which can be soluble (as in EGF) or a membrane-bound protein on another cell (as in the Boss protein ["Bride of sevenless"] presented to the sevenless RTK). The cytoplasmic domains of the RTKs are autophosphorylated once they are dimerized, and this allows them to bind the adaptor protein and to stimulate the Ras G protein. The Ras G protein translocates the Raf protein into the cell membrane, thereby activating it. This can be inhibited by the gap proteins, which can inactivate Ras. The active Raf protein initiates the cascade of phosphorylation that ends in a phosphorylated (activated) transcription factor entering into the nucleus and effecting RNA transcription.

tween Hedgehog and Wingless were conserved between insects and vertebrates in the formation of the limbs. Indeed, the same interactions are used for establishing the segmentation pattern in early *Drosophila* embryos (see Chapter 14) and for establishing compartments in the mammalian brain (see Chapter 7). It has also been shown that numerous DNA-protein interactions regulating specific genes are conserved across divergent species. Thus, the alcohol dehydrogenase gene is controlled in the *Drosophila* fat body by the same set of proteins that governs its expression in the human liver (Figure 23.24; Abel et al., 1992).

Among the earliest known of the homologous pathways was the RTK-Ras signal transduction pathway which has recently been found throughout the animal kingdom, although used for strikingly different functions (see Chapter 3; Figure 23.25). In *Drosophila*, the determination of the photorecep-

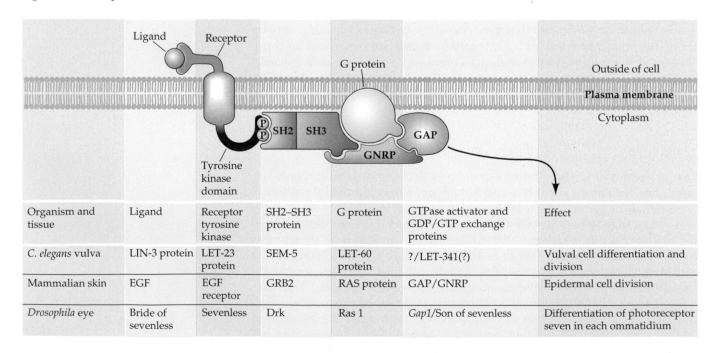

Organism and tissue	Ligand	Receptor tyrosine kinase	SH2–SH3 protein	G protein	GTPase activator and GDP/GTP exchange proteins	Effect
C. elegans vulva	LIN-3 protein	LET-23 protein	SEM-5	LET-60 protein	?/LET-341(?)	Vulval cell differentiation and division
Mammalian skin	EGF	EGF receptor	GRB2	RAS protein	GAP/GNRP	Epidermal cell division
Drosophila eye	Bride of sevenless	Sevenless	Drk	Ras 1	*Gap1*/Son of sevenless	Differentiation of photoreceptor seven in each ommatidium

tor seven is accomplished when the Sevenless protein (on the presumptive photoreceptor 7) binds to the Bride of sevenless protein (Boss) on photoreceptor 8. This interaction activates the tyrosine kinase of the Sevenless protein to phosphorylate itself. The DRK protein then binds to these newly phosphorylated tyrosines through its Src-homology-2 (SH2) region and activates the Son of sevenless (SOS) protein. This protein is a guanosine nucleotide exchanger and exchanges GDP for GTP on the Ras1 G protein. This activates the G protein, enabling it to transmit its signal to the nucleus through the MAP kinase cascade. This same system has been found to exist in the determination of the nematode vulva, the mammalian epidermis, and the *Drosophila* terminal segments. The similarity in these systems is so striking that many of the components are interchangeable between species. The gene for human GRB2 can correct the phenotypic defects of *Sem-5*-deficient nematodes, and the nematode SEM-5 protein can bind to the phosphorylated form of the human EGF receptor (Stern et al., 1993).

Homologous pathways form the basic infrastructure of development. The targets of these pathways can change, depending on the organism. In the ectoderm of one organism, the RTK-Ras pathway may activate the genes responsible for proliferation. In another organism, the same pathway may activate the genes responsible for making a photoreceptor. And in a third organism, the pathway activates the genes needed for constructing a vulva.

Creating new cell types: The basic evolutionary mystery

One of the greatest unsolved questions in evolutionary and developmental biology is, How can organisms evolve a new type of cell? This is an important question, since changes in phyla have been associated with the evolution of new cell types. Hypothetically, new combinations of genes could also create new cell types. However, this remains an unproved hypothesis. Kauffman (1993) has mathematically modeled the generation of new cell types from a random genome consisting of 10,000 genes, each regulated by 2 other genes. In such cases, he finds only 100 stable states of interaction (out of 210,000 possible states). Each of these stable states represents a differentiated cell type. In some cases, the mutation of a regulatory gene is sufficient for the restructuring of the interactions, and a new cell type is created. Most genes, however, remain unaffected by this new arrangement.

The creation of a new cell type is a rare event in nature and can often change the nature of the beast. As Figure 23.26 shows, the vertebrates are thought to have arisen from invertebrates in several steps that involved the formation and modification of new cell types.

As mentioned earlier in this chapter, the neural crest cells were important in the origin of chordates. While we do not know how neural crest cells arose, Holland and colleagues (1996) have provided a fascinating speculation that involves dissociation, duplication and divergence, and co-option. It also involves the vertebrate homologues of the *Drosophila* gene discussed earlier, *Distal-less*. Amphioxus is a protochordate that has notochord, somites, and a hollow neural tube. It lacks a brain and facial structures, and, most importantly, it lacks neural crest cells. Like *Drosophila*, amphioxus has but one copy of the *Distal-less* gene per haploid genome, and as in *Drosophila*, this gene is expressed in the epidermis and central nervous system. However, whereas amphioxus has only one copy of this gene, vertebrates have five or six closely related copies of *Distal-less*, each of which probably originated from a single ancestral gene that resembles the one in amphioxus (Price, 1993; Boncinelli, 1994). These *Distal-less* homologues have found new functions. Some are in the mesoderm, a place where *Distal-less* is not ex-

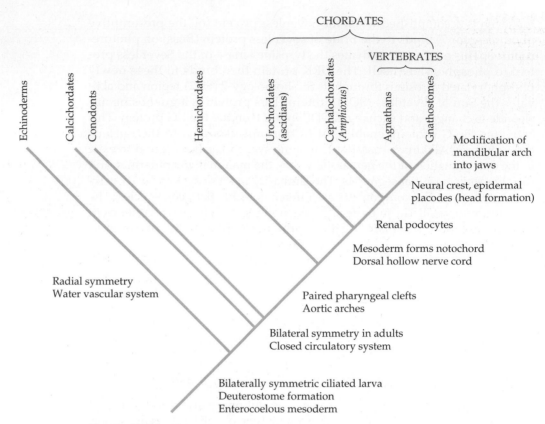

Figure 23.26
Developmental changes in the evolution of invertebrate to vertebrate. The original deuterostome invertebrates were able to form the echinoderms and other organisms that eventually gave rise to the vertebrate lineage. The ability of the mesoderm to form a notochord and its overlying ectoderm to become a neural tube separated the chordates from the remaining invertebrates. The development of neural crest cells and the epidermal placodes that give rise to the sensory nerves of the face distinguish the vertebrates from the protochordates. (After Gans, 1989; Langille and Hall, 1989.)

pressed in amphioxus. Other vertebrate *Distal-less* homologues are expressed in the forebrain, mimicking an expression pattern seen in the anterior of the amphioxus neural tube. This suggests that the vertebrate forebrain is homologous to the anterior neural tube of amphioxus. Still another vertebrate *Distal-less* homologue is expressed in the neural crest cells. Although it remains unproved, it is possible that the new type of *Distal-less* gene could cause the migratory ectodermal cells of amphioxus to evolve into neural crest cells.

A new evolutionary synthesis

In 1922, Walter Garstang declared that ontogeny (individual development) does not recapitulate phylogeny (evolution); it creates phylogeny. The animals that arose late in evolutionary history did not arise through a terminal addition onto an existing embryo. Rather, they arose by mutations that affected the interactions of modules already existing in the *Bauplan* of the organism:

> *A house is not a cottage with an extra story on the top. A house represents a higher grade in the evolution of a residence, but the whole building is altered—foundation, timbers, and roof—even if the bricks are the same.*

Thus, when we say that the contemporary one-toed horse evolved from a five-toed ancestor, we are saying that hereditable changes occurred in the differentiation of limb mesoderm into chondrocytes during embryogenesis in the horse lineage. In this perspective, evolution is the result of hereditary changes affecting development.* This is the case whether the mutation changes the reptilian embryo into a bird or changes the color of *Drosophila* eyes.

That developmental perspective, however, was lost during the 1940s. One of the major events in evolutionary theory has been the "modern synthesis" of evolutionary biology and Mendelian genetics (Mayr and Provine, 1980). One outcome of this hard-won merger is that evolution has been redefined to mean changes in gene frequencies in a population over time."Since evolution is a change in the genetic composition of populations," wrote Dobzhansky (1937), "the mechanisms of evolution constitute problems of population genetics." The developmental approach to evolution was excluded from the synthesis (Hamburger, 1980; Gottlieb, 1992; Dietrich, 1995; Gilbert et al., 1996). It was thought that population genetics could explain macroevolution, so morphology and development were seen to play little role in modern evolutionary theory (Adams, 1991). In other words, macroevolution (the large morphological changes seen between species, classes, and phyla) could be explained by the mechanisms of microevolution, the "differential adaptive values of genotypes or deviations from random mating or both these factors acting together" (Torrey and Feduccia, 1979).

However, this view has had its critics (its heretics, some would say). Perhaps the foremost of these was Richard Goldschmidt. Goldschmidt began his book *The Material Basis of Evolution* (1940) with a challenge to the modern synthesis.

> *I may challenge the adherents of the strictly Darwinian view, which we are discussing here, to try to explain the evolution of the following features by accumulation and selection of small mutants: hair in mammals, feathers in birds, segmentation in arthropods and vertebrates, the transformation of the gill arches in phylogeny including the aortic arches, muscles, nerves, etc.; further, teeth, shells of molluscs, ectoskeletons, compound eyes, blood circulation, alternation of generations, statocysts, ambulacral systems of echinoderms, pedicellaria of the same, cnidocysts, poison apparatus of snakes, whalebone, and finally chemical differences like hemoglobin vs. hemocyanin.*

Goldschmidt claimed that new species did not arise from the mechanisms of microevolution and that population genetics was unable to explain new types of structures that involve several components changing simultaneously. Such macroevolutionary change "requires another evolutionary method than that of sheer accumulation of micromutations." Goldschmidt saw homeotic mutants as "macromutations" that could change one structure into another and possibly create new structures or new combinations of structures. These mutations would not be in the structural genes but in the regulatory genes. A new species, he asserted, would start as a "hopeful monster" (a rather unfortunate phrase having its antecedent in Metchnikoff's prose).

---

*One way of visualizing this is to use a mathematical analogy (Gilbert et al., 1996):
Functional biology = anatomy, physiology, cell biology, gene expression
Developmental biology = δ [functional biology]/δt
Evolutionary biology = δ [developmental biology]/δt

At the same time, Conrad H. Waddington was attempting to find developmental mechanisms for producing such new species. He, too, looked at homeotic mutations in flies as models for drastically new phenotypes, and he formulated the notion of competence transfer ("genetic assimilation," see Chapter 21) to explain certain aspects of morphological evolution. Few scientists paid attention to Goldschmidt or Waddington because they were not writing in the population genetics paradigm of the modern synthesis and their scientific programs were suspect. (Goldschmidt did not believe in Morgan's notion of the gene as a particulate entity, and Waddington's work was misinterpreted as supporting the inheritance of acquired traits.) However, in the 1970s, events in paleontology (the punctuated equilibrium theory), events in society (the Creationists giving the microevolutionary contest to the biologists but contesting macroevolution), and events in molecular biology (notably King and Wilson's 1975 paper showing that chimps and humans have DNA that is greater than 99 percent identical) prompted scientists to consider seriously the view that mutations in regulatory genes can create large changes in morphology.

In the 1990s, the techniques of molecular biology enabled biologists to discover (1) homologous regulatory genes such as *Pax6* that control the development of the same organs thoroughout the animal kingdom, (2) homologous developmental pathways whose functions can change between organisms or between cells of the same organism, and (3) the changing patterns of the homeotic gene expression that allow different parts of the body to have different structures and functions. Such discoveries have converged to form a developmental evolutionary synthesis that incorporates the population genetic approach but which expands evolutionary theory to explain macroevolutionary phenomena as well. The developmental evolutionary synthesis also retains a multiplicity of paradigms. In some instances (such as the creation of neural crest cells), a qualitative change occurs, whereas in other cases (such as the formation of the pocket gopher pouch), quantity becomes quality when a threshold is passed. Signaling the coalescence of this synthesis, "Evolutionary Developmental Biology" has become a separate topic in an encyclopedia of science (Hall, 1996), and *Roux's Archives of Developmental Biology*, one of the oldest journals of experimental embryology, has changed its title to *Development, Genes, and Evolution*.

We are at a remarkable point in our understanding of nature, for a synthesis of developmental genetics with evolutionary biology may transform our appreciation of the mechanisms underlying evolutionary change and animal diversity. Such a synthesis is actually a return to a broader-based evolutionary theory that fragmented at the turn of the past century (Figure 23.27). In the late 1800s, evolutionary biology contained the sciences that we now call evolutionary biology, systematics, ecology, genetics, and development. When Wilhelm Roux (1894) announced the creation of "developmental mechanics," he did not fully break with evolutionary biology. Rather, he stated that "an *ontogenetic* and a *phylogenetic developmental mechanics* are to be perfected." He noted that the developmental mechanics of embryos (the ontogenetic branch) would proceed faster than the phylogenetic studies, but posited that "in consequence of the intimate causal connections between the two, many of the conclusions drawn from the investigation of ontogeny [would] throw light on the phylogenetic processes."

One hundred years later, we are now at the point where we can attend to the second of Roux's developmental mechanics and create a unified theory of evolution.

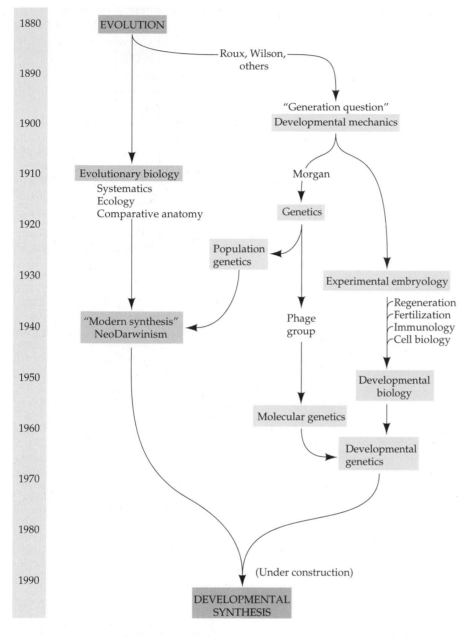

Figure 23.27
Disciplinary road map of the evolutionary side of biology, 1880 to the present. For the sake of clarity, other paths (such as those from genetics to human genetics or from evolution to immunology) have not been shown.

LITERATURE CITED

Abel, T., Bhatt, R. and Maniatis, T. 1992. A Drosophila CREB/ATF transcriptional activator binds to both fat body- and liver-specific regulatory elements. *Genes Dev.* 6: 466–480.

Adams, M. 1991. Soviet perspectives on evolutionary theory. *In* L. Warren and M. Meselson (eds.), *New Perspectives in Evolution.* Liss/Wiley, New York.

Averoff, M. and Cohen, S. M. 1997. Evolutionary origin of insect wings from ancestral gills. *Nature* 385: 627–630.

Ahlberg, P. E. 1997. How to keep a head in order. *Nature* 385: 489–490.

Alberch, P. and Gale, E. 1983. Size dependency during the development of the amphibian foot. Colchicine induced digital loss and reduction. *J. Embyol. Exp. Morphol.* 76: 177–197.

Alberch, P. and Gale, E. 1985. A developmental analysis of an evolutionary trend. Digit reduction in amphibians. *Evolution* 39: 8–23.

Alexandrov, D. A. and Sergievsky, S. O. 1984. A variant of sympatric speciation in snails. *Malacol. Rev.* 17: 147.

Anderson, D. T. 1973. *Embryology and Phylogeny of Annelids and Arthropods.* Pergamon Press, Oxford.

Averoff, M. and Cohen, S. M. 1997. Evolutionary origin of insect wings from ancestral gills. *Nature* 385: 627–630.

Bateson, W. 1894. *Materials for the Study of Variation.* Cambridge University Press, Cambridge.

Berrill, N. J. 1955. *The Origins of the Vertebrates.* Oxford University Press, New York.

Berrill, N. J. 1987. Early chordate evolution. I. Amphioxus, the riddle of the sands. *Int. J. Invert. Repr. Dev.* 11: 1–27.

Bodmer, R. The gene *tinman* is required for specification of the heart and visceral muscles in *Drosophila. Development* 118: 719–729.

Boncinelli, E. 1994. Early CNS development: *Distal-less* related genes and forebrain development. *Curr. Opin. Neurobiol.* 4: 29–36.

Bonner, J. T. 1988. *The Evolution of Complexity*. Princeton University Press, Princeton.

Bowring, S. A., Grotzinger, J. P. Isachsen, C. E., Knoll, A. H., Pelechaty, S. M. and Kolosov, P. 1993. Calibrating rates of early Cambrian evolution. *Science* 261: 1293–1298.

Brusca, R. C. and Brusca, G. J. 1990. *Invertebrates*. Sinauer Associates, Sunderland, MA.

Brylski, P. and Hall, B. K. 1988. Ontogeny of a macroevolutionary phenotype: The external cheek pouches of geomyoid rodents. *Evolution* 42: 391–395.

Burke, A. C. 1989a. Epithelial-mesenchymal interactions in the development of the chelonian Bauplan. *Fortschr. Zool.* 35: 206–209.

Burke, A. C. 1989b. Development of the turtle carapace: Implications for the evolution of a novel Bauplan. *J. Morphol.* 199: 363–378.

Burke, A. C., Nelson, A. C., Morgan, B. A. and Tabin, C. 1995. Hox genes and the evolution of vertebrate axial morphology. *Development* 121: 333–346.

Buss, L. W. 1987. *The Evolution of Individuality*. Princeton University Press, Princeton, NJ.

Carroll, S. B., Gates, J., Keys, D. N., Paddock, S. W., Panganiban, G. E. F., Selegue, J. E. and Williams, J. A. 1994. Pattern formation and eyespot determination in butterfly wings. *Science* 265: 109–114.

Carroll, S. B., Weatherbee, S. D. and Langeland, J. A. 1995. Homeotic genes and the regulation and evolution of insect wing number. *Nature* 375: 58–61.

Chen, J.-Y., Dzik, J., Edgecombe, G. D., Ramsköld, L. and Zhou, G.-Q. 1995. A possible Early Cambrian chordate. *Nature* 377: 720–722.

Chernyak, L. and Tauber, A. I. 1988. The birth of immunology: Metchnikoff, the embryologist. *Cell. Immunol.* 117: 218–233.

Chernyak, L. and Tauber, A. I. 1991. *From Metaphor to Theory: Metchnikoff and the Origin of Immunology*. Oxford University Press, New York.

Chiang, A., O'Connor, M. B., Paro, R., Simon, J. and Bender, W. 1995. Discrete Polycomb-binding sites in each parasegment domain of the bithorax complex. *Development* 121: 1681–1689.

Clack, J. A. 1989. Discovery of the earliest known tetrapod stapes. *Nature* 342: 425–427.

Clark, B. and Murray, J. 1969. Ecological genetics and speciation in land snails of the genus *Partula. Biol. J. Linn. Soc.* 1: 31–42.

Conklin, E. G. 1915. *Heredity and Environment in the Development of Men*. Princeton University Press, Princeton, NJ.

Conway Morris, S. and Whittington, H. B. 1979. The animals of the Burgess Shale. *Sci. Am.* 240(1): 122–133.

Darwin, C. 1859. *The Origin of Species*. John Murray, London.

Dietrich, M. (1995). Richard Goldschmidt's "heresies" and the evolutionary synthesis. *J. Hist. Biol.* 28: 431–461.

Dobzhansky, T. G. 1937. *Genetics and the Origin of Species*. Columbia University Press, New York.

Dobzhansky, T. G. 1955. *Evolution, Genetics, and Man*. Wiley, NY.

Dyke, C. 1988. *The Evolutionary Dynamics of Complex Systems*. Oxford University Press, NY.

Elinson, R. P. 1987. Changes in developmental patterns: Embryos of amphibians with large eggs. *In* R. A. Raff and E. C. Raff (eds.), *Development as an Evolutionary Process*. Alan R. Liss, New York, pp. 1–21.

Gans, C. 1989. Stages in the origin of vertebrates: Analysis by means of scenarios. *Biol. Rev.* 64: 221–268.

Gans, C. and Northcutt, R. G. 1983. Neural crest and the origin of vertebrates: A new head. *Science* 220: 268–274.

Garstang, W. 1922. The theory of recapitulation: a critical restatement of the biogenetic law. *J. Linn. Soc. Zool.* 35: 81–101.

Garstang, W. 1928. Presidential address to the British Association for the Advancement of Science, Section D. Republished in *Larval Forms and Other Zoological Verses*, 1985. University of Chicago Press, Chicago, pp. 77–98.

Gasman, D. 1971. *The Scientific Origins of National Socialsim: Social Darwinism in Ernst Haeckel and the German Monist League*. MacDonald, London.

Gaunt, S. J. 1994, Conservation in the *Hox* code during morphological evolution. *Int. J. Dev. Biol.* 38: 549–552.

Gilbert, S. F. 1996. Cellular dialogues in organogenesis. *In* M. E. Martini-Neri, G. Neri, and J. M. Opitz, (eds.) *Gene Regulation and Fetal Development: Proceedings of the Third International Workshop on Fetal Genetic Pathology, 1993*. Wiley-Liss, NY, pp. 1–12.

Gilbert, S. F., Opitz, J. M. and Raff, R. A. 1996. Resynthesizing evolutionary and developmental biology. *Dev. Biol.* 173: 357–372.

Gish, D. T. 1973. *Evolution? The Fossils Say No!* Creation-Life Publishers, San Diego.

Goldschmidt, R. B. 1940. *The Material Basis of Evolution*. Yale University Press, New Haven.

Goodrich, E. S. 1930. *Studies on the Structure and Development of Vertebrates*, Macmillan, London.

Gottlieb, G. 1992. *Individual Development and Evolution: The Genesis of Novel Behavior*. Oxford University Press, New York.

Gould, S. J. 1977. *Ever Since Darwin*. Norton, New York.

Gould, S. J. 1989. *Wonderful Life*. Norton, New York.

Gould, S. J. 1990. An earful of jaw. *Natural History* 1990(3): 12–23.

Gould, S. J. and Lewontin, R. C. 1979. The spandrels of San Marcos and the Panglossian paradigm: A critique of the adaptationist program. *Proc. R. Soc. London* [B] 205: 581–598.

Grant, P. 1978. *Biology of Developing Systems*. Holt, Rinehart & Winston, New York.

Grindley, J. C., Davidson, D. R. and Hill, R. E. 1995. The role of *Pax-6* in eye and nasal development. *Development* 121: 1433–1442.

Halder, G., Callaerts, P. and Gehring, W. J. 1995. Induction of ectopic eyes by targeted expression of the eyeless gene in *Drosophila. Science* 267: 1788–1792.

Hall, B. K. 1992. *Evolutionary Developmental Biology*. Chapman and Hall, London.

Hall, B. K. 1996. Evolutionary developmental biology. *McGraw-Hill Yearbook of Science and Technology*. McGraw-Hill, NY. pp. 110–112.

Hamburger, V. 1980. Embryology and the Modern Synthesis in evolutionary theory. *In* E. Mayr and W. Provine, (eds.) *The Evolutionary Synthesis: Perspectives on the Unification of Biology* Cambridge University Press, NY, pp. 97–112.

Hampé, A. 1959. Contribution à l'étude du développement et la régulation des déficiences et excédents dans la patte de l'embryon de poulet. *Arch. Anat. Microsc. Morph. Exp.* 48: 347–479.

Hanson, I. M. and seven others. 1994. Mutations at the *PAX6* locus are found in heterogeneous anterior segment malformations including Peter's anomaly. *Nature Genet.* 6: 168–173.

Henry, J. J., Amemiya, S., Wray, G. A. and Raff, R. A. 1989. Early inductive interactions are involved in restricting cell fates of mesomeres in sea urchin embryos. *Dev. Biol.* 136: 140–153.

Hodgson, S. and Saunders, K. 1980. A probable case of the homozygous condition of the *Aniridia* gene. *J. Med. Genet.* 6: 478–480.

Hofmann, A. and Glabe, C. 1994. Bindin, a multifunctional sperm ligand and the evolution of new species. *Semin. Dev. Biol.* 5: 233–242.

Hogan, B., Hirst, E. M. A., Horsburgh, G., Cohen, J., Hetherington, C. M., Fisher, G. and Lyon, M. F. 1986. Small eyes (Sey): A homozygous lethal mutation on chromosome 2 which affects the differentiation of both lens and nasal placodes in the mouse. *J. Embryol. Exp. Morphol.* 97: 95–110.

Holder, N. 1983. Developmental constraints and the evolution of vertebrate limb patterns. *J. Theor. Biol.* 104: 451–471.

Holland, N. D., Panganiban, G., Henyey, E. L. and Holland, L. Z. 1996. Sequence and developmental expression of *AmphiDll*, an amphioxus *Distal-less* gene transcribed in the ectoderm, epidermis, and nervous system: insights into evolution of craniate forebrain and neural crest. *Development* 122: 2911–2920.

Holland, P. W. H. and Garcia-Fernàndez, J. 1996. *Hox* genes and chordate evolution. *Dev. Biol.* 173: 382–395.

Holley, S., Jackson, P. D., Sasai, Y., Lu, B., De Robertis, E. M., Hoffmann, F. M. and Ferguson, E. L. 1995. A conserved system for dorsal-ventral patterning in insects and vertebrates involving *sog* and *chordin*. *Nature* 376: 249–23.

Hopson, J. A. 1966. The origin of the mammalian middle ear. *Am. Zool.* 6: 437–450.

Jefferies, R. P. S. 1986. *The Ancestry of the Vertebrates*. British Museum of Natural History, London.

Jones, M. L. and Gardiner, S. L. 1989. On the early development of the vestimentiferan tube worm *Ridgeia* sp. and observations on the nervous system and trophosome of *Ridgeia* sp. and *Riftia pachyptila*. *Biol. Bull.* 177: 254–276.

Kauffman, S. A. 1993 *Origins of Order*. Oxford University Press, NY.

Kemp, T. S. 1982. *Mammal-Like Reptiles and the Origin of Mammals*. Academic Press, New York.

King, M. C. and Wilson, A. C. 1975. Evolution at two levels in humans, and chimpanzees. *Science* 188: 107–116.

Kline, D., Simoncini, L., Mandel, G., Maue, R. A., Kado, R. T. and Jaffe, L. A. 1988. Fertilization events induced by neurotransmitters after injection of mRNA in *Xenopus* eggs. *Science* 241: 464–467.

Köntges, G. and Lumsden, A. 1996. Rhombencephalic neural crest segmentation is preserved throughout craniofacial ontogeny. *Development* 122: 3229–3242

Kowalevsky, A. 1871. Weitere Studien II. Die Entwicklung der einfachen Ascidien. *Arch. Micr. Anat.* 7: 101–130.

Kukalova-Peck, J. 1978. Origin and evolution of insect wings and their relation to metamorphosis as documented by the fossil record. *J. Morphol.* 156: 53–126.

Kukalova-Peck, J. 1992. The uniramia do not exist: the ground plan of the Ptergota as revealed by Permian Diahanopterodea from Russia (Insecta: Paleodictyopteroidea), *Can. J. Zool.* 70: 236–255.

Langille, R. M. and Hall, B. K. 1989. Developmental processes, developmental sequences and early vertebrate phylogeny. *Biol. Rev.* 64: 73–91.

Larson, A., Wake, D. B., Maxson, L. R. and Highton, R. 1981. A molecular phylogenetic perspective on the origins of morphological novelties in the salamander of the tribe plethodontini (Amphibia, Plethodontidae). *Evolution* 35: 405–422.

Lee, Y.-H., Ota, T. and Vacquier, V. D. 1995. Positive selection is a general phenomenon in the evolution of abalone sperm lysin. *Mol. Biol. Evol.* 12: 231–238.

Lewis, E. B. 1978. A gene complex controlling segmentation in *Drosophila*. *Nature* 276: 565–570.

Lillie, F. R. 1898. Adaptation in cleavage. *Biological Lectures from the Marine Biological Laboratories, Woods Hole, Massachusetts.* Ginn, Boston, pp. 43–67.

Lohnes, D. and seven others. 1994. Function of retinoic acid receptors in development. I. Craniofacial and skeletal abnormalities in RAR double mutants. *Development* 120: 2723–2743.

Lovejoy, A. O. 1964. *The Great Chain of Being*. Harvard University press, Cambridge.

Malicki, J., Cianetti, L. C., Peschle, C. and McGinnis, W. 1992. Human HOX4B regulatory element provides head-specific expression in *Drosophila* embryos. *Nature* 358: 345–347.

Manak, J. R. and Scott, M. P. 1994. A class act: conservation of homeodomain protein functions. *Development 1994 Suppl.* 61–71.

Maderson, P. F. A. 1975. Embryonic tissue interactions as the basis for morphological change in evolution. *Am. Zool.* 15: 315–327

Margulis, L. and Schwartz, K. V. 1988. *The Five Kingdoms*, 2nd Ed. W. H. Freeman, San Francisco.

Mark, M., Rijli, F. M. and Chambon, P. 1995. Alteration of *Hox* gene expression in the branchial region of the head causes homeotic transformations, hindbrain segmentation defects and atavistic changes. *Semin. Dev. Biol.* 6: 275–284.

Mayr, E. 1966. *Animal Species and Evolution*. Harvard University Press, Cambridge.

Mayr, E. and Provine, W. 1980. *The Evolutionary Synthesis: Perspectives on the Unification of Biology*. Harvard University Press, Cambridge, MA.

Metchnikoff, E. 1886. *Embryologische Studien an Medusen. Ein Beitrag zur Genealogie der Primitivorgane.* Vienna.

Metchnikoff, E. 1891. Zakon zhizni. Popovodu nektotorykh proizvedenii gr. L. Tolstogo. *Vest. Evropy* 9: 228–260. Quoted and translated in Chernyak and Tauber, 1990, *From Metaphor to Theory: Metchnikoff and the Origin of Immunology.* Oxford University Press, New York.

Metz, E. C. and Palumbi, S. R. 1996. Positive selection and sequence rearrangements generate extensive polymorphisms in the gamete recognition protein bindin. *Mol. Biol. Evol.* 13: 397–406.

Metz, E. C., Kane, R. E., Yanagimachi, H. and Palumbi, S. R. 1994. Fertilization between closely related sea urchins is blocked by incomppatabilities during sperm-egg attachment and early stages of fusion. *Biol. Bull.* 187: 23–34.

Moyle, W. R., Campbell, R. K., Myers, R. V., Bernard, M. P., Han, Y. and Wang, X. 1994. Co-evolution of ligand-receptor pairs. *Nature* 368: 251–255.

Müller, F. 1864. *Für Darwin*. Engelmann, Leipzig.

Müller, G. B. 1989. Ancestral patterns in bird limb development: A new look at Hampé's experiment. *J. Evol. Biol.* 1: 31–47.

Müller, J., Gaunt, S. and Lawrence, P. 1995. Function of the Polycomb protein is conserved in mice and flies. *Development* 121: 2847–2852.

Ospovat, D. 1981. *The Development of Darwin's Theory*. Cambridge University Press, Cambridge.

Oster, G. F., Shubin, N., Murray, J. D. and Alberch, P. 1988. Evolution and morphogenetic rules: The shape of the vertebrate limb in ontogeny and phylogeny. *Evolution* 42: 862–884.

Owen, R. 1848. *On the Archetype and Homologies of the Vertebrate Skeleton*. London.

Panganiban, G., Nagy, L. and Carroll, S. B. 1994. The role of the *Distal-less* gene in the development and evolution of insect limbs. *Curr. Biol.* 4: 671–675.

Piatigorsky, J. and Wistow, G. 1991. The recruitment of crystallins: New functions precede gene duplication. *Science* 252: 1078–1079.

Popadić, A., Risch, D., Peterson, M., Rogers, B. T. and Kaufman, T. C. 1996. Origin of arthropod mandible. *Nature* 380: 395.

Price, M. 1993. Members of the *Dlx-* and *Nkx2* gene families are regionally expressed in the developing forebrain. *J. Neurobiol.* 24: 1385–1399.

Quiring, R., Walldorf, U., Kloter, U. and Gehring, W. J. 1994. Homology of the *eyeless* gene of Drosophila to the *small eye* gene of mice and *Aniridia* in humans. *Science* 265: 785–789.

Raff, R. A. 1994. Developmental mechanisms in the evolution of animal form: Origins and evolvability of body plans. *In* S. Bengston, (ed.) *Early Life on Earth*. Columbia University Press, NY, pp. 489–500.

Raff, R. A. 1996. *The Shape of Life: Genes, Development, and the Evolution of Animal Form*. University of Chicago Press, Chicago.

Raff, R. A. and Wray, G. A. 1989. Heterochrony: Developmental mechanisms and evolutionary results. *J. Evol. Biol.* 2: 409–434.

Raff, R. A., Wray, G. A. and Henry, J. J. 1991. Implications of radical evolutionary changes in early development for concepts of developmental constraint. *In* L. Warren and M. Meselson (eds.), *New Perspectives in Evolution.* Liss/Wiley, New York.

Reichert, C. B. 1837. Entwicklungsgeschichte der Gehörknöchelchen der sogenannte Meckelsche Forsatz des Hammers. *Müller's Arch. Anat. Phys. wissensch. Med.* 177–188.

Riedl, R. 1978. *Order in Living Systems: A Systems Analysis of Evolution.* John Wiley and Sons, NY.

Rijli, F. M., Mark, M., Lakkaraju, S., Dierich, A., Dollé, P. and Chambon, P. 1993. A homeotic transformation is generated in the rostral branchial region of the head by the disruption of *Hoxa-2*, which acts as a selector gene. *Cell* 75: 1333–1349.

Romer, A. S. 1970. The Chanares (Argentina) Triassic reptile fauna VI. A chiniquodontid cynodont with an incipient squamosal-dentary jaw articulation. *Breviora* 344: 1–18.

Roth, V. L. 1984. On homology. *Biol. J. Linn. Soc.* 22: 13–29.

Roux, W. 1894. The problems, methods, and scope of developmental mechanics. *Biological lectures of the Marine Biology Laboratory, Woods Hole.* Ginn, Boston., pp. 149–190.

Rowe, T. 1995. Quoted in Fischman, J. 1995. Why mammalian ears went on the move. *Science* 270: 1436.

Shaw, A., Lee, Y.-H., Stout, C. D. and Vacquier, V. D. 1994. The species-specificity and structure of abalone sperm lysin. *Semin. Dev. Biol.* 5: 209–215.

Shen, W. and Mardon, G. 1997. Ectopic eye development in *Drosophila* induced by directed *dachshund* expression. *Development* 124: 15–52.

Shu, D., Zhang, X. and Chen, L. 1996. Reinterpretation of *Yunnanozoon* as the earliest known hemichordate. *Nature* 380: 428–430.

Slack, J. M. W. 1983. *From Egg to Embryo: Determinative Events in Early Development.* Cambridge University Press.

Slack, J. M. W., Holland, P. W. H. and Graham, C. F. 1993. The zootype and the phylotypic stage. *Nature* 361: 490–492.

Slijper, E. J. 1962. *Whales.* (Translated by A. J. Pomerans.) Basic Books, New York.

Spemann, H. 1901. Über Correlationen in der Entwicklung des Auges. *Verh. Anat. Ges.* [Vers. Bonn] 15: 61–79.

Stern, M. J. and eight others. 1993. The human *GRB2* and *Drosophila drk* genes can functionally replace the *Caenorhabditis elegans* cell signalling gene *sem-5. Mol. Biol. Cell.* 4: 1175–1188.

Stockard, C. R. 1941. The genetic and endocrine basis for differences in form and behaviour as elucidated by studies of contrasted pure-line dog breeds and their hybrids. *Am. Anat. Memoirs* 19.

Subramanian, V., Meyer, M. I. and Gruss, P. 1995. Disruption of the murine homeobox gene *Cdx1* affects axial skeletal identities by altering the mesodermal expression domains of *Hox* genes. *Cell* 83: 641–653.

Thomson, K. S. 1988. *Morphogenesis and Evolution.* Oxford University Press, New York.

Torrey, T. W. and Feduccia, A. 1979. *Morphogenesis of the Vertebrates.* Wiley, New York.

van der Hoeven, F., Zákány, J. and Duboule, D. 1996. Gene transpositions in the HoxD complex reveal a hierarchy of regulatory controls. *Cell* 85: 1025–1035.

Van Valen, L. M. 1976. Energy and evolution. *Evol. Theor.* 1: 179–229.

von Baer, K. E. 1828. *Entwicklungsgeschichte der Thiere: Beobachtung und Reflexion.* Bornträger, Königsberg.

von Baer, K. E. 1886. *Autobiography of Dr. Karl Ernst von Baer.* (Translated by H. Schneider.) Science History Publications, Canton, MA, pp. 261–262.

Waddington, C. H. 1938. The morphogenetic function of a vestigial organ in the chick. *J. Exp. Biol.* 15: 371–376.

Waddington, C. H. 1940. *Organisers and Genes.* Cambridge University Press, Cambridge.

Waddington, C. H. 1956. *Principles of Embryology.* Allen and Unwin, London.

Waddington, C. H., Needham, J. and Brachet, J. 1936. Studies on the nature of the amphibian organization centre. III. The activation of the evocator. *Proc. R. Soc. London* [B] 120: 173–198.

Wagner, G. P. 1996. Homologues, natural kinds, and the evolution of modularity. *Amer. Zool.* 36: 36–43.

Wake, D. B. and Larson, A. 1987. A multidimensional analysis of an evolving lineage. *Science* 238: 42–48.

Warren, R. W., Nagy, L., Selegue, J., Gates, J. and Carroll, S. 1994. Evolution of homeotic gene regulation and function in flies and butterflies. *Nature* 372: 458–461.

Whittington, H. B. 1985. *The Burgess Shale.* Yale University Press, New Haven.

Wilson, E. B. 1898. Cell lineage and ancestral reminiscence. *Biological Lectures from the Marine Biological Laboratories, Woods Hole, Massachusetts.* Ginn, Boston, pp. 21–42.

Winsor, M. P. 1969. Barnacle larvae in the nineteenth century: A case study in taxonomic theory. *J. Hist. Med. Allied Sci.* 24: 294–309.

Wolpert, L. 1983. Constancy and change in the development and evolution of pattern. *In* B. C. Goodwin, N. Holder and C. C. Wylie (eds.), *Development and Evolution.* Cambridge University Press, Cambridge.

Wray, G. A., Levinton, J. S. and Shapiro, L. H. 1996. Molecular evidence for deep precambrian divergence among metazoan phyla. *Science* 274: 568–573.

Yokouchi, Y., Sakiyama, J.-i., Kameda, T., Iba, H., Suzuki, A., Veno, N. and Kuroiwa, A. 1996. BMP-2/-4 mediate programmed cell death in chicken limb buds. *Development* 122: 3725–3734.

Young, C. M., Vázquez, E., Metaxas, A. and Tyler, P. A. 1996. Embryology of vestimentaran tube worms from deep-sea methane/sulfide seeps. *Nature* 381: 514–516.

Zangerl, R. and Williams, M. E. 1975. New evidence on the nature of the jaw suspension in Paleozoic anacanthus sharks. *Paleontology* 18: 333–341.

Zou, H. and Niswander, L. 1996. Requirement for BMP signaling in interdigital apoptosis and scale formation. *Science* 272: 738–741.

Zuckerkandl, E. 1994. Molecular pathways to parallel evolution. I. Gene nexuses and their morphological correlates. *J. Mol. Evol.* 39: 661–678.

Sources for chapter-opening quotations

Bard, J. 1990. *Morphogenesis: The Cellular and Molecular Processes of Developmental Anatomy*. Cambridge University Press, New York, p. 9.

Cézanne, P. Quoted in J. Gasquet 1921. *Cézanne*. Paris. pp. 79–80.

Claude, A. 1974. The coming of age of the cell. Nobel lecture, reprinted in *Science* 189 (1975): 433–435.

Conrad, J. 1920. *The Rescue: A Romance of the Shallows*. Doubleday, Page and Co., Garden City, NJ 1924, p. 447.

Darwin, C. 1859. *On the Origin of Species*. Reprinted 1958. New American Library, New York, p. 403.

Darwin, C. 1871. *The Descent of Man*. Murray, London, p. 893.

Darwin, E. 1791. Quoted in M. T. Ghiselin, *The Economy of Nature and the Evolution of Sex*. University of California Press 1974, Berkeley, p. 49.

Doyle, A. C. 1891. "A Case of Identity" in The *Adventures of Sherlock Holmes*. Reprinted in *The Complete Sherlock Holmes Treasury*, 1976. Crown, New York, p. 31.

Eliot, T. S. 1936. "The Hollow Men," Part V in *Collected Poems 1909–1962*. Harcourt, Brace and World, New York, pp. 81–82. Copyright T. S. Eliot.

Eliot, T. S. 1942. "Little Gidding" in *Four Quartets*. Harcourt, Brace and Company, New York, 1943, p. 39. Copyright T. S. Eliot.

Fuentes, C. 1989. *Christopher Unborn*. Trans. A. MacAdam. Farrar, Straus, and Giroux, New York, p. 281.

Garstang, W. 1922. The theory of recapitulation: A critical restatement of the biogenetic law. *J. Linn. Soc. Zool.* 35: 81–101.

Geoffroy Saint-Hilaire, E. 1807. Considérations sur les pièces de la tête osseuse de animaux vertebrés, et particulièrment sur celles du crêne des oiseaux. *Ann. Mus Hist. Nat.* 10: 342–343.

Hadorn, E. 1955. *Developmental Genetics and Lethal Factors*. London 1961, p. 105.

Hardin, G. 1968. *Exploring New Ethics for Survival: The Voyage of the Spaceship Beagle*. Viking Press, New York, p. 45.

Harrison, R. G. 1933. Some difficulties of the determination problem. *Am. Nat.* 67: 306–321.

Holub, M. 1990. "From the Intimate Life of Nude Mice" in *The Dimension of the Present Moment*. Trans. D. Habova and D. Young, Faber and Faber, London, p. 38.

Holt, Rinehart & Winston, New York 1949, p. 25.

Just, E. E. 1939. *The Biology of the Cell Surface*. Blakiston, Philadelphia, p. 288.

Kohler, R. E. 1994. Lords of the Fly: *Drosophila* Genetics and the Experimental Life. University of Chicago Press, Chicago, IL, p. 33.

Lessing, G. E. 1778. "Eine Duplik." Reprinted in F. Muncker (ed.), *Sämtliche*, Schriften 13, Göschen, Leipzig, 1897, p. 23.

Levi-Montalcini, R. 1988. *In Praise of Imperfection*. Basic Books, New York, p. 90.

Muller, H. J. 1922. Variation due to change in the individual gene. *Am. Nat.* 56: 32–50.

Ozick, C. 1989. *Metaphor and Memory*. Alfred A. Knopf, New York, p. 111.

Ramón y Cajal, S. 1937. *Recollections of My Life*. Trans. E. H. Craigie and J. Cano. MIT Press, Cambridge, MA. pp. 36–37.

Rostand, J. 1960. *Carnets d'un Biologiste*. Librairie Stock, Paris.

Rushdie, S. 1989. *The Satanic Verses*. Viking, New York, p. 8.

Schultz, J. 1935. Aspects of the relation between genes and development in *Drosophila*. *Am. Nat.* 69: 30–54.

Shelley, M. W. 1817. *Frankenstein, or The Modern Prometheus*. Oxford University Press 1969, p. 36.

Spemann, H. 1943. *Forschung und Leben*. Quoted in T. J. Horder, J. A. Witkowski and C. C. Wylie, *A History of Embryology*, Cambridge University Press 1986, Cambridge, p. 219.

Stern, C. 1936. Genetics and ontogeny. *Am. Nat.* 70: 29–35.

Tennyson, A. 1886. *Idylls of the King*. Macmillan, London 1958, p. 292.

Thomas, L. 1979. "On Embryology" in *The Medusa and the Snail*, Viking Press, New York, p. 157.

Thomson, J. A. 1926. *Heredity*. Putnam, New York, p. 477.

Virchow, R. 1858. *Die Cellularpathologie in ihre Begrundung auf physiologische und pathologische Gewebelehre*. Berlin, p. 493.

Virgil. 37 B.C.E. *Georgics II*, p. 490.

Waddington, C. H. 1956. *Principles of Embryology*. Macmillan, New York, p. 5.

Waddington, C. H. 1957. *The Strategy of the Genes*. Allen and Unwin, London, pp. 154–155.

Weiss, P. 1960. Ross Granville Harrison, 1870–1959. Memorial minute. Rockefeller Inst. Quarterly, p. 6.

Whitehead, A. N. 1934. *Nature and Life*. Cambridge University Press, Cambridge, p. 41.

Whitman, W. 1855. "Song of Myself" in *Leaves of Grass and Selected Prose*. S. Bradley (ed.), Holt, Rinehart & Winston, New York 1949, p. 25.

Whitman, W. 1867. "Inscriptions" in *Leaves of Grass and Selected Prose*. S. Bradley (ed.), Holt, Rinehart & Winston, New York 1949, p. 1.

Williams, C. M. 1959. Hormonal regulation of insect metamorphosis. *In* W. D. McElroy and B. Glass (eds.), *The Chemical Basis of Development*. Johns Hopkins University Press, Baltimore, p. 794.

Wilson, E. B. 1923. *The Physical Basis of Life*. Yale University Press, New Haven, p. 10.

Wolpert, L. 1986. Quoted in J. M. W. Slack, *From Egg to Embryo: Determinative Events in Early Development*. Cambridge University Press, Cambridge, p. 1.

Author Index

Subject Index

In-text definitions of terms are indexed in boldface type.

abdA gene. *see abdominal A* gene
AbdB gene. *see Abdominal B* gene
abdominal A (abdA) gene, 569, 572, 575
Abdominal B (AbdB) gene, 569, 572, 575, 637, 638, 647
abx gene. *see anterobithorax* gene
ac gene. *see achaete* gene
Acanthostega, 726
Accutane, 829
ACE. *see* Adenylation control element
Acetabularia, morphogenesis, 6–10
Acetylation
    in nucleosome disruption, 436–437
    in X chromosome inactivation, 450
Acetylcholine, 147, 148, 279, 291
achaete (ac) gene, 585
Achaete-scute gene complex, 585
Achondroplasia, 109–110, 357
Acidic FGF. *see* Fibroblast growth factor 1
Acini, 684
Acron, 558
Acrosomal process, 123, 129–130
Acrosomal vesicle, 122–123, 129–130, 132
Acrosome, 122–123, 129–130, 132, 857
Acrosome reaction, 129–130, 131, 138
ACTH. *see* Adrenocorticotropic hormone
Actin, 15. *see also* Microfilaments
    in acrosomal process, 130
    in egg microfilaments, 126–127
    in gamete fusion, 139–140
    in meroistic oogenesis, 868
    in microspikes, 277
actin gene, 611
α-actinin, 104
Actinomycin D, 740
Activin, 617, 619
    in epithelial branching, **686**
    in left-right asymmetry, 648–649
activin receptor IIa (cActRIIa), **648–649**
Adaptive enzymes, 47
Adaxial cells, 221
Adenyl cyclase, 621
Adenylation control element (ACE), 484
Adepithelial cells, 749
Adherans junctions, 92
Adhesion. *see* Cell adhesion
Adhesive gradients, axon migration, 314–315
Adolescence, mammary development, 765

Adrenal medulla, 284, 293
Adrenergic neurons, 291
Adrenocorticotropic hormone (ACTH), 407
Adult hemoglobin, 437
Aedes, 808
Aequorin, 144
AER. *see* Apical ectodermal ridge
Affinity. *see* Cell affinity
Aggregation
    in *Dictyostelium,* 22–23
    histotypic, 84
AGM region, 379
AIDS, 688, 788
Air sacs, 383
Albumin, 189, 683
Alcohol, as teratogen, 828, 833–835
Alcohol dehydrogenase-2, 837
Algae, in developmental symbiosis, 808–809
Alisphenoid bone, 906
Allantois, 31, 361, 845
Alligators, sex determination in, 799
Allometry, 891–893
Alternation of generations, 888
Alternative RNA splicing, 466–471
Alveoli, 383
Alzheimer's disease, 334
Amacrine neurons, 281
Ambystoma, 703, 743
Amelia, 830
Ameloblasts, 682
AMH. *see* Anti-Müllerian duct hormone
Amiloride, 151–152
γ-aminobutyric acid, 279
δ-aminolevulinate synthase (DALA synthase), 494
AML1 transcription factor, 379–380
Ammonia
    in *Dictyostelium* differentiation, 28
    excretion by tadpoles, 735
Ammonotelic organisms, 735
Amnion, 31, 186, **243–244, 361**
Amnionte vertebrates, 361
Amniote egg, 31
Amniotic fluid, 244
Amniotic sac, 186
Amoeboflagellates, differentiation, 10–11
Amotrophic lateral sclerosis, 334

Amphibian. *see also* Amphibian metamorphosis; specific types
    activation of embryonic genome, 489
    autonomous development, 603
    bottle cells in, 226–229
    cleavage, 173–174
    cloning experiments, 42–45
    conditional development, 602–603
    egg rearrangement, 156–157
    germ cell migration, 843–844
    mesoderm patterning in, 606–607
    morulae, 173–174
    neurulation, 255
    oogenesis, 861–864
    primary embryonic induction in, 603–605
    progressive determination in, 600–603
    reaggregation experiments, 80, 82–83
    region-specific induction, 621–623
    specification of polarity, 607–609
Amphibian gastrulation, 221–232
Amphibian metamorphosis
    behavioral changes, 740
    general morphological changes, 734–735
    heterochrony and, 743–744
    neuronal changes, 739–740
    tail degeneration, 735–738
    threshold concept, 739
    thyroid hormones and, 735–742
Amphioxus, 169, 888, 911–912
Amphiregulin, 687
Ampulla, 132, 180
Amygdala, 786
Anaerobic respiration, hypertrophic chondrocytes, 356
Analogous structures, 727
Analogy, 727
Anchor cell, 690–691, 692
Andrenomedullary cells, 286
Androgen insensitivity syndrome, 763, 783
Anencephaly, 262
Angelman syndrome, 445
Angioblasts, 367–369
Angiogenesis, 369–370
Angiogenetic clusters, 368
Angiopoietin-1, 369
Animal cap, 229
Animal hemisphere, 156

Digestive tube, 380
derivatives, 382–383
Digital arch, 726
Digits, formation, 724–727
5α-dihydrotestosterone, **783**
Dihydrotestosterone, in sex determination, 783–784
Dioxin, 836
Dipoltene, 852
Direct development, 743–745
Discoidal cleavage, **169,** 188–192
disheveled gene, 565
Disheveled protein, 566
Disruptions, 828
Dissociation, in development, 891–893
Distal tip cell, 853
Distal-less (Dll) gene, 573, 752
in butterfly polyphenism, 815
in evolution, 908–909, 911–912
Distal-less protein, in imaginal disc determination, 750
Divergence, in development, 893–894
Diversity, emergence of phyla and, 888
Dizygotic twins, 186
Dlx3 gene, 715
DMZ. *see* Dorsal marginal zone
DNA. *see also* cDNA; DNA synthesis
accessibility to *trans*-regulators, 432–434
boundary sequences, 454
cloning, 55–57
DNase footprinting, 555
in eukaryotes, 5–6
hormone responsive elements, 420–423
hybridization, 54–55, 58–59
insertion techniques, 69–70
locus control regions, 437–441
methylation, 442–446
mobility shift assay, 414
nuclear matrix and, 451–454
polymerase chain reaction and, 66, 68–69
in prokaryotes, 5–6
recombinant, 56
repair enzymes in meiosis, 852
retinoic acid response elements, 829
sequencing, 59–61
strand separation, 453–454
DNA blots, 58–59
DNA ligase, 56
DNA polymerase, 69
DNA synthesis
in cell cycle, 196
cyclin E and, 200
in fertilization, 151–152
MPF and, 197
in neural tube cells, 270
DNA-bending proteins, 423
DNA-binding domain, 404, 421
DNase footprinting, 555
DNase I, 433–434
DNase I-hypersensitive sites, 434–436, 451
locus control regions and, 440, 441
methylation, 444
Domain, 454
Dopamine, 279
Dorsal, defined, 341
dorsal gene, 577–578
Dorsal lip of the blastopore, 222–224
in initiation of gastrulation, 602
planar signals from, 627

in primary embryonic induction, 603–605
Dorsal marginal zone (DMZ), 627
in vitro migration, 231
Dorsal mesoderm
bone development, 351–358
generation of somite types, 344–347
induction, 610–612
Noggin protein and, 617
in organogenesis, 255
paraxial, 341, 343
skeletal muscle differentiation, 347–351
somite formation, 343–344
Dorsal protein, 577, 585
asymmetric signaling and, 578–581
Drosophila fate map and, 582–585
gradient, 581–585
translocation, 577–578
Dorsal root ganglia, 284, 287
precursors, 292
thalidomide and, 831
Dorsal switch protein 1 (DSP1), 584
dorsalin gene, **264**
Dorsalin protein, 661
Dorsalizing genes, 578
Dorsal-ventral axis
cartesian coordinate model and, 585
central nervous system, 264
determination, 224–225, 238
Dorsal protein in, 577–578
Dorsal protein translocation in, 578–585
in imaginal discs, 751–752
in limb formation, 721–722
in mammals and birds, 647–650
in meroistic oogenesis, 869
similarity to lymphocyte differentiation, 582
specification, 607–610
Dorsolateral hinge points (DLHP), 260
Dosage compensation, 446
Dot blot, 64, 66
"Double assurance," 627, 668
doublesex (dsx) gene, 469–471, 790
in *Drosophila* sex determination, 793–794
Doublesex protein, 404
in sex determination cascades, 794–795
in yolk protein uptake, 870
Down's syndrome, 827
dpp gene. *see decapentaplegic* gene
Drosophila. see also Insect metamorphosis
activation of embryonic genome, 489
anterior-posterior polarity, 545–577
axes determination, 480–481, 585
body plan, 545–546
boundary sequences in, 454
cell fate commitment in, 559–561
cleavage cycle, 196
developmental regulation of MPF kinase, 199
differential poly(A) tail shortening in, 476
dorsal-ventral polarity, 577–585
dosage compensation in, 446
ecdysis in, 756–757
in evolution, 907–909
eye mutations, 688–690
fate map, 582–585
gastrulation, 543–545
germ cells in, 849–850, 855
Hedgehog protein family in, 659
Hom-C genes in, 643, 905

homeodomain proteins, 405
homeotic gene homology, 616, 637–638, 640
homeotic mutations, 570–571
homeotic selector genes, 569–577
imaginal discs, 747–753
integrins and, 104–105
maternal effect genes, 546–559
midblastula transition, 194–195
molting, 747–748
motor neuron specification, 310–312
nervous system formation, 545
neural cells specification, 585
neuroblast spacing, 692–693
neuromeres, 647
nucleosome disruptional complexes, 436
oogenesis, 867–870
organ primordia, 585
parthenogenesis in, 861
periodicity in, 546
retinal photoreceptor formation, 688–690
RTK-Ras pathway in, 910–911
salivary gland specification, 585
segmentation genes, 559–569
Semaphorin II and, 321–322
sex determination, 468–471, 788–795
spermatogenesis, 858–859
transcription factors, 400, 401
Wnt proteins in, 660
X chromosome mutations, 38
DSP1. *see* Dorsal switch protein 1
dsx gene. *see doublesex* gene
Duck
BMP4 expression, 904–905
digit formation, 724–725
Ductus arteriosus, 372
Duplication, in development, 893–894
D/V axis. *see* Dorsal-ventral axis
dwarf mouse, 407
Dwarfism, 658
Dynein, 123–124

ea gene. *see easter* gene
Ear
development, 295
evolution in mammals, 894–896
easter (ea) gene, 581
Easter protein, 581
E-cadherin, 93, 184
in cell migration, 240
metanephrogenic mesenchyme and, 679
in regulation of connexin, 98
Ecdysis, 756–757
Ecdysis-triggering hormone, 756
Ecdysone, 754, 755–756
in polyphenism, 814
in yolk protein uptake, 870
Ecdysone receptor (EcR), 758, 759–761
Ecdysone receptor (EcR) gene, 758, 760, 761
Echinoderm, cleavage, 169
Eclosion, 757
Eclosion hormone, 756
ECM18, 214
EcR. *see* Ecdysone receptor
EcR gene. *see Ecdysone receptor* gene
Ectoderm, 3–4. *see also* Apical ectodermal ridge
amphibian, 222–224, 232
avian, 241–242

ABOUT THE BOOK

Editor: Andrew D. Sinauer

Project Editor: Nan Sinauer

Production Manager: Christopher Small

Electronic Bookbuilder: Janice Holabird

Illustration Program: J/B Woolsey Associates

Book Design: Susan Brown Schmidler

Cover Design: MBDesign

Copy Editor: Janet Greenblatt

Cover Manufacturer: Henry N. Sawyer Company, Inc.

Book Manufacturer: Courier Westford, Inc.